Purity, Spectra and Localisation

This book is an account of a fruitful interaction between algebra, mathematical logic, and category theory. It is possible to associate a topological space to the category of modules over any ring. This space, the Ziegler spectrum, is based on the indecomposable pure-injective modules. Although the Ziegler spectrum arose within the model theory of modules and plays a central role in that subject, this book concentrates on its algebraic aspects and uses.

The central aim is to understand modules and the categories they form through associated structures and dimensions which reflect the complexity of these, and similar, categories. The structures and dimensions considered arise through the application of ideas and methods from model theory and functor category theory. Purity and associated notions are central, localisation is an ever-present theme and various types of spectrum play organising roles.

This book presents a unified account of material which is often presented from very different viewpoints and it clarifies the relationships between these various approaches. It may be used as an introductory graduate-level text, since it provides relevant background material and a wealth of illustrative examples. An extensive index and thorough referencing also make this book an ideal, comprehensive reference.

MIKE PREST is Professor of Pure Mathematics at the University of Manchester.

ENCYCLOPEDIA OF MATHEMATICS AND ITS APPLICATIONS

All the titles listed below can be obtained from good booksellers or from Cambridge University Press. For a complete series listing visit
http://www.cambridge.org/uk/series/sSeries.asp?code=EOM

69 T. Beth, D. Jungnickel, and H. Lenz *Design Theory I, 2nd edn*
70 A. Pietsch and J. Wenzel *Orthonormal Systems for Banach Space Geometry*
71 G. E. Andrews, R. Askey and R. Roy *Special Functions*
72 R. Ticciati *Quantum Field Theory for Mathematicians*
73 M. Stern *Semimodular Lattices*
74 I. Lasiecka and R. Triggiani *Control Theory for Partial Differential Equations I*
75 I. Lasiecka and R. Triggiani *Control Theory for Partial Differential Equations II*
76 A. A. Ivanov *Geometry of Sporadic Groups I*
77 A. Schinzel *Polynomials with Special Regard to Reducibility*
78 H. Lenz, T. Beth, and D. Jungnickel *Design Theory II, 2nd edn*
79 T. Palmer *Banach Algebras and the General Theory of *-Algebras II*
80 O. Stormark *Lie's Structural Approach to PDE Systems*
81 C. F. Dunkl and Y. Xu *Orthogonal Polynomials of Several Variables*
82 J. P. Mayberry *The Foundations of Mathematics in the Theory of Sets*
83 C. Foias, O. Manley, R. Rosa and R. Temam *Navier–Stokes Equations and Turbulence*
84 B. Polster and G. Steinke *Geometries on Surfaces*
85 R. B. Paris and D. Kaminski *Asymptotics and Mellin–Barnes Integrals*
86 R. McEliece *The Theory of Information and Coding, 2nd edn*
87 B. Magurn *Algebraic Introduction to K-Theory*
88 T. Mora *Solving Polynomial Equation Systems I*
89 K. Bichteler *Stochastic Integration with Jumps*
90 M. Lothaire *Algebraic Combinatorics on Words*
91 A. A. Ivanov and S. V. Shpectorov *Geometry of Sporadic Groups II*
92 P. McMullen and E. Schulte *Abstract Regular Polytopes*
93 G. Gierz et al. *Continuous Lattices and Domains*
94 S. Finch *Mathematical Constants*
95 Y. Jabri *The Mountain Pass Theorem*
96 G. Gasper and M. Rahman *Basic Hypergeometric Series, 2nd edn*
97 M. C. Pedicchio and W. Tholen (eds.) *Categorical Foundations*
98 M. E. H. Ismail *Classical and Quantum Orthogonal Polynomials in One Variable*
99 T. Mora *Solving Polynomial Equation Systems II*
100 E. Olivieri and M. Eulália Vares *Large Deviations and Metastability*
101 A. Kushner, V. Lychagin and V. Rubtsov *Contact Geometry and Nonlinear Differential Equations*
102 L. W. Beineke and R. J. Wilson (eds.) with P. J. Cameron *Topics in Algebraic Graph Theory*
103 O. Staffans *Well-Posed Linear Systems*
104 J. M. Lewis, S. Lakshmivarahan and S. Dhall *Dynamic Data Assimilation*
105 M. Lothaire *Applied Combinatorics on Words*
106 A. Markoe *Analytic Tomography*
107 P. A. Martin *Multiple Scattering*
108 R. A. Brualdi *Combinatorial Matrix Classes*
110 M.-J. Lai and L. L. Schumaker *Spline Functions on Triangulations*
111 R. T. Curtis *Symmetric Generation of Groups*
112 H. Salzmann, T. Grundhöfer, H. Hähl and R. Löwen *The Classical Fields*
113 S. Peszat and J. Zabczyk *Stochastic Partial Differential Equations with Lévy Noise*
114 J. Beck *Combinatorial Games*
115 L. Barreira and Y. Pesin *Nonuniform Hyperbolicity*
116 D. Z. Arov and H. Dym *J-Contractive Matrix Valued Functions and Related Topics*
117 R. Glowinski, J-L. Lions and J. He *Exact and Approximate Controllability for Distributed Parameter Systems*
118 A. A. Borovkov and K. A. Borovkov *Asymptotic Analysis of Random Walks*
119 M. Deza and M. Dutour Sikirić *Geometry of Chemical Graphs*
120 T. Nishiura *Absolute Measurable Spaces*
121 M. Prest *Purity, Spectra and Localisation*
122 S. Khrushchev *Orthogonal Polynomials and Continued Fractions: From Euler's Point of View*
123 H. Nagamochi and T. Ibaraki *Algorithmic Aspects of Graph Connectivities*
124 F. W. King *Hilbert Transforms I*
125 F. W. King *Hilbert Transforms II*
126 O. Calin and D.-C. Chang *Sub-Riemannian Geometry*

Purity, Spectra and Localisation

MIKE PREST
University of Manchester

CAMBRIDGE UNIVERSITY PRESS
Cambridge, New York, Melbourne, Madrid, Cape Town, Singapore, São Paulo, Delhi

Cambridge University Press
The Edinburgh Building, Cambridge CB2 8RU, UK

Published in the United States of America by Cambridge University Press, New York

www.cambridge.org
Information on this title: www.cambridge.org/9780521873086

First published 2009

Printed in the United Kingdom at the University Press, Cambridge

A catalogue record for this publication is available from the British Library

ISBN 978-0-521-87308-6 hardback

To Abigail, Stephen and Iain

Contents

Preface

In his paper [726], on the model theory of modules, Ziegler associated a topological space to the category of modules over any ring. The points of this space are certain indecomposable modules and the definition of the topology was in terms of concepts from model theory. This space, now called the Ziegler spectrum, has played a central role in the model theory of modules. More than one might have expected, this space and the ideas surrounding it have turned out to be interesting and useful for purely algebraic reasons. This book is mostly about these algebraic aspects.

The central aim is a better understanding of the category of modules over a ring. Over most rings this category is far too complicated to describe completely so one must be content with aiming to classify the most significant types of modules and to understand more global aspects in just a broad sense, for example by finding some geometric or topological structure that organises some aspect of the category and which reflects the complexity of the category.

By "significant types of modules" one might mean the irreducible representations or the "finite" (finite-dimensional/finitely generated) ones. Here I mean the pure-injective modules. Over many rings this class of modules includes, directly or by proxy, the "finite" ones. There is a decomposition theorem which means that for most purposes we can concentrate on the indecomposable pure-injective modules.

The Ziegler spectrum is one example of an "organising" structure; it is a topological space whose points are the isomorphism classes of indecomposable pure-injectives, and the Cantor–Bendixson analysis of this space does reflect various aspects of complexity of the module category. There are associated structures: the category of functors on finitely presented modules; the lattice of pp conditions; the presheaf of rings of definable scalars. Various dimensions and ranks are defined on these and they are all linked together.

Here I present a cluster of concepts, techniques, results and applications. The inputs are from algebra, model theory and category theory and many of the results and methods are hybrids of these. The way in which these combine here is something which certainly I have found fascinating. The applications are mainly algebraic though not confined to modules since everything works in good enough abelian categories. Again, I have been pleasantly surprised by the extent to which what began in model theory has had applications and ramifications well beyond that subject.

Around 2000 it seemed to me that the central part of the subject had pretty well taken shape, though mainly in the minds of those who were working with it and using it. Much was not written down and there was no unified account so, foolhardily, I decided to write one. This book, which is the result, has far outgrown my original intentions (in length, time, effort, ...). In the category of books it is a pushout of a graduate-level course and a work of reference.

Introduction

The Ziegler spectrum, Zg_R, of a ring R is a topological space. It is defined in terms of the category of R-modules and, although a Ziegler spectrum can be assigned to much more general categories, let us stay with rings and modules at the beginning. The points of Zg_R are certain modules, more precisely they are the isomorphism types of indecomposable pure-injective (also called algebraically compact) right R-modules. Any injective module is pure-injective but usually there are more, indeed a ring is von Neumann regular exactly if there are no other pure-injective modules (2.3.22). If R is an algebra over a field k, then any module which is finite-dimensional as a k-vector space is pure-injective (4.2.6). Every finite module is pure-injective (4.2.6). Another example is the ring of p-adic integers, regarded as a module over any ring between $\mathbb{Z}$ and itself (4.2.8). The pure-injective modules mentioned so far are either "small" or, although large in some sense, have some kind of completeness property. There is something of a general point there but, as it stands, it is too vague: not all "small" modules are pure-injective. For example, the finite-length modules over the first Weyl algebra, $A_1(k)$, over a field k of characteristic zero are not pure-injective (8.2.35). They are small in the sense of being of finite length, but large in that they are infinite-dimensional. Nevertheless each indecomposable finite-length module over the first Weyl algebra has a pure-injective hull (a minimal pure, pure-injective extension, see Section 4.3.3) which is indecomposable. Indeed, associating to a finite-length module its pure-injective hull gives a bijection between the set of (isomorphism types of) indecomposable finite-length modules over $A_1(k)$ and a subset of the Ziegler spectrum of $A_1(k)$ (8.2.39).

Ziegler defined the topology of this space in terms of solution sets to certain types of linear conditions (5.1.21) but there are equivalent definitions: in terms of morphisms between finitely presented modules (5.1.25); also in terms of finitely presented functors (10.2.45). Ziegler showed that understanding this space, in the very best case obtaining a list of points and an explicit description of the topology, is the key to answering most questions about the model theory of modules over the given ring. For that aspect one may consult [495]. Most subsequent advances have been driven more by algebraic than model-theoretic questions though much of what is here can be reformulated to say something about the model theory of modules.

Over some rings there is a complete description of the Ziegler spectrum (see especially Section 5.2 and Chapter 8); for $R = \mathbb{Z}$ the list of points is due to Kaplansky [330], see Section 5.2.1.

A module which is of finite length over its endomorphism ring is pure-injective (4.2.6, 4.4.24) but, unless the ring is right pure-semisimple (conjecturally equivalent to being of finite representation type, see Section 4.5.4), one should expect there to be "large" points of the Ziegler spectrum. Over an artin algebra a precise expression of this is the existence of infinite-dimensional indecomposable pure-injectives (5.3.40) if the ring is not of finite representation type. This is an easy consequence of compactness of the Ziegler spectrum of a ring (5.1.23). Even if one is initially interested in "small" modules, for example, finite-dimensional representations, the "large" modules may appear quite naturally: often the latter parametrise, in some sense or another, natural families of the former (e.g. 5.2.2, Sections 4.5.5 and 15.1.3). Examples of such large parametrising modules are the generic modules of Crawley-Boevey (Section 4.5.5).

A natural context for most of the results here is that of certain, "definable", subcategories of locally finitely presented abelian categories: the latter are, roughly, abelian categories in which objects are determined by their "elements", see Chapter 16. There are reasons for working in the more general context beyond simply wider applicability. Auslander and coworkers, in particular Reiten, showed how, if one is interested in finite-dimensional representations of a finite-dimensional k-algebra, it is extremely useful to move to the, admittedly more abstract, category of k-linear functors from the category of these representations to the category of k-vector spaces. It turns out that, in describing ideas around the Ziegler spectrum, moving to an associated such functor category often clarifies concepts and simplifies arguments (and leads to new results!). By this route one may also dispense with the terminology of model theory, though model theory still provides concepts and techniques, and replace it with the more widely known terminology of categories and functors. In particular, one may define the Ziegler spectrum of a ring as a topology on the (set of isomorphism types of) indecomposable injectives in the corresponding functor category (12.1.17). This topology is dual (Section 5.6) to another topology, on the same set, which one might regard as the (Gabriel–)Zariski spectrum of the functor category (14.1.6).

This equivalence between model-theoretic and functorial methods is best explained by an equivalence, 10.2.30, between the model-theoretic category of "imaginaries" (in the sense of Section B.2) and the category of finitely presented functors. There is some discussion of this equivalence between methods, and see Appendix C, but I have tended to avoid the terminology of model theory, except where I consider it to be particularly efficient or where there is no algebraic equivalent. That is simply because it is less well known, though I hope that one effect of this book will be that some people become a little more comfortable with it. This does mean that those already familiar with model theory might have to work a bit harder than they expected: the terminological adjustments are quite slight, but the conceptual adjustments (the use of functorial methods) may well require more effort. It should be noted that much of the relevant literature does assume familiarity with the most basic ideas from model theory.

As mentioned already, it is possible to give definitions (5.1.1, 5.1.25) of the Ziegler spectrum of a ring purely in terms of its category of modules, that is, without reference to model theory or to any "external" (functor) category. In fact the book begins by taking a "naïve", element-wise, view of modules and gradually, though not monotonically, takes an increasingly "sophisticated" view. I discuss this now.

I believe that there is some advantage in beginning in the (relatively) concrete context of modules and, consequently, in the first part of the book, modules are simply sets with structure and most of the action takes place in the category of modules. Many results are presented, or at least surveyed, in that first part, so a reader may refer to these without having to absorb the possibly unfamiliar functorial point of view.

Nevertheless, it was convincingly demonstrated by Herzog in the early 1990s that the most efficient and natural way to prove Ziegler's results, and many subsequent ones, is to move to the appropriate functor category. Indeed it was already appreciated that work, particularly of Gruson and Jensen, ran, in places, parallel to pre-Ziegler results in the model theory of modules and some of the translation between the two languages (model-theoretic and functorial) was already known.

Furthermore, many applications have been to the representation theory of finite-dimensional algebras, where functorial methods have become quite pervasive. So, at the beginning of Part II, we move to the functor category. In fact, the ground will have been prepared already, in the sense that I call on results from Part II in more than a few proofs in Part I. The main reason for this anticipation, and consequent complication in the structure of logical dependencies in the book, is that I wish to present the functorial proofs of many of the basic results. The original model-theoretic and/or algebraic ("non-functorial") proofs are available elsewhere, whereas the functorial proofs are scattered in the literature and, in many cases, have not appeared.

In the second part of the book modules become certain types of functors on module categories: in the third part they become functors on functor categories. This third part deals with results and questions clustering around relationships between definable categories. Model theory reappears more explicitly in this part because it is, from one point of view, all about interpretability of one category in another.

One could say that Part I is set mostly in the category Mod-R, that Part II is set in the functor category (R-mod, **Ab**) and that Part III is set in the category $\big((R\text{-mod}, \mathbf{Ab})^{\mathrm{fp}}, \mathbf{Ab}\big)$. The category, $\mathrm{Ex}\big((R\text{-mod}, \mathbf{Ab})^{\mathrm{fp}}, \mathbf{Ab}\big)$, of exact functors on the category $(R\text{-mod}, \mathbf{Ab})^{\mathrm{fp}}$ appears and reappears in many forms throughout the book.

A second spectrum also appears. The Zariski spectrum is well known in the context of commutative rings: it is the space of prime ideals endowed with the Zariski topology. It is also possible to define this space, *à la* Gabriel [202], in terms of the category of modules, namely as the set of isomorphism types of indecomposable injectives endowed with a topology which can be defined in terms of morphisms from finitely presented modules. That definition makes sense in the category of modules over any ring, indeed in any locally finitely presented abelian category. Applied to the already-mentioned functor category (R-mod, **Ab**), one obtains what I call the Gabriel–Zariski spectrum. It turns out that this space can also be obtained from the Ziegler spectrum, as the "dual" topology which has, for a basis of open sets, the complements of compact open sets of the Ziegler topology.

Much less has been done with this than with the Ziegler topology and I give it a corresponding amount of space. I do, however, suspect that there is much to discover about it and to do with it. The Gabriel–Zariski spectrum has a much more geometric character than the Ziegler spectrum, in particular it carries a sheaf of rings which generalises the classical structure sheaf: it is yet another "non-commutative geometry".

Chapters 1–5 and 10–12, minus a few sections, form the core exposition. The results in the first group of chapters are set in the category of modules and lead the reader through pp conditions and purity to the definition and properties of the Ziegler spectrum. The methods used in the proofs change gradually; from elementary linear algebra to making use of functor categories. One of my reasons for writing this book, rather than being content with what was already in the literature, was to present the basic theory using these functorial methods since they have lead to proofs which are often much shorter and more natural than the original ones. The second group of chapters introduces those methods, so the reader of Chapters 1–5 must increasingly become the reader of Chapters 10–12.

Beyond this core, further general topics are presented in Chapters 6 and 14 (rings and sheaves of definable scalars, the Gabriel–Zariski topology) and in

Chapters 7 and 13 (dimensions). Chapter 9, on ideals in mod-R, leads naturally to the view of Part II.

Already that gives us a book of some 500 pages, yet there is much more which should be said, not least applications in specific contexts. At least some of that is said in the remaining pages, though often rather briefly. Chapter 8 contains examples and descriptions of Ziegler spectra over various types of ring. Some of the most fruitful development has been in the representation theory of artin algebras (Chapter 15). The theory applies in categories much more general than categories of modules and Chapters 16 and 17 present examples. In these chapters the emphasis is more on setting out the basic ideas and reporting on what has been done, so rather few proofs are given and the reader is referred to the original sources for the full story.

Though the book begins with systems of linear equations, by the time we arrive at Part III we are entering very abstract territory, an additive universe which parallels that of topos theory. Chapter 18 introduces this, though not at great length since this is work in progress and likely not to be in optimal form.

Ziegler's paper was on the model theory of modules and, amongst all this algebraic development, we should not forget the open questions and developments in that subject, so Appendix D is a, very brief, update on the model theory of modules per se.

Beyond this, there is background on model theory in Appendices A and B, as well as general background (Appendix E) and a model theory/functor category theory "dictionary" (Appendix C).

Relationship with the earlier book and other work As to the relationship of this book with my earlier one *Model Theory and Modules* [495], this is, in a sense, a sequel but the emphases of the two books are very different. The earlier book covered model-theoretic aspects of modules and related algebraic topics, and it was written from a primarily model-theoretic standpoint. In this book the viewpoint is algebraic and category-theoretic though it is informed by ideas from model theory. No doubt the model theory proved to be an obstacle for some readers of [495] and perhaps the functor-category theory will play a similar role in this book. But I hope that by introducing the functorial ideas gradually through the first part of this book I will have made the path somewhat easier. Readers who have some familiarity with the contents of [495] will find here new results and fresh directions and they will find that the text reflects a great change in viewpoint and expansion of methods that has taken place in the meantime.

The actual overlap between the books is rather smaller than one might expect, given that they are devoted to the same circle of ideas. Part of the reason for this is that many new ideas and results have been produced in the intervening years.

If that were all, then an update would have sufficed. But one of the reasons for my writing this book is to reflect the fundamental shift in viewpoint, the adoption of a functor-category approach, that has taken place in the area. Although, in this context, the languages of model theory and functor-category theory are, in essence and also in many details, equivalent (indeed, I provide, at Appendix C, a dictionary between them) there have proved to be many conceptual advantages in adopting the latter language. Readers familiar with the arguments of [495] or Ziegler's paper, will see here how complex and sometimes apparently ad-hoc arguments become natural and easy in this alternative language. That is not to say that the insights and techniques of model theory have been abandoned. In fact they inform the whole book, although this might not always be apparent. Some model-theoretic ideas have been explicitly retained, for example the notion of pp-type, because we need this concept and because there is no algebraic name for it. On the other hand, there is no need in this book to treat formulas as objects of a formal language, so I refer to them simply as conditions (which is, in any case, how we think of them).

In the more model-theoretic approach there was a conscious adaptation of ideas from module theory (at least on my part, see, for example, [495, p. 173], and, I would guess, also by Garavaglia, see especially [209]), using the heuristic that pp conditions are generalised ring elements, that pp-types are generalised ideals (in their role as annihilators) and that various arguments involving positive quantifier-free types in injective modules extend to pp-types in pure-injective modules. Moving to the functor category has the effect of turning this analogy into literal generalisation. See, for example, the two proofs of 5.1.3 bearing in mind the heuristic that a pp-type is a generalised right ideal.

Various papers and books contain significant exposition of some of the material included here. Apart from my earlier book, [495], and some papers, [493], [497], [503], [511], there are Rothmaler's, [620], [622], [623], the survey articles [61], [484], [568], [514], [728], a large part of the book, [323], of Jensen and Lenzing, the monograph, [358], of Krause and sections in the books of Facchini, [183], and Puninski, [564]. There is also the more recent [516] which deals with the model theory of definable additive categories.

In this book I have tried to include at least mention of all recent significant developments but, in contrast to the writing of [495], I have tried, with a degree of success, to restrain my tendency to aim to be encylopaedic. There are some topics which I just mention here because, although I would have liked to have said more, I do not have the expertise to say anything more useful that what can easily be found already in the literature. These include: the work of Guil Asensio and Herzog developing a theory of purity in Flat-R (see the end of Section 4.6); recent and continuing developments around cotilting modules and cotorsion theories (see the end of Section 18.2.3); work of Beligiannis and others around triangulated categories and generalisations of these (see Chapter 17).

Thanks, especially, to: Ivo Herzog who, in the early 1990s, showed me how Ziegler's arguments become so much easier and more natural in the functor category; Kevin Burke, Bill Crawley-Boevey and Henning Krause who convinced me in various ways of the naturality and power of the functor category idiom and of the desirability of adopting it.

I am also grateful to a number of colleagues, in particular to Gena Puninski, for comments on a draft of this.

General comments Occasionally I define a concept or prove a result that is defined or proved elsewhere in the book. Oversights excepted, this is to increase the book's usefulness as a reference. For example, if a concept is defined using the functor-category language, or in a very general context, it can be useful to have an alternative definition, possibly in a more particular context, also to hand.

I also play the following mean trick. In Part I, R is a ring and modules are what you think they are. In Part II, I reveal that R was, all the time, a small preadditive category and that what you thought were modules were actually functors. To have been honest right through Part I would have made the book even harder to read. In any case, you have now been warned.

Bibliography As well as containing items which are directly referenced, the bibliography is an "update" on that in [495], so contains a good number of items which do not occur in the bibliography of [495] but which continue or are relevant to some of the themes there. In particular I have tried to be comprehensive as regards including papers which fall within the model theory of modules, even though I have only pointed to the developments there.

Conventions and notations

Conventions A module will be a right module if it matters and if the contrary is not stated.

Tensor product $\otimes$ means, unless indicated otherwise, tensor product over R, $\otimes_R$, where R is the ring of the moment.

By a functor between preadditive categories is meant an additive functor even though "additive" is hardly ever said explicitly.

"Non-commutative" means "not necessarily commutative".

To say that a tuple of elements is **from** M is to say that each entry of the tuple is an element of M (the less accurate "in" in place of "from" may have slipped through in places).

The value of a function f (or functor F) on an element a (or object A or morphism g) is usually denoted fa (respectively FA, Fg) but sometimes, for clarity or to avoid ambiguity, $f(a)$ or $f \cdot a$ (and similarly).

I make no distinction between monomorphisms and embeddings and my use of the terms is determined by a mixture of context and whim: similarly for map, morphism and homomorphism.

In the context of categories I write $\mathcal{A} \simeq \mathcal{B}$ if the categories $\mathcal{A}$ and $\mathcal{B}$ are naturally equivalent. Sometimes the stronger relation of isomorphism holds but we don't need a different symbol for that.

Equality, "=", often is used where $\simeq$ (isomorphism or natural equivalence) would be more correct, especially, but not only, if our choice of copy has not yet been constrained.

I write $l(\overline{a})$ for the **length** (number of entries) of a tuple $\overline{a}$. By default a tuple has length "n" and this explains (I hope all) unheralded appearances of this symbol. If $\overline{a}$ is a tuple, then its typical coordinate/entry is a_i.

Tuples and matrices are intended to match when they need to. So, if $\overline{a}, \overline{b}$ are tuples, then the appearance of "$\overline{a} + \overline{b}$" implies that they are assumed to have the same length, and the expression means the tuple with ith coordinate equal to $a_i + b_i$. Similarly, if $\overline{a}$ and $\overline{b}$ are tuples and H is a matrix, then writing "$(\overline{a}\,\overline{b})H = 0$" implies that the number of rows of H is the length of $\overline{a}$ plus the length of $\overline{b}$. The notations $(\overline{a}\,\overline{b})$, $(\overline{a}, \overline{b})$ and $\overline{ab}$ are all used for the tuple whose entries are the entries of $\overline{a}$ followed by those of $\overline{b}$.

If X is a subset of the R-module M^n and W is a subset of R^n, then by $X \cdot W$ we mean the subgroup of M generated by the $(a_1, \ldots, a_n) \cdot (r_1, \ldots, r_n) = \sum_{i=1}^{n} a_i r_i$ with $(a_1, \ldots, a_n) \in X$ and $(r_1, \ldots, r_n) \in W$. If, on the other hand, $X \subseteq M$ and $W \subseteq R^n$, then $X \cdot W$ will mean the subgroup of M^n generated by the $(ar_1, \ldots, ar_n)$ with $a \in M$ and $(r_1, \ldots, r_n) \in W$. Sometimes the "$\cdot$" will be omitted. So read this as "dot product" if that makes sense, otherwise diagonal product.

For matrices I use notation such as $(r_{ij})_{ij}$ meaning that i indexes the rows, j the columns and the entry in the ith row and jth column is r_{ij}. I also use, I hope self-explanatory, partitioned-matrix notation in various places.

Notation such as $(M_i)_i$ is used for indexed sets of objects and I tend to refer to them (loosely and incorrectly – there may be repetitions) as sets.

If I say that a ring is noetherian I mean right and left noetherian: similarly with other naturally one-sided conditions.

The "radical" of a ring R means the Jacobson radical and this is denoted $J(R)$ or just J. The notation "rad" is used for the more extended (to modules, functors, categories) notion.

If $\mathcal{C}$ is a category, then a class of objects of $\mathcal{C}$ is often identified with the full subcategory which has these as objects.

The notation (co)ker(f) and term (co)kernel are used for both the object and the morphism.

The notation $^{\perp}$ is used for both Hom- and Ext-orthogonality, depending on context.

The ordering on pp-types is by solution sets, so I write $p \geq q$ if $p(M) \geq q(M)$ for every M. Therefore $p \geq q$ iff $p \subseteq q$ (the latter inclusion as sets of pp conditions). This is the opposite convention from that adopted in [495]. There are arguments for both but I think that the convention here makes more sense.

If χ is a condition and $\overline{v}$ is a sequence of variables, then writing $\chi(\overline{v})$ implies that the free variables of χ all occur in $\overline{v}$ but not every component variable of $\overline{v}$ need actually occur in χ.

I will be somewhat loose regarding the distinction between small and skeletally small categories, for instance stating a result for small categories but applying it to skeletally small categories such as mod-R.

The term "tame (representation type)" sometimes includes finite representation type, but not always; the meaning should be clear from the context.

The term "torsion theory" will mean hereditary torsion theory unless explicitly stated otherwise.

Morphisms in categories compose to the left, so fg means do g then f. So, in the category $\mathbb{L}_R^{\mathrm{eq}+}$ associated to the category of right R-modules, pp-defined maps compose to the left. The action of these maps on right R-modules is, however, naturally written on the right, so the ring of definable scalars of a right R-module is the *opposite* of a certain endomorphism ring in the corresponding localisation of this category. I have tried to be consistent in this but there may be places where an ^{op} should be inserted.

The term "preprint" is used to mean papers which are in the public domain but which might or might not be published in the future.

Notations The notation $\mathbb{Z}_n$ is used for the factor group $\mathbb{Z}/n\mathbb{Z}$. The localisation of $\mathbb{Z}$ at p is denoted by $\mathbb{Z}_{(p)}$ and the p-adic integers by $\overline{\mathbb{Z}_{(p)}}$.

End(A) denotes the endomorphism ring of A and Aut(A) its automorphism group. The group, Hom(A, B), of morphisms from A to B is usually abbreviated to (A, B). Similar notation is used in other categories, for example, if $\mathcal{C}, \mathcal{D}$ are additive categories (with $\mathcal{C}$ skeletally small), then $(\mathcal{C}, \mathcal{D})$ denotes the category of additive functors from $\mathcal{C}$ to $\mathcal{D}$.

$R[X_1, \dots, X_n]$ denotes the ring of polynomials with coefficients from R in indeterminates $X_1, \dots, X_n$ which commute with each other and with all the elements of R; $R\langle X_1, \dots, X_n\rangle$ denotes the free associative R-algebra in indeterminates $X_1, \dots, X_n$; "non-commutative polynomials over R" with the elements of R acting centrally (R will be commutative whenever this notation appears).

Annihilators: $\mathrm{ann}_R(A) = \{r \in R : ar = 0\ \forall a \in A\}$, where $A \subseteq M_R$, and the obvious modifications $\mathrm{ann}_R(a)$, $\mathrm{ann}_R(a_1, \dots, a_n)$;
$\mathrm{ann}_M(T) = \{a \in M : at = 0\ \forall t \in T\}$ where $T \subseteq R$.

The **socle**, $\mathrm{soc}(M)$, of a module M is the sum of all its simple submodules.

$C(R)$ denotes the **centre**, $\{r \in R : sr = rs\ \forall s \in R\}$, of the ring R.

pp_R^n is the lattice of equivalence classes of pp conditions in n free variables for right R-modules.

$\mathrm{pp}^n(M)$ is the lattice of subgroups of M^n pp-definable in M; this may be identified with pp_R^n modulo the equivalence relation of having the same value on M.

$\mathrm{pp}^n(\mathcal{X})$ is pp_R^n modulo the equivalence relation of having the same value on every member of $\mathcal{X}$ and this is equal to $\mathrm{pp}^n(M)$ for any M with $\langle M\rangle = \mathcal{X}$.

$\mathrm{pp}^n(X)$ (for X a closed subset of the Ziegler spectrum) is $\mathrm{pp}^n(\mathcal{X})$ if $\mathcal{X}$ is a definable subcategory with $\mathcal{X} \cap \mathrm{Zg}_R = X$.

$\mathbb{L}_R^{\mathrm{eq}+}$, $(\mathbb{L}_R^{\mathrm{eq}+})_M$, $(\mathbb{L}_R^{\mathrm{eq}+})_{\mathcal{X}}$ (also $\mathbb{L}^{\mathrm{eq}+}(\mathcal{X})$), $(\mathbb{L}_R^{\mathrm{eq}+})_X$ denote the categories of pp sorts and pp-definable maps for R-modules and then relativised respectively as above.

$\langle M\rangle$ is the definable subcategory generated by M.

$[\phi]_M$ denotes the the image of $\phi \in \mathrm{pp}_R^n$ in $\mathrm{pp}^n(M)$ (similarly with X or $\mathcal{X}$ in place of M).

$[\phi, \psi]$, $[\phi, \psi]_M$, $[\phi, \psi]_{\mathcal{X}}$, $[\phi, \psi]_X$ (with the assumption $\phi \geq \psi$) denote the interval in the lattice pp_R^n between ϕ and ψ and then the respective relativisations as above.

$\mathrm{Zg}(\mathcal{X})$ is $\mathcal{X} \cap \mathrm{Zg}_R$, equipped with the relative topology, when $\mathcal{X}$ is a definable subcategory of Mod-R.

Latt$(-)$ denotes the lattice of subobjects of $(-)$; the lattice of finitely presented, respectively finitely generated, subobjects is indicated by superscript ^{fg}, resp. ^{fp}, or, if finitely generated = finitely presented, just by ^{f}.

Mod(T) is the class of modules on which every pp-pair in T (a set of pp-pairs) is closed.

pinj_R denotes the class of indecomposable pure-injective right R-modules or, more usually, the set of isomorphism classes of these.

Pinj_R denotes the class (or full subcategory) of all pure-injective right R-modules.

proj-R denotes the class of finitely generated projective right R-modules.

Flat-R denotes the class of all flat right R-modules and R-Flat the class of all flat left R-modules; and similarly, lower case indicates "small" (indecomposable, finitely presented, finitely generated, as appropriate), upper case indicates all.

The choice of subscript $_R$ or notation like -R is not significant (thus $\mathrm{Inj}_R = \mathrm{Inj}\text{-}R = \mathrm{Inj}(\mathrm{Mod}\text{-}R)$). What is significant is the notational difference between proj-$\mathcal{A}$ (or $\mathrm{proj}_{\mathcal{A}}$) and proj($\mathcal{A}$): the former refers to projective right modules *over* the category $\mathcal{A}$, the latter to projective objects *in* the category $\mathcal{A}$.

I will write, for instance, "the set of indecomposable injectives" even though it is the isomorphism types of these which form a set and which is what I really mean.

These are various ways of writing the same thing: $\overline{a} \in \phi(M)$; $M \models \phi(\overline{a})$; $\overline{a}$ satisfies ϕ in M; $\phi(\overline{a})$ is true in M; $\overline{a}$ is a solution of/to ϕ in M. And similarly with a pp-type p in place of the pp condition ϕ.

$N|M$ means N is a direct summand of M.

Add($\mathcal{Y}$), respectively add($\mathcal{Y}$), denotes the closure of the class $\mathcal{Y}$ under direct summands of arbitrary, respectively finite, direct sums.

Here are four notations for the same object: $\mathbb{L}_R^{\mathrm{eq}+}$; $(\mathrm{mod}\text{-}R, \mathbf{Ab})^{\mathrm{fp}}$; $\mathrm{Ab}(R^{\mathrm{op}})$; fun-$R$. Defined as different objects, these small abelian categories turn out to the same (that is, equivalent categories).

Selected notations list

supp(M): support of M
$\mathrm{pp}^M(\overline{a})$: pp-type of $\overline{a}$ in M
$H(M)$: pure-injective hull of M
$H(\overline{a})$: hull of $\overline{a}$
$H(p)$: hull of pp-type p
$E(M)$: injective hull of M
$\langle M \rangle$: definable subcategory generated by M
$\langle \phi \rangle$: pp-type generated by ϕ
$D\phi$: (elementary) dual of pp condition ϕ
dF: dual of functor F
$(M, \overline{a})$: pointed module
$\phi(M)$: solution set of ϕ in M
R_M: ring of definable scalars of M
$(A, -)$: representable functor
$(\star)_{(-)}$: localisation of $\star$ at $-$
(ϕ/ψ), (F), (f): basic Ziegler-open sets
$[\phi/\psi]$ etc.: complement, $(\phi/\psi)^c$, of above
$(-)^{\mathrm{fp}}$: subcategory of finitely presented presented objects
$\overrightarrow{F}$: extension of F to functor commuting with direct limits
$\mathbb{L}_R^{\mathrm{eq}+}$: category of imaginaries/pp sorts and maps
Zg_R, $\mathrm{Zg}(\mathcal{D})$: Ziegler spectrum
Zar_R: rep-Zariski spectrum
(L)Def_R: sheaf/presheaf of definable scalars
pp_R^n: lattice of pp conditions
$\mathrm{Latt}^{\mathrm{f}}$: lattice of "finite" subobjects
fun-R and $\mathrm{fun}^{\mathrm{d}}$-$R$; functor category and dual functor category
w($-$): width
mdim($-$); m-dimension
KGdim($-$), KG($-$): Krull–Gabriel dimension
Udim($-$), UD($-$): uniserial dimension
CB($-$): Cantor–Bendixson rank

Part I

Modules

1

Pp conditions

We start by introducing some elementary but possibly unfamiliar concepts: pp conditions and pp-types. These are used throughout the book. The link between pp conditions and functors is mentioned, though briefly; much more will be made of this later. The main theme of this chapter is the link between pp conditions and finitely presented modules. Free realisations of pp conditions are introduced. Elementary duality, which links pp conditions for right and left modules, is defined. This duality will be one of our main tools.

1.1 Pp conditions

After defining pp conditions and giving examples in Section 1.1.1 it is noted in Section 1.1.2 that these define functors from the category of modules to that of abelian groups. There is a potentially checkable criterion, 1.1.13, for implication of pp conditions, that is, for inclusion of the corresponding functors. In Section 1.1.3 we see that the set of pp conditions in a given number of free variables forms a modular lattice.

1.1.1 Pp-definable subgroups of modules

This section introduces pp (positive primitive) conditions and the sets they define in modules. The concept is illustrated by a variety of examples.

Consider a finite homogeneous system of R-linear equations:

$$\sum_{i=1}^{n} x_i r_{ij} = 0 \qquad j = 1, \ldots, m.$$

Here the x_i are variables, the r_{ij} are elements of a given ring R and this is a system of equations for *right* R-modules.

We also write this as

$$\bigwedge_{j=1}^{m} \sum_{i=1}^{n} x_i r_{ij} = 0.$$

The symbol $\bigwedge$ is taken from mathematical logic and denotes conjunction, that is, "and". This system of equations may be regarded as a single condition, $\theta(\overline{x})$ say, on the tuple $\overline{x} = (x_1, \ldots, x_n)$ of variables.

The **solution set** to $\theta(\overline{x})$ in any right R-module M is

$$\theta(M) = \left\{ \overline{a} = (a_1, \ldots, a_n) \in M^n : \bigwedge_{j=1}^{m} \sum_{i=1}^{n} a_i r_{ij} = 0 \right\}.$$

This is an $\mathrm{End}(M)$-submodule of M^n, where the action of $\mathrm{End}(M)$ on M^n is the diagonal one: $f\overline{a} = (fa_1, \ldots, fa_n)$ for $f \in \mathrm{End}(M)$ and $\overline{a} \in M^n$. It is not in general an R-submodule of M^n.

The simplest examples of such conditions θ are those of the form $xr = 0$ for some $r \in R$. In this case $\theta(M) = \{a \in M : ar = 0\} = \mathrm{ann}_M(r)$, the **annihilator** of r **in** M. Indeed any condition of the type above may be regarded as the generalised annihilator condition $\overline{x}H = 0$, where H is the $n \times m$ matrix $(r_{ij})_{ij}$. Then $\theta(M)$ is just the kernel of the map, $\overline{x} \mapsto \overline{x}H$ from M^n to M^m which is defined by right multiplication by the matrix H; a map of left $\mathrm{End}(M)$-modules.

Such annihilator-type conditions will not be enough: we close under projections to obtain generalised divisibility conditions.

Thus, given a subgroup $\theta(M)$ as defined above, consider its image under projection to the first k, say, coordinates:

$$\left\{ \overline{a} = (a_1, \ldots, a_k) \in M^k : \exists a_{k+1}, \ldots, a_n \in M \text{ such that } \bigwedge_{j=1}^{m} \sum_{i=1}^{n} a_i r_{ij} = 0 \right\}.$$

The condition $\phi(\overline{x}) = \phi(x_1, \ldots, x_k)$, which is such that this projection of $\theta(M)$ is exactly its **solution set**, $\phi(M)$, in M, can be written

$$\exists x_{k+1}, \ldots, x_n \bigwedge_{j=1}^{m} \sum_{i=1}^{n} x_i r_{ij} = 0.$$

This can be abbreviated as

$$\exists \overline{x}' \theta(\overline{x}, \overline{x}') \quad \text{where } \overline{x} = (x_1, \ldots, x_k) \quad \text{and} \quad \overline{x}' = (x_{k+1}, \ldots, x_n).$$

Any condition ϕ of this form is a **pp condition** and any subgroup of M^k of the form $\phi(M)$ is said to be a **pp-definable subgroup** of M or, more accurately, a **subgroup of M^k pp-definable** in M. The terminology "pp", an abbreviation of "positive primitive", is from logic and refers to the formal shape of the condition. The terms **finitely matrizable subgroup** and **subgroup of finite definition** are also used, following Zimmermann, respectively Gruson and Jensen, for what is here called a pp-definable subgroup. Note that $\phi(M)$ is a submodule of ${}_{\mathrm{End}(M)}M^k$, where $\mathrm{End}(M)$ has the diagonal action.

In the above condition ϕ the variables $x_1, \ldots, x_k$ are said to be **free** (to be replaced by values) whereas $x_{k+1}, \ldots, x_n$ are **bound** (by the existential quantifier). We write ϕ or $\phi(x_1, \ldots, x_k)$ depending on whether or not we wish to display the free variables. A condition like θ with no bound variables, that is, a system of linear equations, is a **quantifier-free** pp condition.

The simplest examples of pp conditions which are not annihilator conditions are divisibility conditions of the form $\exists y\,(x = yr)$. The solution set in M to this condition is $Mr = \{mr : m \in M\}$. Any pp condition can be expressed as a generalised divisibility condition: if $\phi(\overline{x})$ is $\exists \overline{y}\,(\overline{x}\,\overline{y})H = 0$, where $(\overline{x}\,\overline{y})$ should be read as the row vector with entries those of $\overline{x}$ followed by those of $\overline{y}$, then, writing H as a block matrix $H = \begin{pmatrix} A \\ -B \end{pmatrix}$, this condition may be rewritten as $\exists \overline{y}(\overline{x}A = \overline{y}B)$, and may be read as $B \mid \overline{x}A$, ("B divides $\overline{x}A$").

Examples 1.1.1. (a) Let $R = \mathbb{Z}$ and let p be a non-zero prime.

Let $M = \mathbb{Z}_p^{(I)}$ be an elementary abelian p-group, where the index set I may be infinite and where $\mathbb{Z}_p$ denotes the group with p elements. The condition $x = x$, which we write in preference to the equivalent $x0 = 0$, defines all of M and the condition $x = 0$ defines the zero subgroup. There are no more pp-definable subgroups because, for any two non-zero elements a, b of M there is $f \in \mathrm{End}(M)$ such that $fa = b$.

If, instead, we take $M = \mathbb{Z}_{p^n}^{(I)}$, for some n, then the pp-definable subgroups of M will be $M > Mp = \mathrm{ann}_M(p^{n-1}) > \cdots > Mp^{n-1} = \mathrm{ann}_M(p) > 0$. Clearly these are pp-definable, by both annihilation and divisibility conditions, and there are no more pp-definable subgroups because there are no more $\mathrm{End}(M)$-submodules.

Taking $M = \mathbb{Z}_{p^2} \oplus \mathbb{Z}_{p^3}$, again we obtain a chain

$$M > \mathbb{Z}_{p^2} \oplus p\mathbb{Z}_{p^3} > p\mathbb{Z}_{p^2} \oplus p^2\mathbb{Z}_{p^3} > p\mathbb{Z}_{p^2} \oplus p\mathbb{Z}_{p^3} > 0 \oplus p^2\mathbb{Z}_{p^3} > 0.$$

One may compute the cyclic $\mathrm{End}(M)$-submodules to see that there are no more pp-definable subgroups.[1]

Take $M = \mathbb{Z}_2 \oplus \mathbb{Z}_3$ for an example where the pp-definable subgroups do not form a chain.

[1] In contrast to the erroneous diagram at [495, p. 22].

In all those examples the pp-definable subgroups were exactly the End(M)-submodules (4.4.25 says why) but that is not general nor even typical. For instance, a left coherent ring R which is not left noetherian has, by 2.3.19, left ideals which are not pp-definable as subgroups of the ring regarded as a right module, R_R, over itself.

(b) If R is any ring and $L = \sum_1^n Rr_i$ is a finitely generated left ideal of R, then L is a pp-definable subgroup of R_R. A defining condition for L is $\exists y_1, \ldots, y_n\, (x = \sum_1^n y_i r_i)$, for which we may alternatively use matrix notation,

$$\exists \overline{y}\, (x\ \overline{y}) \begin{pmatrix} 1 \\ -r_1 \\ \vdots \\ -r_n \end{pmatrix} = 0,$$

or "dot-product" notation, $\exists \overline{y}\, (x = \overline{y} \cdot \overline{r})$.

(c) Let $R = k\langle X, Y : YX - XY = 1\rangle$ be the first Weyl algebra over a field, k, of characteristic zero. Because R is a Dedekind (though not commutative) domain ([441, 5.2.11]), it is the case, see 2.4.10 and 2.4.15, that pp conditions for R-modules are simple combinations of basic annihilation and divisibility conditions.

Since k is in the centre of R, multiplication by any element of k on an R-module M is an R-endomorphism of M. Therefore every pp-definable subgroup of M is a vector space over k.

Every non-zero R-module is infinite-dimensional as a vector space over k: if M is finite-dimensional, with k-basis $a_1, \ldots, a_n$ say, then $\mathrm{ann}_R(M) = \bigcap_1^n \mathrm{ann}_R(a_i)$, is a co-finite-dimensional two-sided ideal of R so, since this ring is simple ([441, 1.3.5]), equals R. It will be shown, 8.2.28, that if M is a finitely generated module over this ring, then every pp-definable subgroup of M is either finite-dimensional or co-finite-dimensional. If, further, M is simple and End(M) $= k$, for example, if M is simple and k is algebraically closed (see 8.2.27(3)), then every finite-dimensional subspace of M is pp-definable. But there will be co-finite-dimensional subspaces of M which are not pp-definable: for example, if k, hence R, is countable, then there are only countably many pp conditions but there are uncountably many co-finite-dimensional subspaces.

Similar remarks apply to Verma modules over $U(\mathrm{sl}_2(k))$, where k is an algebraically closed field of characteristic 0 (§8.2.4).

(d) Over a von Neumann regular ring every pp condition is equivalent to one which is quantifier-free, that is, to a system of linear equations, and this property characterises these rings (2.3.24).

It is easy to produce pp conditions more complicated than those above.

Example 1.1.2. Let k be a field and let R be the string algebra (Section 8.1.3) $k[a, b : ab = 0 = ba = a^2 = b^3]$, the k-path algebra of the Gelfand–Ponomarev quiver $\mathrm{GP}_{2,3}$ (p. 584). Take M to be the string module shown.

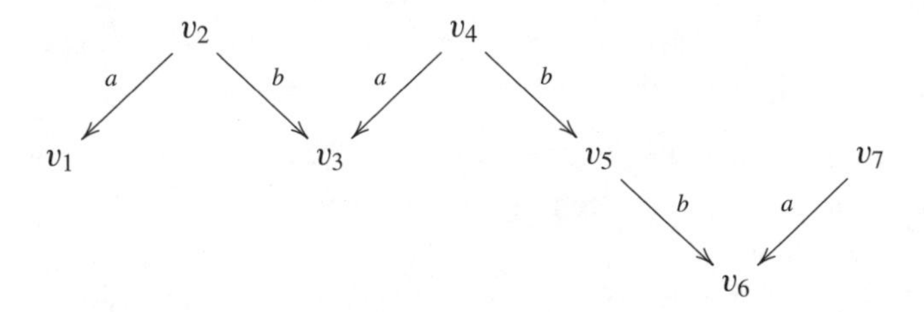

Thus, as a k-vectorspace, M has basis $v_1, \ldots, v_7$ and the actions of a and b are determined by their actions on the v_i, which are as shown. The convention is that the actions not shown are 0. So $v_7 a = v_6$ and $v_7 b = 0$.

There is a natural "pp-description" $\phi(x)$ of the element $v_1 \in M$, namely

$$xa = 0 \wedge \exists y\, (x = ya \wedge \exists z\, (yb = za \wedge \exists w\, (zb^2 = wa \wedge wb = 0))),$$

which rearranges to standard form as

$$\exists y, z, w\, (xa = 0 \wedge x = ya \wedge yb = za \wedge zb^2 = w \wedge wb = 0).$$

Clearly $v_1 \in \phi(M)$ but note that v_6 also satisfies the condition ϕ: directly or by 1.1.7, having noted that $v_2 \mapsto v_7$, $v_4 \mapsto 0$, $v_7 \mapsto 0$ defines an endomorphism of M taking v_1 to v_6. One may check that $\phi(M) = v_1 k \oplus v_6 k$. This subspace, which is $\ker(b^2) \cdot a$, can therefore be defined more simply by the pp condition $\exists y\, (x = ya \wedge yb^2 = 0)$ but one may produce arbitrarily complicated examples along these lines.

Example 1.1.3. Let Q be the quiver $\widetilde{D_4}$ which is as shown.

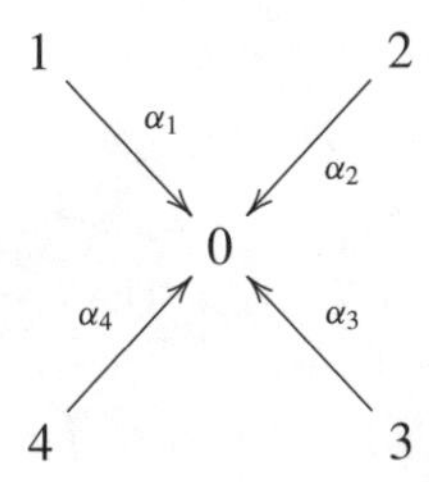

Take R to be any ring and let M be an R-representation (Section 15.1.1) of $\widetilde{D_4}$. Thus M is given by replacing each vertex by an R-module and each arrow by an R-linear map; M may be regarded as a module over the path algebra $R\widetilde{D_4}$. The pp-definable subgroups of M include:

each "component" Me_i, where e_i is the idempotent of the path algebra corresponding to vertex i;

the image of each arrow $\mathrm{im}(\alpha_i) = M\alpha_i$ (α_i may be regarded as an element of $R\widetilde{D_4}$);

any R-submodule of Me_0 obtained from these four images *via* intersection and sum. There is quite a variety of these.

For example, suppose that k is a field, that V is a k-vector space and that f is an endomorphism of V. We can "code up" the structure (V, f) in a k-representation, M_f, of $\widetilde{D_4}$ as follows.

Set $V_0 = V \oplus V$ and let $\alpha_1, \ldots, \alpha_4$ be, respectively, the inclusions of the following subspaces: $V \oplus 0$, $0 \oplus V$, $\Delta = \{(a, a) : a \in V\}$, $\mathrm{Gr}(f) = \{(a, fa) : a \in V\}$. Consider the following pp condition $\rho(x, y)$ with free variables x, y:

$$x = xe_1 \wedge y = ye_1 \wedge \exists u_2, u_4, u_3\, (u_2 = u_2e_2 \wedge u_4 = u_4e_4 \wedge x\alpha_1 + u_2\alpha_2 \\ = u_4\alpha_4 \wedge u_3 = u_3e_3 \wedge y\alpha_1 + u_2\alpha_2 = u_3\alpha_3).$$

Unwinding this condition, one can see that $\rho(a, b)$ holds, that is, $(a, b) \in \rho(M)$, iff $a \in V$ (V identified with $V \oplus 0$) and $b = fa$.

Thus, for example, $\exists x \rho(x, y)$ and $\rho(x, 0)$ define, respectively, the image and kernel of f as subspaces of $V \oplus 0$. The pp condition $\rho_2(x, y)$ which is $\exists z\, (\rho(x, z) \wedge \rho(z, y))$ defines the graph of the action of f^2 on V, etc.

If, in place of $\widetilde{D_4}$ we were to take the "5-subspace quiver" which has an additional vertex and an arrow pointing from it to vertex 0, then we could code up[2] any additional endomorphism, g, of V via its graph as with f above. Thus pp conditions may be used to define the action of any (non-commutative) polynomial in f and g. Therefore the set of pp-definable subgroups is, in the case of this 5-subspace quiver, "wild".

Example 1.1.4. Let $\widetilde{A_1}$ be the quiver shown, the **Kronecker quiver**

$$1 \underset{\beta}{\overset{\alpha}{\rightrightarrows}} 2$$

and let M be any representation (over k). The subgroup, conveniently (and correctly if one thinks in terms of representations rather than modules) denoted $M\beta\alpha^{-1}$, that is, $\{a \in Me_1 : \alpha(a) \in \mathrm{im}(\beta)\}$, is pp-definable. Similarly $M\beta\alpha^{-1}\beta$, $M\beta\alpha^{-1}\beta\alpha^{-1}, \ldots$ are pp-definable subgroups of M. For each $\lambda \in k$, so is the subgroup $\{a \in Me_1 : \beta(a) = \lambda\alpha(a)\}$.

[2] More precisely, interpret, in the sense of Section 18.2.1.

Example 1.1.5. Let D be a division ring. If M is any D-module, then End(M) acts transitively on the non-zero elements of M, so the only pp-definable subgroups of M are 0 and M.

Sometimes we will use the following notation from model theory: if χ is a condition with free variables $\overline{x}$, we write $\chi(\overline{x})$ if we wish to display these variables, and if $\overline{a}$ is a tuple of elements from the module M, then the notation $M \models \chi(\overline{a})$, read as "$M$ satisfies $\chi(\overline{a})$" or "$\overline{a}$ satisfies the condition χ in M", means $\overline{a} \in \chi(M)$, where $\chi(M)$ denotes the solution set of χ in M. We could dispense with this notation but much of the relevant literature makes use of it and we find it an efficient notation.[3]

Proposition 1.1.6. ([731, 6.5]) *Suppose that R is a local ring with (Jacobson) radical J and suppose that $J^2 = 0$. Then the pp-definable subgroups of R_R are R, J and the left ideals of finite length (that is, of finite dimension over the division ring R/J).*

Proof (in part). All the left ideals listed are pp-definable: if r is any non-zero element of J, then J is defined by $xr = 0$ and if $L \leq {}_RR$ is of finite length, generated by $r_1, \dots, r_n$ say, then L is defined by the pp condition $\exists y_1, \dots, y_n \, (x = \sum_1^n y_i r_i)$.

Suppose, for a contradiction, that some infinitely generated left ideal L were pp-definable, say $L = \phi(R)$. Zimmermann, see [731], deals with this by working directly with the matrices involved in the pp condition. Here we give an alternative proof in the case that the radical is split, i.e., $R = J + D \cdot 1$, where $D = R/J$. In that case if $(s_\lambda)_{\lambda \in \Lambda}$ is a D-basis for J, then $X_\lambda \mapsto s_\lambda$ induces an isomorphism $D[X_\lambda \, (\lambda \in \Lambda)] \, / \, \langle (X_\lambda X_\mu)_{\lambda,\mu} \rangle \simeq R$, where $\langle (X_\lambda X_\mu)_{\lambda,\mu} \rangle$ denotes the ideal generated by all the products $X_\lambda X_\mu$. Let σ be any automorphism of the D-vectorspace J and let $\alpha_\sigma : R \to R$ be the map induced by the endomorphism of the polynomial ring which takes X_λ to the linear combination of the X_μ with image σs_λ: clearly α_σ is an automorphism of the ring R. Applying α_σ to the elements of R appearing in the pp condition ϕ gives a pp condition, ϕ^σ say, and clearly $\phi^\sigma(R) = \alpha_\sigma(L)$.

Suppose now that R is countable. Since L is infinite-dimensional it has uncountably many images, $\alpha_\sigma(L)$, as σ varies. But, because R is countable, there are only countably many pp conditions ϕ^σ: a contradiction.

For the general case suppose that $\phi(x)$ is $\exists \overline{y} \, \theta(x, \overline{y})$, where θ is a quantifier-free pp condition (a system of linear equations). Choose a countable subset, T, of L which is linearly independent over D. For each $t \in T$ choose $\overline{s} = (s_1^t, \dots, s_n^t)$ from R such that $\theta(t, \overline{s})$ holds, that is, $(t, \overline{s}) \in \theta(R)$. Let $\Lambda' \subseteq \Lambda$ be such that all

[3] Under our notational conventions, writing $\chi(\overline{a})$ already implies that the length of $\overline{a}$ equals that of $\overline{x}$.

$t \in T$ and all corresponding s_i^t lie in the subring of R generated by 1 together with the s_λ with $\lambda \in \Lambda'$. If necessary put an additional λ with $X_\lambda \notin L$ into Λ' Let D' be a countable sub-division-ring of D such that every $d \in D$ appearing in ϕ, or in any expression of a $t \in T$ or an s_i^t as a linear combination of the s_λ, is in D'. Then $R' = D'[X_\lambda(\lambda \in \Lambda')]/\langle (X_\lambda X_\mu)_{\lambda,\mu}\rangle$ is a countable local ring with radical squared 0 and, by construction, $\phi(R') = L \cap R' < J'$, giving a contradiction.

There is generalisation of this at 4.4.15. □

1.1.2 The functor corresponding to a pp condition

In this section a criterion (1.1.13, 1.1.17) for one pp condition to be stronger than another is established.

Each pp condition ϕ defines a functor, F_ϕ, from the category, Mod-R, of right R-modules to the category, **Ab**, of abelian groups. On objects the action of F_ϕ is $M \mapsto \phi(M)$. If $f : M \to N$ is a morphism in Mod-R, then $F_\phi f$ is defined to be the restriction/corestriction of f to a map from $\phi(M)$ to $\phi(N)$; this is well defined by the following lemma.

Lemma 1.1.7. *If $f : M \to N$ is a morphism and ϕ is a pp condition, then $f(\phi(M)) \leq \phi(N)$.*

Proof. The condition $\phi(\overline{x})$ has the form $\exists \overline{y}\, \theta(\overline{x}, \overline{y})$, where θ is a finite system of R-linear equations. Let $\overline{a} \in \phi(M)$, so there is a tuple $\overline{b}$ from M such that $(\overline{a}, \overline{b}) \in \theta(M)$. Note that f preserves solutions of R-linear equations: $f(\sum_i a_i r_i + \sum_k b_k s_k) = \sum_i f(a_i) r_i + \sum_k f(b_k) s_k$. So $(f\overline{a}, f\overline{b}) \in \theta(N)$, hence $f\overline{a} \in \phi(N)$, as required. □

If the pp condition ϕ has free variables $\overline{x} = (x_1, \ldots, x_n)$, then, observing that $\phi(M) \leq M^n$, we see that F_ϕ is a subfunctor of the nth power of the forgetful functor from Mod-R to **Ab**.

It is an important property of these pp-defined functors that they commute with direct limits, 1.2.31, as well as products, 1.2.3. The full story is given by 10.2.30 (and 18.1.19).

Corollary 1.1.8. *If ϕ is a pp condition and M is any module, then $\phi(M)$ is closed under the (diagonal) action of* End(M)*, that is, $\phi(M)$ is a submodule of* ${}_{\mathrm{End}(M)}M^n$*, where ϕ has n free variables.*

Corollary 1.1.9. *If ϕ is a pp condition with one free variable, then $\phi(R_R)$ is a left ideal of R.*

Corollary 1.1.10. *If M is any module and ϕ is a pp condition with n free variables, then $M \cdot \phi(R) \leq \phi(M) \leq M^n$.*

Proof. By $M \cdot \phi(R)$ is meant $\{\sum_{j=1}^m a_j \overline{r}_j : a_j \in M, \overline{r}_j \in \phi(R), m \geq 1\}$. This is generated by its subgroups $a \cdot \phi(R)$ for $a \in M$, and these are the images of $\phi(R)$ under the morphisms $(r_1, \ldots, r_n) \mapsto (ar_1, \ldots, ar_n)$ from R^n to M^n. □

Example 1.1.11. Suppose that C is a finitely presented module, say $C = \sum_1^n c_i R$ with defining relations $\sum_1^n c_i r_{ij} = 0$, $j = 1, \ldots, m$. That is, there is an exact sequence $R^m \to R^n \to C \to 0$ where the map between the free modules is given by left multiplication, on column vectors, by the matrix $(r_{ij})_{ij}$.

Define θ to be the quantifier-free pp condition $\bigwedge_1^m \sum_1^n x_i r_{ij} = 0$. Then $F_\theta \simeq \mathrm{Hom}_R(C, -)$, as functors from Mod-$R$ to **Ab**. For, if M is any module, then a morphism $f : C \to M$ is determined by the images, $fc_1, \ldots, fc_n$, of $c_1, \ldots, c_n$, and the tuple $(fc_1, \ldots, fc_n)$ satisfies all the equations $\sum_1^n x_i r_{ij} = 0$. Conversely, any n-tuple, $(a_1, \ldots, a_n)$, of elements of M which satisfies θ determines, by sending c_i to a_i, a morphism from C to M. So the functors F_θ and $\mathrm{Hom}_R(C, -)$ (usually we write just $\mathrm{Hom}(C, -)$ or even $(C, -)$) agree on objects and it is easy to see that they agree on morphisms, hence are isomorphic by the natural[4] transformation $\eta : \mathrm{Hom}(C, -) \to F_\theta$ defined by $\eta_M : f \in \mathrm{Hom}(C, M) \mapsto (fc_1, \ldots, fc_n) \in M^n$ at each module M.

Since θ was an arbitrary quantifier-free pp condition this shows that quantifier-free pp conditions correspond exactly to representable functors. A more precise statement is at 10.2.34.

Example 1.1.12. Let $R = \mathbb{Z}$. The torsion functor, $M \mapsto \tau M = \{a \in M : an = 0 \text{ for some } n \in \mathbb{Z}, n \neq 0\}$ is an infinite sum of pp-defined subfunctors of the forgetful functor, namely the $M \mapsto \mathrm{ann}_M(n)$ for $n \in \mathbb{Z}, n \neq 0$, and plausibly is not itself defined by a single pp condition. Indeed, it is not isomorphic to a pp-defined functor: various proofs are possible; for example, if $\tau(-)$ were of the form $F_\phi(-)$, then the class of torsion groups would be definable (by the pair $(x = x)/\phi(x)$) in the sense of Section 3.4.1 but that is clearly in contradiction to 3.4.7.

The next result gives the condition, in terms of the defining matrices, for one pp condition to imply another and hence for two pp conditions to be equivalent in the sense of having identical solution sets in every module (that is, they define the same functor). Our approach follows [564, p. 124 ff.], see also [497, p. 187].

If ϕ and ψ are pp conditions such that $\psi(M) \leq \phi(M)$ in every module M, then we say that ψ **implies** or **is stronger than** ϕ and write $\psi \to \phi$ or, more often,

[4] Once a generating tuple for C has been chosen.

$\psi \le \phi$ (reflecting the ordering of the solution sets). It will be seen later, 1.2.23, that it is enough to check this in every finitely presented module.

We write $\phi \equiv \psi$ and say that these pp conditions are **equivalent** if $\phi \ge \psi$ and $\phi \le \psi$. In all these definitions ϕ and ψ are assumed to have the same number of free variables.[5] By a **pp-pair** ϕ/ψ we mean a pair of pp conditions with $\phi \ge \psi$.

It was observed earlier that every pp condition with free variables $\overline{x}$ can be written in the form $B \mid \overline{x}A$, that is, $\exists \overline{y}\,(\overline{y}B = \overline{x}A)$, for some matrices A, B. The following implications between such divisibility conditions are immediate:

$B \mid \overline{x}A \Rightarrow BC \mid \overline{x}AC$ for any matrix C;
$B \mid \overline{x}A \Rightarrow B_0 \mid \overline{x}A$ if $B = B_1B_0$ for some matrices B_1, B_0;
$B \mid \overline{x}A \Rightarrow B \mid \overline{x}D$ if $A = A_0B + D$ for some matrices A_0, D.

A consequence of the proof of the next lemma is that every implication between pp conditions is obtained by a sequence (of length no more than three) of such implications.

Lemma 1.1.13. ([495, 8.10]) *Let $\phi(\overline{x})$, being $\exists \overline{y}\,(\overline{x}\,\overline{y})H_\phi = 0$, and $\psi(\overline{x})$, being $\exists \overline{z}\,(\overline{x}\,\overline{z})H_\psi = 0$, be pp conditions. Then $\psi \le \phi$ iff there are matrices $G = \begin{pmatrix} G' \\ G'' \end{pmatrix}$ and K such that $\begin{pmatrix} I & G' \\ 0 & G'' \end{pmatrix} H_\phi = H_\psi K$, where I is the $n \times n$ identity matrix, n being the length of $\overline{x}$, and 0 denotes a zero matrix with n columns.*

Proof. Suppose that ψ is $B' \mid \overline{x}A'$ and ϕ is $B \mid \overline{x}A$, so $H_\psi = \begin{pmatrix} A' \\ -B' \end{pmatrix}$ and $H_\phi = \begin{pmatrix} A \\ -B \end{pmatrix}$.

($\Leftarrow$) With an obvious notation, let us write the three types of immediate implication seen above as, respectively, $\begin{pmatrix} A \\ -B \end{pmatrix} \Rightarrow \begin{pmatrix} AC \\ -BC \end{pmatrix}$, $\begin{pmatrix} A \\ -B_1B_0 \end{pmatrix} \Rightarrow \begin{pmatrix} A \\ -B_0 \end{pmatrix}$, $\begin{pmatrix} A_0B + D \\ -B \end{pmatrix} \Rightarrow \begin{pmatrix} D \\ -B \end{pmatrix}$. Suppose that we have the matrix equation in the statement of the result: so $A - G'B = A'K$ and $-G''B = -B'K$. Then $\begin{pmatrix} A' \\ -B' \end{pmatrix} \Rightarrow \begin{pmatrix} A'K \\ -B'K \end{pmatrix}$ (first type) $= \begin{pmatrix} A - G'B \\ -G''B \end{pmatrix} \Rightarrow \begin{pmatrix} A - G'B \\ -B \end{pmatrix}$ (second type) $\Rightarrow \begin{pmatrix} A \\ -B \end{pmatrix}$ (third type), as required.

($\Rightarrow$) Suppose that the matrices A', B' are $n \times m$ and $l \times m$ respectively. Let M be the module freely generated by elements $x_1, \dots, x_n, y_1, \dots, y_l$ subject to the

[5] Rather, to be in the same number of free variables: see the footnote 3 of Chapter 3.

relations $(\overline{x}\,\overline{y})\begin{pmatrix} A' \\ -B' \end{pmatrix} = 0$. That is, M is the cokernel of the morphism from R^m to R^{n+l} given by $\overline{z} \mapsto \begin{pmatrix} A'\overline{z} \\ -B'\overline{z} \end{pmatrix}$ (writing elements of these free modules as column vectors). Let e_i be the ith unit element in R^{n+l} and denote its image in M by a_i if $i \le n$ and by b_i if $n+1 \le i \le n+l$. Clearly $(\overline{a}\,\overline{b})\begin{pmatrix} A' \\ -B' \end{pmatrix} = 0$, so $\overline{a}$ satisfies the condition ψ in M and hence, by assumption, it satisfies ϕ. So there is $\overline{d}$ in M with $(\overline{a}\,\overline{d})\begin{pmatrix} A \\ -B \end{pmatrix} = 0$. Since $\overline{d}$ is from M and $\overline{a}\overline{b}$ generates M, there are matrices G', G'' such that $\overline{d} = (\overline{a}\,\overline{b})\begin{pmatrix} G' \\ G'' \end{pmatrix}$. Therefore $(\overline{a}\,\overline{b})\begin{pmatrix} I & G' \\ 0 & G'' \end{pmatrix}\begin{pmatrix} A \\ -B \end{pmatrix} = 0$ and so, by construction of M, the matrix $\begin{pmatrix} I & G' \\ 0 & G'' \end{pmatrix}\begin{pmatrix} A \\ -B \end{pmatrix}$ belongs to the submodule of R^{n+l} generated by $\begin{pmatrix} A' \\ -B' \end{pmatrix}$ (regarding the latter as a sequence of column vectors). That is, there is a matrix K such that $\begin{pmatrix} I & G' \\ 0 & G'' \end{pmatrix}\begin{pmatrix} A \\ -B \end{pmatrix} = \begin{pmatrix} A' \\ -B' \end{pmatrix} K$, as required. □

Note that the sizes of G and K in the statement of 1.1.13 are determined by those of H_ϕ and H_ψ.

Here are the three basic implications exhibited in matrix form, as in the lemma:

$$\begin{pmatrix} A \\ -B \end{pmatrix} \Rightarrow \begin{pmatrix} AC \\ -BC \end{pmatrix} \text{ since } \begin{pmatrix} AC \\ -BC \end{pmatrix} = \begin{pmatrix} A \\ -B \end{pmatrix} C;$$

$$\begin{pmatrix} A \\ -B_0B_1 \end{pmatrix} \Rightarrow \begin{pmatrix} A \\ -B_1 \end{pmatrix} \text{ since } \begin{pmatrix} I & 0 \\ 0 & B_0 \end{pmatrix}\begin{pmatrix} A \\ -B_1 \end{pmatrix} = \begin{pmatrix} A \\ -B_0B_1 \end{pmatrix};$$

$$\begin{pmatrix} A_0B + D \\ -B \end{pmatrix} \Rightarrow \begin{pmatrix} D \\ -B \end{pmatrix} \text{ since } \begin{pmatrix} I & -A_0 \\ 0 & I \end{pmatrix}\begin{pmatrix} D \\ -B \end{pmatrix} = \begin{pmatrix} A_0B + D \\ -B \end{pmatrix}.$$

Example 1.1.14. Suppose that $\phi(x)$ is $r \mid x$, that is, $\exists y \begin{pmatrix} x & y \end{pmatrix}\begin{pmatrix} 1 \\ r \end{pmatrix} = 0$ and that $\psi(x)$ is $s \mid x$, that is, $\exists y \begin{pmatrix} x & y \end{pmatrix}\begin{pmatrix} 1 \\ s \end{pmatrix} = 0$, for some elements $r, s \in R$. Then, by 1.1.13, $\psi \le \phi$ iff there are matrices K and $G = \begin{pmatrix} G' \\ G'' \end{pmatrix}$ such that $\begin{pmatrix} 1 & G' \\ 0 & G'' \end{pmatrix}\begin{pmatrix} 1 \\ r \end{pmatrix} = \begin{pmatrix} 1 \\ s \end{pmatrix} K$. Clearly G must be 2×1 and K is 1×1, so this equation is actually $\begin{pmatrix} 1 & g' \\ 0 & g'' \end{pmatrix}\begin{pmatrix} 1 \\ r \end{pmatrix} = \begin{pmatrix} 1 \\ s \end{pmatrix}(k)$. Therefore $\psi \leq \phi$ iff there are k, g' and g'' in R such that $k = 1 + g'r$ and $g''r = sk$. It is easy to check that this is equivalent to there being $t \in R$ such that $tr = s$, in other words, to the condition that s is in the left ideal generated by r.

This gave an answer which was obvious anyway. A somewhat less obvious example follows.

Example 1.1.15. Take ϕ to be $\exists y\,(x = yr \wedge ys = 0)$, i.e. $\exists y\,\begin{pmatrix} x & y \end{pmatrix}\begin{pmatrix} 1 & 0 \\ r & s \end{pmatrix} = 0$, and ψ to be $\exists y\,(x = yr' \wedge ys' = 0)$, i.e. $\exists y\,\begin{pmatrix} x & y \end{pmatrix}\begin{pmatrix} 1 & 0 \\ r' & s' \end{pmatrix} = 0$.

Then $\psi \leq \phi$ iff there exist $g', g'', k_{ij} \in R$ such that

$$\begin{pmatrix} 1 & g' \\ 0 & g'' \end{pmatrix}\begin{pmatrix} 1 & 0 \\ r & s \end{pmatrix} = \begin{pmatrix} 1 & 0 \\ r' & s' \end{pmatrix}\begin{pmatrix} k_{11} & k_{12} \\ k_{21} & k_{22} \end{pmatrix},$$

that is,

$$\begin{pmatrix} 1 + g'r & g's \\ g''r & g''s \end{pmatrix} = \begin{pmatrix} k_{11} & k_{12} \\ r'k_{11} + s'k_{21} & r'k_{12} + s'k_{22} \end{pmatrix}.$$

So $k_{11} = 1 + g'r, k_{12} = g's$ and the conditions reduce to: there exist g', g'', k_{21}, k_{22} such that $g''r = r' + r'g'r + s'k_{21}$ and $g''s = r'g's + s'k_{22}$, that is, $g'' - r' - r'g'r \in s'R$ (which implies $r' \in s'R + Rr$) and $g''s - r'g's \in s'R$.

Since g'' is arbitrary, the $-rg'$ can be dropped, so the conditions are equivalent to: there exists g'' such that $g''r - r' \in s'R$ and $g''s \in s'R$. In other words, in $R/s'R$ there is an element c $(= g'' + s'R)$ such that $cs = 0$ and $cr = r' + s'R$. That is, $r' + s'R \in \big(\text{ann}_{(R/s'R)}(s)\big)r$.

Thus $\psi \leq \phi$ iff the image of r' in $R/s'R$ lies in $(\text{ann}_{R'/s'R}(s)) \cdot r$.

Let us check directly that this condition is sufficient: say $t \in R$ with $ts \in s'R$ and $tr - r' \in s'R$ and suppose $a \in M$ satisfies ψ, say $b \in M$ is such that $a = br'$ and $bs' = 0$. Then $r' = tr + s'u$ for some $u \in R$, so $a = br' = btr + bs'u = bt \cdot r$ and $bt \cdot s = bs'u'$ (for some $u' \in R) = 0$. Therefore a does satisfy ϕ in M.

It is perhaps in finding a criterion and in proving necessity that 1.1.13 is most useful.

Rephrased, and with a slight notational change, 1.1.13 becomes the following.

Corollary 1.1.16. *The implication $B' \mid \overline{x}A' \leq B \mid \overline{x}A$ holds iff there are matrices G, H, K such that $A = A'K + GB$ and $HB = B'K$.*

The following is a criterion for a conjunction of pp conditions to imply a pp condition, that is, for $\phi_1 \wedge \cdots \wedge \phi_t \leq \phi$ (the notation "$\wedge$" is explained in the next section).

Corollary 1.1.17. ([564, 10.4]) *The implication $\bigwedge_1^t (B_i \mid \overline{x}A_i) \leq B \mid \overline{x}A$ between pp conditions holds iff there are matrices G, G_i, K_i $(i = 1, \ldots, t)$ such that $A = \sum_1^t A_i K_i + GB$ and $G_i B = B_i K_i$ $(i = 1, \ldots, t)$.*

Proof. Write the pp condition $\bigwedge_1^t B_i \mid \overline{x}A_i$ as

$$\exists \overline{y}_1, \ldots, \overline{y}_t \begin{pmatrix} \overline{x} & \overline{y}_1 & \ldots & \overline{y}_t \end{pmatrix} \begin{pmatrix} A_1 & A_2 & \ldots & A_t \\ -B_1 & 0 & \ldots & 0 \\ 0 & -B_2 & \ldots & 0 \\ \vdots & \vdots & \ddots & \vdots \\ 0 & 0 & \ldots & -B_t \end{pmatrix} = 0.$$

Then 1.1.13 says that this implies the condition $\exists \overline{y} \, (\overline{x} \; \overline{y}) \begin{pmatrix} A \\ -B \end{pmatrix} = 0$ iff a matrix equation of the following form holds, where we have written the matrices G'' and K (in the notation of 1.1.13) in convenient block form and have replaced G' there by G:

$$\begin{pmatrix} I & G \\ 0 & G_1 \\ \vdots & \vdots \\ 0 & G_t \end{pmatrix} \begin{pmatrix} A \\ -B \end{pmatrix} = \begin{pmatrix} A_1 & A_2 & \ldots & A_t \\ -B_1 & 0 & \ldots & 0 \\ 0 & -B_2 & \ldots & 0 \\ \vdots & \vdots & \ddots & \vdots \\ 0 & 0 & \ldots & -B_t \end{pmatrix} \begin{pmatrix} K_1 \\ K_2 \\ \vdots \\ K_t \end{pmatrix}.$$

The statement is then obtained by writing this out as $1 + t$ matrix equations. □

1.1.3 The lattice of pp conditions

The set of equivalence classes of pp conditions in n free variables forms a modular lattice. This will be relativised to a given module in Section 3.2.1. At 10.2.17 it will be shown that the lattice of pp conditions in n free variables is naturally isomorphic to that of finitely generated subfunctors of the nth power, $(R^n, -)$, of the forgetful functor from finitely presented R-modules to abelian groups (the relativised version is 12.3.18).

The set of subgroups of M^n pp-definable in M is closed under finite intersection and finite sum. To see this let $\phi(\overline{x})$, being $\exists \overline{y} \, \theta(\overline{x}, \overline{y})$, and $\phi'(\overline{x})$, being $\exists \overline{y}' \, \theta'(\overline{x}, \overline{y}')$, be two pp conditions in the same free variables. Let $\theta''(\overline{x}, \overline{y}, \overline{y}')$ be the system of equations which is the union of $\theta(\overline{x}, \overline{y})$ and $\theta'(\overline{x}, \overline{y}')$ (after re-naming variables if necessary, so that $\overline{y}$ and $\overline{y}'$ are disjoint). As a condition, this is the conjunction θ and θ', written $\theta(\overline{x}, \overline{y}) \wedge \theta'(\overline{x}, \overline{y}')$, of θ and θ' (with some variables possibly renamed). Then the condition

$$\exists \overline{y}, \overline{y}'\, \theta''(\overline{x}, \overline{y}, \overline{y}')$$

clearly defines $\phi(M) \cap \phi'(M)$ in every module M. We denote this condition $\phi \wedge \phi'$ and call it the **conjunction** of ϕ and ϕ'.

In order to define the **sum** $\phi(M) + \phi'(M)$ of the same two conditions we can use the pp condition

$$\exists \overline{z}, \overline{z}', \overline{y}, \overline{y}' \big(\theta(\overline{z}, \overline{y}) \wedge \theta(\overline{z}', \overline{y}') \wedge \overline{x} = \overline{z} + \overline{z}'\big)$$

(where $\overline{z}$ and $\overline{z}'$ are two "disjoint renamings" of $\overline{x}$). This condition is denoted $\phi + \phi'$. An efficient way of expressing this condition is $\exists \overline{z}, \overline{z}' \big(\phi(\overline{z}) \wedge \phi'(\overline{z}') \wedge \overline{x} = \overline{z} + \overline{z}'\big)$ (as usual, read "$\wedge$" as "and"). We extend the term "pp condition" to expressions such as this and $\phi \wedge \phi'$ which, though not, strictly speaking, pp conditions as originally defined (that is, with all the quantifiers to the left), can be re-expressed in such a form.

Clearly both $\wedge$ and $+$ are well-defined operations on equivalence classes of pp conditions. Also, the implication preordering, $\psi \leq \phi$, on pp conditions induces an ordering on equivalence classes of pp conditions. It is then immediate that the poset of equivalence classes of pp conditions (in, say, n free variables) is a lattice, with $+$ and $\wedge$ being the join and meet operations respectively. We denote this **lattice of pp conditions** by pp_R^n, often dropping the superscript when $n = 1$. We usually identify a point in this lattice – an equivalence class of pp conditions – with any pp condition which represents it. This lattice is modular: one may check directly or use that it is isomorphic to a sublattice of the lattice of subobjects of an object in an abelian category (3.2.1 or 10.2.17).

The lattice of pp conditions is sufficiently complicated that it is not in general easy, or even possible, to represent it, though reasonable pictures can be made in some cases. This lattice has finite length iff the ring has finite representation type (4.5.21). Beyond that there are at least partial descriptions in quite complicated cases see, for example, [566, p. 328] and [564, pp. 178, 192],[6] [640, §§2,3][7] and the pictures implicit in [598].[8] The latter two references deal with functors, but 10.2.17 states the equivalence with pp conditions.

6 Puninski's description of this lattice over certain uniserial rings, see Section 8.2.8.
7 Schröer's description for a certain domestic string algebra.
8 Ringel's description of a maximal chain through this lattice over the dihedral algebra in characteristic 2.

Example 1.1.18. For $R = \mathbb{Z}_2$ the lattice pp_R^1 is a chain with two elements: $x = x > x = 0$ and for $R = \mathbb{Z}_4$ it is the chain $x = x > x2 = 0 > 2|x > x = 0$. For $R = \mathbb{Z}_2 \times \mathbb{Z}_4, = e_1R \oplus e_2R$ say, the lattice is as shown.

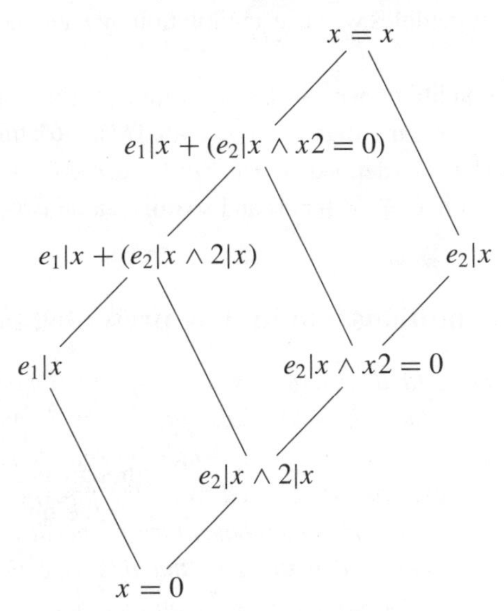

1.1.4 Further examples

Any condition which can be formed from R-linear equations (or pp conditions) by applying conjunction, sum and existential quantification is (equivalent to) a pp condition. One cannot, however, use disjunction ("or") or negation (inequations) and stay within this good class of conditions which define subgroups and which are preserved by morphisms.

Here are a few more examples of properties of elements and tuples which can be expressed using pp conditions. Many more can be seen in the pages that follow.

If I is a finitely generated right ideal, then the condition $xI = 0$ is a pp condition on x: if I is generated by $r_1, \dots, r_m$ take $\phi(x)$ to be $\bigwedge_1^m xr_i = 0$. More generally, if I is a finitely generated submodule of R^n, the condition on n-tuples, $\overline{x}I = 0$, is pp, where $\overline{x}I = \{\overline{x} \cdot \overline{r} : \overline{r} \in I\}$ and where $\overline{x} \cdot \overline{r}$ means $\sum_1^n x_i r_i$.

If L is a finitely generated left ideal, the condition $L \mid x$ is a pp condition (in *right* modules), where, by "L divides x", we mean $x \in ML = \{\sum_i a_i l_i : a_i \in M, l_i \in L\}$, M being the ambient module. For, if L is generated as a left module by $s_1, \dots, s_m$, then $\phi(x)$ may be taken to be $\exists y_1, \dots, y_m \left(x = \sum_1^m y_i s_i\right)$.

More generally, if L is a left submodule of R^n and M is a right module we set $M \cdot L = \{\overline{a} \in M^n :$ there are $b_1, \dots, b_m \in M$ and $\overline{l}_1, \dots, \overline{l}_m \in L$ such that $\overline{a} = \sum_1^m b_j \overline{l}_j\} \leq M^n$. If L is finitely generated, by $\overline{s}_1, \dots, \overline{s}_m$ say, then

$ML = \{\overline{a} \in M^n : \exists b_1, \ldots, b_m \text{ such that } \overline{a} = \sum_1^m b_j \overline{s}_j\}$ and this is clearly a pp-definable subgroup of M^n. Indeed, if $\phi(\overline{x})$ is $\exists y_1, \ldots, y_m \left(\bigwedge_{i=1}^n x_i = \sum_{j=1}^m y_j s_{ij}\right)$, where $\overline{s}_j = (s_{ij})_i$, then $L = \phi(R_R)$ and $\phi(M) = ML = M\phi(R)$ (cf. 1.1.10).

Note that for right modules we have annihilation by *right* ideals, divisibility by *left* ideals.

From given pp conditions we can build up more – for instance those which express: "$x \in \mathrm{ann}_M I \cdot r$", more generally "$x \in \phi(M)r$" (defined by $\exists y\, (\phi(y) \wedge x = yr)$); "$x \in \phi(M)s^{-1}$" (defined by $\phi(xs)$); "$x \in \phi(M)r + \psi(M)s$" (defined by $\exists y, z\, \big(\phi(y) \wedge \psi(z) \wedge x = yr + zs\big)$) and so on.

1.2 Pp conditions and finitely presented modules

The notion of a type is fundamental in model theory: roughly, the type of an element in a structure is the collection of all "first-order"[9] *conditions that it satisfies. In our context we need to know only which pp conditions are satisfied and this restricted set of conditions is referred to as the pp-type of the element. In Section 1.2.1 this is defined and some elementary properties noted. It is shown that a pp-type is finitely generated, in the sense that this collection of pp conditions is generated by a single condition, iff it is the pp-type of an element in a finitely presented module. This leads to the notion of "free realisation" of a pp condition. The important correspondence between inclusions of finitely generated pp-types and morphisms between finitely presented modules is shown. A link between irreducible pp conditions and "strongly indecomposable" finitely presented modules (that is, those with local endomorphism ring) is made in Section 1.2.3.*

1.2.1 Pp-types

Rather than fixing a pp condition and looking at its solution sets in various modules, that is, rather than looking at the functor corresponding to the pp condition, we may fix an element in a module and consider all the pp conditions satisfied by that element. This set of conditions is called the pp-type of the element. This pp-type is thus a collection of pieces of information about that element: information that says which elements of R annihilate the element and which divide it, as well as information which may be expressed through more complicated pp conditions.

More generally, given a tuple $\overline{a} = (a_1, \ldots, a_n)$ of elements from a module M, the **pp-type of $\overline{a}$ in** M, is

$$\mathrm{pp}^M(\overline{a}) = \{\phi : \phi \text{ is a pp condition and } \overline{a} \in \phi(M)\}.$$

[9] This just refers to the fact that quantifiers range over elements, not subsets.

If the length of the tuple $\overline{a}$ is n, then $\mathrm{pp}^M(\overline{a})$ may be referred to as a **pp-n-type**.[10] The superscript ^{M} may be dropped when it is not needed.[11]

Remark 1.2.1. Suppose that $\overline{a}$ is from M and that M is a direct summand of N. Then $\mathrm{pp}^M(\overline{a}) = \mathrm{pp}^N(\overline{a})$.

To see this, apply 1.1.7 to the injection of M into N and also to a projection from N onto M to conclude that $\overline{a}$ satisfies exactly the same pp conditions in M as in N.

For the equality it is enough that M be pure in N, see 2.1.1.

Lemma 1.2.2. *Given n and, for each i in some index set, an n-tuple $\overline{a}_i$ from a module M_i, set $\overline{a} = (\overline{a}_i)_i \in \prod_i M_i^n = (\prod_i M_i)^n = M$, say.*[12] *Then $\mathrm{pp}^M(\overline{a})$ is the intersection of the $\mathrm{pp}^{M_i}(\overline{a}_i)$.*

Proof. If $\overline{a}$ satisfies a pp condition ϕ, say $\phi(\overline{x})$ is $\exists \overline{y}\,(\overline{x}\,\overline{y})H = 0$, choose $\overline{b} = (\overline{b}_i)_i$ from M such that $(\overline{a}\,\overline{b})H = 0$. Then $(\overline{a}_i\,\overline{b}_i)H = 0$ for each i and so each $\overline{a}_i$ satisfies ϕ in M_i. For the converse, reverse the argument, assembling witnesses, $\overline{b}_i$, to the fact that $M_i \models \phi(\overline{a}_i)$, into a witness, $\overline{b} = (\overline{b}_i)_i$, to the fact that $M \models \phi(\overline{a})$. □

Essentially the same argument gives the following.

Lemma 1.2.3. *Given any modules $(M_i)_i$ and any pp condition ϕ one has $\phi(\bigoplus_i M_i) = \bigoplus_i \phi(M_i)$ and $\phi(\prod_i M_i) = \prod_i \phi(M_i)$.*

Remark 1.2.4. An observation which can be useful in computing examples is that if $\overline{a}$ is a tuple from a module M and if f is in the radical of the endomorphism ring of M, then $\mathrm{pp}^M(\overline{a} + f\overline{a}) = \mathrm{pp}^M(\overline{a})$ (because $1 + f$ is invertible).

Example 1.2.5. Let $U = U(\mathrm{sl}_2(k))$ be the universal enveloping algebra of the Lie algebra $\mathrm{sl}_2(k)$, where k is an algebraically closed field of characteristic zero, and let $L(n)$ $(n \geq 0)$ denote the unique $(n+1)$-dimensional simple U-module. Consider $L = \bigoplus_{n \geq 0} L(n)$. We will take our modules to be left modules in this

[10] I remark that not every element of $\overline{a}$ need be involved in any given condition: $x_1 \cdot 0 = 0$ is a condition in $\mathrm{pp}^M(\overline{a})$ whatever the length of the tuple $\overline{a}$ (though then this has to be regarded as a condition in possibly more free variables, not all of which actually appear).

[11] I should make clear that, in the definition above, by a pp condition ϕ we do mean the condition (that is, the functor F_ϕ), not the particular formula used to define it. If one were to define a pp-type to be a set of formulas, then one should first fix a tuple $(x_1, \dots, x_n)$ of distinct free variables and restrict the pp conditions appearing in the definition to those with free variables (from among) $x_1, \dots, x_n$. That artifice results in various side-issues, unimportant but awkward, which are usually (that is, in standard accounts of model theory) ignored or dismissed with a comment. That is part of why, in this book, I have used the term "condition" instead of "formula" (see Appendix E, p. 703 for the distinction). There is further discussion around this in Section 3.2.2.

[12] The identification of $\prod_i M_i^n$ with $(\prod_i M_i)^n$ is, like the identification of the domain of a monomorphism with its image, convenient and usually harmless, and allows us to treat $(\overline{a}_i)_i$ as an n-tuple of elements of $\prod_i M_i$.

example. Notation is as in Section 8.2.3: in particular "x" and "y", along with "h", now have roles as notations for certain elements of sl_2 so we use letters v, w for variables in conditions. For undefined terms see Section 8.2.3.

The condition $\phi_x(v)$, being $xv = 0$, defines the highest weight space, $L(n)_n = \{a \in L(n) : ha = na\}$, of $L(n)$ where we set $M_\lambda = \{a \in M : ha = \lambda a\}$ for $\lambda \in k$. So $\phi_x(L) = \bigoplus_{n\geq 0} L(n)_n$.

For $\lambda \in k$ the condition $\phi_\lambda(v)$, being $hv = \lambda v$, defines the λ-weight space of $L(n)$, so $\phi_\lambda(L) = \bigoplus_{n\geq 0} L(n)_\lambda$.

More interesting are pp conditions like $\phi_{3,0}(v)$, which is

$$(xv = 0) \wedge \exists w, w_0 \big((hw_0 = 0) \wedge (w - w_0 = x^3 w) \\ \wedge \exists w_1 \, (y | w_1 \wedge w_1 + v = w)\big).$$

One may check that the clauses involving w_0 and w say (in $L(n)$) that $w \in \bigoplus_{i\geq 0} L(n)_{6i}$ if n is even and $w = 0$ otherwise. The latter part of the condition says that v is the highest-weight component of w and the very first part says that v is in $L(n)_n$. It follows that $\phi_{3,0}(L(n)) \neq 0$ iff $n \equiv 0 \bmod 6$.

If, in the above condition, we were to use $(x^2 v = 0) \wedge (y|v)$ in place of $xv = 0$ (in order to pick out $L(n)_{n-2}$) and $y^2|w_1$ in place of $y|w_1$, and if we denoted the resulting condition by $\phi_{3,1}$, then $\phi_{3,1}(L(n)) \neq 0$ iff $n \equiv 2 \bmod 6$.

More generally, one may write down similar pp conditions $\phi_{m,i}$, $m \geq 2$, $0 \leq i < m$. These various conditions, when applied to L, pick out a quite complicated collection of pp-definable subgroups of this direct sum.

This example appears in [282, p. 273]: it shows that the finite-dimensional representations of $\mathrm{sl}_2(k)$ together do not have m-dimension (for which see Section 7.2), that is, (7.2.3) L has a densely ordered set of pp-definable subgroups.

Returning to the general case, if p is a pp-n-type and $\overline{a} \in M^n$, then we write $M \models p(\overline{a})$ if[13] $M \models \phi(\overline{a})$ for every $\phi \in p$. The notation allows $\overline{a}$ to satisfy more pp conditions than are in p. Then define the **solution set** of p in M to be $p(M) = \{\overline{a} \in M^n : M \models p(\overline{a})\} = \bigcap_{\phi\in p} \phi(M)$. So this is the solution set to an, in general, infinite set of conditions each of which is the projection of a system of linear equations.

1.2.2 Finitely presented modules and free realisations

Here it is shown that the pp-type of any element, or finite tuple, in a finitely presented module is finitely generated (1.2.6). Conversely, every pp condition has a "free realisation" in a finitely presented module (1.2.14). Free realisations of

[13] Recall that $M \models \phi(\overline{a})$ is just another way of writing $\overline{a} \in \phi(M)$.

the sum (1.2.27) and the conjunction (1.2.28 of pp conditions are identified, as are free realisations of divisibility (1.2.29) and annihilator (1.2.30) conditions.

In the category of finitely presented modules inclusion of pp-types corresponds exactly to existence of morphisms (1.2.9). An analogous result for pure-injective modules is 4.3.9. Pp-definable subgroups of M and End(M)*-submodules of M are compared. A projective presentation of the functor defined by a pp condition is given at 1.2.19. Finally it is noted that pp conditions commute with direct limits (1.2.31).*

This section is loosely based on [495, §§8.2, 8.3].

Clearly any pp-type is closed under conjunction of conditions ($\phi, \psi \in \mathrm{pp}^M(\overline{a})$ implies $\phi \wedge \psi \in \mathrm{pp}^M(\overline{a})$) and under implication ($\psi \in \mathrm{pp}^M(\overline{a})$ and $\psi \leq \phi$ implies $\phi \in \mathrm{pp}^M(\overline{a})$). So $\mathrm{pp}^M(\overline{a})$ may be regarded as a filter in the lattice, pp_R^n (where $n = l(\overline{a})$), of equivalence classes of pp conditions.

If there is a single condition, ϕ_0, in $\mathrm{pp}^M(\overline{a})$ such that $\mathrm{pp}^M(\overline{a}) = \{\phi \in \mathrm{pp}_R^n : \phi \geq \phi_0\}$, then we say that the pp-type of $\overline{a}$ is **finitely generated (by** ϕ_0) and write $\mathrm{pp}^M(\overline{a}) = \langle \phi_0 \rangle$. For any pp condition ϕ_0 define $\langle \phi_0 \rangle = \{\phi : \phi \geq \phi_0\}$. Because the conjuction of finitely many pp conditions is again a pp condition there is no distinction between "finitely generated" and "1-generated".

Warning: for $\mathrm{pp}^M(\overline{a}) = \langle \phi_0 \rangle$ to hold it is *not* enough that $\phi_0 \in \mathrm{pp}^M(\overline{a})$ and $\phi_0(M) \leq \phi(M)$ for every $\phi \in \mathrm{pp}^M(\overline{a})$; rather, it is necessary that for every such ϕ we have $\phi_0 \leq \phi$, meaning $\phi_0(A) \leq \phi(A)$ for *every* module A (by 1.2.23 it is enough to test on finitely presented modules A).

Lemma 1.2.6. *If M is a finitely presented module and $\overline{a}$ is a tuple from M, then the pp-type of $\overline{a}$ in M is finitely generated.*

Proof. From any presentation of M and expression of $\overline{a}$ in terms of those generators we produce a generator for $\mathrm{pp}^M(\overline{a})$.

Suppose that $\overline{b}$ is a finite tuple of elements which together generate M and let $\overline{b}H = 0$, where H is a matrix with entries in the ring, give a finite generating set for the relations on $\overline{b}$. Let G be a matrix such that $\overline{a} = \overline{b}G$. Then the required pp condition is $\exists \overline{y}\ (\overline{x}\ \overline{y}) \begin{pmatrix} I & 0 \\ -G & H \end{pmatrix} = 0$. For certainly $\overline{a}$ satisfies this condition and, since the condition defines exactly the tuple $\overline{a}$ in the module M, every other pp condition satisfied by $\overline{a}$ in M must be a consequence of this one (cf. proof of 1.1.13, alternatively use 1.2.7 and 1.1.7), as required. □

Proposition 1.2.7. *Suppose that M is a finitely presented module and that $\overline{a}$ is a tuple from M with pp-type generated by ϕ:* $\mathrm{pp}^M(\overline{a}) = \langle \phi \rangle$. *Let N be any module and let $\overline{c} \in \phi(N)$. Then there is a morphism from M to N taking $\overline{a}$ to $\overline{c}$.*

Proof. Denote by ϕ' the pp condition, $\exists \overline{y}\ (\overline{x}\ \overline{y}) \begin{pmatrix} I & 0 \\ -G & H \end{pmatrix} = 0$, produced in the proof of 1.2.6. Then ϕ implies (indeed, is equivalent to) ϕ' so $\overline{c}$ satisfies ϕ' in N. Therefore there is $\overline{d}$ in N such that $\overline{d}H = 0$ and $\overline{c} = \overline{d}G$. Since the generators, $\overline{b}$, of M used to produce ϕ' are freely generated subject to $\overline{b}H = 0$, the map $\overline{b} \mapsto \overline{d}$ extends to a well-defined morphism f from M to N which takes $\overline{a} = \overline{b}G$ to $\overline{d}G = \overline{c}$, as required. □

We extend the ordering on pp conditions to pp-types in the obvious way: if p and q both are pp-n-types, then write $q \le p$ (that is, $q \to p$, "q implies p") if $p \subseteq q$. The ordering is that of solution sets: $q \le p$ iff $q(M) \subseteq p(M)$ for every module M. The lattice of pp-types, which contains that of pp conditions via $\phi_0 \mapsto \langle \phi_0 \rangle$, is considered at the end of Section 3.2.1.

The next corollary is immediate from 1.1.7 which, with 1.2.7, gives the result.

Lemma 1.2.8. *If $f : M \to M'$ is any morphism and $\overline{a}$ is a tuple from M, then $\mathrm{pp}^M(\overline{a}) \ge \mathrm{pp}^{M'}(f\overline{a})$.*[14]

Corollary 1.2.9. *Suppose that M is a finitely presented module, that $\overline{a}$ is a tuple from M, that N is any module and that $\overline{b}$ is a tuple from N. Then there is a morphism $f : M \to N$ with $f\overline{a} = \overline{b}$ iff $\mathrm{pp}^M(\overline{a}) \ge \mathrm{pp}^N(\overline{b})$.*

Illustration 1.2.10. Let $r \in R$. By the proof of 1.2.6 the pp-type of r in R is generated by the condition $r|x$. Suppose that f is a morphism from the cyclic right ideal rR to a module M. There is, according to 1.2.9, an extension of f to a morphism from R to M iff $r|x \in \mathrm{pp}^M(fr)$, that is, iff $fr \in Mr$.

The criterion 1.2.9 fails for arbitrary M: take $M = \overline{\mathbb{Z}_{(p)}}$, the p-adic integers regarded as a $\mathbb{Z}$-module, $N = \mathbb{Z}_{(p)}$, the localisation of $\mathbb{Z}$ at p, embedded in M, and take $a = b$ to be any non-zero element of N. Then $\mathrm{pp}^M(a) = \mathrm{pp}^N(b)$ since N is pure in M (2.4.12) but $\mathrm{Hom}_{\mathbb{Z}}(\overline{\mathbb{Z}_{(p)}}, \mathbb{Z}_{(p)}) = 0$. One argument for this last point is to use that $\mathbb{Z} \to \mathbb{Z}_{(p)}$ is an epimorphism of rings so (5.5.1) $\mathrm{Hom}_{\mathbb{Z}}(\overline{\mathbb{Z}_{(p)}}, \mathbb{Z}_{(p)}) = \mathrm{Hom}_{\mathbb{Z}_{(p)}}(\overline{\mathbb{Z}_{(p)}}, \mathbb{Z}_{(p)})$. Now, any non-zero $\mathbb{Z}_{(p)}$-submodule of $\mathbb{Z}_{(p)}$ is isomorphic to $\mathbb{Z}_{(p)}$, hence projective. So existence of a non-zero morphism from $\overline{\mathbb{Z}_{(p)}}$ to $\mathbb{Z}_{(p)}$ would imply that $\mathbb{Z}_{(p)}$ is a direct summand of $\overline{\mathbb{Z}_{(p)}}$ – this is a contradiction since $\overline{\mathbb{Z}_{(p)}}$ is indecomposable as a group. It will be shown (4.3.9) that if we allow M to be arbitrary but require that N be pure-injective, then the conclusion of 1.2.9 is recovered.

[14] If f is such that $\mathrm{pp}^M(\overline{a}) > \mathrm{pp}^{M'}(f\overline{a})$ (strict inclusion), then the usual description of this is that "f is strictly pp-type-increasing on $\overline{a}$". Regarding a pp-type as a collection of pp conditions, this makes sense terminologically but it does rather clash with the ordering (by solution sets) on pp-types, so I will tend to avoid this terminology.

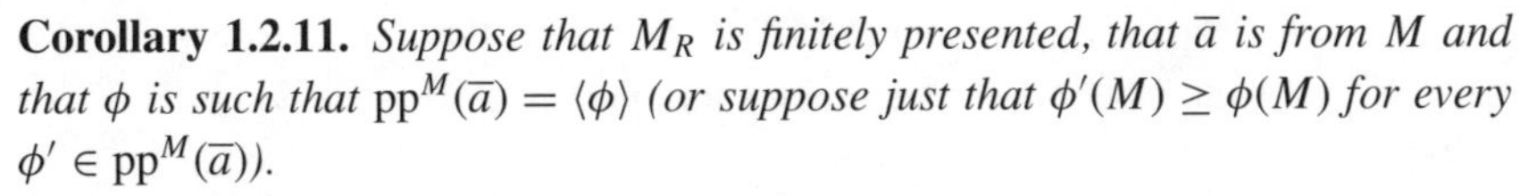

Corollary 1.2.11. *Suppose that M_R is finitely presented, that $\overline{a}$ is from M and that ϕ is such that $\mathrm{pp}^M(\overline{a}) = \langle \phi \rangle$ (or suppose just that $\phi'(M) \geq \phi(M)$ for every $\phi' \in \mathrm{pp}^M(\overline{a})$).*

Then $\phi(M) = \mathrm{End}(M_R) \cdot \overline{a}$.

Proof. By 1.1.7, $\mathrm{End}(M_R) \cdot \overline{a} = \{f\overline{a} : f \in \mathrm{End}(M_R)\} \leq M^{l(\overline{a})} \leq \phi(M)$ and the converse is immediate from 1.2.7 (even with the weaker hypothesis, since that implies $\phi(M) = \phi_0(M)$ where ϕ_0 generates $\mathrm{pp}^M(\overline{a})$). □

Corollary 1.2.12. *Suppose that M_R is finitely presented and let $S = \mathrm{End}(M_R)$. Then every finitely generated S-submodule of M is a pp-definable subgroup of M_R. In particular, if M_R is finitely presented and has the ascending chain condition on S-submodules, then the pp-definable subgroups of M_R are exactly the submodules of ${}_SM$.*

Proof. By 1.1.8 every pp-definable subgroup of M_R is an S-submodule of M. For the converse, let L be a finitely generated submodule of ${}_SM$. Choose $a_1 \in L$ and (1.2.6) let ϕ_1 be a pp condition which generates the pp-type of a_1 in M_R. By 1.2.11, $\phi_1(M) = Sa_1$. If Sa_1 is properly contained in L take $a_2 \in L \setminus Sa_1$ and let ϕ_2 generate the pp-type of a_2 in M_R. Then $(\phi_1 + \phi_2)(M) = \phi_1(M) + \phi_2(M) = Sa_1 + Sa_2$. Continue. By finite generation of L eventually $(\phi_1 + \cdots + \phi_n)(M) = L$, as required.

(The statement and proof also apply to subgroups of M^n pp-definable in M and submodules of M^n under the diagonal S-action.) □

Corollary 1.2.13. *If M is a finitely presented module and $\mathrm{End}(M) = D$ is a division ring, then every finite-dimensional D-subspace of M is a pp-definable subgroup of M_R. More generally, for all n, every finite-dimensional D-subspace of M^n has the form $\phi(M)$ for some pp condition ϕ with n free variables.*

A **free realisation** of a pp condition ϕ is a finitely presented module C and a tuple $\overline{c}$ from C such that $\mathrm{pp}^C(\overline{c}) = \langle \phi \rangle$. So, if $\overline{a}$ is a tuple from the finitely presented module M, then $(M, \overline{a})$ is a free realisation of ϕ, where ϕ generates $\mathrm{pp}^M(\overline{a})$ (1.2.6).

Proposition 1.2.14. ([495, 8.12]) *Every pp condition has a free realisation.*

Proof. Let ψ be a pp condition. The proof of 1.1.13($\Rightarrow$) produced a free realisation, $(M, \overline{a})$ in the notation there, of ψ. For that proof shows that if $\phi \in \mathrm{pp}^M(\overline{a})$, then the right-hand condition of the statement of 1.1.13 is satisfied, so, by the first part of the proof of 1.1.13, $\psi \leq \phi$. □

Thus, to obtain a free realisation $(C_\phi, \overline{c}_\phi)$ of the pp condition $\phi(\overline{x})$, being $\exists \overline{y}\, \theta(\overline{x}, \overline{y})$ with θ a system of linear equations, form the free R-module on generators $\overline{x}\,\overline{y}$, factor by the submodule generated by the linear terms of θ; that gives C_ϕ, then take the image of $\overline{x}$ in this module for $\overline{c}_\phi$. Also see Example 1.2.15 below.

A somewhat different approach to free realisations is in [735, §1]. Also see [22, §7] for related considerations. See 4.3.70 for minimal free realisations. It will be shown, 3.2.6, also 4.3.9, that arbitrary pp-types have "free realisations" in a rather weaker sense.

Given a pp formula, a free realisation is constructed from it by an entirely explicit procedure. We illustrate this with some examples below. Also, 1.2.27, 1.2.28 below give explicit free realisations for the sum and conjunction of pp conditions.

Example 1.2.15. To obtain a free realisation, $(C, \overline{c})$, of the pp condition $\exists \overline{y}\,(\overline{x} = \overline{y}G \wedge \overline{y}H = 0)$, where $H = (s_{ij})_{ij}$ is $k \times l$, take

$$C = \bigoplus_{i=1}^{k} y_i R / \sum_{j=1}^{l} \Big(\sum_{i=1}^{k} y_i s_{ij}\Big) R = \operatorname{coker}(H : R^l \to R^k),$$

where the map is left multiplication (of column vectors) by H. Then set $\overline{c} = \overline{d}G$, where $\overline{d}$ is the image of $\overline{y}$ in C.

In particular, a free realisation of the condition $xr = 0$ is $(R/rR, 1 + rR)$ and a free realisation of the condition $r|x$ is (R, r). See 1.2.29 and 1.2.30 for converses.

Applying the above recipe to the pp condition over $\mathbb{Z}$, which is $\exists y_1, y_2\,(x = y_1 + y_2 \wedge y_1 4 = 0 \wedge y_2 4 = 0)$, that is,

$$\exists \overline{y}\left(x = \overline{y}\begin{pmatrix}1\\1\end{pmatrix} \wedge \overline{y}\begin{pmatrix}4 & 0\\0 & 4\end{pmatrix} = 0\right),$$

illustrates that if a pp condition is not in "simplest" form (if there is such), then the free realisation computed from it may well have "redundant" direct summands.

Example 1.2.16. Consider the butterfly quiver (for quivers with relations, see Section 15.1.1).

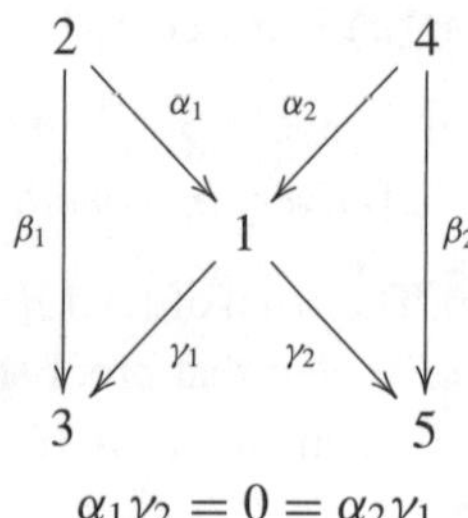

Let R be the path algebra of this quiver over some field. Let $\phi(x)$ be the pp condition $x = xe_2 \wedge x\beta_1 = 0 \wedge \alpha_2 \mid x\alpha_1$. We compute a free realisation by following the proof of 1.1.13($\Rightarrow$).

First, write ϕ in standard form, as $\exists y\,(x \quad y)\begin{pmatrix} 1-e_2 & \beta_1 & \alpha_1 \\ 0 & 0 & -\alpha_2 \end{pmatrix} = 0$. Let C be the cokernel of the morphism $R^3 \to R^2$, given by

$$\overline{z} = \begin{pmatrix} z_1 \\ z_2 \\ z_3 \end{pmatrix} \mapsto \begin{pmatrix} 1-e_2 & \beta_1 & \alpha_1 \\ 0 & 0 & -\alpha_2 \end{pmatrix} \begin{pmatrix} z_1 \\ z_2 \\ z_3 \end{pmatrix} = \begin{pmatrix} (1-e_2)z_1 + \beta_1 z_2 + \alpha_1 z_3 \\ -\alpha_2 z_3 \end{pmatrix}.$$

This is the module with generators a_1, a_2 say (the images of $(1, 0)$ and $(0, 1)$ in R^2) and relations $a_1 = a_1 e_2$, $a_1\beta_1 = 0$, $a_1\alpha_1 = a_2\alpha_2$. Then a_1 in C is a free realisation of ϕ. A k-basis of C is as shown.

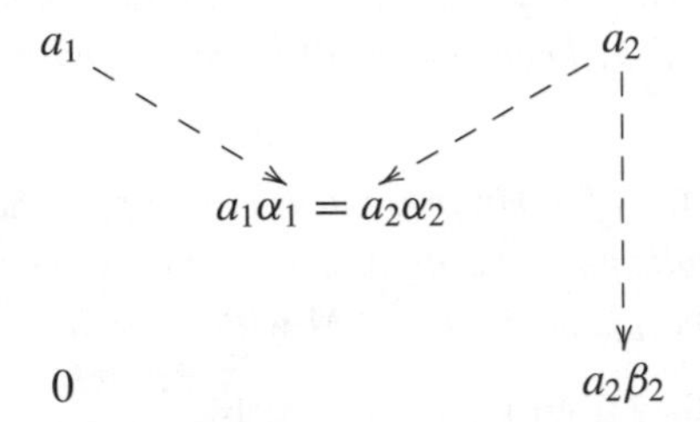

Of course, this is exactly what one would expect from thinking about what the pp condition "says" but our point is that one can compute these in practice by a method which applies even when the condition is too complicated to be comprehended easily. It should be said, though, that this procedure simply yields a finite presentation of a module and a distinguished tuple from it: describing this module in some other, more "structural", way would be a separate task.

Corollary 1.2.17. (to 1.2.9) *If $(C, \overline{c})$ is a free realisation of the pp condition ϕ and if M is any module, then $\overline{a}$ from M satisfies ϕ iff there is a morphism $f : C \to M$ with $f\overline{c} = \overline{a}$.*

That is, $\phi(M) = \{f\overline{c} : f \in (C, M)\}$.

Corollary 1.2.18. (a rephrasing of 1.1.11) *Suppose that $C \in$ mod-R (the category of finitely presented modules) is generated by the finite tuple $\overline{c}$ and that $\mathrm{pp}^C(\overline{c})$ is generated by ϕ. Then (C, M) is naturally isomorphic to $\phi(M)$ for every $M \in$ Mod-R.*

Corollary 1.2.19. ([495, proof of 12.3]) *If ϕ is a pp condition, with free realisation $(C, \overline{c})$, and if M is any module, then the sequence of abelian groups $0 \to (C/\langle\overline{c}\rangle, M) \to (C, M) \to \phi(M) \to 0$ is exact, where $\langle\overline{c}\rangle$ denotes the submodule of C generated by the entries of $\overline{c}$ and where the map is induced by the canonical projection $C \to C/\langle\overline{c}\rangle$.*

Proof. The kernel of the map $f \mapsto f\overline{c}$ (notation as in 1.2.17) consists of those f which are zero on $\overline{c}$, that is, which factor through $C \to C/\langle\overline{c}\rangle$. □

The sequence is natural in M so this gives a presentation of the functor F_ϕ, see 10.2.29.

If $C \in \text{mod-}R$, $\overline{c} \in C^n$ and $M \in \text{Mod-}R$, then the **trace of** $(C, \overline{c})$ **in** M is $\text{tr}_{(C,\overline{c})}(M) = \{f\overline{c} : f \in (C, M)\} \leq M^n$. By 1.2.17 (and 1.2.6) this is $\phi(M)$ where $\text{pp}^M(\overline{c}) = \langle \phi \rangle$. The second statement below is by 1.2.14.

Corollary 1.2.20. (for example, [323, 6.3]) *If $C \in \text{mod-}R$, $\overline{c} \in C^n$ and $M \in \text{Mod-}R$, then $\text{tr}_{(C,\overline{c})}(M)$ is a subgroup of M^n pp-definable in M. Conversely, every subgroup of M^n pp-definable in M has the form $\text{tr}_{(C,\overline{c})}(M)$ for some finitely presented module C and some n-tuple $\overline{c}$ of elements in C.*

Corollary 1.2.21. (to 1.2.9) *Let $(C_\phi, \overline{c}_\phi)$ be a free realisation of ϕ and let $(C_\psi, \overline{c}_\psi)$ be a free realisation of ψ. Then $\phi \geq \psi$ iff there is a morphism from C_ϕ to C_ψ taking $\overline{c}_\phi$ to $\overline{c}_\psi$.*

Corollary 1.2.22. *If C is finitely presented, ϕ is a pp condition and $\overline{c} \in \phi(C)$, then $(C, \overline{c})$ is a free realisation of ϕ iff for every (finitely presented) module M and $\overline{a} \in \phi(M)$ there is a morphism $f : C \to M$ with $f\overline{c} = \overline{a}$.*

Proof. This is immediate from 1.2.17 and 1.2.14. □

Corollary 1.2.23. *If ϕ and ψ are pp conditions, then $\psi \leq \phi$ iff $\psi(M) \leq \phi(M)$ for every finitely presented module M.*

Remark 1.2.24. Let $\phi \geq \psi$ be pp conditions and let $f : C_\phi \to C_\psi$ take $\overline{c}_\phi$ to $\overline{c}_\psi$, where $(C_\phi, \overline{c}_\phi)$ and $(C_\psi, \overline{c}_\psi)$ are free realisations of ϕ, ψ respectively. Then, for any module M, we have

$$\phi(M)/\psi(M) = \text{tr}_{(C_\phi,\overline{c}_\phi)}(M)/\,(f, M)\,\text{tr}_{(C_\psi,\overline{c}_\psi)}(M),$$

the group of images of $\overline{c}_\phi$ in M modulo those which factor along f. Here (f, M) is the map from (C_ψ, M) to (C_ϕ, M) induced by f.

Say that ϕ/ψ is a **pp-pair** if ϕ, ψ are pp conditions with $\phi \geq \psi$. Occasionally we extend the terminology to allow the conditions not to be ordered this way and then, by the pp-pair ϕ/ψ, we mean $\phi/\phi \wedge \psi$. The pp-pair ϕ/ψ is **open in** M if $\phi(M)/\psi(M) \neq 0$, otherwise it is **closed in** (or **on**) M.

Corollary 1.2.25. *If $\phi > \psi$ are pp conditions, then there is a finitely presented module C such that the pair ϕ/ψ is open in C.*

Proof. Take $(C, \overline{c})$ to be a free realisation of ϕ. Then $\overline{c} \in \phi(C) \setminus \psi(C)$. □

Lemma 1.2.26. *If $(C, \overline{c})$ is a free realisation of the pp condition ϕ and if H is a matrix such that $\overline{c}H$ is defined, then $(C, \overline{c}H)$ is a free realisation of the pp condition $\exists \overline{y}\,(\overline{x} = \overline{y}H \wedge \phi(\overline{y}))$.*

Proof. Let $(B, \overline{b})$ be a free realisation of the pp condition $\exists \overline{y}\,(\overline{x} = \overline{y}H \wedge \phi(\overline{y}))$, which we denote by ψ. Then there is $\overline{d}$ in B with $\overline{b} = \overline{d}H$ and $\overline{d} \in \phi(B)$. By 1.2.17 there is a morphism from C to B taking $\overline{c}$ to $\overline{d}$, hence taking $\overline{c}H$ to $\overline{d}H = \overline{b}$. Therefore, by 1.1.7, $\mathrm{pp}^C(\overline{c}H) \geq \mathrm{pp}^B(\overline{b})$.

Since $\overline{c}H$ certainly satisfies ψ in C there is, by 1.2.17, a morphism from B to C taking $\overline{b}$ to $\overline{c}H$ and this gives the other inclusion. That is, $\mathrm{pp}^C(\overline{c}H) = \mathrm{pp}^B(\overline{b})$ is generated by ψ, as required. □

Lemma 1.2.27. *If $\overline{a}$ is from M, with pp-type generated by ϕ, and if $\overline{a}'$ is from M', with pp-type generated by ϕ', then the pp-type of*[15] *$\overline{a} + \overline{a}'$ in $M \oplus M'$ is generated by $\phi + \phi'$.*

Proof. By definition of the condition $\phi + \phi'$ and 1.2.1 certainly $\overline{a} + \overline{a}' \in (\phi + \phi')(M \oplus M')$ so $\mathrm{pp}^{M\oplus M'}(\overline{a} + \overline{a}') \leq \langle \phi + \phi' \rangle$. On the other hand 1.1.7 applied to the projection maps shows that $\mathrm{pp}^{M\oplus M'}(\overline{a} + \overline{a}') \geq \langle \phi \rangle, \langle \phi' \rangle$ and hence (consider the lattice of pp conditions) that $\mathrm{pp}^{M\oplus M'}(\overline{a} + \overline{a}') \geq \langle \phi + \phi' \rangle$, giving the desired equality. □

Note that by 1.2.3 the pp-type of $\overline{a} + \overline{a}'$ in the direct sum $M \oplus M'$ above is just $\mathrm{pp}^M(\overline{a}) \cap \mathrm{pp}^{M'}(\overline{a}')$ (and remember that the ordering of pp-types is that of solution sets, hence opposite to their ordering as sets of conditions).

The following notation is convenient. If $\overline{a}$ is an n-tuple from the module M, then it induces a morphism from the free module (with specified basis) R^n given by $e_i \mapsto a_i$, where e_i denotes the element with 1 in the ith position and zeroes elsewhere. We denote this morphism also by $\overline{a}$.

Lemma 1.2.28. *If $\overline{a}$ is from M, with pp-type generated by ϕ, and if $\overline{a}'$ is from M', with pp-type generated by ϕ', then the pp-type of $g\overline{a} = g'\overline{a}'$ in the pushout M'' shown below is generated by $\phi \wedge \phi'$.*

$$\begin{array}{ccc} R^n & \xrightarrow{\overline{a}} & M \\ {\scriptstyle \overline{a}'}\downarrow & & \downarrow{\scriptstyle g} \\ M' & \xrightarrow[g']{} & M'' \end{array}$$

Proof. By 1.1.7 both ϕ and ϕ' are in the pp-type of $g\overline{a}$ in M, so $\mathrm{pp}^{M''}(g\overline{a}) \leq \langle \phi \wedge \phi' \rangle$. Suppose that $(C, \overline{c})$ is a free realisation of $\phi \wedge \phi'$. If we assume that M and M' are finitely presented, then, by 1.2.7, there are morphisms $f : M \to C$ with $f\overline{a} = \overline{c}$ and $f' : M' \to C$ with $f'\overline{a}' = \overline{c}$. By the pushout property[16] there

[15] By our conventions, writing $\overline{a} + \overline{a}'$ implies that the two tuples have the same length.
[16] A pseudocokernel of the direct sum will suffice, as noted in [144, 3.4, Lemma 1].

is, therefore, $h : M'' \to C$ with $hg = f$ and $hg' = f'$, hence taking $g\overline{a}$ to $\overline{c}$. Therefore $\mathrm{pp}^{M''}(g\overline{a}) \geq \langle \phi \wedge \phi' \rangle$.

In the general case, not assuming that M, M' are finitely presented, replace C by its pure-injective hull (§4.3.3) and use 4.3.9 (and 2.1.1). □

Lemma 1.2.29.[17] *A pp condition is freely realised in a projective module iff it is equivalent to a divisibility condition, that is, to one of the form* $\exists \overline{y}\, (\overline{x} = \overline{y}G)$.

Proof. ($\Rightarrow$) Without loss of generality (by 1.2.1) the projective module is R^m: say $\overline{c} = (c_1, \dots, c_n)$ in R^m is a free realisation of $\phi(x_1, \dots, x_n)$. Let $\overline{e} = (e_1, \dots, e_m)$ where e_i is the ith chosen unit of R^m. Then $\overline{c} = \overline{e}H$ for some $m \times n$ matrix H. So (cf. proof of 1.2.6) $\phi(\overline{x})$ is equivalent to $\exists y_1, \dots, y_m\, (\overline{x} = \overline{y}H)$.

($\Leftarrow$) Say $\phi(\overline{x})$ is $\exists y_1, \dots, y_m\, (\overline{x} = \overline{y}H)$. Set $P = R^m$ and $\overline{c} = \overline{e}H$ (notation as above). Then $(P, \overline{c})$ is a free realisation of ϕ. For if M is any module and $\overline{a} \in \phi(M)$, say $\overline{a} = \overline{b}H$, then sending $\overline{e}$ to $\overline{b}$ gives a well-defined morphism from P to M which takes $\overline{c}$ to $\overline{a}$, so 1.2.22 applies. □

Lemma 1.2.30. *A pp condition is equivalent to an annihilation condition iff it has, for a free realisation, a generating tuple in a finitely presented module.*

Proof. ($\Rightarrow$) The condition $\overline{x}H = 0$ is freely realised by the generating tuple in the module with corresponding generators and relations.

($\Leftarrow$) If $\overline{a}$ generates the finitely presented module A and $(A, \overline{a})$ is a free realisation of the pp condition ϕ, then, since $\mathrm{pp}^A(\overline{a})$ is clearly generated by a condition of the form $\overline{x}H = 0$, this condition is equivalent to ϕ. □

We will need the fact that pp functors commute with direct limits (Appendix E, p. 714).

Proposition 1.2.31. *If ϕ is a pp condition and $M = \varinjlim_\lambda M_\lambda$, then $\phi(M) = \varinjlim_\lambda \phi(M_\lambda)$, the maps between the $\phi_\lambda(M)$ being those induced by the maps between the M_λ.*

Proof. By 1.1.7, each $\phi(M_\lambda)$ maps to $\phi(M)$. Conversely, given a tuple $\overline{a}$ in $\phi(M)$, choose a morphism from a free realisation, $(C, \overline{c})$, of ϕ to $(M, \overline{a})$. Because C is (by definition) finitely presented this morphism will, E.1.12, lift through one of the M_λ, so $\overline{a}$ will be in the image of $\phi(M_\lambda)$. Since the category **Ab** of abelian groups is Grothendieck (Appendix E, p. 708) this is enough. □

[17] Facts about pp conditions, including this, are put to use in [442] in investigating the question of when every projective module is a direct sum of finitely generated modules.

1.2.3 Irreducible pp conditions

Say that a pp condition ϕ is **irreducible** if whenever ψ_1, ψ_2 are pp conditions with $\psi_1, \psi_2 < \phi$ it follows that $\psi_1 + \psi_2 < \phi$. First we show that a pp condition is irreducible if it is freely realised in a finitely presented module with local endomorphism ring. Under some hypotheses the converse is true. For pp-types and pure-injective modules the analogue, 4.3.49, of the next result is an equivalence. A module with local endomorphism ring must be indecomposable but the condition of having local endomorphism ring ("absolute indecomposability") is stronger (just consider $\mathbb{Z}_{\mathbb{Z}}$). We will say that a module is **strongly indecomposable** if it has local endomorphism ring.

We say "irreducible" (the term "indecomposable" is used by some) because pp-types may be regarded as generalised right ideals and a right ideal I is **irreducible**, that is, not the intersection of two larger right ideals, iff R/I is uniform iff the injective hull of R/I is indecomposable. Then the next result and even more so 4.3.49 justify the terminology, particularly when set in the functor category[18] (12.2.3). The result below appears in a very early version of [276]/[277]; it is recast in functorial terms in Section 2 of [276].

Theorem 1.2.32. *Let ϕ be a pp condition. If ϕ has a free realisation in a finitely presented module with local endomorphism ring, then ϕ is irreducible. If every finitely presented module is a direct sum of modules with local endomorphism ring, that is, if R is Krull–Schmidt (Section 4.3.8), then the converse is true.*

Proof. Assume that ϕ has a free realisation $(C, \overline{c})$, where C is a finitely presented strongly indecomposable module. Suppose that $\psi_1, \psi_2 \leq \phi$ are such that $\psi_1 + \psi_2 = \phi$. For $i = 1, 2$, let $(A_i, \overline{a}_i)$ be a free realisation of ψ_i. By 1.2.7 there are morphisms f_i from C to A_i taking $\overline{c}$ to $\overline{a}_i$ for $i = 1, 2$. Set $A = A_1 \oplus A_2$ and $\overline{a} = \overline{a}_1 + \overline{a}_2$ (that is, $(\overline{a}_1, \overline{0}) + (\overline{0}, \overline{a}_2)$). By 1.2.27, $\psi_1 + \psi_2 = \phi$ generates the pp-type of $\overline{a}$ in the finitely presented module A. So, by 1.2.7, there is a morphism $g : A \to C$ taking $\overline{a}$ to $\overline{c}$. Now, $1 - g(f_1 + f_2)$ is an endomorphism of C with non-zero kernel (it sends $\overline{c}$ to 0) so, since $\mathrm{End}(C)$ is local, $gf_1 + gf_2$ is invertible (see Appendix E, p. 716). Again, since $\mathrm{End}(C)$ is local, at least one of gf_1, gf_2, say the former, is invertible. But then $\mathrm{pp}^C(\overline{c}) \geq \mathrm{pp}^{A_1}(\overline{a}_1) \geq \mathrm{pp}^C(g\overline{a}_1) = \mathrm{pp}^C(gf_1\overline{c}) = \mathrm{pp}^C(\overline{c})$, the last since gf_1 is invertible. Therefore ϕ is equivalent to ψ_1, as required.

The proof of the converse becomes quite easy in the relevant functor category, so we refer forward to 12.1.16 for that. □

It follows easily (cf. 1.2.27) from the definition that if ϕ is an irreducible pp condition, if $\mathrm{pp}^M(\overline{a}) = \langle\phi\rangle$ (M any module) and if M decomposes as $M =$

[18] Especially once duality is applied, converting $+$-irreducibility into $\cap$-irreducibility.

$M_1 \oplus M_2$ with projections $\pi_i : M \to M_i$, then $\mathrm{pp}^{M_i}(\pi_i\overline{a}) = \langle\phi\rangle$ for $i = 1$ or $i = 2$.

1.3 Elementary duality of pp conditions

Given any pp condition for right modules one can write down a "dual" condition for left modules. This duality, defined in Section 1.3.1, is shown, in Section 1.3.2, to correspond to "tensor-annihilation". Duality is also realised via suitable hom-dualities in Section 1.3.3. There is a generalisation beyond tensor-annihilation in Section 1.3.4.

Mittag–Leffler modules, those in which the pp-type of every n-tuple is finitely generated, are discussed in Section 1.3.5.

1.3.1 Elementary duality

We show that the lattice of pp conditions for left modules is, via an explicit and natural duality, anti-isomorphic to the lattice of pp conditions for right modules (1.3.1).

The simplest example of a pp condition is an annihilation condition of the form $xr = 0$. The condition for left modules dual to this is the divisibility condition $r \mid x$, that is $\exists y(rx = y)$. More generally, if $\phi(\overline{x})$ is the condition $\exists \overline{y}\,(\overline{x}\;\overline{y})H = 0$, then the **dual** condition, $D\phi(\overline{x})$ (where now $\overline{x}$ is a column vector and will be substituted by elements from *left* modules), is defined to be $\exists \overline{z} \begin{pmatrix} I & H' \\ 0 & H'' \end{pmatrix} \begin{pmatrix} \overline{x} \\ \overline{z} \end{pmatrix} = 0$, where $\begin{pmatrix} H' \\ H'' \end{pmatrix} = H$. This becomes more clearly a duality between annihilation and divisibility if we write ϕ in the form $\exists \overline{y}\,(\overline{x}A = \overline{y}B)$. Then the dual of ϕ is the condition $\exists \overline{z} \begin{pmatrix} I & A \\ 0 & -B \end{pmatrix} \begin{pmatrix} \overline{x} \\ \overline{z} \end{pmatrix} = 0$. Replacing $\overline{z}$ by $-\overline{z}$, this is the pp condition $\exists \overline{z}\,(\overline{x} = A\overline{z} \wedge B\overline{z} = 0)$ and, respectively taking $B = 0$ or $A = I$ one sees that annihilation and divisibility are dual. It will follow from the next result, and it is easy to check directly, cf. the introduction to [620], that any pp condition can be written in this dual form $\exists \overline{z}\,(\overline{x} = \overline{z}A \,\wedge\, \overline{z}B = 0)$. The first diagram below indicates ϕ, the second $D\phi$.

To go from left to right conditions, take a pp condition, $\psi(\overline{x})$, for left modules, of the form $\exists \overline{y}\, H \begin{pmatrix} \overline{x} \\ \overline{z} \end{pmatrix} = 0$, alternatively of the form $\exists \overline{y}\,(A\overline{x} = B\overline{y})$. Then the dual condition, $D\psi$, for right modules is $\exists \overline{z}\, (\overline{x} \quad \overline{z}) \begin{pmatrix} I & 0 \\ H' & H'' \end{pmatrix} = 0$, where $(H' \quad H'') = H$, alternatively $\exists \overline{z}\,(\overline{x} = \overline{z}A \wedge \overline{z}B = 0)$.

Proposition 1.3.1. ([495, 8.21], [494, §2]) *For each $n \geq 1$ the operator D is a duality between the lattice of equivalence classes of pp conditions with n free variables for right modules and the corresponding lattice for left modules. That is, for every pp condition ϕ we have $D^2\phi$ equivalent to ϕ and also $\psi \leq \phi$ iff $D\phi \leq D\psi$.*

Thus $D : (\mathrm{pp}^n_R)^{\mathrm{op}} \simeq \mathrm{pp}^n_{R^{\mathrm{op}}} = {}_R\mathrm{pp}^n$.

Proof. For each of the three basic implications, in the matrix form given after the proof of 1.1.13, we show that a reverse implication holds between the duals.

We have $\begin{pmatrix} I & A \\ 0 & B \end{pmatrix}\begin{pmatrix} I & 0 \\ 0 & C \end{pmatrix} = \begin{pmatrix} I & AC \\ 0 & BC \end{pmatrix}$ so, by 1.1.13 transposed for left modules, we have $\begin{pmatrix} I & AC \\ 0 & BC \end{pmatrix} \Rightarrow \begin{pmatrix} I & A \\ 0 & B \end{pmatrix}$.

We have $\begin{pmatrix} I & A \\ 0 & B_0B_1 \end{pmatrix} = \begin{pmatrix} I & 0 \\ 0 & B_0 \end{pmatrix}\begin{pmatrix} I & A \\ 0 & B_1 \end{pmatrix}$ and hence $\begin{pmatrix} I & A \\ 0 & B_1 \end{pmatrix} \Rightarrow \begin{pmatrix} I & A \\ 0 & B_0B_1 \end{pmatrix}$.

We have $\begin{pmatrix} I & A_0B + D \\ 0 & B \end{pmatrix} = \begin{pmatrix} I & A_0 \\ 0 & I' \end{pmatrix}\begin{pmatrix} I & D \\ 0 & B \end{pmatrix}$ (I' an identity matrix of appropriate size), hence $\begin{pmatrix} I & D \\ 0 & B \end{pmatrix} \Rightarrow \begin{pmatrix} I & A_0B + D \\ 0 & B \end{pmatrix}$.

So reversal of the ordering follows by the discussion around 1.1.13.

To compute the action of D^2, suppose ϕ has matrix $\begin{pmatrix} A \\ -B \end{pmatrix}$; then $D^2\phi$ has matrix $\begin{pmatrix} I & 0 \\ I & A \\ 0 & -B \end{pmatrix}$. From the following equations we deduce, using (the transpose of) 1.1.13, that ϕ and $D^2\phi$ are equivalent.

$$\begin{pmatrix} I & -I & 0 \\ 0 & 0 & -I \end{pmatrix}\begin{pmatrix} I & 0 \\ I & A \\ 0 & -B \end{pmatrix} = \begin{pmatrix} 0 & -A \\ 0 & B \end{pmatrix} = \begin{pmatrix} A \\ -B \end{pmatrix}(0 \quad -I)$$

$$\begin{pmatrix} I & 0 \\ 0 & 0 \\ 0 & -I \end{pmatrix}\begin{pmatrix} A \\ -B \end{pmatrix} = \begin{pmatrix} A \\ 0 \\ B \end{pmatrix} = \begin{pmatrix} I & 0 \\ I & A \\ 0 & -B \end{pmatrix}\begin{pmatrix} A \\ -I \end{pmatrix}$$

A proof for the functorial version of this duality is at 10.3.6 and relative versions can be found at 1.3.15, 3.4.18 and 5.4.12. □

In particular, $x = x$ and $x = 0$ are dual. For another easy example, consider $(xr = 0) \leq (xs = 0)$ iff $(s|x) \leq (r|x)$ for left modules (iff, see 1.1.14, $s \in rR$).

Example 1.3.2. If $\phi(x)$ is $\exists y, z\, \big((x = yr + zr') \wedge (ys = 0) \wedge (zs' = 0)\big)$, then

$$H = \begin{pmatrix} 1 & 0 & 0 \\ -r & s & 0 \\ -r' & 0 & s' \end{pmatrix}$$

so $D\phi(x)$ is

$$\exists y, z, w \begin{pmatrix} 1 & 1 & 0 & 0 \\ 0 & -r & s & 0 \\ 0 & -r' & 0 & s' \end{pmatrix} \begin{pmatrix} x \\ y \\ z \\ w \end{pmatrix} = 0,$$

which is $\exists y, z, w\, \big((x = -y) \wedge (ry = sz) \wedge (r'y = s'w)\big)$, which is equivalent to the condition $s|rx \wedge s'|r'x$.

A certain degree of care is necessary when computing duals of conditions with more than one free variable. Suppose that we are dealing with n-place pp conditions, where $n > 1$ and that we take the condition to be $x_1 = 0$. The dual of this is not $x_1 = x_1$: for the original condition is equivalent to $x_1 = 0 \wedge x_2 = x_2 \wedge \cdots \wedge x_n = x_n$, which, has dual $x_1 = x_1 \wedge x_2 = 0 \wedge \cdots \wedge x_n = 0$. So it is necessary to bear in mind that for each n there is a duality between the lattices of n-place pp conditions for right and left modules and to distinguish between conditions which belong to different lattices but which can be defined by the same formula.

The next result is immediate from 1.3.1.

Corollary 1.3.3. *For any pp conditions, ϕ and ψ, in the same number of free variables we have $D(\phi + \psi) = D\phi \wedge D\psi$ and $D(\phi \wedge \psi) = D\phi + D\psi$.*

For example, take ϕ to be $xr = 0$ and ψ to be $xs = 0$. Then $D\,(xr = 0 + xs = 0)$, that is, $D\,\big(\exists y, z\,(x = y + z \wedge yr = 0 \wedge zs = 0)\big)$ is $r|x \wedge s|x$ (a special case of 1.3.2) and $D(xr = 0 \wedge xs = 0)$ is $r|x + s|x$, that is, $\exists y, z\,(x = ry + sz)$.

Example 1.3.4. Let G be an $n \times m$ matrix over the ring R. Then the dual of the annihilation condition $\overline{x}G = 0$ (that is, $\overline{x}$ annihilates each column of G) is the divisibility condition $G \mid \overline{x}$. Note what this means: for a left module L an n-tuple $\overline{a}$ of elements of L is a solution to $G \mid \overline{x}$ iff $\overline{a} = G\overline{b}$ for some (column) m-tuple $\overline{b}$ from L and this is so iff $\overline{a}$ has the form $\sum_1^m \overline{k}_j \cdot b_j$ where $\overline{k}_1, \ldots, \overline{k}_m$ are the columns of G and the b_j are elements of L.

1.3.2 Elementary duality and tensor product

After giving a matrix criterion for "tensor-annihilation" it is shown (1.3.7, "Herzog's Criterion") that this exactly reflects duality for pp conditions.

If $\overline{a} = (a_1, \ldots, a_n)$ and $\overline{b} = (b_1, \ldots, b_n)$ with the $a_i \in M_R$ and the $b_i \in_R N$, then by $\overline{a} \otimes \overline{b}$ we mean $\sum_{i=1}^n a_i \otimes b_i \in M \otimes_R N$.

Proposition 1.3.5. (see [663, I.8.8]) *Let $\overline{a}$ be a tuple from the right R-module M and let $\overline{b}$ be from the left R-module N. Then $\overline{a} \otimes \overline{b} = 0$ in $M \otimes_R N$ iff there are $\overline{c}$ from M and $\overline{d}$ from N and matrices G, H such that $\begin{pmatrix} \overline{a} & \overline{0} \end{pmatrix} = \overline{c} \begin{pmatrix} G & H \end{pmatrix}$ and $\begin{pmatrix} G & H \end{pmatrix} \begin{pmatrix} \overline{b} \\ \overline{d} \end{pmatrix} = 0$.*

Proof. ($\Leftarrow$) If there are such matrices, then $\overline{a} = \overline{c}G, \overline{0} = \overline{c}H$ and $G\overline{b} + H\overline{d} = 0$. So $\overline{a} \otimes \overline{b} = \overline{c}G \otimes \overline{b} = \overline{c} \otimes G\overline{b} = \overline{c} \otimes (-H\overline{d}) = -\overline{c}H \otimes \overline{d} = 0$.

($\Rightarrow$) Extend $\overline{b}$ to a (possibly infinite) generating sequence $\overline{b}\,\overline{b}'$ for N. So there is an exact sequence $0 \to K \xrightarrow{j} R^{(I)} \xrightarrow{p} N \to 0$ (I indexing the tuple $\overline{b}\,\overline{b}'$) with $p\overline{e} = \overline{b}$ and $p\overline{e}' = \overline{b}'$, where $\overline{e}\,\overline{e}'$ is the tuple $(e_i)_{i \in I}$ and where e_i has 1 in the ith position and zeroes elsewhere.

Tensoring with M gives the exact sequence $M \otimes K \xrightarrow{1_M \otimes j} M \otimes R^{(I)} \xrightarrow{1_M \otimes p} M \otimes N \to 0$.

Since $\overline{a} \otimes \overline{b} = 0$ we have $\overline{a} \otimes \overline{e} \in \ker(1_M \otimes p) = \mathrm{im}(1_M \otimes j)$, say $\overline{a} \otimes \overline{e} = \overline{c} \otimes j(\overline{k})$ with $\overline{k}$ from K. Since $\overline{e}\,\overline{e}'$ generates $R^{(I)}$ there is a matrix $A = (G \;\; H)$ with $j(\overline{k}) = A \begin{pmatrix} \overline{e} \\ \overline{e}' \end{pmatrix}$ (here we remember that tuples from left modules are, by our conventions, column vectors). So $\overline{a} \otimes \overline{e} = \overline{c} \otimes A \begin{pmatrix} \overline{e} \\ \overline{e}' \end{pmatrix} = \overline{c}A \otimes (\overline{e}\,\overline{e}')$ and so the isomorphism $M \otimes R^{(I)} \simeq M^{(I)}$ takes $\overline{a} \otimes \overline{e}$, regarded as $(\overline{a}\,\overline{0}) \otimes (\overline{e}\,\overline{e}')$, to the same image as $\overline{c}A \otimes (\overline{e}\,\overline{e}')$, therefore $(\overline{a}\,\overline{0}) = \overline{c}A$.

The tuples $\overline{b}' = p\overline{e}'$ and $\overline{0}$ could be infinite but A has only finitely many non-zero rows so, reducing A, $\overline{0}$ and $\overline{b}'$ to suitable finite parts, the equations $0 = pj(\overline{k}) = A \begin{pmatrix} \overline{b} \\ p\overline{e}' \end{pmatrix}$ and $(\overline{a}\,\overline{0}) = \overline{c}A$ give the desired solution. □

Corollary 1.3.6. *Suppose that $\overline{a}$ is from M_R and that ${}_RL$ is a finitely generated left R-module, generated by the tuple $\overline{l}$. Then $\overline{a} \otimes \overline{l} = 0$ iff there is $\overline{c}$ from M and a matrix G such that $\overline{a} = \overline{c}G$ and $G\overline{l} = 0$.*

Proof. One direction is obvious. For the other suppose that $\overline{a} \otimes \overline{l} = 0$: by 1.3.5 there are $\overline{c}$ in M, $\overline{d}$ in L and matrices G, H as there. Since $\overline{l}$ generates L there is a matrix K such that $\overline{d} = K\overline{l}$. So $(G + HK)\overline{l} = 0$. Also, since $\overline{a} = \overline{c}G$ and $0 = \overline{c}H$, we have $\overline{c}(G + HK) = \overline{a}$. Replace G by $G + HK$ to obtain the conclusion. □

Theorem 1.3.7. ([274, 3.2], Herzog's Criterion) *Let $\overline{a}$ from M_R and $\overline{b}$ from ${}_RN$ be tuples of the same length. Then $\overline{a} \otimes \overline{b} = 0$ in $M \otimes_R N$ iff there is a pp condition ϕ (for right modules) such that $\overline{a} \in \phi(M)$ and $\overline{b} \in D\phi(N)$, that is, using the model-theoretic notation, $M \models \phi(\overline{a})$ and $N \models D\phi(\overline{b})$.*

Proof. Suppose that $\overline{a} \otimes \overline{b} = 0$ and choose G, H as in 1.3.5. Then $\overline{a} \in \phi(M)$, where ϕ is the condition $\exists \overline{y} \left(\overline{x} + \overline{y}G = 0 \wedge \overline{y}H = 0\right)$. In matrix form this is $\exists \overline{y}\, (\overline{x}\ \overline{y}) \begin{pmatrix} I & 0 \\ G & H \end{pmatrix} = 0$, which is $D\psi$ where ψ is the pp condition (for left R-modules), $\exists \overline{z} \begin{pmatrix} G & H \end{pmatrix} \begin{pmatrix} \overline{x} \\ \overline{z} \end{pmatrix} = 0$. By choice of G, H we have $\overline{b} \in \psi(N)$. By 1.3.1 ψ is equivalent to $D\phi$, as required.

For the converse, suppose that $\phi(\overline{a})$ holds in M and $D\phi(\overline{b})$ holds in N. Without loss of generality, ϕ is of the above, "dual", form (see just before 1.3.1) so the argument reverses. □

For example, if ϕ is $s|xr$, so $D\phi$ is $\exists y\, (x = ry \wedge ys = 0)$, and if $a \in \phi(M_R)$, say $ar = cs$, and if $b \in D\phi({}_RN)$, say $b = rd$ where $sd = 0$, then $a \otimes b = a \otimes rd = ar \otimes d = cs \otimes d = c \otimes sd = 0$. The general case in this direction is essentially the same.

Corollary 1.3.8. *If $(C, \overline{c})$ is a free realisation of ϕ and if $\overline{l}$ is a tuple from the left module L, then $\overline{c} \otimes \overline{l} = 0$ iff $\overline{l} \in D\phi(L)$.*

Corollary 1.3.9. *For any module M and pp condition ϕ in n free variables, $\phi(M) = \{\overline{a} \in M^n : \forall_R L,\ \overline{a} \otimes D\phi(L) = 0\}$.*

Corollary 1.3.10. *Suppose that $\overline{a}$ is a finite tuple from the finitely presented module A and let ϕ be a pp condition which generates $\mathrm{pp}^A(\overline{a})$ (1.2.6). Then $\mathrm{ann}_R(\overline{a}) = D\phi({}_RR)$.*

Example 1.3.11. Suppose that R is a k-algebra, where k is a field, let M be an R-module and set $M^* = \mathrm{Hom}_k(M, k)$ to be its k-dual, with the structure of a left R-module ($rf \cdot a = f(ar)$, $r \in R$, $f \in M^*$).

Let ϕ be $xr = 0$ for some $r \in R$, so $\phi(M) = \mathrm{ann}_M(r)$ and $D\phi(M^*) = rM^*$.

Clearly if $f \in D\phi(M^*)$, then $f \cdot \phi(M) = 0$. Conversely, if $f \in M^*$ and $f \cdot \phi(M) = 0$, then $g'(ar) = fa$ gives a well-defined k-linear map from Mr to k which (by injectivity of k_k) has at least one extension to some $g \in M^*$. Thus $f = rg \in D\phi(M^*)$.

This will be generalised in the next section.

1.3.3 Character modules and duality

It is shown that for any module its lattice of pp-definable subgroups and that of any suitable dual module are anti-isomorphic via duality of pp conditions (1.3.15). It is noted that the embedding of a module into its hom-dual is pure (1.3.16) and that strongly indecomposable finitely presented modules have strongly decomposable duals (1.3.19).

Given a right R-module M, let $S \to \mathrm{End}(M_R)$ be any ring morphism to its endomorphism ring, for instance: the identity of $\mathrm{End}(M_R)$; the canonical morphism from $\mathbb{Z}$ to $\mathrm{End}(M_R)$; the embedding of k into $\mathrm{End}(M_R)$ if R is a k-algebra. Regard M as a left S-module via this map. Let ${}_SE$ be any injective S-module and let $M^{*_E} = \mathrm{Hom}_S({}_SM, {}_SE)$, written just M^* for short, be the "character module" of M with respect to E; we refer to this as a (hom-)dual of M. This has a natural left R-module structure given by $rf.m = f(mr)$ for $f \in M^*, r \in R, m \in M$ (so $(sr)f.m = f(msr) = rf.ms = s.rf.m$). The key lemma is the following.

Lemma 1.3.12. (for example, [528, 1.5], [730, §2(c)]) *Let $\psi(\overline{x})$ be a pp condition for right R-modules with $l(\overline{x}) = n$. Then $\overline{f} \in (M^*)^n$ annihilates $\psi(M)$ iff $\overline{f} \in D\psi(M^*)$.*

Proof. If $\overline{f} \in D\psi(M^*)$ and $\overline{a} \in \psi(M)$, then, by 1.3.7, $\overline{a} \otimes \overline{f} = 0$ in $M \otimes_R M^*$ so certainly the value of $\overline{f}\overline{a}$ (that is, $\sum_i f_i a_i$), which is the image of $\overline{a} \otimes \overline{f}$ under the natural map $M \otimes_R M^* \to E$, is zero.

For the converse, suppose that $\psi(\overline{x})$ is $\exists \overline{y}\, (\overline{x}A = \overline{y}B)$, where the matrix A is $n \times m$ and B is $k \times m$. Then $D\psi$ is $\exists \overline{z}\, \big(\overline{x} = A\overline{z} \wedge B\overline{z} = 0\big)$. Suppose that $\overline{f}$ annihilates $\psi(M)$.

Consider the S-submodule $M^nA = \{\overline{c}A : \overline{c} \in M^n\}$ of M^m. Define $g' : M^nA \to E$ by $g(\overline{c}A) = \overline{f}\overline{c}$. This is well defined since if $\overline{c}A = \overline{c}'A$, then, because $(\overline{c} - \overline{c}')A = 0 = \overline{0}B$, certainly $\overline{c} - \overline{c}' \in \psi(M)$, hence $f\overline{c} = f\overline{c}'$.

Next, consider $M^kB \leq M^m$. Define $g'' : M^kB \to E$ to be the zero map. Note that g' and g'' agree on the intersection of their domains since $M^nA \cap M^kB$ is exactly $\psi(M)A$, on which g' is zero. So $g' + g''$ is defined unambiguously on $M^nA + M^kB$.

By injectivity of E there is an extension of the S-linear map $g' + g''$ to a morphism $\overline{g}$, say, from M^m to E. We may regard $\overline{g}$ as an m-tuple of elements of M^*.

By definition of the R-structure on M^*, for every $\overline{c} \in M^n$ we have $A\overline{g} \cdot \overline{c} = \overline{g}(\overline{c}A) = g'(\overline{c}A) = \overline{f}\overline{c}$, hence $A\overline{g} = \overline{f}$. Also, for every $\overline{d} \in M^k$ we have $B\overline{g} \cdot \overline{d} = \overline{g}(\overline{d}B) = g''(\overline{d}B) = 0$, so $B\overline{g} = 0$. Therefore $\overline{f} \in D\psi(M^*)$, as required. □

Corollary 1.3.13. *If $\psi(M) \leq \phi(M)$, then $D\phi(M^*) \leq D\psi(M^*)$. In particular if $\phi \geq \psi$ is a pp-pair with $\phi(M) = \psi(M)$, then $D\psi(M^*) = D\phi(M^*)$.*

Proof. If $\overline{f} \in D\phi(M^*)$, then, by 1.3.12, $\overline{f} \cdot \phi(M) = 0$ so $\overline{f} \cdot \psi(M) = 0$ and hence, again by 1.3.12, $\overline{f} \in D\psi(M^*)$. □

Corollary 1.3.14. (for example, observed in [277, p. 335]) *If the pp condition ϕ with n free variables defines a submodule, as opposed to just a subgroup, of M^n, then $D\phi$ defines a submodule of $(M^*)^n$, where $M^* = M^{*_E}$, with E as at the beginning of this section.*

Proof. Say ϕ is $\exists \overline{y}\, (\overline{x}\ \overline{y}) \begin{pmatrix} A \\ B \end{pmatrix} = 0$. For $r \in R$ let ϕ_r denote the pp condition $\phi(\overline{x}r)$, that is, $\exists \overline{y}\, (\overline{x}\ \overline{y}) \begin{pmatrix} rA \\ B \end{pmatrix} = 0$ (rA meaning $rI \cdot A$, where I is the appropriate identity matrix): so ϕ_r defines the subgroup of tuples $\overline{a}$ such that $\phi(\overline{a}r)$ holds.

Computing duals one sees that $D\phi_r$ is $\exists \overline{z}\, \big(\overline{x} = -rA\overline{z} \wedge B\overline{z} = 0\big)$ which, setting $\overline{x}'$ to be $A\overline{z}$, is $\exists \overline{x}'\, \big(\overline{x} = r\overline{x}' \wedge D\phi(\overline{x}')\big)$, that is, "$\overline{x} \in rD\phi$".

Since $\phi(M)$ is a submodule of M^n, one has $\phi(M) \leq \phi_r(M)$ so, by 1.3.13, $D\phi_r(M^*) \leq \phi(M^*)$, that is, $rD\phi(M^*) \leq D\phi(M^*)$, as required. □

The converse to 1.3.13 holds if E is large enough.

Theorem 1.3.15. ([730, Lemma 2, Prop. 3], see also [274], [528]) *Suppose that M is a right R-module, that $S \to \mathrm{End}(M_R)$ is a ring morphism and that ${}_SE$ is injective. If, for every pp condition ψ, the S-module $M/\psi(M)$ embeds in a power of E (in particular, if E is an injective cogenerator of S-*Mod*), then for every pp-pair $\phi \geq \psi$ (in n free variables for any n) we have*

$$\phi(M) = \psi(M) \quad \textit{iff} \quad D\psi(M^*) = D\phi(M^*).$$

That is, for every n, $\phi \leftrightarrow D\phi$ induces an anti-isomorphism between the lattice of subgroups of M^n pp-definable in M_R and the lattice of subgroups of M^{*n} pp-definable in ${}_RM^*$.

Proof. Suppose that $\phi > \psi$ is a pp-pair in one free variable and that $\phi(M) > \psi(M)$: choose $a \in \phi(M) \setminus \psi(M)$. By assumption there is an S-linear map $f' : M/\psi(M) \to E$ with $f(a + \psi(M)) \neq 0$. Let $f : M \to E$ be the composition of the natural projection $M \to M/\psi(M)$ with f', so $fa \neq 0$.

Since $f \cdot \psi(M) = 0$, 1.3.12 gives $f \in D\psi(M^*)$ and, since $f \cdot \phi(M) \neq 0$, it is also the case that $f \notin D\phi(M^*)$, so $D\psi(M^*) > D\phi(M^*)$.

The same proof works for pp conditions in n free variables. In fact, by 3.4.10 applied to the definable category generated by M (see just before 3.4.9), the case $n = 1$ implies the general case. □

For example, if $R = \mathbb{Z}$ and $E = \bigoplus_{p\ \mathrm{prime}} \mathbb{Z}_{p^\infty}$ is the minimal injective cogenerator for $\mathbb{Z}$-modules, then the p-adic integers $\overline{\mathbb{Z}_{(p)}}$ and the p-Prüfer group $\mathbb{Z}_{p^\infty}$

correspond under this duality and the result above implies that their lattices of pp-definable subgroups are anti-isomorphic.

Now, given M, S and E as above, suppose that a suitable dualising module also has been chosen for M^* and denote by M^{**} the image of M^* under this second duality (it is shown at 4.3.31 that this double dual is pure-injective).

Corollary 1.3.16. *With notation and assumptions, for both dualities, as in 1.3.15 the canonical map $M \to M^{**}$ is a pure embedding (for pure embeddings see Section 2.1.1).*

Proof. In the criterion 2.1.6 for purity it suffices to consider pp conditions ψ in one free variable. Suppose that $a \in \psi(M^{**})$. Then, by 1.3.12 applied to M^* and since $D^2\psi \leftrightarrow \psi$, $fa = 0$ for every $f \in D\psi(M^*)$. If a were not in $\psi(M)$ there would be, by the proof of 1.3.15, some $f \in M^*$ annihilating $\psi(M)$ hence, by 1.3.12, with $f \in D\psi(M^*)$, but with $fa \neq 0$, a contradiction. □

Lemma 1.3.17. *If M_R is finitely presented, ${}_SN_R$ is any bimodule and ${}_SE$ is injective, then there is a natural isomorphism* $\mathrm{Hom}_S(\mathrm{Hom}_R(M_R, {}_SN_R), {}_SE) \simeq M \otimes_R \mathrm{Hom}_S({}_SN_R, {}_SE)$.

Proof. When $M = R^n$ the left-hand side is $\mathrm{Hom}_S(N^n, E) = (\mathrm{Hom}_S(N, E))^n$, which is naturally isomorphic to the right-hand side.

More generally, given a presentation $R^m \to R^n \to M \to 0$, the sequence $0 \to (M, N) \to (R^n, N) \to (R^m, N)$ is exact and hence, since E is injective, $((R^m, N), E) \to ((R^n, N), E) \to ((M, N), E) \to 0$ is exact. Moreover, since $\otimes$ is right exact, also $R^m \otimes (N, E) \to R^n \otimes (N, E) \to M \otimes (N, E) \to 0$ is exact. Therefore we have the diagram

$$\begin{array}{ccccccc} ((R^m, N), E) & \longrightarrow & ((R^n, N), E) & \longrightarrow & ((M, N), E) & \longrightarrow & 0 \\ \simeq\downarrow & & \simeq\downarrow & & & & \\ R^m \otimes (N, E) & \longrightarrow & R^n \otimes (N, E) & \longrightarrow & M \otimes (N, E) & \longrightarrow & 0 \end{array}$$

with exact rows so (by the Five Lemma for example, [708, 1.3.3]) there is an isomorphism as required. □

Corollary 1.3.18. *If M_R is finitely presented, $S = \mathrm{End}(M_R)$ and if ${}_SE$ is injective, then $M \otimes_R M^{*_E} \simeq E$ as S-modules.*

Proof. Take $M = N$ in 1.3.17, noting that in this case the isomorphisms in the proof are S-linear. □

Lemma 1.3.19. (see [277, p. 336]) *Suppose that M_R is a finitely presented module with $S = \mathrm{End}(M_R)$ local. Let $E = E(S/J(S))$ (the operator E denotes injective* *hull, Appendix E, p. 708). Then* $_RM^* = \mathrm{Hom}_S({}_SM_R, {}_SE)$ *is indecomposable and also has local endomorphism ring* ($\simeq \mathrm{End}({}_SE)$).

Proof. Set $T = \mathrm{End}({}_RM^*)$. Then $T = \mathrm{Hom}_R({}_RM^*, {}_R\mathrm{Hom}_S({}_SM_R, {}_SE)) \simeq \mathrm{Hom}_S({}_SM \otimes_R M^*, {}_SE) \simeq \mathrm{Hom}_S({}_SE, {}_SE)$ (the last by 1.3.18) which is indeed local, being the endomorphism ring of an indecomposable injective (E.1.23). □

See [528, §1] for an alterative "axiomatic" approach to elementary duality between pp conditions and also, [528, §§3,4], for an extension to infinitary conditions. The term "local duality" is used in [164].

1.3.4 Pp conditions in tensor products

There is a generalisation of 1.3.7 under which the equation $\overline{a} \otimes \overline{b} = 0$ *is regarded as "*$\overline{a} \otimes \overline{b}$ *satisfies the pp condition* $x = 0$*".*

Let M_R be a right R-module and let $_RN_S$ be an (R, S)-bimodule. Let $\sigma(\overline{x})$ be a pp condition for right S-modules. Given $\overline{a}$ in M we may consider those $\overline{b}$ in N such that $\overline{a} \otimes \overline{b}$ satisfies σ in $(M \otimes_R N)_S$ (if σ has n free variables where $n > 1$, then $\overline{a}$ and $\overline{b}$ must be partitioned into n subtuples of varying, but matching, lengths). The result proved below implies that if $\overline{a}$ has finitely generated pp-type, generated by a pp condition ϕ, then there is a pp condition, written $(\sigma : \phi)$, for (R, S)-bimodules, such that, for any $_RN_S$, $\overline{a} \otimes \overline{b} \in \sigma(M \otimes_R N_S)$ iff $\overline{b} \in (\sigma : \phi)({}_RN_S)$.

Recall that (R, S)-bimodules are the same as right $R^{\mathrm{op}} \otimes S$-modules, so the notion of a pp condition for bimodules is included in that notion for modules. We use the "natural" notation: a typical element of $R^{\mathrm{op}} \otimes S$ has the form $\sum r_l \otimes s_l$ but we write $\sum r_l x s_l$ rather than $x \sum r_l \otimes s_l$. So a typical pp condition for (R, S)-bimodules is written $\exists y_1, \ldots, y_m \bigwedge_j \left(\sum_i \sum_l r_{ilj} x_i s_{ilj} + \sum_k \sum_h r'_{khj} y_k s'_{khj} = 0\right)$.

Theorem 1.3.20. ([509, 2.1]) *Suppose that* σ *is a pp condition for right* S*-modules and that* ϕ *is a pp condition for right* R*-modules. There is a pp condition,* $(\sigma : \phi)$*, for* (R, S)*-bimodules such that for any right* R*-module* M *and* (R, S)*-bimodule* N*:*

(a) if $\overline{a} \in \phi(M)$ *and* $\overline{b} \in (\sigma : \phi)(N)$*, then* $\overline{a} \otimes \overline{b} \in \sigma((M \otimes_R N)_S)$*;*
(b) if the pp-type of $\overline{a}$ *in* M *is generated by* ϕ*, then for any* $\overline{b}$ *from* N *we have* $\overline{a} \otimes \overline{b} \in \sigma(M \otimes N)$ *iff* $\overline{b} \in (\sigma : \phi)(M)$*;*
(c) if $\overline{a}$ *from* M *and* $\overline{b}$ *from* N *are such that* $\overline{a} \otimes \overline{b} \in \sigma(M \otimes N)$*, then there is some pp condition* ϕ *such that* $\overline{a} \in \phi(M)$ *and* $\overline{b} \in (\sigma : \phi)(N)$*.*

Proof. Fix a free realisation, $\overline{a}$ in C_ϕ, of ϕ. Say C_ϕ is generated by $\overline{c}$ with presentation $\overline{c}G = 0$ and suppose that $\overline{a} = \overline{c}H$, so $\phi(\overline{x})$ is equivalent to $\exists \overline{y} \left(\overline{y}G = 0 \wedge \overline{x} = \overline{y}H\right)$. Note that the pp-type of $\overline{c}$ in C_ϕ is generated by $\overline{y}G = 0$. Suppose that $\sigma(\overline{u})$ is $\exists\overline{v}\,(\overline{v}L = \overline{u}K)$. (So G, H are matrices over R and L, K are matrices over S.)

Let N be an (R, S)-bimodule and let $\overline{b}$ be a tuple from N. Then $C_\phi \otimes N \models \sigma(\overline{a} \otimes \overline{b})$ iff there are $\overline{a}'$ from C_ϕ and $\overline{b}'$ from N such that[19] $(\overline{a}' \otimes \overline{b}')L = (\overline{a} \otimes \overline{b})K$. Now $\overline{c}$ generates C_ϕ, so $\overline{a}' = \overline{c}H'$ for some matrix H'. Hence $\overline{a}' \otimes \overline{b}'L = \overline{c} \otimes (H'\overline{b})L$. Therefore, $\sigma(\overline{a} \otimes \overline{b})$ holds iff there is $\overline{b}''$ from N such that $\overline{c} \otimes \overline{b}''L = \overline{a} \otimes \overline{b}K$, $= \overline{c} \otimes H\overline{b}K$, and that is so iff $\overline{c} \otimes (H\overline{b}K - \overline{b}''L) = 0$ for some $\overline{b}''$. By 1.3.6 and since $(C_\phi, \overline{c})$ is a free realisation of $\overline{y}G = 0$, this is so iff $G|(H\overline{b}K - \overline{b}''L)$ is true in ${}_RN$ for some $\overline{b}''$.

Therefore the condition $(\sigma : \phi)(\overline{u})$ required for (b) is $\exists\overline{w}\; G|(H\overline{u}K - \overline{w}L)$. Statement (a) follows easily by 1.2.17. For the last part, suppose $\overline{a} \otimes \overline{b} \in \sigma(M \otimes N)$, say $\overline{a}' \otimes \overline{b}'L = \overline{a} \otimes \overline{b}K$, hence $\overline{a}'\overline{a} \otimes (-\overline{b}'L)\overline{b}K = 0$. By 1.3.7 there is a pp condition ϕ_0 for right R-modules such that $\overline{a}'\overline{a} \in \phi_0(M)$ and $(-\overline{b}'L)\overline{b}K \in D\phi_0(N)$. Let ϕ be the pp condition $\exists\overline{w}\phi_0(\overline{w}, \overline{x})$, so $\overline{a} \in \phi(M)$. Let $(C, \overline{c})$ be a free realisation of ϕ, say $\overline{d}$ from C is such that $(\overline{d}, \overline{c}) \in \phi_0(C)$. By 1.3.7, $\overline{dc} \otimes (-\overline{b}'L)\overline{b}K = 0$, that is, $\overline{d} \otimes \overline{b}'L = \overline{c} \otimes \overline{b}K$, so $\overline{c} \otimes \overline{b} \in \sigma(C \otimes_R N)$. By part (a) it follows that $\overline{b} \in (\sigma : \phi)$, as required. □

In [509] various elementary properties of the operator $(- : -)$ are given, as is a variety of examples and applications (expressed model-theoretically). For instance, using this one may define the tensor product of closed subsets of the right and left Ziegler spectra, equivalently of definable subcategories of Mod-R and R-Mod.

A somewhat related observation is that, given S and a pp condition σ for S-modules, for any finitely presented R-module C, there is a pair, $\eta \geq \zeta$, of pp conditions for (R, S)-bimodules such that for each $N \in R$-Mod-S we have $\sigma(C \otimes_R N) \simeq \eta(N)/\zeta(N)$. Model-theoretically this is clear since ($- \otimes N$ being right exact) $C \otimes N$ and hence $\sigma(C \otimes N)$ is part of $N^{\mathrm{eq}+}$ (see Section B.2). Alternatively use the characterisation (18.1.20) of pp-pairs on a module category as those functors to **Ab** which commute with direct limits and products, plus the fact that $C \otimes_R -$ is such a functor.

1.3.5 Mittag–Leffler modules

Mittag–Leffler modules, which include pure-projective, in particular finitely presented, modules (1.3.25), are characterised (1.3.22) as those in which every pp-type realised is finitely generated. Every countably generated Mittag–Leffler

[19] Recall that if σ has more than one free variable, then $\overline{a}, \overline{b}, \overline{a}', \overline{b}'$ should be read as tuples of tuples, equivalently as partitioned tuples, and $\overline{a}' \otimes \overline{b}'$ interpreted accordingly.

module is pure-projective (1.3.26) and, though the converse is false, 1.3.27 gives an approximate converse.

A module M is said to be **Mittag–Leffler** if for every collection $(N_i)_i$ of left R-modules, the canonical morphism, $M \otimes (\prod_i N_i) \to \prod_i (M \otimes N_i)$, which is determined by its taking $m \otimes (n_i)_i$ to $(m \otimes n_i)_i$, is injective (recall that $\otimes$ means, by default, $\otimes_R$). These were introduced in [588, pp. 69, 70] and further investigated in [647], [256], [30], [621], [735] for example.

Lemma 1.3.21. *The class of Mittag–Leffler modules is closed under arbitrary direct sums and under direct summands.*

Proof. The first assertion follows since tensor product commutes with direct sums (and since $\bigoplus_j \prod_i M_{ji}$ embeds naturally in $\prod_i \bigoplus_j M_{ji}$). For the second, consider, given a direct summand, M', of M, the commutative diagram shown.

$$\begin{array}{ccc} M \otimes (\prod_i N_i) & \xrightarrow{f} & \prod_i (M \otimes N_i) \\ {\scriptstyle j}\uparrow & & \uparrow \\ M' \otimes (\prod_i N_i) & \xrightarrow[f']{} & \prod_i (M' \otimes N_i) \end{array}$$

□

Theorem 1.3.22. ([621, 2.2]) *A module M is Mittag–Leffler iff for every tuple $\overline{a}$ from M the pp-type $\mathrm{pp}^M(\overline{a})$ is finitely generated.*[20]

Proof. Suppose that $\overline{a} \otimes \overline{b} \in M \otimes (\prod N_i)$ is sent to 0 by $M \otimes (\prod_i N_i) \to \prod_i (M \otimes N_i)$. That is, $\overline{a} \otimes \overline{b}_i = 0$ for each i, where $\overline{b} = (\overline{b}_i)_i$. By 1.3.7 there is, for each i, some pp condition ϕ_i such that $\overline{a}$ satisfies ϕ_i in M and $\overline{b}_i$ satisfies $D\phi_i$ in N_i. If the pp-type of $\overline{a}$ is finitely generated, by ϕ say, then (by 1.3.1 since $\phi \le \phi_i$ so $D\phi_i \le D\phi$) every $\overline{b}_i$ satisfies $D\phi$. Therefore, by 1.2.2, $\overline{b}$ satisfies $D\phi$ and hence, by 1.3.7, $\overline{a} \otimes \overline{b} = 0$.

That gives one direction: for the other, suppose that M is Mittag–Leffler and let $\overline{a}$ be a tuple from M. For each $\phi_i \in \mathrm{pp}^M(\overline{a})$ let $\overline{d}_i$ from $C_i \in R\text{-Mod}$ be a free realisation of $D\phi_i$: so $\overline{a} \otimes \overline{d}_i = 0$ by 1.3.7. Consider $\overline{d} = (\overline{d}_i)_i$ in $\prod_i C_i$. The map from $M \otimes (\prod_i C_i)$ to $\prod_i (M \otimes C_i)$ takes $\overline{a} \otimes \overline{d}$ to $(\overline{a} \otimes \overline{d}_i)_i = 0$. By assumption it must be that $\overline{a} \otimes \overline{d} = 0$ and so, by 1.3.7, there must be $\phi \in \mathrm{pp}^M(\overline{a})$ such that $\overline{d}$, and hence every $\overline{d}_i$, satisfies $D\phi$. Since $(C_i, \overline{d}_i)$ is a free realisation of $D\phi_i$ we deduce that $D\phi \ge D\phi_i$ and so, by 1.3.1, $\phi \le \phi_i$ for every i. It follows that $\mathrm{pp}^M(\overline{a})$ is generated by ϕ, as required. □

The next result is then immediate from 1.2.6.

[20] This has a model-theoretic interpretation: a module is Mittag–Leffler iff it is positively atomic. Also see [384].

Corollary 1.3.23. *Every finitely presented module is Mittag–Leffler.*

In fact the following is true.

Proposition 1.3.24. ([394, Sätze 1,2]) *A module M is finitely presented iff the canonical map $M \otimes (\prod_i N_i) \to \prod_i (M \otimes N_i)$ is an isomorphism for all collections of modules $(N_i)_i$ (and M is finitely generated iff each such map is a surjection).*

Since, 2.1.26, a module is pure-projective iff it is a direct summand of a direct sum of finitely presented modules, the next result follows.

Corollary 1.3.25. *Every pure-projective module is Mittag–Leffler.*

The converse is not true in general, see, for example, [30, Prop. 7], indeed, [29, Thm 5] (also see [621, 2.4]) any locally pure-projective module (in the sense of [29]) is Mittag–Leffler. The following is, however, true.

Proposition 1.3.26. ([588, p. 74, 2.2.2]) *Every countably generated Mittag–Leffler module is pure-projective.*

Proof. Let $a_1, a_2, \dots, a_n, \dots$ be a countable generating set for M. Let $\pi : N \to M$ be any pure epimorphism (see Section 2.1.1): we prove that π splits. Let $\phi_n(x_1, \dots, x_n)$ be a pp condition such that $\mathrm{pp}^M(a_1, \dots, a_n) = \langle \phi_n \rangle$ (this exists by 1.3.22). Suppose, inductively, that we have $b_1, \dots, b_n \in N$ with $\pi b_i = a_i$ and $(b_1, \dots, b_n) \in \phi_n(N)$.

Since $\exists x_{n+1}\, \phi_{n+1}(a_1, \dots, a_n, x_{n+1})$ is true in M (take a_{n+1} for x_{n+1}), then $\exists x_{n+1}\, \phi_{n+1}(x_1, \dots, x_n, x_{n+1}) \in \mathrm{pp}^M(a_1, \dots, a_n)$, so, by choice of ϕ_n, we have $\phi_n(\overline{x}) \le \exists x_{n+1}\, \phi(\overline{x}, x_{n+1})$ and hence, since N satisfies $\phi_n(b_1, \dots, b_n)$, there is some $b''_{n+1} \in N$ such that $(b_1, \dots, b_n, b''_{n+1}) \in \phi_{n+1}(N)$.

Since π is a pure epimorphism there is, by 2.1.14, a tuple $(b'_1, \dots, b'_{n+1})$ from N such that $\pi b'_i = a_i$ $(i = 1, \dots, n+1)$ and $(b'_1, \dots, b'_{n+1}) \in \phi_{n+1}(N)$, so $(b_1 - b'_1, \dots, b''_{n+1} - b'_{n+1}) \in \phi_{n+1}(N)$. Since $b_1 - b'_1, \dots, b_n - b'_n$ all are in $\ker(\pi)$ and since $\ker(\pi)$ is pure in N there is $b \in \ker(\pi)$ with $(b_1 - b'_1, \dots, b_n - b'_n, b) \in \phi_{n+1}(N)$. It follows that $(b_1, \dots, b_n, b'_{n+1} + b) \in \phi_{n+1}(N)$ and this tuple maps under π to $(a_1, \dots, a_n, a_{n+1})$. Set $b_{n+1} = b'_{n+1} + b$.

Define $M \to N$ by $a_i \mapsto b_i$. By construction this gives a well-defined map (if a linear combination of the a_i is 0, then, because this is part of the pp-type of the a_i involved, so is the corresponding linear combination of the b_i) so π is indeed split, as required. □

Theorem 1.3.27. ([588, p. 73, 2.2.1] also see [647, 3.16]) *The following conditions on a module M are equivalent.*

(i) M is Mittag–Leffler;
(ii) every finite subset of M is contained in a pure-projective pure submodule of M;
(iii) every countable subset of M is contained in a pure-projective pure submodule of M.

Proof. (iii)⇒(ii) is trivial.

(ii)⇒(i) Take any tuple $\overline{a}$ from M and let M' be a pure, pure-projective submodule of M which contains $\overline{a}$. Then $\mathrm{pp}^M(\overline{a}) = \mathrm{pp}^{M'}(\overline{a})$ which, by 1.3.25, is finitely generated. By 1.3.22, M is Mittag–Leffler.

(i)⇒(iii) Let X_0 be a countable subset of M. There are just countably many finite tuples, $\overline{a}$, of elements of X_0. By 1.3.22, the pp-type of any such $\overline{a}$ is finitely generated; say $\phi(\overline{x})$, being $\exists \overline{y}\, \theta(\overline{x}, \overline{y})$ with θ quantifier-free, generates $\mathrm{pp}^M(\overline{a})$. Choose and fix some tuple $\overline{b}$ from M such that $\theta(\overline{a}, \overline{b})$ holds in M. Do this for each finite tuple $\overline{a}$ from X_0 and let X_1 be the union of X_0 and all elements appearing in any of the chosen tuples $\overline{b}$.

Repeat this with X_1 in place of X_0 to obtain X_2, and so on. Set $X_\omega = \bigcup_n X_n$. If $\overline{a}$ is a finite tuple from X_ω, then it is already from X_n for some n and so a generator for $\mathrm{pp}^M(\overline{a})$ is, by construction, witnessed in X_{n+1}, hence in X_ω. That is, for every $\overline{a}$ from X_ω we have $\mathrm{pp}^M(\overline{a}) = \mathrm{pp}^{M_\omega}(\overline{a})$, where M_ω is the submodule of M generated by X_ω. It follows, by 2.1.20, that M_ω is pure in M, hence is Mittag–Leffler and so, by 1.3.26, M_ω is pure-projective. □

Generalisations to arbitrary purities can be found in [620, Chpt. 3]. Also see [621], [384], [737].

Gruson and Jensen ([256, 5.8]) prove that a module M is Mittag–Leffler iff the functor $M \otimes - : R\text{-mod} \to \mathbf{Ab}$ (Section 12.1.1) is locally coherent (Section 2.3.3).

2
Purity

This chapter presents basic results on purity and certain kinds of modules (absolutely pure, flat, coherent). Then detailed connections between the shape of pp conditions and direct-sum decomposition of finitely presented modules, especially over RD rings, are established.

2.1 Purity

Sections 2.1.1 and 2.1.3 contain the definition of purity (of embeddings, of exact sequences, of epimorphisms) and basic results on this. Pure-projective modules are introduced in Section 2.1.2: the dual concept of pure-injectivity is the theme of Chapter 4.

Homological dimensions based on pure-exact sequences are described in Section 2.2.

For purity and pure-injectivity in more general contexts see, for example, [86], [622].

2.1.1 Pure-exact sequences

Purity is defined first in terms of pp conditions/solutions of systems of equations; there are other characterisations (2.1.7, 2.1.19, 2.1.28). An exact sequence is pure iff it is a direct limit of split exact sequences (2.1.4). There is a characterisation of pure epimorphisms in terms of lifting of pp conditions (2.1.14). A short exact sequence which is pure is split if its last non-zero term is finitely presented (2.1.18). Every subset of a module is contained in a pure submodule which is not too much bigger (2.1.21).

Suppose that A is a submodule of B and that $a \in A$. If ϕ is a pp condition and $a \in \phi(A)$, then certainly $a \in \phi(B)$ (1.1.7). The converse need not be true, for instance, the non-zero element in $A = \mathbb{Z}_2 \leq \mathbb{Z}_4 = B$ is divisible by 2 in B but not in A. Say that A is **pure** in B if $\phi(A) = A^{l(\overline{x})} \cap \phi(B)$ for every pp condition $\phi = \phi(\overline{x})$. The embedding of A into B is then said to be a **pure embedding**. More generally, a monomorphism $f : A \to B$ is a **pure monomorphism** if fA is a pure submodule of B. (We tend not to distinguish between embeddings and inclusions so, in practice, in this situation we would say that A is pure in B.) The notion is due to Prüfer [533] in the context of abelian groups and the concept for arbitrary rings is due to Cohn [126]. It was further developed by various people, for example, see [660], [661], [699], [192], [588]; also see the notes to Chapter V of [198]. More general notions of purity were also considered by Butler and Horrocks [109]. Purity and associated notions have been developed in very general algebraic systems, for which see the references at the end of Section 4.2.1. Also see the survey [308].

Remark 2.1.1. Since $\phi(A) \subseteq A^n \cap \phi(B)$ is always true (where ϕ has n free variables) we have the following reformulations of the definition: A is pure in B

iff for every tuple $\overline{a}$ from A and pp condition ϕ, if $\overline{a}$ satisfies ϕ in B then $\overline{a}$ satisfies ϕ in A (in the model-theoretic notation, $B \models \phi(\overline{a})$ implies $A \models \phi(\overline{a})$);
iff for every tuple $\overline{a}$ from A one has $\mathrm{pp}^A(\overline{a}) = \mathrm{pp}^B(\overline{a})$;
iff every finite system of R-linear equations with constants from A and a solution from B already has a solution from A.

(For the last, note that solvability of a given system of equations with constants $\overline{a}$ can be expressed by the fact that $\overline{a}$ satisfies a certain pp condition.)

Certainly any split embedding is pure (project a solution in the larger module to the smaller one). An example of a pure, non-split, embedding is the embedding of the localisation $\mathbb{Z}_{(p)}$ of $\mathbb{Z}$ at a prime p into its completion in the p-adic topology, the ring, $\overline{\mathbb{Z}_{(p)}}$, of p-adic integers: that this embedding of $\mathbb{Z}$- (or $\mathbb{Z}_{(p)}$-)modules is pure follows, for example, using the fact, see 2.4.11, that $\mathbb{Z}$ is an RD ring. Since, see 2.1.18, pure embeddings between finitely presented modules are always split, one has to look to infinitely presented modules to obtain such examples. In some sense, see 4.3.18, examples are almost ubiquitous but they may be hard to describe explicitly.

Lemma 2.1.2. *(a) If A is a direct summand of B, then the embedding of A into B is pure.*
(b) Any composition of pure embeddings is a pure embedding.
(c) Any direct limit of pure embeddings is a pure embedding.
(d) Any direct product of pure embeddings is a pure embedding.

Proof. Let $\overline{a}$ be in A. If $\pi : B \to A$ splits the inclusion, then, by 1.1.7, $\text{pp}^A(\overline{a}) \geq \text{pp}^B(\overline{a}) \geq \text{pp}^A(\pi\overline{a} = \overline{a})$, which establishes (a). Part (b) is also immediate from the definition and part (d) is immediate from 1.2.3. As for the third part, take a directed system of pure embeddings: say Λ is a directed index set and for each $\lambda \in \Lambda$ there is a pure embedding, $i_\lambda : A_\lambda \to B_\lambda$, as well as, for each pair $\lambda \leq \mu$, a pair of morphisms $g_{\lambda\mu} : A_\lambda \to A_\mu$, $k_{\lambda\mu} : B_\lambda \to B_\mu$ such that $k_{\lambda\mu} i_\lambda = i_\mu g_{\lambda\mu}$ and such that, for $\lambda \leq \mu \leq \nu$ we have $g_{\mu\nu} g_{\lambda\mu} = g_{\lambda\nu}$ and $h_{\mu\nu} h_{\lambda\mu} = h_{\lambda\nu}$. Let the direct limit be $i : A \to B$. Since direct limit in Grothendieck categories is (left) exact (see Appendix E, p. 707) this is an embedding: the claim is that it is a pure embedding.

Take $\overline{a}$ from A and suppose that $i\overline{a}$ satisfies the pp condition ϕ in B. Say ϕ is $\exists \overline{y}\,(\overline{x}\ \overline{y})H = 0$ and suppose that $\overline{b}$ from B is such that $(i\overline{a}\ \overline{b})H = 0$. Choose λ such that each element in the tuple $\overline{a}$ has a preimage in A_λ and each element in $\overline{b}$ has a preimage in B_λ; so if $\overline{a} = (a_1, \ldots a_n)$, then for each j, $ia_j = k_{\lambda\infty}(i_\lambda a_{j\lambda})$ for some $a_{j\lambda} \in A_\lambda$, where $k_{\lambda\infty}$ denotes the natural map from B_λ to the limit B. Each column of the matrix H gives an equation saying that a certain linear combination of the ia_j and b_k is zero: by the definition of the direct limit, that equation can hold only if some preimage of this linear combination is zero. So, if λ is chosen "large enough", we can suppose that there are preimages, $i_\lambda\overline{a}'$ and $\overline{b}'$ of $i\overline{a}$ and $\overline{b}$ in B_λ satisfying $(i_\lambda\overline{a}'\ \overline{b}')H = 0$. That is, $i_\lambda\overline{a}'$ satisfies ϕ in B_λ and hence, by purity of A_λ in B_λ, $\overline{a}'$ satisfies ϕ in A_λ. Therefore, by 1.1.7, $\overline{a} = g_{\lambda\infty}\overline{a}'$ satisfies ϕ in A, as required. □

A special case of part (c) above arises if, for every λ, there is a pure submodule A_λ of a fixed module B and if this system of submodules is directed upwards (for every λ, μ there is $\nu \geq \lambda, \mu$ with $A_\lambda + A_\mu \leq A_\nu$): then the union = sum of the A_λ is pure in B.

Corollary 2.1.3. *A direct limit of split embeddings is a pure embedding.*

In fact the converse holds. By a **pure-exact sequence** we mean a short exact sequence $0 \to A \xrightarrow{j} B \xrightarrow{\pi} C \to 0$, where j is a pure embedding. Since, in a module category (more generally in a Grothendieck category), every direct limit of short exact sequences is exact, it follows that every direct limit of pure-exact sequences is pure-exact.

Proposition 2.1.4. ([391, 2.3]) *Every pure-exact sequence is a direct limit of split exact sequences of finitely presented modules.*[1] *Hence an exact sequence is pure-exact iff it is a direct limit of split exact sequences.*

[1] This is also true in any locally finitely presented abelian category but, of course, it requires that there be "enough" finitely presented objects so, although purity can be defined in contexts more general than that, see Section 16.1.2, this result does not hold in such generality.

Proof. Suppose that $0 \to A \xrightarrow{k} B \xrightarrow{\pi} C \to 0$ is pure exact and that $\big((C_i)_i, (f_{ij} : C_i \to C_j)_{ij}\big)$ is a directed system of finitely presented modules, with direct limit $\big(C, (f_i : C_i \to C)_i\big)$.

For each i set $B_i = \{(b, c) \in B \oplus C_i : \pi b = f_i c\}$ and let $g_i : B_i \to B$, $\pi_i : B_i \to C_i$ be induced by the projection maps. This is the pullback of π and f_i. Clearly $0 \to A_i = \ker(\pi_i) \to B_i \to C_i \to 0$ is exact. The morphism $f_{ij} : C_i \to C_j$ induces a morphism between the corresponding pullback sequences by defining $g_{ij} : B_i \to B_j$ by $g_{ij}(b, c) = (b, f_{ij}c)$ (which, note, is in B_j) and taking $A_i \to A_j$ to be the restriction.

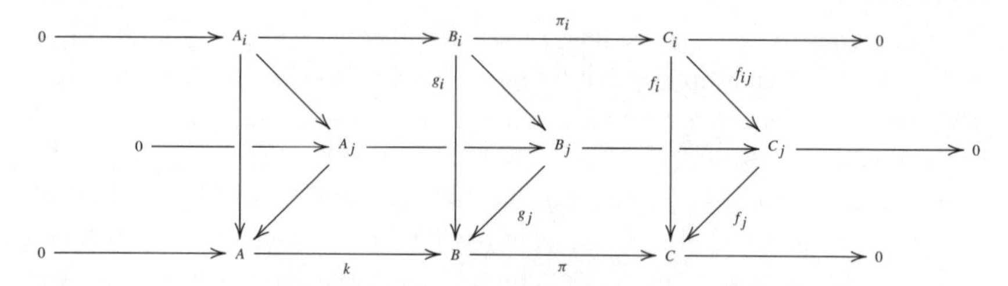

Thus, one may check, the original sequence is a direct limit of the exact sequences $0 \to A_i \to B_i \to C_i \to 0$ (for example, $g_j \cdot g_{ij}(b, c) = g_j(b, f_{ij}c)$). These sequences are pure by 2.1.22, hence are split by 2.1.18. □

It follows from this (alternatively, from 2.1.7 below) that purity of an embedding or an exact sequence is a **Morita-invariant** property, meaning that it is preserved by any equivalence between module categories.

Corollary 2.1.5. ([391, 2.5]) *Any functor between module categories which commutes with direct sums and direct limits preserves pure-exact sequences.*

For instance, the pure-exact, non-split sequence[2] $0 \to \mathbb{Z}_{(p)} \to \overline{\mathbb{Z}_{(p)}} \to \mathbb{Q}^{(2^{\aleph_0})} \to 0$ of $\mathbb{Z}_{(p)}$-modules is a direct limit of split exact sequences of the shape $0 \to \mathbb{Z}_{(p)} \to \mathbb{Z}_{(p)} \oplus \mathbb{Z}^n_{(p)} \to \mathbb{Z}^n_{(p)} \to 0$. The way these exact sequences fit together into a directed system reflects the index set $(2^{\aleph_0})$ for the copies of $\mathbb{Q}$ and the expression of $\mathbb{Q}$ as $\varinjlim \big(\mathbb{Z}_{(p)} \xrightarrow{p\times -} \mathbb{Z}_{(p)} \xrightarrow{p\times -} \cdots\big)$.

Proposition 2.1.6. ([619]) *An inclusion, $A \leq B$, of modules is pure iff $\phi(A) = A \cap \phi(B)$ for every pp condition $\phi(x)$ with one free variable.*

Proof. The proof is by induction on the number, n, of free variables. So suppose that $A \leq B$ is an inclusion such that for every n-place pp condition ϕ we have

[2] The cokernel of the embedding $\mathbb{Z}_{(p)} \to \overline{\mathbb{Z}_{(p)}}$ is torsionfree, divisible and has cardinality $2^{(\aleph_0)}$, hence is $\mathbb{Q}^{(2^{\aleph_0})}$.

$\phi(A) = A^n \cap \phi(B)$. Consider a $(1+n)$-place condition $\phi(x, \overline{y})$ and let $a, \overline{c}$ from A be such that $\phi(a, \overline{c})$ holds in B.

Suppose that $\phi(x, \overline{y})$ is $\exists \overline{z}\, \theta(x, \overline{y}, \overline{z})$ with θ quantifier-free and choose $\overline{b}$ from B such that $\theta(a, \overline{c}, \overline{b})$ holds. Let $\phi'(\overline{y})$ be the condition $\exists x\, \phi(x, \overline{y})$. Then $\overline{c} \in \phi'(B)$, hence, by the induction hypothesis, $\overline{c} \in \phi'(A)$, say $a', \overline{d}$ are from A such that $\theta(a', \overline{c}, \overline{d})$ holds. Therefore $\theta(a - a', \overline{0}, \overline{b} - \overline{d})$ holds so, if $\psi(x)$ is the condition $\exists \overline{z} \theta(x, \overline{0}, \overline{z})$, then $a - a' \in \psi(B) \cap A$, which, by assumption, equals $\psi(A)$. So there is $\overline{e}$ from A such that $\theta(a - a', \overline{0}, \overline{e})$ holds. Combine with $\theta(a', \overline{c}, \overline{d})$ to deduce that $\theta(a, \overline{c}, \overline{e} + \overline{d})$ holds in A. In particular, $\phi(\overline{a}, \overline{c})$ holds in A, as required. □

Proposition 2.1.7. (see [3, p. 85]) *An embedding $f : A \to B$ is pure iff for every morphism $f' : A' \to B'$ in* mod-R *and every $g : A' \to A$ and $h : B' \to B$ such that $hf' = fg$ there is $k : B' \to A$ such that $kf' = g$.*

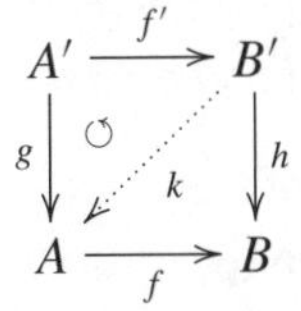

$$\begin{array}{ccc} A' & \xrightarrow{f'} & B' \\ {\scriptstyle g}\downarrow & \circlearrowleft \quad \swarrow {\scriptstyle k} & \downarrow{\scriptstyle h} \\ A & \xrightarrow[f]{} & B \end{array}$$

Proof. Suppose that the embedding f is pure and is in a commutative square as shown. Take a finite tuple, $\overline{b}$, that generates B', with relations given by $\overline{b}G = 0$; since B' is finitely presented, there is such a matrix G. Take a finite generating tuple, $\overline{a}$, for A', and let the matrix H be such that $f'\overline{a} = \overline{b}H$. Then $fg\overline{a} = h\overline{b}H$, so $fg\overline{a}$ satisfies the condition $\exists \overline{y}\, (\overline{y}G = 0 \wedge \overline{x} = \overline{y}H)$ in B. By purity of f, $g\overline{a}$ also satisfies this condition in A, so there is $\overline{c}$ in A such that $g\overline{a} = \overline{c}H$ and $\overline{c}G = 0$. The latter means that sending $\overline{b}$ to $\overline{c}$ extends to a well-defined morphism, k say, from B' to A, with $kf'\overline{a} = g\overline{a}$, hence with $kf' = g$, as required.

For the converse suppose that $B \models \phi(f\overline{a})$, where $\overline{a}$ is from A and ϕ is a pp condition. Let $(C, \overline{c})$ be a free realisation of ϕ. Then, for some h, there is (by 1.2.17) the commutative diagram shown.

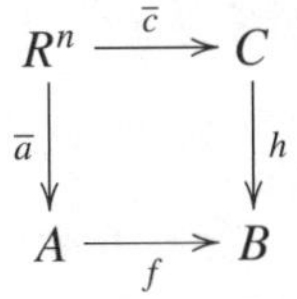

$$\begin{array}{ccc} R^n & \xrightarrow{\overline{c}} & C \\ {\scriptstyle \overline{a}}\downarrow & & \downarrow{\scriptstyle h} \\ A & \xrightarrow[f]{} & B \end{array}$$

By hypothesis there is $k : C \to A$ taking $\overline{c}$ to $\overline{a}$. Since $C \models \phi(\overline{c})$ also $A \models \phi(\overline{a})$, as required. □

Corollary 2.1.8. *Suppose that the diagram shown is commutative, that f is a pure embedding, B' is finitely presented, A' is a finitely generated submodule of B' and i is the inclusion. Then there is $k : B' \to A$ with $ki = h'$.*

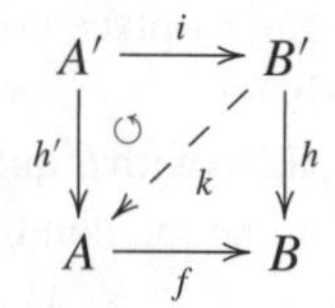

Proof. Choose an epimorphism $\pi : R^n \to A'$, set $f' = i\pi$, $g = h'\pi$ and apply 2.1.7. □

Corollary 2.1.9. *Suppose that $f : A \to B$ is a pure embedding, $h : B' \to B$ is a morphism with B' finitely presented and let $\overline{b}$ be a finite tuple from B'. If $h\overline{b}$ is from the image of A in B, then there is $k : B' \to A$ with $fk\overline{b} = h\overline{b}$.*

Proof. Take A' in 2.1.8 to be the submodule generated by the entries of $\overline{b}$. □

Lemma 2.1.10. *For any collection $(M_i)_i$ of modules the canonical embedding of the direct sum, $\bigoplus_i M_i$, into the direct product, $\prod_i M_i$, is pure.*

Proof. For each finite subset, I', of the index set, the embedding of $\bigoplus_{i \in I'} M_i$ into $\prod_i M_i$ is split. The embedding of $\bigoplus_i M_i$ into $\prod_i M_i$ is the direct limit of these embeddings and, therefore, is pure by 2.1.2(3). (Of course this follows just as easily, and more directly, from the definition.) □

Similarly, and by induction, the submodule of $\prod_i M_i$ consisting of those tuples with less than κ non-zero entries (κ an infinite cardinal number), is pure in $\prod_i M_i$.

Lemma 2.1.11. *Every product of pure embeddings, is pure.*

Proof. If $f_i : A_i \to B_i$ all are pure embeddings, then their product, $(f_i)_i : \prod_i A_i \to \prod_i B_i$, is, on considering coordinates (and using 1.2.3), easily seen to be pure. □

The next lemma is immediate from the definition of purity and 1.1.7.

Lemma 2.1.12. *If $f : M \to N$ and $g : N \to N'$ are morphisms such that the composition gf is a pure embedding, then f is a pure embedding.*

Proposition 2.1.13. *Any pushout of a pure embedding is pure.*

Proof. (Recall, for example, [195, 2.54], that the pushout of an embedding is an embedding.) Consider a pushout diagram

$$\begin{array}{ccc} M & \xrightarrow{f} & N \\ {\scriptstyle g}\downarrow & & \downarrow{\scriptstyle g'} \\ N' & \xrightarrow[f']{} & N'' \end{array}$$

If f is a split embedding, split by $h : N \to M$ say, then, by the pushout property applied to the diagram below, there is a morphism $k : N'' \to N'$ such that $kf' = 1_{N'}$ and $kg' = gh$. In particular, f' is a split embedding.

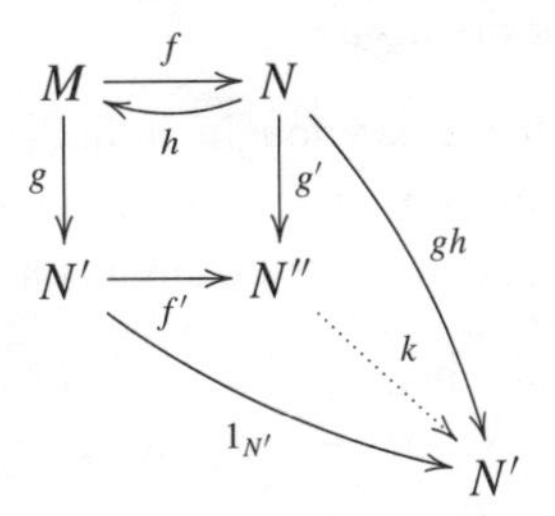

For the general case, if f is pure we may use that it is, 2.1.4, a direct limit of split embeddings in order to complete the proof. Perhaps easier than writing down that proof properly is to give the elementwise proof, which we do now.

Suppose that in the first diagram the morphism f is pure. Suppose that ϕ is a pp condition, say ϕ is $\exists \overline{y}\, (\overline{x}G = \overline{y}H)$, and take $\overline{a}$ from N' with $f'\overline{a} \in \phi(N'')$. Say $\overline{b}''$ from N'' is such that $f\overline{a}'G = \overline{b}''H$. Recall that the pushout, N'', is $N' \oplus N/\{(gc, -fc) : c \in M\}$ so, if $\overline{b}''$ is the image of, say, $(\overline{b}', \overline{b}) \in N' \oplus N$, we have, in $N' \oplus N$, that $(\overline{a}', \overline{0})G = (\overline{b}', \overline{b})H + (g\overline{c}, -f\overline{c})$ for some $\overline{c}$ from M. From one of the resulting equations, $\overline{0} = \overline{b}H - f\overline{c}$, in N one obtains, by purity of M in N, that $\overline{c} = \overline{c}'H$ for some $\overline{c}'$ from M. The other equation gives $\overline{a}'G = \overline{b}'H + g\overline{c} = (\overline{b}' + g\overline{c}')H$, so $\overline{a}' \in \phi(N')$, as required. □

The pullback of a pure embedding need not be pure: for example, the intersection of the direct summands of $\mathbb{Z}_2 \oplus \mathbb{Z}_4 \oplus \mathbb{Z}_2$ generated by $(1, 1, 0)$, respectively $(0, 1, 1)$, is not a pure submodule.

An epimorphism $B \to C$ is a **pure epimorphism** if its kernel is a pure submodule of B.

Proposition 2.1.14. ([619]) *A morphism $\pi : B \to C$ is a pure epimorphism iff for every pp condition ϕ and every $\overline{c} \in \phi(C)$ there is $\overline{b} \in \phi(B)$ such that $\pi\overline{b} = \overline{c}$.*

Proof. First assume that the sequence $0 \to A \xrightarrow{\pi} B \to C \to 0$ is pure-exact. Suppose that $C \models \phi(\overline{c})$, where ϕ is $\exists\, \overline{y}\, (\overline{x}\, \overline{y})H = 0$, say $\overline{d}$ is from C with $(\overline{c}\, \overline{d})H = 0$. Choose any inverse images, $\overline{e}, \overline{b}$, of $\overline{c}, \overline{d}$ respectively. Then $(\overline{e}\, \overline{b})H = \overline{a}$ is from $A = \ker(\pi)$. Purity of A in B, applied to $\psi(\overline{a})$, where $\psi(\overline{x}')$ is the pp condition $\exists \overline{x}, \overline{y}\, \big((\overline{x}\, \overline{y})H = \overline{x}'\big)$, gives that there are $\overline{a}'$ and $\overline{a}''$ in A such that $(\overline{a}'\, \overline{a}'')H = \overline{a}$. Then $(\overline{e} - \overline{a}', \overline{b} - \overline{a}'')H = 0$ so $\overline{e} - \overline{a}'$ is an inverse image of $\overline{c}$ satisfying ϕ, as required.

For the converse assume that solutions lift and take ϕ pp and $\overline{a}$ in A such that $B \models \phi(\overline{a})$. Supposing that ϕ is $\exists \overline{y}\, (\overline{x}\, \overline{y})H = 0$ we have $(\overline{a}\, \overline{b})H = 0$ for some $\overline{b}$

from B. Then $\pi\overline{b}$ satisfies the condition $(\overline{0}\,\overline{y})H = 0$, so, by assumption, there is $\overline{b}'$ from B with $\pi\overline{b}' = \pi\overline{b}$ and $(\overline{0}\,\overline{b}')H = 0$. Then $\overline{b} - \overline{b}'$ is a tuple from $A = \ker(\pi)$ and $(\overline{a}, \overline{b} - \overline{b}')H = 0$, showing that $\overline{a}$ satisfies ϕ in A. We conclude that the sequence is pure-exact, as desired. □

One may check ([619]) that it is enough, as in 2.1.6, to test with ϕ having a single free variable. From the proof of 2.1.14 it is enough to test with $\phi(\overline{x})$ being a quantifier-free condition. Of course, one cannot combine these simplifications.

Remark 2.1.15. For example, by, say, 1.2.31, if $M = \varinjlim M_\lambda$ is a direct limit, then the canonical map from $\bigoplus_\lambda M_\lambda \to M$ is a pure epimorphism ([588, p. 56]).

Corollary 2.1.16. *A morphism $\pi : B \to C$ is a pure epimorphism iff for every finitely presented module D every morphism $g : D \to C$ lifts to a morphism $g' : D \to B$ with $g = \pi g'$.*

Proof. If this condition is satisfied and if ϕ is a pp condition, then let $(C_\phi, \overline{c}_\phi)$ be a free realisation of ϕ (§1.2.2). If $\overline{c} \in \phi(C)$, then there is induced (1.2.17) a morphism $g : C_\phi \to C$ with $g\overline{c}_\phi = \overline{c}$ which, by assumption, factors as $g = \pi g'$ for some $g' : C_\phi \to B$. Set $\overline{b} = g'\overline{c}_\phi$ and note that the criterion of 2.1.14 is thus satisfied.

For the converse, suppose that π is a pure epimorphism and take $g : D \to C$ with D finitely presented, generated by the finite tuple $\overline{d}$, say. By 1.2.6, $\mathrm{pp}^D(\overline{d}) = \langle\phi\rangle$ for some pp condition ϕ. Choose, by 2.1.14, $\overline{b} \in \phi(B)$ such that $\pi\overline{b} = g\overline{d}$. Then $\overline{d} \mapsto \overline{b}$ well defines a map $g' : D \to B$ such that $\pi g' = g$, as required. □

Corollary 2.1.17. *If $0 \to A \to B \xrightarrow{\pi} C \to 0$ is a pure exact sequence and if ϕ/ψ is a pp-pair (see after 1.2.24), then ϕ/ψ is closed in B iff ϕ/ψ is closed in both A and C.*

Proof. For ($\Leftarrow$), if $\overline{b} \in \phi(B)$, then (1.1.7) $\pi\overline{b} \in \phi(C)$, so $\pi\overline{b} \in \psi(C)$. Then, by 2.1.14, there is $\overline{b}' \in \psi(B)$ with $\pi\overline{b}' = \pi\overline{b}$, so $\overline{b} - \overline{b}' \in \ker(\pi) = A$. Thus $\overline{b} - \overline{b}' \in \phi(A) = \psi(A) \le \psi(B)$. Hence $\overline{b} = \overline{b}' + (\overline{b} - \overline{b}') \in \psi(B)$, as required.

The other direction is immediate from the definitions and 2.1.14. □

Corollary 2.1.18. ([391, 2.4]) *If $0 \to A \to B \xrightarrow{\pi} C \to 0$ is pure-exact and C is finitely presented, then this sequence is split.*

In particular any pure embedding in mod-R *splits.*

Proof. Take $g = \mathrm{id}_C$ in 2.1.16. □

Corollary 2.1.19. *If $A \to B$ is an embedding in* mod-R*, then this is a pure embedding iff for every $C \in$* mod-R *the induced map $(B, C) \to (A, C)$ is surjective.*

Lemma 2.1.20. *If $A' \le A$ is generated by a subset S of A' and for each tuple $\overline{a}$ from S we have $\mathrm{pp}^{A'}(\overline{a}) = \mathrm{pp}^A(\overline{a})$, then A' is pure in A.*

Proof. Any tuple $\overline{b}$ from A' can be expressed as $\overline{a}G$ for some tuple $\overline{a}$ from S and matrix G over R. Therefore, if $\phi(\overline{z})$ is a pp condition, then $\phi(\overline{b})$ is equivalent to $\phi(\overline{a}G)$: the latter may be regarded as a pp condition on $\overline{a}$, so holds in A' iff it holds in A. Therefore $\phi(\overline{b})$ holds in A' iff it holds in A, as required. □

A similar statement and proof applies to epimorphisms.

Lemma 2.1.21. ([391, 3.1]) *Let A be a submodule of B. Then A is contained in a pure submodule, B', of B of cardinality no more than* $\max\{\aleph_0, \mathrm{card}(R), \mathrm{card}(A)\}$.

Proof. Let $\phi(x)$, being $\exists y_1, \dots, y_n\, \theta(x, y_1, \dots, y_n)$ with θ quantifier-free, be a pp condition. For each $a \in A \cap \phi(B)$ choose and fix $b_1, \dots, b_n \in B$ such that $\theta(a, b_1, \dots, b_n)$ holds.

Define A_1 to be the submodule generated by A together with all such chosen $b_1, \dots, b_n$ as both ϕ and a vary (over the set of pp conditions in one free variable, resp. over $A \cap \phi(B)$). Note that $\mathrm{card}(A_1) \le \kappa = \max\{\aleph_0, \mathrm{card}(R), \mathrm{card}(A)\}$.

Now repeat this process with A_1 in place of A to obtain A_2. Continue. We obtain $A \le A_1 \le A_2 \le \cdots \le A_n \le \cdots \le B$, where for each n and each ϕ one has $A_n \cap \phi(B) \le \phi(A_{n+1})$ by construction. Let $B' = \bigcup_n A_n$. Take $a \in B' \cap \phi(B)$. Then $a \in A_n$ for some n so, as remarked, $a \in \phi(A_{n+1}) \le \phi(B')$, as required.

(Those who know some model theory will recognise this as a restricted version of the downwards Löwenheim–Skolem Theorem, A.1.10.) □

2.1.2 Pure-projective modules

Modules projective over pure epimorphisms, the pure-projective modules, are exactly the direct summands of direct sums of finitely presented modules (2.1.26). Every module is a pure-epimorphic image of a pure-projective module (2.1.25).

A module C is **pure-projective** if every pure-exact sequence $0 \to A \to B \to C \to 0$ splits. The result below is dual to 2.1.13.

Proposition 2.1.22. *If $\pi : B \to C$ is a pure epimorphism and $g : D \to C$ is any morphism, then in the pullback diagram shown π' is a pure epimorphism.*

$$\begin{array}{ccc} P & \xrightarrow{\pi'} & D \\ \downarrow & & \downarrow g \\ B & \xrightarrow[\pi]{} & C \end{array}$$

Proof. We check the criterion of 2.1.14, so suppose $\overline{d} \in \phi(D)$. Let $(C_\phi, \overline{c})$ be a free realisation of ϕ; then (1.2.17) there is $f : C_\phi \to D$ with $f\overline{c} = \overline{d}$. Suppose that C_ϕ is generated by the finite tuple $\overline{a}$ with generating relations $\overline{a}H = 0$, and, say, $\overline{c} = \overline{a}G$.

We have $C \models gf\overline{a}.H = 0$, so there is, by hypothesis and 2.1.14, $\overline{b}$ from B with $\overline{b}H = 0$ and $\pi\overline{b} = gf\overline{a}$. Denote by h the (well-defined) morphism from C_ϕ to B which takes $\overline{a}$ to $\overline{b}$. Then $\pi h = gf$, so, by the pullback property, there is $k : C_\phi \to P$ with, in particular, $\pi' k = f$, hence with $\pi'(k\overline{a}.G) = \overline{d}$. Also $k\overline{a}G = k\overline{c} \in \phi(P)$, as required.

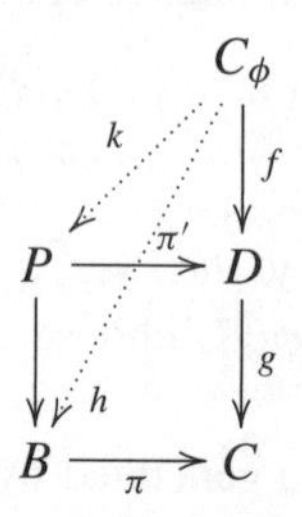

□

Lemma 2.1.23. *A module D is pure-projective iff whenever $\pi : B \to C$ is a pure epimorphism and $g : D \to C$ is any morphism, there is a lift $g' : D \to B$ with $\pi g' = g$.*

Proof. ($\Leftarrow$) Given a pure-exact sequence $0 \to A \to B \to D \to 0$ apply the assumed property of D with $g = \mathrm{id}_D$.

($\Rightarrow$) Given a pure epimorphism $\pi : B \to C$, being part of the pure-exact sequence $0 \to A \to B \to C \to 0$, say, form the pullback exact sequence $0 \to A \to P \to D \to 0$ induced by $g : D \to B$. Then we are in the situation of 2.1.22 and conclude that this second sequence is pure-exact, hence split, which yields the desired lift. □

Corollary 2.1.24. *Any direct summand of a pure-projective module is pure-projective. Any direct sum of pure-projective modules is pure-projective.*

Proposition 2.1.25. *Given any module M there is a pure epimorphism $P \to M$, where P is a direct sum of finitely presented modules (in particular P is pure-projective).*

Proof. Let Λ be the set of isomorphism classes of finitely presented modules and consider $\pi : P = \bigoplus_{A \in \Lambda} A^{(A,M)} \to \pi M$, where the component of π indexed by A, $f \in (A, M)$ is f. Certainly this is an epimorphism (take $A = R$) and the condition of 2.1.14 is satisfied since P has, as a direct summand, a free realisation of each pp condition. □

The dual result, that every module purely embeds in a pure-injective module, also is true (4.3.18), indeed will be central to much of what we do.

Corollary 2.1.26. *A module is pure-projective iff it is a direct summand of a direct sum of finitely presented modules.*

Proof. One half is by 2.1.24 and since, by 2.1.18, finitely presented modules are pure-projective. The other direction is immediate from 2.1.25. □

Over some rings pure-projective modules simply are the direct sums of finitely presented modules. For example, this is true over $\mathbb{Z}$, over Krull–Schmidt rings, over noetherian serial rings: for references see the introduction to [579] where it is shown that this also holds over hereditary noetherian rings. There are, however, examples which make the "direct summand of" in 2.1.26 necessary. For instance, it is shown in [566] that over some uniserial rings there is an indecomposable infinitely generated pure-projective module (the module V in 2.5.8).

Corollary 2.1.27. (for example, [730, Obs. 8]) *Suppose that M is pure-projective. Then every finitely generated* End(M)*-submodule of M is pp-definable.*

That follows easily from 1.2.12 and 2.1.26.

2.1.3 Purity and tensor product

We give another characterisation of purity which is often used as the definition.

Recall that unadorned $\otimes$ means $\otimes_R$.

Proposition 2.1.28. *Let $f : A \to B$ be an embedding of right R-modules. Then f is pure iff for every left R-module L the map $f \otimes 1_L : A \otimes L \to B \otimes L$ of abelian groups is monic. It is enough to test with L ranging over the finitely presented left R-modules.*[3]

Proof. Suppose that L is any left R-module and that $f\overline{a} \otimes \overline{l} = 0$ in $B \otimes L$. By 1.3.5 there are $\overline{b}$ in B, $\overline{k}$ in L and matrices G, H such that $(f\overline{a}\ \overline{0}) = \overline{c}\begin{pmatrix} G & H \end{pmatrix}$ and $\begin{pmatrix} G & H \end{pmatrix}\begin{pmatrix} \overline{l} \\ \overline{k} \end{pmatrix} = 0$. So, in B, $f\overline{a}$ satisfies the pp condition $\exists \overline{y}\,(\overline{x} = \overline{y}G \wedge \overline{y}H = 0)$. If f is a pure embedding, then $\overline{a}$ satisfies this condition also in A; say $\overline{a}'$ is such that $\overline{a} = \overline{a}'G$ and $\overline{a}'H = 0$. Then by 1.3.5, now applied in $A \otimes L$, we have $\overline{a} \otimes \overline{l} = 0$. Thus if f is pure, then $f \otimes 1_L$ is monic for all ${}_RL$.

We show that it is enough to have this for all finitely presented L in order to have the converse. Let $\overline{a}$ in A satisfy the pp condition ϕ in B. Let $(C_{D\phi}, \overline{c}_{D\phi})$ be a free realisation of the dual condition $D\phi$ (recall, Section 1.3.1, 1.2.14, that this includes that $C_{D\phi}$ is finitely presented). By 1.3.7, $f\overline{a} \otimes \overline{c}_{D\phi} = 0$ in $B \otimes C_{D\phi}$. Assuming that $f \otimes 1_{C_{D\phi}}$ is monic it must be that $\overline{a} \otimes \overline{c}_{D\phi} = 0$ in $A \otimes C_{D\phi}$ and hence, by 1.3.8, it follows that $\overline{a}$ satisfies ϕ in A, and f is pure, as required. □

[3] It is also sufficient to test with L ranging over indecomposable pure-injective left R-modules, [274, 8.1], indeed, [274, 8.4], over just those in the dual of the support of A. This follows from 1.3.7 and 5.1.3.

For instance, the fact that $\mathbb{Z} \leq \mathbb{Q}$ is not pure is illustrated by tensoring with $\mathbb{Z}_2 = \mathbb{Z}/2\mathbb{Z}$ (see 12.1.4). Since tensor product is right exact we have the following formulation.

Corollary 2.1.29. *An exact sequence $0 \to A \to B \to C \to 0$ is pure-exact iff, on tensoring with any (finitely generated) left module L, the resulting sequence $0 \to A \otimes L \to B \otimes L \to C \otimes L \to 0$ is exact.*

The importance of this is seen particularly in Section 12.1.1.

Corollary 2.1.30. *If the sequence $0 \to A \to B \to C \to 0$ of right R-modules is pure-exact, then, for every (R, S)-bimodule L, the sequence $0 \to A \otimes_R L \to B \otimes_R L \to C \otimes_R L \to 0$ is a pure-exact sequence of right S-modules.*

Proof. Let K be a left S-module. The sequence $0 \to (A \otimes_R L) \otimes_S K \to (B \otimes_R L)) \otimes_S K \to (C \otimes_R L) \otimes_S K \to 0$ is isomorphic to $0 \to A \otimes_R (L \otimes_S K) \to B \otimes_R (L \otimes_S K) \to C \otimes_R (L \otimes_S K) \to 0$, so, by 2.1.29, is exact. Hence, by 2.1.29, the sequence $0 \to A \otimes_R L \to B \otimes_R L \to C \otimes_R L \to 0$ is pure-exact. $\square$

Corollary 2.1.31. *If $A \to B$ is a pure embedding of R-modules and if $R \to S$ is any morphism of rings, then the embedding $A \otimes_R S_S \to B \otimes_R S_S$ of S-modules is pure. For instance, if R is commutative and P is a prime ideal of R, then the embedding of localisations $A_{(P)} \to B_{(P)}$, that is $A \otimes R_{(P)} \to B \otimes R_{(P)}$, is pure.*

2.2 Pure global dimension

Projective, injective and global dimensions are defined in terms of exact sequences of modules projective or injective with respect to these. One may define dimensions based on particular classes of exact sequences (for example, see [109]), in particular pure-exact sequences. That the resulting dimensions measure something quite different from the usual ones can be seen from the connection (2.2.2) with the cardinality of the ring. We state some results connecting these dimensions to the topics of this book.

Extending the terminology for short exact sequences, say that an exact sequence $\cdots \to M_{n+1} \xrightarrow{f_{n+1}} M_n \xrightarrow{f_n} M_{n-1} \to \cdots$ is **pure-exact** if the image of each f_n is a pure subobject of M_{n-1}. A **pure-projective resolution** of a module M is a pure-exact sequence $\cdots \to P_n \to P_{n-1} \to \cdots \to P_0 \to M \to 0$, where each P_i is pure-projective. By 2.1.25 every module has a pure-projective resolution. The **pure-projective dimension** of M is no more than n, $\text{pprojdim}(M) \leq n$, if there is such a resolution with $P_{n+1} = 0$. If there is no such n, set $\text{pprojdim}(M) = \infty$. Dually, a **pure-injective resolution** of a module M is a pure-exact sequence

$0 \to M \to N_0 \to \cdots \to N_n \to N_{n+1} \to \cdots$, where each N_i is pure-injective. By 4.3.19 every module has a pure-injective resolution. The **pure-injective dimension** of M is no more than n, $\mathrm{pinjdim}(M) \le n$, if there is such a resolution with $N_{n+1} = 0$.

Just as for the injective dimension and the projective dimension of a ring one has $\sup\{\mathrm{pprojdim}(M) : M \in \text{Mod-}R\} = \sup\{\mathrm{pinjdim}(M) : M \in \text{Mod-}R\}$ and this common value is the **pure global dimension**, $\mathrm{pgldim}(R)$, of R. The proof is like that for global dimension and uses the functors $\mathrm{Pext}(-,-)$, where $\mathrm{Pext}^i(M, N)$ is defined to be the cohomology of the ith term of the sequence obtained by forming $\mathrm{Hom}(P_\bullet, N)$, where $P_\bullet$ is a pure-projective resolution of M. One shows that this is the same as the cohomology at the ith term of the sequence obtained by forming $\mathrm{Hom}(M, I_\bullet)$, where $I_\bullet$ is a pure-injective resolution of N and this gives the equality (see, for example, [73, §2.5]). One may also write $\mathrm{rpgldim}(R)$ to emphasise that this refers to the category of right R-modules.

Examples 2.2.1. (a) A ring has right pure global dimension 0 iff every right module is pure-projective iff every right module is pure-injective. These are exactly the right pure-semisimple rings (Section 4.5.1) and the question of whether right pure global dimension 0 is equivalent to left pure global dimension 0 is the long-standing pure-semisimplicity conjecture, 4.5.26, which is that right *or* left pure global dimension 0 implies finite representation type.[4]

(b) If R is von Neumann regular, then every exact sequence is pure-exact (2.3.22) so the pure global dimension and the global dimension are equal.

(c) If k is any field, then the pure global dimension of the polynomial ring $k[T]$ is 1. More generally, this is the pure global dimension of every commutative Dedekind domain (a theorem of Kulikov, see [323, 11.19(ii), 11.32]). Also see 2.2.5(b) and [34] for some other tame algebras.

(d) On the other hand various wild algebras attain the maximum value of pure global dimension compatible with 2.2.2. Suppose that k is a field and let $\kappa = \max(\mathrm{card}(k), \aleph_0)$. Suppose that $\kappa = \aleph_t$ for some $t \in \mathbb{N}$. Then the right and left pure global dimension of each of the rings $k\langle X_1, \dots, X_n\rangle$ $(n \ge 2)$, $k[X_1, \dots, X_n]$ $(n \ge 2)$, $k[X, Y, Z]/(X, Y, Z)^2$ and kQ for any wild quiver Q without relations, is $t + 1$, [33, 3.1, 3.2, 4.1] (also [341, 2.3, 2.4] and 2.2.5 below. Therefore, since whether or not $2^{\aleph_0} = \aleph_t$ depends on the choice of model of set theory, the value of the pure global dimension of the polynomial ring $\mathbb{C}[X, Y]$ is independent of ZFC set theory! (Cf. [475].)

A variety of further examples may be found in the book of Jensen and Lenzing [323, esp. Chpt. 11] and the papers [33], [34].

[4] For rings; it is false for ringoids, [97].

The following surprising theorem is due independently to Kielpinski and Simson and to Gruson and Jensen (for (two) proofs see [323, 7.47, 11.21]), also see [613].

Theorem 2.2.2. ([254, 1.3(A)], [341, 2.2], see [323, 7.47]) *If the ring R has cardinality $\aleph_t$ with $t \in \mathbb{N}$, then* $\operatorname{pgldim}(R) \leq t+1$. *In particular, the pure global dimension of any countable ring which is not pure-semisimple is* 1.

The first statement below follows from the definitions via pure-injective resolutions (and the pure-injective version of Schanuel's Lemma).

Lemma 2.2.3. *Suppose that* $\operatorname{pgldim}(R) = 1$. *Let M be a pure submodule of the pure-injective module N. Then N/M is pure-injective. If $(N_i)_{i \in I}$ is a collection of pure-injective R-modules and $\mathcal{F}$ is any filter on I then the reduced product $\prod_i N_i/\mathcal{F}$ (§3.3.1) is pure-injective.*[5]

The second statement follows since the kernel of the map $\prod_i N_i \mapsto \prod_i N_i/\mathcal{F}$ is, by 3.3.1, pure in $\prod_i N_i$ (which is pure-injective by 4.3.2). For example $N^I/N^{(I)}$ is a reduced product, as are the ultraproducts used in Chapter 3.

Corollary 2.2.4. (see [323, 7.44]) *Suppose that R is a countable ring and that M is a pure submodule of N. If N is pure-injective, then so is N/M. Any reduced product of pure-injective R-modules is pure-injective.*

For finite-dimensional algebras there are connections between pure global dimension and representation type.

Theorem 2.2.5. *(a)* ([33, 4.1]) *Let R be the path algebra, over a field k, of a wild quiver without relations; so R is a wild hereditary algebra. Then* $\operatorname{pgldim}(R)$ *has the maximum value allowed by 2.2.2. The same conclusion holds if k is algebraically closed and R is a wild local or commutative finite-dimensional algebra.*
(b) ([34, 3.3], [468] for the Kronecker quiver) *Let R be the path algebra over a field of an extended Dynkin quiver; so R is a tame hereditary algebra. Then* $\operatorname{pgldim}(R) = 2$ *unless R is countable, in which case the value is* 1.

There are also strong connections between pure global dimension and the derived functors of $\varprojlim$. If M is countably presented, then it is the direct limit of a countable sequence $M_0 \xrightarrow{f_0} M_1 \xrightarrow{f_1} \cdots$ of finitely presented modules. This gives rise to a pure-exact sequence $0 \to \bigoplus_i M_i \xrightarrow{1-(f_i)_i} \bigoplus M_i \to M \to 0$, which is a pure-projective resolution of M, hence the pure-projective dimension of M is 1. More generally, one has the following (see, for example, [323, 7.47] or [73, §2.6]).

[5] Cf. 4.2.17.

Theorem 2.2.6. ([256, p. 1652], [341, 1.4]) *If the module M is $\aleph_t$-presented, then* $\mathrm{pprojdim}(M) \leq t + 1$.

The connection with the derived functors of $\varprojlim$ comes via the fact that if M is written as a direct limit $M = \varinjlim_i M_i$ then for every N there is an induced map $(M, N) \to \varprojlim_i (M_i, N)$. We refer the reader to [320], [256], [73] (also [74]) for details.

Also see, for example, [254], [256] for the connections between the cofinality of the lattice of pp-definable subgroups, the derived functors of $\varprojlim$ and "L-dimension".

2.3 Absolutely pure and flat modules

Absolutely pure modules are characterised in various ways in Section 2.3.1; similarly (dually really) for flat modules in Section 2.3.2. Coherence of modules and rings, and the connection with pp conditions, is described in Section 2.3.3.

Over von Neumann regular rings pp conditions reduce to simple solvability of systems of linear equations (no projections are required). This and some consequences are in Section 2.3.4.

2.3.1 Absolutely pure modules

Absolute purity is a weakening of injectivity. There are characterisations at 2.3.1 and 2.3.2 and another, in terms of annihilation of elementary duals, at 2.3.3.

An R-module M is **absolutely pure** if every embedding $M \to N$ in Mod-R with domain M is pure. A module M is **fp-injective** if for every embedding $i : A \to B$ with finitely presented cokernel and every morphism $f : A \to M$ there is $g : B \to M$ such that $gi = f$.

Proposition 2.3.1. *For any module M the following are equivalent:*

(i) M is absolutely pure;
(ii) M is fp-injective;
(iii) $\mathrm{Ext}^1(C, M) = 0$ *for every finitely presented module*[6] C.

Proof. (i)⇒(iii) Suppose that M is absolutely pure and consider any short exact sequence $0 \to M \to N \to C \to 0$ with C finitely presented, representing an

[6] Conversely, a finitely generated module C is finitely presented iff $\mathrm{Ext}^1(C, M) = 0$ for every absolutely pure module M, [177], [233, 2.1.10].

element of $\mathrm{Ext}^1(C, M)$. By assumption this is pure, hence, by 2.1.18, it is split, as required.

(iii)⇒(ii) Consider any short exact sequence $0 \to A \xrightarrow{i} B \to C \to 0$ with C finitely presented. Consider the beginning of the induced long exact sequence: $0 \to (C, M) \to (B, M) \to (A, M) \to \mathrm{Ext}^1(C, M)$, where the last is zero by assumption. That is, $(i, M) : (B, M) \to (A, M)$ is epi and so every morphism from A to M lifts through i to a morphism from B to M.

(ii)⇒(i) Suppose that M is embedded in N and that $\overline{a}$ from M is such that $\overline{a} \in \phi(N)$, where ϕ is a pp condition. Let $C \xrightarrow{g} N$ be a morphism taking $\overline{c}$ to $\overline{a}$, where $(C, \overline{c})$ is a free realisation of ϕ (1.2.14, 1.2.17). Let $f : C' \to M$ be the restriction/corestriction of g from the submodule, C', of C generated by $\overline{c}$, to M. Note that C/C' is finitely presented (being finitely presented modulo finitely generated, E.1.16) so, by assumption, there is a morphism $C \to M$ lifting f and we deduce, by 1.1.7, that $\overline{a} \in \phi(M)$, as required. □

Clearly every injective module is absolutely pure.

Proposition 2.3.2. *A module is absolutely pure iff it is a pure submodule of an injective module.*

Proof. One direction is by the definition. For the other suppose that $i : M \to E$ is a pure embedding and that E is injective. Let $j : M \to N$ be any embedding. Since E is injective, there is $g : N \to E$ such that $gj = i$. By 2.1.12 j is a pure embedding, as required. □

We will see, 4.4.17, that over a right noetherian ring absolutely pure = injective and, conversely, if every absolutely pure module is injective, then R must be right noetherian (4.4.17). In contrast, if R is von Neumann regular, then, 2.3.22, every module is absolutely pure (but every module is injective only if R is semisimple artinian).

Every absolutely pure module is divisible, where a module M is **divisible** if for every $r \in R$ and inclusion $M \leq N$ we have $Mr = M \cap Nr$ (if R is a domain this is equivalent to $Mr = M$ for every non-zero $r \in R$). The converse is not true: for instance, every preinjective module over the Kronecker quiver (Section 8.1.2) is easily seen to be divisible but need not be absolutely pure (= injective since this algebra is noetherian). For another example, if $R = \mathbb{Z}[X]$ and $Q = \mathbb{Q}(X)$ is its quotient field, then Q/R is divisible but not injective.

If ϕ is a pp condition for right modules, then $D\phi({}_RR)$ is a submodule of the right R-module R^n (where n is the number of free variables of ϕ). Recall that if $\overline{c}$ is an n-tuple, then by $\mathrm{ann}_R(\overline{c})$ we mean $\{\overline{r} \in R^n_R : \overline{c} \cdot \overline{r} (= \sum_i c_i r_i) = 0\}$. The next result is immediate by 1.3.8; precursors are [174, 3.2], [209, Thm 5], [726, 5.8], [323, 6.12].

Proposition 2.3.3. ([527, 1.3]) *A module M is absolutely pure iff for every pp condition ϕ we have* $\phi(M) = \mathrm{ann}_M D\phi(_R R)$.

Proof. Suppose M is absolutely pure. By 1.3.7 certainly $\phi(M) \leq \mathrm{ann}_M D\phi(_R R)$. Let $(C, \overline{c})$ be a free realisation of ϕ and let $\overline{a}$ be from $\mathrm{ann}_M D\phi(R)$ which, by 1.3.10, equals $\mathrm{ann}_M(\mathrm{ann}_R(\overline{c}))$. Then $\overline{c} \mapsto \overline{a}$ well defines a morphism from the submodule, C', of C generated by $\overline{c}$ to M. Since the cokernel of the inclusion $C' \leq C$ is finitely presented, this extends, by 2.3.1, to a morphism from C to M, so $\overline{a} \in \phi(M)$ (by 1.1.7), as required.

For the converse, let ϕ be a pp condition and assume $\phi(M) = \mathrm{ann}_M D\phi(_R R)$. Consider the embedding of M into its injective hull $E(M)$. By the first part, $\phi(E(M)) = \mathrm{ann}_{E(M)} D\phi(_R R)$, so clearly $\phi(M) = M^n \cap \phi(E(M))$ (n the number of free variables of ϕ). Therefore M is pure in $E(M)$ and hence, by 2.3.2, is absolutely pure. □

Corollary 2.3.4. *If M is absolutely pure, then the lattice,* $\mathrm{pp}^1(M)$, *of pp-definable subgroups of M is a quotient of the lattice of right ideals of R.*

Proof. The map $I(\leq R_R) \mapsto \mathrm{ann}_M(I)$ is, by 2.3.3, surjective. □

This gives bounds on the values of various ranks, see Chapter 7, on M in terms of their values on R. Also cf. 2.3.10.

Remark 2.3.5. It is immediate, for example from 2.3.3, that any direct product or direct sum of absolutely pure modules is absolutely pure, as is any pure submodule of an absolutely pure module.[7]

2.3.2 Flat modules

Flat modules are direct limits of projective modules (2.3.6) and have a characterisation in terms of pp conditions (2.3.9). If a flat module occurs at the "end" of a short exact sequence, then that sequence must be pure-exact (2.3.14).

A module M is **flat** if the functor $M \otimes - : R\text{-Mod} \to \mathbf{Ab}$ is exact, that is, since tensor product always is right exact, if whenever $j : L \to N$ is a monomorphism between left R-modules the map $1_M \otimes j : M \otimes L \to M \otimes N$ is monic. It is enough to test with $N = R$ and L any finitely generated left ideal ([663, I.10.6], [626, 2.11.4]).

It is immediate from the definition that any direct summand of a flat module is flat and any direct sum of flat modules is flat. Since $\otimes$ is a left adjoint, hence

[7] Rothmaler [625] examines preservation of pp and other conditions under inverse limits of epimorphisms and shows [625, §6] that certain inverse limits of absolutely pure modules are absolutely pure, and certain inverse limits of flat modules over coherent rings are flat.

commutes with direct limits, every direct limit of flat modules is flat (this also follows from 2.3.9 below). Over arbitrary rings a direct product of flat modules need not be flat (see 2.3.21).

Theorem 2.3.6. *(a) Every projective module is flat, indeed a module is flat iff it is a direct limit of projective modules.*
(b) Every finitely presented flat module is projective (finitely generated is not enough, see, for example, [578, 3.5]).
(c) Every flat right R-module is projective iff R is right perfect (see 3.4.27).

We will give a proof of part (a) at 2.3.13. Part (b) follows from (a) and part (c) is due to Bass [46, Thm P] and can also be found in [187, 22.31A] and, outlined, in [626, Ex. 2.11.7]. It is not the case, see [391, 4.1], that every flat module is a direct limit of projective submodules.

Corollary 2.3.7. *Any direct limit (more generally filtered colimit) of flat modules is flat.*

For, each flat module in a directed or filtered system can be replaced by a directed system of projectives which clearly may be taken to be finitely presented and then these fit together into a single system which, one may check, has the same direct limit as the original one.

For flat modules the criterion of 1.3.5 simplifies.

Proposition 2.3.8. *A right R-module M is flat iff for every left R-module L, every tuple $\overline{a}$ from M and tuple $\overline{l}$ from L we have $\overline{a} \otimes \overline{l} = 0$ iff there is $\overline{c}$ from M and a matrix H such that $\overline{a} = \overline{c}H$ and $H\overline{l} = 0$.*

Proof. Suppose that M satisfies this simplified criterion for a tensor to be zero. Let $j : L \to N$ be a monomorphism of left modules and suppose that $\overline{a} \otimes \overline{l} = 0$ in $M \otimes N$, where $\overline{l}$ is from L. By assumption there is $\overline{c}$ from M and a matrix H such that $\overline{a} = \overline{c}H$ and $H\overline{l} = 0$. By 1.3.5 (or just obviously) it follows that already $\overline{a} \otimes \overline{l} = 0$ in $M \otimes L$, and so $1_M \otimes j$ is monic, as required.

For the converse, suppose that M is flat and suppose that $\overline{a} \otimes \overline{l} = 0$ in $M \otimes L$. By assumption $\overline{a} \otimes \overline{l} = 0$ in $M \otimes L'$, where L' is the submodule of L generated by $\overline{l}$. Then 1.3.6 applies to give the desired conclusion. □

Recall, 1.1.10, that if ϕ is any pp condition and M is any module, then $M \cdot \phi(R) \le \phi(M)$. Note that $\overline{a} \in M^n$ is a member of $M \cdot \phi(R) = \{\sum_{j=1}^{m} c_j \overline{r}_j : c_j \in M, \overline{r}_j \in \phi(R)\}$ iff there is $\overline{c} \in M^m$ for some m and an $m \times n$ matrix H, the rows of which are elements of $\phi(R)$, such that $\overline{a} = \overline{c}H$ (cf. 1.3.4).

Theorem 2.3.9. ([731, 1.3], [617, Prop. 4]) *A module M is flat iff for every pp condition ϕ we have $\phi(M) = M \cdot \phi(R)$ iff for every pp condition ϕ with one free variable we have $\phi(M) = M \cdot \phi(R)$.*

Proof. Suppose that M is flat. The containment $M \cdot \phi(R) \leq \phi(M)$ is 1.1.10, so, for the converse, take $\overline{a} \in \phi(M)$. Let $(D, \overline{d})$ be a free realisation of the dual $D\phi$; by 1.3.7, $\overline{a} \otimes \overline{d} = 0$. By 2.3.8 there is $\overline{c}$ in M and a matrix H such that $\overline{a} = \overline{c}H$ and $H\overline{d} = 0$. From the second equation it follows by 1.3.8 that each row of H is an element of $\phi(R)$. Then, see the comment immediately preceding the theorem, the first matrix equation exhibits $\overline{a}$ as an element of $M \cdot \phi(R)$.

For the converse we check the criterion of 2.3.8. So suppose that $\overline{a} \otimes \overline{l} = 0$ in $M \otimes L$. By 1.3.7 there is ϕ such that $M \models \phi(\overline{a})$ and $L \models D\phi(\overline{l})$. By assumption $\overline{a} = \overline{c}H$ for some matrix H whose rows are in $\phi(R)$ and which, therefore (by 1.3.7), satisfies $H\overline{l} = 0$, as required.

Suppose, finally, that $\phi(M) = M \cdot \phi(R)$ holds for all pp conditions ϕ in $n - 1$ (≥ 1) free variables: we show that it follows for pp conditions in n free variables.

Say $(a, \overline{a}') \in \phi(M)$, where $\phi = \phi(x, \overline{x}')$ with $l(\overline{x}') = n - 1$. Let $\psi(x)$ be the condition $\exists \overline{x}' \phi(x, \overline{x}')$. Then $a \in \psi(M) = M \cdot \psi(R)$ by assumption. So $a = \sum c_i r_i$ for some $c_i \in M$ and $r_i \in \psi(R)$, say $(r_i, \overline{r}'_i) \in \phi(R)$, for each i.

Then $(a, \overline{a}') - \sum c_i(r_i, \overline{r}'_i) = (0, \overline{a}' - \sum c_i \overline{r}'_i)$ and, note, this is in $\phi(M)$. Let $\psi_0(\overline{x}')$ be the pp condition $\phi(0, \overline{x}')$. So $\overline{a}' - \sum c_i \overline{r}'_i \in \psi_0(M)$ which, by the inductive assumption, equals $M \cdot \psi_0(R)$, say $\overline{a}' - \sum c_i \overline{r}'_i = \sum d_j \cdot \overline{s}_j$ with $d_j \in M$ and $\overline{s}_j \in \psi_0(R)$, so $(0, \overline{s}_j) \in \phi(R)$ for each j.

Then $(a, \overline{a}') = \sum c_i(r_i, \overline{r}'_i) + \sum d_j(0, \overline{s}_j) \in M \cdot \phi(R)$, as required. □

The map $L(\leq {}_R R) \mapsto ML$ is, by 2.3.9, surjective so we obtain the next corollary (cf. 2.3.4 and Section 3.4.3).

Corollary 2.3.10. *If M is flat, then the lattice, $\mathrm{pp}^1(M)$, of pp-definable subgroups of M is a quotient of the lattice of left ideals of R.*

Corollary 2.3.11. *Any pure submodule of a flat module is flat.*

Proof. We check the criterion of 2.3.9. Suppose that N is a pure submodule of the flat module M. Let ϕ be a pp condition with one free variable. By 1.1.10, $\phi(N) \geq N \cdot \phi(R)$. If $a \in \phi(N) \leq \phi(M) = M \cdot \phi(R)$, say $a = \sum m_i r_i$ with each $r_i \in \phi(R)$, then, by purity of N in M, $a = \sum n_i r_i$ for some $n_i \in N$, as required for 2.3.9. □

It also follows from 2.1.14 and 2.3.9 that if $0 \to K \to F \to G \to 0$ is pure-exact and F is flat, then G also is flat.

A module M is **torsionfree** if for all $m \in M$ and $r \in R$, if $mr = 0$, then $m = \sum m_i s_i$ for some $s_i \in R$ with $s_i r = 0$. Of course, if R is a domain this reduces to the usual definition, $mr = 0$ implies $r = 0$. Taking $\phi(x)$ to be $xr = 0$ in 2.3.9 gives the following.

Corollary 2.3.12. *Every flat module is torsionfree.*

Corollary 2.3.13. ([391, 1.2], also see [390], and [246]) *Every flat module is a direct limit of projective modules.*

Proof. Let F be a flat module. Like any module (Appendix E, p. 714) it may be represented as a direct limit of finitely presented modules and, cf. the proof of 2.1.25, we may assume that every morphism from a finitely presented module is represented in the directed (or, better, filtered) system. It will be enough to show that the maps from finitely generated projective modules are cofinal in this directed system, meaning that if $f : C \to F$ is any morphism with C finitely presented, then f factorises through a projective module.

Say $\overline{c}$ is a finite generating sequence for C, with relations generated by $\overline{c}G = 0$, where G is an $n \times m$ matrix with entries in R. Since $f\overline{c}$ satisfies the pp condition $\overline{x}G = 0$ there are, by 2.3.9, $\overline{a} \in F^t$ and a $t \times n$ matrix H (for some t) with each row $\overline{r}_k$ satisfying $\overline{r}_k G = 0$ such that $f\overline{c} = \overline{a}H$. For each $k = 1, \dots, t$ sending $\overline{c}$ to $\overline{r}_k$ well defines a map, g, from C to R^n so there is an induced map $C \to (R^n)^t$. Let h be the map from R^{nt} to F^n which takes e_{ki} (in the obvious notation) to a_{ki} (that is, a_k in the ith copy of F). Then $hg\overline{c_i} = h(\sum_{k=1}^t r_{ki}) = \sum_{k=1}^t a_{ki} r_{ki} = (\sum_{k=1}^t a_k r_{ki})_i$ (again, the last subscript i denotes an element in the ith copy of F). Thus, $hg = f$, as required. □

Proposition 2.3.14. (essentially [116, 2.2], see e.g. [663, I.11.1]) *A module F is flat iff every exact sequence of the form $0 \to A \to B \xrightarrow{g} F \to 0$ is pure-exact.*[8]

Proof. ($\Rightarrow$) To show that g is a pure epimorphism, and hence that the sequence is pure-exact, we use the criterion of 2.1.14. So suppose that ϕ is pp and $\overline{c} \in \phi(F)$. Since F is flat, $\phi(F) = F \cdot \phi(R)$ (2.3.9), so $\overline{c} = \sum d_i \overline{r}_i$ for some $d_i \in F$ and $\overline{r}_i \in \phi(R)$. Choose $b_i \in B$ with $gb_i = d_i$ and set $\overline{b} = \sum b_i \overline{r}_i$, so $g\overline{b} = \overline{c}$. Note that $b_i \overline{r}_i \in B \cdot \phi(R) \leq \phi(B)$, hence $\overline{b} \in \phi(B)$, as required.

($\Leftarrow$) Consider a surjection from a free module $\pi : R^{(\kappa)} \to F$. By hypothesis this is a pure epimorphism. We use this to check the criterion, 2.3.9, for flatness. So take ϕ pp and $\overline{a} \in \phi(F)$. By 2.1.14 there is $\overline{r}$ in the free module with $\pi\overline{r} = \overline{a}$ and $\overline{r} \in \phi(R^{(\kappa)})$. So, in fact, $\overline{r} \in \phi(R^t)$ for some integer t. Denote by $\overline{r}_k$ ($k = 1, \dots, t$) the projection of $\overline{r}$ to the kth coordinate. Then $\overline{r}_k \in \phi(R)$ and, if we set $b_k = \pi e_k$, where e_k is the kth unit of R^t, then $\overline{a} = \sum_{k=1}^t b_k \overline{r}_k \in F \cdot \phi(R)$, as required. □

[8] Also note 4.6.1.

Combined with 2.3.11 this implies that any (long) exact sequence of flat objects, in particular any projective resolution of a flat object, is pure-exact and hence that for any flat module F the functors $\mathrm{Pext}^n(F, -)$ (Section 2.2) and $\mathrm{Ext}^n(F, -)$ are naturally isomorphic.

The duality between the characterisations, 2.3.9 and 2.3.3, of flat and absolutely pure, combined with 1.3.15 (and 4.3.29 and 4.3.12) gives a proof of the fact that a module M is flat iff $M^* = \mathrm{Hom}_{\mathbb{Z}}(M, \mathbb{Q}/\mathbb{Z})$ is injective (see, for example, [663, I.10.5]).

2.3.3 Coherent modules and coherent rings

Coherence for modules is a strengthening of finite presentation. The condition on the ring implies that every finitely presented module is coherent (2.3.18) and is also equivalent to finite generation, as ideals on the appropriate side, of its pp-definable subgroups (2.3.19). Coherence of the ring has consequences for absolutely pure (2.3.20) and flat (2.3.21) modules (a stronger result, 3.4.24, will be obtained later).

A module is **locally coherent** if every finitely generated submodule is finitely presented. It is **coherent** if it is locally coherent and is itself finitely generated/presented. A ring R is **right coherent** if R is coherent as a right R-module. For example, any right noetherian ring is right coherent. Both [663], [713] contain the basic results around this notion; see also [17] for the more general setting. We give some of the proofs here.

In terms of elements, a module M is locally coherent iff for every finite tuple $\overline{a}$ from M, $\mathrm{ann}(\overline{a}) = \{\overline{r} \in R^{n=l(\overline{a})} : \sum a_i r_i = 0\}$ is finitely generated as a submodule of R^n: consider the exact sequence $0 \to \mathrm{ann}(\overline{a}) \to R^n \to \langle \overline{a} \rangle \to 0$, where $\langle \overline{a} \rangle$ denotes the module generated by the entries of $\overline{a}$.

Proposition 2.3.15. *A module M is locally coherent iff*

(a) for any $a \in M$ the annihilator, $\mathrm{ann}_R(a)$, is finitely generated and
(b) whenever A and B are finitely generated submodules of M their intersection also is finitely generated.

Proof. If M is locally coherent, then, as already remarked, the first condition must hold. For the second condition, consider the exact sequence $0 \to A \cap B \to A \oplus B \to A + B \to 0$. Since all of A, B and $A + B$ are finitely generated, hence finitely presented, $A \cap B$ must be finitely generated (by E.1.16).

For the converse, assume the conditions (a) and (b): we show that (a) holds with tuples of any finite length in place of a, which will be enough.

So suppose inductively that we have this for $(n-1)$-tuples and take $\overline{a} = (a_1, \ldots, a_n) \in M^n$. Set $A = \sum_1^n a_i R$ and $A' = \sum_1^{n-1} a_i R$ and consider the exact sequence $0 \to a_nR \cap A' \to a_nR \oplus A' \to A \to 0$. By assumption $a_nR \cap A'$ is finitely generated. By induction, a_nR and A', hence $a_nR \oplus A'$, are finitely presented. Therefore A is finitely presented, as required. □

Lemma 2.3.16. *Let* $0 \to A \to B \xrightarrow{\pi} C \to 0$ *be an exact sequence with A and C locally coherent modules. Then B is locally coherent.*

Proof. Take $\overline{b}$ from B and consider $(A : \overline{b}) = \{\overline{r} \in R^{l(\overline{b})} : \sum b_i r_i \in A\} = \mathrm{ann}_R(\pi\overline{b})$. Since C is locally coherent this is finitely generated, generated by $\overline{r}_1, \dots, \overline{r}_t$ say. Then $\overline{b} \cdot (A : \overline{b}) = \langle \overline{b} \rangle \cap A$ is finitely generated, by $\overline{b} \cdot \overline{r}_1, \dots, \overline{b} \cdot \overline{r}_t$, so, since A is locally coherent, is finitely presented. Therefore in the exact sequence $0 \to \langle \overline{b} \rangle \cap A \to \langle \overline{b} \rangle \to \pi\langle \overline{b} \rangle \to 0$ both $\langle \overline{b} \rangle \cap A$ and $\pi\langle \overline{b} \rangle$ are finitely presented. Therefore (E.1.16) $\langle \overline{b} \rangle$ is finitely presented, as required. □

Lemma 2.3.17. *If B is locally coherent and* $A \le B$ *is finitely generated, then* B/A *is locally coherent.*

Proof. Let $\overline{b}$ be a tuple from B. We have an exact sequence $0 \to \langle \overline{b} \rangle \cap A \to \langle \overline{b} \rangle \to (\langle \overline{b} \rangle + A)/A \to 0$. By 2.3.15, $\langle \overline{b} \rangle \cap A$ is finitely generated so, since $\langle \overline{b} \rangle$ is finitely presented, $(\langle \overline{b} \rangle + A)/A$, which is a typical finitely generated submodule of B/A, is finitely presented, as required. □

It follows that the coherent R-modules form an abelian subcategory of Mod-R (cf. 10.2.4). In general the finitely presented modules do not form an abelian subcategory, indeed mod-R is an abelian subcategory of Mod-R iff R is right coherent (E.1.19).

Corollary 2.3.18. ([116, 2.2], see, for example, [663, I.13.1]) *A ring R is right coherent iff for every* $r \in R$, $\mathrm{ann}_R(r)$ *is finitely generated and whenever I and J are finitely generated right ideals of R their intersection* $I \cap J$ *is finitely generated. This holds iff every finitely presented right R-module is coherent.*

Proof. The first equivalence is by 2.3.15, the second is immediate by 2.3.17 (and 2.3.16 for the fact that R^n is coherent if R is). □

Theorem 2.3.19. ([731, 1.3], [617, Prop. 7]) *For any ring R the following are equivalent.*

(i) The ring R is left coherent.
(ii) The pp-definable subgroups of R_R *are precisely the finitely generated left ideals of R.*
(iii) The pp-definable subgroups of R_R^n *are precisely the finitely generated left R-submodules of* R^n.

Proof. (i)⇒(iii) Suppose that $\phi(\overline{x})$, being $\exists\overline{y}\,(\overline{x}\,\overline{y})H = 0$, where H is $k \times m$, is a pp condition for right R-modules. The solution set in R to the system of equations $(\overline{x}\,\overline{y})H = 0$ is the kernel of the morphism from ${}_RR^k$ to ${}_RR^m$ defined by $\overline{z} \mapsto \overline{z}H$. If the columns of H are $\overline{r}_1, \dots, \overline{r}_m$, then this kernel is $\bigcap_{j=1}^m \mathrm{ann}(\overline{r}_i)$ ("ann" in this context meaning left annihilator) which, if R is left coherent, is, by 2.3.15, finitely generated. The projection of this solution set to the first $n(= l(\overline{x}))$ coordinates, that is, $\phi(R)$, is therefore finitely generated, as required.

(iii)⇒(ii) is trivial.

(ii)⇒(i) The left annihilator of any element of R is pp-definable in R_R, as is the intersection of any two finitely generated left ideals, so the conditions in 2.3.18 for left coherence are satisfied. □

Corollary 2.3.20. ([174, §4], also see [495, 15.40] and [323, 6.13]) *If R is right coherent and M_R is absolutely pure, then every pp-definable subgroup of M is definable by a quantifier-free pp condition.*

Proof. Let $\phi(\overline{x})$ be pp and let M_R be absolutely pure. By 2.3.3, $\phi(M) = \mathrm{ann}_M D\phi({}_RR)$. By 2.3.19, $D\phi({}_RR) = \sum_1^m \overline{r}_j R$ say, so $\phi(M) = \bigcap_1^m \mathrm{ann}_M(\overline{r}_j)$ and hence $\phi(M)$ is defined by the quantifier-free condition $\bigwedge_1^m \overline{x} \cdot \overline{r}_j = 0$, that is, $\overline{x}H = 0$, where the $\overline{r}_j$ form the columns of H. □

In fact, if R is right coherent, if M is absolutely pure and if $\phi \geq \psi$ is a pp-pair (Section 1.2.2), then there is (see 11.1.44) a finitely presented module A such that $\phi(M)/\psi(M) \simeq (A, M)$.

Theorem 2.3.21. ([116, 2.1]) *The ring R is left coherent iff the class of flat right R-modules is closed under product.*

Proof. If R is left coherent and $\psi(x)$ is a pp condition for right R-modules, then $\psi(R_R)$ is a finitely generated left ideal (2.3.19), say $\psi(R_R) = \sum_{j=i}^n Rr_i$. Therefore the condition $\psi(R_R) \mid x$ is expressible by a pp condition, namely $\exists y_1, \dots, y_n \left(x = \sum_1^n y_i r_i\right)$. So, if $\psi(x)$ and $\psi(R_R) \mid x$ are equivalent in every component of a product, then, by 1.2.3, they are equivalent in the product. It follows from 2.3.9 that every product of flat right R-modules is flat.

For the converse, and assuming the condition just on copies of R_R, let $\psi(x)$ be a pp condition for right R-modules and let $(r_\lambda)_\lambda$ $(\lambda \in \Lambda)$ be a generating set for the left ideal $\psi(R_R)$. Then $a = (r_\lambda)_\lambda \in \psi(R_R^\Lambda)$ and so, by 2.3.9, $a \in R^\Lambda \cdot \psi(R)$, say $a = \sum_{i=1}^n b_i r_{\lambda_i}$ for some $\lambda_1, \dots, \lambda_n$ and some $b_i \in R^\Lambda$. This implies that $\psi(R) = \sum_{i=1}^n Rr_{\lambda_i}$ is finitely generated, so, by 2.3.19, R is left coherent, as required. □

Clearly there is some kind of duality between absolutely pure and flat modules. One formal expression of this is 3.4.24.

2.3.4 Von Neumann regular rings

Von Neumann regular rings may be characterised in numerous ways (2.3.22), including the simplification of pp conditions to quantifier-free ones (2.3.24). It follows that the lattice of pp conditions in n free variables is isomorphic to the lattice of finitely generated right (equally left) submodules of R^n. That R being von Neumann regular is equivalent to "elimination of quantifiers" for R-modules is the content of 2.3.24.

A ring R is a **von Neumann regular ring** if for every element $r \in R$ there is $s \in R$ such that $r = rsr$. Examples are semisimple artinian rings, endomorphism rings of vector spaces over fields, boolean rings, the ring of continuous functions from the unit interval to $\mathbb{R}$ (with pointwise operations), any factor ring of a von Neumann regular ring. There are many conditions equivalent to von Neumann regularity and we list some of them. Since the defining condition is obviously right/left symmetric, for each "right" condition below there is an equivalent "left" condition, which is not stated.

Theorem 2.3.22. *The following conditions are equivalent for a ring R:*

(i) R is von Neumann regular;
(ii) every cyclic right ideal of R is generated by an idempotent;
(iii) every finitely generated right ideal is generated by an idempotent, that is, is a direct summand of R_R;
(iv) for every n, every finitely generated submodule of R_R^n is a direct summand;
(v) every right R-module is flat;
(vi) every right R-module is absolutely pure, that is, every embedding between R-modules is a pure embedding;
(vii) every exact sequence of R-modules is pure-exact;
(viii) every pure-injective module (see Chapter 4) is injective.

See, for example, [240, 1.1, 1.11, 1.13,], [663, §I.12] for (i)–(v), for (vi) and (vii) use 2.1.3 and for (viii) use, say, 2.3.2 and the fact that every module purely embeds in a pure-injective module, which is 4.3.19. The next result follows quite easily using 2.1.18.

Corollary 2.3.23. *If R is von Neumann regular, then every short exact sequence of finitely presented modules is split, indeed every finitely presented module is projective, R is (right and left) semihereditary and (right and left) coherent.*

We show that von Neumann regular rings are exactly those over which every pp condition is equivalent to one without quantifiers, that is, to a system of equations. This result has appeared, at least in part, in various sources, [731, §4], [207, Lemma 8], [256, 2.3], [617, Prop. 11], with the full result by Hodges, see [710, Thm 1] or [296, A.1.4], and may be read as saying that over von Neumann regular rings pp conditions, pp-types etc. reduce to more familiar "purely algebraic" ideas in that they do not require existential quantifiers (= projections) for their expression. Therefore von Neumann regular rings lie at one extreme.

Theorem 2.3.24. *The ring R is von Neumann regular:*

iff for every pp condition, $\phi(\overline{x})$, for right R-modules there is a matrix G over R such that $\phi(\overline{x})$ is equivalent to the system of equations $\overline{x}G = 0$;
iff for every pp condition, $\phi(\overline{x})$, for right R-modules there is a matrix G over R such that $\phi(\overline{x})$ is equivalent to the divisibility condition $G \mid \overline{x}$.

In each case the matrix G may be taken to be (square and) idempotent.

Proof. Suppose that R is von Neumann regular and let $(C, \overline{c})$ be a free realisation of ϕ. Since, by 2.3.22, C is flat we know, by 2.3.9, that $\phi(C) = C \cdot \phi(R)$. By 2.3.23 and 2.3.19 $\phi(R)$ is a finitely generated left R-submodule of $R^{n=l(\overline{x})}$, hence, by 2.3.22, is a direct summand of ${}_RR^n$, and thus has the form R^nG for some idempotent matrix G. So $\phi(C) = C^nG$ (the set of n-tuples of the form $\overline{d}G$ for some tuple $\overline{d}$ with entries in C) and hence $\overline{c}$ satisfies the condition $G \mid \overline{x}$. Therefore $\phi \leq G \mid \overline{x}$. On the other hand, for any module M, the solution set in M to $G \mid \overline{x}$, that is, M^nG, equals $M \cdot R^nG = M \cdot \phi(R) \leq \phi(M)$ so also $G \mid \overline{x} \leq \phi$, as required.

It was noted in 1.3.4 that the elementary dual to the condition $G \mid \overline{x}$ is the condition $\overline{x}G = 0$ so (by right/left symmetry of the von Neumann regularity condition and 1.3.1) we also have the other form.

For the converse, if we have, on the right or the left, that every pp condition is equivalent to a quantifier-free one (that is, to a system of equations), then clearly, on that side, every embedding between modules is pure, so von Neumann regularity of R follows by 2.3.22. □

Corollary 2.3.25. *If R is von Neumann regular, then the lattice, pp^n_R, of pp conditions in n free variables for right modules is isomorphic to the lattice of finitely generated submodules of the left R-module R^n and is also isomorphic to the lattice of finitely generated submodules of the right R-module R^n. Hence $\mathrm{pp}^n_R \simeq {}_R\mathrm{pp}^n$.*

Proof. This is immediate by 2.3.19 and the observation that if two pp conditions coincide on R, then they coincide on all flat, that is, on all, R-modules (by 2.3.9). The duality follows from 2.3.24. (Observe that for $n = 1$ the isomorphism between the lattices of finitely generated = cyclic right and left ideals is given by $eR \mapsto R(1 - e)$. For $n > 1$, 1 is replaced by the $n \times n$ identity matrix I and the idempotent $e \in R$ is replaced by any idempotent $n \times n$ matrix G.) □

Proposition 2.3.26. ([102, 2.7], [218, p. 10]) *A ring is von Neumann regular iff it is right coherent and every monomorphism between finitely presented right modules is pure.*

Proof. One direction is immediate from 2.3.22. For the other, one may use the fact, see [359, 5.9], that in a locally coherent category every short exact sequence is a direct limit of short exact sequences between finitely presented objects, plus the fact that a direct limit of pure-exact sequences is pure-exact 2.1.2(3). (Burke's proof used the fact, see [495, 16.16] for references, that the theory of modules over a ring has elimination of quantifiers iff the ring is von Neumann regular.) □

See Section 8.2.12 for more on the last condition in the statement of that result.

2.4 Purity and structure of finitely presented modules

In Section 2.4.1 direct sum decomposition of finitely presented modules is related to the expressibility of pp conditions as sums and conjunctions of simpler conditions. A ring is RD if purity reduces to simple divisibility; these rings are discussed in Section 2.4.2. Serial rings are RD and have further simplifications, described in Section 2.4.3, to the structure of finitely presented modules and pp conditions. These results were used by Puninski to obtain examples of "strange" direct sum decomposition properties of some modules over uniserial rings. This is reported in Section 2.5.

2.4.1 Direct sum decomposition of finitely presented modules

The size of the matrix of a pp condition is related to the complexity (in terms of number of generators and relations) of modules in which it can be realised (2.4.1). Decomposition of a pp condition as a sum, or as a conjunction, of simpler conditions is related to its being realised in a module which is a direct sum of modules of a certain complexity (2.4.2). This yields conditions for every finitely presented module to be a direct summand of a direct sum of modules of a given complexity (2.4.6, 2.4.7).

Say that a module M is (k, l)-**presented** if there is an exact sequence $R^l \to R^k \to M \to 0$. We allow k, l to be "∞", by which we mean finite but with no finite bound specified, so (∞, ∞)-presented means finitely presented. We also refer to (1,1)-presented modules as being **cyclically presented**.

It follows from the proof of 1.2.6 that if $\overline{c}$ is a tuple in the (k, l)-presented module M, then there is a pp condition of the form $\exists \overline{y}\, (\overline{x} = \overline{y}G \wedge \overline{y}H = 0)$, with H a $k \times l$ matrix, generating $\mathrm{pp}^M(\overline{c})$. Any pp condition equivalent to one of the form $\exists \overline{y}\, (\overline{x} = \overline{y}G \wedge \overline{y}H = 0)$ with H of size no more than $k \times l$ (without loss of generality, by adding zeroes, H is $k \times l$), we refer to as a k-**gen**, l-**rel** condition. Also in this terminology we allow either k or l to be replaced by "∞" (that is, "finitely").

Lemma 2.4.1. *A pp condition for right R-modules is k-gen, l-rel iff it has a free realisation in a (k, l)-presented module.*

Proof. One direction has just been observed. The other is 1.2.15. □

The dual (Section 1.3.1) of the condition $\exists \overline{y}\, (\overline{x} = \overline{y}G \wedge \overline{y}H = 0)$ is the pp condition $\exists \overline{z}\, (G\overline{x} = H\overline{z})$ for *left* R-modules: we will refer to such as a **dual**-k-**gen**, l-**rel** condition for left R-modules so, transposing, we say that a pp condition for *right* R-modules is **dual**-k-**gen**, l-**rel** if it is equivalent to one of the form $\exists \overline{y}\, (\overline{x}G = \overline{y}H)$, where H is $l \times k$. Note that a k-gen, l-rel condition for *left* modules is one of the form $\exists \overline{z}\, (\overline{x} = G\overline{z} \wedge H\overline{z} = 0)$ where H is $l \times k$.

Proposition 2.4.2. *A pp condition ϕ is equivalent to one of the form $\phi_1 + \cdots + \phi_t$, where each ϕ_i is a k-gen, l-rel condition iff ϕ has a free realisation in a (direct summand of a) direct sum of (k, l)-presented modules.*

Proof. By 1.2.27 and 2.4.1 if each ϕ_i is k-gen, l-rel ($i = 1, \ldots, t$), then $\phi_1 + \cdots + \phi_t$ has a free realisation in a direct sum of (k, l)-presented modules.

For the converse, if ϕ has a free realisation, $(C, \overline{c})$, with C a direct summand of $M = M_1 \oplus \cdots \oplus M_t$, where each M_i is (k, l)-presented, then write $\overline{c} = \overline{c}_1 + \cdots + \overline{c}_t$, where $\overline{c}_i$ is from M_i. Since C is a direct summand of M we have (1.2.1) $\mathrm{pp}^M(\overline{c}) = \mathrm{pp}^C(\overline{c}) = \langle \phi \rangle$ and so, by 1.2.27, ϕ is equivalent to $\phi_1 + \cdots + \phi_t$, where ϕ_i generates $\mathrm{pp}^{M_i}(\overline{c}_i) = \mathrm{pp}^M(\overline{c}_i)$. By 2.4.1 each ϕ_i is k-gen, l-rel, as required. (Regarding the parenthetical phrase, note that $(M, \overline{c})$ is also a free realisation of ϕ.) □

Our terminology is such that the above lemma holds for left and for right modules.

Applying elementary duality (1.3.1) yields the following (which also holds if "left" and "right" are interchanged).

Corollary 2.4.3. *For any ring R the following are equivalent.*

(i) Every finitely presented right R-module is a direct summand of a direct sum of (k, l)-presented modules.

(ii) *Every pp condition for right R-modules is equivalent to one of the form $\phi_1 + \cdots + \phi_t$, where each ϕ_i is a k-gen, l-rel condition.*
(iii) *Every pp condition for left R-modules is equivalent to one of the form $\psi_1 \wedge \cdots \wedge \psi_t$, where each ψ_i is a dual-k-gen, l-rel condition.*

The next result refines the descriptions of pure embeddings (Section 2.1.1) and pure epimorphisms (2.1.14).

Lemma 2.4.4. ([575, 1.5, 2.3]) *Let $0 \to A \to B \xrightarrow{\pi} C \to 0$ be an exact sequence of right R-modules. Then the following are equivalent:*

(i) *for every pp condition ϕ of the form $\phi_1 + \cdots + \phi_t$, where each ϕ_i is k-gen, l-rel, we have $\phi(C) = \pi\phi(B)$;*
(ii) *for every pp condition ψ of the form $\psi_1 \wedge \cdots \wedge \psi_t$, where each ψ_i is dual-l-gen, k-rel, we have $\psi(A) = A^n \cap \psi(B)$ (where n is the number of free variables of ψ).*

Proof. (i)$\Rightarrow$(ii) Suppose that ψ_i is $\exists \overline{y}_i\, (\overline{x}G_i = \overline{y}_i H_i)$, where each H_i is $k \times l$, and take $\overline{a} \in A^n \cap \psi(B)$. Then, for each i, there is $\overline{b}_i$ from B with $\overline{a}G_i = \overline{b}_i H_i$. Since $\pi\overline{b}_i.H_i = \pi(\overline{b}_i H_i) = 0$ there is, by assumption (take $\phi_i(\overline{x})$ to be $\exists \overline{y}\, (\overline{x} = \overline{y} \wedge \overline{y}H_i = 0)$), for each i, some $\overline{b}_i'$ from B with $\overline{b}_i' H_i = \overline{0}$ and $\overline{b}_i - \overline{b}_i'$ from $\ker(\pi) = A$. Then $(\overline{b}_i - \overline{b}_i')H_i = \overline{b}_i H_i = \overline{a}G_i$, so $A \models \exists \overline{y}_i\, (\overline{a}G_i = \overline{y}_i H_i)$ for each i, that is $\overline{a} \in \psi(A)$.

(ii)$\Rightarrow$(i) Suppose that $\overline{c} \in \phi(C)$, say $\overline{c} = \overline{c}_1 + \cdots + \overline{c}_t$ with $\overline{c}_i \in \phi_i(C)$. Suppose that ϕ_i is $\exists \overline{z}_i\, (\overline{x} = \overline{z}_i G_i \wedge \overline{z}_i H_i = 0)$ (where each H_i is $k \times l$), say $\overline{d}_i$ from C is such that $\overline{c}_i = \overline{d}_i G_i$ and $\overline{d}_i H_i = 0$. Choose preimages, $\overline{b}_i$ from B, of the $\overline{d}_i$.

For each i, $\overline{b}_i H_i$ is a tuple from A so, by assumption, there is $\overline{a}_i$ from A such that $\overline{a}_i H = \overline{b}_i H_i$. Then $(\overline{b}_i - \overline{a}_i)H_i = 0$ for each i and the image of $(\overline{b}_i - \overline{a}_i)G_i$ in C is $\overline{d}_i G_i = \overline{c}_i$. Hence $\sum_i (\overline{b}_i - \overline{a}_i)G_i$ is an element of $\phi(B)$ mapping to $\overline{c}$, as required.[9] □

Given a set $\mathcal{H}$ of matrices over R, let $\Psi_{\mathcal{H}}$ be the collection of pp conditions of the form $\exists \overline{y}\, (\overline{x}G = \overline{y}H)$, where G is arbitrary and $H \in \mathcal{H}$ (we assume that $\mathcal{H}$ contains zero matrices so that $\Psi_{\mathcal{H}}$ includes the trivial conditions $\overline{x} = \overline{x}$ and $\overline{x} = \overline{0}$). The next result gives a criterion for every pp condition to be expressible in terms of a certain restricted set of pp conditions.

[9] From the proof one sees that the core statement is: if $0 \to A \to B \xrightarrow{\pi} C \to 0$ is exact and H is any matrix (with n rows), then, if ψ is the condition $\exists \overline{y}\, (\overline{x} = \overline{y}H)$ and ϕ_G is notation for the condition $\exists \overline{z}\, (\overline{x} = \overline{z}G \wedge \overline{z}H = 0)$ (G any matrix with n columns), we have $\psi(A) = A^n \cap \psi(B)$ iff for every G we have $\phi_G(C) = \pi\phi_G(B)$. (The apparent asymmetry is illusory because A is a submodule: we could rephrase this by replacing $\overline{x}$ by $\overline{x}K$ for any K when we form ψ.) For more on this very local notion of purity see [575] and [620]. Also see [554].

Proposition 2.4.5. *Suppose that $\mathcal{H}$ is a set of matrices and that $\Psi_{\mathcal{H}}$, as above, is the corresponding set of pp conditions. Suppose that for every exact sequence $0 \to A \to B \xrightarrow{\pi} C \to 0$ of R-modules this sequence is pure if for every $\psi(\overline{x}) \in \Psi_{\mathcal{H}}$ we have $\psi(A) = A^n \cap \psi(B)$. Then every pp condition for R-modules is equivalent to one of the form $\psi_1 \wedge \cdots \wedge \psi_t$ with the $\psi_i \in \Psi_{\mathcal{H}}$.*

Proof. Let η be any pp condition (with n free variables) for right R-modules and let $(C, \overline{c}_\eta)$ be a free realisation of η. Set $\Psi' = \{\psi \in \Psi_{\mathcal{H}} : \overline{c}_\eta \in \psi(C)\}$ ($\Psi_{\mathcal{H}}$ contains the trivial conditions $\overline{x} = \overline{x}$, so Ψ' is not empty) and let p be the closure of Ψ' under conjunction ($\wedge$) and implication. That is, p is the filter generated by Ψ' in the lattice pp_R^n (Section 1.1.3). At this point we call on 3.2.5, which implies that there is a module M and $\overline{b}$ from M such that $\mathrm{pp}^M(\overline{b}) = p$.

Form the pushout shown.

$$\begin{array}{ccc} R^n & \xrightarrow{\overline{c}_\eta} & C \\ {\scriptstyle \overline{b}}\downarrow & & \downarrow{\scriptstyle g} \\ M & \xrightarrow[f]{} & P \end{array}$$

Note that $f\overline{b} = g\overline{c}_\eta \in \eta(P)$. We claim that for all $\psi(\overline{x}) \in \Psi_{\mathcal{H}}$ we have $\psi(M) = f^{-1}\big(fM^{l(\overline{x})} \cap \psi(P)\big)$. To see this let $\overline{a} \in f^{-1}\big(fM^{l(\overline{x})} \cap \psi(P)\big)$. Say $\psi(\overline{x})$ is $\exists \overline{y}\,(\overline{x}G = \overline{y}H)$, where $H \in \mathcal{H}$, so $f\overline{a}G = \overline{d}H$ for some $\overline{d}$ from P. By the construction of pushout, $P = (M \oplus C)/\{(\overline{b}\overline{r}, -\overline{c}_\eta\overline{r}) : \overline{r} \in R^{l(\overline{b})}\}$ so, representing $\overline{d}$ as the image of a tuple, $(\overline{m}, \overline{c})$, from $M \oplus C$, we have $(\overline{a}G, \overline{0}) = (\overline{m}H, \overline{c}H) + (\overline{b}K, -\overline{c}_\eta K)$ for some matrix K. Projecting to C we have $\overline{c}H = \overline{c}_\eta K$ and so the pp condition $\exists \overline{y}\,(\overline{x}K = \overline{y}H)$ is in Ψ', hence is satisfied by $\overline{b}$, say $\overline{b}K = \overline{b}'H$ with $\overline{b}'$ from M. Then, projecting the above equation to M, we have $\overline{a}G = \overline{m}H + \overline{b}K = \overline{m}H + \overline{b}'H = (\overline{m} + \overline{b}')H$, so we see that $\overline{a}$ lies in $\psi(M)$, as claimed.

It follows that f is monic (take ψ to be $\overline{x} = \overline{0}$) and that the exact sequence beginning with $0 \to M \to P$ satisfies the assumption in the statement of the result, hence f is a pure embedding. Therefore, since $f\overline{b} = g\overline{c}_\eta \in \eta(P)$, also $\overline{b} \in \eta(M)$ and hence $\eta \in p$. That is, there are $\psi_1, \dots, \psi_n \in \Psi'$ such that $\psi_1 \wedge \cdots \wedge \psi_n \leq \eta$. Since, by definition of Ψ', each $\psi_i \geq \eta$, we deduce that η is equivalent to $\psi_1 \wedge \cdots \wedge \psi_n$, as required. □

Theorem 2.4.6. ([575, 2.5]) *For any ring R the following are equivalent, given $k, l \in \mathbb{P} \cup \{\infty\}$:*

(i) every finitely presented right R-module is a direct summand of a direct sum of (k, l)-presented modules;

(ii) every finitely presented left R-module is a direct summand of a direct sum of (l, k)-presented modules;

(iii) *every pp condition ϕ for right R-modules is equivalent to one of the form $\phi_1 + \cdots + \phi_t$, where each ϕ_i is a k-gen, l-rel condition (that is, ϕ_i has the form $\exists \overline{y}\,(\overline{x} = \overline{y}G_i \wedge \overline{y}H_i = 0)$ with H_i a $k \times l$ matrix);*

(iv) *every pp condition ψ for right R-modules is equivalent to one of the form $\psi_1 \wedge \cdots \wedge \psi_s$, where each ψ_j is a dual-l-gen, k-rel condition (that is, ψ_j has the form $\exists \overline{y}\,(\overline{x}G_i = \overline{y}H_i)$ with H_i a $k \times l$ matrix).*

Proof. The pairs (i), (iii) and (ii), (iv) are equivalent by 2.4.3. If we have, say, (iii), then, by 2.1.14, an exact sequence $0 \to A \to B \xrightarrow{\pi} C \to 0$ of right modules is pure iff for every condition ϕ of the form in (iii) we have $\phi(C) = \pi\phi(B)$. It follows by 2.4.4 that such a sequence is pure iff for every condition $\psi(\overline{x})$ of the form in (iv) we have $\psi(A) = A^{l(\overline{x})} \cap \psi(B)$. By 2.4.5 this implies that every pp condition for right modules is equivalent to one of the form in (iv).

The converse, (iv)$\Rightarrow$(iii), follows by applying 2.4.3 to move the hypotheses (iv) to the category of left modules (where it becomes like (iii) but with k, l interchanged), making the same argument there and then using 2.4.3 again to return to the category of right modules. □

Theorem 2.4.7. (Puninski, see [575, 2.9]) *If every finitely presented right R-module is a direct summand of a direct sum of (k, l)-presented modules and also a direct summand of a direct sum of (k', l')-presented modules, then every finitely presented right module is a direct summand of a direct sum of $(\min(k, k'), \min(l, l'))$-presented modules. (Here $k^{(\prime)}$ and $l^{(\prime)}$ are positive integers or "∞".)*

Proof. The first assumption is equivalent, by 2.4.6, to every pp condition for right modules being equivalent to a finite conjunction of dual-l-gen, k-rel conditions. For the duration of this proof let us describe this state of affairs thus: every pp condition is l, k. The second assumption is equivalent to every pp condition being l', k'. We claim that every pp condition is l', k.

To see this, let ϕ be any dual-l'-gen, k'-rel condition, that is, $\phi(\overline{x})$ has the form $\exists \overline{y}\,(\overline{x}G = \overline{y}H)$, where H is $k' \times l'$. By assumption ϕ is l, k, hence equivalent to a conjuction $\phi_1 \wedge \cdots \wedge \phi_t$, where ϕ_i is, say, $\exists \overline{z}\,(\overline{x}G_i = \overline{z}H_i)$ with each H_i being $k \times l$. Since $\bigwedge_1^t \phi_i \leq \phi$ we have, by 1.1.17, that there are matrices C_i, D_i and E such that $G = \sum_1^t G_i D_i + EH$ and $C_i H = H_i D_i$ $(i = 1, \dots, t)$.

Now, the pp condition $H_i \mid \overline{x}G_i$ implies $H_i D_i \mid \overline{x}G_i D_i$, which is equivalent to $C_i H \mid \overline{x}G_i D_i$, implying $H \mid \overline{x}G_i D_i$. Let ψ_i denote the condition $H_i D_i \mid \overline{x}G_i D_i$ and note that $H_i D_i$ is a $k \times l'$ matrix. Also ϕ is equivalent to $\bigwedge_1^t \phi_i$ and this implies $\bigwedge_1^t \psi_i$. This last implies $H \mid \overline{x}\,(\sum_1^t G_i D_i)$, which is equivalent to $H \mid \overline{x}(G - EH)$, which, in turn, is equivalent to $H \mid \overline{x}G$, that is, to ϕ. So ϕ is equivalent to $\bigwedge_1^t \psi_i$: that is, in the temporary terminology, ϕ is l', k. But every pp condition is, by hypothesis and 2.4.6, equivalent to a conjunction of pp conditions such as ϕ and hence every pp condition is l', k.

By symmetry (applied via 2.4.6(i)($\Leftrightarrow$)(ii)) every pp condition is also l, k'. So every pp condition is each of the four possible forms $(-), (-)$ and the result follows. □

Corollary 2.4.8. ([575, 2.13(b)]) *Let R be a simple noetherian ring with right and left Krull dimension at most n. Then every finitely presented module is a direct summand of a direct sum of modules each of which is $(2n+1)$-presented.*

Proof. By [659, 4.4] every indecomposable finitely generated module over such a ring is generated by $2n+1$ elements, so every finitely presented right or left module is $(2n+1, \infty)$-presented. By 2.4.6(i)$\Leftrightarrow$(ii), every finitely presented module is also a direct summand of a direct sum of $(\infty, 2n+1)$-presented modules, and 2.4.7 gives the result. □

For example, this applies to the nth Weyl algebra over a field of characteristic 0 (which is a simple noetherian ring of Krull dimension n [441, 6.6.15]).

Proposition 2.4.9. ([575, 2.16]) *If a module is n-generated and is also k-related (with some possibly different set of generators), then it is $(n, n+k)$-presented.*

Proof. Suppose that M is (n, l)-presented and also (m, k)-presented. Say $\overline{a} \in M^n$ and $\overline{b} \in M^m$ are generating tuples with relations given by $\overline{a}G = 0$ and $\overline{b}H = 0$ respectively, where G is $n \times l$ and H is $m \times k$. Then $\overline{b} = \overline{a}B$ and $\overline{a} = \overline{b}A$ for some $n \times m$ matrix B and $m \times n$ matrix A.

We claim that $\mathrm{pp}^M(\overline{a})$ is generated by the condition $(\overline{x}BA = \overline{x}) \wedge (\overline{x}BH = 0)$. Certainly $\mathrm{pp}^M(\overline{a})$ contains this condition and, if $(C, \overline{c})$ is a free realisation of this condition, which we denote by θ, and if we set $\overline{d} = \overline{c}B$, then $\overline{d}A = \overline{c}$ and $\overline{d}H = 0$, so there is a morphism $M \to C$ determined by sending $\overline{b}$ to $\overline{d}$ and hence taking $\overline{a} = \overline{b}A$ to $\overline{c}$. Therefore $\mathrm{pp}^M(\overline{a}) \geq \mathrm{pp}^M(\overline{c}) = \langle\theta\rangle \geq \mathrm{pp}^M(\overline{a})$, as claimed.

Thus the matrix, $\begin{pmatrix} I - BA & BH \end{pmatrix}$, of relations for $\overline{a}$ gives a presentation for M and is $n \times (n+k)$. □

2.4.2 Purity over Dedekind domains and RD rings

A ring is RD iff pp conditions reduce to simple divisibility statements (2.4.10). The results of the previous section allow a characterisation of being Morita-equivalent to an RD ring (2.4.19).

A ring R is **RD** (for "relatively divisible") if purity reduces to simple divisibility, that is, if the condition for an embedding $A \leq B$ to be pure is simply that for every $r \in R$ we have $Ar = A \cap Br$.

Proposition 2.4.10. *A ring R is RD:*

iff every pp condition $\phi(\overline{x})$ is equivalent to a conjunction of conditions of the form $s \mid x_i r$;
iff every pp condition with one free variable is equivalent to a conjunction of conditions of the form $s \mid xr$.

Proof. Suppose that R is RD. Let $\mathcal{H}$, as in 2.4.5, be the set of matrices of the form $\begin{pmatrix} 1 \\ r \end{pmatrix}$ $(r \in R)$. From that result we deduce that every pp condition $\phi(\overline{x})$ is equivalent to a conjunction of conditions of the form $s \mid x_i r$ (each of which, in particular, involves only one free variable). The third equivalent is a special case of the second and, by 2.1.6, implies that the ring is RD. $\square$

The ring, $\mathbb{Z}$, of integers and the polynomial ring, $k[T]$, over any field k are, by the next result, examples of RD rings.

Corollary 2.4.11. ([703, 2.6], also see [575, 2.6]) *A ring is RD:*

iff every finitely presented right module is a direct summand of a direct sum of cyclically presented modules;
iff the same is true for every left module.

In particular the RD condition on one side implies it on the other.

Proof. This is immediate by 2.4.6 with $k = l = 1$ since any condition of the form $\exists y\, (x = yr)$ is 1×1. $\square$

Example 2.4.12. The module $\mathbb{Z}_{(p)}$ (over $\mathbb{Z}$, or over the ring $\mathbb{Z}_{(p)}$) is pure in its completion, $\overline{\mathbb{Z}_{(p)}}$, in the p-adic topology.

Serial rings also are examples (2.4.21), as ([699, Thm 1]) are **Prüfer rings** – commutative rings each of whose localisations at a prime ideal is a valuation ring. Indeed, a commutative ring is RD iff it is a Prüfer ring ([701, Thm 3], see [575, 4.5]). The resulting simplified criterion for pure-injectivity is given for Prüfer rings in [699, Thm 4]. The next corollary is by 2.4.6(iii)(⇔)(iv).

Corollary 2.4.13. ([551]) *If R is RD, then every pp condition $\phi(\overline{x})$ is equivalent to a sum of conditions of the form $\exists y\, (\overline{x} = y\overline{s} \wedge yr = 0)$, that is, $\exists y\, \big((\bigwedge_{i=1}^n x_i = ys_i) \wedge (yr = 0)\big)$.*

Thus, if M is any R-module, then the subgroups of M^n pp-definable in M are those of the form $\sum_{j=1}^m (\mathrm{ann}_M r_j)\overline{s}_j$ with $r_1, \dots, r_m \in R$ and $\overline{s}_1, \dots, \overline{s}_m \in R^n$.

For another example, the first Weyl algebra, $A_1(k)$, over a field of characteristic 0 is RD (see 2.4.15). An element $r \in A_1(k)$ may be regarded as a differential operator with polynomial coefficients. So the pp-definable subgroups of $A_1(k)$-modules are finite sums of sets obtained by applying differential operators to solution sets of differential equations.

Say that a ring is **right Warfield** if every finitely presented right module is a direct summand of a direct sum of cyclic finitely presented (that is, $(1, \infty)$-) modules.

Corollary 2.4.14. ([551], see [575, 2.10]) *If R is right and left Warfield, then R is RD.*

Proof. If R is left Warfield, then, by 2.4.6, every right R-module is a direct summand of a direct sum of finitely generated, 1-related modules. So 2.4.6, combined with the right Warfield property 2.4.7, gives that every right module is a direct summand of a direct sum of cyclically presented modules, as required by 2.4.11. □

A not necessarily commutative ring is a **Dedekind prime ring** if it is a hereditary noetherian prime (**HNP**) ring[10] and every non-zero one-sided ideal is a projective generator. A **fractional ideal** of a noetherian prime ring R is an R-submodule of the full quotient (simple artinian) ring, D, of R such that $aI \subseteq R$ and $bR \subseteq I$ for some invertible $a, b \in D$. A fractional ideal I of R is **invertible** if there is a fractional ideal, J of R with $IJ = IJ = R$. If every non-zero right ideal of R contains an invertible ideal, then R **has enough invertible ideals**: Dedekind prime rings, Dedekind domains in particular, are examples (see [441, §5.2] for all this).

Corollary 2.4.15. *Every hereditary noetherian prime ring with enough invertible right ideals is RD.*

Proof. For Dedekind prime rings this follows from, for example, [441, §5.7] where it is shown that every finitely generated (= finitely presented since R is right and left noetherian) module on either side is a direct sum of projective and cyclic modules. So R is right and left Warfield, hence RD by 2.4.14. For the general case, see [170, §2], [171, §3]. □

The class of rings appearing in 2.4.15 includes the first Weyl algebra, $A_1(k)$, over any field of characteristic zero since it is a (non-commutative) Dedekind domain. In contrast ([575, 5.1]) since, by 8.2.3, any RD domain must be right and left **semihereditary** (that is, every finitely generated right/left ideal must be projective), it follows that if k is a field of non-zero characteristic, then the algebra $A_1(k)$ is not RD, by [441, 7.5.8]. There is more on RD domains in Section 8.2.1.

[10] A ring is **prime** if, for any ideals I, I', $II' = 0$ implies $I = 0$ or $I' = 0$.

Von Neumann regular rings are RD by default since every embedding between modules over such a ring is pure (see 2.3.22). Serial rings are RD (2.4.21).

Proposition 2.4.16. *Over an RD ring absolutely pure = divisible and flat = torsionfree for (right and left) modules. In particular, any divisible module over a right noetherian RD ring is injective.*

Proof. It was noted after 2.3.2 that, over any ring, absolutely pure implies divisible. It is immediate from the definitions that over an RD ring the converse is true. Then the last statement follows, using 4.4.17.

It was also noted, in 2.3.12, that over any ring flat implies torsionfree. Suppose, for the converse, that R is RD and M_R is torsionfree. We check the criterion of 2.3.9. By 2.4.10 every pp condition with one free variable is a sum of conditions of the form $\exists y\,(x = yr \wedge ys = 0)$. So every pp-definable subgroup of any module is a sum of some of the form $\mathrm{ann}_M s \cdot r$ (with $r, s \in R$). By definition of torsionfree, $\mathrm{ann}_M s = M \cdot \mathrm{ann}_R s$, so clearly the criterion of 2.3.9 is satisfied. □

Lemma 2.4.17. ([575, 6.1]) *Suppose that every finitely presented right R-module is a direct summand of a direct sum of (k, l)-presented modules. Let $n = \max\{k, l\}$. Then $M_n(R)$ is an RD ring.*

Proof. By 2.4.6 every finitely presented right module and every finitely presented left module is a direct summand of a direct sum of n-generated finitely presented modules. So, over $M_n(R)$, every finitely presented right/left module is a direct summand of a direct sum of cyclic finitely presented modules (if $a_1, \dots, a_n$ generate M over R, then $(a_1, \dots, a_n)$ generates M^n over $M_n(R)$, cf. 3.2.4). That is, $M_n(R)$ is right and left Warfield and hence, by 2.4.14, RD. □

For example, see the remark after 2.4.8, the ring of $(2n + 1) \times (2n + 1)$ matrices over the nth Weyl algebra over a field of characteristic 0 is RD. Thus the nth Weyl algebra over a field of characteristic 0 is Morita-equivalent to an RD ring.

Corollary 2.4.18. ([575, 6.2]) *Every ring of finite representation type (for these see Section 4.5.3) is Morita equivalent to an RD ring.*

Corollary 2.4.19. ([575, 6.3]) *A ring is Morita equivalent to an RD ring iff for some integers k, l every finitely presented right module is a direct summand of a direct sum of (k, l)-presented modules.*

Proof. One direction is 2.4.17. For the other suppose that S is Morita equivalent to an RD ring R. Say $R = eM_n(S)e$ for some n and some idempotent $e \in M_n(S)$. So the equivalence Mod-$S \to$ Mod-R takes M_S to $M^n e$. If M_S is finitely presented, then the latter is, by assumption and 2.4.11, a direct summand of a direct sum of cyclic modules. So (cf. proof of 2.4.17) every finitely presented (right or left) S-module is a direct summand of a direct sum of n-generated S-modules. The condition for left modules (2.4.11) translates, by 2.4.6, to every finitely presented right S-module being a direct summand of a direct sum of (∞, n)-presented S-modules. Then 2.4.7 finishes the proof. □

In [136] related ideas are considered, especially over commutative rings.

2.4.3 Pp conditions over serial rings

Serial rings are RD (2.4.21) and there is further simplification to the shape of pp conditions over such rings (2.4.22). The general results of Section 1.1.2 on implications between pp conditions are refined, yielding various criteria which are useful in computations (for example, in proving the results reported in the subsequent section).

A module M is **uniserial** if its lattice of submodules is linearly ordered, equivalently if for all $a, b \in M$ either $a \in bR$ or $b \in aR$. A **serial** module is one which is a direct sum of (possibly infinitely many) uniserial modules. A ring R is **serial** if R is serial as a right module and as a left module. Examples include: valuation rings; the ring of $n \times n$ upper-triangular matrices over any division ring; the ring $\begin{pmatrix} \mathbb{Z}_{(p)} & \mathbb{Q} \\ 0 & \mathbb{Q} \end{pmatrix}$. Many more examples may be found in [564]. See that reference and, for example, [183, Chpt. 5] for background on these rings.

Suppose that R is a serial ring and that $e \in R$ is an idempotent which is **primitive** (or **indecomposable**), that is, if $e = f + g$ with f, g **orthogonal** ($fg = 0 = gf$) idempotents, then $f = 0$ or $g = 0$. Say that a pp condition $\phi(x)$ in one free variable is an e-**condition** if it implies $e \mid x$, that is, if $\phi(M) \leq Me$ for every right R-module M.

In a serial ring, 1 is a sum, $e_1 + \cdots + e_m$, of pairwise orthogonal primitive idempotents (the e_i are unique up to conjugation by an invertible element of R, see [564, 1.19]) so every module is a direct sum, $M = \bigoplus_1^m Me_i$, of corresponding pp-definable subgroups. Note that $R = \bigoplus_1^n e_i R$ is a decomposition of R_R as a direct sum of uniserial modules.

Theorem 2.4.20. ([157, Thm], [702, Thm 3.3] (see [564, 2.3])) *Suppose that R is a serial ring with $1 = \sum e_i$ a representation of 1 as a sum of orthogonal primitive idempotents. Then every finitely presented R-module is a direct sum of cyclically presented modules of the form $e_i R/rR$, where $r \in e_i Re_j$ for some e_i, e_j.*

Corollary 2.4.21. *Every serial ring is RD.*

The corollary follows from 2.4.11 and 2.4.20. The next result is, therefore, a refinement of 2.4.6(i)⇒(iii).

Proposition 2.4.22. ([173, 1.6], [564, 11.1]) *If R is serial then every pp condition in one free variable is equivalent to a sum of pp conditions of the form $(s \mid x) \wedge (xt = 0)$, where $s \in e_i R$ and $t \in Re_j$ for some primitive idempotents e_i, e_j. If ϕ is an e-condition (e a primitive idempotent of R), then we can take $s \in e_i Re$ and $t \in eRe_j$.*

Proof. Suppose that $\phi(x)$ is a pp condition with one free variable and let (C, c) be a free realisation of ϕ. By 2.4.20, $C = M_1 \oplus \cdots \oplus M_t$ with each M_k cyclically presented of the form in that result. Set $c = a_1 + \cdots + a_t$ with $a_k \in M_k$ and let ψ_k be a generator (1.2.6) for $\mathrm{pp}^{M_k}(a_k) = \mathrm{pp}^C(a_k)$ so, by 1.2.27, ϕ is equivalent to the sum $\psi_1 + \cdots + \psi_t$. We show that each ψ_i has the required form.

Consider $M = e_i R / rR$, where e_i is a primitive idempotent and $r \in e_i R$ and let $a = e_i s + rR \in M$, where we can suppose $s \in e_i R$. We claim that $\mathrm{pp}^M(a)$ is generated by the condition $\exists y\, (yr = 0 \wedge x = ys)$.

Certainly a satisfies this condition in M. For the converse, suppose that $b \in N$ (N any module) satisfies the condition, say $d \in N$ with $b = ds$ and $dr = 0$. The morphism $R \to N$ taking 1 to d sends r to $dr = 0$ and also sends $e_i s$ to $de_i \cdot s = ds = b$. Therefore there is an induced morphism $M \to N$ sending a to b. By 1.2.22 this is enough.

Note that if ϕ is an e-condition, then so is each ψ_k and, therefore, each element s as above may be chosen to lie in $e_i Re$.

Continuing with this notation, consider r and s. If $s \in rR$, then, of course, $a = 0$. Otherwise, since r, s both belong to the uniserial module $e_i R$, we have $r \in sR$, say $r = st$. Note that $r = \sum_j re_j \le e_i R$, so, by uniseriality of $e_i R$, there is some j such that $rR = re_j R$ and, in particular, $re_j = r$. Therefore, without loss of generality, $t = te_j$ and also, if $s \in Re$, then without loss of generality $t = ete_j$. Then the pp condition above which generates $\mathrm{pp}^M(a)$ becomes $\exists y\, (yst = 0 \wedge x = ys)$, which is clearly equivalent to $s \mid x \wedge xt = 0$, with s and t of the required form. □

Corollary 2.4.23. ([564, 11.2, 11.3]) *If R is serial and e is a primitive idempotent of R, then the lattice of all e-conditions with one free variable is distributive (Appendix E, p. 713).*

In particular if R is uniserial, then the lattice pp_R^1 (§1.1.3) is distributive.

Proof. By 2.4.22 the lattice of e-conditions with one free variable is generated, as a lattice (that is, under $\wedge$ and $+$), by the e-conditions of the form $s \mid x$ with $s \in Re$ and $xt = 0$ with $t \in eR$. Because e is primitive, so Re and eR are uniserial, both these sets of pp conditions are totally ordered. Therefore the lattice of

e-conditions is a quotient lattice of the lattice freely generated by two chains. By [249, Thm 13] the latter is distributive and, since any quotient of a distributive lattice is distributive (the condition of distributivity being given by equations and, therefore, being preserved by morphisms of lattices), we have the result. □

In fact, see [173, pp. 153–4], [564, 11.4], if R is serial and e is a primitive idempotent of R, then, for every module M which is finitely presented (more generally pure-projective) or pure-injective (for these see Chapter 4), the lattice of $\mathrm{End}(M)$-submodules of Me is distributive. If M is indecomposable, then this lattice is a chain (8.2.50).

The following lemmas are somewhat technical but Puninski uses them to excellent effect in describing the lattice of pp conditions over various sorts of serial ring (see Section 8.2.7).

Lemma 2.4.24. ([572, 3.1]) *Suppose that R is a serial ring, that e is a primitive idempotent of R and that $r, t \in Re$, $s, u \in eR$. Then $(r \mid x \wedge xs = 0) \le (t \mid x + xu = 0)$ iff $Rr \le Rt$ or $uR \le sR$ or $RruR \le RtsR$.*

By 2.4.22, every e-condition in a single free variable is a sum of pp conditions of the form $r \mid x \wedge xs = 0$ with $r \in Re$ and $s \in eR$, equivalently (the dual, 1.3.1, form) it is an intersection of pp conditions of the form $t \mid x + xu = 0$ with $t \in Re$ and $u \in eR$.

Corollary 2.4.25. ([572, 3.2–3.5]) *Suppose that R is serial, that $e^2 = e$ is a primitive idempotent of R and that $r, t \in Re$ and $s, u \in eR$. Then:*

(a) $r \mid x \wedge xs = 0 \le t \mid x$ iff $Rr \le Rt$ or $sR = eR$ or $RrR \le RtsR$;
(b) $r \mid x \wedge xs = 0 \le xu = 0$ iff $uR \le sR$ or $ru = 0$;
(c) $r \mid x \le t \mid x + xu = 0$ iff $Rr \le Rt$ or $ru = 0$;
(d) $xs = 0 \le t \mid x + xu = 0$ iff $Rt = Re$ or $uR \le sR$ or $RtR \le RtsR$.

By the remark before 2.4.25 every implication, $\psi \le \phi$, of e-conditions has the form $\psi = \sum_i (r_i \mid x \wedge xs_i = 0) \le \bigwedge_j \big((t_j \mid x) + (xu_j = 0)\big) = \phi$, and this will hold iff, for every i, j, we have $r_i \mid x \wedge xr_i = 0 \le (t_j \mid x + xu_j = 0)$, and then 2.4.25 applies. That is, we have an effective (relative to information about membership of one- and two-sided ideals of R) way to determine whether $\psi \le \phi$ whenever ϕ, ψ are e-conditions.

Lemma 2.4.26. ([572, 4.1]) *Suppose that R is serial, that e is a primitive idempotent of R, and that $r, r_i \in Re$ and $s, s_i \in eR$ $(i = 1, \dots, n)$. If $\bigwedge_i (r_i s_i \mid xs_i) \le rs \mid xs$, then there are i, j such that $(r_i s_i \mid xs_i) \wedge (r_j s_j \mid xs_j) \le rs \mid xs$.*

Therefore every implication of the form $\bigwedge (r_i s_i \mid xs_i) \le rs \mid xs$ can be obtained as a composition of trivial implications and "fusions", the latter being the term used in [572, §4] for: if $r, t \in Re$ and $s, u \in eR$ with $Rr \leq Rt$ and $sR \leq uR$ and $RrsR \leq RtuR$, then the condition $(rs \mid xs) \wedge (tu \mid xu)$ is equivalent to $ru \mid xu$.

2.5 Pure-projective modules over uniserial rings

This section gives an account, without proofs, of some counterexamples of Puninski on direct sum decomposition of modules over serial rings.

If modules $M_1, \ldots, M_n$ have local endomorphism rings then, by the Krull–Remak–Schmidt–Azumaya Theorem (E.1.24), their direct sum, M, has good decomposition properties: the decomposition of M into indecomposables is essentially unique and it follows from E.1.25 that direct summands of M have complements which are direct sums of some of the M_i.

For direct sums of infinitely many M_i with local endomorphism rings uniqueness of decomposition holds but the statement about direct summands does not. If, however, the M_i are indecomposable injective or pure-injective modules, then one regains the property of complements provided one stays within the relevant class of modules by replacing any infinite direct sum by its injective, resp. pure-injective, hull (see E.1.27, 12.1.6). The general situation is complex and subtle, but there have been recent advances in understanding, see, for example, Facchini's book [183].

For instance, the following question of Matlis ([438, p. 517]) is still open. Let M be a direct sum of indecomposable injectives and let M' be a direct summand of M: is M' also a direct sum of (necessarily injective) indecomposable modules?

Even for serial modules the situation is not straightforward. Problem 10 of [183] asks whether a direct summand of a serial module must be serial. Puninski [565] answered this in the negative by producing an example of a direct summand of a serial module which is not a direct sum of indecomposables, so which cannot be a direct sum of serial modules. This example also provided a negative answer to Problem 11 in [183], which asks (the stronger question) whether every pure-projective over a serial ring is serial.

In some cases there is a positive solution: if the ring is commutative or right noetherian, then a direct summand of a serial module must be serial, see [183, 9.25]. Also, it is enough to consider serial modules which are direct sums of countably many uniserial modules since, [183, 2.49], if N is a direct summand of a serial module, then $N = \bigoplus N_i$ with each N_i a direct summand of a countable direct sum of uniserial modules.

The endomorphism ring of a uniserial module need not be local: there is a counterexample over a ring of Krull dimension 1 at [183, 9.26].[11] If, however, the ring is commutative or right noetherian, then the endomorphism ring of a uniserial module is local (Fuchs and Salce [199, Cor. 3], see [564, 2.11]).

Another open question, (see [183, Problem 9]), is whether an indecomposable direct summand of a serial module must be uniserial. Furthermore, Open Problem 12 in [183] asks whether every direct summand of a serial module must contain an indecomposable direct summand – or can there be superdecomposable direct summands of serial modules?

Another question (see [183, Problem 15]) which has been resolved by Puninski ([566]) is whether there is a uniserial module over a uniserial domain R which is not quasi-small. A module is **quasi-small** if whenever it is a direct summand of a module of the form $\bigoplus_i M_i$ it is a direct summand of the direct sum of finitely many of the M_i. Puninski's counterexample is, in fact, a direct summand of $(R/rR)^{(\aleph_0)}$ for some $r \in R, r \neq 0$, hence pure-projective. So it also answers in the negative the question of Puninski and Tuganbaev [581, 16.32] of whether every pure-projective over a uniserial ring is a direct sum of finitely presented modules.

Puninski's (counter)examples are modules over nearly simple uniserial rings. A ring is **uniserial** if it is right and left uniserial as a module over itself and a uniserial ring R is **nearly simple** if it is not artinian and has exactly three ideals, namely 0, the Jacobson radical $J(R)$ and R (the condition "non-artinian" can be replaced by "$J^2 \neq 0$" because then $J^2 = J$ so J cannot be a finitely generated one-sided ideal). An example from [80, 6.5] of a nearly simple uniserial ring is presented at [566, pp. 325–6].

An important feature of both [566] and [565] is a classification of the pure-projectives: complete in the case of [566, 7.8] (for nearly simple uniserial domains) and partial in [565, §5] (for exceptional uniserial rings).

Over a nearly simple uniserial domain there are three indecomposable modules such that every pure-projective is a direct sum of copies of just these three modules. Finite direct sums of these indecomposables behave well but unique decomposition fails for countably generated pure-projectives: there are some non-trivial identifications of infinite direct sums of the indecomposables ([566, 8.1], cf. 2.5.10).

In [565] this classification of pure-projectives is extended in part to exceptional uniserial rings. A nearly simple uniserial ring is **exceptional** if it is prime but not a domain. The existence of exceptional uniserial rings is non-trivial: the first

[11] Also, although the example from [80], see [564, 2.12], is over a ring of Krull dimension 2, a suitable factor has Krull dimension 1.

example comes from Dubrovin [159]. The three indecomposables seen in the nearly simple case also appear here but it is not the case that every pure-projective is a direct sum of copies of them: Puninski constructed a pure-projective module which is not a direct sum of indecomposables and he conjectures that every pure-projective is a direct sum of copies of these four modules. Since every pure-projective module over a uniserial ring is a direct summand of a serial module this gives a counterexample for Facchini's Problem 10 stated before. Below we indicate the steps to proving these results.

The structure theory of pure-projectives over coherent exceptional uniserial rings was developed further in [567] and the final step made by Příhoda [532, 5.1], using his remarkable theorem on the structure of projectives (arbitrary projectives which are isomorphic modulo their radicals are isomorphic, [532, 2.3]), confirming Puninski's conjectured classification of pure-projective modules over these rings ([567, Conj. 5.9]), see 2.5.10.

Proposition 2.5.1. ([565, 2.1]) *A uniserial ring R is right coherent iff for every $r \in J(R)$ the right annihilator of r is a principal right ideal. In particular a uniserial domain is right and left coherent.*

Every finitely presented module over a uniserial ring R is a direct sum of modules of the form R/r_iR with $r_i \in R$ (2.4.20), so a module is pure-projective iff it is a direct summand of a direct sum of modules of this form.

Proposition 2.5.2. ([566, 4.2, 4.3], [565, 3.5]) *Let R be a nearly simple uniserial ring. Then all modules of the form R/rR with $r \in J(R)$, $r \neq 0$ are isomorphic.*[12]

Proposition 2.5.3. ([566, 7.1]) *Every pure-projective module over a uniserial domain is locally coherent, that is, every finitely generated submodule is finitely presented.*

The following is an analogue of a result of Kaplansky for projective modules (both are consequences of a rather general result, see [183, 2.47]).

Proposition 2.5.4. ([566, 7.2]) *Let R be arbitrary. Then every pure-projective module over R is a direct sum of countably generated (pure-projective) modules.*

Theorem 2.5.5. ([566, 7.8]) *Let R be a nearly simple uniserial domain and let $r \in J(R)$, $r \neq 0$. Then the following are equivalent for an R-module M:*

(i) M is pure-projective;
(ii) $M \simeq R^{(\alpha)} \oplus (R/rR)^{(\beta)} \oplus K^{(\gamma)}$ for some α, β, γ, where K is the unique countably generated locally coherent uniserial torsion module.

[12] Over any uniserial ring $R/rR \simeq R/sR$ iff $RrR = RsR$.

Theorem 2.5.6. ([566, 8.1]) *Let K be a countably generated locally coherent uniserial torsion module over a nearly simple uniserial domain R. Then K is not quasi-small. Moreover, $K \oplus (R/rR)^{(\aleph_0)} \simeq (R/rR)^{(\aleph_0)}$ for every non-zero $r \in J(R)$.*

This shows that it is not the case that every pure-projective module over serial ring is a direct sum of finitely presented modules since K is pure-projective, uniserial (hence indecomposable) but not finitely presented. Puninski asked whether it is possible to obtain the description of the pure-projective modules without use of the methods from model theory. Příhoda subsequently did this in [531, §4].

An analysis, [565, 4.1–4.4], of the implications between pp conditions over an exceptional uniserial ring yields a very explicit description of the lattice of pp conditions (recall, 2.4.21, that serial rings are RD). In particular, pp^1_R consists of two chains, one of which contains all simple annihilator conditions $xr = 0$ and the other contains all simple divisibility conditions $s|x$. All pp conditions are simple combinations of these (a diagram is given at [565, p. 318]).

It is shown, [565, 5.1], that the exceptional uniserial ring arising from Dubrovin's construction is coherent.

Proposition 2.5.7. ([565, 5.2] (cf. 2.5.3)) *For any uniserial ring R and non-zero element $r \in R$, the module R/rR is coherent. Hence every pure-projective module over an exceptional uniserial coherent ring is locally coherent.*

If R is an exceptional uniserial coherent ring, then, [565, 5.8], there is a unique-to-isomorphism countably-infinitely generated locally coherent uniserial module, V, and this module is pure-projective. It is isomorphic to any countably-infinitely generated right ideal of R.

Theorem 2.5.8. ([565, 5.9]) *Let R be an exceptional uniserial coherent ring and let M_R be a pure-projective module such that R_R is not a direct summand of M. Then $M \simeq (R/rR)^{(\beta)} \oplus V^{(\gamma)}$ for some β, γ.*

Theorem 2.5.9. ([565, 6.2]) *Let R be an exceptional uniserial coherent ring. Then there is a pure-projective module W_R such that W is not a direct sum of indecomposable modules.*

Roughly, a chain $(M_0, m_0) \xrightarrow{f_0} (M_1, m_1) \xrightarrow{f_1} \cdots$ of free realisations of certain pp conditions is constructed, with all the f_i embeddings. The module W is their direct limit. The pp-types of the elements m_i in W are computed and it is shown that certain other pp-types have no free realisation in W. This information, with 1.3.22 and 1.3.26, yields that W is pure-projective and also that it does not decompose as a direct sum of indecomposables. The analysis of the lattice of pp conditions (in one free variable) plays a crucial role. As mentioned before 2.5.1, the complete structure of the pure-projective modules over these rings was subsequently established through [567, 5.9ff.] and [532, 5.1].

Theorem 2.5.10. *Suppose that R is an exceptional uniserial coherent ring, let r be a non-zero element of $J(R)$ and let V, W be as above. Then every pure-projective module has the form $R^{(\alpha)} \oplus (R/rR)^{(\beta)} \oplus V^{(\gamma)} \oplus W^{(\delta)}$. Such decompositions are not unique but the identifications are determined by the relations: $V \oplus (R/rR)^{(\omega)} \simeq (R/rR)^{(\omega)}$; $V \oplus W \simeq R^{(\omega)} \oplus (R/rR)$; $W \oplus R^{(\alpha)} \simeq W$ for $\alpha \leq \omega$.*

3

Pp-pairs and definable subcategories

Definable subcategories of module categories form the main theme of this chapter. Before those are defined, reduced products are introduced and used to prove the existence of realisations of pp-types. In the early part of the chapter the opportunity is taken to set up the category of pp-pairs and pp-definable maps.

3.1 Pp conditions and morphisms between pointed modules

An n-**pointed module** is a module M together with a specified n-tuple, $\overline{a} \in M^n$, of elements of M. Our notation for this is, $(M, \overline{a})$. A **morphism** of n-pointed modules, say from $(M, \overline{a})$ to $(N, \overline{b})$, is a morphism $f : M \to N$ of R-modules which takes $\overline{a}$ to $\overline{b}$, that is, $fa_i = b_i$ for $i = 1, \ldots, n$. If we choose and fix a free basis $e_i = (0, \ldots, 1, \ldots, 0)$ ("1" in the ith position, zeroes elsewhere) of R^n we may regard an n-pointed module as a morphism $R^n \to M$; for this morphism determines and is determined by $\overline{a} = (a_1, \ldots, a_n)$ where a_i is the image of e_i. Generalising the identification of the elements of a module M with the morphisms from the free module R_R to M, we use the notation $\overline{a} : R^n \to M$ also to denote this morphism. Thus a morphism from $\overline{a} : R^n \to M$ to $\overline{b} : R^n \to N$ is a morphism $h : M \to N$ with $h\overline{a} = \overline{b}$, so, depending on the point of view, $h\overline{a}$ can be read as composition or evaluation.

We make this independent of choice of basis of R^n and, at the same time, generalise to a definition which makes sense in any locally finitely presented category, by saying that if A is a finitely presented R-module, then an A-**pointed module** is a morphism of the form $A \to M$ for some $M \in$ Mod-R, and a **morphism** from $f : A \to M$ to $g : A \to N$ is a morphism $h : M \to N$ such that $hf = g$.

Say that a morphism f' **factors initially** through f if $f' = gf$ for some morphism g and denote this by writing $f \geq f'$. This gives a preorder on the

A-pointed modules in mod-R, that is, on the collection of morphisms in mod-R with domain A. The associated equivalence relation we denote by $\sim$.

Lemma 3.1.1. $A \xrightarrow{f} B \geq A \xrightarrow{f'} B'$ *iff, in the pushout diagram shown below, g is a split monomorphism.*

$$\begin{array}{ccc} A & \xrightarrow{f} & B \\ {\scriptstyle f'}\downarrow & & \downarrow{\scriptstyle g'} \\ B' & \xrightarrow[g]{} & C \end{array}$$

Proof. The direction ($\Leftarrow$) is clear. For the other, if $f' = hf$, then consider the outer part of the diagram below and use the pushout property to obtain k.

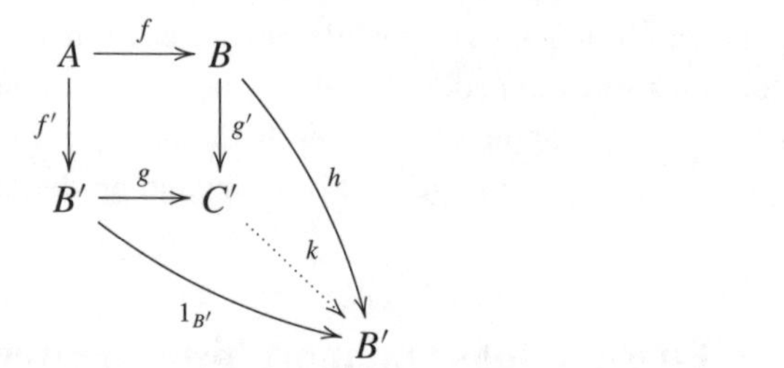

□

Lemma 3.1.2. *Let $A \in$ mod-R. The poset of $\sim$-equivalence classes of morphisms in* mod-R *with domain A is a modular lattice with join given by direct sum and meet given by pushout.*

Proof. Let $A \xrightarrow{f} B$, $A \xrightarrow{f'} B'$ be in mod-R. If $A \xrightarrow{g} C$ with $g \geq f, f'$, then each of f, f' factors initially through g and hence the direct sum $\begin{pmatrix} f \\ f' \end{pmatrix} : A \to B \oplus B'$ of f, f' factors initially through g, so $g \geq \begin{pmatrix} f \\ f' \end{pmatrix}$. Since $\begin{pmatrix} f \\ f' \end{pmatrix}$ is clearly an upper bound for both f and f', it follows that it is, indeed, the join. Dually, if $f, f' \geq A \xrightarrow{h} D$, that is, if h factors initially through each of f and f', then h factors initially through the pushout of f and f', which is thus clearly the meet of f and f'.

Modularity can be checked directly but follows from 3.1.5 below. □

We will refer to this poset as the **lattice of morphisms** in mod-R **with domain** A, or as the **lattice of A-pointed finitely presented modules**. The next result is immediate by 1.2.9.

Lemma 3.1.3. *Let $\overline{a}$ be a finite generating tuple for the finitely presented module A. Let $f : A \to B$ and $g : A \to C$ be morphisms in* mod-R. *Then $f \geq g$ iff* $\mathrm{pp}^B(f\overline{a}) \geq \mathrm{pp}^C(g\overline{a})$.

Taking A to be R^n and $\overline{a}$ to be a free basis, $\overline{e}$, for R^n gives the following.

Corollary 3.1.4. *The lattice, pp_R^n, of pp conditions with n free variables is naturally isomorphic to the lattice of equivalence classes of n-pointed finitely presented modules.*

The isomorphism is independent of the choice of free basis, $\overline{e}$, since such a choice affects neither the pp-type of the image under a morphism nor the equivalence class of the morphism, for if $\overline{e}'$ is another free basis, then there is an automorphism of the module R^n taking $\overline{e}$ to $\overline{e}'$.

The next corollary generalises that above.

Corollary 3.1.5. *Let $\overline{a}$ be any generating (n-)tuple for the finitely presented module A and let ϕ be a pp condition with $\mathrm{pp}^A(\overline{a}) = \langle\phi\rangle$. Then the lattice of A-pointed finitely presented modules is isomorphic to the interval $[\phi, 0] = \{\psi : \phi \geq \psi \geq \overline{x} = 0\}$ in the lattice pp_R^n. The isomorphism is given, in terms of representatives of equivalence classes, by $(A \xrightarrow{f} B) \mapsto \mathrm{pp}^B(f\overline{a})$ and, inversely, $\psi(\in [\phi, 0]) \mapsto (A \xrightarrow{f} C)$, where $(C, f\overline{a})$ is any free realisation of ψ (1.2.21).*

3.2 Pp-pairs

The lattice of pp conditions is considered again in Section 3.2.1. The (abelian) category of pp-pairs and its localisations are introduced in Section 3.2.2. It is also shown that elementary duality lifts to this category.

3.2.1 Lattices of pp conditions

The lattice of pp-definable subgroups of a module is a quotient of that of all pp conditions (3.2.3) and there is a module which has the latter for its lattice of pp-definable subgroups (3.2.1). Pp conditions in n free variables may be regarded as pp conditions in one free variable over the $n \times n$ matrix ring (3.2.4). Any filter of pp conditions in a lattice of pp conditions is the pp-type of some tuple in some module (3.2.5). The direct image of a pp condition along a morphism of rings is defined and its evaluation described (3.2.7).

Given a module M let $\mathrm{pp}^n(M)$ denote the (modular) lattice of subgroups of M^n pp-definable in M. If $n = 1$ we write just $\mathrm{pp}(M)$ and call this the **lattice of pp-definable subgroups of** M.

Proposition 3.2.1. *There is a module M^* such that the natural map from pp_R^n to $\mathrm{pp}^n(M^*)$ is an isomorphism for every n.*

Proof. Take M^* to be the direct sum of one copy of each finitely presented module. By 1.2.25 (and 1.2.3) each pp-pair is open in M^* and this is enough for $\phi \mapsto \phi(M)$ to define an isomorphism from pp_R^n to $\mathrm{pp}^n(M)$. □

Lemma 3.2.2. *If M is a pure submodule of N, then, for each n, the natural map from $\mathrm{pp}^n(N)$ to $\mathrm{pp}^n(M)$ given by $\phi(N) \mapsto \phi(M) (= \phi(N) \cap M^n)$ is a surjection of lattices.*

That is direct from the definitions, the kernel-congruence of the map being determined by the pp-pairs, $\phi \geq \psi$, which are identified ("closed") in M. Similarly, in view of 2.1.14, the map $\phi(N) \mapsto \phi(N/M)$ is, if M is a pure submodule of N, a surjection from $\mathrm{pp}^n(N)$ to $\mathrm{pp}^n(M)$; furthermore, a pp-pair ϕ/ψ is closed in N/M iff the natural embedding $\phi(M)/\psi(M) \to \phi(N)/\psi(N)$ is surjective.

Corollary 3.2.3. *For any module M, the pp-lattice $\mathrm{pp}^n(M)$ is a quotient of the lattice, pp_R^n, of all pp conditions in n free variables*

That is because M is pure in $M \oplus M^*$ where M^* is as in 3.2.1.

Given an interval $[\phi, \psi] = \{\eta : \phi \geq \eta \geq \psi\}$ of pp_R^n we denote by $[\phi, \psi]_M$ the image of this in $\mathrm{pp}^n(M)$, that is, the lattice of those subgroups of M^n which are pp-definable in M and lie between $\phi(M)$ and $\psi(M)$. Sometimes we use the notation $[\eta]_M$, rather than $\eta(M)$, when we are thinking of this typical element as an element of a lattice, rather than a subgroup per se.

Proposition 3.2.4. ([620, 0.1.1]) *For any ring R the lattice, pp_R^n, of pp conditions in n free variables is isomorphic to the lattice, pp_R, of pp conditions in one free variable over the matrix ring $M_n(R)$.*

Proof. First note that an R-module M is n-generated iff the $M_n(R)$-module corresponding to it under the standard Morita equivalence of R and $M_n(R)$ is cyclic for, cf. the proof of 2.4.17, that module is M^n with the action of right multiplication by matrices. It follows that the lattice of finitely presented R^n-pointed R-modules is isomorphic to the lattice of finitely presented $M_n(R)$-pointed $M_n(R)$-modules. Then apply 3.1.4. □

An informal corollary is: any reasonable finiteness condition on the lattice pp_R^1, in particular, one which is defined in terms just of the category of modules and which is, therefore, Morita invariant, extends to pp_R^n for every n.

Given a module M we will say that a pp-pair ϕ/ψ is an M-**minimal pair** if $\phi(M) > \psi(M)$ and if there is no pp-definable subgroup of M strictly between $\phi(M)$ and $\psi(M)$. See Sections 5.3.3 and 12.5 for more on these.

Fix n and consider the pp-type, p, of some n-tuple in some module: the (equivalence classes of) pp conditions in p form a filter in pp_R^n. In fact every filter in pp_R^n arises in this way.

Theorem 3.2.5. *Let p be a filter in the lattice pp_R^n of pp conditions. Then there is a module M and $\overline{a}$ in M with $\mathrm{pp}^M(\overline{a}) = p$.*

Proof. (A model-theoretic proof of this will be given in 3.3.6.) For each $\phi \in p$ choose a free realisation, $(C_\phi, \overline{c}_\phi)$, of ϕ and denote also by $\overline{c}_\phi$ the morphism from R^n to C which takes a chosen and fixed basis, $\overline{e}$, of R^n to $\overline{c}_\phi$. Consider the diagram in the category of R-modules consisting of all these morphisms, $\overline{c}_\phi$, for $\phi \in p$. Since Mod-R is cocomplete this diagram has a colimit (an "infinite pushout") with object M and morphisms $f_\phi : C_\phi \to M$. By construction, these morphisms satisfy $f_\phi \overline{c}_\phi = f_{\phi'} \overline{c}_{\phi'} = \overline{a}$, say, for all ϕ, ϕ' in p. So, by 1.1.7, $\mathrm{pp}^M(\overline{a})$ contains all $\phi \in p$.

If, conversely, $\psi \in \mathrm{pp}^M(\overline{a})$, then, by 1.2.17, there is $f : C \to M$ taking $\overline{c}$ to $\overline{a}$, where $(C, \overline{c})$ is a free realisation of ψ. Now, M is $\bigoplus_{\phi \in p} C_\phi / K$, where K is generated by the entries of all the $\overline{c}_\phi - \overline{c}_{\phi'}$ for $\phi, \phi' \in p$, and M clearly is the direct limit of the finite such pushouts. So f factors through one of these finite pushouts, say that of $\overline{c}_{\phi_1}, \dots, \overline{c}_{\phi_k}$. Thus the image of $\overline{e}$ in this pushout satisfies the condition ψ. By 1.2.28, $\phi_1 \wedge \cdots \wedge \phi_k \leq \psi$ and $\psi \in p$, as required. □

An entirely different construction of such a realisation of p is given in 4.1.5. Also see [420, 4.3].

Corollary 3.2.6. *Given any pp-type p there is a module M and a tuple $\overline{a}$ in M such that for every module N and tuple $\overline{b}$ in N there exists a morphism from M to N taking $\overline{a}$ to $\overline{b}$ iff $p \geq \mathrm{pp}^N(\overline{b})$.*

Proof. Let $M, \overline{a}$ be as in 3.2.5. Suppose that $p \geq \mathrm{pp}^N(\overline{b})$. Then, with notation as there, for each $\phi \in p$ there is, by 1.2.17, a morphism $g_\phi : C_\phi \to N$ with $g_\phi \overline{c}_\phi = \overline{b}$. By definition of M as a pushout there is, therefore, $g : M \to N$ with $g f_\phi = g_\phi$, in particular with $g\overline{a} = \overline{b}$.

The converse is immediate from 1.1.7. □

Now suppose that $f : R \to S$ is a morphism of rings and that ϕ is a pp condition for R-modules, say ϕ is $\exists \overline{y}\, (\overline{x}\ \overline{y}) H = 0$. We define the pp condition $f_*\phi$ for S-modules to be $\exists \overline{y}\, (\overline{x}\ \overline{y}) f H = 0$, where if H is the matrix $(r_{ij})_{ij}$, then fH is the matrix $(f r_{ij})_{ij}$. The next result follows easily.

Lemma 3.2.7. *Let $f : R \to S$ be any morphism of rings, let ϕ be a pp condition for R-modules and let $f_*\phi$ be the corresponding pp condition for S-modules. If*

M_S is an S-module, then $f_*\phi(M) = \phi(M_R)$, that is, $f_*\phi(M) = \phi(f^*M)$, where $f^* : \text{Mod-}S \to \text{Mod-}R$ is the restriction of scalars functor.[1]

The next result is immediate from 1.1.13.

Lemma 3.2.8. *If $f : R \to S$ is a morphism of rings and $\psi \leq \phi$ are pp conditions for R-modules, then $f_*\psi \leq f_*\phi$.*

Therefore f induces, for each n, a homomorphism of posets of equivalence classes of pp conditions, $f_* : \mathrm{pp}^n_R \to \mathrm{pp}^n_S$. Just from the definition of the map and the formal construction of meet and join of pp conditions it is clear that this map does commute with meet and join. This map need be neither injective (take f to be the inclusion of $\mathbb{Z}$ into $\mathbb{Q}$) nor surjective (take the inclusion of $\mathbb{Q}$ into $\mathbb{Q}[T]$).

Lemma 3.2.9. *If $f : R \to S$ is any morphism of rings and $M_S \to N_S$ is a pure embedding, then $M_R \to N_R$ is pure.*

Proof. Let ϕ be a pp condition for R-modules. Then $\phi(M_R) = f_*\phi(M_S)$ (by 3.2.7) $= M^n \cap f_*\phi(N_S)$ (by assumption) $= M^n \cap \phi(N_R)$ (by 3.2.7). □

The **elementary socle series** of a module M is defined ([274, 10.2]) as follows. Set $\mathrm{elsoc}^0(M) = 0$. Then define $\mathrm{elsoc}^1(M)$ to be the sum of all minimal pp-definable subgroups (and to be 0 if there are none), that is, it is the sum of all $\phi(M)$ such that the interval $[\phi(M), 0]$ in the lattice of pp-definable subgroups of M has no more than two points. By 1.1.7 this is an $\mathrm{End}(M)$-submodule. It is also an R-submodule of M since, if $\phi(M)$ is minimal, if $r \in R$ and if $\psi(M)$ is a pp-definable subgroup of $\phi(M)r$ properly contained in $\phi(M)$, then $\phi(M) \cap \psi(M)r^{-1}$ $= \{a \in \phi(M) : ar \in \psi(M)\}$ is properly contained in $\phi(M)$ and is non-zero if $\psi(M)$ is non-zero. The definition is continued inductively, with $\mathrm{elsoc}^{\alpha+1}(M)$ being the sum of all $\phi(M)$ such that $\phi(M) \nleq \mathrm{elsoc}^\alpha(M)$ and which are minimal such. At limit ordinals one takes the sum of the already defined groups.

There is a dual, elementary radical series, which is literally dual in the sense of Section 1.3.1.

This socle series was used by Herzog to get information on the dual (in the sense of Section 5.4) of a Σ-pure-injective module and has been used by

[1] In terms of pp conditions as functors, see 10.2.17, the functor $F_\phi : \text{Mod-}R \to \mathbf{Ab}$ commutes with direct limits (1.2.31) and products (1.2.3), as does the restriction of scalars functor f^*, hence does the composition which is, therefore, determined by its action on mod-S. By 18.1.19, this restriction to mod-S is a finitely presented object of the functor category (mod-S, $\mathbf{Ab}$). Since it is clearly a subfunctor of the nth power of the forgetful functor, 10.2.15 shows that it has the form F_ψ for some pp condition ψ for S-modules. But it was a lot easier to write down ψ, that is, $f_*\phi$, directly!

Kucera to describe the fine structure of some injective modules [381]. These series also make an appearance in [277]. There is probably more to be done with them. Other types of layering and ranks are considered in [380], [377], [378], [115].

We finish by pointing out that the collection of all pp-types in n free variables forms a complete lattice. The join operation is given by intersection as sets, $\bigvee_i p_i = \bigcap_i p_i$, and meet is given by $\bigwedge_i p_i = \{\phi : \phi \geq \phi_{j_1} \wedge \cdots \wedge \phi_{j_n}$ for some $\phi_{j_i} \in p_i\}$. This follows from the anti-isomorphism, 12.2.1, between the poset of pp-types and the, complete modular, lattice of all subfunctors of the nth power of the forgetful functor. Alternatively, show it directly as in [495, §8.1]. Of course there is a version relative to any module or definable subcategory, given by evaluation on a suitable generator of the subcategory, namely one which realises every pp-type. The $\wedge$-compact elements are just the finitely generated pp-types. That is, pp_R^n is the lattice of $\wedge$-compact elements of the lattice of all pp-types. In the relative version the compact elements are those which are generated within the definable subcategory by a single pp condition.

3.2.2 The category of pp-pairs

The category of pp-pairs is defined and shown to be abelian (3.2.10). Evaluation at any module is an exact functor from this category to the category of abelian groups (3.2.11). Elementary duality lifts to an anti-equivalence between the category of pp-pairs for right modules and that for left modules (3.2.12). Each module induces an exact functor from the category of pp-pairs to the corresponding category of pp-pairs which is defined relative to that module (3.2.15).

The use of pairs of pp conditions, rather than just single conditions (that is, pairs with lower term zero) was fundamental to the advances in the model theory of modules made by Ziegler in [726]. Model-theoretically this is related to Shelah's notion of imaginaries, see Section B.2. Herzog [274] realised that these pairs, equivalently intervals in the lattices pp_R^n, can be made the objects of an abelian category, with morphisms being the pp-definable maps. His notation for this category was $(\text{Mod-}R)^{\mathrm{eq}+}$: we will use the notation[2] $\mathbb{L}_R^{\mathrm{eq}+}$.

The objects of the **category $\mathbb{L}_R^{\mathrm{eq}+}$ of pp-pairs** are the pp-pairs ϕ/ψ; recall that means ϕ, ψ are pp conditions with $\phi \geq \psi$. We regard $\phi(x)/\psi(x)$ and $\phi(y)/\psi(y)$ as defining the *same* object. This is discussed after 3.2.11.

[2] The common part of the notation, eq+, refers to the imaginaries construction in Section B.2, where the corresponding enriched language is defined.

A morphism from the pair ϕ/ψ (both with free variables[3] $\overline{x}$) to the pair ϕ'/ψ' (both with free variables $\overline{x}'$) is given by a pp condition[4] $\rho(\overline{x}, \overline{x}')$ which satisfies:

whenever $\rho(\overline{a}, \overline{a}')$ and $\phi(\overline{a})$ hold then so does $\phi'(\overline{a}')$;
whenever $\rho(\overline{a}, \overline{a}')$ and $\psi(\overline{a})$ hold so does $\psi'(\overline{a}')$;
whenever $\phi(\overline{a})$ holds there is $\overline{a}'$ with $\rho(\overline{a}, \overline{a}')$.

Observe that these conditions are those necessary and sufficient for ρ to define a function from $\phi(M)/\psi(M)$ to $\phi'(M)/\psi'(M)$ for every module M, namely the function $\overline{a} + \psi(M) \mapsto \overline{a}' + \psi'(M)$, where $\overline{a} \in \phi(M)$ and M satisfies $\rho(\overline{a}, \overline{a}')$.

The function defined by ρ on any module M is additive since ρ is pp, hence it defines a subgroup of the appropriate power of M. Two such pp conditions are equivalent in the sense defined before 1.1.13 iff they define the same morphism in every module: the morphisms of $\mathbb{L}^{\text{eq}+}$ may be formally defined as the resulting equivalence classes.

That is, the objects of $\mathbb{L}^{\text{eq}+}$ are the pp-pairs and the morphisms are the pp-definable maps between such pairs.

In order to check that a pp condition defines a morphism in every module we may use 1.1.13 because the requirements on ρ all may be written as implications between pp conditions, namely:

$\rho(\overline{x}, \overline{x}') \wedge \phi(\overline{x})$ implies $\phi'(\overline{x}')$;
$\rho(\overline{x}, \overline{x}') \wedge \psi(\overline{x})$ implies $\psi'(\overline{x}')$;
$\phi(\overline{x})$ implies $\exists \overline{x}' \rho(\overline{x}, \overline{x}')$

and so all can be checked, in principle, using 1.1.13. Similarly for equivalence, pp conditions ρ and σ define the same morphism iff $\phi(\overline{x}) \wedge \rho(\overline{x}, \overline{x}') \wedge \sigma(\overline{x}, \overline{x}'')$ implies $\psi'(\overline{x}' - \overline{x}'')$ – again 1.1.13 gives an, in principle, effective criterion.

On the other hand, a "semantic" (referring to solution sets in modules) as opposed to "syntactic" (referring to the form of a condition) check can be made in any module of the sort appearing in 3.2.1, that is, in any module M^* in which every pp-pair is open; for the conditions above assert that certain pp-pairs are closed in every module and this may be tested in such a module M^*.

In the situation above, where a pp condition ρ defines a function from ϕ/ψ to ϕ'/ψ', we will refer to both the function and to any pp condition which defines

[3] Remember that means the free variables of ϕ and ψ are *among* those of $\overline{x}$, so it is important to keep note of which variables a condition is "in". For instance, the condition $x = 0$, regarded as a condition with free variable just x defines the 0 subfunctor of the forgetful functor. But the same condition, regarded as a condition in x and y defines the subfunctor, $M \mapsto \{0\} \times M \leq M^2$, of the square of the forgetful functor.

[4] Variables in $\overline{x}'$ should be renamed if necessary to ensure that no variable occurs in both $\overline{x}$ and $\overline{x}'$, since ρ should define a subgroup of $(-)^{l(\overline{x})} \oplus (-)^{l(\overline{x}')}$.

it as a **pp-definable function**. The term "definable scalars" also will be used for morphisms of this category and for those of its various localisations (12.8.3) especially for the endomorphisms of the image under localisation of the pair $x = x/x = 0$ (Section 6.1.1).

There is a natural additive structure on the category $\mathbb{L}_R^{\mathrm{eq}+}$: given two definable functions, ρ, σ, with the same domain and codomain their sum is defined by the pp condition, $\tau(\overline{x}, \overline{x}')$, which is $\exists \overline{y}, \overline{z}\, \big(\rho(\overline{x}, \overline{y}) \wedge \sigma(\overline{x}, \overline{z}) \wedge \overline{x}' = \overline{y} + \overline{z}\big)$.

Proposition 3.2.10. *The category $\mathbb{L}_R^{\mathrm{eq}+}$ is abelian.*

Proof. The following are all easily checked. The pair ϕ/ϕ (for any ϕ) is a zero object. The direct sum of ϕ/ψ and ϕ'/ψ' is $\big(\phi(\overline{x}) \wedge \phi'(\overline{x}')\big)/\big(\psi(\overline{x}) \wedge \psi'(\overline{x}')\big)$, where the sequences, $\overline{x}, \overline{x}'$, of variables have no common element (rename some variables if need be). If ρ defines a morphism from ϕ/ψ to ϕ'/ψ', then $\ker(\rho)$ is the pair ψ''/ψ, where ψ'' is $\phi(\overline{x}) \wedge \exists \overline{x}' \,\big(\rho(\overline{x}, \overline{x}') \wedge \psi(\overline{x}')\big)$ and $\mathrm{im}(\phi)$ is (isomorphic to) the pair $\phi''/\psi' \wedge \phi''$, where ϕ'' is $\exists \overline{x}\, \big(\phi(\overline{x}) \wedge \rho(\overline{x}, \overline{x}')\big)$. From this it follows easily that every morphism has a kernel and a cokernel, that every monomorphism is the kernel of its cokernel and that every epimorphism is the cokernel of its kernel. I will not write out any more details because one also has a proof via 10.2.30. $\square$

Note that ϕ/ψ is a minimal pair iff, as an object of $\mathbb{L}_R^{\mathrm{eq}+}$, it is **simple** (has no proper subobject apart from zero).

Given any module M there is an obvious notion of "evaluation at M": given by $\phi/\psi \mapsto \phi(M)/\psi(M)$ at each pp-pair ϕ/ψ and given by $\rho \mapsto \rho(M)$ at the morphism defined by the pp condition ρ. Clearly this defines a functor from $\mathbb{L}_R^{\mathrm{eq}+}$ to **Ab** indeed, by 1.1.8, to the category of left $\mathrm{End}(M)$-modules.

Lemma 3.2.11. *Let M be any module. Then evaluation at M is an exact (additive, as always) functor from $\mathbb{L}_R^{\mathrm{eq}+}$ to* **Ab**.

Proof. Any exact sequence in $\mathbb{L}_R^{\mathrm{eq}+}$ is isomorphic to one of the form $0 \to \phi'/\psi \to \phi/\psi \to \phi/\phi' \to 0$ with $\psi \leq \phi' \leq \phi$ and evaluation at M gives a, clearly exact, sequence of abelian groups. $\square$

It turns out, 18.1.4, that the R-modules are precisely the exact functors from $\mathbb{L}_R^{\mathrm{eq}+}$ to **Ab**.

It is worth taking care to clarify the use of variables. First, we cannot regard $\phi(x)/\psi(x)$ and $\phi(y)/\psi(y)$ as different objects of $\mathbb{L}^{\mathrm{eq}+}$: for then no pp condition could define a non-trivial endomorphism of an object. So it is necessary to regard these as literally the same object of $\mathbb{L}^{\mathrm{eq}+}$. (Also see comments near the beginning of Section 1.2.1.)

Second, if $\phi(x)$ is a pp condition, then the same functor $M \mapsto \phi(M)$ is defined by the pp condition $\phi(y)$, which is like ϕ but with every occurrence of the free variable x replaced by y. At least, that is so unless the substitution allows an occurrence of "y" to be captured by a quantifier $\exists y$, resulting in an unintended change of meaning. For example, consider $\phi(x)$ to be $\exists y\,(x2 = 0 \wedge y = 0)$: then $F_{\phi(x)}$ defines $M \mapsto M2$ but $F_{\phi(y)}$, which places no restriction on any free variable, may be regarded as a pp condition on one free variable, and, as such, defines $M \mapsto M$! The usual way to get round this silly effect is to rename quantified variables if necessary.

Next we show that if $\rho(\overline{x}, \overline{y})$ defines a function from ϕ/ψ to ϕ'/ψ', then $D\rho(\overline{x}, \overline{y})$ defines a function from $D\psi'/D\phi'$ to $D\psi/D\phi$. In this way duality of pp conditions (1.3.1) extends to a duality between the categories $\mathbb{L}_R^{\mathrm{eq}+}$ and ${}_R\mathbb{L}^{\mathrm{eq}+}$. For the functor-category version of this duality see 10.3.4.

For a simple example, before we start the proof, take $\rho(x, y)$ to be $xr - y = 0$: the condition which defines multiplication by $r \in R$ and which is a morphism from the pair $x = x/x = 0$ to itself (note the above comments about use of variables). Since $\rho(x, y)$ is $(x \;\; y)\begin{pmatrix} r \\ -1 \end{pmatrix} = 0$, $D\rho(x, y)$ is

$$\exists z \begin{pmatrix} 1 & 0 & r \\ 0 & 1 & -1 \end{pmatrix} \begin{pmatrix} x \\ y \\ z \end{pmatrix} = 0,$$

that is, $\exists z\,(x = -rz \wedge y = z)$, that is, $x = -ry$, which is the function multiplication by $-r$ from the object $x = x/x = 0$ of ${}_R\mathbb{L}^{\mathrm{eq}+}$ to itself.

In order to obtain a functor (since the identity map should be taken to the identity, not multiplication by -1), we then twist, multiplying by -1.

Also, let us note, before we start, that the conditions for $\rho(\overline{x}, \overline{y})$ to define a function from the pair ϕ/ψ to the pair ϕ'/ψ' are:

$\phi(\overline{x}) \le \exists \overline{y}\,\big(\rho(\overline{x}, \overline{y}) \wedge \phi'(\overline{y})\big)$ and
$\rho(\overline{x}, \overline{y}) \wedge \psi(\overline{x}) \;\le\; \psi'(\overline{y})$.

It is the conditions in this form that we will use (for ρ) and check (for $D\rho$) in the proof.

Theorem 3.2.12. ([274, 2.9]) *Elementary duality induces a functor, also denoted D, from $\mathbb{L}_R^{\mathrm{eq}+}$ to ${}_R\mathbb{L}^{\mathrm{eq}+}$ defined on objects by $\phi/\psi \mapsto D\psi/D\phi$ and on morphisms by $\rho \mapsto -D\rho$. This is a contravariant equivalence of categories, $\mathbb{L}_R^{\mathrm{eq}+} \simeq ({}_R\mathbb{L}^{\mathrm{eq}+})^{\mathrm{op}}$ and D^2 is equivalent to the identity.*

Proof. It would take less space to move to the functor category (10.2.29) do everything there (Section 10.3) then move back, or to proceed as in [274, §2], but it is perhaps an interesting exercise to see how to do this directly using the criterion 1.1.13. So suppose that ρ determines a map from ϕ/ψ to ϕ'/ψ'. Take:

$$\rho(\overline{x},\overline{y}) \text{ to be } \exists\overline{u}\ (\overline{x}\ \overline{y}\ \overline{u})\begin{pmatrix} A_1 \\ A_2 \\ B \end{pmatrix} = 0;$$

$$\phi(\overline{x}) \text{ to be } \exists\overline{v}\ (\overline{x}\ \overline{v})\begin{pmatrix} C \\ D \end{pmatrix} = 0;$$

$$\phi'(\overline{y}) \text{ to be } \exists\overline{v}'\ (\overline{y}\ \overline{v}')\begin{pmatrix} C' \\ D' \end{pmatrix} = 0;$$

$$\psi(\overline{x}) \text{ to be } \exists\overline{w}\ (\overline{x}\ \overline{w})\begin{pmatrix} E \\ F \end{pmatrix} = 0;$$

$$\psi'(\overline{y}) \text{ to be } \exists\overline{w}'\ (\overline{y}\ \overline{w}')\begin{pmatrix} E' \\ F' \end{pmatrix} = 0.$$

By assumption

(1) $\phi(\overline{x}) \leq \exists\overline{y}\,\big(\rho(\overline{x},\overline{y}) \wedge \phi'(\overline{y})\big)$ and
(2) $\rho(\overline{x},\overline{y}) \wedge \psi(\overline{x}) \ \leq\ \psi'(\overline{y})$.

That is,

$$(1)'\ \exists\overline{v}\ (\overline{x}\ \overline{v})\begin{pmatrix} C \\ D \end{pmatrix} = 0 \ \leq\ \exists\overline{y},\overline{u},\overline{v}'\ (\overline{x}\ \overline{y}\ \overline{u}\ \overline{v}')\begin{pmatrix} A_1 & 0 \\ A_2 & C' \\ B & 0 \\ 0 & D' \end{pmatrix} = 0 \text{ and}$$

$$(2)'\ \exists\overline{u},\overline{w}\ (\overline{x}\ \overline{y}\ \overline{u}\ \overline{w})\begin{pmatrix} A_1 & E \\ A_2 & 0 \\ B & 0 \\ 0 & F \end{pmatrix} = 0 \ \leq\ \exists\overline{w}'\ (\overline{x}\ \overline{y}\ \overline{w}')\begin{pmatrix} 0 \\ E' \\ F' \end{pmatrix} = 0.$$

These, by 1.1.13, translate to the solvability of the following matrix equations:

$$(1)''\ \begin{pmatrix} I_1 & G'_1 & G'_2 & G'_3 \\ 0 & G''_1 & G''_2 & G''_3 \end{pmatrix}\begin{pmatrix} A_1 & 0 \\ A_2 & C' \\ B & 0 \\ 0 & D' \end{pmatrix} = \begin{pmatrix} C \\ D \end{pmatrix}\begin{pmatrix} K & K' \end{pmatrix},$$

$$(2)''\ \begin{pmatrix} I_1 & 0 & H_1 \\ 0 & I_2 & H_2 \\ 0 & 0 & H_3 \\ 0 & 0 & H_4 \end{pmatrix}\begin{pmatrix} 0 \\ E' \\ F' \end{pmatrix} = \begin{pmatrix} A_1 & E \\ A_2 & 0 \\ B & 0 \\ 0 & F \end{pmatrix}\begin{pmatrix} L \\ L' \end{pmatrix},$$

where I_1 and I_2 are identity matrices matching the lengths of $\overline{x}$, $\overline{y}$ respectively.

Therefore, there are matrices $G_1', G_2', \ldots, L, L'$ such that the following sets of matrix equations hold

$$\begin{aligned}
(1)''' \quad & A_1 + G_1'A_2 + G_2'B = CK \text{ and } G_1'C' + G_3'D' = CK' \\
& G_1''A_2 + G_2''B = DK \text{ and } G_1''C' + G_3''D' = DK' \\
(2)''' \quad & H_1F' = A_1L + EL' \\
& E' + H_2F' = A_2L \\
& H_3F' = BL \\
& H_4F' = FL'.
\end{aligned}$$

The dual pp conditions are as follows:

$$D\rho(\overline{x}, \overline{y}) \text{ is } \exists\overline{z} \begin{pmatrix} I_1 & 0 & A_1 \\ 0 & I_2 & A_2 \\ 0 & 0 & B \end{pmatrix} \begin{pmatrix} \overline{x} \\ \overline{y} \\ \overline{z} \end{pmatrix} = 0,$$

$$D\phi(\overline{x}) \text{ is } \exists\overline{v}_0 \begin{pmatrix} I_1 & C \\ 0 & D \end{pmatrix} \begin{pmatrix} \overline{x} \\ \overline{v}_0 \end{pmatrix} = 0,$$

$$D\phi'(\overline{y}) \text{ is } \exists\overline{v}_0' \begin{pmatrix} I_2 & C' \\ 0 & D' \end{pmatrix} \begin{pmatrix} \overline{y}' \\ \overline{v}_0' \end{pmatrix} = 0,$$

$$D\psi(\overline{x}) \text{ is } \exists\overline{w}_0 \begin{pmatrix} I_1 & E \\ 0 & F \end{pmatrix} \begin{pmatrix} \overline{x} \\ \overline{w}_0 \end{pmatrix} = 0,$$

$$D\psi'(\overline{x}) \text{ is } \exists\overline{w}_0' \begin{pmatrix} I_2 & E' \\ 0 & F' \end{pmatrix} \begin{pmatrix} \overline{y} \\ \overline{w}_0' \end{pmatrix} = 0.$$

Written in non-matrix format, these conditions are:

$$D\rho\colon \exists\overline{z} \begin{cases} \overline{x} = A_1\overline{z} \\ \overline{y} = A_2\overline{z}\,, \\ 0 = B\overline{z} \end{cases}$$

$$D\phi\colon \exists\overline{v}_0 \begin{cases} \overline{x} = C\overline{v}_0 \\ 0 = D\overline{v}_0 \end{cases},$$

$$D\phi'\colon \exists\overline{v}_0' \begin{cases} \overline{y} = C'\overline{v}_0' \\ 0 = D'\overline{v}_0' \end{cases},$$

$$D\psi\colon \exists\overline{w}_0 \begin{cases} \overline{x} = E\overline{w}_0 \\ 0 = F\overline{w}_0 \end{cases},$$

$$D\psi'\colon \exists\overline{w}_0' \begin{cases} \overline{y} = E'\overline{w}_0' \\ 0 = F'\overline{w}_0' \end{cases}.$$

We need to prove

(A) $D\psi'(\overline{y}) \leq \exists\overline{x}\,\big(D\rho(\overline{x}, \overline{y}) \wedge D\psi(\overline{x})\big)$
(B) $D\rho(\overline{x}, \overline{y}) \wedge D\phi'(\overline{y}) \leq D\phi(\overline{x})$.

We do this "formally" (treating the introduced variables as solutions to their respective conditions). So for (A) we assume $\overline{y} = E'\overline{w}_0'$ and $0 = F'\overline{w}_0'$ and then find $\overline{x}$, $\overline{z}$ and $\overline{w}_0$ which make the right hand side of (A) hold for $\overline{y}$.

Using equations (2)‴ we have

$\overline{y} = E'\overline{w}_0' = (A_2L - H_2F')\overline{w}_0' = A_2L\overline{w}_0'$ (since $F'\overline{w}_0' = 0$) so set $\overline{z} = L\overline{w}_0'$ and therefore take $\overline{x} = A_1\overline{z} = A_1L\overline{w}_0'$.

Then $\overline{x} = A_1L\overline{w}_0' = (H_1F' - EL')\overline{w}_0' = -EL'\overline{w}_0'$ so set $\overline{w}_0 = -L'\overline{w}_0$.

Then check:

$B\overline{z} = BL\overline{w}_0' = H_3F'\overline{w}_0' = 0$ and
$F\overline{w}_0 = -FL'\overline{w}_0' = H_4F'\overline{w}_0' = 0$.

Thus (A) is established; that is, $D\rho$ defines a total relation from $D\psi'$ to $D\psi$.

As for (B), we suppose that $\overline{x}, \overline{y}$ are such that there are $\overline{z}$ and $\overline{v}_0'$ such that

$$\begin{cases} \overline{x} = A_1\overline{z} \\ \overline{y} = A_2\overline{z} \\ 0 = B\overline{z} \end{cases} \quad \text{and} \quad \begin{cases} \overline{y} = C'\overline{v}_0' \\ 0 = D'\overline{v}_0' \end{cases}.$$

We need to produce $\overline{v}_0$ such that $\overline{x} = C\overline{v}_0$ and $0 = D\overline{v}_0$. Using equations (1)‴ we have

$\overline{x} = A_1\overline{z} = (CK - G_1'A_2 - G_2'B)\overline{z} = CK\overline{z} - G_1'\overline{y}$ (since $A_2\overline{z} = \overline{y}$ and $B\overline{z} = 0$)
$= CK\overline{z} - G_1'C'\overline{v}_0' = CK\overline{z} - (CK' - G_3'D')\overline{v}_0' = CK\overline{z} - CK'\overline{v}_0'$ (since $D'\overline{v}_0' = 0$)
$= C(K\overline{z} - K'\overline{v}_0')$.
So set $\overline{v}_0 = K\overline{z} - K'\overline{v}_0'$.
Then $D\overline{v}_0 = DK\overline{z} - DK'\overline{v}_0' = (G_1''A_2 + G_2''B)\overline{z} - (G_1''C' + G_3''D')\overline{v}_0' = G_1''A_2\overline{z} - G_1''C'\overline{v}_0' = G_1''\overline{y} - G_1''\overline{y} = 0$ as required.

Thus (B) is established. Thus $D\rho$ (and hence also $-D\rho$) is well defined and functional as a relation from $D\psi'/D\phi'$ to $D\psi/D\phi$.

It remains to check the effect of $(-)D$ on identity morphisms and composition. Any identity morphism is represented by a condition of the form $\overline{x} = \overline{y}$ and this has dual

$$\exists\overline{z} \begin{pmatrix} I & 0 & I \\ 0 & I & -I \end{pmatrix} \begin{pmatrix} \overline{x} \\ \overline{y} \\ \overline{z} \end{pmatrix} = 0,$$

which is equivalent to $\overline{x} = -\overline{y}$, so the negative of the dual is, indeed, the identity functor.

Finally, suppose that $\rho(\overline{x}, \overline{y})$ and $\sigma(\overline{y}, \overline{z})$ define composable functions, the composition, therefore, being defined by $\exists \overline{y} \left(\rho(\overline{x}, \overline{y}) \wedge \sigma(\overline{y}, \overline{z})\right)$, which we denote by $\tau(\overline{x}, \overline{z})$. Say $\rho(\overline{x}, \overline{y})$ is

$$\exists \overline{u} \begin{pmatrix} \overline{x} & \overline{y} & \overline{u} \end{pmatrix} \begin{pmatrix} A_1 \\ A_2 \\ B \end{pmatrix} = 0$$

and $\sigma(\overline{y}, \overline{z})$ is

$$\exists \overline{v} \begin{pmatrix} \overline{y} & \overline{z} & \overline{v} \end{pmatrix} \begin{pmatrix} C_1 \\ C_2 \\ D \end{pmatrix} = 0$$

so that $\tau(\overline{x}, \overline{z})$ is

$$\exists \overline{y}, \overline{u}, \overline{v} \begin{pmatrix} \overline{x} & \overline{z} & \overline{y} & \overline{u} & \overline{v} \end{pmatrix} \begin{pmatrix} A_1 & 0 \\ 0 & C_2 \\ A_2 & C_1 \\ B & 0 \\ 0 & D \end{pmatrix} = 0.$$

Then

$$D\tau(\overline{x}, \overline{z}) \text{ is } \exists \overline{u}', \overline{v}' \begin{pmatrix} I_1 & 0 & A_1 & 0 \\ 0 & I_3 & 0 & C_2 \\ 0 & 0 & A_2 & C_1 \\ 0 & 0 & B & 0 \\ 0 & 0 & 0 & D \end{pmatrix} \begin{pmatrix} \overline{x} \\ \overline{z} \\ \overline{u}' \\ \overline{v}' \end{pmatrix} = 0,$$

$$D\rho(\overline{x}, \overline{y}) \text{ is } \exists \overline{u}'' \begin{pmatrix} I_1 & 0 & A_1 \\ 0 & I_2 & A_2 \\ 0 & 0 & B \end{pmatrix} \begin{pmatrix} \overline{x} \\ \overline{y} \\ \overline{u}'' \end{pmatrix} = 0 \text{ and}$$

$$D\sigma(\overline{y}, \overline{z}) \text{ is } \exists \overline{v}'' \begin{pmatrix} I_2 & 0 & C_1 \\ 0 & I_3 & C_2 \\ 0 & 0 & D \end{pmatrix} \begin{pmatrix} \overline{y} \\ \overline{z} \\ \overline{v}'' \end{pmatrix} = 0.$$

Computing the composition of $D\sigma$ and $D\rho$ we obtain, note, $D\tau(\overline{x}, -\overline{y}) = -D\tau(\overline{x}, \overline{y})$. Therefore the composition of $-D\sigma$ and $-D\rho$ is $-D\tau$, as required. □

Corollary 3.2.13. ([274, 3.4]) *For any pp-pair ϕ/ψ the ring of pp-definable endomorphisms of ϕ/ψ is isomorphic to the ring of pp-definable endomorphisms of $D\psi/D\phi$ (we regard the one ring as having multiplication defined by operating* *from the right,*[5] *the other from the left: otherwise the one ring is isomorphic to the opposite of the other).*

Corollary 3.2.14. *If R is commutative, then $\mathbb{L}_R^{\mathrm{eq}+}$ is isomorphic to its opposite category.*

Proof. By 3.2.12, $\mathbb{L}_R^{\mathrm{eq}+} \simeq (\mathbb{L}_{R^{\mathrm{op}}}^{\mathrm{eq}+})^{\mathrm{op}}$ and if R is commutative, so $R \simeq R^{\mathrm{op}}$, the latter is isomorphic to $(\mathbb{L}_R^{\mathrm{eq}+})^{\mathrm{op}}$. □

Given any module M the constructions of this section may be relativised by working modulo M: we may use the same objects (but it will be equivalent to use pairs from the lattices $\mathrm{pp}^n(M)$ in place of the pp_R^n) and a pp condition ρ will define a function in the relative category iff it defines one just at M, rather than requiring this at all modules (ρ will then, by 3.4.11, define a function on every object in the definable category generated by M); further, ρ and σ will be equivalent if they define the same function on M. Denote the resulting category $(\mathbb{L}_R^{\mathrm{eq}+})_M$. Clearly there is a functor $\mathbb{L}_R^{\mathrm{eq}+} \longrightarrow (\mathbb{L}_R^{\mathrm{eq}+})_M$ which is surjective on objects but which will not in general be full: for example, take $M = \mathbb{Q}_\mathbb{Z}$; there are more pp-definable functions on ($x = x/x = 0$ evaluated at) $\mathbb{Q}_\mathbb{Z}$ than on ($x = x/x = 0$ evaluated at) arbitrary $\mathbb{Z}$-modules.

The following is easily verified.

Proposition 3.2.15. *For any module M the category $(\mathbb{L}_R^{\mathrm{eq}+})_M$ is abelian and the canonical functor $\mathbb{L}_R^{\mathrm{eq}+} \longrightarrow (\mathbb{L}_R^{\mathrm{eq}+})_M$ is exact. Indeed, this is the quotient of $\mathbb{L}_R^{\mathrm{eq}+}$ by the Serre subcategory (§11.1.1) of $\mathbb{L}_R^{\mathrm{eq}+}$ which has objects those ϕ/ψ such that $\phi(M) = \psi(M)$.*

3.3 Reduced products and pp-types

Reduced products are defined and relations to direct limits pointed out in Section 3.3.1. Łoś' Theorem for reduced products is proved in Section 3.3.2. Reduced products are used to realise pp-types in Section 3.3.3.

3.3.1 Reduced products

Reduced products and ultraproducts are defined and expressed as direct limits of products (3.3.1). Direct limits are pure in suitable reduced products (3.3.2).

[5] If $S = \mathrm{End}(\phi/\psi)$, if M is any right module and L is any left module, then $\phi(M)/\psi(M)$ is naturally a right S-module and $D\psi(L)/D\phi(L)$ is naturally a left S-module.

Let M_i $(i \in I)$ be modules and let $\mathcal{F}$ be a **filter** on the index set I. This means that $\mathcal{F}$ is a set of subsets of I, which is non-empty, which is closed upwards ($J \subseteq K \subseteq I$ and $J \in \mathcal{F}$ implies $K \in \mathcal{F}$) and which is closed under finite intersection ($J, K \in \mathcal{F}$ implies $J \cap K \in \mathcal{F}$). We also require any filter to be **proper** in the sense that it does not contain $\emptyset$.

Define an equivalence relation $\sim\ = \sim_{\mathcal{F}}$ on the product $\prod_{i \in I} M_i$ by $(a_i)_i = \overline{a} \sim \overline{b} = (b_i)_i$ iff $\{i \in I : a_i = b_i\} \in \mathcal{F}$ (if we call the subsets of I which lie in $\mathcal{F}$ "large", then $\overline{a} \sim \overline{b}$ iff $\overline{a}$ and $\overline{b}$ agree on a large set of coordinates).

Let $Z = \{\overline{a} \in \prod_i M_i : \overline{a} \sim 0\}$. Since $\mathcal{F}$ is closed under intersection, Z is closed under addition and, since $\mathcal{F}$ is upwards closed, Z is closed under multiplication by elements of R. So we may factor out the submodule Z to obtain $(\prod_{i \in I} M_i)/Z$, which is denoted $\prod_{i \in I} M_i/\mathcal{F}$, or just $\prod_i M_i/\mathcal{F}$, and called the **reduced product** of the M_i **with respect to** $\mathcal{F}$. Let $\pi : \prod_i M_i \to \prod_i M_i/\mathcal{F}$ be the projection, so $\pi\overline{a} = \pi\overline{b}$ iff $\overline{a} \sim \overline{b}$. We use the notation $\overline{a}/\mathcal{F}$ for $\pi\overline{a}$.

This reduced product may also be realised as a direct limit of "partial products", as follows.[6] For $J \in \mathcal{F}$ there is the product $\prod_J M_i = \prod_{i \in J} M_i$, and, if $J \supseteq K$, there is the canonical projection $\pi_{JK} : \prod_J M_i \to \prod_K M_i$. So we have a directed system with objects $(\prod_J M_i)_{J \in \mathcal{F}}$ and morphisms the π_{JK} with $J \supseteq K$. Since $\mathcal{F}$ is closed under intersection this system is directed by $\mathcal{F}$ under reverse inclusion. Let M_∞, with limit maps $\pi_{J\infty} : \prod_J M_i \to M_\infty$, be the direct limit of the system. We claim that this is the reduced product defined above. For locally finitely presented abelian categories, where we can talk about elements (see, for example, Section 10.2.4), the proof of equivalence which is given below applies.

Lemma 3.3.1. *Let M_i $(i \in I)$ be modules and let $\mathcal{F}$ be a filter on I. Then the reduced product $\prod_{i \in I} M_i/\mathcal{F}$ is isomorphic to the direct limit $\varinjlim_{J \in \mathcal{F}^{(\mathrm{op})}} \prod_J M_i$. The exact sequence $0 \to Z \to \prod_I M_i \to \prod_{i \in I} M_i/\mathcal{F} \to 0$ is pure-exact.*

Proof. Note that there is, in fact, a directed system of split exact sequences, $0 \to \prod_{I \setminus J} M_i \to \prod_I M_i \to \prod_J M_i \to 0$ $(J \in \mathcal{F})$ which, since we are working in a Grothendieck category, has exact direct limit, $0 \to \varinjlim_{I \setminus J, J \in \mathcal{F}} \prod_{I \setminus J} M_i \to \prod_I M_i \to \varinjlim_{J \in \mathcal{F}} \prod_J M_i \to 0$. The left-hand term of this exact sequence consists exactly of those elements of $\prod_I M_i$ which are zero on a set of coordinates in $\mathcal{F}$, in other words it is the kernel, Z, of the map from $\prod_i M_i$ to $\prod_i M_i/\mathcal{F}$ described above.

The sequence is pure-exact by 2.1.3. □

[6] This definition makes sense in more general additive categories, even when there is no "elementwise" definition available, see Appendix E, p. 715.

In the case where all component modules are isomorphic to a single module M, the reduced product is written as $M^I/\mathcal{F}$ and referred to as a **reduced power**. There is a **diagonal map** from M to $M^I/\mathcal{F}$ given by $\Delta : a \mapsto (a)_i/\mathcal{F}$. From the definition of the relation $\sim$ on M^I it is immediate that this is an embedding (by 3.3.4 it is a pure embedding).

The terms **ultraproduct** and **ultrapower** are used if $\mathcal{F}$ is an **ultrafilter**, meaning that $\mathcal{F}$ is maximal with respect to inclusion among all proper filters on I, equivalently if for every $J \subseteq I$ either J or $I \setminus J$ is in $\mathcal{F}$.

The following partial converse to 3.3.1 can be found at [621, 3.1] (in fact, it holds in a quite general context, see [622][7]).

Theorem 3.3.2. *If $((M_i)_{i\in I}, (g_{ij})_{ij})$ is a directed system of modules with direct limit $(M, (g_{i\infty})_i)$, then M is a pure submodule of a reduced product (which may be taken to be an ultraproduct) of the M_i.*

Proof. Set $\mathcal{F}_0 = \{\{j : j \geq i_1, \dots, i_n\} : i_1, \dots, i_n \in I, n \geq 1\}$ and let $M^* = \prod M_i/\mathcal{F}$, where $\mathcal{F}$ is any filter containing $\mathcal{F}_0$, equivalently containing all sets of the form $\{j : j \geq i\}$ as i varies over I. Set $\sim \; = \sim_{\mathcal{F}}$.

Define $f : M \to M^*$ as follows: given $a \in M$, choose i such that there is $a_i \in M_i$ with $g_{i\infty}a_i = a$. Set $fa = (a_j)_j/\sim$, where $a_j = g_{ij}a_i$ if $j \geq i$, and $a_j = 0$, say, otherwise.

This map is well defined since if also k is such that there is $b_k \in M_k$ with $g_{k\infty}b_k = a$, then there is $l \geq i, k$ such that $g_{kl}b_k = g_{il}a_i$ (because $g_{k\infty}b_k = g_{i\infty}a_i$) so, for $j \geq l$, one has $g_{ij}a_i = g_{kj}b_k$. Therefore, the tuple $(a_i)_i \in \prod M_i$ defined as above using a_i is equal, on a set of indices belonging to $\mathcal{F}_0$, hence belonging to $\mathcal{F}$, to the corresponding tuple defined using b_k, so they have the same image in M^*.

The map is monic: if $fa = 0$, then, by definition of reduced product, $a_j = 0$ for all $j \in J$ for some $J \in \mathcal{F}$. Let i be as in the definition of fa. Then $J \cap \{j : j \geq i\} \neq \emptyset$ so there is $j \geq i$ with $a_j = 0$, that is, with $g_{ij}a_i = 0$, hence $a = g_{i\infty}a_i = g_{j\infty}g_{ij}a_i = 0$.

Finally, we use 3.3.3 below to show that this embedding is pure. Let $\overline{a} = (a_1, \dots, a_n)$ be from M. Choose i such that there are $b_1, \dots, b_n \in M_i$ with $g_{i\infty}b_t = a_t$ for $t = 1, \dots, n$. Suppose that ϕ is pp and that $(fa_1, \dots, fa_n) \in \phi(M^*)$. Say $fa_t = (c_i^t)_i/\sim$. By 3.3.3, $J' = \{j : (c_j^1, \dots, c_j^n) \in \phi(M_j)\} \in \mathcal{F}$, so, since $\mathcal{F}$ is a filter, $J' \cap \{j : j \geq i\} \neq \emptyset$. Say $j \geq i$ is such that $(c_j^1, \dots, c_j^n) \in \phi(M_j)$. Note that, by construction of the fa_t, we have, without loss of generality, $g_{j\infty}c_j^t = a_t$. So, by 1.1.7, we deduce $(a_1, \dots, a_n) \in \phi(M)$ as required. □

[7] But a corresponding statement for inverse limits is only sometimes true, for example see [389].

The following terminology also will be needed. If $\mathcal{S}$ is a set of subsets of a set I then the filter **generated** by $\mathcal{S}$ is $\langle \mathcal{S} \rangle = \{J \subseteq I : J \supseteq K_1 \cap \cdots \cap K_n \text{ for some } K_i \in \mathcal{S}\}$. Provided $\mathcal{S}$ has the **finite intersection property** (that is, every finite intersection $K_1 \cap \cdots \cap K_n$ with the $K_i \in \mathcal{S}$ is non-empty) this will be a proper filter.

A filter $\mathcal{F}$ is **principal** if it has the form $\{J \subseteq I : J \supseteq I_0\}$ for some $I_0 \subseteq I$. An ultrafilter is principal iff it has the form $\{J \subseteq I : i_0 \in J\}$ for some fixed $i_0 \in I$. In that case it is easy to see that $\prod M_i/\mathcal{F} \simeq M_{i_0}$, so it is **non-principal** ultrafilters which give something new.

There is a degree of non-explicitness inherent in the use of ultraproducts because the existence of non-principal *ultra*filters depends on the axiom of choice: there are no explicit recipes for deciding exactly which subsets of I are to go into an ultrafilter $\mathcal{F}$. So it is typical that some set of subsets of I with the finite intersection property is specified and then one just "chooses" an ultrafilter containing those sets.

This is a natural place to mention the notion of an **ideal** in a partially ordered set (we have defined filters in the power set of a set I but the same definition works in any partially ordered set): this is a subset, $\mathcal{I}$, closed under finite join (for example, union or sum, depending on context) and such that whenever $b \leq a$ and $a \in \mathcal{I}$ then $b \in \mathcal{I}$, and also $0 \in \mathcal{I}$ ($\mathcal{I}$ is non-empty) and $1 \notin \mathcal{I}$ ($\mathcal{I}$ is proper). If R is a commutative von Neumann regular ring, see Section 8.2.11, with $B(R)$ its boolean algebra of idempotents, then there is a bijection between (algebraic) ideals of R and (poset) ideals of $B(R)$, given by $I \mapsto I \cap B(R)$, $\mathcal{I} \mapsto \sum_{e \in \mathcal{I}} eR$.

3.3.2 Pp conditions in reduced products

The solution set of a pp condition in a reduced product is described (3.3.3). The diagonal embedding into a reduced product is pure (3.3.4), as is any reduced product of pure embeddings (3.3.5).

The next result is true in the general model-theoretic context and is essentially Łoś' Theorem (see A.1.7), the fundamental result on the model-theoretic properties of ultraproducts. Because the result below deals only with pp conditions it holds for the more general reduced products.[8]

Proposition 3.3.3. *Let M_i ($i \in I$) be modules and for each i let $\overline{a}_i$ be an n-tuple from M_i. Let $\mathcal{F}$ be a filter on I. Form the reduced product $\prod_i M_i/\mathcal{F}$ and let*

[8] Łoś' Theorem deals with arbitrary conditions, including those with negations, and is correspondingly restricted to ultraproducts.

$\overline{a} = (\overline{a}_i)_i/\mathcal{F}$ be the corresponding n-tuple from $\prod_i M_i/\mathcal{F}$. Let ϕ be a pp condition with n free variables.

Then $\overline{a}/\mathcal{F} \in \phi(\prod_i M_i/\mathcal{F})$ iff $\{i \in I : \overline{a}_i \in \phi(M_i)\} \in \mathcal{F}$.

Proof. Suppose that $J = \{i \in I : M_i \models \phi(\overline{a}_i)\}$ is in $\mathcal{F}$. Set $\overline{a}' = (\overline{a}_i)_{i \in J} \in \prod_J M_i$. By 3.3.1 the reduced product is the direct limit of the $\prod_K M_i$ for $K \in \mathcal{F}$. Clearly $\pi_{J\infty}\overline{a}' = \overline{a}/\mathcal{F}$, where $\pi_{J\infty}$ is the limit map from $\prod_J M_i$ to $\prod_i M_i/\mathcal{F}$. Since each $\overline{a}_i$, for $i \in J$, satisfies ϕ, so does $\overline{a}'$ (1.2.3) and hence, by 1.1.7, so does $\overline{a}/\mathcal{F}$.

For the converse, suppose that $M^* = \prod_i M_i/\mathcal{F} \models \phi(\overline{a}/\mathcal{F})$. Let $(C, \overline{c})$ be a free realisation of ϕ. Then, 1.2.17, there is a morphism, f say, from C to M^* taking $\overline{c}$ to $\overline{a}/\mathcal{F}$. Because C is finitely presented and M^* is the direct limit of the $\prod_{J\in\mathcal{F}} M_i$ this morphism factors through $\prod_J M_i$ for some $J \in \mathcal{F}$, say $f' : C \to \prod_J M_i$ is such that $\pi_{J\infty} f' = f$. Set $\overline{a}' = f'\overline{c}$. Then $\overline{a}'$ satisfies ϕ in $\prod_J M_i$. Since $\pi_{J\infty}\overline{a}' = \overline{a}/\mathcal{F}$ the ith coordinate of $\overline{a}'$ is equal to $\overline{a}_i$ on a subset, J', of J which is still in $\mathcal{F}$. Therefore, for each $i \in J'$, the coordinate $\overline{a}_i$ is in $\phi(M_i)$ as required. □

Corollary 3.3.4. *If M is any module, I is any set and $\mathcal{F}$ is any filter on I, then the diagonal embedding of M into $M^I/\mathcal{F}$ is pure.*

Proof. Suppose that $\overline{a}$ is a tuple from M, that $\Delta\overline{a}$ is its image in $M^I/\mathcal{F}$ under the diagonal map Δ and that $\Delta\overline{a} \in \phi(M^I/\mathcal{F})$ for a given pp condition ϕ. By 3.3.3 it follows immediately that $\overline{a} \in \phi(M)$. □

One may define reduced products of morphisms in the obvious way (or via the category of representations of the quiver $A_2 = \bullet \to \bullet$); the result is a morphism between the corresponding reduced products of objects. Indeed, the reduced product construction, for a given index set I and filter, $\mathcal{F}$, on I, is functorial, that is, defines an additive functor from $(\text{Mod-}R)^I$ to Mod-R. If, for each $i \in I$, we have the morphism $f_i : M_i \to N_i$, then we use the notation $(f_i)_i/\mathcal{F} : \prod M_i/\mathcal{F} \to \prod N_i/\mathcal{F}$ for the resulting morphism.

Proposition 3.3.5. *Any reduced product of pure embeddings is a pure embedding.*

Proof. Let $f = (f_i)_i/\mathcal{F} : A^* = \prod_i A_i/\mathcal{F} \to \prod_i B_i/\mathcal{F} = B^*$ be a reduced product of the pure embeddings $f_i : A_i \to B_i$. Suppose that the image under f of $\overline{a} = (\overline{a}_i)_i/\mathcal{F}$ from A^* satisfies the pp condition ϕ in B^*. Let $g : C \to B^*$ take $\overline{c}$ to $f\overline{a}$, where $(C, \overline{c})$ is a free realisation of ϕ. Using 3.3.1, g factors through some product $\prod_{j\in J} B_j$ with $J \in \mathcal{F}$, say $g = \pi^B_{J\infty} g'$, where $g' : C \to \prod_J B_j$ and $\pi^B_{J\infty} : \prod_J B_j \to \prod B_i$ is the canonical map. Then purity of $\prod_J A_j$ in $\prod_J B_j$ (2.1.11) yields, by 2.1.9, a morphism g'' from C to $\prod_J A_j$ with $\iota g''\overline{c} = g'\overline{c}$ ($\iota : \prod_J A_j \to \prod_J B_j$ being the inclusion). Then $\pi^A_{J\infty} g''\overline{c} = \overline{a}$ so 1.1.7 gives that $\overline{a}$ satisfies ϕ in A^*, as required. □

3.3.3 Realising pp-types in reduced products

A different proof of the next result was given in 3.2.5 and another is given at 4.3.39. The model-theoretic style of proof below will also be used to prove the stronger result 4.1.4.

Theorem 3.3.6. *Let p be a filter in the lattice pp_R^n of pp conditions. Then there is a module M and an n-tuple, $\overline{a}$, in M such that $\mathrm{pp}^M(\overline{a}) = p$.*

Proof. Let I denote the set of finite subsets of p. For each such subset $S = \{\psi_1, \dots, \psi_k\}$ of p let $(C_S, \overline{c}_S)$ be a free realisation of $\psi_1 \wedge \dots \wedge \psi_k$. Consider the set, $\mathcal{S}$, of those subsets of I of the form $\langle\{\psi\}\rangle = \{S \in I : \psi \in S\}$ with $\psi \in p$. Because p is closed under $\wedge$ clearly $\mathcal{S}$ has the finite intersection property, so let $\mathcal{F} = \{J \subseteq I : J \supseteq \langle\{\psi_1\}\rangle \cap \dots \cap \langle\{\psi_k\}\rangle$ for some $\psi_1, \dots, \psi_k \in p\}$ be the filter generated by $\mathcal{S}$.

Form the reduced product $M = \prod_{S \in I} C_S/\mathcal{F}$ and the n-tuple $\overline{c} = (\overline{c}_S)_S/\mathcal{F}$ of elements of M. Given a pp condition $\phi \in p$ the set, $\langle\{\phi\}\rangle$, of finite subsets, S, which contain ϕ and such that, therefore, $\overline{c}_S$ satisfies the condition ϕ, is in the filter $\mathcal{F}$. So, by 3.3.3, $\overline{c} \in \phi(M)$.

On the other hand, if ϕ is a pp condition and $\overline{c} \in \phi(M)$, then, by 3.3.3, the set $J = \{S \in I : \overline{c}_S \in \phi(C_S)\}$ is in $\mathcal{F}$, hence contains some set of the form $\langle\{\psi_1\}\rangle \cap \dots \cap \langle\{\psi_k\}\rangle$ with the $\psi_i \in p$. In particular $S_0 = \{\psi_1, \dots, \psi_k\} \in J$, so $\overline{c}_{S_0} \in \phi(C_{S_0})$. By definition of free realisation, $\psi_1 \wedge \dots \wedge \psi_k \le \phi$, therefore $\phi \in p$, as required. □

3.4 Definable subcategories

Section 3.4.1 introduces definable subcategories and in Section 3.4.2 the duality between definable subcategories of right and left modules is proved. Section 3.4.3 gives examples relating to finiteness conditions (being coherent, noetherian) on the ring. Some relations between definability and covariant finiteness are shown in Section 3.4.4.

3.4.1 Definable subcategories

The definition of definable subcategory is followed by examples and the basic characterisation, 3.4.7, in terms of closure conditions. These subcategories are also closed under pure-injective hulls and pure epimorphisms (3.4.8). Pp conditions with one free variable suffice to define these categories (3.4.10) and the definable subcategory generated by a finite union is identified (3.4.9).

The definable subcategory generated by a module is determined by the pp-pairs closed in that module (3.4.11) and every definable subcategory has a generator (3.4.12).

We will say that a subcategory or subclass of Mod-R is definable if it is specified by the coincidence of various pp conditions. Precisely, let $T = \{\phi_\lambda/\psi_\lambda\}_{\lambda\in\Lambda}$ be a set of pp-pairs (so ϕ_λ, ψ_λ are pp conditions and $\phi_\lambda \geq \psi_\lambda$ for each λ). Define Mod(T) ("Mod" for "model" not "module") to be the full subcategory of Mod-R consisting of those modules M such that $\phi_\lambda(M) = \psi_\lambda(M)$ for every $\lambda \in \Lambda$. This is a typical **definable subcategory** of Mod-R and we say that the object class of Mod(T) is a **definable subclass** of Mod-R.

The terminology comes from model theory: these classes are definable (the terms elementary and axiomatisable are also used), albeit by conditions of a special form.[9]

Example 3.4.1. Suppose that R is a domain. Then the class of torsionfree modules is definable: take the pairs of conditions to be all those of the form $(xr = 0)/(x = 0)$ as r ranges over the non-zero elements of R.

Recall (just before 2.3.12) the definition of torsionfree over more general rings: that if $xr = 0$, then there are n, $y_1, \dots, y_n$ and $s_1, \dots s_n \in \mathrm{ann}_{_RR}(r)$ such that $x = \sum_1^n y_i s_i$. As it stands it is not clear that this is expressible by the closure of a pp-pair or set of pp-pairs (indeed, in general, it is not; an example is at 5.3.62). If, however, each $\mathrm{ann}_{_RR}(r)$ $(r \in R)$ is finitely generated as a left ideal, then torsionfree is a definable property, using pairs $\big(\exists y_1, \dots, y_n\, (x = \sum_1^n y_i s_i)\big)/(x = 0)$, where $\mathrm{ann}_{_RR}(r) = \sum_1^n Rs_i$.

Dually, the class of divisible modules over a domain is always definable: use pp-pairs of the form $(x = x)/(r \mid x)$. Using just some of these we see that if R' is a ring of fractions of R, then the category of R'-modules forms a definable subcategory of Mod-R. Indeed, 5.5.4, this is true whenever $R \to R'$ is an epimorphism of rings (this is obvious for surjections).

It follows from 3.4.7 that the torsion modules, even for $R = \mathbb{Z}$, do not form a definable subcategory of the module category.[10]

Example 3.4.2. If M is a module of finite length over its endomorphism ring, then the modules which are direct summands of direct sums of copies of M form a definable subcategory (4.4.30).

[9] These are examples of classes with "coherent" or "basic" axiomatisations, see, for example, [3], [419] and also [584] for the comparison.

[10] They do, however, form a locally finitely presented category in their own right, hence a definable category in the sense of Chapter 18; the "problem" is the embedding into Mod-$\mathbb{Z}$, cf. Section 16.2.

Example 3.4.3. If Φ is any set of pp conditions, then the intersection of the kernels of the corresponding functors F_ϕ (introduced after 1.1.7) is a definable category, being defined by the set of pairs $\phi/0, \phi \in \Phi$. For example, if A is finitely presented, then, 1.2.18, there is a pp condition ϕ such that for every module M one has $(A, M) \simeq \phi(M)$. Therefore if $\mathcal{A}$ is a collection of finitely presented modules, then $\{M \in \text{Mod-}R : (A, M) = 0 \text{ for all } A \in \mathcal{A}\}$ is a definable subcategory. A large class of examples is considered in [395, §4].

Example 3.4.4. If A is an FP_2 module (that is, there is a projective resolution of A whose first (from the right), second and third projective modules are finitely generated), then there is a pp-pair ϕ/ψ such that for all M_R one has $\phi(M)/\psi(M) \simeq \text{Ext}^1(A, M)$ (10.2.35). So, if $\mathcal{A}$ is a collection of FP_2 modules, then $\mathcal{A}^\perp = \{M : \text{Hom}(A, M) = 0 = \text{Ext}^1(A, M) \text{ for all } A \in \mathcal{A}\}$ is a definable subcategory of Mod-R.

Example 3.4.5. If R is right coherent, then the class of absolutely pure right modules is definable (2.3.3 and 2.3.19) and, dually, the class of flat left modules is definable (2.3.9 and 2.3.19). The definability of either of these classes is equivalent to right coherence of R, see 3.4.24. The conditions for the classes of injective, respectively projective, modules to be definable are much more restrictive (3.4.28).

It will follow from 10.2.31 that if F is any finitely presented functor from mod-R to **Ab** and if $\overrightarrow{F}$ is its canonical extension to Mod-R (the extension which commutes with direct limits, Section 10.2.8), then the kernel of $\overrightarrow{F}$, meaning $\{M \in \text{Mod-}R : \overrightarrow{F}M = 0\}$, is definable as, therefore, will be any intersection of such classes. Conversely, 10.2.32, every definable subcategory has this form.

Example 3.4.6. If $\mathcal{F}$ is the torsionfree class for a hereditary torsion theory on Mod-R, then $\mathcal{F}$ is definable iff the torsion theory is of finite type (11.1.20). If also R is right coherent, more generally, if the torsion theory is elementary (Section 11.1.3), then the localised category $(\text{Mod-}R)_\tau$, regarded as a subcategory of Mod-R, is a definable subcategory (11.1.35, 11.1.21).

Most of the results in this section originate in the model theory of modules and the appearance of reduced products (especially ultraproducts) is a reflection of that. A guiding principle in my choice of proofs, apart from avoiding model theory per se, because the basic techniques are less well known, has been to give "module" proofs where reasonable but to use "functor category" proofs where these are better in some sense (simpler, clearer, more elegant). Hence the appearance of functorial methods in this section. For the reader who has not yet looked at Part II, I summarise the setting for the proof. This summary is extremely brief, and for the definitions and results the reader should look at Sections 12.1.1 and 11.1.2 in particular.

To each right module M is associated the functor $M \otimes_R -$ which takes (finitely presented) left R-modules to abelian groups. The functors of this form are, up to isomorphism, exactly the absolutely pure objects of the category, (R-mod, **Ab**), of all additive functors from R-mod to **Ab** (12.1.6). Identifying M with $M \otimes -$ allows us to regard Mod-R as a full subcategory of (R-mod, **Ab**) (12.1.3). To each definable subcategory, $\mathcal{X}$, of Mod-R there corresponds (Section 12.3) a hereditary torsion theory, $\tau_{D\mathcal{X}}$, of finite type (see Section 11.1.2 for this notion) on (R-mod, **Ab**) and then $\mathcal{X}$, via this embedding into the functor category, becomes the class of absolutely pure $\tau_{D\mathcal{X}}$-torsionfree objects (12.3.2). That is, a definable subclass of Mod-R may be regarded as the intersection of a torsionfree class of functors with Mod-R. To a pp-pair ϕ/ψ corresponds the quotient functor F_ϕ/F_ψ in (mod-R, **Ab**) and the duality (10.3.4) between (mod-R, **Ab**)^{fp} and (R-mod, **Ab**)^{fp} maps this to the pair $F_{D\psi/D\phi}$ in the latter category. The link is the formula $(F_{D\psi/D\phi}, M \otimes -) \simeq \phi(M)/\psi(M)$ (10.3.8). Therefore, defining a subcategory of Mod-R as the class of those modules on which a set, T, of pp-pairs is closed, is equivalent to first specifying a torsion theory on (R-mod, **Ab**) by declaring that the functors $F_{D\psi/D\phi}$ with $\phi/\psi \in T$ are to be torsion, and then defining the class of modules which, via the embedding, are torsionfree for that torsion theory (an object is torsionfree if there are no non-zero morphisms from a torsion object to it, see Chapter 11 for torsion theory).

In the earlier and more model-theoretic literature the classes of modules which are the subject of the next result did appear but, more frequently, and reflecting the subject matter of model theory, more attention was paid to "the class of models of a complete theory of modules closed under products". Such classes are in bijection to definable subclasses: given one of the former, close it under pure submodules to get a definable subcategory; for the other direction, given a definable subcategory, take the subclass of modules M which generate the definable subcategory (in the sense defined after 3.4.8) and which satisfy $M \equiv M^{\aleph_0}$ (for $\equiv$, elementary equivalence, see Appendix A, especially A.1.3). A proof of the equivalence of the model-theoretic and algebraic definitions of these classes (in a more general setting) is given by Rothmaler in [622, 8.1].

Theorem 3.4.7. *The following conditions on a subclass $\mathcal{X}$ of* Mod-R *are equivalent:*

(i) $\mathcal{X}$ is definable;
(ii) $\mathcal{X}$ is closed under direct products, direct limits and pure submodules;
(iii) $\mathcal{X}$ is closed under direct products, reduced products and pure submodules;
(iv) $\mathcal{X}$ is closed under direct products, ultrapowers and pure submodules.

Such a subclass is, in particular, closed under direct sums and direct summands. (Closure under isomorphism should be added where needed.)

Proof. (i)⇒(ii) If a subcategory $\mathcal{X}$ is definable, then, directly from the definitions, it is closed under pure submodules and, by 1.2.3, it is closed under direct products.

For closure under direct limits, let $((M_\lambda)_\lambda, (f_{\lambda\mu})_{\lambda<\mu})$ be a directed system in $\mathcal{X}$, with direct limit M and canonical maps $f_{\lambda\infty} : M_\lambda \to M$ to the limit. Suppose that ϕ/ψ is a pp-pair which is closed in every M_λ. Let $(C, \overline{c})$ be a free realisation of ϕ and let $\overline{a} \in \phi(M)$. By 1.2.17 there is a morphism $g : C \to M$ taking $\overline{c}$ to $\overline{a}$. Since C is finitely presented this morphism factors through some M_λ: there is, for some λ, a morphism $g' : C \to M_\lambda$ such that $g = f_{\lambda\infty}g'$. Now $g'\overline{c} \in \phi(M_\lambda)$, $= \psi(M_\lambda)$ by hypothesis, so, by 1.1.7, $\overline{a} \in \psi(M)$, as required.

By 3.3.1 (ii)⇒(iii), and (iv) is contained in (iii), so it remains to prove (iv)⇒(i).

Let $\mathcal{X} \subseteq \text{Mod-}R$ be closed under direct products, pure submodules and ultrapowers. Let $T = \{\phi/\psi : \forall M \in \mathcal{X},\ \phi(M) = \psi(M)\}$ be the set of pp-pairs which are closed on every object of $\mathcal{X}$. Let $\mathcal{X}' = \text{Mod}(T)$ be the corresponding definable class, so $\mathcal{X} \subseteq \mathcal{X}'$. It must be shown that $\mathcal{X}' \subseteq \mathcal{X}$. The original proof (see, for example, [495, 2.31]) uses general model theory, in particular the result, A.1.6, that two structures are elementarily equivalent iff they have isomorphic ultrapowers, together with pp-elimination of quantifiers for modules (A.1.1) and is immediate from these results.

For the functorial proof, first we quote 3.3.2, which shows (iv)⇒(ii), to see that $\mathcal{X}$ is also closed under direct limits. Then we use torsion theory in the functor category $(R\text{-mod}, \mathbf{Ab})$. So we replace each module M by the corresponding functor $M \otimes -$ and, via this embedding (see 12.1.3 and the paragraphs before the statement of the theorem above), regard $\mathcal{X}$ and $\mathcal{X}'$ as subcategories of $(R\text{-mod}, \mathbf{Ab})$.

By 10.2.31 every finitely presented functor in $(R\text{-mod}, \mathbf{Ab})$ is isomorphic to one of the form $F_{D\psi}/F_{D\phi}$ and since, 10.3.8, $(F_{D\psi}/F_{D\phi}, M \otimes -) \simeq \phi(M)/\psi(M)$, it follows that the set, T, of pp-pairs corresponds to the class, denote it $\mathcal{S}$, of finitely presented functors $F' \in (R\text{-mod}, \mathbf{Ab})$ such that $(F', M \otimes -) = 0$ for every $M \in \mathcal{X}$. Therefore $\mathcal{X}' = \{M' : (F', M \otimes -) = 0 \text{ for all } F' \in \mathcal{S}\}$ is the intersection of the image of Mod-R in $(R\text{-mod}, \mathbf{Ab})$ with the corresponding torsionfree class $\mathcal{F} = \{G \in (R\text{-mod}, \mathbf{Ab}) : (F', G) = 0 \text{ for all } F' \in \mathcal{S}\}$. Since this hereditary torsion theory (§11.1.1) has torsion class generated by finitely presented objects, it is, 11.1.14, of finite type and so $\mathcal{F}$ is, by 11.1.13, the closure of the image of $\mathcal{X}$ in $(R\text{-mod}, \mathbf{Ab})$ under products, direct limits, injective hulls and subobjects. In particular if $M' \in \mathcal{X}'$, then $M' \otimes -$ is in this closure of the image of $\mathcal{X}$.

Now, the functor $M \mapsto M \otimes -$ commutes with products and direct limits (12.1.3) so the image of $\mathcal{X}$ is already closed under these operations. Also, injective hulls in the functor category correspond to pure-injective hulls in Mod-R, 12.1.8, and by 4.3.21 (which does not use this part of 3.4.7) every definable category is already closed under pure-injective hulls, hence its image in the functor category is closed under injective hulls. It remains to note that under this functor, pure embeddings in Mod-R correspond to embeddings in the image of Mod-R (12.1.3), so if $M' \otimes -$ is a subfunctor of $M \otimes -$ with $M \in \mathcal{X}$, then M' is a pure submodule of M, hence already is in $\mathcal{X}$, as required. □

Note that closure under direct limits and reduced products is not enough: regarded as a full subcategory of the category of abelian groups, the category of $\overline{\mathbb{Z}_{(p)}}$-modules is closed under these operations but not under pure subgroups.

It will be shown in 4.3.21 that definable subcategories are also closed under pure-injective hulls. It will also follow, from 3.4.11 and 2.1.17, that the cokernel in Mod-R of any pure embedding between objects of a definable subclass $\mathcal{X}$ is also in $\mathcal{X}$. So "purity in $\mathcal{X}$ is the restriction of purity in Mod-R".[11] We record these points here.

Theorem 3.4.8. *Every definable subcategory of* Mod-R *is closed in* Mod-R *under pure-injective hulls and images of pure epimorphisms*

If $\mathcal{M}$ is a class of modules, then by the definable subcategory **generated** by $\mathcal{M}$ we mean the intersection of all definable subcategories containing $\mathcal{M}$, equivalently, by 3.4.7, the closure of $\mathcal{M}$ under direct products, direct limits and pure submodules, equivalently, again by 3.4.7, the definable subcategory determined by $T_{\mathcal{M}} = \{\phi/\psi : \phi/\psi$ is a pp-pair closed on every $M \in \mathcal{M}\}$. We extend the terminology from classes to single modules, M, in the obvious way. Write $\langle \mathcal{M} \rangle$, respectively $\langle M \rangle$, for the generated definable subcategory.

Note that if a class is already closed under direct products and direct limits, then its closure under pure submodules will be a definable subcategory, that is, it is not then necessary to apply the closure operations again. For, by 2.1.2 and 3.3.5, condition (iii) of 3.4.7 is satisfied.

Proposition 3.4.9. *If $\mathcal{X}_1, \dots, \mathcal{X}_n$ are definable subcategories of* Mod-R*, then the definable subcategory generated by $\mathcal{X}_1 \cup \dots \cup \mathcal{X}_n$ consists of all modules M which can be purely embedded into a module of the form $M_1 \oplus \dots \oplus M_n$ with $M_i \in \mathcal{X}_i$.*

Proof. Let $\mathcal{X}$ be the class of modules M as described. Certainly $\mathcal{X}$ contains each $\mathcal{X}_i$, and any definable subcategory $\mathcal{Y}$ which contains $\mathcal{X}_1 \cup \dots \cup \mathcal{X}_n$ must, by

[11] Though it will not be the case, cf. 2.1.4, that every pure exact sequence of $\mathcal{X}$ is a direct limit of split exact sequences in $\mathcal{X}$ unless the full subcategory $\mathcal{X}$ happens to be finitely accessible – see 16.1.15.

3.4.7, contain $\mathcal{X}$. We check the conditions of 3.4.7(iii) in order to show that $\mathcal{X}$ is definable.

By definition (and 2.1.2(2)) $\mathcal{X}$ is closed under pure submodules.

If, for $\lambda \in \Lambda$, M_λ is pure in $M_{1\lambda} \oplus \cdots \oplus M_{n\lambda}$, then $\prod_\lambda M_\lambda$ is (2.1.11) pure in $\prod_\lambda (M_{1\lambda} \oplus \cdots \oplus M_{n\lambda}) \simeq \prod_\lambda M_{1\lambda} \oplus \cdots \oplus \prod_\lambda M_{n\lambda}$. Each $\prod_\lambda M_{i\lambda}$ is in $\mathcal{X}_i$ by 3.4.7, so $\prod_\lambda M_\lambda$ is in $\mathcal{X}$.

As for closure under reduced products, and retaining the notation, let $\mathcal{F}$ be a filter on Λ. Then there is an embedding $\prod_\lambda M_\lambda/\mathcal{F} \to \prod_\lambda (M_{1\lambda} \oplus \cdots \oplus M_{n\lambda})/\mathcal{F}$ which, by 3.3.5, is pure and, since, as is easily checked, reduced product commutes with finite direct sum, the second term is isomorphic to $\prod_\lambda M_{1\lambda}/\mathcal{F} \oplus \cdots \oplus \prod_\lambda M_{n\lambda}/\mathcal{F}$. By 3.4.7 each $\prod_\lambda M_{i\lambda}/\mathcal{F}$ is in $\mathcal{X}_i$ so $\prod_\lambda M_\lambda/\mathcal{F} \in \mathcal{X}$, as required. □

For an example which shows that this description does not extend to definable subcategories generated by infinitely many definable subcategories, see 3.4.13.

Proposition 3.4.10. *If $\mathcal{X}$ is a definable subcategory of* Mod-R, *if T' is the set of pp-pairs which are closed on every module in $\mathcal{X}$ and if $T_1 = \{\phi/\psi : \phi/\psi \in T'$ and ϕ, ψ have just one free variable$\}$, then each of T' and T_1 defines the subcategory $\mathcal{X}$.*

Proof. That T' defines $\mathcal{X}$ is immediate.

To show that pp conditions in one free variable suffice, first note that a pair ϕ/ψ, where ϕ, ψ have n free variables, determines a subquotient, F_ϕ/F_ψ, of the nth power, $(R, -)^n$, of the forgetful functor. As discussed before and within the proof of 3.4.7, the duals of these functors, as ϕ/ψ ranges over T', determine a torsion theory of finite type on (R-mod, **Ab**) such that $\mathcal{X}$ is the intersection of the image of Mod-R in (R-mod, **Ab**) with the corresponding torsionfree class. But to determine the torsion class it is, 11.1.39, enough to give those finitely generated subquotients of $(R, -)$ which are torsion, that is, by 10.2.30, those subquotients of $(R, -)$ of the form F_ϕ/F_ψ which are torsion. That is, it is enough to give the pp-pairs with one free variable which are closed on $\mathcal{X}$. □

For a non-functorial proof of the above result one may use, say, [495, 2.30] (cf. 2.1.6). The fact that pp conditions with a single variable suffice (for most things) is built right into the foundations of the model-theoretic approach, in that the "invariants statements" which arise from the fundamental pp-elimination of quantifiers theorem, A.1.1 (for the original references and a proof see, for example, [495, 2.13]), are defined in terms of such conditions.

Corollary 3.4.11. *A module M' belongs to the definable subcategory generated by the module M iff every pp-pair closed on M is closed on M'.*

In particular $\langle M \rangle =$ Mod(T), *where $T = \{\phi/\psi : \phi/\psi$ is closed on $M\}$.*

Lemma 3.4.12. *If $\mathcal{X}$ is a definable subcategory of* Mod-R*, then there is a module M such that $\mathcal{X}$ is the definable subcategory generated by M.*

Proof. Take M to be a direct sum of sufficiently many members of $\mathcal{X}$ that, if a pp-pair is open on some member of $\mathcal{X}$, then it is open on M. □

Example 3.4.13. We compute the definable subcategory, $\mathcal{X}$, generated by the set $\{\mathbb{Z}_p : p \text{ prime}\}$ of simple abelian groups. In view of the fact, 5.1.4 below, that every definable subcategory is determined by the indecomposable pure-injective modules in it, it is these which we compute; then, if $M \in \mathcal{X}$, the pure-injective hull of M will, by 7.3.6 and 5.2.6, be the pure-injective hull of a direct sum of copies of those indecomposables: we count that as one kind of reasonable description of $\mathcal{X}$.

Set $M = \bigoplus_p \mathbb{Z}_p$. Noting that $Mp = Mp^2$ for every prime p, we have, by 3.4.11, that the pp-pair $(p \mid x)/(p^2 \mid x)$ is closed on every member of $\mathcal{X}$, so the module, $\overline{\mathbb{Z}_{(p)}}$, of p-adic integers does not lie in $\mathcal{X}$. Dually the pp-pair $(xp^2 = 0)/(xp = 0)$ is closed on M, hence on every member of $\mathcal{X}$, hence the p-Prüfer group, $\mathbb{Z}_{p^\infty}$, does not belong to $\mathcal{X}$. These observations also allow us to exclude all $\mathbb{Z}_{p^n}$ with $n > 1$ from $\mathcal{X}$. Since, by 5.2.3, the simple modules $\mathbb{Z}_p$ are isolated points of the Ziegler spectrum, $\mathrm{Zg}_{\mathbb{Z}}$, of $\mathbb{Z}$ and since, by 5.1.23, this is a compact space, there must be at least one more point of the Ziegler spectrum in the closure of the set of these simple modules, that is, by 5.1.6, there is at least one more indecomposable pure-injective in the definable category, $\mathcal{X}$, generated by these. The only possibility is the torsionfree divisible module $\mathbb{Q}$.

What we have actually done, in terms of the concepts introduced in Chapter 5, is to describe the closed subset of the Ziegler spectrum which corresponds in the sense of 5.1.6 to the definable category $\mathcal{X}$.

In this example we can say a little more. Since, 4.3.22 (which is a continuation of this example), the pure-injective hull of M is the direct product, $H(M) = \prod_p \mathbb{Z}_p$, the modules in $\mathcal{X}$ are precisely those of the form $M' \oplus \mathbb{Q}^{(\kappa)}$, where κ is any cardinal number and where M' purely embeds in a product of copies of the simple abelian groups $\mathbb{Z}_p$.

The same analysis applies to the definable subcategory generated by all the simple modules, over any commutative Dedekind domain, in particular over the ring, $k[T]$, of polynomials in one variable over a field k (cf. 5.2.1).

A similar analysis could be applied to the modules at the mouths of generalised tubes (see Section 15.1.3), in particular (see Section 8.1.2) to the quasi-simple modules over a tame hereditary algebra (for which an argument using representation embeddings (Section 5.5.2) from $k[T]$-modules, instead of a direct argument, could be used).

If M' belongs to the definable subcategory generated by M, then we also write $\text{supp}(M') \subseteq \text{supp}(M)$. The notation refers to the support, $\text{supp}(M)$, of M, which is, see a little after 5.1.6, the closed subset of the Ziegler spectrum consisting of the indecomposable pure-injectives in $\langle M \rangle$.

We extend the notation introduced in Section 3.2.1 by writing $\text{pp}(\mathcal{X})$ for $\text{pp}(M)$ and $[\phi, \psi]_{\mathcal{X}}$ for $[\phi, \psi]_M$, where M is any module generating $\mathcal{X}$. Also, extending the notation in Section 3.2.2, write $(\mathbb{L}_R^{\text{eq}+})_{\mathcal{X}}$ or $\mathbb{L}^{\text{eq}+}(\mathcal{X})$ for $(\mathbb{L}_R^{\text{eq}+})_M$, where M is any module which generates $\mathcal{X}$. Note that if M' is in the definable subcategory generated by M, then there is a natural surjective map of lattices $\text{pp}(M) \to \text{pp}(M')$ and an exact surjective functor of small abelian categories $(\mathbb{L}_R^{\text{eq}+})_M \longrightarrow (\mathbb{L}_R^{\text{eq}+})_{M'}$ (cf. 3.2.2, 3.2.15).

Lemma 3.4.14. *Let $f : R \to S$ be a morphism of rings with corresponding restriction of scalars functor $f^* :$ Mod-$S \to$ Mod-R and let $\mathcal{Y}$ be a definable subcategory of* Mod-S. *Then the closure of $f^*\mathcal{Y}$ under pure submodules is a definable subcategory of* Mod-R.

Since restriction of scalars commutes with products and direct limits, this follows by 3.4.7. Closure under pure submodules is necessary: take S to be the path algebra (Section 15.1.1) over a field k of the quiver $A_2 = (\bullet \to \bullet)$, $\mathcal{Y}$ to be the definable subcategory generated by $k \to k$ (so, by 4.4.30, all objects in $\mathcal{Y}$ are even-dimensional vector spaces) and f to be the inclusion of $R = k$ in S.

Proposition 3.4.15. *Suppose that $\mathcal{X}$ is a definable subcategory of* Mod-R. *Let $\mathcal{X}_\forall$ be the class of submodules of modules in $\mathcal{X}$. Then $\mathcal{X}_\forall$ is a definable subcategory.*

Proof. Certainly $\mathcal{X}_\forall$ is closed under pure subobjects and closure under products is also immediate. By 3.4.7 it will be enough to check closure under reduced products, so suppose $A_i \leq M_i \in \mathcal{X}$ for $i \in I$ and let $\mathcal{F}$ be a filter on I. The proof of 3.3.5 shows that a reduced product of embeddings is an embedding, so we have $\prod_i A_i/\mathcal{F} \leq \prod_i M_i/\mathcal{F} \in \mathcal{X}$, as required. □

3.4.2 Duality and definable subcategories

Duality gives a bijection between definable subcategories of right and left modules which reverses implication of pp conditions (3.4.18). In general, definable subcategories are not closed under inverse limit.

If $\mathcal{X} = \text{Mod}(T)$ is a definable subcategory of Mod-R, then its **dual** $D\mathcal{X}$ is the definable subcategory of R-Mod defined by the set of pp-pairs $DT = \{D\psi/D\phi : \phi/\psi \in T\}$ (for D see Section 1.3.1). This is independent of choice of T, given $\mathcal{X}$.

Lemma 3.4.16. *If T and T' are sets of pp-pairs such that* $\mathrm{Mod}(T) = \mathrm{Mod}(T')$, *then* $\mathrm{Mod}(DT) = \mathrm{Mod}(DT')$. *In particular, if* $\mathcal{X} = \mathrm{Mod}(T)$, *then* $D\mathcal{X} = \{L \in R\text{-Mod} : D\psi(L) = D\phi(L)$ *for every pp-pair* ϕ/ψ *such that* $\phi(M) = \psi(M)$ *for every* $M \in \mathcal{X}\}$.

Proof. By 3.4.12, $\mathcal{X} = \mathrm{Mod}(T) = \mathrm{Mod}(T')$ is the definable subcategory generated by some module M. Without loss of generality T consists of all pairs closed on M. By 1.3.15, it follows that DT equals the set of pairs closed on M^* (any suitable dual in the sense of 1.3.15) and hence $D\mathcal{X} = \langle M^* \rangle$ (by 3.4.11).

If $L \in D\mathcal{X}$, then, since $DT' \subseteq DT$, every pair in DT' is closed on L. Conversely, if every pair in DT' is closed on L, then, by 1.3.15, every pair in $D^2T' = T'$ (by 1.3.1) is closed on L^* (a suitable dual), so, by assumption, every pair in T is closed on L^*, hence every pair in DT is closed on L, so $L \in D\mathcal{X}$, as required. □

Corollary 3.4.17. (of the proof) *For any module M the definable subcategory of* R-Mod *generated by the dual,* $M^* = \mathrm{Hom}_{\mathbb{Z}}(M, \mathbb{Q}/\mathbb{Z})$, *of M is the dual of the definable subcategory of* Mod-R *generated by* M: $\langle M^* \rangle = D(\langle M \rangle)$.

Corollary 3.4.18. *If $\mathcal{X}$ is any definable subcategory of* Mod-R, *then* $D^2\mathcal{X} = \mathcal{X}$ *and, for every* n, $\mathrm{pp}^n(\mathcal{X}) \simeq (\mathrm{pp}^n(D\mathcal{X}))^{\mathrm{op}}$.

Example 3.4.19. If R is a domain, then, as observed already (3.4.1), the class of torsionfree modules is definable, by pairs of the form $(xr = 0)/(x = 0)$. The dual of such a pair is $(x = x)/(r|x)$. Thus the class of torsionfree right modules is dual to the class of divisible left modules. The class of torsionfree divisible right modules is, therefore, dual to the class of torsionfree divisible left modules.

Corollary 3.4.20. *Given a definable subcategory,* $\mathcal{X}$, *of* Mod-R, *a right R-module* $M \in \mathcal{X}$, *a ring morphism,* $S \to \mathrm{End}(M_R)$, *and any injective left S-module E, the "dual" of M with respect to E, that is,* ${}_R(\mathrm{Hom}_S({}_SM_R, {}_SE))$, *is a member of* $D\mathcal{X}$.

Proof. Apply 1.3.13. □

Corollary 3.4.21. *For any module M and duality as in Section 1.3.3 the double dual M^{**} belongs (by 1.3.13) to the definable subcategory generated by M:* $M^{**} \in \langle M \rangle$. *If the duality satisfies the condition of 1.3.15, then* $\langle M^{**} \rangle = \langle M \rangle$.

Example 3.4.22. If $R \to R'$ is an epimorphism of rings, then the dual of Mod-R', considered as a (definable, by 5.5.4) subcategory of Mod-R, is R'-Mod $\subseteq R$-Mod. There are various ways to prove this. For example, using 3.4.17, if M is any module in Mod-$R' \subseteq$ Mod-R, then M^* as in 3.4.17 carries both a left R-module and left R'-module structure, the latter extending the former. So we do have $D(\text{Mod-}R') \subseteq R'$-Mod. Then equality follows by symmetry and since $D^2\mathcal{X} = \mathcal{X}$ for every definable subcategory $\mathcal{X}$.

Suppose that $(M_\lambda)_\lambda$ is an inversely directed system of modules, all belonging to the definable subcategory $\mathcal{X}$, and set $M = \varprojlim M_\lambda$. In general, M need not be in $\mathcal{X}$ (see 3.4.23 below) but, under some assumptions, it will be.

Choose a suitable dualising module, as in 1.3.15, for the modules in $\mathcal{X}$. The inverse system dualises to a direct system, $(M_\lambda^*)_\lambda$, of left R-modules which, by 3.4.20, are in $D\mathcal{X}$, so, by 3.4.7, their direct limit, L say, is in $D\mathcal{X}$. Choose a dualising module for $D\mathcal{X}$. Just from the definition of direct limit $L^* = \varprojlim M_\lambda^{**}$. The canonical pure embeddings (1.3.16) $M_\lambda \to M_\lambda^{**}$ (natural in the M_λ) induce a morphism $f : M = \varprojlim M_\lambda \to \varprojlim M_\lambda^{**} = L^*$, so, since, by 3.4.20, $L^* \in \mathcal{X}$, it remains to see whether f is a pure embedding.

Suppose that $\overline{a}$ from M is such that $f\overline{a} \in \phi(L^*)$. Then, with $\pi'_\lambda : L^* \to M_\lambda^{**}$ denoting the canonical map, $\pi'_\lambda f\overline{a} \in \phi(M_\lambda^{**})$ and so, by purity of $M_\lambda \to M_\lambda^{**}$, we have $\pi_\lambda\overline{a} \in \phi(M_\lambda)$, where $\pi_\lambda : M \to M_\lambda$ denotes the canonical map. If, for some reason, we know that $\phi(\varprojlim M_\lambda) = \varprojlim \phi(M_\lambda)$, then we can finish, deducing purity of f and hence, by 3.4.7, that $M = \varprojlim M_\lambda$ is in $\mathcal{X}$. This will not, however, hold in general.

A context where f is pure, indeed is an isomorphism, is one where the modules M_λ are reflexive ($M_\lambda^{**} = M_\lambda$), for instance if they are finite length modules over an artin algebra and the duality is the natural one (see 5.4.4). A specific example is the closure of a coray of epimorphisms in a generalised tube (see 15.1.10, Section 8.1.2). More generally, modules of finite endolength (Section 4.4.3) are reflexive (5.4.17). Also see the remarks at the end of Section 4.2.3.

Example 3.4.23. Let $R = k[a, b : a^2 = 0 = b^2 = ab]$ and set M_n $(n \geq 1)$ to be the k-vectorspace with basis u, v_i $(i \geq 1)$, v'_i $(1 \leq i \leq n-1)$ with $ua = 0 = ub$, $v_ia = u$ $(i \geq 1)$, $v_ib = 0$ $(i \geq n)$, $v_ib = v'_i$ $(i \leq n-1)$, $v'_ia = 0 = v'_ib$ $(1 \leq i \leq n-1)$. There are natural epimorphisms $\cdots \to M_n \to \cdots \to M_2 \to M_1$ between these modules ($M_n \to M_{n-1}$ is the natural map from M_n to $M_n/v'_{n-1}R$). Let M denote the inverse limit of this system. Note that in each M_n there is the inclusion $\mathrm{im}(a) \leq \mathrm{ann}(b) \cdot a$, whereas, in M, $\mathrm{im}(a)$ is one-dimensional but $\mathrm{ann}(b) \cdot a \leq \mathrm{ann}(a) \cdot a = 0$. So there is a definable subcategory of Mod-R containing all the M_n but not containing their inverse limit.

3.4.3 Further examples of definable subcategories

A ring is right coherent iff the absolutely pure right modules are definable, iff the flat left modules are definable (3.4.24). It is right noetherian iff the injective right modules are definable and is right coherent and left perfect iff the projective left modules are definable (3.4.28). Over von Neumann regular rings the definable subcategories are in bijection with the two-sided ideals of the ring (3.4.29). Over commutative rings a module generates the same definable subcategory as all its localisations together (3.4.31).

Theorem 3.4.24. ([174, 3.16, 3.23], [631, Thm 4], [274, 9.3]) *The ring R is right coherent iff the class of absolutely pure right R-modules is definable iff the class of flat left R-modules is definable. If these conditions are satisfied, then the classes of absolutely pure right modules and flat left modules are dual to each other.*

Proof. If R is right coherent and ψ is any pp condition for left R-modules, then, by 2.3.19, $\psi({}_R R)$ is finitely generated as a right R-module. In that case each requirement $\overline{x} \cdot D\phi({}_R R) = 0$, where ϕ is a pp condition for right R-modules, is a pp condition, so the criterion, 2.3.3, for absolute purity is "definable": a module is absolutely pure iff for every ϕ the pp-pair $(\overline{x} \cdot D\phi({}_R R) = 0)/\phi(\overline{x})$ is closed in that module.

Furthermore, the criterion, 2.3.9, for flatness, namely that $\psi({}_R L) = \psi({}_R R) \cdot L$ for every pp condition ψ for left R-modules, also becomes definable since $\psi({}_R R) \mid x$ is a pp condition whenever $\psi({}_R R)$ is finitely generated as a right ideal (Section 1.1.4).

The converse for flatness is immediate from 2.3.21 and 3.4.7, while that for absolute purity follows from any of a variety of proofs, one which uses reduced products (Section 3.3.1, Appendix E, p. 715).

If R is not right coherent there is, by 2.3.19, a pp condition ϕ for right modules such that $I = D\phi({}_R R)$ is not finitely generated. Let Λ denote the set of finitely generated right ideals, I_λ, contained in I. For each $\lambda \in \Lambda$ let E_λ be the injective hull (Appendix E, p. 708) of R/I_λ and let $a_\lambda \in E_\lambda$ be an element with annihilator exactly I_λ. For each λ set $S_\lambda = \{\mu : I_\mu \geq I_\lambda\}$. This collection $\{S_\lambda\}_\lambda$ has the finite intersection property since $S_\lambda \cap S_\mu \supseteq \{\nu : I_\nu \geq I_\lambda + I_\mu\}$, so let $\mathcal{F}$ be the filter on Λ that it generates. Form the reduced product $E^* = \prod_\lambda E_\lambda/\mathcal{F}$ and consider the element $a = (a_\lambda)_\lambda/\mathcal{F}$. For each $\lambda \in \Lambda$, because $S_\lambda \in \mathcal{F}$, it follows, by 3.3.3, that $aI_\lambda = 0$. Therefore $a \cdot D\phi({}_R R) = 0$. Each E_λ is absolutely pure so, if the absolutely pure modules formed a definable subcategory, then, by 3.4.7, E^* would be absolutely pure and so, by 2.3.3, we would have $a \in \phi(E^*)$. But then, by 3.3.3, there would be (many) λ such that $a_\lambda \in \phi(E_\lambda)$, hence such that $a_\lambda \cdot D\phi({}_R R) = 0$ – which is a contradiction since $\mathrm{ann}_R(a_\lambda) = I_\lambda$.

For the last statement it is enough to observe that the class of absolutely pure modules is defined by the pairs of the form $\big(\overline{x} \cdot D\phi({}_R R) = 0\big) \,/\, \phi(\overline{x})$ and the class of flat modules by those of the form $D\phi(\overline{x}) \,/\, (D\phi({}_R R))|\overline{x}$ (every left pp condition, $\psi(\overline{x})$, has, by 1.3.1, the form $D\phi(\overline{x})$ for some right pp condition ϕ). Clearly such pairs are dual. □

For another view of the statement concerning flat modules see 3.4.41.

Corollary 3.4.25. *The ring R is right coherent iff the class of absolutely pure right modules is closed under direct limits.*

The last part of 3.4.24 is, in a modified form, true over any ring R.

Proposition 3.4.26. ([274, 12.2]) *The dual of the definable subcategory, $\langle R \rangle$, generated by R_R is the definable subcategory generated by the set of injective modules, equivalently, is the definable subcategory, $\langle E \rangle$, where E is an injective cogenerator for R*-Mod.

Proof. [527, 4.1] Suppose that the pp-pair ϕ/ψ is closed on R_R and let E be any injective left R-module. By 2.3.3, $D\psi(E) = \mathrm{ann}_E(\psi(R_R)) = \mathrm{ann}_E(\phi(R_R)) = D\phi(E)$. Thus, $D\psi/D\phi$ is closed on E. If, conversely, $\phi(R) > \psi(R)$, then the quotient, $\phi(R_R)/\psi(R_R)$ is a non-zero left R-module. Let E' be the injective hull of ${}_RR/\psi(R_R)$ and set $a = 1 + \psi(R_R) \in E'$. Then $a \in \mathrm{ann}_{E'}(\psi(R_R)) \setminus \mathrm{ann}_{E'}(\phi(R_R))$ so, again by 2.3.3, $D\psi(E') > D\phi(E')$ which (by 3.4.11) is enough. □

This example is generalised by 10.4.2.

Now we consider the stronger notions of injectivity and projectivity. A ring R is said to be **right perfect** if it satisfies the conditions of the following theorem.

Theorem 3.4.27. ([46, Thm P], also see, for example, [187, 22.29, 22.31A], [663, VIII.5.1]) *The following conditions on a ring R are equivalent:*

(i) every right R-module M has a **projective cover** *(a surjection $\pi : P \to M$ with P projective such that* $\ker(\pi)$ *is an* **inessential** *submodule of P, meaning that if $N \leq P$ and $N + \ker(\pi) = P$ then already $N = P$);*

(ii) $R/J(R)$ is semisimple artinian and $J(R)$ is **right T-nilpotent** *(given any sequence $r_0, r_1, \dots, r_n, \dots$ of elements of $J(R)$ there is n such that $r_n \dots r_1 r_0 = 0$);*

(iii) every flat right R-module is projective;

(iv) R has the descending chain condition on finitely generated left ideals.

Theorem 3.4.28. ([174, 3.19], [631, Thm 5])

(a) A ring R is right noetherian iff the class of injective right R-modules is definable.

(b) A ring R is right coherent and left perfect iff the class of projective left R-modules is definable.

Proof. (a) By 4.4.17 a ring is right noetherian iff every direct sum of injective modules is injective iff every absolutely pure right module is injective. Since right noetherian implies right coherent, 3.4.24 gives one direction and, since every definable subcategory is closed under direct sum, the converse follows.

(b) From the conditions on the ring the condition on the modules is immediate by 3.4.24 and 3.4.27. If, on the other hand, the projective left modules form a definable class, then, by 3.4.7, every power of ${}_RR$ is projective, so, by 2.3.21, R is right coherent. Also, every flat left module, being a direct limit of projective modules, must be projective (again by 3.4.7), so, by 3.4.27, R is left perfect. □

In fact, the injectives form a definable class exactly when they all are $(\Sigma-)$ pure-injective, see 4.4.17. For projective modules the implication goes only one way, see 4.4.22 and 4.4.23. Eklof and Sabbagh also characterised the rings such that the class of free modules is definable, [631, Thm 6].

Now suppose that R is a von Neumann regular ring. Since, 2.3.22, every embedding between R-modules is pure, a subcategory, $\mathcal{X}$, of Mod-R is definable iff it is closed under products, direct limits and submodules. By 3.4.8 and 4.3.12 any such subcategory is also closed under injective hulls. Therefore the definable subcategories of Mod-R are exactly the torsionfree classes for hereditary torsion theories of finite type (see 11.1.13). Since every finitely generated right ideal of R is generated by an idempotent any such class is, by 11.1.14, specified by the filter of idempotents e such that $(R/eR, \mathcal{X}) = 0$, that is, by those idempotents $f = 1 - e$ such that $(fR, \mathcal{X}) = 0$. The two-sided ideal generated by the latter set of idempotents is exactly the relevant torsion submodule of R and, note, is the ideal $\mathrm{ann}_R(\mathcal{X}) = \{r \in R : Mr = 0\ \forall M \in \mathcal{X}\}$. Conversely, if I is a two-sided ideal of R, then $\{M \in \text{Mod-}R : (fR, M) = 0 \text{ for all } f \in I\}$ is a definable subcategory. These maps between definable subcategories and ideals are easily seen to be inverse and to yield the following (for example, see [495, 16.19], [686, 2.1(ii)], [278, 4.6], [282, p. 255]).

Proposition 3.4.29. *Suppose that R is von Neumann regular. There are bijections between:*

(i) definable subcategories of Mod-R*;*
(ii) torsionfree classes for finite type hereditary torsion theories on Mod-R*;*
(iii) two-sided ideals of R.

The next result easily follows.

Proposition 3.4.30. *Suppose that R is von Neumann regular and let $\mathcal{X}$ be a definable subcategory of* Mod-R*. If $I = \mathrm{ann}_R(\mathcal{X})$, then $\mathcal{X} =$* Mod-(R/I)*, the latter being regarded as a full subcategory of* Mod-R *via $R \mapsto R/I$.*

If M is any R-module, then $\langle M \rangle =$ Mod-$(R/\mathrm{ann}_R(M))$.

Now suppose that R is a commutative ring and let $r \in R$ be a non-zero-divisor. If M is any R-module, then $M \otimes_R R[r^{-1}]$ is the direct limit of the diagram

$M \xrightarrow{-\times r} M \xrightarrow{-\times r} \dots$ so this and more general localisations of M are in the definable subcategory generated by M. Garavaglia proved that, together, the localisations of M at maximal ideals of R generate the same definable subcategory as M.

Theorem 3.4.31. ([209, Thm 3]) *If R is commutative and M is any module, then M and $\bigoplus_P M_{(P)}$, where P runs over*[12] maxspec(R), *generate the same definable subcategory of* Mod-R.

The proof uses the fact that localisation "commutes with" pp conditions. Recall (1.1.8) that if R is commutative, then every pp-definable subgroup is actually a submodule.

Lemma 3.4.32.[13] ([207, Lemma 4], [726, 5.5]) *Let R be a commutative ring and let S be a multiplicatively closed subset of R. If M is a module, then denote by $M_S = M \otimes R[S^{-1}]$ the localisation of M at S. Let ϕ be a pp condition. Then for any module M we have $\phi(M_S) = (\phi(M))_S$.*

This is used to show that the canonical map $M \to \prod_{P \in \mathrm{maxspec}(R)} M_{(P)}$ is a pure embedding, indeed these two modules generate the same definable category (in fact they are elementarily equivalent). For proofs see the original references or [495, pp. 49,50]. Similar results can be proved in other contexts, for example ([431, p. 191]) over integral group rings.

Finally, we mention that, as noted after 5.5.5, every category of the form Mod-S, where S is a finitely presented k-algebra (k a field), is a *full* definable subcategory of some module category Mod-R, where R is a finite-dimensional k-algebra.

3.4.4 Covariantly finite subcategories

Covariant finiteness is related to pp conditions (3.4.33), factorisation of morphisms (3.4.36) and definable subcategories.

In this section we describe one source of definable subcategories.

A subcategory, $\mathcal{D}$, of an additive category, $\mathcal{C}$, is said to be **covariantly finite (in $\mathcal{C}$)** if every $C \in \mathcal{C}$ has a **left $\mathcal{D}$-approximation,**[14] meaning a morphism $f : C \to D \in \mathcal{D}$ such that for every $f' : C \to D' \in \mathcal{D}$ there is $h : D \to D'$ with $hf = f'$. The morphism h is not required to be unique, so this is weaker than representability of the functor $(C, -) \restriction \mathcal{D}$, in particular, it is weaker than $\mathcal{D}$ being

[12] The set maxspec(R) of maximal ideals may be replaced by the set Spec(R) of all prime ideals, or any set in between.

[13] For results relating elementary equivalence and localisation see, for instance, [323, p. 112ff.].

[14] A weakly universal arrow from C to the inclusion $\mathcal{D} \longrightarrow \mathcal{C}$, in the terminology of [418, p. 231].

a reflective subcategory. There is the obvious dual notion of **contravariantly finite** subcategory. Also see covers and envelopes in Section 4.6.

Lemma 3.4.33. *A subcategory* $\mathcal{D} \subseteq$ mod-R *is covariantly finite in* mod-R *iff for every pp condition* ϕ *the set* $\{\psi \text{ pp} : \phi \geq \psi$ *and* ψ *has a free realisation in* $\mathcal{D}\}$ *has a maximal element,* $\phi_{\mathcal{D}}$ *say.*

Proof. ($\Rightarrow$) Given ϕ let $(C_\phi, \overline{c}_\phi)$ be a free realisation and let $f : C_\phi \to D \in \mathcal{D}$ be a left $\mathcal{D}$-approximation. Set $\overline{d} = f\overline{c}_\phi$ and let ψ_0 be a pp condition which generates $\mathrm{pp}^D(\overline{d})$ (1.2.6); so $\phi \geq \psi_0$. If $(D', \overline{d}')$ with $D' \in \mathcal{D}$ is a free realisation of some pp condition $\psi \leq \phi$, then, since there is (by 1.2.9) $f' : C_\phi \to D'$ taking $\overline{c}_\phi$ to $\overline{d}'$ and hence there is $h : D \to D'$ with $hf = f'$ so with $h\overline{d} = \overline{d}'$, we deduce (by 1.2.21) that $\psi_0 \geq \psi$. Thus ψ_0 is the required maximal pp condition.

($\Leftarrow$) Given $C \in$ mod-R finitely presented, let $\overline{c}$ be a generating sequence for C and take a generator, η say, for $\mathrm{pp}^C(\overline{c})$. Let $(D, \overline{d})$, with $D \in \mathcal{D}$, be a free realisation of $\eta_{\mathcal{D}}$, the existence of which we assume. Since $\eta \geq \eta_{\mathcal{D}}$ there is $f : C \to D$ taking $\overline{c}$ to $\overline{d}$. Given any morphism $f' : C \to D' \in \mathcal{D}$, take a generator, ψ say, of $\mathrm{pp}^{D'}(f'\overline{c})$. Since $\eta \geq \psi$ (1.1.7), also $\eta_{\mathcal{D}} \geq \psi$, so, by 1.2.17, there is $h : D \to D'$ taking $\overline{d}$ to $f'\overline{c}$. Therefore $hf\overline{c} = f'\overline{c}$ so, since $\overline{c}$ generates C, $hf = f'$, as required. □

Suppose that $\mathcal{D} \subseteq$ Mod-R is closed under finite direct sums. Set $\overrightarrow{\mathcal{D}}$ to be the closure of $\mathcal{D}$ in Mod-R under direct limits and set

$$\mathcal{D}^{\text{fact}} = \{C \in \text{Mod-}R : \text{ for all } f : A \to C \text{ with } A \in \text{mod-}R \text{ there is a } \dots$$
$$\dots \text{ factorisation of } f \text{ through some member of } \mathcal{D}\}.$$

If $A \in$ mod-R and $C \in \overrightarrow{\mathcal{D}}$, say $C = \varinjlim D_\lambda$ with the $D_\lambda \in \mathcal{D}$, then any morphism $A \to C$ factors through some D_λ so $\overrightarrow{\mathcal{D}} \subseteq \mathcal{D}^{\text{fact}}$. One has the following converse. This result (for modules) is due to Lenzing. A particular case appears in [71, §§4,5] and there is a treatment of the general case in [144, 4.1]. In the general case Mod-R is replaced by any finitely accessible category and mod-R by its full subcategory of finitely presented objects.

Proposition 3.4.34. ([395, 2.1]) *If* $\mathcal{D} \subseteq$ mod-R *is closed under finite direct sums, then* $\overrightarrow{\mathcal{D}} = \mathcal{D}^{\text{fact}}$. *For every object* $M \in \overrightarrow{\mathcal{D}}$ *there is a pure epimorphism* $\bigoplus D_i \to M$ *with the* D_i *in* $\mathcal{D}$.

Proof. Let $M \in \mathcal{D}^{\text{fact}}$ and let $\mathcal{I}$ be the (directed by direct sum) system of all morphisms $A \to M$ with A in (some small version of) mod-R. Then $M = \varinjlim \mathcal{I}$. Let $\mathcal{J}$ be the (directed by direct sum) system of all morphisms $D \to M$ with $D \in \mathcal{D}$ (rather, take a set of these: enough, using that $M \in \mathcal{D}^{\text{fact}}$, for the proof to work). By assumption $\mathcal{J}$ is cofinal (Appendix E, p. 714) in $\mathcal{I}$. So $M = \varinjlim \mathcal{J} \in \overrightarrow{\mathcal{D}}$, as required.

Given $M \in \overrightarrow{\mathcal{D}}$ consider $P = \bigoplus D^{(D,M)} \to M$, where D runs over an isomorphism-representative set of objects of $\mathcal{D}$. This (cf. proof of 2.1.25) is a pure epimorphism. For, take a pp condition ϕ and $\overline{a} \in \phi(M)$. Let $(C, \overline{c})$ be a free realisation of ϕ. Then, 1.2.17, there is a morphism $C \to M$ taking $\overline{c}$ to $\overline{a}$ which, by the first part, factors through $P \to M$. Therefore, by 2.1.14, this is a pure epimorphism. □

Lemma 3.4.35. ([395, 2.2]) *If $\mathcal{D}$ is a subcategory of* mod-R *closed under finite direct sums, then $\mathcal{D}^{\text{fact}}$ is closed under pure submodules.*

Proof. Let $M \in \mathcal{D}^{\text{fact}}$ and let $M' \to M$ be a pure embedding: we show $M' \in \mathcal{D}^{\text{fact}}$. So take $f : A \to M'$ with A finitely presented. The composition with $M' \to M$ factors, by the proof of 3.4.34, through $\bigoplus_i D_i \to M$ as in the statement of that result. Since A is finitely presented it follows that there is a commutative square as shown for some $D \in \mathcal{D}$.

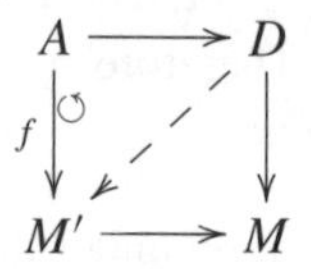

By 2.1.7 there is a factorisation as shown, as required. □

Say that $\mathcal{D} \subseteq \mathcal{C}$ is **covariantly finite with respect to** $\mathcal{C}' \subseteq \mathcal{C}$ if for every $C' \in \mathcal{C}'$ there is $f : C' \to D \in \mathcal{D}$ such that for every $f' : C' \to D' \in \mathcal{D}$ there is $h : D \to D'$ with $hf = f'$.

Proposition 3.4.36. *Let $\mathcal{D}$ be a subcategory of* Mod-R. *Then $\mathcal{D}^{\text{fact}}$ is closed under products if $\mathcal{D}$ is covariantly finite with respect to* mod-R. *The converse also holds if $\mathcal{D}$ is skeletally small or is closed under pure submodules.*

Proof. ($\Leftarrow$) Let $M = \prod M_\lambda$ with $M_\lambda \in \mathcal{D}^{\text{fact}}$ and take $f : A \to M$ with $A \in$ mod-R. Set $f_\lambda : A \to M_\lambda$ to be the composition of f with projection to M_λ. Since $M_\lambda \in \mathcal{D}^{\text{fact}}$ there is a factorisation $f_\lambda = h_\lambda g_\lambda$ for some $g_\lambda : A \to D_\lambda \in \mathcal{D}$ and $h_\lambda : D_\lambda \to M_\lambda$. Putting these together, we have $f = \prod f_\lambda = (\prod h_\lambda)(\prod g_\lambda)$. Let $g : A \to D \in \mathcal{D}$ be a left $\mathcal{D}$-approximation of A. Then, for each λ, there is $k_\lambda : D \to D_\lambda$ with $g_\lambda = k_\lambda g$, hence $\prod g_\lambda = (\prod k_\lambda)g$. So $f = (\prod h_\lambda)(\prod k_\lambda)g$ factors through g. Thus $\prod M_\lambda \in \mathcal{D}^{\text{fact}}$.

($\Rightarrow$) Assume that for every $A \in$ mod-R there is a *set* of morphisms $(f_\lambda : A \to D_\lambda \in \mathcal{D})_\lambda$ which is **jointly initial** for morphisms from A to $\mathcal{D}$ in the sense that every morphism from A to $\mathcal{D}$ factors through at least one of these. By assumption $\prod D_\lambda \in \mathcal{D}^{\text{fact}}$, so $\prod f_\lambda : A \to \prod D_\lambda$ factors through some $f' : A \to D' \in \mathcal{D}$ which is, therefore, a left $\mathcal{D}$-approximation of A (every morphism from A to $\mathcal{D}$ factors through some f_λ and hence through f').

As for conditions under which such a jointly initial set exists: this is clear if $\mathcal{D}$ is skeletally small, so suppose that $\mathcal{D}$ is closed under pure submodules. Let $\kappa = \max\{\text{card}(R), \aleph_0\}$ and consider the set of all morphisms from A to the various $\leq\kappa$-generated members of $\mathcal{D}$. Any morphism from A to a member, D, of $\mathcal{D}$ has image contained in some submodule C of D which is $\leq \kappa$-generated and which, moreover, by 2.1.21, is contained in a pure, still $\leq \kappa$-generated, submodule of D. By assumption such a submodule is also in $\mathcal{D}$, as required. □

The next result is [144, 4.1+4.2], also [355, 4.13] (and [358, Chpt. 3]), stated for modules and is immediate by 3.4.7 together with the results above.

Corollary 3.4.37. *If $\mathcal{D} \subseteq$ mod-R is closed under finite direct sums, then $\mathcal{D}^{\text{fact}} = \overrightarrow{\mathcal{D}}$ is a definable subcategory of* Mod-R *iff $\mathcal{D}$ is covariantly finite in* mod-R.

Proposition 3.4.38. *If $\mathcal{D} \subseteq$ mod-R is closed under finite direct sums and is covariantly finite in* mod-R*, then $\mathcal{D}^{\text{fact}} \subseteq$* Mod-$R$ *is defined by the set of pairs of the form $(\phi/\phi_{\mathcal{D}})$ as ϕ ranges over pp conditions.*

Proof. Let T be the set $\{\phi/\phi_{\mathcal{D}}\}_\phi$ of pp-pairs as ϕ ranges over all pp conditions, the $\phi_{\mathcal{D}}$ being defined by 3.4.33. By 1.2.17 and definition of $\phi_{\mathcal{D}}$ and $\mathcal{D}^{\text{fact}}$ we have $\mathcal{D}^{\text{fact}} \subseteq \text{Mod}(T)$. If, on the other hand, $M \in \text{Mod}(T)$ and $f : A \to M$ is a morphism with A finitely presented, say A is generated by the finite tuple $\overline{a}$ whose pp-type in A is generated by the pp condition θ (1.2.6), then $f\overline{a} \in \theta(M)$ so, by assumption, $f\overline{a} \in \theta_{\mathcal{D}}(M)$. Therefore, since $\overline{a}$ generates A and since $\theta_{\mathcal{D}}$ has a free realisation in $\mathcal{D}$, f factors through $\mathcal{D}$. We deduce that $M \in \mathcal{D}^{\text{fact}}$, as required. □

For covariantly finite subcategories of the category of pure-projective modules, see [293]. For applications to classes of modules with bounded projective dimension, see [13].

There are also some related results in [372] involving subclasses of Mod-R which are not necessarily closed under direct limits, for instance the following.

Proposition 3.4.39. ([372, 2.1,2.2]) *Suppose that the subcategory $\mathcal{D}$ of* Mod-R *is closed under pure submodules. If $\mathcal{D}$ is closed under products, then $\mathcal{D}$ is covariantly finite in* Mod-R*. The converse is true if $\mathcal{D}$ is closed under direct summands.*

Example 3.4.40. Say that a subcategory $\mathcal{D}$ is predecessor-closed if the existence of a non-zero morphism from A to $\mathcal{D}$ implies that A is in $\mathcal{D}$. Such a subcategory is automatically covariantly finite, hence if $\mathcal{D}$ is such a subcategory of mod-R and is also closed under direct sums, then its $\varinjlim$-closure is the definable subcategory of Mod-R that it generates. Examples include, over a tame hereditary algebra, the closure under finite direct sums of the modules in the preprojective component together with any collection of tubes (Section 8.1.2), and, over a tubular canonical algebra, the closure under direct sums of all indecomposable finitely generated modules belonging to the preprojective component $\mathcal{P}$ or a tube in a family $\mathcal{T}_r$ of with index less than $\alpha \in \mathbb{R}$ (using the notation of Section 8.1.4). For the tame hereditary case the corresponding definable subcategories and Ziegler-closed subsets may be described using the results in Sections 8.1.2 and 8.1.4. See [395], [590], [612] for further examples.

Part of 3.4.24 can be obtained as a corollary of 3.4.37.

Corollary 3.4.41. *The following are equivalent for a ring R:*

(i) R is left coherent;
(ii) the category, proj-R, *of finitely generated projective right R-modules is covariantly finite;*
(iii) the category Flat-R *of flat right R-modules is definable.*

Proof. Since, 2.3.6, Flat-$R = \overrightarrow{\text{proj-}R}$, 3.4.37 gives the equivalence of the latter two conditions.

Suppose that R is left coherent and let ϕ be a pp condition for right R-modules. Then, 2.3.19, $\phi(R_R)$ is a finitely generated submodule of ${}_R R^{\overline{x}}$, say it is $\sum_1^n R\overline{r}_i$. Let ϕ' be $\exists y_1, \dots, y_n\, (\overline{x} = \sum y_i \overline{r}_i)$, so $\phi(R) = \phi'(R)$. By 1.2.29, ϕ' is freely realised in a finitely generated projective module. Also, by 2.3.9, $\phi(P) = P \cdot \phi(R) = P \cdot \phi'(R) = \phi'(P)$ for every $P \in \text{proj-}R$, so $\phi' = \phi_{\text{proj-}R}$ and the condition of 3.4.33 holds.

Conversely, if proj-R is covariantly finite, take any pp condition ϕ. By 3.4.33, $\phi_{\text{proj-}R}$ exists and is freely realised in some projective module. By 1.2.29, ϕ' is equivalent to a divisibility condition $\exists y_1, \dots, y_n\, (\overline{x} = \sum y_i \overline{r}_i)$. In particular, $\phi(R) = \phi_{\text{proj-}R}(R) = \sum R\overline{r}_i$, so $\phi(R_R)$ is finitely generated. It follows by 2.3.19 that R is left coherent. □

There are further examples in [144, §5], for instance ([144, 5.2]) if R is a Λ-order in a finite-dimensional K-algebra, where Λ is a commutative Dedekind domain and K is its field of fractions and if latt-R denotes the class of lattices over R, then $\overrightarrow{\text{latt-}R}$ is the class of R-modules which are Λ-torsionfree.

Proposition 3.4.42. *Let $\mathcal{D}$ be a definable subcategory of* Mod-R. *Then $\mathcal{D}$ is covariantly finite in* Mod-R.

Proof. Let $M \in$ Mod-R: we must find a left $\mathcal{D}$-approximation to M. To take the product over all morphisms $M \to D \in \mathcal{D}$ would be to take a product over a proper set: in fact a subset suffices, namely the product, $f = (f_{\lambda\mu})_{\lambda\mu} : M \to \prod_\lambda D_\lambda = D$ as D_λ ranges over all isomorphism classes of members of $\mathcal{D}$ of cardinality $\leq \kappa = \mathrm{card}(M) + \mathrm{card}(R) + \aleph_0$ and $f_{\lambda\mu}$ ranges over all morphisms in (M, D_λ). To check that this works, take $f' : M \to D' \in \mathcal{D}$. The image of f' is contained in a pure submodule, D'' say, of D' of cardinality $\leq \kappa$ (2.1.21). Then f, followed by projection to (a factor D_λ isomorphic to) D'', then inclusion in D', gives the required factorisation. □

4
Pp-types and pure-injectivity

Algebraically compact and pure-injective modules are introduced and shown to be the same. The general structure theory of these modules is presented and the existence of hulls is proved. Then the focus narrows to Σ-pure-injectives and, further, to modules of finite endolength. The connections to representation type and to the pure-semisimplicity conjecture are discussed. In this chapter full use is made of the functorial methods which are developed in Part II.

4.1 Pp-types with parameters

The notion of pp-type is extended to allow parameters (constants) from a particular module. Every pp-type with parameters in a module may be realised in a pure extension of that module; that larger module may be taken to belong to the definable subcategory generated by the first (4.1.4). The pp-type being "realised" may be understood in the strong sense that no non-implied pp conditions are satisfied (4.1.5).

Pp conditions were defined in terms of systems of homogeneous linear equations. By considering inhomogeneous systems the definition may be extended to allow parameters, that is, elements from a particular module, to appear in pp conditions.

Precisely, given a module M, consider an inhomogeneous system, $\overline{x}'H = \overline{a}$, of R-linear equations with the tuple $\overline{a}$ from M, H a matrix over R and $\overline{x}'$ a tuple of variables. Such a system may be represented as $(\overline{x}'\ \overline{a})G = 0$ for some matrix G so let us more generally consider systems written in this form: $(\overline{x}'\ \overline{a})G = 0$, with G arbitrary. For such an inhomogeneous quantifier-free condition (or **quantifier-free condition with parameters**) we use notation such as $\theta(\overline{x}', \overline{a})$. The notation reflects how it may be obtained from the homogeneous system $\theta(\overline{x}', \overline{z})$, that is, $(\overline{x}', \overline{z})G = 0$, by substituting values from M for the variables in $\overline{z}$. The projection of the condition $\theta(\overline{x}', \overline{a})$ which is obtained by quantifying out some of the variables in $\overline{x}'$ results in a condition of the form $\exists \overline{y}(\overline{x}\ \overline{y}\ \overline{a})G = 0$, where $\overline{x}'$ has been split as $\overline{x}\ \overline{y}$. We describe this as a **pp condition with parameters** and use notation such as $\phi(\overline{x}, \overline{a})$ to display the free variables and the parameters. We write $\phi(M, \overline{a})$ for the solution set $\{\overline{c} \in M^{l(\overline{x})} : \exists \overline{b} \in M^{l(\overline{y})} \text{ with } (\overline{c}\ \overline{b}\ \overline{a})\, G = 0\}$ of solutions of $\phi(\overline{x}, \overline{a})$ in M. Replacing $\overline{a}$ by $\overline{0}$ gives a pp condition in our original sense.

Remarks 4.1.1. (a) Suppose that $\phi(\overline{x}, \overline{a})$ is a pp condition with parameters. Then $\phi(M, \overline{a})$ is either empty or is a coset (a **pp-definable coset**) of the pp-definable subgroup $\phi(M, \overline{0})$ in $M^{l(\overline{x})}$. Every coset of a pp-definable subgroup is definable by a pp condition with parameters for, given a pp condition $\phi(\overline{x})$ and tuple $\overline{a}$, the coset $\overline{a} + \phi(M)$ is defined by the condition $\exists \overline{z}\, (\overline{x} = \overline{a} + \overline{z} \wedge \phi(\overline{z}))$ which is (at least can be rewritten as) a pp condition with parameters.

(b) If $\phi(\overline{x}, \overline{a})$ and $\psi(\overline{x}, \overline{b})$ are pp conditions with the same number of free variables and M is any module, then $\phi(M, \overline{a}) \cap \psi(M, \overline{b})$ is either the empty set or a coset of $\phi(\overline{x}, \overline{0}) \wedge \psi(\overline{x}, \overline{0})$.

(c) (additivity of pp conditions) If $\phi(\overline{x}, \overline{y})$ is a pp condition and if $\phi(\overline{a}, \overline{b})$ and $\phi(\overline{a}', \overline{b}')$ hold, then so does $\phi(\overline{a} - \overline{a}', \overline{b} - \overline{b}')$.

All these points are very easily checked.

Example 4.1.2. Consider the localisation, $\mathbb{Z}_{(p)}$, of the ring of integers at a prime p, regarded as a module over itself. By 1.1.9 the pp-definable subgroups of $M = \mathbb{Z}_{(p)}$ are just the ideals Mp^n for $n \geq 0$, together with 0. Allowing parameters in pp conditions enables us to define each coset in the p-branching tree of cosets of the various Mp^n. This example is continued at 4.2.7.

Recall that a pp-type is a possibly infinite system of pp conditions. We may consider pp-types with parameters. Let M be any module. Then a set, p, of pp conditions with n free variables and with parameters from M is a **pp-(n-)type over** M if every finite subset has a solution from M. Of course, pp-types without parameters automatically have $\overline{0}$ as a solution. Such a set of conditions is also referred to as a **pp-type with parameters** (**from** or **in** M). The requirement of finite solvability is exactly that the collection $\{\phi(M, \overline{b}) : \phi(\overline{x}, \overline{b}) \in p\}$ of subsets of $M^{l(\overline{x})}$ has the finite intersection property. We describe that state of affairs by saying that p is **finitely satisfied** (that is, finitely solvable) in M.

Notations such as p, q are used for pp-types with parameters as well as for pp-types without parameters but, whenever we say simply "pp-type", we mean one without parameters unless it is clear from the context that the more general notion is intended. Usually we do not display the parameters of a pp-type in the notation (there may well be infinitely many of them) but sometimes it is convenient to display some of them, for example if they are to be replaced by other parameters.

Therefore notation such as $p(\overline{x}, \overline{a})$ might not be displaying all the (non-zero) parameters of the pp-type p.

Remark 4.1.3. As was the case for pp-types without parameters there is no loss in generality, because the solution set is unchanged, in supposing that p is closed under conjunction ($\phi_1, \dots, \phi_k \in p$ implies $\phi_1 \wedge \dots \wedge \phi_k \in p$) and implication ($\psi \in p$ and $\psi \leq \phi$ implies $\phi \in p$), that is, that p is a filter in the partially ordered set of pp conditions with parameters from M. When convenient, we will take this as included in the definition of pp-type with parameters.

If $\phi(\overline{x}, \overline{b})$ is a pp condition with parameters from M and if $M \leq N$, then it makes sense to evaluate ϕ on N: it makes better sense if M is a pure submodule of N for in that case the solution set, $\phi(M, \overline{b})$, in M is the intersection of $M^{l(\overline{x})}$ with the solution set, $\phi(N, \overline{b})$, of ϕ in N. That follows directly from the definition of pure embedding (§2.1.1). A **solution** for a pp-type p with parameters from M is a tuple, $\overline{a}$, from a module N which contains M as a pure submodule such that $\overline{a}$ satisfies all the conditions in p. We write $p(N)$ for the set, $\bigcap\{\phi(N, \overline{b}) : \phi(\overline{x}, \overline{b}) \in p\}$, of all solutions to p in N; of course this set may well be empty. We refer to the elements of $p(N)$ as **solutions of** (or **for**) p **in** N; the term **realisation** of p also will be used.

Theorem 4.1.4. *Let p be a pp-type with parameters from the module M. Then there is a pure embedding of M into a module, $M^\star$, which contains a solution for p. The module $M^\star$ may be taken to be in the definable subcategory, $\langle M \rangle$, of* Mod-R *generated by M.*

Proof. The proof is a variant of that of 3.3.6. Without loss of generality p is closed under conjunction and implication. Let I be the set of finite subsets of p and, for each such subset, $S = \{\phi_1, \dots, \phi_k\}$, choose a solution, $\overline{a}_S$ say, of $\phi_1 \wedge \dots \wedge \phi_k$ in M; there is a solution because p is finitely satisfied in M. Let $\mathcal{F}$ be the filter on I generated by the sets of the form $\langle\{\phi\}\rangle = \{S \in I : \{\phi\} \subseteq S\}$ as ϕ ranges over p. Let $M^\star = M^I/\mathcal{F}$ be the corresponding reduced product (Section 3.3.1) and let $\overline{a} = (\overline{a}_S)_S/\mathcal{F}$ be the tuple of $M^\star$ manufactured from the tuples $\overline{a}_S$. By 3.4.7, $M^\star$ belongs to $\langle M \rangle$.

Recall, 3.3.4, that the diagonal embedding of M in $M^\star$ is pure. We identify M with its image in $M^\star$: that allows us to regard the parameters appearing in the pp conditions in p as belonging to $M^\star$ and so (clearly) p may be regarded as a pp-type over $M^\star$. Then $\overline{a}$ realises this pp-type. For, if $\phi \in p$, then for each $S \in \langle\{\phi\}\rangle$ we have $\overline{a}_S \in \phi(M)$ and so, by 3.3.3, $\overline{a} \in \phi(M^\star)$, as required. □

This is actually a special case of a (perhaps the) basic result in model theory, the Compactness Theorem (A.1.8), which says that any set of conditions which is finitely satisfied in a structure has a solution in an elementary extension of that structure.

Theorem 4.1.5. *With the notation of 4.1.4 we may further suppose that the solution, $\overline{a}$, for p satisfies the following condition: for every pp condition, $\psi(\overline{x}, \overline{c})$, with parameters from M we have $\overline{a} \in \psi(M^\bigstar, \overline{c})$ iff $\psi(M, \overline{c}) \geq \phi(M, \overline{b})$ for some $\phi(\overline{x}, \overline{b}) \in p$. That is, if we suppose p to be closed under conjunction and implication, then $\overline{a}$ satisfies those pp conditions with parameters from M which are in p, and no more.*

Proof. Let the index set I and filter $\mathcal{F}$ be as in the proof of 4.1.4. However, given $S = \{\phi_1, \ldots, \phi_k\} \in I$, replace M by its power $M_S = M^{\phi(M)}$, where $\phi = \phi_1 \wedge \cdots \wedge \phi_k \in p$ by assumption, and replace the arbitrary solution of ϕ by the tuple $\overline{a}_S = (\overline{d})_{\overline{d} \in \phi(M)} \in M^{\phi(M)}$. The point about $\overline{a}_S$ is that it satisfies a given pp condition $\psi(\overline{x}, \overline{c})$ iff each element $\overline{d} \in \phi(M)$ satisfies that condition, that is, iff $\psi(M, \overline{c}) \geq \phi(M)$.

The embedding of M into each product M_S is pure so, forming the reduced product, $M^\bigstar = \prod_{S \in I} M_S/\mathcal{F}$, of the M_S, we still have that M is a pure submodule of $M^\bigstar$.

Let $\overline{a} = (\overline{a}_S)_{S\in I}/\mathcal{F}$. By 3.3.3 and the definition of $\mathcal{F}$, if $\overline{a} \in \psi(M^\bigstar, \overline{c})$, then for some (in fact many) $S \in I$ we have $\overline{a}_S \in \phi(M)$, where $\phi = \phi(\overline{x}, \overline{b})$, say, is as above. Therefore each $\overline{d} \in \phi(M, \overline{b})$ satisfies $\psi(\overline{x}, \overline{c})$. That is, $\phi(M, \overline{b}) \leq \psi(M, \overline{c})$. So $\psi(\overline{x}, \overline{c}) \in p$, as required. □

Tuples with a prescribed pp-type (over the empty set) were produced by different means in 3.2.5.

4.2 Algebraic compactness

Algebraically compact modules are defined and examples are given in Section 4.2.1. The related notion of linear compactness is considered in Section 4.2.2 and, in Section 4.2.3, algebraically compact modules are characterised as algebraic direct summands of compact topological modules. Algebraic compactness for abelian groups is discussed briefly in Section 4.2.4. Any module has an algebraically compact ultrapower (Section 4.2.5).

4.2.1 Algebraically compact modules

Algebraically compact modules are defined and variants on the definition established (4.2.1, 4.2.2). Examples include injective modules (4.2.4), modules of finite length over their endomorphism rings (4.2.6), the module of p-adic integers (4.2.7).

A module M is **algebraically compact** if every collection of cosets of pp-definable subgroups of M which has the finite intersection property has non-empty intersection. That is, if, for each $\lambda \in \Lambda$, ϕ_λ is a pp condition with one free variable and $a_\lambda \in M$, then $\bigcap_{\lambda\in\Lambda}(a_\lambda + \phi_\lambda(M)) \neq \emptyset$ provided that for every finite $\Lambda' \subseteq \Lambda$ the intersection $\bigcap_{\lambda\in\Lambda'}(a_\lambda + \phi_\lambda(M))$ is non-empty. This definition, which refers to the solvability of infinite systems of inhomogeneous pp conditions, may be rephrased (4.2.1) in terms of solvability of pp-types and also (4.2.2) in terms of solvability of systems of linear equations.

It will be shown (4.3.11) that algebraic compactness and pure-injectivity coincide so, throughout this and the following sections, one may replace "algebraically compact" by "pure-injective".

Lemma 4.2.1. *A module M is algebraically compact iff every pp-1-type with parameters from M has a solution in M iff, for every n, every pp-n-type with parameters from M has a solution from M.*

Proof. Just from the definitions and the remark 4.1.1(a) it is immediate that algebraic compactness is equivalent to the solvability-in-M condition for pp-types with one free variable. So it has to be shown that if the solvability condition holds for 1-types, then it holds for n-types for all n. The proof goes by induction on n. Recall that the notation $M \models \phi(\overline{a})$ ("M satisfies $\phi(\overline{a})$") means $\overline{a} \in \phi(M)$.

Suppose that every pp-n-type with parameters from M has a solution from M. Let $p(x, \overline{y})$, with $l(\overline{y}) = n$, be a pp-$(1+n)$-type over M. Let $q(\overline{y})$ be "$\exists x\, p(x, \overline{y})$", that is,

$$q(\overline{y}) = \{\exists x\, \phi(x, \overline{y}) : \phi(x, \overline{y}) \in p\}.$$

Then, it is claimed, q is a pp-n-type over M. For, given $\phi_1, \dots, \phi_k \in p$ there are $a, \overline{b}$ from M such that $M \models \phi_i(a, \overline{b})$ for $i = 1, \dots, k$, so $M \models \exists x \phi_i(x, \overline{b})$ for $i = 1, \dots, k$. That is, the typical finite subset, $\{\exists x\, \phi_1(x, \overline{y}), \dots, \exists x\, \phi_k(x, \overline{y})\}$, of $q(\overline{y})$ has a solution in M, so q is indeed a pp-n-type with parameters from M.

By the induction hypothesis there is $\overline{c}$ from M satisfying every condition in q. Now let $r(x)$ be "$p(x, \overline{c})$", that is,

$$r(x) = \{\phi(x, \overline{c}) : \phi(x, \overline{y}) \in p\}.$$

Take $\phi_1(x, \overline{y}), \dots, \phi_k(x, \overline{y}) \in p$, so $\{\phi_1(x, \overline{c}), \dots, \phi_k(x, \overline{c})\}$ is a typical finite subset of r. As remarked before there is no loss in generality in assuming that p is closed under finite conjunction so we may suppose that $\phi(x, \overline{y}) = (\phi_1 \wedge \cdots \wedge \phi_k)$ is in p. Therefore $\exists x \phi(x, \overline{y}) \in q$ so, by choice of $\overline{c}$, we have $M \models \exists x \phi(x, \overline{c})$, say $(a', \overline{c}) \in \phi(M)$, that is, $(a', \overline{c}) \in \phi_i(M)$ for $i = 1, \dots, k$.

Thus $\{\phi_1(x, \overline{c}), \ldots, \phi_k(x, \overline{c})\}$ is indeed satisfied in M, so r is a pp-1-type over M.

By assumption there is $a \in M$ satisfying each condition, $\phi(x, \overline{c})$, in r. Then the tuple $(a\,\overline{c})$ is a solution for p in M, as required. □

Proposition 4.2.2. *A module M is algebraically compact iff every system of equations, in possibly infinitely many variables, with parameters from M and which is finitely solvable in M, is solvable in M.*

Proof. (⇒) First we see how to replace one variable by a value. Write $\Theta(\overline{x})$ for a system of equations with parameters from M, where now $\overline{x}$ is a possibly infinite set of variables. Choose and fix one of these variables, say x'. Consider the pp-1-type, p, over M obtained by existentially quantifying out all other variables: if

$$x'r_j + \sum_{i=1}^{k} x_{\lambda_i} s_{ij} = b_j \qquad (j = 1, \ldots, m),$$

with r_j, s_{ij} in R and $b_j \in M$, are finitely many equations in Θ, with all the x_{λ_i} different from x' and from each other and r_j, s_{ij} possibly zero, then put the pp condition

$$\exists y_1, \ldots, y_k \left(\bigwedge_{j=1}^{m} x'r_j + \sum_{i=1}^{k} y_i s_{ij} = b_j \right)$$

into p. Note that p is finitely satisfied in M since the system Θ is finitely satisfied in M, so p is, indeed, a pp-type over M.

By assumption there is a solution, a' say, for p in M. Let $\Theta'(a'/x')$ denote the system of equations of the form $a'r + \sum_1^k x_{\lambda_i} s_i = b$, where $x'r + \sum_1^k x_{\lambda_i} s_i = b$ ranges over equations in Θ, which is obtained by replacing every occurrence of x' by the value a'. By construction of p and choice of a', Θ' is finitely solvable in M. Also Θ' has one fewer free variable than Θ. Using transfinite induction (having ordered the variables by an ordinal) or Zorn's Lemma (applied to the poset of "partially solved" systems derived from Θ) one finishes the argument.

(⇐) Given a pp-type, $p(\overline{x})$, over M, a system of equations is produced in the following way. For each pp condition $\phi \in p$, say ϕ is

$$\exists \overline{y} \left(\bigwedge_{j=1}^{m} \sum_i x_i r_{ij} + \sum_k y_k s_{kj} = b_j \right),$$

with the $r_{ij}, s_{kj} \in R$ and the $b_j \in M$, extract the equations

$$\sum_i x_i r_{ij} + \sum_k y_{\phi k} s_{kj} = b_j \qquad (j = 1, \ldots, m)$$

(note the new subscripts for the ys) having relabelled the previously quantified variables in a way that keeps such variables arising from different conditions in p distinct. Let Θ be the set of all such equations as ϕ ranges over p. So the variables appearing in Θ are the original variables $\overline{x}$ together with, for each condition in p, a finite set of new (and distinct from all the others) free variables.

Since p is finitely solvable in M so, clearly, is Θ. By assumption Θ has a solution in M and, from the construction, this yields a solution of p in M. □

A proof similar to that of 4.2.2(⇒) gives the next result.

Corollary 4.2.3. *Let M be algebraically compact. Suppose that Φ is any set of pp conditions with parameters from M and possibly infinitely many free variables occurring as the conditions range over Φ. If Φ is finitely satisfied in M, then Φ has a solution in M.*

Example 4.2.4. Any injective module is algebraically compact. This will follow immediately from 4.3.11 but we can see it directly here. Let p be a pp-type over an injective module E. Because E is a direct summand of any module which contains it, all we have to do is produce a solution to the pp-type in some extension of E; for then a projection of this solution to E will give a solution in E. But that is what 4.1.4 provides.

Remark 4.2.5. Let p be any pp-type with parameters from M and let $p_0 = \{\phi(\overline{x}, \overline{0}) : \phi(\overline{x}, \overline{b}) \in p\}$ denote the set of pp conditions which is obtained by replacing all parameters appearing in conditions $\phi \in p$ by 0. Then p_0 is a pp-type (closed under intersection and implication if p is) and $p(M)$, if non-empty, is a coset of $p_0(M)$. The same holds for any N containing M as a pure submodule.

This all follows easily from the definitions. For example, if $\overline{a}, \overline{c} \in p(M)$, then for each $\phi(\overline{x}, \overline{b}) \in p$ we have both $\phi(\overline{a}, \overline{b})$ and $\phi(\overline{c}, \overline{b})$ true in M, so, by 4.1.1(c), $\phi(\overline{a} - \overline{c}, \overline{0})$ holds and we deduce that $\overline{a} - \overline{c} \in p_0(M)$ (and the argument more or less reverses).[1]

Proposition 4.2.6. *If a module M has the descending chain condition on* End(M)-*submodules, or just on pp-definable subgroups, then it is algebraically compact.*

Proof. Take any pp-1-type p with parameters in M. Since M has, by 1.1.8, the descending chain condition on pp-definable subgroups it has the dcc on cosets of these. Therefore the filter of sets of the form $\phi(M, \overline{b})$, where $\phi(\overline{x}, \overline{b}) \in p$, has

[1] The pp-type p_0 is the **homogeneous** pp-type **corresponding to** p. This is of importance in, for instance, [495], especially the parts that deal with model-theoretic stability theory, but is not needed much here.

a least element, which is non-empty since p is finitely realised in M. Thus p is realised in M, so M is algebraically compact. □

Therefore every finite module, as well as every module which is finite-dimensional over a field contained in the centre of the ring, is algebraically compact. Modules which satisfy the hypothesis of 4.2.6 are the subject of Section 4.4.2.

Example 4.2.7. The module $\mathbb{Z}_{(p)}$, regarded as a module over $\mathbb{Z}$, or over the ring $\mathbb{Z}_{(p)}$, is not algebraically compact. For, the tree$^{(\mathrm{op})}$ of pp-definable cosets in $\mathbb{Z}_{(p)}$, ordered by inclusion, is p-branching at every node and is of infinite depth. Each of the $2^{\aleph_0}$ infinite branches through this tree gives a filter, indeed a chain, of pp-definable cosets, that is, defines a pp-type with parameters from $\mathbb{Z}_{(p)}$. Clearly the pp-types corresponding to different branches are incompatible, so a solution for one cannot be a solution for any other. Therefore any algebraically compact module containing $\mathbb{Z}_{(p)}$ purely must have cardinality at least $2^{\aleph_0}$: in particular, it cannot be $\mathbb{Z}_{(p)}$.

In fact, by 4.2.8, the completion of $\mathbb{Z}_{(p)}$ in the p-adic topology – the topology which has, for a basis of open neighbourhoods of 0, the subgroups $\mathbb{Z}_{(p)} \cdot p^n$ – namely the p-adic integers, $\overline{\mathbb{Z}_{(p)}}$, is algebraically compact as a $\mathbb{Z}$- (or $\mathbb{Z}_{(p)}$-, or even $\overline{\mathbb{Z}_{(p)}}$-) module and contains $\mathbb{Z}_{(p)}$ as a pure submodule.

In general, one may complete any module M in a *very* roughly similar way, by adding solutions for pp-types with parameters in M, to obtain an algebraically compact module which contains M as a pure submodule. If this is done in a suitably economical way one obtains a minimal such extension, called the pure-injective hull of M, see Section 4.3.3 (however, for a proof along the lines just indicated, see [495, §4.1] rather than that section, where a very different proof is given).

Algebraic compactness was introduced by Mycielski [451] for general algebraic structures and was further investigated by Węglorz [706], Mycielski and Ryll-Nardzewski [452] and Taylor [673], among others.

4.2.2 Linear compactness

Linear compactness is somewhat coarser than algebraic compactness but it is preserved by extension (4.2.10) and inverse limit (4.2.11). A module which is linearly compact over its endomorphism ring is algebraically compact (4.2.8).

A module M is **linearly compact** (with respect to the discrete topology) if every collection of cosets of submodules of M which has the finite intersection property has non-empty intersection.[2] More generally, if M is a topological module (see Section 4.2.3) which is Hausdorff, then it is a **linearly compact** topological module if every collection of cosets of closed submodules with the finite intersection property has non-empty intersection.[3]

Lemma 4.2.8. *If M_R is a module which is linearly compact as a module over its endomorphism ring, then M is an algebraically compact R-module. In particular, any linearly compact module over a commutative ring is algebraically compact.*

Proof. This is immediate because (1.1.8) every pp-definable subgroup of M is an $\mathrm{End}(M_R)$-submodule of M. □

It follows that a commutative linearly compact ring, in particular any complete discrete valuation ring, is algebraically compact as a module over itself. As a partial converse ([699, Prop. 9]), if R is commutative and either noetherian or a valuation ring, then R_R is algebraically compact iff it is linearly compact.

Example 4.2.9. Here is an algebraically compact module which is not linearly compact over its endomorphism ring: $R = k[x_i (i \in \omega) : x_i x_j = 0 \ \forall i, j]$. The module R_R is algebraically compact (even Σ-algebraically compact $= \Sigma$-pure-injective, see 5.3.62). It is not, however, artinian over its endomorphism ring (that is, as a (left) module over itself) so, at least if k is countable, an argument like that in 4.2.7 shows that it cannot be linearly compact.

In general, algebraically compact modules are not closed under extensions, see 8.1.18,[4] but linearly compact modules do have this property (the proof below works for linearly compact topological modules as well). See 4.6.3 for some equivalents to pure-injectives being closed under extension.

Theorem 4.2.10. ([725, Prop. 9]) *If $0 \to A \to B \xrightarrow{\pi} C \to 0$ is an exact sequence with A and C linearly compact, then B is linearly compact.*

Proof. Let $(b_\lambda + B_\lambda)_\lambda$ be a collection of cosets of submodules of B which has the finite intersection property. Without loss of generality this collection is closed under finite intersection. The collection $(\pi b_\lambda + \pi B_\lambda)_\lambda$ of cosets of submodules of C also has the finite intersection property, so, since C is linearly compact, there is $b \in B$ such that $b \in b_\lambda + B_\lambda + A$ for every λ. In particular $(b_\lambda - b + B_\lambda) \cap A \neq \emptyset$ for every λ and, moreover, the collection $((b_\lambda - b + B_\lambda) \cap A)_\lambda$ of cosets of

[2] So, cf. the definition of algebraic compactness: submodules, not subgroups, and no definability requirement.
[3] The concept comes from Lefschetz [392].
[4] Also [716, 3.5.1] for a general criterion.

submodules of A has the finite intersection property so there is $a \in A \cap \bigcap_\lambda (b_\lambda - b + B_\lambda)_\lambda$. Then $a + b \in \bigcap_\lambda (b_\lambda + B_\lambda)$ as required. □

Proposition 4.2.11. *The inverse limit of any inversely directed system of linearly compact topological modules is linearly compact*

The above result is stated as [725, Prop. 4] where, as at [198, p. 70], it is attributed to Lefschetz [392, I; 38, 39].

See [725], [614], [486, § 6.2] for more on linear compactness.

4.2.3 Topological compactness

It is shown that a module is algebraically compact iff it is a direct summand (just as a module) of a topological module which is compact (4.2.13).

Say that an R-module M is **compact** if it may be equipped with a Hausdorff topology in such a way that it becomes a compact **topological module**, that is, the addition $(a, b) \mapsto a + b : M \times M \to M$ and multiplication $(r, m) \mapsto mr : R \times M \to M$ are continuous, where R is given the discrete topology and each product is given the product topology. A straightforward check shows that it is equivalent to require that addition and each scalar multiplication $m \mapsto mr : M \to M$ be continous.

Proposition 4.2.12. (for example, [699, Prop. 7]) *Let M_R be any module and let $S \to \mathrm{End}(M_R)$ be a morphism of rings. Let ${}_SG$ be any compact S-module. Then ${}_SM^* = \mathrm{Hom}({}_SM, {}_SG)$ carries the structure of a compact Hausdorff topological left R-module, the topology being induced from that on G.*

Proof. By Tychonoff's Theorem the product topology on G^M is compact. Recall that the product topology has, for a basis of open sets, those of the form $\times_{a \in M} U_a$, where all U_a are open subsets of G and all but finitely many are equal to G. It is easily checked that G^M, with the natural structure of a left R-module, is a topological module. Regard M^* as a subset of G^M via $f \mapsto (fa)_{a \in M}$: we check that M^* is a closed subset, which will be enough.

Let $(g_a)_a \in G^M \setminus M^*$. Regard this as the function $g : M \to G$ defined by $g(a) = g_a$: by assumption g is not S-linear. So there are $a, b \in M$ and $s \in S$ such that $g_{as+b} \neq sg_a + g_b$. Since ${}_SG$ is Hausdorff there are disjoint open sets U, V with $g_{as+b} \in U$ and $sg_a + g_b \in V$. By the definition of a topological module there are open subsets V_1, V_2 of G such that $sg_a \in V_1$, $g_b \in V_2$ and $+ : V_1 \times V_2 \to V$ and, further, an open subset, V_3, of G such that $g_a \in V_3$ and $s \times - : V_3 \to V_1$. Let W be the open neighbourhood of g which is $\times_{c \in M} U_c$ with $U_a = V_3$, $U_b = V_2$, $U_{sa+b} = U$ and all other $U_c = G$. No element of W is, regarded as a function from M to G, S-linear, completing the proof. □

Corollary 4.2.13. ([699, Thm 2]) *A module is algebraically compact iff it is an algebraic direct summand of a compact module.*

Proof. Given any module M, and taking S and G above to be $\mathbb{Z}$ and the circle group ${}_{\mathbb{Z}}(\mathbb{R}/\mathbb{Z})$ with its usual topology respectively, there is, 1.3.16, a pure embedding $M \to M^{**}$ (notation as above). The module M^{**} is algebraically compact (= pure-injective) by 4.3.29 and, by 4.2.12 applied to M^*, this module is compact. If M is algebraically compact, hence (4.3.11) pure-injective, then this embedding is split, giving one direction.

For the other direction, suppose that N is a compact module. Each finite power N^k also is compact, so the solution set to any system (not necessarily homogeneous) of linear equations is, by continuity of addition and scalar multiplication, a closed subset of the relevant finite power of N. The projection of such a set to the first coordinate is, therefore, also closed. That is, each pp-definable coset in N is closed. Since N is compact it follows that N is algebraically compact. If M is a direct summand of N (now ignoring the topology), then M is also algebraically compact by 4.2.1 (cf. the argument of 4.2.4). □

This gives another proof, cf. 4.2.7, that the ring, $\overline{\mathbb{Z}_{(p)}}$, of p-adic integers is algebraically compact as a module over itself (hence also over $\mathbb{Z}$ and over the localisation $\mathbb{Z}_{(p)}$). Indeed, there is the following.

Theorem 4.2.14. ([699, Prop. 10]) *Let R be a commutative ring and let M be an R-module linearly topologised by declaring the MI, with I a product of powers of maximal ideals, to be a basis of open neighbourhoods of* 0. *Let $\overline{M}$ denote the completion of M in this topology. For each maximal ideal P denote by $\overline{M_{(P)}}$ the completion of M with respect to the P-adic topology. Then the diagonal map $M \to \prod_{P \in \mathrm{Maxspec}(R)} \overline{M_{(P)}}$ induces an isomorphism (of topological modules) $\overline{M} \simeq \prod_{P \in \mathrm{Maxspec}(R)} \overline{M_{(P)}}$.*

Theorem 4.2.15. ([699, Thm 3]) *Suppose that R is a commutative noetherian ring and suppose that M_R is finitely generated. Let $\overline{M}$ denote the completion of M in the topology defined in 4.2.14. Then M is Hausdorff in this topology and the inclusion $M \to \overline{M}$ is a pure-injective hull of M, in particular it is pure.*

It follows that $\overline{\mathbb{Z}_{(p)}}$ is the pure-injective hull of $\mathbb{Z}_{(p)}$ and that $\prod_p \overline{\mathbb{Z}_{(p)}}$, where the product is over all non-zero primes, is the pure-injective hull of $\mathbb{Z}$ (for pure-injective hulls see Section 4.3.3).

Couchot, [135, Thm 12], extends some of this by showing that if R is a commutative valuation ring, then for every finitely generated R-module M, $M \otimes_R H(R)$ is the pure-injective hull of M. Also see [134], [136].

A commutative noetherian ring is pure-injective as a module over itself iff it is a finite product of complete local rings ([323, 11.3]).

There is more in [285] and especially in [284] on pure-injectivity and linear topologies. Let $\mathcal{F}$ be a filter in pp_R, not necessarily proper, so allowing pp_R itself. For any module M declare the sets $\phi(M)$ with $\phi \in \mathcal{F}$ to be a basis of open neighbourhoods of 0 and denote the resulting topological module by $M_{\mathcal{F}}$. Say that $M_{\mathcal{F}}$ is a **finite matrix topology** if every pp-definable subgroup $\psi(M) \leq M$ is closed in this topology. Herzog showed, [284, Props. 4, 5, Thm 7], that the class of modules M such that $M_{\mathcal{F}}$ is a finite matrix topology is closed under products, pure submodules and pure-injective hulls. Also, [284, Thm 7], $M_{\mathcal{F}}$ is a finite matrix topology iff $H(M)_{\mathcal{F}}$ is a finite matrix topology (where $H(M)$ denotes the pure-injective hull of M) iff $H(M)_{\mathcal{F}}$ is Hausdorff.

Say that M is $\mathcal{F}$-**dense** if $M_{\mathcal{F}}$ is a finite matrix topology and if $M_{\mathcal{F}}$ is a dense subset of $H(M)_{\mathcal{F}}$. Then, [284, Thm 13], the assignment $M \mapsto H(M)$ is functorial[5] on the category of $\mathcal{F}$-dense modules and is given by $H(M) = \varprojlim_{\phi \in \mathcal{F}} M/\phi(M)$ ([284, Prop. 11]).

Say that a module is **almost pure-injective** if there is a filter $\mathcal{F}$ for which it is $\mathcal{F}$-dense. Then, [284, Thm 17], if R_R is almost pure-injective, its pure-injective hull $H(R)$ has a natural ring structure which extends that on R and which, Herzog suggests, makes $H(R)$ the pure analogue of the ring of quotients of a non-singular ring.

If R is a hereditary generalised Weyl algebra in the sense of Section 8.2.5 and the filter $\mathcal{F}$ is as in the statement of 8.2.1 then, [284, Ex. 19], every finitely generated R-module is $\mathcal{F}$-dense, so forming pure-injective hulls is functorial from mod-R to Mod-R.

4.2.4 Algebraically compact $\mathbb{Z}$-modules

The term "algebraically compact" was coined by Kaplansky [330] (§17 of the 1954 edition; the later edition has an expanded section on these groups) to describe those abelian groups which are direct summands of compact abelian groups (see 4.2.13). Łoś defined an equivalent notion, [411], [412], and Balcerzyk, [36], noted the equivalence. Maranda, [424], defined the equivalent notion of pure-injectivity for abelian groups (see [198, Notes to Chpt. VII] for further comments and references).

Theorem 4.2.16. ([330, §17] also see [198, 40.1] or [250, Cor. 34]) *An abelian group is algebraically compact iff it has the form $E \oplus \prod_p A_p$, where E is divisible*

[5] The operation of assigning (pure-)injective hulls is seldom functorial.

(equivalently, injective) and, for each prime p, A_p is an abelian group which is complete in its p-adic topology

A decomposition of pure-injective modules which applies over any ring is, in general form, 4.4.2, and, for the special case of PI Dedekind domains, is in Section 5.2.1. Just a little work (see 4.3.22) is required to compare the product decomposition above with the hull-of-a-direct-sum decomposition in Section 5.2.1.

Theorem 4.2.17. ([312] ([37]) for the case $A_i = \mathbb{Z}$ for all i) *For any countable family $(A_i)_{i \in \mathbb{N}}$ of abelian groups the quotient $\prod_i A_i / \bigoplus_i A_i$ is algebraically compact.*

Also see [198, §42] for a proof. This is a case of a far more general result, see 4.2.19 and the comment before that.

This fails for uncountable products: Gerstner, [231, Satz 3], showed that $\mathbb{Z}^\kappa / \mathbb{Z}^{(\kappa)}$ is not algebraically compact if κ is uncountable. Also see the next section. For positive results referring to "filter quotients", see [194] and references therein.

4.2.5 Algebraic compactness of ultraproducts

Every module has an algebraically compact ultrapower (4.2.18): we refer the reader to model theory texts for a proof (of a much more general fact) but discuss what is involved.

Theorem 4.2.18. ([628, Thm 1]) *Let R be a ring. Then there is an index set I and an ultrafilter $\mathcal{U}$ on I such that for every $M \in$ Mod-R the ultraproduct $M^I/\mathcal{U}$ is algebraically compact. Indeed, $(-) \mapsto (-)^I/\mathcal{U}$ defines a functor from* Mod-R *to the subcategory of algebraically compact objects and takes pure-exact sequences to split exact sequences.*

The proof of this result uses a theorem, on saturation of ultraproducts, from model theory. To give the complete proof here would be quite a diversion, so I give just an outline.[6]

Say that a structure (for example, an R-module) M is a κ**-saturated structure**, where κ is an infinite cardinal, if every set, Φ, of formulas (see Appendix A) with parameters from M and of cardinality strictly less than κ, and which is finitely satisfied in M has a solution in M (so this generalises the definition of algebraic compactness). Let μ be the cardinality of the set of all formulas without parameters (for the kind of structures being considered). The theorem that we quote says that,

[6] The proof of 4.3.21 is also relevant.

given a set I of cardinality $\kappa > \mu$, there is a suitable ("ω_1-incomplete, card(I)-good") ultrafilter, $\mathcal{U}$, on I such that every ultraproduct, $\prod_{i\in I} M_i/\mathcal{U}$ of structures is κ-saturated. For this see, for example, [114, 6.1.8] or other model theory text such as [296], [435]. It remains to note that if $\kappa > \mu = \max\{\mathrm{card}(R), \aleph_0\}$, then any κ-saturated module is algebraically compact. For, every set of pp conditions which is finitely satisfied in a module, is essentially a filter, of cosets of pp-definable subgroups of which there are no more than μ. Therefore, if M is a κ-saturated module, then every set of pp conditions which is finitely satisfied in M must have a solution in M, so M is algebraically compact ([628, Thm 1]).

Therefore, for every $\kappa > \mathrm{card}(R) + \aleph_0$ if I is a set of cardinality κ, then there is an ultrafilter, $\mathcal{U}$, on I such that for every R-module M the ultraproduct $M^I/\mathcal{U}$ is algebraically compact.

If R is countable, then the hypotheses needed on the ultrafilter simplify and one has the following simpler statement, which is given a direct proof at [323, 7.49].

Theorem 4.2.19. *Suppose that R is countable. If M_i $(i \in I)$ is any infinite family of R-modules and $\mathcal{F}$ is a filter on I which is ω_1-incomplete, meaning that some countable subset of $\mathcal{F}$ has empty intersection, then $\prod M_i/\mathcal{F}$ is algebraically compact.*

(To see that such an ultrafilter exists, split the set I into a countable disjoint union $I = \bigcup_n I_n$ and take the complements $I \setminus I_n$ of these: this set has the finite intersection property, so extends to an ultrafilter which, by construction, is ω_1-incomplete.)

Recall, 2.2.3, that if $\mathrm{pgldim}(R) \leq 1$ (for example, if R is countable), then every ultraproduct of algebraically compact modules is algebraically compact.

By way of contrast, one has, for example, the following, which is [323, 8.40]. Let $R = \mathbb{C}[X]$ and let $\mathcal{U}$ be any ultrafilter on a countable set I. Then the module $R^I/\mathcal{U}$ is not algebraically compact. Also see [388].

4.3 Pure-injectivity

Pure-injectivity is introduced in Section 4.3.1. That algebraic compactness is equivalent to pure-injectivity is the content of Section 4.3.2. Pure-injective hulls are defined and shown to exist, via the existence of injective hulls in the functor category, in Section 4.3.3. In the same section the relation between pp-pairs on a module and on its pure-injective hull is shown to be closed; in particular definable subcategories are closed under pure-injective hulls. Suitable hom-duals of modules are shown to be pure-injective in Section 4.3.4. A rather more subtle notion of hull, which was important in the development of the model theory of *modules, is discussed in Section 4.3.5, where it is also noted that there is just a set of isomorphism types of indecomposable pure-injectives. Section 4.3.6 starts with the Jacobson radical of the endomorphism ring of an indecomposable pure-injective. The notion of irreducibility for pp-types is analogous to $\cap$-irreducibility for one-sided ideals and that is equivalent to indecomposability of the injective hull. One of Ziegler's key results was the equivalence of irreducibility of a pp-type to indecomposability of the hull. The functorial approach gives a short proof which is presented in this section.*

Section 4.3.7 considers pure-injective hulls of strongly indecomposable finitely presented modules. A ring is Krull–Schmidt if every finitely presented module is a direct sum of strongly indecomposable modules: these are considered in Section 4.3.8. Hulls of quotient pp-types are defined in Section 4.3.9 and used to describe indecomposable summands of pure-injective modules in Section 4.3.10.

4.3.1 Pure-injective modules

Pure-injectivity is defined and a criterion (4.3.6) for pure-injectivity proved. Often it is easy to see using this criterion that certain functors preserve pure-injectivity. It is shown that inclusion of pp-types reflects the existence of morphisms when the codomain is pure-injective (4.3.9).

An R-module N is **pure-injective** if, given any pure embedding $f : A \to B$ in Mod-R, every morphism $g : A \to N$ lifts through f: there exists $h : B \to N$ such that $hf = g$.

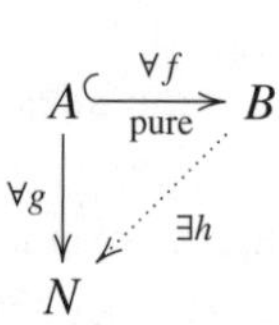

$$\begin{array}{ccc} A & \xrightarrow[\text{pure}]{\forall f} & B \\ {\scriptstyle \forall g}\downarrow & \swarrow {\scriptstyle \exists h} & \\ N & & \end{array}$$

Proposition 4.3.1. *A module N is pure-injective iff every pure embedding $N \to M$ with domain N is split.*

Proof. ($\Rightarrow$) Apply the definition of pure-injective with g, as in the definition, the identity map, 1_N.

($\Leftarrow$) Given a pure embedding $f : A \to B$ and a morphism $g : A \to N$ form the pushout shown.

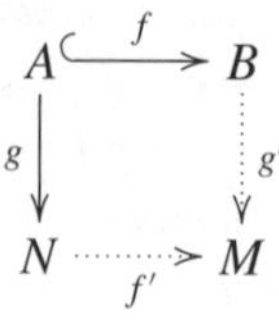

$$\begin{array}{ccc} A & \xrightarrow{f} & B \\ {\scriptstyle g}\downarrow & & \downarrow{\scriptstyle g'} \\ N & \xrightarrow[f']{} & M \end{array}$$

By 2.1.13, f' is a pure embedding, so it is split, by $k : M \to N$ say. Set $h = kg'$ to obtain a map with $hf = g$, as required. □

Any injective module is pure-injective because such a module has the lifting property over all embeddings, pure or not. Since, as we will see (4.3.11), pure-injectivity and algebraic compactness coincide for modules, some more examples are provided by 4.2.6 and 4.2.7. A module over a von Neumann regular ring is pure-injective iff it is injective because, 2.3.22, over such a ring all embeddings are pure.

Lemma 4.3.2. *Any direct product of pure-injective modules is pure-injective. Any direct summand of a pure-injective module is pure-injective.*

Proof. Suppose that the modules N_i $(i \in I)$ all are pure-injective. If $f : A \to B$ is pure and $g : A \to \prod_i N_i$, then, with the $\pi_i : \prod_j N_j \to N_i$ being the projections, each $\pi_i g$ factors through f, say $h_i : B \to N_i$ is such that $h_i f = \pi_i g$. The h_i together yield a morphism $h : B \to \prod_i N_i$ with $\pi_i h = h_i$ for all i, hence with $hf = g$.

If $N = N' \oplus N''$ is pure-injective, $f : A \to B$ is a pure embedding and $g : A \to N'$ is any morphism, then, with $j : N' \to N$ denoting the inclusion, there is $h : B \to N$ such that $hf = jg$. Let $\pi : N \to N'$ be the projection with kernel N''. Then $g = \pi j g = \pi h f$, as required. □

Example 4.3.3. An infinite direct sum of pure-injective modules need not be pure-injective. For instance $M = \bigoplus_n \mathbb{Z}_{2^n}$ is not pure-injective even though each component of the direct sum is (by 4.2.6). For, the embedding of M into $M' = \prod_n \mathbb{Z}_{2^n}$ is pure by 2.1.10 but is not split.

To see that it does not split suppose, for a contradiction, that $\pi : M' \to M$ does split the inclusion. Set $a = (2^{n-1}_{2^n})_n \in M'$ and, for each n, set $a_n = (1_2, 2_{2^2}, \dots, 2^{n-1}_{2^n}, 0, \dots) \in M$. For each n, $2^n \mid a - a_n$, from which $2^n \mid \pi a - a_n$ follows. But, for some n, the nth coordinate of πa is 0 hence $2^n \mid 2^{n-1}_{2^n}$ – contradiction.

Example 4.3.4. ([729, Prop. 1] (which proves a more general result and a variety of other preservation results)) If R is a ring such that R_R is pure-injective, then, for any index set I, so is the power series ring $R[[X_i : i \in I]]$ as a right module over itself. In particular, if k is a field, then $k[[X]]$ is pure-injective as a module over itself.

Example 4.3.5. ([732, Thm 1]) The group ring RG is pure-injective as a right module over itself iff the same is true for R and G is finite.

Given any module, M, and index set, I, the summation map, Σ, from the direct sum $M^{(I)}$ to M is defined by $\Sigma(a_i)_i = \sum_i a_i$. Equivalently, it is defined as the unique (by definition of coproduct) map such that for each inclusion $j_i : M \to M^{(I)}$ (of M to the ith component) we have $\Sigma \cdot j_i = 1_M$. Note, for future reference, that this definition of Σ makes sense in any additive category with coproducts.

The following criterion for pure-injectivity is due to Jensen and Lenzing.

Theorem 4.3.6. ([323, 7.1]) *A module M is pure-injective iff for every index set, I, the summation map $\Sigma : M^{(I)} \to M$, given by $(x_i)_i \mapsto \sum_i x_i$, factors through the natural embedding of $M^{(I)}$ into the corresponding direct product M^I.*

Proof. Necessity of the condition is by definition of pure-injectivity and 2.1.10. For the converse we establish algebraic compactness of M: by 4.3.11 below that is equivalent to pure-injectivity of M.

If the ring is uncountable (at least if M has uncountably many pp-definable subgroups) the idea can be somewhat obscured, so first we treat the case where M has just countably many pp-definable subgroups.

Let I be the set of pp-definable subgroups of M. Denote by $f : M^I \to M$ a morphism which extends the summation map $\Sigma : M^{(I)} \to M$. Suppose that $p(x)$ is a pp-type with parameters from M: we must produce a solution to p in M. Choose a countable cofinal set $a_1 + \phi_1(x) \geq a_2 + \phi_2(x) \geq \cdots \geq a_n + \phi_n(x) \geq \cdots$ in p and consider the element $c = (a_1, a_2 - a_1, a_3 - a_2, \ldots, a_n - a_{n-1}, \ldots) \in M^I$. We claim that fc satisfies every condition in p.

Since $a_n + \phi_n(M) = a_{n+1} + \phi_n(M)$, we have $a_{n+1} - a_n \in \phi_n(M)$ and so $a_{m+1} - a_m \in \phi_n(M) \geq \phi_m(M)$ for all $m \geq n$. Therefore, writing $c = c_{\leq n} + c_{>n}$, where $c_{\leq n} = (a_1, \ldots, a_n - a_{n-1}, 0, \ldots)$, and using that $fc_{\leq n} = \Sigma c_{\leq n}$, we have $fc = a_1 + (a_2 - a_1) + \cdots + (a_n - a_{n-1}) + fc_{>n} = a_n + fc_{>n}$. By the observation just made and 1.2.3, $c_{>n} \in \phi_n(M)$, so $fc \in a_n + \phi_n(M)$. By cofinality of the $a_n + \phi_n(M)$ in p we deduce that fc satisfies every condition in p.

We treat the general case by transfinite induction on a cardinal κ, where the induction hypothesis is that for every cardinal $\lambda < \kappa$ every pp-type p with parameters from M which has a cofinal subset of cardinality $\leq \lambda$ has a solution in M.

Let $f : M^\kappa \to M$ be a morphism extending the summation map from $M^{(\kappa)}$ to M. We use the standard convention that κ is identified with the set of all ordinals strictly less than κ: $\kappa = \{\alpha : \alpha < \kappa\}$. Suppose that $\{a_\alpha + \phi_\alpha(x) : \alpha < \kappa\}$ is a cofinal set in p (p as above).

For $\alpha < \kappa$ set $\Phi_\alpha = \{a_\beta + \phi_\beta(x) : \beta < \alpha\}$. By our induction hypothesis (and since every ordinal strictly less than κ has strictly less than κ ordinals below it) Φ_α has a solution, b_α say, in M. Note that for each $\beta < \alpha$ we have $b_\alpha - a_\beta \in \phi_\beta(M)$.

In particular for

$$\delta \leq \gamma < \gamma'$$

it is the case that

$$b_{\gamma'} - b_\gamma \in \phi_\delta(M). \qquad (*)$$

Every tuple $c = (c_\alpha)_\alpha \in M^\kappa$ has, for each α, an expression $c = c_{<\alpha} + c_{=\alpha} + c_{>\alpha}$ in the obvious notation (so $(c_{<\alpha})_\beta = c_\beta$ if $\beta < \alpha$, and $= 0$ otherwise, and $(c_{=\alpha})_\beta = c_\alpha$ if $\beta = \alpha$, and $= 0$ otherwise). Take $c = (a_0, b_1 - a_0, b_2 - b_1, \ldots, b_{\omega+1} - fc_{<\omega}, b_{\omega+2} - b_{\omega+1}, \ldots, b_{\lambda+1} - fc_{<\lambda}, b_{\lambda+2} - b_{\lambda+1}, \ldots)$, explicitly $c_\alpha = b_{\alpha+1} - fc_{<\alpha}$ if α is a limit ordinal and $c_\alpha = b_{\alpha+1} - b_\alpha$ otherwise (apart from $\alpha = 0, 1$).[7] We claim that fc is a solution for p.

It is enough to show that $fc \in a_\alpha + \phi_\alpha(M)$ for each $\alpha < \kappa$. Clearly this is true for finite α (as in the first part of the proof) so suppose $\alpha \geq \omega$. Write $\alpha = \lambda + n$, where λ is a limit ordinal and $n \geq 0$ is an integer. Then $c = c_{<\lambda} + c_{=\lambda} + c_{=\lambda+1} + \cdots + c_{=\lambda+n} + c_{>\alpha}$, so $fc = fc_{<\lambda} + b_{\lambda+1} - fc_{<\lambda} + b_{\lambda+2} - b_{\lambda+1} + \cdots + b_{\lambda+n+1} - b_{\lambda+n} + fc_{>\alpha} = b_{\alpha+1} + fc_{>\alpha}$. Now, $b_{\alpha+1} \in a_\alpha + \phi_\alpha(M)$ by choice of $b_{\alpha+1}$. Furthermore, each coordinate of $c_{>\alpha}$ of the form $b_{\gamma+1} - b_\gamma$ is in $\phi_\alpha(M)$ by $(*)$ above.

Also, for every limit ordinal γ and $\beta < \gamma$ we have $fc_{<\gamma} \in b_\beta + \phi_\beta(M)$: we prove this by induction on γ, having fixed β. Set $\beta = \gamma' + m$, where γ' is a limit ordinal and $m \geq 0$ is an integer. Then $fc_{<\gamma} = fc_{<\gamma'} + b_{\gamma'+1} - fc_{<\gamma'} + \cdots + b_\beta - b_{\beta-1} + c'$ say, $= b_\beta + c'$. Note that each coordinate of c' is (by induction on γ and $(*)$) in $\phi_\beta(M)$ so $fc_{<\gamma} \in b_\beta + \phi_\beta(M)$ as required. Furthermore, this coset equals $b_{\gamma+1} + \phi_\beta(M)$ by $(*)$.

Therefore each coordinate of $c_{>\alpha}$ of the form $b_{\gamma+1} - fc_{<\gamma}$ also lies in $\phi_\alpha(M)$. Therefore $fc_{>\alpha} \in \phi_\alpha(M)$, so $fc \in b_{\alpha+1} + \phi_\alpha(M) = a_\alpha + \phi_\alpha(M)$, as required. □

Lemma 4.3.7. *If $f : R \to S$ is any morphism of rings and N_S is pure-injective, then N, regarded as an R-module, is pure-injective.*

This is immediate by 4.3.6, but also easy, hence more direct, from any of the other equivalents to pure-injectivity/algebraic compactness.[8]

[7] So the definition of the coordinates of c has to be by induction on α because of the "self-reference" at the limit clause.

[8] In particular the lemma is also immediate once we have 4.3.11, since it follows directly from the definition of algebraic compactness (as opposed to that of pure-injectivity) that this is an "internal" property of a module, meaning that it depends only on the structure of the module, not (cf. the comments after 4.3.11 on injectivity) on the context, that is, the category in which it lies.

An additive functor $F : \mathcal{C} \longrightarrow \mathcal{D}$ between module or other additive categories which have arbitrary direct sums is said to **commute with** or **preserve** direct sums if whenever C_i $(i \in I)$ are objects of $\mathcal{C}$, with direct sum (that is, coproduct) $\bigoplus_i C_i$, and canonical injections $j_k : C_k \to \bigoplus_i C_i$, then the map from $\bigoplus_i FC_i$ to $F(\bigoplus_i C_i)$ induced by the $Fj_k : FC_k \to F(\bigoplus_i C_i)$, is an isomorphism, and similarly for direct products, direct limits etc.

A functor $F : \mathcal{C} \longrightarrow \mathcal{D}$ **preserves** a property of objects if whenever C has that property (as an object of $\mathcal{C}$), then so does FC (as an object of $\mathcal{D}$), and F **reflects** a property if whenever FC has that property, then so does C. The terminology is extended in the obvious way to properties of morphisms.

Corollary 4.3.8. ([323, 7.35]) *If F : Mod-$R \longrightarrow$ Mod-S is an additive functor which commutes with direct sums and direct products, then F preserves pure-injectivity: if N is a pure-injective R-module, then FN is a pure-injective S-module.*

Proof. Suppose that N_R is pure-injective. Then, 4.3.6, for any index set I there is $g : N^I \to N$ such that $gj = \Sigma$, where $j : N^{(I)} \to N^I$ is the inclusion and $\Sigma : N^{(I)} \to N$ is the summation map. So $FgFj = F\Sigma$. Note that $F\Sigma : F(N^{(I)}) = (FN)^{(I)} \to N$ is the summation map – because this map is characterised by each inclusion followed by it being the identity map. Also $Fj : F(N^{(I)}) = (FN)^{(I)} \to (FN)^I = F(N^I)$ is the canonical map, being induced by the coordinate inclusions. Therefore, by 4.3.6, FN is pure-injective. □

The following property of pure-injective modules, cf. 1.2.9, has been central in the model-theoretic-influenced line of development. Recall that $\mathrm{pp}^M(\overline{a}) \geq \mathrm{pp}^N(\overline{b})$ means exactly that $\overline{a} \in \phi(M)$ implies $\overline{b} \in \phi(N)$ for all pp conditions ϕ.

Theorem 4.3.9. *If a tuple $\overline{a}$ is from the module M and the tuple $\overline{b}$ is from the pure-injective module N, then there is a morphism $f : M \to N$ taking $\overline{a}$ to $\overline{b}$ iff $\mathrm{pp}^M(\overline{a}) \geq \mathrm{pp}^N(\overline{b})$.*

Proof. This may be proved by an argument in the style of that for 4.3.11 (see [495, 2.8]). The proof we give here uses the functorial methods of Part II and is actually very simple.

One direction is immediate from 1.1.7.

For the converse, if $p = \mathrm{pp}^M(\overline{a}) \geq \mathrm{pp}^N(\overline{b}) = q$, then $F_{Dp} \leq F_{Dq}$ (12.2.1) so there is the canonical morphism $h : (R^n \otimes -)/F_{Dp} \to (R^n \otimes -)/F_{Dq}$, where

$n = l(\overline{a}) = l(\overline{b})$. By 12.2.5, $(\overline{a} \otimes -) : (R^n \otimes -) \to (M \otimes -)$ factorises as shown, and $(R^n \otimes -)/F_{Dp} \to (M \otimes -)$ is an embedding. Similarly for $\overline{b} \otimes -$.[9]

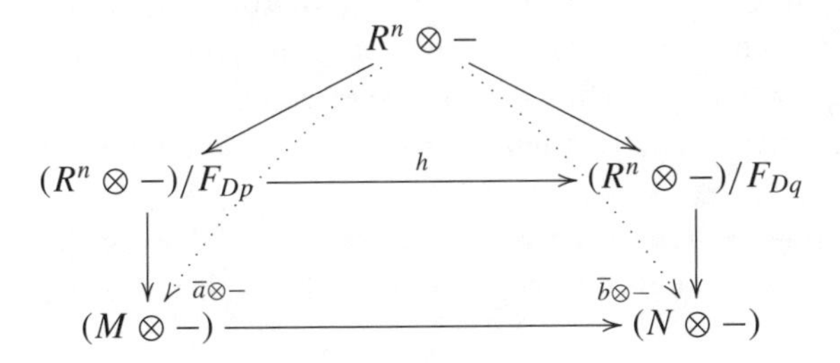

The composition $(R^n \otimes -)/F_{Dp} \to (R^n \otimes -)/F_{Dq} \to (N \otimes -)$ lifts, by injectivity of $N \otimes -$ (12.1.6), through $(R^n \otimes -)/F_{Dp} \to (M \otimes -)$. Since the embedding, $M \mapsto M \otimes -$, of Mod-R into the functor category is full, 12.1.3, there is $f : M \to N$ such that $(f \otimes -)(\overline{a} \otimes -) = (\overline{b} \otimes -)$, that is such that $f\overline{a} = \overline{b}$, as required. □

Corollary 4.3.10. *Let $\overline{a}$ be a tuple from the pure-injective module N and set $p = \mathrm{pp}^N(\overline{a})$. Then $p(N) = S \cdot \overline{a}$ where $S = \mathrm{End}(N_R)$.*

This should be compared with 1.2.11.

4.3.2 Algebraically compact = pure-injective

For the next result see Węglorz [706, 2.3], Zimmermann [731, 2.1], Warfield [699, Thm 2] (and references discussed after this last).

Theorem 4.3.11. *An R-module is pure-injective iff it is algebraically compact.*

Proof. Suppose that N is pure-injective and that p is a pp-type with parameters from N. By 4.1.4 there is a pure extension, $N^\star$, of N and $\overline{a}$ from $N^\star$ satisfying (every condition in) p. Since N is pure-injective this pure embedding is split. Projecting $\overline{a}$ to N gives a solution to p from N so, by 4.2.1, N is algebraically compact.

For the other direction suppose that N is algebraically compact, let $f : A \to B$ be a pure embedding and let $g : A \to N$ be any morphism. For each element $b \in B \setminus A$ introduce a variable x_b. For each linear relation of the form $\sum b_i r_i = a$ that holds in B, where the b_i are in $B \setminus A$, the r_i in R and a in A, form the

[9] The morphisms $\overline{a} \otimes -$ and $\overline{b} \otimes -$ are shown stippled in the diagram only for visual clarity: they are part of the initial data.

equation $\sum x_{b_i} r_i = ga$. Let Θ be the set of all such equations – a set of equations with parameters from N.

A solution to Θ in N will allow us to lift the morphism g through f: if $c_b \in N$ is the value assigned by the solution to x_b, then mapping $b(\in B \setminus A)$ to c_b will give, by construction of Θ, an R-linear map extending g.

Since N is algebraically compact there will be a solution to Θ provided that every finite subset has a solution (4.2.2). Let $\sum_{i=1}^n x_{b_i} r_{ij} = ga_j$ $(j = 1, \dots, m)$ be a finite subset of Θ (we can have a common set of variables by allowing 0 as a coefficient), so the relations $\sum b_i r_{ij} = a_j$ $(j = 1, \dots, m)$ hold in B. That is,

$$B \models \exists y_1, \dots, y_n \bigwedge_{j=1}^{m} \left(\sum_i y_i r_{ij} = a_j \right),$$

so, by purity of f, there are $a'_1, \dots, a'_n$ in A with $\sum_i a'_i r_{ij} = a_j$ for $j = 1, \dots, m$. Then $\sum g a'_i . r_{ij} = ga_j$ $(j = 1, \dots, m)$ in N and Θ is indeed finitely solvable in N, as required. □

It follows that the notion of pure-injectivity is (equivalent to one which is) "internal": it does not depend on the surrounding category. This is in marked contrast to the notion of injectivity. For example, if $R = kG$, the group ring of a finite group G over a field k, and if S is the **trivial** R-module (the vector space k with each element of G acting as the identity; a pure-injective module by 4.2.6), then the full subcategory of Mod-R whose objects are the direct sums of copies of S is equivalent to the category, Mod-k, of k-vectorspaces and, as an object of this category, S is injective but, in general, it is not an injective object of Mod-R.

Rather paradoxically, in view of the above comment, we make heavy use of the embedding, $M \mapsto M \otimes -$, of Mod-R into the functor category (R-mod, **Ab**) under which a module is pure-injective iff its image in the functor category is injective (12.1.6).

4.3.3 Pure-injective hulls

Every module has a pure-injective hull (a minimal pure-injective pure extension) (4.3.18). We give a functor-category proof which emphasises the analogy with existence of injective hulls (indeed, which is deduced from that). For absolutely pure modules the pure-injective hull is the injective hull (4.3.12).

Definable subcategories are closed under pure-injective hulls (4.3.21), indeed a pp-pair is open on the hull iff it is open on the module (4.3.23). This leads to more general statements that if a pp-pair is "small" on a module, then it is small on the pure-injective hull.

A **pure-injective hull** for a module M is a pure embedding $M \to N$ with N pure-injective and N minimal such, in the sense that there is no factorisation of this map through any direct summand of N.

Lemma 4.3.12. ([627, Props. 2, 3], also see [662, 2.5]) *A module is injective iff it is absolutely pure and pure-injective.*[10] *If M is absolutely pure, then $M \to N$ is a pure-injective hull of M iff it is an injective hull of M.*

Proof. All but one point, that a pure-injective hull $M \to N$ of an absolutely pure modules M must be an injective hull, are direct from the definition. For this, since the embedding of M into an injective hull $j : M \to E(M)$ is pure, there is $g : E(M) \to N$ whose composition with j is the embedding of M into N. Since M is essential in $E(M)$ (see Appendix E, p. 708) g must be an embedding so (a copy of) $E(M)$ lies between M and N and this, by minimality of N, must equal N. □

A pure embedding $j : M \to N$ is said to be **pure-essential** if whenever $f : N \to N'$ is a morphism such that fj is a pure embedding, then f also must be a pure embedding. It follows from the results below that a pure-injective hull is a maximal pure-essential extension.[11]

In this section it is shown that every module has a pure-injective hull. We freely use functorial methods and the full embedding of Mod-R into the functor category (R-mod, **Ab**), which, on objects, is $M \mapsto M \otimes -$. For alternative proofs see, for example, [109, §§14,21], [340, Thm 3], [660, 4.5, 9.1], [699, Prop. 6], [647, 4.1], [726, 3.6], [495, §§4.1, 4.2] (the last two prove the existence of hulls in the more general sense of Section 4.3.5: that is due originally to Fisher, in the unpublished [190, 7.18], and a more direct proof was given by Garavaglia in, the also unpublished(!), [209, Lemma 2]).

By 12.1.3, a morphism $j : M \to N$ is a pure embedding in Mod-R iff $(j \otimes -) : (M \otimes -) \to (N \otimes -)$ is a monomorphism in (R-mod, **Ab**). Recall (Appendix E, p. 708) the definition of essential embedding: just drop the word "pure" throughout the above definition of pure-essential.

10 It follows from 2.3.3 that an absolutely pure module M is injective iff it has the defining property for algebraic compactness (Section 4.2.1) just for cosets of annihilators of finite sets of elements of R.

11 Curiously, another, inequivalent, definition, given in [699, p. 709] but going back as far as [340, p. 130], has been used in the literature. The gap between this and other definitions was noticed, for example, it was commented on by Sabbagh, and Gómez Pardo and Guil Asensio [239] pointed out an example showing that the definitions differ; in particular, this second notion is not transitive. Its definition is that whenever L is a non-zero subobject of N such that $M \cap L = 0$, then $M \simeq (M + L)/L \to N/L$ is not a pure embedding, that is, if fj (as above) is a pure embedding, then f must be an embedding.

Lemma 4.3.13. *An embedding $j : M \to N$ is pure-essential iff the morphism $(j \otimes -) : (M \otimes -) \to (N \otimes -)$ is an essential embedding.*

Proof. Suppose that the embedding $j \otimes -$ of functors is essential and let $f : N \to N'$ be a morphism such that fj is a pure monomorphism hence, 12.1.3, such that $(f \otimes -)(j \otimes -)$ is monic. Then, by assumption, $f \otimes -$ is monic so, again by 12.1.3, f is a pure monomorphism.

For the converse, suppose that j is pure-essential and suppose that $\alpha : (N \otimes -) \to F$ is a morphism in $(R\text{-mod}, \mathbf{Ab})$ such that $\alpha.(j \otimes -)$ is monic. Let α' be the composition of α with an embedding of F into an injective hull which, by 12.1.6, may be taken to have the form $N' \otimes -$ for some $N' \in \text{Mod-}R$. Because the embedding to the functor category is full (12.1.3), $\alpha' = f \otimes -$ for some $f : N \to N'$. Since $(fj) \otimes -$ is monic, fj is a pure monomorphism, so, by assumption, f is a pure monomorphism. Therefore $f \otimes -$ is monic and so, therefore, is α, as required. □

Directly from this and the fact, 12.1.6, that a module N is pure-injective iff the functor $N \otimes -$ is injective, we have the next result.

Corollary 4.3.14. *A morphism $j : M \to N$ is a pure-injective hull of M iff $(j \otimes -) : (M \otimes -) \to (N \otimes -)$ is an injective hull in the functor category $(R\text{-mod}, \mathbf{Ab})$.*

Corollary 4.3.15. *(a) A composition of pure-essential embeddings is pure-essential.*
(b) If $M \to N$ is a pure-essential embedding and $M \leq M' \leq N$ with M' pure in N then both $M' \to N$ and $M \to M'$ are pure-essential.

Proof. (a) and the first part of (b) are immediate from the definitions (alternatively from the corresponding results for essential embeddings and 4.3.13). For the second part of (b), given $f : M' \to M''$ such that the composition $M \to M' \to M''$ is pure, form the pushout of f and $M' \to N$ and note that, by 2.1.13, the pushout of $M' \to N$ is pure. Then use that $M \to N$ is pure-essential to deduce that the pushout of f is pure. Then use 2.1.12. Alternatively, move to the functor category, where the proof becomes immediate (if A is essential in C and $A \leq B \leq C$ then, trivially, A is essential in B). □

Proposition 4.3.16. *Let $M \to N$ be a pure embedding with N pure-injective. Then $M \to N$ is a pure-injective hull of M iff M is pure-essential in N.*

Proof. If M were not pure-essential in N, then, by 4.3.13, the embedding of $M \otimes -$ into the injective (by 12.1.6) functor $N \otimes -$ would not be essential, hence would factor through a proper direct summand, $N' \otimes -$ say (necessarily of this form by 12.1.6), of $N \otimes -$. Then N' would be a proper direct summand of N containing M, contradicting the definition of pure-injective hull.

If, for the converse, $M \to N$ factors through a direct summand, M', of N, then the composition, $M \to N \to M'$, with the projection of N to M', is pure so, since $M \to N$ is pure-essential, $N \to M'$ is a (pure) embedding. That is, $M' = N$ as required. □

Proposition 4.3.17. *If $j : M \to N$ is a pure-injective hull of M and $f : M \to N'$ is a pure embedding with N' pure-injective, then there is $g : N \to N'$ with $gj = f$. Any morphism g with $gj = f$ must be a pure, hence split, embedding.*

Proof. Moving to the functor category, the hypothesis is that $(j \otimes -) : (M \otimes -) \to (N \otimes -)$ is an injective hull and $(f \otimes -) : (M \otimes -) \to (N' \otimes -)$ is an embedding into an injective. Since $N' \otimes -$ is injective, there is $\alpha : (N \otimes -) \to (N' \otimes -)$ such that $\alpha(j \otimes -) = (f \otimes -)$. Since the embedding of Mod-R into (R-mod, **Ab**) is full, α has the form $g \otimes -$ for some $g : N \to N'$, so $gj = f$. Furthermore, $\alpha = g \otimes -$ is monic, because its restriction to $M \otimes -$ is monic and $M \otimes -$ is essential in $N \otimes -$, so, by 12.1.3, g must be pure, as required. □

Anticipating the next result, we write $H(M)$ for any module N such that $M \to N$ is a pure-injective hull of M. Strictly speaking, it is the embedding which is the pure-injective hull but usually the context means that we do not have to be so explicitly precise.

For references for the next theorem see the comments before 4.3.13.

Theorem 4.3.18. (Existence and uniqueness of pure-injective hull) *Every module M has a pure-injective hull which is unique to isomorphism over M: if $j : M \to N$ and $j' : M \to N'$ are pure-injective hulls of M, then there is an isomorphism $f : N \to N'$ such that $fj = j'$.*

Proof. Let $(M \otimes -) \to E$ be an injective hull; the existence theorem is E.1.8. By 12.1.6, $E \simeq N \otimes -$ for some pure-injective module N and the embedding is $j \otimes -$ for some pure embedding $j : M \to N$. By 4.3.14 this is a pure-injective hull of M.

If $j' : M \to N'$ is another pure-injective hull, then $(j' \otimes -) : (M \otimes -) \to (N' \otimes -)$ is an injective hull, so, by uniqueness of injective hull (which we quote but which is easy from the definitions), there is an isomorphism, $f \otimes -$ say, from $N \otimes -$ to $N' \otimes -$ such that $(f \otimes -)(j \otimes -) = (j' \otimes -)$, hence f is an isomorphism with $fj = j'$ as required. □

Corollary 4.3.19. *For every module M there is a pure embedding $M \to N$ with N pure-injective.*

This corollary also follows from 4.2.18 and 4.3.11.

Remark 4.3.20. It follows that every endomorphism of a module lifts to at least one endomorphism of its pure-injective hull. By pure-essentiality of the hull over the module any automorphism of M will lift to an automorphism of the hull.

Theorem 4.3.21. ([628, Cor 4 to Thm 4]) *If $\mathcal{X}$ is a definable subcategory of* Mod-R *and if $M \in \mathcal{X}$, then the pure-injective hull of M is in $\mathcal{X}$.*

Proof. We outline a proof which uses the reduced product construction (Section 3.3.1). Alternatively we could quote 4.2.18. A simpler proof is given as 12.3.4; the functorial/torsion-theoretic approach there could be used to develop a rather simpler though conceptually more sophisticated treatment of definable subcategories based on 12.3.2.

Here we use the fact (4.3.11) that algebraically compact = pure-injective. In the proof of 4.1.4 we started with a module M and a pp-type p with parameters from M, then produced a pure extension, $M^\star$, of M which was a reduced power of M containing a realisation of (that is, solution for) p. By 3.4.7 if $M \in \mathcal{X}$, then $M^\star \in \mathcal{X}$. We can repeat this process, transfinitely, and obtain a pure, algebraically compact, extension of M, as follows.

Use the ordinals less than some ordinal λ to enumerate all pp-types (by 4.2.1 1-types will be enough) with parameters from M: $p_1, p_2, \dots, p_\alpha, \dots (\alpha < \lambda)$. Use the (proof of) 4.1.4 to produce a pure extension, M_1 say, of M which realises p_1, then, again, to produce a pure extension, M_2, of M_1 and hence of M, that also realises p_2, etc. At limit ordinals one takes the union (= the direct limit so, by 3.4.7, we remain within the class $\mathcal{X}$): the extension of M is still pure by 2.1.2(3). In this way we produce $M^1 \in \mathcal{X}$ (the superscript is an index, not a power!) with M pure in M^1 and such that M^1 realises every pp-(1-)type with parameters from M.

But M^1 might not be algebraically compact: M^1 need not realise every pp-type with parameters from M^1. So we repeat the process, with M^1 in place of M, to get M^2 in place of M^1. And so on, transfinitely.

This, potentially rather extravagant, process does stop – eventually M catches up with M^1 – for the following reason (also relevant to the proof of 4.2.18). There are at most $\kappa = \mathrm{card}(R) + \aleph_0$ pp conditions for R-modules so any pp-type is determined by specifying at most this number of cosets, hence needs at most this number of parameters to specify it. In our second transfinite process, eventually we take the union at some limit ordinal γ which has cofinality strictly greater than κ. That means that for any pp-type, p, with parameters over the union, M^γ, all the $\leq\kappa$ parameters which are needed to specify p appear already in some M^β with $\beta < \gamma$ and therefore there is a realisation of p in $M^{\beta+1} \leq M^\gamma$. It follows, by

4.2.1, that M^γ is algebraically compact, hence pure-injective. The pure-injective hull of M is a direct summand of M^γ and so is also in $\mathcal{X}$. □

Example 4.3.22. Let $M = \bigoplus_p \overline{\mathbb{Z}_p}$ (it follows from 4.2.7 that $\overline{\mathbb{Z}_p} = H(\mathbb{Z}_p)$). Then $H(M) = \prod_p \overline{\mathbb{Z}_p}$. For we know that $H(M)$ is a summand of this product since by 4.3.2 the product is pure-injective and M is pure in it. By 3.4.7 any summand must be in the definable category generated by the modules $\overline{\mathbb{Z}_p}$ (p prime). By the argument of 3.4.13 the only possible complement of $H(M)$ in $\prod_p \overline{\mathbb{Z}_p}$ is a direct sum of copies of $\mathbb{Q}$, but the projection of such a divisible abelian group to any coordinate would be divisible, hence 0, as required. (Also see 5.2.7.)

Corollary 4.3.23. ([628, Cor. 4 to Thm 4]) *For every module M and pp-pair ϕ/ψ, ϕ/ψ is open on $H(M)$ iff it is open on M, that is, $\phi(M) > \psi(M)$ iff $\phi(H(M)) > \psi(H(M))$.*

Proof. By 4.3.21 applied to the definable subcategory generated by M, and 3.4.11, if a pp-pair is open on $H(M)$, then it is open on M. The converse is true because M is pure in $H(M)$. □

Recall (Section 3.2.1) that $\mathrm{pp}^n(M)$ denotes the lattice of subgroups of M^n pp-definable in M.

Lemma 4.3.24. *For every module M and for every integer n, the natural map of lattices from $\mathrm{pp}^n(H(M))$ to $\mathrm{pp}^n(M)$ is an isomorphism.*

Proof. Since M is pure in $H(M)$ there is a natural surjection from $\mathrm{pp}(H(M))$ to $\mathrm{pp}(M)$ (3.2.2). By 4.3.23 this map is injective. □

The actual statement of Sabbagh's result [628, Cor. 4 to Thm 4] is that every module is elementarily equivalent to, indeed is an elementary substructure of (Appendix A), its pure-injective hull. The result below is an immediate consequence of that (as are 4.3.21 and 4.3.23).

Proposition 4.3.25. *Let M be any module and let ϕ/ψ be a pp-pair such that $\phi(M)/\psi(M)$ is finite. Then $\phi(H(M))/\psi(H(M))$ and $\phi(M)/\psi(M)$ have the same cardinality, where $H(M)$ denotes the pure-injective hull of M.*

Proof. We use the formula $\phi(M)/\psi(M) \simeq (F_{D\psi/D\phi}, M \otimes -)$ (10.3.8) and the fact (12.1.6) that for every module M the functor $M \otimes -$ is an absolutely pure object of the functor category (R-mod, **Ab**).

Take any morphism $\alpha : F_{D\psi/D\phi} \to (H(M) \otimes -)$. Since $M \otimes -$ is essential in $H(M) \otimes -$ (4.3.13) there is $F' \leq F_{D\psi/D\phi}$ with $\alpha F' \neq 0$ and $\alpha F' \leq (M \otimes -)$. We may take F' to be finitely generated and, therefore (10.2.3), finitely presented.

Since $M \otimes -$ is absolutely pure in (R-mod, **Ab**) there is, by 2.1.7, a morphism $\alpha' : F_{D\psi/D\phi} \to (M \otimes -)$ which agrees with α on F'.

If $\alpha \neq \alpha'$, then we repeat the argument with $\alpha - \alpha'$ in place of α to obtain a morphism $\alpha'' \in (F_{D\psi/D\phi}, M \otimes -)$ which agrees with $\alpha - \alpha'$ on some F'', which may be taken to contain F', by construction, properly.

We continue this process (with $\alpha - \alpha' - \alpha''$ replacing $\alpha - \alpha'$, etc.), which must terminate since the morphisms $\alpha' + \alpha'' + \cdots$ agree with α on a strictly increasing sequence of subobjects of $F_{D\psi/D\phi}$ and since $(F_{D\psi/D\phi}, M \otimes -)$ is finite. The process can only terminate when $\alpha = \alpha' + \alpha'' + \cdots \in (F_{D\psi/D\phi}, M \otimes -)$. Hence the embedding of $(F_{D\psi/D\phi}, M \otimes -)$ into $(F_{D\psi/D\phi}, H(M) \otimes -)$ induced by $(M \otimes -) \to (H(M) \otimes -)$ is surjective, as required. □

The proof shows that the groups $\phi(H(M))/\psi(H(M))$ and $\phi(M)/\psi(M)$ are isomorphic. It also adapts to give the following.

Proposition 4.3.26. ([247], see [522, 3.4]) *Let M be a module over a k-algebra R, where k is any field. Let ϕ/ψ be any pp-pair. If $\phi(M)/\psi(M)$ is finite-dimensional over k, then $\phi(H(M))/\psi(H(M))$ has the same dimension. In particular, if $\phi(M)$ is finite-dimensional, then $\phi(H(M)) = \phi(M)$.*

More generally, the conclusion holds for any module M over any ring provided the End(M)*-module structure on $\phi(M)/\psi(M)$ makes $\phi(M)/\psi(M)$ a finite-dimensional vector space over some field k.*

Proof. We use the notation of the proof of 4.3.25. The strictly increasing chain of functors $F' < F'' < \cdots$ annihilated by the various differences between the αs induces a strictly decreasing chain of k-subspaces, $\mathrm{ann}(F') = \{\beta : \beta F' = 0\} > \mathrm{ann}(F'') > \cdots$, of $\mathrm{Hom}(F_{D\psi/D\phi}, M \otimes -) + k \cdot \alpha \leq \mathrm{Hom}(F_{D\psi/D\phi}, H(M) \otimes -)$ (identifying $\mathrm{Hom}(F_{D\psi/D\phi}, M \otimes -)$ with its image in $\mathrm{Hom}(F_{D\psi/D\phi}, H(M) \otimes -)$). Finite dimensionality of $\phi(M)/\psi(M) \simeq \mathrm{Hom}(F_{D\psi/D\phi}, M \otimes -)$, applied at the end of the proof of 4.3.25, yields the conclusion. □

Essentially the same proof gives the next result which, given a different proof, is used in the classification results in [166] (see Section 8.2.9).

Proposition 4.3.27. ([166, 2.1]) *Let R be a local commutative ring with maximal ideal J and suppose that J is finitely generated. Let M be any R-module and let $H(M)$ denote its pure-injective hull. If M/MJ is finite-dimensional over R/J, then the embedding of M/MJ into $H(M)/H(M)J$ is an isomorphism.*

Proof. Since J is finitely generated (as a left ideal) there is a pp condition $\psi(x)$ such that $\psi(M) = MJ$ for every module M. Take $\phi(x)$ to be $x = x$, so $\phi(M) = M$. Since R/J is a field and M/MJ an R/J-vectorspace, 4.3.26 applies to give the result. (Note that $MJ = H(M)J \cap M$) by purity of M in $H(M)$ and pp-definability of $(-)J$, so M/MJ does embed in $H(M)/H(M)J$.) □

The three results above, and somewhat related results in [495, §4.6]), were originally proved by (various and) very different arguments. Here is another example of a variant on a result of Granger from [247]. Again, the functorial proof is very natural. As an alternative route to such results one may use (the proof of) [284, Lemma 6] which shows that if M is any module, if ϕ/ψ is a pp-pair and if $a \in \phi(H(M)) \setminus \psi(H(M))$, then there is ψ' with $\phi > \psi' \geq \psi$ and $a \notin \psi'(H(M))$ such that $a \in M + \psi'(H(M))$.

Proposition 4.3.28. *Suppose ϕ/ψ is an M-minimal pair. Then $(\phi/\psi)(H(M)/M) = 0$, where $H(M)$ is the pure-injective hull*[12] *of M.*

Proof. It is enough to show that the natural inclusion $\phi/\psi(M) \to \phi/\psi(H(M))$ is surjective. We give a functorial proof of this fact.

Consider $F = F_{D\psi}/F_{D\phi}$ – a finitely presented functor in (mod-R, **Ab**) (10.2.30). By 12.1.6 we have an exact sequence of functors $0 \to (M \otimes -) \to (H(M) \otimes -) \to (H(M)/M \otimes -) \to 0$. Apply $(F, -)$ to obtain the long exact sequence $0 \to (F, M \otimes -) \to (F, H(M) \otimes -) \to (F, H(M)/M \otimes -) \to \mathrm{Ext}^1(F, M \otimes -) \to \cdots$. Also by 12.1.6 the functor $M \otimes -$ is absolutely pure = fp-injective, hence $\mathrm{Ext}^1(F, M \otimes -) = 0$ and so, since $(F_{D\psi}/F_{D\phi}, M \otimes -) \simeq (\phi/\psi)(M)$ (10.3.8) etc., we deduce that $(\phi/\psi)(H(M)/M) = 0$ iff $(\phi/\psi)(M) \to (\phi/\psi)(H(M))$ is surjective. This required no assumption on ϕ/ψ.

Now, the assumption that ϕ/ψ is a minimal pair is exactly that F_ϕ/F_ψ, hence $F_{D\psi}/F_{D\phi}$, is a simple functor. Since $M \otimes -$ is essential in $H(M) \otimes -$ (4.3.14) any morphism from $F_{D\psi}/F_{D\phi}$ to $H(M) \otimes -$ must factor through $M \otimes -$, so $(F, M \otimes -) \to (F, H(M) \otimes -)$ is indeed surjective, as required. □

4.3.4 Pure-injective extensions via duality

The hom-dual of a module is pure-injective (4.3.29). The double dual is in the same definable subcategory as the module, so this gives another proof that definable subcategories are closed under pure-injective hulls (4.3.32).

Proposition 4.3.29. *If ${}_SM_R$ is an (S, R)-bimodule and ${}_SE$ is an injective left S-module, then $M^* = \mathrm{Hom}_S({}_SM_R, {}_SE)$ is a pure-injective left R-module.*

[12] I also mention that, on the way to proving that n-cotilting modules are pure-injective, Šťovíček proved ([665, Thm 11]) that if A is any module such that ${}^\perp A = \{M \in \text{Mod-}R : \mathrm{Ext}^1(M, A) = 0\}$ is closed under pure submodules and products, then it is also closed under pure epimorphic images and for any $M \in {}^\perp A$ also $H(M)/M \in {}^\perp A$.

Proof. We prove that M^* is algebraically compact, hence pure-injective. Let $\big(f_\lambda + \phi_\lambda(M^*)\big)_{\lambda\in\Lambda}$ be a collection of cosets of pp-definable subgroups of M^* with the finite intersection property. We must produce an element of M^* in their intersection $\bigcap_\lambda \big(f_\lambda + \phi_\lambda(M^*)\big)$, that is, an element $f \in M^*$ with $f - f_\lambda \in \phi_\lambda(M^*)$ for every λ, that is, by 1.3.12, with $f \restriction D\phi_\lambda(M) = f_\lambda \restriction D\phi_\lambda(M)$.

Since $\big(f_\lambda + \phi_\lambda(M^*)\big) \cap \big(f_\mu + \phi_\mu(M^*)\big)$ is non-empty it follows that $f_\lambda \restriction D\phi_\lambda(M)$ and $f_\mu \restriction D\phi_\mu(M)$ agree on $D\phi_\lambda(M) \cap D\phi_\mu(M)$ so have a common extension to $D\phi_\lambda(M) + D\phi_\mu(M)$ $(= D(\phi_\lambda \wedge \phi_\mu)(M))$. Hence the set of specifications $f \restriction D\phi_\lambda(M) = f_\lambda \restriction D\phi_\lambda(M)$ $(\lambda \in \Lambda)$ is consistent and defines a function from $\sum_{\lambda\in\Lambda} D\phi_\lambda(M)$ to E. Since E is injective and $\sum_\lambda D\phi_\lambda(M)$ is an S-submodule of M, such a function extends to a function from M to E, yielding $f \in M^*$ with $f \in \bigcap_\lambda \big(f_\lambda + \phi_\lambda(M^*)\big)$, as required. □

In the next proposition we assume that A, B, C all are (S, R)-bimodules and that ${}_SE$ is injective. For example, take $S = \mathbb{Z}$ and $E = \bigoplus\{\mathbb{Z}_{p^\infty} : p \text{ prime}\} = \mathbb{Q}/\mathbb{Z}$ to be a minimal injective cogenerator for $\mathbb{Z}$-Mod. As usual, set $(-)^* = \mathrm{Hom}_S({}_S(-)_R, {}_SE)$.

Proposition 4.3.30. *If the exact sequence* $0 \to A \xrightarrow{i} B \xrightarrow{\pi} C \to 0$ *is pure-exact, then the dual sequence* $0 \to C^* \xrightarrow{\pi^*} B^* \xrightarrow{i^*} A^* \to 0$ *is split. The converse is true if E is an injective cogenerator for S-*Mod.

Proof. Let $\phi(x)$ be a pp condition for *left* R-modules and suppose $f \in C^*$ is such that $\pi^* f \in \phi(B^*)$. By 1.3.12, $\pi^* f \cdot D\phi(B) = 0$. Let $c \in D\phi(C)$. Since the original sequence is pure-exact, there is, by 2.1.14, $b \in D\phi(B)$ with $\pi b = c$. Then $fc = f\pi b = \pi^* f \cdot b = 0$. So $f \cdot D\phi(C) = 0$ and we deduce from 1.3.12 that $f \in \phi(C^*)$, as required for purity (by 2.1.6 or by replacing the elements by tuples, as one prefers).

Assume now that the dual sequence is pure, hence, by what has just been shown, that its double dual is pure. By 4.3.29 the sequence is, therefore, split. If E is an injective cogenerator there is an embedding (see 1.3.16) of the original sequence into its double dual, as shown.

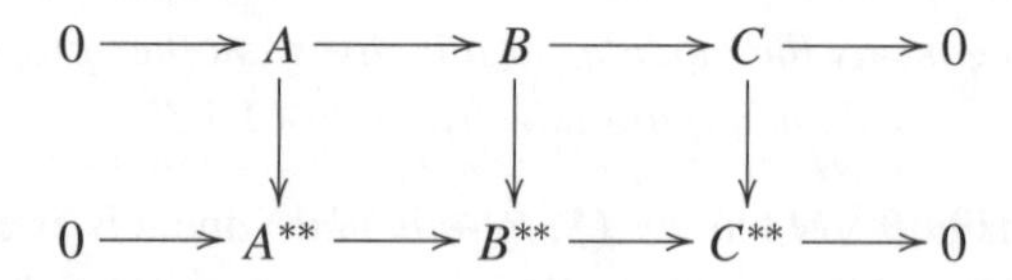

By 1.3.16 $A \to A^{**}$ is pure so the composition $A \to B^{**}$ is pure and hence, by 2.1.12, $A \to B$ is pure, as required. □

It was seen at 1.3.16 that the embedding of M into M^{**} is pure so we have the following.

Corollary 4.3.31. *Let ${}_SM_R$ be an (S, R)-bimodule and let ${}_SE$ be an injective cogenerator for S-*Mod. *Then the embedding of M into M^{**} given by $m \mapsto (f \mapsto fm)$ $(m \in M,\ f \in M^{**})$ is a pure embedding into a pure-injective module.*

In particular, the pure-injective hull of M is a direct summand of M^{**}. This double dual also lies in the definable subcategory generated by M (by two applications of 1.3.13).

Corollary 4.3.32. *Let ${}_SM_R$ be an (S, R)-bimodule and let ${}_SE$ be an injective cogenerator for S-*Mod. *Then M^{**} belongs to the definable subcategory generated by M.*

See [285] for a teasing apart of the map $M \mapsto M^{**}$ including conditions [285, §6] where one gets exactly the pure-injective hull and more information in the case where $M = R_R$.

4.3.5 Hulls of pp-types

It is also possible to define the notion of hull, not just of a module, but of any subset A of a module M. This is a, unique-to-isomorphism (4.3.35), minimal direct summand of the pure-injective hull of M which contains A (4.3.33). It depends on the pp-type of A in M but this apparent subtlety of the notion is dispelled once it is realised that this is simply injective hull in the functor category.

It is also shown that there is just a set of indecomposable pure-injective R-modules up to isomorphism (4.3.38) and yet another proof of existence of realisations of pp-types is given (4.3.39).

The notion of hull introduced in this section is due to Fisher, [190, §3]; Garvaglia presents Fisher's results, with somewhat different proofs, in [209, §1]. These manuscripts were circulated but neither was published. Published accounts are in [726, §3] and [495, §§4.1, 4.2]. All these proofs were essentially model-theoretic; Fisher's was actually set in the rather general context of categories of what he called "abelian structures". The functorial account is somewhat simpler.

Let $\overline{a}$ be an n-tuple from a module M and set $p = \mathrm{pp}^M(\overline{a})$ to be its pp-type (Section 1.2.1). Regarded as a morphism from R^n to M, $\overline{a}$ induces a morphism between functors, $(\overline{a} \otimes -) : (R^n \otimes -) \to (M \otimes -)$, and, corresponding to the embedding of $M \otimes -$ into its injective hull $H(M) \otimes -$ (12.1.8), there is a morphism, which we also denote by $\overline{a} \otimes -$, from $R^n \otimes -$ to $H(M) \otimes -$, namely that induced by regarding $\overline{a}$ as a tuple from $H(M)$. The injective hull of $\mathrm{im}(\overline{a} \otimes -)$ is a direct summand of $H(M) \otimes -$, hence (12.1.6) has the form $H^M(\overline{a}) \otimes -$ for some pure-injective module, let us denote it $H^M(\overline{a})$, which is a direct summand of $H(M)$. If M already is pure-injective, and so contains $H^M(\overline{a})$ as a direct summand, then we refer to $H^M(\overline{a})$ as the **hull of $\overline{a}$ in M**. It is unique up to automorphism of M fixing $\overline{a}$ because injective hulls have the corresponding property.

If $A \subseteq M$ is any subset of the module M, then the same construction may be made using $\left(\sum_{a\in A} \mathrm{im}(a \otimes -)\right) \leq (H(M) \otimes -)$ and we write $H^M(A)$. Of course, if A is a submodule, then we just use the image of $A \otimes -$ in $M \otimes -$. Let us say all this formally.

Proposition 4.3.33. *If N is pure-injective and A is any subset of N, then there is a direct summand, N', of N such that N' contains A and such that no proper direct summand of N' contains A. Such a module, N', is unique up to isomorphism over A and is a copy of the hull of A in N.*

Proof. If A is finite, the case we use most, arrange it as a tuple $\overline{a}$. The morphism $\overline{a} : R^n \to N$ gives rise to $(\overline{a} \otimes -) : (R^n \otimes -) \to (N \otimes -)$. Let N' be such that $N' \otimes -$ is a copy of the injective hull of the image of this morphism, contained in $N \otimes -$. Since $N' \otimes -$ is a direct summand of $N \otimes -$ so is N' a direct summand of N' (12.1.3). Any direct summand of N containing $\overline{a}$ would give a direct summand of $N \otimes -$ containing the image of $\overline{a} \otimes -$, so N' is minimal. Similarly if A is an arbitrary subset or submodule of M. The last statement is by uniqueness to isomorphism of injective hulls (Appendix E, p. 709). □

This hull depends not just on the isomorphism type of the submodule generated by $\overline{a}$ but also on how $\overline{a}$ sits in M: take $R = \mathbb{Z}$, $M = \mathbb{Z}_2 \oplus \mathbb{Z}_4$, $a = (1, 0)$, $a' = (0, 2)$ to illlustrate the point. More precisely, it is determined by $\mathrm{pp}^M(\overline{a})$. Indeed, it is shown in 12.2.5 that $\ker(\overline{a} \otimes -) = F_{Dp} = \sum\{F_{D\phi} : \phi \in p\}$, so $H^M(\overline{a})$ is determined by $p = \mathrm{pp}^M(\overline{a})$. We may, therefore, denote this module by $H(p)$ and call it the **hull** of the pp-type p. For future reference we extract the following statement.

Remark 4.3.34. If p is a pp-type and $H(p)$ its hull, then $(H(p) \otimes -) = E\big((R^n \otimes -)/F_{Dp}\big)$.

Proposition 4.3.35. *Let M, M' be any modules and let $H(M)$, $H(M')$ be their respective pure-injective hulls. Suppose that $\overline{a}$ from M and $\overline{a}'$ from M' have the same pp-type $p = \mathrm{pp}^M(\overline{a}) = \mathrm{pp}^{M'}(\overline{a}')$. Then there are direct summands N of $H(M)$ and N' of $H(M')$, with $\overline{a}$ from N and $\overline{a}'$ from N', and an isomorphism $f : N \to N'$ such that $f\overline{a} = \overline{a}'$. These summands may be taken to be copies of $H(p)$.*

Proof. Moving to the functor category, the assumption on $\overline{a}$ and $\overline{a}'$ is that $\mathrm{im}(\overline{a} \otimes -) \simeq \mathrm{im}(\overline{a}' \otimes -)$ so, by the corresponding result for injective objects, there are direct summands E of $H(M) \otimes -$ and E' of $H(M)' \otimes -$, containing those respective images, and which are isomorphic by a morphism α such that the composition

$\alpha.(\overline{a} \otimes -)$ is equal to $\overline{a}' \otimes -$. Translating back to the module category gives the statement. □

Corollary 4.3.36. *If N is a pure-injective module, if $\overline{a} = (a_1, \dots, a_n)$ is from N and $p = \mathrm{pp}^N(\overline{a})$, then there is a copy of the hull, $H(p)$, of p which contains all the a_i and which is a direct summand of N.*

Given the module $H(p)$, if $\overline{b}$ is any tuple, of any length, with entries from $H(p)$, then $H(p)$ contains, as a direct summand, the hull of $\overline{b}$. In particular, if $H(p)$ is indecomposable and $\overline{b} \neq \overline{0}$, then $H(p)$ is the hull of $\overline{b}$.

Corollary 4.3.37. ([726, 4.1]) *If N is an indecomposable pure-injective module, then there is a pp-type p such that $N \simeq H(p)$. Indeed, if $\overline{a} \neq \overline{0}$ is from N, then $N = H(\mathrm{pp}^N(\overline{a}))$.*

The next result is stated[13] explicitly as [726, 4.2(10] and follows already from [209, Lemma 2].

Corollary 4.3.38. *There is just a set of indecomposable pure-injective R-modules up to isomorphism, indeed there are at most $2^{\mathrm{card}(R)+\aleph_0}$.*

Proof. There is just a set of pp-types (without parameters) and, by 4.3.37, every indecomposable pure-injective is isomorphic to the hull of a pp-type. More precisely there are $\mathrm{card}(R) + \aleph_0$ pp conditions therefore no more than $2^{\mathrm{card}(R)+\aleph_0}$ pp-types for R-modules. □

This approach gives an alternative route to showing that pp-types are always realised, cf. 3.3.6.

Theorem 4.3.39. *Let p be a pp-n-type. Then there is a (pure-injective) module N and an n-tuple, $\overline{a}$, from N such that $\mathrm{pp}^N(\overline{a}) = p$.*

Proof. Let E be the injective hull of the functor $(R^n \otimes -)/F_{Dp}$, where $F_{Dp} = \sum\{F_{D\phi} : \phi \in p\}$. This has the form $N \otimes -$ for some pure-injective module N. Then the projection from $R^n \otimes -$ to $(R^n \otimes -)/F_{Dp}$ followed by inclusion into $N \otimes -$ has the form $\overline{a} \otimes -$ for some tuple $\overline{a}$ in N since, 12.1.3, $\big((R^n \otimes -), (N \otimes -)\big) \simeq (R^n, N)$. We have, cf. the proof of 12.2.1, $\phi \in p$ iff $F_{D\phi} \leq F_{Dp}$ and this is so iff the morphism $\overline{a}$ factors through $(R^n \otimes -) \to (R^n \otimes -)/F_{D\phi}$. By 10.3.9 this is the case iff $\overline{a} \in \phi(N)$. Therefore the pp-type of $\overline{a}$ in N is p. □

This is an appropriate place to mention the notion of pp-essential, which was defined in [495, p. 70].[14] Though I do not make much use of it here, the functorial

[13] Rather, mis-stated; the word "compact", meaning pure-injective, is omitted.
[14] The equivalent notion, "small over", was defined by Ziegler [726, 3.7].

approach clarifies the notion, which is the following. Suppose that M is a module and let $A \leq C \leq M$. Then the embedding of A in C is **pp-essential** (**relative to**) M if whenever $f : M \to M'$ is a morphism which preserves the pp-type of A, then also it preserves the pp-type of C. To say that f preserves the pp-type of A, $\mathrm{pp}^M(A) = \mathrm{pp}^{M'}(fA)$, means that $\mathrm{pp}^M(\overline{a}) = \mathrm{pp}^{M'}(f\overline{a})$ for every finite tuple $\overline{a}$ from A.

Lemma 4.3.40. *Suppose that $A \leq C \leq M$. Then A is pp-essential in C relative to M iff the image of $A \otimes -$ in $M \otimes -$ is essential in the image of $C \otimes -$ in $M \otimes -$. (Images refer to those under the maps induced by the inclusions.)*

Proof. Let $i : A \to C$, $j : C \to M$ be the inclusions. So the second condition is that $\mathrm{im}((ji) \otimes -)$ is an essential subobject of $\mathrm{im}(j \otimes -)$, both contained in $M \otimes -$.

Let $f : M \to M'$. The condition that f preserve the pp-type of A is, by 12.2.5, exactly that the induced map $\mathrm{im}((ji) \otimes -) \to \mathrm{im}((fji) \otimes -)$ be a monomorphism, so the result follows quite directly from the definition of essential embedding and the fact, 12.1.3, that the tensor embedding from Mod-R to (R-mod, **Ab**) is full. □

Corollary 4.3.41. (cf. [495, 4.12 ff.]) *Suppose that $A \leq C \leq M$ and that A is pp-essential in C relative to M. Let $g : A \to N$ be such that $\mathrm{pp}^N(gA) = \mathrm{pp}^M(A)$ and suppose that N is pure-injective. Then there is an extension of g to a morphism $h : C \to N$ and, for every such extension, one has $\mathrm{pp}^N(hC) = \mathrm{pp}^M(C)$. In particular if $C = M$ then hM is pure in N.*

Proof. Moving to the functor category, and with notation, as above, this is simply the statement that, since $N \otimes -$ is injective, any morphism from $\mathrm{im}((ji) \otimes -)$ to $N \otimes -$ extends to one from $\mathrm{im}(j \otimes -)$, which, by 4.3.40, must be monic if the morphism from $\mathrm{im}((ji) \otimes -)$ is. □

The submodule A appearing in 4.3.41 could as well be replaced by a generating tuple $\overline{a}$ of elements. Note that, by 4.3.40 and 4.3.14, $\overline{a}$ is pp-essential in M iff the pure-injective hull, $H(M)$, of M is a copy of the hull, $H(\overline{a})$, of $\overline{a}$.

Lemma 4.3.42. *Suppose that $\overline{a}$ is a tuple from the module M and that $\overline{a}$ is pp-essential in M. If $f : M \to M'$ is any morphism such that $\mathrm{pp}^{M'}(f\overline{a}) = \mathrm{pp}^M(\overline{a})$, then f is a pure embedding.*

Proof. Consider the composition of f with the embedding of M' into its pure-injective hull. By pure-injectivity of that module, this factors through the embedding of M into its pure-injective hull. By the hypothesis and 4.3.41 the morphism from $H(M)$ to $H(M')$ is pure, so the original morphism f is pure. □

4.3.6 Indecomposable pure-injectives and irreducible pp-types

Every indecomposable pure-injective module has local endomorphism ring (4.3.43) and is a module over a certain localisation of the ring: that obtained by inverting the central elements which act automorphically on the module (4.3.44). The Jacobson radical of an indecomposable pure-injective consists of the endomorphisms which strictly increase pp-types (4.3.45). Indecomposable pure-injectives are isomorphic iff they realise a common non-zero pp-type (4.3.47).

A pp-type is irreducible iff it is realised in an indecomposable pure-injective module (4.3.49). A technical-seeming, but useful, way of producing irreducible pp-types, hence indecomposable pure-injectives, is 4.3.52. Every minimal pp-pair in an indecomposable pure-injective is one-dimensional over the top of the endomorphism ring (4.3.53).

Restriction of scalars certainly need not preserve indecomposability of pure-injectives but indecomposable summands of restrictions of pure-injectives are summands of restrictions of indecomposables (4.3.54).

In our terminology we normally exclude the zero module when using the term "indecomposable". Just occasionally it is allowed to slip in when this makes for neater formulations.

First we look at the effect of morphisms on pp-types. The first result is from [190, 7.14], [729, Thm 9], [256, 1.3], [726, 4.3].

Theorem 4.3.43. *Every indecomposable pure-injective module has local endomorphism ring.*

Proof.[15] If N is an indecomposable pure-injective, then $N \otimes -$ is an indecomposable injective in (R-mod, **Ab**) (12.1.6), hence has local endomorphism ring (E.1.23). Fullness of the embedding of Mod-R into (R-mod, **Ab**), 12.1.3, gives $\mathrm{End}(N) \simeq \mathrm{End}(N \otimes -)$. This proof, which is direct from fullness of the embedding of Gruson and Jensen, [254], is much simpler than the proofs which are set in Mod-R. □

If $P \subseteq C(R)$ is a prime ideal contained in the centre, $C(R)$, of R, then the elements of $C(R) \setminus P$ may be inverted, giving the central localisation of R at P: $R_{(P)} = \{rs^{-1} : s \in C(R) \setminus P\}$. There is the canonical map $R \to R_{(P)}$ given

[15] This proof, and the preprint version of [256], was pointed out to me by Peter Vámos around 1980.

by $r \mapsto r.1^{-1}$. Note that if $r \in C(R)$, then multiplication by r is an R-linear endomorphism of any R-module.

Corollary 4.3.44. ([726, 5.4], [495, 2.$\mathbb{Z}$.8]) *Let N be an indecomposable pure-injective module over the ring R. Let*

$$P = \{r \in C(R) : \textit{multiplication by } r \textit{ is not a bijection on } N\}$$
$$= C(R) \cap \mathrm{JEnd}(N)$$

where $\mathrm{JEnd}(N)$ *denotes the Jacobson radical of the endomorphism ring of N.*

Then P is a prime ideal of the centre of R and N is naturally a module over the localisation, $R_{(P)}$, of R obtained by inverting all central elements not in P.[16]

Proof. The fact that $P = C(R) \cap \mathrm{JEnd}(N)$, hence is an ideal of $C(R)$, is by 4.3.43 and clearly it is prime. Since each element of $C(R) \setminus P$ acts as an isomorphism on N the R-module structure on N extends canonically to an $R_{(P)}$-module structure. □

The propositions which follow appeared variously in work of Garavaglia, Prest and Ziegler on the model theory of modules. See [495], especially Chapter 4 there, for references ([495, p. 84] for the next result).

Proposition 4.3.45. *Suppose that N is an indecomposable pure-injective module and let $f \in \mathrm{End}(N)$. Then the following are equivalent:*

(i) $f \in \mathrm{JEnd}(N)$;
(ii) there is $a \in N$ *such that* $\mathrm{pp}^N(a) > \mathrm{pp}^N(fa)$;
(iii) for all non-zero tuples $\overline{a}$ *from N we have* $\mathrm{pp}^N(\overline{a}) > \mathrm{pp}^N(f\overline{a})$.

Proof. (i)⇒(iii) Let $\overline{a} = (a_1, \dots, a_n)$ be a tuple from N with a_1, say, non-zero. If $\mathrm{pp}^N(\overline{a}) = \mathrm{pp}^N(f\overline{a})$, then there is, by 4.3.9, $g \in \mathrm{End}(N)$ such that $gf\overline{a} = \overline{a}$, in particular, $gfa_1 = a_1$. But if $f \in \mathrm{JEnd}(N)$, then, since $\mathrm{End}(N)$ is local (4.3.43), $1 - gf$ would be an automorphism of N, a contradiction.

(iii)⇒(ii) is trivial and (ii)⇒(i) is immediate since if $f \in \mathrm{End}(N) \setminus \mathrm{JEnd}(N)$, then, since $\mathrm{End}(N)$ is local, f is an automorphism of N, so $\mathrm{pp}^N(a) = \mathrm{pp}^N(fa)$ for all $a \in N$. □

Corollary 4.3.46. ([495, 4.12] for references) *If N, N' are pure-injective modules with N indecomposable, if $\overline{a}$ is a non-zero tuple from N and $\overline{b}$ is a tuple from N'*

[16] Sometimes one gets more: for example, it is shown in [428, 2.1] that if R is a Dedekind domain of characteristic 0, if G is a finite group and if $N \in \mathrm{Zg}_R$ is R-torsionfree, then either N is a simple QG-module, where Q is the quotient field of R, or there is a maximal prime P such that N is a (pure-injective) module over $\overline{R_{(P)}}G$, where $\overline{R_{(P)}}$ is the completion of R at P.

with $\mathrm{pp}^N(\overline{a}) = \mathrm{pp}^{N'}(\overline{b})$, *then* N *is isomorphic to a direct summand of* N' *by a map taking* $\overline{a}$ *to* $\overline{b}$.

This is immediate by 4.3.35, or just say: by 4.3.9 there are morphisms $f : N \to N'$ and $g : N' \to N$ taking $\overline{a}$ to $\overline{b}$ and $\overline{b}$ to $\overline{a}$ respectively; by 4.3.45 the composition gf is an isomorphism, as required.

Corollary 4.3.47. *If* M, N *are indecomposable pure-injective modules and if* $\overline{a}$ *from* M *and* $\overline{b}$ *from* N *are such that* $\mathrm{pp}^M(\overline{a}) = \mathrm{pp}^N(\overline{b})$, *then there is an isomorphism from* M *to* N *taking* $\overline{a}$ *to* $\overline{b}$.

A somewhat stronger form of 4.3.46 is the following.

Proposition 4.3.48. *Suppose that* N *is an indecomposable pure-injective and let* $f : N \to M$ *be a morphism such that there is some non-zero* $a \in N$ *with* $\mathrm{pp}^M(fa) = \mathrm{pp}^N(a)$. *Then* f *embeds* N *as a direct summand of* M, *in particular, for every tuple* $\overline{c}$ *from* N *we have* $\mathrm{pp}^N(f\overline{c}) = \mathrm{pp}^M(\overline{c})$.

Proof. Consider the induced morphism $f \otimes - : (N \otimes -) \to (M \otimes -)$. By 12.2.5 this is monic when restricted to the image of $a \otimes - : (R \otimes -) \to (N \otimes -)$, so, since $N \otimes -$ is an indecomposable, hence uniform, injective functor, $f \otimes -$ is monic and so $f(N \otimes -)$ is direct summand of $M \otimes -$. Therefore by fullness of the tensor embedding (12.1.3), fN is a direct summand of M, as required. □

Now we concentrate on indecomposable pure-injectives. Note that such a module has cardinality bounded above by $2^{\mathrm{card}(R)+\aleph_0}$ ([726, 4.2] or, more generally, [628]). A pp-n-type p is **irreducible** if whenever ψ_1, ψ_2 are pp conditions (in n free variables) not in p there is some $\phi \in p$ such that $\phi \wedge \psi_1 + \phi \wedge \psi_2 \notin p$.

The next result played a fundamental role in [726]; the original proof looks very different but is essentially the same as that given here.

Theorem 4.3.49. (Ziegler's Criterion, [726, 4.4]) *Let* p *be a pp-type. Then the hull,* $H(p)$, *of* p *is indecomposable iff* p *is irreducible.*

Proof. By 12.2.3, p is irreducible iff the functor $(R^n \otimes -)/F_{Dp}$ is uniform, equivalently, iff the injective hull, $E\big((R^n \otimes -)/F_{Dp}\big) = H(p) \otimes -$, is indecomposable. □

The next construction is an analogue/extension of one in ring theory where pp conditions are replaced by elements of a ring and pp-types by right ideals, see [495, 4.33] for references. For a functorial version of this result see 12.2.7.

Lemma 4.3.50. *Let* Φ, Ψ *be sets of pp conditions (all with free variables from among* $x_1, \dots, x_n$*) with* Φ *closed under finite conjunction,* Ψ *closed under finite* *sum and such that for no $\phi \in \Phi$ and $\psi \in \Psi$ do we have $\phi \leq \psi$. Let p be a pp-type maximal with respect to $\Phi \subseteq p$ and $p \cap \Psi = \emptyset$. Then p is irreducible (hence, by 4.3.49, the hull of p, $H(p)$, is indecomposable).*

Proof. Suppose that θ_1, θ_2 are pp conditions not in p. By maximality of p there are $\eta_1, \eta_2 \in p$ and $\psi_1, \psi_2 \in \Psi$ such that $\eta_i \wedge \theta_i \leq \psi_i$ $(i = 1, 2)$. Then $\eta_1 \wedge \theta_1 + \eta_2 \wedge \theta_2 \leq \psi_1 + \psi_2 \in \Psi$ is not in p, as required. □

Recall (Section 3.2.1) that the lattice, $\mathrm{pp}(M)$, of pp-definable subgroups of M is isomorphic to the lattice of pp conditions modulo equivalence when evaluated in M.

Lemma 4.3.51. *Let M be any module and suppose that Φ is a filter in $\mathrm{pp}^n(M)$. Let $p \subseteq \mathrm{pp}^n_R$ be the full inverse image of Φ under the natural surjection $\mathrm{pp}^n_R \to \mathrm{pp}^n(M)$. Then $H(p)$ belongs to the definable subcategory generated by M.*

Proof. We apply 4.1.5 (without parameters, so really just 3.3.6 with the extra argument of 4.1.4) to $p = \{\phi \in \mathrm{pp}^n_R : \phi(M) \in \Phi\}$ to obtain $M^\bigstar \in \langle M \rangle$ which contains a tuple $\overline{a}$ with $\mathrm{pp}^{M^\bigstar}(\overline{a}) = p$. By 4.3.21 it may be supposed that $M^\bigstar$ is pure-injective, hence (4.3.36) that $H(p)$ is a direct summand of $M^\bigstar$. Thus $H(p) \in \langle M \rangle$. □

With this we have a relative version of 4.3.50.

Corollary 4.3.52. *Let M be any module and let Φ, Ψ be subsets of $\mathrm{pp}^n(M)$ with Φ closed under finite conjunction and Ψ closed under finite sum and such that for every ϕ with $[\phi]_M \in \Phi$ and ψ with $[\psi]_M \in \Psi$ we have $\phi(M) \not\leq \psi(M)$. Let Φ' be a subset of $\mathrm{pp}^n(M)$ maximal with respect to $\Phi \subseteq \Phi'$ and $\Phi' \cap \Psi = \emptyset$. Let p be the full inverse image of Φ' in pp^n_R. Then p is irreducible and $H(p)$ is in the definable subcategory generated by M.*

Proof. Irreducibility of p follows easily from 4.3.50. The second statement follows from 4.3.51. The model-theoretic version of the proof can be found at [495, 4.33] and the functorial version is 12.2.7. □

Recall (Section 3.2.1) that $\phi \geq \psi$ is an N-minimal pair if $\phi(N) > \psi(N)$ and there is no pp-definable subgroup of N strictly between $\phi(N)$ and $\psi(N)$.

Proposition 4.3.53. ([495, 4.5.3] for the case where N is Σ-pure-injective) *Suppose that N is an indecomposable pure-injective module, let $S = \mathrm{End}(N)$ and let $D = S/J(S)$ be the "top" of the local (4.3.43) ring S. If ϕ/ψ is an N-minimal pair, then $\phi(N)/\psi(N)$ is a one-dimensional vector space over the division ring D.*

Proof. The proof in [495, 4.5.3] for the case where N is Σ-pure-injective uses Ziegler's irreducibility criterion (4.3.49). The functorial proof is simpler and more general but does need more than we have hitherto quoted from Part II. We need the isomorphism $\phi(N)/\psi(N) \simeq (F_{D\psi/D\phi}, N \otimes -)$ (10.3.8) under which, one may check, the S-module structure on $\phi(N)/\psi(N)$ becomes the left action (post-composition) of $S = \text{End}(N) \simeq \text{End}(N \otimes -)$ on $(F_{D\psi/D\phi}, N \otimes -)$. We also need that N-minimality of ϕ/ψ translates, 12.5.3, to simplicity (as an object of the localised category) of the image, $(F_{D\psi/D\phi})_{D\tau}$, of $F_{D\psi/D\phi}$ under the localisation of $(R\text{-mod}, \mathbf{Ab})$ at the finite type torsion theory $D\tau$ cogenerated by the injective object $N \otimes -$ (see, for example, 11.1.29). Now $N \otimes -$ is (by definition of $D\tau$) $D\tau$-torsionfree, and is injective, hence (11.1.5) $(N \otimes -)_{D\tau}$ is isomorphic to $(N \otimes -)$ and so, applying the localisation functor $(-)_{D\tau}$, we have $(F_{D\psi/D\phi}, N \otimes -) \simeq ((F_{D\psi/D\phi})_{D\tau}, (N \otimes -)_{D\tau})$. Since $(F_{D\psi/D\phi})_{D\tau}$ is simple, so $(N \otimes -)_{D\tau}$ must be its injective hull, it follows that the division ring D is the endomorphism ring of $(F_{D\psi/D\phi})_{D\tau}$ and that $(F_{D\psi/D\phi}, N \otimes -)$ is one-dimensional over D, as required. (For a fuller summary of what has been used here, see the beginning of Section 5.3.1.) □

The next result is useful (see Section 5.5.2) in comparing Ziegler spectra over different rings. Also see the more general 18.2.24.

Proposition 4.3.54. ([500, Prop. 2]) *Let $f : R \to S$ be a morphism of rings and let $f^* : \text{Mod-}S \longrightarrow \text{Mod-}R$ be the induced restriction of scalars functor. Suppose that N is an indecomposable pure-injective R-module such that N is a direct summand of f^*M for some S-module M. Then there is an indecomposable pure-injective S-module N', which may be taken in the definable subcategory of* Mod-S *generated by M, such that N is a direct summand of f^*N'.*

Proof. Recall from Section 3.2.1 that if ϕ is a pp condition for R-modules, then $f_*\phi$ is the pp condition for S-modules obtained by replacing each element of R by its image under f. Choose $a \in N \setminus 0$ and let $p = \text{pp}^N(a)$. Let f_*p denote the corresponding set of pp conditions for S-modules (this is closed under conjunction by construction) and consider the image, p', of this set under the quotient map $\text{pp}_S \to \text{pp}(M_S)$. For any $\phi \in \text{pp}_R$ denote the image of $f_*\phi$ in $\text{pp}(M_S)$ by ϕ_M.

Since $N \mid f^*M$, for every $\phi \in p$ and for every $\psi \in \text{pp}_R$ with $\psi \notin p$ it follows that $\psi_M \not\geq \phi_M$. Therefore we may apply 4.3.52 and obtain an irreducible pp-type, q, for S-modules with $f_*^{-1}q = p$ (the construction gives $f_*^{-1}q \supseteq p$, and by construction $f_*^{-1}q$ can contain no $\psi \in \text{pp}_R \setminus p$) and with $N' = H(q)$ in the definable subcategory of Mod-S generated by M. To see that $N \mid f^*N'$ take $b \in N'$ with $\text{pp}^{N'}(b) = q$. Then $\text{pp}^{f^*N'}(b) = f_*^{-1}q = p$, so the hull of b within f^*N' (4.3.36) is a copy of N. □

4.3.7 Pure-injective hulls of finitely presented modules

The pure-injective hull of a finitely presented module with local endomorphism ring is indecomposable; furthermore, that hull determines the finitely presented module up to isomorphism (4.3.55). These results apply in particular to finitely presented modules of finite length.

The next result was proved in [495, 11.24, 11.27] for indecomposable finitely presented modules of finite length, in [549, Thm 1] for the case $A = R$, and in [275, p. 157], also see [278, p. 535], in the general form below. The functorial proof that we give is from the last reference.

Theorem 4.3.55. *If A is a finitely presented module with local endomorphism ring, then the pure-injective hull, $H(A)$, of A is indecomposable. If B is a finitely presented module with $H(B) \simeq H(A)$, then $B \simeq A$.*

Proof. We use the fact, 10.2.36, that if A is finitely presented, then the functor $A \otimes -$ lies in the subcategory, $(R\text{-mod}, \mathbf{Ab})^{\mathrm{fp}}$, of finitely presented functors in $(R\text{-mod}, \mathbf{Ab})$. By 12.1.8, $H(A) \otimes -$ is the injective hull in $(R\text{-mod}, \mathbf{Ab})$ of $A \otimes -$. By 12.1.13, $A \otimes -$ is an injective object of $(R\text{-mod}, \mathbf{Ab})^{\mathrm{fp}}$, so, since, 12.1.3, $\mathrm{End}(A \otimes -) \simeq \mathrm{End}(A)$, it is indecomposable, hence uniform (in the category of finitely presented functors,[17] hence in the full functor category), so $H(A) \otimes -$ is uniform, hence indecomposable. Therefore, 12.1.5, $H(A)$ is indecomposable, as required.

For the second part, since $(H(B) \otimes -) \simeq (H(A) \otimes -)$ is uniform, so is $B \otimes -$. Therefore, up to isomorphism, $A \otimes -$ and $B \otimes -$ have a non-zero common subobject, which we may take to be finitely generated, hence, by 10.2.2, finitely presented. Therefore, by 12.1.13, each of $A \otimes -$ and $B \otimes -$ is an (injective containing, therefore, since indecomposable, an) injective hull of this object in $(R\text{-mod}, \mathbf{Ab})^{\mathrm{fp}}$ and so, injective hulls being unique to isomorphism when they do exist, $(A \otimes -) \simeq (B \otimes -)$ hence $A \simeq B$, as required. □

Recall that we also use the terminology "strongly indecomposable" for a module with local endomorphism ring.

Corollary 4.3.56. *Suppose that A and B are strongly indecomposable finitely presented modules. If there are $\overline{a}$ from A and $\overline{b}$ from B with $\mathrm{pp}^A(\overline{a}) = \mathrm{pp}^B(\overline{b})$, then $A \simeq B$.*

[17] Some care is needed since the category of finitely presented functors need not have injective hulls, as opposed to enough injectives. The argument is that, if $A \otimes -$ is not uniform, say $F, G \leq A \otimes -$ are non-zero with $F \cap G = 0$, then the map from $F + G$ to $A \otimes -$, which is the identity on F and 0 on G, extends to an endomorphism f of $A \otimes -$, but neither f nor $1 - f$ is invertible, contradicting that the endomorphism ring of $A \otimes -$ is local.

Proof. By 4.3.55, $H(A)$ and $H(B)$ are indecomposable so 4.3.47 gives $H(A) \simeq H(B)$, hence, by 4.3.55, $A \simeq B$. □

Proposition 4.3.57. (Fitting's Lemma) *If M is a module of finite length and $f \in \mathrm{End}(M)$, then $M = \mathrm{im}(f^n) \oplus \ker(f^n)$ for some n.*

Proof. Since M is of finite length an endomorphism is a monomorphism iff it is an epimorphism iff it is an automorphism. Moreover, since M is of finite length the increasing, respectively decreasing, sequence of kernels, respectively images, of the powers of f stabilise. Say $\mathrm{im}(f^n) = \mathrm{im}(f^{n+1}) = \cdots$ and $\ker(f^n) = \ker(f^{n+1}) = \cdots$. Let $a \in M$. Then $f^n(a) = f^{2n}b$ for some b, so $a = (a - f^n b) + f^n b \in \ker(f^n) + \mathrm{im}(f^n)$. Also f^n acts as an epimorphism on $\mathrm{im}(f^n)$ so, since $\mathrm{im}(f^n)$ has finite length, $\mathrm{im}(f^n) \cap \ker(f^n) = 0$. □

Corollary 4.3.58. *If M is an indecomposable module of finite length, then* $\mathrm{End}(M)$ *is local.*

Proof. Let $f \in \mathrm{End}(M)$. By Fitting's Lemma, there is n such that either $\mathrm{im}(f^n) = M$ – in which case f is epi so, because M is of finite length, an isomorphism – or $\ker(f^n) = M$ – in which case f is nilpotent, hence in $\mathrm{JEnd}(M)$. This is enough (see Appendix E, p. 716). □

Corollary 4.3.59. (essentially [495, 11.24]) *If M is a finitely presented indecomposable module of finite length, then its pure-injective hull, $H(M)$, is indecomposable.*

Corollary 4.3.60. (essentially [495, 11.27]) *Suppose that M and N are finitely presented indecomposable modules of finite length. If $H(M) \simeq H(N)$, equivalently if there is $\overline{a}$ from M and $\overline{b}$ from N such that $\mathrm{pp}^M(\overline{a}) = \mathrm{pp}^N(\overline{b})$, then $M \simeq N$.*

Remark 4.3.61. Note that if A is any module with local endomorphism ring, $\overline{a}$ is a non-zero tuple from A and $f \in \mathrm{End}(A)$ with $f\overline{a} = \overline{a}$, then $f \in \mathrm{Aut}(A)$. For, if not, then $1 - f \in \mathrm{Aut}(A)$ – but $(1 - f)\overline{a} = 0$.

Lemma 4.3.62. *Suppose that A is finitely presented with local endomorphism ring and let $\overline{a}$ be from A with $\mathrm{pp}^A(\overline{a})$ generated by the pp condition ϕ (1.2.6). If $\phi > \psi_1, \psi_2$ (in $\mathrm{pp}_R^{l(\overline{a})}$), then $\phi(A) > \psi_1(A) + \psi_2(A)$, in particular $\phi > \psi_1 + \psi_2$.*

Proof. Since $\mathrm{pp}^A(\overline{a}) = \langle \phi \rangle$, from $\phi > \psi_1, \psi_2$ we deduce $\phi(A) > \psi_1(A), \psi_2(A)$. If the sum $\psi_1(A) + \psi_2(A)$ were equal to $\phi(A)$, then there would be $\overline{b}_i$ from $\psi_i(A)$ $(i = 1, 2)$ with $\overline{a} = \overline{b}_1 + \overline{b}_2$.

Now $\phi \in \mathrm{pp}^A(\overline{b}_i)$, so, by 1.2.7, there is $f_i \in \mathrm{End}(A)$ with $f_i\overline{a} = \overline{b}_i$ $(i = 1, 2)$. But then $f_1 + f_2$ fixes $\overline{a}$ hence, by 4.3.61, $f_1 + f_2 \in \mathrm{Aut}(A)$. Since $\mathrm{End}(A)$ is local either f_1 or f_2 is an automorphism – a contradiction because each strictly increases the pp-type of $\overline{a}$. □

We remark here, see 10.2.27 for a proof, that a finitely presented module M has local endomorphism ring iff the representable functor $(M, -)$ (from mod-R to **Ab**) has a unique maximal proper subfunctor, that is, is a local functor, where we say that an object of an additive category is **local** if it has a unique maximal proper subobject.

The next result is a corollary of 4.3.42.

Corollary 4.3.63. *Suppose that C is a finitely presented module with local endomorphism ring and let $\overline{a}$ be a non-zero tuple from C. If a morphism $C \to M$ ($M \in$ Mod-R) preserves the pp-type of $\overline{a}$, then it is a pure embedding.*

4.3.8 Krull–Schmidt rings

Every pp condition over a Krull–Schmidt ring gives a functor on finitely presented modules which is a sum of local functors (4.3.64); indeed this characterises these rings, as does the condition that every finitely generated functor have a projective cover (4.3.69). From results of Warfield on a duality between finitely presented left and right modules over semiperfect rings, it follows that a ring is right Krull–Schmidt iff it is left Krull–Schmidt (4.3.68). Every pp condition has a minimal free realisation iff the ring is Krull–Schmidt (4.3.70). Over these rings there is a bijection between indecomposable finitely presented modules and simple functors on finitely presented modules (4.3.71). Pure-projective modules over these rings are just the direct sums of finitely presented modules (4.3.72).

Following [19, §2] we will say that an additive category $\mathcal{A}$ is **Krull–Schmidt** if every indecomposable object of $\mathcal{A}$ has a local endomorphism ring and if every object of $\mathcal{A}$ is a finite direct sum of indecomposable objects. Therefore (E.1.24) every object of $\mathcal{A}$ has an essentially unique decomposition as a direct sum of indecomposable objects. We will also follow [276, p. 162], in saying that a ring R is **Krull–Schmidt** if mod-R is Krull–Schmidt, that is, if every finitely presented right R-module is a direct sum of indecomposable modules with local endomorphism rings. As defined, this is "right Krull–Schmidt" but it turns out, 4.3.68, that the notion is right/left symmetric.

For example, any right artinian ring is Krull–Schmidt (by 4.3.58), as is any serial ring which is commutative or right noetherian (see comments near the beginning of Section 2.5 and 2.4.20). By 4.3.55, if R is a Krull–Schmidt ring, then the pure-injective hull, $H(M)$, of each indecomposable finitely presented module, M, is indecomposable so, by the Krull–Remak–Schmidt–Azumaya Theorem (E.1.24 and 4.3.55), $H(M) \simeq H(N)$ iff $M \simeq N$ for $M, N \in$ mod-R.

Lemma 4.3.64. ([276, §2]) *Suppose that R is a Krull–Schmidt ring and let ϕ be a pp condition. Then F_ϕ is a sum of local functors (Section 10.2.3).*

Proof. Let $(C, \overline{c})$ be a free realisation of ϕ (Section 1.2.2). By assumption $C = C_1 \oplus \cdots \oplus C_k$, where each C_i has local endomorphism ring. Decompose $\overline{c} = \overline{c}_1 + \cdots + \overline{c}_k$ with $\overline{c}_i$ from C_i accordingly. Let ϕ_i be a generator for $\mathrm{pp}^{C_i}(\overline{c}_i)$ (1.2.6). By 1.2.27, $\phi = \phi_1 + \cdots + \phi_k$ (at least, up to equivalence of pp conditions), that is, $F_\phi = F_{\phi_1} + \cdots + F_{\phi_k}$. By 10.2.28 each F_{ϕ_i} is local, as required. □

A ring R is **semiperfect** (Bass, [46]) if R is **semilocal** (that is, R/J is semisimple artinian) and if **idempotents lift modulo** the radical J (that is, each idempotent of the ring R/J is the image, modulo J, of an idempotent of R).

Theorem 4.3.65. (see, for example, [663, §VIII.4], [183, 3.6]) *The following conditions on a ring R are equivalent:*

(i) R is semiperfect;
(ii) R_R is a direct sum of local projectives (that is, projectives with simple top);
(iii) every finitely generated module has a projective cover;
(iv) every simple module has a projective cover.

Therefore (since $\mathrm{End}(eR) \simeq eRe$ if $e^2 = e$) any Krull–Schmidt ring is semiperfect, hence every finitely generated module over such a ring has a projective cover.

In order to state some results of Warfield on modules over semiperfect rings we need the definition of **Auslander–Bridger duality** [24]. Let R be any ring and let $M \in \text{mod-}R$. Take a presentation $P_1 \xrightarrow{f} P_0 \to M \to 0$ of M with P_0, P_1 finitely generated projectives. For any right R-module A set $A^\vee$ to be the left R-module $A^\vee = \mathrm{Hom}_R(A_R, {}_RR_R)$ and similarly for morphisms. Define the **Auslander–Bridger dual** of M, let us denote it $D_{AB}M$, by exactness of the sequence $P_0^\vee \xrightarrow{f^\vee} P_1^\vee \to D_{AB}M \to 0$. A different choice of presentation of M may well result in a different value for $D_{AB}M$: this dual is, in general, defined only up to stable equivalence (that is, to adding or removing projective direct summands). If, however, objects of mod-R have minimal projective resolutions in Mod-R, then one obtains a well-defined dual module. This is developed in [703, §2].

Lemma 4.3.66. ([703, 2.3]) *If R is semiperfect, if $M \in$ mod-R has no non-zero projective direct summands and if $P_1 \xrightarrow{f} P_0 \to M \to 0$ is a minimal projective presentation of M, then $P_0^\vee \xrightarrow{f^\vee} P_1^\vee \to D_{AB}M \to 0$ is a minimal projective presentation of $D_{AB}M$, and $D_{AB}M$ has no projective direct summands.*

Theorem 4.3.67. ([703, 2.4]) *Suppose that R is semiperfect. Then Auslander–Bridger duality defines a bijection between the isomorphism classes of finitely* *presented right R-modules with no non-zero projective direct summands and the isomorphism classes of finitely presented left R-modules with no non-zero projective direct summands. This bijection commutes with direct sum decompositions and, if M is indecomposable with local endomorphism ring, then so is $D_{AB}M$.*

Corollary 4.3.68. *If R is a right Krull–Schmidt ring, then it is also left Krull–Schmidt.*

Theorem 4.3.69. (see [276, 2.3]) *For any ring R the following are equivalent:*

(i) R is Krull–Schmidt;
(ii) every finitely generated functor in (mod-R, **Ab**) *has a projective cover;*
(iii) every finitely generated projective object of (mod-R, **Ab**) *is a direct sum of local functors;*

Proof. The equivalence of (ii) and (iii) is just the equivalence (iii)⇔(ii) of 4.3.65 in a more general setting (see [647, 5.6]). The equivalence (i)⇔(iii) is direct from 10.2.27. □

Corollary 4.3.70. *Every pp condition for right R-modules has a minimal free realisation (in the sense of Section 10.2.3) iff R is Krull–Schmidt iff every pp condition for left R-modules has a minimal free realisation.*

Proof. This is immediate from 10.2.25 and 4.3.69 (and 4.3.68). □

Theorem 4.3.71. ([19, 2.3]) *Suppose that R is a Krull–Schmidt ring. Then $M \mapsto S_M = (M, -)/\mathrm{rad}(M, -)$ is a bijection between isomorphism classes of finitely presented indecomposable right R-modules and simple functors in* (mod-R, **Ab**). *Given a simple functor $S \in$* (mod-R, **Ab**), *then $S = S_M$, where M is the unique indecomposable module in* mod-R *with $SM \neq 0$.*

Proof. By 10.2.27 to every indecomposable finitely presented module M there corresponds the local projective functor $(M, -)$ and to that corresponds its top, the simple functor $(M, -)/\mathrm{rad}(M, -)$. Conversely, if S is a simple functor, then, by 4.3.68, it has a projective cover, necessarily indecomposable, so (10.1.14) that projective has the form $(M, -)$ for some indecomposable finitely presented R-module M. Since there is a non-zero morphism $(M, -) \to S$ the Yoneda Lemma (10.1.7) gives $SM \neq 0$. Again by the Yoneda Lemma, if N is an indecomposable module with $SN \neq 0$, then $((N, -), S) \neq 0$, so, since $(N, -)$ is local, S must be the top, $(N, -)/\mathrm{rad}(N, -)$, of $(N, -)$, therefore (local projectives with the same top must be isomorphic) $N \simeq M$. □

Those indecomposable finitely presented modules M which are such that $\mathrm{rad}(M, -)$ is finitely generated will correspond, via $M \mapsto H(M)$, exactly to the points of the Ziegler spectrum of R which are isolated by a minimal pair (see the comment after 5.3.31). Furthermore, 5.3.42, the hulls of finitely presented indecomposable modules over a Krull–Schmidt ring form a dense subset of the spectrum.

The next result follows from the Crawley–Jonsson–Warfield Theorem (E.1.26).

Proposition 4.3.72. *If R is a Krull–Schmidt ring, then an R-module is pure-projective iff it is a direct sum of finitely presented modules.*

For other rings over which this is true see [579] as well as the comments after 2.1.26.

Puninski investigates the Krull–Schmidt condition for serial rings in [559].

4.3.9 Linking and quotients of pp-types

Linking of elements of a module via a pp formula was important in developing the model theory of modules. This, and the results around it, can be described alternatively in terms of the functor category: we give an example at 4.3.74. Another model-theoretic notion which can be described very differently in terms of functors is that of a tuple modulo some pp-definable equivalence relation – an "imaginary" in the sense of Section B.2. These also have hulls (4.3.75), generalising those considered in Section 4.3.5.

Say that tuples $\overline{a}$ and $\overline{b}$ (not necessarily of the same length) from a module M are **linked (in** M**)** if there is a pp condition $\phi(\overline{x}, \overline{y})$ such that $\phi(\overline{a}, \overline{b})$ holds in M but $\phi(\overline{a}, \overline{0})$, hence also $\phi(\overline{0}, \overline{b})$, does not hold in M. If so, then the pp condition ϕ **links** $\overline{a}$ and $\overline{b}$ and $\overline{a}$ and $\overline{b}$ are **linked** (by ϕ).

Example 4.3.73. Let $R = K[u, v : u^2 = v^2 = uv = 0]$, where K is a field. Take $M = R, M' = J(R) = uK + vK, a = u, b = v$. Then a and b are linked in M. To see this, let $\phi(x, y)$ be $\exists z\, (zu = a \wedge zv = b)$: taking $z = 1$ we see that $\phi(a, b)$ holds in M. On the other hand, $\phi(a, 0)$ does not hold in M since $\mathrm{ann}_M(u) = \mathrm{ann}_M(v)$. Yet, a and b are not linked in M' since they lie in complementary direct summands (then projecting any pp relation $\psi(a, b)$ gives $\psi(a, 0)$).

The next result is due to Fisher [190, §3] and can be found in Garavaglia's [209, Lemma 1] (also see [495, §4], [726, 3.8, 3.10]). The proofs in those places are quite different from the functor-category proof below.

Proposition 4.3.74. *Let N be a pure-injective module and suppose that there is a finite tuple $\overline{a}$ from N such that N is the hull of $\overline{a}$. Let $\overline{b}$ be a non-zero (that is, not all coordinates zero) tuple of any length in N. Then $\overline{a}$ and $\overline{b}$ are linked in N.*

Proof. Since the image of $(\overline{a} \otimes -) : (R^n \otimes -) \to (N \otimes -)$ is essential in $N \otimes -$ (cf. proof of 4.3.33) and the image of $(\overline{b} \otimes -) : (R^m \otimes -) \to (N \otimes -)$ is non-zero, the intersection of $\mathrm{im}(\overline{a} \otimes -)$ and $\mathrm{im}(\overline{b} \otimes -)$ is non-zero. Therefore the object P in the pullback diagram shown is non-zero, so has a finitely generated subobject, F' say, which has non-zero image in $N \otimes -$.

$$\begin{array}{ccc} P & \longrightarrow & (R^n \otimes -) \\ \downarrow & & \downarrow {\scriptstyle \overline{a}\otimes -} \\ (R^m \otimes -) & \underset{\overline{b}\otimes -}{\longrightarrow} & (N \otimes -) \end{array}$$

By definition of pullback, P is the kernel of the morphism $\big((\overline{a} \otimes -), -(\overline{b} \otimes -)\big) : (R^n \otimes -) \oplus (R^m \otimes -) \to (N \otimes -)$ which corresponds to the tuple $(\overline{a}, -\overline{b}) \in R^{n+m}$. So F', being a finitely generated subobject of $(R^{n+m} \otimes -) \simeq ({}_R R^{n+m}, -)$, has, by 10.2.16, the form $F_{D\phi'}$ for some ϕ' (a pp condition for right R-modules with $n + m$ free variables). Since $F_{D\phi'}$ is contained in the kernel of $\big((\overline{a} \otimes -), -(\overline{b} \otimes -)\big)$ 10.3.9 implies $(\overline{a}, -\overline{b}) \in \phi'(N)$. Define $\phi(\overline{x}, \overline{y})$ to be $\phi'(\overline{x}, -\overline{y})$. It remains to be seen that $(\overline{a}, \overline{0}) \notin \phi^{(')}(N)$.

If it were the case that $(\overline{a}, \overline{0}) \in \phi'(N)$, then, again from 10.3.9, $F' = F_{D\phi'}$ would be contained in the pullback functor $(\leq (R^{n+m} \otimes -))$ obtained by replacing $\overline{b} \otimes -$ with the zero map from $R^m \otimes -$ to $N \otimes -$ and this would imply that $\mathrm{im}(a \otimes -) = 0$, hence that $\overline{a} = 0$ – contradiction, as required.[18] □

For M any module, $\overline{a}$ a tuple from M and ψ a pp condition (with $n = l(\overline{a})$ free variables) we denote by $\overline{a}_\psi$ the image of $\overline{a}$ under the natural projection $M^n \to M^n/\psi(M)$ (of groups or, 1.1.8, left $\mathrm{End}(M)$-modules).

We want to be able to refer to the pp-type of $\overline{a}_\psi$: there are two approaches. The model-theoretic one is to view $\overline{a}_\psi$ as being a tuple in a sort attached to M (see Section B.1.1). One introduces an enriched language with variables which take values in these extra sorts; in particular, one can define the pp-type of such a tuple. It fits better with the methods used in this book if we define the pp-type of $\overline{a}_\psi$ to be a set of functors. It does follow from 10.2.30 that these approaches are entirely equivalent.

So we define the **pp-type** of $\overline{a}_\psi$ in M to be

$$\mathrm{pp}^M(\overline{a}_\psi) = \{\phi_\psi : \overline{a}_\psi \in (\phi(M) + \psi(M))/\psi(M) = (F_{\phi+\psi}/F_\psi)(M)\}.$$

The model-theoretic meaning given to "ϕ_ψ" is that it is a pp condition in a language enriched by the quotient sort $(-)^{l(\overline{a})}/\psi(-)$; the functor-theoretic meaning is that it is simply the functor $(F_\phi + F_\psi)/F\psi$. Therefore (by 10.2.30), in the latter view

[18] It is not necessary that N be the hull of $\overline{a}$ (in particular, pure-injective), just that $\mathrm{im}(\overline{a})$ be essential in $N \otimes -$, in other words (4.3.40) that $\overline{a}$ be pp-essential in N.

(the one we take here), $\mathrm{pp}^M(\overline{a}_\psi)$ is the set of finitely presented functors F such that $\overline{a}_\psi \in FM$.

If $p = \mathrm{pp}^M(\overline{a})$, we write p_ψ for $\mathrm{pp}^M(\overline{a}_\psi)$. This is well defined since it is just $\{(F_\phi + F_\psi)/F_\psi : \phi \in p\}$. We may refer to $\overline{a}_\psi$ and p_ψ as "$\overline{a}$ mod ψ" and "p mod ψ" respectively. Denote by $p_\psi(M)$ the set of realisations of p_ψ in M, that is, $p_\psi(M) = (p(M) + \psi(M))/\psi(M)$.

Proposition 4.3.75. ([382, §4]) *Let N be a pure-injective module, let $\overline{a}$ be a tuple from N of length n and let ψ be a pp condition. Then there is a unique-to-isomorphism pure-injective direct summand, $H = H(\overline{a}_\psi)$, of N which contains $\overline{a}_\psi$ ("contains" in the sense that $\overline{a}_\psi \in H^n/\psi(H)$) and is minimal such. If N' is any pure-injective module, if $\overline{a}'$ from N' is such that $\overline{a}'_\psi$ has the same pp-type as $\overline{a}_\psi$ and if $H(\overline{a}'_\psi)$ is similarly chosen, then there is an isomorphism from $H(\overline{a}_\psi)$ to $H(\overline{a}'_\psi)$ taking $\overline{a}_\psi$ to $\overline{a}'_\psi$.*

Proof. The proof is similar to that for ordinary hulls, see Section 4.3.5. What has to be observed first is that, by 10.3.8, $N^n/\psi(N) \simeq (F_{D\psi}, N \otimes -)$ and, from this point of view, $\overline{a}_\psi$ is simply the restriction of $(\overline{a} \otimes -) : (R^n \otimes -) \to (N \otimes -)$ to $F_{D\psi} \le (R^n \otimes -)$. So $H(\overline{a}_\psi)$ is the module $H \le N$ such that $H \otimes -$ is the injective hull of the image, $(\overline{a} \otimes -) \cdot F_{D\psi}$, of this restricted morphism. $\square$

We refer to $H(\overline{a}_\psi)$ above as the **hull** of $\overline{a}_\psi$, or of p_ψ, where p_ψ is the pp-type of $\overline{a}_\psi$, writing $H(\overline{a}_\psi)$, respectively $H(p_\psi)$.

By 12.2.5, F_{Dp}, for p a pp-n-type, is the kernel of $(\overline{a} \otimes -) : (R^n \otimes -) \to (M \otimes -)$, where $\overline{a}$ from M has pp-type p. Similarly we may write F_{Dp_ψ} for the kernel, $F_{D\psi} \cap F_{Dp}$, of the restricted morphism and, for any module M, one has $(F_{D\psi}/(F_{D\psi} \cap F_{Dp}), M \otimes -) \simeq (p(M) + \psi(M))/\psi(M) = p_\psi(M)$ (see 12.2.4). Note, from the proof of 4.3.75, that $(H(p_\psi) \otimes -) \simeq E(F_{D\psi}/(F_{D\psi} \cap F_{Dp})) \simeq E\big((F_{Dp} + F_{D\psi})/F_{Dp}\big)$.

Proposition 4.3.76. ([102, 5.7], also cf. [382, 4.2, 4.3]) *Suppose that p is a pp-type and that ψ and θ are pp conditions. Then $\psi \le \theta$ implies $H(p_\theta) \mid H(p_\psi)$.*

Proof. This is immediate from the above discussion and the observation that $\psi \le \theta$ implies $D\theta \le D\psi$, so $(F_{Dp} + F_{D\theta})/F_{Dp} \le (F_{Dp} + F_{D\psi})/F_{Dp}$, hence $(H(p_\theta) \otimes -) = E\big((F_{Dp} + F_{D\theta})/F_{Dp}\big) \mid E\big((F_{Dp} + F_{D\psi})/F_{Dp}\big) = (H(p_\psi) \otimes -)$. $\square$

4.3.10 Irreducible pp-types and indecomposable direct summands

Direct summands of the hull of a pp-type correspond to quotients of that pp-type by "large" conditions (4.3.79), equivalently to uniform factors of a functor corresponding to the pp-type.

The results in this section come mainly from [726, §7]. The "functorial" proofs are due to Burke ([101], see [102, §5]). The non-functorial proofs, see [495, 9.16], are significantly harder.

Say that a pp condition $\psi \notin p$ is **large in** the pp-type p if, for all $\theta_1, \theta_2 \geq \psi$, if $\theta_1 \notin p$ and $\theta_2 \notin p$, then there is $\phi \in p$ with $\theta_1 \wedge \phi + \theta_2 \wedge \phi \notin p$. So p is irreducible (§4.3.6) iff $\overline{x} = 0$ is large in p.

Lemma 4.3.77. *The pp condition ψ is large in the pp-type p iff the functor $(F_{D\psi} + F_{Dp})/F_{Dp}$ is uniform iff p_ψ is irreducible.*

Proof. Suppose that ψ is large in p and let $(F_{D\theta_i} + F_{Dp})/F_{Dp}$ $(i = 1, 2)$ be non-zero finitely generated subfunctors of $(F_{D\psi} + F_{Dp})/F_{Dp}$. That is, translating between the model-theoretic and functorial languages (see, for example, Section 12.2), suppose θ_1, θ_2 are as in the definition of large: then for some $\phi \in p$ as in the definition of large we have $(F_{D\theta_1} + F_{D\phi}) \cap (F_{D\theta_2} + F_{D\phi}) \nleq F_{Dp}$, hence $\big((F_{D\theta_1} + F_{Dp})/F_{Dp}\big) \cap \big((F_{D\theta_2} + F_{Dp})/F_{Dp}\big) \neq 0$, as required.

Conversely, if $(F_{D\psi} + F_{Dp})/F_{Dp}$ is uniform and neither of $\theta_1, \theta_2 \geq \psi$ is in p, then both functors $(F_{D\theta_i} + F_{Dp})/F_{Dp}$ are non-zero, so, since $F_{Dp} = \sum_{\overrightarrow{\phi \in p}} F_{D\phi}$, there are $\phi_i \in p$ with $F_{D\theta_i} + F_{D\phi_i} \nleq F_{Dp}$ $(i = 1, 2)$, so, setting $\phi = \phi_1 \wedge \phi_2 \in p$ (hence $F_{D\phi} = F_{D\phi_1} + F_{D\phi_2}$) we have, by uniformity, $(F_{D\theta_1} + F_{D\phi}) \cap (F_{D\theta_2} + F_{D\phi}) \nleq F_{Dp}$, thus (by 1.3.1) $\theta_1 \wedge \phi + \theta_2 \wedge \phi \notin p$.

The equivalence with the last statement follows from the discussion just after 4.3.75. □

Proposition 4.3.78. ([726, 7.6]) *If p is a pp-type and N is an indecomposable direct summand of $H(p)$, then there is a pp condition ψ such that $N \simeq H(p_\psi)$.*

Proof. Say $H(p) = H(\overline{a})$, where $\overline{a}$ has pp-type p, so, see Section 4.3.5, $(H(p) \otimes -) \simeq E(\mathrm{im}(\overline{a} \otimes -))$. Since $N \mid H(p)$ also $(N \otimes -) \mid (H(p) \otimes -)$.

Since $\mathrm{im}(\overline{a} \otimes -)$ is essential in $H(p) \otimes -$, there is a non-zero finitely generated subfunctor F of $R^n \otimes -$ such that $(\overline{a} \otimes -)(F)$ is non-zero and is contained in $(N \otimes -)$ and hence is essential in $(N \otimes -)$ because $N \otimes -$ is an indecomposable injective. Say $F = F_{D\psi}$ (10.2.16). By the discussion after 4.3.75, $N \simeq H(\overline{a}_\psi)$, as required. □

Theorem 4.3.79. ([726, 7.6]) *(a) The indecomposable direct summands of a module of the form $N = H(\overline{a})$ are precisely the $H(\overline{a}_\psi)$ for ψ a pp condition which is large in $p = \mathrm{pp}^N(\overline{a})$.*

(b) The indecomposable summands of a module of the form $H(p)$ are, up to isomorphism, precisely the $H(p_\psi)$ for ψ a pp condition which is large in p.

Proof. This follows by 4.3.77, 4.3.78 and, for the stronger statement (a), their proofs. □

The proof of the next result is similar to that of 4.3.78.

Proposition 4.3.80. ([102, 5.10]) *If p is a pp-type and ψ is a pp condition, then, if $N \mid H(p_\psi)$, there is a pp condition θ with $\psi \le \theta$ and $H(p_\theta) \mid N$. In particular if N is indecomposable, then $H(p_\theta) \simeq N$.*

For more on the form of direct summands of hulls of pp-types, including representations related to that in 5.3.47(ii), see [101, §§6.3, 6.4].

4.4 Structure of pure-injective modules

The basic structure theorem for pure-injective modules is in Section 4.4.1. Σ-pure-injective modules are characterised in Section 4.4.2 and shown to be direct sums of indecomposables. Section 4.4.3 considers modules satisfying the stronger condition of being of finite length over their endomorphism rings. Crawley-Boevey's correspondence between "characters" and modules of finite endolength is outlined in Section 4.4.4. Modules with the ascending chain condition on pp-definable subgroups are considered, though rather briefly, in Section 4.4.5.

4.4.1 Decomposition of pure-injective modules

Every pure-injective is the direct sum of a "discrete" part, which is the pure-injective hull of a direct sum of indecomposable summands, and a "superdecomposable" part without any indecomposable direct summands (4.4.2). Over some rings there are no superdecomposable pure-injectives. An example of a superdecomposable pure-injective module is given at 4.4.3.

Lemma 4.4.1. *If N is an indecomposable pure-injective and N is purely embedded in, hence a direct summand of, $\bigoplus_i M_i$, where the M_i are arbitrary, then N is a direct summand of M_i for some i.*

Proof. This is part of the proof of the Krull–Remak–Schmidt–Azumaya Theorem (see Appendix E, p. 717); all that is used is that N has local endomorphism ring and is a direct summand of $\bigoplus_i M_i$.

Let $j_i : M_i \to M = \bigoplus_j M_j$ and $\pi_i : M \to M_i$ be the canonical inclusions and projections, so $\sum_i j_i\pi_i = 1_M$ (if the index set is infinite interpret this equality as: for every $a \in M$, $\sum_i j_i\pi_i a = a$). Let $j : N \to M$ be the inclusion and let $\pi : M \to N$ be a projection, so $\pi j = 1_N$.

Then $1_N = \pi 1_M j = \sum_i (\pi j_i)(\pi_i j)$. Since End($N$) is local not all the $(\pi j_i)(\pi_i j)$ can lie in JEnd(N) (if the sum is infinite just apply it to any non-zero element of N to see this), that is, for some i, $(\pi j_i)(\pi_i j)$ is an automorphism of N, so, for some automorphism g of N, the morphism $(g\pi j_i)(\pi_i j)$ is the identity of N. Therefore $\pi_i j : N \to M_i$ is monic and is split, as required. □

Say that a pure-injective module is **superdecomposable** ("continuous" in the terminology of [495]) if it has no (non-zero) indecomposable summands. By the result below, a pure-injective module with no superdecomposable direct summand is the pure-injective hull of a direct sum of indecomposable pure-injectives (for such pure-injective modules the term "discrete" was used in [495]).

Theorem 4.4.2. ([190, 7.21], also see references within the proof) *Let N be a pure-injective module. Then $N \simeq H(\bigoplus_\lambda N_\lambda) \oplus N_c$, where each N_λ is indecomposable pure-injective and where N_c is a superdecomposable pure-injective. The modules N_λ, together with their multiplicities, as well as the module N_c, are determined up to isomorphism by N.*

Proof. This result has been given a variety of proofs (see [726, §6], [181, Cor. 3]). The simplest, from [256, §1], is to use the embedding (§12.1.1) of Mod-R into (R-mod, **Ab**), which turns the pure-injective modules into the injective functors, and then to quote the corresponding result for injectives. That result, stated as E.1.9, may be found in, for example, [486, Exer. 5.5 to §5.3], also [663, V.6.8]. □

For instance, $\prod_p \mathbb{Z}_p = H(\bigoplus_p \overline{\mathbb{Z}_p})$ (4.3.22).

In general, neither N_c nor a complement to N_c in N is unique (as opposed to being unique up to isomorphism), see [610, p. 25ff.].

Although the zero module satisfies, by default, the criterion for superdecomposability, if we say, for example, that a ring has a superdecomposable pure-injective then, of course, we mean a non-zero such module.

Example 4.4.3. Let R be a domain which is not right Ore, equivalently not uniform as a right R-module. For example, let $R = k\langle X, Y\rangle$ be the free associative algebra ("ring of non-commutative polynomials") generated by two indeterminates over a field k. This is not an Ore domain, for instance the right ideals generated by X and by Y have no element in common except 0. Since every non-zero right ideal contains a copy of R_R there is no non-zero uniform right ideal. It follows, see E.1.7, that the injective hull $E(R_R)$ has no indecomposable direct summand, so is an example of a superdecomposable (pure-)injective.

Further examples may be obtained from this one (see [323, p. 184 ff.]) via suitable functors between module categories (for example, strict representation embeddings, Section 5.5.2): for instance every strictly wild algebra has a superdecomposable pure-injective module (7.3.27) which can be realised as the image of the above injective module under a strict representation embedding.

The existence of superdecomposable pure-injectives over various types of rings is the topic of Section 7.3.1.

4.4.2 Σ-pure-injective modules

A module is Σ-pure-injective if every direct sum of copies of it is pure-injective. It is enough to check for countably many copies (4.4.8). A module is Σ-pure-injective iff it has the descending chain condition on pp-definable subgroups (4.4.5, 4.4.6). Every countable pure-injective module is Σ-pure-injective (4.4.9), as is every countable-dimensional pure-injective module over an algebra (4.4.10).

The definable subcategory generated by a Σ-pure-injective module consists entirely of Σ-pure-injectives (4.4.12). Every Σ-pure-injective is a direct sum of indecomposable modules (4.4.19).

The condition that every injective right module be Σ(-pure)-injective is equivalent to every absolutely pure module being injective and this is equivalent to the ring being right noetherian (4.4.17). The situation for flat modules is not entirely dual (4.4.22, 4.4.23). A module is Σ-pure-injective iff the corresponding tensor functor is Σ-injective (4.4.21).

A module M is **Σ-pure-injective** (= **Σ-algebraically compact**) if the direct sum of any (infinite) number of copies of M is pure-injective. Any Σ-pure-injective module must, in particular, be pure-injective. In the model-theoretic literature the notion of a "totally transcendental" module is defined in an entirely different way but is equivalent (see, for example, [495, 3.2]).[19]

Example 4.4.4. Any injective module over a right noetherian ring is Σ-pure-injective because, see 4.4.17, over a right noetherian ring, every direct sum of injectives is injective.

Say that M has the **descending chain condition (dcc) on pp-definable subgroups** if for every descending sequence $M \geq \phi_0(M) \geq \phi_1(M) \geq \cdots$ of pp-definable subgroups of M there is n such that $\phi_n(M) = \phi_{n+1}(M) = \cdots$ (that is, $\phi_n = \phi_{n+k}$ for all $k \geq 1$). The next result appears as [255, Thm], [731, 3.4], [417, Lemma 3] + [207, Lemma 5]= [208, Thm 1].

Theorem 4.4.5. *A module M is Σ-pure-injective iff M has the descending chain condition on pp-definable subgroups.*

[19] See, for example, [443], [208], [40, §8]; the last shows that a countable ring is of finite representation type iff there are only countably many countable modules up to isomorphism.

Proof. Choose an infinite cardinal number κ such that $\kappa \geq \mathrm{card}(M)$ and $\kappa^{\aleph_0} > \kappa$ (the latter will hold if the cofinality of κ is $> \aleph_0$: see your favourite book on set theory (I suggest [176]) for this and other facts that we use from the arithmetic of infinite numbers). If M does not have the descending chain condition on pp-definable subgroups, then choose pp conditions ϕ_i in one free variable such that $\phi_0(M) > \phi_1(M) > \cdots$. Then $\phi_0(M^{(\kappa)}) > \phi_1(M^{(\kappa)}) > \cdots$ (1.2.3). The cosets of these subgroups of $M^{(\kappa)}$ form a κ-branching tree of depth $\aleph_0$, hence there are $\kappa^{\aleph_0}$ branches (that is, nested chains of cosets) of this tree. Since $M^{(\kappa)}$ is pure-injective (= algebraically compact by 4.3.11) each of these branches must have non-empty intersection, so $\mathrm{card}(M^{(\kappa)}) \geq \kappa^{\aleph_0}$. But the direct sum, $M^{(\kappa)}$, has cardinality no more than κ, a contradiction.[20]

For the converse, if M has the dcc on pp-definable subgroups, then, by 1.2.3, so does $M^{(I)}$ for any set I so, by 4.2.6, $M^{(I)}$ is algebraically compact, hence pure-injective. □

For example ([731, 2.6]), any module which is of finite length or just artinian over its endomorphism ring is Σ-pure-injective (4.2.6).

Corollary 4.4.6. *If M has the dcc on pp-definable subgroups, then, for every n, M has the dcc on subgroups of M^n pp-definable in M.*

Proof. This could be proved directly but, noting that the first part of the proof of 4.4.5 applies just as well to subgroups of M^n pp-definable in M, it is immediate from the equivalence there. □

Corollary 4.4.7. *If $M$is Σ-pure-injective and $\overline{a}$ is a tuple from M with $p = \mathrm{pp}^M(\overline{a})$, then there is a pp condition ϕ such that $p(M) = \phi(M) = \mathrm{End}(M_R) \cdot \overline{a}$.*

Proof. The first equality is by the previous two results: if $\phi_0 \in p$ is such that $\phi_0(M) > p(M)$, then take $\phi_1 \in p$ such that $\phi_1(M) \not\geq \phi_0(M)$; if $(\phi_0 \wedge \phi_1)(M) = \phi_0(M) \cap \phi_1(M) > p(M)$, then take $\phi_2 \in p$ such that $\phi_2(M) \not\geq (\phi_0 \wedge \phi_1)(M)$; etc. The second equality is by 4.3.10. □

Beware that this does *not* imply $p = \langle \phi \rangle$. Take, for instance, $M = \mathbb{Q}$, $a = 1$: then ϕ in 4.4.7 may be taken to be $x = x$ which certainly does not generate (the divisibility conditions in) $\mathrm{pp}^{\mathbb{Q}}(1)$.

It is enough, in the definition of Σ-pure-injective, to check that $M^{(\aleph_0)}$ is pure-injective.

Theorem 4.4.8. ([731, 3.4], [207, (proof of) Lemma 6]) *A module M is Σ-pure-injective iff $M^{(\aleph_0)}$ is pure-injective.*

[20] If that was a bit brief see the proof of 4.4.8 below. There is also a relevant diagram at [495, p. 57] and a similar one at [491, p. 513].

Proof. For the non-trivial direction suppose that $M^{(\aleph_0)}$ is pure-injective but that M has an infinite properly descending chain, $\phi_1(M) > \phi_2(M) > \cdots$, of pp-definable subgroups. Set $N = M^{(\aleph_0)} = \bigoplus_{n\geq 1} M_n$, where $M_n \simeq M$ for all n. For each i choose $a_i \in \phi_i(M_i) \setminus \phi_{i+1}(M_i)$, so also $a_i \in \phi_i(N) \setminus \phi_{i+1}(N)$.

Given n and any function $\eta : \{1, \ldots, n\} \to \{0, 1\}$, define the element a_η to be $\sum\{a_i : i \in \{1, \ldots, n\}, \eta i = 1\}$. Then the cosets $a_\eta + \phi_{n+1}(N)$, as η ranges over the 2^n such functions, are disjoint (if i is minimal such that $\eta i \neq \eta' i$, then clearly a_η and $a_{\eta'}$ belong to different cosets of $\phi_{i+1}(N)$).

So, using the countably many parameters a_η (as n, η vary), we can describe, by sets of pp conditions, all cosets of an infinite 2-branching tree$^{(\mathrm{op})}$ of cosets of the $\phi_n(N)$. Any function ζ from the set of positive integers to $\{0, 1\}$ describes an infinite branch of this tree, namely the nested sequence of cosets $(a_{\zeta\restriction\{1,\ldots,n\}} + \phi_{n+1}(N))$. Since N is pure-injective, there is, for each such ζ, an element, a_ζ, in $\bigcap_n (a_{\zeta\restriction\{1,\ldots,n\}} + \phi_{n+1}(N))$. Note, also regarding the argument for 4.4.5 above and 4.4.9 below, that if $\zeta \neq \zeta'$ then $a_\zeta \neq a_{\zeta'}$.

Let ζ be such that there are infinitely many i with $\zeta i = 1$. Since a_ζ belongs to the direct sum $\bigoplus_n M_n$ one has $a_\zeta = \sum_{i=1}^n c_i$, with $c_i \in M_i$, for some n. Let $m > n$ be such that $\zeta m = 1$. Observe that $a_\zeta - a_{\zeta\restriction\{1,\ldots,m\}} \in \phi_{m+1}(N)$. But this is impossible, since the projection of $a_{\zeta\restriction\{1,\ldots,m\}}$ to M_m is a_m, which is not in $\phi_{m+1}(N)$, and the projection of a_ζ to M_m is zero. □

Corollary 4.4.9. ([208, Footnote 2], [732, Prop. 3]) *Every countable pure-injective module is Σ-pure-injective.*

Proof. If M is pure-injective and has an infinite descending chain of pp-definable subgroups then, by the argument of 4.4.8, $M^{(\aleph_0)}$ realises $2^{\aleph_0}$ different pp-types with parameters. Therefore $\mathrm{card}(M) \geq 2^{\aleph_0}$.[21] □

Corollary 4.4.10. *If R is a k-algebra, where k is a field, then every pure-injective module which is countable-dimensional over k is Σ-pure-injective.*

Proof. We argue as before but also ensure that we choose linearly independent coset representatives at each stage. This will give us $2^{\aleph_0}$ linearly independent elements, the required contradiction. In more detail (though changing the argument slightly), suppose there is an infinite descending chain $M = \phi_0(M) > \phi_1(M) > \cdots$. Without loss of generality, $\dim_k(\phi_i(M)/\phi_{i+1}(M)) \geq 2$ for each i. Choose a k-subspace of $M/\phi_1(M)$ of the form $V_0 \oplus V_1$ with $V_0, V_1 \neq 0$ and let V_i' be the full inverse image of V_i in M. Choose $a_i \in V_i' \setminus \phi_1(M)$, so a_0, a_1 are linearly independent modulo $\phi_1(M)$.

[21] So $\mathrm{card}(M) < 2^{\aleph_0}$ can replace countability.

Then choose a subspace $V_{00} \oplus V_{01} \oplus V_{10} \oplus V_{11}$ of $M/\phi_2(M)$ with each $V_{ij} \neq 0$ such that, if V'_{ij} denotes the full inverse image of V_{ij}, then $V'_{ij} \leq V_i$. Choose $a_{ij} \in V'_{ij} \setminus \phi_2(M)$ such that $a_{ij} - a_i \in \phi_1(M)$, so $a_{00}, a_{01}, a_{10}, a_{11}$ are linearly independent modulo $\phi_2(M)$ and $a_{ij} + \phi_1(M) = a_i + \phi_1(M)$. Continue in this way.

Then the a_η produced as in 4.4.8 are linearly independent, since, given $a_{\eta_1}, \ldots$ a_{η_k}, we may choose i such that the a_{η_i} lie in different cosets modulo $\phi_i(M)$, hence are linearly independent modulo $\phi_i(M)$, hence are linearly independent. So $\dim_k M \geq 2^{\aleph_0}$, as required. □

Proposition 4.4.11. ([417, Lemma 1]) *If M and N are Σ-pure-injective modules, then their direct sum $M \oplus N$ is Σ-pure-injective.*

Proof. If $\phi_0 \geq \phi_1 \geq \cdots$ is a descending chain of pp conditions, then, by the dcc on pp-definable subgroups, eventually $\phi_i(M) = \phi_{i+1}(M) = \cdots$ and $\phi_i(N) = \phi_{i+1}(N) = \cdots$, so, since $\phi_i(M \oplus N) = \phi_i(M) \oplus \phi_i(N)$, it follows by 4.4.5 that $M \oplus N$ is Σ-pure-injective. □

Proposition 4.4.12. *If M is Σ-pure-injective and M' is a module belonging to the definable subcategory of* Mod-R *generated by M, then M' is Σ-pure-injective. In particular this is so for pure submodules of M.*

Proof. Any pp-pair which is open on M' must be open on M (3.4.11). Since M has the dcc on pp-definable subgroups, so, therefore, has M'. By 4.4.5, M' is Σ-pure-injective. □

Corollary 4.4.13. *If M is Σ-pure-injective and if M' is a pure submodule of M, then M' is a direct summand of M.*

Example 4.4.14. If R is a local ring with radical J and $J^2 = 0$, then R is Σ-pure-injective ([731, 6.5]). This follows directly from 4.4.5 and the description, 1.1.6, of the pp-definable subgroups of R_R. Any ring which is Σ-pure-injective as a module over itself must be semiprimary (for example, see [323, 11.1])

An example of a ring which is pure injective as a module over itself on one side but not on the other is given in [729, §4]. An example of such a ring which is also artinian on one side is given in [732, §2].

There is a generalisation of the example above.

Proposition 4.4.15. ([729, Thm 5]) *Let S be a ring which, as a right module over itself, is Σ-pure-injective. Let $(X_\lambda)_\lambda$ be a set (of any cardinality) of indeterminates and let $k \geq 1$. Then the ring $R = S[(X_\lambda)_\lambda]/\langle (X_\lambda)_\lambda \rangle^k$ is Σ-pure-injective as a right module over itself.*

The pp-definable subgroups of R_R are the left ideals of the form $L_0 + JL_1 + \cdots + J^i L_i + \cdots + J^{k-1}L_{k-1}$ where $L_0, \ldots, L_{k-1}$ are finitely generated left ideals of R and where $J = \langle (x_\lambda)_\lambda \rangle$ (x_λ being the image of X_λ in R) is the radical of R.

For a related result see the comments after 4.6.3.

A module E is **Σ-injective** if every direct sum $E^{(I)}$ of copies of E is injective.

Lemma 4.4.16. (see [185, Prop. 3] for (i)⇔(iii)[22]) *The following are equivalent for a module E:*

(i) E is Σ-injective;
(ii) E is absolutely pure and Σ-pure-injective;
(iii) $E^{(\aleph_0)}$ is injective.

Proof. The equivalence of (i) and (ii) follows from the fact that $E^{(\aleph_0)}$ is, for any injective module E, absolutely pure, (2.3.5), together with 4.3.12. Then use 4.4.8 for (iii). □

Theorem 4.4.17. *The following are equivalent for a ring R:*

(i) R is right noetherian;
(ii) every absolutely pure right R-module is injective;
(iii) every direct sum of injective right R-modules is injective;
(iv) every injective right R-module is Σ-injective.

Proof. (i)⇒(ii) If $\phi(M) > \psi(M)$ are pp-definable subgroups of the absolutely pure module M, then, by 2.3.3, the inclusion $D\psi({}_RR) \leq D\phi({}_RR)$ of right ideals is proper. So the ascending chain condition on right ideals of R yields the descending chain condition on pp-definable subgroups of M. Therefore, by 4.4.5, M is (Σ-)pure-injective hence, by 4.4.16, (Σ-)injective.

(ii)⇒(iii) Any direct sum of (injective hence) absolutely pure modules is absolutely pure (as remarked at 2.3.5) so this follows.

(iii)⇒(iv) This is immediate from the definition of Σ-injective.

(iv)⇒(i) If E_R is an injective cogenerator for Mod-R and $I_1 \leq I_2 \leq \cdots \leq I_n \leq \cdots$ is an ascending chain of finitely generated right ideals, then $\text{ann}_E(I_1) \geq \text{ann}_E(I_2) \geq \cdots \geq \text{ann}_E(I_n) \geq \cdots$ is a descending chain of pp-definable subgroups of E so, if E is Σ-pure-injective, then, by 4.4.5, this chain of pp-definable subgroups must stabilise. Since E is an injective cogenerator, the chain of right ideals

[22] Part (i)⇔(iii) appears as [188, 0.1] where it is attributed variously to Cartan, Eilenberg, Bass and Papp and it also appears as [195, Exercise 6.D].

also must stabilise: if $I_n < I_{n+1}$, then in some module, and hence in E, there is an element which annihilates I_n but not I_{n+1}. Therefore R is right noetherian. $\square$

Another equivalent condition is, 3.4.28, that the injective right R-modules form a definable subcategory of Mod-R.

More generally, if M is an absolutely pure module with the dcc on annihilators, then, [31, Satz 1], M is Σ-injective; this follows from 2.3.3 and 4.4.5.

Example 4.4.18. Although, 4.4.11, any finite direct sum of Σ-pure-injective modules must be Σ-pure-injective, Example 4.3.3 shows that an infinite direct sum of Σ-pure-injective modules need not even be pure-injective. In fact, the argument of 4.4.8 adapts to show that if such a direct sum is not Σ-pure-injective, then it cannot be pure-injective. There are more general considerations in [311].

The next result appears variously in [255, Thm], [731, p. 1100], [208, Lemma 1, Thm 4] (and in [195, Exercise 6.D] for injectives over noetherian rings). The decomposition is unique by 4.3.43 and E.1.24.

Theorem 4.4.19. *If M is Σ-pure-injective, then M is a direct sum of indecomposable (Σ-pure-injective) modules.*

Proof. This proof is from [208]. Alternatively one could use the embedding (12.1.6) into the functor category (see 4.4.21 below) and then quote the corresponding result for Σ-injective objects in Grothendieck categories ([438, 2.5, 2.7] generalised).

Let $a \in M$, $a \neq 0$. Let N be a pure submodule of M which does not contain a and is maximal such: this exists by Zorn's Lemma (recall that the union of a chain of pure submodules is pure, 2.1.2(3)). By 4.4.13, N is a direct summand of M, say $M = N_0 \oplus N$. We claim that N_0 is indecomposable.

If $N_0 = N_1 \oplus N_2$, then, since $(N_0 \oplus N) \cap (N_1 \oplus N) = N$ we have, say, $a \notin N_1 \oplus N$, so, by maximality of N, $N_1 = 0$, as required for the claim.

Now let $\mathcal{P}$ be the family of sets, $\{N_i\}_{i \in I}$, of submodules of M such that each N_i is indecomposable and such that the sum $\sum_i N_i$ is direct and is pure in M. The union of any chain of such families is, note, still in $\mathcal{P}$, so, by Zorn's Lemma, there is a maximal such family: let $N = \bigoplus_i N_i$ be the (direct) sum of its members.

Because N is pure in M it is, by 4.4.13, a direct summand of M, say $M = N \oplus N'$. By 4.4.12 N' is Σ-pure-injective, so the first part of the proof applies to give that N', if non-zero, has an indecomposable direct summand, clearly contradicting maximality of the family $\{N_i\}_i$. We deduce that $M = N$, as required. $\square$

There is a kind of converse. For (slightly different) routes to the proof see [323, 8.1] and [308, Thm 10].

Theorem 4.4.20. *For any module M the following are equivalent:*

(i) M is Σ-pure-injective;
(ii) every power M^I of M is a direct sum of indecomposable modules with local endomorphism ring;
(iii) there is a cardinal κ such that every power M^I of M is a direct sum of modules of cardinality $\leq \kappa$.

Lemma 4.4.21. *A module M is Σ-pure-injective iff the functor $M \otimes -$ is a Σ-injective object of* (R-mod, **Ab**).

Proof. This follows from 12.1.6 and 12.1.3. □

For projectivity only part of the analogue of 4.4.17 is true. The situation is rather different than for injectivity. Recall, 3.4.28, that definability of the class of projective left modules is equivalent to R being right coherent and left perfect. For the next result see [208, Rmk 5 after Lemma 4], also [630, Prop. 8] for when R is countable.

Proposition 4.4.22. *If R is right coherent and left perfect, then every flat left module is (Σ-)pure-injective.*

Proof. If R is right coherent and left perfect and if ψ is a pp condition for left R-modules, then, by 2.3.19, $\psi({}_RR)$ is a finitely generated right ideal, therefore, by 2.3.9 and 3.4.27, every flat module has the dcc on pp-definable subgroups, that is, by 4.4.5, every flat module is Σ-pure-injective. □

That the converse is false is shown by the next example.

Example 4.4.23. (see [617, 3.3]) If R is Σ-pure-injective as a module over itself, then every flat left module must be Σ-pure-injective (by 4.4.12, since, by 2.3.13, every flat module is in the definable subcategory generated by R) so any non-right-coherent such ring will provide a counterexample to the converse of 4.4.22. By 4.4.14 if R is a local ring with $J(R)^2 = 0$, then R_R is Σ-pure-injective but such a ring need not be coherent: for an example, the ring $k[(x_i)_{i \geq 0} : x_i x_j = 0 \ \forall i, j]$ is not coherent since it plainly fails the annihilator condition of 2.3.18.

4.4.3 Modules of finite endolength

Modules of finite endolength are Σ-pure-injective (4.4.24) and every endo-submodule of such a module is pp-definable (4.4.25). The definable subcategory generated by a module M of finite endolength consists entirely of modules of finite endolength (4.4.27); indeed each module in that definable subcategory is *a direct sum of copies of direct summands of M (4.4.30). If a module is indecomposable and of finite endolength, then each of its direct powers is a direct sum of copies of it (4.4.28). A module of finite endolength which is also either artinian or noetherian is of finite length (4.4.34).*

A module M is of **finite endolength** if, when regarded as a module over its endomorphism ring, it has finite length. Write $\mathrm{el}(M)$ for the length of the module ${}_{\mathrm{End}(M)}M$. The first result is immediate by 4.4.5 and 1.1.8.

Corollary 4.4.24. *Every module of finite endolength is Σ-pure-injective.*

Proposition 4.4.25. *If M is of finite endolength, then every $\mathrm{End}(M)$-submodule of M is pp-definable in M_R. Hence $\mathrm{el}(M)$ is the length of the lattice, $\mathrm{pp}(M)$, of pp-definable subgroups of M.*

In particular, a module is of finite endolength iff the lattice of pp-definable subgroups of M has finite length.

Proof. Set $S = \mathrm{End}(M)$ and let $a \in M$. Since M has the dcc on pp-definable subgroups, there is (4.4.7) $\phi \in \mathrm{pp}^M(a)$ such that $Sa = \phi(M)$. Since M is of finite endolength, every S-submodule is finitely generated. Consider $\sum_1^k Sa_i$: if ϕ_i is such that $\phi_i(M) = Sa_i$, then $\sum_1^k Sa_i = (\sum_1^k \phi_i)(M)$ is pp-definable. □

Lemma 4.4.26. *If M and N are modules of finite endolength, then their direct sum $M \oplus N$ is of finite endolength.*

Proof. Any $\mathrm{End}(M \oplus N)$-submodule, L, of $M \oplus N$ contains its projections, $\pi_M L$ and $\pi_N L$ to M, N respectively, hence $L = \pi_M L \oplus \pi_N L$ (direct sum of abelian groups). Note that $\pi_M L$ is an $\mathrm{End}(M)$-submodule of M and similarly for $\pi_N L$.

Given any chain of $\mathrm{End}(M \oplus N)$-submodules of $M \oplus N$, then, projecting, we obtain two chains, each of which, by assumption, must be finite. By the first observation that can happen only if the original chain is finite, as required. □

It is the case, 4.4.31, that if M and N as above have no direct summand in common, then $\mathrm{el}(M \oplus N) = \mathrm{el}(M) + \mathrm{el}(N)$. The proof of 4.4.12 shows the following.

Proposition 4.4.27. *If M is of finite endolength and if M' is a module in the definable subcategory of* Mod-R *generated by M, then M' is of finite endolength with $\mathrm{el}(M') \leq \mathrm{el}(M)$. In particular, this is so for pure submodules (equivalently, by 4.4.13, direct summands) of M.*

Theorem 4.4.28. ([209, Thm 13], [253, Lemma 4]) *Suppose that M is an indecomposable module of finite endolength. Then for every index set I the direct power M^I is a direct sum of copies of M.*

Proof. Since M^I is, by 1.2.3, of finite endolength hence Σ-pure-injective, it follows, by 4.4.19, that M^I is a direct sum of indecomposable modules. Let N be any indecomposable summand of M^I: we show that $N \simeq M$.

Choose a non-zero element $a \in N$. So $N = H(a)$ (4.3.37). Let $p = \mathrm{pp}^N(a) = \mathrm{pp}^{M^I}(a)$. By 4.4.7 there is a pp condition ϕ such that $\phi(M^I) = p(M^I)$.

Since M^I has finite endolength and $a \neq 0$, there is a pp condition ψ such that $\psi(M^I)$ is properly contained in $\phi(M^I)$ and is maximal such. Since, by 4.3.49, p is irreducible, $\psi(M^I)$ must contain every pp-definable subgroup properly contained in $\phi(M^I)$. To see this, suppose not; then $\phi(M^I) = \psi(M^I) + \psi'(M^I)$ for some ψ' with $\psi'(M^I) < \phi(M^I)$. Since $a \notin \psi(M^I)$, $\psi'(M^I)$ neither ψ nor ψ' is in p so, by irreducibility of p, there is $\phi' \in p$ with $\phi' \wedge \psi + \phi' \wedge \psi' \notin p$. But $\phi'(M^I) \geq \phi(M^I)$ by choice of ϕ, so $(\phi' \wedge \psi + \phi' \wedge \psi')(M^I) = \psi(M^I) + \psi'(M^I) = \phi(M^I)$ which contains a, a contradiction.

Now, since $\phi(M^I) > \psi(M^I)$, there must be some $c \in \phi(M) \setminus \psi(M)$. We claim that both $\mathrm{pp}^M(c)$ $(= \mathrm{pp}^{M^I}(c))$ and $\mathrm{pp}^N(a)$ consist exactly of those pp conditions ϕ' such that $\phi'(M) \geq \phi(M)$. If $\phi'(M) \geq \phi(M)$, then certainly c and a are in $\phi'(M)$. If $\phi'(M) \ngeq \phi(M)$, then $\phi'(M) \cap \phi(M) \leq \psi(M)$, so, since neither c nor a lies in $\psi(M)$, neither is in $\phi'(M)$. We conclude that $\mathrm{pp}^M(c) = \mathrm{pp}^N(a)$ so, by 4.3.47, $M \simeq N$, as required. □

This is expanded in 4.4.30 below.

Lemma 4.4.29. *Suppose that M is of finite endolength. Then $M = M_1^{(\kappa_1)} \oplus \cdots \oplus M_k^{(\kappa_k)}$ for some indecomposable modules $M_1, \ldots, M_k$ of finite endolength and some index sets κ_i.*

Proof. We already know, from 4.4.19, that M is a direct sum of indecomposables, so it must be shown that, up to isomorphism, only finitely many indecomposables occur.

This follows rather directly from the methods of Part II (Section 12.5) since the localisation of $R \otimes -$ at the finite type torsion theory cogenerated by $M \otimes -$ will be of finite length, hence will have non-zero morphisms to only finitely many indecomposable injectives. Alternatively use 5.3.6. Or proceed as follows.

Suppose that ϕ is such that $\phi(M)$ is minimal among non-zero pp-definable subgroups of M. Then, if M_1, M_2 are indecomposable summands of M with $\phi(M_1) \neq 0$ and $\phi(M_2) \neq 0$, it must be that $M_1 \simeq M_2$. For, pick $a_i \in \phi(M_i)$ with $a_i \neq 0$. Clearly $\mathrm{pp}^M(a_1) = \{\phi' : \phi'(M) \geq \phi(M)\} = \mathrm{pp}^M(a_2)$ so $M_1 \simeq M_2$ by 4.3.47. Note, furthermore, that $\phi(M)$ is a simple $\mathrm{End}(M)$-submodule of M (by 4.4.25), so there can be only finitely many distinct such $\phi(M)$. With respect to a decomposition of M as a direct sum of indecomposables M_λ, remove all those indecomposables M_λ with $\phi(M_\lambda) \neq 0$ for some ϕ such that $\phi(M)$ is minimal.

By what has been noted already we have removed copies of only finitely many indecomposables. Let M' be what remains. Now repeat this process with M' in place of M, noting that if ϕ is such that $\phi(M')$ is minimal, then the length of $\phi(M)$ as an $\mathrm{End}(M)$-submodule is at least 2.

Continue in this way. Since M has finite endolength the process stops at some stage and we deduce that M has only finitely many indecomposable summands up to isomorphism. □

Corollary 4.4.30. *Suppose that M is a module of finite endolength and let $M_1, \dots, M_k$ be, up to isomorphism, the distinct indecomposable direct summands of M. Then every module M' in the definable subcategory generated by M is a direct sum of copies of $M_1, \dots, M_k$ (and conversely). In particular this applies to all modules of the form M^I.*

Proof. One may move to the functor category and make use of 12.3.2, or base a short argument on 5.3.6. An alternative proof proceeds as follows.

If M' is in the definable subcategory generated by M, consider $M \oplus M'$. By 4.4.27 and 4.4.26 this has finite endolength. Choose a non-zero element, $a \in N$, where N is any indecomposable direct summand of M'. Now go through the proof of 4.4.28 with $M \oplus M'$ in place of M^I, bearing 3.4.11 in mind, and check that it shows that N is isomorphic to a direct summand of M. □

Lemma 4.4.31. *If M, N are modules of finite endolength and have no non-zero direct summand in common, then* $\mathrm{el}(M \oplus N) = \mathrm{el}(M) + \mathrm{el}(N)$.

Proof. Say $M = M_1^{(\lambda_1)} \oplus \cdots \oplus M_s^{(\lambda_s)}$ and $N = N_1^{(\kappa_1)} \oplus \cdots \oplus N_t^{(\kappa_t)}$ with the M_i, N_j all indecomposable and pairwise non-isomorphic (4.4.29). By 4.4.25, $\mathrm{el}(M \oplus N)$ is the length of the lattice $\mathrm{pp}(M \oplus N)$ which, by 5.3.6 (and cf. the proof of 4.4.11), is the sum of those of $\mathrm{pp}(M)$ and $\mathrm{pp}(N)$ for, by that result, there is no shared proper interval in the lattice of pp conditions. □

The next result is immediate from 1.3.15 and 4.4.25.

Proposition 4.4.32. *If M is a module of finite endolength and E is a duality module for M as in 1.3.15, then* $\mathrm{el}(M^*) = \mathrm{el}(M)$.

Example 4.4.33. ([139, 1.3], [417, Thm 1] for $R = \mathbb{Z}$) Let R be a semiprime right and left Goldie ring, with semisimple artinian quotient ring Q, and suppose that M is a Σ-pure-injective R-module. Then M is the sum of a divisible R-module and a module which is annihilated by a regular element of R. If M is of finite endolength, then the divisible summand is the restriction to R of a Q-module. (For definitions and results used here see [441, 2.3.1].)

For, since R satisfies the left Ore condition, the pp-definable subgroups of M of the form Mc with $c \in R$ regular are directed downwards: given $c, d \in R$ regular, there are $r, s \in R$ with s regular and $rc = sd$ also, therefore, regular, and then $Mc \cap Md \geq M(sd)$. So there is a minimal such subgroup, Mc say. Using the Ore condition it is easy to check that Mc is a divisible submodule of M, that $\mathrm{ann}_M(c)$ is a submodule of M and that $M = Mc \oplus \mathrm{ann}_M(c)$, showing the first statement.

A torsion divisible module N will have an infinite ascending chain of pp-definable subgroups. For, take $c \in R$ regular with $\mathrm{ann}_N(c) \neq 0$ and consider $\mathrm{ann}_N(c) \leq \mathrm{ann}_N(c^2) \leq \cdots$. If $\mathrm{ann}_N(c^n) = \mathrm{ann}_N(c^{n+1}) = \cdots$, then take $a \in \mathrm{ann}_N(c^n)$. Since N is divisible $a = bc^n$ for some $b \in N$, so $bc^{2n} = ac^n = 0$, hence $0 = bc^n = a$. Thus in the case that M is of finite endolength, the divisible summand must be torsionfree, hence a Q-module.

It then follows fairly directly ([139, 1.4]) that if R is a non-artinian simple noetherian ring, then the only R-modules of finite endolength are the direct sums of copies of the simple modules over the (simple artinian) quotient ring of R.

Also if R is a non-artinian hereditary noetherian prime ring, then, see [139, 1.4], every R-module of finite endolength is a direct sum of R-modules of finite length and copies of the simple module over the simple artinian quotient ring of R. Conversely, every finite direct sum of such modules is of finite endolength.

Proposition 4.4.34. ([141, 4.6]) *If N_R is of finite endolength and N_R has either the acc or dcc on submodules, then N_R is of finite length.*

Proof. This follows from a generalisation (also at [141, 4.6]) of a theorem of Lenagan (see [441, 4.1.6] or [243, 7.10]). The statement is that if a bimodule has finite length on one side and the acc on the other, then it has finite length on both. □

4.4.4 Characters

Modules of finite endolength give rise to Sylvester rank functions (4.4.36) and thence to characters; indeed, there is a natural bijection between irreducible characters and indecomposable modules of finite endolength (4.4.37).

In this section we only point out some concepts and, mostly without proofs, results.

In [141] Crawley-Boevey defines a **character** of mod-R to be a function $\chi : \text{mod-}R \longrightarrow \mathbb{N}$ satisfying the following conditions:

(1) $\chi(A \oplus B) = \chi(A) \oplus \chi(B)$; and
(2) for every exact sequence of the form $A \to B \to C \to 0$ in mod-R one has $\chi(C) \leq \chi(B) \leq \chi(A) + \chi(C)$.

This is an adaptation of the notion of a **Sylvester module rank function**, see [638, pp. 96, 97], which is a function, ρ, from mod-R to $(1/n)\mathbb{Z}$, for some n, which satisfies the two conditions given and is normalised so that $\rho(R) = 1$. There is a bijection, [638, 7.12], between Sylvester rank functions and certain equivalence classes of ring homomorphisms from R to simple artinian rings. Note that if $R \to M_n(D)$, where D is a division ring, is such a homomorphism, then $M_n(D)$, regarded as an R-module, is of finite endolength.

Lemma 4.4.35. ([141, §5]) *If N is a right module of finite endolength, then the function $\chi_N : R\text{-mod} \to \mathbb{N}$ given by $\chi_N(L) = \text{length}_{\text{End}(N)}(N \otimes L)$ is a character of R-mod.*

Proof. Note that $\text{length}_{\text{End}(N)}(N \otimes L)$ is finite if ${}_RL$ is finitely generated. For, if ${}_RR^m \to L$ is an epimorphism, then so is $N \otimes R^m \to N \otimes L$ and $N \otimes R^m \simeq N^m$ has finite length as a left End(N)-module.

The conditions for a character are satisfied because $N \otimes -$ commutes with direct sum and is right exact. □

It is the case that every character of R-mod arises in this way. This result, stated below as 4.4.39, is due to Crawley-Boevey. A different proof was given by Burke, [101, Chpt. 5], [103, §3], and is one which allows some generalisation beyond modules of finite endolength. See also Herzog's [278, §9]. Crawley-Boevey's proof involves extending a character to finitely presented, then arbitrary, functors in (R-mod, **Ab**).[23] Burke's proof works directly with localised lattices of pp conditions. We give some indication of the latter. Following [638, p. 97] say that a **Sylvester matrix rank function** for R is a function, ρ, from the set of all finite rectangular matrices with entries from R to $(1/n)\mathbb{Z}$ (for some n) such that, for all matrices H_i (of appropriate sizes):

(1) $\rho(I_1) = 1$ (I_1 denotes the 1×1 identity matrix);

(2) $\rho\begin{pmatrix} H_1 & 0 \\ 0 & H_2 \end{pmatrix} = \rho(H_1) + \rho(H_2)$;

(3) $\rho\begin{pmatrix} H_1 & 0 \\ H_3 & H_2 \end{pmatrix} \geq \rho(H_1) + \rho(H_2)$;

(4) $\rho(H_1 H_2) = \min\{\rho(H_1), \rho(H_2)\}$.

Let N be any module of finite endolength. By 4.4.25 el(N) is the length of the lattice $\text{pp}^1(N)$ and so is also the length of the opposite lattice, which, by 1.3.15, has the form $\text{pp}^1(DN)$ for a certain left module DN. Define a function on matrices

[23] Since $N \otimes L$ with L finitely presented may be thought of as the value of a sort of $\mathbb{L}_R^{\text{eq+}}$ at N, that is, as part of the many-sorted structure $N^{\text{eq+}}$, the first step may be thought of as extending a character from part to all of $N^{\text{eq+}}$.

as follows. Given an $m \times n$ matrix H set $\psi_H(\overline{x})$ to be the pp condition (for left modules, with n free variables) $H\overline{x} = 0$, so $D\psi_H(\overline{x})$ is the pp condition $H|\overline{x}$ for right modules. Set $\rho_N(H) = \mathrm{length}\big([D\psi_N(\overline{x}), (\overline{x} = \overline{0})]_N\big)/\mathrm{el}(N)$. Recall that $[\sigma, \tau]_N$ is the notation for an interval in the lattice $\mathrm{pp}(N)$ (§3.2.1).

Proposition 4.4.36. ([101, 5.1.1], [103, 3.1]) *Let N be a module of finite endolength. Then ρ_N is a Sylvester matrix rank function.*

From a Sylvester matrix rank function ρ one may define a Sylvester module rank function ρ' on R-mod as follows. Let L be a finitely presented left module and take a presentation, say $R^m \to R^n \to L \to 0$, of L, where the morphism from R^m to R^n is given by $\overline{r} \mapsto \overline{r}H$, where H is an $m \times n$ matrix with entries in R. Define $\rho'(L) = n - \rho(H)$. This is independent of presentation of L and is a Sylvester module rank function, [638, 7.2]. Therefore, if N_R is a module of finite endolength, then a character on R-mod is obtained from N by setting $\xi_N = \rho'_N \cdot \mathrm{el}(N)$. Explicitly, if $L \in R$-mod is as above, then $\xi_N(L) = \big(n - \mathrm{length}\big([D\psi_N(\overline{x}), (\overline{x} = \overline{0})]_N\big)/\mathrm{el}(N)\big) \cdot \mathrm{el}(N) = \mathrm{length}\,(\mathrm{pp}^n(N)) - \mathrm{length}$ $([D\psi_N(\overline{x}), (\overline{x} = \overline{0})]_N)$ and this is equal to $\mathrm{length}\big([(\overline{x} = \overline{x}), D\psi_N(\overline{x})]_N\big)$. The next result, which follows fairly directly from 1.3.7 (or 10.3.8), is then used.

Lemma 4.4.37. ([101, 5.1.2], [103, 3.3]) *Let M be a right R-module. Let $L =$ $\mathrm{coker}(\overline{r} \mapsto \overline{r}H : {}_RR^m \to {}_RR^n)$, where H is an $m \times n$ matrix over R, be a finitely presented left R-module. Let $\psi(\overline{x})$ be the pp condition $H\overline{x} = 0$ for left modules (so L is generated by a free realisation of ψ, see 1.2.30). Then $L \otimes_R M \simeq M^n/D\psi(M)$ as* $\mathrm{End}(M)$*-modules. If M is of finite endolength, then every* $\mathrm{End}(M)$*-submodule of $L \otimes_R M$ has the form $D\phi(M)/D\psi(M)$ for some $\phi \le \psi$.*

It follows immediately from this and the definition of χ_N that $\chi_N = \xi_N$. It also follows (for example, from [103, 3.5]) that if N is an R-module of finite endolength and M is a direct summand of N, then $\chi_M(L) \le \chi_N(L)$ for every $L \in R$-mod. Furthermore, the definable subcategory of Mod-R generated by M is equal to that generated by N iff $\chi_M = \chi_N$.

There is a dual way to define characters, namely, if K is a left R-module of finite endolength, then set $\chi^K(L) = \mathrm{length}_{\mathrm{End}(K)}\big(\mathrm{Hom}({}_RL, {}_RK)\big)$. Then χ^K is a character of R-mod. The relation with duality of Ziegler spectra follows with the aid of 5.4.15.

Lemma 4.4.38. ([141, §6], [103, 5.4]) *Let N be an indecomposable right R-module of finite endolength. Then $\chi^{DN} = \chi_N$, where DN is the dual module of N in the sense of 5.4.17.*

Now let χ be a character for left R-modules. From this we produce a definable subcategory of R-Mod, as follows. Consider a pp-pair $\phi < \phi'$ for left R-modules.

There are, by 10.2.14, projective resolutions of the functors F_ϕ and $F_{\phi'}$ (without loss of generality, with common first projective functor): $0 \to (B,-) \to (A,-) \to F_\phi \to 0$ and $0 \to (B',-) \to (A,-) \to F_{\phi'} \to 0$. Let T_χ be the collection of all pp-pairs ϕ'/ϕ with $\chi(B) = \chi(B')$. Let $\mathcal{X}_\chi$ be the subcategory of R-Mod defined (see Section 3.4.1) by T_χ. Denote by X_χ and pp_χ the corresponding closed subset of the Ziegler spectrum (Section 5.1) and the corresponding quotient lattice, $\mathrm{pp}(\mathcal{X}_\chi)$, of pp_R (see just before 3.4.14). It may be shown ([101, 5.2.2], [103, 4.5]) that all this is well defined.

Then it is shown that pp_χ has finite length and that if L is a left module, necessarily of finite endolength, which generates the definable subcategory $\mathcal{X}_\chi$, then $\chi = \chi_{DL} = \chi^L$.

Thus, also see [141, §5], [278, §§8,9], every character arises from a module of finite endolength.

Examples which, with manageable computations, can be used to illustrate these processes are provided by finite-dimensional representations of quivers.[24]

Crawley-Boevey proved, [141, 5.1], that if $N_1, \dots, N_n$ are non-isomorphic indecomposable R-modules of finite endolength, then the characters $\chi_{N_1}, \dots, \chi_{N_n}$ are linearly independent over $\mathbb{Z}$ and he established unique decomposition.

Theorem 4.4.39. ([141, 5.2], [101, 5.3.2] for an alternative proof) *Every character χ of R-*mod *has a unique decomposition of the form $a_1\chi_{N_1} + \cdots + a_n\chi_{N_n}$ with the N_i non-isomorphic indecomposable right R-modules of finite endolength and the a_i positive integers. The definable subcategory, $\mathcal{X}_\chi$, of R-*Mod *corresponding to χ is generated by $DN_1, \dots, DN_n$.*

There are further general results and examples (in the context of artin algebras) in [634].

Burke [103] extends these ideas by allowing ordinal-valued characters and is able to extend much of this to Σ-pure-injective modules. Also see [144, pp. 1662–3] for extension to more general (than module) categories.

By 4.4.39 there is a bijection between irreducible (with respect to the above decomposition) characters and indecomposable modules of finite endolength, hence, 12.1.9, between irreducible characters and indecomposable injective objects of finite endolength in the, locally coherent, functor category (R-mod, **Ab**). Crawley-Boevey extended this to locally finitely presented categories, obtaining ([143]) a bijection between indecomposable injectives of finite endolength and irreducible additive functions. Both 5.1.24 and 2.3.3 can (just) be discerned in the proofs.

[24] For example, if $R = kA_3$, see 5.1.2, then, since this is small and of finite representation type, everything, even the functor category/category of pp-pairs, can be computed explicitly and easily.

4.4.5 The ascending chain condition on pp-definable subgroups

There is a characterisation of modules with the acc on pp-definable subgroups (4.4.41) and then some examples of constructions which preserve this property.

There is a well-developed theory of modules with the descending chain condition on pp-definable subgroups, see Section 4.4.2ff., but the ascending chain condition (**acc**) has been much less studied, despite the duality between (right) modules with the one condition and (left) modules with the other: the dual in the sense of Section 1.3.3 of any Σ-pure-injective module has, by 1.3.15 and 4.4.5, the acc on pp-definable subgroups.

If M is a module with the acc on pp-definable subgroups, then its pure-injective hull $H(M)$ also, by 4.3.24, has this property. By 7.2.3 and 7.1.5 it has m-dimension, so, by 5.4.15, $H(M)$ is a direct summand of a module of the form $\mathrm{Hom}_S({}_RN_S, E_S)$, where N is a Σ-pure-injective left R-module, $S = \mathrm{End}(N)$ and E is a minimal injective cogenerator for Mod-S. Also, by 7.3.6, $H(M)$ is the pure-injective hull of a direct sum of indecomposable pure-injective modules.

Any module M with the acc on $\mathrm{End}(M)$-submodules certainly satisfies the acc on pp-definable subgroups (by 1.1.8). In particular, any left noetherian ring, regarded as a right module over itself, satisfies this condition. By 2.1.27 a pure-projective module has the acc on pp-definable subgroups iff it has the acc on $\mathrm{End}(M)$-submodules.

Example 4.4.40. Even if $M \in$ Mod-R is noetherian over its endomorphism ring it need not be the case that every $\mathrm{End}(M)$-submodule of M is a pp-definable subgroup of M_R. Take $R = \mathbb{Z}_{(p)}$, $M = \mathbb{Z}_{(p)} \oplus \overline{\mathbb{Z}_{(p)}}$. Since $\mathrm{Hom}(\overline{\mathbb{Z}_{(p)}}, \mathbb{Z}_{(p)}) = 0$ (for example, by 2.2.4, using 2.3.14), $0 \oplus \overline{\mathbb{Z}_{(p)}}$ is an $\mathrm{End}(M)$-submodule but it is not a pp-definable subgroup, for example, since $\mathrm{pp}^M(1, 0) = \mathrm{pp}^M(0, 1)$, which is so since an automorphism of the pure submodule $\mathbb{Z}_{(p)} \oplus \mathbb{Z}_{(p)}$ interchanges $(0, 1)$ and $(1, 0)$.

The next result, characterising modules with the acc on pp-definable subgroups, may be compared with those in Section 1.3.5.

Theorem 4.4.41. ([735, 2.5]) *For any module M the following are equivalent:*

(i) M has the ascending chain condition on pp-definable subgroups;
(ii) the canonical morphism $\mu : M^I \otimes_R L \to (M \otimes_R L)^I$ is injective for all left modules ${}_RL$ and index sets I;
(iii) as (ii) but with I countably infinite.

Proof. (i)$\Rightarrow$(ii) Let $\overline{a} = (a_1, \dots, a_n) \in (M^I)^n$ and $\overline{l} = (l_1, \dots, l_n) \in L^n$ with, say, $a_j = (a_{ji})_{i \in I}$ $(j = 1, \dots, n)$. Set $\overline{a}_i = (a_{1i}, \dots, a_{ni})$. If $\mu(\overline{a} \otimes \overline{l}) = 0$, then each $\overline{a}_i \otimes \overline{l}$ is 0, so, by Herzog's criterion (1.3.7), there is a pp condition ϕ_i such that

$\overline{a}_i \in \phi_i(M)$ and $\overline{l} \in D\phi_i(L)$. By assumption there is ϕ such that $\phi(M) = \sum_i \phi_i(M)$, $= \phi_{i_1}(M) + \cdots + \phi_{i_t}(M)$ say. By 1.3.1 $D\phi(L) = D\phi_{i_1}(L) \cap \cdots \cap D\phi_{i_t}(L)$, so $\overline{l} \in D\phi(L)$. By 1.2.3 $\overline{a} \in \phi(M^I)$, so, again by 1.3.7, $\overline{a} \otimes \overline{l} = 0$, as required.

(ii)⇒(iii) is trivial, so suppose that (iii) holds. Assume, for a contradiction that $\phi_1(M) < \phi_2(M) < \cdots < \phi_i(M) < \cdots$ is a strictly increasing sequence of pp-definable subgroups of M. Choose $a_i \in \phi_i(M) \setminus \phi_{i-1}(M)$. Without loss of generality (replace each ϕ_i by $\phi_1 + \cdots + \phi_i$) $\phi_1 \leq \phi_2 \leq \cdots \leq \phi_i \leq \cdots$ in the lattice, pp_R, of pp conditions (not just in the lattice pp(M)). Let q be the pp-type generated (under conjunction and implication) by the $D\phi_i$, that is, $\{\psi \in {}_R\mathrm{pp}; \psi \geq D\phi_i$ for some $i\}$, and, 3.2.5, let L and $l \in L$ be such that $\mathrm{pp}^L(l) = q$. Since $l \in D\phi_i(L)$, each $a_i \otimes l = 0$ so, by assumption, $(a_i)_i \otimes l = 0$. Therefore, by 1.3.7, there is a pp condition ψ such that $(a_i)_i \in \psi(M^{\aleph_0})$ and $l \in D\psi(L)$. Since $\mathrm{pp}^L(l) = q$, the latter implies $D\psi \geq D\phi_i$ for some i, hence, by 1.3.1, $\psi \leq \phi_i$. But then $a_j \in \phi_i(M)$ for all j, contradiction. □

The dual of 4.4.6, that is, the fact that the dcc on pp-definable subgroups of M is equivalent to the dcc on subgroups of M^n pp-definable in M for all n, follows by that result, 1.3.1, and 1.3.15 (for instance).

Zimmermann proved that the acc on pp-definable subgroups is stable under polynomial extension (an extension of the Hilbert Basis Theorem).

Theorem 4.4.42. ([736, 1.4, 1.5]) *A module M_R has the ascending chain condition on pp-definable subgroups iff for every finite tuple $\overline{X}$ of indeterminates the $R[\overline{X}]$-module $M[\overline{X}]$ has the ascending chain condition on pp-definable subgroups.*

Corollary 4.4.43. ([736, 1.6]) *If $R[\overline{X}] \to S$ is a ring homomorphism with S finitely generated as a left $R[\overline{X}]$-module and if M_R has the ascending chain condition on pp-definable subgroups, then so does the right S-module $M \otimes_R S$.*

In [736] there is also a corresponding variant on the Artin–Rees Lemma and extension of the Krull intersection theorem.

4.5 Representation type and pure-injective modules

Equivalents to, and consequences of, pure-semisimplicity are given in Section 4.5.1. Using duality, strong restrictions are obtained in Section 4.5.2 on the finitely presented right and left modules over right pure-semisimple rings. Finite representation type and the connection with pure-semisimplicity are described in Section 4.5.3. It is not known whether pure-semisimplicity implies finite representation type: some cases where it is known to be true are seen in Section 4.5.4. An indecomposable module of finite endolength is said to be generic if it is not "small".

In some contexts generic modules specialise to families of "small" modules; this is described in Section 4.5.5 and linked to representation type of algebras.

4.5.1 Pure-semisimple rings

A ring is right pure-semisimple if every right module over it is (Σ-)pure-injective. In fact, apparently much weaker conditions are enough (4.5.7).

Right pure-semisimplicity of a ring is equivalent to the lattice of pp conditions for right modules having the descending chain condition (4.5.1), hence to the acc on pp conditions for left modules (4.5.3). Right pure-semisimple rings are right artinian, every indecomposable module is of finite length and every right module is a direct sum of these indecomposables (4.5.4). There is a characterisation in terms of chains of morphisms between indecomposables (4.5.6). The condition is also equivalent to every non-zero functor having a simple subfunctor (4.5.8).

A ring R is **right pure-semisimple** if every right R-module is pure-injective, in which case every right R-module is, by the definition, Σ-pure-injective. The consequences of this are quite extreme (in fact, conjecturally even more so, see 4.5.26 below).

Pure-semisimple rings have been studied from a variety of points of view. As a result, it can be difficult to assign credit to (a particular version of) some of the earlier results. Therefore I have not attempted to do so here, but at least some of the relevant references can be found in [491, §2] and [495, p. 229].

Theorem 4.5.1. *A ring R is right pure-semisimple*

iff the lattice, pp_R, of pp conditions with one free variable for right modules has the dcc

iff for each n, the lattice, pp^n_R has the dcc.

Proof. There is, by 3.2.1, a module M such that each lattice, $\mathrm{pp}^n(M)$, of subgroups of M^n pp-definable in M is isomorphic to pp^n_R. If R is right pure-semisimple, then, by 4.4.5, each of these lattices has the dcc.

Conversely, if N is any module, then the lattice $\mathrm{pp}(N)$ is a quotient of pp^1_R (3.2.3), so, if the latter has the dcc, so does $\mathrm{pp}(N)$ and, therefore, by 4.4.5, N is (Σ-)pure-injective and R is indeed right pure-semisimple. □

Corollary 4.5.2. *A ring R is right pure-semisimple iff every pp-type for right R-modules is finitely generated.*

Proof. Suppose that R is right pure-semisimple. Take $p \in \mathrm{pp}^n_R$ and choose $\phi_1 \in p$. If ϕ_1 does not generate p, there is $\phi_2 \in p$ with $\phi_1 > \phi_2$. Repeat with ϕ_2 in place of ϕ_1; and so on. By the dcc (4.5.1) eventually we have ϕ_n with $\langle \phi_n \rangle = p$.

For the converse, if R is not right pure-semisimple, then, by 4.5.1, there is an infinite descending chain $\phi_0 > \phi_1 > \cdots > \phi_i > \cdots$ in pp_R. Clearly $\{\phi : \phi \geq \phi_i$ for some $i\}$ is a pp-type which is not finitely generated. □

So, for such rings, the lattice of pp-types (end of Section 3.2.1) coincides with the lattice of pp conditions.

Applying elementary duality, 1.3.1, one obtains the following.

Corollary 4.5.3. ([494, 3.1]) *A ring R is right pure-semisimple iff each lattice, ${}_R\mathrm{pp}^n$, of pp conditions for left modules has the acc iff ${}_R\mathrm{pp}$ has the acc.*

Theorem 4.5.4. *If R is right pure-semisimple, then R is right artinian, every right R-module is a direct sum of indecomposable modules with local endomorphism ring and every indecomposable right R-module is of finite length.*

Proof. If E is an injective right R-module, then $E^{(\aleph_0)}$ is absolutely pure (2.3.5). By assumption it is pure-injective, so it is injective by 2.3.2. Therefore (4.4.17) R is right noetherian.

Every finitely generated left ideal of R is right pp-definable so R has the dcc on finitely generated left ideals, that is, R is right perfect. But any right perfect right noetherian ring is (see [663, VIII.5.2]) right artinian.

Every right R-module is a direct sum of indecomposables because, 4.4.19, this is true of every Σ-pure-injective module.

Let N be an indecomposable right R-module and take $a \in N$, $a \neq 0$. Let ϕ be a pp condition which generates $p = \mathrm{pp}^N(a)$ (4.5.2). Say $\phi(x)$ is $\exists \overline{y}\, (x\ \overline{y})H = 0$. Choose $\overline{b}$ from N such that $(a\ \overline{b})H = 0$ and let M be the submodule of N generated by a and the entries of $\overline{b}$: so M is finitely generated, hence of finite length.

Note that $a \in \phi(M)$ so $\mathrm{pp}^M(a) = p$. The modules M, N are, by assumption, pure-injective, so, by 4.3.46, there is a direct summand of M isomorphic to N (or note that, by 4.3.41, M must be a direct summand of, hence equal to, N). We deduce that N is of finite length, as required. □

Proposition 4.5.5. *If R is right pure-semisimple, then for every chain $M_0 \xrightarrow{f_1} M_1 \xrightarrow{f_2} \cdots \to M_n \to \cdots$ of non-isomorphisms between indecomposable modules there is n such that $f_n \ldots f_1 = 0$.*

Proof. By 4.5.4 each M_i is finitely presented. Let $\overline{c}$ be a finite generating tuple for M_0. Say $\mathrm{pp}^{M_0}(\overline{c}) = \langle \phi_0 \rangle$ (1.2.6 or 4.5.2). For each i let ϕ_i be a generator for $\mathrm{pp}^{M_i}(f_i \ldots f_1\overline{c})$. Then (1.1.7), $\phi_0 \geq \phi_1 \geq \cdots \geq \phi_n \geq \cdots$ and, by 1.2.9 and the fact that no f_i is split, these inclusions are proper, until $\phi_n(\overline{x})$ is equivalent to $\overline{x} = \overline{0}$. By 4.5.1 such an n exists. □

Under some further assumptions on the ring the above condition is also sufficient for right pure-semisimplicity: for artin algebras this is [19, §3], which is strengthened in the next result.

Proposition 4.5.6. ([276, 3.2]) *Let R be a Krull–Schmidt ring. Then R is right pure-semisimple iff for every chain $M \xrightarrow{f_1} M_1 \xrightarrow{f_2} \cdots \to M_n \to \cdots$ of non-isomorphisms between indecomposable finitely presented modules there is n such that $f_n \dots f_1 = 0$.*

Proof. For the direction converse to 4.5.5 suppose that R is not right pure-semisimple. By 4.5.1 the lattice pp_R^1 does not have the dcc – let us say "has infinite depth". Set ϕ_0 to be $x = x$. By 4.3.64, F_{ϕ_0} is a (finite) sum of local functors: certainly at least one of these, F_{ψ_0} say, is such that the interval $[F_{\psi_0}, x = 0]$ in pp_R^1 has infinite depth. So there is $F_{\phi_1} < F_{\psi_1}$ such that $[\phi_1, x = 0]$ has infinite depth. Write F_{ϕ_1} as a sum of local functors and continue, to obtain an infinite properly descending chain $\phi_0 > \phi_1 > \cdots$ with the F_{ϕ_i} local. Let (C_i, c_i) be a free realisation (Section 1.2.2) of F_{ϕ_i}: we may choose this so that $(c_i, -) : (C_i, -) \to F_{\phi_i}$ is a projective cover (by 4.3.69 and 10.1.14) and then, by 10.2.27 and since F_{ϕ_i}, hence $(C_i, -)$, is local, C_i is indecomposable. By 1.2.9 there are morphisms $f_i : C_i \to C_{i+1}$ taking c_i to c_{i+1}, and no composition, $f_n \dots f_0$, of these is zero. □

Various apparently quite weak conditions of the form "the structure of arbitrary R-modules is not as complicated as it might be" turn out to be equivalent to pure-semisimplicity. See, for example, [495, §11.1], [729, §4] (that (ii)⇒(i), hence the equivalence of (i) and (ii), can be proved along the lines of 4.4.8).

Theorem 4.5.7. *The following conditions on a ring R are equivalent:*

(i) R is right pure-semisimple;
(ii) every right R-module is a direct sum of indecomposable modules;
(iii) there is a cardinal number κ such that every right R-module is a direct sum of modules, each of which is of cardinality less than κ;
(iv) there is a cardinal number κ such that every right R-module is a pure submodule of a direct sum of modules each of cardinality less than κ.

Theorem 4.5.8. ([19, §1]) *A ring R is right pure-semisimple iff every non-zero functor in* (mod-R, **Ab**) *has a simple subfunctor.*

A proof of this result using pp conditions is at [495, 12.18].

4.5.2 Finite length modules over pure-semisimple rings

After stating the Harada–Sai Lemma (4.5.10) it is shown that from a right pure-semisimple ring one may produce another which is also left artinian and such that the respective categories of right and left finitely presented modules are dual (4.5.12). It follows that over a right pure-semisimple ring there are only finitely many indecomposable modules of each finite length (4.5.13).

Every finitely presented left module over a right pure-semisimple ring is of finite endolength (4.5.14) and duality induces a bijection between indecomposable finitely presented left modules and isolated points of the right Ziegler spectrum (4.5.15). It follows that each indecomposable finitely presented left module is also an isolated point of its spectrum (4.5.16) and each isolated point of the right spectrum is of finite endolength (4.5.17).

Lemma 4.5.9. ([494, 3.4]) *Suppose that $\phi_1, \dots, \phi_k$ are pp conditions, with n free variables, and that $\phi_1 + \cdots + \phi_k \in p$, where p is a pp-type which is realised by a non-zero tuple in the module M. Then there is a pp-type q, not the type of the zero tuple, realised in M such that there is i with $\phi_i \in q$.*

Proof. Let $(C_i, \overline{c}_i)$ be a free realisation of ϕ_i. By 1.2.27 $\overline{c} = \sum_1^k \overline{c}_i$ in $C = \bigoplus_1^k C_i$ is a free realisation of $\phi_1 + \cdots + \phi_n$. By 1.2.17 there is a morphism $f : C \to M$ taking $\overline{c}$ to $\overline{a}$, where $\overline{a}$ from M realises p. For some i, $f\overline{c}_i$ is non-zero. Let $q = \mathrm{pp}^M(f\overline{c}_i)$. □

Proposition 4.5.10. (Harada–Sai Lemma) ([267, 1.2], see, for example [27, VI.1.3]) *Let $N_0 \to N_1 \to \cdots \to N_t$ be a sequence of non-isomorphisms between indecomposable modules each of which is of length no greater than n. If the composition of these morphisms is non-zero, then $t \le 2^n$.*

Theorem 4.5.11. ([494, 3.6]) *If R is right pure-semisimple, then, for each n, there are only finitely many indecomposable finitely presented left (sic) R-modules of length n, up to isomorphism.*

Proof. Fix $n \ge 1$ and let L_λ $(\lambda \in \Lambda)$ be the set of distinct isomorphism types of finitely presented indecomposable left R-modules of length n. For each λ set $\Phi_\lambda = \{\phi : \langle\phi\rangle = \mathrm{pp}^{L_\lambda}(a) \text{ for some } a \in L_\lambda, a \ne 0\}$ (the fact that each of the pp-types $\mathrm{pp}^{L_\lambda}(a)$ is finitely generated is 1.2.6). By 4.3.60, $\Phi_\lambda \cap \Phi_\mu = \emptyset$ if $\lambda \ne \mu$. Set $\Phi = \bigcup_\lambda \Phi_\lambda$.

Suppose that $\phi_0 \ge \phi_1 \ge \cdots$ is a decreasing chain in Φ. By 1.2.17 there is a corresponding chain of morphisms $L_{\lambda_0} \xrightarrow{f_0} L_{\lambda_1} \xrightarrow{f_1} \cdots$ with $f_i(a_i) = a_{i+1}$, where the $a_i \in L_{\lambda_i}$ are such that $\mathrm{pp}^{L_{\lambda_i}}(a_i) = \langle\phi_i\rangle$. By the Harada–Sai Lemma, 4.5.10, some composition $f_n \cdots f_1 f_0$ is zero, hence $a_{n+1} = 0$. Therefore every such chain is finite.

It follows that there is a minimal element, ϕ_0 say, of Φ. Let λ_0 be such that $\phi_0 \in \Phi_{\lambda_0}$. Next choose a minimal element, ϕ_1 say, of $\Phi \setminus \Phi_{\lambda_0}$ (the above argument applies to any subset of Φ). Supposing that Λ is infinite, continue in this way, choosing ϕ_k to be a minimal element of $\Phi \setminus (\Phi_{\lambda_0} \cup \cdots \cup \Phi_{\lambda_{k-1}})$.

Consider the increasing chain $\phi_0 \leq \phi_0 + \phi_1 \leq \cdots$. By 4.5.3 this chain must terminate, so at some stage $\phi_{k+1} \leq \phi_0 + \cdots + \phi_k$. By 4.5.9 there is $\phi \in \Phi_{\lambda_{k+1}}$ and $i \in \{0, \ldots, k\}$ with $\phi \leq \phi_i$ so, since $\Phi_{\lambda_{k+1}} \cap \Phi_{\lambda_i} = \emptyset$, with $\phi < \phi_i$. But this contradicts the choice of ϕ_i as being minimal.

Therefore Λ must be finite, as required. □

Theorem 4.5.12. ([648, Thm], also [649, 2.4]) *Suppose that R is right pure-semisimple and that E is the minimal injective cogenerator for* Mod-R. *Let $S = \mathrm{End}(E)$. Then S is right pure-semisimple and left artinian and ${}_SM \mapsto \mathrm{Hom}_S({}_SM, {}_SE_R)$ is a duality between the categories of finitely presented left S-modules and the category of finitely presented right R-modules.*

This shows ([648, p. 256]) that for the pure-semisimple conjecture it is enough to consider rings which are artinian on both sides.

Therefore, if R is right pure-semisimple and if S is as above, then there are, by 4.5.11, only finitely many finitely presented left S-modules of each finite length. The duality preserves the length of a module so the desired conclusion about *right* R-modules is obtained.

Corollary 4.5.13. *If R is right pure-semisimple, then there are, up to isomorphism, only finitely many indecomposable right R-modules of length n.*

This was obtained in [494, 3.8]. A proof using somewhat different methods appears in [730, Cor. 10] and a proof using still other methods in [712, 3.2]; [718] gives a partial result.

By 4.3.43 a right pure-semisimple ring is right and, by 4.3.68, left, Krull–Schmidt so every finitely presented left R-module also is a direct sum of modules with local endomorphism ring. In fact, the following is true.

Theorem 4.5.14. ([277, 2.3]) *Every finitely presented left module over a right pure-semisimple ring is of finite endolength.*

Proof. As remarked above, every finitely presented left module is a direct sum of indecomposable modules with local endomorphism rings so it is enough to consider the case where ${}_RL$ is indecomposable and finitely presented.

By 4.5.3, ${}_RL$ has the ascending chain condition on pp-definable subgroups, so, setting $S = \mathrm{End}({}_RL)$, by 1.2.12 L_S is noetherian, hence each of its non-zero submodules has at least one maximal proper submodule. Therefore the non-zero terms of the series $L_S \geq \mathrm{rad}(L_S) \geq \mathrm{rad}^2(L_S) \geq \cdots \geq \mathrm{rad}^n(L_S) \geq \cdots$ form a properly descending chain, where, for $K_S \leq L_S$ we set $\mathrm{rad}(K) = \bigcap\{K' : K'$ is a maximal proper S-submodule of $K\}$ and $\mathrm{rad}^{i+1}(K) = \mathrm{rad}(\mathrm{rad}^i(K))$. Since L_S is noetherian, each factor $\mathrm{rad}^i(L)/\mathrm{rad}^{i+1}(L)$ is a finitely generated module over $S/\mathrm{rad}(S)$, so, if $\mathrm{rad}^n(L) = 0$ for some n, then L_S will have finite length, hence ${}_RL$ will have finite endolength.

By 1.2.12, $\mathrm{rad}^n(L_S) = \phi_n({}_RL)$ for some pp condition ϕ_n, so, by 1.3.15, there is an increasing sequence $D\phi_1(L^*) \leq D\phi_2(L^*) \leq \cdots \leq D\phi_n(L^*) \leq \cdots$ of pp-definable subgroups of $L^*_R = \mathrm{Hom}_S\big({}_RL_S, E(S/\mathrm{rad}(S))_S\big)$, strictly increasing where the original sequence is strictly decreasing. Since each $\phi_n(L)$ is a submodule of ${}_RL$, each $D\phi_n(L^*)$ is a submodule of L^*_R (1.3.14). But L^*_R is finitely generated, by 1.3.19 and 4.5.4 or just directly since an epimorphism $R^n \to L$ dualises to a monomorphism $L^* \to R^n$ and R is right artinian, 4.5.4. Hence L^*_R is of finite length. Therefore these sequences eventually stabilise, and $\mathrm{rad}^n(L_S) = 0$ for some n, as required. □

In particular, by 4.4.24, every module in R-mod is pure-injective.

In the remainder of this section, and in the next, we call freely on concepts and results from Chapter 5.

Suppose that R is right pure-semisimple and let L be an indecomposable finitely presented left R-module. By 4.5.3, L has the acc on pp-definable subgroups, hence (7.2.5) has m-dimension. Of course, this also follows from 4.5.14. Therefore (5.4.14) L is reflexive, so the elementary dual, DL (Section 5.4), of L is, by 5.4.15 and 1.3.19, the indecomposable finite length right R-module $L^*_R = \mathrm{Hom}_S({}_RL_S, E_S)$, where E_S is the minimal injective cogenerator for Mod-S seen in the proof just above.

Theorem 4.5.15. ([277, 2.5] (also [276, 3.5])) *Suppose that R is right pure-semisimple. Then elementary duality induces a bijection, $L(\simeq D^2L) \leftrightarrow DL$, between the set, R-ind, of isomorphism classes of finitely presented indecomposable left R-modules and the isolated points of the right Ziegler spectrum Zg_R.*

The fact that the duals of the indecomposable finitely presented left modules are the isolated points of Zg_R follows by 5.3.47, and the fact that, since R_R is right pure-semisimple, neg-isolation is equivalent to isolation in Zg_R. Indeed it is clear, 5.3.17, that isolated points are isolated by minimal pairs so, by 5.4.12, the dual of an isolated point is isolated.

Corollary 4.5.16. *Suppose that R is right pure-semisimple. If L is a finitely presented indecomposable left R-module, then L is an isolated point of ${}_R\mathrm{Zg}$.*

Thus we have a bijection between the isolated points of ${}_R\mathrm{Zg}$ and of Zg_R, the former being exactly the indecomposable finitely presented left R-modules. By 5.4.17 this bijection preserves endolengths.

Corollary 4.5.17. ([276, 3.6]) *If R is right pure-semisimple, then every isolated point of Zg_R is of finite endolength.*

Later results will give the following further information. Since pp_R has mdim $< \infty$ also $\mathrm{mdim}({}_R\mathrm{pp}) = \mathrm{mdim}(\mathrm{pp}_R) < \infty$ (7.2.4) and, since the isolation condition holds on each side (5.3.17), it follows (5.3.60) that mdim $=$ CB ($=$ KGdim) have the same values on the right and left; indeed 5.4.20, ${}_R\mathrm{Zg} \simeq \mathrm{Zg}_R$ and, by the same result, the Cantor–Bendixson analysis on ${}_R\mathrm{Zg}$ and Zg_R go in step, in particular there is a homeomorphism between corresponding closed sets at each stage. It also follows, by 14.1.8, that ${}_R\mathrm{Zar} \simeq \mathrm{Zar}_R$ and so $({}_R\mathrm{LDef}, {}_R\mathrm{Zar}) \simeq (\mathrm{LDef}_R, \mathrm{Zar}_R)$.

4.5.3 Rings of finite representation type

A ring is of finite representation type iff the lattice of pp conditions has finite length (4.5.21) equivalently iff every module has finite endolength (4.5.22). It follows that the property is two-sided (4.5.24) and that finite representation type is equivalent to the conjunction of right and left pure-semisimplicity (4.5.25).

The ring R is of right **finite representation type** if every right R-module is a direct sum of indecomposable modules and there are, up to isomorphism, only finitely many indecomposable modules. By 4.5.7 such a ring is right pure-semisimple. The condition is, in fact, two-sided (4.5.24).

Examples 4.5.18. Any field, more generally any semisimple artinian ring, is of finite representation type since every module is a direct sum of simple modules and there are only finitely many of these up to isomorphism. The rings $\mathbb{Z}_n$ and $k[X]/\langle X^n \rangle$ are of finite representation type, as is any factor of a commutative Dedekind domain by a non-zero ideal. If k is a field of characteristic p and G is a finite group, then the group ring kG is of finite representation type iff G has cyclic p-Sylow subgroups (for example, [27, §VI.3]). A commutative ring is of finite representation type iff it is an artinian principal ideal ring [704, 1.2]. The path algebra, over a field, of a connected quiver without relations is of finite representation type iff the quiver is of Dynkin type [203] (for example, see [27, §VIII.5]). For example, the path algebra of the quiver A_n is the ring of upper-triangular $n \times n$ matrices with entries in k; the indecomposable representations and maps between these are easily found and displayed in the corresponding Auslander–Reiten quiver (see, for example, just below and 15.1.5). If R_1 and R_2 are rings of finite representation type, then so is their product $R_1 \times R_2$, as is any factor ring of R_1, as is the codomain of any ring epimorphism with domain R_1, as is any ring Morita equivalent to R_1.

Example 4.5.19. For an example in more detail, take R to be the path algebra of the quiver $A_3 = 1 \xrightarrow{\alpha} 2 \xrightarrow{\beta} 3$. One may check (also see [27, VIII.6.3])) that there are just six indecomposable right R-modules. They are, with their dimension vectors: the projective modules $P_3 = (001) = S_3$, $P_2 = (011)$ and $P_1 = (111) = I_3$; the last is also injective and the other indecomposable injectives are $I_2 = (110)$ and $I_1 = (100) = S_1$; the remaining simple module is $S_2 = (010)$. The labelling is such that S_i is the simple module associated to vertex i, P_i is its projective cover and I_i its injective hull.

Every R-module is a direct sum of copies of these. This example is continued at 5.1.2, 5.1.9 and 15.1.5.

Proposition 4.5.20.[25] ([20, 2.4]) *A ring R is of finite representation type iff it is right pure-semisimple and whenever N_R is an indecomposable direct summand of a product, $\prod_\lambda N_\lambda$, of indecomposable R-modules N_λ, then $N \simeq N_\lambda$ for some λ.*

Proof. If R is right pure-semisimple and has the stated property, then, by 5.3.48 and 5.3.45, every point of Zg_R is isolated. Since, 5.1.23, this space is compact, it must, therefore, be finite, so R is of finite representation type.

Suppose, for the converse, that R is of finite representation type. If N is a direct summand of $\prod_\lambda N_\lambda$, then (3.4.7) N is in the definable subcategory generated by the N_λ, so (5.1.1) N is in the Ziegler-closure of $\{N_\lambda\}_\lambda$. By 5.3.26 the topology on the Ziegler spectrum of R is trivial, so, in fact, $N \in \{N_\lambda\}_\lambda$, as required. □

Corollary 4.5.21. ([491, 3.9]) *A ring R is of (right) finite representation type iff the lattice, pp_R, of pp conditions (in one free variable) for right R-modules has finite length.*

Proof. Suppose that R is of finite representation type with $N_1, \dots, N_k$ being the indecomposable right R-modules up to isomorphism. Again we use the fact, proved later (5.3.26), that if R is of finite representation type, then Zg_R is a discrete space. Since, therefore, each N_i is isolated, it follows, by 5.3.14, that each of these must have finite endolength and so also, therefore, does $M = N_1 \oplus \cdots \oplus N_k$. Since the definable subcategory generated by M is the whole of Mod-R, it follows (by 3.4.11, alternatively 4.4.27) that $\mathrm{pp}_R \simeq \mathrm{pp}(M)$ has finite length.

[25] The right artinian rings for which this condition on summands of products holds are characterised in [8, 4.2]. An example of a right artinian ring for which the condition fails is given in [1], also see the discussion in [473, §1].

If, conversely, pp_R has finite length, then, by 4.5.1, R is right pure-semisimple and, by 5.3.8, there are just finitely many indecomposable modules, so R is of finite representation type.

That asking for the condition on pp_R^1 implies it on pp_R^n can be argued directly but also follows from 4.4.6 and 1.3.1. □

Since, 3.2.1, there is an R-module M with $\mathrm{pp}_R \simeq \mathrm{pp}(M)$ the next result follows.

Corollary 4.5.22. *A ring R is of (right) finite representation type iff every R-module has finite endolength.*

This has been improved by Dung and García, who prove the following, among other results.

Theorem 4.5.23. ([163, 3.4]) *If R is right pure-semisimple and every indecomposable right R-module has finite endolength, then R is of finite representation type.*

It is shown later than if R has finite representation type, then its Ziegler spectrum is a finite discrete space, 5.3.26. Also, if R is not of finite representation type, then there is an infinite-dimensional indecomposable pure-injective module, 5.3.40.

Theorem 4.5.24. ([153, 1.1] also [19, 3.6]) *A ring is of right finite representation type iff it is of left finite representation type.*

Proof. ([495, 8.24]) By 1.3.1 the condition of finite length on the lattice of pp conditions is right/left symmetric so this is immediate by 4.5.21. □

The next result was found independently by a number of people, see [200, p. 433] for some references. This proof using elementary duality comes from [495, 8.25].

Corollary 4.5.25. *The ring R is of finite representation type iff R is both right and left pure-semisimple.*

Proof. The direction ($\Rightarrow$) was noted at the beginning of the section. For the other direction, if R is left pure-semisimple, then ${}_R\mathrm{pp}$ has the dcc, so, by 1.3.1, pp_R has the acc. Since the latter also has the dcc if R is right pure-semisimple, this modular lattice must be of finite length and so R is of finite representation type by 4.5.21. □

What is unknown is whether right pure-semisimplicity implies finite representation type!

4.5.4 The pure-semisimplicity conjecture

If R is right pure-semisimple and has a self-duality, then it is of finite representation type (4.5.29). If R is right pure-semisimple and is a PI ring, then it is of finite representation type (4.5.30). It is not yet known whether every right pure-semisimple ring is of finite representation type.

Conjecture 4.5.26. (Pure-Semisimplicity Conjecture) If R is right pure-semisimple, then R is of finite representation type.

For artin algebras this was proved by Auslander [20]; a short proof is given later (5.3.41). Simson showed ([649, 3.3]) that for hereditary rings it would be enough to prove that every hereditary right pure-semisimple ring is left artinian. Herzog showed ([276, 6.9]) that if there is a counterexample to the pure-semisimplicity conjecture then there is a hereditary counterexample. So for the general conjecture it would be enough to prove that every hereditary right pure-semisimple ring is left artinian.

Herzog showed, 4.5.29 below, that if there is a duality between the categories of right and left R-modules and R is right pure-semisimple, then R is of finite representation type. We indicate what goes into the proof.

The notions of preinjective and preprojective module appear in Section 8.1.2 in the context of artin algebras. One may extend the terminology by saying that an indecomposable module $N \in \text{mod-}R$ is **preinjective** if there is a finitely presented module M such that N is not a direct summand of M and such that every embedding $f : N \to K$ with $K \in \text{mod-}R$ is split provided K has no direct summand in common with M; that is, N is injective in the subcategory of mod-R which is obtained by "removing" add(M). Dually say that N is **preprojective** if there is a finitely presented module M such that N is not a direct summand of M and such that every epimorphism $g : K \to N$, where K has no direct summand in common with M, is split.[26]

Suppose that R is right pure-semisimple. Since the lattice, ${}_R\text{pp}$, of pp conditions for left modules has the ascending chain condition (4.5.3) there is a pp condition ϕ_0 in one free variable such that the interval $[\phi_0, x = 0]$ in ${}_R\text{pp}$ has finite length and which is maximal such. By 5.3.6 the open set $(\phi_0, x = 0)$ has just finitely many points. Herzog, [276, 4.2], showed that if R is right pure-semisimple and left artinian and if $N \in R\text{-mod}$ is indecomposable, then N is preinjective iff $\phi_0(N) \neq 0$. Hence, over such a ring there are only finitely many preinjective left R-modules.

[26] It should be remarked that there are other ways of extending those notions beyond the context of modules over hereditary algebras.

Next, note that if there is a duality $R'\text{-mod} \simeq (\text{mod-}R)^{\text{op}}$, then a finitely presented left R'-module is preinjective iff its dual is preprojective in mod-R. If R is right pure-semisimple, then there is such a duality (see [276, §3]) with, [648, Thm], R' also right pure-semisimple and left artinian. The next result then follows

Corollary 4.5.27. ([276, 4.3]) *If R is right pure-semisimple, then there are only finitely many preprojective right R-modules.*

In fact, [276, 4.4], if R is right pure-semisimple and ϕ is a pp condition, in one free variable, for right R-modules such that the interval $[x = x, \phi]$ has finite length, then if $L \in R\text{-mod}$ is indecomposable with $L > \phi(L)$ then L is preprojective. It is then shown, [276, 4.5], that if R is right pure-semisimple and the indecomposable $N \in R\text{-mod}$ is preinjective, then the elementary dual, DN, is preprojective and is an isolated point of Zg_R. Therefore, if $\text{i}_l(R)$ is the number of preinjective left R-modules and if $\text{p}_r(R)$ is the number of preprojective right R-modules, then $\text{i}_l(R) \leq \text{p}_r(R)$.

Theorem 4.5.28. ([276, 5.2]) *Suppose that R is right pure-semisimple and left artinian.*[27] *Then R is of finite representation type iff* $\text{i}_l(R) = \text{p}_r(R)$.

Say that R has a **self-duality** if there is a duality $R\text{-mod} \simeq (\text{mod-}R)^{\text{op}}$. For example, every artin algebra has a self-duality (15.1.3). Now use the duality between preinjectives and preprojectives to deduce the next theorem.

Theorem 4.5.29. ([276, 5.3]) *If R is right pure-semisimple and has a self-duality, then R is of finite representation type.*

In [276, §§5,6] there are various corollaries including the next theorem, and also some consequent constraints on counterexamples to the pure-semisimplicity conjecture. Also see papers [651], [653] of Simson, again playing off right and left modules, and reducing to hard questions about division rings, in pursuit of a counterexample.

Theorem 4.5.30. ([276, 6.7]) *If R is right or left pure-semisimple and satisfies a polynomial identity, then R is of finite representation type.*

Other proofs have been given subsequently by Krause [346] and Schmidmeier [636].

In particular, this covers the case of artin algebras (see 5.3.41) since such a ring is finitely generated as a module over a commutative subring, hence can be realised as a subring of a matrix ring over a commutative ring, hence, see for example, [441, 13.1.13], is PI.

[27] The left artinian assumption can be dropped, see [161, 5.14].

Theorem 4.5.31. ([276, 6.9], the first part [649, 3.3]) *If there is a counterexample to the Pure-semisimplicity Conjecture, then there is a counterexample which is right and left hereditary, not left artinian and whose Jacobson radical is a minimal ideal.*

Indeed, [649, 3.3], if there is a counterexample, then there is one of the form $\begin{pmatrix} D & {}_DM_E \\ 0 & E \end{pmatrix}$, where D and E are division rings and M is a (D, E)-bimodule.

4.5.5 Generic modules and representation type

An indecomposable module is said to be generic if it is of finite endolength and not small. For noetherian algebras it is clear what "not small" should mean (4.5.32). Representation types of algebras are defined and existence of generic modules is related to these. In particular, if an artin algebra has infinitely many modules of endolength n, then there is a generic module of endolength at most n (4.5.40) and over every finite-dimensional algebra of infinite representation type, over an algebraically closed field, there is a generic module (4.5.42). Moreover, for such algebras, domestic representation type is equivalent to there being only finitely many generic modules (4.5.44).

Following Crawley-Boevey [139], say that a module over an artin algebra is **generic** if it is indecomposable, of finite endolength (Section 4.4.3) and is not of finite length. Over more general rings it may be more appropriate to use "not finitely presented" as the way to say that the module is not small. So, for a general ring, we will say that a module is **generic** if it is indecomposable, of finite endolength and is not finitely presented.[28]

For instance, $\mathbb{Q}$ is generic as a $\mathbb{Z}$-module but not as a $\mathbb{Q}$-module.

The appropriateness of the term "generic" is illustrated by the results in this section and by those outlined in Sections 8.1.2, 8.1.3, 8.1.4, where, for various sorts of ring, generic modules are generic points (in an algebraic-geometric sense) in the dual-Ziegler ="Zariski" topology (Section 14.3).

Proposition 4.5.32. ([141, 7.1, Prop.]) *Suppose that R is a* **noetherian algebra***; that is, the centre, $C(R)$, of R is noetherian and R is finitely generated as a module over its centre. Let N be an indecomposable R-module of finite endolength. Then:*

(i) N is of finite length iff
(ii) N is finitely presented iff
(iii) N is the source of a left almost split map in Mod-R *(see Section 5.3.3).*

[28] Ringel [610, p. 35] suggests dropping the exclusion clause but the current terminology is quite convenient.

These conditions on N are equivalent to the same conditions on the elementary dual of N (for that see 5.4.17).

Note that the equivalence of (i) and (ii) is by 4.4.34.

Examples 4.5.33. If R is a prime, non-artinian right Goldie ring, with simple artinian quotient ring Q, then the unique generic right R-module is the simple Q-module (regarded as an R-module) and the endolength of this module equals the uniform dimension of R ([139, 1.4], see 4.4.33).

If R is noetherian or a PI ring, then the generic modules of endolength 1 are in natural bijection with the ideals I of R such that R/I is a domain but not a division ring (see [139, 1.5]).

Perhaps this is the place to point to Cohn's notion, see [128], of spectrum of a ring R, being the set of equivalence classes of epimorphisms from R to simple artinian rings. This is described for certain artin algebras in [600]. See also [129], [638], [634].

Say that R is an **artin algebra** if the centre, $C(R)$, of R is artinian and R is finitely generated as a module over $C(R)$. If $C \subseteq C(R)$ is artinian and R_C is finitely generated, then say that R is an **artin C-algebra**. For such an algebra every finitely generated module is of finite length, both as an R-module and as a $C(R)$-module. Since every pp-definable subgroup of an R-module M is a $C(R)$-submodule of M (1.1.8) the following is immediate.

Remark 4.5.34. If R is an artin algebra, then every finitely generated R-module is of finite endolength, in particular (4.4.24) is pure-injective.

Suppose that R is a finite-dimensional algebra over a field k. Then R is said to be of **wild representation type** (or simply, is **wild**) if there is a representation embedding (see Section 5.5.2) from the category of modules over $k\langle X, Y\rangle$ to the category of modules over R. Often a definition making reference only to the categories of finite-dimensional modules is used; indeed, it is not really clear what the definition of "wild" should be outside the context of finite-dimensional algebras over algebraically closed fields (that case is clear-cut, see 4.5.41 below, and [139], [141]). The terminology is extended to other types of ring R, and to more general categories, see, for example, [650], via "if S is wild and there is a representation embedding from S-modules to R-modules then R is wild".

If R is a finite-dimensional algebra over a field k, then R is said to be of **tame representation type** if, for every $d \geq 1$ there are finitely many $(k[X], R)$-bimodules $B_1, \dots, B_n$ such that all but finitely many indecomposable R-modules of dimension d have the form $(S \otimes_{k[X]} B_i)_R$ for some i and for some simple $k[X]$-module S (see, for example, [156] for equivalent variants on this definition). If

$n = n(d)$ is bounded as d increases, then R is of **domestic** representation type. Again the terminology can be extended (to some degree informally) to other classes of rings. The next definition gives a plausibly good extension of the concept to a wide class of rings.

A noetherian algebra R is **generically tame** if for every d there are only finitely many generic modules of endolength d ([141]); R is **generically wild** if there is a generic module G whose endomorphism ring is not PI equivalently, [141, 7.4, proof of Cor.], the top, $\mathrm{End}(G)/\mathrm{JEnd}(G)$, of the endomorphism ring of G is a division ring which is infinite-dimensional over its centre. By 4.4.32 a ring is generically tame on the right iff it is so on the left. Crawley-Boevey asked whether the same is true of the property of generic wildness.

If R is a finite-dimensional algebra over an algebraically closed field, then R is exactly one of finite representation type, generically tame or generically wild, [141, 7.4, Cor.]. Crawley-Boevey made the conjecture ([139, p. 159]) that any noetherian algebra, if not of finite representation type, is either generically tame or generically wild and not both.

Question 4.5.35. Is it true that every noetherian algebra, if not of finite representation type, is either generically tame or generically wild and not both?

A finite-dimensional k-algebra R is **strictly wild** if there is a strict representation embedding (see Section 5.5.2) from the category of $k\langle x, y\rangle$-modules to the category of R-modules. For example, any wild hereditary finite-dimensional algebra is strictly wild, see [92, §6] and [141, 8.4] (though they use different definitions of strictly wild).

An artin algebra R has **strongly unbounded representation type** if for infinitely many d there exist infinitely many indecomposables of length d; by a result, [657, 3.3], of Smalø, it is enough to have this for one d.

Theorem 4.5.36. ([141, 7.3 Thm]) *If R is an artin algebra, then R has a generic module iff there are infinitely many indecomposable modules of some fixed finite endolength iff, provided all simple R-modules are infinite, R has strongly unbounded representation type.*

Question 4.5.37. ([139]) Does every artin algebra which is not of finite representation type have a generic module?

We can ask for more general rings R whether not being pure-semisimple implies the existence of a generic R-module. Without restriction the answer is, as Crawley-Boevey pointed out, no. For suppose that R does not have invariant basis number, say $R_R^n \simeq R_R^m$ for some integers $n \neq m$; then for every module M we have $M^n \simeq M \otimes_R R^n \simeq M^m$ as $\mathrm{End}(M)$-modules. If M were of finite endolength, then this would be impossible by essential uniqueness of decomposition of such modules. For an example of such a ring take $R = \text{End}_k(V)$, where V is an infinite-dimensional vector space over a field k; then $R_R \simeq R_R \oplus R_R$. Also see Section 15.1.3, [158] and further results in [357] for existence of generic modules over artin algebras.

Theorem 4.5.38. ([141, 9.6 Thm.]) *Suppose that R is an artin C-algebra, where C is a commutative artinian local ring with maximal ideal J. Consider the following conditions:*

(i) there exists d such that there are infinitely many indecomposable modules of length d over C;
(ii) there exists l such that there are infinitely many indecomposable modules of endolength l and finite length over C;
(iii) there is a $((C/J)[X], R)$-bimodule which is indecomposable over R and is of finite length over $(C/J)[X]$;
(iv) there is a generic R-module.

Then (ii), (iii), (iv) are equivalent and, if $k = C/I$ is infinite, these are equivalent to (i).

Corollary 4.5.39. *Suppose that R is an artin algebra. Then there is a generic R-module iff, for some integer n, there are infinitely many non-isomorphic finitely presented modules of endolength n.*

For an example which shows that the equivalence fails for general artinian rings see [139, 1.8].

Herzog saw a wonderfully simple proof ([278, 9.6]) for one direction of this equivalence.

Corollary 4.5.40. *Let R be an artin algebra. If there are infinitely many (indecomposable) modules of endolength n, then there is a generic module of endolength $\leq n$.*

Proof. By 5.1.16 the indecomposable modules of endolength $\leq n$ form a closed subset of the Ziegler spectrum of R. By assumption this is an infinite set. By compactness of the spectrum, 5.1.23, there must be a non-isolated point. By 5.3.37 that point cannot be finitely generated, as required. □

For finite-dimensional algebras over an algebraically closed field one may say more. Note that endolength = length for modules over a finite-dimensional k-algebra where every simple module is one-dimensional over k, in particular if k is algebraically closed.

Theorem 4.5.41. ([139, 4.4]) *Suppose that R is a finite-dimensional algebra over an algebraically closed field k. Then the following conditions are equivalent:*

(i) R is of tame representation type;
(ii) for every d, R has only finitely many generic modules of endolength d, that is, R is generically tame;
(iii) for every generic R-module G, $\mathrm{End}(G)/J\mathrm{End}(G) \simeq k(X)$, the field of rational functions in one indeterminate;
(iv) R is not of wild representation type.

It is also shown, [139, 4.6, 4.7], that if R is a finite-dimensional algebra over an algebraically closed field and if G is a generic R-module, then $\mathrm{End}(G)$ is split over its radical, that is, there is a ring morphism $\mathrm{End}(G)/J\mathrm{End}(G) \to \mathrm{End}(G)$ which splits the projection from $\mathrm{End}(G)$ to $\mathrm{End}(G)/\mathrm{JEnd}(G)$, and any two splittings are conjugate. Also if $\mathrm{End}(G)$ is commutative, then $\mathrm{JEnd}(G) = 0$.

Corollary 4.5.42. ([139, 4.5]) *If R is a finite-dimensional algebra over an algebraically closed field, then R has infinite representation type iff there is a generic R-module.*

More precisely, [139, 4.4, 5.2–5.5], if R is a tame finite-dimensional algebra over an algebraically closed field k and if G is a generic R-module, then there is, for some non-zero polynomial $f \in k[X]$, a $(k[X, f^{-1}], R)$-bimodule M^G which is free, of rank equal to the endolength of G, such that $k(X) \otimes_{k[X,f^{-1}]} M^G_R \simeq G_R$. Moreover, the functor $- \otimes_{k[X,f^{-1}]} M^G_R : \text{Mod-}k[X, f^{-1}] \longrightarrow \text{Mod-}R$ reflects isomorphism and preserves indecomposability and almost split sequences (in particular, is a representation embedding in the sense of Section 5.5.2). Furthermore, for every d, all but finitely many (indecomposable) R-modules of k-dimension d have the form $(k[X, f^{-1}]/\langle g\rangle) \otimes M^G_R$ for some polynomial $g \neq 0$ and some M^G_R.

Set $\mu_R(d)$ to be the smallest l such that l bimodules as above suffice to cover, in the sense of the previous paragraph, all but finitely many R-modules of dimension d. Also set $g_R(d)$ to be the number of generic R-modules of endolength d.

Theorem 4.5.43. ([139, 5.6]) *If R is a tame algebra over an algebraically closed field then $\mu_R(n) = \sum_{d|n} g_R(d)$.*

Corollary 4.5.44. ([139, 5.7]) *Suppose that R is a finite-dimensional algebra over an algebraically closed field. Then:*

(i) $\mu_R(n)$ is bounded as $n \to \infty$ iff
(ii) R has finitely many generic modules iff
(iii) R is of domestic representation type.

See Krause [363] for more on generic tameness, in particular a characterisation of tameness which, making use of the idempotent ideals of Section 9.2, is purely in terms of the category mod-R. Geiss and Krause, [228, Thm A, 8.3], showed that generic tameness is preserved by derived equivalence between finite-dimensional algebras over an algebraically closed field, as is generic domesticity. Also see [358, Chpt. 10], [353], [354], [51], [611], [738].

Krause showed, [350, §3], that if R, S are artin algebras, then an equivalence F between the stable module categories $\underline{\text{mod}}$-R and $\underline{\text{mod}}$-S (see Section 17.1) induces a bijection between the generic R- and S-modules. Hence, to the extent that representation type can be characterised in terms of generic modules, it is preserved by stable equivalence. In the same paper he extended the Auslander–Reiten translate (see Section 15.1.2) to arbitrary modules over an artin algebra, shows ([350, 5.12]) that this preserves pure-injectivity, Σ-pure-injectivity and being of finite endolength and, [350, 5.15], that it induces a homeomorphism from the non-projective points of the Ziegler spectrum (see Chapter 5) to the non-injective points of that space. If R is a tame finite-dimensional algebra over an algebraically closed field, then this operation fixes every generic point ([350, 5.16]). Krause and Zwara, [375, Thm] showed that, given an equivalence of stable module categories over artin algebras, endolength is preserved up to multiplication by a constant which depends only on F and they deduced that the rate of growth of $\mu_{(-)}(n)$ (μ is defined above) is preserved. Lenzing, [399], gave a streamlined proof of some of these results.

4.6 Cotorsion, flat and pure-injective modules

Relations between flat, cotorsion and pure-injective modules are outlined. For instance, the pure-injective hull of every flat module is flat iff the flat cover of every pure-injective module is pure-injective (4.6.2). The class of pure-injective modules is closed under extension iff every cotorsion module is pure-injective (4.6.3).

Recall, 2.3.14, that a module F is flat iff every exact sequence of the form $0 \to A \to B \xrightarrow{g} F \to 0$ is pure-exact. A module A is **cotorsion** if $\text{Ext}(F, A) = 0$ for every flat module F. So every pure-injective module is cotorsion. The modules F such that $\text{Ext}(F, A) = 0$ for every cotorsion module A are exactly the flat modules, in fact, testing with pure-injective modules is enough.

Proposition 4.6.1. ([716, 3.4.1]) *If a module F is such that* $\text{Ext}(F, N) = 0$ *for every pure-injective module N, then F must be flat.*

Proof. Let L be any left R-module. The right R-module $\mathrm{Hom}_{\mathbb{Z}}({}_RL_{\mathbb{Z}}, \mathbb{Q}/\mathbb{Z})$ is, by 4.3.29, pure-injective so we have

$$0 = \mathrm{Ext}^1_R\big(F, \mathrm{Hom}_{\mathbb{Z}}({}_RL_{\mathbb{Z}}, \mathbb{Q}/\mathbb{Z})\big) \simeq \mathrm{Hom}_{\mathbb{Z}}\big(\mathrm{Tor}_1(L, F), \mathbb{Q}/\mathbb{Z}\big)$$

(a derived version of the Hom/$\otimes$ adjointness, see, e.g., [69, 2.8.5]). Thus $\mathrm{Tor}_1(L, F) = 0$ for every left R-module L and this implies (for example, [708, Ex. 3.2.1]) that F is flat. □

The projective/injective and flat/absolutely pure dualites are quite extensive but they are not perfect. For instance, although every module has an injective hull it is only over (semi)perfect rings that every (finitely generated) module has a projective cover (see 4.3.65).

Every module does, however, have a flat cover: that this should be so was conjectured by Enochs but was proved only much later, see [81, Thm 3]. There is a general theory of hulls and covers, see, for example, [716, Chpt. 1], and the basic definitions are as follows.

Let $\mathcal{F}$ be any class of objects in a category $\mathcal{C}$ and let C be a object of $\mathcal{C}$. Then an $\mathcal{F}$-**precover** of C is a morphism $f : F \to C$ with $F \in \mathcal{F}$ such that every morphism $f' : F' \to C$ with $F' \in \mathcal{F}$ lifts through f (that is, there is $g : F' \to F$ with $fg = f'$) and this is an $\mathcal{F}$-**cover** if for every $h \in \mathrm{End}(F)$ if $fh = f$ then $h \in \mathrm{Aut}(F)$. Dualising gives the notions of an $\mathcal{F}$-**preenvelope** and an $\mathcal{F}$-**envelope** (or $\mathcal{F}$-**hull**). Notice that every object of $\mathcal{C}$ having an $\mathcal{F}$-precover is equivalent to $\mathcal{F}$ being contravariantly finite in $\mathcal{C}$ (Section 3.4.4). Taking $\mathcal{F}$ to be the class of projective modules gives the notion of projective cover and taking $\mathcal{F}$ to be the class of flat modules gives the notion of flat cover.

If $f : F \to M$ is a flat cover, then $\ker(f)$ is cotorsion, [178, 2.2]. Also the class of cotorsion modules is closed under extension, [716, 3.1.2], so the flat cover of any cotorsion module is cotorsion. In particular, the flat cover of a pure-injective module is cotorsion, though not necessarily pure-injective. Indeed Rothmaler proved the following.

Theorem 4.6.2. ([624, 2.3]) *For any ring R the following are equivalent:*

(i) the pure-injective hull of every flat right R-module is flat;
(ii) the flat cover of every pure-injective right R-module is pure-injective;
(iii) the flat cover of every cotorsion right R-module is pure-injective;
(iv) every flat cotorsion right R-module is pure-injective;
(v) for every flat right R-module F with pure-injective hull, $H(F)$, the quotient $H(F)/F$ is flat.

Compare the last condition with the following result of Xu.

Theorem 4.6.3. ([716, §3.5]) *For any ring R the following are equivalent:*

(i) the class of pure-injective right R-modules is closed under extension;
(ii) for every right R-module M, with pure-injective hull $H(M)$, the quotient $H(M)/M$ is flat;
(iii) every cotorsion right R-module is pure-injective.

Existence of cotorsion envelopes is equivalent to existence of flat covers, [716, 3.4.6], so it is now known that every module does have a cotorsion envelope (as does every sheaf of modules, [179], and complex of modules, [5]). The embedding of any module into its cotorsion envelope is pure since, [716, 3.4.2], the cokernel of this embedding is flat. The conditions of 4.6.2 are also equivalent to the cotorsion envelope of every flat module being pure-injective, and hence to the coincidence of cotorsion envelopes and pure-injective hulls for all flat modules ([624, 2.9]).

For countable rings, the conditions of 4.6.2 are equivalent to the pure-injective hull of R being flat, see [527, 5.3]. In that paper the failure of a flat module to have flat pure-injective hull is linked (characterised in the case of countable rings) with failure of a certain chain condition on pp-definable subgroups.

It is also shown in [624, 3.2] that R_R is Σ-pure-injective iff R is right perfect and satisfies the conditions of 4.6.2. (Compare with the result of Zimmermann [731, 6.4] that if R_R is Σ-pure-injective, then R is semiprimary with the acc on right annihilators.)

Note the following result somewhat analogous to 4.6.2.

Theorem 4.6.4. ([118]) *Let R be commutative noetherian. Then the following are equivalent:*

(i) the injective hull of every flat module is flat;
(ii) the flat cover of every injective module is injective;
(iii) the injective hull of R is flat;
(iv) the localisation $R_{(P)}$ is quasi-Frobenius for every prime $P \in \mathrm{Ass}(R)$ (for the latter see, for example, [439, p. 39]).

Guil Asensio and Herzog have been pursuing a programme of developing the theory of purity in Flat-R, the full subcategory of flat modules. Since this category is closed under direct limits (2.3.7) and is generated by the finitely generated $=$ finitely presented projective modules (2.3.6), Flat-R is a finitely accessible category (Section 16.1.1, 16.1.3 in particular). It is not in general locally finitely presented, indeed, by 16.1.5 and 2.3.21, it is locally finitely presented iff R is left coherent. So, see 16.1.6, it will not in general be an abelian category.

Nevertheless, as a finitely accessible category, Flat-R has a theory of purity (see 16.1.15) but when this category does not have products this theory of purity will not simply be the restriction of that of Mod-R. Indeed, by [283, Lemma 3] the pure-injective objects of Flat-R are exactly the cotorsion modules. The fact ([81]) that every module has a flat cover, and hence that every module has a cotorsion envelope, implies that every flat module has a pure-injective hull in the sense of the category Flat-R. In this development of the theory around purity and pure-injectivity in categories Flat-R cotorsion modules therefore replace the pure-injective modules.

In [257] Guil Asensio and Herzog extended, a result [729, Thm 9] (see 12.1.10) of Zimmermann–Huisgen and Zimmermann for rings which are pure-injective as modules over themselves, showing that if R is cotorsion as a module over itself then $R/J(R)$ is von Neumann regular, self-injective and idempotents lift modulo $J(R)$ ([257, Cor. 4, Thm 6, Cor. 9]). These rings are more general than those which are pure-injective over themselves, indeed, any right perfect ring is even Σ-cotorsion, and conversely ([258, Thm, Cor. 10]), where a module M is **Σ-cotorsion** if every direct sum of copies of M is cotorsion. This property is developed as an analogue of the Σ-pure-injectivity condition (Section 4.4.2). For instance, every flat Σ-cotorsion module is a direct sum of modules with local endomorphism ring ([258, Prop. 7]), cf. 4.4.19, and in [260] characterisations, though not as clear-cut as for Σ-pure-injectivity, in terms of chain conditions of certain pp-definable subgroups are obtained, cf. 4.4.5. In [259, Cor. 10] it is proved that every flat cotorsion module has a pure embedding into a product of indecomposable flat cotorsion modules, cf. 5.3.53.

Also see Section 18.2.3.

5

The Ziegler spectrum

The Ziegler spectrum is defined in a variety of equivalent ways. Isolation of points in closed subsets proves to be a central issue. There is a duality between the right and left spectra over any ring. Some morphisms between rings induce continuous maps between spectra. Even more use is made of functorial techniques.

5.1 The Ziegler spectrum

Section 5.1.1 introduces the Ziegler spectrum and has proofs of the basic results. Closed points are considered. Examples and some special cases are given. It is proved in Section 5.1.2 that the topology has a basis of compact[1] *open sets and that the Ziegler spectrum over any ring is compact. There is a description of the topology in terms of morphisms between finitely presented modules in Section 5.1.3.*

What is now called the Ziegler spectrum is a topological space which was introduced by Ziegler in his landmark paper [726]. This space played a central role in that paper, which was concerned with the model theory of modules. For example the "support" of an elementary (= axiomatisable) class of modules is a closed subset of the spectrum and the correspondence between classes of modules and closed sets is bijective (5.1.6) when restricted to those axiomatisable classes which are closed under direct sums and direct summands (these have been called definable classes in Section 3.4.1). Most broad questions about the model theory of modules can be phrased as questions about the points and topology of this space and can be tackled, and often answered, in those terms. For instance, the use of Cantor–Bendixson rank (Section 5.3.6) clarified many existing results in the model theory of modules and opened up new avenues of research.

[1] "Quasicompact" in the terminology of some.

Since then it has been realised that there are many motives which have nothing to do with model theory for studying this space. Two important themes are links between the complexity of the module category and the complexity of this space and the ways in which single ("large") points of this space parameterise families of ("small") modules.[2]

5.1.1 The Ziegler spectrum via definable subcategories

The Ziegler spectrum is defined via definable subcategories (5.1.1). Every proper pp-pair defines a non-empty open set (5.1.3) and the topology may be defined by pp-pairs (5.1.8). Each definable subcategory is generated by the indecomposable pure-injectives in it (5.1.4). The Ziegler spectrum is a Morita invariant (5.1.17).

Over a right coherent ring closed sets of indecomposable injectives correspond to finite-type torsionfree classes (5.1.11).

Indecomposable modules of finite endolength are closed points and the converse is true if R is countable (5.1.12). The indecomposable modules of endolength $\leq n$ form a closed subset of the Ziegler spectrum (5.1.16).

Let R be any ring. The (right) Ziegler spectrum, Zg_R, of R is a topological space, the underlying set of which is the set of isomorphism types of indecomposable pure-injective right R-modules: we use pinj_R to denote this set. These isomorphism types really do form a set, as opposed to a proper class (4.3.38). Normally we do not count the zero module among the indecomposable modules, so 0 is not (officially) in this set but from time to time we allow it to slip in for ease of statement.

There are quite a few equivalent ways of defining the topology. Here is the first.

If $\mathcal{X}$ is any definable subcategory of Mod-R (Section 3.4.1),[3] then, abusing notation slightly, we let $\mathcal{X} \cap \mathrm{pinj}_R$ denote the set of isomorphism classes of indecomposable pure-injectives in $\mathcal{X}$. It has been seen already, 4.3.21, that $\mathcal{X}$ is closed under pure-injective hulls and it will be shown below, 5.1.5, that if $\mathcal{X} \neq \{0\}$, then there is an indecomposable pure-injective in $\mathcal{X}$, indeed distinct definable subcategories are distinguished by their intersections with pinj_R.

Define a topology on pinj_R by declaring the sets of the form $\mathcal{X} \cap \mathrm{pinj}_R$ with $\mathcal{X}$ definable to be closed sets. The resulting space we denote by Zg_R and call the **Ziegler spectrum** for right R-modules (or the right Ziegler spectrum of R). The Ziegler spectrum for left R-modules is denoted ${}_R\mathrm{Zg}$ or $\mathrm{Zg}_{R^{\mathrm{op}}}$. We will often use the notation Zg_R instead of pinj_R for the underlying set of this topology (but

[2] This, as a reason for studying infinite-dimensional modules over finite-dimensional algebras, is proposed and developed in [600, §3].

[3] Recall that these are defined by declaring certain pp-pairs to be closed.

the latter notation remains useful when we discuss other topologies on this set, for which see Sections 5.3.7, 5.6, 14.1.3).

Theorem 5.1.1. *The closed sets of the Ziegler topology are exactly those of the form $\mathcal{X} \cap \mathrm{Zg}_R$ with $\mathcal{X} \subseteq$ Mod-R a definable subcategory.*

Proof. Closure of sets of this form under finite union and arbitrary intersection must be checked.

Let $\mathcal{X}$ be the definable subcategory of Mod-R generated by the union of the definable subcategories $\mathcal{X}_1, \ldots, \mathcal{X}_n$. Clearly $X_1 \cup \cdots \cup X_n \subseteq X$, where $X_i = \mathcal{X}_i \cap \mathrm{Zg}_R$ and $X = \mathcal{X} \cap \mathrm{Zg}_R$. If $N \in X$, then, by 3.4.9, N is pure in $M_1 \oplus \cdots \oplus M_n$ for some $M_i \in \mathcal{X}_i$. But then, by 4.4.1, N is a direct summand of one of the M_i and hence is in $\mathcal{X}_i$. We conclude that $X = X_1 \cup \cdots \cup X_n$.

To show closure under arbitrary intersection take, for $\lambda \in \Lambda$, $\mathcal{X}_\lambda$ to be the definable subcateory defined by a set, T_λ, of pp-pairs:[4] $\mathcal{X}_\lambda = \mathrm{Mod}(T_\lambda)$. Let Y be the subset of Zg_R defined by $T = \bigcup_\lambda T_\lambda$. If N belongs to Y, then N belongs to each X_λ (since all pairs in T_λ are closed on N). Conversely, if $N \in \bigcap_\lambda X_\lambda$, then all pairs in T are closed on N, so $N \in Y$. Thus $\bigcap_\lambda X_\lambda = \mathrm{Mod}(T) \cap \mathrm{Zg}_R$, as required. □

Example 5.1.2. Let R be the path algebra of the quiver $A_3 = 1 \xrightarrow{\alpha} 2 \xrightarrow{\beta} 3$. This ring of finite representation type has six indecomposable modules (see 4.5.19), all pure-injective by 4.2.6, so Zg_R has exactly six points. Each indecomposable module N is of finite length over k, hence of finite endolength. Therefore, by 4.4.30, the definable subcategory that N generates consists exactly of the direct sums, $N^{(\kappa)}$, $\kappa \geq 0$. So $\{N\}$ is closed. Therefore Zg_R has the discrete topology. The argument applies to any ring of finite representation type (5.3.26).

This example is used below (5.1.9) to illustrate another description of the topology.

Theorem 5.1.3. ([726, 4.8]) *Suppose that ϕ/ψ is a proper (that is, $\phi > \psi$) pp-pair. Then there is an indecomposable pure-injective module N with $\phi(N) > \psi(N)$. If M is any module with $\phi(M) > \psi(M)$, then there is an indecomposable pure-injective with this property in the definable subcategory of* Mod-R *generated by M.*

Proof. We give two proofs. The first is more or less the original one. The second is the arguably more conceptual functorial proof (which was described to me in the early 1990s by Herzog). In fact they begin in the same way but use techniques

[4] Notation as in Section 3.4.1.

from different areas to finish the argument. Comparing these illustrates to some extent the model-theory/functor category dictionary (Appendix C).

First proof: By Zorn's lemma there is a pp-type[5] p which contains ϕ, does not contain ψ and is maximal such. By 4.1.4 there is a module, $M^\star$, in the definable subcategory generated by M and there is $\overline{a}$ from M^* such that $\mathrm{pp}^{M^\star}(\overline{a}) = p$. We claim that p is irreducible and hence, by Ziegler's criterion, 4.3.49, that the hull, $H(p)$, of p is indecomposable.

To see this, let ψ_1 be a pp condition not in p. Then the pp-type generated by p and ψ_1, that is, $\{\phi' : \phi' \geq \phi_1 \wedge \psi_1 \text{ for some } \phi_1 \in p\}$, must, by maximality of p, contain ψ, so there is $\phi_1 \in p$ with $\phi_1 \wedge \psi_1 \leq \psi$. Therefore if ψ_1, ψ_2 are pp conditions not in p, then there are $\phi_1, \phi_2 \in p$ with $\phi_i \wedge \psi_i \leq \psi$, hence with $\phi_1 \wedge \psi_1 + \phi_2 \wedge \psi_2 \leq \psi$, in particular with $\phi_1 \wedge \psi_1 + \phi_2 \wedge \psi_2 \notin p$. So p is indeed irreducible and $H(p)$ is indecomposable.

Replacing $M^\star$ by its pure-injective hull (4.3.21) we have, by 4.3.36, that $H^{M^\star}(\overline{a}) = H(p)$ is a direct summand of $M^\star$. Because $\phi \in p$ and $\psi \notin p$ it is the case that $a \in \phi\big(H(p)\big)\backslash\psi\big(H(p)\big)$ so $H(p) \in (\phi/\psi)$ (and $H(p)$ is indecomposable and in the definable subcategory generated by M), as required.

Second proof: This uses the functorial methods of Part II, see the summary paragraphs just before 3.4.7. Suppose that ϕ and ψ have n free variables. So $F_{D\phi} < F_{D\psi} \leq (R^n \otimes -)$. Since $F_{D\psi}$ is a finitely generated functor (10.2.29) there is a maximal proper subfunctor, F', of $F_{D\psi}$ which contains $F_{D\phi}$. Then $S = F_{D\psi}/F'$ is a simple functor, hence has indecomposable injective hull, E, say. Since E is injective, the inclusion of S into E extends to a morphism, α say, from $(R^n \otimes -)/F'$ to E. Now, 12.1.9, $E \simeq N \otimes -$ for some indecomposable pure-injective R-module N, so the composition of the projection $(R^n \otimes -) \to (R^n \otimes -)/F'$ followed by α has the form $\overline{a} \otimes -$ for some $\overline{a} : R^n \to N$, that is, for some n-tuple $\overline{a}$ from N. Since $F_{D\phi} \leq \ker(\overline{a} \otimes -)$, 10.3.9 implies $\overline{a} \in \phi(N)$ and, since $F_{D_\psi} \nleq \ker(\overline{a} \otimes -)$, the same result gives $\overline{a} \notin \psi(N)$. Thus we have $N \in (\phi/\psi)$, as required.

To get the second statement in this proof, observe that the hypothesis $\phi(M) > \psi(M)$ implies, 10.3.8, that there is a non-zero morphism from $F_{D\psi}/F_{D\phi}$ to $M \otimes -$. Choose F' at the outset of the proof to contain the kernel of this morphism and to be maximal proper τ-closed in $F_{D\psi}$, where τ is the hereditary finite type torsion theory cogenerated by $E(M \otimes -)$ (see Section 12.3 and 11.1.12(iii) or 11.1.14 which shows that there is a maximal proper τ-closed subfunctor). Then the localisation at τ of the functor $F_{D\psi}/F'$ is simple and τ-torsionfree. Hence its injective hull E, which has the form $N \otimes -$ for some pure-injective N, is torsionfree, so, by 12.3.2, N is in the definable subcategory of Mod-R, generated by M.

[5] That is, a set of pp conditions such that $\phi', \phi'' \in p \Rightarrow \phi' \wedge \phi'' \in p$ and $\phi' \in p, \phi' \leq \phi'' \Rightarrow \phi'' \in p$, see Section 1.2.1.

If, in this second proof, we set $p = \mathrm{pp}^N(\overline{a})$, then this is as p appearing in the first proof modulo the fact that in each case there is a choice to be made: of pp-type in the first proof; of morphism from $R^n \otimes -$ to E in the second. Also $F' = F_{Dp} \cap F_{D\psi}$. These really are two renditions of the same proof. □

Corollary 5.1.4. *If $\mathcal{X}$ is a definable subcategory of* Mod-R, *then $\mathcal{X}$ is generated as such by the indecomposable pure-injectives in it.*

Proof. Let T be the set of pp-pairs closed on $\mathcal{X}$. For each ϕ/ψ not in T choose, by 5.1.3, $N \in \mathcal{X} \cap \mathrm{pinj}_R$ with $\phi(N) > \psi(N)$. Let $\mathcal{X}'$ be the definable subcategory of Mod-R generated by all these indecomposable pure-injectives. So $\mathcal{X}' \subseteq \mathcal{X}$. But also every pp-pair open on some member of $\mathcal{X}$ is, by construction, open on some member of $\mathcal{X}'$ so, by the definition of definable subcategory, $\mathcal{X}' \supseteq \mathcal{X}$, as required.

(The functorial translation of this is that every hereditary torsion theory of finite type on the functor category is determined by the *indecomposable* torsionfree injectives – see 11.1.29.) □

Corollary 5.1.5. ([726, 4.7, 4.10]) *If $\mathcal{X} \neq 0$ is a definable subcategory of* Mod-R, *then $\mathcal{X} \cap \mathrm{Zg}_R \neq \emptyset$. If $\mathcal{X}, \mathcal{X}' \subseteq$* Mod-$R$ *are definable, then $\mathcal{X} = \mathcal{X}'$ iff $\mathcal{X} \cap \mathrm{Zg}_R = \mathcal{X}' \cap \mathrm{Zg}_R$.*

Corollary 5.1.6. (essentially [726, 4.10]) *There is a bijection between definable subcategories of* Mod-R *and closed subsets of* Zg_R *given by $\mathcal{X} \mapsto \mathcal{X} \cap \mathrm{Zg}_R$ and $X \mapsto$ the definable subcategory generated by X, equivalently by $\bigoplus_{N \in X} N$.*

From this and 5.1.1 we deduce an alternative description of the topology.

Corollary 5.1.7. *The closed subsets of* Zg_R *are exactly those of the form $[T] = \{N \in \mathrm{Zg}_R : \phi(N) = \psi(N)$ for all $\phi/\psi \in T\}$, where T is an arbitrary set of pp-pairs (by 3.4.10 pairs of pp conditions with one free variable suffice).*

For any pp-pair, ϕ/ψ, set $(\phi/\psi) = \{N \in \mathrm{Zg}_R : \phi(N) > \psi(N)\}$. Also write $[\phi/\psi] = (\phi/\psi)^c$ for the complement, so a typical closed set has the form $\bigcap_\lambda [\phi_\lambda/\psi_\lambda]$ and a typical open set the form $[T]^c = \bigcup\{(\phi/\psi) : \phi/\psi \in T\}$.

Corollary 5.1.8. *A basis of open sets for the Ziegler topology consists of the $(\phi/\psi) = \{N \in \mathrm{Zg}_R : \phi(N) > \psi(N)\}$ as ϕ/ψ ranges over pp-pairs (with one free variable).*

In Ziegler's original definition the (ϕ/ψ) were declared to be open and then it was a non-trivial issue, [726, 4.9], to show that they form a basis for a topology.[6] Ziegler proved that the closed sets correspond to definable subcategories,

[6] A direct proof that the (ϕ/ψ) form a basis – that if $N \in (\phi_1/\psi_1) \cap (\phi_2/\psi_2)$, then there is (ϕ/ψ) such that $N \in (\phi/\psi) \subseteq (\phi_1/\psi_1) \cap (\phi_2/\psi_2)$ – may be found at 5.1.20 in the next subsection.

[726, 4.10]. In fact, being interested in the model theory of modules, Ziegler used the somewhat more general elementary, rather than definable, classes and the connection with closed sets was not to take the intersection of the class with pinj_R but rather to consider the "support" of an elementary class, meaning all indecomposable pure-injectives which occur as direct summands of members of that class.

Example 5.1.9. Returning to the example where R is the path algebra of A_3 (see 5.1.2), isolating pairs for the six points may be taken as follows (one may choose other isolating pairs, for instance $\big((\beta \mid x)\,/\,(\alpha\beta \mid x)\big)$ for P_2).

$$\{P_3\} = \big((x = xe_3)\,/\,(\beta \mid x)\big)$$
$$\{P_2\} = \big((x = xe_2)\,/\,(\alpha \mid x + x\beta = 0)\big)$$
$$\{P_1\} = \big((x = xe_1)\,/\,(x\alpha\beta = 0)\big)$$
$$\{I_2\} = \big((\alpha \mid x \wedge x\beta = 0)\,/\,(x = 0)\big)$$
$$\{I_1\} = \big((x = xe_1 \wedge x\alpha = 0)\,/\,(x = 0)\big)$$
$$\{S_2\} = \big((x = xe_2 \wedge x\beta = 0)\,/\,(\alpha \mid x)\big)$$

(Here we use the convention that (ϕ/ψ) means $(\phi/\phi \wedge \psi)$ in case $\phi \ngeq \psi$.)

If M is any module then, by 5.1.6, the definable subcategory, $\langle M \rangle$, generated by M is also generated by the set $\langle M \rangle \cap \mathrm{Zg}_R$ of indecomposable pure-injectives. So the next result follows by 3.4.11.

Corollary 5.1.10. *Suppose that $\phi > \psi$ and $\phi' > \psi'$ are pp-pairs. If $(\phi/\psi) \subseteq (\phi'/\psi')$, then, for every module M, if $\phi(M) > \psi(M)$, then $\phi'(M) > \psi'(M)$ (strict inclusions).*

We give some examples.

The notation inj_R is used to denote the set, a subset of pinj_R, of isomorphism classes of indecomposable injective R-modules. By 3.4.24 and 4.3.12, inj_R is a closed subset of Zg_R if R is right coherent, and the converse is [527, 4.4]. The remainder of the next result is essentially contained in [487, 2.5]. For torsion theories (of finite type) see Section 11.1.1 (resp. Section 11.1.2).

Theorem 5.1.11. *If R is any right coherent ring, then a subset of inj_R is Ziegler-closed iff it has the form $\mathcal{F} \cap \mathrm{inj}_R$, where $\mathcal{F}$ is the torsionfree class for some hereditary torsion theory of finite type on* Mod-R.

For any ring R, inj_R is Ziegler-closed if R is right coherent.

Proof. A torsionfree class, $\mathcal{F}$, is a definable subcategory of Mod-R iff the corresponding torsion theory is of finite type (11.1.20), so any subset of inj_R of the form given will be Ziegler-closed.

For the converse, if $X \subseteq \mathrm{inj}_R$ is a closed subset of Zg_R, then it has the form $\mathcal{X} \cap \mathrm{Zg}_R$ for some definable subcategory, $\mathcal{X}$, of Mod-R. Since R is right coherent so, 3.4.24, the class, Abs_R, of absolutely pure right R-modules is definable, we may replace $\mathcal{X}$ by the definable subcategory $\mathcal{Y} = \mathcal{X} \cap \mathrm{Abs}_R$. Let $\mathcal{F}$ be the class of submodules of members of $\mathcal{Y}$. Note that $\mathcal{F} \cap \mathrm{inj}_R = X$. Then, by 3.4.7, $\mathcal{F}$ is closed under submodules, products and direct limits (since direct limits of embeddings are, see Appendix E, p. 707, embeddings). It is also, by 4.3.21, closed under injective hulls. That is, 11.1.1 and 11.1.12, the hereditary torsion theory cogenerated by $\mathcal{F}$ is of finite type, as required.

The second statement was noted just above. $\square$

For example, the Ziegler-closed subsets of $\mathrm{inj}_{\mathbb{Z}}$ are, since $\mathbb{Z}$ is noetherian so (11.1.14) every hereditary torsion theory is of finite type, and by 11.1.1, those of the form $\{\mathbb{Z}_{p^\infty} : p \in \mathcal{P}\} \cup \{\mathbb{Q}\}$, where $\mathcal{P}$ is any set of non-zero prime integers.

An example of a non-coherent ring, so with inj_R not closed, is at 14.4.7.

The **support**, $\mathrm{supp}(M)$, of a module M is defined to be the Ziegler-closed set $\langle M \rangle \cap \mathrm{Zg}_R$, where $\langle M \rangle$ denotes the definable subcategory of Mod-R generated by M. Note that 3.4.11 may be read as a criterion for inclusion of supports in terms of pp-pairs: $\mathrm{supp}(M') \subseteq \mathrm{supp}(M)$ iff every pp-pair closed on M is closed on M'.

Thus to every module is associated a closed subset of the spectrum, and two modules are associated to the same closed set[7] iff they generate the same definable subcategory of Mod-R. If R is such that each group $\phi(-)/\psi(-)$ is trivial or infinite, this reduces simply to M and M' being elementarily equivalent (A.1.2).

Theorem 5.1.12. *If N is an indecomposable module of finite endolength, then N is a closed point of Zg_R. The converse is true if* $\mathrm{pp}(N)$ *(Section 3.2.1) is countable (in particular it is true if R is countable).*

Proof. By 4.4.30 the definable subcategory generated by N consists just of the direct sums of copies of N, so the corresponding closed subset, the closure of N, is just $\{N\}$.

For the partial converse we refer forward to 5.3.21 and 5.3.22 applied with $X = \{N\}$ and $[\phi, \psi] = [x = x, x = 0]$. $\square$

Question 5.1.13. Is it true that over any ring a closed point of the Ziegler spectrum has finite endolength?

The converse to 5.1.12 is true over some particular kinds of ring. For serial rings this is 8.2.59. For rings wih Krull–Gabriel dimension (= m-dimension) $< \infty$

[7] In model-theoretic terms, $\mathrm{supp}(M) = \mathrm{supp}(M')$ iff $M^{(\aleph_0)}$ and $M'^{(\aleph_0)}$ are elementarily equivalent (A.1.3). Krause, for example, [347], describes such modules as being "purely equivalent".

(Sections 7.2, 13.2.1) or, more generally rings with width $< \infty$ (Section 7.3.1), hence with no superdecomposable pure-injective modules (7.3.5), it follows by 5.3.24 and 5.3.22.

The next corollary is immediate from 4.4.30. The one after that follows from 4.4.19 and 5.1.6.

Corollary 5.1.14. *If N is of finite endolength, then* $\mathrm{supp}(N)$ *is the finite set of isomorphism types of indecomposable summands of N.*

Corollary 5.1.15. *If N is Σ-pure-injective, then* $\mathrm{supp}(N)$ *is the closure in* Zg_R *of the set of indecomposable direct summands of N.*

That fails for general modules N. For instance, if N is a superdecomposable pure-injective (for example, see 4.4.3), then it has, by definition, no indecomposable direct summands but, by 5.1.4, there is at least one indecomposable pure-injective in the definable subcategory generated by N and hence in the support of N.

Example 5.1.16. (Herzog, [278, p. 554]) Let R be any ring and let $n \geq 0$. Then the indecomposable modules of endolength at most n form a closed subset of Zg_R.

To see this, let $\phi_0 > \phi_1 > \cdots > \phi_{n+1}$ be any length $n+1$ chain of pp conditions with one free variable. Each set (ϕ_i/ϕ_{i+1}) is open so, therefore, is the set $\bigcap_{i=0}^{n}(\phi_i/\phi_{i+1})$. Note that any point of this open set must have endolength at least $n+1$. And any point of endolength at least $n+1$ (including those without finite endolength) clearly belongs to at least one such open set. Therefore the union of these open sets over all such chains of pp conditions is exactly the complement of the set of points of endolength at most n, as required.

Proposition 5.1.17. *If R and S are* **Morita equivalent** *(that is, their categories of (right) modules are equivalent), say α : Mod-$R \longrightarrow$ Mod-S is an equivalence, then their Ziegler spectra are homeomorphic via α.*

Proof. Since α preserves endomorphism rings it takes indecomposable R-modules to indecomposable S-modules. It also takes finitely presented objects to finitely presented objects because this property is described in terms of the category (the corresponding covariant representable functor should commute with direct limits, see Appendix E, p. 709). Therefore it preserves purity and pure-injectivity (by the characterisations 2.1.4 or 2.1.7, respectively by the definition). That is enough to see that it induces a bijection of sets between Zg_R and ${}_R\mathrm{Zg}$. The definition of the topology in terms of definable subcategories shows that the topology is preserved as well (the closure conditions, 3.4.7, for definable subcategories are clearly Morita invariant). □

Remark 5.1.18. If $R = R_1 \times R_2$ is a product of rings, with $1_R = 1_{R_1} + 1_{R_2}$ the corresponding decomposition of 1 into orthogonal commuting idempotents, then Zg_R is the disjoint union of copies of Zg_{R_1} and Zg_{R_2}, namely the clopen sets $\big((x = x1_{R_1})/(x = 0)\big)$ and $\big((x = x1_{R_2})/(x = 0)\big)$, which are Mod-$R_1 \cap \mathrm{Zg}_R$ and Mod-$R_2 \cap \mathrm{Zg}_R$.

5.1.2 Ziegler spectra via pp-pairs: proofs

It is proved that pp-pairs define a basis for the topology (5.1.20) and that each of these basic open sets is compact (5.1.22). Since the whole space is given by such a pair the Ziegler spectrum of a ring is compact. A neighbourhood basis of an indecomposable may be described in terms of any type of which it is the hull (5.1.21).

First we give the functorial proof, promised in the previous section, of the fact that the sets $(\phi/\psi) = \{N \in \mathrm{pinj}_R : \phi(N) > \psi(N)\}$ form a basis of Zg_R. Then we show that a basis of open neighbourhoods of $N \in \mathrm{Zg}_R$ can be found from any pp-type p for which $N = H(p)$. Ziegler's original proofs are (at least, look) very different. The proofs here are straightforward in the sense that, at each stage, the next step is fairly obvious, given the situation. The original proofs, because they did not sit inside a well-developed framework, required more ingenuity.

Lemma 5.1.19. *Suppose that $G \leq A \oplus B$ are objects of an abelian category. Then $\pi_A G/(G \cap A) \simeq \pi_B G/(G \cap B)$, where π_A, π_B denote the projections from $A \oplus B$ to A, respectively B.*

Proof. $\pi_A G/G \cap A \simeq \big(G/G \cap B\big)/\big((G \cap A + G \cap B)/G \cap B\big) \simeq G/(G \cap A + G \cap B)$ which is isomorphic to $\pi_B G/G \cap B$ by symmetry. □

In Ziegler's paper this (see [726, 8.9]) was much used, in the form where G is the subgroup defined by a pp condition $\phi(\overline{x}, \overline{y})$, $\pi_A G$ is the subgroup defined by $\exists \overline{y}\, \phi(\overline{x}, \overline{y})$ (so $G \cap A$ is defined by $\phi(\overline{x}, \overline{0})$) and π_B the subgroup defined by $\exists \overline{x}\, \phi(\overline{x}, \overline{y})$. Then the conclusion is "$\exists \overline{y}\, \phi(\overline{x}, \overline{y}) / \phi(\overline{x}, \overline{0}) \simeq \exists \overline{x}\, \phi(\overline{x}, \overline{y}) / \phi(\overline{0}, \overline{y})$", that is, the corresponding quotients of pp functors are isomorphic.[8] In the proof below we use this to find isomorphic non-zero finitely presented subquotients of two finitely generated functors.

Theorem 5.1.20. ([726, 4.9]) *Suppose that $N \in (\phi_1/\psi_1) \cap (\phi_2/\psi_2)$ in Zg_R. Then there is a pp-pair ϕ/ψ such that $N \in (\phi/\psi) \subseteq (\phi_1/\psi_1) \cap (\phi_2/\psi_2)$.*

[8] Ziegler, [726, 8.9], suggested the name "Goursat's Theorem" for this on the basis of the pictures one might draw to illustrate this lemma and the result of that name in complex analysis.

Proof. By 10.3.8 the assumption means that there are non-zero homomorphisms $\alpha_1 : F_{D\psi_1}/F_{D\phi_1} \to (N \otimes -)$ and $\alpha_2 : F_{D\psi_2}/F_{D\phi_2} \to (N \otimes -)$. Since N is an indecomposable pure-injective module, $N \otimes -$ is an indecomposable injective functor (12.1.6), hence is uniform, hence $\mathrm{im}(\alpha_1) \cap \mathrm{im}(\alpha_2) \neq 0$. It follows that the object P in the pullback diagram shown is non-zero.

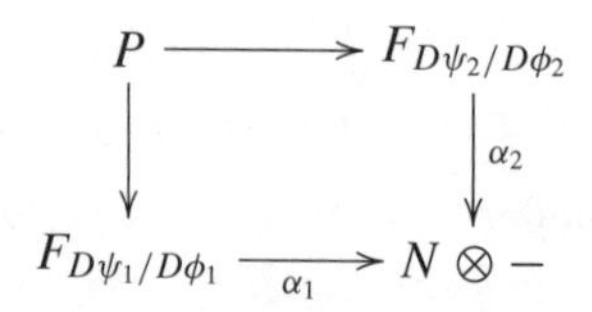

Choose a finitely generated subfunctor, G, of P with non-zero image in $N \otimes -$. Since G is a subfunctor of $F_{D\psi_1/D\phi_1} \oplus F_{D\psi_2/D\phi_2}$ (Appendix E, p. 707) it is, by 10.2.3, finitely presented, hence so are $G \cap (F_{D\psi_1/D\phi_1})$ and $G \cap (F_{D\psi_2/D\phi_2})$ (by 10.2.7). Therefore the corresponding isomorphic quotients appearing in 5.1.19 are finitely presented subquotients of $F_{D\psi_1/D\phi_1}$, resp. $F_{D\psi_2/D\phi_2}$, so this common quotient is isomorphic to $F_{D\psi'/D\phi'}$ for some pp-pair ϕ'/ψ' (10.2.31). Note that this functor is mapped to $N \otimes -$ with non-zero image since $\left(G \cap F_{D\psi_1/D\phi_1}\right) \oplus \left(G \cap F_{D\psi_1/D\phi_2}\right)$ is sent to 0 in $N \otimes -$.

Therefore, by 10.3.8, $(\phi'/\psi')(N) \simeq (F_{D\psi'/D\phi'}, N \otimes -) \neq 0$ and, since $F_{D\psi'/D\phi'}$ is a subquotient of $F_{D\psi_i/D\phi_i}$ $(i = 1, 2)$, we also have $(\phi'/\psi') \subseteq (\phi_1/\psi_1) \cap (\phi_2/\psi_2)$ by 12.1.19, as required.

(For a model-theoretic, as opposed to "functorial", proof of this and the next two results see the original paper or [495, 4.66].) □

Note that, by the proof, the pair (ϕ/ψ) in the statement may be chosen such that $F_{D\psi/D\phi}$ is isomorphic to a subquotient of both $F_{D\psi_1/D\phi_1}$ and $F_{D\psi_2/D\phi_2}$; equivalently, by 1.3.1, $F_{\phi/\psi}$ is isomorphic to a subquotient of both F_{ϕ_1/ψ_1} and F_{ϕ_2/ψ_2}. This form of the statement is given at [278, 3.13]; also note 12.1.19.

Theorem 5.1.21. ([726, 4.9]) *Let $N = H(p)$ be a point of Zg_R so p is an irreducible pp-type (4.3.49). Then the sets (ϕ/ψ) for $\phi > \psi$ with $\phi \in p$ and $\psi \notin p$ form a basis of open neighbourhoods of N.*

Proof. Let $\overline{a}$ from N be such that $\mathrm{pp}^N(\overline{a}) = p$. Consider the corresponding (non-zero) morphism $(\overline{a} \otimes -) : (R^n \otimes -) \to (N \otimes -)$. Suppose that $N \in (\phi_1/\psi_1)$, so, 10.3.8, there is a non-zero morphism $\alpha : F_{D\psi_1/D\phi_1} \to (N \otimes -)$. Proceed as in the proof of 5.1.20, forming the pullback and obtaining the common subquotient $F_{D\psi'/D\phi'}$, of $F_{D\psi_1/D\phi_1}$ and $R^n \otimes -$ as there. We may choose the pp conditions ϕ' and ψ' so that, with notation as before, $F_{D\phi'} = G \cap (R^n \otimes -)$ and $F_{D\psi'} = \pi_{(R^n \otimes -)}G$.

Also as in the proof of 5.1.20, we have $N \in (\phi'/\psi') \subseteq (\phi_1/\psi_1)$. It remains only to show that $\phi' \in p$ and $\psi' \notin p$. For that we use 12.2.5 which says that the kernel of $\overline{a} \otimes -$ is $F_{Dp} = \sum_{\phi \in p} F_{D\phi} \leq (R^n \otimes -)$. Therefore, since the image of $F_{D\psi'/D\phi'}$ in $N \otimes -$ is non-zero, certainly $F_{D\psi'} \nleq F_{Dp}$, that is, $\psi' \notin p$. On the other hand, since (with the previous notation) $F_{D\phi'} = G \cap (R^n \otimes -)$ is sent to zero by $\overline{a} \otimes -$, also $F_{D\phi'} \leq F_{Dp}$, hence $\phi' \in p$, as required. □

Theorem 5.1.22. ([726, 4.9]) *The compact open sets of* Zg_R *are exactly the* (ϕ/ψ) *with* ϕ/ψ *a pp-pair (with an arbitrary number of free variables).*

Proof. Each such set is compact: if $(\phi/\psi) = \bigcup_\lambda (\phi_\lambda/\psi_\lambda)$, then, by 12.3.19, $F_{\phi/\psi}$ belongs to the Serre subcategory of $(\text{mod-}R, \mathbf{Ab})^{\text{fp}}$ generated by the $F_{\phi_\lambda/\psi_\lambda}$, so, necessarily (11.2.1), it belongs to the Serre subcategory generated by just finitely many of them. Therefore, by the same result, (ϕ/ψ) is the union of the corresponding finitely many open subsets.

Since sets of this kind form a basis of the topology, an open set is compact exactly if it is a finite union of such sets. But $(\phi_1/\psi_1) \cup \cdots \cup (\phi_k/\psi_k) = (\phi/\psi)$, where ϕ is $\phi_1(\overline{x}_1) \wedge \cdots \wedge \phi_k(\overline{x}_k)$ and ψ is $\psi_1(\overline{x}_1) \wedge \cdots \wedge \psi_k(\overline{x}_k)$, where the sequences, $\overline{x}_i$, of free variables should be taken to be disjoint. □

An arguably simpler proof of compactness of (ϕ/ψ) is as follows. By 5.1.6 the equality $(\phi/\psi) = \bigcup_\lambda (\phi_\lambda/\psi_\lambda)$, that is, $[\phi/\psi] = \bigcap_\lambda [\phi_\lambda/\psi_\lambda]$, is equivalent to $\mathcal{X} = \bigcap_\lambda \mathcal{X}_\lambda$, where $\mathcal{X}$, respectively $\mathcal{X}_\lambda$, is the definable subcategory of Mod-R defined by the pp-pair ϕ/ψ, resp. $\phi_\lambda/\psi_\lambda$. That is, a module M satisfies $\forall \overline{x}\, (\phi(\overline{x}) \to \psi(\overline{x}))$ iff it satisfies $\{\forall \overline{x}\, (\phi_\lambda(\overline{x}) \to \psi_\lambda(\overline{x}))\}_\lambda$. By the Compactness Theorem of logic (in the form stated after A.1.8) this set may be replaced by a finite one and hence $\mathcal{X} = \mathcal{X}_{\lambda_1} \cap \cdots \cap \mathcal{X}_{\lambda_k}$ for some $\lambda_1, \ldots, \lambda_k$ and hence $(\phi/\psi) = (\phi_{\lambda_1}/\psi_{\lambda_1}) \cup \cdots \cup (\phi_{\lambda_k}/\psi_{\lambda_k})$.

Taking (ϕ/ψ) to be $(x = x/x = 0)$ in 5.1.22 gives compactness of the whole space.

Corollary 5.1.23. *For every ring* R *the Ziegler spectrum* Zg_R *is compact.*

This corollary is one of the few results which do not hold if the ring R is replaced by a small preadditive category, $\mathcal{A}$ and, correspondingly, Mod-R is replaced by the functor category $(\mathcal{A}^{\text{op}}, \mathbf{Ab})$ (see the end of Section 10.2.4).

Proposition 5.1.24. ([726, 7.10]) *Suppose that* p, q *are irreducible pp-types with* $H(p)$ *not isomorphic to* $H(q)$ *and let* ϕ/ψ *be a pp-pair such that* $\phi \in p, q$ *and* $\psi \notin p, \psi \notin q$. *Then there is a pp condition* θ *with* $\phi > \theta > \psi$ *and either* $(\theta \notin p$ *and* $\theta \in q)$ *or* $(\theta \notin q$ *and* $\theta \in p)$.

Proof. Choose $a \in H(p)$ and $b \in H(q)$ with pp-types p and q respectively. Consider $(a \otimes -) : (R \otimes -) \to (H(p) \otimes -)$ and $(b \otimes -) : (R \otimes -) \to (H(q) \otimes -)$. Set $K' = \ker(a \otimes -)$ and $K'' = \ker(b \otimes -)$. By 12.2.5, both $F' = F_{D\psi}/(F_{D\psi} \cap K')$ and $F'' = F_{D\psi}/(F_{D\psi} \cap K'')$ are non-zero, F' embeds in $H(p) \otimes -$ and F''

embeds in $H(q) \otimes -$. These injective functors are, by assumption, not isomorphic, so $F' \neq F''$, that is, $F_{D\psi} \cap K' \neq F_{D\psi} \cap K''$. So there is a finitely presented functor F with, say, $F \leq F_{D\psi} \cap K'$ but $F \nleq F_{D\psi} \cap K''$. Since both $F_{D\psi} \cap K'$ and $F_{D\psi} \cap K''$ contain $F_{D\phi}$ we may suppose that $F \geq F_{D\phi}$. By 10.2.16, $F = F_{D\theta}$ for some pp condition θ. Note that $\phi \geq \theta > \psi$. Then, in this case, we have (by 12.2.5) $a \in \theta(H(p))$ but $b \notin \theta(H(q))$ and hence $\theta \notin q$ and $\theta \in p$ is as required. □

5.1.3 Ziegler spectra via morphisms

Let $f : A \to B$ be a morphism in mod-R. For any module M this induces a morphism of abelian groups $(f, M) : (B, M) \to (A, M)$. Set

$$(f) = \{N \in \mathrm{Zg}_R : (f, N) : (B, N) \to (A, N) \text{ is not onto}\}$$

to be the set of points, N, of the spectrum for which there is a morphism from A to N which does not factor through f. The fact that these sets also give a basis for the Ziegler spectrum comes from Crawley-Boevey's exposition [145].

Theorem 5.1.25. *The sets of the form (f) where f ranges over* mod-R *form a basis of open sets for* Zg_R*, indeed these are exactly the compact open sets of the space.*[9]

Proof. Let $f : A \to B$ be a morphism in mod-R. Let $\overline{a}$ be a finite generating tuple for A, with presentation $\overline{a}H = 0$ (H a matrix over R). Let ψ be a pp condition generating the pp-type of $f\overline{a}$ in B (1.2.6). Set $\phi(x)$ to be $\overline{x}H = 0$: then $(f) = (\phi/\psi)$. For if $N \in (f)$, say $g : A \to N$ does not factor through f, then $g\overline{a} \in \phi(N)$ (1.1.7) but $g\overline{a} \notin \psi(N)$ since, otherwise, there would, by 1.2.9, be a morphism from B to N taking $f\overline{a}$ to $g\overline{a}$, so g would factor through f. Conversely, if $N \in (\phi/\psi)$, say $\overline{c} \in \phi(N)\backslash\psi(N)$, then, by 1.2.9, there is a morphism from A to N taking $\overline{a}$ to $\overline{c}$ and this morphism, again by 1.2.9, cannot factor through f.

To see that every open set (ϕ/ψ) has the form (f) for some morphism f in mod-R, choose a free realisation $(C_\phi, \overline{c}_\phi)$ of ϕ. Suppose that $\overline{d}$ is a finite generating tuple for C_ϕ and let $\overline{d}H = 0$, with H a matrix over R be a presentation of the finitely presented module C_ϕ. Let K be a matrix such that $\overline{c}_\phi = \overline{d}K$. Let $(C', \overline{d'})$ be a free realisation of the pp condition $\overline{y}H = 0 \wedge \psi(\overline{y}K)$. Since $(\overline{y}H = 0) \geq (\overline{y}H = 0 \wedge \psi(\overline{y}K))$ there is a morphism $f : C_\phi \to C'$ taking $\overline{d}$ to $\overline{d'}$. We claim that $(\phi/\psi) = (f)$.

[9] This description, of the spectrum and also of finitely presented functors, is taken as basic in [358], where the property $N \notin (f)$ is described as N being "f-injective". Krause develops the idea of inverting collections of such maps, as in universal localisation, in [358, §12.3].

If $N \in (\phi/\psi)$, say $\overline{a} \in \phi(N) \setminus \psi(N)$, then there is, by 1.2.17, a morphism g from C_ϕ to N taking $\overline{c}_\phi$ to $\overline{a}$. If this were to factor through f, then we would have $\overline{a} = g\overline{d}K$ in $\psi(N)$, a contradiction. So $N \in (f)$. Conversely if $N \in (f)$, say $g : C_\phi \to N$ does not factor through f, then set $\overline{a} = g\overline{c}_\phi$. If we had $\overline{a} \in \psi(N)$, then we would have $g\overline{d}H = 0 \wedge \psi(g\overline{d}K)$ so, by 1.2.17, there would be $g' : C' \to N$ with $g'\overline{d'} = g\overline{d}$, hence with $g'f = g$, contrary to assumption. Therefore $\overline{a} \in \phi(N)\setminus\psi(N)$, as required. □

The argument towards the end of this proof comes from [101, 3.1.7] (where it is used to prove 10.2.14(b)); it shows that every pp-pair ϕ/ψ is isomorphic in the category, $\mathbb{L}_R^{\mathrm{eq}+}$, of pp pairs to one of the form θ/η with θ quantifier-free. Indeed if, as in the proof above, $\overline{y}H = 0 \wedge \overline{x} = \overline{y}K$ "presents" a free realisation of ϕ (cf. proof of 1.2.6), then the quotients $(\overline{y}H = 0) \,/\, (\overline{y}H = 0 \wedge \psi(\overline{y}K))$ and $\phi(\overline{x})/\psi(\overline{x})$ are easily seen to be (pp-definably) isomorphic in every module M. Replacing the category of pp pairs by the equivalent (10.2.30) category of finitely presented functors, this becomes the fact that every subquotient of an object within this category is a factor of a representable functor.

5.2 Examples

5.2.1 The Ziegler spectrum of a Dedekind domain

The list of indecomposable pure-injectives over a commutative Dedekind domain is given (5.2.2): in fact it is enough to assume that the ring is a Dedekind domain which satisfies a polynomial identity. The proof goes via localisation to discrete valuation domains (5.2.1). This is followed by a complete description of the topology: in general terms at 5.2.3 and in detail after that. The intersection of two compact open sets need not be compact (5.2.5). Corollaries of the general description are that the m-dimension is 2 and that there is no superdecomposable pure-injective (5.2.6). The pure-injective hull of the ring is computed (5.2.7).

The ring, $\mathbb{Z}$, of integers and the ring, $k[X]$, of polynomials in one indeterminate over a field are the archetypal examples of Dedekind domains. There is, however, no need to confine ourselves to the commutative case, indeed, for some applications we need non-commutative Dedekind domains. Here are the definitions: for more detail consult, say, [441, §5.2].

A **Dedekind domain** is a **domain** (that is, a ring without zero-divisors) which is right and left noetherian, **hereditary** (every right ideal is a projective module, equivalently, for example, [441, 5.4.3], in the presence of the right and left noetherian assumption, every left ideal is projective) and which has no **idempotent ideals** (two-sided ideals I such that $I^2 = I$).

For a commutative domain, being Dedekind is equivalent to being hereditary ([663, 4.8]) and this is equivalent to every localisation at a maximal ideal being a (commutative) discrete valuation domain (for example, [331, Thm 96]).

In the non-commutative situation there is a wide gap between the case where R is **PI**, meaning that R satisfies some polynomial identity (that is, there is some non-zero non-commutative polynomial in $\mathbb{Z}\langle X_1, \ldots, X_n \rangle$ for some n, which is zero when evaluated at all n-tuples of R) and the case where R is not PI. In this section we deal only with the, relatively similar to commutative, PI case. In this case the elements of R may be represented as matrices over some commutative field. More precisely, R is a maximal hereditary order (see [441, §5.3]) in a **central simple algebra** (the ring of $n \times n$ matrices over some division ring which is finite-dimensional as a vector space over its centre).

The points First consider the case where R is local. Then R is a (**non-commutative**) **discrete valuation domain**, meaning a domain which has a unique maximal ideal, P say, such that the ideals of R are precisely $R > P > P^2 > \cdots > P^n > \cdots > 0$. Then any element $\pi \in P \setminus P^2$ generates P as a right ideal and as a left ideal. The PI condition is equivalent to the division ring R/P being finite-dimensional over its centre (which is a field).

The next two results appear as 1.3 and 1.6 of [505] but the proof is just as in the commutative case, which is [726, 5.1, 5.2], also see [172, §5], [323, 8.56]. The pure-injective modules over these rings were also described in [436, §§4,5].

Theorem 5.2.1. *Let R be a PI discrete valuation domain with maximal ideal P. The points of* Zg_R *are the following:*

(a) the indecomposable modules, R/P^n, of finite length, for $n \geq 1$;
(b) the completion, $\overline{R} = \varprojlim_n R/P^n$, of R in the P-adic topology;
(c) the Prüfer module $R_{P^\infty} = E(R/P)$;
(d) the quotient division ring, $Q = Q(R)$, of R.

Of these, all but $\overline{R}$ are Σ-pure-injective, the R/P^n are of finite endolength, as is Q, whereas R_{P^∞} is of infinite length over its endomorphism ring.

Proof. First it must be checked that all these modules are indecomposable and pure-injective. Injective modules are pure-injective (by the definitions) as are modules of finite endolength (4.4.24), as is the completion of R at P (4.2.15). The other assertions about these modules are easily checked. So it remains to show that there are no more points. The proof we give is that of [726]. For a "homological" proof see [323, p. 205 ff.]. For abelian groups this is in [330, §17] (second edition), see also [198, §40].

Let N be an indecomposable pure-injective module and choose any non-zero element $a \in N$. Since R is PI it may be supposed, [441, 13.7.9], that π, a generator for the maximal ideal P, is in the centre of R. Set $h(a) = \sup\{n : \pi^n \mid a\}$ (a non-negative integer or ∞). The annihilator, $\mathrm{ann}_R(a)$, of a is a power of P or 0 and, since $\mathrm{ann}(a) = P^{m+1}$ implies that $a\pi^m \neq 0$ and $a\pi^m \cdot \pi = 0$, we may restrict to the case where $\mathrm{ann}(a)$ is P or 0. We consider the various possibilities for $h(a)$ and $\mathrm{ann}(a)$.

Case (i) $h(a) = n$, $\mathrm{ann}(a) = P$. Say $a = b\pi^n$. Then $bR \simeq R/P^{n+1}$ is pure-injective. It is also the case that bR is pure in N. For, by 2.4.15, it is enough to check pp conditions of the simple form in 2.4.10. So suppose that $br \in Ns$, say $br = cs$, where $r = \pi^k u$ and $s = \pi^l t$ with u, t units of R. Since $b\pi^{n+1} = 0$, we may suppose $k \leq n$. Then $a = b\pi^k u.\pi^{n-k}u^{-1} = c\pi^l t\pi^{n-k}u^{-1}$ so $h(a) = n$ implies $n - k + l \leq n$, that is, $l \leq k$. Hence $Br \in bRs$, as required. Therefore $N \simeq bR \simeq R/P^{n+1}$.

Case (ii) $h(a) = n$, $\mathrm{ann}(a) = 0$. Say $a = b\pi^n$. Then $br = 0$ implies $ar = 0$, so $r = 0$. Thus $bR \simeq R$. Furthermore, bR is pure in N: otherwise there would be an equality of the form $b\pi^k u = c\pi^l t$ with $l > k$, hence a non-zero (since $h(a) = n$) torsion element $bu - c\pi^{l-k}t$, so we could move to Case (i) or (iii) (and, indeed, a contradiction). Therefore N is isomorphic to the pure-injective hull of R which, 4.2.15 (cf. 4.2.7), is just the completion of R at P.

Case (iii) $h(a) = \infty$, $\mathrm{ann}(a) = P$. For each $n \geq 1$ there is some element $b_n \in N$ with $a = b_n\pi^n$. Therefore the set $\{x_0 = a\} \cup \{x_i\pi = x_{i-1} : i \geq 1\}$ of pp conditions (with the parameter a from N and with infinitely many free variables) is finitely satisfied in N, so, by assumption and 4.2.2, has a solution in N: say there are $b_i \in N$ $(i \geq 1)$ with $b_1\pi = a$, $b_2\pi = b_1, \dots$. It is easy to check that these generate a copy of the injective module $E(R/P)$ contained in, hence a direct summand of, hence equal to, N.

Case (iv) $h(a) = \infty$, $\mathrm{ann}(a) = 0$. By Cases (i)–(iii) it may be assumed that every non-zero element of N satisfies these two conditions. Let $q \in Q \setminus R$, say $q = \pi^n u$ where $n < 0$ and u is a unit of R. Because $h(a) = \infty$, there is $b \in N$ such that $a = b\pi^{-n}$. Since N is torsionfree this b is unique, so we may set $aq = bu$. In this way we can define a map, $a \mapsto aq$ $(q \in Q)$, which is easily checked to be an R-homomorphism, from Q to N. Thus there is a copy of the injective module Q embedded in N so $N \simeq Q$.

□

Now let R be any PI Dedekind domain. If $N \in \mathrm{Zg}_R$, then, 4.3.44, the elements of the centre, $C(R)$, of R which do not act as automorphisms of N form a prime ideal and N is a module over the corresponding localisation of R. Because R is an Azumaya algebra [589, 22.4], there is a bijection, [441, 13.7.9], between the ideals of R and those of the centre of R, given by intersection, $I \mapsto I \cap C(R)$. Hence this localisation is a discrete valuation domain. So, with the above result, this yields the following.

Theorem 5.2.2. ([436, §§4, 5], [505, 1.6]) *Let R be a PI Dedekind domain which is not a division ring. The points of Zg_R are as follows:*

(a) the indecomposable modules, R/P^n, of finite length, for $n \geq 1$ and for P a maximal ideal of R;
(b) the completion, $\overline{R_P} = \varprojlim_n R/P^n$, of R in the P-adic topology for P a non-zero, equivalently maximal, prime of R, we call these **adic** *modules;*
(c) the **Prüfer** *modules $R_{P^\infty} = E(R/P)$ as P ranges over the non-zero primes of R;*
(d) the quotient division ring, $Q = Q(R)$, of R.

For the commutative case this can be found in [330] and in [726, 5.2] and, with a functor-category proof, in [323, 8.56].

Since any Dedekind prime ring is Morita equivalent to a Dedekind domain ([441, 5.2.12]), the result applies equally to the more general class of rings, by 5.1.17 and since the statement of the theorem is essentially Morita-invariant ("division ring" in (d) must be replaced by "simple artinian ring"). The results here are extended to PI hereditary noetherian prime rings at 8.1.13.

The topology

Theorem 5.2.3. ([505, 1.4, 1.6], [726, 5.2] for the commutative case) *Let R be a PI Dedekind domain. The points of Zg_R of the form R/P^n, where P is a non-zero prime of R, are the isolated points. If these are all removed, then the adic and Prüfer modules are the isolated points in what remains. The closed points are the finite length points and the quotient division ring Q.*

Remark 5.2.4. In the terminology which will be introduced in Section 5.3.6, the modules of finite length are the points of Cantor–Bendixson rank 0; the adic and Prüfer points are those of rank 1 and the quotient division ring, Q, of R is the unique point of rank 2.

For each point of Zg_R a basis of open neighbourhoods will be written down. With this list to hand one may simply check the assertions of 5.2.3.

Since each two-sided ideal, I, of R is finitely generated, the conditions $xI = 0$ and $I \mid x$ (the latter meaning $x \in MI$, where M is the module to which x belongs) are pp conditions: recall (Section 1.1.4) that if $I = \sum a_i R$, then $xI = 0$

is rendered by $\bigwedge_i xa_i = 0$ and if $I = \sum Rb_j$, then the condition $I \mid x$ is rendered by $\exists y_1, \ldots, y_m\, (x = \sum_j y_j b_j)$.

R/P^n : This point is isolated by the pair ϕ/ψ, where ϕ is $(P^{n-1} \mid x) \wedge (xP = 0)$ and ψ is $(P^n \mid x) \wedge (xP = 0)$: for, if $N \in (\phi/\psi)$, then there is an element of $\mathrm{ann}_N(P)$ which is divisible by P^{n-1} but not by P^n. By inspection, R/P^n is the only such module among those listed in 5.2.2.

$\overline{R_P}$: The set $\big((x = x) / (P \mid x)\big)$ containing just this point together with the points of finite length associated to the same prime is open, so a neighbourhood basis is given by the sets of the form $\big((P^n \mid x) / (P^{n+1} \mid x)\big) = \{\overline{R_P}\} \cup \{R/P^m : m \geq n+1\}$. No open neighbourhood of $\overline{R_P}$ can omit infinitely many points of the form R/P^n because then the, closed, complement would contain their inverse limit $\overline{R_P}$ – see the discussion before 3.4.23

R_{P^∞}: The set $((xP = 0)/(x = 0))$ containing just this point together with the points of finite length associated to P is open, so a neighbourhood basis is given by the sets of the form $\big((xP^{n+1} = 0) / (xP^n = 0)\big) = \{R_{P^\infty}\} \cup \{R/P^m : m \geq n+1\}$. No open neighbourhood of R_{P^∞} can omit infinitely many points R/P^n because then the complement would contain their direct limit, R_{R^∞}, so could not, by 3.4.7, be closed.

Q: For every prime P, the module Q is a direct summand of the product of infinitely many copies of R_{P^∞}. Also Q is a direct summand of the direct limit of the directed system $\overline{R_P} \xrightarrow{\pi\times-} \overline{R_P} \xrightarrow{\pi\times-} \cdots$ so, by 3.4.7, Q is in the Ziegler-closure of each point of infinite length. Therefore, every open neighbourhood of Q contains every point of infinite-length and hence, by compactness of the spectrum and isolation of the finite length points, can exclude only finitely many (finite-length) points. A neighbourhood basis is given by those sets of the form $\big((x = x) / (xI = 0)\big) = \mathrm{Zg}_R \setminus \{R/P_1^t, \ldots, R/P_k^t : t \leq m\}$, where $P_1, \ldots, P_k$ are any distinct non-zero primes, m is any integer with $m \geq 1$ and $I = (P_1 \times \cdots \times P_k)^m$. This set may also be defined by the dual condition $\big((I \mid x) / (x = 0)\big)$.

The finite-length points and Q are of finite endolength, so, by 5.1.12, are closed points of Zg_R. As noted already, Q is in the closure of every point apart from these, therefore there are no more closed points.

Remark 5.2.5. The intersection of the basic, hence (5.1.22) compact, open sets $\big((x = x) / (P \mid x)\big)$ and $\big((xP = 0) / (x = 0)\big)$ is the non-compact set consisting of the finite-length points associated to the prime P.

The information in 5.2.3 makes it easy to determine all the definable subcategories of Mod-R for R a PI Dedekind domain since such a subcategory $\mathcal{X}$ is determined (5.1.6) by its intersection, X, with Zg_R. We describe the possibilities for X. That is, we describe all the closed subsets of Zg_R.

Suppose first that X contains no points of finite length. Since the finite-length points are isolated, the subset Zg'_R of Zg_R which remains when the isolated points are removed is closed and, by 5.2.3, every point of that space except Q is isolated. So for X we may take any subset of $\mathrm{Zg}'_R \setminus \{Q\}$, together with Q (which has to be there because it is in the closure of every point of Zg'_R).

Next, if X contains only finitely many finite-length points, then X is the disjoint union $X_0 \cup X_1$ of $X_1 = X \cap \mathrm{Zg}'_R$ and $X_0 = X \setminus X_1$ with both X_0 and X_1 closed. For X_1 we may take any closed set as above and for X_0 any finite set of finite-length (hence closed) points.

The last case is that X contains infinitely many finite-length points. Since the relative topology on $\mathrm{Zg}_R \setminus \mathrm{Zg}'_R$ is discrete (every finite-length point is open) the set X being closed imposes no restriction on X_0 (X_0 as above) which may, therefore, be any subset of $\mathrm{Zg}_R \setminus \mathrm{Zg}'_R$. Given a non-zero prime P of R, let us say that the points R/P^n, $\overline{R_P}$ and R_{P^∞} are those "associated" to P. If the prime P is associated to infinitely many points R/P^n of $X_0 = X \setminus \mathrm{Zg}'_R$, then, by the proof of 5.2.3, both $\overline{R_P}$ and R_{P^∞}, and hence also Q, must be in X. Otherwise we may have neither, one of, or both $\overline{R_P}$, R_{P^∞} in X, without restriction and with no effect on choices for other primes. In any case, since X_0 is infinite, Q must be in X (by compactness of Zg_R). All restrictions have been described.

The dimensions appearing in the next result are defined in Sections 7.2 and 13.2.

Corollary 5.2.6. *Let R be a PI Dedekind domain. Then* $\mathrm{mdim}(\mathrm{pp}_R) = \mathrm{KG}(R) = 2$. *There is no superdecomposable pure-injective R-module. Hence every pure-injective R-module is the pure-injective hull of a direct sum of indecomposable pure-injective modules.*

Proof. The analysis of the topology shows that the equivalent conditions of 5.3.16 are satisfied. That the Cantor–Bendixson rank (Section 5.3.6) of Zg_R equals 2 has been noted above, so the first statement follows by 5.3.60. From that and 7.3.5 the second statement follows. Then the third is a consequence of the general structure theorem 4.4.2. □

Garavaglia showed, [209, Thm 3], for principal ideal domains, that the "elementary Krull dimension" is 2. This dimension, see 7.1.3, is somewhat coarser, though coexistent, with m-dimension.

Example 5.2.7. We compute the pure-injective hull of $\mathbb{Z}$, more generally, the pure-injective hull of a PI Dedekind domain, R, regarded as a module over itself. Here is one of various possible proofs.

There can be no non-zero divisible factor of $H(R)$, for if $H(R) = Q' \oplus H'$, where Q' is divisible, then (consider the pp-type of 1_R) the embedding of R into $H(R)$ followed by projection to H' must be a pure embedding, so, since the embedding of R into $H(R)$ is pure-essential (Section 4.3.3), $Q' = 0$. By 5.2.6 every direct summand of $H(R)$ is indecomposable. Since every factor must be torsionfree and cannot be divisible it remains to determine how often each adic factor $\overline{R_{(P)}} = H(R_{(P)})$ occurs. Certainly each must occur since the pp-pair $(x = x)\,/\,(P|x)$ is open on R, hence on its pure extension $H(R)$ and, by the proof of 5.2.3, that pair isolates $\overline{R_{(P)}}$ among torsionfree points of Zg_R. By 4.3.26 each occurs just once. Therefore $H(R) = H\big(\bigoplus_P \overline{R_{(P)}}\big)$. An alternative representation, proved just as in 4.3.22, is $H(R) = \prod_P \overline{R_{(P)}}$.

5.2.2 Spectra over RD rings

The special form of pp conditions over RD rings (Section 2.4.2) yields a simpler description of a basis of open neighbourhoods (5.2.8).

Over an RD ring every pp condition in one free variable is equivalent to a conjunction of conditions of the form $s \mid xr$ (2.4.10) and, 2.4.13, also is equivalent to a sum of conditions of the form $\exists y\,(x = yt \wedge yu = 0)$ (r, s, t, u all in R). Since pairs of pp conditions in one free variable are sufficient to give a basis for the Ziegler spectrum (5.1.8) it may be shown that the fairly simple form of pp conditions defining a basis for Zg_R, where R is a Dedekind domain (see the proof of 5.2.3), extends to RD rings in general. Also see Section 8.2.7, in particular 8.2.57, for other forms of a basis.

Proposition 5.2.8. ([521, 3.5]) *Suppose that R is RD. Then there is a basis of Zg_R consisting of sets of the form*

$$\big(\exists y\,(x = yt \wedge yu = 0)\,/\,(\exists y\,(x = yt \wedge yu = 0) \wedge (s \mid xr))\big)$$

with $r, s, t, u \in R$.

Proof. Let ϕ/ψ be a pair of pp conditions in one free variable. By 2.4.10 ψ is equivalent to a condition of the form $\bigwedge_{i=1}^{n} s_i \mid xr_i$ and, by 2.4.13, ϕ is equivalent to a condition of the form $\sum_{j=1}^{m} \exists y\,(x = yt_j \wedge yu_j = 0)$. Set ψ_i to be $s_i \mid xr_i$ and ϕ_j to be $\exists y\,(x = yt_j \wedge yu_j = 0)$.

First note that $(\phi/\psi) = \bigcup_j (\phi_j/\phi_j \wedge \psi)$: the inclusion $\supseteq$ is immediate because $\phi_j/\phi_j \wedge \psi \simeq (\phi_j + \psi)/\psi \leq \phi/\psi$ and, for the converse, if $a \in \phi(N) \setminus \psi(N)$, then $a = \sum_j a_j$ with $a_j \in \phi_j$ so it cannot be that all a_j lie in ψ.

Next note that $(\phi_j/\phi_j \wedge \psi) = \bigcup_i (\phi_j/\phi_j \wedge \psi_i)$: again, the inclusion $\supseteq$ is immediate because $\phi_j/\phi_j \wedge \psi_i$ is a quotient of $\phi_j/\phi_j \wedge \psi$ and, for the converse, if $a \in \phi_j(N) \setminus \psi(N)$, then there is some i such that $a \in \phi_j(N) \setminus \psi_i(N)$, as required. □

Remark 5.2.9. The RD condition is two-sided (2.4.11), so, applying duality $(\phi/\psi) \mapsto (D\psi/D\phi)$ (1.3.1) to a basis of this form for left modules, one obtains a basis for Zg_R of the dual form: $\big((u \mid xt + \exists y\,(x = yr \wedge ys = 0))\,/\,(u \mid xt)\big)$: but clearly this is no different from the basis in 5.2.8, which is, therefore, of a "self-dual" form. The same will be true of the alternative form of basis given for serial rings in 8.2.57.

Extracting the general point, we have the following.

Lemma 5.2.10. ([521, 3.4]) *Suppose that the ring R is such that there are two sets Φ, Ψ of pp conditions with the property that every pp condition in one free variable may be written as a finite sum of elements of Φ and also as a finite conjunction of elements of Ψ. Then a basis of Zg_R is given by those sets of the form $(\phi/\phi \wedge \psi)$ with $\phi \in \Phi$ and $\psi \in \Psi$.*

For, as in the proof of 5.2.8, one has $\big((\sum_i \phi_i)/\psi\big) = \bigcup_i (\phi_i/\phi_i \wedge \psi)$ and then $\big(\phi/(\bigwedge_j \psi_j)\big) = \bigcup_j (\phi/\phi \wedge \psi_j)$.

5.2.3 Other remarks

Proposition 5.2.11. *Suppose that R is a von Neumann regular ring (Section 2.3.4). Then Zg_R is the set, inj_R, of isomorphism types of indecomposable injective modules, endowed with the topology which has a basis of open sets of the form*

$$(xe = 0\,/\,x = 0) = \{E \in \mathrm{inj}_R : \mathrm{ann}_E(e) \neq 0\} = ((1-e)|x\,/\,x = 0)$$

as e ranges over idempotents of R. An alternative basis of open sets consists of those of the form

$$(x = x\,/\,e|x) = \{E \in \mathrm{inj}_R : \mathrm{ann}_E(1-e) \neq E\} = (x = x\,/\,x(1-e) = 0)$$

as e ranges over idempotents of R.

Proof. By the comment before 4.3.2 the points of the space are as described. Every finitely presented module is projective, so the functor category (R-mod, **Ab**) is equivalent to R-Mod (10.2.38). Take a basic open set (ϕ/ψ). Then, under this equivalence of categories, the functor $F_{D\psi/D\phi}$ corresponds to a finitely generated projective module which, if these conditions have just one free variable, is of the form eR for some $e^2 = e \in R$. By 10.3.8, for any module M, $\phi(M)/\psi(M) \simeq (F_{D\psi/D\phi}, M \otimes -) \simeq (eR \otimes -, M \otimes -)$ (by 10.2.39) $\simeq (eR, M) \simeq Me = \mathrm{ann}_M(1-e)$, from which the second, hence also the first, form of basis is derived.

Alternatively one may start from 2.3.24, which implies that the typical pp condition with one free variable has the form $xe = 0$ for some $e^2 = e \in R$, and then show that pairs $(xe = 0 / xe = 0 \wedge xf = 0)$ simplify to these forms (for example, see [495, 16.18]). □

Remark 5.2.12. It is easy to check that if $f = (f_i)_i : \bigoplus_i A_i \to B$ is a morphism in mod-R, then, with the notation of Section 5.1.3, $(f) = \bigcup_i (f_i)$, so if R is a Krull–Schmidt ring, then its Ziegler spectrum has a basis of the form (f) with f having indecomposable domain (and, if R is right coherent, f either a monomorphism or epimorphism by 5.2.13 below).

Lemma 5.2.13. *If $f = f''f'$ is a factorisation in* mod-R *with f' an epimorphism, then $(f) = (f') \cup (f'')$.*

Therefore, if R is right coherent, then the sets (f) with f either a monomorphism or epimorphism in mod-R *form a basis of open sets of* Zg_R.

Proof. Say $A \xrightarrow{f'} A' \xrightarrow{f''} B$. Let $N \in (f'')$, say $g : A' \to N$ does not factor through f''. Since f' is epi it follows directly that gf' cannot factor through f either, hence $N \in (f)$. Next, suppose $N \in (f')$, say $h : A \to N$ does not factor through f'; since f' is epi, $\ker(g) \not\geq \ker(f')$, so it cannot factor through $f = f''f'$. The fact that $(f) \subseteq (f') \cup (f'')$ is equally direct.

For the second statement consider the epi-mono factorisation of an arbitrary morphism in mod-R (the assumption of right coherence is needed for the image to be in mod-R). □

The example described before 8.2.45 is a non-discrete Ziegler spectrum with just three points and Cantor–Bendixson rank 1. Even the implications of having just one point in the spectrum are not clear – see 8.2.86.

5.3 Isolation, density and Cantor–Bendixson rank

In this section increasing use is made of the functor categories from Part II, indeed many results are most naturally stated making reference to relevant functors. Section 5.3.1 begins with a summary of the main points used, then continues to show the relation between isolation, respectively isolation by a minimal pair, and simple functors, resp. finitely presented simple functors. Some consequences for supports of Σ-pure-injectives are derived.

The isolation condition, that any isolated point (in a closed set) be isolated by a minimal pair, is the topic of Section 5.3.2. A number of consequences flow from satisfying this condition, the condition is known to hold in many circumstances, and there is no known example where it fails to hold. Results connecting isolation by a minimal pair and existence of almost split morphisms are in Section 5.3.3. The question of when the set of isolated points is dense in the spectrum is considered in Section 5.3.4, with some consequences for artin algebras.

In Section 5.3.5 the points corresponding to injective hulls of arbitrary simple functors, the neg-isolated pure-injectives, are identified. Also, elementary cogenerators are defined and existence of these is proved.

In Section 5.3.6 it is shown how the Cantor–Bendixson analysis of the Ziegler spectrum (which proceeds by removing isolated points) correlates with the analysis of the lattice of pp conditions (which proceeds by collapsing intervals of finite length).

Burke's full support topology is introduced in Section 5.3.7 and linked to type-definable subcategories and neg-isolation.

5.3.1 Isolated points and minimal pairs

Points of the Ziegler spectrum, that is, indecomposable pure-injective modules, are in bijection with indecomposable injective functors. Isolation of a point implies that the corresponding injective functor is the injective hull of a simple functor (5.3.1), though not conversely. Isolation in a closed subset implies the same condition in a localisation of the functor category (5.3.1). Isolation by a minimal pair (absolute or relative to a closed subset) is equivalent to the simple functor being finitely presented (5.3.5) and the functor is computed directly from an isolating pair (5.3.2). For finitely generated pp-types, isolation of the hull is equivalent to isolation by a minimal pair (5.3.12).

Any finitely presented simple injective module is isolated by a minimal pair (5.3.9).

Indecomposable pure-injectives which share a minimal pair are isomorphic (5.3.6). If pp-types have isomorphic hulls and one has a minimal pair, then so does the other (5.3.11).

If the lattice of pp conditions relative to a closed set is of finite length, then that closed set is finite, discrete, and consists of modules of finite endolength (5.3.8). Any Σ-pure-injective module of infinite endolength has a non-isolated point in its support (5.3.14).

A finitely presented module with indecomposable hull which is isolated by a minimal pair embeds purely into any module in which that pair is open (5.3.15).

In this section considerable use is made of the functor category and localisation (Chapters 10–12). In order to be able to read the proofs in the remainder of this chapter one will need some knowledge of the contents of those chapters. Proofs which do not use the functorial methods, which rather use model-theoretic methods, may be found in the literature (for example, in [726], [495], [274], [275]). In particular, localisation corresponds to working relative to closed subsets of the spectrum; essentially, to working relative to a complete theory of modules.

Here is a brief summary of what is used: for more detail see Section 12.3 in particular. Let X be a closed subset of Zg_R and let $\mathcal{X}$ be the corresponding (5.1.6) definable subcategory. Let $\tau = \tau_{DX}$ be the hereditary torsion theory on $(R\text{-mod}, \mathbf{Ab})$ cogenerated (11.1.1) by the injective functors (12.1.6) $N \otimes -$ for $N \in X$; equivalently τ is the hereditary torsion theory of finite type maximal, in the sense of inclusion of torsion classes, such that all $N \otimes -$ are torsionfree, where N runs over a dense subset of X, see 12.3.7. Let $Q_\tau : (R\text{-mod}, \mathbf{Ab}) \longrightarrow (R\text{-mod}, \mathbf{Ab})_\tau$ be the corresponding (11.1.5) localisation functor; we also write F_τ, and in this case, F_{DX}, for $Q_\tau F$. Then $M \otimes -$ is τ-torsionfree iff $\mathrm{supp}(M) \subseteq X$ (see 12.3.2) and, in that case, $Q_\tau(M \otimes -) \simeq (M \otimes -)$ (11.1.5) is an absolutely pure object of $(R\text{-mod}, \mathbf{Ab})_\tau$ (12.3.3). In particular the indecomposable injective objects of the localised functor category $(R\text{-mod}, \mathbf{Ab})_\tau$ are exactly those isomorphic to some $N \otimes -$ with $N \in X$.

Given a pp-pair, ϕ/ψ, the corresponding functor, $F_{D\psi/D\phi}$, in $(R\text{-mod}, \mathbf{Ab})^{\mathrm{fp}}$ (10.2.30 and Section 10.3) is τ-torsion, hence, 11.1.5, becomes zero in the localised category iff ϕ/ψ is closed on $\mathcal{X}$ (12.3.17). Equivalently, by definition of the topology, this is so iff that pair is closed on every member of (any dense subset of) X. Indeed, for any pair ϕ/ψ the lattice $[\phi, \psi]_{\mathcal{X}}$ of $\mathcal{X}$-equivalence classes (see after 3.4.13) of pp conditions in the interval $[\phi, \psi]$ is isomorphic to the lattice of finitely generated (= finitely presented, 12.3.21) subfunctors of $Q_\tau(F_{D\psi/D\psi})$, see 12.3.18. In particular, $\mathrm{pp}^n(X) \simeq \mathrm{Latt}^{\mathrm{f}}\big((R^n \otimes -)_\tau\big)$: the lattice of pp conditions modulo equivalence on $\mathcal{X}$ (we write $\mathrm{pp}^n(X) = \mathrm{pp}^n(\mathcal{X})$) is isomorphic to the lattice of finitely generated (= finitely presented) subfunctors of the localised nth power of the forgetful functor.

As is the case for $(R\text{-mod}, \mathbf{Ab})$, the subcategory of finitely presented functors in $(R\text{-mod}, \mathbf{Ab})_\tau$ is naturally isomorphic, 12.3.20, to the localised category of pp-imaginaries $(\mathbb{L}_R^{\mathrm{eq}+})_X$ (this is defined near the end of Section 3.2.2). In particular every finitely presented functor in the localised category is a localisation of a finitely presented functor (11.1.33), that is, is isomorphic to one of the form $Q_\tau(F_{D\psi/D\phi}) \in ((R\text{-mod}, \mathbf{Ab})_\tau)^{\mathrm{fp}} \simeq ((R\text{-mod}, \mathbf{Ab})^{\mathrm{fp}})_\tau$.

At least initially, proofs will be given in rather more detail than strictly necessary in order to help the reader find the relevant ideas and results from those later chapters. Model-theoretic versions of some of the results and proofs are in [495, §§9.3, 9.4].

Theorem 5.3.1. *If a point $N \in \mathrm{Zg}_R$ is isolated, then $N \otimes -$ is the injective hull of a simple functor. More generally, if X is a closed subset of Zg_R and if $N \in X$*

is an isolated point of X, then $N \otimes -$ is the injective hull of a (finitely generated) functor F such that the localisation, F_{DX}, of F at the torsion theory corresponding to X is simple. In this case $N \otimes -$, regarded as an object of the localised category $(R\text{-mod}, \mathbf{Ab})_{DX}$, is the injective hull of a simple object.

Proof. Say $(\phi/\psi) \cap X = \{N\}$. By 12.3.15, N is the only point of X such that $((F_{D\psi/D\phi})_{DN}, N \otimes -) \neq 0$. By the adjunction of 11.1.5 and the fact that $(N \otimes -) \simeq (N \otimes -)_{DX}$ (by definition of τ_{DX} and, for example, 12.3.3) N is the only point of X such that $(F_{D\psi/D\phi}, N \otimes -) \neq 0$. Choose $f : F_{D\psi/D\phi} \to (N \otimes -)$ non-zero.

Choose a maximal subfunctor, G say, of $F_{D\psi/D\phi}$ which contains $\ker(f)$ and which is not τ_{DX}-dense in $F_{D\psi/D\phi}$; since $F_{D\psi/D\phi}$ is finitely generated and since τ_{DX} is of finite type this is possible by confinality, 11.1.14, of finitely generated τ_{DX}-dense subobjects of $F_{D\psi/D\phi}$. Then G is τ_{DX}-closed in $F_{D\psi/D\phi}$, that is, $(F_{D\psi/D\phi})/G$ is τ_{DX}-torsionfree since the torsion class is closed under extension. Since the localisation functor is exact (11.1.5) and every G' with $G < G' < F_{D\psi/D\phi}$ is such that $(F_{D\psi/D\phi})/G'$ is torsion, the localisation $\big((F_{D\psi/D\phi})/G\big)_{DX}$ is a simple object of the quotient category $(R\text{-mod}, \mathbf{Ab})_{DX}$.

The injective hull of this simple object S is an indecomposable injective object of $(R\text{-mod}, \mathbf{Ab})_{DX}$, hence has the form $N' \otimes -$ for some $N' \in X$ (12.3.3) and, since S is (by exactness of localisation) a factor of $(F_{D\psi/D\phi})_{DX}$, by the first comments we must have $N' = N$. By construction, the restriction of the embedding of $S = \big((F_{D\psi/D\phi})/G\big)_{DX}$ to $(F_{D\psi/D\phi})/G$ is monic, so $N \otimes -$ is the injective hull of an object as described. □

The module N, necessarily pure-injective, is said to be neg-isolated if $N \otimes -$ is the injective hull of a simple functor. Conditions equivalent to this are given at 5.3.47. Neg-isolated pure-injectives are discussed in Section 5.3.5; they are the isolated points of a finer topology, for which see Section 5.3.7.

If N is finitely presented and $N \otimes -$ is the injective hull of a simple functor, then, by 10.2.36, that functor must be finitely presented and hence, 5.3.3, N will be isolated. In general, however, if N is such that $N \otimes -$ is the injective hull of a simple functor, then N need not be an isolated point of Zg_R. For example, there are rings, such as the first Weyl algebra, where the Ziegler spectrum has no isolated points (8.2.34), but certainly there are simple functors in the functor category: any non-zero finitely generated functor has, by Zorn's Lemma, a maximal proper subfunctor, hence a simple quotient. On the other hand, there are rings, see 5.3.43, where the isolated points are exactly those corresponding to injective hulls of simple functors. It is always the case that the injective hull of a *finitely presented* simple functor is isolated (5.3.3 below).

If X is a closed subset of Zg_R, then say that a pp-pair ϕ/ψ is an X**-minimal pair** if for some (equivalently, every) module M with $\mathrm{supp}(M) = X$ it is an M-minimal pair. Recall from Section 3.2.1 that this means that $\phi(M) > \psi(M)$ and there is no proper pp-definable subgroup of M strictly between $\phi(M)$ and $\psi(M)$. From this it follows (3.4.11) that for every M' with $\mathrm{supp}(M') \subseteq X$, equivalently with $M' \in \mathcal{X}$, where $\mathcal{X}$ is the definable subcategory corresponding to X, either $\phi(M') = \psi(M')$ or ϕ/ψ is an M'-minimal pair. If $X = \mathrm{Zg}_R$ (so $\mathcal{X} = \text{Mod-}R$) we simply refer to a **minimal pair**. We also use terminology such as $\mathcal{X}$**-minimal pair**.

Say that a pp-type p **contains** or **has** a (−)-**minimal pair** if there is a (−)-minimal pair ϕ/ψ with $\phi \in p$ and $\psi \notin p$.

Theorem 5.3.2. ([726, 8.12]) *If X is a closed subset of Zg_R and if ϕ/ψ is an X-minimal pair, then $(\phi/\psi) \cap X = \{N\}$ for some N, which is, therefore, an isolated point of X. Furthermore, the functor F in the statement of 5.3.1 may be taken to be $F_{D\psi}/F_{D\phi}$ so $N \otimes -$ is the injective hull of the simple finitely presented functor $(F_{D\psi}/F_{D\phi})_{DX}$.*

Proof. By 12.3.18 the hypothesis that ϕ/ψ is an X-minimal pair is equivalent to the localisation $(F_{D\psi}/F_{D\phi})_{DX}$ of $F_{D\psi}/F_{D\phi}$ being a simple finitely presented functor of $(R\text{-mod}, \mathbf{Ab})_{DX}$. Here $_{DX}$ denotes localisation at the hereditary torsion theory, τ_{DX}, of finite type on $(R\text{-mod}, \mathbf{Ab})$ corresponding to the closed subset X (as described in the introduction to this section).

Also, by 12.3.12, the points of X correspond bijectively to the (isomorphism classes of) indecomposable injectives of $(R\text{-mod}, \mathbf{Ab})_{DX}$.

Since $S = (F_{D\psi}/F_{D\phi})_{DX}$ is simple, there is a unique indecomposable injective E of $(R\text{-mod}, \mathbf{Ab})_{DX}$ such that $(S, E) \neq 0$ (namely the injective hull of S), that is, by 12.1.9, there is a unique point N of X such that $(S, N \otimes -) \neq 0$. If N' is any point of X and if $(F_{D\psi}/F_{D\phi}, N' \otimes -) \neq 0$, then also $(S, N' \otimes -) \neq 0$ by the adjunction in 11.1.5, as in the proof of 5.3.1. We conclude that N is the unique point, N', of X such that $(F_{D\psi}/F_{D\phi}, N' \otimes -) \neq 0$, that is, (10.3.8), such that $\phi(N)/\psi(N) \neq 0$, as required for the first statement, and the second statement also has been proved. □

Corollary 5.3.3. *If N is an indecomposable pure-injective such that $N \otimes -$ is the injective hull of a simple finitely presented functor in $(R\text{-mod}, \mathbf{Ab})$, then N is an isolated point of Zg_R. More generally, if X is a closed subset of Zg_R and if $N \otimes -$ is the injective hull of a τ_{DX}-torsionfree functor F, the image of which in the localised category $(R\text{-mod}, \mathbf{Ab})_{DX}$ is simple and finitely presented, then $N \in X$ and N is an isolated point of X with $(F) \cap X = \{N\}$.*

Proof. For the definition of the basic open set (F) see Section 12.1.3. Since F is τ_{DX}-torsionfree so is its injective hull $N \otimes -$ (since F is essential in $N \otimes -$ and since every subobject of a torsionfree object is torsionfree), so, by 12.3.5, $N \in X$. The remainder follows by (the proof of) 5.3.2. □

In the situation of 5.3.2, respectively 5.3.3, say that N is **isolated in X by the X-minimal pair** ϕ/ψ, resp. by the functor F. In the case that $X = \mathrm{Zg}_R$, say simply that N is **isolated by a minimal pair**. It is an open question whether an isolated point must be isolated by a minimal pair. That is, if N is isolated, hence (5.3.1) $N \otimes -$ is the injective hull of a simple functor, is that simple functor finitely presented? There are many cases where this is known to be so (see the next section). For example, the isolated points of the Ziegler spectrum of a PI Dedekind domain were described in Section 5.2.1 and clearly are isolated by minimal pairs. We will see in 5.3.37 that the isolated points of Zg_R for R an artin algebra are exactly the finite-length points, and the existence of isolating minimal pairs is just the existence of almost split sequences. But in general this is open.

Question 5.3.4. Is every isolated point of Zg_R isolated by a minimal pair? More generally, if X is a closed subset of the Ziegler spectrum is every isolated point of X isolated by an X-minimal pair?

Corollary 5.3.5. *A point $N \in \mathrm{Zg}_R$ is isolated by a minimal pair iff $N \otimes -$ is the injective hull of a finitely presented simple functor in* (R-mod, **Ab**). *More generally, if X is a closed subset of Zg_R, then $N \in X$ is isolated in X by an X-minimal pair iff $N \otimes -$ is the injective hull of a finitely presented functor F such that the localisation, F_{DX}, of F at the torsion theory corresponding to X is simple.*

Corollary 5.3.6. (see [495, 9.3]) *Suppose that X is a closed subset of Zg_R, that $N, N' \in X$ and that ϕ/ψ is an X-minimal pp-pair which is open in both N and N'. Then $N \simeq N'$.*

Since $N, N' \in (\phi/\psi) \cap X$ this follows from 5.3.2.

Lemma 5.3.7. *If N is an isolated point in the closed subset X of Zg_R and is also isolated by a Y-minimal pair within some closed subset Y of X, then N is isolated in X by an X-minimal pair.*

Proof. Suppose that N is isolated in X by (ϕ/ψ) and in Y by the Y-minimal pair (ϕ'/ψ'). Choose by 5.1.20 a pp-pair ϕ''/ψ'' with $N \in (\phi''/\psi'') \subseteq (\phi/\psi) \cap (\phi'/\psi')$. Then this pair isolates N in X and, as remarked after the proof of 5.1.20, the functor $F_{D\psi''/D\phi''}$ may be taken to be a subquotient of $F_{D\psi'/D\phi'}$. Since the localisation of the latter at τ_{DY} is simple, so is the localisation of the former. But since (ϕ''/ψ'') isolates N in X, that is, $\big((F_{D\psi''/D\phi''})_{DX}, N' \otimes -\big) = 0$ for every $N' \in X \setminus \{N\}$, all subquotients of the functor $(F_{D\psi''/D\phi''})_{DX}$ must be τ_{DY}-torsionfree, hence that functor must already be simple, as required. □

Observe that an indecomposable pure-injective N is isolated by a minimal pair within some closed subset of the Ziegler spectrum iff it is isolated by a minimal pair within its own closure; equivalently, there is a pp-pair $\phi > \psi$ such that $\phi(N) > \psi(N)$ and such that there is no pp-definable subgroup of N properly between these (cf. 5.3.16). By 5.3.5 it is equivalent that the localisation $(R\text{-mod}, \mathbf{Ab})_{DN}$ contain some finitely presented simple object. Such modules are reflexive in the sense of Section 5.4, see 5.4.12.

If X is a closed subset of Zg_R, we extend the notation from near the end of Section 3.4.1 by writing $\mathrm{pp}(X)$ (resp. $\mathrm{pp}^n(X)$) for the lattice $\mathrm{pp}(\mathcal{X})$ ($\mathrm{pp}^n(\mathcal{X})$) of pp conditions (with n free variables) modulo X, where $\mathcal{X}$ is the definable subcategory of Mod-R corresponding to X. If M is any module with $\mathrm{supp}(M) = X$, then $\mathrm{pp}(X)$ ($\mathrm{pp}^n(X)$) is isomorphic to the lattice of pp-definable subgroups of M (subgroups of M^n pp-definable in M).

Corollary 5.3.8. *Suppose that $X \subseteq \mathrm{Zg}_R$ is closed. If $\mathrm{pp}(X)$ is of finite length (say t), then X is a finite discrete set, of (no more than t) modules of finite endolength.*

Proof. Take a maximal chain in $\mathrm{pp}(X)$. If $N \in X$, then at least one of the simple intervals of this chain is an N-minimal pair; then apply 5.3.6 and 5.3.2. □

Example 5.3.9. ([521, 3.3]) Suppose that E_R is an indecomposable injective which is simple and finitely presented. Then E is isolated by a minimal pair in Zg_R.

To see this, let $a \in E$ be non-zero and let $I = \mathrm{ann}_R(a)$. Since E is finitely presented, I is a finitely generated right ideal. Let ϕ be the pp condition $xI = 0$ (Section 1.1.4) and let ψ be the condition $x = 0$. If N is any (indecomposable pure-injective) module such that $\phi(N) > \psi(N)$, that is, such that $\mathrm{Hom}(E, N) \neq 0$, then, since E is simple, E embeds in N and so, being injective, E is a direct summand of N.

Thus $(\phi/\psi) = \{E\}$.

Proposition 5.3.10. *Let N be an indecomposable pure-injective and set $S =$ $\mathrm{End}(N)$. Let ϕ/ψ be a pp-pair. Then ϕ/ψ is an N-minimal pair iff $\phi(N)/\psi(N)$ is a simple S-module.*

Proof. Since pp-definable subgroups are S-modules, if $\phi(N)/\psi(N)$ is a simple S-module, then ϕ/ψ is an N-minimal pair. Suppose, conversely, that ϕ/ψ is an N-minimal pair: we show that $\phi(N)/\psi(N)$ is a simple S-module.

Choose $\overline{a} \in \phi(N) \setminus \psi(N)$: we show that $S\overline{a} + \psi(N) = \phi(N)$. So let $\overline{b} \in \phi(N)$. Set $p = \mathrm{pp}^N(\overline{a})$. If $\phi' \in p$ and $\phi \geq \phi'$, then $\phi'(N) + \psi(N) = \phi(N)$ because

$\phi' \not\leq \psi$ (since $\psi \notin p$) and the pair ϕ/ψ is minimal. Therefore the set

$$\Phi(\overline{u}, \overline{v}) = \{\phi'(\overline{u}) \wedge \psi(\overline{v}) \wedge \overline{u} + \overline{v} = \overline{b} : \phi' \in p,\ \phi' \leq \phi\}$$

of pp conditions with parameter $\overline{b}$ is finitely satisfied in N. Since N is pure-injective (= algebraically compact) there is, by 4.2.1, a solution, $(\overline{a}', \overline{c})$, for Φ in N. So $\mathrm{pp}^N(\overline{a}') \supseteq \mathrm{pp}^N(\overline{a})$, $\overline{c} \in \psi(N)$ and $\overline{a}' + \overline{c} = \overline{b}$. By 4.3.9 there is $s \in S$ with $s\overline{a} = \overline{a}'$. Thus $\overline{b} \in S\overline{a} + \psi(N)$, as claimed. Since $\overline{a}$ was arbitrary in $\phi(N) \setminus \psi(N)$ the quotient $\phi(N)/\psi(N)$ is indeed a simple S-module.

(Let me also outline an alternative, functorial, proof (not really different from the proof above). First note that $(F_{D\psi/D\phi}, N \otimes -) \simeq \phi(N)/\psi(N)$ (10.3.8). Also, the localisation of $F_{D\psi/D\phi}$ at the finite-type torsion theory cogenerated by $N \otimes -$ is a simple object of the localised category: if $\psi \leq \phi' \leq \phi$, then evaluation on N identifies ϕ' with either ψ or ϕ, that is, by the above formula, either $F_{D\psi/D\phi'}$ or $F_{D\phi'/D\phi}$ is torsion. Moreover, $((F_{D\psi/D\phi})_{DN}, N \otimes -) \simeq (F_{D\psi/D\phi}, N \otimes -) \neq 0$ (by the adjunction in 11.1.5), so, if f is any morphism from $(F_{D\psi/D\phi})_{DN}$ to $N \otimes -$, then the image of f is just the socle, T say, of $N \otimes -$. Note that the endomorphism ring of T is S/J. Also every morphism from $(F_{D\psi/D\phi})_{DN}$ to $N \otimes -$ is f followed by an endomorphism of T. Thus $\phi(N)/\psi(N) \simeq ((F_{D\psi/D\phi})_{DN}, N \otimes -)$ is a cyclic module over S/J, as required.) □

Proposition 5.3.11. ([726, 8.10], see [495, 9.12]) *Suppose that X is a closed subset of* Zg_R *and that $N \in X$. Let p and q be pp-types with $H(p) \simeq N \simeq H(q)$. If p has an X-minimal pair, then so does q.*

Proof. Suppose $\overline{a}$ from N realises p and $\overline{b}$ from N realises q. By 4.3.34 and 12.2.5 the morphism $(\overline{a} \otimes -) : (R^{l(\overline{a})} \otimes -) \to (N \otimes -)$ has kernel F_{Dp}, $\overline{b} \otimes -$ has kernel F_{Dq} and $N \otimes -$ is the (indecomposable) injective hull of both $\mathrm{im}(\overline{a} \otimes -)$ and $\mathrm{im}(\overline{b} \otimes -)$. By assumption (and cf. the proof of 5.3.2) there is $\psi \notin p$ such that the localisation of $(F_{D\psi} + F_{Dp})/F_{Dp}$ $(\leq N \otimes -)$ at τ_{DX} is a simple finitely presented object of $(R\text{-mod}, \mathbf{Ab})_{DX}$.

The intersection of $(F_{D\psi} + F_{Dp})/F_{Dp}$ with $\mathrm{im}(\overline{b} \otimes -) \simeq (R^{l(\overline{b})} \otimes -)/F_{Dq}$ is non-zero, so contains a non-zero finitely generated subobject, necessarily (by 10.2.16) of the form $(F_{D\psi'} + F_{Dq})/F_{Dq}$ for some pp condition ψ' with $\psi' \notin q$. The localisation $\big((F_{D\psi'} + F_{Dq})/F_{Dpq}\big)_{DX}$ is a subfunctor of, hence equal to, $\big((F_{D\psi} + F_{Dp})/F_{Dp}\big)_{DX}$. We have $(F_{D\psi'} + F_{Dq})/F_{Dq} \simeq F_{D\psi'}/(F_{D\psi'} \cap F_{Dq}) = F_{D\psi'}/(F_{D\psi'} \cap \sum_{\phi' \in q} F_{D\phi'})$ (by E.1.3) $= \varinjlim_{\phi' \in q} F_{D\psi'}/(F_{D\psi'} \cap F_{D\phi'})$. Since localisation is a left adjoint (11.1.5), it commutes with direct limits, so $\big((F_{D\psi'} + F_{Dq})/F_{Dq}\big)_{DX} \simeq \varinjlim_{\phi' \in q} \big(F_{D\psi'}/(F_{D\psi'} \cap F_{D\phi'})\big)_{DX}$.

Since the simple functor on the left-hand side is finitely presented, we deduce $\big((F_{D\psi'} + F_{Dq})/F_{Dq}\big)_{DX} \simeq \big(F_{D\psi'}/(F_{D\psi'} \cap F_{D\phi'})\big)_{DX}$ for some $\phi' \in q$. Therefore

$D\psi'/(D\psi' \wedge D\phi')$ is a DX-minimal pair, hence (12.3.18, 1.3.1) $(\psi' + \phi')/\psi'$, equivalently $\phi'/(\phi' \wedge \psi')$ is an X-minimal pair with $\phi \in q$ and $\phi' \wedge \psi' \notin q$, as required. □

Of course, if $N = H(p)$ and p has an X-minimal pair then that pair is open in N, hence isolates N in X.

Proposition 5.3.12. ([495, 9.26]) *Let $N = H(p) \in \mathrm{Zg}_R$ and suppose that p is a finitely generated pp-type.*

Then N is an isolated point of Zg_R iff p contains a minimal pair.

More generally if $N \in X$, with X a closed subset of Zg_R, then N is an isolated point of X iff p contains an X-minimal pair, which then isolates N in X. For this one needs only the weaker hypothesis on p that there is ϕ such that ϕ generates p modulo X in the sense that $p = \{\phi' : \phi'_X \geq \phi_X\}$, where subscript $_X$ denotes image in pp(X).

Proof. One direction is by 5.3.2. For the converse, suppose that N is an isolated point of X. Let F be as in 5.3.1. In particular $S = F_{DX}$ is a simple object of the localised category and $N \otimes -$ is the injective hull of S.

Let ϕ generate p modulo X as stated. This, by 12.3.16, is exactly the assertion that $(F_{Dp})_{DX} = (F_{D\phi})_{DX}$. Therefore $(F_{Dp})_{DX}$ is a finitely generated object of $(R\text{-mod}, \mathbf{Ab})_{DX}$, so (by 11.1.33 and E.1.16) $\big((R^n, -)/F_{Dp}\big)_{DX} = (R^n, -)_{DX}/(F_{Dp})_{DX}$ is, by exactness of localisation (11.1.5), finitely presented. By 12.2.5 $N \otimes -$ is the injective hull of this functor so the coherent (11.1.34) functor $((R^n, -)/F_{Dp})_{DX}$ contains S, which is, therefore, finitely presented. By 5.3.5 (and 11.1.33) N is isolated in X by a minimal pair, as required. □

With 1.2.6 this give the next result.

Corollary 5.3.13. ([276, 2.5]) *If N is an isolated point of Zg_R and is the pure-injective hull of a finitely presented module, then N is isolated by a minimal pair.*

In particular every finitely presented isolated point of Zg_R is isolated by a minimal pair.

For more on these modules see Section 5.3.3, in particular the characterisation 5.3.31.

In fact in 5.3.12 and 5.3.13 the condition that N be isolated can be weakened to $N \otimes -$ being the injective hull of a simple functor (in the relevant localised category), that is, to N being neg-isolated. For, if $a \in N, a \neq 0$ has finitely generated pp-type p, then the finitely presented, therefore coherent, functor $(R \otimes -)/F_{Dp}$ embeds, via $a \otimes -$, in $N \otimes -$ (12.2.5). By assumption this contains a simple, hence finitely presented simple, functor, as required.

Corollary 5.3.14. (essentially [478, 6.7], see [495, 6.28]) *Suppose that M is a Σ-pure-injective module of infinite endolength. Then there is a non-isolated point of* supp(M)*; that is, there is a point of* supp(M) *which is not isolated in the relative topology on that set.*

Proof. In 4.3.52, and working in the lattice pp(M), take Φ to be $\{x = x\}$ and Ψ to consist of all the pp conditions ψ such that the interval $[\psi, 0]_M$ in the lattice pp(M) has finite length. By assumption $\Phi \cap \Psi = \emptyset$. Let p be an irreducible pp-type constructed as in 4.3.52 and set $N = H(p) \in \text{supp}(M)$. By 5.3.12 it will be enough to show that p contains no M-minimal pair. If it did contain such a pair, then clearly there would be $\phi \in p$ and $\eta \notin p$ with $[\phi, \eta]_M$ a simple (that is, two-point) interval. Since $\eta \notin p$ there is, by construction of p (that is, maximality, see 4.3.52), $\phi' \in p$ such that $\phi' \wedge \eta \in \Psi$ and hence such that $\phi \wedge \phi' \wedge \eta \in \Psi$. But then the interval $[(\phi \wedge \phi') \wedge \eta, 0]_M$ is of finite length and also (by modularity of the lattice) $[\phi \wedge \phi', \eta \wedge \phi']_M$ is a two-point interval, so $[\phi \wedge \phi', 0]_M$ would also be of finite length, hence $\phi \wedge \phi' \in \Psi \cap p$, a contradiction.

That was the original proof. The functorial version is as follows. Consider the localisation $(R \otimes -)_{DM}$ at the finite type torsion theory on (R-mod, **Ab**) corresponding to M. Since M is Σ-pure-injective, this object is noetherian, by 4.4.5 and duality, 12.3.18. Since pp(M) has infinite length, $(R \otimes -)_{DM}$ has infinite length. It follows easily that there is a uniform quotient of $(R \otimes -)_{DM}$ which has no simple subobject. That is what the construction above gives: take the quotient of $(R \otimes -)_{DM}$ by a subfunctor which is maximal with respect to containing no finitely generated subfunctor of finite colength; there is a similar argument in the proof of 5.3.22. So (E.1.7) there is an indecomposable injective which is not the injective hull of a simple functor, therefore, (5.3.1), there is an indecomposable pure-injective which is not isolated. □

Proposition 5.3.15. *Let R be any ring and let X be a closed subset of* Zg_R. *Suppose that A is a finitely presented R-module with* supp$(A) \subseteq X$ *and such that the pure-injective hull, $H(A)$, is indecomposable and is isolated in X by an X-minimal pair, ϕ/ψ. Let M be any module with* supp$(M) \subseteq X$. *If $\phi(M)/\psi(M) \neq 0$, then there is a pure embedding of A into M.*

Proof. We work in the localised category of functors $(R\text{-mod}, \mathbf{Ab})_{DX}$. For notational simplicity we drop the subscripts indicating localisation but all functors should be understood as belonging to this category. Thus, for example, by $A \otimes -$ we mean the image of the usual tensor functor in this category.

Let $F = (F_{D\psi/D\phi})_{DX}$. The assumption that (ϕ/ψ) is an X-minimal pair is exactly the assumption that F is a simple functor, and the assumption that $H(A)$ is isolated in X by this pair is exactly the assumption that $H(A) \otimes - (= (H(A) \otimes -)_{DX}$

since supp(A) $\subseteq X$) is the injective hull of F (5.3.5). From $\phi(M)/\psi(M) \neq 0$ it follows that there is a non-zero, hence monic, map $f : F \to (M \otimes -)$ $(= (M \otimes -)_{DX})$. Similarly (or since $A \otimes -$ is essential in $H(A) \otimes -$, 4.3.14), there is an inclusion $j : F \to (A \otimes -)$. By 12.3.3, $M \otimes -$ is absolutely pure in the functor category $(R\text{-mod}, \mathbf{Ab})_{DX}$ and, since A is finitely presented, $A \otimes -$ also is finitely presented (10.2.36), in the original category, hence also in the localisation. Therefore (2.3.1 and E.1.16) f factors through j, say via $(h \otimes -) : (A \otimes -) \to (M \otimes -)$, where $h : A \to M$ (12.1.3). Since the simple functor F is essential in $A \otimes -$ and f is monic, ker$(h \otimes -) = 0$. Hence, 12.1.6, h is a pure embedding of A into M, as required. □

5.3.2 The isolation condition

The isolation condition is the assertion that every isolated point is isolated by a minimal pair. It is sufficient to test this on the closures of points (5.3.16). Assumptions which imply that the isolation condition holds include: existence of m-dimension (5.3.17); countability of the ring (5.3.21); non-existence of superdecomposable pure-injectives (5.3.24); distributivity of the lattice of pp conditions (5.3.28).

Under a countability assumption, existence of an interval with no minimal pair implies that there are continuum many indecomposable pure-injectives (5.3.18) and conversely (5.3.19).

If the isolation condition holds, then every closed point is of finite endolength (5.3.23). If the isolation condition holds and the Ziegler spectrum is a discrete space, then the ring is of finite representation type; the converse holds unconditionally (5.3.26).

If a point of the Ziegler spectum belongs to an open set of the form (ϕ/ψ), where ϕ/ψ is a minimal pair, then, by 5.3.2, that point is isolated, by that pair, but it is not known whether the converse is true. Under a variety of hypotheses it is known that an isolated point is isolated by a minimal pair but determining whether or not this is true in complete generality may well be hard (for instance, see 8.2.85)

So we adopt this condition as a useful hypothesis which holds in many cases. The **isolation condition** on R is that for every closed subset, X, of Zg_R and every isolated point, N, of X, meaning $N \in X$ and N is isolated in the relative topology on X, there is a pp-pair ϕ/ψ which is X-minimal such that $(\phi/\psi) \cap X = \{N\}$. Say that the isolation condition[10] **holds for** X if one has this condition for all

[10] In [495, §10.4] I used the rather random notation ($\wedge$) for this condition.

closed subsets of X. In terms of functors, the isolation condition is, by 5.3.1 and 5.3.2, that if the injective hull of a simple functor is isolated (in a closed set), then that simple functor is finitely presented (in the corresponding localisation).

If X is a closed subset of Zg_R, then, cf. the definition of $[\phi, \psi]_M$ near the beginning of the previous section and in Section 3.4.1, we let $[\phi, \psi]_X$ denote the quotient of the interval $[\phi, \psi]$ in the lattice of pp conditions obtained by identifying ϕ' and ψ' iff $\phi'(M) = \psi'(M)$ for some, equivalently all, M with $\mathrm{supp}(M) = X$.

Proposition 5.3.16. *Let X be a closed subset of Zg_R. Then the following are equivalent:*

(i) the isolation condition holds for X;
(ii) each $N \in X$ which is isolated in some closed subset of X is isolated by a minimal pair in its closure, supp(N)*;*
(iii) each $N \in X$ which is isolated in some closed subset of X is isolated by a Y-minimal pair in some closed subset, Y, of X.

Proof. Let $X_N = \mathrm{supp}(N)$ denote the Ziegler-closure of N.

(i)⇒(iii) is by the definition.

(iii)⇒(ii) is immediate: if $\{N\} = (\phi/\psi) \cap Y$ and ϕ/ψ is Y-minimal, then, since $\mathrm{supp}(N) \subseteq Y$, ϕ/ψ is supp(N)-minimal.

(ii)⇒(i) Suppose $\{N\} = (\phi/\psi) \cap Y$, where Y is some closed subset of X. By assumption $\{N\} = (\phi'/\psi') \cap \mathrm{supp}(N)$ for some supp(N)-minimal pair ψ'/ψ'. Thus, if $\overline{a}' \in \phi'(N) \setminus \psi'(N)$, then there is the supp(N)-minimal pair ϕ'/ψ' in $\mathrm{pp}^N(\overline{a}')$. Hence, if $\overline{a} \in \phi(N) \setminus \psi(N)$, there is, by 5.3.11, a supp(N)-minimal pair, ϕ''/ψ'' say, in $\mathrm{pp}^N(\overline{a})$. It follows, by modularity of the lattice $\mathrm{pp}^{l(\overline{a})}(N)$, that $\phi \wedge \phi'' + \psi > \phi \wedge \psi'' + \psi$ is either a supp(N)-minimal pair or else $\phi \wedge \phi'' + \psi = \phi \wedge \psi'' + \psi$. The latter cannot be so: for, $\overline{a} \in (\phi \wedge \phi'')(N) = \phi(N) \cap \phi''(N)$ and $\overline{a}$ is in neither $(\phi \wedge \psi'')(N)$ nor $\psi(N)$, so, by 4.3.49, there is $\phi_0 \in \mathrm{pp}^N(\overline{a})$ such that $\overline{a} \notin (\phi_0 \wedge \phi \wedge \psi'' + \phi_0 \wedge \psi)(N)$. It follows that the pair $\phi_0 \wedge \phi \wedge \phi'' + \phi_0 \wedge \psi > \phi_0 \wedge \phi \wedge \psi'' + \phi_0 \wedge \psi$ is supp(N)-minimal.

We claim that this pair is also Y-minimal. For, if $N' \in Y \setminus \{N\}$, then, by hypothesis, $\phi(N') = \psi(N')$, so it follows that $(\phi_0 \wedge \phi \wedge \phi'' + \phi_0 \wedge \psi)(N') = (\phi_0 \wedge \phi \wedge \psi'' + \phi_0 \wedge \psi)(N')$, for both terms lie between $\phi_0 \wedge \phi$ and $\phi_0 \wedge \psi$ and these are equal in N'. Therefore N is the only point of Y on which this pair $(\phi_0 \wedge \phi \wedge \phi'' + \phi_0 \wedge \psi)/(\phi_0 \wedge \phi \wedge \psi'' + \phi_0 \wedge \psi)$ is open, so, since there is no θ with $\phi_0 \wedge \phi \wedge \phi'' + \phi_0 \wedge \psi > \theta > \phi_0 \wedge \phi \wedge \psi'' + \phi_0 \wedge \psi$ in N, there is no θ with $\phi_0 \wedge \phi \wedge \phi'' + \phi_0 \wedge \psi > \theta > \phi_0 \wedge \phi \wedge \psi'' + \phi_0 \wedge \psi$ on *any* point with support in Y. Thus we have a Y-minimal pair isolating N in Y, completing the proof.

The proof just given gives a good idea of the kind of argument, using manipulations in the lattice of pp conditions, frequently used in Ziegler's original proofs in

[726]. For contrast we give a functorial proof which replaces these manipulations by ones which involve subobjects and quotient objects. These are not esssentially different but they make more use of algebra in the "structural" sense.

With notation as at the beginning of the first proof, let $S' = (F_{D\psi'/D\phi'})_{DN}$ be the localisation of $F_{D\psi'/D\phi'}$ at the torsion theory, τ_{DN}, corresponding to the closed subset supp(N). By 5.3.2 $N \otimes -$ is the injective hull of S', which is a finitely presented simple object of this localised functor category. Taking "X" in 12.3.15 to be supp(N), we deduce that there is a non-zero morphism $f : F = (F_{D\psi/D\phi})_{DN} \to (N \otimes -)$. Since S' is simple and essential in $N \otimes -$, it is contained in the image of f so choose a finitely generated (in $(R\text{-mod}, \mathbf{Ab})_{DN}$) subfunctor, F', of F which maps onto S'. Let $K = \ker(f \restriction F')$: a finitely generated object of this localised functor category by E.1.16. By 12.3.20 F'/K may be taken to be the localisation at τ_{DN} of a functor of the form $F_{D\psi''/D\phi''}$ for some pp-pair ϕ''/ψ''. (Since F'/K is a subquotient of $(F_{D\psi/D\phi})_{DN}$ we may suppose, by 11.1.6, that $F_{D\psi''/D\phi''}$ is isomorphic to a subquotient of $F_{D\psi/D\phi}$; compare the proof above.)

We claim that the localisation, F'', of $F_{D\psi''/D\phi''}$ at τ_{DY} is a simple object of $(R\text{-mod}, \text{Ab})_{DY}$. For otherwise there would be a finitely generated F_1 with $0 < F_1 < F''$. Let σ denote the torsion theory on $(R\text{-mod}, \mathbf{Ab})_{DY}$ which localises $(R\text{-mod}, \text{Ab})_{DY}$ to $(R\text{-mod}, \text{Ab})_{DN}$. If each of F_1 and F''/F_1 were σ-torsionfree, then the localisation, F', of F'' at σ would have length at least 2, a contradiction. Since the $(N' \otimes -) \simeq (N' \otimes -)_{DY}$ for $N' \in Y$ cogenerate the localised functor category $(R\text{-mod}, \text{Ab})_{DY}$ and, 12.3.13, the $N' \otimes -$ with $N' \in X_N$ cogenerate σ, this would give a non-zero morphism from either F_1 or F''/F_1 to some $N' \otimes -$ with $N' \in Y \setminus \{N\}$. By injectivity of $N' \otimes -$ this would give a non-zero morphism from $(F_{D\psi/D\phi})_{DY}$ to $N' \otimes -$. But by 12.3.15 we have $((F_{D\psi/D\phi})_{DY}, N' \otimes -) = 0$, the required contradiction.

Thus $F'' = (F_{D\psi''/D\phi''})_{DY}$ is a simple functor and so ϕ''/ψ'' is a Y-minimal pair, as required. □

For the dimension, m-dimension, referred to in the following results, see Section 7.2.

The next result is completed at 5.3.19.

Proposition 5.3.17. (see [495, 10.16]) *If* mdim(pp(X)) $< \infty$, *then X satisfies the isolation condition.*

Proof. Suppose that Y is a closed subset of X and N is an isolated point of Y, say $\{N\} = Y \cap (\phi/\psi)$. Since $[\phi, \psi]_Y$ is a subquotient of pp(X), mdim$[\phi, \psi]_Y < \infty$ (7.1.2) so it must be (7.2.5) that $[\phi, \psi]_Y$ contains a simple subinterval, that is, a Y-minimal pair ϕ'/ψ', and clearly $(\phi'/\psi') \cap Y = \{N\}$. □

Theorem 5.3.18. ([726, 8.3]) *Suppose that X is a closed subset of* Zg_R *such that* $\mathrm{pp}^n(X)$ *is countable for every n.*

If $[\phi, \psi]_X$ is a non-trivial interval which contains no minimal pair, then $(\phi/\psi) \cap X$ has more than one point, indeed it contains at least $2^{\aleph_0}$ points.

More generally, if $\mathrm{mdim}[\phi, \psi]_X = \infty$, *then there are at least $2^{\aleph_0}$ points in $(\phi/\psi) \cap X$.*

Proof. The proof we give is the functorial rendition of that appearing in Ziegler's paper. Some of the points of correspondence are indicated in footnotes. For simplicity of notation we give the proof for the case $X = \mathrm{Zg}_R$. The general case is obtained by carrying out the same proof in the localisation, $(R\text{-mod}, \mathbf{Ab})_{DX}$, of the functor category which corresponds to X, so each functor F appearing should be replaced by its localisation F_{DX}.

Because the interval $[\phi, \psi]$ has no minimal pair or, more generally, if its m-dimension is ∞ it must, by 7.2.5, contain a densely ordered chain, say $\{\phi_r\}_{r \in \mathbb{Q} \cap (0,1)}$ with $\phi_s < \phi_r$ if $s < r$. Consider the corresponding finitely generated subfunctors $F_r = F_{D\phi_r}$ of $R^n \otimes -$, where n is the common number of free variables of ϕ and ψ. By 1.3.1, $F_s > F_r$ iff $s < r$.

For each real number $\alpha \in (0, 1)$ a subfunctor G_α of $R^n \otimes -$ will be defined such that $G_\alpha \geq \sum_{r \geq \alpha} F_r$; such that for all $s < \alpha$, $F_s \not\leq G_\alpha$; and which is maximal subject to these requirements.

First note that for any such G_α, the quotient $(R^n \otimes -)/G_\alpha$ is uniform, hence the injective hull $E\big((R^n \otimes -)/G_\alpha\big)$ is indecomposable. For, if $G, G' > G_\alpha$, then, by maximality of G_α, there are $s, s' < \alpha$ such that $G \geq F_s$ and $G' \geq F_{s'}$ and hence such that $G \cap G' \geq F_{\min\{s,s'\}}$ – in particular $G \cap G' > G_\alpha$.

Our first use of the countability hypothesis is to constrain and link the choices of the G_α. Enumerate the finitely generated subfunctors of $R^n \otimes -$ as, say, H_n ($n \geq 1$) (by hypothesis and 10.2.16 there are just countably many). Define each G_α inductively as follows: $G_\alpha^0 = \sum_{r \geq \alpha} F_r$ and, having defined G_α^n, set $G_\alpha^{n+1} = G_\alpha^n$ if $G_\alpha^n + H_n \geq F_s$ for some $s < \alpha$ and $G_\alpha^{n+1} = G_\alpha^n + H_n$ otherwise. That is, to construct a maximal subfunctor containing $\sum_{r \geq \alpha} F_r$ and not containing any F_s with $s < \alpha$, we add H_1 if it is consistent to do so, then add H_2 if consistent, etc. Clearly $G_\alpha = \sum_n G_\alpha^n$ is as described above.

By 12.2.1, $G_\alpha = F_{Dp_\alpha}$ for some pp-type p_α; so $p_r \in p_\alpha$ iff $r \geq \alpha$. By 12.1.9 there is $N_\alpha \in \mathrm{Zg}_R$ such that $(N_\alpha \otimes -) \simeq E((R^n \otimes -)/G_\alpha)$, $= E_\alpha$ say. By 12.2.5 if $\overline{a}_\alpha$ from N_α is such that $(\overline{a}_\alpha \otimes -) : (R^n \otimes -) \to (N_\alpha \otimes -)$ is the composition $(R^n \otimes -) \to (R^n \otimes -)/G_\alpha \to E_\alpha$, then[11] $\mathrm{pp}^{N_\alpha}(\overline{a}_\alpha) = p_\alpha$.

[11] Our p_α is Ziegler's q_α ([726, p. 189]); his inductive starting point, p_α, was the starting point, G_α^0, of our induction.

Suppose that there were $\alpha \in (0, 1)$ such that there exist uncountably many β with $E_\beta \simeq E_\alpha$. Fix such an α; then for each such β, if we form the pullback shown:

$$\begin{array}{ccc} P & \xrightarrow{g_\alpha} & (R^n \otimes -) \\ {\scriptstyle g_\beta}\downarrow & & \downarrow{\scriptstyle \overline{a}_\alpha \otimes -} \\ (R^n \otimes -) & \xrightarrow[\overline{a}_\beta \otimes -]{} & E_\beta \simeq E_\alpha \end{array}$$

Then, since $E_\beta \simeq E_\alpha$ is uniform, the image of P in G_α is non-zero. Note that $P \leq (R^n \otimes -) \oplus (R^n \otimes -)$ by construction of pullback. Choose a finitely generated subfunctor, P_β say, of P with non-zero image in E_α, hence with $g_\beta P_\beta \nleq G_\beta = \ker(\overline{a}_\beta \otimes -)$.

By hypothesis there are only countably many possibilities for P_β as β varies so there is some $P' \leq (R^{2n} \otimes -)$ such that $P_\beta = P'$ for infinitely many β. Since g_β is just restriction of the projection $(R^{2n} \otimes -) \to (R^n \otimes -)$ to the first n coordinates we may identify the $g_\beta \upharpoonright P_\beta = P'$ for these[12] β. Since $g_\beta P'$ is a finitely generated subfunctor of $R^n \otimes -$ it is equal to H_n for some n. As noted already, $H_n \nleq G_\beta$ for all β with $P_\beta = P'$. Choose $\beta < \beta'$ with $P_\beta = P' = P_{\beta'}$ and such that for $i = 1, \dots, n$ we have $H_i \leq G^n_\beta$ iff $H_i \leq G^n_{\beta'}$ (possible since there are infinitely many β to choose from).

Consider $K = \ker\big((\overline{a}_{\beta'} - \overline{a}_\beta) \otimes -\big) \leq (R^n \otimes -)$: this contains all F_r with $r \geq \beta'$ and no F_s with $s < \beta'$; for otherwise it would contain F_s for some s with $\beta < s < \beta'$ and then $F_s \leq \ker(\overline{a}_\beta \otimes -)$ and $F_s \leq K$ would yield $F_s \leq \ker(\overline{a}_{\beta'} \otimes -)$ – contrary to definition of $G_{\beta'}$. In particular K contains $G^0_{\beta'}$ and, by choice of β, β', for each $i = 1, \dots, n$, if $G^n_{\beta'}$ contains H_i so also does G^n_β, and hence so does K. Therefore, by definition of $G^n_{\beta'}$, $K \geq G^n_{\beta'}$. It follows that K cannot contain H_n since otherwise it would contain $G^n_{\beta'} + H_n$, hence, since $H_n \nleq G_{\beta'}$, it would, by definition of $G^{n+1}_{\beta'}$, contain F_s for some $s < \beta'$, contrary to what was shown above. On the other hand, $\ker\big((\overline{a}_\beta, -\overline{a}_\alpha) \otimes -\big)$ and $\ker\big((\overline{a}_{\beta'}, -\overline{a}_\alpha) \otimes -\big)$ both contain P' (by construction), hence so does the kernel, $\ker\big((\overline{a}_{\beta'} - \overline{a}_\beta, 0) \otimes -\big)$, of the difference of these two maps. Projecting along $g_\beta (= g_{\beta'})$, we deduce $K = \ker\big((\overline{a}_{\beta'} - \overline{a}_\beta) \otimes -\big) \geq g_\beta P' = H_n$, a contradiction, as required.

The final statement of the theorem is covered by the comment made at the beginning of the proof but perhaps it is also worth noting that, by 7.1.6 applied to m-dimension, if $\text{mdim}[\phi, \psi] = \infty$, then there is some closed subset Y of X

[12] Our P' corresponds to Ziegler's pp condition $\chi_\alpha(x, y)$ ([726, 8.3]). For, the dual of this condition defines a finitely generated subfunctor of $R^{2n} \otimes -$ (the argument there has $n = 1$) which is in the kernel of the map $(a \otimes -, a_\alpha \otimes -)$, our $(\overline{a}_\alpha \otimes -, -\overline{a}_\beta \otimes -)$, from $(R \otimes -) \oplus (R \otimes -)$. For, by 10.3.6, $\chi_\alpha(a, a_\alpha)$ means exactly that $F_{D\chi_\alpha} \leq \ker(a \otimes -, a_\alpha \otimes -)$. The condition $\neg\chi_\alpha(0, a_\alpha)$ corresponds (by 10.3.6) exactly to our condition that $g_\beta P' \nleq \ker(\overline{a}_\beta \otimes -)$. See "linking" in the model-theory/functor-category dictionary, Appendix C.

such that $[\phi, \psi]_Y$ has no minimal pair. Then one may finish by observing that $(\phi/\psi) \cap Y \subseteq (\phi/\psi) \cap X$. □

The proof above looks considerably longer than that in [726, 8.3] or [495, 10.15], partly because more detail is given but also, to some extent, because this is an instance where the model-theoretic machinery and notation is rather efficient.

Regarding the countability hypothesis, what is used is that both $\mathrm{pp}^n(X)$ and $\mathrm{pp}^{2n}(X)$ are countable.[13] The former does not imply the latter; for example take R to be an uncountable field, so pp^1_R is the two-point lattice but pp^2_R contains all the inequivalent pp conditions $x = yr$, $r \in R$.

For the converse we need the Cantor–Bendixson analysis of the space (Section 5.3.6).

Theorem 5.3.19. ([726, 8.1,8.4]) *Suppose that X is a closed subset of Zg_R such that $\mathrm{pp}^n(X)$ is countable for each n. Then the following conditions are equivalent:*

(i) $\mathrm{pp}^1(X)$ *contains a densely ordered subset;*
(ii) for some n, $\mathrm{pp}^n(X)$ *contains a densely ordered subset;*
(iii) for every n, $\mathrm{pp}^n(X)$ *contains a densely ordered subset;*
(iv) $\mathrm{card}(X) = 2^{\aleph_0}$;
(v) X is uncountable.

Proof. (i)⇒(ii) is trivial, (ii)⇒(iii) is 7.2.5, (iii)⇒(iv) is by 5.3.18 and (iv)⇔(v) is trivial.

If (i) does not hold, then, by 7.2.3, $\mathrm{mdim}(\mathrm{pp}^1(X)) < \infty$. So, by 5.3.17, X satisfies the isolation condition and by 5.3.60 the Cantor–Bendixson rank of X is $<\infty$. Therefore each point, N, of X is an isolated point of some Cantor–Bendixson derivative $X^{(\alpha)}$ so there is a pair ϕ_N/ψ_N such that $(\phi_N/\psi_N) \cap X^{(\alpha)} = \{N\}$. Clearly different points cannot be isolated in this sense by the same pair ϕ/ψ. There are only countably many pairs ϕ/ψ, hence only countably many points of X. □

If R is countable,[14] then, since in any given pp condition only finitely many elements of R appear, there are only countably many pp conditions; we deduce the following. For Krull–Gabriel dimension, KG(R), see Section 13.2.

[13] A number of results from Ziegler's paper require a countability hypothesis in one direction. Subsequently very little has been done – for a partial exception see [298] – towards removing this hypothesis, or showing that it is necessary, although sometimes it has been bypassed.

[14] In fact, it is enough to have countability modulo the relation on the lattice of pp conditions which identifies $\phi \geq \psi$ if the width of the interval $[\phi, \psi]$ is $< \infty$ (for this and its connection with superdecomposability see Section 7.3.1). In terms of functors, collapsing by this equivalence relation is equivalent to localising at the torsion theory whose torsion class is generated by all the quotients $F_{D\psi/D\phi}$ where $\mathrm{w}[\phi, \psi] < \infty$. If the resulting lattice is countable, then the proof and result are valid ([495, 10.13] for modules, [104, 5.5] for the general case).

Corollary 5.3.20. ([726, 8.1]) *If R is a countable ring, then Zg_R is countable iff* $\mathrm{mdim}(\mathrm{pp}_R) < \infty$ *iff* $\mathrm{KG}(R) < \infty$.

Corollary 5.3.21. *If R is countable, then it satisfies the isolation condition.*

Proposition 5.3.22. ([495, 10.17]) *Assume that the isolation condition holds for the closed subset X of Zg_R. Suppose that $(\phi/\psi) \cap X$ has just one point. Then $[\phi, \psi]_X$ is an interval of finite length. In particular, if X has just one point, then that point must be of finite endolength.*

Proof. The proof is essentially the same as that in [495, 10.17], though set in the functorial context.

As usual, we work in the localisation $(R\text{-mod}, \mathbf{Ab})_{DX}$ of $(R\text{-mod}, \mathbf{Ab})$ at the torsion theory corresponding to X.

Let N be the unique point of $(\phi/\psi) \cap X$. Then, by 12.3.15 and 12.3.3, $N \otimes -$ is the unique indecomposable injective of $(R\text{-mod}, \mathbf{Ab})_{DX}$ such that $(F, N \otimes -) \neq 0$, where $F = (F_{D\psi}/F_{D\phi})_{DX}$. By assumption there is a pp-pair ϕ'/ψ' such that $(\phi'/\psi') \cap X = \{N\}$ and such that $[\phi', \psi']_{DX}$ is a simple interval, hence such that $S = (F_{D\psi'}/F_{D\phi'})_{DX}$ is a simple functor in the localised category. Because $(S, N \otimes -) \neq 0$ it follows that $N \otimes -$ is the injective hull of S.

Suppose, for a contradiction, that $[\phi, \psi]_{DX}$ has infinite length, so, by 12.3.18, F is of infinite length. Let F' be a subfunctor of F which is maximal with respect to containing no finitely generated subfunctor, F_0, of F with F/F_0 of finite length (this exists by Zorn's Lemma).

We claim that F/F' is uniform. Take subfunctors F_1, F_2 of F with $F' < F_i$ for $i = 1, 2$. By maximality of F' there are finitely generated functors F_1', F_2' with $F_i' \le F_i$ and F/F_i' of finite length ($i = 1, 2$). Then $F_1' \cap F_2'$ is a finitely generated (by 12.3.11 and 2.3.15, cf. 10.2.7) subfunctor of F with $F/(F_1' \cap F_2')$ of finite length. So $F_1' \cap F_2' \nleq F'$ and hence $F < (F_1' \cap F_2') + F' \le F_1 \cap F_2$, as required.

Therefore the injective hull, E, of F/F' is indecomposable and, since $(F, E) \neq 0$, it must be that $E \simeq N \otimes -$. It follows that F/F' contains a copy of the finitely presented simple functor S. Let F'' with $F' < F'' \le F$ be such that $F''/F' \simeq S$. By maximality of F', the functor F'' contains a finitely generated subfunctor, F_0, such that the quotient F/F_0 is of finite length. By choice of F', certainly $F_0 \nleq F'$, so, since F''/F' is simple, we have $F_0/F' \cap F_0 \simeq F''/F' \simeq S$. Since S is finitely presented, $F' \cap F_0 = \ker(F_0 \to S)$ is (by E.1.16) finitely generated. But also $F' \cap F_0 \le F'$ and $F/F' \cap F_0$ is of finite length – contradicting our choice of F', as required. □

Corollary 5.3.23. *Assume that the isolation condition holds for the closed subset X of Zg_R. Then every closed point in X is of finite endolength.*

Recall, 5.1.12, that the converse is always true.

Theorem 5.3.24. ([726, 8.11]) *If there is no superdecomposable pure-injective R-module, then R satisfies the isolation condition.*

Proof. As was done in the case of 5.3.18 we prove that the isolation condition holds for $X = \mathrm{Zg}_R$, the general case being established by the same argument carried out in the corresponding localisation of (R-mod, **Ab**) (also see 5.3.25 below).

Suppose that the interval $[\phi, \psi]$ has no minimal pair. It will be shown that (ϕ/ψ) contains at least two non-isomorphic points. One is easy to find: let G be a maximal proper subfunctor of $F_{D\psi}$ containing $F_{D\phi}$. Then $F_{D\psi}/G$ is simple, so its injective hull, $N \otimes -$ say, yields, by 12.2.4, a point $N \in (\phi/\psi)$.

Next, since $[\phi, \psi]$ has no minimal pair, there is a densely ordered chain, $(\phi_r)_{r \in \mathbb{Q} \cap (0,1)}$, of pp conditions between ϕ and ψ, with $\phi_r > \phi_s$ when $r > s$. Therefore there is a densely ordered chain $(F_r = F_{D\phi_r})_r$, with $F_r < F_s$ if $r > s$, lying between $F_{D\psi}$ and $F_{D\psi}$.

Let H be maximal among subfunctors of $F_{D\psi}$ containing $F_{D\phi}$ and with the property that $H + F_r < H + F_s$ whenever $r > s$ (existence is by Zorn's Lemma and since, 10.2.29, the F_s are finitely generated). Now, $E(F_{D\psi}/H) \simeq N' \otimes -$ for some pure-injective module N' (by 12.1.6). By hypothesis, there are no superdecomposable pure-injectives, so N' has an indecomposable direct summand, N_0, say. Since $N_0 \otimes -$ is essential in $N' \otimes -$ and is injective it follows that $(F_{D\psi/D\phi}, N_0 \otimes -) \neq 0$; so $N_0 \in (\phi/\psi)$. Suppose, for a contradiction, that $N_0 \simeq N$. It follows that $F_{D\psi}/H$ has a simple subfunctor, say H'/H. By maximality of H there are r, t with $r > t$ and $H' + F_r = H' + F_t$. Then, if s is such that $r < s < t$, all of the sums $H' + (H + F_t)$, $H' + (H + F_s)$, $H' + (H + F_r)$ coincide. By simplicity of H'/H, the intersections $H' \cap (H + F_t)$, $H' \cap (H + F_s)$, $H' \cap (H + F_r)$ can take at most two values, H and H'. This contradicts modularity of the lattice of subfunctors (see Appendix E, p. 712).

Therefore N_0, N are distinct points in (ϕ/ψ), as required. □

Hence if there is no superdecomposable pure-injective and if the open set (ϕ/ψ) contains just a single point, then that point is isolated by a minimal pair (ϕ'/ψ') (with $\phi \geq \phi' > \psi' \geq \psi$). Ziegler proved a local version: if N is in the closed subset X of Zg_R and if N is a direct summand of *every* pure-injective module which generates the definable subcategory, $\mathcal{X}$, corresponding to X, then N is isolated by a minimal pair ([726, 9.3(2)]).

The generalisation of 5.3.24 to arbitrary closed sets also follows from 12.5.8 applied to the corresponding localisation of the functor category.

Corollary 5.3.25. *If X is a closed subset of the Ziegler spectrum and if there is no superdecomposable pure-injective in the corresponding definable subcategory, then X satisfies the isolation condition.*[15]

Proposition 5.3.26. *Let R be any ring. If R is of finite representation type (§4.5.3), then Zg_R is discrete. The converse is true if R satisfies the isolation condition.*

Proof. If R is of finite representation type, then, by 4.5.22, every indecomposable module is of finite endolength, hence, by 5.1.12, is a closed point of Zg_R. There are only finitely many such points in Zg_R, so this space has the discrete topology.

Suppose, conversely, that Zg_R is discrete, hence, being compact (5.1.23), is finite, and suppose that R satisfies the isolation condition. By 5.3.23 every point of Zg_R has finite endolength. Therefore Mod-R is generated as a definable category by finitely many modules of finite endolength, so, by 4.4.27, every R-module has finite endolength and, therefore, by 4.5.22, R is of finite representation type. □

Question 5.3.27. Suppose that Zg_R is a discrete space. Does it follow that R is of finite representation type? (By 5.1.23 Zg_R is finite. The example given before 8.2.45 shows that a finite Ziegler spectrum need not be a discrete space.)

In [570, 3.3] Puninski showed that, for a definable subcategory, $\mathcal{X}$, of Mod-R over any ring R the following are equivalent:

(i) the lattice, pp($\mathcal{X}$), of pp conditions modulo $\mathcal{X}$ is distributive;
(ii) every pure-injective module in $\mathcal{X}$ is **distributive** (that is, has distributive lattice of submodules) as a module over its endomorphism ring;
(iii) every point in $\mathcal{X} \cap \mathrm{Zg}_R$ is uniserial as a module over its endomorphism ring, equivalently its lattice of pp-definable subgroups is a chain.

The next result then follows from 7.3.16.

Theorem 5.3.28. ([570, 3.4]) *Let R be any ring and let X be a closed subset of Zg_R. If the lattice, $\mathrm{pp}^1(X)$, of pp conditions modulo X is distributive, then X satisfies the isolation condition.*

Corollary 5.3.29. ([570, 3.5]) *Every commutative Prüfer ring satisfies the isolation condition.*

I remark that it is not known whether the isolation condition is preserved by right/left duality (see 5.4.10).

[15] For a model-theoretic, as opposed to "functorial", proof see the original paper of Ziegler or [495, 10.16].

5.3.3 Minimal pairs and left almost split maps

For a strongly indecomposable finitely presented module, that is, a finitely presented module with local endomorphism ring, isolation of its hull by a minimal pair is equivalent to the module being the domain of a left almost split morphism (5.3.31). It follows that if R is an artin algebra, then every indecomposable finitely generated module is an isolated point of the Ziegler spectrum (5.3.33).

A morphism $f : A \to B$ in mod-R (or a more general category $\mathcal{C}$) is **left almost split** (in mod-R, resp. in $\mathcal{C}$)) if it is not a split embedding and if every morphism $g : A \to C$ in mod-R (resp. $\mathcal{C}$) which is not a split embedding factors through, f (see, for example, [27, p. 137]). These occur as "left-hand portions" of almost split sequences, examples of which can be seen in Section 15.1.1.

Lemma 5.3.30. (cf. 4.3.45) *Suppose that A is a finitely presented module with local endomorphism ring. A morphism $f : A \to B$ in* mod-R *is not a split embedding iff there is $\overline{a} \neq \overline{0}$ from A such that* $\mathrm{pp}^B(f\overline{a}) > \mathrm{pp}^A(\overline{a})$ *iff for every non-zero tuple $\overline{a}$ from A we have* $\mathrm{pp}^B(f\overline{a}) > \mathrm{pp}^A(\overline{a})$.

Proof. Suppose that f is not a split embedding and let $\overline{a} \neq \overline{0}$ be from A. If $\mathrm{pp}^B(f\overline{a}) = \mathrm{pp}^A(\overline{a})$, then, by 1.2.9, there is $g : B \to A$ with $gf\overline{a} = \overline{a}$ hence with $1 - gf$ not invertible. Since End(A) is local gf must be invertible and so f must be monic and is split by $(gf)^{-1}g$, a contradiction.

For the (strong) converse, suppose that $\overline{a} \neq \overline{0}$ from A is such that $\mathrm{pp}^B(f\overline{a}) > \mathrm{pp}^A(\overline{a})$. If g is any morphism from B to A, then $\mathrm{pp}^A(gf\overline{a}) \geq \mathrm{pp}^B(f\overline{a}) > \mathrm{pp}^A(\overline{a})$ so g cannot split f, as required. □

The next result should also be compared with 5.3.47.

Theorem 5.3.31. ([141, 2.3], see also [521, 3.7] and cf. [19, 2.7], [278, 7.7]) *Suppose that A is a finitely presented module with local endomorphism ring, so, by 4.3.55, $H(A)$ is indecomposable. Then the following conditions are equivalent:*

(i) $H(A)$ is isolated in Zg_R *(by a minimal pair);*

(ii) there is a morphism $f : A \to B$ in mod-R *which is not a split monomorphism and which is such that every morphism $g : A \to M$ in* Mod-R *which is not a pure embedding factors through f;*

(iii) A is the domain of a left almost split morphism in mod-R, *that is, as (ii) but with* mod-R *in place of* Mod-R.

If these conditions hold, then f above is either a non-split monomorphism or is an epimorphism with simple essential kernel: the latter case occurs exactly if A is absolutely pure.

Proof. The parenthetical comment in (i) is by 5.3.13.

(iii)⇒(i) Let $\overline{a}$ be a generating tuple for A, with $\overline{x}H = 0$ being the generating relations, and take ϕ to be the condition $\overline{x}H = 0$. Let ψ be a pp condition which generates $\mathrm{pp}^B(f\overline{a})$ (1.2.6). We claim that ϕ/ψ is a minimal pair.

First note that $\phi > \psi$ – otherwise, by 1.2.9, there would be $g : B \to A$ with $gf\overline{a} = \overline{a}$, contradicting, by 5.3.30, that f is not a split monomorphism.

Next, if ϕ' is a pp condition with $\phi > \phi'$, then, taking $(D, \overline{d})$ to be a free realisation (1.2.14) of ϕ', there is, by 1.2.9, a morphism $g : A \to D$ taking $\overline{a}$ to $\overline{d}$. Since $\phi > \phi'$, g is not a pure embedding, so, by hypothesis, there is $h : B \to D$ with $hf = g$, hence such that $hf\overline{a} = \overline{d}$. In particular $\psi \geq \phi'$. Thus ϕ/ψ is a minimal pair which is open on A, so, since A is pure in $H(A)$, which is open on $H(A)$. Therefore (5.3.2), ϕ/ψ isolates $H(A)$, as required.

(i)⇒(ii) Suppose that ϕ/ψ is a minimal pair isolating $H(A)$. By 4.3.23, $\phi(A) > \psi(A)$. Let $\overline{a} \in \phi(A) \setminus \psi(A)$ and (1.2.6) let ϕ generate $\mathrm{pp}^A(\overline{a})$. So $\phi' \leq \phi$ and $\phi' \not\leq \psi'$, hence $\phi'/\phi' \wedge \psi$ is, by modularity, a minimal pair, so, without loss of generality, we may suppose that ϕ generates the pp-type of $\overline{a}$. By 5.3.11 we may replace our original choice of $\overline{a}$ (and ϕ, ψ) by a generating tuple and reach the same conclusion. So from now we assume that $\overline{a}$ generates A (that is, its entries together do). Therefore, by 4.3.62, for any ψ', if $\psi' < \phi$ it must be that $\psi' \leq \psi$.

Let $(B, \overline{b})$ be a free realisation of ψ. By 1.2.9 there is a morphism $f : A \to B$ taking $\overline{a}$ to $\overline{b}$ and, by 5.3.30, f is not a split monomorphism. If $g : A \to M$ is any morphism in Mod-R which is not a pure embedding, then $\mathrm{pp}^A(\overline{a}) > \mathrm{pp}^M(g\overline{a})$ (4.3.63), so, by the above, $g\overline{a}$ satisfies ψ in M. Therefore, by 1.2.9, there is $h : B \to M$ with $h\overline{b} = g\overline{a}$, hence, since $\overline{a}$ generates A, with $hf = g$. This establishes the equivalence of the conditions since trivially (ii)⇒(iii).

For the second part (cf., for example, [27, 1.13]), suppose that f as in (iii) is not monic. Let $K' \leq \ker(f)$ with K' finitely generated and non-zero. Then f factors through the natural map $f' : A \to A/K'$, so, by assumption on f, f' factors through f, hence $\ker(f) = K'$. Therefore $\ker(f) = K'$, hence $\ker(f)$ is simple. If there were a non-zero submodule L of A with $L \cap \ker(f) = 0$, then the projection $g : A \to A/L$ also would have to factor through $f : A \to A/\ker(f)$, which is impossible. Therefore $\ker(f)$ is essential in A.

Finally, if f is an epimorphism, then there can be no non-pure embedding of A into any module – otherwise we could take such an embedding for $g : A \to M$ in (ii). Therefore A is absolutely pure. □

Note that in the proof of (i)⇒(ii) we proved, with notation as there, that F_ψ is the *unique* maximal proper subfunctor of F_ϕ, that is, F_ϕ is a local functor in the sense of Section 10.2.3. Indeed, if we take $\overline{a}$ as in (i)⇒(ii) and regard it as a morphism from R^n to A, then it induces an epimorphism $(A, -) \to F_\phi \leq (R^n, -)$. Since

$(A, -)$ is local (10.2.27), it then follows that F_ϕ is local. (Also see the comment after 4.3.71.)

Slightly modifying the argument above, one derives the result that if R is Krull–Schmidt, then mod-R has left almost split maps, in the sense that every indecomposable finitely presented module is the domain of a left almost split map, iff for every indecomposable $A \in$ mod-R the radical (see Section 10.2.3) of the functor $(A, -)$ is finitely generated, that is, iff every simple functor in (mod-R, **Ab**) is finitely presented ([19, 2.7]). Note the correspondence between Auslander's "universally minimal" elements and the minimal pairs that appear here.

Corollary 5.3.32. *Suppose that A is a finitely presented module with local endomorphism ring. Then $H(A)$ is isolated (by a minimal pair) iff there is a left almost split map in* mod-R *with domain A.*

If R is an artin algebra, then every indecomposable module of finite length has local endomorphism ring (4.3.58) and is the domain of a left almost split map (see 15.1.4), which is an embedding unless the module is an injective (= absolutely pure since R is noetherian, see 3.4.28) with simple socle. So the next result is, by 5.3.32, a direct consequence of the existence of left almost split maps for modules over artin algebras. The converse is also true and is proved in the next section as 5.3.37.

Corollary 5.3.33. ([495, 13.1]) *Let R be an artin algebra. Then every indecomposable finitely generated module is a point of* Zg_R *and is isolated by a minimal pair.*

Corollary 5.3.34. *Suppose that R is an artin algebra and that the interval* $[\phi, \psi]$ *in* pp_R^n *is of infinite length. Then the open set* (ϕ/ψ) *contains an infinite-dimensional point of* Zg_R.

Proof. (cf. 5.3.57) Otherwise (ϕ/ψ) contains only finitely generated, hence isolated, points, so, since it is compact (5.1.22), it contains only finitely many points, $N_1, \dots, N_t$ say. Each of these is of finite endolength, so, since the interval $[\phi, \psi]$ is of infinite length, there is a pp-pair ϕ'/ψ' with $\phi \geq \phi' > \psi' \geq \psi$, hence with $(\phi'/\psi') \subseteq (\phi/\psi)$, which is closed on each of $N_1, \dots, N_t$. This pair must, therefore, be closed on every point of Zg_R, contradicting 5.1.3. □

5.3.4 Density of (hulls of) finitely presented modules

Over an artin algebra the set of finite-length points is dense in the Ziegler spectrum (5.3.36); this follows from the fact that over these rings every pure-injective module is a direct summand of a direct product of finite-length modules (5.3.35). Therefore the indecomposable modules of finite length are exactly the isolated points (5.3.37). It follows that if there are only finitely many indecomposable modules of finite length, then this artin algebra is actually of finite representation type (5.3.38). On the other hand, if there are infinitely many modules of finite length, then there must be a infinitely generated indecomposable pure-injective module (5.3.40). As a result, one deduces Auslander's result that any pure-semisimple artin algebra is of finite representation type (5.3.41).

More generally, if the ring is Krull–Schmidt, then the set of hulls of strongly indecomposable finitely presented modules is dense in the spectrum (5.3.42). With the further condition that radicals of corresponding representable functors are finitely generated one deduces that these points are also isolated, by minimal pairs (5.3.43).

The first result seems to have been noticed by various people, see the references to [495, 13.2] for at least some of them.[16]

Theorem 5.3.35. *Suppose that R is an artin algebra. Then every pure-injective R-module is a direct summand of a direct product of modules of finite length. More generally, every module is a pure submodule of a direct product of modules of finite length.*

Proof. Let $E = E\big(C(R)/J(C(R)\big)$ be the injective hull of the $C(R)$-module $C(R)/J(C(R))$, where $C(R)$ denotes the centre of R. Then $(-, E_R)$, respectively $(-, {}_RE)$, is a functor from right modules to left modules, resp. vice versa, which restricts to a duality between the categories of finitely presented right and finitely presented left modules (15.1.3). Write M^* for $\mathrm{Hom}_R(M, E)$ (see Section 1.3.3).

Let M be any right R-module. By 2.1.25 there is a pure epimorphism $\pi : L \to M^*$, where L is a direct sum of finitely presented left R-modules. Then $\pi^* : M^{**} \to L^*$ is a pure monomorphism (by 4.3.30) in Mod-R. Recall, 1.3.16, that M purely embeds in M^{**}, hence in L^*. But L^* is a direct product of duals of finitely presented modules, that is, a direct product of finitely presented modules. In particular if M is pure-injective, then M is a direct summand of a direct product of modules of finite length. □

We deduce that over any artin algebra the points of finite length are dense in the Ziegler spectrum.

Corollary 5.3.36. ([495, 13.3]) *Suppose that R is an artin algebra. Then the closure of the set of finite length points in Zg_R is the whole of Zg_R.*

[16] [134] should be added.

Proof. By 5.3.35 and 3.4.7 the definable subcategory generated by the indecomposable finite length modules is the whole of Mod-R, so, by 5.1.6, the Ziegler-closure of this set of points is the whole of Zg_R. □

A more general result is 5.3.42 below. Also 5.3.43 generalises the next result.

Corollary 5.3.37. ([495, 13.4]) *Suppose that R is an artin algebra. Then the isolated points of Zg_R are exactly the indecomposable modules of finite length.*

Proof. By 5.3.33 every indecomposable module of finite length is an isolated point of Zg_R. Every point of Zg_R is in the closure of these isolated points, so, if isolated, must already be among them. □

More generally, [357, 4.7, 4.8], if R is a dualising ring in Krause's sense, then an indecomposable module of finite endolength is isolated in the spectrum iff it is finitely presented, and this condition characterises dualising rings.

The definition of finite representation type given in Section 4.5.3 is not the original one, which, in the context of artin algebras, was that there are just finitely many indecomposable modules of finite length, up to isomorphism. We give a proof of the result that for artin algebras this is equivalent to the apparently stronger definition that we used.

Theorem 5.3.38. ([670, 9.5]) *Suppose that R is an artin algebra which has only finitely many indecomposable modules of finite length. Then R is of finite representation type and every module is a direct sum of copies of these indecomposables.*

Proof. By 5.3.36 these finitely many points are dense in Zg_R. Being of finite endolength, each of them is, by 5.1.12, a closed point, so Zg_R consists of these points and no others. By 5.1.6 the definable subcategory generated by these points is the whole of Mod-R, so, by 4.4.30, every module is a direct sum of copies of these. In particular, there are no more indecomposable modules. □

Corollary 5.3.39. *An artin algebra is of finite representation type iff Zg_R is finite.*

The above is not an equivalence for arbitrary rings; see 8.2.84.

Compactness of the Ziegler spectrum gives an easy proof of the following result of Auslander.

Theorem 5.3.40. ([20, Thm A]) *Suppose that R is an artin algebra. If there are infinitely many indecomposable modules of finite length, then there is an infinitely generated indecomposable (pure-injective) R-module.*

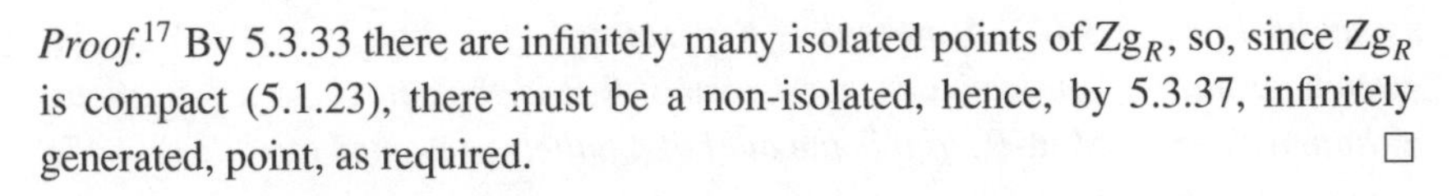

Proof.[17] By 5.3.33 there are infinitely many isolated points of Zg_R, so, since Zg_R is compact (5.1.23), there must be a non-isolated, hence, by 5.3.37, infinitely generated, point, as required. □

Corollary 5.3.41. ([20, Thm A]) *Every pure-semisimple artin algebra is of finite representation type.*

Proof. Since, 4.5.4, every indecomposable module over a pure-semisimple ring is of finite length, this is immediate from 5.3.40. □

Theorem 5.3.42. ([278, 5.4]) *Suppose that R is a Krull–Schmidt ring (Section 4.3.8). Then the set of points of the form $H(A)$ with A finitely presented and indecomposable is dense in Zg_R.*

Proof. Every non-empty basic open set has, by 10.2.45, the form (F), where F is a non-zero finitely presented functor in (mod-R, **Ab**). By 4.3.69, F has a projective cover P. By the same result, P has a local, in particular an indecomposable, direct summand, say $(A, -)$, where $A \in$ mod-R (10.1.14). Now, with notation as in Section 10.2.8, $\overrightarrow{F}A = FA \simeq ((A, -), F)$ (by 10.1.7) $\neq 0$, so, by 10.2.46, $\overrightarrow{F}(H(A)) \neq 0$, that is, $H(A) \in (F)$, as required. □

Corollary 5.3.43. *Suppose that R is a Krull–Schmidt ring and suppose that for every indecomposable finitely presented right R-module A the radical of the functor $(A, -)$ is finitely generated; that is, see comments after 5.3.31, suppose that* mod-R *has left almost split maps. Then the isolated points of Zg_R are exactly those of the form $H(A)$, where A is indecomposable and finitely presented. Also, every isolated point is isolated by a minimal pair.*

Proof. Continuing the proof of 5.3.42, our assumption implies that the functor $S = (A, -)/\mathrm{rad}(A, -)$ is finitely presented. Hence (S) is a Ziegler-open set (10.2.45) and, by 5.3.5, $H(A)$ is the unique point in it. □

An example where hulls of "small" (in this case meaning finite-length) modules are dense in the spectrum but where, in contrast with the situation above, there are no isolated points at all is provided by the first Weyl algebra over a field of characteristic 0 (and similar rings), see 8.2.30.

5.3.5 Neg-isolated points and elementary cogenerators

An indecomposable pure-injective is neg-isolated iff the corresponding functor is the injective hull of a simple functor (5.3.45). In particular any isolated point

[17] Another proof is in [720].

is neg-isolated (5.3.44). A pp-type is neg-isolated iff its hull is (5.3.46). An indecomposable pure-injective is neg-isolated iff it is the source of a left almost split morphism in Mod-R*, iff it is the dual of a finitely presented module (5.3.47). The neg-isolation property is equivalent to not being produced non-trivially as a summand of a product (5.3.48).*

A module is an elementary cogenerator if the definable subcategory it generates can be generated using just products and pure submodules. This is equivalent to every neg-isolated point in its support being a direct summand (5.3.50). Every definable category has an elementary cogenerator (5.3.52) and every Σ*-pure-injective is an elementary cogenerator (5.3.54).*

Say that a pp-type p in n free variables is **neg-isolated (by** ψ**)** if there is a pp condition $\psi \in \mathrm{pp}_R^n$ which is not in p and such that p is maximal not containing ψ. This is equivalent to the functor $(F_{D\psi} + F_{Dp})/F_{Dp}$ being simple, see 12.5.5. More generally, if X is a closed subset of Zg_R, then say that the pp-n-type p is X-**neg-isolated (by** ψ**)** if there is a pp condition ψ such that there is no pp condition in p which is X-equivalent to ψ and if p is, as a set of pp conditions, maximal such. That is, p is the inverse image under the natural map $\mathrm{pp}_R^n \to \mathrm{pp}^n(X)$ of a set of elements of $\mathrm{pp}^n(X)$ which does not include the image of ψ and which is maximal such.[18]

Any such type is, by 4.3.52, irreducible, so the hull, $H(p)$, of p is indecomposable. Indeed, by 12.5.5 and 12.2.4, $H(p) \otimes -$ is the injective hull of the functor $(F_{D\psi} + F_{Dp})/F_{Dp}$ and the localisation of this functor at the torsion theory corresponding to X is simple.

The terminology is extended to points of the spectrum by saying that $N \in X$ is X-**neg-isolated (by** ψ**)** if $N = H(p)$ for some X-neg-isolated pp-type p. In 5.3.46 it is shown that, in this case, every pp-type realised in N is neg-isolated.

Corollary 5.3.44. ([495, 9.25]) *If an indecomposable pure-injective R-module is an isolated point of X, then it is X-neg-isolated.*

That is just 5.3.1.

By 12.5.5 a pp-type p is X-neg-isolated (by ψ) iff the localised functor $\big((R^n \otimes -)/F_{Dp}\big)_{DX}$ has simple essential socle, namely $\big((F_{D\psi} + F_{Dp})/F_{Dp}\big)_{DX}$. If we were to start with a pp condition ψ, then to obtain a pp-type which is X-neg-isolated by ψ we take a maximal proper subfunctor, G', of $F_{D\psi}$ which is τ_{DX}-closed, then choose a maximal subfunctor, G, of $R^n \otimes -$ such $G \cap F_{D\psi} =$

[18] I should perhaps remark that the terminology "p is isolated", which appears in [495], for example, is from model theory (it refers to isolation in a certain space of types) and does not refer to isolation in the Ziegler spectrum (though if an irreducible type is isolated in the model-theoretic sense, then, [495, 9.26], its hull will be isolated in the spectrum).

G' (G also will be τ_{DX}-closed) and then let p be the pp-type such that $F_{Dp} = G$ (12.2.1). The localisation of $F_{D\psi}/(F_{D\psi} \cap F_{Dp}) \simeq (F_{D\psi} + F_{Dp})/F_{Dp}$ at τ_{DX} will then be a simple object of the quotient category $(R\text{-mod}, \mathbf{Ab})_{DX}$ and every simple object of this category has such a form as ψ and p vary. We state some of this in a form for future reference (cf. 5.3.1).

Remark 5.3.45. An indecomposable pure-injective N is neg-isolated iff $N \otimes -$ is the injective hull of a simple functor. More generally, and precisely, if N is X-neg-isolated, where X is a closed subset of Zg_R, say $N = H(p)$, where p is a pp-type which is X-neg-isolated by the pp condition ψ, then the image of $N \otimes -$ in the localisation $(R\text{-mod}, \mathbf{Ab})_{DX}$ is the injective hull of a simple object, S, of this localised category and $S \simeq ((F_{D\psi} + F_{Dp})/F_{Dp})_{DX}$.

As is noted after 12.5.5, if p is a neg-isolated finitely generated pp-type, then F_{Dp} will be finitely generated, hence S above will be finitely presented and so $H(p)$ will be isolated by a minimal pair in Zg_R; the same conclusion holds in the relative case if p is finitely generated relative to X in the sense of 5.3.12.

From the fact above, namely that N is X-neg-isolated iff $N \otimes -$ is the injective hull of a functor whose localisation at the torsion theory corresponding to X is simple, we get easy proofs of various results which were proved originally by model-theoretic arguments.

Proposition 5.3.46. (Ziegler, see [495, 9.24]) *If X is a closed subset of Zg_R and $N = H(p)$ for some X-neg-isolated pp-type p, then, if $N \simeq H(q)$, also q is X-neg-isolated.*

Proof. (This proof of the result is from [101, 6.5.1].) Let $\overline{a}$ from N be such that $\mathrm{pp}^N(\overline{a}) = q$. By 12.2.5 $\overline{a} : R^n \to N$ induces an embedding of $(R^n \otimes -)/F_{Dq}$ into $N \otimes -$. If N is X-neg-isolated, then, as remarked above, $\mathrm{soc}(N \otimes -)_{DX}$ is simple and essential in $N \otimes -$, hence $((R^n \otimes -)/F_{Dq})_{DX}$ has a simple essential socle, which, by the remarks above, implies that q is X-neg-isolated. □

The next result should be compared with 5.3.31.

Theorem 5.3.47. ([141, 2.3]) *Suppose that N is an indecomposable pure-injective R-module. Then the following conditions are equivalent:*

(i) *$N \otimes -$ is the injective hull of a simple functor in $(R\text{-mod}, \mathbf{Ab})$, that is, (5.3.45), N is neg-isolated;*
(ii) *$N \simeq \mathrm{Hom}_S({}_RL_S, E_S)$, where L is a finitely presented left R-module, $S = \mathrm{End}(L)$ and E is the injective envelope of a simple left S-module;*
(iii) *N is the source of a left almost split morphism in Mod-R.*

Proof. (iii)⇒(i) Suppose that $f : N \to M$ is a left almost split morphism in Mod-R. Choose any non-zero element $a \in N$ and let $p = \mathrm{pp}^N(a)$. Set $q = \mathrm{pp}^M(fa)$ and choose $\psi \in q \setminus p$. We claim that p is neg-isolated by ψ (which, by 5.3.45, will be enough).

If it were not, then there would be a pp-type p' strictly containing p but not containing ψ. Let $c \in N'$, with N' pure-injective, realise p' (4.3.39). By 4.3.9 there is a morphism $h : N \to N'$ taking a to c which, since p' strictly contains p, cannot be a split monomorphism. Therefore there is a morphism $g : M \to N'$ with $gfa = ha$. Since fa, hence gfa satisfies ψ, this is a contradiction.

(i)⇒(iii) Let a be any non-zero element of N. By 5.3.45 the pp-type of a in N is neg-isolated, by ψ say. Let (C, c) be a free realisation of ψ and define M to be the pushout shown and let b denote the common image of a and c in M.

$$\begin{array}{ccc} R^n & \xrightarrow{c} & C \\ {\scriptstyle a}\downarrow & & \downarrow \\ N & \xrightarrow[f]{} & M \end{array}$$

Certainly f is not a split embedding, since it strictly increases the pp-type of a. And, if $h : N \to M'$ is not a split embedding, it cannot be a pure monomorphism, since N is pure-injective, so it increases the pp-type of some, hence (4.3.48) every, element of N. In particular the pp-type of ha contains ψ, so, by 1.2.17, there is a morphism from C to M' taking c to ha. Since M is the pushout we deduce that there is a morphism from M to M' taking b to ha, as required.

(i)⇒(ii) The equivalence of (i) and (ii) may be proved, as in [141], using the description of Auslander's correspondence between maximal ideals in endomorphism rings of finitely presented modules and simple quotients of representable functors ([21]). Our proofs here are not really different but are said in terms of pp-types.

Let $a \in N$ be non-zero and suppose that $p = \mathrm{pp}^N(a)$ is neg-isolated by ψ. Let $({}_RL, c)$ be a free realisation of $D\psi$ and set $S = \mathrm{End}(S)$. Consider the S-linear map $a \otimes_R 1_L : {}_RL_S \to (N \otimes_R L)_S$ given by $l \mapsto a \otimes l$. Since c freely realises $D\psi$ and $\psi \notin p$, we have $a \otimes c \neq 0$ by 1.3.8. Let $\mathcal{M} = \{s \in S : cs \in \Sigma_{\phi \in p} D\phi(L)\}$. Noting that the sum is directed (for p is closed under $\wedge$), it is clear that $\mathcal{M}$ is a right ideal of S. We claim that $\mathcal{M}$ is a maximal right ideal.

Let $t \in S \setminus \mathcal{M}$, so $ct \notin F_{Dp}(L) = \sum_{\phi \in p} D\phi(L)$. Since ${}_RL$ is finitely presented, there is (1.2.6) a pp condition η such that $\mathrm{pp}^L(ct) = \langle \eta \rangle$. Suppose first that $\eta = D\psi$. Then (1.2.9) there is $t' \in S$ such that $ct.t' = c$. Thus we have $1_S = tt' + (1 - tt') \in tS + \mathcal{M}$, as required. Suppose, on the other hand, that $\eta < D\psi$. Since $t \notin \mathcal{M}$, we have, for all $\phi \in p$, $\eta \not\leq D\phi$, that is, $D\eta \notin p$, so, by maximality of p with respect to not containing ψ, there is $\phi \in p$ with $\phi \wedge D\eta \leq \psi$, that is, by 1.3.1,

$D\phi + \eta \geq D\psi$, so, by modularity, $(D\psi \wedge D\phi) + \eta = D\psi \wedge (D\phi + \eta) = D\psi$. Therefore $c = c_1 + c_2$ for some $c_1 \in (D\psi \wedge D\phi)(L)$ and $c_2 \in \eta(L)$. By 1.2.17, there are $t_1, t_2 \in S$ with $ct_1 = c_1$, $ct.t_2 = c_2$. Note that $t_1 \in \mathcal{M}$. Thus $c = c(t_1 + tt_2)$ and $1_S = (1 - (t_1 + tt_2)) + (t_1 + tt_2) \in \mathcal{M} + (\mathcal{M} + tS)$, as required.

By 1.3.7 $a \otimes c\mathcal{M} = 0$, so $(a \otimes c)(S/\mathcal{M})$ is a simple S-module. Let E be its injective hull. By injectivity of E there is a morphism $f : (N \otimes L)_S \to E_S$ which extends the embedding of $(a \otimes c)S/\mathcal{M}$ in E. Set $N' = \mathrm{Hom}_S({}_RL_S, E_S)$ and also set $g \in N'$ to be the composition $f(a \otimes 1_L)$.

We claim that $\mathrm{pp}^{N'}(g) = p$. Since $g(c) \neq 0$, 1.3.8 implies that $\psi \notin \mathrm{pp}^{N'}(g)$. Let $\phi \in p$. If $b \in D\phi(L)$, then $gb = f(a \otimes b) = f0 = 0$, so, by 1.3.12, $g \in \phi(N')$. Thus $\mathrm{pp}^{N'}(g) = p = \mathrm{pp}^N(a)$, so, by 4.3.46, N is a direct summand of N'.

It remains to show that N' is indecomposable. So consider, $\mathrm{End}(N'_R) = \mathrm{Hom}_R(N'_R, \mathrm{Hom}_S({}_RL_S, E_S)) \simeq \mathrm{Hom}_S(N' \otimes_R L_S, E_S) = \mathrm{Hom}_S(\mathrm{Hom}_S({}_RL_S, E_S) \otimes_R L, E_S)$ which, by 1.3.17^{op} (with L for M and N) and since ${}_RL$ is finitely presented, is isomorphic to $\mathrm{Hom}_S(\mathrm{Hom}_S(\mathrm{Hom}_R({}_RL, {}_RL_S)_S, E_S), E_S) \simeq \mathrm{Hom}_S(\mathrm{Hom}_S(S_S, E_S), E_S) \simeq \mathrm{Hom}_S(E_S, E_S)$, which is a local ring (E.1.23); in particular N' is indecomposable, as required.

(ii)⇒(i) Finally, suppose that $N \simeq \mathrm{Hom}_S({}_RL_S, E_S)$ as in the statement. The argument at the end of the previous part shows that N is indecomposable. Choose any non-zero element $g \in N$: it must be shown that the pp-type, p say, of g in N is neg-isolated. Since the image of g is non-zero it contains the simple submodule T of E, so there is $c \in L$ such that $gc \in T$ and $gc \neq 0$. By 1.2.6, the pp-type of c in L is generated by a pp condition which, by 1.3.1, may be taken to have the form $D\psi$ for some pp condition ψ for right R-modules. Certainly $\psi \notin p$, by 1.3.7, since the bilinear map $N \otimes_R L \to E$ given by $f \otimes l \mapsto f(l)$ takes $g \otimes c$ to $gc \neq 0$. We claim that p is neg-isolated by ψ.

Let $\phi \in p$ and suppose $b \in D\phi(L)$. By 1.3.7, $gb = 0$. Thus $F_{Dp}(L) = \sum_{\phi \in p} D\phi(L) \leq \ker(g)$. Conversely, if $b \in \ker(g)$ has pp-type in L generated by η, then, by 1.3.12, $D\eta \in p$, hence $b \in F_{Dp}(L)$.

Now suppose $\psi' \notin p$. Again by 1.3.12 $gD\psi'(L) \neq 0$, so, because E has simple essential socle, $c \in gD\psi'(L)$. Thus $c = d + \ker(g)$ for some $d \in D\psi'(L)$, say $c = d + b$ with, say, $b \in D\phi(L)$, where $\phi \in p$. Thus $c \in D\psi'(L) + D\phi(L)$, so, since $D\psi$ generates the pp-type of c, $D\psi' + D\phi \geq D\psi$, so, by 1.3.1, $\psi' \wedge \phi \leq \psi$. Thus the pp-type p is indeed maximal with respect to not containing ψ, that is, is neg-isolated, as required. □

Also see [357, §§3,4] for the condition that N be the source of a *minimal* left almost split morphism and also for the case when N is of finite endolength. In particular, [357, 4.2], if R is Krull–Schmidt, then every neg-isolated point is the source of a minimal left almost split morphism in Mod-R.

Beligiannis proves the analogue of this result in compactly generated triangulated categories, [64, 7.9].

Proposition 5.3.48. ([495, 9.29], cf. [20, 2.3, 3.2]) *Let X be a Ziegler-closed set. A pp-type p is X-neg-isolated iff whenever $H(p)$ is a direct summand of a product, $\prod_\lambda N_\lambda$, of indecomposable pure-injectives in X, then already $H(p) \simeq N_\lambda$ for some λ.*

Proof. We give the proof for $X = \mathrm{Zg}_R$: the general case is just the same but carried out in the localised functor category $(R\text{-mod}, \mathbf{Ab})_{DX}$. If N is neg-isolated, then, 5.3.45, $N \otimes -$ is the injective hull of a simple functor, S say. If N is a direct summand of a product $\prod_\lambda N_\lambda$, then, by 12.1.3, $N \otimes -$ is a direct summand of $\prod_\lambda (N_\lambda \otimes -)$, so S embeds in the latter. Therefore there is a non-zero map, hence an embedding, from S to one of the direct summands $N_\lambda \otimes -$, which must therefore be isomorphic to $N \otimes -$. So $N_\lambda \simeq N$, as required.

For the converse, let $a \in N$ be non-zero and set $p = \mathrm{pp}^N(a)$. For each pp condition $\psi \notin p$ let p_ψ be a pp-type which contains p, does not contain ψ and is maximal such. By the definitions $N_\psi = H(p_\psi)$ is neg-isolated. Consider the element $a' = (a_\psi)_\psi \in N^* = \prod_\psi N_\psi$, where $a_\psi \in N_\psi$ has pp-type p_ψ, and where ψ runs over all pp conditions not in p. By 1.2.3 $\mathrm{pp}^{N^*}(a') = p$, so, 4.3.36, N is a direct summand of N^*. By assumption N is isomorphic to one of the N_ψ, therefore, by 5.3.46, N is neg-isolated, as required. □

Recall that the support, $\mathrm{supp}(M)$, of a module M is $\langle M \rangle \cap \mathrm{Zg}_R$.

Corollary 5.3.49. *Suppose that X is a closed subset of Zg_R such that there is no superdecomposable pure-injective module with support contained in X. Suppose that $N \in X$ is X-neg-isolated and let M be a pure-injective module with* $\mathrm{supp}(M) = X$. *Then N is a direct summand of M.*

In particular if N is any indecomposable pure-injective which is neg-isolated and if the lattice $\mathrm{pp}(N)$ *has width $<\infty$, then N is a direct summand of every pure-injective module with support equal to that of N.*

Proof. By 4.4.2 M is the pure-injective hull of a direct sum of indecomposable pure-injectives, so clearly M is a direct summand of the product of those indecomposables and then 5.3.48 applies. For the second statement note 7.3.5. □

A pure-injective module M is said to be an **elementary cogenerator** ([495, §9.4]) if every module in the definable subcategory $\langle M \rangle$ of Mod-R generated by M embeds purely in a product of copies of M. Say that a definable subcategory $\mathcal{X}$ has an elementary cogenerator if it is the definable category generated by some elementary cogenerator: we will see below that every definable subcategory has an elementary cogenerator.

Theorem 5.3.50. ([495, 9.32]) *A pure-injective module M is an elementary cogenerator iff it has, as a direct summand, a copy of every neg-isolated point of its support,* supp(M)*, that is, iff M realises every M-neg-isolated pp-type.*

Proof. By 12.5.7, M is an elementary cogenerator iff $M \otimes -$ is a cogenerator of the hereditary torsion theory τ_{DX} corresponding to $X = \text{supp}(M)$, equivalently, by 11.1.5, if $M \otimes -$ is an injective cogenerator for the localised category $(R\text{-mod}, \mathbf{Ab})_{DX}$. If it is such a cogenerator, then certainly $M \otimes -$ must embed every simple object of that category, hence the injective hull of each simple object of $(R\text{-mod}, \mathbf{Ab})_{DX}$ must be a direct summand of $M \otimes -$. Therefore, by 5.3.45, $M \otimes -$ has, as a direct summand, a copy of $N \otimes -$ for each neg-isolated point N of X.

The converse is also true because if $\mathcal{C}$ is any locally finitely generated abelian category (such as $(R\text{-mod}, \mathbf{Ab})_{DX}$), then an injective object which embeds every simple object is an injective cogenerator (see E.1.11). □

Example 5.3.51. Let $R = \mathbb{Z}$. Then $\mathbb{Z}_{p^\infty}$ is an elementary cogenerator (directly or by 5.3.54 below), and $\overline{\mathbb{Z}_{(p)}}$ is not, because $\mathbb{Q}$ is a point (indeed, the only other point) of its Ziegler-closure but does not embed in any power of $\overline{\mathbb{Z}_{(p)}}$.

Corollary 5.3.52. ([495, 9.36]) *Every definable subcategory has an elementary cogenerator.*

Proof. As commented at the end of the proof of 5.3.50, $(R\text{-mod}, \mathbf{Ab})_{DX}$ (where X is the support of the definable subcategory) has an injective cogenerator, so, by that proof, the definable subcategory has an elementary cogenerator, namely any pure-injective module which realises every X-neg-isolated pp-type. □

For Mod-R this is also [181, Thm 1].

Corollary 5.3.53. *Every module embeds purely in a direct product of indecomposable pure-injective modules.*

Corollary 5.3.54. ([495, 9.33]) *Every Σ-pure-injective module is an elementary cogenerator.*

Proof. Suppose that M is Σ-pure-injective. By 5.3.50 it is enough to show that, setting $X = \text{supp}(M)$, every X-neg-isolated point of X is a direct summand of M. Let $N = H(p)$ with p being X-neg-isolated, by ψ say, be such a point. By 4.4.5 there is a pp condition ϕ such that $p(N) = \phi(N)$, hence such that $\phi/\phi \wedge \psi$ is an X-minimal pair which therefore (5.3.2) isolates $H(p)$ in X. Since, by 5.1.15, X is the Ziegler-closure of the indecomposable direct summands of M, $H(p)$, being isolated, must already be such a summand, as required. □

If M is Σ-pure-injective and contains, as a direct summand, a copy of every point of its support, supp(M), then every direct product of copies of M is a direct summand of a direct sum of copies of M. Krause and Saorín ([372]) refer to such modules as "product-complete".

5.3.6 Cantor–Bendixson analysis of the spectrum

The Cantor–Bendixson analysis of a space is described, then applied to the Ziegler spectrum. Assume the isolation condition. Then the pp intervals collapsed when restricting to the first derivative are exactly those of finite length (5.3.57). There is an inductive extension stated in terms of m-dimension of the lattice of pp conditions (5.3.59). Assuming the isolation condition, the Cantor–Bendixson rank of the spectrum equals the m-dimension of the lattice of pp conditions (5.3.60) and then it follows that the right and left Ziegler spectra have the same Cantor–Bendixson rank (5.3.61).

The Cantor–Bendixson analysis of a space layers the space according to how far points are from being isolated. This analysis, applied to the Ziegler spectrum, fits well with various algebraic properties of the points and properties (especially measures of "complexity") of the module category.

Let T be a topological space. A point $p \in T$ is **isolated** if $\{p\}$ is an open set: thus p cannot be a limit of other points. Such a point is assigned **Cantor–Bendixson rank** 0: $\mathrm{CB}(p) = 0$. Denote by T' the set of non-isolated points: this is a closed subset of T and is called the **first Cantor–Bendixson derivative** of T. Consider T' with the induced topology: there may well be isolated points of this space, so we continue the process. That is, set $T^{(0)} = T$, $T^{(1)} = T'$ and, having defined $T^{(\alpha)}$, set $T^{(\alpha+1)} = T^{(\alpha)\prime}$. There is no reason to confine this process to finite steps, so at limit ordinals, λ, put $T^{(\lambda)} = \bigcap\{T^{(\alpha)} : \alpha < \lambda\}$; again a closed subset of T. Set $\mathrm{CB}(p) = \alpha$ if $p \in T^{(\alpha)} \setminus T^{(\alpha+1)}$ and, in that case, say that p has **Cantor–Bendixson rank** α. Define $T^{(\infty)} = \bigcap\{T^{(\alpha)} : \alpha \text{ is an ordinal}\}$. Since T is a set there is α such that $T^{(\infty)} = T^{(\alpha)}$.

If T is compact, then each $T^{(\alpha)}$, being a closed subset, will be compact.

It may happen that at some stage we reach the empty set, $T^{(\alpha)} = \emptyset$. If so, take the minimal such α (recall that every non-empty set of ordinals has a least member). If T is compact and if λ is a limit ordinal, then it cannot be that $\bigcap_{\beta<\lambda} T^{(\beta)} = \emptyset$ unless one of the $T^{(\beta)}$ with $\beta < \lambda$ is already empty. Therefore if T is compact this minimal α is not a limit ordinal, hence $\alpha - 1$ exists. In any case, whether or not T is compact, if there is α such that $T^{(\alpha)} = \emptyset$ and $T^{(\alpha-1)} \neq \emptyset$ say that T has **Cantor–Bendixson rank** $\alpha - 1$ and write $\mathrm{CB}(T) = \alpha - 1$. Also in this case say that T **has** Cantor–Bendixson rank, **CB rank** for short and note that $\mathrm{CB}(T)$ is the maximum rank of points of T. For instance $\mathrm{CB}(T) = 0$ exactly if every point is isolated.

The other possibility is that $T^{(\infty)} \neq \emptyset$, in which case we say that the Cantor–Bendixson rank of T is "∞" or "**undefined**" and set $\mathrm{CB}(T) = \infty$. Also set the CB rank of any point in $T^{(\infty)}$ to be ∞.

Lemma 5.3.55. *If the space T is compact and has Cantor–Bendixson rank then there are just finitely many points of maximal Cantor–Bendixson rank.*

Proof. Say $CB(T) = \alpha$. Since T is compact, so is $T^{(\alpha)}$, which, since it has no non-isolated points, must be finite. □

Lemma 5.3.56. *If T has Cantor–Bendixson rank, then the isolated points are dense in T.*

Proof. Otherwise there would be a non-empty open set U with no isolated points. By induction on α, $U \subseteq T^{(\alpha)}$ for every α, so $T^{\infty} = \bigcap_\alpha T(\alpha) \neq \emptyset$. □

Recall that if X is a closed subset of Zg_R, then $\mathrm{pp}^n(X)$ denotes the quotient lattice of pp^n_R consisting of X-equivalence classes of pp conditions. Also if $X' \subseteq X$ is a closed subset, then there is the natural epimorphism of lattices $\pi : \mathrm{pp}^n(X) \to \mathrm{pp}^n(X')$ which takes the X-equivalence class of a pp condition to its X'-equivalence class. Note that the kernel (see Section 7.1) of π consists of those pairs (ϕ, ψ) which have the same X'-equivalence class, equivalently such that the pair $(\phi + \psi)/(\phi \wedge \psi)$ is closed on every point of X'.

Proposition 5.3.57. *Assume that the isolation condition holds for the closed subset X of Zg_R. If X' is the first Cantor–Bendixson derivative of X, then for every n the kernel of the epimorphism of lattices $\pi : \mathrm{pp}^n(X) \to \mathrm{pp}^n(X')$ is the set of pairs, ϕ, ψ, of pp conditions such that the interval $[\phi + \psi, \phi \wedge \psi]_X$ has finite l ength.*

Proof. If ϕ/ψ is an X-minimal pair, that is, if the interval $[\phi, \psi]_X$ is simple, then (5.3.2) $(\phi/\psi) \cap X = \{N\}$ for some, isolated, point N. In particular, $\phi(N') = \psi(N')$ for every non-isolated point of X. Therefore $[\phi, \psi]_{X'}$ is trivial as, therefore, is $[\phi_1, \psi_1]_X$ whenever $[\phi_1, \psi_1]_X$ has finite length.

For the converse, suppose that $[\phi, \psi]_X$ has infinite length. It must be shown that $(\phi/\psi) \cap X' \neq \emptyset$.

If $(\phi/\psi) \cap X' = \emptyset$, then each point of $(\phi/\psi) \cap X$ is isolated in X, so, by compactness of (ϕ/ψ) (5.1.22), $(\phi/\psi) \cap X$ is finite, say $\{N_1, \dots, N_t\}$ with each N_i an isolated point of X.

If $t = 1$, then $[\phi, \psi]_X$ is finite by 5.3.22.

If $t > 1$, set Y_i to be the closed set $X' \setminus \{N_j : j \neq i\}$. Then $(\phi/\psi) \cap Y_i = \{N_i\}$, so, by 5.3.22, $[\phi, \psi]_{Y_i}$ is of finite length, l_i say. It follows that $[\phi, \psi]_X$ has length no more than $l = l_1 + \cdots + l_t$ for, if $\phi = \phi_0 > \phi_1 > \cdots > \phi_{l+1} = \psi$, then for

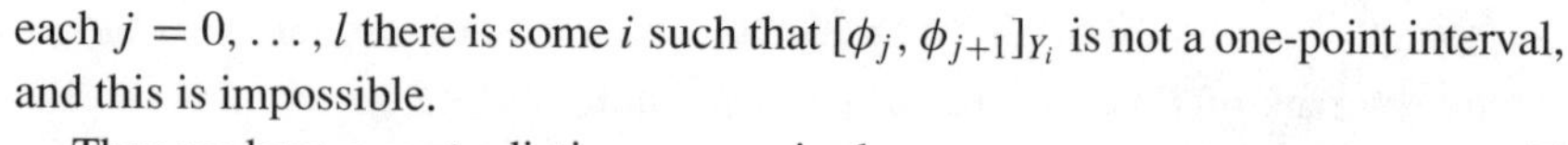

each $j = 0, \ldots, l$ there is some i such that $[\phi_j, \phi_{j+1}]_{Y_i}$ is not a one-point interval, and this is impossible.

Thus we have a contradiction, as required. □

Corollary 5.3.58. *Let M be any module and set $X = \mathrm{supp}(M)$. Suppose that the isolation condition holds for X. Then every indecomposable direct summand of $H(H(M)/M)$ lies in the Cantor–Bendixson derivative X'.*

Proof. This follows directly from 4.3.28 and 4.3.23. □

We extend 5.3.57 by induction.

Proposition 5.3.59. ([495, 10.19], [726, 8.6]) *Assume that the isolation condition holds for the closed subset X of Zg_R. If α is an ordinal and $X^{(\alpha)}$ denotes the αth Cantor–Bendixson derivative of X, then the kernel of the epimorphism of lattices $\mathrm{pp}^n(X) \to \mathrm{pp}^n(X^{(\alpha)})$ is the set of pairs ϕ, ψ such that $\mathrm{mdim}[\phi + \psi, \phi \wedge \psi] < \alpha$ (for m-dimension see Section 7.2).*

Proof. It is enough to consider the case where $\phi \geq \psi$. Assume that we have the statement for an ordinal α. Clearly the interval $[\phi, \psi]$ is collapsed by $\mathrm{pp}^n(X) \to \mathrm{pp}^n(X^{(\alpha+1)})$ iff $[\phi, \psi]_{X^{(\alpha)}}$ is collapsed by $\mathrm{pp}^n(X^{(\alpha)}) \to \mathrm{pp}^n(X^{(\alpha+1)} = X^{(\alpha)\prime})$ and, by 5.3.57, this is so iff $[\phi, \psi]_{X^{(\alpha)}}$ has finite length. Now, that is easily seen to be so iff there is a chain $\phi = \phi_0 > \phi_1 > \cdots > \phi_l = \psi$ such that each interval $[\phi_i, \phi_{i+1}]_{X^{(\alpha)}}$ $(i = 0, \ldots, l-1)$ is simple, hence satisfies the condition that for every ϕ' with $\phi_i \geq \phi' \geq \phi_{i+1}$ either $[\phi_i, \phi']_{X^{(\alpha)}}$ or $[\phi', \phi_{i+1}]_{X^{(\alpha)}}$ is trivial. By induction (and 7.1.2) this is the condition that either $\mathrm{mdim}[\phi_i, \phi'] < \alpha$ or $\mathrm{mdim}[\phi', \phi_{i+1}] < \alpha$. Just from the definition of m-dimension this is equivalent to each $[\phi_i, \phi_{i+1}]$, and hence $[\phi, \psi]$, being of m-dimension $< \alpha + 1$.

That deals with successor ordinals. If λ is a limit ordinal and $[\phi, \psi]_{X^{(\lambda)}}$ is trivial, then $(\phi/\psi) \cap X^{(\lambda)} = \emptyset$, so, by compactness of (ϕ/ψ) (and since $X^{(\lambda)} = \bigcap_{\alpha<\lambda} X^{(\alpha)}$), we have $(\phi/\psi) \cap X^{(\alpha)} = \emptyset$ for some $\alpha < \lambda$. So $\mathrm{mdim}[\phi, \psi] < \alpha < \lambda$, as required. □

For the dimension KG (Krull–Gabriel dimension), which appears in the next result, see Section 13.2.1.

Corollary 5.3.60. ([495, 10.19], [726, 8.6]) *Assume that the isolation condition holds for the closed subset X of Zg_R. Then $\mathrm{CB}(X) = \mathrm{mdim}(\mathrm{pp}(X))$. In particular, if $\mathrm{CB}(X) < \infty$, then the finitely many points of maximal CB-rank all are of finite endolength.*

Therefore, if the isolation condition holds for Zg_R, for example (5.3.17) if $\mathrm{mdim}(\mathrm{pp}_R) < \infty$, that is, if $\mathrm{KG}(R) < \infty$, then $\mathrm{CB}(\mathrm{Zg}_R) = \mathrm{mdim}(\mathrm{pp}_R) = \mathrm{KG}(R)$.

Although $\mathrm{mdim}(\mathrm{pp}_R) = \mathrm{KG}(R)$ (13.2.2) $= \mathrm{mdim}(\mathrm{pp}_{R^{\mathrm{op}}})$ (7.2.4) $= \mathrm{KGdim}(R^{\mathrm{op}})$ it is not known whether or not $\mathrm{CB}({}_R\mathrm{Zg}) = \mathrm{CB}(\mathrm{Zg}_R)$ always holds, nor whether the isolation condition on Zg_R implies the isolation condition on ${}_R\mathrm{Zg}$.

Corollary 5.3.61. *If R satisfies the isolation condition, then* $\mathrm{CB}(\mathrm{Zg}_R) = \mathrm{CB}({}_R\mathrm{Zg})$.

The next example illustrates the above analysis of a closed subset of the spectrum, by using it to determine the definable subcategory generated by a particular module.

Example 5.3.62. Let $R = k[X_i\ (i \geq 0)]/\langle (X_i)_{i\geq 0}\rangle^2 = k[x_i\ (i \geq 0) : x_i x_j = 0\ \forall i, j]$, where k is a field. Consider the module R_R. The pp-definable subgroups of R_R are R, $J = \langle (x_i)_{i\geq 0}\rangle$ and the ideals of finite k-dimension 4.4.15. The module R_R is Σ-pure-injective (4.4.5), hence has the dcc on pp-definable subgroups. Let $\mathcal{X}$ be the definable subcategory of Mod-R generated by R_R and set $X = \mathcal{X} \cap \mathrm{Zg}_R$. Since R_R is indecomposable it is the unique isolated point of X (by 5.3.17 and 5.3.6 for example). By 5.3.57 the lattice of pp conditions for the first CB derivative, X', of X is the two-point lattice, so (again by 5.3.6) there is a unique non-isolated point of X. The obvious candidate is the simple module R/J: how can we check this? One possibility is the following.

Consider $M = R^{\mathbb{N}}$ and the element $a = (x_i)_i$ in M. The submodule N of M generated by a is isomorphic to R/J, so it will be enough to show that the inclusion of N in M is pure, for then, since N is pure-injective, N must be a point of X. The pp-definable subgroups of M are defined by the same pp conditions as those of R (1.2.3), so are, apart from M and 0, defined by the following conditions, where we use v for our free variable ("x" being notationally too close to the elements of R): ϕ_J, which is $vx_0 = 0$, which defines $J^{\mathbb{N}}$ and which satisfies $\phi_J(N) = N = N \cap \phi_J(M)$; for each element $a \in J$ the condition ϕ_a, which is $a|v$ and which satisfies $\phi_a(N) = 0 = N \cap \phi_a(M)$; finite sums ϕ of the ϕ_a, which must, therefore, also satisfy $\phi(N) = N \cap \phi(M)$. Thus N is pure in M, hence $X = \{R_R, R/J\}$ and so, by 4.4.19, $\mathcal{X}$ consists of arbitrary direct sums of copies of these modules.

5.3.7 The full support topology

Allowing open sets to be defined by pp-types rather than just pp conditions gives a finer topology on the set of indecomposable pure-injectives (5.3.65). This may also be defined in terms of subcategories closed under products, pure-submodules and pure-injective hulls (5.3.63), equivalently (Section 12.7) torsion theories in the functor category which are not necessarily of finite type. As for the Ziegler topology, realised types give neighbourhood bases of points (5.3.66).

The isolated points of this full support topology are exactly the neg-isolated modules (5.3.67). Unlike the Ziegler topology, the space is not compact (5.3.68).

In [101], see [102], Burke introduced a topology on the set, pinj_R, of isomorphism classes of indecomposable pure-injective R-modules which is finer than the Ziegler topology. As is the case for the Ziegler topology there are many equivalent ways to define this topology, which Burke called the **full support** topology.[19] We give a definition which parallels that which was used to define the Ziegler topology (Section 5.1.1). Burke pointed out ([102, p. 20]) that this topology appears in [486, 5.3.9].

Say that a subcategory or subclass of Mod-R is **type-definable** if there is a set, $\{p_\lambda/\psi_\lambda\}_{\lambda\in\Lambda}$, of pairs with p_λ a pp-type and ψ_λ a pp condition not in p_λ, such that, for all $M \in$ Mod-R, $M \in \mathcal{X}$ iff $p_\lambda(M) = \psi_\lambda(M) \cup p_\lambda(M)$ for all λ. The next result is comparable to 3.4.7. For example the absolutely pure modules always form a type-definable class (by 2.3.5, 4.3.12), cf. 3.4.24; see [528] for more examples. Note that every definable subcategory of Mod-R is type-definable.

Theorem 5.3.63. *Suppose that $\mathcal{X}$ is a subclass of* Mod-R *which is closed under pure-injective hulls. Then $\mathcal{X}$ is type-definable iff it is also closed under products and pure submodules.*

Proof. (Recall that $p(M) = \bigcap_{\phi\in p} \phi(M)$.) Clearly any class of modules defined by conditions of the form $p(M) \subseteq \psi(M)$ is closed under products (1.2.3) and pure submodules.

For the converse, let $\mathcal{X} \subseteq$ Mod-R be closed under products, pure submodules and pure injective hulls. Let τ be the smallest hereditary torsion theory (not necessarily of finite type!) on (R-mod, **Ab**) such that all $M \otimes -$ with $M \in \mathcal{X}$ are torsionfree (cf. the beginning of Section 12.3). We claim that $\mathcal{X}$ consists exactly of those $M \in$ Mod-R such that $M \otimes -$ is τ-torsionfree. Since $\mathcal{X}$ is closed under pure-injective hulls, it follows by 4.3.14 that the class $\{M \otimes - : M \in \mathcal{X}\}$ is closed under injective hulls. Since, 12.1.3, $M \mapsto (M \otimes -)$ preserves products, this class is closed under products. If $M \otimes -$ is τ-torsionfree, then $M \otimes -$ embeds in some τ-torsionfree injective $N \otimes -$ with, by the first observation, $N \in \mathcal{X}$, so, by 12.1.3, the corresponding morphism $M \to N$ is a pure embedding, hence $M \in \mathcal{X}$. By 11.1.1 our claim is established.

It follows by 11.1.11 that τ is determined by the τ-torsion finitely generated objects of (R-Mod, **Ab**). By 12.2.2 these are exactly the functors of the form $(F_{D\psi} + F_{Dp})/F_{Dp}$, where p is a pp-type and ψ is a pp condition. Let Φ be the set

[19] Originally, in view of the definition below and the comparison with the Ziegler, that is, formulas-over-formulas, topology, he named it the types-over-formulas topology. The name "full support" reflects the functorial version of the definition, for which see Section 12.7.

of all pairs p, ψ, let us write p/ψ, where $(F_{D\psi} + F_{Dp})/F_{Dp}$ is τ-torsion. By what was shown above, $\mathcal{X}$ consists exactly of those M such that $((F_{D\psi} + F_{Dp})/F_{Dp}, M \otimes -) = 0$ for all $p/\psi \in \Phi$, that is, by 12.2.4, such that $p(M)/p(M) \cap \psi(M) = 0$ for all $p/\psi \in \Phi$, as required. □

The assumption that $\mathcal{X}$ be closed under pure-injective hulls is necessary: there is an example of Piron ([482, Chpt. 4], see [496, p. 169])) of a module M, pp-type p and pp condition ψ such that $\psi(M) \geq p(M)$ but $\psi(H(M)) \not\geq p(H(M))$.

Consider those subsets Z of pinj_R of the form $Z = \mathcal{Z} \cap \mathrm{pinj}_R$, where $\mathcal{Z}$ is a type-definable subcategory of Mod-R.

Theorem 5.3.64. ([101, §7.4], [102, 6.10]) *The subsets of the form $\mathcal{Z} \cap \mathrm{pinj}_R$ for $\mathcal{Z}$ a type-definable subcategory of* Mod-R *closed under pure injective hulls are the closed sets of a topology on* pinj_R.

Proof. This is 12.7.1 via 12.1.9 and 12.2.2 as in the proof of 5.3.63. □

In contrast to the Ziegler topology (5.1.23), this space, for which the notation fs_R is introduced in Section 12.7, need not be compact (see 5.3.68 below).

We comment that, in contrast to the situation (5.1.6) for the Ziegler topology, 5.3.64 need not induce a bijection between type-definable classes closed under pure-injective hulls and fs_R-closed subsets of pinj_R, see 12.7.5, although there will be a bijection if there are no superdecomposable pure-injectives since then every hereditary torsion theory is, by 11.1.1 and 4.4.2 (for injective objects), determined by the indecomposable torsionfree injectives. That is, a type-definable subcategory, even one closed under pure-injective hulls, is not necessarily determined by its intersection with pinj_R.

Burke's original definition of this topology was that having, for a basis of open sets, those of the form $(p/\psi) = \{N \in \mathrm{pinj}_R : \exists \overline{a} \text{ from } N \text{ with } \mathrm{pp}^N(\overline{a}) \supseteq p \text{ and } \psi \notin \mathrm{pp}^N(\overline{a})\}$, where p is a pp-type and ψ a pp condition not in p.

Corollary 5.3.65. ([102, §7]) *The sets of the form (p/ψ) for p a pp-type and ψ a pp condition not in p form a basis of open sets for the full support topology. In order to have a basis it is sufficient to take only those (p/ψ) with p irreducible and p and ψ with just one free variable.*

Proof. The first statement is by 5.3.63 and 5.3.64.

The second statement follows from 12.2.1 and injecting 11.1.39 into the proof of 5.3.63 at the point where we describe the finitely generated torsion functors which determine the torsion theory. □

The proof of the next result is like that of 5.1.21, only slightly simpler since we need just a finitely generated (rather than finitely presented) functor.

Corollary 5.3.66. *If $N = H(p) \in \text{pinj}_R$, then the sets of the form (p/ψ) with $\psi \notin p$ form a basis of open neighbourhoods of N in the full support topology.*

The next result follows by 12.5.6 and 12.7.4.

Proposition 5.3.67. ([101], see [102, 8.2, 6.11]) *Let X be a Ziegler-closed subset of* pinj_R. *An indecomposable pure-injective $N \in X$ is isolated in the full support topology restricted to X iff it is a neg-isolated point of X (which, 5.3.45, is so iff $N \otimes -$ is the injective hull of a simple functor in the corresponding localised category* $(R\text{-mod}, \mathbf{Ab})_{DX}$*). The* fs_R*-isolated points are dense in* pinj_R *when this is given the full support topology.*

Example 5.3.68. ([101, 7.1.15]) Let $R = \mathbb{Z}$. Each Prüfer group $\mathbb{Z}_{p^\infty}$ is neg-isolated, hence is isolated in the full support topology, and no other infinitely generated points are neg-isolated (see 5.3.51) so the isolated points in the $\text{fs}_{\mathbb{Z}}$ topology are the points of finite length and the Prüfer points. If p is the pp-type of any non-zero element of $\mathbb{Q}$, then the $\text{fs}_{\mathbb{Z}}$-open set $(p(x)/x = 0)$ contains $\mathbb{Q}$ and no adic point. Hence $\mathbb{Q}$, as well as all adic points, are of Cantor–Bendixson rank 1 with respect to this topology (which is, therefore, not compact).

In [102, §7] it is pointed out that a result from [723] takes the following form in this terminology.

Theorem 5.3.69. ([723, 3.2, 4.1]) *Suppose that R is a hereditary finite-dimensional algebra:*

(a) R is of tame representation type iff the type-definable subcategory of Mod-R *generated by the preinjective modules is definable;*
(b) R is of finite representation type iff the type-definable subcategory of Mod-R *generated by the preprojective modules is definable.*

There is a little more on this topology, results on which are developed in [101, Chpt. 7], in Sections 12.7 and 14.4.

5.4 Duality of spectra

For any ring the right and left Ziegler topologies are homeomorphic as locales (5.4.1) and for countable rings they are homeomorphic once topologically equivalent points have been identified (5.4.7). This uses the fact that over countable rings every irreducible closed set has a generic point (5.4.6); it is not known whether this is true without restriction.

Points which have a minimal pair with respect to any closed set are reflexive (5.4.12), meaning that they have a well-defined dual in the 'opposite' space. If m-dimension is defined, then every point is reflexive (5.4.14) and the right and left Ziegler spectra are homeomorphic at the level of points (5.4.20). The dual of a reflexive point may be found via Hom-duality (5.4.17).

Herzog discovered that the duality (Section 1.3.1) between the lattices of pp conditions for right and for left modules, lifts to give a homeomorphism (or, at least, something close to one) between the right and left Ziegler spectra of a ring.

In many cases we really do have a homeomorphism: a bijection between the points of the right and left spectra which is a homeomorphism of the spaces. In all cases we have a homeomorphism "at the level of topology", see 5.4.1. No counterexample is currently known to there being a literal homeomorphism in all cases.

The Ziegler spectrum may contain points which are topologically indistinguishable (an extreme example is given at 8.2.93). This is one possible obstruction to the right and left spectra being homeomorphic in general since a possibility which has not been ruled out is that there is a ring, R, and a point, N, in Zg_R such that there are κ points topologically indistinguishable from N whereas its "dual" in ${}_R\mathrm{Zg}$ has $\mu \neq \kappa$ points topologically indistinguishable from it. Indeed, it is not at all clear whether one should expect "natural" bijections between clusters of topologically indistinguishable points. In any case, we can identify topologically indistinguishable points and then ask whether the resulting quotient spaces are homeomorphic.

Even after identifying topologically indistinguishable points, however, it is not known whether there is a homeomorphism between the quotient spaces, although if R is countable (5.4.7) and over many other rings, by 5.4.20, there is. But what is always true is that there is an isomorphism between the topologies of Zg_R and ${}_R\mathrm{Zg}$. By the *topology* we mean the lattice, indeed complete Heyting algebra, of open subsets.

If T is any topological space, then its lattice of open sets is a **complete Heyting algebra**, meaning that, as well as being a lattice under the operations of $\cup$ and $\cap$, it has arbitrary joins $\bigcup$ and $\cap$ distributes over $\bigcup$: $X \cap \bigcup_\lambda Y_\lambda = \bigcup_\lambda (X \cap Y_\lambda)$. There is the obvious category of complete Heyting algebras and a **locale** is an object of the opposite category, that is, a complete Heyting algebra but where a morphism from the locale X to the locale Y is a morphism from the complete Heyting algebra Y to the complete Heyting algebra X (then, if locales are regarded as generalised topological spaces, the arrows go in the "right" direction).

We will say that two spaces are **homeomorphic at the level of topology** (or that we have an **isomorphism of topologies**) if their algebras of open sets are isomorphic by a bijection which respects intersections and arbitrary unions.

Theorem 5.4.1. ([274, 4.4]) *For any ring R the right and left Ziegler spectra of R are homeomorphic at the level of topology, the map being defined by taking the basic open set (ϕ/ψ) to $(D\psi/D\phi)$.*

Proof. First, this map is well defined: suppose $(\phi/\psi) = (\phi'/\psi')$. By 1.3.15, $N \in (D\psi/D\phi)$ iff $\phi(N^*) > \psi(N^*)$, where the module duality $*$ is as there. By 5.1.3 this is so iff $(\langle N^*\rangle \cap \mathrm{Zg}_R) \cap (\phi/\psi) \neq \emptyset$, where $\langle N^*\rangle$ denotes the definable subcategory of Mod-R generated by N^*, and that is true iff $(\langle N^*\rangle \cap \mathrm{Zg}_R) \cap (\phi'/\psi') \neq \emptyset$, hence iff $N \in (D\psi'/D\psi')$.

Therefore we may set $D(\phi/\psi) = (D\psi/D\phi)$. Since $D^2\phi$ is equivalent to ϕ (1.3.1) it follows that this map is a bijection on the given basis.

Next, the map D on open sets is order-preserving. For suppose that $(\phi/\psi) \subseteq (\phi'/\psi')$. By 1.3.15, $N \in (D\psi/D\phi)$ iff $\phi(N^*) > \psi(N^*)$ and this, by 5.1.10, implies $\phi'(N^*) > \psi(N^*)$, so $N \in (D\psi'/D\phi')$ as required.

The map D, being bijective and order-preserving on these basic open sets, extends uniquely to a bijection, which we also denote by D, on the finite unions and intersections of such sets. It remains to extend D to arbitrary unions of open sets.

Set $DU = \bigcup_\lambda (D\psi_\lambda/D\phi_\lambda)$ whenever $U = \bigcup_\lambda (\phi_\lambda/\psi_\lambda)$: it must be shown that this is well defined. So suppose that also $U = \bigcup_\mu (\phi'_\mu/\psi'_\mu)$. It will be sufficient, by symmetry, to show that $(D\psi'_\mu/D\phi'_\mu) \subseteq \bigcup_\lambda (D\psi_\lambda/D\phi_\lambda)$. Since $(\phi'_\mu/\psi'_\mu) \subseteq \bigcup_\lambda (\phi_\lambda/\psi_\lambda)$ compactness of the first set, 5.1.22, gives $(\phi'_\mu/\psi'_\mu) \subseteq (\phi_{\lambda_1}/\psi_{\lambda_1}) \cup \cdots \cup (\phi_{\lambda_n}/\psi_{\lambda_n})$ for some $\lambda_1, \dots, \lambda_n$. Then, by what has been shown already, $(D\psi'_\mu/D\phi'_\mu) = D(\phi'_\mu/\psi'_\mu) \subseteq D\big((\phi_{\lambda_1}/\psi_{\lambda_1}) \cup \cdots \cup (\phi_{\lambda_n}/\psi_{\lambda_n})\big) = (D\psi_{\lambda_1}/D\phi_{\lambda_1}) \cup \cdots \cup (D\psi_{\lambda_n}/D\phi_{\lambda_n})$, as required. □

As in the proof above, we denote by DU the image of the open set U in the "opposite" Ziegler spectrum. We also extend this notation to closed subsets, X, of Zg_R by setting $DX = (D(X^c))^c$, where ^{c} denotes set-theoretic complement. In line with our existing use of the term, we refer to this duality (on the various types of objects on which it is defined) as **elementary duality**.

Corollary 5.4.2. *For any ring R the ($\bigcap$-complete) lattices of closed subsets of Zg_R and ${}_R\mathrm{Zg}$ are isomorphic by the map D.*

Example 5.4.3. If R is commutative, then the equivalence between Mod-R and R-Mod induces a homeomorphism $\mathrm{Zg}_R \simeq {}_R\mathrm{Zg}$ which will not, in general, coincide with this duality. The composition of these gives a homeomorphism of Zg_R to itself which need not be the identity. For example, if $R = \mathbb{Z}$, then elementary duality interchanges the p-adic $\overline{\mathbb{Z}_{(p)}}$ and p-Prüfer $\mathbb{Z}_{p^\infty}$ groups and leaves points of other types fixed. This may be seen by direct computation (many proofs are possible), using the information in Section 5.2.1. For example, if P is a prime ideal, then once the isolated points have been removed the P-Prüfer point is isolated in what remains by $(xP = 0 / x = 0)$; elementary duality transforms this to the open set (for left modules) $(x = x / P \mid x)$, which isolates the P-adic point within the first Cantor–Bendixson derivative. Here we are using the observation that the Cantor–Bendixson process is preserved by the duality of 5.4.1 provided the isolation condition holds (without this assumption one cannot be sure that the dual, $(D\psi/D\phi)$, of a pair (ϕ/ψ) which isolates a point on the one side contains just one point).

Example 5.4.4. If R is an artin algebra, then there is a duality mod-$R \simeq (R\text{-mod})^{\mathrm{op}}$ given by $M \mapsto \mathrm{Hom}_C({}_CM_R, {}_CE)$, where E is the injective hull of the top, C/JC, of the centre, $C = C(R)$, of R (15.1.3). If R is a finite-dimensional k-algebra, then one may use $M \mapsto \mathrm{Hom}_k({}_kM_R, k)$ instead. Then 1.3.15 applies to show that this is the restriction of the above duality to the isolated points (5.3.37) of Zg_R and ${}_R\mathrm{Zg}$.

Example 5.4.5. (see [278, p. 530]) If R is a commutative noetherian local ring with maximal ideal J and if R is complete in the J-adic topology, then, [438], the duality $M \mapsto \mathrm{Hom}_R(M, E(R/J))$ gives a duality, which agrees with elementary duality, between the category of finitely generated R-modules and the category of artinian R-modules. Note that by 4.2.15 and 4.4.5 both these categories consist of pure-injective modules.

A closed subset X of a topological space is said to be **irreducible** if whenever $X = Y \cup Z$ with Y and Z both closed, then either $X = Y$ or $X = Z$, equivalently, if the intersection of any finitely many non-empty open subsets of X is non-empty. A **generic point** of a closed set X is a point whose closure equals X. Note that if a closed set has a generic point then that set is irreducible.

Proposition 5.4.6. ([274, 4.7]) *If X is an irreducible closed subset of Zg_R such that X has a countable basis of open sets in the relative topology, then X has a generic point.*

Proof. Suppose that $\{X \cap (\phi_n/\psi_n)\}_{n\geq 1}$ is a countable basis of non-empty open sets for the relative topology on X. By 5.1.8 it may be supposed that all these pp conditions ϕ_n, ψ_n have just one free variable. As in a number of previous proofs, we work in the localisation $(R\text{-mod}, \mathbf{Ab})_{DX}$ of the functor category, and write F_{DX} for the image in $(R\text{-mod}, \mathbf{Ab})_{DX}$ of $F \in (R\text{-mod}, \mathbf{Ab})$.

Define a sequence of pairs $F_n > F_n'$ of finitely generated subfunctors of $(R^n \otimes -)_{DX}$ as follows. Set $F_1 = (F_{D\psi_1})_{DX} > (F_{D\phi_1})_{DX} = F_1'$; the inclusion is proper since $X \cap (\phi_1/\psi_1)$ is non-empty and by 12.3.15. Suppose inductively that $F_1 \geq F_2 \geq \cdots \geq F_n > F_n' \geq \cdots \geq F_1'$ have been defined and that the (non-empty) open subset (F_n/F'_n) of X (in the notation of 12.3.14) is contained in each $X \cap (\phi_i/\psi_i)$, $i = 1, \dots, n$.

Since X is irreducible, $(F_n/F'_n) \cap (\phi_{n+1}/\psi_{n+1}) \neq \emptyset$, so, by the argument of 5.1.20 applied in the localised category, F_n/F'_n and $(F_{D\psi_{n+1}}/F_{D\phi_{n+1}})_{DX}$ have, up to isomorphism, a common non-zero finitely presented subquotient, G_{n+1} say, with $(G_{n+1}) \subseteq (F_n/F'_n) \cap (\phi_{n+1}/\psi_{n+1})$. Therefore there are finitely generated F_{n+1} and F'_{n+1} with $F_n \geq F_{n+1} > F'_{n+1} \geq F'_n$ and with $F_{n+1}/F'_{n+1} \simeq G_{n+1}$, hence, by 12.1.20, with $(F_{n+1}/F'_{n+1}) \subseteq X \cap (\phi_{n+1}/\psi_{n+1})$. Thus the induction continues.

Now we apply 12.2.7 with G there being $\sum_n DF_n$ and H there being $\sum_n F'_n$: note that the conditions for that result are satisfied. We obtain $H' \geq \sum F_n$ as there such that $((R^n \otimes -)/H')_{DX}$ is uniform, hence such that $E((R^n \otimes -)/H')_{DX}$ is indecomposable injective, hence of the form $N \otimes -$ for some $N \in X$. Also by construction we have that for every n $(F_n/F'_n, N \otimes -) \neq 0$, hence $N \in \bigcap_n (F_n/F'_n)$ belongs to every open subset of X, as required. □

Given points x, y of any topological space T, set $x \approx y$ if x and y belong to exactly the same open sets (that is, each is in the closure of the other) and say that x and y are **topologically indistinguishable**. Write $T/\approx$ for the quotient set of T by this equivalence relation, endowed with the quotient topology. For examples where this is not equality see 8.2.74 and Section 8.2.12.

Theorem 5.4.7. ([274, 4.9]) *If X is a closed subset of Zg_R and has a countable basis of open sets, then $X/\approx$ is homeomorphic to $DX/\approx$. In particular, if R is countable, then $\mathrm{Zg}_R/\approx$ is homeomorphic to ${}_R\mathrm{Zg}/\approx$.*

Proof. Let $N \in X$ and denote by Z_N its closure, which is irreducible. The dual closed subset, DZ_N, of ${}_R\mathrm{Zg}$ also is irreducible. For, if $(D\psi/D\phi) \cap DZ_N \cap (D\psi'/D\phi') = \emptyset$, then $(D\psi/D\phi) \cap (D\psi'/D\phi') \subseteq (DZ_N)^c$, so, by 5.4.1, $(\phi/\psi) \cap (\phi'/\psi') \subseteq Z_N^c$. Therefore, by irreducibility of Z_N, either $(\phi/\psi) \cap Z_N$ or $(\phi/'\psi') \cap Z_N$ is empty, say the former, so $(\phi/\psi) \subseteq Z_N^c$. By 5.4.1 again, $(D\psi/D\phi) \subseteq (DZ_N)^c$, so $(D\psi/D\phi) \cap DZ_N = \emptyset$.

Let N' be a generic point of DZ_N: this exists by 5.4.6 (by 5.4.1 the countability hypothesis dualises). Although N' might not be unique, all generic points of an irreducible closed set are $\approx$-equivalent so $(N/\approx) \mapsto (N'/\approx)$ is a well-defined map from $\mathrm{Zg}_R/\approx$ to ${}_R\mathrm{Zg}/\approx$ which, by right/left symmetry, is a bijection. Also, if $N \in (\phi/\psi)$, then $(\phi/\psi) \cap Z_N \neq \emptyset$ so $(D\psi/D\phi) \cap DZ_N \neq \emptyset$, hence $N' \in (D\psi/D\phi)$, so this map is clearly a homeomorphism. □

Question 5.4.8. For every ring are the right and left Ziegler spectra homeomorphic?

Question 5.4.9. For every ring do the right and left Ziegler spectra have the same Cantor–Bendixson rank?

Question 5.4.10. Assume that the isolation condition holds for the closed subset X of Zg_R. Does the isolation condition hold for $DX \subseteq {}_R\mathrm{Zg}$?

Proposition 5.4.11. *Assume that the closed subset X of Zg_R and the dual subset DX of ${}_R\mathrm{Zg}$ both satisfy the isolation condition. Then* $\mathrm{CB}(DX) = \mathrm{CB}(X)$.

This is immediate from 5.3.60 and 7.2.4.

Ideally one wants a pointwise homeomorphism, hence a duality at the level of points. For some points we have this. In [274] Herzog showed that any point of Zg_R which opens, and hence is relatively isolated by, an X-minimal pair (Section 5.3.2) for some closed subset X, has a dual point in Zg_R. Krause, in [347], [348], [358, §4.3], gave another treatment of these (calling them "simply reflexive"). In [275, 3.2] Herzog extended this to certain neg-isolated (Section 5.3.5) points, namely those which are the hull of a neg-isolated pp-type p which satisfies the condition that, if p is neg-isolated by ψ, then there is $\phi \in p$ such that the functor $F_{D\psi}/F_{D\phi}$ is uniform (or, dually, such that F_ϕ/F_ψ is uniform).

We follow [275] and [347] in extending the terminology from [274] and say that a point N of Zg_R is **reflexive** if there is a unique point, which we will refer to as the **elementary dual**,[20] DN, of N, in ${}_R\mathrm{Zg}$ such that for all pp-pairs ϕ/ψ one has $N \in (\phi/\psi)$ iff $DN \in (D\psi/D\phi)$ and which is itself reflexive (with dual necessarily N). That is, a point N is reflexive if each closed set, $\mathrm{supp}(N) \subseteq \mathrm{Zg}_R$ and $D\mathrm{supp}(N) \subseteq {}_R\mathrm{Zg}$ has a unique generic point, namely N in the first and the module we denote by DN in the second. Thus the modules termed "reflexive" in [274] ("simply reflexive" in [358]) and those referred to as "critical" and "cocritical" in [275] all are reflexive in this use of the term. Essentially by definition, if N is reflexive, then DN is reflexive and $D^2N = N$. Also note that if every point of Zg_R is reflexive, then there are no topologically indistinguishable points in that space.

In "functorial" terms, the condition is (see 12.3.7): N is reflexive iff each of the finite type hereditary torsion theory on (R-mod, **Ab**) cogenerated by $N \otimes -$ and the dual of this torsion theory on (mod-R, **Ab**) is cogenerated by a single and unique indecomposable injective functor ($N \otimes -$ in the first case and $- \otimes DN$ in the second).

Theorem 5.4.12. ([274, 4.10]) *If X is a closed subset of Zg_R and $N \in \mathrm{Zg}_R$ is isolated in X by an X-minimal pair, then N is reflexive and DN is isolated in DX by a DX-minimal pair.*

Specifically, if $N \in X$ is isolated in X by the X-minimal pair ϕ/ψ, then DN is isolated in DX by the DX-minimal pair $D\psi/D\phi$.

[20] Krause, [347], used the terminology "purely opposed" to describe the relation between N and DN.

Elementary duality gives, for such N and for each n, an anti-isomorphism: $\mathrm{pp}^n(N) \simeq (\mathrm{pp}^n(DN))^{\mathrm{op}}$.

Proof. By 1.3.1, since ϕ/ψ is an X-minimal pair, $D\psi/D\phi$ is a DX-minimal pair, so, by 5.3.6, there is a unique point, let us denote it by DN, in DX on which $D\psi/D\phi$ is open. Thus $(D\psi/D\phi) \cap DX = \{DN\}$. The same comments apply with supp(N) in place of X but, since the open subset $(\phi/\psi) = \{N\}$ is dense in supp(N), it follows by 5.4.1 that $(D\psi/D\phi)$ is dense in $D\,\mathrm{supp}(N)$. Thus DN is the unique generic point of $D\,\mathrm{supp}(N)$, as required.

The last statement is by 1.3.1, 5.1.7 and the definition of $D\,\mathrm{supp}(N)$. □

The duality between lattices of pp-definable subgroups extends to a duality, 12.3.20, between the respective functor categories/categories of pp-imaginaries.[21]

Corollary 5.4.13. *If N is neg-isolated and if the isolation condition holds (for the closure of N), then N is reflexive.*

Corollary 5.4.14. *If X is a closed subset of* Zg_R *and* $\mathrm{mdim}(X) < \infty$, *then every point of X is reflexive.*

Proof. By 5.3.60 the Cantor–Bendixson rank of X is defined. Therefore every point of X is isolated in some closed subset (one of the Cantor–Bendixson derivatives) of X, so, by 5.4.12, is reflexive. □

For instance, if R is a PI Dedekind domain (Section 5.2.1), then every point is reflexive, with the dual of $(R/P^n)_R$ being ${}_R(R/P^n)$, $\overline{R_{(P)}}_R$ and ${}_R(R_{P^\infty})$ dual to each other and Q_R dual to ${}_RQ$. If R is commutative and von Neumann regular, then ([274, §5]) if N_R is reflexive its dual is, as a consequence of 3.4.30, ${}_RN$. For an example where a module and its dual seem very different, see 8.2.40.

Herzog proves [274, 9.2], without any assumption of coherence (cf. 3.4.24), that if $X \subseteq \mathrm{Zg}_R$ is a closed subset consisting of reflexive modules all of which are flat, then all the points of DX are injective modules.

Corollary 5.4.15. *Suppose that N is an indecomposable pure-injective module with an N-minimal pair. Then DN is a direct summand of the left R-module* $\mathrm{Hom}_S({}_SN_R, {}_SE)$, *where* $S = \mathrm{End}(N)$ *and E is the minimal injective cogenerator of* S-Mod.

If N is of finite endolength, then $\mathrm{Hom}_S({}_SN_R, {}_SE)$ *is a direct sum of copies of DN.*

If N is finitely presented and of finite endolength, then $\mathrm{Hom}_S({}_SN_R, {}_SE) \simeq DN$.

[21] In contrast, there is in general little connection between the endomorphism rings of N and DN, in particular they need not be (anti-)isomorphic, see, for example, [347, §6].

Proof. By 1.3.15, $\text{supp}(\text{Hom}_S({}_SN_R,\ {}_SE)) = D\,\text{supp}(N)$ and $\text{Hom}_S({}_SN_R,\ {}_SE)$ is pure-injective by 4.3.29. If ϕ/ψ is an N-minimal pair, then $D\psi/D\phi$ is DN-minimal and is open on $\text{Hom}_S({}_SN_R,\ {}_SE)$, so, by 5.3.49 (rather, the proof of 5.3.48), DN is a direct summand of $\text{Hom}_S({}_SN_R,\ {}_SE)$.

If N is of finite endolength, then (5.1.12) it is the only point of supp(N), hence DN is the only point of the dual of this closed set. By 1.3.1 DN also is of finite endolength, so is a direct sum of indecomposables (necessarily copies of DN) by 4.4.29.

The last statement follows by (the proof of) 5.3.47. □

More generally, the last statement holds for N finitely presented and Σ-pure-injective ([274, 10.5, 10.6, 11.2]) and, without the assumption of finite presentation, there is a similar description of the dual at [274, 10.5]. Also see [274, 12.3] and [276, 2.6] for more results along these lines. The next result is proved by an argument like that used in 5.3.47.

Proposition 5.4.16. ([278, p. 538]) *Suppose that M is finitely presented and has local endomorphism ring. If the pure-injective hull $H(M)$ of M is reflexive, then $D(H(M)) = \text{Hom}_S({}_SM_R,\ {}_SE)$, where $S = \text{End}(H(M))$ and E is the minimal injective cogenerator for S-*Mod.

Corollary 5.4.17. *If N is an indecomposable module of finite endolength, then N is reflexive and the endolengths of N and its dual DN are equal.*

The last part follows from 4.4.32.

Corollary 5.4.18. *If every point of the closed subset X of Zg_R is reflexive, then X and DX are homeomorphic via elementary duality (5.4.1).*

That is immediate from 5.4.1 and the definition of reflexive. The next result is immediate from 5.4.14 and 5.4.18.

Corollary 5.4.19. *If $X \subseteq \text{Zg}_R$ is closed and* $\text{mdim}(X) < \infty$, *then X and DX are homeomorphic via elementary duality.*

Corollary 5.4.20. ([274, p. 52]) *If* $\text{KGdim}(R) < \infty$, *equivalently (13.2.1) if* $\text{mdim}(\text{pp}_R) < \infty$, *then elementary duality induces a homeomorphism* $\text{Zg}_R \simeq {}_R\text{Zg}$.

5.5 Maps between spectra

Epimorphisms of rings induce homeomorphic embeddings of spectra (Section 5.5.1), as do representation embeddings (Section 5.5.2).

5.5.1 Epimorphisms of rings

An epimorphism of rings induces a homeomorphic embedding of spectra (5.5.3) and the image of the restriction of scalars functor is a definable subcategory (5.5.4). In particular this applies to universal localisations.

A morphism $f : R \to S$ of rings is an **epimorphism** (in the category of rings), if whenever $g, h : S \to T$ are ring morphisms, if $gf = hf$ then $g = h$. That is, the image of R in S is "large enough" to detect whether or not two morphisms from S differ. For instance, $\mathbb{Z} \to \mathbb{Q}$ is an epimorphism of rings, though not surjective.

Theorem 5.5.1. ([646, 1.3]) *A morphism $R \to S$ of rings is an epimorphism iff the induced restriction of scalars functor from* Mod-S *to* Mod-R *is full, that is, if for all S-modules M, N the natural embedding of groups* $\mathrm{Hom}_S(M, N) \to \mathrm{Hom}_R(M, N)$ *is a bijection. This is so iff the natural map $S \otimes_R S \to S$ determined by $s \otimes t \mapsto st$ is an isomorphism.*

In particular, if $R \to S$ is an epimorphism, then Mod-S may be regarded as a full subcategory of Mod-R.

Corollary 5.5.2. *If $f : R \to S$ is an epimorphism of rings and M_S is indecomposable, then M_R is indecomposable.*

Proof. If $\mathrm{End}(M_R)$ contains a non-zero, non-identity idempotent, then, by 5.5.1, so must $\mathrm{End}(M_S)$. □

Theorem 5.5.3. ([501, Thm 1, Cor. 9], also see [323, 8.62]) *Suppose that $f : R \to S$ is an epimorphism of rings.*

(a) If N is a pure-injective (indecomposable) S-module, then, as an R-module, N is pure-injective (and indecomposable).
(b) If $\mathcal{X}$ is a definable subcategory of Mod-S*, then $\mathcal{X}$, regarded as a class of R-modules, is a definable subcategory of* Mod-R.

Hence there is an induced homeomorphic embedding of Zg_S *as a closed subset of* Zg_R.

Proof. The first statement is immediate by 5.5.2 and 4.3.7. For the second statement, the first two of the characterising closure conditions in 3.4.7(ii) are indifferent as to whether tested in Mod-R or Mod-S. For the third, closure under pure submodules, we refer forward, to 6.1.11 (the proof could be given here but fits better in Chapter 6). Concerning the last statement, by the first part, there is a map of Ziegler spectra (as sets) which is injective because f is an epimorphism and, because the topology is determined by the intersections of the spectra with definable subsets, 5.1.6, this map is, by the second part, a homeomorphic embedding. □

Corollary 5.5.4. *If $R \to S$ is an epimorphism of rings, then* Mod-S *is a definable subcategory of* Mod-R.

It will be seen later, 6.1.9, that every pp condition for S-modules is equivalent, in every S-module, to one which involves scalars from R only.

Rings of fractions in the sense of Ore are obtained by inverting suitable sets of elements, see, for example, [441, §2.1] or [663, Chpt. II], and the canonical map from a ring to its ring of fractions is an example of an epimorphism of rings (see, for example, [663, p. 230]). More generally, so are universal localisations in the sense of Cohn, [127], also see [423], [638], [130].

Theorem 5.5.5. ([638, 4.1 and pp. 56–7]) *Let R be any ring and let Σ be a set of morphisms between finitely generated projective R-modules. Then there is a morphism of rings $R \to R_\Sigma$, where possibly $0 = 1$ in R_Σ, such that for each morphism $H : P \to Q$ in Σ the morphism $H \otimes 1_{R_\Sigma} : P \otimes_R R_\Sigma \to Q \otimes_R R_\Sigma$ is invertible, and $R \to R_\Sigma$ is universal such. That is, if $f' : R \to S$ is any morphism of rings such that for every $H \in \Sigma$ the map $H \otimes 1_S$ is invertible, then there is a unique morphism $g : R_\Sigma \to S$ such that $f' = gf$.*

The morphism $R \to R_\Sigma$ is an epimorphism of rings.

The morphism $R \to R_\Sigma$ is the **universal localisation** of R at Σ.

For example, if S is a finitely presented k-algebra, then it is easy to see that there is a finite-dimensional k-algebra R such that Mod-S is a definable subcategory of Mod-R. But in [464] it is shown that there is a finite-dimensional k-algebra R and a ring epimorphism $R \to S'$, where S' is Morita equivalent to S.

The next result will be used in connection with tame hereditary algebras (Section 8.1.2).

Theorem 5.5.6. (Dicks and Bergman, see [638, 4.9]) *If R is a right hereditary ring, then every universal localisation, R_Σ, of R also is right hereditary.*

There is a characterisation of epimorphisms in terms of localised functor categories at 12.8.7.

There is the following related result of Jensen and Lenzing (also see Sections 18.2 and 18.2.5 for more on functors which commute with direct limits and products).

Proposition 5.5.7. ([323, 8.61]) *Suppose that F :* Mod-$S \longrightarrow$ Mod-R *is a functor which commutes with direct limits and products and is full. Suppose also that for each S-module M the kernel of the induced ring morphism* $\mathrm{End}(M) \to \mathrm{End}(FM)$ *is contained in the Jacobson radical of* $\mathrm{End}(M)$. *Then:*

(a) F preserves pure-exact sequences;
(b) if $M, M' \in$ Mod-S, *then $M' \in \langle M \rangle$ iff $FM' \in \langle FM \rangle$;*

(c) *a module N_S is an (indecomposable) pure-injective iff $(FN)_R$ is an (indecomposable) pure-injective;*
(d) *if N_S is pure-injective, then N has an indecomposable direct summand iff FN does.*

In this book the space of indecomposable pure-injective modules is interpreted as a kind of spectrum of a ring. There are relations between this and various other general notions of spectra, some of which involve ring epimorphisms, see [600], [129], [638], [615], [616] for instance.

5.5.2 Representation embeddings

Representation embeddings between module categories induce homeomorphic embeddings of Ziegler spectra, with closed image (5.5.9). Suitable tensor functors induce continuous maps (5.5.10).

Let R and S be finite-dimensional algebras over a field k. It is a result of Watts, [705], and Eilenberg, [169], that a functor from Mod-S to Mod-R is k-linear, **exact** (takes exact sequences to exact sequences) and preserves products and direct sums iff it is of the form $- \otimes_S B_R$ for some (S, R)-bimodule B on which k acts centrally and with B finitely generated as an S-module.[22] Such a functor F is said to be a **representation embedding** if F preserves indecomposability (if M is indecomposable so is FM) and **reflects isomorphism** (if $FM \simeq FN$, then $M \simeq N$). It follows directly that ${}_SB$ must in this case be a projective generator, finitely generated if, as is often the case, the definition is given for such a functor from mod-S to mod-R.

Example 5.5.8. Let k be a field and consider the Kronecker quiver $\widetilde{A_1}$ (1.1.4) $1 \underset{\beta}{\overset{\alpha}{\rightrightarrows}} 2$ There are two obvious representation embeddings from Mod-$k[T]$ to Mod-R. The first, F_1, is defined on objects by taking the $k[T]$-module M to the representation of $\widetilde{A_1}$ which has a copy of the vector space M at each vertex, has α an isomorphism (write it as 1 and think of it as an identification of the two copies) and β is multiplication by T (that is, if $\alpha a = b$, then $\beta a = bT$). The second, F_2, reverses the roles of α and β. Both have the obvious action on morphisms.

It is easy to check that these are representation embeddings, indeed are strict representation embeddings, that is, are full. The bimodule, tensoring with which

[22] If the requirement that finitely generated modules be taken to finitely generated modules is dropped, then even tame algebras become "wild", see, for example, [90], [91], [599, p. 407], [94], [609].

is F_1, is simply the image of $k[T]$ under F_1, equipped with the left $k[T]$-action (so this is free of rank 2 over $k[T]$).

This definition may be extended to arbitrary rings by saying that a functor from Mod-S to Mod-R (R, S any rings) is a **representation embedding** if it is equivalent to one of the form $- \otimes_S B_R$, where ${}_SB$ is a finitely generated projective generator of Mod-S, and if the functor preserves indecomposability and reflects isomorphism.

Since ${}_SB$ is a finitely generated projective generator the ring $S' = \mathrm{End}({}_SB)$ is Morita equivalent to S (for example, [663, IV.10.7]). Indeed, the functor F factorises as the Morita equivalence $- \otimes_S B_{S'}$ followed by the restriction of scalars functor from Mod-S' to Mod-R via the natural map from R to S'. Since Morita equivalence induces a homeomorphism of Ziegler spectra, 5.1.17, we may assume (and we do in the proofs below) that ${}_SB_R = {}_SS_R$ and that the representation embedding is simply restriction of scalars.

Theorem 5.5.9. ([500, Thm 7]) *Suppose that there is a representation embedding of* Mod-S *into* Mod-R. *Then this functor induces a homeomorphic embedding of* Zg_S *as a closed subset of* Zg_R.

Proof. By the comments above we may suppose that the representation embedding has the form f^*, that is, restriction of scalars, for some $f : R \to S$. Let $N' \in \mathrm{Zg}_S$. By assumption, N'_R is indecomposable and, by 4.3.7, it is pure-injective. So that gives the map $\mathrm{Zg}_S \to \mathrm{Zg}_R$, which, again by assumption, is injective.

Suppose that $X \subseteq \mathrm{Zg}_S$ is closed, corresponding (5.1.6) to the definable subcategory $\mathcal{X}$ of Mod-S, and let N be in the closure of X as a subset of Zg_R. By 3.4.14, N is isomorphic to a direct summand of a module of the form f^*M, where $M_S \in \mathcal{X}$. By 4.3.54 there is $N' \in X$ such that $N \mid N'_R$ and hence, since by assumption we have a representation embedding and N'_S is indecomposable, $N = N'_R \in X$. So the induced map, which we may also denote $f^* : \mathrm{Zg}_S \to \mathrm{Zg}_R$ is a closed map.

Finally, to see that this map is continuous, let (ϕ/ψ) be a basic open subset of Zg_R. It is immediate from 3.2.7 that $(f_*\phi/f_*\psi) = (\phi/\psi) \cap f^*\mathrm{Zg}_S$, as required. $\square$

For example, each of the functors F_1, F_2 in 5.5.8 induces an embedding of the Ziegler spectrum of $k[T]$, which is described in Section 5.2.1, into that of $k\widetilde{A}_1$. In particular each provides certain infinite-dimensional points of $\mathrm{Zg}_{k\widetilde{A}_1}$. Indeed, it turns out, see Section 8.1.2 (8.1.8 especially), that every infinite-dimensional point of the spectrum of $k\widetilde{A}_1$ is in the image of one or other of these functors. (Note that the images have a huge overlap, since F_1 and F_2 are essentially two affine lines covering the projective line over k, see Section 8.1.2.) Coverings like this are used in that section to describe the Ziegler spectra of arbitrary tame hereditary algebras.

With modified assumptions one may obtain part of the conclusion; the proof of the next result uses the $(- : -)$ operation on pp conditions from Section 1.3.4.

Proposition 5.5.10. ([500, Thm 9]) *Suppose that R, S are any rings. Let ${}_S B_R$ be a bimodule which is of finite endolength as a bimodule and with ${}_S B$ finitely presented. Suppose that $- \otimes_S B_R : \text{Mod-}S \longrightarrow \text{Mod-}R$ sends indecomposable pure-injectives to indecomposable pure-injectives, hence induces a map from Zg_S to Zg_R. Then this induced map is continuous.*

If one assumes just that a functor from one module category to another commutes with direct limits and products, then, although nothing as strong as 5.5.9 will be true, one can say something about preservation and reflection of definable subcategories, see 18.2.4.

A **strict representation embedding** is a representation embedding where the functor $- \otimes_S B_R$ is full. This is equivalent, in the notation established earlier, to the restriction of scalars functor from Mod-$S'(= \text{End}({}_S B))$ to Mod-R being full, that is, 5.5.1, to the inclusion $R \to S'$ being an epimorphism of rings. So this case is also covered by the results of Section 5.5.1. In particular, by 5.5.3 and 5.1.17, a strict representation embedding from Mod-S to Mod-R induces a homeomorphic inclusion of Zg_S as a closed subset of Zg_R.

The next statement follows (for the first part see the proof of 5.5.9, alternatively apply 1.3.23 and 4.3.8).

Proposition 5.5.11. *If $F : \text{Mod-}S \to \text{Mod-}R$ is a representation embedding, then F preserves pure-injectivity. If F is a strict representation embedding, then it preserves pure-injective hulls.*

We also point out that a somewhat less restrictive notion of representation embedding appears in [607] and [263].

5.6 The dual-Ziegler topology

A topological space is **spectral**, [294], if it is T_0 (that is, there are no topologically indistinguishable points), compact, if the compact open sets are closed under finite intersection and form a basis for the topology, and if every irreducible closed set has a generic point. Given a spectral space, Hochster [294] defined a new topology on the same set by taking as a basis of open sets the complements of the compact open sets. This **dual** topology is again spectral and its dual is the original space [294, Prop. 8].

Now, the Ziegler spectrum is not spectral: it fails on at least two counts since it need not be T_0 (see Section 8.2.12 for the extreme case) and since an intersection of compact open sets need not be compact (5.2.5). If R is countable, then, 5.4.6, any irreducible closed set does have a generic point but it is not known whether this is true in general.

Nevertheless, we may follow Hochster's definition and define the **dual Ziegler** topology to be that which has, for a basis of open sets, the complements, $[\phi/\psi]$, of the compact open sets (5.1.22). Note that this does give a basis for a topology since $[\phi/\psi] \cap [\phi'/\psi'] = [\phi \times \phi'/\psi \times \psi']$, where $(\phi \times \phi')(\overline{x}, \overline{x}')$ is the pp condition $\phi(\overline{x}) \wedge \phi'(\overline{x}')$ where we ensure, by renaming variables if necessary, that the sets $\overline{x}, \overline{x}'$ of variables are disjoint (this pair defines the direct sum of the objects ϕ/ψ, ϕ'/ψ' in $\mathbb{L}_R^{\mathrm{eq}+}$, see 3.2.10).

A spectral space may be recovered from its dual [294, Prop. 8] but this fails here. It is, in general, not possible to recover the Ziegler topology from the dual-Ziegler topology, see Section 14.3.5.

In Chapter 14 it will be shown that this topology coincides with what may reasonably be regarded as the classical Zariski topology lifted to the functor category. It will also be shown that rings of definable scalars, see Chapter 6, give a presheaf over this dual-Ziegler topology which generalises the classical Zariski structure sheaf.

For some examples see 14.1.2 and Section 14.3.

6

Rings of definable scalars

Rings of pp-definable scalars, and the more general pp-type-definable scalars, are defined and their basic properties developed. Section 12.8 gives another way of arriving at these rings.

6.1 Rings of definable scalars

Rings of definable scalars are defined in Section 6.1.1. In Section 6.1.2 it is shown that every element of an epimorphic extension of a ring is definable (in its actions on modules) over that ring. An example is given to show that the notion of localisation implicit in rings of definable scalars does not always yield an epimorphism of rings.

Classical localisations are shown in Section 6.1.3 to be examples of rings of definable scalars. Duality preserves rings of definable scalars (Section 6.1.4). The rings of definable scalars of the points of the spectrum of a PI Dedekind domain are computed in Section 6.1.5.

In Section 6.1.6 we allow scalars defined by pp-types, that is, by infinite sets of pp conditions. These are compared with rings of definable scalars and used to show that rings of definable scalars can be realised as biendomorphism rings.

6.1.1 Actions defined by pp conditions

Scalars defined by pp conditions are those which extend across definable subcategories (6.1.1). If a closed subset contains the support of R_R, then its ring of definable scalars is R (6.1.5). The ring of definable scalars is just a part of the category of pp-pairs (6.1.7).

Let $\mathcal{X}$ be a class of R-modules. The elements of R act as scalars on the modules in $\mathcal{X}$ but, on these particular modules, other scalars may act. For instance, if $R = \mathbb{Z}$ and if $\mathcal{X}$ is the class of torsionfree divisible modules, then there is, on each member of $\mathcal{X}$, a natural action of the ring $\mathbb{Q}$ which extends the $\mathbb{Z}$-action. We require such actions ("scalars") to commute with all R-linear maps between these modules, in particular with the action of End(M) for each $M \in \mathcal{X}$, so they should act as biendomorphisms on each module in $\mathcal{X}$. But we shall also require that our scalars be pp-definable from the R-action. In general, not every biendomorphism will be a definable scalar.

For example, the biendomorphism ring of the Prüfer group $\mathbb{Z}_{p^\infty}$, regarded as a $\mathbb{Z}$-module, is the ring of p-adic integers, which is uncountable. But the ring $\mathbb{Z}$ is countable and, therefore, there are only countably many pp conditions with which to define scalars, so there can be only countably many *definable* scalars (though see Section 6.1.6 below for the more general notion where we allow pp-*type-definable* scalars). In fact the definable scalars in this example are exactly the elements of the localisation $\mathbb{Z}_{(p)}$: this can be shown directly but follows immediately from 6.1.8, and also from 6.1.17, alternatively from 6.1.8 and 6.1.23.

Now we make this precise; also see Section 3.2.2 where the more general notion of pp-definable map appears. Let $\rho(x, y)$ be a pp condition with two free variables. Then ρ defines a relation, the solution set of ρ, $\rho(M) \subseteq M^2$, on every module M. This relation is **total** on M if for every $a \in M$ there exists some $b \in M$ such that $(a, b) \in \rho(M)$ and this relation is **functional** on M if for every $a \in M$ there is at most one $b \in M$ such that $(a, b) \in \rho(M)$. Since ρ is a pp condition it follows (4.1.1(c)) that the latter requirement is equivalent to $(0, b) \in \rho(M)$ implying $b = 0$. If ρ is both total and functional on M, then, clearly, it defines a function from M to M by $a \mapsto b$ if $(a, b) \in \rho(M)$. By 4.1.1 this function is additive and by 1.1.7 this function commutes with every endomorphism of M. In this situation we say that ρ **defines a scalar on** M and that the corresponding biendomorphism of M is a **definable scalar of** M. Of course, distinct pp conditions may well define the same subset of M^2 hence define the same scalar.

More generally a definable scalar of $\mathcal{X} \subseteq$ Mod-R is given by a pp condition which defines a scalar on each member of $\mathcal{X}$. So a **definable scalar of** $\mathcal{X}$ may formally be defined to be the class of such a pp condition under the equivalence relation of defining the same function on every member of $\mathcal{X}$.

For any module M the set of definable scalars of M forms a subring, which we denote by R_M, of Biend(M): the sum of the scalars defined by ρ and σ is defined by the pp condition $\exists u, v\, (\rho(x, u) \wedge \sigma(x, v) \wedge y = u + v)$ and their product $\rho\sigma$ (composition to the right) is defined by $\exists z\, (\rho(x, z) \wedge \sigma(z, y))$. We call R_M the **ring of definable scalars** of M. This definition goes back at least to Garavaglia, [209, §6].

Lemma 6.1.1. *If ρ defines a scalar on M, then it does so on the definable subcategory generated by M. Hence, if $\mathcal{X}$ is a definable subcategory and if M is any module which generates $\mathcal{X}$, then the ring of definable scalars of M is identical to the ring of definable scalars of $\mathcal{X}$. In particular, the ring of definable scalars of a module M and of its pure-injective hull, $H(M)$, coincide.*

Proof. The conditions for ρ to define a scalar on a module M' are that both pp-pairs $(x = x) / \exists y\rho(x, y)$ and $\rho(0, y) / (y = 0)$ be closed on M'. If these are closed on M, then, by 3.4.11, if M' is in the definable subcategory generated by M they will be closed on M', as required. The last assertion follows by 4.3.23. □

It makes sense, therefore, to extend the terminology and notation to closed subsets of the Ziegler spectrum, so set $R_M = R_{\mathcal{X}} = R_X$ if M and $\mathcal{X}$ are as above and if $X = \mathcal{X} \cap \mathrm{Zg}_R$ (see 5.1.6).

Remarks 6.1.2. (a) The canonical map from R to $\mathrm{Biend}(M)$ factors through R_M, via the map multiplication by r, that is, the map which takes $r \in R$ to the scalar defined by the pp condition $xr - y = 0$. Strictly speaking, the ring of definable scalars is this R-algebra $R \to R_M$, rather than just the ring R_M.

(b) If $X \subseteq Y$ is an inclusion of closed subsets of Zg_R, then there is a corresponding morphism $R_Y \to R_X$ of R-algebras: if a pp condition defines a scalar on every module supported on Y (that is, in the definable subcategory $\mathcal{Y}$ which is such that $\mathcal{Y} \cap \mathrm{Zg}_R = Y$) certainly it defines a scalar on every module supported on X. (There is a presheaf of rings here, though over the dual topology, see Section 14.1.4.)

(c) The ring of definable scalars of the module R_R is $R(= \mathrm{Biend}(R_R))$ itself.

Proposition 6.1.3. *Suppose that $X \subseteq \mathrm{Zg}_R$ is closed and that $Y \subseteq X$ is dense in X. Then the canonical map $R_X \to \prod_{N \in Y} R_N$ is an embedding.*

Proof. (For a sheaf-theoretic gloss on this, see 14.1.13.) For $N \in Y$ there is the canonical morphism $R_X \to R_N$ (by 6.1.2(2)): combining these gives the canonical map referred to in the statement.

If $s \in R_X$ is defined on (modules with support contained in) X by the pp condition $\rho_S(x, y)$ and is non-zero, then there is some M with $\mathrm{supp}(M) \subseteq X$ and $Ms \neq 0$, that is, there are $a, b \in M$ with $(a, b) \in \rho_s(M)$ (that is, $b = as$) and $b \neq 0$. So, by 5.1.3, the open set $\big(\exists x\, \rho_s(x, y) / y = 0\big)$ has non-empty intersection with X, hence, since Y is dense in X, $\big(\exists x\, \rho_s(x, y) / y = 0\big) \cap Y \neq \emptyset$. Take N in the latter set: certainly $Ns \neq 0$ so ρ_s defines a non-zero element of R_N, as required. □

Lemma 6.1.4. *The ring of definable scalars of Zg_R is R.*

Proof. Suppose that $\rho(x, y)$ defines a scalar on every right module M. Say ρ is $\exists \overline{z}\,(x\ y\ \overline{z})H = 0$, where the matrix H has entries from R. There is some $s \in R$ such that $\rho(1, s)$ holds in R, say $(1\ s\ \overline{r})H = 0$ with $\overline{r}$ from R. Then, for every module M and $a \in M$, we have $a(1\ s\ \overline{r})H = 0$, that is, $(a\ as\ a\overline{r})H = 0$. Hence $\rho(a, as)$ holds in M, so the action defined by ρ is just multiplication by s. □

Corollary 6.1.5. (of proof of 6.1.4) *Suppose that X is a closed subset of* Zg_R *and that R belongs to the corresponding definable subcategory of* Mod-R. *Then the ring of definable scalars of X is R.*

Example 6.1.6. Let $R = \mathbb{Z}$ and consider $M = \mathbb{Z}_{(2)}$ – the localisation of $\mathbb{Z}$ at the prime 2. Since any endomorphism of M as a $\mathbb{Z}$-module is an endomorphism of M as a $\mathbb{Z}_{(2)}$-module (for example, by 5.5.1, because $\mathbb{Z} \to \mathbb{Z}_{(2)}$ is an epimorphism of rings) certainly $\mathrm{End}(M) = \mathbb{Z}_{(2)}$, hence $\mathrm{Biend}(M) = \mathbb{Z}_{(2)}$. But clearly the action of every element of $\mathbb{Z}_{(2)}$ on $\mathbb{Z}_{(2)}$ is pp-definable in terms of the $\mathbb{Z}$-action. (For instance, the action of $\frac{2}{3}$ is defined by the pp condition $x2 - y3 = 0$ which, at least on $\mathbb{Z}_{(2)}$, does define a function.) Therefore the ring of definable scalars of the $\mathbb{Z}$-module $\mathbb{Z}_{(2)}$ is exactly $\mathbb{Z}_{(2)}$ – and coincides with the biendomorphism ring.

This is in contrast to the example, $\mathbb{Z}_{2^\infty}$, considered at the beginning of the chapter, where the ring of definable scalars is strictly smaller than the biendomorphism ring.

The rings of definable scalars of points of Zg_R turn out, 14.1.12, to be the stalks of a sheaf of rings over this set of points endowed with the dual-Ziegler = rep-Zariski topology.

Remark 6.1.7. The category of pp-definable sorts and pp-definable maps for a module M was defined at the end of Section 3.2.2. From the definitions it is immediate that the ring R_M of definable scalars of M is, if it acts on the right, the opposite to the endomorphism ring of the "home sort", that is, the object $x = x/x = 0$ in $(\mathbb{L}_R^{\mathrm{eq}+})_M$.

More generally, if ϕ/ψ is a pp-pair (hence also an object in $\mathbb{L}_R^{\mathrm{eq}+}$) and M is a module, then one may define the **ring,** $R_M^{\phi/\psi}$**, of definable scalars of M in sort** ϕ/ψ to be the ring of pp-definable (in M) maps (acting on the right) from $\phi(M)/\psi(M)$ to itself, equivalently, the opposite of the endomorphism ring of the image of ϕ/ψ in $(\mathbb{L}_R^{\mathrm{eq}+})_M$ (see 3.2.15).

If two rings are Morita equivalent, then their definable subcategories are in natural bijection. It is shown at [358, 11.5] that the corresponding rings of definable scalars also are Morita equivalent.

6.1.2 Rings of definable scalars and epimorphisms

If $R \to S$ is an epimorphism of rings, then the ring of definable scalars of the image of the restriction of scalars functor is exactly S (6.1.8) and every pp condition over S is equivalent to one over R (6.1.9). For any morphism $f : R \to S$ of rings the subring of S-actions which are pp-definable over R is contained in the dominion of f (6.1.12).

An example is given of a ring of definable scalars which is not an epimorphic extension (6.1.13).

There is a more technical result concerning some conditions under which the lattice of pp conditions is complemented, yielding that the ring of definable scalars is an epimorphism to a von Neumann regular ring (6.1.15, 6.1.16).

Theorem 6.1.8. ([501, Thm 1]) *If $f : R \to S$ is an epimorphism of rings and if* Mod-S *is regarded as a (definable, by 5.5.4) subcategory of* Mod-R*, then the ring of definable scalars of this class is S, regarded as an R-algebra via f. In particular, if S is regarded as an R-module via f, then the ring of definable scalars of S_R is exactly S.*

Proof. By (the proof of) 6.1.4 the ring of definable scalars of Mod-$S \subseteq$ Mod-R (equally of S_R) can be no greater than S: the issue is whether every multiplication by an element of S is definable over R. So let $s \in S$. By 5.5.1, $1 \otimes_R s - s \otimes_R 1 = 0$, that is, $(1, s) \otimes_R (-s, 1) = 0$. So, by 1.3.7, there is a pp condition, $\rho = \rho_s$, over R such that $\rho(1, s)$ holds in S_R and $D\rho(-s, 1)$ holds in ${}_RS$. We claim that $\rho(x, y)$ defines multiplication by s in every S-module M.

Let $a \in M$. As in the proof of 6.1.4 we deduce that $\rho(a, as)$ holds in M, so ρ defines a total relation on each S-module. To check that this relation is functional, suppose that $(0, b) \in \rho(M)$. Since $(-s, 1) \in D\rho({}_RS)$, hence $(-s, 1) \in D\rho({}_SS)$, it follows by 1.3.7 that in $M_S \simeq M \otimes_S S_S$ we have $(0, b) \otimes (-s, 1) = 0$, that is, $b = 0$, as required. □

Therefore, if $R \to S$ is an epimorphism of rings, then the action of S on any S-module is pp-definable in terms of the R-action.[1] This is extended in the next result. Recall (just before 3.2.8) that if $f : R \to S$ is any morphism of rings and ϕ is a pp condition for R-modules, then $f_*\phi$ is the corresponding pp condition for S-modules which is obtained by replacing each element $r \in R$ appearing in a pp condition by its image fr.

[1] Of course, the converse also is true: if every S-action is R-definable, then the restriction of scalars functor from Mod-S to Mod-R is full so $R \to S$ is an epimorphism.

Proposition 6.1.9. *If $f : R \to S$ is an epimorphism of rings and ϕ is any pp condition for S-modules, then there is a pp condition, ϕ', for R-modules such that $\phi = f_*\phi'$.*

Proof. The point is that any S-linear equation $\sum_1^n x_i s_i = 0$ can be replaced by a pp condition over R, namely, $\exists z_1, \dots, z_n \left(\bigwedge_1^n \rho_i(x_i, z_i) \wedge \sum_1^n z_i = 0\right)$, where $\rho_i = \rho_{s_i}$ is, as in the proof of 6.1.8, a pp condition over R which defines the scalar, multiplication by s_i, on S-modules when they are regarded as R-modules. Clearly one can use this to produce the required condition ϕ'. □

It is easy to see, [501, Cor. 4], that this operation commutes with elementary duality of pp conditions.

Corollary 6.1.10. *Suppose that $R \to S$ is an epimorphism of rings. If $\iota : M \to N$ is an embedding of S-modules which, regarded as an embedding of R-modules, is pure, then ι is pure.*

Corollary 6.1.11. *Suppose that $R \to S$ is an epimorphism of rings, that N is an S-module and that M is a pure R-submodule of N_R. Then M is a (pure, by 6.1.10) S-submodule of N.*

Proof. Let $a \in M$, let $s \in S$ and let ρ_s be as in the proof of 6.1.8. Since ρ_s is total on N, $N \models \exists y\, \rho_s(a, y)$. Since M is pure in N_R, $M \models \exists y\, \rho_s(a, y)$, say $b \in M$ is such that $(a, b) \in \rho_s(M)$. So both (a, b) and (a, as) are in $\rho_s(N)$ hence, since ρ_s is functional, $as = b \in M$, as required. □

This completes the proof of 5.5.3 and, in particular, (the parenthetical statement in 6.1.8) we see that if $R \to S$ is an epimorphism of rings, then Mod-S is a definable subcategory of Mod-R, and S-Mod is a definable subcategory of R-Mod, 5.5.4.

For an extension of this to the categories, $\mathbb{L}^{\mathrm{eq}+}$, of pp-pairs, see 12.8.7. For an alternative approach see [358, §11.5].

If $f : R \to S$ is any morphism of rings, then the **dominion** of f is the set of elements $s \in S$ such that, for every pair $g, h : S \to T$ of ring morphisms, if $hf = gf$, then $hs = gs$. Clearly the dominion of f is a subring of S and f is an epimorphism iff its dominion is S (see [663, §XI.1]). Also denote by R_f the subring of S consisting of those elements $s \in S$ such that the action of s on every S-module is pp-definable in terms of the fR-action. More precisely, $s \in R_f$ if there is a pp condition $\rho \in \mathrm{pp}_R^2$ such that for every M_S and $a, b \in M$, $(a, b) \in f_*\rho(M)$ iff $b = as$.

Corollary 6.1.12. *Let $f : R \to S$ be any morphism of rings. Then R_f is contained in the dominion of f.*

Proof. If $s \in R_f$, then the action of s on any S-module, in particular on T_S (as in the definition), is determined by the fR-action so s is in the dominion of f. More precisely, both $(1_T, gs) \in f_*\rho(T_R)$ and $(1_T, hs) \in f_*\rho(T_R)$, the R-module structures on T induced by g and h being the same, so $gs = hs$. □

Example 6.1.13. ([498, A1.9]) If $f : R \to R_X$ is a ring of definable scalars it need not be the case that f is an epimorphism of rings.

It will be proved below (6.1.20) that if M is a module of finite endolength, then all its biendomorphisms are definable scalars. So it is necessary only to exhibit a finite-dimensional module, M, over a k-algebra R (k a field) with the natural map from R to $\mathrm{Biend}(M)$ not an epimorphism of rings.

Take R to be the algebra $k[a, b, c : ab = ac = bc = 0 = a^2 = b^2 = c^2]$, where k is any field. For the module take the four-dimensional indecomposable "string module" (for these see Section 8.1.3) M with k-basis $x, xa = yb, y, yc$.

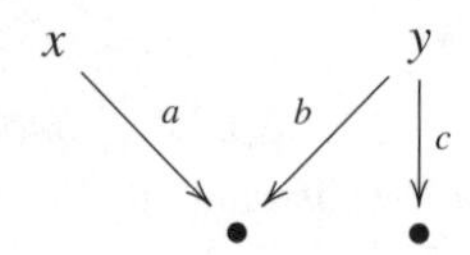

Since M is indecomposable and of finite length its endomorphism ring, S, is local (4.3.58), and clearly $S/J(S) \simeq k$. In the radical, $J(S)$, of S there are the (images of the) nilpotent elements a, b, c of R (which, note, is commutative so has a natural map to $\mathrm{End}(M)$) and there is also the element – let us call it d – which sends x to yc and sends y to 0. A one-line computation shows that the image of x under any element of S must be contained in $xk \oplus xak \oplus yck$ and similarly for y. So we have a k-basis, $\{1, a, b, c, d\}$ for S.

Therefore $S \simeq R[d : d^2 = 0 = ad = bd = cd]$. Since this ring is commutative and contains R it must also be the biendomorphism ring of M.

Now, consider the canonical embedding of R into S followed by composition with: on the one hand, the identity morphism $S \to S$; on the other, the endomorphism of S which sends d to 0 and fixes R. Since the compositions are equal, it follows that $R \to S = R_N$ is not an epimorphism.

If $\mathcal{X}$ is a definable subcategory and if $R \to R_{\mathcal{X}}$ is an epimorphism of rings it need not be the case that $\mathcal{X} = \text{Mod-}R_{\mathcal{X}}$ (clearly(!), take $\mathcal{X} = \langle R_R \rangle$, noting 6.1.5). But it is true for von Neumann regular rings (Section 2.3.4).

Corollary 6.1.14. (to 3.4.30) *Suppose that R is a von Neumann regular ring and let $\mathcal{X}$ be a definable subcategory of* Mod-R. *Then $R \to R_{\mathcal{X}}$ is an epimorphism of rings, indeed $R_{\mathcal{X}} = R/I$ for some two-sided ideal, I, of R, and the inclusion $\mathcal{X} \longrightarrow$* Mod-$R$ *is just the functor* Mod-$R_{\mathcal{X}} \longrightarrow$ Mod-R *induced by $R \to R_{\mathcal{X}}$.*

The following result of Herzog is proved, using the functorial approach, at 12.8.10: if M is such that every pp-definable subgroup of M has a pp-definable complement, then its ring of definable scalars R_M is von Neumann regular and the canonical map $R \to R_M$ is an epimorphism of rings. The first statement can can also be seen directly as follows. Let $\rho(x, y)$ be a pp condition defining a scalar, $s \in R_M$, on M and consider the image of s: this is defined by the pp condition $\phi(x)$, which is $\exists x\, \rho(x, y)$, so, by assumption, it has a pp-definable complement, say $M = \phi(M) \oplus \phi'(M)$ (as a group). Let $e \in R_M$ be the scalar defined by the condition $\sigma(x, y)$, which is $\exists v\, (x = y + v \wedge \phi(y) \wedge \phi'(v))$, that is, e is projection to the image of s, hence is an idempotent with $s = se$. It is also easy to write down a pp condition which defines $t \in R_M$ such that $ts = e$ (t takes $x \in M$ to the unique element in some chosen pp-definable complement of $\ker(t)$ which is mapped by s to xe). Thus $R_M e = R_M s$, so (2.3.22) R_M is von Neumann regular.

Lemma 6.1.15. ([406, 3.5]) *Suppose that M is finitely presented, that* $\mathrm{End}(M) = k$ *is a division ring, that every pp-definable subgroup of M is finite-dimensional or cofinite-dimensional and that $\bigcap\{\phi(M) : \phi(M)$ is cofinite-dimensional $\} = 0$. Then* $\mathrm{pp}^1(M)$ *is complemented.*

Proof. Suppose that ϕ is a pp condition such that $\phi(M)$ is finite-dimensional over k. Choose ψ such that $\psi(M)$ is cofinite-dimensional (that is, $\dim_k(M/\psi(M)) < \infty$) and $\phi(M) \cap \psi(M) = 0$. Such exists: otherwise, choose ψ with $\psi(M)$ cofinite-dimensional and with $\dim_k(\phi(M) \cap \psi(M))$ minimal. By hypothesis there is ψ' with $\psi'(M)$ cofinite-dimensional and $\psi'(M) \not\geq \psi(M) \cap \phi(M)$. Then $\psi \wedge \psi'$ is cofinite-dimensional and its intersection with $\phi(M)$ is strictly smaller than $\phi(M) \cap \psi(M)$, a contradiction.

Now choose any k-vectorspace complement, W, of $\psi(M)$ in M with $W \geq \phi(M)$ and choose a complement, W', for $\phi(M)$ within W. By 1.2.13, $W' = \phi_0(M)$ for some ϕ_0. Then $\psi(M) \oplus \phi_0(M)$ is a pp-definable complement for $\phi(M)$.

On the other hand, if $\phi(M)$ is cofinite-dimensional, then, by 1.2.13, it has a pp-definable complement. □

With 12.8.10 the next result follows. It applies, for example, to finite length modules over the first Weyl algebra over a field of characteristic 0 (see 8.2.41).

Corollary 6.1.16. *Suppose that M is finitely presented, that* $\mathrm{End}(M) = k$ *is a division ring, that every pp-definable subgroup of M is finite- or cofinite-dimensional and that $\bigcap\{\phi(M) : \phi(M)$ is cofinite-dimensional $\} = 0$. Then the ring, R_M, of definable scalars of M is von Neumann regular and $R \to R_M$ is an epimorphism of rings.*

Burke, see [106, §3], extended the notion of definable scalars by allowing actions defined by possibly infinite sets of pp conditions and, among other things, used the resulting ring of pp-type-definable scalars to give simplified proofs of some results on rings of (ordinary) definable scalars, see Section 6.1.6. Rings of type-definable scalars are associated to the type-definable classes discussed in connection with the full support topology of Section 5.3.7, just as rings of definable scalars are associated to definable classes of modules.

6.1.3 Rings of definable scalars and localisation

Finite-type torsion-theoretic localisations, including those which yield classical rings of quotients, give rings of definable scalars of cogenerating injectives (6.1.17).

Rings of definable scalars can be realised as biendomorphism rings of elementary cogenerators (6.1.18). The ring of definable scalars of a Σ-pure-injective module which is finitely generated over its endomorphism ring coincides with its biendomorphism ring (6.1.19); in particular this applies to modules of finite endolength (6.1.20); there is a similar result for finitely presented modules (6.1.21).

The ring morphism $\mathbb{Z} \to \mathbb{Z}_{(2)}$ is covered by 6.1.8 but is also a special case of another general result which concerns the notion of localisation discussed in Section 11.1.1, applied to the module category Mod-R.

Let E be an injective R-module. Then E cogenerates a hereditary torsion theory, the torsion modules being those M with $(M, E) = 0$ and the torsionfree modules being those which embed in some power of E (11.1.1). For instance, if $R = \mathbb{Z}$ and if $E = \mathbb{Q}$, then we get the usual notions of torsion and torsionfree for abelian groups. If $E = \mathbb{Q} \oplus \mathbb{Z}_{2^\infty}$ or $= \mathbb{Z}_{2^\infty}$, then we get the notion of torsion appropriate to the example $\mathbb{Z} \to \mathbb{Z}_{(2)}$.

Corresponding to any hereditary torsion theory τ on Mod-R there is a localisation functor on Mod-R (11.1.5) and the image of R under this functor carries a natural ring structure (for example, [663, §IX.1]). Furthermore, there is a canonical ring morphism from R to this **ring of quotients** or **localisation** of R. In the notation of 11.1.5 this is the canonical map $R \to Q_\tau R = R_\tau$. In the two examples above the rings of quotients are $\mathbb{Z} \to \mathbb{Q}$ and $\mathbb{Z} \to \mathbb{Z}_{(2)}$ respectively. A torsion theory is said to be of finite type if it is determined by the finitely *presented* torsion objects (11.1.14).

Theorem 6.1.17. ([498, A1.7]) *Let E be an injective R-module and suppose that the hereditary torsion theory cogenerated by E is of finite type (for example, see 11.1.14, suppose that R is right noetherian). Then the corresponding localisation of R coincides with the ring of definable scalars of E.*

Proof. Let τ denote the torsion theory cogenerated by E and let R_τ denote the localisation of R at τ.

First we show that R_τ is naturally embedded in R_E as an R-algebra. Let $s \in R_\tau$. By the construction of R_τ (see [663, IX.2.2]) and 11.1.14, there is a finitely generated right ideal I which is τ-dense in R and with $sI \le R$. Say $b_1, \dots, b_n$ generate I. Define $\rho_s(x, y)$ to be $\bigwedge_{i=1}^n (x.sb_i = yb_i)$ – a pp condition over R because $sb_i \in R$.

Let $e \in E$. Clearly $(e, es) \in \rho_s(E)$. For any $e' \in E$, $(0, e') \in \rho_s(E)$ implies $\bigwedge_i (0 = e'b_i)$, hence $e'I = 0$. But E is τ-torsionfree and R/I is τ-torsion, so $e' = 0$. Therefore ρ_s is total and functional on E so the map $s \mapsto \rho_s$ defines an embedding from R_τ to R_E. It is an embedding because if ρ_s defines the zero map on E, then $E \cdot sI = 0$ so, since E is injective and R/I, hence sR/sI, is τ-torsion, $E \cdot sR = 0$. But E cogenerates τ, in particular R_τ embeds in a power of E, hence s must equal 0.

For the converse suppose that $\rho(x, y)$ defines an element of R_E. By 2.3.3 $\rho(E) = \mathrm{ann}_E\, D\rho({}_RR)$. Let $\{(r_i, t_i)\}_i$ be a set, possibly infinite, of generators for $D\rho({}_RR) \le R_R^2$ and set $I = \sum_i t_i R$. We have $(a, b) \in \rho(E)$ iff $ar_i + bt_i = 0$ for each i and, since $\rho(0, E) = 0$, if $b \in E$ satisfies $bI = 0$ then $b = 0$. If I were not τ-dense in R, then there would be a τ-torsionfree R-module M with a non-zero element annihilated by I. But M embeds in some product of copies of E, contradicting the previous observation. Hence I is τ-dense in R, so, by 11.1.14, there is a finite subset, (r_i, t_i), $i = 1, \dots, n$ after reindexing, of the chosen generating set for I such that $I' = \sum_1^n t_i R$ is τ-dense in R. Let $\rho'(x, y)$ be the pp condition $\bigwedge_i^n xr_i + yt_i = 0$.

We claim that $\rho(x, y)$ is equivalent to (defines the same function as) $\rho'(x, y)$ in E. Certainly $(a, b) \in \rho(E)$ implies $(a, b) \in \rho'(E)$. For the reverse implication suppose that $(a, b) \in \rho'(E)$. Since ρ defines a scalar on E we also have $(a, b') \in \rho(E)$, hence $(a, b') \in \rho'(E)$, for some $b' \in E$. Therefore $(0, b - b') \in \rho'(E)$, that is, $(b - b')I' = 0$, so $b = b'$ because E is τ-torsionfree and I' is τ-dense in R. Thus the claim is established.

Now suppose that M is any τ-torsionfree, τ-injective R-module. Since M is τ-torsionfree it embeds in a power, E^κ say, of E, so, since (6.1.1) ρ' defines a scalar on E^κ, if $m \in M$, there is $e \in E^\kappa$ with $(m, e) \in \rho'(E^\kappa)$. That is, $et_i = -mr_i$ for $i = 1, \dots, n$, so $eI' \le M$. Since M is τ-injective and R/I', hence eR/eI', is τ-torsion, the inclusion of eI' in M extends to a morphism $f : eR \to M$. Since $(fe - e)I' = 0$ and E^κ is τ-torsionfree it follows that $e = fe \in M$. Therefore ρ' defines a scalar on M.

In particular, there is an element $s \in R_\tau$ such that $\rho'(1, s)$ holds in R_τ (which certainly is τ-injective and τ-torsionfree). Every τ-torsionfree, τ-injective R-module is naturally an R_τ-module, so the proof of 6.1.4 applies to show that ρ'

defines multiplication by s on every such module, in particular on E. Since ρ and ρ' define the same function on E, ρ defines multiplication by s on E. Hence the inclusion of R_τ in R_E is surjective, as required. □

In fact every ring of definable scalars really is a localisation at a hereditary torsion theory of finite type, but in the functor category (12.8.2).

It is the case, see [663, IX.3.3], that any ring of quotients, $R \to R_\tau$, with respect to a hereditary torsion theory τ may be obtained as the biendomorphism ring of a suitably large power of a cogenerating injective for τ. There is an analogous result, 12.8.4, for rings of definable scalars, indeed this result for rings of definable scalars follows from the generalisation of the result just mentioned to the functor-category setting (see Section 12.8) and may be phrased as follows.

Proposition 6.1.18. ([498, A.4.1], [106, 4.3], [358, 11.7]) *Let X be a closed subset of Zg_R and let M be such that $\mathrm{supp}(M) = X$. Suppose also that M is an elementary cogenerator (see 5.3.52) for the corresponding definable subcategory $\mathcal{X}$. Then there is an index set I such that the associated ring of definable scalars, R_X, is the biendomorphism ring of M^I.*

A non-functorial proof uses type-definable scalars: see 6.1.34.

Proposition 6.1.19. ([498, A.1.5], see [106, 3.6]) *If M is a Σ-pure-injective module which is finitely generated over its endomorphism ring, then the ring of definable scalars of M coincides with its biendomorphism ring. (Also compare 6.1.31 below.)*

Proof. Let $g \in \mathrm{Biend}(M_R)$: it must be shown that the action of g on M is pp-definable in M_R. Set $S = \mathrm{End}(M_R)$ and suppose that $a_1, \dots, a_k \in M$ are such that ${}_SM = \sum_1^k Sa_i$. Then g is determined by its action on $\overline{a} = (a_1, \dots, a_k)$, so consider $\overline{a}g$. Since M is Σ-pure-injective, there is, 4.4.7, a pp condition $\phi(\overline{x}, \overline{y})$ which defines the solution set in M^{2k} of $\mathrm{pp}^M(\overline{a}, \overline{a}g)$, so $\phi(M) = S \cdot (\overline{a}, \overline{a}g) \leq M^{2k}$ (diagonal action of S). Consider the pp condition $\rho(u, v)$ which is

$$\exists x_1, \dots, x_k\, \exists y_1, \dots, y_k \left(u = \sum_1^k x_i \wedge v = \sum_1^k y_i \wedge \overline{\phi}(\overline{x}, \overline{y})\right),$$

where $\overline{\phi}(x_1, \dots, x_k, y_1, \dots, y_k)$ is $\bigwedge_{i=1}^k \phi_i(x_i, y_i)$, where $\phi_i(x_i, y_i)$ is

$$\bigwedge_i \exists z_{i1}, \dots, \hat{z}_{ii}, \dots, z_{ik}, w_{i1}, \dots, \hat{w}_{ii}, \dots, w_{ik}$$

$$\phi(z_{i1}, \dots, x_i, \dots, z_{ik}, w_{i1}, \dots, y_i, \dots, w_{ik}).$$

Note that, by choice of ϕ, if $a, b \in M$, then $(a, b) \in \phi_i(M)$ iff there is $s \in S$ with $sa_i = a$ and $sa_i g = b$, in particular, for each i and s, $(sa_i, sa_i g_i) \in \phi_i(M)$. We claim that ρ defines the action of g in M.

First, $\rho(u, v)$ defines a total relation from u to v: given $c \in M$ we have $c = \sum_1^k s_i a_i$ for some $s_i \in S$, hence $cg = \sum_1^k s_i a_i g$. In particular, $(c, cg) \in M$. As commented above, $\bigwedge_{i=1}^k \phi_i(s_i a_i, s_i a_i g_i)$ holds. Therefore $(c, cg) \in \rho(M)$.

It remains to show that ρ is functional, so suppose $(0, d) \in \rho(M)$. Then there are $c_i, d_i \in M$ such that $0 = \sum_1^k c_i, d = \sum_1^k d_i$ and such that $(c_i d_i) \in \phi_i(M)$ for each i. As commented above, it follows that there are $s_i \in S$ with $s_i a_i = c_i$ and $s_i a_i g = d_i$ for $i = 1, \ldots, k$. So $d = \sum d_i = \sum s_i a_i g = (\sum s_i a_i)g$ and $0 = \sum c_i = \sum s_i a_i$, from which we deduce $d = 0$, as required. □

Because a module of finite endolength certainly is finitely generated over its endomorphism ring and is Σ-pure-injective (4.4.24) one has the following corollary.

Corollary 6.1.20. *If M is a module of finite endolength, then the ring of definable scalars of M coincides with its biendomorphism ring.*

For another route to this result see 6.1.33. For yet another see [347, 2.10]. The argument from the proof of 6.1.19 applies, using 1.2.6 and 1.2.7 in place of 4.4.7, to give the following.

Proposition 6.1.21. ([498, A.1.5′]) *If M is a finitely presented module which is finitely generated over its endomorphism ring, then its ring of definable scalars coincides with its biendomorphism ring.*

6.1.4 Duality and rings of definable scalars

Dual closed subsets have (anti-)isomorphic rings of definable scalars (6.1.22); similarly for reflexive indecomposable pure-injective modules (6.1.23).

Theorem 6.1.22. ([274, 6.3]) *Let X be a closed subset of Zg_R and let DX be the dual (5.4.2) closed subset of ${}_R\mathrm{Zg}$. Then $R_X \simeq R_{DX}$, where we regard R_X as operating from the right and R_{DX} from the left.*

Proof. For $X = \mathrm{Zg}_R$ this is by 3.2.13 with ϕ/ψ there being $(x = x)/(x = 0)$. The localised, to X, result is contained in 12.3.20. □

Corollary 6.1.23. *Let N be an indecomposable pure-injective which is reflexive (Section 5.4), so there is a corresponding dual point, DN, of ${}_R\mathrm{Zg}$. Then $R_N \simeq R_{DN}$.*

Proof. If X is the Ziegler-closure of N, then, by definition, DX is the Ziegler-closure of DN so this is immediate from 6.1.22. □

For example, the 2-Prüfer and 2-adic abelian groups, $\mathbb{Z}_{2^\infty}$, $\overline{\mathbb{Z}_{(2)}}$ are dual and both have the localisation $\mathbb{Z}_{(2)}$ for ring of definable scalars.

For another example, if R is right coherent, then the class of absolutely pure right modules is dual to the class of flat left modules (3.4.24). The ring of definable scalars of the latter is, by 6.1.5, the ring R itself. We deduce from this that if E is an injective cogenerator of Mod-R, then $R_E \simeq R$. But, actually, by 3.4.26 the fact that $R_E \simeq R$ is true for any ring R.

6.1.5 Rings of definable scalars over a PI Dedekind domain

Let R be a (non-commutative) PI Dedekind domain. The points of Zg_R were described in Section 5.2.1. We compute the ring of definable scalars at each of these points.

R/P^n: This is a module of finite length over its endomorphism ring, so, by 6.1.20, its ring of definable scalars coincides with its biendomorphism ring, which is just the ring R/P^n.

R_{P^∞}: By 6.1.17 this has, for its ring of definable scalars, the localisation of R at the torsion theory cogenerated by this injective module, that is, the ring $R_{(P)}$.

$\overline{R_P}$: Since $\overline{R_P}$ is the elementary dual module (in the sense of Section 5.4) to R_{P^∞}, by 6.1.22 and the previous case this module also has the localisation $R_{(P)}$ for its ring of definable scalars.

Q: This is a module of finite length over its endomorphism ring (since R is PI the division ring Q is finite-dimensional over its centre) so, by 6.1.20, its ring of definable scalars is its biendomorphism ring, which is just Q.

6.1.6 Rings of type-definable scalars

Generic elements, those with most free pp-type, are produced (6.1.24) and used to define the ring of type-definable scalars of a pure-injective module. These scalars extend to direct summands of products (6.1.27), though not, in general, to the definable subcategory generated by the module (example after that result). They coincide with the definable scalars if the module is Σ-pure-injective (6.1.28) or, more generally, is an elementary cogenerator (6.1.34). The ring of definable scalars of a pure-injective module which contains a generic element, hence, 6.1.25, is cyclic over its endomorphism ring, equals its biendomorphism ring (6.1.29).

Similarly if the pure-injective is finitely generated over its endomorphism ring (6.1.31).

For a module of finite endolength the ring of definable scalars, the ring of type-definable scalars and the biendomorphism ring all coincide (6.1.33).

If we allow "scalars" defined by infinite sets of pp conditions ("pp-type-definable scalars"), then, in general, we obtain a ring which is larger than the ring of definable scalars. For example, if $M = \mathbb{Z}_{(2)}$, then, 6.1.6, its ring of definable scalars is $\mathbb{Z}_{(2)}$ but it will follow from 6.1.29 that the ring of type-definable scalars is the whole biendomorphism ring $\overline{\mathbb{Z}_{(2)}}$ since $1 \in \overline{\mathbb{Z}_{(2)}}$ is generic in the terminology introduced below. Even the ring of type-definable scalars may be a proper subring of the biendomorphism ring: as M varies, the size of the ring of type-definable scalars is limited by the number of pp-types, which is $\leq 2^{\text{card}(R)+\aleph_0}$ so is limited by the size of the ring, whereas the size of the biendomorphism ring of M is not. On the other hand, by 6.1.29, if one takes M to be a "large enough" pure-injective module, then the two rings (biendomorphism, and type-definable scalars) do coincide.

In order to develop, as in [106], a theory of type-definable scalars it is necessary that pp-types have solutions, therefore (in view of 4.2.1) we work in the context of pure-injective modules.

Partly for convenience, partly for variety, we give the definition of the ring of type-definable scalars in the following way, rather than directly mimicking the development for definable scalars.

If M is any module, say that an n-tuple, $\overline{c}$, of elements of M^I, where I is some index set, is **generic** (for M) if for every n-tuple, $\overline{a}$, of elements of M (*sic*) one has[2] $\text{pp}^{M^I}(\overline{c}) \geq \text{pp}^M(\overline{a})$. By 1.2.2 this is equivalent to $\text{pp}^{M^I}(\overline{c}) \geq \text{pp}^{M^I}(\overline{b})$ for every tuple $\overline{b}$ from M^I. So a tuple generic for M will also be generic for M^I and we may simply refer to a "generic" tuple.[3]

Lemma 6.1.24. *If M is any module and n is any integer, then there is a set I such that M^I contains a generic n-tuple. The pp-type of any generic n-tuple contains exactly those pp conditions, ϕ, with n free variables such that $\phi(M) = M^n$. In particular, there is a unique pp-n-type of a generic for M and every n-tuple with this pp-type is generic.*

Proof. For each n-tuple, $\overline{b}$, in M take a copy, $M_{\overline{b}}$, of M. The tuple $\overline{c} = (\overline{b})_{\overline{b}}$ in $M^{M^n} \simeq (M^M)^n$ is, by construction and 1.2.2, the required pp-type. If ϕ is any pp condition which defines a proper subgroup of M^n, then there is some tuple,

[2] Recall that the ordering on pp-types is $p \geq q$ iff $p \subseteq q$ as sets of pp conditions.
[3] This use of the term "generic" comes from model theory, see, for example, [290].

$\overline{b}$, from M not satisfying ϕ, therefore no generic n-tuple satisfies ϕ. The last statement is obvious. □

Lemma 6.1.25. *If N is pure-injective and I is any set, then an n-tuple $\overline{c}$ from N^I is generic iff $(N^I)^n$ is cyclic as a module over* End(N^I)*, generated by $\overline{c}$. (Recall that the action of* End(N^I) *on tuples is the diagonal one.)*

Proof. If $\overline{c}$ is generic and $\overline{a}$ is any n-tuple from N^I, then since $\mathrm{pp}^{N^I}(\overline{c}) \geq \mathrm{pp}^{N^I}(\overline{a})$ there is, by 4.3.9, $f \in \mathrm{End}(N^I)$ with $f\overline{c} = \overline{a}$, so $(N^I)^n$ is indeed generated by $\overline{c}$ over its endomorphism ring. If, conversely, $(N^I)^n = \mathrm{End}(N^I) \cdot \overline{c}$, then for every tuple $\overline{a}$ in N^I there is $f \in \mathrm{End}(N^I)$ with $f\overline{c} = \overline{a}$, so (by 1.2.8) $\mathrm{pp}^{N^I}(\overline{c}) \geq \mathrm{pp}^{N^I}(\overline{a})$ and $\overline{c}$ is generic. □

Lemma 6.1.26. *If N is pure-injective and if, as a left* End(N)*-module, $N^n = \sum_{i\in I} \mathrm{End}(N) \cdot \overline{c}_i$, then $\overline{c} = (\overline{c}_i)_i \in (N^n)^I \simeq (N^I)^n$ is a generic n-tuple.*

Proof. Take $\overline{a}$ from N^n. By assumption $\overline{a} = \sum f_i\overline{c}_i$, where all but finitely many $f_i \in \mathrm{End}(N)$ are zero. Define a morphism from N^I to N to be application of the f_i, followed by the summation map to N. This takes $\overline{c}$ to $\overline{a}$, so $\mathrm{pp}^{N^I}(\overline{c}) \geq \mathrm{pp}^N(\overline{a})$, as required. □

A **(pp-)type-definable scalar** of a pure-injective module N is given by a pp-type $p(x, y)$ (with no non-zero parameters) such that for each $a \in N$ there is a unique $b \in N$ such that $p(a, b)$ holds. Recall the notation for this: $N \models p(a, b)$ means that $p(a, b)$ holds in N, that is, $(a, b) \in \phi(N)$ for every $\phi \in p$. As with pp-definable scalars, different pp-types might define the same function on N; the scalar is the corresponding function (or equivalence class of pp-types).

It is enough to define these scalars generically. To see this, choose I such that N^I contains a generic element $a = (a_i)_{i\in I}$ say. Suppose that $p(x, y)$ is a pp-type such that there is a unique $b \in N^I$ with $N \models p(a, b)$. Then p **defines a scalar on** N, meaning that for each $c \in N$ there is a unique $d \in N$ with $(c, d) \in p(N)$. For, if $a' \in N$, then since a is generic there is, by 4.3.9, $f : N^I \to N$ with $fa' = a$, so from $(a, b) \in p(N^I)$ we deduce $(a', fb) \in p(N)$ (1.2.8). The relation on N defined by p is functional as well as total since from $(0, d) \in p(N)$ one obtains $p(a, b + d')$, where $d' = (d, 0, \dots)$, therefore, by uniqueness of b, $d = 0$. Conversely if p defines a scalar on N, then applying it coordinate-wise it defines a scalar on N^I so determines, and is determined by, the resulting action on a generic.

Lemma 6.1.27. *If a pp-type $p(x, y)$ defines a scalar on N, then it defines a scalar on every direct summand of every power of N.*

Proof. Given $(a_i)_{i\in I} \in N^I$, for each i, take $b_i \in N$ such that $N \models p(a_i, b_i)$. By 1.2.3 $N^I \models p\big((a_i)_i, (b_i)_i\big)$. If $N^I \models p(0, (b_i)_i)$, then for each i we have

$N \models p(0, b_i)$, so $b_i = 0$. If M is a direct summand of N^I and $a = (a_i)_i \in M$, then $N \models p\big((a_i)_i, b = (b_i)_i\big)$ as above, so projecting with π_M to M gives $M \models p(a, \pi_M b)$, and $M \models p(0, d)$ implies $N \models p(0, d)$, therefore $d = 0$ (so necessarily $\pi_M b = b$). □

In order to define addition and multiplication of type-definable scalars, it is, as shown above, enough to define (and check) the action on any generic element.

Suppose, therefore, that $p(x, y)$ and $q(x, y)$ define scalars on N. Let $b, c \in N^I$ be such that $p(a, b)$ and $q(a, c)$ hold, where a is a generic as before. Define the pp-type $(p + q)(x, y)$ to be $\mathrm{pp}^{N^I}(a, b + c)$; one can check that this is generated by all conditions of the form $\exists w, z \, \big(\phi(x, w) \wedge \phi'(x, z) \wedge y = w + z\big)$, where $\phi \in p$ and $\phi' \in q$. If $(p + q)(0, e)$ holds for some $e \in N^I$, then, by 4.3.9, there is $f \in \mathrm{End}(N^I)$ with $fa = 0$ and $f(b + c) = e$. Now, from $p(a, b)$ we deduce $p(fa, fb)$, so, since $p(0, N^I) = 0$, $fb = 0$. Similarly $fc = 0$. Therefore $e = 0$. Thus $p + q$ defines a scalar at the generic a, hence defines a scalar on N.

As for multiplication, suppose that $p(x, y)$ and $q(x, y)$ define scalars on N, say $p(a, b)$ and $q(a, c)$ hold where $a \in N^I$ is generic and $b, c \in N^I$. Because a is generic, there is $g \in \mathrm{End}(N^I)$ such that $b = ga$, so $q(a, c)$ yields $q(b, gc)$. Define the composition, p then q, to be the scalar defined by $r = \mathrm{pp}^{N^I}(a, gc)$. As before, it must be checked that $r(0, d)$ implies $d = 0$. If $r(0, d)$ holds, then there is $f \in \mathrm{End}(N^I)$ with $fa = 0$ and $fgc = d$. Now, since $p(a, b)$ holds and $fa = 0$, also $fb = 0$ and then, since $q(fb, fgc)$ holds, also $fgc = 0$, as required (and one may check that r is generated by conditions of the form $\exists z \, \big(\phi(x, z) \wedge \phi'(z, y)\big)$ where $\phi \in p$ and $\phi' \in q$).

Thus to any pure-injective N is associated its **ring**, R_N^∞, of **(pp-)type-definable scalars**, defined via a suitably large power of N. Note that there is a natural inclusion $R_N \subseteq R_N^\infty$; the definable scalars correspond to those type-definable scalars which are determined by a pp-type which, modulo N, is finitely generated.

In contrast with rings of definable scalars (6.1.1), the ring of type-definable scalars of a pure-injective module N need not act on every module in the support of, or in the definable subcategory generated by, N. For instance, by 6.1.31 below, the ring of type-definable scalars of $\overline{\mathbb{Z}_{(p)}}$ is $\overline{\mathbb{Z}_{(p)}}$ but the module $\mathbb{Q}$, which is in the definable subcategory of Mod-$\mathbb{Z}$ generated by $\overline{\mathbb{Z}_{(p)}}$, is not even a $\overline{\mathbb{Z}_{(p)}}$-module (any torsionfree $\overline{\mathbb{Z}_{(p)}}$-module must be uncountable). This illustrates that the topology which fits with type-definable scalars is the full support topology.

Lemma 6.1.28. *If N is Σ-pure-injective, then $R_N = R_N^\infty$.*

Proof. If N is Σ-pure-injective, then, 4.4.7, every pp-type, $p(x, y)$, is equivalent in N to a single pp condition, $\phi(x, y)$, so every type-definable scalar is a definable scalar. □

Observe that for any module M and any index set I there is a natural map $\text{Biend}(M^I) \to \text{Biend}(M)$. For any element of $\text{Biend}(M^I)$ must commute with all coordinate projections and permutations of coordinates in M^I, therefore takes any diagonal element $(a)_i \in M^I$ $(a \in M)$ to another diagonal element, so induces an action on M which clearly commutes with $\text{End}(M)$.

An example where $\text{Biend}(M^{\aleph_0}) \neq \text{Biend}(M)$ is given by $N = \mathbb{Z}_{2^\infty}$. By the (proof of the) next result $\text{Biend}(\mathbb{Z}_{2^\infty}^{\aleph_0}) = R^\infty_{\mathbb{Z}_{2^\infty}}$ and, by 6.1.28, this equals $R_{\mathbb{Z}_{2^\infty}}$ which, as seen at the beginning of this chapter, is the ring $\mathbb{Z}_{(2)}$. On the other hand, $\text{End}(\mathbb{Z}_{2^\infty})$ is the commutative ring $\overline{\mathbb{Z}_{(2)}}$ which also, therefore, equals $\text{Biend}(\mathbb{Z}_{2^\infty})$.

Proposition 6.1.29. ([106, 3.4]) *Suppose that N is pure-injective and that N^I contains a generic element for N. Then $R^\infty_N = \text{Biend}(N^I)$ $(\subseteq \text{Biend}(N))$. In particular if N already contains a generic element then $R^\infty_N = \text{Biend}(N)$.*

Proof. Certainly $R^\infty_N \subseteq \text{Biend}(N^I)$ (by 6.1.27). For the converse, let $g \in \text{Biend}(N^I)$ and consider $p = \text{pp}^{N^I}(a, ag)$, where $a \in N^I$ is generic. We claim that p defines a scalar. If $c \in N^I$, then there is $f \in \text{End}(N^I)$ with $fa = c$, so $p(c, fag)$ holds. Thus the relation defined on N^I by p is total. Also, if $(0, d) \in p(N^I)$, then, by 4.3.9, there is $f \in \text{End}(N^I)$ with $fa = 0$ and $fag = d$, hence with $d = 0$, as required. □

In view of 6.1.24 we deduce the following.

Corollary 6.1.30. ([106, 3.5]) *For any pure-injective module N there is a set I such that $R^\infty_N = \text{Biend}(N^I)$.*

Proposition 6.1.31. *Suppose that N is a pure-injective module which is finitely generated over its endomorphism ring. Then $R^\infty_N = \text{Biend}(N)$.*

Proof. If $N = \sum_1^k Sa_i$, where $S = \text{End}(N)$, then $a = (a_1, \ldots, a_k) \in N^k$ is a generic for N, so $R^\infty_N = \text{Biend}(N^k)$ (by 6.1.29), which, since k is a natural number, is easily seen to be equal to $\text{Biend}(N)$ (cf. the discussion before 6.1.29 or use that $M_k(S)$ is Morita equivalent to S). □

Example 6.1.32. By 6.1.23 (and 5.4.3) the $\mathbb{Z}$-modules $\mathbb{Z}_{p^\infty}$ and $\overline{\mathbb{Z}_{(p)}}$ (equivalently, 6.1.1, $\mathbb{Z}_{(p)}$) have the same ring of definable scalars, $\mathbb{Z}_{(p)}$, as noted at the beginning of this chapter. In contrast, the ring of type-definable scalars of $\mathbb{Z}_{p^\infty}$ is, by 6.1.28, $\mathbb{Z}_{(p)}$ but that of $\overline{\mathbb{Z}_{(p)}}$ is, by 6.1.31, $\overline{\mathbb{Z}_{(p)}}$.

Combining this with 6.1.28 one obtains the following (which also gives a different proof of 6.1.20).

Corollary 6.1.33. ([106, 3.6]) *If N is of finite endolength, then the ring of definable scalars of N coincides with the biendomorphism ring of N and with the ring of type-definable scalars of N.*

The next result is from [498, A4.1].[4]

Proposition 6.1.34. *If N is an elementary cogenerator (§5.3.5), then $R_N = R_N^\infty$.*

Proof. Suppose that the pp-type $p(x, y)$ defines a scalar on N. Then $p(0, N) = 0$, that is, $\bigcap_{\phi(x,y)\in p} \phi(0, N) = 0$. We claim that there is $\phi \in p$ such that $\phi(0, N) = 0$.

If this were not so, then $\{\phi(0, y) : \phi(x, y) \in p\}$ would extend to a pp-type, $q(y)$ say, which is maximal with respect to not containing the condition $y = 0$. That is, q is N-neg-isolated by the condition $y = 0$. By 5.3.50 there is $b \in N$ with $\mathrm{pp}^N(b) = q$. Thus $b \in p(0, N)$ and $b \neq 0$, contradiction.

So there is $\phi(x, y) \in p$ such that $\phi(0, N) = 0$. We claim that for all $c, d \in N$, $(c, d) \in p(N)$ iff $(c, d) \in \phi(N)$. The direction ($\Rightarrow$) follows since $\phi \in p$. For the other direction, since p defines a scalar on N there is $d' \in N$ such that $(c, d') \in p(N)$ and hence such that $(c, d') \in \phi(N)$. But then $(0, d - d') \in \phi(0, N)$ so $d = d'$, as required. That is, the action defined by p is also defined by ϕ, as required. □

Corollary 6.1.35. *If N is an elementary cogenerator then there is a set I such that $R_N = R_N^\infty = \mathrm{Biend}(N^I)$.*

Somewhat different proofs of this are given at [498, A4.1] and at [358, 11.7].

Example 6.1.36. We compute the ring of definable scalars for the closed subset of $\mathrm{Zg}_\mathbb{Z}$ consisting of the simple modules $\{\mathbb{Z}_p : p \text{ prime}\}$ together with $\mathbb{Q}$. By 3.4.13 we may take N as in 6.1.35 to have the form $\prod_p \mathbb{Z}_p^{\kappa_p} \oplus \mathbb{Q}^\lambda$ for some cardinals κ_p, λ, where p runs over the non-zero primes of $\mathbb{Z}$. Write this module as $\prod_p N_p$ where now p runs over $\mathrm{Spec}(\mathbb{Z})$ including 0. Let $S = \mathrm{End}(N)$, so $S = (\prod_p N_p, \prod_p N_p) \simeq \prod_p (N_p, \prod_q N_q) = \prod_p (N_p, N_p)$ is just a product, $S = \prod_p S_p$ of endomorphism rings of vector spaces over the various fields $\mathbb{F}_p$, where $\mathbb{F}_0$ means $\mathbb{Q}$. Regarded as an S-module, M decomposes as a product, $\prod_p {}_{S_p}(N_p)$ of simple modules N_p over the S_p. The endomorphism ring of N_p as an S_p-module is just $\mathbb{F}_p$. So $B = \mathrm{End}_S N$ equals $\prod_p \mathbb{F}_p$, that is, in our more usual notation, the ring of definable scalars is the von Neumann regular ring $\left(\prod_p \mathbb{Z}_p\right) \times \mathbb{Q}$, the ring morphism from the original ring $\mathbb{Z}$ to this being the obvious one.

[4] It should have appeared in print in [106] but was somehow omitted.

7

M-dimension and width

A general framework for defining dimensions on lattices is described. The special cases m-dimension and width are defined; the former is related to the number of points of the spectrum, the latter to existence of superdecomposable pure-injectives. These dimensions can be recognised in the complexity of morphisms between finitely presented modules. Related dimensions in categories are in Chapter 13.

7.1 Dimensions on lattices

A general construction of dimensions on lattices is given, based on inductively collapsing lattices of certain kinds. After some lemmas we consider Krull dimension, which is based on collapsing intervals with the descending chain condition (7.1.4).

Recall (Appendix E, p. 712) that a lattice is a poset (a partially ordered set) in which every pair of elements, a, b, has a unique least upper bound, or join, denoted $a + b$ and a unique greatest lower bound, or meet, denoted $a \wedge b$. Recall also that a lattice is modular if it satisfies the identity $a \wedge (c + b) = a \wedge c + b$ whenever $a \geq b$.

Let $\mathcal{L}$ be a class of modular lattices which is closed under sublattices and quotient lattices. The examples in which we will be particularly interested are, first, the class consisting of the two-point lattice and the trivial lattice and, second, the class of all **chains** (totally ordered posets). Let L be any modular lattice with 0 (bottom element) and 1 (top element). In our applications L will be a lattice of pp conditions (or pp-definable subgroups) or the lattice of finitely generated subfunctors of some coherent functor.

Consider the congruence $\sim = \sim_{\mathcal{L}}$ which is defined to be the smallest congruence on L which collapses all intervals of L which belong to $\mathcal{L}$. That is, $\sim$ is the smallest equivalence relation on L which is such that if $a > b$ are in L and the interval $[a, b] \in \mathcal{L}$ then $a \sim b$, and which is a congruence, meaning that if $a \sim b$ then[1] for any $c \in L$ also $a + c \sim b + c$ and $a \wedge c \sim b \wedge c$.

Lemma 7.1.1. ([495, 10.3]) *Let $\mathcal{L}$ be a class of modular lattices closed under sublattices and quotient lattices, let L be a modular lattice and let $\sim = \sim_{\mathcal{L}}$ be the corresponding congruence on L. Let $a, b \in L$. Then $a \sim b$ iff there is a finite chain $a + b = c_n \geq c_{n-1} \geq \cdots \geq c_0 = a \wedge b$ such that each interval $[c_i, c_{i-1}]$ belongs to $\mathcal{L}$.*

Proof. If there is such a chain, then certainly $a + b \sim a \wedge b$, hence $a \sim b$. For the converse, we define a relation $\equiv$ on L by $a \equiv b$ if there is such a chain between $a + b$ and $a \wedge b$, show that this is a congruence, which, since it contains, and is contained in, $\sim$, must, therefore, coincide with $\sim$. The details, which call on modularity at every turn, are straightforward.

For example, if $a \equiv b$ and $c \in \mathcal{L}$, then, with notation as given, it has to be shown that $a \wedge c \equiv b \wedge c$, so we have to produce a suitable chain between $a \wedge c + b \wedge c$ and $a \wedge b \wedge c$. Note that $(a + b) \wedge c \geq a \wedge c + b \wedge c$ and there is the chain, $(a + b) \wedge c = c_n \wedge c \geq \cdots \geq c_0 \wedge c = a \wedge b \wedge c$, with each interval $[c_i \wedge c, c_{i-1} \wedge c]$ being a quotient of the corresponding interval $[c_i, c_{i-1}]$, hence in $\mathcal{L}$. (The map is $d \in [c_i, c_{i-1}] \mapsto d \wedge c$: this is surjective, since if $e \in [c_i \wedge c, c_{i-1} \wedge c]$ then $e + c_{i-1} \in [c_i, c_{i-1}]$ and $(e + c_{i-1}) \wedge c = (c_i \wedge c \wedge e + c_{i-1}) \wedge c = c_i \wedge c \wedge e + c_{i-1} \wedge c$ (by modularity) $= e$.) Similarly one checks that $a + c \equiv b + c$ and that $\equiv$ is an equivalence relation. □

Following [495, §10.2] set $L_0 = L$ and form the quotient $L \to L_1 = L/\sim$. Since $\sim$ is a congruence, L_1 is a modular lattice. One may repeat this process with L_1 in place of L, inductively setting $L_{n+1} = (L_n)_1$. The process may be continued transfinitely as follows. Let $\pi_n : L \to L_n$ denote the composite quotient map and let $\sim_n$ denote the **kernel** of π_n, that is, $\{(a, b) \in L \times L : \pi_n a = \pi_n b\}$. Clearly $\sim = \sim_1 \subseteq \cdots \subseteq \sim_n \subseteq \cdots$. Define $\sim_\omega$ to be the union of the $\sim_n$ with $n < \omega$ and set $L_\omega = L/\sim_\omega$. Continue: that is, having defined L_α and $\sim_\alpha = \ker(\pi_\alpha : L \to L_\alpha)$, define $L_{\alpha+1} = (L_\alpha)_1$ and $\sim_{\alpha+1} = \ker(\pi_\alpha : L \to L_\alpha \to L_{\alpha+1})$ and, if λ is a limit ordinal, set $\sim_\lambda = \bigcup_{\alpha<\lambda} \sim_\alpha$ and $\pi_\lambda : L \to L_\lambda = L/\sim_\lambda$.

Let α be the least ordinal such that L_α is the **trivial** (that is, one-point) **lattice**, if such an α exists. If the top and bottom of L are identified at some limit stage, then they must already have been identified at a previous stage so α is not a limit ordinal. Hence we may set $\mathcal{L}$-dim$(L) = \alpha - 1$ and say that the **$\mathcal{L}$-dimension** of

[1] A congruence is the kernel of a morphism: the set of pairs with the same image.

L is $\alpha - 1$. For example, by 7.1.1, L has finite length iff $\mathcal{L}$-dim$(L) = 0$, where $\mathcal{L}$ consists of just the two-point and one-point lattices.

If there is no such ordinal α, then set $\mathcal{L}$-dim$(L) = \infty$ and say that the $\mathcal{L}$-dimension of L is ∞ or "undefined".

Lemma 7.1.2. *If L' is a subquotient of the lattice L, then $\mathcal{L}$-dim$(L') \leq \mathcal{L}$-dim(L).*

Proof. If $a' > b'$ are in L', then choose $a > b$ in L with $\pi a = a'$ and $\pi b = b'$, where π is a partial morphism from L onto L' (a **morphism** of lattices is one which preserves $+$ and $\wedge$, a **partial morphism** is one defined on a sublattice). If $a \sim b$, then, by 7.1.1 and since $\mathcal{L}$ is closed under sublattices and quotients, $a' \sim b'$ in L'. Similarly (by a transfinite induction) for each congruence $\sim_\alpha$. □

Example 7.1.3. ([591]) Let $\mathcal{L}$ consist of all lattices with the **descending chain condition**: $L \in \mathcal{L}$ iff for every descending sequence $a_1 \geq a_2 \geq \cdots \geq a_i \geq \cdots$ of points of L there is some n such that $a_n = a_m$ for all $m \geq n$. The corresponding dimension is called **Krull dimension** and is denoted Kdim$(-)$. It was defined in [591] (as the "deviation" of a poset) but usually has been applied to the lattice of all submodules of a module and it leads to a generalisation of the classical notion of Krull dimension of a commutative ring ([245, 3.5] or see, say, [626, 3.5.51]).

A modular lattice has Krull dimension 0 iff it has the descending chain condition. Inductively, a modular lattice L has Krull dimension α iff it does not have Krull dimension strictly less than α and if, for every descending sequence $a_1 \geq a_2 \geq \cdots \geq a_i \geq \cdots$ of points of L, there is some n such that the Krull dimension of the interval $[a_i, a_{i+1}]$ is strictly less than α for all $i \geq n$. This is the usual definition of Krull dimension: it is easy to check that it is equivalent to that defined by the inductive process above.

This dimension was applied to the lattice of pp-definable subgroups of a module by Garavaglia: he called the result the **elementary Krull dimension** of the module and, in a long, influential, though never published, paper, [209], he derived many consequences of a module having this dimension and he also computed this dimension for various kinds of modules.[2]

Proposition 7.1.4. *Let L be a modular lattice. Then the following are equivalent:*

(i) Kdim$(L) < \infty$;
(ii) L has no non-trivial densely ordered subchain;[3]
(iii) the poset $(\mathbb{Q}, <)$ does not embed in L.

[2] This paper of Garavaglia was a major inspiration and precursor for the work that transformed the model theory of modules around 1980.
[3] By "subchain" I mean a subposet which is a chain.

Proof. A poset P is **densely ordered** if for every pair, $a < b$, of comparable distinct points of P there is a point, c, lying strictly between them: $a < c < b$. Any non-trivial densely ordered poset embeds the poset $(\mathbb{Q}, <)$: $(\mathbb{Q}, <)$ is isomorphic, as a partially ordered set, to the poset of dyadic fractions (those of the form $n/2^k$) lying in the interval $(0, 1) \subseteq \mathbb{Q}$ so one may define an embedding by induction in the obvious way (see below).

Clearly $\mathbb{Q}$ has no non-trivial subinterval with the descending chain condition, hence $\mathbb{Q} = \mathbb{Q}_0 = \cdots = \mathbb{Q}_\alpha$ for all α. Therefore $\mathrm{Kdim}(\mathbb{Q}) = \infty$ and so, if L embeds $\mathbb{Q}$ (or any other non-trivial densely ordered poset), then $\mathrm{Kdim}(L) = \infty$ by 7.1.2.

For the converse, suppose that $\mathrm{Kdim}(L) = \infty$. Then there is α such that $L_\alpha = L_{\alpha+1}$, hence such that $L_\alpha = L_\beta$ for every $\beta > \alpha$ and L_α is not the trivial lattice. Let $a < b \in L_\alpha$. Then there must be some point $c \in L_\alpha$ lying strictly between a and b, otherwise the congruence $\sim$ would not be the identity on L_α contradicting that $L_\alpha = L_{\alpha+1}$. Therefore L_α is densely ordered and non-trivial, hence contains a copy of $\mathbb{Q}$. We must pull this back to a copy in L. For this it is more convenient to use, in place of $\mathbb{Q}$, the poset of dyadic fractions lying in the closed interval $[0, 1] \cap \mathbb{Q}$ because we use an inductive argument to ensure that the inverse images in L that we choose are comparable. Suppose, then, that $a_i \in L_\alpha$ ($i = n/2^m$, $m \geq 1$, $0 \leq n \leq 2^m$) are ordered as are their indices. Let $\pi : L \to L_\alpha$ denote the projection from L to $L_\alpha = L/\sim_\alpha$.

Choose $b_1 \in L$ with $\pi b_1 = a_1$. Choose $c_0 \in L$ with $\pi c_0 = a_0$ and set $b_0 = c_0 \wedge b_1$. Then $b_0 < b_1$ and $\pi(b_0) = \pi(c_0) \wedge \pi(b_1) = a_0$.

Suppose, inductively on m, that we have chosen $b_{n/2^m} \in L$ for $n = 0, \ldots, 2^m$ with $\pi(b_{n/2^m}) = a_{n/2^m}$ and that these are ordered as are their indices. Let $n \in 0, \ldots, 2^m - 1$ and set $j = (2n+1)/2^{m+1}$. Set $i = n/2^m, i' = (n+1)/2^m$. Choose $c_j \in L$ such that $\pi c_j = a_j$. Set $b_j = b_i + c_j \wedge b_{i'}$ Because the lattice L is modular and $b_i < b_{i'}$ we have $b_i + (c_j \wedge b_{i'}) = (b_i + c_j) \wedge b_{i'}$ so b_j is an element lying between b_i and $b_{i'}$, and $\pi(b_j) = \pi(b_i) + \pi(c_j) \wedge \pi(b_{i'}) = a_i + a_j \wedge a_{i'} = a_j$.

This construction yields a densely ordered chain in L, as required. □

Corollary 7.1.5. *If L is a modular lattice with either the ascending or the descending chain condition, then* $\mathrm{Kdim}(L) < \infty$.

Proof. For the dcc this is by definition. For the acc it is immediate from 7.1.4 since if L' is a subposet of L and if $a > b$ are in L', then there is a maximal element strictly between a and b, so L' cannot be densely ordered. □

The characterisation, 7.1.4, of having Krull dimension generalises.

Proposition 7.1.6. ([495, 10.5]) *The $\mathcal{L}$-dimension of a lattice L is undefined iff there is some non-trivial subquotient of L which contains no non-trivial interval in $\mathcal{L}$.*

In particular, if L is a modular lattice with $\mathcal{L}$-dim$(L) < \infty$, *then for every* $a > b$ *in L there are* $a', b' \in L$ *with* $a \geq a' > b' \geq b$ *such that the interval* $[a', b']$ *is in* $\mathcal{L}$.

Proof. If L' is a subquotient of L with no non-trivial interval in $\mathcal{L}$, then $(L')_1 = L$, so if L' is non-trivial, then $\mathcal{L}$-dim$(L') = \infty$ and, therefore, by 7.1.2, $\mathcal{L}$-dim$(L) = \infty$.

For the converse, if $\mathcal{L}$-dim$(L) = \infty$, choose α such that $L_\alpha = L_{\alpha+1}$. Then the congruence $\sim$ on L_α is trivial so L_α contains no interval in $\mathcal{L}$.

By the first statement, if $\mathcal{L}$-dim$(L) < \infty$, then every non-trivial interval $[a, b]$ in L contains a non-trivial subinterval in $\mathcal{L}$. □

7.2 M-dimension

The dimension based on collapsing intervals of finite length, equivalently simple intervals, is defined and shown to coexist with Krull dimension (7.2.3), then applied to lattices of pp-definable subgroups (7.2.5). This dimension, m-dimension, is preserved by duality (7.2.7). This dimension has value 0 *for all modules exactly if the ring is of finite representation type (7.2.8).*

Let $\mathcal{L}$ be the class consisting of two-point and trivial (one-point) lattices. Apply the definitions of the previous section to obtain the dimension for modular lattices based on collapsing intervals in $\mathcal{L}$. We call this **m-dimension**,[4] denoting it mdim$(-)$. This dimension appears in [726, p. 191] as a refinement of Garavaglia's elementary Krull dimension (see 7.1.3) and is developed also in [495, §10.2]. The next results are special cases of 7.1.6 and 7.1.1 respectively.

Lemma 7.2.1. *If L is a lattice with* mdim$(L) < \infty$, *then L contains a minimal pair (that is, there are* $a > b$ *in L such that there is no* $c \in L$ *with* $a > c > b$*).*

Lemma 7.2.2. *Let L be a modular lattice. Let* $a, b \in L$*. Then a and b are identified by the m-dimension congruence on L iff the interval* $[a + b, a \wedge b]$ *is of finite length. In particular, if* $a \geq b$, *then* mdim$[a, b] = 0$ *iff* $[a, b]$ *is of finite length.*

Proposition 7.2.3. *Let L be a modular lattice.*

(i) mdim$(L^{\mathrm{op}}) = $ mdim(L)
(ii) Kdim$(L) \leq$ mdim(L)
(iii) Kdim$(L) = \infty$ *iff* mdim$(L) = \infty$ *iff L contains a densely ordered subset.*

Proof. (i) By 7.2.2 it is clear that two points are identified under $\sim$ in L iff they are identified under $\sim$ in L^{op} and this continues through the inductive definition.

[4] The "m" stands for minimal (collapsing) or maximal (-valued dimension), as one prefers.

The point is, of course, that, in contrast to Krull dimension, the basic class, $\mathcal{L}$, of lattices is closed under taking opposite lattices. The same will be the case for the dimension that we define in Section 7.3.1.

(ii) This is clear since the basic class of lattices used to define m-dimension is contained in the class used to define Krull dimension. Again, this is a general point.

(iii) If the m-dimension congruence, $\sim$, on a lattice L is just equality, then L has no two-point intervals and hence L, if non-trivial, is densely ordered. So the argument of 7.1.4 for Krull dimension applies to show that if $\text{m-dim}(L) = \infty$, then L contains a densely ordered chain and so, by that result, $\text{Kdim}(L) = \infty$. □

Corollary 7.2.4. *For any ring R and $n \geq 1$,* $\text{mdim}(\text{pp}^n_R) = \text{mdim}({}_R\text{pp}^n)$.

The corollary is immediate by 1.3.1.

Corollary 7.2.5. *For any ring R and closed subset, X, of Zg_R the following conditions are equivalent:*

(i) $\text{pp}^1(X)$ *does not contain a densely ordered subposet;*
(ii) $\text{mdim}(\text{pp}^1(X)) < \infty$*;*
(iii) $\text{Kdim}(\text{pp}^1(X)) < \infty$*;*
(i)^n–(iii)^n the same conditions for $\text{pp}^n(X)$ *(for any n);*
(i)^{op}–(iii)^{op} the same conditions for $\text{pp}^1(DX)$*, where $DX \subseteq {}_R\text{Zg}$ is the closed subset of ${}_R\text{Zg}$ dual (5.4.2) to X.*

In particular, if these conditions are satisfied, then every non-trivial interval of each $\text{pp}^n(X)$ *contains a two-point subinterval.*

Proof. Equivalence of the first three conditions is immediate from 7.2.3. Equivalence of these with (i)^n–(iii)^n is immediate (cf. proof of 7.3.8) once one makes the identification, 13.2.1, of m-dimension with Krull–Gabriel dimension, the identification, 12.3.18, of $\text{pp}^n(X)$ with the lattice of finitely generated subfunctors of the localisation $(R^n, -)_X$ and then uses the definition of Krull–Gabriel dimension via localisation (Section 13.2.1) and the fact that $(R^n, -)_X \simeq (R, -)^n_X$ is torsion iff $(R, -)_X$ is torsion. A proof which works directly with the lattices $\text{pp}^n(X)$ and projections between them is also possible.

Equivalence with the last three conditions is immediate from the definitions, 1.3.1, 1.3.15 and 7.2.3.

The last statement is 7.2.1. □

The notation and terminology is extended to modules, definable subcategories and closed subsets of the Ziegler spectrum via the corresponding lattices of pp conditions. In particular, if M is a module we write $\text{mdim}(M)$ for $\text{mdim}(\text{pp}(M))$.

It will be seen in Section 7.3.1 (7.3.6 in particular) that if mdim $< \infty$, then every pure-injective is the pure-injective hull of a direct sum of indecomposable pure-injectives.

Example 7.2.6. If N is Σ-pure-injective, then mdim$(N) < \infty$: by 4.4.5, pp(N) has the dcc, that is, Kdim(pp(N)) = 0s, so, by 7.2.5, mdim(pp(N)) $< \infty$. Therefore, by 1.3.15 any elementary dual of a Σ-pure-injective module has m-dimension. Thus, or directly by 2.3.10, every flat module over a left noetherian ring has m-dimension.

Corollary 7.2.7. *If $\mathcal{X}$ is any definable subcategory of* Mod-R, *then* mdim$(\mathcal{X})$ = mdim$(D\mathcal{X})$, *where $D\mathcal{X}$ is the subcategory of* R-Mod *dual (Section 3.4.2) to $\mathcal{X}$. Similarly, if $N \in \mathrm{Zg}_R$ is reflexive (that is, has an elementary dual in the sense of Section 5.4), then N and its dual have the same m-dimension.*

Proof. This is immediate from 7.2.3(i) together with 3.4.18, respectively 5.4.12. □

The following is a rephrasing of 4.5.21 (KG(R) denotes the Krull–Gabriel dimension of R, see Section 13.2.1).

Proposition 7.2.8. *A ring R is of finite representation type iff* mdim$(\mathrm{pp}_R) = 0$ *iff* KG$(R) = 0$.

For another dimension ("2dim"), which coexists with m-dimension, see 7.3.17.

For a summary of what is known on the values of m-dimension = Krull–Gabriel dimension, see Section 13.2.1.

7.2.1 Calculating m-dimension

The m-dimension of the lattice freely generated by a pair of chains is the Cantor sum of their m-dimensions (7.2.9).

Let L_1, L_2 be chains with top 1_i and bottom 0_i $(i = 1, 2)$. Following Puninski ([570, §4]) denote by $L_1 \otimes L_2$ the modular lattice freely generated by L_1 and L_2 subject to identifying 1_1 with 1_2 and 0_1 with 0_2 (so if L_2 has just two elements, then $L_1 \otimes L_2 \simeq L_1$). By [249, Thm 13] this is a distributive lattice. Every element of $L_1 \otimes L_2$ has a unique representation of the form $(a_n \wedge b_1) + (a_{n-1} \wedge b_2) + \cdots + (a_1 \wedge b_n)$ with $a_1 < \cdots < a_n = a_n \wedge (b_1 + a_{n-1}) \wedge + \cdots + (b_{n-1} + a_1) \wedge b_n$ with $a_1 < \cdots < a_n$ in L_1 and $b_1 < \cdots < b_n$ in L_2. In [570] (also [571, §3]) Puninski discusses this and gives a graphical way of representing elements of $L_1 \otimes L_2$ as elements of the algebra generated, under $\cup$ and $\cap$, by certain rectangles in the first quadrant of the "$L_1 \times L_2$-plane".

Proposition 7.2.9. ([570, 4.2]) *Suppose that* L_1 *and* L_2 *are non-trivial chains with 0 and 1. Then* $\mathrm{mdim}(L_1 \otimes L_2) = \mathrm{mdim}L_1 \oplus \mathrm{mdim}L_2$, *where* $\oplus$ *denotes the Cantor sum of ordinals (Appendix E, p. 713).*[5]

Remark 7.2.10. ([570, 4.3]) The same holds for the modular lattice generated by L_1 and L_2 without identification of tops and bottoms.

The result above has been very useful in computing m-dimension in particular examples and classes of examples, see [570, §5], [524, §§5, 7]. The same techniques have been used to compute width, see Section 7.3, 7.3.17 in particular. Calculations are much simplified in intervals of pp_R which are distributive and such intervals quite often can be found.

7.2.2 Factorisable systems of morphisms and m-dimension

There is a factorisable system of morphisms between finitely presented modules iff there is a densely ordered chain of pp conditions (7.2.12) iff the m-dimension, equivalently the Krull–Gabriel dimension, is ∞ *(7.2.13). For countable rings this is equivalent to there being continuum many indecomposable pure-injectives (7.2.14). Examples of factorisable systems are given (7.2.11, 7.2.15). There is a result on preservation of factorisable systems under functors (7.2.16).*

In 3.1.4 the connection induced by 1.2.9, between the ordering on pp conditions in pp_R^n and the ordering on n-pointed finitely presented modules was made precise. From that one can see that any "finiteness condition" on the lattice of pp conditions can be re-expressed as a condition on morphisms between finitely presented modules. In this section, following [504], this is done for the condition of having m-dimension.

A collection $\{M_\alpha, \overline{a}_\alpha, f_{\alpha\beta} : \alpha < \beta,\ \alpha, \beta \in \mathbb{Q}^{\geq 0}\}$ of finitely presented R-modules M_α, n-tuples (fixed n), $\overline{a}_\alpha \in M_\alpha^n$ from them, and morphisms between them, $f_{\alpha\beta} : M_\alpha \to M_\beta$, is a **factorisable system** if ([504]):

- $f_{\beta\gamma} f_{\alpha\beta} = f_{\alpha\gamma}$ whenever $\alpha < \beta < \gamma$;
- $f_{0\alpha}(\overline{a}_0) = \overline{a}_\alpha$ for all $\alpha > 0$ (so this is a $\mathbb{Q}^{\geq 0}$-directed system in the category of n-pointed finitely presented R-modules);
- for each $\alpha < \beta$ there is no $g : M_\beta \to M_\alpha$ such that $g f_{\alpha\beta}(\overline{a}_\alpha) = \overline{a}_\alpha$ (by 1.2.9 this is precisely the condition that each morphism $f_{\alpha\beta}$ strictly increase the pp-type of $\overline{a}_\alpha$).

[5] Note that this is very different from calculating the m-dimension of a product of lattices: it is easy to check (inductively, noting that the product of two intervals of finite length has finite length) that $\mathrm{mdim}(L_1 \times L_2) = \max\{\mathrm{mdim}(L_1), \mathrm{mdim}(L_2)\}$.

If the module M_0 is dropped from a factorisable system as above, then the second condition should be replaced by:

- $f_{\alpha\beta}(\overline{a}_\alpha) = \overline{a}_\beta$ for all $\alpha < \beta$.

The resulting notion is equivalent, as is that obtained by using any countable densely ordered set of indices, such as $\mathbb{Q} \cap [0, 1]$, $\mathbb{Q} \cap (0, 1)$ or the dyadic fractions in $[0, 1]$. For most purposes we may take $n = 1$.

Example 7.2.11. Let R be a commutative von Neumann regular ring with atomless boolean algebra, $B(R)$, of idempotents (see Section 8.2.11). Then from these idempotents one obtains a factorisable system, indexed by $(\mathbb{Q} \cap [0, 1])^{(\mathrm{op})}$, of finitely presented (projective) modules as follows.

Choose $e_1^2 = e_1 \in R$, $e_1 \neq 0$. Since $B(R)$ is atomless there is an idempotent $e_{\frac{1}{2}}$ with $e_1 R > e_{\frac{1}{2}} R > 0$. Consider the non-zero idempotents $e_{\frac{1}{2}}$ and $e_1 - e_{\frac{1}{2}}$. Again since $B(R)$ is atomless there are idempotents $e_{\frac{1}{4}}$ and $e'_{\frac{1}{4}}$ with $e_{\frac{1}{2}} > e_{\frac{1}{4}} > 0$ and $(e_1 - e_{\frac{1}{2}})R > e'_{\frac{1}{4}} R > 0$. Set $e_{\frac{3}{4}} = e_{\frac{1}{2}} + e'_{\frac{1}{4}}$ (idempotent since $e_{\frac{1}{2}} e'_{\frac{1}{4}} = 0$). Continue in the obvious way. For each dyadic fraction, r, set $P_r = e_r R$, $a_r = e_r$ and let f_{rs}, for $r > s$, be the morphism $P_r \to P_s$ determined by $e_r \mapsto e_s$. There is no morphism taking e_s back to e_r because $e_s(e_{1-s}) = 0$ whereas $e_r(e_{1-s}) \neq 0$.

Clearly a factorisable system is just a densely ordered chain in the lattice of equivalence classes of n-pointed modules for some n, so one has the following (which is given a direct proof from 1.2.9 in [504]).

Theorem 7.2.12. ([504, 0.3]) *For any ring R the following conditions are equivalent.*

(i) There is a densely ordered chain of pp conditions in n free variables for some n.
(ii) There is a factorisable system in mod-R.

Corollary 7.2.13. *For any ring R the following conditions are equivalent:*

(i) $\mathrm{mdim}(\mathrm{pp}_R) < \infty$;
(ii) $\mathrm{KG}(R) < \infty$;
(iii) there is no factorisable system in mod-R.

The corollary follows by 7.2.5 and 13.2.1.

The next result, which is immediate from 7.2.12 and 5.3.19, illustrates the link, via pp conditions and pp-types, between complexity of the category of finitely presented modules and complexity of the set/space of indecomposable pure-injective modules.

Corollary 7.2.14. *A countable ring has a factorisable system of morphisms between finitely presented modules iff it has continuum many indecomposable pure-injectives.*

This result, combined with specific arguments such as at 7.2.15 below, was used to show, around the mid-1980s (see the introduction to [504]), that over many finite-dimensional algebras there are far more indecomposable pure-injective modules than those which are easily visible.

There is a connection between existence of factorisable systems of morphisms in mod-R and the transfinite radical of mod-R (9.1.9) though the connection is not completely understood (see 9.1.11).

In situations where one has a good knowledge of the category of finitely presented modules (such as over many finite-dimensional algebras), the (non-) existence of a factorisable system can be straightforward to check.

Example 7.2.15. ([504, pp. 449–50]) Given any field k let $\mathrm{GP}_{m,n}$ denote the **Gelfand–Ponomarev algebra** ([230], see Section 15.1.1) $k[a, b : ab = 0 = ba = a^m = b^n]$ which, for $m, n \geq 2, m + n \geq 5$, is of tame but non-domestic representation type. The indecomposable finite-dimensional representations of this quiver are string and band modules of the sort discussed in Section 8.1.3. A factorisable system of maps between string modules over $\mathrm{GP}_{2,3}$ may be obtained as follows.

Let w be any finite word such that wb is defined (refer to Section 8.1.3, also [110] for terminology and notation). Consider the "obvious" morphism of string modules $f : M(wba^{-1}) \to M(wb^2a^{-1})$ – the morphism which annihilates the image of a then makes the generator divisible by b. Notice that this has a factorisation (where the morphisms are the "obvious" ones):

$$M(wba^{-1}) \to M(wba^{-1}ba^{-1}) \to M(wba^{-1}b^2a^{-1}) \to M(wba^{-1}b^2a^{-1}ba^{-1}) \\ \to M(wba^{-1}b^2a^{-1}b^2a^{-1}) \to M(wb^2a^{-1}).$$

Writing this as

$$A \to B \to C \to D \to E \to F$$

note that the morphisms $B \to C$ and $D \to E$ have the same form as f (for different "w").

So define $M_0 = M(ba^{-1})$ and let a_0 be any non-zero element in M_0b, say a basis element at the "left-hand end" according to the conventions of Section 8.1.3. Set $M_1 = M(b^2a^{-1})$ and let $f_{01} : M_0 \to M_1$ denote the "obvious" map (as above). Taking these to be A, respectively, F, in the above factorisation, let $M_{1/2} = C$ and let $f_{0(1/2)}$ and $f_{(1/2)1}$ be the compositions $A \to C$ and $C \to F$. Now repeat with $A \to C$, resp. $C \to F$ (factorising $B \to C$, resp. $D \to E$).

Proceeding by induction, one obtains a factorisable system of morphisms indexed by the dyadic fractions in [0, 1], as required: the third condition in the definition of factorisable system is satisfied because each of these modules M_α is indecomposable, [230], (see Section 8.1.3) and none of the morphisms is an isomorphism, indeed, by [110, §3], M_α is not isomorphic to M_β for $\alpha \neq \beta$.

A picture is worth many words:

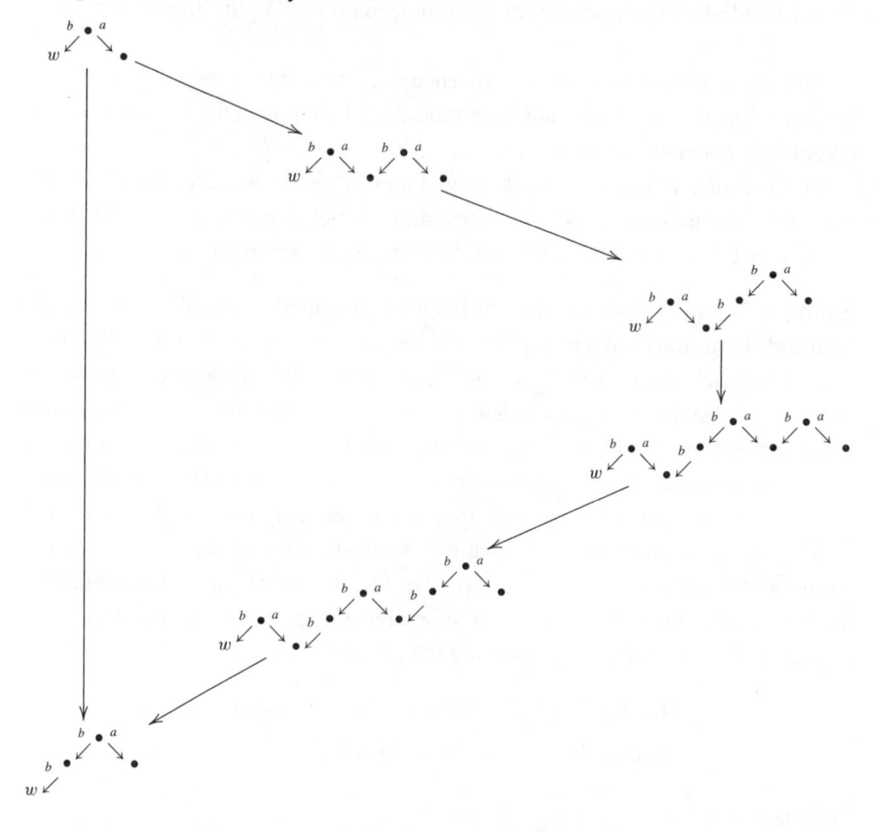

If R is any non-domestic string algebra, then, as has been noted by various people (for example, [640, 5.5]), a similar construction of a factorisable system may be made, either directly or by using a transfer result such as 7.2.16. Other algebras with factorisable systems (of non-isomorphic indecomposable modules) are the tubular canonical algebras: the existence of such a system is immediate from Ringel's description of mod-R ([602, 5.2.4], see Section 8.1.4).

There are two possible obstructions to a functor carrying a factorisable system to a factorisable system. One is the possibility that we lose track of the distinguished n-tuples: for these are just morphisms $R^n \to M_\alpha$ and the functor might be defined on a subcategory of mod-R which does not include R^n. But all that we really want from the $f_{\alpha\beta}$ is that they are strictly pp-type-increasing and that there is some tuple whose images through the system are non-zero. Since any subsystem will do, the latter only needs a single $f_{\alpha\beta}$ with $\alpha < \beta$ and some a_α with $f_{\alpha\beta}a_\alpha \neq 0$. The main problem is the possibility that (components of) morphisms in the system become isomorphisms or, more to the point, the morphisms $f_{\alpha\beta}$ may be sent to morphisms which are not strictly pp-type-increasing in the image category.

With some assumptions then, as illustrated by the following transfer result, this possibility may be avoided. Recall the notion of a representation embedding from Section 5.5.2.

Proposition 7.2.16. ([504, 0.2]) *Let R, S be artin algebras and let $\mathcal{C}$ be a full subcategory of* mod-S.

(a) Suppose that F is a representation embedding from $\mathcal{C}$ to mod-R. *Then the image of any factorisable system of mutually non-isomorphic indecomposable modules in $\mathcal{C}$ is a factorisable system in* mod-R.
(b) Suppose that F is a representation embedding from $\mathcal{C}$ to mod-R. *Suppose that there is a factorisable system of indecomposable modules in $\mathcal{C}$. Then there is a factorisable system of indecomposable modules in* mod-R.
(c) Suppose that F is an equivalence of $\mathcal{C}$ with a full subcategory of mod-R. *Then the image of any factorisable system in $\mathcal{C}$ contains (in the sense of the proof below) a factorisable system in* mod-R.

Proof. In each part, suppose that $\{M_\alpha, \overline{a}_\alpha, f_{\alpha\beta} : \alpha < \beta, \alpha, \beta \in \mathbb{Q}^{\geq 0}\}$ is the factorisable system in $\mathcal{C}$.

(a) A representation embedding preserves both indecomposability and non-isomorphism, so the images $N_\alpha = FM_\alpha$ are pairwise non-isomorphic indecomposable finitely presented R-modules. No $g_{\alpha\beta} = Ff_{\alpha\beta}$ can be zero since F is right exact (being given by tensor product, Section 5.5.2) and preserves non-isomorphism. Let $\overline{c}_0$ be a finite generating set for N_0 and, for each α, set $\overline{c}_\alpha = f_{0\alpha}(\overline{c}_0)$. Then, we claim, $\{N_\alpha, \overline{c}_\alpha, g_{\alpha\beta} : \alpha < \beta, \alpha, \beta \in \mathbb{Q}^{\geq 0}\}$ is a factorisable system. The only point which is not quite immediate is that the morphisms $g_{\alpha\beta}$ are strictly pp-type increasing. To see this, suppose that for some $\alpha < \beta$ there is h with $hg_{\alpha\beta}(\overline{c}_\alpha) = \overline{c}_\alpha$. Then (by 4.3.61) $hg_{\alpha\beta}$ is an automorphism of N_α, hence $g_{\alpha\beta}$ is a split embedding, hence is an isomorphism, a contradiction, as required.

(b) Given a factorisable system of indecomposable modules, each indecomposable, being of finite endolength, can occur only finitely many times in the system (or use the Harada–Sai Lemma, 4.5.10). Therefore, from a factorisable system of indecomposable modules of finite length, one may inductively construct a new factorisable system in which each indecomposable occurs at most once (at each stage, just pass over any indecomposable which has already been put into the new system). Then apply part (a) to deduce the result.

(c) Since every element of mod-S is a finite direct sum of indecomposable modules, each M_α may be replaced by a direct summand which is minimal containing $\overline{a}_\alpha$ (that is, a hull of $\overline{a}_\alpha$ in the sense of Section 4.3.5). More precisely, let $M_\alpha = M'_\alpha \oplus M''_\alpha$, where M'_α is a minimal direct summand of M_α containing $\overline{a}_\alpha$. Let $f'_{\alpha\beta}$ be the component of $f_{\alpha\beta}$ between M'_α and M'_β. Then $\{M'_\alpha, \overline{a}_\alpha, f'_{\alpha\beta} : \alpha < \beta, \alpha, \beta \in \mathbb{Q}^{\geq 0}\}$ is clearly a factorisable system.

Let $N_\alpha, g_{\alpha\beta}$ denote the images under F of $M'_\alpha, f'_{\alpha\beta}$. Let $\overline{c}_0$ be a finite generating set for N_0 and set $\overline{c}_\alpha = g_{0\alpha}(\overline{c}_0)$. Notice that N_α is minimal containing $\overline{c}_\alpha$: otherwise $g_{0\alpha}$ would factorise through a proper direct summand of N_α, so, by fullness, the same would be true of $f'_{0\alpha}$ and M'_α and this would contradict the completeness of our initial pruning. As in part (a), there is only one non-trivial point in checking that we have a factorisable system. So suppose there were some α, β and h with $hg_{\alpha\beta}(\overline{c}_\alpha) = \overline{c}_\alpha$. Since N_α is the minimal direct summand of N_α containing $\overline{c}_\alpha$, and similarly for N_β and $\overline{c}_\beta$, Fitting's Lemma (4.3.57) yields that $g_{\alpha\beta}$ is an isomorphism hence, by fullness, that $f'_{\alpha\beta}$ is an isomorphism, a contradiction, as required. □

In particular, for artin algebras, a strict representation embedding (Section 5.5.2) carries factorisable systems to factorisable systems and any representation embedding carries factorisable systems of non-isomorphic indecomposable modules to factorisable systems.

Question 7.2.17. Are domestic algebras characterised among finite-dimensional algebras of infinite representation type by not having a dense chain of pp conditions/morphisms?

This is very much related to conjectures of Prest, Ringel and Schröer concerning the transfinite powers of the radical of the category mod-R and the Krull–Gabriel dimension of a ring, see Section 9.1.4.

Example 7.2.18. Let R be the path algebra over a field k of the quiver $1 \rightrightarrows 2$ (three arrows). There is a strict representation embedding from the category Mod-$k\langle X, Y\rangle$ to Mod-R defined by sending the $k\langle X, Y\rangle$-module M to the representation with M_k at each of the two vertices, taking the first arrow to be an isomorphism, regarded as identifying the two copies of M, and taking the second and third arrows to be multiplication by X and Y respectively. Conversely, the definable subcategory of R-modules consisting of those representations where the first arrow is an isomorphism is equivalent to Mod-$k\langle X, Y\rangle$.

Recall that the path algebra of a quiver is of finite or tame representation type provided the quiver[6] is a union of Dynkin and extended Dynkin (= Euclidean) diagrams. All other quivers without relations have wild representation type: it is enough to show this for a set of minimal such quivers, minimal in the sense that any other quiver contains one of these. The quiver above is one of a small set of minimal wild quivers and the others in that set can be found, together with proofs of their wildness, in Brenner's [92].

It was shown in [492] that the functors used by Brenner are interpretation functors in the sense of Section 18.2.1 and, from that, undecidability of the theory of modules over wild hereditary algebras follows. For example, representations of the 5-subspace quiver code up arbitrary $k\langle X, Y\rangle$-modules, see 1.1.3. That completed one line of results leading from Baur's papers, [47], [48], though only one: it is still an open question whether every wild quiver (with relations) has undecidable theory of modules; a representation embedding seems too weak to give an interpretation functor, though, see Section 5.5.2, a strict representation embedding suffices.

7.3 Width

Width and its relation to existence of superdecomposable pure-injectives is described in Section 7.3.1. A summary of what is known about this dimension over various kinds of rings is given in Section 7.3.2.

7.3.1 Width and superdecomposable pure-injectives

The width, and its variant, breadth, of a lattice is defined and a "densely-splitting" test lattice identified (7.3.1).

If the lattice of pp-definable subgroups of a pure-injective module contains a chain, then it has an indecomposable direct summand on which that chain does not collapse (7.3.2). A superdecomposable pure-injective has no intervals in its lattice of pp-definable subgroups with width (7.3.4). If a ring has Krull–Gabriel dimension (equivalently m-dimension), more generally if the lattice of pp conditions has width, then there is no superdecomposable pure-injective over that ring (7.3.5). The converse is true if the ring is countable (7.3.9) but unknown in general. Hence, for a countable ring, there is a superdecomposable pure-injective right module iff there is a left such module (7.3.12).

If every point has an open neighbourhood given by a pp-pair with width, then the isolation condition holds (7.3.16).

[6] Rather, the underlying undirected graph.

The lattice freely generated by a pair of densely ordered chains does not have width (7.3.17).

Let $\mathcal{L}$ be the class of lattices which are chains, that is, totally ordered. We apply the construction of Section 7.1 and call the resulting notion of dimension **breadth**. In [726, p. 183] Ziegler defined what he called the "width" of a lattice; as with Krull dimension and m-dimension (7.2.3), width and breadth grow at different rates but the one is defined iff the other is ([495, 10.7], also for example, [577, §3]) and in this section what we are interested in is whether or not this dimension is defined. The actual value of breadth as defined here is relevant to uniserial dimension (Section 13.4), and also appears in Section 8.2.8.

Let L be a modular lattice. Say that L is **wide** if, given any two distinct comparable points, $\phi > \psi$, in L, there exist incomparable points, $\theta_1, \theta_2 \in L$, with $\phi > \theta_i > \psi$, $i = 1, 2$. That is, a lattice is wide iff no non-trivial interval is a chain. In this context these lattices play the role that densely ordered sets did in the case of m-dimension (7.3.1, cf. 7.2.3). It will be convenient to broaden the scope of the definition, so let P be any subposet of a modular lattice L. Say that P is **wide** if, given any two distinct comparable points $\phi > \psi$ in P, there exist incomparable points $\theta_1, \theta_2 \in P$ with $\phi > \theta_i > \psi$, $i = 1, 2$, and there exist $\theta_{11}, \theta_{12}, \theta_{21}, \theta_{22} \in P$ such that $\theta_1 + \theta_2 \geq \theta_{11} > \theta_1 > \theta_{12} \geq \theta_1 \wedge \theta_2$ and $\theta_1 + \theta_2 \geq \theta_{21} > \theta_2 > \theta_{22} \geq \theta_1 \wedge \theta_2$ (see the diagram shown where double lines could, but need not be, equalities and single lines show strict inclusion).

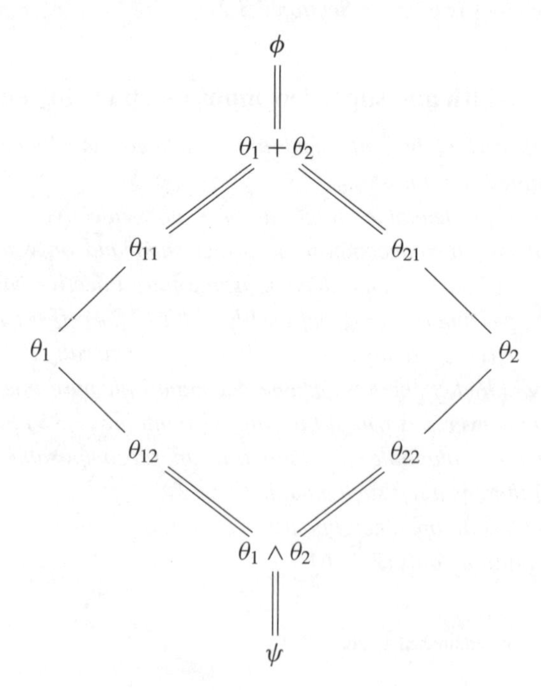

Here the join $+$ and the meet $\wedge$ refer to the operations in L. If P is itself a sublattice, then the definition simplifies since then we may take $\theta_{i1} = \theta_1 + \theta_2$ and $\theta_{i2} = \theta_1 \wedge \theta_2$.

Theorem 7.3.1. *The breadth of a modular lattice L is defined (that is, $< \infty$) iff L contains no wide poset iff L has no wide subquotient.*

Proof. In view of 7.1.6, and noting that distinct points of a wide subposet can be identified by the congruence used to define breadth, what has to be shown is that if L has a wide subquotient then L has a wide subposet. Suppose then that K, say, is a wide sublattice in a quotient lattice of L. Let L' be the full inverse image of K in L; note that L' is a sublattice of L and let $\pi : L' \to K$ be the projection to K. We construct, inductively, a wide subposet of L'.

Choose distinct comparable points $\phi > \psi$ in K, so there exist incomparable points $\theta_1, \theta_2 \in K$ with $\phi > \theta_i > \psi$ $i = 1, 2$. Choose points θ_{ij} in K so as to obtain a diagram as above. Choose inverse images, ϕ', ψ', θ_i', θ_{ij}' in L' of these points. Replacing ϕ' by $\phi' + \psi'$, which, note, has the same image in L as ϕ', we may suppose that $\phi' > \psi'$. Then replacing θ_1' by $\phi' \wedge (\theta_1' + \psi') = (\phi' \wedge \theta_1') + \psi'$, we may suppose that $\phi' \geq \theta_1' \geq \psi'$ (and, of course, these will be strict inclusions). And so on, to obtain a set, P_0 say, of points of L' arranged as in the diagram above.

Inductively, we suppose that we have constructed a finite poset, P_n, in L' such that, for every comparable pair of points $\eta > \zeta$ in P_n we have $\pi\eta > \pi\zeta$. We construct P_{n+1} by repeating the construction which resulted in P_0, between every pair $\eta > \zeta$ in P_n which is such that there is no member of P_n strictly between $\eta > \zeta$, of course retaining the choices η and ζ (as "ϕ' and ψ'" above). Thus $P_n \subseteq P_{n+1}$ and between every pair of neighbouring comparable points of P_n we have inserted a diagram as above. The union $\bigcup_n P_n$ will be a wide subposet of L, as required. □

Theorem 7.3.2. ([726, §7]) *Suppose that N is pure-injective, that ϕ, ψ are pp conditions with $\phi(N) > \psi(N)$ and that the pp-definable subgroups of N lying between $\phi(N)$ and $\psi(N)$ form a chain. Then there is an indecomposable direct summand, N_0, of N (with $\phi(N_0) > \psi(N_0)$).*

Proof. Choose $a \in \phi(N) \setminus \psi(N)$. Define corresponding "upper and lower cuts" in the interval $[\phi(N), \psi(N)]$ by $U = \{\phi' : \phi(N) \geq \phi'(N) \geq \psi(N) \text{ and } a \in \phi'(N)\}$ and $L = \{\psi' : \phi(N) \geq \psi'(N) \geq \psi(N) \text{ and } a \notin \psi'(N)\}$. Clearly, if $\psi' \in L$ and $\phi' \in U$, then $\phi' > \psi'$. Let p be a pp-type maximal with respect to $p \supseteq U$ and $p \cap L = \emptyset$. By 4.3.50, p is irreducible.

Consider the following set of pp conditions, in free variables x and y and with parameter a (for pp conditions with parameters see Section 4.1): $\Phi(x, y) = p(x) \cup \{\psi(y)\} \cup \{a = x + y\}$. Any finite subset, Φ', of Φ is contained in one of the form $\{\phi(x), \phi_1(x), \ldots, \phi_n(x), \psi(y), a = x + y\}$. Set $\phi_0 = \phi \wedge \phi_1 \wedge \cdots \wedge \phi_n$. Since $\phi_0 \in p$ and $\phi(N) \geq \phi_0(N) + \psi(N) \geq \psi(N)$, one has, by construction, $a \in \phi_0(N) + \psi(N)$, say $a = b_0 + c_0$ with $b_0 \in \phi_0(N)$ and $c_0 \in \psi(N)$. This shows (take b_0 for x, c_0 for y) that Φ' is satisfied in N. Thus Φ is finitely satisfied in N, hence (4.2.3), since N is pure-injective, Φ has a solution, (b, c) say, in N.

The claim is that $\mathrm{pp}^N(b) = p$ and hence that the hull of b (§4.3.5), which, by 4.3.49, is indecomposable, is a direct summand of N.

Certainly $p \subseteq \mathrm{pp}^N(b)$. For a contradiction, suppose that $b \in \psi'(N)$ for some pp condition ψ' with $\psi' \notin p$. By maximality of p there is $\phi' \in p$, without loss of generality $\phi' \leq \phi$, with $\phi' \wedge \psi' \leq \psi''$ for some $\psi'' \in L$, hence with $\phi' \wedge \psi' + \psi \in L$. But then we would have $a = b + c \in (\phi' \wedge \psi' + \psi)(N)$ – contrary to definition of L, as required. □

Corollary 7.3.3. *Suppose that N is pure-injective and that there is a pp-pair, ϕ/ψ, with $\phi(N) > \psi(N)$ and such that the interval $[\phi(N), \psi(N)]$ in the lattice of pp-definable subgroups of N has m-dimension:* $\mathrm{mdim}[\phi(N), \psi(N)] < \infty$. *Then there is an indecomposable direct summand of N. In particular, N is not superdecomposable.*

Proof. By 7.2.1 the interval contains a minimal pair, so this follows by 7.3.2. □

Say that the **width** (equivalently **breadth**) of a module M is **undefined** if this is so for the lattice of pp-definable subgroups of M and, in that case, write $\mathrm{w}(M) = \infty$.

Corollary 7.3.4. *If N is a superdecomposable pure-injective module, then* $\mathrm{w}(N) = \infty$, *indeed every non-trivial interval in the lattice of pp-definable subgroups of N has width ∞.*

Proof. Otherwise there would be pp conditions (in any such interval) ϕ, ψ with $\phi(N) > \psi(N)$ and the interval $[\phi, \psi]_N$ a chain. But then, by 7.3.2, N would not be superdecomposable. □

Corollary 7.3.5. *If* $\mathrm{w}(\mathrm{pp}_R) < \infty$, *in particular if* $\mathrm{mdim}(\mathrm{pp}_R) < \infty$ *(equivalently, by 13.2.2, if* $\mathrm{KG}(R) < \infty$*), then there are no superdecomposable pure-injective R-modules and hence, if N is pure-injective, then N is (by 4.4.2) the pure-injective hull of a direct sum of indecomposable pure-injective modules.*

Corollary 7.3.6. *If* $\mathrm{mdim}(\mathrm{pp}(M)) < \infty$, *then every pure-injective module with support contained in* $\mathrm{supp}(M)$ *is the pure-injective hull of a direct sum of indecomposable pure-injective modules.*

These results are due to Garavaglia, [209, Thm 1] (for m-dimension), and Ziegler, [726, 7.1] (for width).

If M is any module which has Krull dimension when regarded as a module over its endomorphism ring, then, by 1.1.8 and 7.2.3, it has m-dimension. In particular, by 7.3.6, its pure-injective hull is the pure-injective hull of a direct sum of indecomposable modules. A particular case is the following.

Corollary 7.3.7. *Suppose that the ring R has left Krull dimension. Then its pure-injective hull, $H(R_R)$, is a direct sum of indecomposable modules.*

Lemma 7.3.8. *The following conditions are equivalent for any ring R:*

(i) $\mathrm{w}(\mathrm{pp}^n_R) = \infty$ *for some n;*
(ii) $\mathrm{w}(\mathrm{pp}^n_R) = \infty$ *for every n.*

Proof. This could be done lattice-theoretically but we use the fact, see Section 13.4, especially 13.4.3, that the width is ∞ iff the uniserial dimension is ∞. Now, the latter dimension is defined via a sequence of localisations of the functor category and, just from the definition of torsion class, one has that $(R, -)$ is torsion, for a given torsion theory, iff every $(R^n, -)$ is torsion. Thus the uniserial dimension, hence (13.4.3) the breadth, of these functors (as n varies) are all the same, as required. □

If R is countable, then there is a converse to 7.3.4.

Theorem 7.3.9. ([726, 7.8(2)]) *Suppose that the lattice of pp-definable subgroups of N is countable. Then there is a superdecomposable pure-injective module, N', with* $\mathrm{supp}(N') \subseteq \mathrm{supp}(N)$ *iff* $\mathrm{w}(N) = \infty$.

A proof (which is a reworked version of Ziegler's) in terms of pp conditions can be found at [495, 10.13]. In this book we give a functorial proof which, because it is rather involved, probably sits better in Part II, see 12.6.1 and 12.6.2.

Question 7.3.10. Suppose that $\mathcal{X}$ is a definable category such that $\mathrm{w}(\mathrm{pp}^n(\mathcal{X})) = \infty$ for some n. Is there a superdecomposable pure-injective module in $\mathcal{X}$? In particular, if $\mathrm{w}(\mathrm{pp}_R) = \infty$ is there a superdecomposable pure-injective R-module?

The answer is positive for serial rings, 7.3.26.

In summary, we have the following.

Theorem 7.3.11. ([726, 7.8]) *Let R be any ring and M any R-module. Let* $\mathrm{pp}(M)$ *denote the lattice of pp-definable subgroups of M.*

(a) If there is a superdecomposable pure-injective module with support contained in $\mathrm{supp}(M)$*, then* $\mathrm{pp}(M)$ *has a wide subposet.*

(b) *If* pp(M) *is countable (in particular, if* R *is countable) and* pp(M) *contains, or has as a subquotient, a wide poset, then there is a superdecomposable pure-injective module* N *with* $\mathrm{supp}(N) \subseteq \mathrm{supp}(M)$.

In particular, if there is a superdecomposable pure-injective module over R*, then each lattice,* pp^n_R*, of pp conditions in n free variables over* R *has width* $= \infty$*. If* R *is countable, or just if, for some* n*,* pp^n_R *is countable and* $\mathrm{w}(\mathrm{pp}^n_R) = \infty$*, then there is a superdecomposable pure-injective* R*-module.*

Corollary 7.3.12. *If there is a superdecomposable pure-injective right* R*-module, then the lattice of pp conditions for left modules has width* ∞*, so if* R *is countable, then there is a superdecomposable left* R*-module.*

Question 7.3.13. If there is a superdecomposable right R-module is there a superdecomposable left R-module? Of course, if the answer to 7.3.10 is positive, then so will be the answer to this.

Just as for m-dimension in Section 7.2.2, one may reformulate the condition of not having width in terms of the lattice of pointed finitely presented modules. In [499] I attempted, not entirely successfully, to give a usable formulation in terms of "splittable systems" of finitely presented modules. This is done better in [577, 5.4].

A **wide poset of** A**-pointed modules** is a set of A-pointed modules such that the corresponding set of equivalence classes (equivalence in the sense of Section 3.1) is wide.

Corollary 7.3.14. *The following conditions for any ring* R *are equivalent.*

(i) *There is wide poset of pp conditions in one free variable over* R.
(ii) *There is a wide poset of (one-)pointed finitely presented modules.*
(iii) *There is a wide poset of* A*-pointed finitely presented modules for some nonzero finitely presented module* A.

Proof. This follows from 3.1.4, together with the observation that A-pointed modules yield R^n-pointed modules and, for instance, 3.2.4. □

Lemma 7.3.15. *Let* R, S *be any rings and let* $\mathcal{C}$ *be a full subcategory of the category of* S*-modules. Suppose that* F *is a full embedding from* $\mathcal{C}$ *to* Mod-R*. If* P *is a wide poset of* A*-pointed modules in* $\mathcal{C}$*, then* FP *is a wide poset of* FA*-pointed modules in* Mod-R.

Proof. By FP we mean the collection of FA-pointed modules $FA \to FM$, where $A \to M$ is in P. The result follows directly from the definitions using the fact that F is full. □

Proposition 7.3.16. ([570, 3.1]) *Let X be a closed subset of* Zg_R, *where R is any ring. Suppose that every point $N \in X$ is contained in a basic open set (ϕ/ψ) such that the quotient interval $[\phi, \psi]_N$ has width. Then the isolation condition (Section 5.3.2) holds for X.*

Proof. The proof in [570] is lattice-theoretic: here I give a functorial proof. Choose $N \in X$. Let X_N denote the Ziegler-closure of N and let subscript ${}_{DN}$ denote localisation at the hereditary finite type torsion theory, τ_{DX_N}, on $(R\text{-mod}, \mathbf{Ab})$ corresponding to N.

Since $[\phi, \psi]_N$ has width it has, by 7.1.6, some subinterval, $[\phi', \psi']_N$ which is a chain. Then the functor $(F_{D\psi'/D\phi'})_N$ is uniserial and, since $\{N\} = (\phi'/\psi') \cap X_N$ (since the latter is contained in $(\phi/\psi) \cap X_N$), N is the only point, N', of X_N such that $\big((F_{D\psi'/D\phi'})_N, N' \otimes -\big) \neq 0$.

Let S_1 be the simple functor $(F_{D\psi'})_N / \sum_{\phi' \geq \phi'' > \psi'} (F_{D\phi''})_N$ (that is, the top of $(F_{D\psi'/D\phi'})_N$). The injective hull in $(R\text{-mod}, \mathbf{Ab})_N$ of S_1 has, by 12.3.5, the form $N_1 \otimes -$ for some $N_1 \in X_N$ and, since clearly $N_1 \in (\phi'/\psi')$, it must be that $N \simeq N_1$.

If there were no minimal pair in $[\phi', \psi']_N$, then we could construct a cut $[\phi', \psi']_N = U \cup L$, where, for $\phi'' \in U$ and $\psi'' \in L$, one has $\phi'' > \psi''$ and where there is no minimal element of U nor maximal element of L. Consider the functor $F_2 = (F_{D\psi'})_N / \sum_{\phi'' \in U} (F_{D\phi''})_N$. This is uniserial, hence uniform, and, since L has no maximal element, has zero socle. The injective hull (in the localised functor category) of F_2 has the form $N_2 \otimes -$ for some $N_2 \in X_N$ and, since clearly $N_2 \in (\phi'/\psi')$, it must be that $N \simeq N_2$.

Since $N_1 \otimes -$ has non-zero socle yet $N_2 \otimes -$ has zero socle, this is impossible. We conclude that there is an N-minimal pair between ϕ' and ψ', so N is isolated by a minimal pair in X_N. By 5.3.16 that is enough to deduce that the isolation condition holds for X. □

In [571] Puninski defined another dimension coexistent with m-dimension. He denoted it $2\dim(L)$. Its definition is not covered by the general one at the beginning of this chapter and is as follows.

Set $2\dim(L) = 0$ if L is the one-point lattice.
Inductively, set $2\dim(L) = \alpha$ if $2\dim(L) \not< \alpha$ and if for every $a \in L$ either $2\dim[1, a] < \alpha$ or $2\dim[a, 0] < \alpha$.

He proves the following results (the first is straightforward). For $L_1 \otimes L_2$ see Section 7.2.1.

Theorem 7.3.17. *(a)* $2\dim(L) < \infty$ *iff L does not contain a densely ordered subchain, hence, 7.2.3, iff* $\mathrm{mdim}(L) = \infty$.

(b) *([571, 3.1]) Suppose that* L_1, L_2 *are chains with* 0 *and* 1 *and that* $\text{2dim}(L_1) = \alpha$ *and* $\text{2dim}(L_2) = \beta$. *Then* $\text{w}(L) \geq \min\{\alpha, \beta\}$.

(c) *([571, 3.2]) Suppose that* L_1, L_2 *are chains, both containing a densely ordered subchain. Then* $\text{w}(L_1 \otimes L_2) = \infty$.

The third part is used to deduce that various rings – non-domestic string algebras ([571, 4.1]) and almost all non-domestic group algebras ([577, 4.11]) – have width undefined, hence, if countable, have superdecomposable pure-injectives. Subsequently Puninski proved the following, bypassing, in some circumstances, the need of a countability hypothesis (note that the hypothesis is stronger than saying that the free product of the lattices embeds in the interval).

Theorem 7.3.18. ([574, 2.3]) *Suppose that an interval in the lattice* pp_R *is freely generated by two chains, each of which contains a densely ordered subchain. Then there is a superdecomposable pure-injective* R*-module.*

By elementary duality (1.3.1) the hypothesis is right/left-symmetric and so, therefore, is the conclusion.

7.3.2 Existence of superdecomposable pure-injectives

For von Neumann regular rings m-dimension is defined iff width is defined iff the ring is semiartinian (7.3.20).

For serial rings there is a superdecomposable pure-injective iff the width is undefined, with no restriction on the cardinality of the ring (7.3.26), hence there is a superdecomposable right pure-injective module iff there is a superdecomposable left pure-injective (7.3.22). A commutative valuation ring (that is, a commutative ring the ideals of which are linearly ordered) has value group containing the rationals iff it has Krull dimension ∞ *iff there is a superdecomposable pure-injective (7.3.23). If the lattice of two-sided ideals of a serial ring has Krull dimension, then there is no superdecomposable pure-injective (7.3.24).*

Every strictly wild finite-dimensional algebra has a superdecomposable pure-injective (7.3.27). It is not known whether this is true for arbitrary wild algebras. Every non-domestic finite-dimensional string algebra has width ∞ *(7.3.28), so, if countable, has a superdecomposable pure-injective (7.3.29). Sometimes such a module may be constructed directly, bypassing the countability assumption.*

If G is a finite, non-trivial, group, then there is a superdecomposable pure-injective over the integral group ring (7.3.30).

Various generalised Weyl algebras have superdecomposable pure-injectives (7.3.34).

In this section we describe briefly what is currently known regarding existence of superdecomposable pure-injective modules over various kinds of rings. Of course, by 7.3.5, rings whose lattice of pp conditions has m-dimension have no superdecomposable pure-injectives, so, for non-existence, also see Section 13.2.1.

von Neumann regular rings

Example 7.3.19. Let $R = \mathrm{End}(V)/\mathrm{soc}(\mathrm{End}(V))$, where V is a countably-infinite-dimensional vector space over a field, so the socle of $\mathrm{End}(V)$ consists of those endomorphisms of finite rank. Then R is a von Neumann regular ring ([240, 1.23]). Let $r \in \mathrm{End}(V) \setminus \mathrm{soc}(\mathrm{End}(V))$ and let e be an idempotent of R such that $e\mathrm{End}(V) = r\mathrm{End}(V)$ (2.3.22). Fix a vector space decomposition $\mathrm{im}(e) = V_1 \oplus V_2$ with V_1, V_2 both infinite-dimensional and let f_i be projection from V to V_i with kernel $\ker(e) \oplus V_{3-i}$. Then, using $\overline{(-)}$ to denote images in R, we have $\overline{e}R = \overline{f_1}R \oplus \overline{f_2}R$ and both $\overline{f_1}R$ and $\overline{f_2}R$ are non-zero. Thus R has no non-zero uniform submodule, hence its injective hull (= pure-injective hull by 2.3.22) is superdecomposable. (R itself is not injective, see [240, 9.34].)

Fisher obtained, for boolean rings, a necessary and sufficient criterion for existence of a superdecomposable pure-injective. This was extended to commutative von Neumann regular rings in [495, 16.26]. The general case is due to Trlifaj. In fact, for von Neumann regular rings, all the dimensions, with the possible exception of the Cantor–Bendixson rank of the spectrum, coincide. For more on that, and on the isolation condition for these rings, see Section 8.2.10.

Theorem 7.3.20. ([684, §1], [495, 16.26] for commutative rings, [190, 7.31] for boolean rings, [495, 16.23] for m-dimension) *Suppose that R is a von Neumann regular ring. Then the following are equivalent:*

(i) $\mathrm{mdim}(\mathrm{pp}_R) < \infty$ *(that is,* $\mathrm{KG}(R) < \infty$*);*
(ii) $\mathrm{w}(\mathrm{pp}_R) < \infty$*;*
(iii) $\mathrm{w}(R_R) < \infty$*;*
(iv) R is semiartinian (equivalently, 13.2.12, seminoetherian);
(v) there is no superdecomposable (pure-)injective R-module.

If R is also commutative, then there are the further equivalent conditions:

(vi) the lattice, $B(R)$, of idempotents of R is superatomic (§8.2.11).

If R is commutative and countable then another equivalent is:

(vi) $\mathrm{spec}(R)$ *is countable.*

The equivalence (ii)⇔(iii) is by 2.3.25 and see 13.2.2 for (i) and 5.3.19 for (vi).

Serial rings The next theorem has a variety of previously proved results as corollaries.

Theorem 7.3.21. ([572, 5.2]) *Suppose that R is a serial ring, with $1 = \sum_1^n e_i$ a decomposition of 1 as a sum of primitive orthogonal idempotents. There is a superdecomposable pure-injective right R-module iff there exists an idempotent $e \in R$ (which may be taken to be one of the e_i) and embeddings of linearly preordered sets $f : (\mathbb{Q}, \leq) \to Re$ and $g : (\mathbb{Q}, \leq) \to eR$ (the preordering on Re being $r \leq s$ iff $Rr \leq Rs$ and similarly for eR) such that for all $q < q'$ in $\mathbb{Q}$ one has $Rfq' < Rfq \cdot gq'R < gqR$ and $Rfq' \cdot gq'R < Rfq \cdot gqR$.*

Since the condition is right/left symmetric one has the following.

Corollary 7.3.22. ([572, 6.1]) *Suppose that R is serial. Then there is a superdecomposable pure-injective right R-module iff there is a superdecomposable pure-injective left R-module.*

It was stated in [726, p. 184, Ex. (1)], and repeated in [495, p. 226] that there are no superdecomposable pure-injectives over valuation domains. That assertion was based on too simplistic a picture of the lattice of pp conditions over such a ring. Puninski pointed out that this was false and gave the exact conditions under which there is a superdecomposable pure-injective.

Theorem 7.3.23.[7] ([550, 4.1, 4.2], see [564, 12.12]) *Suppose that R is a commutative valuation ring. Then there is a superdecomposable pure-injective R-module iff* $\mathrm{Kdim}(R) = \infty$. *If R is a valuation domain, then it is equivalent that the value group of R contain, as a partially ordered subset, a copy of* $(\mathbb{Q}, \leq)$.

In the non-commutative case there are less complete results. For example, 8.2.60, if R is serial and has (right) Krull dimension, then pp_R has m-dimension so, 7.3.5, there is no superdecomposable pure-injective R-module. There is a stronger result.

Theorem 7.3.24. ([572, 6.2]) *If R is serial such that the lattice of two-sided ideals has Krull dimension, then there is no superdecomposable pure-injective R-module.*

Question 7.3.25. ([572, 6.3]) Is the above condition on two-sided ideals, namely that its Krull dimension be defined, also necessary for non-existence of superdecomposable pure-injective R-modules?

Theorem 7.3.26. ([572, 6.4]) *If R is serial, then there is a superdecomposable pure-injective R-module iff the width of* pp_R, $\mathrm{w}(\mathrm{pp}_R)$, *is undefined.*

[7] There is a similar statement for general commutative valuation rings.

That is, for serial rings, the countability hypothesis is not necessary for the conclusion of 7.3.9.

In [569] the first stages of a structure theory for superdecomposable pure-injective modules over commutative valuation domains are developed but little is known in general about the structure of superdecomposable modules.

Wild finite-dimensional algebras Recall that a k-algebra R (k a field) is strictly wild if there is a strict, that is, full, representation embedding (Section 5.5.2) from the category of $k\langle X, Y\rangle$-modules to the category of R-modules.

Proposition 7.3.27. *If R is a strictly wild algebra, then there is a superdecomposable pure-injective R-module. There is also a wide poset of pointed modules in the category,* fd-R, *of finite-dimensional R-modules.*

Proof. There is, 4.4.3, a superdecomposable pure-injective module over $k\langle X, Y\rangle$. By 4.3.8 the image of this module under a representation embedding is pure-injective and, since the functor is full, the image also will be superdecomposable.

It then follows by 7.3.4 and 7.3.1 that the lattice pp_R contains a wide poset. So, by 3.1.4, there is a wide poset of pointed finitely presented modules. If R is not itself finite-dimensional, then to get a wide poset of finite-dimensional modules we can modify the argument as follows.

First observe that kQ, where Q is the quiver $1 \rightrightarrows 2$, is strictly wild (7.2.18): the functor which takes a $k\langle X, Y\rangle$-module M to the representation $M \overset{1}{\underset{X}{\rightrightarrows}} M$ (with a third arrow Y) of Q gives a full embedding. Let $\mathcal{X}$ denote the image of this embedding – the definable subcategory of Mod-kQ consisting of those representations where the "top" map is an isomorphism. Since kQ is a finite-dimensional algebra it follows from what has been observed already that there is a wide poset of pointed finite-dimensional kQ-modules and, with a bit more care, working within $\mathcal{X}$, we can see that there is a wide poset of finite-dimensional modules inside $\mathcal{X}$. Next note that there is a full embedding[8] from $\mathcal{X}$ to Mod-$k\langle X, Y\rangle$: from a representation $M_1 \overset{\gamma}{\underset{\alpha}{\rightrightarrows}} M_2$ (with a third arrow β) where γ is an isomorphism form the $k\langle X, Y\rangle$-module with underlying vector space M_1 and with the action of X being $\gamma^{-1}\alpha$ and that of Y being $\gamma^{-1}\beta$. Thus we now we have a wide poset of finite-dimensional $k\langle X, Y\rangle$-modules which maps, by

[8] This argument can be shortcut by appealing to [92, Thm 3] which gives a strict representation embedding from the category of (finitely generated) kQ-modules to the category of (finite-dimensional) $k\langle X, Y\rangle$-modules.

fullness of the embedding, to a wide poset of R-modules which, by definition of representation embedding, are finite-dimensional. □

For example this applies to wild hereditary algebras finite-dimensional over a field, see [141, 8.4].

The proof above shows that there is a wide poset of pointed finite-dimensional $k\langle X, Y\rangle$-modules ([521, 7.1]). It is not clear whether this is true for every wild algebra and hence whether every (countable) wild algebra has a superdecomposable pure-injective.[9] The argument of [496] gives width $= \infty$ for algebras which are wild in a particular sense of realising algebras as endomorphism rings. The kinds of argument used by Puninsky in [574] for 7.3.18, also see arguments used in [577] for m-dimension, can be used with particular classes of algebras and perhaps could give a general proof (avoiding 7.3.9, hence without the need to assume countability). A good notion of wide poset of pointed modules[10] that can be transferred via representation embeddings might work. But, at the moment, the general wild case is open, both for width $= \infty$ and for existence of superdecomposable pure-injectives.

The ring $\mathbb{F}_2[X, Y, Z]/\langle X, Y, Z\rangle^2$ has 16 elements and, being strictly wild, has a superdecomposable pure-injective. No smaller ring does [576, 5.10] (also cf. [493, p. 256][11]).

String algebras In [495, p. 350] I conjectured that a tame algebra has no superdecomposable pure-injective modules; see also [323, p. 219]. In the context of string algebras subsequent investigations by Point, Prest, Ringel and others seemed to support this and the general impression was that it would turn out to be so (see the introduction to [571]) until Puninski proved quite the opposite.

Theorem 7.3.28. ([571, 4.1]) *Suppose that R is a non-domestic finite-dimensional string algebra. Then* $\mathrm{w}(\mathrm{pp}_R) = \infty$.

As mentioned earlier (see 7.3.17) the proof involves analysing the lattice of pp conditions for R-modules (so continues from Ringel's analysis of these, in the form of finitely presented functors, from [598]).

Corollary 7.3.29. ([571, 4.2]) *Suppose that R is a non-domestic finite-dimensional string algebra over a countable field. Then there is a superdecomposable pure-injective R-module.*

[9] A possible proof was outlined at [495, 13.7] but it is not clear to me now that this can be made to work.

[10] Or "splittable system" in the sense of [499].

[11] Though the analysis there is not complete.

The corollary follows from 7.3.9. Puninski later constructed, [574], a superdecomposable pure-injective directly, that is, without going via width and the consequent need for countability; this was over a specific string algebra but there is a rather general result (7.3.18) which applies to many examples.

Integral group rings The proof of the next theorem makes use of the previous two results and depends heavily on 7.3.17 and 7.3.18. The results of Levy in [403], [404] and reduction to the case where G has prime order are also important ingredients.

Theorem 7.3.30. ([576, 5.9]) *Let G be a finite non-trivial group. Then there is a superdecomposable pure-injective $\mathbb{Z}G$-module.*

On the way to this theorem Puninski, Puninskaya and Toffalori proved that if R is a commutative ring, if G is a finite group and H is a subgroup such that the width of the lattice pp_{RH} of pp conditions over the group ring RH is undefined, then so is the width of pp_{RG}, cf. 15.5.3. As a kind of converse they have the following.

Proposition 7.3.31. ([576, 2.5]) *Let R be a countable commutative Dedekind domain of characteristic* 0 *and let G be a finite group. If there is a superdecomposable pure-injective RG-module, then, for some prime p dividing the order of G, there is a superdecomposable pure-injective $RS(p)$-module, where $S(p)$ is a p-Sylow subgroup of G.*

They also show, [576, 2.6], that if R is commutative and countable, if G is finite and if H is a subgroup of G such that the index of H in G is invertible in R, then there is a superdecomposable pure-injective module over RH if there is one over RG. The proof uses pullback rings (Section 8.2.9, note that $\mathbb{Z}C_p$ is a pullback of Dedekind domains) and string modules (Section 8.1.3) over these.

Theorem 7.3.32. ([576, 5.8]) *Suppose that R is countable and is the pullback of two commutative Dedekind domains neither of which is a field. Then there is a superdecomposable pure-injective R-module.*

In fact, [576, 5.6], if R is the pullback of two commutative artinian valuation rings of lengths at least 2 and 3, then the lattice pp_R of pp conditions does not have width.

Generalised Weyl algebras The assumption of the next result is essentially that there is a representation of the, wild, 5-subspace quiver (see 1.1.3) in the Ext-quiver of the ring R. The construction referred to in the statement comes from a paper of Klingler and Levy, [342, 2.11.4] (it is also discussed at [521, p. 288]).

Lemma 7.3.33. ([521, 6.1, 7.5]) *Let R be a right noetherian ring which is a k-algebra over a field k and over which there are simple nonisomorphic modules $T, T_1, \dots, T_5$ such that* $\mathrm{End}(T) \simeq \mathrm{End}(T_i) \simeq k$ *and* $\mathrm{Ext}(T_i, T) \neq 0$ *for each i. There is a construction, $\mathcal{F}$, on finite-dimensional $k\langle X, Y\rangle$-modules which, applied to any wide poset of A-pointed modules in* fd-$k\langle X, Y\rangle$ *(where A is some finite-dimensional $k\langle X, Y\rangle$-module), yields a wide poset of $\mathcal{F}A$-pointed modules in* mod-R.

Corollary 7.3.34. ([521, 7.6]) *If R is as in 7.3.33 and is countable, then there is a superdecomposable pure-injective R-module. In particular, if k is a countable field, then this applies where R is any of $A_1(k)$ (the first Weyl algebra), $B_1(k)$ (a certain localisation of $A_1(k)$, see before 8.2.34), $A_1(q)$ (the quantum Weyl algebra) and $Usl_2(k)$ (see Section 8.2.3).*

8
Examples

This chapter is devoted to reporting on what is known about Ziegler spectra of various types of ring. In some cases results have merely been collected together and stated; in other cases outline or sample proofs are given.

Over some rings, for example, tame hereditary artin algebras and some string algebras, a complete description of the spectrum has been obtained; over others, for example, generalised Weyl algebras, there are not even any isolated points and little is said about the overall structure of the space. In this, and other, "wild" cases, for example, over the Lie algebra $\mathrm{sl}_2(k)$ and over general pullback rings, it is possible to say something about parts of the space. Over certain rings, the topology on the spectrum is trivial despite there being many points.

8.1 Spectra of artin algebras

We begin, in Section 8.1.1, with some observations and open questions. Section 8.1.2 is devoted to describing the Ziegler spectra of tame hereditary finite-dimensional algebras; Section 8.1.3 the spectra of some domestic string algebras. Pure-injectives over the canonical algebras are discussed briefly in Section 8.1.4.

8.1.1 Points of the spectrum

Representations of finite-dimensional algebras have been a rich source of examples, conjectures and results in the application of model theory to modules. After abelian groups, and modules over commutative Dedekind domains, already quite well explored by the early 1970s (for example, [669], [172], also [330], [198]), this was the context where the potential of the interaction was recognised.

Throughout this section R denotes an artin algebra. For unexplained notation and terminology see Chapter 15. We recall and make some general points, then turn to specific (classes of) examples.

Over any artin algebra the isolated points of Zg_R are the indecomposable finitely generated = finitely presented = finite length modules (5.3.37) and, together these are dense in the space (5.3.36). Here is a quick application of this.

Theorem 8.1.1. ([20, Thm B]) *Suppose that R is an artin algebra and that A is a module of finite length such that there are infinitely many indecomposable modules B of finite length such that $(A, B) \neq 0$. Then there is an infinite-dimensional indecomposable pure-injective module N such that $(A, N) \neq 0$.*

Proof. Since A is finitely presented, there is, by 1.2.18, a pp condition ϕ such that $(A, B) \neq 0$ iff $B \in (\phi)$, where by the latter we mean the basic open subset $(\phi(\overline{x})/\overline{x} = \overline{0})$ of the Ziegler spectrum. By 5.1.22 this set is compact and by assumption it contains infinitely many isolated points, hence it must contain a non-isolated point which, by 5.3.37, is infinite-dimensional, as required. □

The result and proof are valid for somewhat more general rings, cf. 5.3.43.

Over an artin algebra, every finitely generated module, being of finite length over the centre of R, hence of finite length over its endomorphism ring, is a closed point of Zg_R by 5.1.12. Even for artin algebras it is not known whether the converse is true (by 5.1.12 it is true if R is countable).

Question 8.1.2. (specialising 5.1.13) If R is an artin algebra and N is a closed point of the Ziegler spectrum, must N be of finite endolength?

Finite length points are also open and closed in the dual, rep-Zariski, topology (Section 14.1.3).

Lemma 8.1.3. *Let N be an indecomposable finitely generated module over the artin algebra R. Then N is both open and closed in the rep-Zariski topology.*

Proof. The point N is Ziegler-isolated (5.3.33) say $\{N\} = (\phi/\psi)$ for some pp conditions $\psi \leq \phi$ and this set is, by definition, a Zariski-closed set.

Since N is also Ziegler-closed, $\{N\}^c$ is both open and closed, hence (5.1.23) compact, in the Ziegler topology. Therefore (5.1.22) it has the form (ϕ'/ψ') for some pp conditions $\phi' > \psi'$. So $\{N\} = [\phi'/\psi']$ is Zariski-closed, as required. □

Proposition 8.1.4. *(a) Let R be a countable artin algebra and let $N \in \mathrm{Zar}_R$ be rep-Zariski-isolated. Then N is of finite endolength.*
(b) Let R be a finite-dimensional algebra over a countable algebraically closed field and let $N \in \mathrm{Zar}_R$ be rep-Zariski-isolated. Then N is of finite length.

Proof. (a) Say $\{N\} = [\phi/\psi]$. Then N is Ziegler-closed, hence, by 5.1.12 and countability, is of finite endolength.

(b) If N is not of finite length, then it is a generic point in the sense (see Section 4.5.5) of Crawley-Boevey, so, see the comments after 4.5.42, there is a Dedekind domain D with infinitely many primes and a representation embedding from Mod-D to Mod-R with N being the image of the generic D-module Q (= the quotient field of D). By 5.5.9 this induces a homeomorphic embedding of the rep-Zariski spectrum of D into that of R. Since Q is not isolated in Zar_D, N cannot be isolated in Zar_R. □

Question 8.1.5. Let R be an artin algebra (not necessarily countable). Suppose that the indecomposable pure-injective R-module N is isolated in the rep-Zariski topology. Is N of finite endolength?

Recall also that over an artin algebra a module is pure-injective iff it is a direct summand of a direct product of finitely generated modules, 5.3.35, and, 4.3.72, a module is pure-projective iff it is a direct sum of finite length modules.

8.1.2 Spectra of tame hereditary algebras

The Auslander–Reiten quivers of the path algebras of some extended Dynkin quivers are presented. Then the points of the Ziegler spectrum over an arbitrary tame hereditary finite-dimensional algebra are described, as is the topology. The topology may be obtained using universal localisation (see 8.1.6) and a theorem (8.1.7) of Crawley-Boevey. Over any hereditary finite-dimensional algebra every infinite-dimensional point is in the closure of the regular points (8.1.8). These results lead to a description of all the points of the spectrum, and the topology, in the tame case (8.1.11) and there is an analogous result for PI HNP rings (8.1.13). In particular the CB-rank, m-dimension and Krull–Gabriel dimension all have value 2, or 0 if the ring is of finite representation type, (8.1.12). Then a more direct route, taken by Ringel, to the topology is indicated (8.1.14). A proof of this is also outlined in [145, §3].

An example of an extension of pure-injective modules which is not pure-injective is given (8.1.18).

Throughout this section k is a field and R is a finite-dimensional tame hereditary k-algebra which is also **connected**, that is, indecomposable as a ring. Without loss of generality we may replace R by a basic artin algebra to which it is Morita equivalent. The points and topology of the Ziegler spectrum Zg_R will be described. This description is the result of a sequence of papers: [599], [469], [50], [224], [492], [227], see also [495, §13.3], [323, Chpt. 8], then both [505] and [608] dealing with the most general case, though by very different methods.[1] The method of [505] uses the idea of directly computing one or more basic cases such as the Kronecker quiver $\widetilde{A_1}$, then using these to "cover" more complicated ones. That method of proceeding goes back to Donovan and Freislich's description, [155], of the finite-dimensional representations of extended Dynkin quivers. The method of [608] is much more direct and, assuming the description of the points, goes straight to the analysis of what a closed set can be. The introduction to each of [505] and [608] gives comments of the history, some of which is repeated here.

The Auslander–Reiten quivers (Section 15.1.2) of the rings considered are well known and now we recall the broad description (see, for example, [601, esp. pp. 188–9], [602], [27, §VIII.7]).

There are infinitely many tubes (Section 15.1.3) and all but finitely many, in fact all but at most three, are homogeneous, that is have rank 1, so have a single quasi-simple module at the mouth. The quasi-simple modules may be regarded as parametrised by a "weighted projective line" over k – the projective line $\mathbb{P}^1(k)$ over k with just finitely many points, corresponding to the non-homogenous tubes, having weight > 1 – see [396], [225]. A module is said to be **regular** if each of its indecomposable summands lies in a tube.

Apart from the tubes there are just two components of the Auslander–Reiten quiver: a **preprojective** component which consists of the indecomposable projective modules, P, together with all their Auslander–Reiten translates, $\tau^{-n}P$ ($n \geq 0$); and, dually, a **preinjective** component, consisting of the indecomposable injective modules, E, together with their translates, $\tau^n E$ ($n \geq 0$). A module is **preprojective**, respectively **preinjective** if each of its indecomposable summands lies in the preprojective, resp. preinjective, component. Let $\mathbb{P}$, respectively $\mathbb{I}$, denote the set of indecomposable preprojective, resp. preinjective, modules.

The Auslander–Reiten quiver of one of these algebras is usually pictured with $\mathbb{P}$ to the left, the tubes in the middle, and $\mathbb{I}$ on the right; then no non-zero morphisms go from right to left: if P, R and E are preprojective, regular and preinjective modules respectively, then $\mathrm{Hom}(R, P) = \mathrm{Hom}(E, P) = \mathrm{Hom}(E, R) = 0$. Also, given any morphism, f, from P to E and any tube, $\mathbf{T}$, there is a factorisation of f through a direct sum of modules in $\mathbf{T}$.

[1] Also see [600, §6], where a similar picture is arrived at by a different route.

The Kronecker quiver $\widetilde{A_1}$ (1.1.4) shown below will be used as a running (ur-)example.

$$1 \underset{\beta}{\overset{\alpha}{\rightrightarrows}} 2$$

The k-representations of this quiver, equivalently modules over its path algebra $k\widetilde{A_1}$, are described at 15.1.1(b). The preprojective component of the Auslander–Reiten quiver of $k\widetilde{A_1}$ is as follows, where $P_{2n+e} = \tau^{-n} P_e, e = 1, 2, n \geq 0$ and τ^{-1} denotes the inverse of the Auslander–Reiten translate.

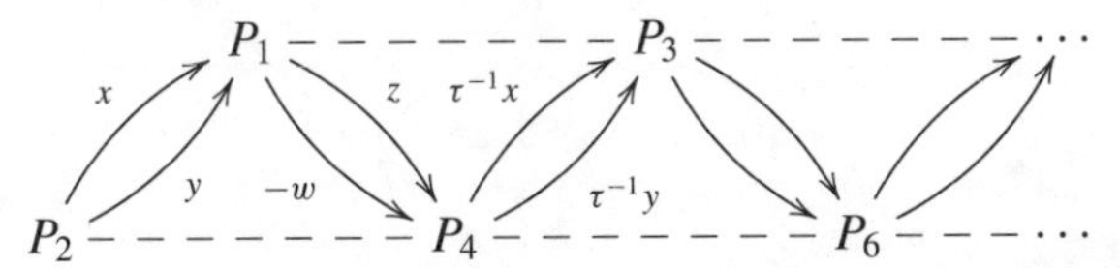

Here x denotes the morphism which maps the non-zero element of P_2 to $e_1\alpha \in P_1$ (using the notation of 15.1.1(b)); y instead takes it to $e_1\beta$. The map w, respectively z, is the embedding of P_1 into P_4 which takes $e_1\alpha$, resp. $e_1\beta$, to the "centre" of P_4 (draw P_4 as a string module to make sense of this).[2] A possibly clearer diagram results if we regard this as having been obtained by identifying all copies of the same module in the next diagram, in which modules are shown by their dimension vectors. Dashed lines show almost split (= Auslander–Reiten) sequences.

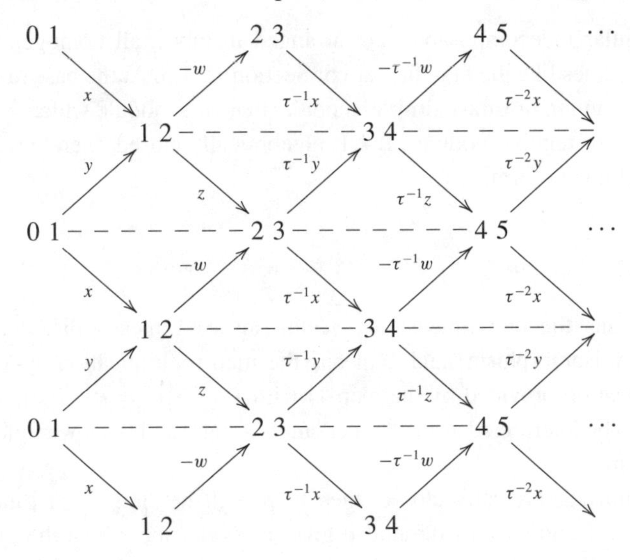

[2] Note that the indecomposable projective modules at the left-hand edge form a copy of the quiver $\widetilde{A_1}$ and that the iterates of τ^{-1}, applied to this left-hand "edge", gives all the modules in this component of the AR-quiver.

The preinjective component has the following shape.

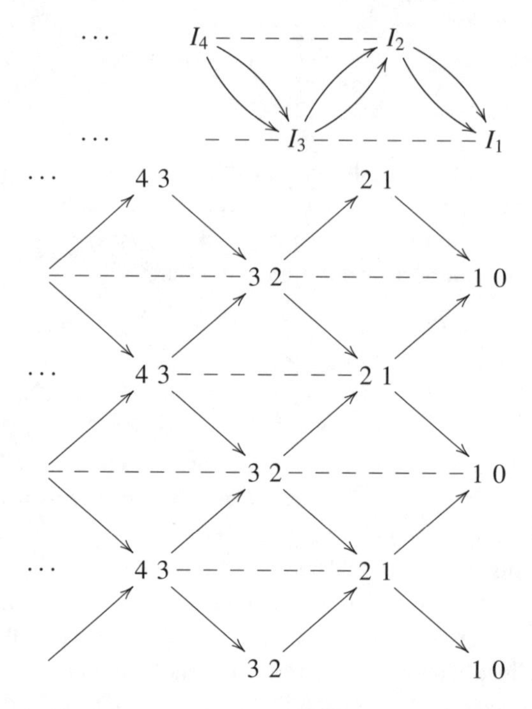

The regular indecomposables are arranged in tubes, all homogeneous, thus half-infinite lines like the first diagram of Section 15.1.3. At the base of each tube is a simple regular, or **quasi-simple**, module, that is, a module which is simple in the category of regular modules. If k is algebraically closed, then these modules have the following form.

$$S_\lambda = k \underset{\lambda}{\overset{1}{\rightrightarrows}} k$$

Here k means the one-dimensional vector space, 1 means the identity map (that is, any isomorphism) and λ means the map multiplication-by-λ (modulo the identification provided by the map 1) with $\lambda \in \mathbb{P}_1(k) = k \cup \{\infty\}$. The value $\lambda = \infty$ is to be interpreted as the "upper" map being 0 and the lower one being an isomorphism.

If k is not algebraically closed, then there will be simple $k[T]$-modules of dimension > 1 and each of these also gives a regular module at the mouth of a tube: if V with T acting invertibly or nilpotently is a simple $k[T]$-module, then the regular module obtained from this replaces each copy of k above with a copy of V and λ by the action of T.

To see non-homogeneous tubes one may consider, for example, our second running example in this section, $\widetilde{A}_3$ with the orientation shown.

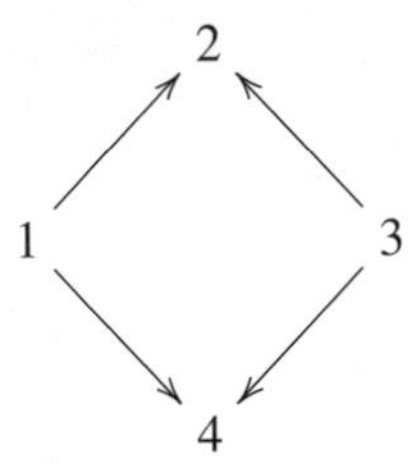

The beginning of the preprojective component for $\widetilde{A}_3$ is shown below; the top and bottom row should be identified.

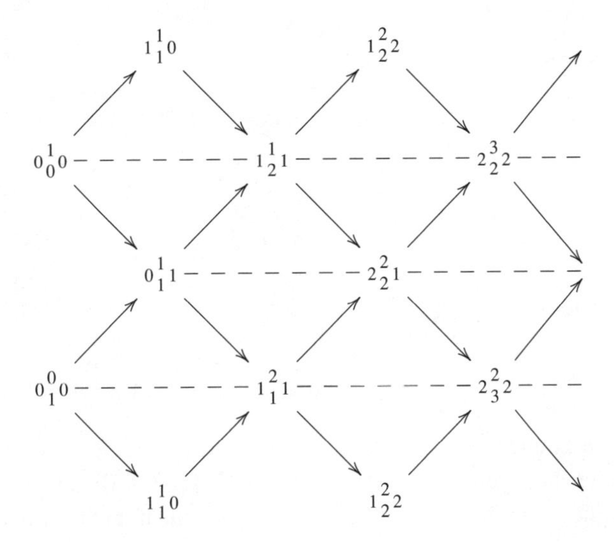

The preinjective component is easily computed from the indecomposable injectives, which have dimensions $1\,{}^{0}_{0}\,0$, $0\,{}^{0}_{0}\,1$, $1\,{}^{0}_{1}\,1$, $1\,{}^{1}_{0}\,1$, and the obvious morphisms between them.

The quasi-simple representations at the mouths of homogeneous tubes have the form $V\,{}^{V}_{V}\,V$ where three of the maps may be taken to be the identity of the vectorspace V and the fourth is a k-linear endomorphism T of V such that V with the action T is a simple $k[T]$-module.

There are two non-homogeneous tubes, each of rank 2: the diagrams below, with identification of identical modules, show the first few layers. The four non-isomorphic indecomposable modules with dimension 1 at each vertex are distinguished by a small dot showing the single map which is zero.

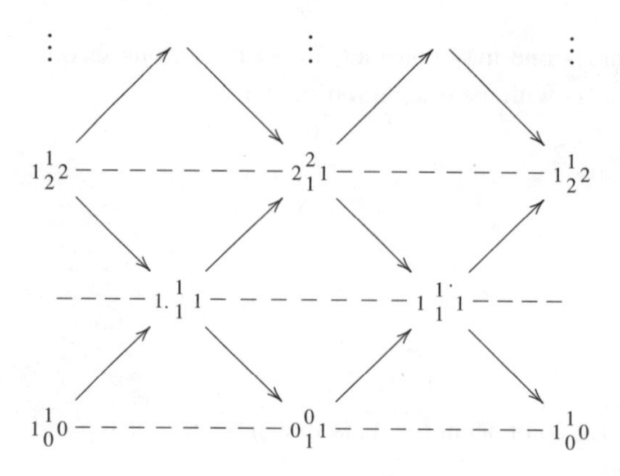

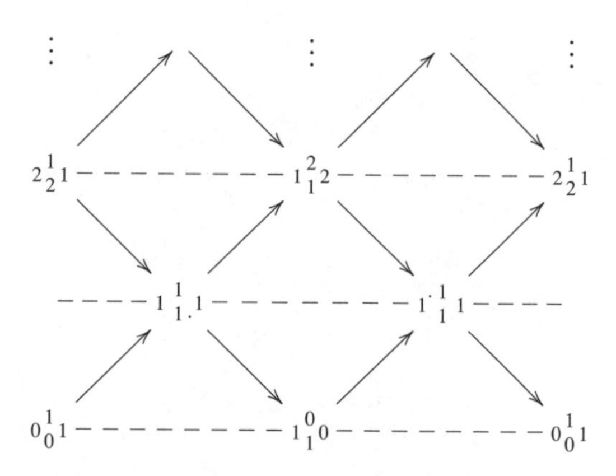

Both these are examples of the rank 2 picture in Section 15.1.3.

We mention here that, following [319, §A.1], [140, §3], a very similar description applies to finite length modules over what one might regard as "infinite-dimensional, tame hereditary algebras", namely those hereditary noetherian prime (HNP) rings which satisfy a polynomial identity (see, for example, Section 14.3.3). One example is at 14.3.8. The precise statement is that if R is a PI HNP ring and if fl-R denotes the category of finite-length R-modules, then fl-R has almost split sequences, an Auslander–Reiten translate and an Auslander–Reiten quiver which consists of tubes, parametrised by maxspec(R), all but finitely many of which are homogeneous. This is pointed out in [140].

The points The points, at least in retrospect, are not so difficult to find and the difficulty here is to prove that the provisional list is, in fact, complete.

Every indecomposable finite-dimensional module, being pure-injective, is a point of the spectrum. One class of infinite-dimensional points is that of "Prüfer" points – these were first discussed in [599]. Then there are the "adic" points, which were described in [492] (and also implicitly in [224], also see [469]). Finally, there is the so-called generic point which also appears in [599, p. 386 ff.]. All these points were described in [492] but the completeness of the list is proved there in detail only over some extended Dynkin quivers. A proof that the list is complete can be extracted, see [323, p. 207ff.] or [35, §4], from Geigle's paper [224, 3.7, 4.6] once one moves the problem into the functor category. Geisler gives a direct proof of completeness of this list in [227].

Prüfer and adic modules are discussed in a more general context in Section 15.1.3. In brief, let S be a quasi-simple module, that is, a regular module at the mouth of a tube. There is a well-defined ray of monomorphisms beginning with S and moving up through the tube. The direct limit of this ray, M_∞ in the notation of that section, is the associated Prüfer module, let us denote it S^∞. Dually, the inverse limit of the unique coray of epimorphisms in the tube ending at S, the module $\widehat{M}$ in the notation of Section 15.1.3, is the adic module associated to S: denote it $\widehat{S}$. Finally, there is the generic module, Q. Each of these modules is indecomposable pure-injective, the generic module obtained from one quasi-simple is the same as that obtained from any other,[3] and the various Prüfer and adic modules are pairwise non-isomorphic.

For example, for $R = k\widetilde{A}_1$ (k a field, algebraically closed for simplicity) the finite-dimensional points are the preprojective, regular and preinjective indecomposable modules seen above. The infinite-dimensional points of Zg_R are as follows.

The (homogeneous) tube with mouth S_λ, and $\lambda \neq \infty$, has a ray of monomorphisms with direct limit (that is, union) S_λ^∞.

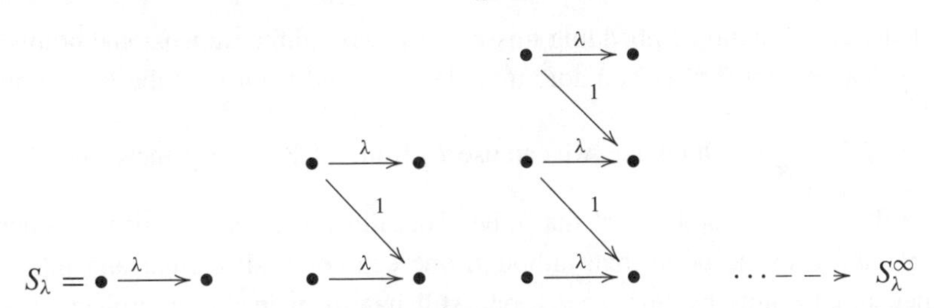

The arrows within each module show the action of β (the "lower" arrow), the action of α being an isomorphism regarded as the identity. Thus in, say, the rank 2, four-dimensional, module, the action of β on the top left basis vector is to take it to λ times the top right vector, plus the lower left vector.

The ray of epimorphisms of that tube has inverse limit $\widehat{S_\lambda}$. The open-circle basis vectors span the kernel of the relevant epimorphism.

[3] The fact that there is just one generic follows from 4.5.43, as well as from the analysis of the topology.

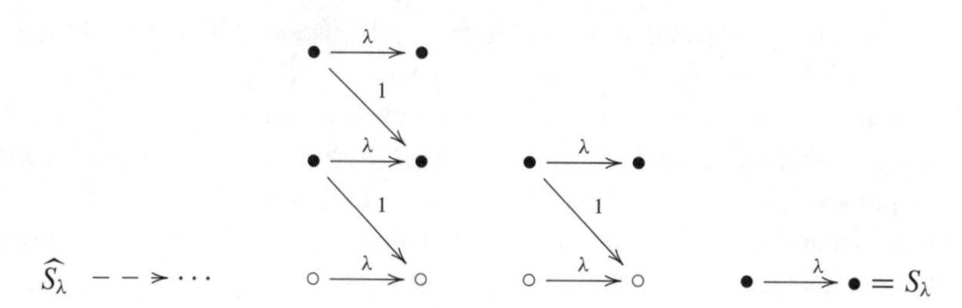

Note that these diagrams do make sense for $\lambda = 0$ and, in that case, yield a string module in the sense of Section 8.1.3. The case $\lambda = \infty$ is as $\lambda = 0$ but with the roles of α and β interchanged.

The generic module may be described via one of the representation embeddings of $k[T]$-modules into $k\widetilde{A}_1$-modules; consider, say, F_1, in the notation of 5.5.8. Clearly F_1 carries the simple $k[T]$-module $k[T]/(T - \lambda)$ to S_λ ($\lambda \neq \infty$). Moreover, one may see directly (or since it is an interpretation functor in the sense of Section 18.2.1) that F_1 commutes with direct limits and, again directly or using duality (Section 1.3.3), that F_1 commutes with inverse limits. It follows that S_λ^∞, respectively $\widehat{S_\lambda}$, is the image of the λ-Prüfer, resp. λ-adic, $k[T]$-module. Furthermore, the image of the remaining infinite-dimensional point $k(T)$ of $\mathrm{Zg}_{k[T]}$ (5.2.2) is, by the general results of this section (one can also argue directly), the generic point Q, of the Ziegler spectrum of $k\widetilde{A}_1$. Thus Q is the representation $k(T) \overset{1}{\underset{T}{\rightrightarrows}} k(T)$.

This representation embedding misses only two infinite-dimensional points of $\mathrm{Zg}_{k\widetilde{A}_1}$, namely the Prüfer and adic modules obtained from the tube with mouth $S_\infty = k \overset{0}{\underset{1}{\rightrightarrows}} k$. Of course, we can use F_2 from 5.5.8 to cover these and all the finite-dimensional modules of that tube. This illustrates the idea of the general proof, and also the point that, although one can cover all regular and infinite-dimensional points by this means, one still has to fit in the preprojective and preinjective modules when describing the topology.

What $k\widetilde{A}_1$ does not illustrate is the reduction process, via universal localisation, described below; for that we will use $\widetilde{A}_3$.

The topology The first description of the topology was given in [492]. The arguments there were rather ad hoc and only certain cases were dealt with completely, the arguments for other cases being indicated in more or less detail. This line of argument was completed, and at the same time some errors in [492] were corrected, by Geisler [227]. His long and complex proofs are contained in his aptly titled section, *Blut, Schweiz und Tränen.*

Later, uniform styles of proof, which also cover the general hereditary artin algebra case (so species over diagrams, such as $\begin{pmatrix} \mathbb{R} & \mathbb{H} \\ 0 & \mathbb{H} \end{pmatrix}$, see [154] for these), and which require considerably less effort, were given in [505] and [608]. The methods used are quite different. I outline that of [505] and indicate that of [608].

The Topology via coverings Suppose in this part that R is a connected tame hereditary finite-dimensional algebra or, *mutatis mutandis*, a PI hereditary noetherian prime ring (there is more on these in Section 8.2.2). The latter rings are actually integral to this approach, as will be seen below. Let $\mathcal{S}$ be any set of quasi-simple modules: recall that these are the indecomposable modules occurring at the mouths of tubes in the Auslander–Reiten quiver. By a **clique** of quasi-simple modules is meant a set of these consisting of the finitely many which occur at the mouth of a single tube. So all but finitely many cliques are singletons. The terminology is taken from noetherian ring theory where the term is used to refer to certain sets of primes, via its application to PI hereditary noetherian prime rings, see [319, §A2 esp.]. Let $\Sigma_{\mathcal{S}}$ denote the set of morphisms between finitely generated projective R-modules with cokernel a member of $\mathcal{S}$. Denote by $R_{\mathcal{S}}$ the universal localisation (see 5.5.5) of R at $\Sigma_{\mathcal{S}}$.

Theorem 8.1.6. ([638, 4.9, 4.7, 5.1], [140, §2]) *Let R be an indecomposable hereditary finite-dimensional algebra and let $\mathcal{S}$ be any set of quasi-simple regular R-modules. Then $R_{\mathcal{S}}$ is hereditary and the canonical map $R \to R_{\mathcal{S}}$ is an epimorphism of rings. Furthermore, for any $R_{\mathcal{S}}$-modules M, N one has:*

$$\mathrm{Ext}^1_{R_{\mathcal{S}}}(M, N) = \mathrm{Ext}^1_R(M, N) \quad (\textit{as well as } \mathrm{Hom}_{R_{\mathcal{S}}}(M, N) = \mathrm{Hom}_R(M, N)).$$

The image of the restriction of scalars functor from Mod-$R_{\mathcal{S}}$ *to* Mod-R *consists of those modules M which satisfy the condition: for all $S \in \mathcal{S}$,* $\mathrm{Hom}(S, M) = 0 = \mathrm{Ext}^1(S, M)$.

For example, let $R = k\widetilde{A_1}$ and choose the clique to be that consisting of S_∞ only. The minimal projective presentation of S_∞ is $0 \to P_2 \xrightarrow{x} P_1 \to S_\infty \to 0$ (notation as on the previous pages) so the corresponding universal localisation of R will be that obtained by formally inverting the morphism x (essentially, inverting α). It is at least plausible that the effect of this is to identify the two vertices of the quiver $\widetilde{A_1}$, hence to obtain the quiver with one vertex and one arrow (see [639] for relevant arguments). A precise argument is to use that the universal localisation of a hereditary ring is an epimorphism of rings (5.5.5) and that the full subcategory of Mod-$k\widetilde{A_1}$ consisting of modules which have x "acting invertibly", is equivalent to the image of the representation embedding F_1 from 5.5.8 which was mentioned above, hence is equivalent to Mod-$k[T]$ (see [639, p. 163]). We note that since $k\widetilde{A_1}$ embeds in this universal localisation (see [140, 2.2]), the latter cannot be simply $k[T]$: one may check that it is the ring $M_2(k[T])$ of 2×2 matrices over $k[T]$. Hence the central simple algebra of the next result is, in this case, $M_2(k(T))$. This example illustrates case (d) below.

Theorem 8.1.7. ([140, 4.2] extending [35, 5.8]) *Let R be an indecomposable finite-dimensional tame hereditary k-algebra and let $\mathcal{S}$ be a set of (isomorphism types of) quasi-simple modules.*

(a) If there is no clique entirely contained in $\mathcal{S}$ (in which case $\mathcal{S}$ is finite, consisting of modules from inhomogeneous tubes), then $R_{\mathcal{S}}$ is a tame hereditary finite-dimensional algebra with card$(\mathcal{S})$ *fewer simple modules than R.*
(b) If $\mathcal{S}$ contains all quasi-simple R-modules, then $R_{\mathcal{S}}$ is a central simple algebra, A, with centre infinite-dimensional over k.
(c) If neither (a) nor (b) but the complement of $\mathcal{S}$ contains at least two quasi-simple modules in the same clique, then $R_{\mathcal{S}}$ is a non-maximal hereditary order in A.
(d) Otherwise $R_{\mathcal{S}}$ is a maximal order in A.

Maximal orders are ([441, 13.9.16]) the Dedekind prime rings which satisfy a polynomial identity.[4] Hence the spectrum of such a ring is described by the results of Section 5.2.1.

An example illustrating case (a) is to take $R = k\widetilde{A_3}$ with orientation as before (page 331). This can be realised as the matrix ring

$$\begin{pmatrix} k & k & 0 & k \\ 0 & k & 0 & 0 \\ 0 & k & k & k \\ 0 & 0 & 0 & k \end{pmatrix}.$$

Take $\mathcal{S} = \{S_{32}\}$, where S_{32} is the quasi-simple module of dimension $0\,{}^{1}_{0}\,1$. The effect of the corresponding localisation is to make S_{32} zero. Referring back to the diagrams showing the two tubes of rank 2 over $k\widetilde{A_3}$, this has the effect of identifying the modules with dimension types $1\,{}^{0}_{1}\,0$ and $1\,{}^{1}_{1}\,1$, hence of collapsing this rank-2 tube to a homogenous tube over $\widetilde{A_2}$ equipped with the following orientation.

[4] They are PI since they are subrings of matrix rings over fields and any matrix ring over a commutative ring satisfies a polynomial identity, see, for example, [441, 13.3.3] or [626, 6.1.17].

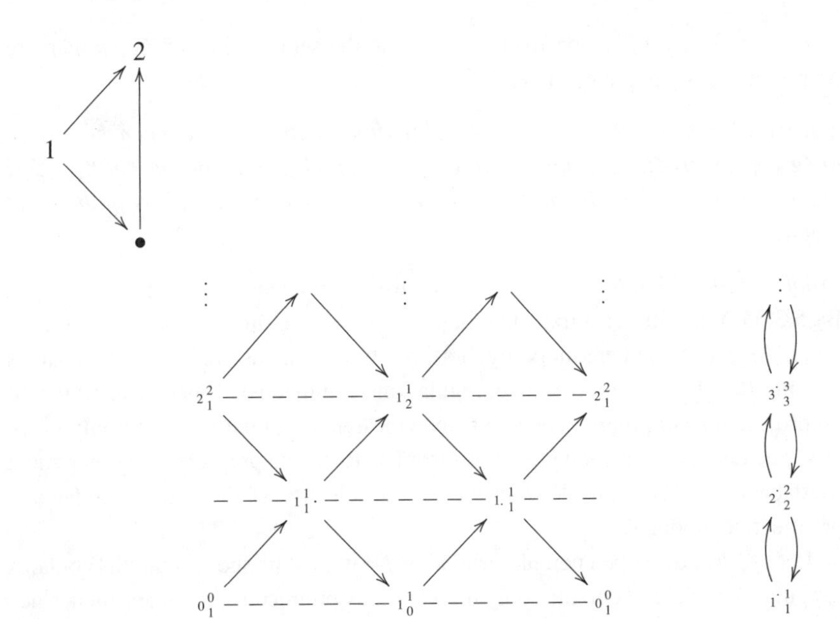

The minimal projective presentation of S_{32} is $0 \to 0\,{}^{0}_{1}\,0 \to 0\,{}^{1}_{1}\,1 \to 0\,{}^{1}_{0}\,1 = S_{32} \to 0$, so the localisation identifies P_4 and P_3. Note how the rank-2 tube above is obtained from the corresponding one for $\widetilde{A_3}$ by identifying the vertices 3 and 4. The effect also may be seen in terms of the matrix representations of the path algebras: think of the identification of vertices 3 and 4 as introducing the matrix unit e_{43}, to obtain

$$\begin{pmatrix} k & k & k & k \\ 0 & k & 0 & 0 \\ 0 & k & k & k \\ 0 & k & k & k \end{pmatrix},$$

then collapse the resulting 2×2 matrix ring formed by $\{e_{33}, e_{34}, e_{43}, e_{44}\}$ to obtain the matrix representation

$$\begin{pmatrix} k & k & k \\ 0 & k & 0 \\ 0 & k & k \end{pmatrix}$$

of the path algebra of $\widetilde{A_2}$.

The next result is essential to this approach since it says that all infinite-dimensional points of the spectrum are in the closure of the regular modules; therefore covering the regular modules by finitely many closed sets is enough to capture all infinite-dimensional points. For tame hereditary algebras, this is from [492, §2], see [495, 13.6]; as was noted in [227, 2.2.1], the proof applies in the general case. Let $X_{\mathbb{P}}$, respectively $X_{\mathbb{I}}$, denote the set of points of Zg_R which are preprojective, resp. preinjective.

Theorem 8.1.8. *Let R be a hereditary finite-dimensional algebra. Then the closure in Zg_R of the set of indecomposable regular modules is the complement of $X_{\mathbb{P}} \cup X_{\mathbb{I}}$; in particular every infinite-dimensional point of Zg_R is in the closure of the set of regular points.*

Proof. (Outline) Let N be an infinite-dimensional indecomposable pure-injective. By 5.3.35 N is a direct summand of a product of finite-dimensional indecomposables. As over tame hereditary algebras, the finite-dimensional indecomposables fall into the classes, preprojective, regular and preinjective. Since submodules of preprojectives are preprojective and since these have a relative projectivity property, one deduces that each projection from N to such a preprojective component must be zero. Therefore N is in the Ziegler-closure of the set of regular and preinjective modules.

Let (ϕ/ψ) be an open neighbourhood of N; it must be shown that this contains a regular module. If it contains only finitely many preinjective modules, then, since these are closed points, we may take a smaller neighbourhood which omits them and which therefore, since N is not isolated (5.3.37), contains a regular module. So we need consider only the case that (ϕ/ψ) contains infinitely many preinjective modules.

A key observation is that if M is a module over any ring R and if ϕ is a pp condition, then there is an integer n such that if $\overline{a} \in \phi(M)$ then there is a finitely generated submodule M' of M, generated by no more than n elements, such that $\overline{a} \in \phi(M')$. For, if $\phi(\overline{x})$ is $\exists \overline{y}\, \theta(\overline{x}, \overline{y})$, where θ is quantifier-free (thus a conjunction of equations) and if $(\overline{a}, \overline{b}) \in \theta(M)$, then M' may be taken to be the module generated by the entries of $\overline{a}$ and $\overline{b}$, so n may be taken to be the total number of distinct variables appearing in θ.

Choose a preinjective module $M \in (\phi/\psi)$, say $\overline{a} \in \phi(M) \setminus \psi(M)$. It is the case that there are only finitely many indecomposable preinjectives of any given dimension, so M may be chosen to have dimension strictly greater than n. Therefore there is a proper submodule M' of M containing $\overline{a}$ and such that $\overline{a} \in \phi(M')$. Writing M' as a direct sum of indecomposables, one sees that there is an indecomposable summand, M'' say, of M' which lies in (ϕ/ψ). By a relative injectivity property of preinjectives, M'', if preinjective, would have to be a direct summand of M, which it is not. So M'' either is regular, as wanted, or is preprojective.

If M'' is preprojective, then we appeal to another property, that any morphism from a preprojective to a preinjective module, in particular the inclusion of M'' in M, must factor through a direct sum of indecomposable regular modules, and thus we see that there must be a regular module in (ϕ/ψ), as required.

The special properties of preprojective, regular and preinjective modules that we used above may be found in various of the references on representations of algebras and [227] gives direct proofs. □

We need a slight generalisation of this, the proof of which may be found at [505, 2.8] (it could also be derived from the results of Section 15.1.3). Say that a regular module M is **isotropic** if the multiplicities, as factors in a "regular composition series" of M, of the quasi-simple modules at the mouth of the tube containing M are all the same. So these are the modules found at "heights" in the tube which are integer multiples of n, where n is the number of quasi-simples at the mouth of the tube containing M.

Proposition 8.1.9. ([505, 2.8]) *Let R be a tame hereditary finite-dimensional algebra. Then every infinite-dimensional point of Zg_R is in the closure of the set of isotropic regular modules.*

The idea now is to cover Zg_R by already known Ziegler spectra, with that of $k[T]$, or a finite localisation thereof, as the ultimate (working backwards to simplicity) case. Without loss of generality, R is indecomposable. First we deal with the case where R has no inhomogeneous tubes, hence ([154, §6]) is of type $\widetilde{A_1}$, $\widetilde{A_{11}}$ or $\widetilde{A_{12}}$. Choosing two different localisations (corresponding to covering the projective line by two affine lines), we obtain a covering of the regular points of Zg_R, hence also (by 8.1.8) we cover the infinite-dimensional points of the spectrum of R, by spectra of two maximal hereditary orders, and those spectra are known (Section 5.2.1).

Theorem 8.1.10. ([505, 2.5]) *Let R be a tame hereditary finite-dimensional algebra of type $\widetilde{A_1}$, $\widetilde{A_{11}}$ or $\widetilde{A_{12}}$. Let S_1, S_2 be non-isomorphic quasi-simple modules and let $R \to R_i = R_{S_i}$ be the corresponding universal localisations. Set X_i to be the induced homeomorphic image of Zg_{R_i}, regarded (5.5.3) as a closed subset of Zg_R. Then $\mathrm{Zg}_R = X_{\mathbb{P}} \cup X_1 \cup X_2 \cup X_{\mathbb{I}}$.*

This allows a description of all the points and, with a bit more work (see [505, 2.13]) to take account of the preprojective and preinjective points, the complete topology.

Next we use the fact that all other cases can be obtained by localising to algebras of the above form, hence their spectra can be covered by homeomorphic copies of the spectra appearing in the result above.

Theorem 8.1.11. ([505, 2.9]) *Let R be an indecomposable tame hereditary finite-dimensional algebra. Then there is a finite number of finite sets, $\mathcal{S}_1, \dots, \mathcal{S}_t$, of quasi-simple modules such that if $R \to R_i = R_{\mathcal{S}_i}$ are the corresponding universal* *localisations and if* $X_i \subseteq \mathrm{Zg}_R$ *denotes the image of* Zg_{R_i} *under the induced map, then:*

(a) each R_i *is a maximal hereditary order in the central simple algebra of 8.1.7;*
(b) each X_i *is a closed subset of* Zg_R *and* $X_1 \cup \cdots \cup X_t$ *contains all the infinite-length points of* Zg_R.

Corollary 8.1.12. ([505, 2.11]; [492] for some cases, [227, 8.1.1] for path algebras of extended Dynkin quivers) *Let R be a tame hereditary finite-dimensional algebra. Then the Cantor–Bendixson rank of the Ziegler spectrum,* Zg_R, *is* 2, *the points of rank* 1 *being the adic and Prüfer points.*

The overlaps of the X_i are easily determined and the part of the topology involving the preprojective and preinjective modules can be described in a uniform way ([505, p. 158ff]).

For instance, there are two non-homogeneous tubes, each of rank 2, over $\widetilde{A_3}$ with the orientation given earlier. Labelling the four quasi-simple modules in these tubes as $\{S_{12}, S_{34}\}, \{S_{14}, S_{32}\}$, we may choose the cliques $\mathcal{S} = \{S_{12}, S_{14}\}$ and $\mathcal{S}' = \{S_{34}, S_{32}\}$ and localise at each of these to the case, up to Morita equivalence, of $\widetilde{A_1}$. The resulting representation embeddings from $\widetilde{A_1}$-modules will suffice to cover all the infinite-dimensional points: thus four representation embeddings from $k[T]$-modules will suffice.

In particular, one has the following description ([505, 2.14]) of neighbourhood bases of the infinite-dimensional points; recall that the finite-dimensional points are isolated. For any quasi-simple S denote by $\mathbb{M}(S)$ the set of (indecomposable regular) modules in the ray of monomorphisms beginning at S and denote by $\mathbb{E}(S)$ the set of modules in the coray of epimorphisms ending at S. Recall that $\mathbb{P}$, respectively $\mathbb{I}$, denotes the set of indecomposable preprojective, resp. preinjective, modules.

- If S is quasi-simple, then a basis of open neighbourhoods of the corresponding adic point, $\widehat{S}$, consists of those sets of the form

$$\{\widehat{S}\} \cup \Big(\big(\{P \in \mathbb{P} : (P, S) \neq 0\} \cup \mathbb{E}(S)\big) \setminus F\Big),$$

where F is any finite set.

- If S is quasi-simple, then a basis of open neighbourhoods of the corresponding Prüfer point, S^∞, consists of those sets of the form

$$\{S^\infty\} \cup \Big(\big(\mathbb{M}(S) \cup \{I \in \mathbb{I} : (S, I) \neq 0\}\big) \setminus F\Big),$$

where F is any finite set.

- A basis of open neighbourhoods of the generic point Q consists of those cofinite subsets of Zg_R which contain Q.

If the tube containing S is homogeneous, then for every preprojective module P, $(P, S) \neq 0$ and for every preinjective module I, $(S, I) \neq 0$ (because every morphism from $\mathbb{P}$ to $\mathbb{I}$ factors through any given tube) so the description simplifies slightly in that case.

The proofs of the theorems above also yield the following results, which extend those of Section 5.2.1.

Theorem 8.1.13. ([505, 3.2, 3.3]) *Let R be a non-artinian PI hereditary noetherian prime ring, that is, a hereditary order in a central simple algebra A, say. Then the points of Zg_R are:*

(a) the indecomposable torsion modules;
(b) for each simple R-module S, the S-adic and S-Prüfer module;
(c) A.

The description of the topology is just as in the Dedekind case (Section 5.2.1). In particular every indecomposable torsion module is an isolated point of Zg_R and together these are dense in Zg_R.

Marubayashi also considers pure-injectives over such rings in [437].

Such an explicit description of the Ziegler spectrum of a ring leads, in principle, to decidability of the theory of modules over that ring ([726, 9.4], [495, Chpt. 17]): in practice rather more work may be required. In [580] this is explained and carefully carried through for the the Kronecker algebra and then combined with the method, from [323], of reduction modulo the radical to obtain decidability of the theory of modules for certain finite "tame" commutative rings.

The topology directly Ringel's description of the closed sets is as follows. The approach is to determine directly when a closed set contains a particular kind of infinite-dimensional point of the spectrum.

Theorem 8.1.14. ([608, Thm A]) *Let R be connected tame hereditary finite-dimensional algebra. Then a subset X of Zg_R is closed iff the following conditions are satisfied.*

(a) If S is a quasi-simple regular module and if there are infinitely many finite-dimensional modules $M \in X$ with $(M, S) \neq 0$, then $\widehat{S} \in X$.
(b) If S is a quasi-simple regular module and if there are infinitely many finite-dimensional modules $M \in X$ with $(S, M) \neq 0$, then $S^\infty \in X$.

(c) *If there are infinitely many finite-dimensional modules in X or if there is at least one infinite-dimensional point in X, then the generic module is in X.*

In both descriptions, 8.1.14 and after 8.1.12, one may see that the supports (the modules not in the kernel) of the functors $(S, -)$ and $(-, S) \simeq \mathrm{Ext}^1(\tau^{-1}S, -)$ (for τ see Section 15.1.2) are essential ingredients of the description. Explicit description of these sets is discussed in [608, §6].

Some results which are proved on the way to this description are the following.

Proposition 8.1.15. ([608, Prop. 4]) *Let R be a connected tame hereditary finite-dimensional algebra and let S be a quasi-simple regular module. Let $f : \widehat{S} \to \widehat{S}$ be any injective, but not surjective, endomorphism of the corresponding adic module. Then the direct limit of the chain of morphisms $\widehat{S} \xrightarrow{f} \widehat{S} \xrightarrow{f} \dots$ is a direct sum of copies of the generic module.*

(Compare that result with those in Section 15.1.3.)

Proposition 8.1.16. ([608, Prop. 5]) *Let R be a connected tame hereditary finite-dimensional algebra. Let Y be an infinite set of indecomposable regular modules with no two having the same quasisimple submodule. Then $\prod_{M \in Y} M / \bigoplus_{M \in Y} M$ is a direct sum of copies of the generic module.*

(Compare with 3.4.13.)

Proposition 8.1.17. ([608, Prop. 6]) *Let R be a tame hereditary algebra and let X be a closed subset of Zg_R. If X contains infinitely many preprojective modules, then it contains at least one adic module associated to each tube. If X contains infinitely many preinjective modules, then it contains at least one Prüfer module associated with each tube.*

The following special case of 3.3.1 is used ([608, p. 111]). Let R be any ring and let M_i $(i \geq 0)$ be a countable sequence of modules. Set $M_{\geq i} = \prod_{j \geq i} M_j$. Then the direct limit of the canonical epimorphisms $M_1 \to M_2 \to \cdots$ is $\prod_i M_i / \bigoplus_i M_i$.

To finish this section we note the following example.

Example 8.1.18. An extension of a pure-injective module by a pure-injective module need not be pure-injective. This example, which uses some of the modules seen above, is taken from [605, p. 436]. Let M be the module $k[T] \underset{T}{\overset{1}{\rightrightarrows}} k[T]$ over the Kronecker algebra (1.1.4). Then its (one-dimensional) submodule $0 \underset{0}{\overset{0}{\rightrightarrows}} k$ certainly is pure-injective, as is the corresponding factor, which is the

Prüfer module S^∞, where S is the quasi-simple module $k \underset{1}{\overset{0}{\rightrightarrows}} k$; the factor has k-basis x_n, y_n $(n \geq 0)$, where $x_n e_1 = x_n, y_n e_2 = y_n, x_0\beta = 0, x_n\beta = y_{n-1}$ $(n \geq 1)$, $x_n\alpha = y_n$. But M is not itself pure-injective (by 4.4.10) since it is countable-dimensional but not Σ-pure-injective. To see the latter, note that the k-linear endomorphism, "$\alpha\beta^{-1}$", of M is well defined and pp-definable. The images of its powers $(\alpha\beta^{-1})^n$ form an infinite properly descending chain of pp-definable subgroups, so, by 4.4.5, M is not Σ-pure-injective. In fact, since M is the image of the $k[T]$-module $k[T]$ under one of the obvious (strict) representation embeddings from Mod-$k[T]$ to Mod-$k\widetilde{A}_1$, see 5.5.8, the pure-injective hull of M is the image of the pure-injective hull of $k[T]$ by 5.5.11. That is, 5.2.7, the direct product of the P-adic modules $\overline{k[T]_{(P)}}$ as P varies over non-zero primes of $k[T]$. Therefore $\overline{M}$ is the direct product of the S'-adic modules as S' varies over the quasi-simple $k\widetilde{A}_1$-modules with the exception of S above (which is, note, the only quasi-simple not in the image of this representation embedding).

In [472, §3] Okoh determines, for modules over tame hereditary algebras, exactly when an extension of a pure-injective by a pure-injective is again pure-injective.

8.1.3 Spectra of some string algebras

The indecomposable finite-dimensional representations of string algebras – string and band modules – are described. Then the Ziegler spectra of the domestic string algebras Λ_n are given (8.1.19), rather briefly, the reader being referred elsewhere for more detail. The string algebra Λ_n has Krull–Gabriel dimension $n + 1$ (8.1.20).

Let Q be a finite quiver with relations and let R be the corresponding path algebra (Section 15.1.1) over a field k. This algebra is a **string algebra** if the following conditions on the quiver hold: there are only finitely many paths in Q which do not have, as a subpath, one of the relations, so R is finite-dimensional; each vertex is the source of at most two arrows and the target of at most two arrows; at each vertex i, for each arrow α with target i there is at most one arrow β with source i such that $\alpha\beta$ is not a relator (that is, 0); dually, at each vertex i, for each arrow β with source i there is at most one arrow α with target i such that $\alpha\beta$ is not a relator. Examples of such quivers are at 1.1.2, 1.1.4, 1.2.16.

A "string" is a walk through the quiver which allows traversing, as well as arrows, their formal inverses, but which includes no relators or inverses of relators. The description of such a walk is a **word** or **string** (we use the terms interchangeably), meaning, a sequence of (names of) arrows ("direct letters") and inverses of arrows ("inverse letters") which does not contain any relator or inverse relator. For some purposes, especially those here, it is useful to allow infinite, possibly doubly infinite, words but usually we will add "infinite" explicitly, so simply "string" will mean finite string.

For example, let $Q = \mathrm{GP}_{23}$ (1.1.2): it has a single vertex and two arrows, α, β, and relations $\alpha\beta = 0, \beta\alpha = 0, \alpha^2 = 0, \beta^3 = 0$. Examples of words are: 1 (notation such as 1_i is used for the empty word representing a path/string of length 0 at vertex i); α; β^{-2}; $\beta\alpha^{-1}$; $\beta\alpha^{-1}\beta^2\alpha^{-1}$. And, for instance, α^{-2}, $\beta^2\alpha$ are non-examples.

To each string w one may associate a "string module" $M(w)$. Also, given a string w which represents a "cyclic" walk, meaning that it starts and ends at the same vertex, is irreducible (that is, is not a proper power) and no power of which contains a relator$^{\pm 1}$, and given an indecomposable $k[T]$-module A, one may define from these data a "band" module $M(w, A)$. Informal definitions are given below and formal definitions can be found in, for example, [110], [605, §1] or [641, §2]. In the first reference it is shown that string and band modules are indecomposable and constitute all the indecomposable finite-dimensional modules over a string algebra. These modules are also non-isomorphic apart from the obvious identifications: that $M(w) \simeq M(w^{-1})$ and similarly for band modules, also allowing for cyclic permutations of w. In the second reference the modules corresponding to infinite words are described. Here we give examples which, we hope, make the meaning, and our conventions for depicting such modules, clear.[5]

Continuing with $R = k(\mathrm{GP}_{23})$, here are some small examples. The first two diagrams represent $M(\beta\alpha^{-1})$ and $M(\alpha\beta^{-1})$ respectively: note that these are isomorphic. The third represents $M(\beta)$, the last two $M(\alpha^{-1}) \simeq M(\alpha)$. $M(1)$ is the one-dimensional module with both α and β acting as zero.

In such diagrams each • represents the generator of a one-dimensional k-vectorspace and the absence of an arrow means the zero action. Thus, for instance, the third diagram represents a two-dimensional module with a generator v such that $v\alpha = 0$ and with $v\beta^2 = 0$.

Another example, over the same algebra, can be seen at 1.1.2. The string module there is, with marginally different notation, $M(ab^{-1}ab^{-2}a)$. The example, over a different string algebra, at 1.2.16 is $M(\alpha_1^{-1}\alpha_2\beta_2^{-1})$.

[5] Since we are not going into much detail here we are allowing some sloppiness regarding right/left conventions, so if you feel that some words would more appropriately be replaced by their inverses, then feel free to make that replacement.

Thus, associated to each word/string is a **string module**, which is formed by placing a one-dimensional vectorspace at each vertex on the walk and defining the action of the path algebra as indicated. A **band** is a string of length at least 1 which starts and ends at the same point, which is such that every power of this string also is a string and which is not a proper power of a proper sub-string. To each band is associated a family of modules. First, choose any indecomposable finite-dimensional $k[T]$-module, A, and place its underlying vector space at each vertex along the walk, except at the terminus (where there will already be at least one copy). In a string module each occurrence of an arrow on the walk acts as an isomorphism between the two one-dimensional spaces at its source and target. In a band module, each occurrence of an arrow, except the last, is defined to act as an isomorphism ("the identity") from the relevant copy of A at its source to the relevant copy of A at its target. The action of the last arrow, from the final to the first copy of A, is defined to be the action of T on A (modulo the identification of all these copies of A). Precise definitions are in the references cited and the basic results, including the fact that every finite-dimensional indecomposable module is a string or a band module, are in [110].

For example, over the Kronecker quiver, the preprojective and preinjective modules (see Section 8.1.2) are string modules and the regular modules (those in tubes) are band modules. The modules in two tubes, those where either β or α acts as 0 ("$\lambda = 0, \infty$"), are obtained through both processes.

All string algebras are tame [110] and a string algebra is domestic (Section 4.5.5) iff there are just finitely many bands (see [605, §11]).

Here we outline the description of the Ziegler spectrum of the algebras Λ_n shown below. The details are in [107]. Since there is also an expository account in [511] (which uses an opposite convention for depicting string modules), and one may see [610] for discussion of the kinds of infinite-dimensional modules involved, we are quite brief here. Here is Λ_2.

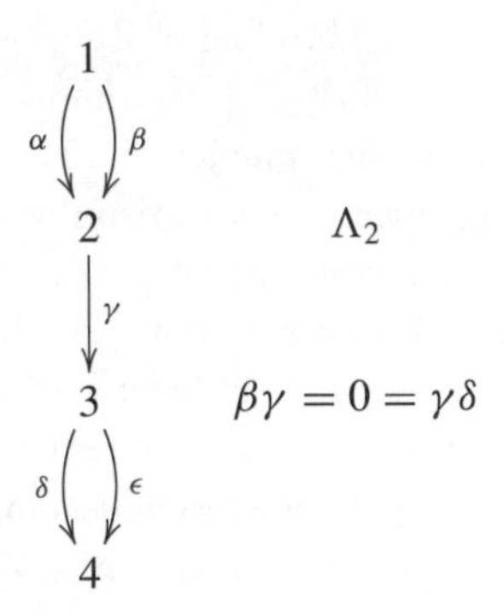

Λ_2

$\beta\gamma = 0 = \gamma\delta$

More generally, we have Λ_n.

$$1 \underset{\alpha_1}{\overset{\beta_1}{\rightrightarrows}} 2 \xrightarrow{\gamma_1} 3 \underset{\alpha_2}{\overset{\beta_2}{\rightrightarrows}} \quad \cdots \quad \underset{\alpha_{n-1}}{\overset{\beta_{n-1}}{\rightrightarrows}} 2n-2 \xrightarrow{\gamma_{n-1}} 2n-1 \underset{\alpha_n}{\overset{\beta_n}{\rightrightarrows}} 2n$$

$$\beta_i\gamma_i = 0 = \gamma_i\alpha_{i+1}$$

The main result, said for Λ_2, is that, apart from the finite-dimensional, isolated, points of the spectrum, there are the Prüfer, adic and generic modules associated with each of the two copies of $\widetilde{A_1}$ contained in Λ_2 (these can be obtained via representation embeddings, as in the previous section), plus a small number of modules which "link" the two copies of $\widetilde{A_1}$. These last modules correspond to certain doubly infinite, eventually periodic, words which are like Prüfer modules to one side and like adic modules to the other. To such a word one associates a corresponding "string" module, using the direct sum of the one-dimensional spaces on the Prüfer side and the direct product of the one-dimensional modules on the adic side. An example is discussed below. The definition and basic properties of such types of modules over general string algebras may be found in [605].

One ingredient of the proof is a partial analysis of finitely presented functors which arise from strings. Such functors are used in [230] and, much more explicitly, in [598]: each paper gives a complete classification of the finite-dimensional representations of certain string algebras. The same comment applies to the classification, [110], in the general case. Subsequent investigation of infinite-dimensional pure-injective representations, for example, [44], [524], [523], [571], [573], [574], [605], continues this line of analysis, but often using the equivalent (10.2.30) language of pp conditions.

We give the line of the proof.

Each quiver Λ_n has n obvious Kronecker subquivers. The idea is very much like that used in [505], and described in Section 8.1.2, for the description of Zg_R for R a tame hereditary algebra. Namely, each of the n obvious representation embeddings from the category of $k\widetilde{A_1}$-modules to the category of Λ_n-modules induces (5.5.9) a homeomorphic embedding of $\mathrm{Zg}_{k\widetilde{A_1}}$ as a closed subset of Zg_R. The union of these closed subsets of $\mathrm{Zg}_{k\Lambda_n}$ is thus known since the overlaps can be computed, so it remains to analyse the complement. This case is more complicated than the tame hereditary case since, in comparison with 8.1.8, that complement does contain infinite-dimensional points of $\mathrm{Zg}_{k\Lambda_n}$.

This necessitates a closer study of certain functors which are defined by strings. Suppose that F is a finitely generated subfunctor of $(R, -)$, where $R = k\Lambda_n$, such that, on finite-dimensional modules, $FN = 0$ for every indecomposable band module N and $\dim_k(FN) \le 1$ for every indecomposable string module N (cf. Section 13.3). It follows ([107, 1.1]) that if $N \in \mathrm{Zg}_R$ is such that $FN \neq 0$, then N, which we know, by 5.3.35, to be a direct summand of a direct product of finite-dimensional modules, is actually a direct summand of a direct product of string modules; moreover, in this case $\dim_k(FN) = 1$ ([107, 1.5]).

As already remarked, apart from the adic, Prüfer and two generic modules which are (see 15.1.12) in the closure of the tubes, that is, which are in the image of one of the representation embeddings from the category of Kronecker modules, there are certain infinite string modules, one of which is shown below.

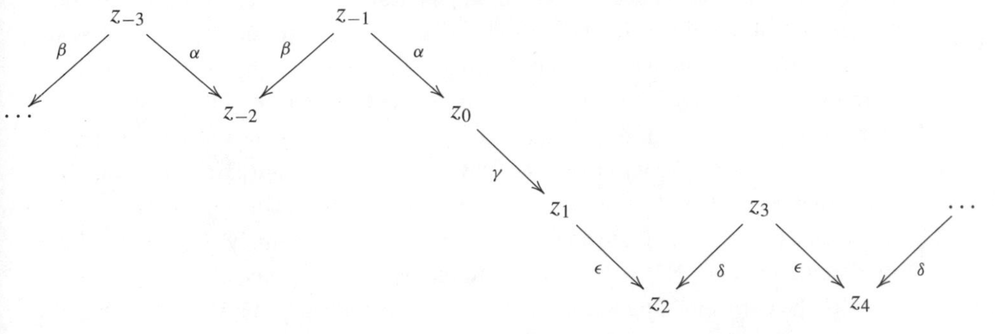

There are two notable endomorphisms of this module. Namely, there is that which sends $Me_3 \oplus Me_4$ to 0 and which "shifts $Me_1 \oplus Me_2$ one place to the left"; the labelling of vertices, hence of idempotents, of Λ_2 is as in the diagram. This is an example of what Ringel terms ([605]) an "expanding" endomorphism. A pp condition, $\rho(x, y)$, which defines this map is

$$\exists x_1, \ldots, x_4 \Big(\bigwedge_{i=1}^{4} x_i = x_i e_i \wedge x = x_1 + \cdots + x_4 \wedge \\ \exists y_1, y_2 \, (y_1\alpha = x_1\beta \wedge y_2\alpha = x_2 \wedge y = y_1 + y_2\beta)\Big).$$

One has to check that this condition does well define a function, so show that the various choices for y_1 and y_2 in the above formula do all yield the same result, but that is easy. The action of this map, regarded as acting on the right, could be described as: $\beta\alpha^{-1}$ on Me_1, $\alpha^{-1}\beta$ on Me_2 and 0 on $Me_3 \oplus Me_4$. This map is, therefore, a biendomorphism as well as an endomorphism, hence is a definable scalar in the sense of Chapter 6.[6] The other notable (bi)endomorphism is that which sends $Me_1 \oplus Me_2$ to 0 and "shifts $Me_3 \oplus Me_4$ one place to the left". On $Me_3 \oplus Me_4$ this is surjective but has a kernel, hence it is a "contracting endomorphism" in Ringel's terminology (compare with the map defined first, which is injective but not surjective on $Me_1 \oplus Me_2$). It follows from Ringel's results ([605, p. 424])

[6] In [605] the actions of such maps and their powers are described in terms of a $k[T]$-action on the infinite string module $\bigoplus_i z_i k$ and a $k[[T]]$-action on the corresponding product module $\prod_i z_i k$.

that the pure-injective hull of this module[7] is $\prod_{i\le 0} z_i k \oplus \bigoplus_{i>0} z_i k$ and it may be checked that this module is indecomposable.

The notation used in [107] for this module is ${}^{\infty}\Pi\gamma\Sigma^{\infty}$. This is, over Λ_2, the only two-sided infinite word. Over Λ_n, $n \ge 3$ there are more possibilities for the central part of a two-sided infinite word. In the case of Λ_2, only γ is possible but there are infinitely many possibilities if $n \ge 3$, since a string may cycle around an intermediate Kronecker quiver a finite, but arbitrarily large, number of times before moving on. Indeed it is exactly that "finite, but arbitrarily large," which results in the Krull–Gabriel dimension / Cantor–Bendixson rank increasing as n increases.

There are also "one-sided infinite" strings and corresponding indecomposable pure-injective modules. Notation for these has the form ${}^{\infty}\Pi_i\star$ (if the infinite side is expanding) or $\star\Sigma_i^{\infty}$ (if the infinite side is contracting), where $\star$ is a finite string and where i indexes the Kronecker subquiver where the infinite part of the string is supported. A related notation is used for the infinite-dimensional points of $\mathrm{Zg}_{k\Lambda_n}$ associated to tubes supported on the ith Kronecker quiver, namely $\Pi_{i,\lambda}$ for the adic module (Section 15.1.3) corresponding to the tube with parameter $\lambda \in \mathbb{P}^1(k)$ and $\Sigma_{i,\lambda}$ for the corresponding Prüfer module. The parametrisation is chosen so that it is those with parameter ∞ that are different from the others, in the sense that they "can connect to the next Kronecker subquiver". The generic module associated to the ith Kronecker subquiver is denoted G_i.

In summary, there are the following infinite-dimensional indecomposable pure-injectives over Λ_n; the finite-dimensional ones are the indecomposable string and band modules described earlier. The symbol $\star$ denotes a finite string.

One-sided infinite:

${}^{\infty}\Pi_i\gamma_i\star$ $(i = 1, \ldots, n)$;

$\star\gamma_{j-1}\Sigma_j^{\infty}$ $(j = 1, \ldots, n)$.

Two-sided infinite:

${}^{\infty}\Pi_i\gamma_i \star \Sigma_j^{\infty}$ $(i < j)$.

Modules associated to bands:

$\Pi_{i,\lambda}$, $\Sigma_{i,\lambda}$ $(i = 1, \ldots, n, \lambda \in \mathbb{P}^1(k))$;

G_i $(i = 1, \ldots, n)$.

These are all the obvious candidates for infinite-dimensional points of the Ziegler spectrum. The main work is to prove that there are no more points and, also, to describe the topology completely by giving a neighbourhood basis of open sets at each point. That is carried out in [107] with full details for Λ_2 and some details in the general case.

[7] Note that any finite shift in where to break from product to direct sum gives the same module.

In particular, the Cantor–Bendixson ranks of the infinite-dimensional points of $\mathrm{Zg}_{k\Lambda_n}$ are as follows; recall, 5.3.33, that the finite-dimensional points all are isolated, that is, of rank 0.

Theorem 8.1.19. ([107, 4.1]) *The Cantor–Bendixson ranks of the infinite-dimensional points of* $\mathrm{Zg}_{k\Lambda_n}$ *are as follows.*

$\mathrm{CB}({}^{\infty}\Pi_i\gamma_i\star) = i$
$\mathrm{CB}(\star\gamma_{j-1}\Sigma_j^{\infty}) = (n+1) - j$
$\mathrm{CB}({}^{\infty}\Pi_i\gamma_i \star \Sigma_j^{\infty}) = i + (n+1) - j$
$\mathrm{CB}(\Pi_{i,\lambda}) = (n+1) - i$ *for* $\lambda \neq \infty$
$\mathrm{CB}(\Pi_{i,\infty}) = i$
$\mathrm{CB}(\Sigma_{i,\lambda}) = i$ *for* $\lambda \neq \infty$
$\mathrm{CB}(\Sigma_{i,\infty}) = (n+1) - i$
$\mathrm{CB}(G_i) = n+1$
In particular $\mathrm{CB}({}_{\Lambda_n}\mathrm{Zg}) = n+1$.

From the details, which yield a complete description of the topology, one may see that the isolation condition (Section 5.3.2) holds: in each of the closed subsets of the Ziegler spectrum which are the successive Cantor–Bendixson derivatives (Section 5.3.6) of $\mathrm{Zg}_{k\Lambda_n}$ each isolated point is isolated by a minimal pair, and that is enough for the isolation condition by 5.3.16. Therefore 5.3.60 yields the following, which was also obtained at the same time and independently by Schröer, who used the lattice of possibly transfinite factorisations of morphisms, hence analysed the transfinite radical series (Section 9.1.2) of mod-R.

Corollary 8.1.20. ([107, 2.3, §4], [641, Thm 14]) *The Krull–Gabriel dimension, hence also the m-dimension, of the path algebra of the quiver* Λ_n *(*$n \geq 2$*) is* $n+1$.

An essential ingredient in all this is the description of pure-injective hulls associated to infinite strings. Ringel shows in [605] that over a domestic string algebra every infinite string is, in each direction that it is infinite, eventually "expanding" or "contracting": these notions have combinatorial descriptions but mean roughly "adic-like" and "Prüfer-like" respectively. Thus a doubly infinite word might be expanding on each side, contracting on each side or expanding on one side and contracting on the other. From any infinite word Ringel built a corresponding infinite-dimensional module, using a direct sum of one-dimensional spaces where there is a half-infinite contracting word and a direct product of one-dimensional spaces where there is a half-infinite expanding word. See above for examples. In [605] he shows that the modules he builds are pure-injective and states that they are indecomposable (this can be checked directly). For the results described above it was also necessary to consider the "other" modules: direct sum modules formed from expanding words and direct product modules formed from contracting words. Burke, in the preprint [105] which builds on [605] and [44], describes the pure-injective hulls of these modules ([105, §§3,4]).

Question 8.1.21. Is every infinite-dimensional pure-injective module over a domestic string algebra a module, Prüfer, adic or generic, arising from a band or else a module of the sort built by Ringel from a singly or doubly infinite string?

Certainly it is the case, [523, 5.4], over any finite-dimensional string algebra, that different one-sided strings yield non-isomorphic indecomposable pure-injectives. Hence, [523, 6.1], any non-domestic such algebra has at least $2^{\aleph_0}$ points in its Ziegler spectrum.

Question 8.1.22. If R is a domestic string algebra is the Krull–Gabriel dimension of R equal to $n+2$, where n is the maximal length of a path in the "bridge quiver"? This is defined in [640], see [642, §4] (for further examples see [524, §3]); it is a quiver with the finitely many (since R is domestic) bands as vertices and the paths between them for arrows.

It is shown in [529, 1.3] that a string algebra R is of domestic representation type iff every finitely presented serial functor in (mod-R, **Ab**) has m-dimension iff every such functor has finite m-dimension.

The analysis of the spectrum over these algebras also shows (see Section 14.3.5) that the Ziegler spectrum of a ring cannot, in general, be recovered from the dual-Ziegler (= rep-Zariski, see Section 14.1.3) spectrum (Section 5.6).

We do not give more details of what is in the papers [44], [605], [105], nor in later papers such as [524], [523], [573], [574] on string modules, partly because setting up notation takes a fair amount of space, partly because the proofs, as opposed to the statements, are quite technical but also because it is clear that we have by no means reached the end of the story here, neither for domestic nor for non-domestic string algebras, and many questions are open.

8.1.4 Spectra of canonical algebras

In this section we are very brief, partly to avoid having to give all the relevant definitions, partly since the investigation of the Ziegler spectrum of these algebras is still very much work in progress. In fact we refer mainly to a single example which is, however, an exemplar of an important class of algebras.

Let R be the k-path algebra (k any field) of the following quiver with the relations $\alpha_2^1\alpha_1^1+\alpha_2^2\alpha_1^2+\alpha_2^3\alpha_1^3=0$ and $\alpha_2^2\alpha_1^2+\lambda\alpha_2^3\alpha_1^3+\alpha_2^4\alpha_1^4=0$ for some $\lambda\in k$, $\lambda\neq 0,1$.

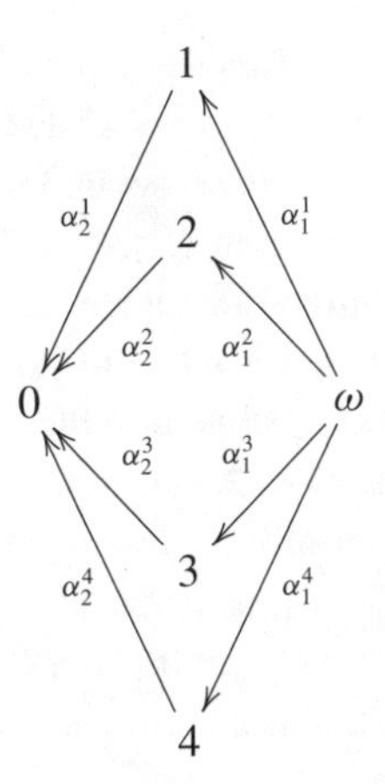

This is a **canonical algebra of type** (2, 2, 2, 2).

Let $\mathcal{P}$ be the set of isomorphism types of finite-dimensional indecomposable R-modules with all α_i^s injective but not all α_i^s bijective. Let $\mathcal{Q}$ be the set of isomorphism types of finite-dimensional indecomposable R-modules with all α_i^s surjective but not all α_i^s bijective. Let $\mathcal{T}$ be the set of remaining isomorphism types of finite-dimensional indecomposable R-modules. Then, [602, p. 162], $\mathcal{T}$ is a sincere stable tubular family of type (2, 2, 2, 2) separating $\mathcal{P}$ from $\mathcal{Q}$. See that reference or any of [603], [590], [612] for what this means (and, altogether, for more details). It does imply that $\mathcal{T}$ is a collection of tubes and any morphism from a module in $\mathcal{P}$ to one in $\mathcal{Q}$ factors through any given one of these tubes. Furthermore, see [602, p. 273 ff.], the tubes are arranged into infinitely many families of tubes, $\mathcal{T}_r$, with $r \in \mathbb{Q}^{\geq 0} \cup \{\infty\}$ with each family $\mathcal{T}_r$, like those seen in Section 8.1.2, a family of tubes indexed by a (weighted) projective line. Also, if $r < s < t$, then any morphism from a module appearing in $\mathcal{T}_r$ to a module appearing in $\mathcal{T}_t$ factors through any tube in $\mathcal{T}_s$. It follows immediately from the definition (Section 7.2.2) that there is a factorisable system of morphisms in rad(mod-R), hence, 7.2.13, that $\mathrm{KG}(R) = \infty$ and (Section 9.1.3) that $\mathrm{rad}_R^\infty \neq 0$. Harland and Puninsky have shown independently there are superdecomposable pure-injectives over these rings, at least in the countable case.

Question 8.1.23. Is there an example of a tame algebra which does not have Krull–Gabriel dimension but which does have uniserial dimension, hence which has no superdecomposable pure-injective?

The finite-dimensional points are described by the classification of finite-dimensional modules over R, [602], also see [225] for a geometric view of these representations.[8]

[8] And [602, p. 277] or [400, §10.8] for a "picture".

Less is known about the infinite-dimensional points. Lenzing, [397, 5.1], proved that the only generic modules are those associated with the families $\mathcal{T}_r$ of tubes. One also has a Prüfer and adic module associated with every quasi-simple module at the mouth of each tube, as in Section 8.1.2. By the remarks above and 7.2.14 it is clear that there are many infinite-dimensional indecomposable pure-injectives beyond these. The obvious place for these to "live" is at cuts of the index set $\mathbb{Q}^{\geq 0} \cup \{\infty\}$ for families of tubes, that is, at indices in $\mathbb{R}^{\geq 0} \cup \{\infty\}$. Constructions of points of Zg_R "living at" irrational such indices are given in [512] and at the end of [590]. Reiten and Ringel proved, amongst many other results relating to the structure of Mod-R for such rings, that all the remaining indecomposable pure-injectives can be found at cuts ([590, Thm 6]). However, it is, so far, quite unclear what the structure of these is and, indeed, how many live at each cut.

8.2 Further examples

8.2.1 Ore and RD domains

The lattice of pp conditions over a right Ore domain is, by reference to its division ring of quotients, partitioned into a filter and a complementary ideal. Divisibility conditions are cofinal in the filter, annihilation conditions in the ideal (8.2.1). An RD domain is left Ore iff it is right Ore (8.2.5); it is not known whether every RD domain is Ore. If a faithful indecomposable module over an RD domain is Σ-pure-injective, then it is Σ-injective (8.2.7).

Recall that R is a right **Ore domain** if R is a domain which is uniform as a right module over itself, that is, for all non-zero elements a, b of R, $aR \cap bR \neq 0$. For example, any right noetherian domain is right Ore but, for instance, the free associative algebra $k\langle X, Y\rangle$ over a field k is not. If R is a right Ore domain, then there is a right division ring, D, of quotients of R, which is actually the injective hull of R equipped with a ring structure (for example, see [243, 6.13] or [663, II.3.8]) and every element of D has the form rs^{-1} for some $r, s \in R$. The only pp-definable subgroups of D as an R-module are D and 0, so, if ϕ is any pp condition (in one free variable) over R, then $\phi(D) = D$ or $\phi(D) = 0$. This partitions pp_R^1 into a filter of "high" conditions and an ideal (in the sense of posets, see Section 3.3.1) of "low" pp conditions. This partition has been used in [290] and [289]. If R is also left Ore, then the right and left quotient division rings of R coincide (for example, [243, p. 95] or [441, 2.1.4]).

Proposition 8.2.1. *Suppose that R is a right Ore domain with right division ring of quotients D. Set $\mathcal{F} = \{\phi : \phi(D) = D\}$, $\mathcal{I} = \{\phi : \phi(D) = 0\}$. Then $\mathcal{F}$ is a filter* *and $\mathcal{I}$ is an ideal in the lattice, pp_R^1, of pp conditions, with one free variable, over R and $\mathcal{F} \cup \mathcal{I} = \mathrm{pp}_R^1$.*

A pp condition ϕ is in $\mathcal{I}$ iff there is some non-zero $s \in R$ with $\phi(x) \leq xs = 0$ and if R is also left Ore, then a pp condition ϕ is in $\mathcal{F}$ iff there is some non-zero $r \in R$ with $r|x \leq \phi(x)$.

Indeed, if R is right and left Ore, then, for every pp condition $\phi(x)$ for right modules, $\phi \in \mathcal{F}$ for right modules iff $D\phi \in \mathcal{I}$ for left modules.

Proof. The first statements are obvious. Given $\phi \in \mathcal{I}$, suppose that there were no pp condition of the form $xs = 0$ $(s \neq 0)$ above it. This implies that if $c \in C$ is a free realisation of ϕ, then $cR \simeq R_R$ since $cs \neq 0$ for every $s \in R$, $s \neq 0$. Therefore the injective hull, $E(C)$, of C decomposes as $E(cR) \oplus E'$ for some E', and $E(cR) \simeq D_R$. But then $c \in \phi(C) \leq \phi(D)$ so it cannot be that $\phi(D) = 0$.

If R is right and left Ore and $\phi(x)$ is a pp condition for right modules, then $\phi(x)$ is equivalent in D to $x = x$ and hence (for example, 1.3.15) $D\phi(x)$ is equivalent in the dual module, which is ${}_RD$, of D_R to $x = 0$, so the last statement follows.

Also in the case that R is right and left Ore, if $\phi \in \mathcal{F}$, then, by the first paragraph, $D\phi(x) \leq rx = 0$ for some non-zero $r \in R$, hence, 1.3.15, $\phi(x) \geq r|x$. □

Therefore, if $\phi \in \mathcal{F}$, then there is $r \in R$, $r \neq 0$ such that $\phi(M) \geq Mr$ for every module M and if $\phi \in \mathcal{I}$ there is $s \in R$, $s \neq 0$ such that $\phi(M) \leq \mathrm{ann}_M(s)$ for every module M.

The result above can be used to prove the following which, for commutative domains, was shown by Herzog and Puninskaya.

Proposition 8.2.2. ([289, 3.3], see [522, 8.3]) *Let M be a divisible module over a right and left Ore domain R. Then the ring of definable scalars R_M is the subring $\{sr^{-1} : \mathrm{ann}_M(r) \leq \mathrm{ann}_M(s),\ r, s \in R,\ r \neq 0\}$ of the quotient division ring of R.*

Also see [534], [535].

Next we consider RD domains (see Section 2.4.2 for these).

Proposition 8.2.3. ([575, 5.1]) *An RD domain is right and left semihereditary, hence right and left coherent.*

Proof. By [713, 39.13] a ring R is right semihereditary iff every factor module of an absolutely pure module is absolutely pure. Over an RD ring absolutely pure = divisible (2.4.16) and a factor of a divisible module is clearly divisible.

That any right semihereditary domain is right coherent is immediate from 2.3.18. □

Proposition 8.2.4. ([575, 5.2]) *If R is a domain, then R is right Ore iff there is a non-zero torsionfree absolutely pure right R-module.*

Proof. If R is right Ore, then one may take its right quotient ring for a module as in the statement. For the other direction, let $r, s \in R$ be non-zero. It must be shown that $rR \cap sR \neq 0$, that is, that the pp-pair $(r|sx) / (x = 0)$ is open on ${}_R R$. Now that, by 3.4.26 and 3.4.18, is so iff the dual pair $(x = x) / (\exists y\, (x = ys \wedge yr = 0)$ is open on some injective right R-module. If M is non-zero and torsionfree, then certainly this pair is open on M, so if M is also absolutely pure this pair is open on its injective hull (4.3.23), as required. □

Corollary 8.2.5. ([575, 5.3]) *An RD domain is left Ore iff it is right Ore.*

Proof. Since, 8.2.3, R is right and left coherent this is immediate from 8.2.4 and 3.4.24. □

Question 8.2.6. ([575, p. 2153]) Is every RD domain Ore?

A ring is **right Bezout** if every finitely generated right ideal is cyclic. It is the case ([575, 5.5]) that an RD domain is right Bezout iff it is left Bezout. It is not known whether every right and left Bezout domain is RD. It is shown in [575, 5.7] that this is so if the ring has the ascending chain condition on cyclic right or left ideals.

Proposition 8.2.7. ([575, 5.10]) *Suppose that M is a faithful indecomposable module over an RD Ore domain R.*

(a) If M is pure-injective but not injective, then there is an infinite properly descending chain of pp-definable subgroups of the form $Mr_1 > Mr_2 > \cdots$ with the $r_i \in R$.
(b) If M is not divisible, then there is an infinite properly descending chain of pp-definable subgroups of the form $Mr_1 > Mr_2 > \cdots$ with the $r_i \in R$.
(c) If M is Σ-pure-injective, then M is Σ-injective.

Proof. (a) By the left Ore condition the set of subgroups of M of the form Mr, $r \neq 0$ is downwards-directed, so, if there is no such infinite descending chain, the set $M' = \bigcap\{Mr : r \in R, r \neq 0\}$ is non-zero. The argument for 8.2.11 applies to show that M is injective.

(b) By 2.4.16, M is not absolutely pure so, by 4.3.12, $H(M)$ is not injective. By part (a) $H(M)$ has an infinite descending chain of the form stated, so, by 4.3.23, the same is true of M.

(c) This is now immediate from part (b) and 4.4.5. □

Corollary 8.2.8. *Every Σ-pure-injective module over a simple RD-domain is Σ-injective.*

This applies, for example, to the first Weyl algebra over a field of characteristic 0 since this is a simple RD-domain (comment after 2.4.15).

8.2.2 Spectra over HNP rings

Over a hereditary noetherian domain with enough invertible ideals the set of hulls of indecomposable finite length modules is dense in the Ziegler spectrum provided there is no simple injective left module (8.2.9). The hull of an indecomposable finite-length module over an HNP ring is non-isolated in the spectrum provided the module has non-split extensions with infinitely many simple modules (8.2.12). Conditions under which the spectrum of an HNP domain has no isolated points are given (8.2.14).

The Ziegler spectrum over a PI hereditary noetherian prime ring is described in 8.1.13. In this section, though the ring will be assumed to be HNP, there is no assumption that it is PI so it may well be (see Section 7.3.2) that the Ziegler spectrum is "wild", with no general (known or likely) complete description of the points and topology of the space. The results in this section show that in many cases the hulls of finite-length modules are dense but that there are few or no isolated points. Although the proofs here follow those of [521] rather closely, we include them since they do provide some good illustrations of "pp-techniques".

Examples to bear in mind are $\begin{pmatrix} \mathbb{Z} & 2\mathbb{Z} \\ \mathbb{Z} & \mathbb{Z} \end{pmatrix}$ (see 14.3.8), which is PI, and the first Weyl algebra $A_1(k) = k\langle x, y : yx - xy - 1\rangle$ which is PI if the field k has characteristic $p > 0$ and is not PI if $\mathrm{char}(k) = 0$.

Recall (2.4.15) that hereditary noetherian prime rings with enough invertible ideals, such as both examples above (see [171, §5]), are RD.

Lemma 8.2.9. ([521, 3.6]) *Let R be a hereditary noetherian domain, which is not a field, with enough invertible ideals. Then the following conditions are equivalent:*

(i) the set of pure-injective hulls of finite-length points is dense in Zg_R*;*
(ii) there are no simple injective left R-modules.

Proof. (i)⇒(ii). Otherwise let $M = R/L$ be a simple injective left module and choose a set, $r_1, \dots, r_n$, of generators for the left ideal L. By 5.3.9 the pp-pair $(Lx = 0) / (x = 0)$ is a minimal pair. By elementary duality (1.3.1) the pair $(x = x) / (L \mid x)$ also is minimal and, since $R \neq L$, this pair is open on the right module R_R. Thus the unique (by 5.3.2), hence isolated, pure-injective indecomposable module on which this pair opens is a direct summand of the pure-injective hull, $H(R)$, of R, hence is torsionfree, hence cannot be the pure-injective hull of a finite length point (since every finite length module is torsion).

(ii)$\Rightarrow$(i). Let (ϕ/ψ) be a non-empty basic open set. It will be enough to find a finitely generated torsion module M with $\phi(M) > \psi(M)$: for then M is (see [441, 5.7.4]) a direct sum of indecomposable finite-length modules, so (ϕ/ψ) must open on at least one of these and hence, 4.3.23, it must open on the (indecomposable by 4.3.59) pure-injective hull of that module.

By 5.2.8 it may be supposed that ϕ has the form $\exists y\,(x = yr \wedge ys = 0)$, for some $r, s \in R$, and that ψ has the form $\phi(x) \wedge t' \mid xt$, for some $t', t \in R$. If $s \neq 0$, then R/sR is a cyclic torsion module and the pair $(R/sR, r + sR)$ is easily seen (use 1.2.15) to be a free realisation (Section 1.2.2) of ϕ. Therefore, since $(r + sR) \notin \psi(R/sR)$ – otherwise ψ would not be strictly contained in ϕ (in the lattice of pp conditions) – the module $M = R/sR$ is as desired.

Otherwise $s = 0$ and, as above and since (R, r) is clearly a free realisation of ϕ, $rt \notin Rt'$. In particular, $rt \neq 0$, hence $rt \notin I$ for some nonzero right ideal I; otherwise rtR would be a simple module, hence R would be an artinian ring, contrary to assumption. If $t' = 0$, then (ϕ/ψ) is open on R/I, which we may take for M. Otherwise $t' \neq 0$. Since R/Rt' is not an injective left module and R is a noetherian RD domain it follows by 8.2.11 below that for some $s' \in R$, $s' \neq 0$ the image of rt in R/Rt' is not divisible by s'. Thus $rt \notin s'R + Rt'$, hence t' does not divide rt in $R/s'R$, so (ϕ/ψ) is open on $R/s'R$. □

Lemma 8.2.10. *Suppose that M is a right module over the left Ore domain R. Then $M' = \bigcap\{Mr : r \neq 0, r \in R\}$ is a submodule of M.*

Proof. Take $a \in M'$ and $t \in R$ with $t \neq 0$: we prove that $at \in M'$. Let $r \in R, r \neq 0$. By the left Ore condition there are $r', t' \in R$ such that $r'r = t't \neq 0$. Since $a \in M'$ there is $b \in M$ with $a = bt'$. Then $at = bt't = (br')r \in Mr$, as required. □

If R is a left Ore ring, meaning that the set of non-zero divisors satisfies the left Ore condition, then there is a similar statement and proof, with the modification that $M' = \bigcap\{Mr : r \text{ is a non-zero-divisor in } R\}$.

Corollary 8.2.11. *Suppose that R is a right and left noetherian RD domain and that M is a finitely presented right R-module with local endomorphism ring. If $\bigcap\{Mr : r \neq 0, r \in R\} \neq 0$, then M is injective.*

Proof. (cf. argument for 8.2.7(a)) Consider the embedding of M into its injective hull, $E(M)$. Suppose there is $a \in \bigcap\{Mr : r \neq 0, r \in R\}$, $a \neq 0$. Since, 2.4.10, every pp condition is a conjunction of conditions of the form $s|xr$, it follows by 8.2.10 that $\mathrm{pp}^M(a) = \mathrm{pp}^{E(M)}(a)$. Therefore, by 4.3.35, the pure-injective hull, $H(M)$, of M has a non-zero, necessarily injective, direct summand in common with $E(M)$. Since, 4.3.55, $H(M)$ is already indecomposable, it must be injective and hence, since M is pure in $H(M)$ (therefore M is absolutely pure) and R is right noetherian, M itself is injective by 4.4.17. □

Proposition 8.2.12. ([521, 3.8]) *Let R be a hereditary noetherian prime ring. Suppose that M is an indecomposable module of finite length and suppose that there are infinitely many simple modules S such that* $\mathrm{Ext}(S, M) \neq 0$. *Then $H(M)$ is not isolated by a minimal pair in* Zg_R.

Proof. By 4.3.59, $H(M)$ is a point of Zg_R. Since, by assumption, M has non-trivial extensions, M is not absolutely pure so, by 5.3.31, if $H(M)$ is isolated by a minimal pair, then there is a left almost split monomorphism $f : M \to N$ with $N \in \text{mod-}R$. Set $C = \mathrm{coker}(f)$. Being a finitely presented module over an HNP ring, N has a decomposition $N = N' \oplus N''$ with N' torsion and N'' torsionfree (see [441, 5.7.4]). Since $f(M) \subseteq N'$, we may suppose that N is torsion, so both N and C are of finite length (see [441, 5.4.5]). Let S be a simple module with $\mathrm{Ext}^1_R(S, M) \neq 0$, say $0 \to M \to X \to S \to 0$ is a non-split exact sequence. Then there is the following commutative diagram:

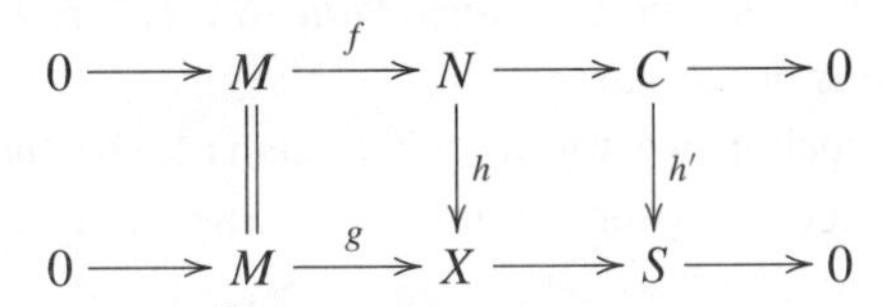

where h exists since g is not a split embedding and with the induced map h' being non-zero since f is not split. But then there are infinitely many simple modules S with $\mathrm{Hom}(C, S) \neq 0$, contradicting that C is of finite length. □

Corollary 8.2.13. ([521, 3.9]) *Let R be a hereditary noetherian prime ring such that for every simple module T there are infinitely many simple modules S with* $\mathrm{Ext}(S, T) \neq 0$. *Then no point of Zg_R of the form $H(M)$ with M of finite length is isolated by a minimal pair.*

Proof. Let M be an indecomposable module of finite length. By 8.2.12 it is enough to prove that $\mathrm{Ext}(S, M) \neq 0$ for infinitely many simple modules S. We prove this by induction on the length of M, with the base case being our assumption. If M is not simple, then choose a simple submodule T and consider the short exact sequence $0 \to T \to M \to M' \to 0$. By induction there are infinitely many S such that $\mathrm{Ext}(S, M') \neq 0$. For every such S we have, since R is hereditary, the exact sequence $0 \to \mathrm{Hom}(S, T) \to \mathrm{Hom}(S, M) \to \mathrm{Hom}(S, M') \to \mathrm{Ext}(S, T) \to \mathrm{Ext}(S, M) \to \mathrm{Ext}(S, M') \to 0$ so $\mathrm{Ext}(S, M) \neq 0$, as desired. □

Theorem 8.2.14. ([521, 3.11]) *Let R be a hereditary noetherian domain with enough invertible ideals and with no injective simple left module. Suppose that for* *every simple right module T there are infinitely many simple modules S such that* $\mathrm{Ext}(S, T) \neq 0$. *Then there are no isolated points in* Zg_R.

Proof. If N were an isolated point of Zg_R, then, by 8.2.9, N would have the form $H(M)$ for some module, M, of finite length. Since M is finitely presented, $H(M)$ is isolated by a minimal pair (5.3.13), contradicting 8.2.13. □

This applies to a wide class of generalised Weyl algebras (see 8.2.27 and 8.2.30(d) below).

8.2.3 Pseudofinite representations of sl_2

The ring of definable scalars for the set of finite-dimensional representations of the Lie algebra sl_2 *is a von Neumann regular ring and is an epimorphic extension of* sl_2 *(8.2.15). The proof of this result of Herzog is outlined. The key is to show that the lattice of pp conditions is complemented (8.2.22) and, for this, a certain anti-automorphism of the lattice of pp conditions relative to the set of finite-dimensional representations is important (8.2.17). Related algebraic and model-theoretic results are discussed.*

Throughout this section we follow what seems to be the usual convention and deal with *left* modules over the universal enveloping algebra of the Lie algebra sl_2.

Let k be a field of characteristic zero, which we also assume to be algebraically closed, though certainly not all of what we say needs this. Let L be the Lie algebra, $\mathrm{sl}_2 = \mathrm{sl}_2(k)$, of 2×2 matrices over k with trace 0. A k-basis of L is given by $x = \begin{pmatrix} 0 & 1 \\ 0 & 0 \end{pmatrix}$ $y = \begin{pmatrix} 0 & 0 \\ 1 & 0 \end{pmatrix}$ $h = \begin{pmatrix} 1 & 0 \\ 0 & -1 \end{pmatrix}$. The Lie bracket, which in terms of the algebra of matrices is $[u, v] = uv - vu$, is given by $[h, x] = 2x$, $[h, y] = -2y$, $[x, y] = h$. For basic definitions on Lie algebras see, for example, [113], [201] or [313]. The category of representations of any Lie algebra L is equivalent to the category left modules over its universal enveloping algebra $U(L)$, which, in the case $L = \mathrm{sl}_2(k)$, is $k\langle x, y, h\rangle / \langle hx - xh = 2x, hy - yh = -2y, xy - yx = h\rangle$. We will not distinguish between representations of L and left $U(L)$-modules.

Every finite-dimensional module over $U = U(\mathrm{sl}_2)$ is a direct sum of simple modules and, for every natural number n, there is, up to isomorphism, exactly one of these, $L(n)$, of dimension $n + 1$. As a k-vectorspace $L(n)$ has a basis $m_{-n}, m_{-n+2}, \ldots, m_{n-2}, m_n$ of eigenvectors of h with the action of sl_2 (and hence of $U(\mathrm{sl}_2)$) determined by $hm_k = km_k$, $xm_{n-2k} = k(n - k + 1)m_{n-2k+2}$ (with m_i interpreted as 0 when $i > n$) and $ym_k = m_{k-2}$ (with a similar comment if $k - 2 < -n$). See, for instance, [201, §11.1] for more on these.

That completely describes the finite-dimensional representations but there are many infinite-dimensional simple modules: some of these are considered in the next section.

Let $\mathcal{X}_{\text{fd}}$ denote the definable subcategory of U-Mod generated by the set of (simple) finite-dimensional representations and let X_{fd} denote the corresponding subset, $\mathcal{X}_{\text{fd}} \cap {}_U\text{Zg}$, of the (left) Ziegler spectrum of U. Denote by U' the ring of definable scalars (Section 6.1.1) of this closed set. Herzog shows in [282], see below, that U' is von Neumann regular and that the canonical map $U \to U'$ is an epimorphism of rings.

Herzog puts the following gloss on the results of [282]. Suppose that we wished to study just the finite-dimensional representations of sl_2. We might try to simplify the context by factoring U by the ideal which is the common annihilator, $\bigcap_n \text{ann}_U(L(n))$, of the finite-dimensional representations. This intersection is, however, zero (for example, [152]), so nothing is gained by this. What we can do instead is to factor out the finitely presented functors which annihilate all the $L(n)$; since the functor $M \mapsto rM$ for $r \in U$ is finitely presented, this really is an extension of the first idea. That is, we factor the free abelian category (Section 10.2.7), $\text{Ab}(U) = (U\text{-mod}, \mathbf{Ab})^{\text{fp}}$, by the Serre subcategory (11.1.40) $\mathcal{S}_{\text{fd}} = \{F \in (U\text{-mod}, \mathbf{Ab})^{\text{fp}} : F\,L(n) = 0 \text{ for all } n \in \mathbb{N}\}$. Then U' is, 12.8.2, the endomorphism ring of the corresponding localisation, $(U, -)_{\mathcal{S}_{\text{fd}}}$, of the forgetful functor and that is, note, the image of U under the canonical map (Section 10.2.7) from U to $\text{Ab}(U)$ followed by "factoring" by the common annihilator, $\mathcal{S}_{\text{fd}}$, of the $L(n)$.

The fact that $U \to U'$ is an epimorphism of rings, hence that U'-Mod is, by 5.5.4, a full, definable subcategory of U-Mod, means that we are indeed just concentrating on a part of the representation theory of U, namely that part determined, in some sense, by the finite-dimensional representations of U. Note also that there are no other simple U-modules which are modules over U': for if ${}_US$ is simple, $a \in S$ is non-zero and $K = \text{ann}_R(a)$, then the pp-pair[9] $(Kv = 0 \,/\, v = 0)$ is open on S and on no other simple module (the condition $Kv = 0$ is pp because U is noetherian), so S is not in the closure, X_{fd}, of the $L(n)$ if it is not already one of the $L(n)$.

Theorem 8.2.15. ([282, Cor. 27]) *Let $U = U(\text{sl}_2(k))$, where k is any algebraically closed field of characteristic zero, and let $U \to U'$ be the ring of definable scalars associated to the definable subcategory, $\mathcal{X}_{\text{fd}}$, generated by the finite-dimensional*

[9] In this section and the next our typical notation for variables in pp conditions is u, v, etc. since x and y are already in use.

*representations $L(n)$, $n \geq 0$. Then U' is a von Neumann regular ring and the morphism $U \to U'$ is an epimorphism of rings. Furthermore, $\mathcal{X}_{\mathrm{fd}} = U'$-*Mod.

The proof is indicated below. The last part is by 6.1.14. By [282, Prop. 39] the $L(n)$ are, therefore, together dense in ${}_{U'}\mathrm{Zg}$ and it is easy to check that each is isolated in that subspace of ${}_U\mathrm{Zg}$; explicitly, $L(n)$ is isolated by $\big((xv = 0 \wedge hv = nv) \,/\, v = 0\big)$.

Proposition 8.2.16. ([282, p. 273]) *With notation as above the m-dimension of* ${}_{U'}\mathrm{pp}$ *is undefined. Hence* $\mathrm{KG}(U') = \infty$ *and there is a superdecomposable injective* U'*-module. The same therefore holds for* $U(\mathrm{sl}_2(k))$.

The first statement is proved using the pp-definable subgroups seen in 1.2.5. The other statements then follow by 13.2.1, for Krull–Gabriel dimension, and 7.3.20, for superdecomposability.

Since each $L(n)$ has trivial (that is, k) endomorphism ring, its ring of definable scalars is, by 6.1.20, the matrix ring $M_{n+1}(k)$, so, by 6.1.3, U' embeds in the product, $\prod_{n\in\mathbb{N}} M_{n+1}(k)$, which is also von Neumann regular. But U' is not simply this product: for example, if k is countable, then so is U, hence so is U' (it is an epimorphic image of U so, by 6.1.8, every element of it is definable by a pp condition over U, of which there are only countably many); in contrast, the product has cardinality that of the continuum.

Regarding the proof, in view of 12.8.10 it must be shown that every finitely generated subfunctor of $(U, -)$ has a "complement modulo $\mathcal{S}_{\mathrm{fd}}$".

This complement is computed quite explicitly using the anti-isomorphism α from U to U^{op} which is defined by interchanging x and y and fixing h. This is used to define a map from the set of pp conditions in one free variable to itself, as follows.

Given a pp condition $\phi(v)$, say ϕ is $\exists \overline{w} \begin{pmatrix} A & B \end{pmatrix} \begin{pmatrix} v \\ \overline{w} \end{pmatrix} = 0$, define $\phi^-(v)$ to be the pp condition $\exists \overline{z} \begin{pmatrix} 1 & \alpha A^{\mathrm{T}} \\ 0 & \alpha B^{\mathrm{T}} \end{pmatrix} \begin{pmatrix} v \\ \overline{z} \end{pmatrix} = 0$ (the superscript denotes transpose).

Beware that this operation is not well defined on equivalence classes of pp conditions: one can produce simple pp conditions, ϕ, ψ, which are equivalent on all U-modules such that ϕ^- and ψ^- are not equivalent. The operation is, however, well defined on equivalence classes of pp conditions modulo $\mathcal{X}_{\mathrm{fd}}$.

Lemma 8.2.17. ([282, Prop 11]) *If ϕ and ψ are pp conditions and $n \in \mathbb{N}$ are such that $\phi(L(n)) = \psi(L(n))$, then $\phi^-(L(n)) = \psi^-(L(n))$.*

Indeed $\phi \mapsto \phi^-$ induces a well-defined anti-isomorphism on the lattice $\mathrm{pp}(\mathcal{X}_{\mathrm{fd}})$. Said in terms of functors, the operation $F_{\mathcal{S}_{\mathrm{fd}}} \mapsto (F^-)_{\mathcal{S}_{\mathrm{fd}}}$, is a well-defined operation on $\mathrm{Latt}^{\mathrm{f}}\big((U, -)_{\mathcal{S}_{\mathrm{fd}}}\big) = \mathrm{Latt}^{\mathrm{f}}\big(U', -\big)$.

Herzog says that a pp condition in one free variable is a **weight condition relative to** the $L(n)$ if for each n one has $h\phi(L(n)) \leq \phi L(n)$, that is, iff $\phi(L(n))$ is a sum of eigenspaces "weight spaces" for h.

Proposition 8.2.18. ([282, Prop. 13]) *If ϕ is a weight condition relative to the $L(n)$, then so is ϕ^- and $(U,-)_{\mathcal{S}_{\mathrm{fd}}} = (F_\phi)_{\mathcal{S}_{\mathrm{fd}}} \oplus (F_{\phi^-})_{\mathcal{S}_{\mathrm{fd}}}$.*

Example 8.2.19. Let F be the functor which picks out the next-to-highest weight/eigenspace, so is defined by $M \mapsto \mathrm{ann}_M(x^2) \cap \mathrm{im}(y)$, that is, by the pp condition

$$\exists w \begin{pmatrix} x^2 & 0 \\ 1 & -y \end{pmatrix} \begin{pmatrix} v \\ w \end{pmatrix} = 0.$$

Then ϕ^- is the condition

$$\exists z_1, z_2 \begin{pmatrix} 1 & y^2 & 1 \\ 0 & 0 & -x \end{pmatrix} \begin{pmatrix} v \\ z_1 \\ z_2 \end{pmatrix} = 0,$$

that is $\exists z_1, z_2 \left(v + y^2 z_1 + z_2 = 0 \wedge x z_2 = 0\right)$, which is just the condition $v \in \mathrm{im}(y^2) + \mathrm{ann}(x)$. Clearly this does define a complement (which is a sum of weight spaces) for ϕ in each $L(n)$.

Say that ϕ is **uniformly bounded relative to** the $L(n)$ $(n \in \mathbb{N})$ if there is a natural number N such that $\dim_k \phi(L(n)) \leq N$ for every n.

Let D denote the quotient division ring of (the right and left noetherian, hence Ore, domain) U. By 8.2.1 the lattice ${}_U\mathrm{pp}^1$ is the disjoint union of the filter $\mathcal{F} = \{\phi : \phi(D) = D\}$ and the ideal $\mathcal{I} = \{\phi : \phi(D) = 0\}$.

Proposition 8.2.20. ([282, Cor. 24]) *The ideal $\mathcal{I}$ consists exactly of the pp conditions which are uniformly bounded relative to the $L(n)$ $(n \in \mathbb{N})$.*

Lemma 8.2.21. ([282, Lemma 25]) *If ϕ is a weight condition and is uniformly bounded relative to the $L(n)$, then the interval $[\phi(x), x = 0]$ in $\mathrm{pp}(\mathcal{X}_{\mathrm{fd}})$ is complemented.*

Theorem 8.2.22. ([282, Thm 26]) *The lattice of finitely generated subobjects of $(U,-)_{\mathcal{S}_{\mathrm{fd}}}$, that is, of $(U',-) = (U_{\mathcal{X}_{\mathrm{fd}}},-)$, is complemented.*

Proof. (sketch) Set $\mathcal{S} = \mathcal{S}_{\mathrm{fd}}$. Let G be a finitely generated subfunctor of $(U,-)_{\mathcal{S}}$: by 12.3.18 $G = (F_\phi)_{\mathcal{S}}$ for some pp condition $\phi \in {}_U\mathrm{pp}$. If ϕ is in the filter $\mathcal{F}$, then Herzog showed that there is a uniformly bounded weight condition ψ such that $G + (F_\psi)_{\mathcal{S}} = (U,-)_{\mathcal{S}}$. By 8.2.21 $G \cap (F_\psi)_{\mathcal{S}}$ has a complement in the interval $[(F_\psi)_{\mathcal{S}}, 0]$ of $\mathrm{Latt}^{\mathrm{f}}(U',-) = \mathrm{Latt}^{\mathrm{f}}\left((U,-)_{\mathcal{S}}\right)$ and this then serves as a complement for G in $\mathrm{Latt}^{\mathrm{f}}\left((U,-)_{\mathcal{S}}\right)$.

If, on the other hand, ϕ is in the ideal $\mathcal{I}$, so, by 8.2.20, ϕ is a weight condition, then, by 8.2.18, ϕ^- is in $\mathcal{F}$ so, by the first paragraph, ϕ_S^- has a complement, say $(U, -)_S = (F_{\phi^-})_S \oplus (F_\psi)_S$. Apply the operation $(^-)$ to obtain $(U, -)_S = (F_\phi)_S \oplus (F_{\psi^-})_S$. Hence $(F_\phi)_S$ does have a complement, as required. □

Herzog's paper [282] contains a great deal more information on the structure of the ring $U' = U_{\mathcal{X}_{\text{fd}}}$ and on its set of simple modules, some of which is summarised now.

As already stated, the modules $L(n)$ are isolated points of ${}_{U'}\text{Zg}$ and together are dense in that space. The $L(n)$ are exactly the finite-dimensional simple U'-modules (see, for example, the remark after [282, Thm 47]) and also (using [282, Thm 14]) are exactly the finitely presented simple U'-modules: there are many more, indeed continuum many [282, Thm 41], simple U'-modules, none of which is simple as a U-module. One of these is the quotient division ring, D, of U, which is U'/I, where $I = \sum_{\phi \in \mathcal{I}} \phi(U')$ ($\mathcal{I}$ as above); this ideal is also the ideal of U' generated by the **highest-weight idempotent** e_x, [282, Thm 30]. The latter is defined to be the idempotent of the ring, U', of definable scalars, corresponding to the pp condition $xv = 0$ which picks out the highest-weight space on each $L(n)$. For any left U'-module M, the U-torsion submodule is a U'-module, namely IM, [282, Prop. 32].

Every indecomposable (pure-)injective U'-module is the injective hull $E = E(M)$ of a simple U'-module M [282, Cor. 35] and, if $M \neq D$, then $\text{End}(E) = \text{End}(M)$ is a commutative field, [282, Cor. 34, Prop. 44], which, if M is infinite-dimensional, is a proper extension of k, [282, p. 273]. It follows, by 5.2.11, that the Ziegler spectrum of U' is in natural bijection with the set of simple U'-modules. The map $E \mapsto \text{ann}_{U'} E$ defines a homeomorphism, [282, Prop. 43], from ${}_{U'}\text{Zg}$ to the space of primitive ideals of U', where the latter is equipped with the **Jacobson–Zariski topology** – that for which the closed sets are those of the form $V(J) = \{I \text{ a primitive ideal} : J \leq I\}$, where J ranges over ideals of U'.

If M is an infinite-dimensional simple U'-module, not equal to D, then the Ziegler-closure of $E(M)$ is $\{E(M), D\}$ [282, Thm 47]. So the closed points of ${}_{U'}\text{Zg}$ are those of finite endolength, which are the $L(n)$ and D. If P is any primitive ideal which is not maximal, then $P < I$, [282, Thm 47], and the natural map $U \to U'/P$ is an inclusion, [282, p. 277]. The ring U'/P is hereditary, [282, Prop. 51], and, by the description of Ziegler-closures of points, its Ziegler spectrum has just two points, $E(U'/P)$ and D.

Herzog also developed a local theory of pseudoweights, applicable to the simple U'-modules other than D and showed that the structure theory has many similarities to that of the $L(n)$, [282, p. 279 ff.].

Concerning the model theory of U'-modules, Herzog also showed that two U'-modules generate the same definable subcategory iff they have the same annihilator in U', [282, Cor. 62]. Therefore, since k is infinite, this is the criterion for elementary equivalence (see A.1.3) as U'-modules, equivalently, as U-modules. There is an explicit, and simple, axiomatisation, [282, Thm 63], of the **pseudo-finite-dimensional** representations of sl_2, meaning those modules, finite- or infinite-dimensional, which are models of the common theory of finite-dimensional representations. It is also shown that the pseudo-finite-dimensional modules are exactly those elementarily equivalent to, that is, which generate the same definable subcategory as, a direct sum of simple U'-modules none of which is isomorphic to D, [282, Prop. 65].

L'Innocente and Macintyre ([405]) have used these results and some close connections with questions about integer points on curves, aiming to prove, at least modulo some conjectures in diophantine geometry, the decidability of the theory of finite-dimensional representations of $\mathrm{sl}_2(k)$.

The finite-dimensional representations of $\mathrm{sl}_2(k)$ may be packed together into a single representation of $\mathrm{sl}_2(k)$ on the coordinate ring $k[X, Y]$ of the affine plane. Herzog and L'Innocente ([288]) have extended some of the results of [282] to the quantum group U_q, which is the coordinate ring of the quantum plane $k_q[X, Y]$ (when $\mathrm{char}(k) \neq 2$) and have proved in particular that the relevant lattice of pp-definable subgroups is complemented, cf. 8.2.22, and that, therefore, the ring of definable scalars of the finite-dimensional representations is von Neumann regular and a ring-epimorphic extension of U_q, cf. 8.2.15.

8.2.4 Verma modules over sl_2

It is shown that the ring of definable scalars of any Verma module over sl_2 is von Neumann regular and is an epimorphic extension of the enveloping algebra (8.2.24). The Ziegler-closure of such a module is computed and shown to have rank 1 (8.2.26).

One may ask whether there are results analogous to those in the previous section if one replaces the finite-dimensional representations by the Verma modules over $\mathrm{sl}_2(k)$ (k still algebraically closed of characteristic 0). It turns out that this links to the generalised Weyl algebras discussed in Section 8.2.5. We retain the notation of the previous section, in particular $U = U(\mathrm{sl}_2(k))$. The results here are taken from [406].

For $\lambda \in k$ let $M = M(\lambda) = U(\mathrm{sl}_2)/I_\lambda$, where $I_\lambda = Ux + U(h - \lambda)$, be the corresponding **Verma module**. This is the U-module generated by an element m_λ (the image of $1 \in U$) with relations $xm_\lambda = 0$, $hm_\lambda = \lambda m_\lambda$. Then a k-basis for $M(\lambda)$ is $m_\lambda, m_{\lambda-2} = ym_\lambda, \ldots, m_{\lambda-2i} = y^i m_\lambda, \ldots$ $(i \geq 0)$ and one may check that $hm_{\lambda-2i} = (\lambda - 2i)m_{\lambda-2i}$. Therefore, setting $M_\mu = \{m \in M : hm = \mu m\}$ to be the μ-eigenspace for the action of h on M, one has $M = \bigoplus_{i \geq 0} M_{\lambda-2i}$ with each $M_{\lambda-2i}$ one-dimensional. The action of x is easily computed to be $xm_{\lambda-2i} = i(\lambda - i + 1)m_{\lambda-2i+2}$ (with $m_{\lambda+2}$ being interpreted as 0).

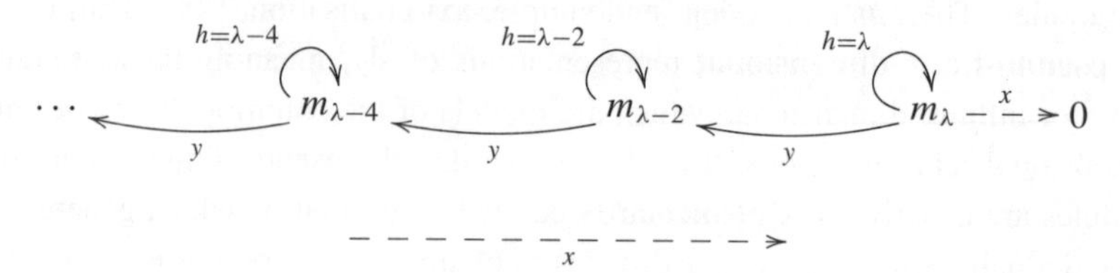

If λ is not a natural number, then $M(\lambda)$ is a simple module, also denoted $L(\lambda)$, and if λ is a natural number, n, then $M(n)$ is a module of length two with composition factors $L(n)$ and $L(-n-2)$ (see, for example, [314, p. 75]).

By 1.2.13 and since the endomorphism ring of each $M(\lambda)$ is k, every finite-dimensional subspace of $M = M(\lambda)$ is pp-definable. Furthermore, every cofinite-dimensional subspace, W, of M which contains some $M_{\leq\lambda-2i} = \bigoplus_{s=i}^{\infty} M_{\lambda-2s}$ is pp-definable. To see this, first note that $M_{\leq\lambda-2i}$ itself is pp-definable (by $y^i | v$). Then write $W = M_{\leq\lambda-2i} \oplus V$, where $V \leq M_{>\lambda-2i}$. Thus W is the sum of two pp-definable subgroups, hence is pp-definable.[10]

Remark 8.2.23. Let $M(\lambda)$, $\lambda \notin \mathbb{N}$ be a Verma module over $\mathrm{sl}_2(k)$ (k an algebraically closed field of characteristic 0). Then $M(\lambda)$ has neither the ascending nor descending chain condition on pp-definable subgroups. The images of successive powers of y give an infinite descending chain and the annihilators of successive powers of x give an infinite ascending chain. Since $M(\lambda)$ is countable-dimensional and is not, since the dcc fails, Σ-pure-injective 4.4.5, it is not pure-injective (by 4.4.10).

Since these simple modules $M(\lambda)$ are infinite-dimensional it makes sense to ask about the Ziegler-closure and ring of definable scalars, not just for the set of all the $M(\lambda)$ ($\lambda \notin \mathbb{N}$), but also for each individual module. A key observation is that the simple Verma module $M(\lambda)$ ($\lambda \notin \mathbb{N}$) is a module over a generalised Weyl algebra in the sense of Bavula (see Section 8.2.5 for these), as follows.

Set $\mu = \lambda^2 + 2\lambda$ and let $C = 2xy + 2yx + h^2$ be the Casimir element of U. Computation shows that the actions of x, y and h on $M(\lambda)$ satisfy the relations: $xy = \frac{1}{4}(\mu - (h-2)^2 - 2(h-2))$ and $yx = \frac{1}{4}(\mu - h^2 - 2h)$, as well as those, $xh = (h-2)x$ and $yh = (h+2)h$, coming from U itself. Furthermore, computing the action of C on $m_{\lambda-2i}$ one see that $C = \mu$ on $M(\lambda)$.

[10] Not every pp-definable cofinite-dimensional subspace contains some $M_{\leq\lambda-2i}$: consider, for instance, the image of $1 + y$. Also, of course, 'most' cofinite-dimensional subspaces are not pp-definable, since there are too many of them – for instance, if k is countable, then they are uncountable in number, whereas there are only countably many pp conditions with which to define subspaces.

Therefore $M(\lambda)$ is a faithful module over the primitive factor ring $B_\mu = U/\langle C - \mu\rangle$ of U (see [152, 8.4.3]).[11] Observe that $\lambda^2 + 2\lambda = \nu^2 + 2\nu$ iff $\nu = \lambda$ or $\nu = -\lambda - 2$ so $M(-\lambda - 2)$ also is a B_μ-module. We check that the rings B_μ are generalised Weyl algebras (§8.2.5).

Denote by σ the automorphism of the polynomial ring $k[h]$ given by $\sigma(h) = h - 2$ and let $a(h)$ be the polynomial $(-h^2 - 2h + \mu)/4$. Then

$$B_\mu = k[h]\langle x, y : xh = \sigma(h)x,\ yh = \sigma^{-1}(h)y,\ xy = a(h-2),\ yx = a(h)\rangle$$

or, making the change of variable $H = h/2$,

$$B_\mu = k[H]\langle x, y : xH = \sigma(H)x,\ yH = \sigma^{-1}(h)y,\ xy = a(H-1),\ yx = a(H)\rangle,$$

where now σ is given by $\sigma(H) = H - 1$ and $a(H) = -H^2 - H + \mu/4$. So this is even a generalised Weyl algebra of the special form, where the automorphism σ is given by $\sigma(H) = H - 1$.

By 8.2.27, B_μ is hereditary and simple iff the roots of $H^2 + H - \mu/4$ do not differ by an integer. The roots of this polynomial are $\lambda/2$ and $-(\lambda/2) - 1$, so the difference is $-\lambda - 1$, which is an integer iff $\lambda \in \mathbb{Z}$. Thus B_μ is hereditary and simple if(f) $\lambda \notin \mathbb{Z}$.

Therefore every Verma module is a simple module over a generalised Weyl algebra with $\sigma(H) = H - 1$. Every Verma module has endomorphism ring k (consider the, commuting, actions on m_λ of an endomorphism and h). More generally, [52, Thm 4] (see 8.2.27), Hom and Ext groups between finite length modules over generalised Weyl algebras with $\sigma(H) = H - 1$ are finite-dimensional over the base field k.

So the results of Section 8.2.5 apply. In particular one obtains the following since the ring $B_\mu = U/\mathrm{ann}_U(M)$ as above satisfies the conditions of 8.2.41.

Corollary 8.2.24. ([406, 4.1]) *Let $M = M(\lambda)$ $(\lambda \in k \setminus \mathbb{N})$ be a Verma module for $\mathrm{sl}_2(k)$ (k algebraically closed of characteristic 0). Then its ring of definable scalars, R_M, is a von Neumann regular ring and the natural map $R \to R_M$ is an epimorphism of rings.*

Corollary 8.2.25. ([406, 5.1]) *Let $M = M(\lambda)$ $(\lambda \in k \setminus \mathbb{N})$ be a Verma module over $U = U(\mathrm{sl}_2(k))$ (k algebraically closed of characteristic 0). Then the Ziegler-closure of the pure-injective hull, $H(M)$, of M consists of $H(M)$ and D_λ, where D_λ is the quotient division ring of $B_\mu = U/\mathrm{ann}_U(M)$. This closed set is homeomorphic to the Ziegler spectrum, Zg_{R_M}, of the ring of definable scalars of M.*

[11] It is the case, [152, 8.4.4], that, as λ varies, these are all the primitive factor rings of U.

Proof. The first statement follows by 8.2.38 and the second follows by 5.5.3 and 3.4.30 since $R \to R_M$ is an epimorphism to a von Neumann regular ring. □

It also follows from 8.2.37 that $H(M(\lambda))/M(\lambda) \simeq D_\lambda^{(\kappa)}$ for some infinite cardinal κ.

Corollary 8.2.26. ([406, 5.2]) *Let* $M = M(\lambda)$ $(\lambda \in k \setminus \mathbb{N})$ *be a Verma module over* $\mathrm{sl}_2(k)$. *Then* $\mathrm{CB}(\mathrm{Zg}_{R_M}) = \mathrm{KG}(R_M) = \mathrm{mdim}(M_R) = 1$.

Proof. That the m-dimension of M is 1 is immediate from the fact, 8.2.28, that every pp-definable subgroup is of finite or cofinite height in the lattice of pp-definable subgroups of M. Therefore, by 5.3.60, or just directly by 8.2.25, Zg_{R_M} has Cantor–Bendixson rank 1. The fact that this equals the Krull–Gabriel dimension of R_M is then a consequence of $R \to R_M$ being an epimorphism, so the class of R_M-modules is the definable closure of M in the category of U-modules. □

8.2.5 The spectrum of the first Weyl algebra and related rings

Generalised Weyl algebras are defined and their properties summarised (8.2.27). Then we restrict to a subclass which contains many important examples. The pp-definable subgroups of any finite length module over one of these algebras are either finite-dimensional or cofinite-dimensional and the m-dimension of such a module is 1 *(8.2.28). Every simple module has non-split extensions with infinitely many simple modules (8.2.29) and one may conclude, for example, that if such a ring is simple and hereditary, then there is no isolated point in the Ziegler spectrum and the category of finite length modules does not have any almost split sequence (8.2.30). In particular, this applies to the first Weyl algebra.*

Over the algebras considered, no infinite-dimensional finite-length module is pure-injective (8.2.35) and the quotient of the hull by the module is a direct sum of copies of the division ring of quotients (8.2.37). Thus the Ziegler-closure of such a module consists of its hull and this ring of quotients (8.2.38). Furthermore, the map from indecomposable finite length modules to their hulls is bijective (8.2.39).

The elementary dual of the hull of a simple module is considered: it is not a module of the same type (8.2.40).

The ring of definable scalars of a finite-length module over a hereditary such algebra is von Neumann regular and is an epimorphic extension of the algebra (8.2.41).

Further results are given on closures of "tubes" (8.2.42) and on the pure-injective hull of the ring (8.2.43). The ring of definable scalars of the injective hull of a simple module and the corresponding flat module are identified (8.2.44).

Let k be a field. The **first Weyl algebra** over k is the algebra $fA_1(k) = k\langle x, y : yx - xy = 1\rangle$. This has a natural realisation as the ring of polynomial differential operators on the coordinate ring, $k[x]$, of the affine line: regard $x \in A_1(k)$ as the operator multiplication by x and y as the operator $\partial/\partial x$ on $k[x]$. If the characteristic of k is 0, then $A_1(k)$ is a simple noetherian hereditary domain. If the characteristic of k is non-zero, then the resulting ring is very different, for example it is a finitely generated module over its centre (for example, [441, 6.6.14]), hence satisfies a polynomial identity and so is covered by the results of Section 8.1.2. In this section it is the Ziegler spectrum of $A_1(k)$, where $\mathrm{char}(k) = 0$, that we investigate.

There are a number of important algebras with properties similar to those of $A_1(k)$. These were given a uniform treatment by Bavula ([52], also see [55]) and are referred to as "generalised Weyl algebras".[12]

In order to define these **generalised Weyl algebras** start with the polynomial ring $k[H]$, where k is a field. Choose an automorphism, σ, of the ring $k[H]$ and a non-constant polynomial $a = a(H) \in k[H]$. From these data define the algebra[13] $k\langle x, y : yx = a,\ xy = \sigma(a),\ by = y\sigma(b),\ bx = x\sigma^{-1}(b),\ \ b \in k[H]\rangle$. For most of the results it is assumed that k is algebraically closed and has characteristic 0. Indeed, many of our results will be for the special case where σ is given, perhaps after a change of variable, by $\sigma(H) = H - 1$; these algebras were considered by Hodges ([295], also see [54]). Generalised Weyl algebras of this form include the first Weyl algebra itself (set $H = yx$ and take $a = H$) and all non-semisimple-artinian primitive factor rings of the universal enveloping algebra of $\mathrm{sl}_2(k)$ (see before 8.2.24).

The action of σ on $k[H]$ induces an action on the maximal ideals of $k[H]$. Since k is algebraically closed, each of these has the form $\langle H - \nu\rangle$ for a unique $\nu \in k$. Thus sigma induces an action on the elements of k. In particular, if $\sigma(H) = H - 1$, then the orbit of $\nu \in k$ under this action is $\nu + \mathbb{Z}$.

The first result summarises properties of these generalised Weyl algebras.

Theorem 8.2.27. *Let R be a generalised Weyl algebra over an algebraically closed field of characteristic 0 and with $\sigma(H) = H - 1$.*

(a) ([52, 2.1]) *R is a noetherian domain with Krull dimension 1. Every proper factor ring of R is finite-dimensional over k.*
(b) ([52, 3.2, Thm 5]) *R is hereditary iff $a(H)$ (in the notation of the definition) has no multiple roots and every orbit contains at most one root of $a(H)$; in*

[12] Here we will consider only those like the first, rather than the nth, $n > 1$, Weyl algebra.
[13] In fact Bavula's definition is more general.

this case R is a **simple ring** (there is no proper, non-zero, two-sided ideal) and every non-zero module is infinite-dimensional and faithful.

(c) ([52, Thm 4 (also Thm 6)], generalising [440, p. 321]) *If M_R is of finite length then for every non-zero $r \in R$, both* $\mathrm{Hom}(R/rR, M)$ *and* $\mathrm{Ext}_R(R/rR, M)$ *are finite-dimensional over k.*

Throughout the remainder of this section we assume that **every generalised Weyl algebra discussed is over an algebraically closed field of characteristic** 0.

Theorem 8.2.28. ([575, 5.12], [522, 3.3]) *Let R be a generalised Weyl algebra with $\sigma(H) = H - 1$ and let M be an R-module of finite length. Then every pp-definable subgroup of M either is finite-dimensional or is co-finite-dimensional in M. If M is infinite-dimensional, then there is no finite bound on the values of dimension which occur in the first case. If R is hereditary and M is not injective, then there is no finite bound on the values of codimension which occur. Hence in either of these cases* $\mathrm{mdim}(M) = 1$.

Every finite-dimensional pp-definable subgroup of M is contained in one of the form $\mathrm{ann}_M(r)$ *and every cofinite-dimensional pp-definable subgroup contains one of the form Mr.*

Proof. Given a pp-definable subgroup, $\phi(M)$, of M, since R is a noetherian domain, there is, by 8.2.1, $r \in R, r \neq 0$ with either $Mr \leq \phi(M)$ or $\phi(M) \leq \mathrm{ann}_M(r)$. Now, as k-vectorspaces, $M/Mr \simeq \mathrm{Ext}_R(R/rR, M)$ and $\mathrm{ann}_M(r) \simeq \mathrm{Hom}(R/rR, M)$, both of which, by 8.2.27, are finite-dimensional. So the first statement follows.

For the second assertion, as M is torsion (and R is Ore), any finite collection of elements from M can be annihilated by a single non-zero element $r \in R$. This shows that arbitrarily large finite values of dimension occur.

The third statement follows from 8.2.7 and the last part from 8.2.1. □

In order to prove some basic results about the Ziegler spectrum of the first Weyl algebra it was realised that the following result was needed: Bavula provided a proof, [53], (which is also given in [521, 4.1]). His result is somewhat more general than what is stated here.

Theorem 8.2.29. ([53, 1.1]) *Let R be a generalised Weyl algebra with $\sigma(H) = H - 1$. Then for every infinite-dimensional simple R-module M there are infinitely many simple modules S such that* $\mathrm{Ext}(S, M) \neq 0$ *and there are infinitely many simple modules T such that* $\mathrm{Ext}(M, T) \neq 0$.

Corollary 8.2.30. ([521, 4.2, 4.3]) *Let R be a generalised Weyl algebra with $\sigma(H) = H - 1$.*

(a) *No infinite-dimensional simple module is the source of a left almost split morphism (see Section 5.3.3) in the category of finite length R-modules.*
(b) *If R is a simple ring, then no simple module is the source of a left almost split morphism in the category of finite length R-modules.*
(c) *If R is a hereditary, hence (8.2.27(2)) simple, ring, then no finitely generated R-module is injective.*
(d) *If R is hereditary, then there is no isolated point in Zg_R. In particular the category of finite length R-modules does not have any almost split sequence.*

Proof. By 8.2.29, for every infinite-dimensional simple module, M, there are infinitely many simple modules S such that $\mathrm{Ext}(S, M) \neq 0$. By 5.3.31 and the proof of 8.2.12 the first, hence also the second, statement follows.

For the third, suppose M were a finitely generated injective R-module. Since R is noetherian M would have a maximal proper submodule, hence a simple factor module, N, which is injective (because R is hereditary). This is in contradiction to 8.2.29.

For the final part we can, in view of 8.2.27, apply 8.2.14, then 5.3.31. □

In particular, the Ziegler spectrum over the first Weyl algebra over an algebraically closed field of characteristic 0 has no isolated points, though the set of pure-injective hulls of finite length points is dense (8.2.9).

In fact, the details of 8.2.29 (see [522, 3.1]) give the following (cf. part (c) above).

Corollary 8.2.31. *Let R be a generalised Weyl algebra with $\sigma(H) = H - 1$. Then no finite length module, apart from 0, is divisible.*

Lemma 8.2.32. *Suppose that R is a generalised Weyl algebra with $\sigma(H) = H - 1$. Let M be an R-module of finite length. Then $\bigcap\{Mr : r \in R, r \neq 0\} = 0$.*

Proof. By 8.2.10, $V = \bigcap\{Mr : r \in R, r \neq 0\}$ is a submodule of M. If V is nonzero, then $M = V$ is divisible, contradicting 8.2.31. □

Remark 8.2.33. If R as in 8.2.32 is hereditary, hence (see 2.4.15) RD, then, by 8.2.11, the conclusion of that result holds for any module M of finite length.

Denote by $B_1(k)$ the localisation of $A_1(k)$ with respect to the Ore set $k[H] \setminus \{0\}$. Then (see [441, 1.3.9]) $B_1(k)$ is a simple left and right principal ideal domain and, by 5.5.3, the Ziegler spectrum of $B_1(k)$ is a closed subset of that of $A_1(k)$.

Corollary 8.2.34. ([521, 4.5]) *Let k be a field of characteristic 0. Then there is no isolated point in the Ziegler spectrum of the first Weyl algebra, $A_1(k)$, nor in that of its localisation $B_1(k)$.*

Proof. Since $A_1(k)$ is a simple hereditary generalised Weyl algebra, 8.2.30 applies. Also, B_1 is a Dedekind domain with no injective simple module by [440, Cor. 4.2]. By [440, 4.1, 4.3] the other hypotheses of 8.2.14 hold, which yields the conclusion. □

Corollary 8.2.35. ([522, 3.2]) *If R is a generalised Weyl algebra with $\sigma(H) = H - 1$ and M is an R-module of finite length which is infinite-dimensional over the base field, then M is not pure-injective.*

Proof. By 8.2.7, M would have to be injective if it were pure-injective but then that would contradict 8.2.30(c). □

Taking $\sigma(H) = H - 1$ and $a(H) = H(H - 1)$ gives a non-hereditary generalised Weyl algebra (which arises from the ring of differential operators on the projective line, see [137, 2.7]) for which ([521, 5.5]) no simple module is the domain of a left almost split morphism in the category of finite length modules, cf. 8.2.30(d). This algebra has global dimension 2, is a factor ring of the universal enveloping algebra $U\mathrm{sl}_2(k)$ and can be regarded as the ring of differential operators on $\mathbb{P}^1(k)$ (see [137]).

Question 8.2.36. ([521, 5.6]) Suppose that R is a generalised Weyl algebra with $\sigma(H) = H - 1$. Is it true, without the extra assumptions of 8.2.30(d) that there are no almost split sequences in the category of finite length R-modules? Is it true that there are no isolated points in Zg_R?

The question about almost split sequences has been answered positively by Puninski, [561, Thm 1], in the case that each orbit of the induced action on k is infinite and no orbit contains more than three distinct roots of $a(H)$. In consequence, no simple module over $\mathrm{sl}_2(k)$ (k algebraically closed of characteristic 0) begins a left almost split sequence in the category of finite length modules over $U(\mathrm{sl}_2(k))$ ([561, p. 612]). If one could replace the category of finite-length modules by mod-$U\mathrm{sl}_2(k)$, then it would follow that for no indecomposable finite-length module M over $U = U(\mathrm{sl}_2(k))$ (k algebraically closed of characteristic 0) is its pure-injective hull $H(M)$ isolated in Zg_R, but whether or not this is true seems to be open.

Now we take a single module of finite length and consider its pure-injective hull and its Ziegler-closure.

Proposition 8.2.37. ([522, 3.5]) *Suppose that R is a generalised Weyl algebra with $\sigma(H) = H - 1$. Let M be an R-module of finite length which is infinite-dimensional over the base field. Then M is the torsion submodule of its (indecomposable by 4.3.59) pure-injective hull, $H(M)$, and $H(M)/M \simeq D^{(I)}$ for some infinite set I, where D is the division ring of fractions of R.*

Proof. If M is finite-dimensional over k, hence pure-injective (4.2.6), there is nothing to prove so suppose that M is infinite-dimensional. Given $r \in R$, $r \neq 0$ apply 4.3.26 to the condition $xr = 0$ to obtain, in view of 8.2.28, the first assertion, hence that $H(M)/M$ is torsionfree. That $H(M)/M$ is divisible follows similarly on considering the pair of conditions $x = x \,/\, r|x$.

For the fact that I is infinite: any countable-dimensional pure-injective is Σ-pure-injective (4.4.10) which, by 8.2.28 and 4.4.5, M, hence $H(M)$ (4.3.24), is not. □

Proposition 8.2.38. ([522, 3.7]) *Suppose that R is a generalised Weyl algebra with $\sigma(H) = H - 1$. Let M be an indecomposable R-module of finite length which is infinite-dimensional over the base field. Then the Ziegler-closure of $H(M)$ is $\{H(M), D\}$, where D denotes the division ring of quotients of R.*

Proof. By the description, 8.2.28, of the pp-definable subgroups of M clearly $\mathrm{mdim}(M) = 1$, so, by 5.3.17, M satisfies the isolation condition. The module $H(M)$ is isolated in its closure, X say, by the open set $\big((xr = 0)/(x = 0)\big)$, where r is any element of R such that $\mathrm{ann}_M(r)$ is non-zero, and there is no other isolated point in X. Once $H(M)$ is removed from X the lattice of pp conditions for what remains is described by 5.3.57: by 8.2.28 this is a two-point lattice so, by 5.3.2, $X \setminus H(M)$ has just one point, necessarily D. □

By 4.3.55 we have the following.

Corollary 8.2.39. *Suppose that R is a generalised Weyl algebra with $\sigma(H) = H - 1$. The map $M \mapsto H(M)$ from the set of (isomorphism classes of) indecomposable finite length R-modules to Zg_R is an injection.*

If M is an indecomposable R-module of finite length, then since, by 8.2.28, $\mathrm{mdim}(M) \leq \infty$, it follows by 5.4.14 that $H(M)$ is reflexive, that is, has an elementary dual, $D(H(M)) \in {}_R\mathrm{Zg}$.

Proposition 8.2.40. ([522, 3.9]) *Let R be a generalised Weyl algebra with $\sigma(H) = H - 1$ and let M be a simple infinite-dimensional R-module. Then $D(H(M))$ is an indecomposable pure-injective left R-module with infinitely generated socle. If R is hereditary, then $D(H(M))$ is the pure-injective hull of its torsion submodule.*

In particular $D(H(M))$ is not simply the pure-injective hull of the holonomic dual (for this see below) of M.

Proposition 8.2.41. *Suppose that R is a generalised Weyl algebra with $\sigma(H) = H - 1$. Let M be a simple R-module. Then the ring, R_M, of definable scalars of M is von Neumann regular and the natural map $R \to R_M$ is an epimorphism. If R is hereditary, then this is true for any R-module of finite length.*

Proof. We apply 6.1.16. The finite/cofinite hypothesis is satisfied, by 8.2.28. The intersection condition is by 8.2.32 and 8.2.28 and, in the hereditary case, the comment after that. □

Continuing with R as above we describe some further results from [522]. Let $\mathcal{O}$ be a non-degenerate orbit (that is, one which contains no root of $a(H)$, in the notation of the definition) of the action described before 8.2.27. Given $\lambda, \mu \in \mathcal{O}$ one has $R/(H-\lambda)R \simeq R/(H-\mu)R$ and this module is simple ([55, §4]), let us denote it by $S_{\mathcal{O}}$. It may be checked that $\mathrm{Ext}(S_{\mathcal{O}}, S_{\mathcal{O}})$ is one-dimensional over k and the category of finite-length modules with all simple factors isomorphic to $S_{\mathcal{O}}$ is just like that corresponding to a homogeneous tube in the sense of Section 15.1.3: there is, up to isomorphism, a single indecomposable module of each finite length; each of these is uniserial; there are the obvious almost split sequences (see [522, §4]). Not surprisingly, there are modules corresponding to the analogues of adic and Prüfer modules seen in Section 8.1.2, but there are differences.

Certainly the direct limit of the obvious inclusions, between the indecomposable modules of finite length built from $S_{\mathcal{O}}$, is an artinian uniserial module but it is not pure-injective [522, 5.2]. Nevertheless, the pure-injective hull of this "Prüfer" module is indecomposable, [522, 5.3], and is in the Ziegler-closure of these finite length indecomposables.

Dually, the inverse limit of the obvious surjections is not pure-injective, [522, 6.2]. At least, if the ring is hereditary, then this "adic" module is in the Ziegler-closure of the finite length indecomposable modules built from $S_{\mathcal{O}}$, [522, 6.3]. The pure-injective hull of this "adic" module is far from indecomposable: it has as a direct summand a copy of every indecomposable flat pure-injective module (see below for these) over B, the localisation of R at the non-zero polynomials in $k[H]$ ([522, 7.3]), plus the elementary dual of the injective hull of the corresponding simple left R-module.

Theorem 8.2.42. ([522, 7.1, 7.2]) *Let R be a hereditary generalised Weyl algebra with $\sigma(H) = H - 1$ and let $\mathcal{O}$ be a non-degenerate orbit of the action induced by σ on k. Let $S = R/(H-\lambda)R$ (any $\lambda \in \mathcal{O}$) be the corresponding simple module. Let S_n denote the uniserial module of length n which has all composition factors isomorphic to S. Then the Ziegler-closure of the set, $\{S_n\}_n$ of these is as follows.*

(a) The isolated points are the S_n ($n \geq 1$).
(b) The points of Cantor–Bendixson rank 1 are: the pure-injective hull of the Prüfer module $\varinjlim_n S_n$; the flat module which is the elementary dual of the injective hull, $E(R/R(H-\lambda))$, of the corresponding simple left R-module;

every indecomposable flat pure-injective module over the localisation, B, of R at the non-zero elements of $k[H]$.

(c) There is one point of rank 2, namely the quotient division ring D of R.

Furthermore, the m-dimension of the lattice of pp conditions for the corresponding definable subcategory is 2.

Finally in this section, let R be a hereditary generalised Weyl algebra with $\sigma(H) = H - 1$. Since R is noetherian the classes of injective (= divisible) right modules and of flat left modules are definable (3.4.24) and (7.2.6) have m-dimension. Denote the corresponding closed subsets of Zg_R by inj_R and, say, X_{flat}.

The indecomposable injective modules are D, the quotient division ring of R, and the injective hulls of the simple R-modules. If $S \simeq R/I$ is simple, then $E(S)$ is isolated in inj_R by the open set $\big((xI = 0)\,/\,(x = 0)\big)$. Thus inj_R has infinitely many isolated points and just one non-isolated point, D. Therefore, by 5.3.60, the definable subcategory of injective modules has m-dimension 1. It follows by 7.2.7 that the dual definable subcategory of flat modules also has m-dimension 1, and (5.4.19) D is the only non-isolated point of X_{flat}.

The indecomposable pure-injective flat module which is dual to $E(S)$, where $S = R/I$, is the unique such module, denote it N_S, which satisfies $\mathrm{Ext}(S^*, N) \neq 0$ [522, §8]. Here S^* is the **holonomic dual**, $S^* = \mathrm{Ext}(S, R)$ (a left R-module via the left R-module structure on R). For each simple module S let R_S denote the (flat epimorphic) localisation of R obtained by making every other simple module torsion: then R_S is maximal in the collection of rings between R and its full quotient ring D (see [241] for these localisations).

Proposition 8.2.43. ([522, 8.1, 8.2]) *Suppose that R is a hereditary generalised Weyl algebra with $\sigma(H) = H - 1$. Then $H(R) = H\big(\bigoplus_S N_S^{(\kappa_S)}\big)$, where S runs over the simple right R-modules and $\kappa_S \geq 1$ for every S.*

The pure-injective hull, $H(R_S)$, of the localisation of R at S is the pure-injective hull of a direct sum of copies of N_S.

The values of the κ_S were not determined in [522]. (In particular are they all equal to 1?) The next result is a special case of 8.2.2.

Proposition 8.2.44. ([522, 8.3]) *Suppose that R is a hereditary generalised Weyl algebra with $\sigma(H) = H - 1$. Let S be a simple right R-module. Then the ring of definable scalars of $E(S)$ is R_S. This is also the ring of definable scalars of the left module N_S and this is the subring $\{tr^{-1} : r, t \in R$ and $\mathrm{ann}_S(r) \leq \mathrm{ann}_S(t)\}$ of D.*

We remark that by [285] (see towards the end of Section 4.2.3) $H(R)$ and all the $H(R_S)$ carry natural ring structures which extend the ring structures on R, respectively on the R_S.

Finally, there are results in [521, §§6, 7] (cf. 7.3.34) on wildness, existence of superdecomposable pure-injectives and undecidability (interpretability of the word problem for groups) over various of these algebras.

8.2.6 Spectra of V-rings and differential polynomial rings

The Ziegler spectrum of a right and left principal ideal domain which is also a V-ring is described in terms of the simple modules; it has rank 1 *(8.2.45).*

The spectrum of a type of twisted polynomial ring is rather similar to that of a tame hereditary algebra; its rank is 2 *(8.2.47).*

Let K be a field and let d be a **derivation** on K: that is, $d : K \to K$ is an additive map which satisfies $d(ab) = a \cdot db + da \cdot b$. Let R be the corresponding differential polynomial ring: this is the ring of polynomials, with (non-central) coefficients from K, in an indeterminate X, and relations $aX = X \cdot da$ $(a \in K)$. By varying K and d one obtains a variety of interesting examples.

For instance, suppose that (K, d) is a universal field with derivation (see, for example, [186, p. 361]). Then (see [186, 7.42]) R is a right and left principal ideal domain and is also a **V-ring**: a ring over which all simple modules are injective. Indeed this ring has a unique simple R-module, $S = R/\langle X \rangle$, so, by the next result, the Ziegler spectrum of R consists of just three points: the injective module S which, since R is noetherian is actually Σ-pure-injective (4.4.17); the indecomposable pure-injective flat (by 3.4.24) module N_S such that $H(R) \simeq H(N_S^{(I)})$ for some I; the quotient division ring of R. Also by the next result, the first two points are isolated and $\mathrm{KG}(R) = \mathrm{CB}(\mathrm{Zg}_R) = 1$. (For these various dimensions, see Sections 5.3.6, 7.2, 13.2.1.)

Proposition 8.2.45. ([521, 5.2]) *Suppose that R is a right and left principal ideal domain which is a V-ring but not a division ring. Then* $\mathrm{CB}(\mathrm{Zg}_R) = 1$, *indeed* $\mathrm{KG}(R) = \mathrm{mdim}(\mathrm{pp}_R) = 1$. *The isolated points of* Zg_R *are the simple modules and the indecomposable direct summands of the pure-injective hull, $H(R)$, of R (these being the duals of the simple left modules). The only non-isolated point is the injective hull (= quotient division ring), $E(R)$, of R.*

Proof. If $M = R/rR$ is a simple module, then its hull is, by 5.3.9, isolated by the minimal pair $(xr = 0)/(x = 0)$. Since R/rR is simple, if $r = st$, then either s or t is a unit (that is, r is irreducible), so the left module R/Rr also is simple, and is isolated by $(rx = 0)/(x = 0)$. Since this is a minimal pair, so, by 1.3.1, is the dual pair, $(x = x)/(r|x)$, for right modules. By 5.3.2 this pair is open in a unique indecomposable pure-injective module. But this pair is open on R, hence (4.3.23) it is open on the pure-injective hull of R. By 7.3.3 the decomposition (4.4.2) of the pure-injective hull of R has an indecomposable summand on which this pair is open. Therefore the point of Zg_R which is isolated by $(x = x)/(r|x)$ is a direct summand of the pure-injective hull of R; in particular, it is torsionfree.[14]

Let $N \in \mathrm{Zg}_R$. If N contains a torsion element, say $a \in N$, $a \neq 0$ with $ar = 0$ for some non-zero $r \in R$, then aR is of finite length so contains a simple module. But every simple R-module is injective so N must be simple. Suppose, on the other hand, that N is torsionfree and is not divisible, say $a \in N \setminus Nr$ for some $r \in R$. The element r may be taken to be such that R/rR is simple: for, in a (one-sided) principal ideal ring, every element r is a product, $s_1 \dots s_n$, of irreducible elements (for example, [130, 1.3.5]) so there is some i such that $N > Ns_i$. Then the pair $(x = x)\,/\,(r|x)$ is open on N so, as seen above, N is a direct summand of $H(R)$ and is isolated.

The remaining case is that N is torsionfree divisible, hence injective, so $N \simeq E(R)$ (which is the module structure on the quotient division ring of R).

It remains to show that $E(R)$ is not an isolated point. So suppose, for a contradiction, that $M = E(R)$ is isolated by a pair, ϕ/ψ say. Since R is a hereditary noetherian domain, it is an RD ring (2.4.15), so, by 5.2.8, it may be supposed that ϕ has the form $\exists y\,(x = yr \wedge ys = 0)$ and that ψ has the form $\phi \wedge u|xt$ for some $r, s, t, u \in R$. Since M is divisible the condition $u|xt$ would be automatically satisfied if u were non-zero, so, since $\phi(M) \neq \psi(M)$, ψ must be the condition $\phi \wedge xt = 0$. Also, because M is torsionfree, $s = 0$. Therefore ϕ is $r|x$ and ψ is $r|x \wedge xt = 0$. So the pair ϕ/ψ opens in R, hence opens in one of the indecomposable summands of $H(R)$. Therefore it will be enough to show that no direct summand of $H(R)$ is injective.

Suppose that $H(R) = M \oplus D^{(\kappa)}$, where D is the quotient division ring of R and D is not a direct summand of M. Set $1 = (a, d)$ accordingly. As already noted, R is RD, so, since R_R, hence $H(R)$, is torsionfree, the pp-type of any element of $H(R)$ is determined by the divisibility conditions that it satisfies. Therefore $\mathrm{pp}^{H(R)}(1) = \mathrm{pp}^M(a)$. By uniqueness of hulls of elements (Section 4.3.5), and since $H(R) = H(1)$ (for 1 generates R), it follows that $H(R) \simeq M$, so this has no direct summand isomorphic to D.

[14] In fact, because R, being left noetherian, has left Krull dimension, the m-dimension of $\mathrm{pp}(R)$ is defined. Therefore (7.3.5) the pure-injective hull of R is the pure-injective hull of a direct sum of indecomposable modules. Indeed, by the end of the proof it will have been shown that the m-dimension of pp_R is defined, so, by 7.3.5, there is no superdecomposable pure-injective R-module.

Finally, note that all three points of the spectrum have m-dimension; so, therefore, has their direct sum. Therefore the lattice pp_R has m-dimension. By 5.3.60, $KG(R) = \mathrm{mdim}(\mathrm{pp}_R) = \mathrm{CB}(\mathrm{Zg}_R) = 1$. □

In particular, the example described before this result is a ring with finite, non-discrete Ziegler spectrum.

Example 8.2.46. ([521, 5.4]) Let k be a field of characteristic 0 and let $K = k((X))$ be the field of Laurent series, equipped with the derivation d/dX. Let $R = K[Y, d/dX]$ be the corresponding differential polynomial ring. If M is an R-module of finite length its pure-injective hull is isolated in Zg_R by a minimal pair. For, by [733, §2] (also see [734]) the category of R-modules of finite length has almost split sequences and so 5.3.31 applies.

Puninski [563] also investigated the Ziegler spectrum of, and the finite length modules over, rings of the form $R = k[X, \delta]$, where k is a field of characteristic 0, δ is a derivation on k whose **field of constants** (those $a \in k$ with $\delta a = 0$) is algebraically closed and where it is assumed that the category of finite-length modules has almost split sequences (for conditions giving this see [734]). This ring is a noetherian principal ideal domain and 5.3.31 and 8.2.9 apply to give, [563, 3.2], that the isolated points of Zg_R are exactly the pure-injective hulls of the indecomposable finite-length modules. The general shape of the spectrum turns out to be similar to that over a Dedekind domain (Section 5.2.1) (or over a tame hereditary finite-dimensional algebra, Section 8.1.2). In particular the Cantor–Bendixson rank is 2.

Theorem 8.2.47. ([563, 5.7, 5.8]) *Let k be a field of characteristic 0, let δ be a derivation on k with algebraically closed field of constants. Suppose that the category of finite length modules over the differential polynomial ring $R = k[X, \delta]$ has almost split sequences. Then the Cantor–Bendixson rank of Zg_R is 2.*

The points of rank 0 are the the pure-injective hulls of finite length indecomposable modules.

The points of rank 1 are the Prüfer and adic-type modules arising from tubes of finite length modules (cf. Section 15.1.3). The Prüfer modules all are injective and the adic modules all arise as as direct summands of the pure-injective hull of R_R.

The single point of rank 2 is the quotient division ring of R.

Puninski also conjectured, [563, 5.9], for rings $R = k[X, \delta]$ where k is a field of characteristic 0, that the Cantor–Bendixson rank of Zg_R is 1 (cf. 5.2.11), 2 (cf. 8.2.47), or ∞ (that is, R is wild).

Question 8.2.48. What are the possible values of Cantor–Bendixson rank of Zg_R (and m-dimension) for R a differential polynomial ring over a field of characteristic 0?

In [562], which uses the results of [563], Puninsky is concerned with modules over $\mathcal{D}_n = \mathcal{O}_n[\partial/\partial x_1, \ldots, \partial/\partial x_n]$, the ring of differential operators in n variables over $\mathcal{O}_n = k[[x_1, \ldots, x_n]]$, the ring of formal power series in n variables over k, an algebraically closed field of characteristic 0.

Let P be a prime ideal of $\mathcal{O}_n$ of coheight 1, so $\mathcal{O}_n/P$ is the coordinate ring of a curve. It is shown that every indecomposable $\mathcal{D}_n$-module of finite length and with support on P is uniserial and its composition factors are either all isomorphic or they alternate between two simples [562, §5]; that is, they are organised in tubes which are homogeneous or of rank 2, (see Sections 15.1.1, 8.1.2). In particular, the Cantor–Bendixson rank of the Ziegler spectrum is 2; in particular (7.3.5) there are no superdecomposable pure-injectives. The detailed statements, [562, 4.2] for $\mathcal{D}_1$ and, more generally, [562, 5.3], for modules supported at a prime as above, are similar to 8.2.47 above.

8.2.7 Spectra of serial rings

Indecomposable pure-injectives over serial rings are "locally uniserial" over their endomorphism rings (8.2.50). A simple criterion for irreducibility of pp-types living at an idempotent is given (8.2.51). All irreducible pp-types arise from specifying a suitable set of divisibility and annihilation conditions (8.2.53) and sets of conditions which give rise to the same point of the spectrum are characterised (8.2.55). There is a simplified form of (local and global) bases of open sets (8.2.57).

Every serial ring satisfies the isolation condition (8.2.58).

A serial ring has Krull–Gabriel dimension iff it has Krull dimension as a module over itself (8.2.60), in particular there are no topologically indistinguishable points in its Ziegler spectrum (8.2.61).

The model theory of modules over serial rings was investigated by Eklof and Herzog [173] and by Puninski [557]. In both these papers a particularly nice basis of the Ziegler topology was found and general characterisations of indecomposable pure-injectives in terms of the right and left ideals of R were given. Subsequently Puninski, [558], [560], in the commutative case, and Reynders, [593], and Puninski, [570], in the general case, have investigated m-dimension(= Krull–Gabriel dimension) and Cantor–Bendixson rank over serial rings. Then there are further results of Puninski for uniserial rings and valuation rings in particular; these are reported in the next section.

Examples 8.2.49. These examples, which illustrate some of the results in this section, are taken from [592, §4.1], which builds on [558, §3]. For more information and detail see those references.

(a) Let $R = \begin{pmatrix} \mathbb{Z}_{(p)} & \mathbb{Q} \\ 0 & \mathbb{Q} \end{pmatrix}$. Then R is right but not left noetherian. The complete set of orthogonal primitive idempotents is $e = \begin{pmatrix} 1 & 0 \\ 0 & 0 \end{pmatrix}$, $f = \begin{pmatrix} 0 & 0 \\ 0 & 1 \end{pmatrix}$. The chain of right ideals contained in eR is

$$0 < I_q < \cdots < I_n < I_{n-1} < \cdots < I_2 < I_1 < eR,$$

where $I_q = \begin{pmatrix} 0 & \mathbb{Q} \\ 0 & 0 \end{pmatrix}$ and $I_n = \begin{pmatrix} p^n\mathbb{Z}_{(p)} & \mathbb{Q} \\ 0 & 0 \end{pmatrix}$. The chain of left ideals contained in Re is

$$0 < \cdots < J_m < J_{m-1} < \cdots < J_2 < J_1 < Re,$$

where $J_m = \begin{pmatrix} p^m\mathbb{Z}_{(p)} & 0 \\ 0 & 0 \end{pmatrix}$. The only right ideal properly contained in fR is the zero ideal. On the other hand, the left ideals contained in Rf are as follows:

$$0 < \cdots < L_n < L_{n-1} < \cdots < L_1 < L_0 < L_{-1} < \cdots < L_{-n} < \cdots < L_q < Rf,$$

where $L_q = \begin{pmatrix} 0 & \mathbb{Q} \\ 0 & 0 \end{pmatrix}$ and $L_t = \begin{pmatrix} 0 & p^t\mathbb{Z}_{(p)} \\ 0 & 0 \end{pmatrix}$ for $t \in \mathbb{Z}$.

(b) Another example, which is right and left noetherian and which is also used as an illustrative example of an HNP ring (in Sections 8.2.2, 14.3.3), is $R = \begin{pmatrix} \mathbb{Z}_{(p)} & p\mathbb{Z}_{(p)} \\ \mathbb{Z}_{(p)} & \mathbb{Z}_{(p)} \end{pmatrix}$.

Recall (Section 2.4.3) that if e is a primitive idempotent of R, then a pp condition $\phi(x)$ is said to be an e-condition if for every module M_R, $\phi(M) \leq Me$. This terminology is extended to pp-types by saying that a pp-type p is an e-**type** if $p(M) \leq Me$ for every module M, that is, iff $e|x \in p$.

Theorem 8.2.50. ([173, pp. 153–4], also see [564, 11.4]) *Suppose that R is a serial ring. Let $N \in \mathrm{Zg}_R$ and let e be a primitive idempotent of R such that $Ne \neq 0$. Then the lattice of pp-definable subgroups of N which are contained in Ne is a chain, indeed Ne is uniserial as an* End(N)*-module.*

Proof. By 2.4.23 the lattice of all e-conditions is distributive and so, by 3.2.3, the lattice of pp-definable subgroups of N contained in Ne is distributive. Consider Ne as a module over $S = \mathrm{End}(N)$, a local ring by 4.3.43. By a result of Stephenson, [664, 1.6], proving that $Sa \cap (Sb + Sc) = (Sa \cap Sb) + (Sa \cap Sc)$ for all $a, b, c \in$

Ne will be enough to show that the lattice of all submodules of ${}_SNe$ is distributive and hence, since S is local, that it is uniserial.

Clearly $Sa \cap (Sb + Sc) \supseteq (Sa \cap Sb) + (Sa \cap Sc)$.

For the converse, let $m \in Sa \cap (Sb + Sc)$. Let p, q, r be respectively the pp-types of a, b, c in N, so, by 4.3.10, $Sa = p(N)(= p(Ne))$, $Sb = q(N)$, $Sc = r(N)$. Let $\Phi(x, y)$ be the set of pp conditions with parameter m which is the union of $q(x) \cup p(x)$, $r(y) \cup p(y)$ and $m = x + y$: we claim that this is finitely satisfied (Section 4.1) in N. Since pp-types are closed under $\wedge$, it is enough to show that $\psi(x)$, $\phi_1(x)$, $\rho(y)$, $\phi_2(y)$, $m = x + y$, where $\phi_1, \phi_2 \in p$, $\psi \in q$, $\rho \in r$, is satisfied in Ne. We can replace each ϕ_i by $\phi = \phi_1 \wedge \phi_2$. By assumption, $m \in \phi(N)$ and $m \in \psi(N) + \rho(N)$. By distributivity $\phi(N) \cap (\psi(N) + \rho(N)) = \phi(N) \cap \psi(N) + \phi(N) + \rho(N)$, so there are elements $b', c' \in Ne$ with $m = b' + c'$ and $b' \in \phi(N) \cap \psi(N)$, $c' \in \phi(N) \cap \rho(N)$, as desired.

Thus Φ is finitely satisfied in N hence, 4.2.3, Φ is satisfied in N: there are $m' \in p(N) \cap q(N)$ and $m'' \in p(N) \cap r(N)$ such that $m = m' + m''$, as required. □

Ziegler's criterion, 4.3.49, for irreducible pp-types takes the following equivalent form for e-types over serial rings. We prove just one direction and, in general, refer to the works cited for proofs of most of the results stated in this and the next section.

Proposition 8.2.51. ([173, p. 162], also see [564, 11.6]) *Suppose that R is serial, that e is an primitive idempotent and p is an e-type. Then p is irreducible iff for all $r \in eR$ and $s \in Re$ if $sr \mid xr \in p$ then either $s \mid x \in p$ or $xr = 0 \in p$.*

Proof. We prove just the direction ($\Rightarrow$). So suppose that p is an irreducible e-type, say $p = \mathrm{pp}^N(a)$ with $a = ae \in N$ and N indecomposable pure-injective. If $sr \mid xr \in p$, then $ar = bsr$ for some $b \in N$, so $a = (a - bs) + bs \in \phi(N) + \psi(N)$, where $\phi(y)$ is $yr = 0$ and $\psi(z)$ is $s \mid z$. Since ${}_SNe$ $(S = \mathrm{End}(N))$ is uniserial, either $\phi(N) \leq \psi(N)$ or $\psi(N) \leq \phi(N)$, so either $a \in \phi(N)$ or $a \in \psi(N)$, that is, $ar = 0$ or $s \mid a$, as required. □

Corollary 8.2.52. ([557, Prop. 2], also see [564, 11,7]) *A pp-type p over a uniserial ring R is irreducible iff for all $r, s \in R$ if $sr \mid xr \in p$, then either $s \mid x \in p$ or $xr = 0 \in p$.*

Let M be any R-module, let e be a primitive idempotent of R and let $a \in Me$. Set $p = \mathrm{pp}^M(a)$. The pp-type p contains the information of which ring elements annihilate a and which divide a. Since R is RD, that is, by 2.4.10, quite close to complete information about p. Set $I_a = \{r \in eR : ar = 0\} = \{r \in eR : xr = 0 \in p\}$: this determines the annihilator, $\mathrm{ann}_R(a)$, which is just $(1 - e)R \oplus I_a$. Also set $J_a = \{s \in Re : s \nmid a\} = \{s \in Re : \text{“}s \mid x\text{''} \notin p\}$. If M is indecomposable pure-injective, then, by 8.2.50, the groups of the form Mse are linearly ordered so J_a is a left ideal of R. In this case it turns out, see below, that the pair I_a, J_a determines the pp-type, p, of a.

For J a left ideal contained in Re set $J^* = Re \setminus J$.

Proposition 8.2.53. ([173, 2.7], [557], also see [564, 11.8]) *Suppose that R is serial and that e is a primitive idempotent of R. Let $I \leq eR$ and $J \leq Re$. Suppose that there is at least one pp-type containing the set*

$$\{s \mid x : s \in Re \setminus J\} \cup \{xr = 0 : r \in I\}$$

of pp conditions and not containing any pp condition in the set

$$\{xt = 0 : t \in eR \setminus I\} \cup \{u|x : u \in J\}.$$

Then there is a exactly one such pp-type, denote it $p_{J^*/I}$, and it is an irreducible e-type. If $N \in \mathrm{Zg}_R$ and $a \in N$ has pp-type $p_{J^*/I}$, then $I_a = I$ and $J_a = J$. Every irreducible e-type has the form $p_{J^*/I}$ for some $I \leq eR$, $J \leq Re$.

Remark 8.2.54. If there is a pp-type as in 8.2.53, then say that the pair (I, J) is **admissible**. Necessary and sufficient conditions on I and J for admissibility are given in [173, 2.1]. An equivalent condition was given by Puninski, see [564, 11.9]: for all $r \in I, t \in eR \setminus I, u \in J, s \in Re \setminus J$ one has $st \neq ut + sr$.

This condition implies ([564, 11.10]) that if $t \in eR \setminus I$ and $s \in Re \setminus J$, then $st \neq 0$. Puninski showed ([558], see [564, 11.13, 2.13]) that this is equivalent to the above condition iff the endomorphism ring of every indecomposable finitely presented module is local, that is, iff R is a Krull–Schmidt ring (Section 4.3.8). That is the case, see [564, 1.26], iff R satisfies the condition that whenever $r \in e_i Re_j$ where e_i and e_j are primitive idempotents of R, then either $e_i Rr \leq rR$ or $rRe_j \leq Rr$.

Thus all indecomposable pure-injectives can be obtained from pairs J^*/I satisfying conditions as above. The condition for two such pairs to give the same point of the Ziegler spectrum also has been determined.

Proposition 8.2.55. ([173, 2.10], [557], also see [564, 11.11]) *Suppose that J_1^*/I_1 and J_2^*/I_2 are respectively e_1- and e_2-pairs satisfying the conditions of 8.2.53. Then $H(p_{J_1^*/I_1}) \simeq H(p_{J_2^*/I_2})$ iff either there exists $r_{12} \in e_1Re_2, r_{12} \neq 0$, such that $I_1 = r_{12}I_2$ and $J_1 r_{12} = J_2$ or there is $r_{21} \in e_2Re_1, r_{21} \neq 0$, such that $r_{21}I_1 = I_2$ and $J_1 = J_2 r_{21}$.*

Therefore, if R is a serial ring, then the points of Zg_R correspond to admissible pairs J^*/I modulo the equivalence relation of 8.2.55.

Example 8.2.56. Referring to Example 8.2.49(a), the equivalence classes of sets of pairs are as follows. Given by each is the Cantor–Bendixson rank of the corresponding point(s); these are computed using the description of neighbourhood bases given in 8.2.57. Details of this and also for Example 8.2.49(b) can be found in [592, §4.1].

$\{(0,0)\}$ (rank 1);
$\{(0, J_m) : m \geq 1\} \cup \{(0, L_t) : t \in \mathbb{Z}\}$ (rank 0);
$\{(I_n, 0) : n \geq 1\}$ (rank 1);
$\{(I_n, J_m) : n, m \geq 1, n+m = k\}$, one for each $k > 1$ (rank 0);
$\{(I_q, J_m) : m \geq 1\}$ (rank 1);
$\{(I_q, 0)\}$ (rank 2).

Proposition 8.2.57. ([593, 2.1, 2.2]) *Suppose that R is a serial ring and let $N \in \mathrm{Zg}_R$. Suppose that e is a primitive idempotent such that $Ne \neq 0$ and let $a \in Ne$. Then a neighbourhood basis of N is given by the open sets of the form: $\big((xr = 0) \wedge (s|x) \ / \ (xt = 0) + (u|x)\big)$ with $r \in I_a$, $s \in J_a^*$, $t \in I^*$, $u \in J$.*

Therefore the open sets of the form $\big((xr = 0) \wedge (s|x) \ / \ (xt = 0) + (u|x)\big)$, with e a primitive idempotent, $r, t \in eR$ and $s, u \in Re$, form a basis of open sets of Zg_R.

Reynders used this to give conditions for a point of a closed subset, X, of Zg_R to be isolated and also for such a point to be isolated by an X-minimal pair. He used these to show, [593, 3.3], that if R is a serial ring with Krull dimension, then R satisfies the isolation condition (§5.3.2). That was proved without any restriction by Puninski.

Theorem 8.2.58. ([570, 3.2], [593, 3.3] if R has Krull dimension) *If R is serial, then R satisfies the isolation condition.*

Proof. Let $N \in \mathrm{Zg}_R$ and choose $a \in N$, $a \neq 0$ and an idecomposable idempotent e such that $ae \neq 0$. By 8.2.50 the interval $[e|x, \ x = 0]$ in $\mathrm{pp}(N)$ is a chain, in particular has width so, by 7.3.16, R satisfies the isolation condition. □

Corollary 8.2.59. *Suppose that R is a serial ring. Then every closed point N of Zg_R is of finite endolength.*

That follows by 5.3.23.

The m-dimension (= Krull–Gabriel dimension) over some classes of serial rings has been calculated (or estimated). In particular this dimension is $< \infty$ iff the Krull dimension of the ring is $< \infty$. The notation $\oplus$ below refers to the Cantor sum (Appendix E, p. 713) of ordinals.[15] The proofs in the first two papers use transfinite powers of the Jacobson radical and results of Müller and Singh, [450]: that in the last paper uses the lattice techniques indicated in Section 7.2.1.

[15] In the first two references, where upper bounds are given in the case of commutative valuation domains and serial rings with Krull dimension respectively, the sum, rather than Cantor sum, appears but this is corrected in [570].

Theorem 8.2.60. ([560, 3.6], [593, 3.11], [570, 3.6, 5.3]) *Let R be a serial ring. Then* $\mathrm{mdim}(\mathrm{pp}_R) = \mathrm{KG}(R)$ *exists iff* $\mathrm{Kdim}(R_R)$ *exists and, in that case* $\mathrm{KG}(R) \leq \mathrm{Kdim}(R_R) \oplus \mathrm{Kdim}(R_R)$.

The bound is attained in some cases, see Section 13.2.1. The next result then follows by 5.4.14.

Corollary 8.2.61. *If R is a serial ring with Krull dimension, then there are no topologically indistinguishable points of Zg_R: if each of $N, N' \in \mathrm{Zg}_R$ is in the closure of the other then $N = N'$.*

8.2.8 Spectra of uniserial rings

A module over a uniserial ring has indecomposable pure-injective hull iff its pp-definable subgroups are linearly ordered and every two non-zero elements of M are linked by a pp condition (8.2.64). A faithful indecomposable pure-injective over a uniserial ring either is injective or is a direct summand of the pure-injective hull of a faithful right ideal (8.2.66). Any uniserial module over a commutative ring has indecomposable pure-injective hull (8.2.67).

Nearly simple uniserial domains (Section 2.5) were used by Puninski as a source of counterexamples. He also proved that over these rings there are no superdecomposable pure-injectives (8.2.72) and their Ziegler spectra are described (8.2.74) and shown to have Cantor–Bendixson rank ∞ (8.2.75). Compare this with the commutative case 7.3.23.

Over exceptional uniserial rings the indecomposable pure-projective modules are described (8.2.76), as is the spectrum (8.2.77).

Various results of the previous section are restated, in simplified form, for uniserial rings.

The first result follows from 8.2.50, the second from 8.2.51.

Proposition 8.2.62. *Over a uniserial ring every indecomposable pure-injective module is uniserial as a module over its endomorphism ring: in particular the lattice of pp-definable subgroups of such a module is a chain.*

Proposition 8.2.63. ([581, 17.7]) *A pp-type p over a uniserial ring R is irreducible iff for every $a, b \in J(R)$ with $a \neq 0, b \neq 0$ if $ab|xb \in p$ then either $a|x \in p$ or $xb = 0 \in p$.*

If N is an indecomposable pure-injective module over any ring, then, 4.3.74, every two non-zero elements of N are linked (in the sense of Section 4.3.9) by some pp condition. In general the converse is not true but over uniserial rings, there is the following.

Proposition 8.2.64. ([557, Prop. 3], see [564, 11.22]) *If M is a module over a uniserial ring, then the pure-injective hull of M is indecomposable iff M is pp-uniserial and every two non-zero elements of M are linked by a pp condition.*

If $a \in N, a \neq 0$ with N an indecomposable pure-injective over a uniserial ring, then, see the previous section, we set $I_a = \text{ann}_R(a)$ and $J_a = \{s \in R : a \notin Ns\}$. Then 8.2.55 reduces to the following.

Corollary 8.2.65. *Suppose that N, N' are indecomposable pure-injective modules over a uniserial ring. Let $a \in N$, $a' \in N'$ be non-zero. If $N \simeq N'$, then there is $r \in R, r \neq 0$ such that either: $rI_a = I_{a'}$ and $J_a = J_{a'}r$; or $I_a = rI_{a'}$ and $J_a r = J_{a'}$.*

Proposition 8.2.66. ([581, 17.17], also cf. [173, 4.18]) *Let M be a faithful indecomposable pure-injective module over a uniserial ring R. Then M either is injective or is a direct summand of the pure-injective hull of some faithful right ideal.*

With stronger conditions, such as commutativity, on R considerably more can be said (see, for example, [182], [199], [173, §§3,4]). For example, there is the following result of Facchini.

Theorem 8.2.67. ([182, 5.1]) *If R is a commutative ring and M is a uniserial R-module, then $H(M)$ is indecomposable.*

Eklof and Herzog asked whether the converse is true for a commutative valuation ring and obtain some partial results ([173, §4]).

Question 8.2.68. If R is a commutative valuation ring is every indecomposable pure-injective module the pure-injective hull of a uniserial module?

For commutative valuation domains there is the following description of isolated points.

Proposition 8.2.69. ([560, 3.1]) *Suppose that R is a commutative valuation domain and let $J(R)$ denote its Jacobson radical.*

If $J(R)$ is not a principal ideal, then Zg_R has no isolated points.

If $J(R)$ is a principal ideal, then the isolated points are those given by pairs (I, J), where I, J both are principal, and each of these is isolated by a minimal pair. Also the isolated points are dense.

Suppose that R is uniserial. Let L_r, respectively L_l, denote the poset (a chain) of all right, resp. left, ideals of R. Each non-zero element $r \in R$ acts on L_r by $I \mapsto rI$. Let L_r/R denote the quotient by the equivalence relation generated by $I \sim rI$, for $I \in L_r$, $r \in R$, $r \neq 0$. Note that this is a factor *set*: the order is far from being preserved. For instance, all non-zero principal (= finitely generated) ideals of R are identified under this relation. Clearly $|L_r/R| \geq 2$ iff R is not right noetherian.

Proposition 8.2.70. ([566, 3.1]) *If R is a uniserial domain, then there is a natural bijection between the sets L_r/R and L_l/R.*

This is used below.

There is an example at [160, 3.8] with the chain of principal non-zero proper right ideals having order-type that of $(\mathbb{R}, \leq)$, with L_r/R having just two points and the Ziegler spectrum consisting of just six points, including two pairs of topologically indistinguishable points. The unique isolated point of Zg_R in that example is R/J; then there are the topologically indistinguishable $E(R/J)$ and $E(R/rR)$, where r is any non-zero element of J; dually, the topologically indistinguishable $H(R)$ and $H(J)$; then the flat and injective point $E(R_R)$.

Spectra of nearly simple uniserial domains In [566, 5.1, 5.2] the following classification of indecomposable pure-injectives, N, over a nearly simple uniserial domain, R, is obtained:

(a) N is injective, $N = E(R/I)$ for some right ideal, I, of R, and then $E(R/I) \simeq E(R/J)$ iff there is some non-zero $r \in R$ with $rI = J$ or $I = rJ$; or

(b) $N = R/J(R)$; or

(c) $N = H(I)$ for some right ideal I of R, and then $H(I) \simeq H(J)$ iff there is some non-zero $r \in R$ with $rI = J$ or $I = rJ$.

That is, the admissible pairs, in the sense of Section 8.2.7, for R are $(J(R), J(R))$, $(0, 0)$ and those of the form $(I, 0)$ and $(0, J)$.

Therefore the cardinality of Zg_R is finite or equals that of L_r/R (= that of L_l/R).

It follows, [566, 5.3], that if R is a countable nearly simple uniserial domain, then Zg_R has continuum many points.

Puninski gave an essentially complete description of the lattice, pp^1_R, of pp conditions over a nearly simple uniserial domain. The general form of the pp conditions is known by 2.4.22 since such a ring is RD; in this case it simplifies further, allowing one to determine when one pp condition implies another.

Proposition 8.2.71. ([566, 5.4, 5.5]) *Let R be a nearly simple uniserial domain.*

(a) Let $r, s, t, u \in R$ with $r \in J(R)$ and $s, t \neq 0$. Then $r|x \wedge xs = 0 \leq t|x + xu = 0$.

(b) Let $s, t, u \in R$ with $u \in J(R)$ and $s, t \neq 0$. Then $xs = 0 \leq tu|xu$.

Proof. (a) Let $\psi(x)$, respectively $\phi(x)$, be the pp condition $r|x \wedge xs = 0$, resp. $t|x + xu = 0$. If it were not the case that $\psi \leq \phi$, then there would, by 5.1.3, be some indecomposable pure-injective module N and $a \in \psi(N) \setminus \phi(N)$. Referring to the above classification of indecomposable pure-injectives, since $as = 0$, N is not torsionfree so cannot (by 4.3.23) be the hull of a non-zero right ideal. Since $r \mid a$ and $r \in J(R)$, it cannot be that $N \simeq R/J(R)$. Therefore it must be that N is injective, hence divisible, contradicting that $t \nmid a$.

(b) First note that by a short computation, over any ring, $t|x + xu = 0$ is equivalent to $tu \mid xu$. So applying elementary duality (1.3.1) and right/left symmetry, the statement of (b) becomes a special case of (a). □

The following diagram from [566, 5.6] illustrates the lattice of pp-conditions over a nearly simple uniserial domain. In particular the lattice has m-dimension $= \infty$ and has width $= 2$. Since it has width there are, by 7.3.5, no superdecomposable pure-injectives.

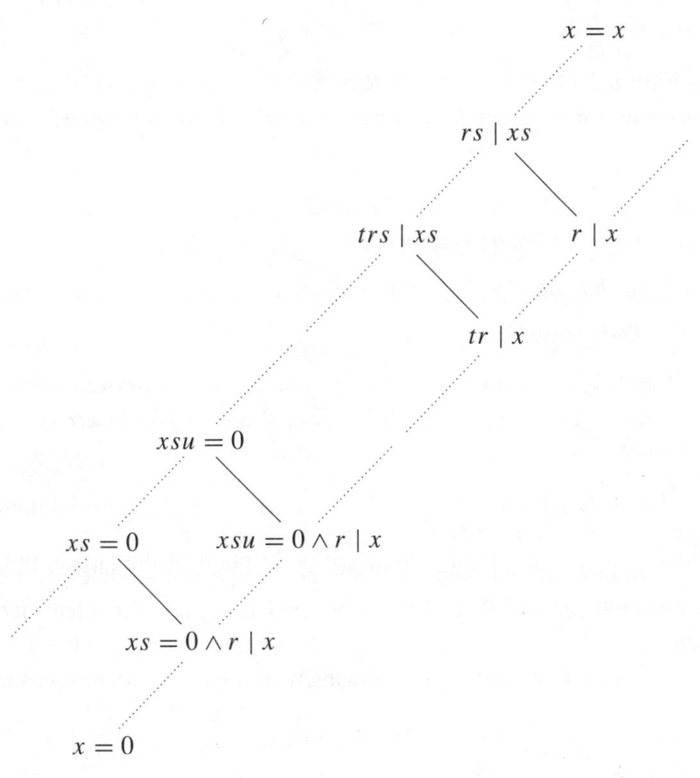

Proposition 8.2.72. ([566, 6.3]) *Over a nearly simple uniserial domain there is no superdecomposable pure-injective module.*

This contrasts with modules over a commutative valuation domain where, 7.3.23, there is a superdecomposable pure-injective module iff the ring does not have Krull dimension.

The Ziegler topology is also given a fairly complete description. For I a proper right ideal of the uniserial domain R, define the **coarseness**, $c(I)$, of I, to be $\{r \in R : \exists s \in R \setminus I \ sr \in I\}$. So $c(I)$ is a right ideal of R. For example, $c(0) = 0$, $c(rR) = J(R)$ for $r \in J(R)$ with $r \neq 0$, and $c(J(R)) = J(R)$. Also $c(rI) = c(I)$ if $r \neq 0$; thus coarseness is defined on equivalence classes in L_r/R.

Proposition 8.2.73. ([566, 5.7]) *A basis for the Ziegler topology over a nearly simple uniserial domain is given by sets of the following forms:*

(a) $\big((xs = 0) / (r|x)\big) = \big((rs|xs) / (r|x)\big) = \{R/J(R)\}$ *for* $r, s \in J(R)$, $r, s \neq 0$;
(b) $\big((xsu = 0) / (xs = 0)\big) = \big((r|x \wedge xsu = 0) / (r|x \wedge xs = 0)\big) = \{E(R/I) : u \in c(I)\}$ *for* $r, s, u \in J(R)$ *non-zero;*
(c) $\big((r|x) / (tr|x)\big) = \big((rs|xt) / (crs|xs)\big) = \{H(I) : t \in c(I)\}$ *for* $r, s, t \in J(R)$ *non-zero.*

Proposition 8.2.74. ([566, 5.8]) *If R is a nearly simple uniserial domain, then the following is a complete list of pairs of topologically indistinguishable points in* Zg_R*:*

(a) $E(R/I)$ *and* $E(R/J)$*, where* $c(I) = c(J)$*;*
(b) $H(I)$ *and* $H(J)$*, where* $c(I) = c(J)$*.*

In particular, for non-zero $r \in J(R)$, the points $E(R/rR)$ and $E(R/J(R))$ are topologically indistinguishable.

Proposition 8.2.75. ([566, 5.9]) *Let R be a nearly simple uniserial domain. The unique isolated point in Zg_R is $R/J(R)$. Once this is removed there is no isolated point in what remains and so* $\mathrm{CB}(\mathrm{Zg}_R) = \infty$.

It is also shown, [566, 6.1], that if R is a nearly simple uniserial domain and $r \in J(R)$, $r \neq 0$, then $H(R/rR) \simeq E(R/rR) \oplus R/J(R)$.

There are just two indecomposable finitely presented modules over a nearly simple uniserial domain: R_R and R/rR, where r is any non-zero element of $J(R)$. The latter does not, however, have local endomorphism ring and this is reflected in, for example, the category mod-R having no left almost split morphisms ([566, 6.4]).

Spectra of exceptional uniserial rings It was an open question whether a direct summand of a serial module must be serial. Puninski gave a counterexample in [565], using modules over exceptional uniserial rings. As with the results above, an analysis of the lattice of pp conditions is central to the paper. Examples of results obtained are the following.

Proposition 8.2.76. ([565, 3.6]) *The indecomposable pure-injective modules over a coherent exceptional uniserial ring R are: the indecomposable injectives $E(R/I)$ for some right ideal I of R; $H(J)$, which is not injective; the division ring R/J, which is Σ-pure-injective.*

Proposition 8.2.77. ([565, 4.6]) *If R is a coherent exceptional uniserial ring, then R/J and $H(J)$ are the isolated points of Zg_R. The remaining points all are injective (hence, [565, 4.5], also flat) and none is isolated in the first Cantor–Bendixson derivative Zg_R^1. In particular $\mathrm{CB}(\mathrm{Zg}_R) = \infty$.*

The counterexample, [565, 6.2], to the conjecture on serial modules is constructed as a direct limit of a sequence of embeddings between free realisations of certain pp conditions. It is shown that every tuple in the resulting module has finitely generated pp-type, that is, 1.3.22, it is Mittag–Leffler and so, since it is countably generated, it is, by 1.3.26, pure-projective.

Puninski gives a fine analysis of the pure-projective modules over such a ring in [567] which, with [532], yields their complete classification. Also see [160].

8.2.9 Spectra of pullback rings

The description of the Ziegler spectrum over a pullback of Dedekind domains seems to be at least as hard as, perhaps is essentially equivalent to, the problem over an arbitrary string algebra (Section 8.1.3) but some points have been described: the separated ones (8.2.80) and those with finite-dimensional top (8.2.83).

If R_1, R_2 are two commutative discrete valuation domains with maximal ideals P_1, P_2 respectively and if there is an isomorphism between their residue fields, $R_1/P_1 \simeq \overline{R} \simeq R_2/P_2$, then, following Levy [402], one may form their pullback in the category of (commutative) rings, denote it by R. So $R = \{(r_1, r_2) : r_i \in R_i \text{ and } \pi_1 r_1 = \pi_2 r_2\}$, where $\pi_i : R_i \to \overline{R}$ is the projection to the common residue field $\overline{R}$. An example of a ring obtained in this way, see below, is the algebra $k[x, y : xy = 0]_{(x,y)}$, which is the infinite-dimensional version of the Gelfand–Ponomarev algebras $\mathrm{GP}_{n,m} = k[x, y : xy = 0 = x^n = y^m]$ seen in Section 15.1.1. The latter are, for $m, n \geq 2$, $m + n \geq 5$, tame, non-domestic, string algebras and there are strong connections, developed in [14], between the classification of finite-dimensional modules over such string algebras (Section 8.1.3) and Levy's classification ([402]) of finitely generated modules over pullback rings. Indeed, there is a more than passing resemblance between the modules discussed below and the infinite string modules discussed in [605] and Section 8.1.3. If R_1 and R_2

are commutative Dedekind domains with a common factor field $R_1/P_1 \simeq R_2/P_2$, then their pullback over this field can be formed in the same way. An example is the integral group ring, $\mathbb{Z}[C_p]$, of the cyclic group, C_p, of order p (a prime): take $R_1 = \mathbb{Z}$ and $R_2 = \mathbb{Z}[\eta]$, where η is a primitive pth root of unity, let π_1 be the canonical projection to $\mathbb{Z}_p$ and let π_2 be the projection to $\mathbb{Z}_p$ with kernel generated by $1 - \eta$.

Example 8.2.78. Let $R_1 = k[X]_{(X)}$, $R_2 = k[Y]_{(Y)}$, where k is a field. Set $P_1 = \langle X \rangle$, $P_2 = \langle Y \rangle$. Then the corresponding pullback ring consists of the pairs $(p, q) \in k[X]_{(X)} \times k[Y]_{(Y)}$, where the constant terms of p and q are equal. This may be seen to be isomorphic to $k[x, y : xy = 0]_{(x,y)}$. The ring R has three prime ideals, $P_1 = \langle X \rangle$, $P_2 = \langle Y \rangle$ and the maximal prime $P = P_1 \oplus P_2$.

As commented above, the complete description of the Ziegler spectrum for such rings would include that for the tame, non-domestic, Gelfand–Ponomarev algebras, and that seems some way from being solved. The structure theory for finitely generated modules over pullback rings, and generalisations, was worked out by Levy, [402], [403], [404].

Let R be a pullback ring as above. Given an R_1-module S_1 and an R_2-module S_2, the direct sum, $S_1 \oplus S_2$, has the structure of an R-module by setting $(s_1, s_2)(r_1, r_2) = (s_1r_1, s_2r_2)$. Any submodule of an R-module of this form is said to be **separated**. By [402, 2.9] a module S is separated iff $SP_1 \cap SP_2 = 0$. Since P_1, P_2 are finitely generated, the separated modules form a definable subcategory of Mod-R, being defined, see Sections 1.1.4 and 3.4.1, by closure of the pp-pair $(P_1|x \wedge P_2|x) / x = 0$. The corresponding points of the Ziegler spectrum Zg_R, that is, the separated indecomposable pure-injectives, were identified by Toffalori and his results were used by Ebrahimi Atani to classify all those points, N, of Zg_R with finite-dimensional **top**, that is, with N/NP finite-dimensional over $\overline{R}$, where $P = P_1 \oplus P_2 = J(R)$. Here we give a rough description of the results. In our statements of their results we continue to assume that R_1, R_2 are local: that is not necessary (for the construction of the pullback ring or for the results) but it is the main case.

Any separated module can be represented as the pullback of an R_1-module, S_1, and an R_2-module, S_2, over a common top $\overline{S} \simeq S_1/S_1P_1 \simeq S_2/S_2P_2$. We denote such a pullback as $(S_1 \to \overline{S} \longleftarrow S_2)$ and, if S is the module which is the pullback of this diagram, then we write $S = (S/SP_1 \to S/SP \longleftarrow S/SP_2)$.

Proposition 8.2.79. ([166, 2.4, 2.5]) *Let R be the pullback of two discrete valuation domains R_1, R_2. Let $S = (S_1 \to \overline{S} \longleftarrow S_2)$ be a separated R-module. If S is pure-injective, then so is each of S_1, S_2. If $\overline{S}$ is finite-dimensional over $\overline{R}$, then the converse is true.*

If $S = (S_1 \to \overline{S} \longleftarrow S_2)$ is separated, with pure-injective hull $H = (H_1 \to \overline{H} \longleftarrow H_2)$, and if $\overline{S}$ is finite-dimensional over $\overline{R}$, then H_i is the pure-injective hull of the R_i-module S_i.

Theorem 8.2.80. ([678, §4], [166, 2.9]) *Let R be the pullback of two discrete valuation domains, R_1, R_2. The indecomposable pure-injective separated R-modules are exactly those of the form $S = (S_1 \to \overline{S} \longleftarrow S_2)$ where:*

(a) $\overline{S} = 0$, one of S_1, S_2 is zero and the other is one of the two injective modules in the Ziegler spectrum of R_{3-i} (for those see 5.2.1);
(b) $\overline{S} = \overline{R}$ and each S_i is either a finite length module in Zg_{R_i} or is the adic point of Zg_{R_i} (again, see 5.2.1 for these).

Although the relative topology on the closed set of separated indecomposable pure-injectives is not described in those references it would be a straightforward exercise to do so on the basis of description (Section 5.2.1) of the spectra of the R_i.

In order to investigate non-separated pure-injectives, use is made of the separated representation of an R-module. If M is any R-module, then a **separated representation** of M is an epimorphism, $S \to M$ with S separated and such that, in any factorisation $S \xrightarrow{f} S' \to M$ of this morphism with S' separated, f must be monic. Equivalently, S is separated, the kernel, K, of the epimorphism $S \to M$ is contained in SP, and $K \cap SP_i = 0$ for $i = 1, 2$. Every module has a separated representation [402, 2.8].

Example 8.2.81. Let R be as in 8.2.78. Let M be the three-dimensional module with k-basis $a, b, ax = by$, where $ay = 0 = bx$; this may be regarded, in the terminology of Section 8.1.3, as a string module over the Gelfand–Ponomarev factor ring, $\mathrm{GP}_{2,2}$, of R. The separated representation of M, is the obvious map from the separated module $S = \big(k[X]/\langle x^2\rangle\big) \oplus \big(k[Y]/\langle y^2\rangle\big)$ with kernel $\langle x - y\rangle$.

Proposition 8.2.82. ([678, §3], also see [166, 2.6]) *Let R be the pullback of two discrete valuation domains. If M is a pure-injective R-module and $S \to M$ is a separated representation of M, then S is pure-injective.*

This allows one to use the description of the indecomposable pure-injective separated modules to investigate arbitary indecomposable pure-injective R-modules. Roughly, the latter, at least those with finite-dimensional top, are obtained by glueing together finitely many modules of the sort described in 8.2.80. Locally the glueing involves only two modules and the glueing is to identify the "right (that is, R_1) socle" of one module with the "left (that is, R_2) socle" of the other. This glueing is a key ingredient in Levy's description of the finitely generated modules. In particular, the possible infinite-length components (Prüfer modules from 8.2.80(a) and adic modules from 8.2.80(b)) can occur only at the "ends" of a chain of glueings. Rather than give all the definitions and a complete list which, to give in detail would be quite lengthy and can, in any case, be found in Ebrahimi Atani's paper [166, §3], we illustrate with an example (modified from [166, p. 4057]). The similarity with the modules seen in Section 8.1.3 has been remarked upon already.

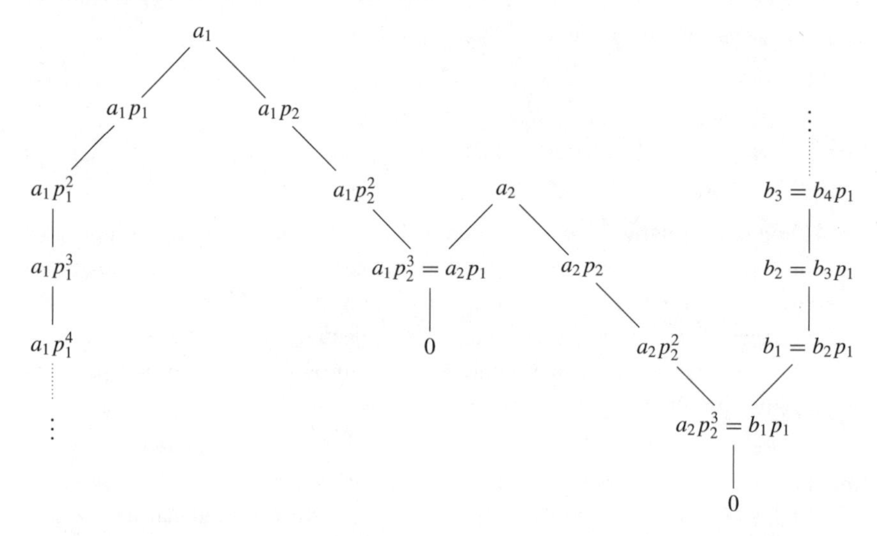

I remark that the restriction to indecomposable pure-injectives having finite-dimensional top is used in a number of ways, for example, 4.3.27 then applies; it is far from clear that such methods could be used to obtain the classification of all the points.

Theorem 8.2.83. ([166, 3.5]) *Let R be the pullback of two discrete valuation domains. Then the points of Zg_R with finite-dimensional top are explicitly described, being obtained by the glueing process indicated above.*

One must show that all the modules in the list obtained are indecomposable pure-injective but the main work goes into showing that every point with finite-dimensional top has this form described. The proof is quite lengthy.

8.2.10 Spectra of von Neumann regular rings

We begin with an example of a non-discrete Ziegler spectrum with just two points (8.2.84). Trlifaj showed that the isolation condition being true for von Neumann regular rings is equivalent to the non-existence of a simple, non-artinian von Neumann regular ring with a one-point spectrum (8.2.85). He also showed that, for von Neumann regular rings, the isolation condition is implied by any of: having cardinality less than the continuum; having all primitive factor rings artinian; being semiartinian (8.2.87).

Simple criteria for isolation and for isolation by a minimal pair are given (8.2.88). If all primitive factor rings are artinian, then the spectrum is Hausdorff iff the ring is biregular (8.2.89).

There is also a brief discussion of strongly uniform modules and the Γ-invariant which measures the extent to which the submodule lattice of a module fails to be complemented.

Recall, 2.3.24, that over a von Neumann regular ring every pp condition is equivalent to a system of equations, so everything becomes "purely algebraic": no existential quantifiers are needed for pp conditions, every exact sequence is pure-exact (2.3.22), the functor category is equivalent to the module category (10.2.38), etc. In particular the Ziegler spectrum is the set of isomorphism classes of indecomposable injective modules and, 5.2.11, a basis of open sets is given by those of the form $\{N \in \mathrm{inj}_R : Ne \neq 0\}$ as e ranges over idempotents of R.

Example 8.2.84. ([495, p. 328]) Let $S = \mathrm{End}_k(V)$, where V is an $\aleph_0$-dimensional vector space over a field k, and let R be its subring $\mathrm{soc}(S) + 1 \cdot k$. It may be checked that R is von Neumann regular and clearly it is semiartinian. By [242, 6.19] every simple module is injective. Since R is semiartinian it follows (see [684, §2]) that every indecomposable injective R-module is the injective hull of a simple module. Therefore Zg_R is the set of simple right R-modules. One of these is $S_1 = R/\big(J(R) = \mathrm{soc}(R)\big)$, a Σ-injective module, by 5.3.60. The only other, S_0, is isomorphic to any simple right ideal; these are all of the form eR, where e is projection to a one-dimensional subspace and all these right ideals are isomorphic. Therefore Zg_R has just two points, with S_0 isolated and S_1 non-isolated. Since $\mathrm{CB}(\mathrm{Zg}_R) = 1$ it follows by 7.3.20 (although this could be checked directly using 2.3.25) that $\mathrm{KG}(R) = 1$.

It has been seen already in 7.3.20 that width and m-dimension coexist for von Neumann regular rings, that is, the one is $< \infty$ iff the other is, and this is so iff the ring is semiartinian. Moreover, in this case $\mathrm{CB}(\mathrm{Zg}_R) < \infty$ and equals $\mathrm{KG}(R)$, 5.3.60. The converse, whether $\mathrm{CB}(\mathrm{Zg}_R) < \infty$ implies $\mathrm{KG}(R) < \infty$, is, however, open: in particular, it is not known whether every von Neumann regular ring satisfies the isolation condition. Indeed, Trlifaj showed the following.

Theorem 8.2.85. ([684, 3.5]) *There exists a von Neumann regular ring R which does not satisfy the isolation condition iff there exists a von Neumann regular ring R' which is simple, non-artinian and with $\mid \mathrm{Zg}_{R'} \mid = 1$.*

Question 8.2.86. Does every von Neumann regular ring satisfy the isolation condition? Equivalently is there a simple, non-artinian von Neumann regular ring with just one indecomposable injective module up to isomorphism?

There are some partial results.

Theorem 8.2.87. ([684, 3.6]) *The isolation condition holds for a von Neumann regular ring R which satisfies any of the following conditions:*

(a) $|R| < 2^{\aleph_0}$ *or;*
(b) all primitive factor rings of R are artinian or;
(c) R is semiartinian.

There is the following characterisation (from [684, 3.2]) of isolation in the Ziegler spectrum over regular rings (also see 10.2.40).

Proposition 8.2.88. *Suppose that R is von Neumann regular and let X be a closed subset of Zg_R. Let $I = \{r \in R : \forall N \in X\ Nr = 0\}$. Let $N \in X$.*

(a) N is an isolated point of X iff there is $e = e^2 \in R/I$ such that $Ne \neq 0$ and N is the only such point of X.
(b) N is isolated by an X-minimal pair iff there is $e = e^2 \in R/I$ such that $e \cdot R/I$ is an indecomposable right R/I-module, such that $Ne \neq 0$ and such that N is the only such point of X.

If R is a von Neumann regular ring with all primitive factor rings artinian, then the indecomposable injective R-modules are, [684, 2.2], exactly the simple modules and, [686, 2.2(i)], Zg_R is homeomorphic to the maximal ideal space of R via $S \mapsto \mathrm{ann}_R(S)$. In particular, Zg_R is a T_1 space. Furthermore, one has the following.

Theorem 8.2.89. ([686, 2.2(ii)]) *If R is von Neumann regular with all primitive factor rings artinian then the following conditions are equivalent:*

(i) Zg_R is Hausdorff;
(ii) Zg_R is a normal space;
(iii) Zg_R is totally disconnected;
(iv) R is a **biregular** *ring: for every $x \in R$, the ideal RxR is generated by a central idempotent.*

There is more information in [686] about the relationship between maxspec(R) and Zg_R for von Neumann regular, rings.

If R is von Neuman regular, then, without further assumptions, there may well be indecomposable injective R-modules with zero socle. Indeed, as pointed out in [686, 2.3], Goodearl showed [242, 3.8] that for each ordinal $\alpha > 1$ there is a prime unit-regular ring with a faithful cyclic module U such that the lattice of submodules of U is anti-isomorphic to the chain of ordinals β with $0 \le \beta \le \alpha$. Hence if α is an infinite limit ordinal, then U is uniform with zero socle and so the injective hull of U is an indecomposable injective with zero socle. This has been developed as follows.

Say that a module M is **strongly uniform** if there is a strictly descending sequence $(U_\alpha)_{\alpha<\kappa}$ of non-zero submodules of M which is cofinal in the sense that every non-zero submodule of M' contains some U_α and, if $\lambda < \kappa$ is a limit ordinal, then $U_\lambda = \bigcap_{\alpha<\lambda} U_\alpha$ (in particular this intersection is non-zero). If M has countably many submodules, then, clearly, this is equivalent to M being uniform but, [685, §2], if k is a field, then $k[X]$, as a module over itself, is uniform but is strongly uniform iff k is countable.

To a strongly uniform module M and sequence $(U_\alpha)_{\alpha<\kappa}$ as above with κ chosen minimal, is associated the set, $\mathcal{E}(M)$, of those ordinals $\alpha < \kappa$ such that, for some $\beta > \alpha$, U_α is not complemented over U_β, meaning that there is no $M' \le M$ with $U_\alpha + M' = M$ and $U_\alpha \cap M = U_\beta$. The set of ordinals resulting from a different choice of cofinal sequence of submodules U_α will agree with $\mathcal{E}(M)$ on a closed unbounded set of ordinals $< \kappa$, and the **Γ-invariant** of M is defined to be the equivalence class of $\mathcal{E}(M)$ in the boolean algebra, $B(\kappa)$, of subsets of κ modulo the filter generated by the closed unbounded subsets of κ. So this, and the related 'Γ^*-invariant' (see the references for definitions and more details) measures the extent to which the lattice of submodules of M fails to be complemented and may be taken to be a reasonable measure of complexity, especially over a von Neumann regular ring.

Trlifaj and Shelah proved a variety of realisation results. For example, given any field F, any invariant $i \in B(\aleph_1)$ can be realised as the Γ^*-invariant of an indecomposable injective module over a **locally matricial** F-algebra, that is, an F-algebra which is a direct limit of finite direct products of full matrix rings over F (such a ring is von Neumann regular). In [645, 4.3] it is shown that if μ is an uncountable cardinal and if F is any field, then there is an F-algebra R such that for every uncountable regular cardinal $\kappa < \mu$ and any $i \in B(\kappa)$ there is a strongly uniform injective R-module (hence point of Zg_R) with Γ^*-invariant equal to i.

8.2.11 Commutative von Neumann regular rings

For commutative von Neumann regular rings the Pierce and Ziegler spectra coincide (8.2.91), as do most dimensions (at least, one is defined iff the other is) (8.2.92).

A **boolean** ring is one which satisfies the identities $x + x = 0$ and $x^2 = x$ – that is, a boolean algebra, regarded as a ring. For any commutative von Neumann regular ring the set, $B(R)$, of its idempotents forms a boolean subring.

Theorem 8.2.90. (for example, [242, pp. 83, 90ff.]) *Suppose that R is a commutative von Neumann regular ring. Then the idempotent elements of R form a boolean algebra, $B(R)$, with join given by $e + f - ef$ and meet given by ef for $e, f \in B(R)$. Every ideal is generated by the idempotents which it contains, so the ideals of R correspond bijectively to the ideals (in the sense of partial orders, see Section 3.3.1) of the boolean algebra $B(R)$. In particular the simple R-modules are in bijection with the maximal ideals, hence also with the ultrafilters on $B(R)$.*

Explicitly, to a maximal ideal, I, of R there corresponds the ultrafilter $\{e \in B(R) : e \notin I\}(= \{e \in B(R) : 1 - e \in I\})$. The set of ultrafilters of $B(R)$ is topologised by taking for a basis of open sets the $O_e = \{I \in \text{maxspec}(R) : e \notin I\}$. Note that these sets are clopen since the complement of O_e is O_{1-e}. This is the **Stone space** of R and, as noted already, it may be identified (to an ultrafilter associate the ideal generated by its complement in $B(R)$) with the space, maxspec(R), of maximal ideals of R, and then it is called the **Pierce spectrum** of R.

To any R-module M is associated a sheaf over maxspec(R), which is defined by assigning to the basic open set O_e the module M/eM and then sheafifying this presheaf defined on a basis. The stalk of this sheaf at a maximal ideal I is M/MI. Taking $M = R$ one obtains a sheaf of rings, the **Pierce sheaf** of R, and the sheaf built from the module M is a module over this ringed space (see Appendix E, p. 716). Indeed, [477], the category of R-modules is equivalent to the category of modules over this ringed space. A boolean ring R is just a commutative regular ring where each factor R/I is the two-element field.

There is a natural bijection between Zg_R and the Pierce spectrum (seen in the comment before 8.2.89) and one has the following, which follows easily from the various definitions.

Theorem 8.2.91. (for example, [495, p. 327]) *Let R be a commutative von Neumann regular ring. Then the Ziegler and Pierce spectra are naturally homeomorphic and the Ziegler and Gabriel–Zariski spectra on the set* inj_R *coincide. Furthermore, the Pierce sheaf is naturally isomorphic with the sheaf of locally definable scalars (Section 14.1.4).*

A boolean ring is **atomic** if every non-zero ideal contains a simple ideal and is **superatomic** if every factor ring is atomic. The terminology may be extended to general commutative regular rings via their boolean subrings of idempotent elements and then the condition is equivalent to R being semiartinian.

Theorem 8.2.92. ([209, Thm 4], see [495, 16.25]) *Suppose that R is a commutative von Neumann regular ring. Then the following conditions are equivalent:*

(i) $\mathrm{mdim}(\mathrm{pp}_R) = \alpha < \infty$;
(ii) $\mathrm{KG}(R) < \infty$;
(iii) $\mathrm{CB}(\mathrm{spec}(R)) = \alpha < \infty$;
(iv) *$B(R)$ is superatomic, with* $\mathrm{mdim}(B(R)) = \alpha$.

If R is countable, then these conditions are satisfied for some α iff $\mathrm{spec}(R)$ *is countable.*

8.2.12 Indiscrete rings and almost regular rings

A ring is indiscrete, that is, the Ziegler topology is trivial, iff it is simple and all finitely presented right and left modules are absolutely pure (8.2.95). Simple von Neumann regular rings are indiscrete (8.2.93) and it was an open question whether there were any other indiscrete rings. For instance any right or left coherent indiscrete ring is von Neumann regular (8.2.96). Examples, indeed a general method for producing examples of indiscrete rings which are not von Neumann regular, were found by Prest, Rothmaler and Ziegler (8.2.97).

Equivalents, for arbitary rings, to every finitely presented module being absolutely pure, "almost regularity", are given (8.2.98) and the almost regular group rings characterised as those where the group is locally finite and the order of every finite subgroup is invertible in the coefficient ring, which is also almost regular (8.2.99).

Say that a ring R is **indiscrete** if the topology on the Ziegler spectrum of R is the indiscrete one – the only open subsets are the empty set and the whole space. Equivalently, by the definition of the topology on Zg_R, Mod-R has no non-trivial, proper, definable subcategories. By 5.4.1 this condition for modules on the one side implies it for the other. An indiscrete ring must be simple since, if I is a non-zero, proper, two-sided ideal of R, then the class of modules M such that $MI = 0$ is a non-zero, proper, definable subcategory of Mod-R.

Proposition 8.2.93. ([690, Thm 1], [527, 2.1]) *Any simple von Neumann regular ring is indiscrete. If R is finite, then it is indiscrete iff it is simple artinian.*

Proof. The first statement is immediate from 3.4.29. The second statement follows, for example, from the observation above. □

Example 8.2.94. Let $S = \text{End}_k(V)$, where V is a countably-infinite-dimensional vector space over a field k. The socle of S consists of all the endomorphisms with finite-dimensional image. Then it is easy to see that $R = S/\text{soc}(S)$ is a simple von Neumann regular ring, hence is indiscrete. By the downwards Löwenheim–Skolem Theorem (A.1.10) it follows easily that there are countable such rings; for a specific example see [240, 5.6].

Theorem 8.2.95. ([527, 2.4]) *A ring R is indiscrete iff it is a simple ring and all its left and right finitely presented modules are absolutely pure.*[16]

Proof. We prove one half to give some flavour of the arguments.

That an indiscrete ring must be simple has been noted already. Suppose that $A \le M$ are R-modules with A finitely presented: we check the definition (Section 2.1.1) of purity. So suppose that $\overline{a}$ is a tuple from A, that ψ is pp and that $\overline{a} \in \psi(M)$; by 1.3.7, $\overline{a} \cdot D\psi({}_RR) = 0$. By 1.2.6 there is a pp condition ϕ which generates the pp-type of $\overline{a}$ in A and, by 1.3.10, $\text{ann}_R(\overline{a}) = D\phi({}_RR)$. Therefore $D\psi({}_RR) \le D\phi({}_RR)$. By indiscreteness of R, $D\psi(L) \le D\phi(L)$ for every left R-module L. That is, $D\psi \le D\phi$ hence, by 1.3.1, $\psi \ge \phi$ and so $\overline{a} \in \psi(A)$, as required. □

Proposition 8.2.96. ([527, 2.14]) *Any left or right coherent indiscrete ring is von Neumann regular.*

This is immediate from 3.4.24 since that result implies that either the flat or absolutely pure modules form a definable class, which must be all of Mod-R, which is enough by 2.3.22.

In fact, essential coherence is sufficient for the conclusion, where R is **essentially coherent** if every finitely generated right submodule $I \le R^n$ has a presentation $0 \to K \to R^m \to I \to 0$ with K containing an essentially finitely generated submodule.

For some time it was an open question ([618]) whether there are non-von-Neumann-regular indiscrete rings. Such rings, indeed a general method of constructing them as twisted direct limits of non-von-Neumann-regular rings of finite representation type, was given in [527].

Example 8.2.97. ([527, §2.2]) There are indiscrete rings which are not von Neumann regular. I indicate the method, referring to the paper for details. The method is based on an example of Menal and Raphael [445]. I also mention that the construction has been used by Hovey, Lockridge and Puninski in [305].

[16] It is sufficient, if R is infinite and simple, that every finitely presented right module be absolutely pure and that R be absolutely pure as a left module over itself.

Let k be an infinite field and let R be any finite-dimensional k-algebra of finite representation type. Let N be a direct sum of one copy of each indecomposable right R-module. Let $\tau_1 : R \to M_t(k)$, where $t = \dim_k(N)$, be a representation of the action of R on N and define $\tau : R \to M_{t+1}(k)$ by $r \mapsto \begin{pmatrix} \tau_1(r) & 0 \\ 0 & r \end{pmatrix}$. This embedding is now repeated, so each entry, r, in any matrix in $M_{t+1}(k)$ is replaced by $\tau(r)$, and this is continued inductively through the integers. Thus one obtains a diagram $R \to M_{t+1}(k) \to M_{(t+1)^2}(k) \to \cdots$ in the category of rings; let R_τ be the direct limit of this diagram. It is shown that R_τ is indiscrete and that if R is not von Neumann regular, then neither will be R_τ. A variation of the construction, indicated in [527], starts with any finite-dimensional algebra over a countable field, and gradually adds in the, possibly infinitely many, isomorphism types of finite-dimensional indecomposable modules,

Garkusha and Generalov gave a homological characterisation of indiscrete rings, [218]. Say that a ring is **almost regular** if it satisfies the equivalent conditions of the next result. Regarding the terminology there: a module M is **fp-flat** if for every monomorphism $K \to L$, with K and L finitely presented left modules, the map $M \otimes K \to M \otimes L$ is monic, and a module M is **fp-mono-injective** (they said just "fp-injective" but that term is often used in place of "absolutely pure") if for every monomorphism $A \to B$ between finitely presented modules, the induced map $(B, M) \to (A, M)$ is surjective (compare with fp-injectivity = absolutely purity where one requires this of every morphism between finitely presented modules, not just monomorphisms).

Theorem 8.2.98. ([218, 3.1]) *The following conditions on a ring R are equivalent:*

(i) Every short exact sequence of finitely presented right and left modules splits.
(ii) Every right module is fp-flat and fp-mono-injective.
(iii) Every left and every right finitely presented module is fp-flat.
(iv) Every left and every right finitely presented module is absolutely pure.

Various other equivalent conditions are given in [218, 3.1]. Also see [214], [213] for related results.

The class of almost regular rings includes the von Neumann regular rings and the statement of 8.2.95 becomes ([527, 2.4], also see [218, 3.3]): a ring is indiscrete iff it is simple and almost regular.

Other examples of almost regular rings are given by the next result; also see [213, §3] for results along the same lines.

Theorem 8.2.99. ([215, Thm 1]) *A group ring RG is almost regular iff R is almost regular, G is locally finite and the order of every finite subgroup of G is invertible in R.*

It is noted in [215] that if R is indiscrete and G is non-trivial, then RG is an example of an almost regular ring which is not indiscrete; also if R is not von Neumann regular, then neither is RG.

9

Ideals in mod-R

In this short chapter, which links into the functorial view of Part II, the subcategory of finitely presented modules is treated as a ring with many objects and a connection between fp-idempotent ideals of this ring and Serre subcategories of the functor category is made. The transfinite powers of the radical are defined and related to the value of m-dimension.

9.1 The radical of mod-R

The radical of the category of finitely presented modules is introduced in Section 9.1.1, and its transfinite powers in Section 9.1.2. In the latter section and in Section 9.1.3 there are results particular for artin algebras. In Section 9.1.4 the transfinite nilpotence of this radical is related to Krull–Gabriel/m-dimension.

9.1.1 Ideals in mod-R

The radical of a Krull–Schmidt category is generated by non-isomorphisms between indecomposable objects (9.1.2).

An **ideal** of mod-R, more generally, of any preadditive category $\mathcal{A}$, is a collection, $\mathcal{I}$, of morphisms in $\mathcal{A}$ which contains all zero morphisms, is closed under addition and is such that for all $f \in \mathcal{I}$ and all $g \in \mathcal{A}$ each composition gf, fg, if defined, is in $\mathcal{I}$. The ideal $\mathcal{I}$ is **generated** by a collection $(g_i)_i$ of morphisms if $\mathcal{I} = \{\sum_i h_i g_i f_i : f_i, h_i \in \mathcal{A}\}$.

There is also the notion of a **left ideal**, meaning a class, $\mathcal{L}$, of morphisms containing all zero morphisms and closed under addition and postcomposition: if $f \in \mathcal{L}$, then, for all g, $gf \in \mathcal{L}$. Dually, a **right ideal** should be closed under precomposition. If $\mathcal{I}$ is a right or left ideal of mod-R set $\mathcal{I}(A, B) = \mathcal{I} \cap (A, B)$ to be the subgroup of (A, B) consisting of those morphisms which are also in

$\mathcal{I}$. Given a morphism, $f : A \to B$, in $\mathcal{A}$ denote by $\mathcal{A}f$ the left ideal **generated** by f: $\mathcal{A}f = \{gf : g \in (B, C),\ C \in \mathcal{A}\} \cup \{0 \in (C, D) : C, D \in \mathcal{A}\}$ and similarly denote by $f\mathcal{A}$ the right ideal generated by f.

Example 9.1.1. The **radical**, rad($\mathcal{A}$), of a preadditive category $\mathcal{A}$ ([337]) is defined by $\mathrm{rad}(A, B) = \{f \in (A, B) : \forall g \in (B, A),\ 1_A - gf \in \mathrm{Aut}(A)\}$. If $\mathcal{A}$ is a ring, regarded as a category with one object (see Section 10.1.1), then this is just the Jacobson radical of the ring.

For other examples of ideals of a category see Section 15.3 and, though in a different context, the construction of the homotopy category and the construction of the stable module category in Section 17.1.

Proposition 9.1.2. *If $\mathcal{A}$ is a Krull–Schmidt category (Section 4.3.8) then* rad$\mathcal{A}$ *is generated by the non-isomorphisms between indecomposable objects of $\mathcal{A}$.*

Proof. Every object of $\mathcal{A}$ is a direct sum of indecomposables and any map, f, between two such objects can be represented by a matrix $(f_{ij})_{ij}$, where $f_{ji} : A_i \to B_j$. Then apply 9.1.4 below. □

Clearly products (defined just as in rings, that is, $\mathcal{I}\mathcal{J} = \{\sum f_i g_i : f_i \in \mathcal{I},\ g_i \in \mathcal{J}\}$) and arbitrary intersections, as well as arbitrary sums, of ideals are ideals. Also, if $\mathcal{I}$ is any ideal, then $\mathcal{I}^n$ is the ideal of morphisms generated by all compositions of n morphisms from $\mathcal{I}$, and $\mathcal{I}^n \cdot \mathcal{I}^m = \mathcal{I}^{n+m}$. For example, the finite powers, $\mathrm{rad}^n(\mathcal{A})$, and first infinite power, $\mathrm{rad}^\omega(\mathcal{A}) = \bigcap_{n\geq 1} \mathrm{rad}^n(\mathcal{A})$, of the radical are ideals.

Remark 9.1.3. If $\mathcal{I}$ and $\mathcal{J}$ are ideals of $\mathcal{A}$, where $\mathcal{A}$ has finite direct sums, then $\mathcal{I}\mathcal{J} = \{fg : f \in \mathcal{I},\ g \in \mathcal{J}\}$. All that has to be checked is that the set of products fg is closed under addition. But, given morphisms $A \xrightarrow{g} B \xrightarrow{f} C$ and $A \xrightarrow{h} D \xrightarrow{k} B$ we have $fg + kh = \begin{pmatrix} f & k \end{pmatrix} \begin{pmatrix} g \\ h \end{pmatrix}$, where $\begin{pmatrix} g \\ h \end{pmatrix} : A \to C \oplus D$ and $\begin{pmatrix} f & k \end{pmatrix} : C \oplus D \to B$ are the obvious maps.

Lemma 9.1.4. *Let $\mathcal{I}$ be an ideal of the category $\mathcal{A}$, where $\mathcal{A}$ has finite direct sums. Let $f = (f_{ij})_{ij} : A = \bigoplus_{j=1}^m \to \bigoplus_{i=1}^n B_i = B$ be in $\mathcal{A}$. Then $f \in \mathcal{I}$ iff $f_{ij} \in \mathcal{I}$ for all i, j.*

Proof. Since $f_{ij} = \pi_i f \iota_j$ (where π_i and ι_j are the obvious projection and inclusion) the direction ($\Rightarrow$) is immediate. The converse follows since f is a sum of morphisms $F_{ij} : A \to B$, where F_{ij} has just one non-zero entry, f_{ij}, and clearly $F_{ij} \in \mathcal{I}$. □

A special case is that $\mathcal{I}$ is **generated by objects**, that is, by identity morphisms: $\mathcal{I} = \mathcal{I}_{\mathcal{M}} = \{f : f \text{ factors through } M \text{ for some } M \in \mathcal{M}\}$, where $\mathcal{M}$ is a subclass of $\mathcal{A}$. For example, the stable module category of a quasi-Frobenius ring is obtained from its module category by factoring out the ideal of morphisms which factor through a projective object (see Section 17.1). In general, if $\mathcal{I}$ is an ideal of a category $\mathcal{A}$, then one may form the category $\mathcal{A}/\mathcal{I}$ which has the same objects as $\mathcal{A}$ and $\mathrm{Hom}_{\mathcal{A}/\mathcal{I}}(A, B) = (A, B)/\mathcal{I}(A, B)$. The result is again a preadditive category but, in contrast to forming the quotient by a localising (11.1.5) or Serre subcategory (11.1.40), if $\mathcal{A}$ is abelian $\mathcal{A}/\mathcal{I}$ need not be; in particular the stable module category of a non-trivial group ring is not abelian, see Section 17.1.

Another example, which is considered in Section 15.3, is given by $\mathcal{A} = \text{Mod-}R$, where R is an artin algebra and $\mathcal{I} =$ the ideal of those morphisms which factor through finitely presented modules.

Note that rad(mod-R) contains no non-zero identity morphisms, so is quite the opposite from an ideal generated by objects. The same is true for rad^{ω} but sometimes, 15.3.8, this ideal can be characterised in a weak sense by a class of objects: namely for R an artin algebra, it turns out to be the ideal of those morphisms which factor through some class of objects in the larger category Mod-R. In general, if $\mathcal{A}$ is a full subcategory of $\mathcal{B}$ and if $\mathcal{M}$ is a class of objects of $\mathcal{B}$, then $\{f \in \mathcal{A} : f \text{ factors through some } M \in \mathcal{M}\}$ is an ideal if $\mathcal{M}$ is closed under finite direct sums.

9.1.2 The transfinite radical of mod-R

Over an artin algebra the irreducible morphisms generate the radical (9.1.5) and morphisms which factor through an infinite-length indecomposable module are in the ω-power of the radical (9.1.6). Morphisms in this ω-radical split off irreducible morphisms (9.1.7) and this determines the definition of the transfinite powers.

Let R be a ring (the definition will also make sense with mod-R replaced by any preadditive category). Define, following [504, p. 453], a radical series for the category mod-R as follows[1] (also cf. [345]). Let $\mathrm{rad} = \mathrm{rad}^1$ denote the radical (9.1.1) of the category mod-R. Define the finite powers rad^n as in the previous section. If λ is a limit ordinal, define $\mathrm{rad}^{\lambda} = \bigcap\{\mathrm{rad}^{\nu} : \nu < \lambda\}$ and if ν is a non-limit infinite ordinal, so uniquely of the form $\nu = \lambda + n$ for some limit ordinal λ and natural number $n \geq 1$, set $\mathrm{rad}^{\nu} = (\mathrm{rad}^{\lambda})^{n+1}$. Finally set $\mathrm{rad}^{\infty} = \bigcap\{\mathrm{rad}^{\nu} : \nu \text{ an ordinal}\}$ and $\mathrm{rad}^0 = \text{mod-}R$. This allows us to assign a rank to

[1] For this definition applied to ideals of a ring see [352].

every morphism f in mod-R: $\text{rk}(f) = \nu$ if $f \in \text{rad}^\nu \setminus \text{rad}^{\nu+1}$ and $\text{rk}(f) = \infty$ if $f \in \text{rad}^\infty$. Call rad^ω the **ω-radical** and rad^∞ the **transfinite radical**.

Note that if $f : A \to B$ is in rad, then f strictly increases the pp-type of every non-zero element (and tuple) of A: $\text{pp}^A(\overline{a}) > \text{pp}^B(\overline{b})$ provided $\overline{a} \neq 0$. This is immediate from the definition and 1.2.9.

Lemma 9.1.5. *If R is an artin algebra, then* rad(mod-R) *is generated as a left ideal (alternatively as a right ideal) of* mod-R *by the irreducible morphisms between indecomposables.*

Proof. Let $f : A \to B$ be non-invertible with $A, B \in$ mod-R indecomposable. If A is not injective, then (15.1.4) there is an almost split sequence $0 \to A \xrightarrow{g} A' \to A'' \to 0$ in mod-R and a factorisation, $f = hg$, for some $h : A' \to B$. Express g as $(g_1, \ldots, g_n)$, where $A' = A_1 \oplus \cdots \oplus A_n$ with each A_i indecomposable and g_i is g followed by projection to A_i. Each g_i is irreducible (see after 15.1.4) and $f = \sum hg_i$. If A is injective, then $f = f'p$ for some f', where the natural map $p : A \to A/\text{soc}(A)$ is irreducible.

The parenthetical comment has a similar proof. □

Lemma 9.1.6. *Suppose that R is an artin algebra and that $f : A \to B$ in* mod-R *factors through some module which has no direct summand of finite length. Then $f \in \text{rad}^\omega$.*

Proof. Say $f = (A \xrightarrow{g} M \xrightarrow{h} B)$ with M indecomposable of infinite length. By 9.1.4 it will be enough to prove the result for the case where A and B both are indecomposable. Suppose, for a contradiction, that $f \in \text{rad}^n \setminus \text{rad}^{n+1}$ and suppose that f has been chosen among all such counterexamples so that $n \geq 1$ is minimal. Let $0 \to A \xrightarrow{k} A' \to A'' \to 0$ be an almost split sequence or, in the case that A is injective, take k to be the canonical map $p : A \to A' = A/\text{soc}(A)$. Since g is not a split embedding, $g = g'k$ for some $g' : A' \to M$, so $f = hg'k$. Since $k \in$ rad and $f \notin \text{rad}^{n+1}$, we have $hg' \notin \text{rad}^n$ yet, since hg' factors through M, this contradicts the choice of n. Hence $f \in \bigcap_n \text{rad}^n = \text{rad}^\omega$. □

For a converse see 15.3.8.

The next result, which was observed by various people (for example, [502, 5.5], [640, p. 82], [358, 8.10]) shows that a plausible definition for the transfinite powers of rad, namely "$\text{rad}^{\omega+1} = \text{rad}^\omega \cdot \text{rad}$", is not useful, so explains the definition chosen.

Proposition 9.1.7. *Let R be an artin algebra and suppose that $A \xrightarrow{f} B \in \text{rad}^\omega$, with A, B indecomposable, is non-zero. Then there are morphisms $h, h' \in \text{rad}^\omega$ and $g, g' \in$ rad *such that* $hg = f = g'h'$.*

Proof. Let $0 \to A \xrightarrow{g} A' \to A'' \to 0$ be an almost split sequence (or take g to be $A \to A/\mathrm{soc}(A)$ if A is injective). For each n there is, by 9.1.3, a factorisation $f = f_n \dots f_1$ with $f_i \in \mathrm{rad}$. Then there is h with $f_1 = hg$, hence $f = f_n \dots f_2 hg$. So $k_{n-1} = f_n \dots f_2 h \in \mathrm{rad}^{n-1} \cap (A', B)$. Therefore, for each m, $\mathrm{rad}^m \cap (A', B) \neq 0$. But (A', B) is a module of finite length over the centre, $C(R)$, of R and each $\mathrm{rad}^m \cap (A', B)$ is a $C(R)$-submodule. Therefore, for some n, $\mathrm{rad}^{n-1} \cap (A', B) = \mathrm{rad}^{\omega}(A', B)$, hence $f = k_{n-1}g$ is a factorisation with $k_{n-1} \in \mathrm{rad}^{\omega}$ and $g \in \mathrm{rad}$.

A dual argument (or duality, mod-$R \simeq (R\text{-mod})^{\mathrm{op}}$, 15.1.3) gives the other kind of factorisation. □

We mention that the notation rad^{∞} was, and still sometimes is, used for what we have denoted rad^{ω}.

Kerner and Skowroński, [339, 1.7], showed, that if rad^{ω} is nilpotent, that is, if $\mathrm{rad}^{\omega+n} = 0$ for some $n \in \mathbb{N}$, then R must be of tame representation type. Also, [339, 1.8], R is of finite representation type iff $\mathrm{rad}^{\omega} = 0$ (in which case it must be that $\mathrm{rad}^n = 0$ for some n); indeed, $\mathrm{rad}^{\omega+1} = 0$ is enough to imply finite type, [124, Thm]. The introduction to [125] discusses related results. That finite representation type implies $\mathrm{rad}^n = 0$ for some n follows quickly from the Harada–Sai Lemma, 4.5.10. For the other direction, if R is not of finite type, then there is, by 5.3.40, an infinitely generated indecomposable pure-injective module N and, 9.1.6, any morphism between finitely presented modules which factors through N must belong to the ω-radical. To get a non-zero such morphism take any simple module which is a subquotient of N and obtain a morphism from its projective cover to its injective hull which factors through N (see the end of the proof of 9.1.12). For further results connecting representation type with nilpotence of rad, and for a number of open questions, see [339], [656], [654], [642] and references therein. Also see [351].

9.1.3 Powers of the radical and factorisation of morphisms

Over an artin algebra there is a factorisable system of morphisms in the radical of mod-R *iff the transfinite radical of* mod-R *is non-zero (9.1.9).*

The rank, $\mathrm{rk}(f)$, of a morphism was defined in the previous section. For factorisable systems of morphisms, see Section 7.2.2.

Lemma 9.1.8. ([504, 0.7]) *Suppose that $f = hg$ are morphisms in* mod-R. *If* $\mathrm{rk}(g), \mathrm{rk}(h) \geq \nu \geq 1$ *then* $\mathrm{rk}(f) > \nu$.

Proof. Since $\mathrm{rad} = \mathrm{rad}(\text{mod-}R)$ is two-sided so is each rad^{μ}, so we need consider only the case that $\mathrm{rk}(g) = \nu = \mathrm{rk}(h)$. For ν finite the result is clear by induction.

Suppose that $\nu = \lambda + n$, where λ is a limit and $n \geq 0$ is a natural number. Then $f \in (\text{rad}^\lambda)^{n+1}(\text{rad}^\lambda)^{n+1} = (\text{rad}^\lambda)^{2n+2}$ so $\text{rk}(f) \geq \lambda + 2n + 1 > \nu$, as required. □

Proposition 9.1.9. ([504, 0.6]) *Let R be an artin algebra. Then the following are equivalent:*

(i) there is a factorisable system of morphisms in rad;
(ii) the transfinite radical, rad^∞, *of* mod-R, *is non-zero;*
(iii) there is a factorisable system of morphisms in rad^∞.

Proof. (i) ⇒ (ii) We use notation as in the definition in Section 7.2.2 of a factorisable system. By assumption each $f_{\alpha\beta}$ has rank ≥ 1. Suppose that some $f_{\alpha\beta}$ is not in the transfinite radical, so has a rank: choose $f_{\alpha\beta}$ with minimal rank, ν, say. Take γ with $\alpha < \gamma < \beta$, so $f_{\alpha\beta} = f_{\gamma\beta} f_{\alpha\gamma}$. Since each of $f_{\gamma\beta}$ and $f_{\alpha\gamma}$ has rank ≥ 1 we have an immediate contradiction to 9.1.8.

(ii) ⇒ (iii) Since mod-R is skeletally small, there is an ordinal ν, which may be assumed least such, with $\text{rad}^\nu = \text{rad}^\infty$. Since $\text{rad}^\infty \subseteq (\text{rad}^\nu)^2$ certainly $(\text{rad}^\infty)^2 = \text{rad}^\infty$. Thus every morphism f in rad^∞ is a finite sum $\sum h_i g_i$ with the morphisms g_i,h_i in rad^∞. Set $g = \sum g_i$ and $h = \sum h_i$ (that is, the domain of g is the domain of the g_i and the codomain of g is the direct sum of the codomains of the g_i and similarly for h). Thus, f can be written as a composition, hg, of morphisms which are in rad^∞. Continuing in the obvious way (factorise g and $h, \dots$) we obtain a system of morphisms which form a factorisable system, the third condition of the definition in Section 7.2.2 being satisfied since no component of an element in rad(mod-R) can be invertible).

(iii) ⇒ (i) This is immediate from the definitions. □

Corollary 9.1.10. *Let R be an artin algebra. If there is a factorisable system of morphisms between indecomposable R-modules, then the transfinite radical* rad^∞ *of* mod-R *is non-zero.*

Question 9.1.11. ([504, p. 454]) Suppose that there is a factorisable system in mod-R: does it follow that $\text{rad}^\infty \neq 0$? That is, if there is a factorisable system, is there one in which no component of any of the morphisms is an isomorphism?

9.1.4 The transfinite radical and Krull–Gabriel/m-dimension

If Krull–Gabriel dimension is defined, then the radical of mod-R *is transfinitely nilpotent (9.1.12); the converse is open. There are many conjectures on the relation between Krull–Gabriel dimension, nilpotence of this radical and representation type of artin algebras.*

Recall, 13.2.2, that Krull–Gabriel dimension (of the relevant functor category) and m-dimension (of the relevant lattice of pp conditions) coincide.

In view of the links (Section 7.2.2) between factorisation of morphisms and m-dimension, equivalently Krull–Gabriel dimension, it is clear that there should be a relation between the (transfinite) index of nilpotence of rad-R and these dimensions. One direction is not difficult to prove (see [358, 8.14]).

Proposition 9.1.12. *If* $\mathrm{KG}(R) = \alpha$, *then* $\mathrm{rad}_R^{\omega\alpha+n} = 0$ *for some natural number* n.

Proof. We prove the following more general fact. Suppose that $A \xrightarrow{f} B$ is a morphism in mod-R. Choose a generating tuple $\overline{a}$ of A, let ϕ, respectively ψ, generate $\mathrm{pp}^A(\overline{a})$, resp. $\mathrm{pp}^B(f\overline{a})$. Suppose that $\mathrm{mdim}[\phi, \psi] = \beta$. Then $\mathrm{rk}(f) \leq \omega\beta + n$ for some n. The proof is by induction on β.

Case $\mathrm{mdim}[\phi, \psi] = 0$: Then the interval $[\phi, \psi]$ has finite length, k say. If $f \in \mathrm{rad}^{k+1}$, then $f = f_{k+1} \dots f_1$ with $f_i : A_{i-1} \to A_i$ in rad(mod-R): by definition f is a sum of such terms but, as in 9.1.3 and the proof of 9.1.9, a sum can be replaced by a single term.

Set $\overline{a}_0 = \overline{a}$, $\overline{a}_1 = f_1\overline{a}_0, \dots, \overline{a}_i = f_i\overline{a}_{i-1}, \dots$; so $\overline{b} = \overline{a}_{k+1}$. Let ϕ_i be a pp condition which generates the pp-type of $\overline{a}_i$ in A_i (1.2.6). Then $\phi = \phi_0 > \phi_1 > \dots > \phi_{k+1} = \psi$, since if $\phi_i = \phi_{i+1}$ then, by 1.2.7, there would be $g : A_{i+1} \to A_i$ with $gf_{i+1}\overline{a}_i = \overline{a}_i$, contradicting that $f_{i+1} \in \mathrm{rad}$. But then this strictly descending chain contradicts that the length of $[\phi, \psi]$ is k, as required.

Case $\mathrm{mdim}[\phi, \psi] = \beta + 1$: Let $\sim_\beta$ be the congruence on the interval $[\phi, \psi]$ which identifies points $\phi' \geq \psi'$ with $\mathrm{mdim}[\phi', \psi'] \leq \beta$. By definition of m-dimension, $[\phi, \psi]/\sim_\beta$ has finite length, k say. If we had $\mathrm{rk}(f) > \omega(\beta + 1) + k$, say $f = f_{k+1} \dots f_1$ with $\mathrm{rk}(f_i) \geq \omega(\beta + 1)$, then, with notation and argument as above, we obtain a chain $\phi = \phi_0 > \phi_1 > \dots > \phi_{k+1} = \psi$ with, by induction, each interval $[\phi_i, \phi_{i+1}]$ having m-dimension not less than or equal to β (since otherwise $\mathrm{rk}(f_i) = \omega\beta + n < \omega(\beta + 1)$), a contradiction as before.

Case $\mathrm{mdim}[\phi, \psi] = \lambda$, a limit ordinal: Recall, from the definition of m-dimension, that this means that, modulo $\bigcup_{\beta<\lambda} \sim_\beta$, the interval $[\phi, \psi]$ has finite length, so the above argument applies also in this case, and we have established our initial claim.

So now suppose that $\mathrm{KG}(R) = \beta$, that is, (13.2.2) $\mathrm{mdim}[x = x, x = 0] = \beta$ and, more generally, $\mathrm{mdim}[\overline{x} = \overline{x}, \overline{x} = 0] = \beta$ for $\overline{x} = (x_1, \dots, x_n)$. Let $f : A \to B$ be a morphism in mod-R, let $\overline{a} = (a_1, \dots, a_n)$ generate A and take $\phi = \phi(\overline{x})$ such that $\mathrm{pp}^A(\overline{a}) = \langle\phi\rangle$ and ψ such that $\mathrm{pp}^B(f\overline{a}) = \langle\psi\rangle$. Then $[\phi, \psi]$ is a subinterval of $[\overline{x} = \overline{x}, \overline{x} = 0]$, hence $\mathrm{mdim}[\phi, \psi] \leq \beta$ and so, by what has been shown, $\mathrm{rk}(f) \leq \omega\beta + n$ for some n. This shows that $\mathrm{rad}^{\omega(\beta+1)} = 0$.

To reduce to $\mathrm{rad}^{\omega\beta+n}=0$ for some n we need the following argument from [640], also see [358, 7.2]. Let $P \to A$ be a projective cover and choose an indecomposable direct summand of P, the projective cover, P_S of the simple module S say, such that the composition $P_S \to P \to A \xrightarrow{f} B$ is non-zero. Since $S = P_S/JP_S$ is a subquotient of B there is a further morphism $B \to E(S)$ to the injective hull of S such that the composition $f_S : P_S \to E_S$ with the above map is non-zero. So this morphism is in the ideal of mod-R generated by f, hence has rank $\geq \mathrm{rk}(f)$. There are, up to isomorphism, only finitely many such morphisms f_S, since there are only finitely many simple modules. Each of these morphisms has rank $\leq \omega\beta + n_S$ for some n_S. So, if $n = \max\{n_S\}_S$, then for every morphism f in mod-R we have $\mathrm{rk}(f) \leq \omega\beta + n$, as required. □

It seems not unreasonable to conjecture the converse but, like 9.1.11, this is completely open.

Question 9.1.13. Suppose that $\mathrm{rad}_R^{\infty}=0$; does it follow that $\mathrm{KG}(R)<\infty$? More precisely, if $\mathrm{rad}_R^{\omega\alpha+n}=0$ for some $n \in \mathbb{N}$, does it follow that $\mathrm{KG}(R) \leq \alpha$?

An early conjecture of Prest [495, p. 350], which is still open, is the following.

Conjecture 9.1.14. Suppose that R is a tame, non-domestic, finite-dimensional algebra. Then $\mathrm{KG}(R)=\infty$.

As has been noted at 7.3.29, the other part of that original conjecture, that there are no superdecomposable pure-injectives over tame algebras, was shown to be false by Puninski.

Another conjecture from [495] also proved to be based on too limited a range of examples: this was that if an algebra is of domestic representation type (Section 4.5.5), then its Krull–Gabriel dimension is 2. The examples of Burke and Prest and Schröer (Section 8.1.3) show that, for any natural number $n \geq 2$ there is a domestic string algebra with Krull–Gabriel dimension equal to n (recall, 7.2.8, that $n=0$ corresponds to finite representation type and, by 15.4.5, $n=1$ is impossible). But there are the following modified conjectures, variously due to Prest, Ringel and Schröer (see [529, 1.5], [643, p. 420]). They can be expressed in terms of Krull–Gabriel dimension or nilpotence of rad.

Conjecture 9.1.15. Suppose that R is a finite-dimensional algebra. Then the following are equivalent:

(i) R is of domestic representation type;
(ii) $\mathrm{KG}(R)<\infty$;
(iii) $\mathrm{KG}(R)<\omega$.

Saying this more precisely in terms of the transfinite powers of the radical of mod-R (see the references mentioned above and also [643, p. 423], [358, §8.2]), we have the following.

Conjecture 9.1.16. Suppose that R is a finite-dimensional algebra. Then the following are equivalent, where rad = rad(mod-R):

(i) R is of domestic representation type;
(ii) $\mathrm{rad}^{\infty} = 0$;
(iii) $\mathrm{rad}^{\omega^2} = 0$.

Furthermore, $\mathrm{KG}(R) = n$ iff $\mathrm{rad}^{\omega(n-1)} \neq 0$ and $\mathrm{rad}^{\omega n} = 0$.

Schröer, [642, Thm 2], has proved the second of these conjectures to be correct for special biserial algebras (this includes the class of string algebras), indeed, [642, Thm 3], he has shown that if n is the maximal length of a path in the "bridge quiver", then $\mathrm{rad}^{\omega(n+1)} \neq 0$ and $\mathrm{rad}^{\omega(n+2)} = 0$. The bridge quiver is a graph with vertices the (indecomposable) bands (in the sense of Section 8.1.3) and with a directed edge from band w to band w' if there is a string u such that wuw' also is a string.

The conjecture stated in terms of Krull–Gabriel dimension is still open, but, for string algebras with a single band of type $\widetilde{A}$ (meaning that no vertex occurs twice), Prest and Puninski ([524, 7.7]) have proved that the Krull–Gabriel dimension is equal to 2 or 3, depending on the shape of the bridge quiver, as conjectured.

Schröer, [640, §6], see [642, Thm 1], gave for each ordinal α of the form $\omega n + t$ with $t \geq 1$, except $\omega + 1$ (which by [124, Thm] cannot occur), an example of an algebra, obtained by modifying the algebras Λ_n seen in Section 8.1.3, with rad having index of nilpotence equal to α.

A useful point, used in the proof of 9.1.12 and made in [640, §2] (also see [358, 8.9]), is that if $f \in$ mod-R is non-zero, then for some simple module S the essentially unique non-zero morphism f_S from the projective cover of S, via S, to the injective hull of S is in the ideal generated by f, being of the form $f_S = hfg$ for some $g, h \in$ mod. So the index of nilpotence of rad_R is equal to the maximal rank of an f_S and hence, since there are just finitely many simple modules over an artin algebra, it follows that this index is not a limit ordinal.

For tame hereditary artin algebras the link between Krull–Gabriel dimension and factorisation of morphisms, including detailed information on the latter, is given in [224, esp. §5].

9.2 Fp-idempotent ideals

An ideal of mod-R *is idempotent iff its functorial annihilator is closed under extensions (9.2.1) and there is an analogous "finite" result (9.2.2).*

A right ideal, I, of a ring R is **idempotent** if $I^2 = I$. This property can be detected in the category of modules by the fact that I is idempotent iff $\mathrm{Ann}(I) = \{M \in \text{Mod-}R : MI = 0\}$ is closed under extensions. To see this: if $0 \to L \to M \to N \to 0$ is exact with $L, N \in \mathrm{Ann}(I)$, then clearly $MI^2 = 0$, so idempotence of I gives closure of $\mathrm{Ann}(I)$ under extension. Conversely, consider the exact sequence $0 \to I/I^2 \to R/I^2 \to R/I \to 0$: both I/I^2 and R/I are in $\mathrm{Ann}(I)$ but R/I^2 is in $\mathrm{Ann}(I)$ iff $I^2 = I$.

"The same" proof works for any one-sided ideal $\mathcal{I}$ of mod-R, where $\mathcal{I}$ is **idempotent** if $\mathcal{I}^2 = \mathcal{I}$. By the **annihilator** of $\mathcal{I}$ we mean $\mathrm{Ann}(\mathcal{I}) = \{F \in (\text{mod-}R, \mathbf{Ab}) : Fh = 0 \text{ for all } h \in \mathcal{I}\}$. Note that the annihilator of a one-sided ideal equals the annihilator of the two-sided ideal that it generates, so in this definition $\mathcal{I}$ can be one- or two-sided.

Proposition 9.2.1. (see [358, C.7]) *An ideal* $\mathcal{I}$ *of* mod-R *is idempotent iff* $\mathrm{Ann}(\mathcal{I}) = \{F \in (\text{mod-}R, \mathbf{Ab}) : Fh = 0$ *for all* $h \in \mathcal{I}\}$ *is closed under extensions.*

Proof. ($\Rightarrow$) Suppose first that $0 \to F \xrightarrow{\alpha} G \xrightarrow{\beta} H \to 0$ is an exact sequence with $F, H \in \mathrm{Ann}(\mathcal{I})$. First it is shown that $G\mathcal{I}^2 = 0$. It is enough to check that if $A \xrightarrow{f} B \xrightarrow{g} C$ is exact with $f, g \in \mathcal{I}$, then $G(gf) = 0$. Consider the diagram.

$$\begin{array}{ccccccccc}
0 & \longrightarrow & FA & \xrightarrow{\alpha_A} & GA & \xrightarrow{\beta_A} & HA & \longrightarrow & 0 \\
 & & \downarrow{\scriptstyle 0=Ff} & & \downarrow{\scriptstyle Gf} & & \downarrow{\scriptstyle Hf=0} & & \\
0 & \longrightarrow & FB & \xrightarrow{\alpha_B} & GB & \xrightarrow{\beta_B} & HB & \longrightarrow & 0 \\
 & & \downarrow{\scriptstyle 0=Fg} & & \downarrow{\scriptstyle Gg} & & \downarrow{\scriptstyle Hg=0} & & \\
0 & \longrightarrow & FC & \xrightarrow{\alpha_C} & GC & \xrightarrow{\beta_C} & HC & \longrightarrow & 0
\end{array}$$

Since $\beta_B Gf = 0$, there is $\gamma : GA \to FB$ with $\alpha_B \gamma = Gf$, so $GgGf = \alpha_C Fg\gamma = 0$, as required. Therefore if $\mathcal{I}^2 = \mathcal{I}$ then $\mathrm{Ann}(\mathcal{I})$ is closed under extensions.

($\Leftarrow$) For the converse, if $\mathcal{I}^2 \neq \mathcal{I}$, then choose A, B in mod-R such that there is $f \in \mathcal{I}(A, B) \setminus \mathcal{I}^2(A, B)$ and consider the functors $\mathcal{I}^2(A, -) \leq \mathcal{I}(A, -) \leq (A, -)$, where $\mathcal{I}^2(A, -)$, for example, is given on objects by $C \mapsto \mathcal{I}^2(A, C)$. There is an exact sequence

$$0 \to \mathcal{I}(A, -)/\mathcal{I}^2(A, -) \to (A, -)/\mathcal{I}(A, -) \to (A, -)/\mathcal{I}(A, -) \to 0,$$

where each "end" functor annihilates $\mathcal{I}$: for example, if $h : C \to D$ is in mod-R, then $(A, h) : (A, C) \to (A, D)$ takes $g : A \to C$ to hg, which is in $\mathcal{I}(A, D)$, so if $h \in \mathcal{I}$, then $\big((A, -)/\mathcal{I}(A, -)\big)h = 0$. But $(A, f)/\mathcal{I}^2(A, f) : (A, A)/\mathcal{I}^2(A, A) \to (A, B)/\mathcal{I}^2(A, B)$ takes 1_A to $(f + \mathcal{I}^2(A, B))/\mathcal{I}^2(A, B) \neq 0$, so $(A, -)/\mathcal{I}(A, -) \notin \mathrm{Ann}(\mathcal{I})$. □

Set $\mathrm{ann}(\mathcal{I}) = (\text{mod-}R, \mathbf{Ab})^{\mathrm{fp}} \cap \mathrm{Ann}(\mathcal{I})$ to be the category/class of finitely presented functors which annihilate $\mathcal{I}$.

Proposition 9.2.2. *If $\mathcal{I}$ and $\mathcal{I}^2$ are finitely generated left ideals of* mod-R, *then $\mathcal{I}$ is idempotent iff* $\mathrm{ann}(\mathcal{I})$ *is closed under extensions.*

Proof. Say $f : A \to B$ generates $\mathcal{I}$, so, for $C, D \in$ mod-R, we have $\mathcal{I}(C, D) = 0$ if $C \neq A$ and $\mathcal{I}(A, D) = \{gf : g \in (B, D)\}$. Then the functor $\mathcal{I}(A, -)$ is isomorphic to the functor $\mathrm{im}\big((B, -) \xrightarrow{(f,-)} (A, -)\big)$ which is finitely generated (even finitely presented, see 10.2.2).

If also $\mathcal{I}^2$ is finitely generated, then so is the functor $\mathcal{I}^2(A, -)$, hence all the functors appearing in the exact sequence in ($\Leftarrow$) of the proof of 9.2.1 are finitely presented, so that proof applies to yield the conclusion of this result. □

Motivated by the above, Krause said, [358, Appx C and Chpt. 5], that an ideal $\mathcal{I}$ of mod-R is **fp-idempotent** if $\mathrm{ann}(\mathcal{I}) = \{F \in (\text{mod-}R, \mathbf{Ab})^{\mathrm{fp}} : F\mathcal{I} = 0\}$ is closed under extensions, that is, see 11.2.6, iff $\mathrm{ann}(\mathcal{I})$ is a Serre subcategory of $(\text{mod-}R, \mathbf{Ab})^{\mathrm{fp}}$. By the results above, idempotent ideals are fp-idempotent and for ideals which are, together with their squares, finitely generated, there is no difference.

At least over artin algebras, there is a natural bijection, 15.3.4, between fp-idempotent ideals of mod-R and Serre subcategories of $(\text{mod-}R, \mathbf{Ab})^{\mathrm{fp}}$, and hence between these and all the other data listed at 12.4.1.

Examples 9.2.3. (a) Let R be an artin algebra. Then rad^ω is an fp-idempotent ideal of mod-R, with $\mathrm{ann}(\mathrm{rad}^\omega) = \mathcal{S}_0$, the Serre subcategory of finite length functors in $(\text{mod-}R, \mathbf{Ab})^{\mathrm{fp}}$.

First, without any assumption on the ring R, let $h(: A \to B) \in \mathrm{rad}^\omega$ and let $F \in \mathcal{S}_0$. By 10.2.30 $F \simeq F_\phi/F_\psi$ for some pp-pair ϕ/ψ, and there is a chain $\phi = \phi_0 > \phi_1 > \cdots > \phi_n = \psi$ of pp conditions with each ϕ_i/ϕ_{i+1} a minimal pair (that is, with the corresponding functor $F_{\phi_i}/F_{\phi_{i+1}}$ simple). Since $h \in \mathrm{rad}^n$ there is a chain $A = A_0 \xrightarrow{h_1} A_1 \xrightarrow{h_2} \cdots \xrightarrow{h_n} A_n = B$ with $h = h_n \cdots h_2 h_1$ and each $h_i \in \mathrm{rad}$. Since each h_i strictly increases the pp-type of every non-zero tuple in A_{i-1} (by 1.2.9) if $\overline{a} \in \phi_0(A)$, then, by induction, $h_i \ldots h_1\overline{a} \in \phi_i(A_i)$ and hence $F_\phi h : \phi(A) \to \phi(B)$ has image contained in $\psi(B)$, that is, $F_\phi h = 0$, showing that $\mathcal{S}_0 \subseteq \mathrm{ann}(\mathrm{rad}^\omega)$.

Suppose, for the converse, that R is an artin algebra and that $F, = F_\phi/F_\psi$ say, is a finitely presented functor of infinite length. By 10.2.1 there is a projective presentation $(B, -) \xrightarrow{(h,-)} (A, -) \to F \to 0$ for some $h : A \to B$ in mod-R. Thus ϕ may be taken to be of the form $\overline{x}H = \overline{0}$, where $\overline{a}$ is some generating tuple of A with $\overline{a}H = \overline{0}$ being a presentation of A, and then ψ is a generator of $\mathrm{pp}^B(h\overline{a})$. By 10.2.11 the lattice of finitely generated subfunctors of F is isomorphic to the lattice of pp conditions in the interval $[\phi, \psi]$, so, by assumption, the latter is of infinite length. It follows by 5.3.34 that there is an infinitely generated point $N \in (\phi/\psi) \subseteq \mathrm{Zg}_R$; so $\overrightarrow{F} N \neq 0$. Therefore, by 15.3.2, there is a morphism f of mod-R which factors through N and such that $Ff \neq 0$. By 9.1.6, $f \in \mathrm{rad}^\omega$ and so $F \notin \mathrm{ann}(\mathrm{rad}^\omega)$, as required.

(b) Suppose that R is an artin algebra and that M is a module of finite endolength. Consider the ideal $[M]$ of morphisms in mod-R which factor through a finite direct sum of copies of M, equivalently, by 4.4.30, through a module in the definable subcategory generated by M; this definable subcategory will be that which corresponds in the sense of 15.3.4 to $[M]$. As noted at [363, p. 396], if M has a summand of finite length, then $[M]$ will contain an idempotent morphism; otherwise, by 4.3.57, $[M]$ will be nilpotent.

These ideals are considered further in Sections 11.2.1 and 15.3.

Krause used these ideals to give an "internal" definition of tameness ([363]).

Appendix A Model theory

This is a quick introduction to certain basic concepts of model theory as well as to some particular results from the model theory of modules. These are not really needed for this book, despite their occasional mention, but many of the papers cited in this book assume some acquaintance with these basics. For a somewhat lengthier introduction to model theory see [513]. That article considers general structures whereas here the definitions are given in the context of modules. More details, including proofs of results stated, can be found in, for instance, [114], [296], [435], [623].

A.1 Model theory of modules

The main results are: pp-elimination of quantifiers for modules (A.1.1); a criterion for elementary equivalence of modules (A.1.2, also A.1.6); the relation between generating the same definable subcategory and elementary equivalence (A.1.3); a criterion for elementary substructure (A.1.4); Łoś' Theorem on ultraproducts (A.1.7); the Compactness Theorem (A.1.8); the downwards and upwards Löwenheim–Skolem Theorems (A.1.10, A.1.12).

Let R be a ring. The usual language, $\mathcal{L}_R$, for R-modules has for its **atomic** (that is, simplest) **formulas** those of the form $\sum_{i=1}^{n} x_i r_i = 0$, where the x_i are **variables** (indeterminates) and the r_i are elements of the ring. Formally, each element $r \in R$ is represented in the language by a function symbol with which to represent "multiplication by r" but, rather than write $f_r(x)$, we write the more natural xr or, in the case of left modules, rx. The symbol 0 is formally added to the language as a constant symbol and there is a binary function symbol to represent addition in a module, with iterated addition abbreviated, as usual, by a summation symbol. An expression such as $\sum_i x_i r_i$ is a **term**.

All other formulas are built up, in particular but obvious ways, from the atomic formulas using the following operators:

$\neg$ for negation, but we write $\sum_i x_i r_i \neq 0$ rather than $\neg(\sum_i x_i r_i = 0)$;
$\wedge$ ("and") for conjunction, we also use $\bigwedge$ for iterated $\wedge$ as in, for example,
$\bigwedge_{j=1}^{m} \sum_i x_i r_{ij} = 0$ which is an abbreviation for the system of equations
$(\sum_i x_i r_{i1} = 0) \wedge \cdots \wedge (\sum_i x_i r_{im} = 0)$;
$\vee$ ("or") for disjunction.

Formulas built up from the atomic formulas using these operations are the **quantifier-free** formulas. One also needs to use parentheses in order to avoid ambiguity: "(A and B) or C" is not logically equivalent to "A and (B or C)".

The formulas are built up from the quantifier-free formulas by prefixing them with **existential quantifiers** $\exists x$ ("there exists an x such that") and **universal quantifiers** $\forall x$ ("for all x").

Any expression built up from the atomic formulas using any combination of these operations is called a **formula** and any such expression is logically equivalent to one which has the form: a string of quantifiers, followed by a quantifier-free formula. For example, $\forall x \, (x = 0 \vee \neg\exists y(x = yr))$ is logically equivalent to $\forall x \forall y \, ((x = 0) \vee (x \neq yr))$. A formula which can be built up from the atomic formulas using only conjunction and existential quantification is called a **positive primitive** or **pp** formula (what we have referred to as a "pp condition").

"Built up" is precisely rendered as follows: any atomic formula is a formula; if ϕ is a formula, then so are $\neg\phi$, $\forall x\phi$ and $\exists x\phi$, where x is any variable; if ϕ and ψ are formulas, then so are $\phi \wedge \psi$ and $\phi \vee \psi$. We can introduce $\rightarrow$ ("implies"), as in $\phi \rightarrow \psi$, as a separate operator or as an abbreviation for $(\neg\phi) \vee \psi$.

If ϕ is a formula, then the expression $\phi(x_1, \dots, x_n)$ is used to mean that the free variables of ϕ occur among $x_1, \dots, x_n$; it is convenient not to insist that each of these actually appears (free) in ϕ. The notion "free variable in a formula" is defined inductively as follows. The **free variables** of a quantifier-free formula are all those variables which appear in it. If ϕ is a formula, then the **free variables** of $\exists x\phi$ are those of ϕ except x, any free occurrences of which in ϕ now occur **bound** or within the **scope of** the existential quantifier $\exists x$. Similarly for $\forall x$. In view of the inductive definition of how formulas are built up we should add the clauses: the free variables of $\neg\phi$ are precisely those of ϕ; the free variables of $\phi \wedge \psi$ and of $\phi \vee \psi$ are those occurring free in either ϕ or ψ. A formula with no free variables is a **sentence**.

Given a language – roughly, a certain choice of basic function, relation and constants symbols – a structure M for that language is a set with corresponding functions, relations and distinguished elements. If σ is a sentence of that language, then we write $M \models \sigma$ ("M **satisfies** σ" or "σ holds in M") if σ is true in M. For the precise definition of this relation, as well as the definition of being a structure for a language, see any book on model theory, for example [114], [296]; in the cases that we consider, the meaning should be clear. More generally, if $\phi = \phi(x_1, \dots, x_n)$ is a formula of the language and if $a_1, \dots, a_n \in M$, then write $M \models \phi(a_1, \dots, a_n)$ if, when every free occurrence of x_i in ϕ is replaced by a_i $(i = 1, \dots, n)$, the resulting statement $\phi(a_1, \dots, a_n)$ is true in M. Again, the meaning should be clear in cases that we consider but for a proper definition consult a model theory text. This is usually regarded as a condition on the tuple $\overline{a} = (a_1, \dots, a_n)$ from M so $M \models \phi(\overline{a})$ is often read as "$\overline{a}$ **satisfies** ϕ (in M)". In this book I have used the term "condition" instead of "formula" in order to emphasise their role rather than their formal structure.

A **theory** is simply a set, T, of sentences and one says that M **satisfies**, or **is a model of**, T, writing $M \models T$, if for every $\sigma \in T$ we have $M \models \sigma$ (also read as "M satisfies σ" or "σ is true in M"). For example, "the theory of R-modules" may be taken to mean some set of axioms for R-modules, written formally. So, if $rs = t$ in R, then $\forall x\big((xr)s = xt\big)$ would be one of those axioms. Often little distinction is made between a theory and its deductive closure, meaning the set of all sentences that follow from it. For instance, one would not normally include $\forall x\,(x+x=0 \to x=0)$ among the axioms for modules because it is a consequence of those axioms but it does belong to "the theory of R-modules", that is, to the set of all sentences which follow from the axioms for R-modules, equivalently the set of all sentences which are true in every R-module. We denote this $\mathrm{Th}(\text{Mod-}R) = \{\sigma \text{ a sentence} : M \models \sigma \text{ for every } M \in \text{Mod-}R\}$. More generally, the theory of a subclass, $\mathcal{D}$, of Mod-R is $\mathrm{Th}(\mathcal{D}) = \{\sigma : M \models \sigma \text{ for all } M \in \mathcal{D}\}$ $(\supseteq \mathrm{Th}(\text{Mod-}R))$. If $\mathcal{D}$ is a singleton, $\mathcal{D} = \{M\}$, then one writes $\mathrm{Th}(M)$. In this last case the theory, T, is **complete** in the sense that for every sentence σ either $\sigma \in T$ or $\neg\sigma \in T$. The theory of all R-modules is not complete: for instance, neither the sentence $\forall x(x=0)$ nor its negation $\neg\forall x(x=0)$, equivalently $\exists x(x \neq 0)$, is in Th(Mod-R).

If T is any set of sentences in the language of R-modules, then the class of **models** of T is $\mathrm{Mod}(T) = \{M \in \text{Mod-}R : M \models T\}$. By the definitions $\mathrm{Th}(\mathrm{Mod}(T))$ is the deductive closure of T. A class of the form $\mathrm{Mod}(T)$ for some set T of sentences is said to be **elementary**, or **axiomatisable** or **definable**. However, this use of the term "definable" is not quite as used in this book,[1] in that an elementary class of modules need not be closed under products: consider, for example, $R = \mathbb{Z}$ and let $T = \{\sigma\}$, where σ is $\forall x, y, z\,(x = y \vee x = z \vee y = z)$. Nor is a general elementary class closed under pure submodules: take $T = \{\exists x(x \neq 0)\}$. By 3.4.7 the "definable" is equivalent to "elementary, plus closed under products and pure submodules". If, for instance, R is an algebra over an infinite central subfield, then, by A.1.3, every elementary subclass of Mod-R will be closed under products. This is a consequence of a fundamental theorem (due to Baur and others) in the model theory of modules which says, see just below, that, although one may write down arbitrarily complicated formulas for modules, they are all equivalent to reasonably simple ones, namely those ("invariants statements") which refer to indices of pp-definable subgroups in each other. Note that if $\phi \geq \psi$ are pp conditions in the same free variables, then "$\mathrm{card}(\phi/\psi) \geq m$" is expressed by the sentence $\exists \overline{x}_1, \dots, \overline{x}_m \left(\bigwedge_1^m \phi(\overline{x}_j) \wedge \bigwedge_{i,j=1, i\neq j}^m \neg\psi(\overline{x}_i - \overline{x}_j)\right)$.

Theorem A.1.1. (pp-elimination of quantifiers) (Baur, Monk, . . . , see [495, p. 36] for references) *Let R be any ring.*

(a) *If σ is a sentence in the language of R-modules, then there is a finite boolean combination, τ, of sentences of the form* $\mathrm{card}(\phi/\psi) \geq m$, *where ϕ, ψ are pp conditions and m is a positive integer, such that σ is equivalent to τ in the sense that for every R-module $M \models \sigma$ iff $M \models \tau$.*

(b) *If χ is any formula (not necessarily pp) in the language of R-modules, then there is a sentence, τ, and a finite boolean combination, η, of pp conditions (that is pp formulas) such that for every module M and tuple $\overline{a}$ from M (matching the free variables of χ) we have $M \models \chi(\overline{a})$ iff both $M \models \tau$ and $M \models \eta(\overline{a})$. In particular, the solution set to χ in every module M is a finite boolean combination (that is, using $\cap$, $\cup$, ^{c}) of pp-definable subgroups. If non-zero constants are allowed in χ, then the solution set will be a finite boolean combination of cosets of pp-definable subgroups.*

The relation of **elementary equivalence** of structures is defined as $M \equiv N$ iff for every sentence σ one has $M \models \sigma$ iff $N \models \sigma$, that is, iff $\mathrm{Th}(M) = \mathrm{Th}(N)$. From A.1.1 one obtains

[1] This is Crawley-Boevey's use from [145], which I have adopted, see Section 3.4.1.

the following criterion, which appears explicity as [207, Thm 2], for elementary equivalence of modules.

Theorem A.1.2. *For any R-modules M_1 and M_2 one has $M_1 \equiv M_2$ iff for all pp-pairs $\phi > \psi$ (in one free variable),* $\mathrm{card}\big(\phi(M_1)/\psi(M_1)\big) = \mathrm{card}\big(\phi(M_2)/\psi(M_2)\big)$ *if at least one of these is finite.*

From this the next result follows directly.

Theorem A.1.3. *For any R-modules M_1 and M_2 the following are equivalent:*

(i) M_1 and M_2 generate the same definable subcategory of Mod-R;
(ii) $\mathrm{supp}(M_1) = \mathrm{supp}(M_2)$;
(iii) for all pp-pairs $\phi > \psi$ (in one free variable) $\phi(M_1) > \psi(M_1)$ iff $\phi(M_2) > \psi(M_2)$;
(iv) $M_1^{\aleph_0} \equiv M_2^{\aleph_0}$.

If M is a structure for a language and M' is a substructure (the definition is fairly obvious), then M' is an **elementary substructure** of M, written $M' \prec M$ (and then M is an **elementary extension** of M') if for every formula $\phi(x_1, \ldots, x_n)$ and elements $a_1, \ldots, a_n$ of M', one has $M' \models \phi(a_1, \ldots, a_n)$ iff $M \models \phi(a_1, \ldots, a_n)$. Roughly, the properties of elements and tuples from M' are the same whether viewed in M' or in M; at least those properties which can be expressed by formulas of the language. This implies $M' \equiv M$ (take ϕ to have no free variables) but is stronger. For instance, the abelian group, $2\mathbb{Z}$, of even integers is isomorphic to the group, $\mathbb{Z}$, of integers so these certainly are elementarily equivalent, but the inclusion of $2\mathbb{Z}$ in $\mathbb{Z}$ is not elementary: take $\phi(x)$ to be $\exists y\, (x = y + y)$ and $a = 2$. Sabbagh characterised elementary substructure in modules in terms of purity (now a direct consequence of A.1.3).

Theorem A.1.4. ([629, Thm 2]) *Let $M' \leq M$ be modules. Then $M' \prec M$ iff $M' \equiv M$ and M' is a pure submodule of M.*

Corollary A.1.5. *A subcategory of* Mod-R *is definable (in the sense of Section 3.4.1) if it is the full subcategory on an axiomatisable class of modules which is closed under finite direct sums and direct summands.*

There is the following completely general result relating ultrapowers (Section 3.3.1) and elementary equivalence. The structures M and N in the statement should, of course, be assumed to be of the same general kind (so that isomorphism between them makes sense).

Theorem A.1.6. (Keisler and Shelah, see [114, 6.1.15]) *Two structures M, N are elementarily equivalent iff they have isomorphic ultrapowers, that is, iff $M^I/\mathcal{F} \simeq N^J/\mathcal{G}$ for some index sets I, J and ultrafilters $\mathcal{F}, \mathcal{G}$ on them.*

It is an old result of model theory that every structure M is elementarily equivalent to each of its ultrapowers: $M \equiv M^I/\mathcal{F}$. This is immediate by Łoś' Theorem, which is just as 3.3.3 but with arbitrary formulas in place of pp formulas and with ultrapowers in place of reduced powers (ultrapowers are needed to take care of any negations and disjunctions in arbitrary formulas). A proof can be found in just about any model theory textbook.

Theorem A.1.7. (Łoś' Theorem, [410]) *If M_i $(i \in I)$ are structures for a language, if $\mathcal{F}$ is an ultrafilter on I, if $\phi(\overline{x})$ is a formula in the language and if $\overline{a} = (a^1, \ldots, a^n)$, $a^j =$*

$(a_i^j)_i/\mathcal{F}$, is a tuple from $M^\star = \prod_I M_i/\mathcal{F}$, then $M^\star \models \phi(\overline{a})$ iff $\{i \in I : M_i \models \phi(a_i^1, \ldots, a_i^n)\} \in \mathcal{F}$.

The Compactness Theorem is central in model theory. It has not been used explicitly here though its proof has (3.3.6 is a special case) and some uses of it have been replaced by compactness of the Ziegler spectrum (5.1.23). Its statement is the following.

Theorem A.1.8. (Compactness Theorem) *(a) Let T be a set of sentences of a language. If every finite subset of T has a model, then T has a model; that is, if for every $T' \subseteq T$ with T' finite, there is a structure M with $M \models T'$ then there is M with $M \models T$.*

(b) Let M be a structure. Suppose that Φ is a set of formulas with parameters from M such that, for every finite subset, Φ', of Φ there is a solution for Φ' in M, that is, a tuple from M which satisfies all the formulas in Φ'. Then there is an elementary extension of M which contains a solution for Φ.

Both say, roughly, that if a situation can be described by infinitely many conditions and if these conditions are finitely compatible, then the situation can be realised. The second part (which is a consequence of the first) implies that every type (like a pp-type but not restricted to pp formulas) can be realised in an elementary extension. A third expression of this result is that if T is a set of sentences and if σ is a sentence such that every structure which satisfies T satisfies σ, then there is a finite subset T' of T such that every structure which satisfies T' must satisfy σ. That is, if T implies σ, then some finite subset of T implies σ.

Occasionally one sees the terminology "syntax/semantics", especially with reference to the distinction between these. Roughly, "syntactic" refers to formal manipulation of the strings of symbols which are formulas; "semantic" refers to arguments which go via their meanings, or solution sets of conditions, in structures.

Example A.1.9. As an illustration of the model-theoretic way of saying things, I give the model-theoretic proof of 3.4.7(iv)$\Rightarrow$(i).

With notation as there, we are supposing that $\mathcal{X} \subseteq \text{Mod-}R$ is closed under direct products, pure submodules and ultrapowers: set $T = \{\phi/\psi : \phi(M) = \psi(M) \text{ for all } M \in \mathcal{X}\}$ and set $\mathcal{X}' = \text{Mod}(T)$. Then $\mathcal{X} \subseteq \mathcal{X}'$ and we must show the converse.

For each pp-pair ϕ'/ψ' which is open on some member of $\mathcal{X}$, choose such a member $M = M_{\phi'/\psi'}$. Replacing M by $M^{\aleph_0}$ if necessary, we may suppose that $\phi'(M)/\psi'(M)$ is infinite. Let M_1 be the direct sum of the $M_{\phi'/\psi'}$ over all such pairs ϕ'/ψ'.

Let $N \in \mathcal{X}'$. By construction and 1.2.3, M_1 and $M_1 \oplus N$ satisfy the condition of A.1.2, hence are elementarily equivalent. Therefore, by A.1.6, they have isomorphic ultrapowers: $M_1^I/\mathcal{F} \simeq (M_1 \oplus N)^J/\mathcal{G}$ for some index sets I, J and ultrafilters $\mathcal{F}, \mathcal{G}$. By assumption, $M_1^I/\mathcal{F} \in \mathcal{C}$. By 3.3.4, $M_1 \oplus N$ is, therefore, a pure submodule of $M_1^I/\mathcal{F}$, hence so is N. Therefore $N \in \mathcal{C}$, as required.

Just occasionally I refer to the Löwenheim–Skolem Theorem: in fact there are two (they can be found in more or less any model theory text). The downwards one says that one can find "small" substructures.

Theorem A.1.10. (Downwards Löwenheim–Skolem Theorem) *If M is a structure for a language $\mathcal{L}$ and $A \subseteq M$, then there is an elementary substructure, M', of M with $A \subseteq M'$ and* $\text{card}(M') = \max(\text{card}(\mathcal{L}), \text{card}(A))$.

The **cardinality** of a language is the number of formulas in it. For instance, if R is a ring of cardinality κ, then the cardinality of the language of R-modules is $\max(\kappa, \aleph_0)$.

Corollary A.1.11. (of 3.4.11 and A.1.3) *If M is an R-module and $A \subseteq M$, then there is a pure submodule $M' \leq M$ which generates the same definable subcategory as M, with $A \subseteq M'$ and* $\mathrm{card}(M') = \max\big(\mathrm{card}(A), \mathrm{card}(R), \aleph_0\big)$.

The upwards theorem just says that there are arbitrarily large elementary extensions.

Theorem A.1.12. (Upwards Löwenheim–Skolem Theorem) *If M is an infinite L-structure and κ is any cardinal with $\kappa \geq \max\{\mathrm{card}(M), \mathrm{card}(L)\}$, then there is an elementary extension of M of cardinality κ.*

Part II

Functors

Context Our attention so far has been centred on categories of modules, though much use has been made of "functorial" techniques. This part begins with an exposition of the general background for, and then the details of, these methods. Henceforth we will work much of the time in categories of functors and localisations of these. We begin with a chapter which introduces additive functor categories; we take the view that functors are just generalised modules and can largely be understood as such. That is followed by a chapter on torsion-theoretic localisation of categories. In our context this is the functorial analogue of restricting to a closed subset of the Ziegler spectrum, equivalently of restricting to a definable subcategory.

Having made the effort to step into this more abstract world, we can increase the reward by widening the applicability of the results that we prove. For, there are many categories of structures other than categories of modules to which the results of Part I apply. Examples include categories of presheaves and sheaves of modules, many subcategories of categories of modules (for a familiar example, the category of torsion abelian groups), categories of comodules, functor categories themselves. Therefore, instead of starting with the category of modules over a ring, from the outset we work in a more general context. But there is no harm in sticking with the modules case if one wishes to do so and I have tried to write this part in a way that allows the reader to ignore the extra generality if he or she so desires.

The correct generality for our results is definable categories. But all these are well-behaved subcategories of locally finitely presented categories, so we may deal with the latter and then relativise. But locally finitely presented categories are themselves well-behaved subcategories of functor categories, so it is in the context of functor categories that our results are, in the first instance, developed, then we explain how to relativise. But a functor category is really just a module category, admittedly over a ring that might not have a 1.

For background material we refer mainly to [195] for additive and abelian categories, [663] and [486] for localisation. The texts [84] and [332] also cover a great deal of relevant background material.

10
Finitely presented functors

A fresh entry-point to the book, this chapter begins with the generalisation of rings to small preadditive categories and of modules to functors on such categories. The basic results on additive functor categories are presented and the category of finitely presented functors is shown to be equivalent to the category of pp-pairs and pp-definable maps. This, in turn, is equivalent to the free abelian category. A good part of the "dictionary" between "pp" and functorial concepts is given.

10.1 Functor categories

The generalisation of rings to small preadditive categories and modules to functors on such categories is explained and illustrated in Section 10.1.1. Section 10.1.2 presents the Yoneda embedding of a category into a functor category via *representable functors. The generating set of representable functors is discussed in Section 10.1.3.*

Some short introductions to the key category of finitely presented functors can be found in, for instance, [69], [183], [323].

10.1.1 Functors and modules

This section contains basic definitions and examples and shows the pointwise nature of some concepts for functors.

We have seen a variety of functors – homological functors and functors defined by pp conditions – from categories of modules to the category, **Ab**, of abelian groups. We have also seen, 1.2.31, that pp functors commute with direct limits, so, since every module is a direct limit of finitely presented modules (E.1.20), these functors are determined by their action on the category of finitely presented modules. Consider the category of all such functors: let (mod-R, **Ab**) denote the category of additive functors from the category, mod-R, of finitely presented right R-modules to **Ab**. The objects are the functors and the morphisms from a functor F to a functor G are the natural transformations from F to G.

This is the context within which much of our work will now be set.

Here is a wider context within which we may work, with no more effort.

First we replace the ring R by something more general, namely a **small preadditive category** $\mathcal{A}$. Recall (see Appendix E, p. 703) that "small" means that the collection of objects, and hence morphisms, of $\mathcal{A}$ is a set;[1] "preadditive" means that, for each pair, A, B, of objects of $\mathcal{A}$, the set, $\mathcal{A}(A, B)$, which we write as just (A, B), of morphisms from A to B is equipped with an abelian group structure such that composition is bilinear: $f(g+h) = fg + fh$ and $(g+h)f = gf + hf$ when these are defined.

For instance, if R is a ring, then consider the category which has just one object, $*$, and which has $(*, *) = R$ with the addition on R giving the abelian group structure on $(*, *)$ and the multiplication on R giving the composition (so, according to our convention for composition of morphisms, rs means do s then r). This allows us to regard the ring R as a small preadditive category with just one object. Conversely, to give such a category is really just to give a ring, namely the ring of endomorphisms of the sole object.

If $\mathcal{A}$ has only finitely many objects, $A_1, \dots, A_n$, set $R = \bigoplus_{i,j}(A_i, A_j)$. Define addition and multiplication (that is, composition) on R pointwise, with the convention that if the domain of f is not equal to the codomain of g, where $f, g \in \bigcup_{i,j}(A_i, A_j)$, then the product fg is zero. It is easy to check that R is a ring and that $1 = e_1 + \cdots + e_n$, where e_i is the identity map of A_i, is a decomposition of $1 \in R$ into a sum of orthogonal idempotents. This ring[2] R codes up almost all the information contained in the category $\mathcal{A}$ (admittedly the objects A_i might not be recoverable from R, but that really does not matter for anything we do here).

Example 10.1.1. For a simple example, let k be any field and suppose that the preadditive category $\mathcal{A}$ has two objects, A_1, A_2 such that $\mathrm{End}(A_1) \simeq k \simeq \mathrm{End}(A_2)$, $(A_1, A_2) \simeq k$, $(A_2, A_1) = 0$. Set $R = \bigoplus_{i,j}(A_1, A_j)$, most naturally represented as $R = \begin{pmatrix} (A_1, A_1) & (A_2, A_1) \\ (A_1, A_2) & (A_2, A_2) \end{pmatrix} = \begin{pmatrix} k & 0 \\ k & k \end{pmatrix}$, which would be the endomorphism ring, acting on the left, of the object $A_1 \oplus A_2$ if it existed in $\mathcal{A}$. Then the category of $\mathcal{A}$-modules (see below) and that of R-modules are equivalent in a very direct way, which one should write down if this is unfamiliar. (Whether right or left modules depends on the conventions you adopt for composition of morphisms.)

[1] In fact, it will be enough if $\mathcal{A}$ is skeletally small, meaning that the isomorphism types of objects of $\mathcal{A}$ form a set.

[2] This ring has a natural representation as the ring of matrices $(f_{ij})_{ij}$ with $f_{ij} \in (A_j, A_i)$.

So, to get something genuinely new, the category $\mathcal{A}$ must have infinitely many objects. Even in that case one may set $R = \bigoplus_{i,j}(x_i, x_j)$ as before to obtain a ring-without-1, but with enough idempotents/local identities, namely finite sums of the e_i. Small preadditive categories are sometimes referred to as **ringoids** or **rings with many objects**. This point of view is pursued seriously in, for instance, [448].

Example 10.1.2. Let A_∞ be the quiver

$$1 \longrightarrow 2 \longrightarrow 3 \longrightarrow \cdots,$$

let k be any field and let $\mathcal{A}$ be the **k-path category** of this quiver. That is, the objects of $\mathcal{A}$ are the vertices of the quiver and the arrows of $\mathcal{A}$ are freely generated as a k-category, that is under composition and forming well-defined k-linear combinations, by the arrows of the quiver which, in this case, means that $\dim_k(i, j) = 1$ if $i \le j$ and $(i, j) = 0$ if $i > j$.

In order to obtain a convenient description of the corresponding ring-without-identity/-with-many-objects R, set $e_{ii} = \mathrm{id}_i$ (the identity arrow for i) for $i \ge 1$ and let $e_{i+1,i}$ represent the arrow from vertex i to vertex $i + 1$. A basis for R as a k-vectorspace is given by the e_{ji} for $i \le j$, where, inductively, we define $e_{j+1,i} = e_{j+1,j}e_{ji}$. Clearly these multiply just like matrix units, so we may regard R as the ring(oid) of those infinitely extending (to the right and down) lower-triangular matrices over k with only finitely many non-zero entries.

Our conventions for representations of quivers (arrows acting on the right) and composition of morphisms (to the left) clash at this point: the path algebra, rather "algebroid", kA_∞ of this quiver (see the beginning of Chapter 15) is the opposite ring-with-many-objects, that is, the ringoid of infinite, to the right and down, upper-triangular matrices with finitely many non-zero entries.

A **left $\mathcal{A}$-module** is a functor (additive, as always) from $\mathcal{A}$ to **Ab**. For instance in the one-object $(*)$ case, obtained from a ring $R = \mathrm{End}(*)$, a functor from $\mathcal{A}$ to **Ab** is determined by the image of $*$, an abelian group, let us denote it by M, together with, for each $r \in R = (*, *)$, an endomorphism of this abelian group M. If we denote the action of this endomorphism as $m \mapsto rm$, then, by functoriality, $(sr)m = s(rm)$. Thus a functor from this category, which we may as well write as R, to **Ab** is a left R-module. It is easy to see that, conversely, every left R-module gives rise to a functor from this one-point category to **Ab** and that, furthermore, the natural transformations between functors are exactly the R-linear maps between modules. Therefore, $(R, \mathbf{Ab}) \simeq R\text{-Mod}$, and similarly in the case that $\mathcal{A}$ is finite and R is the ring constructed above from $\mathcal{A}$. Even for arbitrary preadditive $\mathcal{A}$ one has such a natural equivalence provided one takes care to define the category R-Mod correctly in the case that R does not have a 1 (for example, see [713, Chpt. 10]).

It makes good sense, therefore, to write $\mathcal{A}$-Mod for $(\mathcal{A}, \mathbf{Ab})$ when $\mathcal{A}$ is a small preadditive category, especially since this reminds us that functors are just generalised modules. So, for instance, if R is a ring we could write (mod-R)-Mod instead of (mod-R, $\mathbf{Ab}$) (but I prefer the second notation).

In this last example one might object that $\mathcal{A}$ is supposed to be a small category, whereas the objects of mod-R form a proper class. The category mod-R is, however, skeletally small (the isomorphism types of objects form a set) and that is good enough to avoid any set-theoretic difficulties which might otherwise arise.[3] When we write (mod-R, $\mathbf{Ab}$) we have in mind a small (that is, a set), but not necessarily skeletal (which means just one copy of each object), version of mod-R.

Theorem 10.1.3. *If $\mathcal{A}$ is a skeletally small preadditive category, then the functor category* $\mathcal{A}$-Mod $= (\mathcal{A}, \mathbf{Ab})$ *is abelian, indeed Grothendieck.*

This can be found at, for example, [486, 3.4.2].

Indeed, there is a generating set of finitely generated projective objects which we describe at 10.1.13.

One may check that the category, Mod-R, of *right* R-modules is equivalent to the category of *contravariant* functors to $\mathbf{Ab}$ from the one-point category (corresponding to) R. More generally, and bearing in mind that contravariant functors from a category are covariant functors from its opposite, we often write Mod-$\mathcal{A}$ for $(\mathcal{A}^{\mathrm{op}}, \mathbf{Ab})$.

In a case such as Example 10.1.2, where the morphism sets are k-vectorspaces and composition is k-bilinear, it is natural to consider, instead of the category of additive functors to $\mathbf{Ab}$, the naturally equivalent category of k-linear functors to the category k-Mod $\simeq$ Mod-k of k-vectorspaces. It is easily checked that the forgetful functor from Mod-k to $\mathbf{Ab}$ induces a natural equivalence of functor categories $(\mathcal{A}\text{-mod}, \mathbf{Ab}) \simeq (\mathcal{A}\text{-mod}, \text{Mod-}k)$.

Example 10.1.4. (10.1.2 continued) A left module over the path category, $\mathcal{A}$, of the quiver A_∞ is essentially the same thing as a left module over the corresponding (matrix) ring-without-1, equivalently a representation (see Chapter 15) of the quiver A_∞, equivalently a right module over the path algebra kA_∞. In each case, the essential data are: for each i, a k-vectorspace V_i; for each arrow $i \to i+1$, a k-linear map $T_i : V_i \to V_{i+1}$. More formally, there are easily described equivalences between the categories of these various kinds of objects.

[3] In contrast, the morphisms (natural transformations) between two objects of (Mod-R, $\mathbf{Ab}$) can form a proper class so some care must be taken when dealing with that category, as in Section 18.1.4, towards the end of that section.

We hope that the reader doubtful about functor categories will realise that, when one is trying to make sense of what follows, there is no harm in imagining that "functor" means "module", that "subfunctor" means "submodule" and that "natural transformation between functors" means "linear map between modules". In fact, the Mitchell Embedding Theorem (see [195, Chpt. 4 and 7.34]) says that each small abelian category embeds, via a functor which preserves exact sequences, as a full subcategory of the module category over some ring so, at least locally, one may take the above imaginings literally. See comments in, for example, [195], [416], [448] for reassurance.

What does require rather more effort is the reconciliation of different contexts, that is, the translation from explanation in one context to another. For instance, given a ring R, it will be seen, 12.1.3, that there is a full embedding of the category Mod-R of modules into a certain functor category. This embedding replaces a module M by the functor $M \otimes_R -$ and many concepts that apply to modules, for instance the notion of an element of a module, have a corresponding translation to some concept in the functor category, and vice versa. A proof that involves moving to this functor category will also require various hypotheses and conclusions to be moved back and forth between the contexts and hence some familiarity with the translations.

As remarked already, sometimes the target category **Ab** will be replaced by some other category, usually abelian, like the category Mod-k, but sometimes just preadditive. So, in discussing the basic definitions, let us deal with the general case and assume, at least at the outset, only that $\mathcal{C}, \mathcal{D}$ are preadditive categories, normally with $\mathcal{C}$ (skeletally) small, and that $F : \mathcal{C} \longrightarrow \mathcal{D}$ is a functor (additive under our conventions). Thinking of F as a generalised module we are about to define the notion of submodule. A **subfunctor** of F is simply a subobject of F in the functor category $(\mathcal{C}, \mathcal{D})$: that is, a functor G and a natural transformation $\tau : G \to F$ which is **monic** in $(\mathcal{C}, \mathcal{D})$, meaning that if $\alpha, \beta : H \rightrightarrows G$ are natural transformations such that $\tau\alpha = \tau\beta$, then $\alpha = \beta$, equivalently if $H \xrightarrow{\gamma} G$ is such that $\tau\gamma = 0$, then $\gamma = 0$. To say that τ is a **natural transformation** is to say that for each object $C \in \mathcal{C}$ a morphism $\tau_C : GC \to FC$ is given and these all cohere in the sense that if $C \xrightarrow{f} C'$ is a morphism of $\mathcal{C}$, then the diagram below commutes.

$$\begin{array}{ccc} GC & \xrightarrow{\tau_C} & FC \\ {\scriptstyle Gf}\downarrow & & \downarrow{\scriptstyle Ff} \\ GC' & \xrightarrow[\tau_{C'}]{} & FC' \end{array}$$

If $\mathcal{D}$ has kernels, then, setting $HC = \ker(\tau_C)$ and defining the obvious action of H (that is, restriction) on morphisms, defines a subfunctor $H \xrightarrow{\gamma} G$, where γ_C is the inclusion of HC in GC, such that $\tau\gamma = 0$. Hence, if τ is, as above, assumed to be monic, then $H = 0$; therefore each component τ_C is a monomorphism.

The following lemmas are easily proved, see, for example, [195, p. 110]. The first says that if $\mathcal{D}$ is abelian (this is defined at Appendix E, p. 706) then subfunctors and quotient functors are described "pointwise".

Lemma 10.1.5. *Suppose that $\mathcal{C}$ is a preadditive category, that $\mathcal{D}$ is an abelian category and let $\tau : G \to F$ be a morphism in $(\mathcal{C}, \mathcal{D})$.*

(i) τ is monic iff for every object $C \in \mathcal{C}$ the component $\tau_C : GC \to FC$ at C is a monomorphism in $\mathcal{D}$;
(ii) τ is epi iff for every object $C \in \mathcal{C}$ the component $\tau_C : GC \to FC$ at C is an epimorphism in $\mathcal{D}$.

The image and kernel of a morphism between functors in $(\mathcal{C}, \mathcal{D})$, where $\mathcal{D}$ is abelian, are also given locally: if $\tau : G \to F$ is such a morphism, then $\ker(\tau)$ is the subfunctor of G given at $C \in \mathcal{C}$ by $\ker(\tau) \cdot C = \ker(\tau_C) \leq GC$ and $\mathrm{im}(\tau) \leq F$ is given on objects by $\mathrm{im}(\tau) \cdot C = \mathrm{im}(\tau_C) \leq FC$. In each case the action of the functor on morphisms is the "obvious" one.

Lemma 10.1.6. *Suppose that $\mathcal{C}$ is preadditive and $\mathcal{D}$ is abelian. The sequence $0 \to F' \xrightarrow{\iota} F \xrightarrow{\pi} F'' \to 0$ of functors in $(\mathcal{C}, \mathcal{D})$ is exact iff for every $C \in \mathcal{C}$ the sequence $0 \to F'C \xrightarrow{\iota_C} FC \xrightarrow{\pi_C} F''C \to 0$ is an exact sequence in $\mathcal{D}$.*

In particular, in the situation of 10.1.6, if $F' \leq F$ is an inclusion of functors, then the quotient F/F' is given on objects by $(F/F')C = FC/F'C$. Also the direct sum, $F \oplus G$, of two functors is given pointwise.

Various categorical constructions which apply in module categories apply equally in functor categories. We describe direct limits concretely for future reference. So let $\big((F_\lambda)_\lambda, (\gamma_{\lambda\mu} : F_\lambda \to F_\mu)_{\lambda \leq \mu}\big)$ be a directed system (see Appendix E, p. 714) of functors in $(\mathcal{C}, \mathcal{D})$, where now we assume that $\mathcal{D}$ is Grothendieck abelian; for example, a module category or a functor category. In particular $\mathcal{D}$ has direct limits and these are exact. Hence, E.1.5, the functor category also is Grothendieck abelian. Let $\big(F, (\gamma_{\lambda\infty} : F_\lambda \to F)_\lambda\big)$ be the direct limit of this directed system in $(\mathcal{C}, \mathcal{D})$. For any object C of $\mathcal{C}$ there is a directed system $\big((F_\lambda C)_\lambda, ((\gamma_{\lambda\mu})_C : F_\lambda C \to F_\mu C)_{\lambda \leq \mu}\big)$ in $\mathcal{D}$. Then $\big(FC, ((\gamma_{\lambda\infty})_C : F_\lambda C \to FC)_\lambda\big)$ is, one may check, the direct limit of this system. As for the action of F on morphisms, note that a morphism from C to C' gives rise to a morphism (the obvious definition) between the corresponding directed systems in $\mathcal{D}$ and hence, using the definition of direct limit, to a morphism between their corresponding direct limits FC and FC'. It is more useful to write down than to read the details and they are left to the reader.

10.1.2 The Yoneda embedding

The Yoneda lemma (10.1.7) is fundamental. A proof can be found in almost any book on category theory.

If C is an object of the category $\mathcal{C}$, then there is the corresponding **representable functor** $(C, -) : \mathcal{C} \longrightarrow \mathbf{Set}$ (or to $\mathbf{Ab}$ if $\mathcal{C}$ is preadditive) defined on objects by $D \mapsto (C, D)$ and on morphisms by $f : D \to E$ maps to (C, f), where $(C, f) : (C, D) \to (C, E)$ is defined by $(C, f)g = fg$ for $g \in (C, D)$. There is also the corresponding contravariant representable functor $(-, C)$ defined on objects by $D \mapsto (D, C)$ and on morphisms in the obvious way.

Lemma 10.1.7. (Yoneda Lemma) *Let $\mathcal{C}$ be any preadditive category. Take $C \in \mathcal{C}$ and $F \in (\mathcal{C}, \mathbf{Ab})$. Then there is a natural identification between the group, $((C, -), F)$ of natural transformations from the representable fuctor $(C, -)$ to F and the value group, FC, of F at C.*

Naturality includes that if $\eta : F \to G$ is a natural transformation, then $((C, -), \eta) : ((C, -), F) \to ((C, -), G)$ is identified with the map $\eta_C : FC \to GC$.

Similarly, if $G \in (\mathcal{C}^{\mathrm{op}}, \mathbf{Ab})$ then $((-, C), G) \simeq GC$.

Proof. We sketch the proof since we will, at some points, need the definition of the isomorphism $\theta = \theta_{C,F} : ((C, -), F) \simeq FC$. Let $\tau \in ((C, -), F)$ and consider the component of τ at C, that is, $\tau_C : (C, C) \to FC$. Define $\theta(\tau) = \tau_C(1_C)$. This element $\tau_C 1_C$ determines τ as follows: the component of τ at D, $\tau_D : (C, D) \to FD$, is defined by $\tau_D f = Ff \cdot (\tau_C 1_C)$. We leave to the reader the straightforward check that this works.

The first use of the term "natural" in the statement refers to the dependence of θ on C and F. See, for example, [195, 5.32, 5.34] for a precise statement and proof. □

The Yoneda Lemma (theorem and proof) is completely general (the basic version is for arbitrary categories with functors to **Set**). In particular, it also applies to the category, $(\mathcal{C}, \text{Mod-}k)$, of k-linear functors if $\mathcal{C}$ is a k-linear category.

It follows that if $(C, -)$ and $(D, -)$ are representable functors, then the group, $((C, -), (D, -))$, of natural transformations from $(C, -)$ to $(D, -)$ is naturally equivalent to the value of $(D, -)$ at C, that is, to (D, C). In fact, it is immediate from 10.1.7 that we have a full contravariant embedding of $\mathcal{C}$ to $(\mathcal{C}, \mathbf{Ab})$, called the **Yoneda embedding**.

Corollary 10.1.8. (see, for example, [195, 5.32]) *Given any preadditive category $\mathcal{C}$ the functor from $\mathcal{C}^{\mathrm{op}}$ to $(\mathcal{C}, \mathbf{Ab})$ given on objects by $C \mapsto (C, -)$ and on* *morphisms by sending $f : C \to D$ to $(f, -) : (D, -) \to (C, -)$, where $(f, -)$ is given by $(f, -)g = gf$ whenever $g \in (D, E)$, is a full and faithful embedding.*

Dually, the (covariant) Yoneda embedding of $\mathcal{C}$ into $(\mathcal{C}^{\text{op}}, \mathbf{Ab})$, given on objects by $C \mapsto (-, C)$ and on morphisms in the obvious way, is a full and faithful embedding.

Example 10.1.9. Let R be a ring, regarded as a preadditive category with one object. The Yoneda embedding from R^{op} to $(R, \mathbf{Ab}) = R\text{-Mod}$ is given by taking the object $*$ of R to the functor $(*, -)$. This functor takes $*$ to $(*, *) = R$ and takes $r \in R = (*, *)$ to the map $s \mapsto rs$ $(s \in R)$: $(*, *) \to (*, *)$ – this is just the left module ${}_RR$.

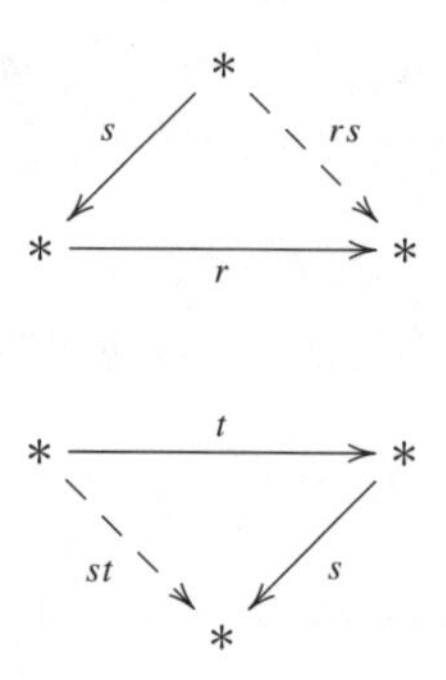

The action of this Yoneda embedding on morphisms is to take $t \in R = (*, *)$ to the endomorphism of ${}_RR$ which takes $s \in {}_RR$ to st. The fact that the Yoneda embedding is full and faithful is just the assertion that $\text{End}({}_RR) = R^{\text{op}}$ (that is, the right action of R).

Of course, the dual Yoneda embedding from R (as a category) to $(R^{\text{op}}, \mathbf{Ab}) = \text{Mod-}R$ takes $*$ to the right module R_R and $R = \text{End}(*)$ isomorphically to $R = \text{End}(R_R)$.

Corollary 10.1.10. *Suppose that G is a subfunctor of the representable functor $(C, -)$. If GC contains an automorphism of C, then $G = (C, -)$.*

Proof. Say $f \in GC$ is an automorphism of C. With the notation of the proof of 10.1.7, let $\tau : (C, -) \to G$ be the Yoneda-correspondent, $\theta_{C,G}^{-1} f$, of f. The composition $\iota\tau$, where ι is the inclusion of G in $(C, -)$, is easily checked to be $(f, -) : (C, -) \to (C, -)$. Precomposed with $(f^{-1}, -)$, this gives the identity of $(C, -)$. Therefore ι is an epimorphism (as well as a monomorphism), hence is the identity of $(C, -)$, as required. $\square$

The representable functor $(A, -)$ is generated by the identity $1_A : A \to A$ in the sense (see [22, p. 171]) that a **generator** for a subfunctor, F', of a functor

$F : \mathcal{C} \longrightarrow \mathbf{Ab}$ is an element $b \in FB$ for some $B \in \mathcal{C}$ such that F' is the image of the (Yoneda-)corresponding map $(B, -) \to F$.

Lemma 10.1.11. (for example, [195, §3.1, 3.21]) *If $A \xrightarrow{f} B \xrightarrow{g} C \to 0$ is an exact sequence in* mod-R, *then the induced sequence $0 \to (C, -) \xrightarrow{(g,-)} (B, -) \xrightarrow{(f,-)} (A, -)$ is exact in* (mod-R, $\mathbf{Ab}$).

If $A \xrightarrow{f} B \xrightarrow{g} C$ is a sequence in mod-R *such that $(C, -) \xrightarrow{(g,-)} (B, -) \xrightarrow{(f,-)} (A, -)$ is exact in* (mod-R, $\mathbf{Ab}$), *then the original sequence is exact. In particular a morphism h of $\mathcal{C}$ is an epimorphism, iff $(h, -)$ is a monomorphism, and if $(h, -)$ is an epimorphism, then h is a monomorphism.*

In the above any additive category with cokernels, in particular any abelian category, may replace mod-R. *(Indeed, what is needed is that g is a pseudocokernel, see Appendix E, p. 704, of f.)*

10.1.3 Representable functors and projective objects in functor categories

Representable functors are finitely generated projectives (10.1.12) and together are generating (10.1.13).

For definitions and basic properties of finitely generated and finitely presented objects see Appendix E, p. 709.

Lemma 10.1.12. *If A is an object of the small preadditive category $\mathcal{C}$, then the representable functor $(A, -)$ is a finitely generated projective object of $(\mathcal{C}, \mathbf{Ab})$. In particular, it is finitely presented.*

Proof. If $(A, -) = \sum_{\overrightarrow{\lambda}} F_\lambda$ is a directed sum of subfunctors, then $1_A \in (A, A) = \sum_{\overrightarrow{\lambda}} F_\lambda A$, so $1_A \in F_\lambda A$ for some λ (strictly, is the image of an element of $F_\lambda A$ under the inclusion $F_\lambda A \to (A, A)$). By the Yoneda Lemma (10.1.7) this gives a morphism $(A, -) \to F_\lambda$ which, composed with the inclusion $F_\lambda \to (A, -)$, is $1_{(A,-)}$; for, that is the endomorphism of $(A, -)$ which Yoneda-corresponds to 1_A. Therefore the inclusion $F_\lambda \to (A, -)$ is an epimorphism, as required.

For projectivity, take an epimorphism $\pi : F \to G$ of functors and a morphism $\alpha : (A, -) \to G$ which, by Yoneda, corresponds to an element $a \in GA$. Since $\pi_A : FA \to GA$ is epi, there is $b \in FA$ mapping to a. Suppose that $\beta \in ((A, -), F)$ Yoneda-corresponds to b. Then $\pi\beta = \alpha$ by the Yoneda Lemma.

The last point is general and follows easily from the definitions in Appendix E, p. 710. $\square$

Proposition 10.1.13. (for example, [195, p. 119]) *If $\mathcal{C}$ is a skeletally small preadditive category, then the representable functors generate the functor category*

$(\mathcal{C}, \mathbf{Ab})$ *in the sense that for every functor* $F : \mathcal{C} \longrightarrow \mathbf{Ab}$ *there is an epimorphism* $\bigoplus_i (A_i, -) \to F$ *for some* $A_i \in \mathcal{C}$. *A functor* F *is finitely generated iff this direct sum may be taken to be finite.*

Proof. For each isomorphism class of objects of $\mathcal{C}$ take a representative A. Define the morphism $\bigoplus_A (A, -)^{(FA)} \to F$ to have component at $a \in FA$ the morphism $f_a : (A, -) \to F$ which Yoneda-corresponds to a. This map is surjective, essentially by definition.

For the second statement, we have $F = \sum_i \operatorname{im}(f_i)$, where f_i is the ith component map of $\bigoplus_i (A_i, -) \to F$, so F finitely generated implies that F is a sum of finitely many of these, therefore finitely many of the direct summands will do. The converse follows (by E.1.13) since, by 10.1.12, each $(A_i, -)$ is finitely generated. □

Say that **idempotents split** in the preadditive category $\mathcal{C}$ if for every $A \in \mathcal{C}$ each idempotent $e = e^2 \in \operatorname{End}(A)$ has a kernel and the canonical map $\ker(e) \oplus \ker(1 - e) \to A$ is an isomorphism (also see Appendix E, p. 704).

Corollary 10.1.14. *If* $\mathcal{C}$ *is a small preadditive category, then the finitely generated projective objects of* $(\mathcal{C}, \mathbf{Ab})$ *are the direct summands of finite direct sums of representable functors. If* $\mathcal{C}$ *has split idempotents and has finite direct sums, then these are precisely the representable functors.*

Proof. The first statement follows from 10.1.12 and 10.1.13.

For the second, if $\mathcal{C}$ has finite direct sums, then $(A_i, -) \oplus (A_j, -) \simeq (A_i \oplus A_j, -)$, so, if F is finitely generated and projective, F is, without loss of generality, a direct summand of a functor of the form $(A, -)$. Let $\pi : (A, -) \to F$ split the inclusion and let $f \in \operatorname{End}(A, -)$ be π followed by the inclusion: so $f^2 = f$. The Yoneda embedding is full and faithful, so there is $e \in \operatorname{End}(A)$ with $(e, -) = f$, hence with $e^2 = e$. By assumption $A = \operatorname{im}(e) \oplus \ker(e)$ and then it follows quickly that $F \simeq (\operatorname{im} e, -)$ so F is representable. □

In particular, the finitely generated projectives of $(\text{mod-}R, \mathbf{Ab})$ are the functors $(A, -)$ with $A \in \text{mod-}R$ and, together, these generate the functor category.

Of course, all the above applies if we replace $\mathcal{C}$ by $\mathcal{C}^{\text{op}}$. In particular, the representable functors $(-, A)$ yield (closing under direct summands of finite direct sums) the finitely generated projective objects of $(\mathcal{C}^{\text{op}}, \mathbf{Ab})$. The injective objects of the functor category $(R\text{-mod}, \mathbf{Ab})$ are described in 12.1.6.

Example 10.1.15. We continue Example 10.1.2. We write $\alpha_{i,i+1}$ for the arrow of the quiver going from vertex i to vertex $i + 1$, more generally, α_{ij} $(i \le j)$ for compositions of these in the path category $\mathcal{A}$. So the morphism α_{ij} of $\mathcal{A}$

corresponds to e_{ji} in the ring R built from the quiver (in 10.1.2). Clearly an additive functor from $\mathcal{A}$ to Mod-k is equivalent to a left R-module, which, in turn, is equivalent to a right module over the path algebra kA_∞, which, in turn, is equivalent to a k-representation of the quiver. We describe the simple, projective and injective objects of this category.

The simple representations (that is, modules) are the S_i, $i \geq 1$, where S_i has dimension 1 at vertex i and is zero elsewhere. For each vertex i there is the indecomposable projective object P_i which is one-dimensional at each $j \geq i$ and zero elsewhere. This is the representable functor $(i, -)$ and one can see that the embeddings $P_1 \leftarrow P_2 \leftarrow \cdots$, of each indecomposable projective as the radical of the next, show $\mathcal{A}^{\mathrm{op}}$ Yoneda-embedded into the functor category $(\mathcal{A}, \text{Mod-}k)$. Clearly P_i is the projective cover of the simple object S_i. Dually, the indecomposable injective representations are the $E_i = E(S_i)$, which have dimension 1 at each vertex $j \leq i$ and zero elsewhere, plus one more, $E_\infty = P_1$, which has dimension 1 at each vertex.

One may compute the corresponding objects for the category of contravariant functors, equivalently covariant functors for the opposite quiver – the quiver with the same vertices but all arrows reversed.

There is more on this example at 14.1.2.

10.2 Finitely presented functors in (mod-R, Ab)

Local coherence of the functor category is proved in Section 10.2.1. In Section 10.2.2 the projective dimension of a finitely presented functor is shown to correspond to how it may be presented by pp conditions. Minimal free realisations and projective covers are shown to correspond in Section 10.2.3. Section 10.2.4 describes how the notion of pp condition is extended to apply to modules over rings with many objects, i.e. to additive functors.

The equivalence of the category of "pp-imaginaries" with the category of finitely presented functors, a key result, is proved in Section 10.2.5. In Section 10.2.6 conditions are given under which various homological functors are finitely presented, hence pp-definable. The free abelian category on a small preadditive category is introduced in Section 10.2.7 and shown to be equivalent to the category of finitely presented functors/pp-pairs.

The extension of functors from finitely presented to arbitrary modules is described in Section 10.2.8.

The article [270] gives an interesting exposition of some of the ideas in this and the next section.

10.2.1 Local coherence of (mod-R, **Ab**)

Every finitely presented functor is the cokernel of a representable morphism (10.2.1). Every finitely generated subfunctor of a finitely presented functor is finitely presented (10.2.3). The finitely presented functors form an abelian subcategory (10.2.4). The projective dimensions of finitely presented functors are bounded above by 2 *(10.2.5) and projective dimension* ≤ 1 *is equivalent to embeddability in a representable functor (10.2.6). Free realisations give surjections from representable functors (10.2.8).*

Throughout R denotes a small preadditive category, for example, but not necessarily, a ring.

Although our original interest was (and remains) $(R^{\mathrm{op}}, \mathbf{Ab}) = \text{Mod-}R$ and $(R, \mathbf{Ab}) = R\text{-Mod}$ it proves to be very useful to move to the richer context of (mod-R, **Ab**) and (R-mod, **Ab**). This move, see Chapter 12, will *not* be via the Yoneda embedding, 10.1.8, which would take us to a different category (see, for example, 16.1.3). Here we prepare the ground by relating pp conditions and finitely presented functors.

This idea of studying the representations of a ring by investigating the representations of its (finitely presented) representations is just the notion of investigating an object via its representations and, since the work of Auslander and Reiten (for example, [17], [25], [22]), has been used very successfully in the representation theory of finite-dimensional algebras.

Let F be a finitely generated functor in (mod-R, **Ab**). Noting that mod-R is closed under finite direct sums there is, by 10.1.13, some $A \in \text{mod-}R$ and an epimorphism $(A, -) \to F$. If F is finitely presented, then the kernel of this epimorphism will be finitely generated, so there will be $B \in \text{mod-}R$ and an exact sequence $(B, -) \to (A, -) \to F \to 0$. By the Yoneda Lemma every morphism from $(B, -)$ to $(A, -)$ is induced by a morphism $f : A \to B$. This gives the following description.

Lemma 10.2.1. *Let* $F \in (\text{mod-}R, \mathbf{Ab})$ *be finitely presented. Then there is a morphism* $f : A \to B$ *in* mod-R *such that* $F \simeq \text{coker}\big((f, -) : (B, -) \to (A, -)\big)$. *Conversely, any functor of this form,* coker$(f, -)$ *for some* $f \in \text{mod-}R$*, is finitely presented.*

Of course, given F, there are many choices for f above.

Given $f : A \to B$ in mod-R write $F_f = \text{coker}(f, -)$. Then $F_f M$ ($M \in \text{Mod-}R$) has the following description: $F_f M = (A, M)/\text{im}(f, M)$ is the abelian group, (A, M), of morphisms from A to M factored by the subgroup consisting of those which factor initially through f. The next result shows that each finitely generated subfunctor of $(A, -)$ is determined by the property of morphisms factoring through some specified morphism.

Lemma 10.2.2. *Suppose that* $A \in$ mod-R *and that* G *is a subfunctor of* $(A, -)$. *Then* G *is finitely generated iff* G *is finitely presented iff* G *has the form* $\mathrm{im}(f, -)$ *for some* $f : A \to B$ *in* mod-R.

Proof. Any functor $(B, -)$ with $B \in$ mod-R is finitely generated, by 10.1.12, so any functor of the form $\mathrm{im}(f, -)$ is finitely generated. Conversely, if G is a finitely generated subfunctor of $(A, -)$, then it is the image of a representable functor $(B, -) \to G$ (by 10.1.13). Compose this with the inclusion $G \to (A, -)$ to obtain a morphism $(B, -) \to (A, -)$ which, by Yoneda, 10.1.7, has the form $(f, -)$ for some $A \xrightarrow{f} B$. It follows that $G = \mathrm{im}(f, -)$.

It remains to show that any such functor is finitely presented. Retaining the notation, let $g : B \to C$ be the cokernel of f, so $A \xrightarrow{f} B \xrightarrow{g} C \to 0$ is exact (by E.1.16, C is finitely presented). Since, for every module D, the functor $(-, D)$ converts this to an exact sequence $0 \to (C, D) \to (B, D) \to (A, D)$ (10.1.11), we obtain an exact sequence of functors $0 \to (C, -) \to (B, -) \xrightarrow{(f,-)} (A, -)$. Therefore $\ker\big((B, -) \to G\big)$ is $(C, -) \to (B, -)$ and hence G is finitely presented, by E.1.15. □

The following strengthening of 10.2.2, the fact that every finitely presented functor in (mod-R, **Ab**) is coherent (cf. Section 2.3.3), is a key property, local coherence, of the functor category. See 16.1.12 for a more general result.[4]

Corollary 10.2.3. *Every finitely generated subfunctor of a finitely presented functor in* (mod-R, **Ab**) *is finitely presented. That is,* (mod-R, **Ab**) *is locally coherent.*

Proof. Suppose $H \leq G$ with G finitely presented and H finitely generated. By 10.2.1 there is $B \in$ mod-R and an epimorphism $\pi : (B, -) \to G$. Since G is finitely presented, $\ker(\pi)$ is finitely generated (E.1.16), so if $H' = \pi^{-1}H$, then H' is finitely generated, being an extension of a finitely generated by a finitely generated object, hence is finitely presented by 10.2.2. By E.1.16 $H \simeq H'/\ker(\pi)$ is finitely presented. □

Denote by $(\text{mod-}R, \mathbf{Ab})^{\mathrm{fp}}$ the full subcategory of finitely presented functors from mod-R to **Ab**. We show that this is an abelian subcategory (this is a general fact, E.1.19).

Proposition 10.2.4. $(\text{mod-}R, \mathbf{Ab})^{\mathrm{fp}}$ *is an abelian subcategory of* (mod-R, **Ab**)*: that is, as well as the subcategory being an abelian category, a sequence* $0 \to F \to G \to H \to 0$ *which is exact in the smaller category is also exact in the larger category.*

[4] Some authors therefore refer to "coherent", rather than "finitely presented", functors in this context.

Proof. By [195, 3.41] (where the term "exact subcategory" is used) it is enough to check that the kernel and cokernel of every morphism $\alpha : G \to F$ in $(\text{mod-}R, \mathbf{Ab})^{\text{fp}}$ are also in this subcategory.

The cokernel of α is a finitely presented object factored by a finitely generated one, hence is finitely presented (E.1.16).

The image of α is a finitely generated, hence, by 10.2.3, finitely presented, subfunctor of F. Therefore (E.1.16) the kernel of α is a finitely generated, therefore finitely presented, subfunctor of G, as required. □

The following notation is used by some authors (but is not used here): $D(R)$ for $(R\text{-mod}, \mathbf{Ab})$ (so $(\text{mod-}R, \mathbf{Ab})$ is $D(R^{\text{op}})$), and $C(R)$ for $(R\text{-mod}, \mathbf{Ab})^{\text{fp}} = D(R)^{\text{fp}}$.

Corollary 10.2.5. [17, §2] *Every finitely generated subfunctor of a representable functor in* $(\text{mod-}R, \mathbf{Ab})$ *has projective dimension less than or equal to 1. Every finitely presented functor has projective dimension less than or equal to 2.*

Proof. The first statement follows from the end of the proof of 10.2.2, along with 10.1.12. Also, since the representable functors form a generating set of finitely generated projective functors (10.1.13) the second statement follows; for every finitely presented functor F therefore has a projective presentation $0 \to (C, -) \to (B, -) \to (A, -) \to F \to 0$. □

Corollary 10.2.6. *A finitely presented functor in* $(\text{mod-}R, \mathbf{Ab})$ *has projective dimension* ≤ 1 *iff it embeds in a representable functor.*

Proof. For the direction not covered by 10.2.5, suppose that the sequence $0 \to (B, -) \xrightarrow{(f,-)} (A, -) \to F \to 0$ induced by $f : A \to B$ is exact. So (10.1.11) f is an epimorphism. Then the sequence $0 \to \ker(f) \to A \to B \to 0$ is exact. Now, $\ker(f)$ is a finitely generated (but not necessarily finitely presented) R-module, so there is an epimorphism $C \to \ker(f)$ with $C \in \text{mod-}R$. From the exact sequence $C \to A \to B \to 0$ one obtains (10.1.11) the exact sequence $0 \to (B, -) \to (A, -) \to (C, -)$ and so F, being $\text{coker}((B, -) \to (A, -))$, embeds in $(C, -)$, as required. □

If R is a ring, one could take C in the proof above to be R^n for some n and deduce that every finitely presented functor of projective dimension ≤ 1 embeds in a direct sum of copies of $(R, -)$. Indeed, if $A \in \text{mod-}R$, then there is an epimorphism $R^m \to A$, hence a monomorphism $(A, -) \to (R^m, -) = (R, -)^m$.

The next result follows directly from 10.2.3 as in 2.3.15

Corollary 10.2.7. *If* $F \in (\text{mod-}R, \mathbf{Ab})$ *is finitely presented and* $G, H \leq F$ *are finitely generated, then their intersection* $G \cap H$ *is finitely generated.*

Let $F \in (\text{mod-}R, \mathbf{Ab})$. Denote by $\text{Latt}(F)$ the lattice of subfunctors of F. If F is finitely presented, then, by 10.2.7, the subposet consisting of finitely generated = finitely presented subfunctors also is a lattice, which we denote by $\text{Latt}^{\text{f}}(F)$.

The next result is stated first for the case where R is a ring and then, slightly reformulated, in the general case.

Proposition 10.2.8. *Suppose that ϕ is a pp condition with n free variables, so $F_\phi \leq (R^n, -)$. Choose any epimorphism $(C, -) \to F_\phi$ with C finitely presented and let $\overline{c}$ from C be such that the composition $(C, -) \to F_\phi \to (R^n, -)$ is $(\overline{c}, -)$. Then $(C, \overline{c})$ is a free realisation of ϕ.*

Conversely, if $(C, \overline{c})$ is a free realisation of the pp condition ϕ and if $(\overline{c}, -) : (C, -) \to (R^n, -)$ is the morphism induced[5] *by $\overline{c} : R^n \to C$ then $F_\phi = \text{im}(\overline{c}, -)$. Every morphism from a representable functor to a finite power of $(R, -)$ arises in this way.*

Proof. For any module M the component of $(C, -) \to F_\phi$ at M, that is, $(C, M) \to \phi(M)$, is given by evaluation at $\overline{c}$ and is an epimorphism. Therefore, for any $\overline{a} \in \phi(M)$ there exists $f : C \to M$ with $f\overline{a} = \overline{c}$, so, by 1.2.22, $(C, \overline{c})$ is a free realisation of ϕ, that is, ϕ generates the pp-type of $\overline{c}$ in C, as required.

Reversing this argument gives the converse, and the last statement follows by Yoneda, 10.1.7, which implies that every morphism $(C, -) \to (R^n, -)$ has the form $(\overline{c}, -)$ for some $\overline{c} : R^n \to C$. □

Let $C \in \text{mod-}R$ and let $c_i \in C$ $(i = 1, \dots, n)$ be elements of C. In the case where R is a ring, c_i may be thought of as literally an element of the set C. In the general case, where R is a small preadditive category, by an element, c_i, of C we mean a morphism from some object G_i in R, rather from its Yoneda image $(-, G_i)$ in Mod-R, to C, in which case we also say that c_i is an element of C of sort G_i (this point of view is discussed in Section 10.2.4). There is no conflict with the usual view of elements in the case where R is a ring because we can identify an element c of the (ordinary) module C with that morphism from the projective module $R_R = (-, R)$ to C which is defined by taking $1 \in R$ to $c \in C$.

The n-tuple $\overline{c} = (c_1, \dots, c_n)$ (c_i as above) may be regarded as a morphism from $H = (-, G_1) \oplus \cdots \oplus (-, G_n) = (-, G_1 \oplus \cdots \oplus G_n)$ to C. The corresponding morphism in the functor category (mod-R, **Ab**) is $(C, -) \xrightarrow{(\overline{c}, -)} (H, -)$. By 10.2.2 the image of $(\overline{c}, -)$ is finitely presented. We use $F_{C,\overline{c}}$ to denote this image. Since C is finitely presented, there is, by 1.2.6, a pp condition $\phi(\overline{x})$ which generates the pp-type of $\overline{c}$ in C. Recall that F_ϕ denotes the functor, a subfunctor of $(H, -)$,

[5] Here $\overline{c}$ is thought of equally as an n-tuple from C and as a morphism from R^n to C.

which takes an R-module M to $\phi(M) \leq G_1(M) \oplus \cdots \oplus G_n(M) = (H, -)(M)$. Here is the formulation, in this generality, of the first part of 10.2.8.

Proposition 10.2.9. *Let $C \in$ mod-R and let $\overline{c}$ be a tuple of elements of C. Let $\phi(\overline{x})$ be a pp condition which generates the pp-type, $\mathrm{pp}^C(\overline{c})$, of $\overline{c}$ in C. Then $F_\phi = F_{C,\overline{c}}$.*

Proof. Consider the morphism $(\overline{c}, -) : (C, -) \to (H, -)$ as above; in the case of a ring, H is just $R^{l(\overline{c})}$. Evaluating at a module M, we have $(\overline{c}, M) : (C, M) \to (H, M)$, which is just evaluation, $f \mapsto f\overline{c}$, at $\overline{c}$. That is, $F_{C,\overline{c}}f = f\overline{c}$, regarded as an element of (H, M) ($= M^n$ if $H = R^n$), so, by 1.2.17, $F_{C,\overline{c}}$ coincides with $F_\phi \leq (H, -)$. □

The converse, cf. 10.2.8, is also easy to formulate and prove.

Note that the submodule generated by an element (or tuple) in this generalised sense is simply its image if it is regarded as a morphism.

Let $A \xrightarrow{h} B$ be a morphism in mod-R and let $\overline{a}$ be a tuple of elements of A. Let $\phi(\overline{x})$ be a pp condition which generates $\mathrm{pp}^A(\overline{a})$ and let $\psi(\overline{x})$ be a generator for $\mathrm{pp}^B(h\overline{a})$. Then the functor $F_{\phi/\psi} = F_\phi/F_\psi$ is a measure of the "drop" in the lattice, $\mathrm{pp}_R^{l(\overline{c})}$, of pp conditions from ϕ to ψ, hence, as $\overline{a}$ varies, of the "length" (in a sense related to concepts in Sections 9.1.2 and 7.2) of the morphism h.

Proposition 10.2.10. *Let $A \xrightarrow{h} B$ be a morphism in* mod-R*, let $\overline{a}$ be a tuple of elements of A, let ϕ, respectively ψ, generate $\mathrm{pp}^A(\overline{a})$, respectively $\mathrm{pp}^B(h\overline{a})$. Then there is an induced epimorphism $F_h \to F_{\phi/\psi}$. This is an isomorphism iff the submodule generated by $\overline{a}$ contains* $\ker(h)$*; in particular, this is so if $\overline{a}$ generates A. In any case there is an inclusion of Ziegler-open sets $(F_{\phi/\psi}) \subseteq (F_h)$.*

Proof. By 10.2.8 there is a diagram as shown, with the vertical solid arrows epimorphisms.

$$\begin{array}{ccccccccc} & & (B,-) & \xrightarrow{(h,-)} & (A,-) & \longrightarrow & F_h & \longrightarrow & 0 \\ & & \downarrow{\scriptstyle (h\overline{a},-)} & & \downarrow{\scriptstyle (\overline{a},-)} & & \vdots & & \\ 0 & \longrightarrow & F_\psi & \longrightarrow & F_\phi & \longrightarrow & F_{\phi/\psi} & \longrightarrow & 0 \end{array}$$

Therefore there is the morphism from F_h to $F_{\phi/\psi}$ as shown, necessarily an epimorphism.

If $\overline{a}$ generates A, then the morphism $\overline{a}$ (from $R^{l(\overline{a})}$ to A if R is a ring) is an epimorphism, so, by 10.1.11, $(\overline{a}, -)$ is monic, hence an isomorphism, and it follows that $F_h \simeq F_{\phi/\psi}$. More generally, the map is an isomorphism iff $\mathrm{im}(h, -) \geq \ker(\overline{a}, -)$ and this is the condition that any morphism $A \xrightarrow{g} C$ with $g\overline{a} = 0$ initially factor through h; equivalently it is the condition that the canonical epimorphism $A \to A/\langle \overline{a} \rangle$ initially factor through h, that is, $\ker(h) \leq \langle \overline{a} \rangle$.

Finally, if $F \to G$ is any epimorphism of finitely presented functors and if $\overrightarrow{G} N \neq 0$ (for $\overrightarrow{G}$ see Section 10.2.8), then $\overrightarrow{F} N \neq 0$, so the final statement follows from the description of the spectrum given in 10.2.45. □

Corollary 10.2.11. *Let $h : A \to B$ be in* mod-R*; choose a generating tuple $\overline{a}$ of A, let ϕ generate the pp-type of $\overline{a}$ in A and let ψ generate* $\mathrm{pp}^B(h\overline{a})$. *Then the lattice of finitely generated subfunctors of $F_h \simeq F_\phi / F_\psi$ is isomorphic to the interval $[\phi, \psi]$ in $\mathrm{pp}_R^{l(\overline{a})}$, by the map which takes η with $\phi \geq \eta \geq \psi$ to F_η / F_ψ.*

Corollary 10.2.12. *Let $h : A \to B$ be a morphism in* mod-R. *Then h is irreducible iff the functor F_h is simple.*

Proof. (Irreducible morphisms are defined after 15.1.4.) In view of the previous corollary, and with notation as there, it must be shown that h is irreducible iff ϕ/ψ is a minimal pair.

If $\phi > \eta > \psi$, then let $(C, \overline{c})$ be a free realisation of η. Then, by 1.2.21, there are $f : A \to C$ and $g : C \to B$ with $f\overline{a} = \overline{c}$ and $g\overline{c} = \overline{b}$. Since $\overline{a}$ generates A it follows that $gf = h$. By strictness of the inclusions of pp conditions and 1.2.21 again, f is not a split monomorphism nor g a split epimorphism, so h is not irreducible.

If ϕ/ψ is a minimal pair, so there is no pp condition η strictly between them, suppose that $h = gf$, where $f : A \to C$ and $g : C \to B$ are in mod-R. Choose a pp condition η with $\langle \eta \rangle = \mathrm{pp}^C(f\overline{a})$. Then either η is equivalent (in $\mathrm{pp}_R^{l(\overline{a})}$) to ϕ or to ψ. If the former, then there is, by 1.2.7, $f' : C \to A$ with $f'f\overline{a} = \overline{a}$, hence with $f'f = 1_A$ and f is a split monomorphism. If the latter, then there is $g' : B \to C$ with $g' \cdot h\overline{a} = f\overline{a}$, hence with $gg' \cdot h\overline{a} = h\overline{a}$. We may assume that B has no proper direct summand containing $h\overline{a}$ so it follows (for example, from 4.3.57) that gg' is an automorphism of B, hence that g is a split epimorphism. Thus h is indeed an irreducible morphism, as required. □

10.2.2 Projective dimension of finitely presented functors

Systems of equations correspond to representable functors (10.2.13), general pp conditions to functors of projective dimension ≤ 1 (10.2.14). The lattice of pp conditions in n free variables may be identified with the lattice of finitely generated subfunctors of the nth power of the representable functor (10.2.17).

The global dimension of the category of finitely presented functors is 2 *unless the ring is von Neumann regular, in which case it is* 0 *and, in that case, the functor category is equivalent to the category of finitely presented modules (10.2.38).*

The pp conditions corresponding to finitely generated subfunctors of a representable functor are identified (10.2.23).

In this section we write as if R is a ring, rather than a more general small preadditive category, but everything works in the more general situation (using what is set up in Section 10.2.4).

Proposition 10.2.13. ([102, 2.1]) *The functor F_ϕ is representable iff ϕ is equivalent to a system of equations.*

Proof. The direction ($\Leftarrow$) follows from 1.2.30 and 1.2.18. The converse was shown at 1.1.11. □

It has been seen already (10.2.5) that every finitely presented functor F has projective dimension, $\mathrm{pdim}(F)$, no more than 2. The other two parts of the next result appear in [102, 2.4].

Proposition 10.2.14. *Let F be a finitely presented functor. Then:*

(a) $\mathrm{pdim}(F) \leq 2$;
(b) $\mathrm{pdim}(F) \leq 1$ *iff* $F \simeq F_\phi$ *for some pp condition* ϕ;
(c) $\mathrm{pdim}(F) = 0$ *iff* $F \simeq F_\theta$ *where* θ *is a system of equations.*

Proof. The last part is 10.2.13 (plus 10.1.14).

For part (b), just by its definition, every functor of the form F_ϕ is a subfunctor of a representable functor (and is finitely presented by 1.2.19, see 10.2.29), so, by 10.2.5, has projective dimension at most 1.

For the converse, if $\mathrm{pdim}(F) \leq 1$, then, by 10.2.6, $F \leq (A, -)$ for some A. Choose an epimorphism $R^n \to A$, hence an embedding $(A, -) \to (R^n, -)$, giving $F \leq (R^n, -)$. Also choose an epimorphism $(C, -) \to F$ with $C \in \text{mod-}R$. The composition $(C, -) \to (R^n, -)$ is induced by a morphism $R^n \to C$, that is, by an n-tuple $\overline{c}$ from C. Let ϕ generate $\mathrm{pp}^C(\overline{c})$. By 10.2.8, $\mathrm{im}(\overline{c}, -)$, that is, F, is just F_ϕ, as required. □

Corollary 10.2.15. *A finitely presented functor is isomorphic to one of the form F_ϕ iff it embeds in a representable functor iff it embeds in $(R^n, -)$ for some n.*

In particular every finitely presented projective functor embeds in $(R^n, -)$ for some n.

Corollary 10.2.16. *Every finitely generated subfunctor of $(R^n, -)$ has the form F_ϕ for some pp condition ϕ.*

Corollary 10.2.17. *For any ring R the lattice, pp_R^n, of pp conditions with n free variables is isomorphic to the lattice, $\mathrm{Latt}^{\mathrm{f}}(R^n, -)$, of finitely generated subfunctors of the nth power of the forgetful functor.*

A more general result is at 10.2.33. Also the lattice of *all* subfunctors of $(R_R^n, -)$ is, by 12.2.1, anti-isomorphic to the lattice of pp-n-types for *left* R-modules.

This corollary also can be said in terms of the pre-ordering of morphisms by factorisation which was discussed in Section 3.1.

Recall that for morphisms f, f' with the same domain we write $f' \leq f$ if f' factors initially through f and we say that f and f' are equivalent, writing $f \sim f'$, if $f \leq f' \leq f$.

Lemma 10.2.18. *Let* $A \xrightarrow{f} B$, $A \xrightarrow{f'} B'$ *be in* mod-R. *Then* $f' \leq f$ *iff* $\mathrm{im}(f', -) \leq \mathrm{im}(f, -)$ *as subfunctors of* $(A, -)$ *and* $f \sim f'$ *iff* $\mathrm{im}(f, -) = \mathrm{im}(f', -)$.

Proof. If $f' = gf$, then $\mathrm{im}(f', -) = \mathrm{im}(gf, -) = (f, -) \cdot \mathrm{im}(g, -) \leq \mathrm{im}(f, -)$. Conversely, if $\mathrm{im}(f', -) \leq \mathrm{im}(f, -)$, then $\mathrm{im}(f', B') \leq \mathrm{im}(f, B')$, so $f' = (f', B')1_{B'} = (f, B')g = fg$ for some $g \in (B, B')$. □

Corollary 10.2.19. *Let* $A \in$ mod-R. *The following lattices are naturally isomorphic:*

(i) the lattice, $\mathrm{Latt}^{\mathrm{f}}(A, -)$, *of finitely generated = finitely presented subfunctors of* $(A, -)$*;*

(ii) the lattice of (equivalence classes of) morphisms in mod-R *with domain* A, *under the pre-ordering* $f \geq f'$ *iff* $f' = gf$ *for some* g*;*

(iii) the lattice of (equivalence classes of) pp conditions ψ *with* $\psi \leq \phi$, *where* ϕ *generates the pp-type of any chosen and fixed finite set of generators for* A.

That follows from 10.2.2 and 3.1.5. In the case $A = R^n$ the third lattice is simply the lattice, pp_R^n, of pp conditions in n free variables.

Note, by the way, that the lattice $\mathrm{Latt}^{\mathrm{f}}(A, -)$ of finitely generated subfunctors of $(A, -)$ has nothing to do with the lattice $\mathrm{pp}(A)$ of pp-definable subgroups of A. For instance, take $A = R_R$, where R is a finite-dimensional algebra. Then $\mathrm{pp}(A)$ has finite length (by 1.1.8) but $\mathrm{Latt}^{\mathrm{f}}(A, -)$ need not even have width (7.3.28).

We make one further remark on this ordering of morphisms.

By 10.2.2 every finitely presented subobject of $(A, -)$ has the form $\mathrm{im}(f, -)$ for some morphism $f \in$ mod-R with domain A, so a typical subquotient of $(A, -)$ is given by a pair $A \xrightarrow{f} B \geq A \xrightarrow{f'} B'$ of morphisms in mod-R. To this pair corresponds the subquotient $\mathrm{im}(f, -)/\mathrm{im}(f', -)$, the value of which at $C \in$ mod-R is the group of morphisms from A to C which factor initially through f modulo those which factor initially through f'. We may represent such a subquotient as a quotient of a representable functor as follows.

Given $A \xrightarrow{f} B \geq A \xrightarrow{f'} B'$ in mod-R, say $g : B \to B'$ is such that $gf = f'$, we have the exact sequence $A \xrightarrow{f} B \xrightarrow{\pi} B/fA \to 0$. Consider $(\pi \;\; g)^{\top} : B \to$

$(B/fA) \oplus B'$, where $^\top$ denotes transpose. Then $\mathrm{im}(f,-)/\mathrm{im}(f',-) \simeq (B,-)/\mathrm{im}((\pi\ g)^\top,-) = (B,-)/\big(\mathrm{im}(\pi,-) + \mathrm{im}(g,-)\big)$.

To see this, let $\pi = \mathrm{coker}(f)$; so we have an exact sequence of functors $0 \to (B/fA,-) \xrightarrow{(\pi,-)} (B,-) \xrightarrow{(f,-)} (A,-) \to F_f \to 0$ and $(B,-)/\mathrm{im}(\pi,-) \simeq \mathrm{im}(f,-)$. Choose $B \xrightarrow{g} B'$ such that $f' = gf$. Then consider the diagram with exact rows, where $\pi' : B' \to B'/gB$ is the canonical map.

$$\begin{array}{ccccccccc}
0 & \longrightarrow & (B'/gB,-) & \xrightarrow{(\pi',-)} & (B',-) & \longrightarrow & \mathrm{im}(f',-) & \longrightarrow & 0 \\
 & & \Big\downarrow{\scriptstyle (g',-)} & & \Big\downarrow{\scriptstyle (g,-)} & & \Big\downarrow & & \\
 & & (B/fA,-) & \xrightarrow[(\pi,-)]{} & (B,-) & \longrightarrow & \mathrm{im}(f,-) & \longrightarrow & 0 \\
 & & & & & & \Big\downarrow & & \\
 & & & & & & \mathrm{im}(f,-)/\mathrm{im}(f',-) & \longrightarrow & 0 \\
 & & & & & & \Big\downarrow & & \\
 & & & & & & 0 & &
\end{array}$$

There is g' with $g'\pi = \pi'g$ since $\pi'gf = 0$.

This gives a projective presentation $(B/fA,-) \oplus (B',-) \xrightarrow{\big((\pi,-)\ (g,-)\big)} (B,-) \to \mathrm{im}(f,-)/\mathrm{im}(f',-) \to 0$ as required.

The next result comes from [17, p. 205].[6]

Proposition 10.2.20. *A ring R is von Neumann regular iff* $\mathrm{gldim}(\text{mod-}R, \mathbf{Ab})^{\mathrm{fp}} = 0$. *Otherwise* $\mathrm{gldim}(\text{mod-}R, \mathbf{Ab})^{\mathrm{fp}} = 2$.

Proof. (from [102, 2.8]) By 10.2.14 it is enough to show that if $\mathrm{gldim}(\text{mod-}R, \mathbf{Ab})^{\mathrm{fp}} \le 1$, then $\mathrm{gldim}(\text{mod-}R, \mathbf{Ab})^{\mathrm{fp}} = 0$.

If $f : A \to B$ is a morphism in mod-R, then F_f has projective dimension ≤ 1, so, by 10.2.6, F_f embeds in a representable functor, hence there is an exact sequence $(B,-) \xrightarrow{(f,-)} (A,-) \xrightarrow{(g,-)} (D,-)$ for some $g : D \to A$ in mod-R. It follows from 10.1.11 that $D \xrightarrow{g} A \xrightarrow{f} B$ is exact. Therefore mod-R has kernels, so, by 2.3.18, R is right coherent; for this shows that $\ker(f)$ is finitely generated and, since $\mathrm{im}(f)$ is a typical finitely generated subobject of B, it follows that every finitely generated submodule of a finitely presented module is finitely presented. Furthermore, if $f : A \to B$ is monic, then $g = 0$, hence $(f,-)$ is epi and so $f : A \to B$ is split. Thus every embedding between finitely presented modules is pure (hence split). By 2.3.26, R is, therefore, von Neumann regular (2.1.19).

[6] In fact, that paper deals with functor categories where the domain is abelian, which mod-R need not be, but much applies more generally.

If R is von Neumann regular, let $F = F_f$, where $f : A \to B$ is in mod-R, be a typical finitely presented functor (10.2.1). Let $A'' = \mathrm{im}(f)$. Since R is coherent (2.3.23) A'' is finitely presented, so $A' = \ker(f)$ is finitely generated. By 2.3.22 the embedding of A' in A is pure, hence (2.1.18) split. So $F_f \simeq (A', -)$ has projective dimension 0 (10.1.12), as required. □

Since regularity is a right/left symmetric condition one has the immediate corollary.

Corollary 10.2.21. gldim(mod-R, **Ab**)^{fp} = gldim(mod-R, **Ab**)^{fp} *for every ring* R.

Combined with 10.2.14, 10.2.20 gives the following model-theoretic corollary, for which there are many proofs (see [495, 16.16] for references). Also see 10.2.40 for a stronger result.

Corollary 10.2.22. *A ring* R *is von Neumann regular iff every pp condition for* R*-modules is equivalent to a system of linear equations.*

We remark that, in contrast to the projective dimension of finitely presented functors, the projective dimension of arbitrary functors in (mod-R, **Ab**) is not bounded by 2. For example, one has, see [256, 10.1], that for any ring R $\mathrm{pgldim}(R) \leq \mathrm{gldim}(\text{mod-}R, \mathbf{Ab}) \leq \mathrm{pgldim}(R) + 2$ (for pure global dimension see Section 2.2).

The next result describes the finitely generated subfunctors of an arbitrary representable functor in terms of pp conditions, cf. 10.2.16. Since, see the comment after 10.2.6, $(C, -)$ embeds in some $(R^n, -)$, it follows by 10.2.16 that the finitely generated subfunctors of $(C, -)$ are given by pp conditions. The first statement follows from 10.2.16 and the second follows easily from 4.3.62.

Proposition 10.2.23. *Let* $C \in$ mod-R. *Take a finite generating tuple* $\overline{c}$ *from* C *and suppose that* $\overline{c}H = 0$ *(where* H *is a matrix over* R*) is a presentation of* C*. Denote by* $\theta(\overline{x})$ *the pp condition* $\overline{x}H = 0$*, so, 1.1.11,* $(C, -) \simeq F_\theta$*. Then the finitely generated subfunctors of* $(C, -)$ *are exactly the* F_ϕ *with* $\phi \leq \theta$*.*

If C *is indecomposable with local endomorphism ring, then the radical of* $(C, -)$ *is* $\sum\{F_\phi : \phi < \theta\}$*.*

10.2.3 Minimal free realisations and local functors

Minimal free realisations (where they exist) correspond to projective covers of the corresponding functors (10.2.25). A finitely presented module has local endomorphism ring iff the corresponding representable functor is local (10.2.27).

This section is based on [276, §2].

Given a pp condition ϕ and a finitely presented module C there is an epimorphism $(C, -) \to F_\phi$ iff some tuple from C is a free realisation (Section 1.2.2) of ϕ (10.2.8).

Say that a finitely presented module C is **minimal over** $\overline{c}$ if every $f \in \mathrm{End}(C)$ which fixes the pp-type of $\overline{c}$, that is, such that $\mathrm{pp}^C(f\overline{c}) = \mathrm{pp}^C(\overline{c})$, is an automorphism of C and, in this case say that $(C, \overline{c})$ is a **minimal** free realisation of ϕ, where ϕ generates $\mathrm{pp}^C(\overline{c})$. This implies, in particular, that there is no proper direct summand of C which contains $\overline{c}$.

Lemma 10.2.24. *Suppose that C is a finitely presented indecomposable module. If* $\mathrm{End}(C)$ *is local, then C is minimal over every non-zero tuple,* $\overline{c}$*, from it and the radical,* $J\mathrm{End}(C)$*, of the endomorphism ring of C is* $\{f \in \mathrm{End}(C) : \mathrm{pp}^C(\overline{c}) > \mathrm{pp}^C(f\overline{c})\}$ *– the set of endomorphisms which strictly increase the pp-type of* $\overline{c}$*.*

Proof. If $\mathrm{End}(C)$ is local and if $f : C \to C$ is such that $\mathrm{pp}^C(f\overline{c}) = \mathrm{pp}^C(\overline{c})$, then, by 1.2.9, there is $g \in \mathrm{End}(C)$ with $gf\overline{c} = \overline{c}$. Then $1 - gf$ has non-zero kernel, so, since $\mathrm{End}(C)$ is local, gf and therefore f must be an automorphism.

The description of the radical follows, since if f strictly increases the pp-type of $\overline{c}$, then it cannot be invertible. □

Proposition 10.2.25. *Let* $(C, \overline{c})$ *be a free realisation of the pp condition* ϕ*. Then* $(C, \overline{c})$ *is a minimal free realisation of* ϕ *iff the epimorphism* $(\overline{c}, -) : (C, -) \to F_\phi$ *induced by* $\overline{c}$ *is a projective cover.*

Proof. Suppose that $(\overline{c}, -)$ is a projective cover and let $f \in \mathrm{End}(C)$ be such that $\mathrm{pp}^C(f\overline{c}) = \mathrm{pp}^C(\overline{c})$. Then $(f\overline{c}, -) = (\overline{c}, -)(f, -)$ is an epimorphism (by 10.2.8). By projectivity of $(C, -)$ there is g as shown making the diagram commute.

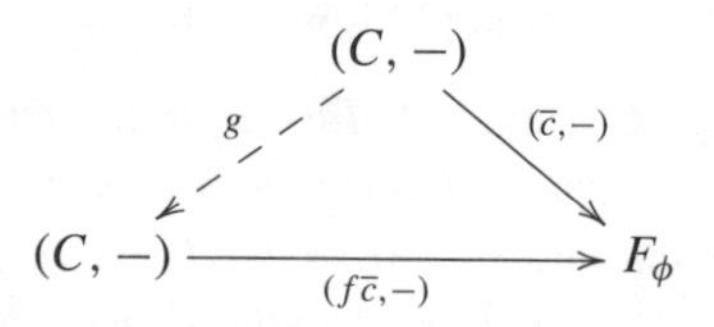

Thus $(\overline{c}, -)(f, -)g = (\overline{c}, -)$, so (from the definition, see Section 4.6, of projective cover), $(f, -)$, hence (10.1.11) f, is an isomorphism.

For the converse suppose that the free realisation is minimal. It must be shown that the kernel, K, of $(C, -) \to F_\phi$ is inessential. Suppose, therefore, that G is a subfunctor of $(C, -)$ with $K + G = (C, -)$. For any $A \in \text{mod-}R$ one has $KA = \{k \in (C, A) : k\overline{c} = 0\}$. In particular, this is so for $A = C$, so $k + g = 1_C$ for some $k \in KC$ and $g \in GC$. Since $k\overline{c} = 0$ it must be that g fixes $\overline{c}$, so, by assumption, g is an automorphism of C. It follows by 10.1.10 that $G = (C, -)$, as required. □

In consequence every pp condition over a Krull–Schmidt ring has a minimal free realisation (4.3.70).

Corollary 10.2.26. ([276, 2.2]) *Suppose that the finitely presented indecomposable module C is minimal over $\overline{c}$. Let ϕ be a generator of $\mathrm{pp}^C(\overline{c})$ and suppose that $(A, \overline{a})$ is a free realisation of ϕ. Then there is a split monomorphism $C \to A$ taking $\overline{c}$ to $\overline{a}$.*

Proof. Since $\mathrm{pp}^C(\overline{c}) = \mathrm{pp}^A(\overline{a})$ there are, by 1.2.9, morphisms $f : C \to A$, $g : A \to C$ with $f\overline{c} = \overline{a}$, $g\overline{a} = \overline{c}$. By 10.2.24 the composition gf is in Aut(C), so f is a split monomorphism, as required. □

A functor G is **local** if it has a unique maximal proper subfunctor. If $C \in$ mod-R, then the **radical**, rad$(C, -)$, of the functor $(C, -)$ is the intersection of all the maximal proper subfunctors of $(C, -)$. By [448, 4.1] this is given by $\mathrm{rad}(C, -).A = \{f \in (C, A) : \forall g \in (A, C), 1 - gf \text{ is invertible}\}$ (this agrees with the definition at 9.1.1).

Proposition 10.2.27. ([276, p. 161]) *If $C \in$ mod-R, then C has local endomorphism ring iff $(C, -) \in$ (mod-R, **Ab**) is a local functor. In that case the radical* rad$(C, -)$ *of* $(C, -)$, *which is given by* $\mathrm{rad}(C, -) \cdot A = \mathrm{rad}(C, A)$ *in the notation of 9.1.1, consists of those $f \in (C, A)$ such that f is strictly pp-type-increasing on any, equivalently every, non-zero tuple from C.*

Proof. If C has local endomorphism ring, then the description of the radical follows directly from the minimality of C over any non-zero tuple from it and 1.2.9.

If C has local endomorphism ring and F G is a proper subfunctor of $(C, -)$, then, by 10.1.10, $FC \neq (C, C)$, so, since End(C) is local, $FC \leq J\mathrm{End}(C)$. Therefore, from the description given after the definition, F must be contained in the radical of $(C, -)$.

For the converse, suppose that $(C, -)$ is local. Let $\overline{c} : R^n \to C$ be a finite generating tuple for C and let ϕ be a pp condition generating $\mathrm{pp}^C(\overline{c})$. Then $(C, -) = F_\phi$. If End(C) were not local, say $f_1, f_2 \in \mathrm{End}(C)$ are non-isomorphisms with $f_1 + f_2 = 1_C$, then choose a generator, ψ_i, for $\mathrm{pp}^C(f_i\overline{c})$ $(i = 1, 2)$. By 1.2.9, $\psi_i < \phi$, that is, $F_{\psi_i} < F_\phi$, and so, by assumption, $F_{\psi_1} + F_{\psi_2} < F_\phi = (C, -)$. But that contradicts that $\overline{c} = f_1\overline{c} + f_2\overline{c} \in (F_{\psi_1} + F_{\psi_2})(C)$. □

Corollary 10.2.28. *Suppose that $C \in$ mod-R has local endomorphism ring and let $\overline{c}$ be a non-zero tuple from C. Let ϕ be a pp condition which generates $\mathrm{pp}^C(\overline{c})$. Then the corresponding functor, F_ϕ, is local.*

Proof. As in 10.2.8, $\overline{c}$ induces an epimorphism $(C, -) \to F_\phi$. By 10.2.27, $(C, -)$ is local. It follows that the same is true for every non-zero image of $(C, -)$. □

Also note 12.1.16.

10.2.4 Pp conditions over rings with many objects

In this chapter, indeed in most of Part II, R is assumed to be a small preadditive category, not necessarily a ring. Most of the time we can pretend that it is a ring but some issues do require explication. One of these is what is meant by a pp condition and by the solutions to such a condition. The key is to accept that an element of a module M is "really just" a morphism from the projective generator[7] *R_R to M. That leads to the definition of an element of an object in any functor category.*[8]

Let R be a small preadditive category. First Yoneda-embed R (10.1.8) into the category, Mod-$R = (R^{\mathrm{op}}, \mathbf{Ab})$, of right R-modules. For simplicity (but not always clarity) we sometimes identify R with its image, thus we identify an object $G \in R$ with $(-, G) \in$ Mod-R. In the case of a ring this is where we "regard the ring as a (right) module" (10.1.9). Then, if M is any right R-module and G is an object of R, an **element** of M **of sort** G, or a G**-element** of M, more accurately a $(-, G)$-element of M, is a morphism $a : (-, G) \to M$ in Mod-R. That is, the elements of M of sort G are the natural transformations from $(-, G)$ to M, hence, by the Yoneda Lemma (10.1.7), they are exactly the elements of $M(G)$.

This accepted, multiplication by elements (that is, morphisms) of R becomes precomposition with maps in the Yoneda-image of R: if $r : G \to H$ is a morphism of R, then there is the induced map $(-, r) : (-, G) \to (-, H)$ which, given a right R-module M, induces $((-, r), M) : ((-, H), M) \to ((-, G), M)$, that is, $M(r) : M(H) \to M(G)$. In this way, and note the contravariance, a morphism $r : G \to H$ in R induces a map, "multiplication by r", from the abelian group of H-elements of M to the group of G-elements of M. If b is an H-element of M (so $b : (-, H) \to M$), then we write br for the composition, that is, the corresponding G-element of M. Thus we have a formalism which looks just like, and may be manipulated just as, that in the case of a ring.

Turning to pp conditions, since there might now be more than one sort, we restrict each variable to stand for elements of a *fixed* sort and sometimes use subscripts attached to variables to indicate sorts. For instance, in the pp condition $xr = 0$, where $r : G \to H$ in R, the variable x stands for elements of sort H (so

[7] Quoting from [195, p. 100]: "This fact, that projective generators are as good as elements, was a part of the folkore of the subject from the beginning."

[8] Indeed, in any finitely accessible category, see Section 16.1.1.

we could write $x_H r = 0$ to emphasise this) and the resulting term, xr, will stand for certain elements of sort G. Therefore, if M is an R-module and $\phi(x)$ is the condition $xr = 0$, then $\phi(M)$ is the group of morphisms from H to M which, precomposed with r, give the zero morphism from G to M.

For another example, take r as above and consider the pp condition $\psi(y)$, which is $r \mid y$ or, showing more detail, $\exists x_H\, (y_G = x_H r)$. Then $\psi(M)$ is the group of morphisms from G to M which factor initially through r.

Note also that a term $\sum_i x_i r_i$ is well defined iff all the r_i have the same domain (the x_i may well be variables of different sorts).

Modulo these elaborations, the definitions, general results and proofs in Part I go through with almost no change. Such few differences as there are arise from the fact that there might be no global $1 \in R$ and even that has limited effect because variables are sorted, so each pp condition refers to only finitely many sorts, for which the corresponding sum of idempotents/identities provides a 1. One difference, cf. 5.1.23, which should be pointed out is that the Ziegler spectrum, Zg_R, will not necessarily be compact if R has infinitely many non-isomorphic objects:[9] although the basic open subsets will be compact, for the proof of 5.1.23 still applies, there need be no pp-pair ϕ/ψ such that $\mathrm{Zg}_R = (\phi/\psi)$. For instance, let R be the k-path category of the quiver A_∞ (10.1.2) – equivalently the category with objects the finitely generated projective representations of this quiver (see 10.1.15). Since every finitely generated functor is a quotient of a functor $(P, -)$ for some finitely generated projective P and since for any P there are certainly non-zero representations M such that $(P, M) = 0$, it follows that there is no pp-pair $\phi > \psi$ such that the open set (ϕ/ψ) is the whole spectrum.

This discussion is extended from functor categories to arbitrary locally finitely presented categories in Section 11.1.5.

10.2.5 Finitely presented functors = pp-pairs

The category of pp-pairs is shown to be equivalent to the category of finitely presented functors on finitely presented modules (10.2.30).

Remark 10.2.29. We have seen in 10.2.8 that, if ϕ is a pp condition, then the corresponding functor F_ϕ is equivalent to the functor $\mathrm{im}\big((\overline{c}, -) : (C, -) \to (R^n, -)\big)$, where $(C, \overline{c})$ is a free realisation of ϕ. In particular, F_ϕ is finitely generated, hence, by 10.2.2, finitely presented. Therefore, by E.1.16, for any pp pair ϕ/ψ the functor $F_{\phi/\psi} = F_\phi/F_\psi$ is finitely presented. We will see that the converse also is true.

[9] Of course, it *might* still be compact, say if R is proj-S for a ring S.

An explicit projective presentation of F_ϕ is seen in 1.2.19: $0 \to (C_\phi/\langle \overline{c}_\phi \rangle, -) \to (C_\phi, -) \to F_\phi \to 0$, where $(C_\phi, \overline{c}_\phi)$ is any free realisation of ϕ.

The next result is the key to the translation between the model-theoretic and functorial approaches.

Theorem 10.2.30. ([101, 3.2.5]) *Let R be a skeletally small preadditive category. Then the category,* $(\text{mod-}R, \mathbf{Ab})^{\text{fp}}$, *of finitely presented functors is equivalent to the category,* $\mathbb{L}_R^{\text{eq}+}$, *of pp-pairs (Section 3.2.2).*

Proof. Let F be a finitely presented functor, say $f : A \to B$ is a morphism in mod-R such that the sequence $(B, -) \xrightarrow{(f,-)} (A, -) \to F \to 0$ is exact (10.2.1). Take a finite generating tuple $\overline{a}$ for A and choose a generator, ϕ, for $\text{pp}^A(\overline{a})$ (1.2.6). Let ψ be a generator for the pp-type of $f\overline{a}$ in B. Then, since $FM = F_f M$ is the group of morphisms from A to M modulo those which factor through f, it follows, see 1.2.24, that $F \simeq F_{\phi/\psi}$. We saw just above that every functor of the latter form is finitely presented.

It remains to show that pp-definable functions between pp pairs correspond to natural transformations between the corresponding functors. So (see Section 3.2.2) let the pp condition ρ define a morphism $\phi/\psi \to \phi'/\psi'$: then for each (finitely presented) module M we have an additive function from $\phi(M)/\psi(M)$ to $\phi'(M)/\psi'(M)$ defined by taking $\overline{a} + \psi(M) \in \phi(M)/\psi(M)$ to the coset $\overline{a}' + \psi'(M)$, where $\overline{a}'$ is the unique, modulo $\psi'(M)$, tuple such that $(\overline{a}, \overline{a}') \in \rho(M)$. If $f : M \to M'$ is in Mod-R, then (1.1.7) $(\overline{a}, \overline{a}') \in \rho(M)$ implies $(f\overline{a}, f\overline{a}') \in \rho(M')$, so this defines a natural transformation, $\tilde{\rho}$ say, from $F_{\phi/\psi}$ to $F_{\phi'/\psi'}$.

For the converse, let $\tau : F_{\phi/\psi} \to F_{\phi'/\psi'}$ be a natural transformation. Considering the component at C, where $(C, \overline{c})$ is a free realisation of ϕ, let $\overline{c}'$ from C be such that $\overline{c}' + \psi'(C) = \tau_C(\overline{c} + \psi(C))$ and let $\rho(\overline{c}, \overline{c}')$ be a generator for the pp-type of $(\overline{c}, \overline{c}')$ in C. We claim that ρ defines a morphism from ϕ/ψ to ϕ'/ψ' and that τ is the natural transformation, $\tilde{\rho}$, corresponding to ρ.

First we show that ρ is functional; the argument in the next paragraph shows that it is total. So suppose that M is a module and that $(\overline{b}, \overline{b}') \in \rho(M)$ with $\overline{b} \in \psi(M)$: it must be shown that $\overline{b}' \in \psi'(M)$. By 1.2.17 there is a morphism $f : C \to M$ taking $(\overline{c}, \overline{c}')$ to $(\overline{b}, \overline{b}')$. Note that, for example, $F_{\phi/\psi} f$ is simply the map induced (1.1.7) between the relevant subquotients. We have $\tau_M \cdot F_{\phi/\psi} f = F_{\phi'/\psi'} f \cdot \tau_C$, so since $\tau_M(\overline{b} + \psi(M)) = \overline{0} + \psi'(M)$ it follows that $0 = f\tau_C(\overline{c} + \psi(M)) = F_{\phi'/\psi'} f(\overline{c}' + \psi'(M)) = \overline{b}' + \psi'(M)$, as required.

Next, given $M \in \text{mod-}R$ and $\overline{d} \in \phi(M)$ there is, by 1.2.17, a morphism $f : C \to M$ with $f\overline{c} = \overline{d}$. Then $\tau_M . F_{\phi/\psi} f = F_{\phi'/\psi'} f \cdot \tau_C$, so $\tau_M(\overline{d} + \psi(M)) = \tau_M(f\overline{c} + \psi(M)) = \tau_M \cdot F_{\phi/\psi} f(\overline{c} + \psi(C)) = F_{\phi'/\psi'} f \cdot \tau_C(\overline{c} + \psi(C)) = F_{\phi'/\psi'} f \cdot \tilde{\rho}_C(\overline{c} + \psi(C)) = \tilde{\rho}_M \cdot F_{\phi/\psi} f \cdot (\overline{c} + \psi(C))$ (since $\tilde{\rho}$ is a natural transformation)

$= \tilde{\rho}_M(f\overline{c} + \psi(M)) = \tilde{\rho}_M(\overline{d} + \psi(M))$. Therefore $\tau_M = \tilde{\rho}_M$ for every $M \in$ mod-R, so $\tau = \tilde{\rho}$, as required. □

A relative version of this is proved at 12.3.20.

Corollary 10.2.31. *Every finitely presented functor in* (mod-R, **Ab**) *is isomorphic to one of the form $F_{\phi/\psi} = F_\phi/F_\psi$ for some pp-pair ϕ/ψ, and every functor of this form is finitely presented.*

Sometimes we refer to a functor of the form $F_{\phi/\psi}$ as a "pp-functor".

In the next result the notation $\overrightarrow{F}$ is used for the extension of a functor on mod-R to one on Mod-R which commutes with direct limits (Section 10.2.8).

Corollary 10.2.32. *A subcategory $\mathcal{X}$ of* Mod-R *is definable iff there is a set $(F_\lambda)_\lambda$ of finitely presented functors from* mod-R *to* **Ab** *such that $\mathcal{X} = \{M \in \text{Mod-}R : \overrightarrow{F_\lambda} M = 0\ \forall \lambda\}$.*

Corollary 10.2.33. *For any small preadditive category R and objects $G_1, \dots, G_n$ of R the lattice of pp conditions with n free variables $x_1, \dots, x_n$, where x_i is a variable of sort G_i, is isomorphic to the lattice* $\mathrm{Latt}^{\mathrm{f}}\big((G_1, -) \oplus \cdots \oplus (G_n, -)\big)$ *of finitely generated subfunctors of the functor $(G_1, -) \oplus \cdots \oplus (G_n, -)$ in* (mod-R, **Ab**)*; here G_i is identified with its image, $(-, G_i) : R^{\mathrm{op}} \longrightarrow$* **Ab**, *in* mod-$R$.

This follows directly from the definitions, 10.2.30 and 10.2.2. A particular case is 10.2.17 (and a slightly more direct proof as there could be given).

A functor in (mod-R, **Ab**)^{fp} is representable iff it is isomorphic to one of the form F_θ with θ a quantifier-free pp condition (see 1.1.11). By the Yoneda Lemma the full subcategory of representable functors is equivalent to mod-R, so we have the following corollary.

Corollary 10.2.34. *Let R be a small preadditive category. Then the full subcategory of $\mathbb{L}_R^{\mathrm{eq}+}$ whose objects are the functors F_θ with θ quantifier-free pp conditions is equivalent to* mod-R. *In particular, if $\theta' \leq \theta$ are quantifier-free pp conditions, then $F_\theta/F_{\theta'} \simeq F_{\theta''}$ for some quantifier-free pp θ''. Moreover, every pp-defined morphism $F_\theta \to F_{\theta'}$, where θ and θ' are quantifier-free, can be defined by a quantifier-free pp condition.*

Proof. The first statement is direct from 10.2.30, 1.1.11 and the Yoneda Lemma. The second follows since mod-R is closed under cokernels in Mod-R. For the last, if $F_\theta \simeq (C_\theta, -)$ and $F_{\theta'} \simeq (C_{\theta'}, -)$ (isomorphisms chosen as at 1.1.11) then it is clear that any morphism $g : C_\theta \to C_{\theta'}$ corresponds to a quantifer-free-defined relation between $\overline{c}_\theta$ and $\overline{c}_{\theta'}$ (notation again as at 1.1.11) of the form $(\overline{x}_\theta, \overline{x}_{\theta'} H)$ where the matrix H is such that $g\overline{c}_\theta = \overline{c}_{\theta'} H$. □

Thus mod-R is "the quantifier-free part" of $\mathbb{L}_R^{\mathrm{eq}+}$.

10.2.6 Examples of finitely presented functors

The functor $\mathrm{Ext}^n_R(M,-)$ *is finitely presented if* M *is* FP_{n+1} *(10.2.35). The functor* $\mathrm{Tor}^R_n(L,-)$ *is finitely presented if* L *is* FP_{n+1} *(10.2.36).*

In this section, following [503, pp. 211–12], we obtain conditions under which various homological functors are definable by pp conditions. By 10.2.31 if $F : \text{Mod-}R \longrightarrow \mathbf{Ab}$ is such that $F \restriction \text{mod-}R$ is finitely presented, then $F \restriction \text{mod-}R$ is isomorphic to a functor of the form $F_{\phi/\psi} = F_\phi/F_\psi$ for some pp-pair ϕ/ψ. If, moreover, F commutes with direct limits, then F is determined by its restriction to mod-R and will be isomorphic to F_ϕ/F_ψ on all modules (see 10.2.43).[10]

We begin with representable functors $(M,-)$, where $M \in \text{Mod-}R$. If M is finitely presented, then, by 10.1.12, $(M,-) \restriction \text{mod-}R$ is finitely presented. The converse is false: if $R = \mathbb{Z}$ and $M = \mathbb{Q}$, then $(\mathbb{Q},-) \restriction \text{mod-}R$ is finitely presented (being the zero functor) but $\mathbb{Q}_\mathbb{Z}$ is not finitely presented. Each functor on mod-R has a unique extension to a functor on Mod-R which commutes with direct limits (Section 10.2.8), so, since the functor $(M,-)$ on Mod-R commutes with direct limits iff M is finitely presented (this is the definition of "finitely presented" – see Appendix E, p. 709), it is exactly when M is finitely presented that $(M,-)$ is this extension to Mod-R of its restriction to mod-R. In that case the restriction of $(M,-)$ is equivalent, by 10.2.30, to a pp functor $F_{\phi/\psi}$, so, since both $(M,-)$ and $F_{\phi/\psi}$ commute with direct limits, these functors also agree on Mod-R. Explicitly this functor $(M,-)$ is $F_{\phi/(\overline{x}=0)}$, where ϕ is given by the pp condition $\overline{x}H = 0$, where $\overline{a} = (a_1, \dots, a_n)$ is a finite generating set for M with presentation $\overline{a}H = 0$ (1.1.11).

Next, consider the functor $\mathrm{Ext}^1(M,-)$, where $M \in \text{Mod-}R$. If M is $\mathbf{FP}_2$, meaning that there is a projective presentation $\cdots \to P_2 \to P_1 \to P_0 \to M \to 0$ with all of P_0, P_1, P_2 finitely generated, then, setting $K = \mathrm{im}(P_1 \to P_0)$ (finitely presented since P_2 is finitely generated), we have an exact sequence $0 \to K \to P_0 \to M \to 0$, hence a long exact sequence $0 \to (M,-) \to (P_0,-) \to (K,-) \to \mathrm{Ext}^1(M,-) \to \mathrm{Ext}^1(P_0,-) = 0$ (0 since P_0 is projective). Therefore $\mathrm{Ext}^1(M,-) \simeq (K,-)/(P_0,-)$ is a quotient of functors which are finitely presented when restricted to mod-R. Thus M being FP_2 implies that $\mathrm{Ext}^1(M,-) \restriction \text{mod-}R$ is finitely presented. Furthermore, by [96, Thm 2], M being FP_2 is exactly the condition needed for $\mathrm{Ext}^1(M,-)$ to commute with direct limits. Therefore if M is FP_2, then $\mathrm{Ext}^1(M,-)$, as a functor on Mod-R, has the form[11] $F_{\phi/\psi}$. An explicit representation of $\mathrm{Ext}^1(M,-)$ in the form F_ϕ/F_ψ can be obtained via the argument above, modulo an explicit presentation of M and a first syzygy K.

[10] In that case we may say, in the terminology of Section B.2, that F assigns in a functorial way, to every module M, a certain sort in $M^{\mathrm{eq}+}$.

[11] The converse is, of course, false: if M is any infinitely generated projective module, then $\mathrm{Ext}(M,-)$ is 0 but M is not FP_2.

This argument generalises to $\mathrm{Ext}^n(M, -)$. With notation as above, if M is $\mathbf{FP}_{n+1}$, meaning that it has a projective resolution the first $n+2$ terms of which are finitely generated, then its first syzygy K is FP_n, so, by induction, $\mathrm{Ext}^{n-1}(K, -) \restriction$ mod-R is finitely presented. As part of the usual long exact sequence we have, for $n \geq 2$,

$$0 = \mathrm{Ext}^{n-1}(P_0, -) \to \mathrm{Ext}^{n-1}(K, -) \to \mathrm{Ext}^n(M, -) \to \mathrm{Ext}^n(P_0, -) = 0,$$

so $\mathrm{Ext}^n(M, -) \simeq \mathrm{Ext}^{n-1}(K, -)$ is, when restricted to mod-R, finitely presented. Furthermore, [96, Thm 2] says that $\mathrm{Ext}^n(M, -)$ commutes with direct limits provided M is FP_{n+1}.

We have shown the following from [503, pp. 211–12] (some of which is in [17]). Also see [323, p. 264].

Theorem 10.2.35. *Let* $M \in$ Mod-R.

(a) *The functor* $(M, -) :$ mod-$R \longrightarrow$ **Ab** *is finitely presented if* M *is finitely presented.*
(b) *The functor* $\mathrm{Ext}^1(M, -) :$ mod-$R \longrightarrow$ **Ab** *is finitely presented if* M *is* FP_2.
(c) *More generally, the functor* $\mathrm{Ext}^n(M, -) :$ mod-$R \longrightarrow$ **Ab** *is finitely presented if* M *is* FP_{n+1}.

Next, consider tensor product. If $L \in R$-Mod is finitely presented, say $P_1 \to P_0 \to L \to 0$ is exact with P_0, P_1 finitely generated projective modules, then there is the sequence $(- \otimes P_1) \to (- \otimes P_0) \to (- \otimes L) \to 0$ which is exact by right exactness of tensor product. Since P_0 is a direct summand of a finitely generated free module, $- \otimes P_0$ is, for some n, a direct summand of $(- \otimes R^n) \simeq (R_R, -)^n$, hence $(- \otimes P_0) \restriction$ mod-R is finitely presented, as is $(- \otimes P_1) \restriction$ mod-R. Therefore $(- \otimes L) \restriction$ mod-R is finitely presented (by E.1.16). Furthermore, for any module L the functor $- \otimes L$, being a left adjoint, commutes with direct limits.

For the converse ([17, 6.1], also [256, 3.3]), suppose that $(- \otimes L) :$ mod-$R \longrightarrow$ **Ab** is finitely presented. Let $R^{(\kappa)} \to L$ be an epimorphism. Then $(R \otimes -)^{(\kappa)} \simeq (R^{(\kappa)} \otimes -) \to (L \otimes -)$ is an epimorphism, so, since $L \otimes -$ is finitely generated, there is (10.1.13) an epimorphism (since $\otimes$ is right exact) $(R \otimes -)^n \to (L \otimes -)$, hence an epimorphism $R^n \to L$. Therefore L is finitely generated. Let $R^{(\lambda)} \to R^n \to L \to 0$ be a presentation of L. This gives rise to the exact sequence $(R \otimes -)^{(\lambda)} \to (R^n \otimes -) \to (L \otimes -) \to 0$, so, since $L \otimes -$ is finitely presented, by E.1.16 there is an exact sequence $(R \otimes -)^m \to (R^n \otimes -) \to (L \otimes -) \to 0$, hence an exact sequence $R^m \to R^n \to L \to 0$ and L is finitely presented, as required.

Finally, consider the derived functors of tensor product. Let $L \in R$-mod be FP_2, so, as argued before, there is a short exact sequence $0 \to K \to P_0 \to L \to 0$

with P_0 finitely generated projective and K finitely presented. This gives the long exact sequence $0 = \mathrm{Tor}_1(P_0, -) \to \mathrm{Tor}_1(L, -) \to (- \otimes K) \to (- \otimes P_0) \to (- \otimes L) \to 0$. Since all of L, P_0, K are finitely presented, so also, by the above, are all the corresponding tensor functors. Therefore, $\mathrm{Tor}_1(L, -)$, being the kernel of $(- \otimes K) \to (- \otimes P_0)$, is finitely generated hence, by 10.2.3, is finitely presented. Since every functor $\mathrm{Tor}_n(L, -)$ commutes with direct limits (for example, [112, VI.1.3]) $\mathrm{Tor}_1(L, -)$ is, therefore, a pp functor if L is FP_2.

The case of $\mathrm{Tor}_n(L, -)$ follows as in the case of Ext^n: we have, as part of the long exact sequence of homology, for $n \geq 2$, $0 = \mathrm{Tor}_n(P_0, -) \to \mathrm{Tor}_n(L, -) \to \mathrm{Tor}_{n-1}(K, -) \to \mathrm{Tor}_{n-1}(P_0, -) = 0$, so, if L is FP_{n+1}, hence K is FP_n, then we deduce, by induction, that $\mathrm{Tor}_n(L, -)$ is finitely presented.

Theorem 10.2.36. ([17, 6.1] for (a), [503, pp. 211, 212] for (b), (c), also see [711, 3.2]) *Let $L \in R$-Mod.*

*(a) The functor $(- \otimes_R L)$: mod-$R \longrightarrow$ **Ab** is finitely presented iff L is finitely presented.*
*(b) The functor $\mathrm{Tor}_1(L, -)$: mod-$R \longrightarrow$ **Ab** is finitely presented if L is FP_2.*
*(c) More generally, the functor $\mathrm{Tor}_n(L, -)$: mod-$R \longrightarrow$ **Ab** is finitely presented if L is FP_{n+1}.*

Note that by the symmetry of $\otimes$ and its derived functors Tor_n, the statements also apply with the module L and argument "$-$" interchanged.

By [256, 5.8] the functor $- \otimes_R L$ is a locally coherent object of the functor category (that is, every finitely generated subfunctor is finitely presented) iff L is Mittag–Leffler (Section 1.3.5).

10.2.7 Free abelian categories

The category of finitely presented functors is the free abelian category on a ring (or small preadditive category) (10.2.37). A ring is von Neumann regular iff this category is equivalent to the category of finitely presented modules (10.2.38). Modules over von Neumann regular rings have elimination of imaginaries (10.2.40).

Let R be a small preadditive category. A **free abelian category** on R is an abelian category, $\mathcal{A}$, together with an additive functor, $R \longrightarrow \mathcal{A}$, such that every additive functor from R to an abelian category $\mathcal{B}$ factors through $R \longrightarrow \mathcal{A}$ via an exact functor, which is unique to natural equivalence, from $\mathcal{A}$ to $\mathcal{B}$.

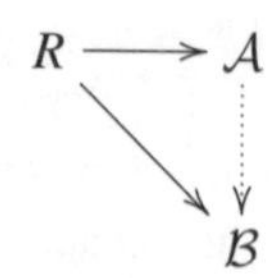

If such a category exists (and, by the next result, it does), then it is unique up to natural equivalence over R. We denote it by Ab(R). The definition of and proof of existence of this category is due to Freyd, [196, 4.1]. The identification of this category with the category of finitely presented functors on finitely presented left modules is stated (for R a ring) in [253, Lemma 1] and a proof is given in [359, 2.10]. Also see Adelman's direct construction ([4]) of the free abelian category.[12] Also see [462, §5.1]. For a very general construction see [65].

Given the small preadditive category R there is the contravariant Yoneda embedding (10.1.8) into R-Mod, given on objects G of R by $G \mapsto (G, -)$ and then another contravariant Yoneda embedding from R-mod to (R-mod, **Ab**) given by $M \mapsto (M, -)$. Denote the composite, which takes G to $((G, -), -)$ by Y^2.

Theorem 10.2.37. *Let R be a small preadditive category. Then the composite Yoneda embedding $Y^2 : R \longrightarrow (R\text{-mod}, \mathbf{Ab})^{\mathrm{fp}}$ given on objects by $G \mapsto ((G, -), -)$ is the free abelian category on R.*

Proof. First, an additive functor $f : R \longrightarrow \mathcal{A}$, where $\mathcal{A}$ is abelian, is extended to a left exact functor, f', from $(R\text{-mod})^{\mathrm{op}}$ (that is, a right exact functor from R-mod) by sending $L \in R$-mod to $\ker(fG \xrightarrow{f\alpha} fH)$, where $G \xrightarrow{\alpha} H$ in R is such that $(H, -) \xrightarrow{(\alpha,-)} (G, -) \to L \to 0$ is exact in R-mod; one may check independence of presentation and extend in the obvious way to an action on morphisms, checking that it also is well defined.

Then f' is extended to a right exact (and, one checks, exact, in addition to well-defined) functor, f'', from $(R\text{-mod}, \mathbf{Ab})^{\mathrm{fp}}$ to $\mathcal{A}$ by sending $F \in (R\text{-mod}, \mathbf{Ab})^{\mathrm{fp}}$ to $\operatorname{coker}(f'M^{\circ} \xrightarrow{f'g^{\circ}} f'L^{\circ})$, where $L \xrightarrow{g} M$ in R-mod is such that $(M, -) \xrightarrow{(g,-)} (L, -) \to F \to 0$ is exact in $(R\text{-mod}, \mathbf{Ab})^{\mathrm{fp}}$ and where superscript $^{\circ}$ indicates objects and morphisms in the opposite category.

Writing out all the details is left to the reader. □

If R is a ring, then Y^2 takes R to the free left module ${}_RR$ and then to the functor $({}_RR, -)$. In terms of pp conditions this has taken R to the pair $x = x/x = 0$ for left modules. In the case that R has many objects the composite Yoneda embedding takes an object, G, of R to, with notation as in Section 10.2.4, the pair $(x_G = x_G)/(x_G = 0)$ (again for left modules).

It follows directly from the definition that $\mathrm{Ab}(R) \simeq (\mathrm{Ab}(R^{\mathrm{op}}))^{\mathrm{op}}$. An explicit description of this duality will be given in Section 10.3. In particular (10.3.4) $\mathrm{Ab}(R^{\mathrm{op}}) \simeq (\text{mod-}R, \mathbf{Ab})^{\mathrm{fp}} \simeq \mathbb{L}_R^{\mathrm{eq}+}$.

[12] The relevance of this category was pointed out to me by Herzog.

A right R-module is a functor $M : R^{\mathrm{op}} \longrightarrow \mathbf{Ab}$, so extends uniquely to an exact functor, which we denote $M^{\mathrm{eq}+} : \mathrm{Ab}(R^{\mathrm{op}}) \longrightarrow \mathbf{Ab}$. The notation is explained in Section B.2, where this is defined in a very different way. This functor is simply evaluation: $M^{\mathrm{eq}+}(F) = \overrightarrow{F}(M)$, where $\overrightarrow{F}$ is the unique extension of F to a functor on Mod-R which commutes with direct limits (Section 10.2.8). For, certainly the functor "evaluation at M" is exact and it takes the image, $(R_R, -)$, of R^{op} in $\mathrm{Ab}(R^{\mathrm{op}})$ to M, so, by uniqueness, this is the extension to $\mathrm{Ab}(R^{\mathrm{op}})$.

If R is von Neumann regular, then it follows from 2.3.22 that every finitely presented module is projective, so mod-$R \simeq$ proj-R and similarly for left modules. Therefore (R-mod, $\mathbf{Ab}$) $\simeq$ (R-proj, $\mathbf{Ab}$) $\simeq$ R-Mod (the second equivalence is easy to prove, see, for example, [448, 3.1]).

Proposition 10.2.38. *A ring R is von Neumann regular iff* (R-mod, $\mathbf{Ab}$) $\simeq$ R-Mod *iff* $\mathrm{Ab}(R) \simeq R$-mod, *where the first equivalence is the Yoneda embedding.*

Proof. From the discussion above and 10.2.37 all that remains is to show that the last condition implies the first. But that is immediate from 10.2.20 since R-mod is the category of projective objects in the category of finitely presented functors, which, therefore, has global dimension 0. □

Corollary 10.2.39. (see [282, p. 255]) *Suppose that R is von Neumann regular. Then the functor which on objects is $P_R \mapsto P \otimes_R -$ is an equivalence* proj-$R \simeq$ $\mathrm{Ab}(R)$ *and the duality* $\mathrm{Ab}(R) \simeq (\mathrm{Ab}(R^{\mathrm{op}}))^{\mathrm{op}}$ *is, regarded as a duality* proj-$R \simeq$ (R-proj)^{op}, *the functor which on objects is $P \mapsto P^*$, where* $P^* = \mathrm{Hom}_R(P_R, R_R)$.

For the last statement one may, for instance, show that $P \otimes - \simeq (P^*, -)$ using 12.1.2.

A further corollary is the following, which, model-theoretically, is elimination of pp-imaginaries (Section B.2). It is immediate from 10.2.14 and 10.2.20 and is generalised in 11.1.44.

Corollary 10.2.40. ([498, A5.3], [282, Prop. 9]) *The ring R is von Neumann regular iff for every pp-pair ϕ/ψ there is a pp condition ξ such that $F_{\phi/\psi} \simeq F_\xi$ in $\mathbb{L}_R^{\mathrm{eq}+}$. In this case ξ may be taken to be quantifier-free, that is, a system of linear equations.*

10.2.8 Extending functors along direct limits

Every functor on finitely presented modules has a unique, well-defined, $\varinjlim$-commuting extension to a functor on the category of all modules. This extension process preserves projective presentations (10.2.41) and annihilation of definable *subcategories (10.2.44). The topology on the Ziegler spectrum may be defined directly using these extended finitely presented functors (10.2.45).*

Every functor $F \in (\text{mod-}R, \mathbf{Ab})$ has a natural extension to a functor from Mod-R to **Ab** which commutes with direct limits. Let $M \in \text{Mod-}R$: then M is a direct limit of objects in mod-R, say $M = \varinjlim M_\lambda$ with $\big((M_\lambda)_\lambda, (g_{\lambda\mu})_{\lambda<\mu}\big)$ a directed system in mod-R. Applying F gives a directed system $\big((FM_\lambda)_\lambda, (Fg_{\lambda\mu})_{\lambda<\mu}\big)$ in **Ab**. This has a direct limit, the object of which we define to be $\overrightarrow{F} M$. It is straightforward to check that the result, $\overrightarrow{F} M$, does not depend on the choice of representation of M as a direct limit and the definition of the extension of F to morphisms between arbitrary modules is similar.

In particular one has the following.

Proposition 10.2.41. ([20, pp. 4–5]) *Let $F \in (\text{mod-}R, \mathbf{Ab})$ have projective presentation $\bigoplus_{j\in J}(B_j, -) \to \bigoplus_{i\in I}(A_i, -) \to F \to 0$ with $A_i, B_j \in \text{mod-}R$. This induces an exact sequence in* (Mod-R, **Ab**)*: $\bigoplus_{j\in J}(B_j, -) \to \bigoplus_{i\in I}(A_i, -) \to \overrightarrow{F} \to 0$, where the representable functors are now read as being in* (Mod-R, **Ab**) *and where $\overrightarrow{F}$ is the unique extension of F to a functor on* Mod-R *which commutes with direct limits.*

Corollary 10.2.42. *If $f : A \to B$ is a morphism in* mod-R*, then the extension, $\overrightarrow{F}_f$, of F_f (10.2.1) to* Mod-R *is the cokernel of $(f, -) : (B, -) \to (A, -)$, where now the representable functors are to be read as functors from* Mod-R *to* **Ab**.

The next result is immediate from 10.2.30.

Proposition 10.2.43. *Suppose that F is a functor from* Mod-R *to* **Ab** *which commutes with direct limits and which is such that the restriction of F to* mod-R *is a finitely presented object of* (mod-R, **Ab**)*. Then $F \simeq F_{\phi/\psi}$ for some pp conditions $\phi \geq \psi$. In particular, the kernel, $\{M \in \text{Mod-}R : FM = 0\}$, of F is a definable subcategory (Section 3.4.1) of* Mod-R.

The functors of the form $\overrightarrow{F}$ sometimes are termed “coherent functors” (also “p-functors” for those which are subfunctors of the forgetful functor). It is easy to check that these are exactly the functors from Mod-R to **Ab** which commute with direct limits and products. A much more general result is true, see 18.1.19.

Corollary 10.2.44. *Let $\mathcal{M} \subseteq \text{Mod-}R$ and let $F \in (\text{mod-}R, \mathbf{Ab})^{\text{fp}}$. If $\overrightarrow{F}\mathcal{M} = 0$, then $\overrightarrow{F}\langle\mathcal{M}\rangle = 0$, where $\langle\mathcal{M}\rangle$ denotes the definable subcategory generated by $\mathcal{M}$.*

Proof. This is immediate from 10.2.43 but it is easy to prove directly (that is, without using 10.2.30), as follows. Let $f : A \to B$ in mod-R be such that $(B, -) \xrightarrow{(f,-)} (A, -) \to \overrightarrow{F} \to 0$ is exact.

Suppose that M is a pure submodule of N and suppose that $\overrightarrow{F}N = 0$, so $(f, N) : (B, N) \to (A, N)$ is an epimorphism. Let $g \in (A, M)$. Then $ig \in (A, N)$, where $i : M \to N$ is the inclusion, so there is $h : B \to N$ with $hf = ig$. By 2.1.7 there is $k : B \to M$ such that $kf = g$. So $(f, M) : (B, M) \to (A, M)$ also is an epimorphism and, therefore, $\overrightarrow{F}M = 0$.

Next, suppose that $\overrightarrow{F}N_\lambda = 0$ $(\lambda \in \Lambda)$. So each $(f, N_\lambda) : (B, N_\lambda) \to (A, N_\lambda)$ is an epimorphism. Let $g \in (A, \prod N_\lambda)$. If $\pi_\mu : \prod N_\lambda \to N_\mu$ is the projection, then, for each μ, there is $h_\mu : B \to N_\mu$ such that $h_\mu f = \pi_\mu g$. Then $h = (h_\lambda)_\lambda \in (B, \prod N_\lambda)$ satisfies $hf = g$, so $\overrightarrow{F}(\prod N_\lambda) = \mathrm{coker}(f, \prod N_\lambda) = 0$.

Finally, the class of modules annihilated by $\overrightarrow{F}$ is closed under direct limits because $\overrightarrow{F}$ commutes with direct limits.

The conclusion therefore follows by 3.4.7. □

Although we have introduced a notation, $\overrightarrow{F}$, for the extension of $F \in (\text{mod-}R, \mathbf{Ab})$ to Mod-R, it is convenient to use the notation F for both and in many places we shall do that (in particular, usually we do not distinguish notationally between a pp functor F_ϕ and its restriction to mod-R).

Corollary 10.2.45. *Let R be any small preadditive category. The basic open sets of the Ziegler spectrum, Zg_R, are those of the form*[13] $(F) = \{N \in \mathrm{Zg}_R : \overrightarrow{F}N \neq 0\}$ *as F ranges over* $(\text{mod-}R, \mathbf{Ab})^{\mathrm{fp}}$.

Proof. This is immediate by 5.1.8 and 10.2.31. □

The next result is, by 10.2.30, 4.3.23 rephrased. Recall (Section 4.3.3) that $H(M)$ denotes the pure-injective hull of M.

Corollary 10.2.46. *For any module M and functor $F \in (\text{mod-}R, \mathbf{Ab})^{\mathrm{fp}}$, $\overrightarrow{F}(M) \neq 0$ iff $\overrightarrow{F}(H(M)) \neq 0$.*

10.3 Duality of finitely presented functors

The category of finitely presented functors on finitely presented right modules and the corresponding category for left modules are dual (10.3.4), with the duality interchanging Hom and ⊗ (10.3.1). The connection between this duality and elementary duality of pp conditions is given (10.3.6). It is noted that the latter duality gives an anti-isomorphism between the lattices of finitely generated subfunctors of the forgetful functors on right and left modules (10.3.7).

[13] Following [22, p. 138] one might call this the "support" of F.

A key formula, expressing the evaluation of a pp-pair on a module as a group of natural transformations between corresponding functors, 10.3.8, is proved. This formula will be extended and generalised later as will be the identification of the "annihilator" of a free realisation as a certain functor (10.3.9).

A ring is von Neumann regular iff every finitely presented functor is injective in the category of finitely presented functors (10.3.10).

The fact that the two categories, $(R\text{-mod}, \mathbf{Ab})^{\text{fp}}$ and $(\text{mod-}R, \mathbf{Ab})^{\text{fp}}$, are dual, though not the description of the duality, follows directly from 10.2.37, since, as remarked after that result, it is an immediate consequence of the definition of free abelian category that $\text{Ab}(R^{\text{op}}) \simeq (\text{Ab}(R))^{\text{op}}$. The explicit definition of the duality comes from [23, §7] and [256, 5.6] (for the case where R is a ring) and is as follows.

Let $F \in (R\text{-mod}, \mathbf{Ab})^{\text{fp}}$. Define the functor $dF \in (\text{mod-}R, \mathbf{Ab})$ by $dF \cdot A = (F, A \otimes -)$ for $A \in \text{mod-}R$; recall, 10.2.36, that $A \otimes -$ is a finitely presented functor if A is a finitely presented module. If $g : A \to B$ is in mod-R, then $dF \cdot g : (F, A \otimes -) \to (F, B \otimes -)$ is defined to be $(F, g \otimes -)$, so is given by $dF \cdot g \cdot \tau = (g \otimes -)\tau$. Thus dF, the **dual** of F is given by the representable functor $(F, -)$ restricted to the image of the embedding (12.1.3) $A \mapsto A \otimes -$, of mod-R into $(R\text{-mod}, \mathbf{Ab})^{\text{fp}}$. The fact that dF is a finitely presented functor is shown below (10.3.3).

The definition of d on morphisms is as follows: given $f : F \to G$ in $(R\text{-mod}, \mathbf{Ab})^{\text{fp}}$ define $df : dG \to dF$ to be $(f, -)$, that is, if $M \in \text{mod-}R$, then the component of df at M is given by taking $\tau \in (G, M \otimes -)$ to $\tau f \in (F, M \otimes -)$.

Dually, if $F \in (\text{mod-}R, \mathbf{Ab})^{\text{fp}}$, then we use the same notation for the functor $dF \in (R\text{-mod}, \mathbf{Ab})$ given on objects by $dF \cdot L = (F, - \otimes L)$.

Example 10.3.1. If $A \in \text{mod-}R$ and $L \in R\text{-mod}$, then $d(A, -) \simeq A \otimes -$ and $d(- \otimes L) \simeq (L, -)$. For the first, $d(A, -) \cdot K = \big((A, -), - \otimes K\big)$, which (10.1.7) is naturally isomorphic to $A \otimes K = (A \otimes -)K$ and, for the second, $d(- \otimes L) \cdot K = (- \otimes L, - \otimes K)$, which is naturally isomorphic to $(L, K) = (L, -)K$.

Lemma 10.3.2. *d is a contravariant exact functor from $(R\text{-mod}, \mathbf{Ab})^{\text{fp}}$ to $(\text{mod-}R, \mathbf{Ab})^{\text{fp}}$.*

Proof. Suppose that $0 \to H \to F \to G \to 0$ is an exact sequence in $(R\text{-mod}, \mathbf{Ab})^{\text{fp}}$. For any $A \in \text{mod-}R$ the sequence $0 \to (G, A \otimes -) \to (F, A \otimes -) \to (H, A \otimes -) \to \text{Ext}^1(G, A \otimes -)$ is exact. By 12.1.6, $A \otimes -$ is an absolutely pure functor, hence (see 2.3.1), since G is finitely presented, $\text{Ext}^1(G, A \otimes -) = 0$, so d is indeed exact. That dF is finitely presented is shown next. □

Proposition 10.3.3. *Let R be a small preadditive category. If $F \in (R\text{-mod}, \mathbf{Ab})^{\text{fp}}$, then $dF \in (\text{mod-}R, \mathbf{Ab})^{\text{fp}}$.*

Proof. Suppose that $(L, -) \xrightarrow{(f,-)} (K, -) \to F \to 0$ is a projective presentation of F with $K \xrightarrow{f} L \in R$-mod (10.2.1). Apply d to obtain the, exact by 10.3.2, sequence $0 \to dF \to d(K, -) \to d(L, -)$, that is, by 10.3.1, $0 \to dF \to (- \otimes K) \to (- \otimes L)$. Since K, L are finitely presented so are the functors $- \otimes K$ and $- \otimes L$ (10.2.36), so the image of $- \otimes K$, being finitely generated, is finitely presented (10.2.3). Therefore (E.1.16) the kernel, dF, of this map is finitely generated, hence (10.2.3 again) finitely presented, as required. □

Theorem 10.3.4. *Let R be a small preadditive category. Then the functor $d : ((R\text{-mod}, \mathbf{Ab})^{\mathrm{fp}})^{\mathrm{op}} \longrightarrow (\text{mod-}R, \mathbf{Ab})^{\mathrm{fp}}$ is an equivalence of categories. If we use d also to denote the corresponding functor from $((\text{mod-}R, \mathbf{Ab})^{\mathrm{fp}})^{\mathrm{op}}$ to $(R\text{-mod}, \mathbf{Ab})^{\mathrm{fp}}$, then d^2 is naturally equivalent to the identity functor on $(R\text{-mod}, \mathbf{Ab})^{\mathrm{fp}}$.*

Proof. Applying d to a presentation $(L, -) \to (K, -) \to F \to 0$ of $F \in (R\text{-mod}, \mathbf{Ab})^{\mathrm{fp}}$ gives the exact sequence $0 \to dF \to (- \otimes K) \to (- \otimes L)$ in $(\text{mod-}R, \mathbf{Ab})^{\mathrm{fp}}$, by the proof of 10.3.3. Applying d again gives, by 10.3.2 and 10.3.1, the exact sequence $(L, -) \to (K, -) \to d^2F \to 0$. From this and what has been shown already the assertions follow. □

The fact that this Auslander–Gruson–Jensen duality coincides with that ([274, 2.9]) of Herzog, once the relevant categories have been identified (10.2.30), was shown in [101, §3.4] and [528, §2].

Proposition 10.3.5. *If M is any right R-module and if F is any functor in $(\text{mod-}R, \mathbf{Ab})^{\mathrm{fp}}$, then there is a natural isomorphism $(dF, M \otimes -) \simeq FM$ (more accurately, $\overrightarrow{F}M$).*

Proof. Suppose that $(B, -) \to (A, -) \to F \to 0$ is a projective presentation of F in $(\text{mod-}R, \mathbf{Ab})^{\mathrm{fp}}$. By 10.3.2 and 10.3.1 this gives the exact sequence $0 \to dF \to (A \otimes -) \to (B \otimes -)$ in $(R\text{-mod}, \mathbf{Ab})^{\mathrm{fp}}$. In turn this gives the exact sequence $(B \otimes -, M \otimes -) \to (A \otimes -, M \otimes -) \to (dF, M \otimes -) \to 0$ (exactness at the last place because, 12.1.6, $M \otimes -$ is absolutely pure), that is, $(B, M) \to (A, M) \to (dF, M \otimes -) \to 0$. Since $(B, M) \to (A, M) \to FM \to 0$ also is exact, we deduce $(dF, M \otimes -) \simeq FM$. □

Recall from Section 1.3.1 the definition of the dual, $D\phi$, of a pp condition ϕ.

Corollary 10.3.6. *If ϕ is a pp condition in n free variables, then $F_{D\phi} = \ker\big(({}_RR^n, -) \to dF_\phi\big)$, where this epimorphism is the dual of the inclusion $F_\phi \to (R^n_R, -)$.*

Proof. Let $(C, \overline{c})$ be a free realisation of ϕ. By 10.2.8 there is the following commutative diagram in $(\text{mod-}R, \mathbf{Ab})^{\text{fp}}$,

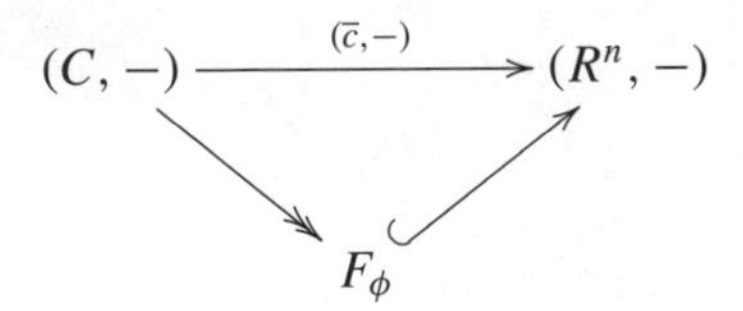

so, 10.3.2 and 10.3.1, there is the following commutative diagram in $(R\text{-mod}, \mathbf{Ab})^{\text{fp}}$

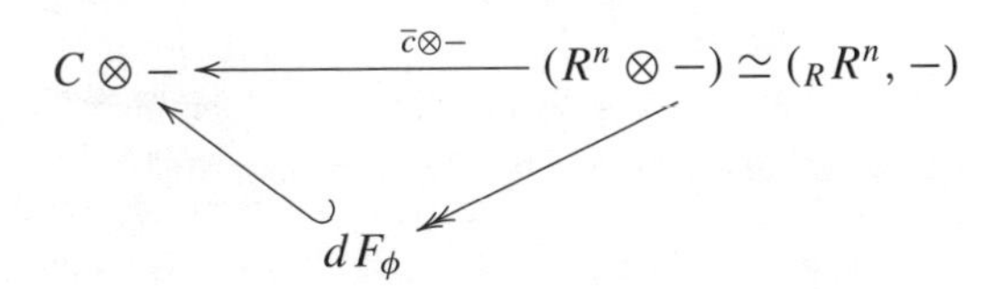

By 1.3.8, $\ker(\overline{c} \otimes -) = F_{D\phi}$, that is, the sequence

$$0 \to F_{D\phi} \to ({}_RR^n, -) \to dF_\phi \to 0$$

is exact, as required. $\square$

If F is a finitely generated subfunctor of $(R^n_R, -)$, then $F = F_\phi$ for some pp condition ϕ in n free variables (10.2.15) so we may set $DF = F_{D\phi}$, a finitely generated subfunctor of $({}_RR^n, -)$, and refer to this as the **elementary dual** of F. This is, by the above result, the kernel of the map from $({}_RR^n, -)$ to the functorial dual, dF, of F.

Corollary 10.3.7. (of 1.3.1) *For each n the duality D induces an anti-isomorphism* $\text{Latt}^{\text{f}}(R^n_R, -) \simeq \big(\text{Latt}^{\text{f}}({}_RR^n, -)\big)^{\text{op}}$ *between lattices of finitely generated subfunctors.*

The following very useful result appears in [497, p. 193], also [101, 3.4.2] and, in part, [144, 3.4, Lemma 2].

Corollary 10.3.8. *If $\phi \geq \psi$ is a pp-pair for right R-modules and M is a right R-module, then there is a natural isomorphism* $(F_{D\psi/D\phi}, M \otimes -) \simeq \phi(M)/\psi(M)$ *as left* $\text{End}(M)$*-modules.*

Proof. In view of 10.3.5 and 10.3.4 it will be enough to show that $d(F_{D\psi/D\phi}) = F_{\phi/\psi}$ $(= F_\phi/F_\psi)$. The exact sequence $0 \to F_{D\phi} \to (R^n \otimes -) \to dF_\phi \to 0$ of 10.3.6 dualises to $0 \to d^2F_\phi = F_\phi \to (R^n, -) \to dF_{D\phi} \to 0$ and similarly for ψ. The exact sequence $0 \to F_{D\phi} \to F_{D\psi} \to F_{D\psi/D\phi} \to 0$ dualises to $0 \to d(F_{D\psi/D\phi}) \to dF_{D\psi} \to dF_{D\phi} \to 0$.

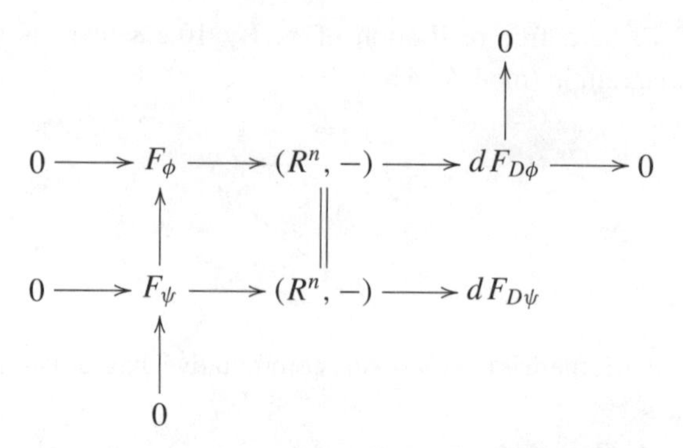

Therefore $d(F_{D\psi/D\phi}) \simeq F_\phi/F_\psi$, as required, the isomorphism commuting with the End(M)-structures by naturality of the isomorphism in 10.3.5. □

Generalisations of this result are at 12.2.4 and 12.3.15. The next result also has a generalisation at 12.2.5.

Corollary 10.3.9. *Let $\overline{a}$ be an n-tuple from the module M. Then the morphism $(\overline{a} \otimes -) : (R^n \otimes -) \to (M \otimes -)$ factors through $(R^n \otimes -)/F_{D\phi}$ iff $\overline{a} \in \phi(M)$. In particular, if $(M, \overline{a})$ is a free realisation of ϕ, then $(R^n \otimes -)/F_{D\phi}$ embeds in $(M \otimes -)$.*

Proof. Let $(C, \overline{c})$ be a free realisation of ϕ. Then $\overline{a} \in \phi(M)$ iff $\overline{a} : R^n \to M$ factors through $\overline{c} : R^n \to C$ (1.2.17) iff $(\overline{a} \otimes -) : (R^n \otimes -) \to (M \otimes -)$ factors through $(\overline{c} \otimes -) : (R^n \otimes -) \to (C \otimes -)$. This happens iff, see the proof of 10.3.6, $\overline{a} \otimes -$ factors through $(R^n \otimes -)/F_{D\phi}$. □

Corollary 10.3.10. (for example, [102, 2.7]) *A ring R is von Neumann regular iff every object of* $(R\text{-mod}, \mathbf{Ab})^{\mathrm{fp}}$ *is injective in that category.*

Proof. If R is von Neumann regular so, 10.2.20, gldim(mod-R, $\mathbf{Ab}$) $= 0$, then every finitely presented functor is, by 10.1.14, representable, hence projective. By the duality, d, between $(\text{mod-}R, \mathbf{Ab})^{\mathrm{fp}}$ and $(R\text{-mod}, \mathbf{Ab})^{\mathrm{fp}}$ (10.3.4) every functor in $(R\text{-mod}, \mathbf{Ab})^{\mathrm{fp}}$ is injective in $(R\text{-mod}, \mathbf{Ab})^{\mathrm{fp}}$. The argument reverses to give the equivalence. □

10.4 Finitistic global dimension

Following Bass ([46]) the right **finitistic dimension**[14] of a ring R is defined to be $\mathrm{Fd}(R) = \sup\{\mathrm{pd}(M) : M \in \text{Mod-}R,\ \mathrm{pd}(M) < \infty\}$: the supremum of the

[14] The left finitistic dimension may differ from this, see, for example, [321, 2.2].

values of projective dimensions of those right R-modules which do have projective dimension defined. The Finitistic Dimension Conjecture is: if R is an artin algebra, then $\mathrm{Fd}(R) < \infty$. Krause pointed out a connection with the Ziegler spectrum and this was generalised by Wilkins.

Proposition 10.4.1. ([356, Lemma 5]) *Suppose that R is right and left artinian. Then the class of modules of projective dimension $\leq n$ is definable.*

Proof. We use the flat dimension of modules: this is defined as is projective dimension (see Appendix E, p. 705) but using resolutions by flat, rather than by just projective, modules. Since R is left artinian so (see 2.3.6(c)) flat = projective for right modules, this coincides with projective dimension. For any ring and module M, the flat dimension of M is $\leq n$ iff $\mathrm{Tor}_{n+1}(M, -) = 0$ (for example, [708, 4.1.10]). Since $\mathrm{Tor}_n(M, -)$ commutes with direct limits (for example, [112, VI.1.3]) it is enough that $\mathrm{Tor}_{n+1}(M, -) \upharpoonright \text{mod-}R = 0$. By (the comment after) 10.2.36 and since R is right noetherian, it follows that the class of modules of projective dimension $\leq n$ is definable, being the intersection of the kernels of the finitely presented functors $\mathrm{Tor}_{n+1}(-, L)$ as L ranges over mod-R. □

Let fp-injdim(M) denote the fp-injective dimension of the module M: this is defined like injective dimension but using fp-injective = absolutely pure modules, see, for example, [323, pp. 240, 377]. For a right noetherian ring this is, by 4.4.17, just injective dimension. Flat dimension, see just above, also called weak dimension, will be denoted flatdim and the corresponding observation, that for right perfect rings this equals projective dimension, is by 3.4.27.

Proposition 10.4.2. ([711, (proof of) 3.3]) *Suppose that R is left coherent. For each $n \geq 1$ the subcategories $\mathcal{X}_n = \{L \in R\text{-Mod} : \text{fp-injdim}(L) \leq n\}$ and $\mathcal{Y}_n = \{M \in \text{Mod-}R : \text{flatdim}(M) \leq n\}$ are definable subcategories of R-Mod and Mod-R respectively and $\mathcal{X}_n$ and $\mathcal{Y}_n$ are dual in the sense of Section 3.4.2.*

Corollary 10.4.3. ([711, p. 175], [356, Prop. 6] for R right and left artinian) *Suppose that R is left coherent and right perfect and let M be any right R-module. Then* $\mathrm{pd}(M) = \sup\{\mathrm{pd}(N) : N \in \mathrm{supp}(M)\}$.

Corollary 10.4.4. ([711, 3.4], generalising [356, Thm 1]) *Suppose that R is left coherent. Then* $\mathrm{Fd}(R) = \sup\{\text{flatdim}(N) : N \in \mathrm{Zg}_R,\ \text{flatdim}(N) < \infty\} = \sup\{\text{fp-injdim}(L) : L \in {}_R\mathrm{Zg},\ \text{fp-injdim}(L) < \infty\}$. *In particular, if R is also right perfect, then* $\mathrm{Fd}(R) = \sup\{\mathrm{pd}(N) : N \in \mathrm{Zg}_R,\ \mathrm{pd}(N) < \infty\}$.

Since the projective dimension of a direct sum of modules is the supremum of the projective dimension of the summands, it follows that $\mathrm{Fd}(R) < \infty$ iff the class of modules with finite projective dimension is definable. That condition certainly implies that the class of indecomposable pure-injective modules with finite projective dimension forms a closed subset of Zg_R and Krause suggested the conjecture that the converse might be true.

Question 10.4.5. Let R be right and left artinian and suppose that the set of indecomposable pure-injective modules with finite projective dimension is a closed subset of Zg_R. Does it follow that $\mathrm{Fd}(R) < \infty$?

Krause also pointed out [356, Cor. 9] that if the set of finitely presented indecomposable pure-injectives with finite projective dimension is dense in the set of all indecomposable pure-injectives with finite projective dimension, then $\mathrm{Fd}(R) = \mathrm{fd}(R)$, where the latter is defined as $\mathrm{Fd}(R)$ but taking the supremum of projective dimension over only the finitely presented modules M which have projective dimension $< \infty$.

For more on the general topic see the survey [309].

We also point to [373] for work, related to that in Section 4.6, in another direction. There Krause and Solberg show that if R is right artinian, then for each $n \geq 0$ there is a pure-injective module P_n such that an R-module M has $\mathrm{pd}(M) \leq n$ iff M is a direct summand of a module M' which has a filtration $M' = M'_t > M'_{t-1} > \cdots > M'_0 = 0$ for some integer t with each factor M_i/M_{i-1} being a direct product of copies of P_n. Also see [658].

11

Serre subcategories and localisation

The chapter begins with a summary of the basic definitions and results on localisation in general abelian categories then gradually moves to finite type localisation in locally coherent categories. The latter is essential for the functorial approach to results which involve relativising to closed subsets of the Ziegler spectrum, that is, to relativising to definable subcategories of the module category. Both localisation of large (Grothendieck) categories at hereditary torsion theories and localisation of possibly small (abelian) categories at Serre subcategories are discussed.

11.1 Localisation in Grothendieck categories

Section 11.1.1 gives the definitions and basic results around localisation at hereditary torsion theories. Finite-type localisation, especially in locally finitely generated categories, is the subject of Section 11.1.2 and elementary localisation, especially in locally finitely presented categories, is the subject of Section 11.1.3. Still stronger results hold for finite-type localisation in locally coherent categories (Section 11.1.4).

In Section 11.1.5 the effect of localisation on pp conditions is described.

11.1.1 Localisation

The localisation process is described and the main summative results stated: 11.1.1 on equivalent data defining a localisation; 11.1.3 characterising localisation functors; 11.1.5 on the effect of localisation. The ordering on localisations is formulated in equivalent ways in 11.1.9, while 11.1.10 expresses a composition of localisations as a single localisation. In locally finitely generated categories a localisation is determined by a Gabriel topology on a generating set (11.1.11).

Torsion-theoretic localisation of abelian categories is one of the main techniques used in this book. In our context this is the functor-theoretic equivalent of working within a definable subcategory or over a closed subset of the Ziegler spectrum. If $\mathcal{X}$ is a definable subcategory of Mod-R, then there is, in general, no functor from Mod-R to $\mathcal{X}$ which gives a "best $\mathcal{X}$-approximation" of each module, but there is such a functor from each of (R-mod, **Ab**) and (mod-R, **Ab**) to a corresponding category of functors on $\mathcal{X}$ (see 12.3.20). This does, via the embedding (12.1.3) of Mod-R into (R-mod, **Ab**), give a "best $\mathcal{X}$-approximation" of any module, but the result is a functor, not a module.

Torsion-theoretic localisation is a generalisation of Ore localisation, which is itself a generalisation of the classical construction of commutative rings of fractions. It is an additive version of the topos-theoretic notion of localisation which was introduced by Grothendieck in algebraic geometry. The main lines were laid down in the thesis of Gabriel [202], and there were many subsequent developments, see, for example, [486, Intro. to Chpt. 4]. We present the basic definitions and results and then focus on torsion theories of finite type. At the beginning, we make somewhat imprecise comments which, we hope, will convey the ideas in the absence of many of the details. For more detail, see [663], [486]. In their emphases these texts are somewhat complementary. The former reference is perhaps more useful since at most places we use only hereditary localisation in Grothendieck categories (that book also has useful introductions to many of the topics discussed in this book). But we also discuss the related process of factoring by a Serre subcategory, which applies to the category of finitely presented functors for instance. That can be found in [486]. In this book "torsion theory" will, by default, mean hereditary torsion theory unless it is clearly stated otherwise. Another account of a good part of the material here can be found at [212].

The rough idea is as follows. Start with an abelian category $\mathcal{C}$ (a module category, a category of sheaves of modules, a functor category, the category of finitely presented functors, . . .). For some purpose we might wish to force certain objects of $\mathcal{C}$ to become 0: these, and all objects which must consequently become zero, will be called the torsion objects. The category obtained by forcing these to be zero will be called a quotient category, or localisation, of $\mathcal{C}$ and the "projection map" from $\mathcal{C}$ to its localisation will be called the localisation functor. Under some assumptions this quotient category can be identified with a subcategory of $\mathcal{C}$. This process might seem similar to that of forcing elements of an additive algebraic structure to be zero but the direct category-theoretic analogue of that is factoring by an ideal of morphisms of $\mathcal{C}$ (see Section 9.1.1) and that process does not usually result in an abelian category (see Section 17.1).

In contrast, the process of localisation, as well as making some morphisms zero, adds new morphisms, in particular, inverses for certain morphisms. Indeed, if the torsion objects are to become zero, then certainly all morphisms to and from them should become zero but, also, all extensions by them should become isomorphisms. In the context of hereditary torsion-theoretic localisation in a Grothendieck category it turns out that this is essentially all that is needed, in the sense that, given any object, if we factor out its torsion subobject and then make the result "divisible with respect to torsion" (injective over morphisms with torsion cokernel), then this determines the localisation functor from $\mathcal{C}$ to the quotient category. In particular, the full subcategory on the resulting objects – those objects D of $\mathcal{C}$ such that $(C, D) = 0$ for all torsion objects C (say then that D is torsionfree) and also such that $\mathrm{Ext}^1(C, D) = 0$ for all torsion objects C (say then that D is relatively injective) – is the quotient category, or is equivalent to it, depending on how one chooses to define the localisation functor.

Now we give the precise definitions. We concentrate on torsion-theoretic localisation; localising at Serre subcategories is discussed later in this section and in Section 11.1.4.

Throughout this chapter, unless specified otherwise, $\mathcal{C}$ denotes an abelian category which is Grothendieck (see Appendix E, p. 707).

A **torsion subcategory** (or subclass) of $\mathcal{C}$ is a full subcategory, $\mathcal{T}$, which is closed in $\mathcal{C}$ under epimorphic images, extensions and arbitrary direct sums. Here we say that a full subcategory, or subclass, $\mathcal{D}$ of $\mathcal{C}$ is **closed under extensions** if for every short exact sequence $0 \to A \to B \to C \to 0$ of $\mathcal{C}$ if $A, C \in \mathcal{D}$, then $B \in \mathcal{D}$. A torsion class is a **hereditary torsion class**, also called a **localising subcategory** if it is also closed in $\mathcal{C}$ under subobjects.

Having defined a notion of torsion, an object will be torsionfree exactly if it contains no non-zero torsion subobject: the conditions characterising such classes are the following. A subclass $\mathcal{F}$ of $\mathcal{C}$ is a **torsionfree class** if it is closed in $\mathcal{C}$ under subobjects, extensions and arbitrary products. Such a class is a **hereditary torsionfree class** if it is also closed under injective hulls. In order to check that a class is a hereditary torsionfree class it is necessary only to check closure under subobjects, products and injective hulls: closure under extensions then follows. Indeed, given an exact sequence $0 \to A \to B \to C \to 0$ with $A, C \in \mathcal{F}$, one may form the pushout sequence $0 \to E(A) \to X \to C \to 0$ where $E(A)$ is the injective hull (Appendix E, p. 708) of A; since $E(A)$ is injective $X \simeq E(A) \oplus C \in \mathcal{F}$, so $B(\leq X) \in \mathcal{F}$. It follows that a torsionfree class for a hereditary torsion theory is determined by the class of torsionfree injective objects.

Given any subclass $\mathcal{A}$ of $\mathcal{C}$ define $\mathcal{A}^{\perp} = \{C \in \mathcal{C} : (\mathcal{A}, C) = 0\}$, where $(\mathcal{A}, C) = 0$ means $(A, C) = 0$ for all $A \in \mathcal{A}$ and, dually, set ${}^{\perp}\mathcal{A} = \{C \in \mathcal{C} : (C, \mathcal{A}) = 0\}$. If $\mathcal{T}$ is a torsion class, then $\mathcal{T}^{\perp}$ is a torsionfree class and $\mathcal{T} = {}^{\perp}(\mathcal{T}^{\perp})$; dually for torsionfree classes. Also, the torsion class is hereditary iff the corresponding torsionfree class is hereditary. A pair $(\mathcal{T}, \mathcal{F})$ of corresponding classes is a **torsion**

theory; a **hereditary torsion theory** if the classes are hereditary. The objects in $\mathcal{T}$ are **torsion**, those in $\mathcal{F}$ are **torsionfree** (for that particular torsion theory). See [663, Chpt. VI] for details and proofs.

Associated to any hereditary torsion theory is the functor, τ, from $\mathcal{C}$ to itself which assigns to every object its largest torsion subobject: $\tau C = \sum\{C' : C' \leq C, \ C' \in \mathcal{T}\} = \max\{C' : C' \leq C, \ C' \in \mathcal{T}\}$. The action of τ on morphisms is simply restriction/corestriction. One can check that τ is a left exact subfunctor of the identity functor on $\mathcal{C}$ (and every such functor arises in this way [663, VI.3.1]). The functor τ is the associated **torsion radical** or **torsion functor**. Since each of $\tau, \mathcal{T}, \mathcal{F}$ determines the others, if the one is given we may use it as a subscript for the other associated objects (for example, $\mathcal{T}_\tau$, $\tau_{\mathcal{T}}$). Let us summarise all this, and more, for ease of reference.

Theorem 11.1.1. (for example, [663, Chpt. VI]) *Let $\mathcal{C}$ be a Grothendieck additive category. A hereditary torsion theory on $\mathcal{C}$ is determined by the following equivalent data:*

(i) a (hereditary torsion) subclass $\mathcal{T}$ closed under quotients, extensions, direct sums and subobjects;

(ii) a (hereditary torsionfree) subclass $\mathcal{F}$ closed under subobjects, products and injective hulls;

(iii) a pair (a hereditary torsion theory) $(\mathcal{T}, \mathcal{F})$ with $(T, F) = 0$ for every $T \in \mathcal{T}$ and $F \in \mathcal{F}$, which is maximal with respect to this condition, and which is such that $\mathcal{T}$ is closed under subobjects, equivalently $\mathcal{F}$ is closed under injective hulls;

(iv) a class $\mathcal{E}$ of injective objects closed under products and direct summands (then $\mathcal{F} = \{C : C \leq E$ for some $E \in \mathcal{E}\}$, see 11.1.7);

(v) a left exact subfunctor (torsion functor) τ of the identity functor on $\mathcal{C}$.

The connections between these are as described above or are simple consequences such as $\mathcal{F} = \{C : \tau C = 0\}$, $\mathcal{T} = \{C : \tau C = C\}$. A further equivalent datum when $\mathcal{C}$ is locally finitely generated is a Gabriel filter/topology (11.1.11).

Suppose that $\mathcal{T}, \mathcal{F}, \tau$ define the same hereditary torsion theory.

If $C \in \mathcal{C}$, then $C' \leq C$ is **τ-dense in** C if the factor C/C' belongs to $\mathcal{T}$ (so, in the quotient category, C' and C will be identified). Let $\mathcal{U}_\tau(C)$ denote the set of τ-dense subobjects of C: it may be checked that this is a filter (which satisfies certain further conditions, cf. 11.1.11 below). An object $D \in \mathcal{C}$ is **τ-injective** if it is injective over all τ-dense embeddings: if C' is τ-dense in C, then any morphism from C' to D extends to one from C to D which, note, will be unique if D is also τ-torsionfree. One may see why objects of the quotient category must be

τ-injective as well as τ-torsionfree: for they do not see any difference between C' and C if C' is τ-dense in C.

Associated with any hereditary torsion theory is a **localisation functor** which may be regarded as a functor from $\mathcal{C}$ to a full subcategory of $\mathcal{C}$. An alternative construction of this functor is described later. The localisation functor is defined on objects by taking $C \in \mathcal{C}$ to $C_\tau = \pi^{-1}\big(\tau\big(E(C_1)/C_1\big)\big)$, where $C_1 = C/\tau C$ and where $\pi : E(C_1) \to E(C_1)/C_1$ is the natural projection. That is, given an object C of $\mathcal{C}$, first factor out the torsion part, τC, of C to get its largest torsionfree quotient $C_1 = C/\tau C$. Then form the largest extension, with no trivial subextensions, of C_1 by torsion objects. This object exists and can be found lying between C_1 and its injective hull: it is called the τ**-injective hull** of C_1 and is denoted $E_\tau(C_1)$. It is the full inverse image in $E(C_1)$ of the torsion subobject of $E(C_1)/C_1$. On morphisms, the action of the functor is natural: a morphism from C to D induces one from $C/\tau C$ to $D/\tau D$ and this extends uniquely to one between the respective τ-injective hulls of $C/\tau C$ and $D/\tau D$. The following points follow easily from the construction.

Lemma 11.1.2. *Suppose that τ is a hereditary torsion theory on a Grothendieck additive category $\mathcal{C}$.*

(a) If E is a τ-torsionfree injective object, then $E \simeq E_\tau$.
(b) If $C' \leq C$, then C' is τ-dense in C iff $(C')_\tau = C_\tau$.
(c) For every object $C \in \mathcal{C}$, $C/\tau C$ is τ-dense in C_τ.

If $\mathcal{C}$ is an arbitrary abelian category, then $\mathcal{S}$ is a **Serre subcategory**, if whenever $0 \to A \to B \to C \to 0$ is an exact sequence in $\mathcal{C}$, then $B \in \mathcal{S}$ iff $A, C \in \mathcal{S}$. There is then a "localisation functor" from $\mathcal{C}$ to a category of fractions, see 11.1.40 (and, for example, [486, §4.3] for details). If more conditions are imposed on $\mathcal{C}$ and $\mathcal{T}$ (see [486, §4.4. ff.], [226, §2]), then we reach the particularly pleasant situation that we described first, where $\mathcal{C}$ is Grothendieck and $\mathcal{T}$ is a hereditary torsion class (in particular is closed under arbitrary direct sums). In that case the category of fractions/quotient category $\mathcal{C}_\tau$ may be realised as a full subcategory of $\mathcal{C}$: let $\mathcal{C}_\tau$ denote the full subcategory of $\mathcal{C}$ whose objects are those of the form C_τ and denote by Q_τ the localisation functor described above, $C \mapsto C_\tau$, from $\mathcal{C}$ to $\mathcal{C}_\tau$. The latter is referred to as the **localisation** or **quotient category** of $\mathcal{C}$ at τ.

The functor Q_τ is exact (see the references for 11.1.5: beware, however, that [486] uses the term "localisation functor" for the composition of our localisation functor with the inclusion of $\mathcal{C}_\tau$ in $\mathcal{C}$ and that, in general, is not exact). Thus the objects of $\mathcal{C}_\tau$ are exactly the τ-torsionfree, τ-injective objects of $\mathcal{C}$. We use $Q_\tau C$ and $Q_\tau\mathcal{C}$ interchangably with C_τ and $\mathcal{C}_\tau$ respectively (notation such as $\mathcal{C}/\mathcal{T}$ is also used for quotient categories). The functor Q_τ acts on objects and morphisms of $\mathcal{C}_\tau$ as the identity functor (at least, after the, non-exact, identification of the localisation with a subcategory of $\mathcal{C}$).

The inclusion, i, of $\mathcal{C}_\tau = Q_\tau \mathcal{C}$ in $\mathcal{C}$ is right adjoint to Q_τ: $(C, iQ_\tau D) \simeq (Q_\tau C, Q_\tau D)$ for $C, D \in \mathcal{C}$. So i is left exact but, although $\mathcal{C}_\tau$ is a full subcategory of $\mathcal{C}$, the inclusion functor is not in general right exact. If $0 \to D' \to D \to D'' \to 0$ is an exact sequence in $\mathcal{C}_\tau$, then, regarded as a sequence in $\mathcal{C}$, the quotient D/D', though certainly τ-torsionfree, need not be τ-injective; the corresponding exact sequence in $\mathcal{C}$ would replace D'' by the, perhaps proper, submodule D/D' (of which D'' would be the τ-injective hull). The functor Q_τ is analogous with, indeed can be seen as, sheafification of presheaves with respect to a certain Grothendieck-type topology (cf. 11.1.11) and then this corresponds to the fact that the inclusion of the category of sheaves in the category of presheaves is not right exact. Also the inclusion functor does not, for instance, commute with infinite direct sums. In general, the direct sum of objects in $\mathcal{C}_\tau$, regarded as embedded in $\mathcal{C}$, is obtained by taking their direct sum in $\mathcal{C}$ and then taking the τ-injective hull of that.

An alternative construction, see [663, §IX.1], of the localisation $\mathcal{C}_\tau$ is to take the same objects as $\mathcal{C}$ but to define morphisms by $\mathcal{C}_\tau(C, D) = \varinjlim_{C'} \varinjlim_{D_0} (C', D/D_0)$, where C' ranges over τ-dense subobjects of C (directed by intersection) and D_0 ranges over τ-torsion subobjects of D (directed by sum). In this category, an enlargement of $\mathcal{C}$ by morphisms, every object C is isomorphic to its τ-injective hull C_τ. One may discern the construction given earlier mirrored in this direct limit of morphisms.[1]

Theorem 11.1.3. (see [486, 4.4.9]) *Suppose that $\mathcal{C}$ and $\mathcal{C}'$ are abelian categories, that $\mathcal{C}$ is Grothendieck and that $Q : \mathcal{C} \longrightarrow \mathcal{C}'$ is an exact functor. Suppose also that Q has a full and faithful right adjoint. Then* $\ker(Q) = \{T \in \mathcal{C} : QC = 0\}$ *is a hereditary torsion subcategory of $\mathcal{C}$ and the right adjoint of Q induces an equivalence between $\mathcal{C}'$ and the localisation of $\mathcal{C}$ at $\mathcal{T}$.*

Example 11.1.4. Take $\mathcal{C} = \mathbf{Ab}$ and take "torsion" to mean 2-torsion by declaring an abelian group M to be torsion if every element of M is annihilated by a power of 2.

Then the objects of the quotient category are the abelian groups D such that D has no 2-torsion and such that D has no non-trivial extension by a 2-torsion module, meaning that if $D \leq M$ and if M/D is 2-torsion then the inclusion of D in M is split.

[1] This second construction is the additive analogue of the ++ construction of sheaves with respect to a Grothendieck topology of a site whereas the first is analogous to the Lawvere–Tierney sheafification process, see [419, §III.5, §V.3].

To get from an arbitrary abelian group, C, to the corresponding object of the quotient category, first factor out the 2-torsion subgroup, τC, the subgroup consisting of all elements of C which are annihilated by some power of 2; set $C_1 = C/\tau C$ – a 2-torsion-free group. Let $E(C_1)$ denote the injective hull of C_1 (this will be a direct sum of ($p \neq 2$)-Prüfer groups and copies of $\mathbb{Q}$) and consider the factor group $E(C_1)/C_1$; consider the 2-torsion subgroup $\tau(E(C_1)/C_1)$ (clearly there will be no contribution from the Prüfer components) and let C_2 be its full inverse image in $E(C_1)$ – so $C_1 \leq C_2 \leq E(C_1)$ and $C_2/C_1 = \tau(E(C_1)/C_1)$ (thus C_2 is obtained by "making C_1 fully divisible by powers of 2"). The localisation functor takes C to C_2. For instance, if $C = \mathbb{Z}$ then $C_2 = \mathbb{Z}[\frac{1}{2}]$. Indeed this localisation functor is easily seen to be equivalent to tensoring with $\mathbb{Z}[\frac{1}{2}]$.

We remark that, as well as localisation in the non-Grothendieck situation (for which see [486, Chpt. 4]), there is a related process of forming orthogonal categories – orthogonal with respect to Hom and various of its derived functors Ext^i – see [226] (also [3] in the non-additive context). In the context of triangulated categories there are also analogues, touched on in Section 17.4.

Theorem 11.1.5. (see [663, §§IX.1, X.1], [486, 4.3.8, 4.3.11, §4.4, 4.6.2]) *Let τ be a hereditary torsion theory on the Grothendieck category $\mathcal{C}$. Then the localised category $\mathcal{C}_\tau$ also is Grothendieck, the localisation functor $Q_\tau : \mathcal{C} \longrightarrow \mathcal{C}_\tau$ is exact and, if $F : \mathcal{C} \longrightarrow \mathcal{C}'$ is any exact functor to a Grothendieck category $\mathcal{C}'$ such that F commutes with direct limits and $\ker(F) \supseteq \mathcal{T}_\tau$, then F factors uniquely through Q_τ.*

The localisation functor $Q_\tau : \mathcal{C} \longrightarrow \mathcal{C}_\tau$ has a right adjoint, namely the inclusion, i, of $\mathcal{C}_\tau$ in $\mathcal{C}$: $\mathcal{C}(C, iD) \simeq \mathcal{C}_\tau(C_\tau, D)$ for every $C \in \mathcal{C}$ and $D \in \mathcal{C}_\tau$. The image of i is, up to natural equivalence, the full subcategory of τ-torsionfree, τ-injective objects of $\mathcal{C}$.

For any τ-torsionfree, τ-injective object C, one has $Q_\tau C \simeq C$ and the injective objects of $(i)\mathcal{C}_\tau$ are exactly the τ-torsionfree injective objects of $\mathcal{C}$.

If $\mathcal{G}$ is a generating set for $\mathcal{C}$, then $Q_\tau \mathcal{G}$ is a generating set for $\mathcal{C}_\tau$.

The last statement follows since the localisation functor is (right) exact and, being a left adjoint, commutes with arbitrary direct sums, indeed with arbitrary direct limits.

On occasion we use the following.

Lemma 11.1.6. *If τ is a hereditary torsion theory on $\mathcal{C}$, if $B \in \mathcal{C}$ and if $0 \to A' \to B_\tau \to C' \to 0$ is an exact sequence in $\mathcal{C}_\tau$, then there is an exact sequence $0 \to A \to B \to C \to 0$ in $\mathcal{C}$ such that $0 \to A_\tau \to B_\tau \to C_\tau \to 0$ is isomorphic to the first sequence in $\mathcal{C}_\tau$.*

To find the sequence $0 \to A \to B \to C \to 0$, apply the adjoint i of Q_τ, note the canonical map $B \to iB_\tau$, obtain A as the pullback and C as the cokernel of $A \to B$:

$$\begin{array}{ccccccccc} 0 & \dashrightarrow & A & \dashrightarrow & B & \cdots\!\!> & C & \cdots\!\!> & 0 \\ & & \downarrow & & \downarrow & & & & \\ 0 & \longrightarrow & iA' & \longrightarrow & iB_\tau & \longrightarrow & iC' & \longrightarrow & 0 \end{array}$$

Given a class $\mathcal{E}$ of injective objects in the Grothendieck category $\mathcal{C}$ denote by $\mathrm{cog}(\mathcal{E})$ the torsion theory **cogenerated** by $\mathcal{E}$: this is the torsion theory with torsion class $\{C : (C, \mathcal{E}) = 0\}$. Since $\mathcal{E}$ is a class of injective objects this torsion class is closed under subobjects, that is, the torsion theory cogenerated by $\mathcal{E}$ is hereditary. The torsionfree class consists of those objects which are **cogenerated** by $\mathcal{E}$, that is, which embed in some product of copies of members of $\mathcal{E}$. These assertions and the next result, restating part of 11.1.1, follow directly from the definitions.

Proposition 11.1.7. *There is a natural bijection between hereditary torsion theories, τ, on the Grothendieck category $\mathcal{C}$ and the classes, $\mathcal{E}$, of injective objects of $\mathcal{C}$, which are closed under products and direct summands. The bijection is given by $\mathcal{E} \mapsto \mathrm{cog}(\mathcal{E})$ and $\tau \mapsto$ the class of τ-torsionfree injectives.*

In particular, every hereditary torsion theory on $\mathcal{C}$ is determined by the class of torsionfree injective objects.

Torsion classes are closed under direct sums and quotients, hence under direct limits.[2] We record this for future reference. The second statement below is immediate from the fact that a hereditary torsion class is closed under subobjects, plus the fact that in a locally finitely generated category every object is the direct limit of the inclusions of its finitely generated subobjects (Appendix E, p. 712).

Lemma 11.1.8. *Every torsion class $\mathcal{T}$ is closed under direct limits. If the category $\mathcal{C}$ is locally finitely generated and $\mathcal{T}$ is hereditary, then $\mathcal{T}$ is the closure under direct limits of the finitely generated objects in $\mathcal{T}$.*

We order torsion theories by inclusion of their torsion classes: set $\tau \le \sigma$ if $\mathcal{T}_\tau \subseteq \mathcal{T}_\sigma$. The following equivalents are direct from the definitions (part (iv) by 11.1.5).

Lemma 11.1.9. *Let τ and σ be hereditary torsion theories on the Grothendieck category $\mathcal{C}$. The following are equivalent:*

[2] Or, just directly, if $\mathcal{T} = {}^{\perp}\mathcal{F}$ and if $T = \varinjlim_\lambda T_\lambda$ with $T_\lambda \in \mathcal{T}$, then, for every $F \in \mathcal{F}$, $(T, F) = (\varinjlim T_\lambda, F) = \varinjlim (T_\lambda, F) = 0$, so $T \in \mathcal{T}$.

(i) $\tau \leq \sigma$ *(that is,* $\mathcal{T}_\tau \subseteq \mathcal{T}_\sigma$*);*
(ii) $\mathcal{F}_\tau \supseteq \mathcal{F}_\sigma$*;*
(iii) $\tau C \leq \sigma C$ *for every* $C \in \mathcal{C}$*;*
(iv) the localisation $Q_\sigma : \mathcal{C} \longrightarrow \mathcal{C}_\sigma$ *factors through* $Q_\tau : \mathcal{C} \longrightarrow \mathcal{C}_\tau$*.*

In (iv) the functor from $\mathcal{C}_\tau$ *to* $\mathcal{C}_\sigma$ *is exact (for example, since, see 11.1.10, it is a localisation of* $\mathcal{C}_\tau$*).*

If τ is a hereditary torsion theory on $\mathcal{C}$, then (11.1.5) $\mathcal{C}_\tau$ is again Grothendieck so we may consider hereditary torsion theories on this category. In fact, every hereditary torsion class on $\mathcal{C}_\tau$ is the image, under the localisation functor Q_τ, of one on $\mathcal{C}$. Using that the localisation functor is exact and a left adjoint one may easily show the next result.

Proposition 11.1.10. *Let* τ *be a hereditary torsion theory on the Grothendieck category* $\mathcal{C}$ *and let* $Q_\tau : \mathcal{C} \longrightarrow \mathcal{C}_\tau$ *be the corresponding localisation functor. Let* σ *be a hereditary torsion theory on* $\mathcal{C}_\tau$ *and consider* $Q_\tau^{-1}(\mathcal{T}_\sigma) = \{C \in \mathcal{C} : Q_\tau C \in \mathcal{T}_\sigma\}$*. Then* $Q_\tau^{-1}(\mathcal{T}_\sigma)$ *is a hereditary torsion class on* $\mathcal{C}$ *and, if we denote the associated torsion theory by* $\tau^{-1}\sigma$*, then* $Q_{\tau^{-1}\sigma} = Q_\sigma Q_\tau : \mathcal{C} \longrightarrow \mathcal{C}_{\tau^{-1}\sigma} = (\mathcal{C}_\tau)_\sigma$*.*

If τ_i $(i \in I)$ *is an increasing chain of hereditary torsion theories (so* $\mathcal{T}_i \subseteq \mathcal{T}_j$ *for* $i \leq j$ *in* I*), then the union* τ*, the torsion theory with torsion class generated by* $\bigcup_i \mathcal{T}_i$*, is hereditary and* $\mathcal{C}_\tau = \varinjlim_{i\in I} \mathcal{C}_{\tau_i}$*.*

We may take the above limit of categories to be the direct limit in the (2–) category of Grothendieck categories and exact functors between them; since all the categories appearing can be taken, see 11.1.5, to be subcategories of $\mathcal{C}$ we do not run into set-theoretic difficulties.

If $\mathcal{C}$ is locally finitely generated (Appendix E, p. 712), then any hereditary torsion theory on $\mathcal{C}$ may be defined in terms of data concerning just the generators. For, by 11.1.8, it will be enough to specify the finitely generated torsion objects, hence to specify which quotients of the generators are torsion. Recall that $\mathcal{U}_\tau(C) = \{C' \leq C : C/C' \in \mathcal{T}_\tau\}$ is the filter of τ-dense subobjects of an object C. The next result is from [202, §V.2] (for modules over a ring) and can be found at, for example, [663, §VI.5]; the general case can be found in [486, §4.9], [212, §2], [415, §2]. It shows clearly that this is an abelian analogue of the notion of Grothendieck topology (for which see [419], for instance).

Proposition 11.1.11. *Suppose that* $\mathcal{C}$ *is a locally finitely generated Grothendieck category with* $\mathcal{G}$ *a generating set of finitely generated objects. Let* τ *be a hereditary torsion theory on* $\mathcal{C}$*. Then* τ *is determined by the filters* $\mathcal{U}_\tau(G)$ $G \in \mathcal{G}$*. These* *filters satisfy the following conditions (additive versions of those for Grothendieck topologies):*

(a) for every $f : G \to H$ *with* $G, H \in \mathcal{G}$ *and every* $H' \in \mathcal{U}_\tau(H)$ *we have* $f^{-1}H' \in \mathcal{U}_\tau(G)$*;*
(b) if $H \in \mathcal{G}$ *and* $H' \leq H$ *is such that there is* $H'' \in \mathcal{U}_\tau(H)$ *such that for all* $G \in \mathcal{G}$ *and* $f : G \to H''$ *we have* $f^{-1}H' \in \mathcal{U}_\tau(G)$*, then* $H' \in \mathcal{U}_\tau(H)$*.*

Conversely, every such collection of filters determines a hereditary torsion theory[3] *on* $\mathcal{C}$*.*

In the case $\mathcal{C} = \text{Mod-}R$ and $\mathcal{G} = \{R\}$ the conditions on the filter $\mathcal{U}_\tau(R)$, which is then called a **Gabriel filter** or, since it can be seen as a basis of open neighbourhoods of 0, a **Gabriel topology**, become:

(1) for every $r \in R$ and $I \in \mathcal{U}_\tau(R)$ the right ideal $(I : r) = \{s \in R : rs \in I\}$ is in $\mathcal{U}_\tau(R)$ (the morphism f appearing in 11.1.11 is multiplication by r);
(2) if I is a right ideal such that there exists $J \in \mathcal{U}_\tau(R)$ such that for every $r \in J$ we have $(I : r) \in \mathcal{U}_\tau(R)$, then $I \in \mathcal{U}_\tau(R)$.

Condition (2) is what makes the corresponding torsion class closed under extension.

11.1.2 Finite-type localisation in locally finitely generated categories

Finite-type localisation is characterised in various ways, in general (11.1.12) and in locally finitely generated categories (11.1.14). The composition of finite-type localisations is of finite type, as is a direct limit of finite-type localisations (11.1.18). The localisation of a locally finitely generated category at a finite-type torsion theory is again locally finitely generated, by the localisations of the finitely generated objects (11.1.15).

A localisation is of finite type iff the torsionfree objects form a definable subcategory (11.1.20).

A detailed account of some of this can also be found in [212, §§5,7].

We will be particularly interested in those hereditary torsion theories which are determined by the class of finitely *presented* torsion objects, cf. 11.1.8.

A hereditary torsion theory is **of finite type** if it satisfies the equivalent conditions of the next result.[4] The results in this section are based on [278, §2.1], [349, §2] and [498, §A3] but most go back further (though not always in this generality) see, for example, [486, 4.15.2].

[3] For a similar result in the non-additive setting see [88, 1.5].
[4] Krause [349], [358] defines this just in the context of locally coherent categories, using a stronger condition which, in such categories, is equivalent, as will be seen in Section 11.1.4.

Proposition 11.1.12. *Let $\mathcal{C}$ be a Grothendieck additive category. The following conditions on a torsion functor $\tau : \mathcal{C} \longrightarrow \mathcal{C}$ are equivalent:*

(i) τ commutes with direct limits;
(ii) the inclusion functor $i : \mathcal{C}_\tau \longrightarrow \mathcal{C}$ commutes with direct limits of monomorphisms;
(iii) the torsionfree class $\mathcal{F}$ is closed under direct limits.

Proof. (i)⇒(iii) If $F = \varinjlim_\lambda F_\lambda$ with $F_\lambda \in \mathcal{F}$ then $\tau F = \tau \varinjlim F_\lambda = \varinjlim \tau F_\lambda = 0$, so $F \in \mathcal{F}$.

(iii)⇒(i) Suppose that $C = \varinjlim C_\lambda$. The directed system of objects C_λ induces, note, a directed system of exact sequences $0 \to \tau C_\lambda \to C_\lambda \to C_\lambda/\tau C_\lambda \to 0$. Since $\mathcal{C}$ is Grothendieck, $\varinjlim$ is exact, so these induce an exact sequence $0 \to \varinjlim \tau C_\lambda \to C \to \varinjlim(C_\lambda/\tau C_\lambda) \to 0$. By hypothesis $\varinjlim(C_\lambda/\tau C_\lambda) \in \mathcal{F}$, so $\varinjlim \tau C_\lambda \geq \tau C$. The other inclusion is by 11.1.8.

(iii)⇒(ii) Suppose $N = \sum_{\to} N_\lambda$ in $\mathcal{C}_\tau$ and let $N' = \sum_{\to} iN_\lambda$ in $\mathcal{C}$. Since (11.1.5) i is left exact $N' \leq iN$. The limit of the directed system of exact sequences $0 \to iN_\lambda \to iN \to iN/iN_\lambda \to 0$ is $0 \to N' \to iN = N \to \varinjlim(iN/iN_\lambda) \to 0$. Now, each iN/iN_λ is in $\mathcal{F}$ (since i is left exact, iN/iN_λ is contained in $i(N/N_\lambda) \in \mathcal{F}$), so, by hypothesis, $\varinjlim(iN/iN_\lambda)$ is in $\mathcal{F}$, therefore is 0 since it is also torsion, being isomorphic to iN/N' (that is torsion since, note, it is in the kernel of Q_τ).

(ii)⇒(iii) Suppose that $\big((F_\lambda)_\lambda, (f_{\lambda\mu} : F_\lambda \to F_\mu)_{\lambda<\mu}\big)$ is a directed system of objects of $\mathcal{F}$ with direct limit F. If $F \notin \mathcal{F}$, then, replacing each F_λ by $f_{\lambda\infty}^{-1}(\tau F)$ (where $f_{\lambda\infty} : F_\lambda \to F$ is the map to the direct limit), we may suppose that $F = \varinjlim_\lambda F_\lambda$ is torsion.

Choose λ such that $f_{\lambda\infty} \neq 0$. For each $\mu \geq \lambda$ let $G_\mu = \ker(f_{\lambda\mu})$ and let $G = \ker f_{\lambda\infty}$, so $G = \sum_{\mu>\lambda} G_\mu$. Note that for every $\mu > \lambda$, F_λ/G_μ is torsionfree and F_λ/G is torsion and non-zero. Therefore (using 11.1.5) each $Q_\tau G_\mu$ is a proper subobject of $Q_\tau F_\lambda$, whereas $Q_\tau G = Q_\tau F_\lambda$. Since Q_τ, being a left adjoint, commutes with direct limits, $\sum_{\mu>\lambda} Q_\tau G_\mu = Q_\tau \sum_\mu G_\mu = Q_\tau G = Q_\tau F_\lambda$. By assumption $iQ_\tau F_\lambda = \sum_{\mu>\lambda} iQ_\tau G_\mu$. Therefore F_λ (which, being torsionfree, is a subobject of $iQ_\tau F_\lambda$) is equal to $\sum_{\mu>\lambda} F_\lambda \cap iQ_\tau G_\mu$. Since $F_\lambda/G_\mu \in \mathcal{F}$ it follows that $F_\lambda \cap iQ_\tau G_\mu = G_\mu$ (for the quotient $iQ_\tau G_\mu/G_\mu$ is torsion). Therefore $F_\lambda = \sum_{\mu>\lambda} F_\lambda \cap iQ_\tau G_\mu = \sum_\mu G_\mu = G$, contradiction, as required. □

In particular, if τ is of finite type, then the composition $iQ_\tau : \mathcal{C} \longrightarrow \mathcal{C}$ commutes with direct sums since Q_τ does. If $\mathcal{C}$ is locally coherent, then the second condition can be extended to arbitrary direct limits (11.1.23, 11.1.32).

Corollary 11.1.13. *If $\mathcal{C}$ is a Grothendieck abelian category and $\mathcal{F} \subseteq \mathcal{C}$, then $\mathcal{F}$ is the torsionfree class for a hereditary torsion theory of finite type on $\mathcal{C}$ iff $\mathcal{F}$ is closed direct limits as well as products, injective hulls and subobjects.*

In locally finitely generated categories one has the following additional (to 11.1.12) characterisations of torsion theories of finite type.

Proposition 11.1.14. *Let $\mathcal{C}$ be a locally finitely generated Grothendieck additive category, generated by $\mathcal{G} \subseteq \mathcal{C}^{\mathrm{fg}}$, and let τ be a hereditary torsion theory on $\mathcal{C}$. Then the following are equivalent:*

(i) τ is of finite type;

(ii) for every finitely generated object, F, of $\mathcal{C}$ the filter $\mathcal{U}_\tau F$ has a cofinal set of finitely generated objects, that is, every element of $\mathcal{U}_\tau(F)$ contains a finitely generated object which is also in $\mathcal{U}_\tau(F)$;

(iii) for every $G \in \mathcal{G}$, the filter $\mathcal{U}_\tau(G)$ has a cofinal set of finitely generated objects.

If $\mathcal{C}$ is locally finitely presented, by $\mathcal{G} \subseteq \mathcal{C}^{\mathrm{fp}}$, then a fourth equivalent condition is:

(iv) the torsion class $\mathcal{T}_\tau$ is generated as such by the class of finitely presented torsion objects.

Proof. (i)$\Rightarrow$(ii) Suppose that F is finitely generated. Let $F' \in \mathcal{U}_\tau(F)$. Since $\mathcal{C}$ is locally finitely generated, F' is the directed sum, $\sum_\lambda F_\lambda$, of its finitely generated subobjects. Since F' is τ-dense in F, 11.1.2(2) gives $Q_\tau F = Q_\tau F' = \sum_\lambda Q_\tau F_\lambda$, so, since $Q_\tau F$ is finitely generated (11.1.16), we have $Q_\tau F = Q_\tau F_\lambda$ for some λ, hence (by 11.1.2 again) F_λ is in $\mathcal{U}_\tau(F)$, as required.

(iii) is a special case of (ii).

(iii)$\Rightarrow$(i) If $C = \varinjlim C_\lambda$, then, as remarked in the proof of 11.1.12(iii)$\Rightarrow$(i), $\varinjlim \tau C_\lambda \leq \tau C$. Suppose, for a contradiction, that the inclusion is proper. Then there is λ such that $\mathrm{im}(f_{\lambda\infty}) \cap \varinjlim \tau C_\lambda$ is strictly contained in $\mathrm{im}(f_{\lambda\infty}) \cap \tau C$, where $f_{\lambda\infty}$ denotes the limit map $C_\lambda \to C$. Since $\mathcal{C}$ is locally finitely generated, there is an epimorphism $\bigoplus G_i \to C_\lambda$ with the $G_i \in \mathcal{G}$, hence there is some $G \in \mathcal{G}$ and morphism $g : G \to C_\lambda$ such that $f_{\lambda\infty}g.G \cap \varinjlim \tau C_\lambda < f_{\lambda\infty}g.G \cap \tau C$.

Since $f_{\lambda\infty}g.G$ is torsion, there is, by assumption, some finitely generated $K_0 \in \mathcal{U}_\tau G$ such that $f_{\lambda\infty}g$ factors through G/K_0, say $f_{\lambda\infty}g = g'\pi$ with $g' : G/K_0 \to C$ and π the projection $G \to G/K_0$. Since K_0, hence gK_0, is finitely generated and $f_{\lambda\infty}(gK_0) = 0$, there is $\mu \geq \lambda$ such that $f_{\lambda\mu}(gK_0) = 0$, hence such that g' factors through $f_{\mu\infty}$, say $g' = f_{\mu\infty}g''$. But $\mathrm{im}(g'') \leq \tau C_\mu$ and so $f_{\lambda\infty}g \cdot G \leq \varinjlim_\nu \tau C_\nu$, a contradiction, as required.

(iii)$\Rightarrow$(iv) Let $T \in \mathcal{T}$ and let $f : \bigoplus_i G_i \to T$ be an epimorphism with the $G_i \in \mathcal{G}$, say $f = (f_i)_i$. Set $G_i' = \ker(f_i)$. Since T, hence $\mathrm{im}(f_i)$, is torsion, $G_i' \in \mathcal{U}(G_i)$. By (iii) there is some finitely generated $G_i'' \in \mathcal{U}(G_i)$ with $G_i'' \leq G_i'$. Then f factors through $\bigoplus_i (G_i/G_i'')$ and, note, each G_i/G_i'' is finitely presented and torsion.

(iv)⇒(i) Suppose that $C = \varinjlim C_\lambda$ is as in the proof of (iii)⇒(i), with $\varinjlim \tau C_\lambda$ properly contained in τC. By assumption, there is some finitely presented $T \in \mathcal{T}_\tau$ and morphism $f : T \to \tau C$ with $\mathrm{im}(f) \nleq \varinjlim \tau C_\lambda$. Consider the composition of f with the inclusion of τC into $C = \varinjlim C_\lambda$. Since T is finitely presented this factors through some C_λ hence, since every image of T is torsion, through some τC_λ, a contradiction, as required. □

Proposition 11.1.15. (generalising [663, p. 271 Exer. 1.3]) *Let $\mathcal{C}$ be a locally finitely generated Grothendieck additive category, generated by $\mathcal{G} \subseteq \mathcal{C}^{\mathrm{fg}}$ and let τ be a hereditary torsion theory of finite type on $\mathcal{C}$. Then the quotient category $\mathcal{C}_\tau$ is locally finitely generated, with the $Q_\tau G$ for $G \in \mathcal{G}$ forming a generating set of finitely generated objects.*

Proof. The $Q_\tau G$ are generating, since, if $C \in \mathcal{C}$, then there is an epimorphism $\bigoplus_i G_i \to C$, with each $G_i \in \mathcal{G}$, which, since Q_τ is exact, yields an epimorphism $\bigoplus_i Q_\tau G_i = Q_\tau(\bigoplus_i G_i) \to Q_\tau C$.

If $Q_\tau G = \sum_{\overrightarrow{\lambda}} F_\lambda$, then $i Q_\tau G = i(\sum_{\overrightarrow{\lambda}} F_\lambda) = \sum_{\overrightarrow{\lambda}} i F_\lambda$ (by 11.1.12(ii)). In particular, since $G/\tau G \leq Q_\tau G$, $G/\tau G \leq \sum_{\overrightarrow{\lambda}} i F_\lambda$ so, since G, hence $G/\tau G$, is finitely generated, $G/\tau G \leq i F_\lambda$ for some λ. Since $G/\tau G$ is τ-dense in $Q_\tau G$ (11.1.2(3)) we have $Q_\tau(G/\tau G) \leq Q_\tau i F_\lambda \leq Q_\tau G$, that is, $F_\lambda = Q_\tau i F_\lambda = Q_\tau G$, as required. □

From the proof above the following points may be extracted.

Corollary 11.1.16. *Let τ be a hereditary torsion theory on the Grothendieck additive category $\mathcal{C}$.*

(a) If τ is of finite type and if $C \in \mathcal{C}$ is finitely generated, then $Q_\tau C$ is finitely generated (as an object of $\mathcal{C}_\tau$!).
(b) If $\mathcal{G}$ is a generating family for $\mathcal{C}$, then $Q_\tau \mathcal{G}$ is a generating family for $\mathcal{C}_\tau = Q_\tau \mathcal{C}$.

In (a) above it need not be the case that $i Q_\tau C$ (where C is finitely generated) is finitely generated: consider, for example, the localisation of Mod-$\mathbb{Z}$ which takes $C = \mathbb{Z}$ to $Q_\tau C = \mathbb{Q}$.

Example 11.1.17. Let X be a non-noetherian topological space and let $\mathcal{O}_X$ be a sheaf of rings over X. Suppose that $U \subseteq X$ is open and not compact. Then, 16.3.2, $j_0 \mathcal{O}_U$ is a finitely generated, even finitely presented, object of the category PreMod-$\mathcal{O}_X$ of presheaves but $j_! \mathcal{O}_X$, which is its localisation at the sheafification torsion theory (Section 16.3.2), is not a finitely generated object of Mod-$\mathcal{O}_X$ (16.3.21). Thus, even though the localised category, in this case the category of sheaves as a localisation of the category of presheaves, might be locally finitely generated (for example, 16.3.17, if X has a basis of compact open sets), the localisations of finitely generated objects need not be finitely generated.

Proposition 11.1.18. *If τ is a torsion theory of finite type on the Grothendieck category $\mathcal{C}$ and if σ is a torsion theory of finite type on $\mathcal{C}_\tau$, then the pullback $\tau^{-1}\sigma$ (defined at 11.1.10) is a torsion theory of finite type on $\mathcal{C}$.*

The union of any chain of torsion theories of finite type is again of finite type.

Proof. Let $F = \varinjlim F_\lambda$ in $\mathcal{C}$ with each F_λ being $\tau^{-1}\sigma$-torsionfree. Each $Q_\tau F_\lambda$ is σ-torsionfree: by 11.1.5, $\sigma Q_\tau F_\lambda = Q_\tau T_\lambda$ for some $T_\lambda \in \mathcal{C}$ and the inclusion $Q_\tau T_\lambda \to Q_\tau F_\lambda$ has the form $Q_\tau f_\lambda$ for some $f_\lambda : T'_\lambda \to F_\lambda$ with T'_λ τ-dense in T_λ. By definition of $\tau^{-1}\sigma$, T'_λ is $\tau^{-1}\sigma$-torsion, so $f_\lambda = 0$, hence $\sigma Q_\tau F_\lambda = 0$. So, since $Q_\tau F = \varinjlim Q_\tau F_\lambda$, by 11.1.12 applied to σ we have $\sigma(Q_\tau F) = 0$. Therefore, if $F_0 = (\tau^{-1}\sigma)(F)$ is the $\tau^{-1}\sigma$-torsion subobject of F, then $Q_\tau F_0 \le \sigma(Q_\tau F) = 0$, so $F_0 \le \tau F = \varinjlim \tau F_\lambda$ (by 11.1.12) $= 0$. Thus condition (iii) of 11.1.12 has been checked for $\tau^{-1}\sigma$, which is, therefore, of finite type.

For the second part, it is easily checked that an object is torsionfree for a union $\tau = \bigcup \tau_i$ (see 11.1.10) iff it is torsionfree for each τ_i, so condition (iii) of 11.1.12 holds for the union if it holds for the τ_i. □

Let τ be a hereditary torsion theory of finite type on the Grothendieck category $\mathcal{C}$. An object $F \in \mathcal{C}$ is τ-**finitely generated** if it contains a τ-dense finitely generated subobject. This implies that $Q_\tau F$ is a finitely generated object of $\mathcal{C}_\tau$, since, if $F' \in \mathcal{U}_\tau(F)$, then (11.1.2) $Q_\tau F' \simeq Q_\tau F$ and, by 11.1.16, since F' is finitely generated so must be $Q_\tau F'$. The converse also holds, at least if we assume $\mathcal{C}$ to be locally finitely generated.

Lemma 11.1.19. *Suppose that τ is a hereditary torsion theory of finite type on the locally finitely generated Grothendieck category $\mathcal{C}$. An object C of $\mathcal{C}$ is τ-finitely generated iff its localisation, $Q_\tau C$, is finitely generated in $\mathcal{C}_\tau$.*

Proof. One direction was discussed just above. For the other, suppose that $Q_\tau C$ is finitely generated. Since $\mathcal{C}$ is locally finitely generated, $C = \sum_{\rightarrow} C_\lambda$ with C_λ ranging over the finitely generated subobjects of C. Since Q_τ commutes with direct limits (by 11.1.5), $Q_\tau C = \sum Q_\tau C_\lambda$, so $Q_\tau C = Q_\tau C_\lambda$ for some λ. By 11.1.2(2), C_λ must be τ-dense in C, as required. □

The next result is proved for module categories in [487, 2.4] and in [344, Thm 1]. For definable subcategories, see Section 3.4.1; they are defined there in the context of module categories but everything works in this generality, with the same proofs.[5]

[5] Or, if one prefers, first replace the ring by a small preadditive category (then the proofs need almost no modification) and then relativise using 11.1.25.

Theorem 11.1.20. ([488, 3.3]) *Let $\mathcal{C}$ be Grothendieck and locally finitely generated. Let τ be a hereditary torsion theory on $\mathcal{C}$. Then τ is of finite type iff $\mathcal{F} = \mathcal{F}_\tau$ is a definable subcategory of $\mathcal{C}$.*

Proof. Let $\mathcal{G} \subseteq \mathcal{C}^{\mathrm{fg}}$ generate $\mathcal{C}$.

By 11.1.12 and since $\mathcal{F}$ is closed under direct products and arbitrary subobjects, the direction ($\Rightarrow$) is immediate from 3.4.7.

For the converse, suppose that $\mathcal{F}$ is definable and suppose, for a contradiction, that there is $G \in \mathcal{G}$ and $I \in \mathcal{U}_\tau(G)$ which contains no finitely generated τ-dense subobject. For each finitely generated subobject $J \leq I$ denote by $\overline{J}$ the **τ-closure** of J in G, that is, the subobject $\overline{J}$ of G such that $\overline{J}/J = \tau(G/J)$, note that $G\overline{J} \in \mathcal{F}$. The finitely generated subobjects of I form a directed system, $(J_\lambda)_\lambda$, with $I = \sum_{\overrightarrow{\lambda}} J_\lambda$. Since $J_1 \subseteq J_2$ implies $\overline{J}_1 \subseteq \overline{J}_2$, the $\overline{J}_\lambda$ also form a directed system and their sum, K, is a proper subobject of G; otherwise, since G is finitely generated, $G = \overline{J}_\lambda$ for some λ, but then J_λ would be τ-dense in G and contained in I, a contradiction. Exactness of direct limits gives $\varinjlim_\lambda G/\overline{J}_\lambda = G/K \neq 0$, so, since $\mathcal{F}$ is definable, hence (3.4.7) closed under direct limits, $G/K \in \mathcal{F}$. But $K \geq I$ so G/K is τ-torsion, a contradiction as required.

(For variety, I also give a direct proof of the direction ($\Rightarrow$) which gives an "explicit" defining set of pp-pairs, in the case where $\mathcal{C}$ is locally finitely presented, by $\mathcal{G} \subseteq \mathcal{C}^{\mathrm{fp}}$.

An object $C \in \mathcal{C}$ is in $\mathcal{F}$ iff every morphism, g, from $G \in \mathcal{G}$ to C, with $\ker(g) \in \mathcal{U}_\tau(G)$ is zero (see the comment just before 11.1.11). If τ is of finite type, then it is enough, by 11.1.11, to check that, for every $G \in \mathcal{G}$ and finitely generated $I \in \mathcal{U}_\tau(G)$, if $g : G \to C$ has $\ker(g) \geq I$, then $g = 0$. Since I is finitely generated the condition $gI = 0$ is expressed by the pp condition $xI = 0$ where x is a variable of sort G; recall (Section 10.2.4) that such a variable ranges over morphisms from G. To write this pp condition more precisely, take an epimorphism $G_1 \oplus \dots \oplus G_n \to I$ and denote the composition of this with the inclusion of I into G by $r = (r_1, \dots, r_n)$, with components $r_i : G_i \to G$. Note that $r_1, \dots, r_n$ are 'scalars' that can be used in pp conditions. Then $xI = 0$ is precisely rendered by $\bigwedge_{i=1}^n xr_i = 0$. Therefore C satisfies the above condition for G, I iff $\phi(C) = 0$ where $\phi = \phi_{G,I}(x)$ is the pp condition $xI = 0$. As G, I vary, the resulting set of pp-pairs, $\phi(x) / x = 0$, serves to define $\mathcal{F}$.) □

11.1.3 Elementary localisation and locally finitely presented categories

Elementary localisation in locally finitely generated categories is characterised (11.1.21, 11.1.23). The Gabriel–Popescu Theorem is stated (11.1.24) and locally finitely presented categories are shown to be elementary localisations, as well as definable subcategories, of functor categories (11.1.27, 11.1.25); conversely, elementary localisations of locally finitely presented categories are locally finitely presented, with the subcategory of finitely presented objects "preserved" by the localisation (11.1.26).

In a locally finitely presented category, finite-type localisations are distinguished by, indeed cogenerated by, the indecomposable torsionfree injectives (11.1.28, 11.1.29). Elementary localisations are closed under composition and direct limits (11.1.30).

The absolutely pure objects in the localisation of a locally finitely presented category at an elementary torsion theory are the torsionfree absolutely pure objects of the original category (11.1.31).

Suppose that $\mathcal{C}$ is a locally finitely presented abelian, and hence (16.1.11) Grothendieck, category. Following [489],[6] say that a hereditary torsion theory τ on $\mathcal{C}$ is an **elementary** torsion theory if it is of finite type and if, for every morphism $f : G \to F$ in the category, $\mathcal{C}^{\text{fp}}$, of finitely presented objects of $\mathcal{C}$, if $\text{im}(f)$ is τ-dense in F then $\ker(f)$ is τ-finitely generated. The term elementary is used because this is exactly the condition one needs for the localised category $\mathcal{C}_\tau$ to be definable, equivalently "elementary" in the model-theoretic sense, when regarded as a subcategory of $\mathcal{C}$, see 11.1.21 and 11.1.27 below.

The next result is proved for module categories in [487, 2.5].

Theorem 11.1.21. *Suppose that $\mathcal{C}$ is a locally finitely presented abelian category, generated by $\mathcal{G} \subseteq \mathcal{C}^{\text{fp}}$. Let τ be a hereditary torsion theory of finite type on $\mathcal{C}$. Then the following are equivalent:*

- *(i) the class, $\mathcal{I}$, of τ-torsionfree, τ-injective objects (that is, $i\mathcal{C}_\tau = iQ_\tau\mathcal{C}$ in the notation of 11.1.5) is a definable subclass of $\mathcal{C}$ (loosely, "$\mathcal{C}_\tau$ is a definable subcategory of $\mathcal{C}$");*
- *(ii) for every $f : G_1 \oplus \cdots \oplus G_n \to G$ with $G_i, G \in \mathcal{G}$ and $\text{im}(f)$ τ-dense in G, $\ker(f)$ is τ-finitely generated;*
- *(iii) τ is an elementary torsion theory.*

Proof. (ii)$\Rightarrow$(i) By 11.1.20 the class, $\mathcal{F}$, of τ-torsionfree objects is definable. It is straightforward to check that an object C is τ-injective iff for every $G \in \mathcal{G}$, $I \in \mathcal{U}_\tau(G)$ and morphism $h : I \to C$ there is an extension of h to a morphism from G to C (use that $\mathcal{G}$ is generating, that pullbacks of dense embeddings are dense and a Zorn's Lemma argument over partial extensions). Because τ is of finite type, if C is torsionfree, then it is enough to check this condition when I is

[6] In [489], in the definition of elementary torsion theory, a single generator G replaces $G_1 \oplus \cdots \oplus G_n$ (or a general finitely presented object as in the definition given here) but it is the above (correct) form of the definition which is actually used in that paper.

finitely generated: for, if $I_0 \leq I$ is finitely generated and τ-dense in I (equivalently in G) and if $h \upharpoonright I_0$ extends to $g : G \to C$, then $\mathrm{im}(h - g \upharpoonright I)$ is torsion, being an image of I/I_0, hence, since a subobject of C, must be zero.

So, in order to show that $iQ_\tau\mathcal{C}$ is definable, it will be enough to show that "h is a morphism from I (to an object of $\mathcal{F}$)" is expressible using pp conditions.

Let $I \in \mathcal{U}_\tau(G)$, with $G \in \mathcal{G}$, be finitely generated. Choose an epimorphism $G_1 \oplus \cdots \oplus G_n \to I$ with $G_i \in \mathcal{G}$ and let r denote the composition with the inclusion of I into G, say $r = (r_1, \dots, r_n)$ with $r_i : G_i \to G$. By assumption there is a finitely generated, τ-dense subobject, K', of $\ker(r)$: say $H_1 \oplus \cdots \oplus H_m \to K'$ with $H_j \in \mathcal{G}$ is an epimorphism. Let $s = (s_{ji})_{ji}$ with $s_{ji} : H_j \to G_i$ be the composition with the inclusion of K' into $G_1 \oplus \cdots \oplus G_n$. Then a morphism from I to $C \in \mathcal{F}$ is given by morphisms $g_1, \dots, g_n$ (with $g_i : G_i \to C$) satisfying $\bigwedge_{j=1}^m \sum_{i=1}^n g_i s_{ji} = 0$. So if $\phi(x_1, \dots, x_n)$ is the pp condition $\bigwedge_{j=1}^m \sum_{i=1}^n x_i s_{ji} = 0$ (where x_i has sort G_i) and $\psi(x_1, \dots, x_n)$ is the pp condition $\exists x \, (\bigwedge_{i=1}^n x_i = x r_i)$ (where x has sort G), then a τ-torsionfree object C satisfies the injectivity condition with respect to the inclusion $I \leq G$ iff $\phi(C) \leq \psi(C)$, that is, iff the pair ϕ/ψ is closed on C. Therefore the set of these pairs ϕ/ψ, as G varies over $\mathcal{G}$ and I varies over finitely generated objects in $\mathcal{U}_\tau(G)$, defines the class of τ-torsionfree, τ-injective objects.

(i)$\Rightarrow$(ii) Assume that the class, $iQ_\tau\mathcal{C}$, of τ-torsionfree, τ-injectives is definable and suppose, for a contradiction, that there is $f : G_1 \oplus \cdots \oplus G_n \to G$ with $G, G_i \in \mathcal{G}$, $\mathrm{im}(f)$ τ-dense in G and $K = \ker(f)$ not τ-finitely generated.

Set $K = \sum_{\overrightarrow{\lambda}} K_\lambda$ where K_λ is the (directed) set of finitely generated subobjects of K and let $\overline{K_\lambda}$ be the τ-closure (see the proof of 11.1.20) of K_λ in $G' = G_1 \oplus \cdots \oplus G_n$. Set Q_λ to be the τ-injective hull (Section 11.1.1) of (the torsionfree object) $G'/\overline{K_\lambda}$, that is, $Q_\lambda = iQ_\tau(G'/\overline{K_\lambda})$. The directed system $\big((G'/\overline{K_\lambda})_\lambda, (\pi_{\lambda\mu})_{K_\lambda \leq K_\mu}\big)$, where $\pi_{\lambda\mu}$ is the natural projection from $G'/\overline{K_\lambda}$ to $G'/\overline{K_\mu}$, gives rise, by functoriality of iQ_τ, to a directed system $\mathcal{Q} = \big((Q_\lambda)_\lambda, (q_{\lambda\mu} = iQ_\tau\pi_{\lambda\mu})_{K_\lambda \leq K_\mu}\big)$ with limit, Q, say. Each Q_λ is in $\mathcal{I}$, so, by assumption, $Q \in \mathcal{I}$. There is the obvious natural map from the directed system $\mathcal{P} = \big((G'/K_\lambda)_\lambda, (\pi'_{\lambda\mu} : G'/K_\lambda \to G'/K_\mu)_{K_\lambda \leq K_\mu}\big)$ to $\mathcal{Q}$, hence an induced map, h say, from $\varinjlim \mathcal{P} = G'/\sum_\lambda K_\lambda = G'/K = I$ to $Q = \varinjlim \mathcal{Q}$. Since $Q \in \mathcal{I}$, there is an extension of h to a morphism $g : G \to Q$. But G is finitely presented, so g, and hence h, factors through one of the maps, $Q_\lambda \to Q$, to the limit. But that implies $K \leq \overline{K_\lambda}$, contradicting that K is not τ-finitely generated.

Equivalence with (iii) follows since condition (i) is independent of the choice of generating subset $\mathcal{G} \subseteq \mathcal{C}^{\mathrm{fp}}$ (although pp conditions make reference to the chosen generating set, the characterisation, 3.4.7, of definable subcategories does not). $\square$

The next lemma will be used in the proof of 16.3.12.

Lemma 11.1.22. *Suppose that $\mathcal{C}$ is a locally finitely presented abelian category, generated by $\mathcal{G} \subseteq \mathcal{C}^{\mathrm{fp}}$. Let τ be a hereditary torsion theory of finite type on $\mathcal{C}$. If for every $G \in \mathcal{G}$ and τ-dense subobject $F \leq G$ there is a finitely presented $F' \leq F$ which is τ-dense in F, then τ is elementary.*

Proof. The second condition of 11.1.21 is checked. With notation and assumption as there, set $H = G_1 \oplus \cdots \oplus G_n$ and $F = \mathrm{im}(f)$: so F is a τ-dense subobject of G. By assumption there is $F' \leq F$ which is τ-dense and finitely presented. Since F' is finitely generated, there is a finitely generated subobject H' of H with $fH' = F'$. Set $K = \ker(f)$. Since F' is τ-dense in G, $H' + K$ is τ-dense in H, so, since τ is of finite type (and by 11.1.14), there is a finitely generated $H_1 \leq H' + K$ which is τ-dense in H. Without loss of generality we may suppose $H_1 \geq H'$, so $fH_1 = F'$. Since F' is finitely presented, the kernel $K_1 = K \cap H_1$ of $f \restriction H_1$ is, by E.1.17, finitely generated and, since H_1 is τ-dense in $H_1 + K$, K_1 is τ-dense in K, as required. □

Proposition 11.1.23. *A hereditary torsion theory τ on a locally finitely presented abelian category $\mathcal{C}$ is elementary iff the inclusion $i : \mathcal{C}_\tau \longrightarrow \mathcal{C}$ commutes with direct limits.*

Proof. If τ is elementary, then $i\mathcal{C}_\tau$ is, by 11.1.21, a definable subcategory of $\mathcal{C}$, so, by 3.4.7, as a subcategory of $\mathcal{C}$, it is closed under direct limits. Therefore the inclusion functor commutes with direct limits.

Suppose, conversely, that i commutes with direct limits, hence $i\mathcal{C}_\tau$ is closed under direct limits in $\mathcal{C}$. The class $i\mathcal{C}_\tau$ is closed under products because each of $\mathcal{F}_\tau$ and the class of τ-injectives (like any injectivity class) is closed under products. Furthermore, $i\mathcal{C}_\tau$ is closed under pure substructures. For, if B is a pure subobject of $C \in i\mathcal{C}_\tau$, then certainly B is τ-torsionfree: we prove that B is also τ-injective.

It is enough (see the proof of 11.1.21(ii)⇒(i)) to consider the case that G is finitely presented, $F \in \mathcal{U}_\tau(G)$ and $f : F \to B$. It is necessary to produce an extension of f to a morphism from G to B. By 11.1.12, τ is of finite type, so there is $F' \in \mathcal{U}_\tau(G)$ with F' finitely generated and $F' \leq F$. Consider the diagram shown where g exists by τ-injectivity of C.

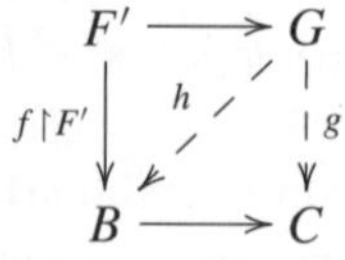

$$\begin{array}{ccc} F' & \longrightarrow & G \\ {\scriptstyle f\restriction F'}\downarrow & \overset{h}{\swarrow} & \downarrow{\scriptstyle g} \\ B & \longrightarrow & C \end{array}$$

By 2.1.8 there is a morphism $h : G \to B$ with $h \restriction F' = f \restriction F'$. Since F' is τ-dense in F and B is torsionfree, this map h in fact extends f (as in the proof of 11.1.21(ii)⇒(i)), so B is τ-torsionfree and τ-injective, as required.

Therefore, by 3.4.7, $i\mathcal{C}_\tau$ is a definable subcategory of $\mathcal{C}$, so, by 11.1.21, τ is elementary. □

Henceforth we will often identify $\mathcal{C}_\tau$ and $i\mathcal{C}_\tau$ (but bear in mind that i does not in general commute with arbitrary colimits).

We will show that every locally finitely presented abelian (hence, 16.1.11, Grothendieck) category $\mathcal{C}$ can be represented as a localisation of a "module" (that is, functor) category at some elementary torsion theory. It is clear already that $\mathcal{C}$ can be embedded as a full subcategory of some functor category. For, if $\mathcal{G}$ is a generating subcategory of $\mathcal{C}^{\mathrm{fp}}$, then there is the full embedding (10.1.7) of $\mathcal{C}$ into Mod-$\mathcal{G} = (\mathcal{G}^{\mathrm{op}}, \mathbf{Ab})$ given on objects by $C \mapsto (-, C) \upharpoonright \mathcal{G}$. The image of this embedding will be identified as a certain localisation of Mod-$\mathcal{G}$.

First we state the Gabriel–Popescu Theorem. This formulation, also see, for example, [212, §4], [415] and references therein, is taken from [488, §I.3]. We do not give the proof since it is a rather direct generalisation of that for the usual formulation (see [486, 3.7.9, 4.14.2], [663, X.4.1]).

Theorem 11.1.24. ([489, 1.1], see also [217, §3]) (Gabriel-Popescu Theorem) *Let $\mathcal{C}$ be an additive Grothendieck category with a small generating subcategory $\mathcal{G}$. Then there is a hereditary torsion theory, τ, on the category* Mod-$\mathcal{G} = (\mathcal{G}^{\mathrm{op}}, \mathbf{Ab})$ *such that $\mathcal{C}$ is equivalent to the corresponding localisation* (Mod-$\mathcal{G})_\tau$.

Theorem 11.1.25. ([489, 2.2]) *Let $\mathcal{C}$ be a locally finitely presented abelian category, let $\mathcal{G} \subseteq \mathcal{C}^{\mathrm{fp}}$ be generating and let $j : \mathcal{C} \longrightarrow$ Mod-$\mathcal{G}$ be the Yoneda embedding $C \mapsto (-, C) \upharpoonright \mathcal{G}$. Set $\mathcal{T} = \{M \in \text{Mod-}\mathcal{G} : (M, j\mathcal{C}) = 0\}$. Then $\mathcal{T}$ is a hereditary torsion class and the corresponding torsion theory, τ, is elementary. Furthermore, $j\mathcal{C}$ may be identified with the localisation* (Mod-$\mathcal{G})_\tau$.

We only outline the argument for this since the proof refers to the details of the proof of the previous result. This comment applies in particular to the proof that $\mathcal{T}$ is a hereditary torsion class and that $j\mathcal{C}$ is equivalent to the localisation (Mod-$\mathcal{G})_\tau$ (see [489] for more detail). So here we prove just that τ is elementary. First we show that τ is of finite type.

So, let $G \in \mathcal{G}$ (identify $\mathcal{G}$ with its image in Mod-$\mathcal{G}$, hence G with $(-, G)$). Let $G' \in \mathcal{U}_\tau(G)$. Then $G' = \sum_{\rightarrow} G_\lambda$, where $\{G_\lambda\}_\lambda$ is the directed set of finitely generated subobjects of G', so $Q_\tau G' = \sum_{\rightarrow} Q_\tau G_\lambda$. Since $Q_\tau G' \simeq Q_\tau G$ (that is, $= Q_\tau(-, G) = G$) is finitely generated, there is λ such that $Q_\tau G' = Q_\tau G_\lambda$. Then (11.1.2) G_λ must be τ-dense in G', as required to show that τ is of finite type.

Next, take a morphism $f : G_1 \oplus \cdots \oplus G_n \to G$ with τ-dense image $F \in \mathcal{U}_\tau(G)$, so $Q_\tau F = Q_\tau G$ and $Q_\tau f : Q_\tau G_1 \oplus \cdots \oplus Q_\tau G_n \to Q_\tau G$ is an epimorphism. Since $Q_\tau G$ is finitely presented, $\ker(Q_\tau f)$ is finitely generated (E.1.16).

Since Q_τ is exact, $Q_\tau(\ker(f)) = \ker(Q_\tau f)$, so, arguing as above, it follows that $\ker(f)$ is τ-finitely generated and τ is indeed elementary.

The next result can be found for functor categories as [488, 4.4], see [489, 2.1], and, for the general case, as [278, 2.16], [349, 2.6], [498, A3.15].

Theorem 11.1.26. *Suppose that τ is an elementary torsion theory on the locally finitely presented category $\mathcal{C}$ and let $\mathcal{G} \subseteq \mathcal{C}^{\mathrm{fp}}$ be generating. Then the localised category $\mathcal{C}_\tau$ is locally finitely presented and $\mathcal{G}_\tau \subseteq (\mathcal{C}_\tau)^{\mathrm{fp}}$ is generating.*

Proof. By 11.1.15, $\mathcal{G}_\tau$ generates $\mathcal{C}_\tau$ and consists of finitely generated objects of $\mathcal{C}_\tau$. It must be shown that if $G \in \mathcal{G}$, then $Q_\tau G \in (\mathcal{C}_\tau)^{\mathrm{fp}}$.

Let $(D_\lambda)_\lambda$ be a directed system in $\mathcal{C}_\tau$ with direct limit $D = \varinjlim_\lambda D_\lambda$. Take $g : Q_\tau G \to D$. By 11.1.23, $\varinjlim_\lambda iD_\lambda = iD$. Let $h : G \to iQ_\tau G$ be the canonical map (by adjointness, but explicitly $G \to G/\tau G \to E_\tau(G/\tau G) = iQ_\tau G$). Since G is finitely presented, there is λ and $k : G \to iD_\lambda$ such that $if_{\lambda\infty} \cdot k = ig \cdot h$, where $f_{\lambda\infty} : D_\lambda \to D$ is the canonical map to the limit. Since iD_λ is τ-torsionfree and τ-injective, k factors (uniquely) as $k = k'h$ for some $k' : iQ_\tau G \to iD_\lambda$ and (i is full) $k' = ig'$ for some $g' : Q_\tau G \to D_\lambda$ (the correspondent to k under the adjunction $(G, iD_\lambda) \simeq (Q_\tau G, D_\lambda)$ of 11.1.5). Since $if_{\lambda\infty} \cdot ig'$ and ig agree on the image of h and iD is τ-torsion-free they are equal, hence $f_{\lambda\infty}g' = g$, as required. □

Corollary 11.1.27. ([489, 2.3]) *The following are equivalent for an additive Grothendieck category $\mathcal{C}$:*

(i) $\mathcal{C}$ is locally finitely presented;
(ii) $\mathcal{C}$ is equivalent to a localisation of a functor category at an elementary torsion theory;
(iii) there is a small preadditive category $\mathcal{G}$ and a localisation Mod-$\mathcal{G} \longrightarrow \mathcal{C}$ *such that $i\mathcal{C}$ is a definable subcategory of* Mod-$\mathcal{G}$, *where $i : \mathcal{C} \longrightarrow$* Mod-$\mathcal{G}$ *is the left adjoint of the localisation functor.*

Proof. The equivalence (i)⇔(ii) is by 11.1.25 and 11.1.26. Then, since i identifies the localised category with the full subcategory of τ-torsionfree, τ-injective objects, the equivalence with (iii) is by 11.1.25 and 11.1.21. □

Proposition 11.1.28. *Suppose that $\mathcal{C}$ is a locally finitely presented abelian category and let τ, τ' be hereditary torsion theories of finite type on $\mathcal{C}$. Then $\tau = \tau'$ iff for every indecomposable injective $E \in \mathcal{C}$, E is τ-torsionfree if and only if E is τ'-torsionfree.*

Proof. If $\tau \neq \tau'$, say there is a τ-torsion object F which is not τ'-torsion, then, by 11.1.14, it may be supposed that F is finitely presented. By Zorn's Lemma, using that $\mathcal{U}_{\tau'}(F)$ has cofinal finitely generated members (11.1.14), there is a subobject G of F maximal with respect to not being τ'-dense in F. Then F/G is uniform, because the intersection of two τ'-dense subobjects of F is still τ'-dense, so has indecomposable injective hull, E, say. Also F/G is τ'-torsionfree: otherwise there would be G' with $F \geq G' > G$ and G τ'-dense in G', so, since G' must be τ'-dense in F and $\mathcal{T}_{\tau'}$ is closed under extensions, G would also be τ'-dense in F, a contradiction. Since F/G is essential in E (and τ is hereditary) E also is τ'-torsionfree but is not τ-torsionfree (since F, hence F/G, is τ-torsion), as required. □

The sets of indecomposable τ-torsionfree injectives which correspond to torsion theories of finite type are described in 12.3.5 for functor categories, in terms of the Ziegler spectrum.

Proposition 11.1.29. *Let $\mathcal{C}$ be a locally finitely presented abelian category and suppose that τ is a hereditary torsion theory of finite type on $\mathcal{C}$. Then τ is cogenerated by the set of τ-torsionfree indecomposable injectives.*

Proof. Let τ'' be the torsion theory cogenerated by the set of τ-torsionfree indecomposable injectives: so

$$\mathcal{T}_{\tau''} = \{C \in \mathcal{C} : (C, E) = 0 \text{ for all indecomposable injective } E \in \mathcal{F}_\tau\}$$

and clearly $\mathcal{T}_\tau \subseteq \mathcal{T}_{\tau''}$. Let $T \in \mathcal{T}_{\tau''} \setminus \mathcal{T}_\tau$ if there is such. Then there is a non-zero morphism $f : T \to E$ with E τ-torsionfree. It will be enough, for a contradiction, to produce a non-zero morphism from T to an indecomposable τ-torsionfree injective.

Choose $T' \leq T$ finitely generated with $fT' \neq 0$ and let $g : T'' \to T'$ be an epimorphism with T'' finitely presented; this exists since $\mathcal{C}$ is locally finitely presented (E.1.15). As in the proof of 11.1.28, there is $T_0 \leq T''$ containing $\ker(g)$ and maximal with respect to not being τ-dense in T'' (note that $\ker(g)$ is not τ-dense in T'' because E is τ-torsionfree). As in that proof, $T'/gT_0 \simeq T''/T_0$ is uniform. Then, also as there, $E(T''/T_0)$ is an indecomposable τ-torsionfree injective and the natural map $T' \to E(T''/T_0)$ extends to a non-zero map from T to $E(T''/T_0)$, the required contradiction. □

Proposition 11.1.30. *If τ is an elementary torsion theory on the locally finitely presented abelian category $\mathcal{C}$ and if σ is an elementary torsion theory on $\mathcal{C}_\tau$, then the pullback, $\tau^{-1}\sigma$ (see 11.1.10), is an elementary torsion theory on $\mathcal{C}$.*

The union of any chain of elementary torsion theories is again elementary.

Proof. By 11.1.21, $\mathcal{C}_\tau$ is a definable subcategory of $\mathcal{C}$ and $(\mathcal{C}_\tau)_\sigma$ is a definable subcategory of $\mathcal{C}_\tau$. It follows that $(\mathcal{C}_\tau)_\sigma = \mathcal{C}_{\sigma^{-1}\tau}$ is a definable subcategory of $\mathcal{C}$

(by 3.4.7, or the general fact (cf. beginning of Section 18.1.1) that a definable subcategory of a definable subcategory of $\mathcal{C}$ is a definable subcategory of $\mathcal{C}$). By 11.1.21 and 11.1.18, it follows that $\sigma^{-1}\tau$ is an elementary torsion theory on $\mathcal{C}$.

For the second statement, if $\tau = \bigcup \tau_i$, then already, by 11.1.18, τ is of finite type. Suppose that $f = G_1 \oplus \cdots \oplus G_n \to G$ with G and the G_i finitely presented, has τ-dense image. Directly from the definition of $\bigcup \tau_i$, we have $G/\mathrm{im}(f) = \tau(G/\mathrm{im}(f)) = \sum \tau_i(G/\mathrm{im}(f))$, so, since $G/\mathrm{im}(f)$ is finitely generated, $\mathrm{im}(f)$ is already τ_i-dense in G for some i. Therefore $\ker(f)$ is τ_i-finitely generated, so certainly is τ-finitely generated, as required. □

Proposition 11.1.31. *Suppose that $\mathcal{C}$ is a locally finitely presented abelian category and that τ is an elementary torsion theory on $\mathcal{C}$. Let C be a τ-torsionfree absolutely pure object of $\mathcal{C}$. Then $C = iQ_\tau C$ and is an absolutely pure object of $\mathcal{C}_\tau$. Conversely, every absolutely pure object of $\mathcal{C}_\tau$ has the form $Q_\tau C$ for some absolutely pure object C of $\mathcal{C}$. Hence* $\mathrm{Abs}(\mathcal{C}_\tau)$, *respectively* $\mathrm{Inj}(\mathcal{C}_\tau)$, *is identified by i with the class, $\mathcal{F} \cap \mathrm{Abs}(\mathcal{C})$, resp. $\mathcal{F} \cap \mathrm{Inj}(\mathcal{C})$, of absolutely pure, resp. injective, τ-torsionfree objects of $\mathcal{C}$.*

Proof. The fact that C is τ-injective, hence equal to its localisation, is direct from the characterisation 2.3.1(iii) and the fact that τ is of finite type (and 11.1.14). For, if $G' \leq G \in \mathcal{C}^{\mathrm{fp}}$ is τ-dense, then there is some finitely generated $G_0 \in \mathcal{U}_\tau G$ with $G_0 \leq G'$. Since C is absolutely pure and G/G_0 is finitely presented the restriction to G_0 of any morphism $f : G' \to C$ extends to a morphism $G \to C$ and, since C is τ-torsionfree, this extension must agree with f (the image of the difference is torsion).

Purity in $\mathcal{C}_\tau$ is, by 16.1.17 and 11.1.21, the restriction of purity in $\mathcal{C}$, so absolute purity of C in the subcategory $i\mathcal{C}_\tau$ is immediate (one may also prove this directly). Note that the statement regarding injectivity is part of 11.1.5. □

Observe that if $C \to D$ is a pure embedding in the locally finitely presented category $\mathcal{C}$ and if τ is elementary, then both the localised morphism $C_\tau \to D_\tau$ and its image in $\mathcal{C}$ are pure embedding (for example, by 11.1.23, 11.1.5 and 2.1.4).

11.1.4 Finite-type localisation in locally coherent categories

The localisation of a locally coherent category at an elementary = finite-type torsion theory is again locally coherent, with finitely presented objects the localisations of finitely presented objects (11.1.33). The localisation is a definable subcategory (11.1.35).

In a locally coherent category a finite-type localisation is determined by the Serre subcategory of finitely presented objects which are torsion (11.1.36), indeed there is a bijection between Serre subcategories of finite type objects and finite-type localisations (11.1.38). A finite-type torsion theory on a functor category is determined by the finitely presented subquotients of the forgetful functor which are torsion (11.1.39).

Each torsion class contains a largest finite-type torsion class (11.1.37).

A summary result on localising at Serre subcategories is stated (11.1.40). In locally coherent categories, these are the restrictions of finite-type localisations (11.1.42).

Most of the results of this section come variously from [278], [349], [498].

Remark 11.1.32. Suppose that $\mathcal{C}$ is locally coherent. A hereditary torsion theory τ is elementary iff it is of finite type: since $\mathcal{C}$ is locally coherent the kernel of any morphism between finitely presented objects is finitely generated.

Theorem 11.1.33. ([498, A3.16, 3.17], [278, 2.16], [349, 2.6]) *Suppose that $\mathcal{C}$ is a locally coherent abelian category with $\mathcal{G}$ as a generating set of finitely presented objects. Let τ be an elementary torsion theory on $\mathcal{C}$. Then the localisation, $\mathcal{C}_\tau$, of $\mathcal{C}$ at τ is again locally coherent, with $\{Q_\tau G : G \in \mathcal{G}\}$ a generating set of finitely presented objects, and has, for its collection of finitely presented objects, the objects of the form $Q_\tau C$, where C is in $\mathcal{C}^{\mathrm{fp}}$: $(\mathcal{C}_\tau)^{\mathrm{fp}} = (\mathcal{C}^{\mathrm{fp}})_\tau$.*

Proof. Since $\mathcal{C}$ is locally coherent, $\mathcal{C}^{\mathrm{fp}}$ is an abelian subcategory (E.1.19) so, since the localisation functor is exact (11.1.5), $(\mathcal{C}^{\mathrm{fp}})_\tau$ is an abelian subcategory of $\mathcal{C}_\tau$. By 11.1.26, $(\mathcal{C}^{\mathrm{fp}})_\tau$ consists of finitely presented objects and is generating. So, to show that $(\mathcal{C}^{\mathrm{fp}})_\tau = (\mathcal{C}_\tau)^{\mathrm{fp}}$, it will be enough to show that the localisation functor Q_τ restricted to $\mathcal{C}^{\mathrm{fp}}$ is full. But if $A, B \in \mathcal{C}^{\mathrm{fp}}$, then (for example, by the second description of the localisation process in Section 11.1.1) any morphism from $Q_\tau A$ to $Q_\tau B$ is the image of one from a τ-dense subobject, A', of A to a quotient, $B' = B/T$, of B' where $T \in \mathcal{T}$. Because τ is of finite type, A' may, by 11.1.14, be chosen to be finitely generated, hence, since $\mathcal{C}$ is locally coherent, finitely presented. Then any morphism $A' \to B/T$ factors through, so may be replaced by, a morphism from B/T' for some finitely generated $T' \leq T$ (so also $B/T' \in \mathcal{C}^{\mathrm{fp}}$), as required. That $\mathcal{C}_\tau$ is locally coherent now follows from E.1.19. □

We extract the main point for easy reference.

Corollary 11.1.34. *If $\mathcal{C}$ is a locally coherent abelian category and τ is a finite-type hereditary torsion theory on $\mathcal{C}$, then $\mathcal{C}_\tau$ is locally coherent and has, for its class of finitely presented objects, $(\mathcal{C}^{\mathrm{fp}})_\tau = \{C_\tau : C \in \mathcal{C}^{\mathrm{fp}}\}$.*

The next corollary is immediate by 11.1.21.

Corollary 11.1.35. ([487, 2.3]) *Suppose that $\mathcal{C}$ is a locally coherent abelian category and that τ is a finite-type torsion theory on $\mathcal{C}$. Then $\mathcal{C}_\tau$, regarded as a subcategory of $\mathcal{C}$, is a definable subcategory.*

Recall that a full subcategory, $\mathcal{S}$, of an abelian category $\mathcal{B}$ is a Serre subcategory if, given any short exact sequence $0 \to A \to B \to C \to 0$ in $\mathcal{B}$, one has $B \in \mathcal{S}$ iff $A, C \in \mathcal{S}$.

If $\mathcal{S} \subseteq \mathcal{C}$, where $\mathcal{C}$ has direct limits, then denote by $\overrightarrow{\mathcal{S}}$ the closure of $\mathcal{S}$ under direct limits in $\mathcal{C}$.

Proposition 11.1.36. ([278, 2.5], [349, 2.8]) *Suppose that $\mathcal{C}$ is a locally coherent abelian category. If $\mathcal{S}$ is a Serre subcategory of $\mathcal{C}^{\mathrm{fp}}$, then $\overrightarrow{\mathcal{S}}$ is the hereditary torsion class generated in $\mathcal{C}$ by $\mathcal{S}$ and $\mathcal{C}^{\mathrm{fp}} \cap \overrightarrow{\mathcal{S}} = \mathcal{S}$. This hereditary torsion theory generated by $\mathcal{S}$ is of finite type.*

If $\mathcal{T}$ is any hereditary torsion class in $\mathcal{C}$, then $\mathcal{T} \cap \mathcal{C}^{\mathrm{fp}}$ is a Serre subcategory of $\mathcal{C}^{\mathrm{fp}}$ and $\overrightarrow{\mathcal{T} \cap \mathcal{C}^{\mathrm{fp}}} \subseteq \mathcal{T}$, with equality iff $\mathcal{T}$ is of finite type.

Proof. To check that $\overrightarrow{\mathcal{S}}$ is a hereditary torsion class, suppose that $C = \varinjlim A_\lambda$ with the $A_\lambda \in \mathcal{S}$ and $g_\lambda : A_\lambda \to C$ being the limit maps. Let $C' \leq C$. Then $C' = \varinjlim g_\lambda^{-1}C'$ and each $g_\lambda^{-1}C'$ is the direct limit of its finitely generated subobjects, which, since they are subobjects of A_λ and $\mathcal{C}$ is locally coherent, are finitely presented, hence in $\mathcal{S}$. Therefore $C' \in \overrightarrow{\mathcal{S}}$. Also, if $f : C \to C''$ is an epimorphism, let $B_\lambda = \ker(fg_\lambda)$. Then $C'' = \varinjlim(A_\lambda/B_\lambda)$ and each A_λ/B_λ is the direct limit of the $A_\lambda/B_{\lambda\mu}$ where $B_{\lambda\mu}$ ranges over the finitely generated subobjects of B_λ. Each $A_\lambda/B_{\lambda\mu}$ is a finitely presented quotient of A_λ, hence is in $\mathcal{S}$. Therefore $C'' \in \overrightarrow{\mathcal{S}}$.

To see that $\overrightarrow{\mathcal{S}}$ is closed under extensions, suppose also that $D = \varinjlim B_\mu$ with $B_\mu \in \mathcal{S}$ and consider an exact sequence $0 \to C \to E \to D \to 0$. Given μ, form the pullback exact sequence shown.

$$\begin{array}{ccccccccc} 0 & \longrightarrow & C & \longrightarrow & P_\mu & \longrightarrow & B_\mu & \longrightarrow & 0 \\ & & \| & & \downarrow & & \downarrow & & \\ 0 & \longrightarrow & C & \longrightarrow & E & \longrightarrow & D & \longrightarrow & 0 \end{array}$$

Given $g_{\mu\mu'} : B_\mu \to B_{\mu'}$ in the directed system for D there is the diagram below.

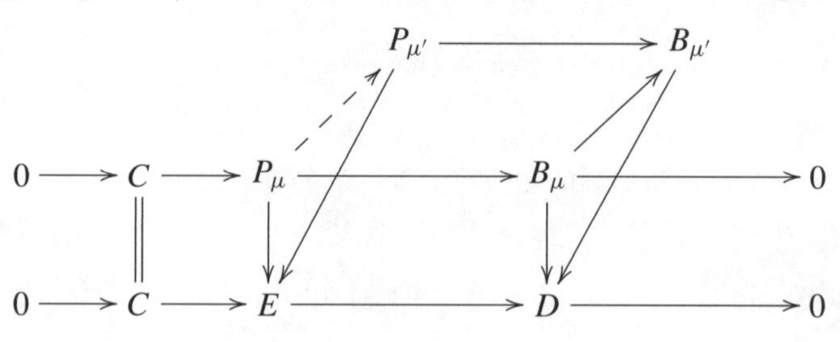

By the pullback property (of $P_{\mu'}$) there is a unique morphism $P_\mu \to P_{\mu'}$ making the top square and left-hand triangle commute. Thus we obtain a directed system of pullback exact sequences, one for each μ. By exactness of direct limits, $E = \varinjlim P_\mu$, so it is sufficient to consider the case where D is already in $\mathcal{S}$. Since $\vec{\mathcal{C}}$ is locally coherent, D is FP_2 (see Section 10.2.6) so, by [96, Thm 2], $\mathrm{Ext}^1(D, -)$ commutes with direct limits, in particular $\mathrm{Ext}^1(D, C = \varinjlim A_\lambda) = \varinjlim \mathrm{Ext}^1(D, A_\lambda)$. So the exact sequence $0 \to C \to E \to D \to 0$ is a direct limit of exact sequences with middle terms in $\mathcal{S}$ (since $D, A_\lambda \in \mathcal{S}$), hence E is indeed in $\vec{\mathcal{S}}$.

Since $\vec{\mathcal{S}}$ certainly is closed under direct sums, the first point has been proved.

Next, $\mathcal{C}^{\mathrm{fp}} \cap \vec{\mathcal{S}} \supseteq \mathcal{S}$ by construction and the reverse inclusion is immediate from the definition of finitely presented.

That τ is of finite type is immediate from 11.1.14.

Now suppose that $\mathcal{T}$ is a hereditary torsion class. Just from the definitions, $\mathcal{T} \cap \mathcal{C}^{\mathrm{fp}}$ is a Serre subcategory of $\mathcal{C}^{\mathrm{fp}}$ and, since $\mathcal{T}$ is closed under direct limits, $\mathcal{T} \supseteq \overrightarrow{\mathcal{T} \cap \mathcal{C}^{\mathrm{fp}}}$ (and the latter is of finite type). By 11.1.14 there is equality iff $\mathcal{T}$ is of finite type. □

In fact, a hereditary torsion class in a locally coherent category is itself a locally coherent abelian category, see [278, 2.7].

Corollary 11.1.37. *If $\mathcal{C}$ is locally coherent and $\mathcal{T}$ is a hereditary torsion class in $\mathcal{C}$, then the torsion class $\overrightarrow{\mathcal{T} \cap \mathcal{C}^{\mathrm{fp}}}$ is the largest hereditary torsion class of finite type contained in $\mathcal{T}$ ([488, 5.3.7]).*

Corollary 11.1.38. ([278, 2.8], [349, 2.10]) *Let $\mathcal{C}$ be a locally coherent category. Then there is a bijection between Serre subcategories, $\mathcal{S}$, of $\mathcal{C}^{\mathrm{fp}}$ and hereditary torsion classes, $\mathcal{T}$, of finite type on $\mathcal{C}$, given by: $\mathcal{S} \mapsto \vec{\mathcal{S}}$ and $\mathcal{T} \mapsto \mathcal{T} \cap \mathcal{C}^{\mathrm{fp}}$.*

Lemma 11.1.39. *If τ is a hereditary torsion theory of finite type on* (R-mod, **Ab**)*, then τ is determined by the torsion finitely presented subquotients of $(R_R \otimes -) = ({}_RR, -)$.*

Proof. Recall, 10.2.3, that (R-mod, **Ab**) is locally coherent. By 11.1.14 and 10.1.13 the factors $({}_RL, -)/F$ with $F \in \mathcal{U}_\tau({}_RL, -)$ and F finitely generated determine τ. An epimorphism $R^n \to L$ induces an embedding $(L, -) \to (R^n, -) \simeq (R^n \otimes -)$, so such a functor $({}_RL, -)/F$ is a subquotient of $(R^n \otimes -)$. Every subquotient of $(R^n \otimes -) \simeq (R \otimes -)^n$ is a finite extension of subquotients of $R \otimes -$ (see below) so the statement follows. □

The fact that every subquotient of $(R, -)^n$ is an iterated extension of subquotients of $(R, -)$ is proved by induction on n, as follows.

First note that, if $F \le (R^n, -)$ is finitely generated, then there is an exact sequence $0 \to K \to F \to \pi F \to 0$, where $\pi : (R, -)^n \to (R, -)$ is projection to (say) the first coordinate. The image πF is finitely generated hence (10.2.3) finitely presented, hence K ($\le (R, -)^{n-1}$) is finitely generated, hence finitely presented. Then, given $F' \le F \le (R^n, -)$ with F, F' finitely generated – so F/F' is, up to isomorphism, a typical finitely presented subquotient of $(R^n, -)$ – consider the diagram shown.

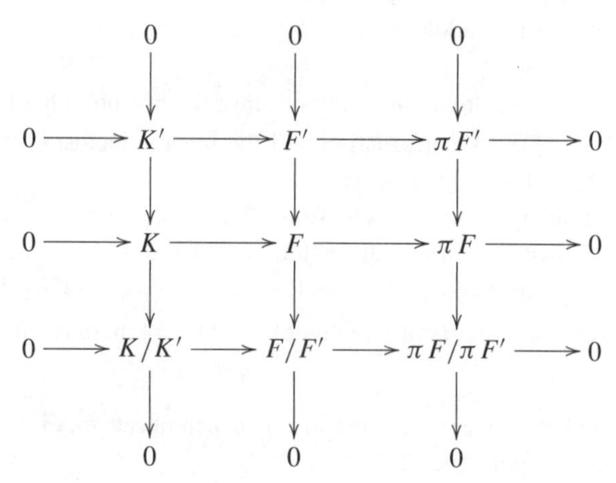

The nine ($= 3 \times 3$) lemma (for example, [195, 2.65], [708, 1.3.2]) gives exactness of the last row. By induction, K/K' is an iterated extension of subquotients of $(R, -)$, hence the same is true of F/F'.

This result 11.1.39, bearing in mind 10.2.17 and 10.2.30, is why many results require a hypothesis only on pp conditions with one free variable (because the hypothesis automatically extends to pp conditions in any finite number of free variables).

If $\mathcal{B}$ is an abelian category and $\mathcal{S}$ is a Serre subcategory, then, as mentioned in Section 11.1.1, there is a notion of localisation, $\mathcal{B} \longrightarrow \mathcal{B}/\mathcal{S}$ of $\mathcal{B}$ at $\mathcal{S}$ (see [486, 4.3.3]). This may be compared with 11.1.3 and 11.1.5.

Theorem 11.1.40. ([486, pp. 172–4]) *Suppose that $\mathcal{B}$ is an abelian category and that $\mathcal{S}$ is a Serre subcategory of $\mathcal{B}$. The quotient category $\mathcal{B}/\mathcal{S}$ is abelian and the localisation functor $F : \mathcal{B} \longrightarrow \mathcal{B}/\mathcal{S}$ is exact. The kernel of F in the sense of* $\ker(F) = \{B \in \mathcal{B} : FB = 0\}$ *is $\mathcal{S}$. Every exact functor $G : \mathcal{B} \longrightarrow \mathcal{B}'$ from $\mathcal{B}$ to an abelian category $\mathcal{B}'$ with* $\ker(G) \supseteq \mathcal{S}$ *factors uniquely through $F : \mathcal{B} \longrightarrow \mathcal{B}/\mathcal{S}$ via an exact and faithful functor.*

Example 11.1.41. Suppose that R is a right coherent ring, so mod-R is abelian (E.1.19). Let Ab(R) be the free abelian category on R (10.2.7). The functor from R, regarded as preadditive category, to mod-R which takes R to the projective module R_R extends to an exact functor from Ab(R) to mod-R. This is the quotient of Ab(R) by the Serre subcategory consisting of those finitely presented functors (note 10.2.37) $F : R\text{-mod} \to \mathbf{Ab}$ which are 0 on ${}_RR$ (the image of R in Ab(R)). This result, that $(R\text{-mod}, \mathbf{Ab})^{\text{fp}}/\{F : F({}_RR) = 0\} \simeq \text{mod-}R$ for a right coherent ring, goes back to Auslander ([17, §3], which is more general, with any small abelian category in place of mod-R). Lenzing gives this result a central role in [398], a survey paper which also touches on a good number of the topics in this book.

Corollary 11.1.42. *Suppose that $\mathcal{C}$ is locally coherent and that τ is a hereditary torsion theory on $\mathcal{C}$ of finite type. Then the restriction of the corresponding localisation Q_τ to $\mathcal{C}^{\text{fp}}$ is the localisation of $\mathcal{C}^{\text{fp}}$ at the Serre subcategory $\mathcal{T}_\tau \cap \mathcal{C}^{\text{fp}}$.*

11.1.5 Pp conditions in locally finitely presented categories

The effect of localisation on lattices of pp conditions is the natural one (11.1.43).

In a locally coherent category absolutely pure objects have both elimination of quantifiers and elimination of imaginaries (11.1.44).

The definition of pp condition was extended from module categories to functor categories in Section 10.2.4. Here we make some remarks on how to extend to general locally finitely presented abelian categories. We are fairly brief here since there is a more extended discussion of all this in Appendix B.

Let $\mathcal{C}$ be a locally finitely presented abelian category, generated by a small full subcategory $\mathcal{G} \subseteq \mathcal{C}^{\text{fp}}$. If some $\mathcal{G}$ is only skeletally small, then replace it by a small version. We can set up the notion of pp condition as in Section 10.2.4, with a sort for each object of $\mathcal{G}$ and a unary function symbol for each morphism in $\mathcal{G}$. This, of course, is the language we set up for Mod-$\mathcal{G}$ and the only difference is in the ordering of pp conditions for $\mathcal{C}$ as opposed to Mod-$\mathcal{G}$: the ordering – $\psi \leq \phi$ iff $\psi(M) \leq \phi(M)$ for every M – will depend on whether we take $M \in \text{Mod-}\mathcal{G}$ or $M \in \mathcal{C}$, with the lattice obtained from the latter being a quotient of that obtained from the former (cf. 3.2.15). For, by 11.1.25 and 11.1.21, $\mathcal{C}$ is both a localisation of and a definable subcategory of Mod-$\mathcal{G}$, that is, is definable using pairs of such pp conditions, so we may simply regard pp conditions, pp-types etc. for $\mathcal{C}$ as being those for Mod-$\mathcal{G}$ relativised to $\mathcal{C}$.

It is the case, see Section 18.1.4, that "the definable structure" on $\mathcal{C}$, in particular pp conditions as certain types of functors on $\mathcal{C}$, can be expressed just in terms of $\mathcal{C}$, as opposed to being defined in terms of some particular representation of $\mathcal{C}$ as a localisation of/definable subcategory of a functor category. Indeed, most associated data are independent of choice of $\mathcal{G}$ (the default choice for generating set $\mathcal{G}$ will be $\mathcal{C}^{\text{fp}}$): if $\mathcal{G}$ and $\mathcal{H}$ both are generating sets of finitely presented objects of $\mathcal{C}$, then any pp condition in the language based on $\mathcal{G}$ can be expressed as a pp condition in the language based on $\mathcal{H}$; this is discussed more in Appendix B.

Thus it is straightforward to extend everything to this context.

For instance, the Ziegler spectrum, $\mathrm{Zg}(\mathcal{C})$, of $\mathcal{C}$ will be a closed subspace of $\mathrm{Zg}_{\mathcal{G}}$, where $\mathcal{G} \subseteq \mathcal{C}^{\mathrm{fp}}$ is any choice of generating set. A basis of open sets for $\mathrm{Zg}(\mathcal{C})$ consists of the sets $(f) = \{N \in \mathrm{Zg}(\mathcal{C}) : (f, N) : (B, N) \to (A, N) \text{ is not epi}\}$ as $f : A \to B$ ranges over morphisms in $\mathcal{C}^{\mathrm{fp}}$. One may prove this via the representation of $\mathcal{C}$ as a localisation of Mod-$\mathcal{G}$ or just prove it directly, more or less as in the case of the category of modules over a ring.

We will not formulate, let alone prove, even the main results in this generality, but we give a couple of examples. Here is one which generalises 10.2.17 and 12.3.18. The proof is just like that in the ring case (one checks that the results leading to 12.3.18 work with $((-, G), -)$ in place of $(R_R, -)$).

Proposition 11.1.43. *Let $\mathcal{C}$ be a locally finitely presented abelian category, generated by $\mathcal{G} \subseteq \mathcal{C}^{\mathrm{fp}}$. Let $\tau = \tau_{\mathcal{C}}$ be the hereditary torsion theory of finite type on* (mod-$\mathcal{G}$, **Ab**) *corresponding (in the sense of 12.4.1) to $\mathcal{C} \subseteq$* Mod-$\mathcal{G}$. *Let $G \in \mathcal{G}$. Then the lattice of pp conditions in $\mathcal{C}$ with (specified) free variable of sort G is isomorphic to the lattice of finitely generated (= finitely presented) subfunctors of the localisation, $(G, -)_\tau$, of $(G, -) \in$* (mod-$\mathcal{G}$, **Ab**) *at τ.*

The first part of the next example of such a result may be compared with 2.3.3 and generalises 10.2.40.

Theorem 11.1.44. *Let $\mathcal{C}$ be a locally coherent abelian category.*

For each pp condition, ϕ, for objects of $\mathcal{C}$ there is a finitely presented object, A, of $\mathcal{C}$ such that $\phi(M) \simeq (A, M)$ for every absolutely pure, in particular for every injective, object M of $\mathcal{C}$.

For each pp-pair $\phi \geq \psi$ there is a finitely presented object A' of $\mathcal{C}$ such that $\phi(M)/\psi(M) \simeq (A', M)$ for every absolutely pure object M of $\mathcal{C}$.

Proof. Let $(C, \overline{c})$ be a free realisation of ϕ and set A to be the subobject of C generated by $\overline{c}$. Given any absolutely pure object M and $\overline{m} \in \phi(M)$ there is $f : C \to M$ taking $\overline{c}$ to $\overline{m}$. The restriction of f to A is an element of (A, M). Conversely, every morphism $g : A \to M$ extends, since M is absolutely pure and by 2.3.1, to a morphism $g : C \to M$ and hence gives an element, $g\overline{c}$, of $\phi(M)$. The correspondence is clearly a bijection since $\overline{c}$ generates A.

Continuing with the above notation, let $(D, \overline{d})$ be a free realisation of ψ and set B to be the subobject generated by $\overline{d}$. Since $\psi \leq \phi$, there is a morphism from C to D taking $\overline{c}$ to $\overline{d}$: this restricts to an epimorphism, h say, from A to B. Let $A' = \ker(h)$. Apply $(-, M)$ to the exact sequence $0 \to A' \to A \to B \to 0$

to obtain the exact sequence $0 \to (B, M) \to (A, M) \to (A', M) \to \mathrm{Ext}^1(B, M)$, where the last is zero (since B is a finitely generated subobject of $C \in \mathcal{C}^{\mathrm{fp}}$, $\mathcal{C}$ is locally coherent and M is absolutely pure). So $(A', M) \simeq (A, M)/(B, M) \simeq \phi(M)/\psi(M)$ (by the first part), as required. □

The above result has a strong model-theoretic significance. The first statement is "positive elimination of quantifiers" for absolutely pure objects, meaning that for every pp condition there is a quantifier-free pp condition to which it is equivalent in every absolutely pure object. The second statement is "pp-elimination of imaginaries", meaning that every sort in the category of pp-sorts is definably isomorphic, in every absolutely pure object, to a subsort of a finite product of "representable sorts". Note that one can go from the first to the second by the observation that $\overline{a} \mapsto \overline{a}G$ is a pp-definable bijection from the sort $(\overline{x} = \overline{x})/(\overline{x}G = 0)$ to a subsort of a finite product of representable sorts.

Absolutely pure objects are not as special as they might seem since the embedding of $\mathcal{C}$ into $((\mathcal{C}^{\mathrm{fp}})^{\mathrm{op}}, \mathbf{Ab})$ takes the objects of $\mathcal{C}$ to precisely the absolutely pure objects of this functor category, see 12.1.6 (and 18.1.6 for the general case). This is used by Baer ([31, §2], also see [144, 3.4, Lemma 2]) to present pp-definable subgroups of modules as essentially quantifier-free definable when one moves to the functor category.

There is also a duality of pp conditions, which is discussed in Section 16.1.3, after the "dual" category of $\mathcal{C}$ has been defined – this is the category which is to $\mathcal{C}$ as R-Mod is to Mod-R.

11.2 Serre subcategories and ideals

Annihilation of ideals of mod-R *by finitely presented functors is considered in Section 11.2.1 and the bijection induced by duality between Serre subcategories (of the "right" and "left" functor categories) is described in Section 11.2.2.*

Remark 11.2.1. Note that, just from the definition regarded as a set of closure conditions, if an object belongs to the Serre subcategory generated by a set $\mathcal{G}$ of objects, then it belongs to the Serre subcategory generated by a finite subset of $\mathcal{G}$.

11.2.1 Annihilators of ideals of mod-R

There is a criterion for the cokernel functor of a morphism to annihilate a morphism (11.2.3), hence for such a functor to annihilate an ideal of mod-R *(11.2.5). The Galois connection established by annihilation, between ideals of* mod-R *and sets of finitely presented functors, is investigated, with particular attention paid to fp-idempotent ideals and Serre subcategories (11.2.7, 11.2.8). Over artin algebras there is a neater correspondence, see Section 15.3.*

It has been remarked already, in Section 10.1.1, that a category such as mod-R, at least a small version of this, may be regarded as a ring without 1. Ideals of such "rings" have been considered in Section 9.1.1. Recall that if $\mathcal{L}$ is an ideal (left, right or two-sided) of mod-R and $A, B \in$ mod-R, then $\mathcal{L}(A, B)$ denotes the group of those morphisms from A to B which are in $\mathcal{L}$. Such an ideal also defines, for each $A \in$ mod-R, a subfunctor, $\mathcal{L}(A, -)$, of $(A, -)$ given on objects by $\mathcal{L}(A, -) \cdot B = \mathcal{L}(A, B)$.

Recall also that if $f : A \to B$ is a morphism in mod-R, then the left ideal generated by f is

$$(\text{mod-}R)f = \{gf : g \in \text{mod-}R \text{ such that } gf \text{ is defined}\}$$
$$\cup \{0 \in (C, D) : C, D \in \text{mod-}R\}.$$

The first statement below is obvious, the second restates part of 10.2.18.

Lemma 11.2.2. *(a) Let* $A \xrightarrow{f} B \in$ mod-R. *Then* $\big((\text{mod-}R)f\big)(A, -) = \text{im}(f, -)$.

(b) Let $A \xrightarrow{f} B$, $A \xrightarrow{g} C$ *be in* mod-R. *Then* $g \in \big((\text{mod-}R)f\big)(A, C)$ *iff* $\text{im}(f, -) \geq \text{im}(g, -)$.

Suppose that $\mathcal{I}$ is an ideal of mod-R. The **annihilator** of $\mathcal{I}$ is

$$\text{ann}(\mathcal{I}) = \{F \in (\text{mod-}R, \mathbf{Ab}) : F\mathcal{I} = 0, \text{ i.e. } Fg = 0 \ \forall g \in \mathcal{I}\}.$$

The following is a useful criterion for a finitely presented functor to annihilate a morphism. Recall, 10.2.1, that if $A \xrightarrow{f} B$ is in mod-R, then $F_f = \text{coker}\big((B, -) \xrightarrow{(f,-)} (A, -)\big)$ is a typical finitely presented functor on mod-R.

Lemma 11.2.3. ([358, Appx. C]) *Let* $A \xrightarrow{f} B, C \xrightarrow{h} D$ *be in* mod-R. *Then* $F_f h = 0$ *iff for all* $A \xrightarrow{g} C$ *in* mod-R *there is* $B \xrightarrow{g'} D$ *in* mod-R *such that* $hg = g'f$, *that is, iff for every* $A \xrightarrow{g} C$ *in* mod-R *one has* $hg \leq f$.

In particular, for any $C \in$ mod-R, $F_f C = 0$ *iff every morphism from* A *to* C *factors initially through* f.

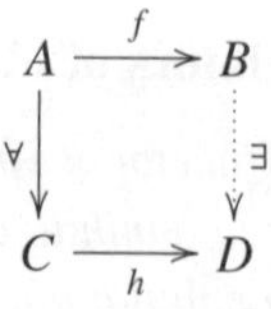

$$\begin{array}{ccc} A & \xrightarrow{f} & B \\ \forall \downarrow & & \vdots\, \exists \\ C & \xrightarrow[h]{} & D \end{array}$$

Proof. Clearly $F_fh : F_fC = \big((A, C)/\mathrm{im}(f, C)\big) \to \big((A, D)/\mathrm{im}(f, D)\big) = F_fD$ is the morphism induced by composition with h (see the diagram). So $F_fh = 0$ iff $\mathrm{im}(A, h) \le \mathrm{im}(f, D)$, as required.

$$\begin{array}{ccccccc}
(B, C) & \xrightarrow{(f,C)} & (A, C) & \longrightarrow & F_fC & \longrightarrow & 0 \\
\Big\downarrow{\scriptstyle (B,h)} & & \Big\downarrow{\scriptstyle (A,h)} & & \Big\downarrow{\scriptstyle F_fh} & & \\
(B, D) & \xrightarrow[(f,D)]{} & (A, D) & \longrightarrow & F_fD & \longrightarrow & 0
\end{array}$$

For the second statement take $h = 1_C$. □

Remark 11.2.4. Recall, 10.2.42, that if $f : A \to B$ is in mod-R, then there is a natural extension $\overrightarrow{F}_f$ of F_f to a functor in (Mod-R, **Ab**) which is also is defined by the exact sequence $(B, -) \xrightarrow{(f,-)} (A, -) \to \overrightarrow{F}_f \to 0$ but where now the representable functors are to be regarded as functors in the category (Mod-R, **Ab**). Clearly the proof of 11.2.3 applies in (Mod-R, **Ab**), so the analogue of 11.2.3 holds with Mod-R in place of mod-R and $\overrightarrow{F}_f$ in place of F_f.

Corollary 11.2.5. *Let $A \xrightarrow{f} B$ be in* mod-R *and let $\mathcal{I}$ be an ideal of* mod-R. *Then $F_f \in \mathrm{ann}(\mathcal{I})$ iff $\mathcal{I}(A, -) \le \mathrm{im}(f, -)$.*

Proof. Because $\mathcal{I}$ is a (left) ideal the criterion of 11.2.3 reduces to the assertion that $F_f\mathcal{I} = 0$ iff every morphism in $\mathcal{I}$ with domain A factors initially through f, and this is, as noted in 11.2.2, simply the condition $\mathcal{I}(A, -) \le \mathrm{im}(f, -)$. □

Remark 11.2.6. If $\mathcal{I}$ is an ideal of mod-R, then $\mathrm{ann}(\mathcal{I})$ is closed under subfunctors and epimorphic images. For, if $0 \to F' \xrightarrow{j} F$ is exact and if $F \in \mathrm{ann}(\mathcal{I})$, then given $A \xrightarrow{h} B \in \mathcal{I}$ we have $j_B \cdot F'h = Fh \cdot j_A = 0$. Since j_B, the component at B of the natural transformation j is monic, $F'h = 0$. Similarly for images.

Consider the converse process: given a class of functors $\mathcal{A} \subseteq (\text{mod-}R, \mathbf{Ab})^{\mathrm{fp}}$, let $\mathrm{ann}(\mathcal{A}) = \{h \in \text{mod-}R : Fh = 0 \text{ for all } F \in \mathcal{A}\}$ be the annihilator of $\mathcal{A}$ in mod-R. Clearly $\mathrm{ann}(\mathcal{A})$ is a two-sided ideal of mod-R (for $F(ghk) = FgFhFk$). An object, A, of mod-R is annihilated by a functor F iff the corresponding identity morphism, 1_A, is in $\mathrm{ann}(F)$, but, in general, knowing the objects annihilated by a functor is much less than knowing the morphisms annihilated by that functor.

Following [358, §5], for each $F \in (\text{mod-}R, \mathbf{Ab})^{\mathrm{fp}}$ and $\mathcal{A} \subseteq (\text{mod-}R, \mathbf{Ab})^{\mathrm{fp}}$ set

$$r_{\mathcal{A}}F = \bigcap\{F' \le F : F' \text{ is finitely generated and } F/F' \in \mathcal{A}\}.$$

In particular, by 10.2.2,

$$r_{\mathcal{A}}(A, -) = \bigcap\{\mathrm{im}(f, -) : A \xrightarrow{f} B \text{ is in mod-}R \text{ and } F_f \in \mathcal{A}\}.$$

It follows, bearing in mind that $g \in \mathrm{im}(f, -)$ iff g factors initially through f, that $\bigcup\{r_{\mathcal{A}}(A, B) : A, B \in \text{mod-}R\}$ is a left ideal of mod-R. We abuse notation somewhat by writing $r_{\mathcal{A}}$ also for this left ideal.

Proposition 11.2.7. ([515, 2.8]) *Let $\mathcal{I}$ be an ideal of* mod-R *and set* $\mathcal{A} = \mathrm{ann}(\mathcal{I})$ *and* $\overline{\mathcal{I}} = \mathrm{ann}(\mathcal{A})$. *Then $\overline{\mathcal{I}}$ is an ideal of* mod-R *with $\mathcal{I} \le \overline{\mathcal{I}} \le r_{\mathcal{A}}$ and $\overline{\mathcal{I}}$ is the largest two-sided ideal of* mod-R *containing $\mathcal{I}$ and contained in the left ideal $r_{\mathcal{A}}$.*

Proof. Clearly $\mathcal{I} \le \overline{\mathcal{I}}$. If $h \in \overline{\mathcal{I}}(A, C)$ and if $A \xrightarrow{f} B$ in mod-R is such that $F_f \in \mathcal{A}$, then $F_f h = 0$, so, by 11.2.3, h factors initially through f, that is, $h \in \mathrm{im}(f, -) \cdot C$. Therefore $\overline{\mathcal{I}}(A, -) \le \mathrm{im}(f, -)$ for all such f, that is, $\overline{\mathcal{I}}(A, -) \le r_{\mathcal{A}}(A, -)$.

There is a largest ideal, $\mathcal{I}'$ say, as described (because any sum of ideals is an ideal), so $\overline{\mathcal{I}}$ is contained in $\mathcal{I}'$. Let $C \xrightarrow{h} D \in \mathcal{I}'$ and let $A \xrightarrow{f} B$ be in mod-R with $F_f \in \mathcal{A}$. Let $A \xrightarrow{g} C$ be in mod-R. Since $\mathcal{I}'$ is assumed to be an ideal, $hg \in \mathcal{I}'(A, D) \le r_{\mathcal{A}}(A, D)$, so $hg \in \mathrm{im}(f, -) \cdot D$. Hence hg factors initially through f for every such g. Therefore, by 11.2.3, $F_f h = 0$ and $h \in \overline{\mathcal{I}}$, as required. □

Note that $\mathcal{A} = \mathrm{ann}(\overline{\mathcal{I}})$, where notation is as above.

If $\mathcal{I}$ is an ideal of mod-R, then, in view of 11.2.6, $\mathrm{ann}(\mathcal{I})$ is a Serre subcatgory of (mod-R, $\mathbf{Ab}$)^{fp} exactly if it is closed under extensions, that is, exactly if $\mathcal{I}$ is fp-idempotent (Section 9.2). In this case we will also use the notation $\mathcal{S}_{\mathcal{I}}$ for $\mathrm{ann}(\mathcal{I})$.

Proposition 11.2.8. ([515, 2.10], cf. [358, 5.4(3)]) *Let R be any preadditive category and suppose that $\mathcal{S}$ is a Serre subcategory of* (mod-R, $\mathbf{Ab}$)^{fp}. *Then $r_{\mathcal{S}}$ is a two-sided ideal of* mod-R. *In particular, if $\mathcal{I}$ is any fp-idempotent ideal of* mod-R, *then* $\overline{\mathcal{I}} = r_{\mathrm{ann}(\mathcal{I})}$, *where* $\overline{\mathcal{I}} = \mathrm{ann}(\mathrm{ann}(\mathcal{I}))$.

Proof. As noted before 11.2.7, $r_{\mathcal{S}}$ is a left ideal of mod-R.

Let $F \in$ (mod-R, $\mathbf{Ab}$)^{fp} and let G be a finitely generated subfunctor of F. We claim that $r_{\mathcal{S}}G \le r_{\mathcal{S}}F$.

To see this, let $F' \le F$ be $\mathcal{S}$-dense in F (that is, $F/F' \in \mathcal{S}$, see 11.1.36, which applies by 10.2.3)) and finitely generated. Then $G/(F' \cap G)$ is a subfunctor of F/F', so $F' \cap G$ is $\mathcal{S}$-dense in G. Since $F' \cap G$ is also finitely generated (because, 10.2.3, F is a coherent functor) $r_{\mathcal{S}}G \le F' \cap G \le F'$. Therefore $r_{\mathcal{S}}G \le F'$ for all such F' and hence $r_{\mathcal{S}}G \le r_{\mathcal{S}}F = \bigcap\{F' \le F : F'$ is finitely generated and $\mathcal{S}$-dense in $F\}$.

Now let $A \xrightarrow{f} B \in r_{\mathcal{S}}(A, B)$ and let $D \xrightarrow{g} A$ be in mod-R. We show $fg \in r_{\mathcal{S}}(D, B)$ by proving that $\mathrm{im}(fg, -) \le r_{\mathcal{S}}(D, -)$, which, since $fg = (fg, -)1_B$, will be enough. Since $f \in r_{\mathcal{S}}$ certainly $\mathcal{L}_f \le r_{\mathcal{S}}$, so, by 11.2.2(1), $\mathrm{im}(f, -) \le r_{\mathcal{S}}(A, -)$. Let $F' \le \mathrm{im}(g, -)$ be finitely generated and $\mathcal{S}$-dense in $\mathrm{im}(g, -)$. Then its full inverse image under $(g, -)$ is $\mathcal{S}$-dense in $(A, -)$, therefore contains a finitely generated functor which is $\mathcal{S}$-dense in $(A, -)$ (since, see 11.1.14, the torsion theory generated by $\mathcal{S}$ is of finite type), hence contains $r_{\mathcal{S}}(A, -)$ and hence contains $\mathrm{im}(f, -)$. Therefore $(g, -) \cdot \mathrm{im}(f, -) \leq F'$. This is so for all such F', so $\mathrm{im}(fg, -) = (g, -) \cdot \mathrm{im}(f, -) \leq r_{\mathcal{S}}(\mathrm{im}(g, -)) \leq r_{\mathcal{S}}(D, -)$ by the first part of the proof, as required. □

This will be used in Section 15.3.

11.2.2 Duality of Serre subcategories

There is a bijection between Serre subcategories of finitely presented functors on the right and on the left (11.2.9), hence also between finite type localisations (11.2.10).

Let $\mathcal{S}$ be a subcategory of $(R\text{-mod}, \mathbf{Ab})^{\mathrm{fp}}$. Define $d\mathcal{S} = \{dF : F \in \mathcal{S}\}$. The next result is, modulo 10.2.30, [274, 2.9].

Theorem 11.2.9. *Let R be a small preadditive category. If $\mathcal{S}$ is a Serre subcategory of* $(\text{mod-}R, \mathbf{Ab})^{\mathrm{fp}}$*, then $d\mathcal{S}$ is a Serre subcategory of* $(R\text{-mod}, \mathbf{Ab})^{\mathrm{fp}}$ *and $d^2\mathcal{S} = \mathcal{S}$. Therefore there is an inclusion-preserving bijection between Serre subcategories of* $(R\text{-mod}, \mathbf{Ab})^{\mathrm{fp}}$ *and* $(\text{mod-}R, \mathbf{Ab})^{\mathrm{fp}}$.

Proof. Suppose that $0 \to F \to G \to H \to 0$ is an exact sequence in $(R\text{-mod}, \mathbf{Ab})^{\mathrm{fp}}$. By 10.3.4, $0 \to dF \to dG \to dH \to 0$ is an exact sequence in $(\text{mod-}R, \mathbf{Ab})^{\mathrm{fp}}$ and every exact sequence in this category is of this form. If $G \in d\mathcal{S}$, then (10.3.4) $dG \in \mathcal{S}$, so, since $\mathcal{S}$ is Serre, $dF, dH \in \mathcal{S}$. Therefore, recall $d^2 = 1$, $F, H \in d\mathcal{S}$. If $F, H \in d\mathcal{S}$, then $dF, dH \in \mathcal{S}$, so, since $\mathcal{S}$ is Serre, $dG \in \mathcal{S}$. Therefore $G \in d\mathcal{S}$. The remaining statements follow from the fact, 10.3.4, that d is a duality. □

If $\mathcal{S}$ is a Serre subcategory of $(\text{mod-}R, \mathbf{Ab})^{\mathrm{fp}}$, then $d\mathcal{S}$ is the Serre subcategory **dual to** $\mathcal{S}$. The next result follows from 11.1.38. The next result is, in an equivalent form, [274, 4.4], and, in this form, is [278, 5.5].

Corollary 11.2.10. *For any small preadditive category R there is a bijection between torsion theories of finite type on* $(\text{mod-}R, \mathbf{Ab})$ *and torsion theories of finite type on* $(R\text{-mod}, \mathbf{Ab})$.

This bijection between Serre subcategories, equivalently torsion theories of finite type, can be extended to arbitrary locally coherent categories in place of the locally coherent functor categories above, for example, use the discussion before 18.1.2.

12
The Ziegler spectrum and injective functors

This could be seen as the technical heart of the book: the category of modules is embedded in a functor category, pure-exact sequences become simply exact sequences, pp conditions and modules now become the same kind of object. Restriction to closed subsets/definable subcategories is exactly finite-type localisation. Rings of definable scalars are shown to be endomorphism rings of localised functors.

12.1 Making modules functors

The category of modules is embedded into the functor category and the main properties of this embedding established in Section 12.1.1. The injective objects of the subcategory of finitely presented functors are identified in Section 12.1.2. The embedding allows the Ziegler spectrum to be realised as a topology on the set of indecomposable injective functors (Section 12.1.3).

12.1.1 The tensor embedding

Tensor product over rings with many objects is defined and illustrated. Then the embedding of the category of right modules into the category of functors on finitely presented left modules via tensor product is defined. This embedding is full, commutes with direct limits and products, and is left adjoint to evaluation at R (12.1.3). It takes pure-exact sequences to (pure-)exact sequences (12.1.3, 12.1.6). It takes modules to the absolutely pure functors, pure-injective modules to the injective functors (12.1.6) and pure-injective hulls to injective hulls (12.1.8). The absolutely pure objects of the functor category are precisely the right exact functors (12.1.7).

In consequence there is a bijection between points of the right Ziegler spectrum and isomorphism classes of indecomposable injective functors on finitely presented left modules (12.1.9).

The image of the embedding is a definable subcategory of the functor category (12.1.12).

As things stand we have, on the one hand, a module category Mod-R and, on the other hand, an associated functor category. We describe now how to move the whole module category into a functor category, with the result that modules and functors on finitely presented modules become objects of the same kind. In fact, it is not quite so simple because right modules become functors on finitely presented *left* modules. But duality (Section 10.3) gives a strong link with functors on finitely presented right modules. This embedding, which was defined by Gruson and Jensen ([254], but see especially [256] for its development) is central to the functorial approach to purity and the Ziegler spectrum. We remark that there is also the usual, but quite different, contravariant embedding of the module category into the functor category via representable functors (10.1.8); there are some comments on this at the end of the section.

We continue to use R to denote a small preadditive category, not necessarily a ring, although we often simplify notation and argument in a non-essential way by dealing with the case where R is a ring. The embedding functor uses tensor product, so first we describe the tensor product of R-modules. That is (see Section 10.1.1), we describe $M \otimes_R L$, where L is a covariant functor and M is a contravariant functor from the category R to **Ab**.

The functor $M \otimes_R -$ from R-Mod to **Ab** is described as follows. On representable (covariant) functors $(G, -) \in R$-Mod, where G is an object of R, that is, on projectives from the standard generating set of R-Mod, set $(M \otimes_R -)(G, -) = MG$. If $G \xrightarrow{r} H$ is a morphism in R, so $(H, -) \xrightarrow{(r,-)} (G, -)$ is a typical morphism between representables, set $(M \otimes_R -)(r, -) = Mr\ (:MH \to MG)$ (remember that M is contravariant on R). The requirement that $M \otimes_R -$ commute with arbitrary direct sums and be right exact then determines the action of $M \otimes_R -$ completely, because every $L \in R$-Mod is an epimorphic image of a direct sum of representable functors, 10.1.13. In more detail, if $\bigoplus_i (G_i, -) \to L \to 0$ is exact in R-Mod, then $M \otimes_R L$ is defined by exactness of $\bigoplus_i MG_i \to M \otimes_R L \to 0$. Similarly for the action on morphisms in R-Mod: use that $\bigoplus_i (G_i, -)$ is (10.1.12) projective to lift a morphism $g : L \to K$ to a morphism from $\bigoplus_i (G_i, -)$ to a projective similarly mapping onto K and then evaluate at M. Note that by using such presentations we obtain a formulation in terms of elements (or "elements", meaning morphisms from a representable functor), namely $(M \otimes g)(\sum_i a_i \otimes l_i) = \sum_i a_i \otimes g(l_i)$.

The functor $- \otimes_R L$ is defined similarly. We work through an example to make this a bit more concrete.

Example 12.1.1. Let $\mathcal{A}$ be the path category of the quiver A_2, which is $1 \xrightarrow{\alpha} 2$, over a field k. So $\mathcal{A}$ has two objects, label them 1 and 2, and each of End(1), End(2), Hom(1, 2) is one-dimensional over k. A left $\mathcal{A}$-module L is $L1 \xrightarrow{L\alpha} L2$, where each of $L1$, $L2$ is a k-vectorspace and $L\alpha$ is a k-linear transformation. A right $\mathcal{A}$-module M has the form $M1 \xleftarrow{M\alpha} M2$. The indecomposable projectives of $\mathcal{A}$-Mod are, as representable functors, $(1, -)$ and $(2, -)$, and in the above form, $k \xrightarrow{1} k$, and $0 \to k$.

Given $M \in \text{Mod-}\mathcal{A}$ the definition of $\otimes$ gives $(1, -) \otimes M = M1$ and $(2, -) \otimes M = M2$. Since the quiver A_2 is of finite representation type, with just the three obvious indecomposable representations, every $L \in \mathcal{A}\text{-Mod}$ has the form $(1, -)^{(\kappa)} \oplus (2, -)^{(\lambda)} \oplus ((1, -)/(2, -))^{(\mu)}$ for some cardinals κ, λ, μ. So, in order to compute the action of $- \otimes M$ on arbitrary left $\mathcal{A}$-modules, it remains to compute $((1, -)/(2, -)) \otimes M$.

Consider the projective presentation

$$(2, -) \xrightarrow{(\alpha, -)} (1, -) \to L = ((1, -)/(2, -)) \to 0.$$

Since tensor is right exact, this gives the exact sequence

$$(2, -) \otimes M \xrightarrow{(\alpha, -)\otimes 1_M} (1, -) \otimes M \to L \otimes M \to 0,$$

that is,

$$M2 \xrightarrow{M\alpha} M1 \to M1/\text{im}(M\alpha) \to 0$$

and we see that $((1, -)/(2, -)) \otimes M \simeq M1/\text{im}(M\alpha)$.

Since one may construe these as modules over the path algebra, with $\mathcal{A}$-Mod corresponding to the category of right modules over the path algebra (see 10.1.2), one could continue and compare this with the usual description of tensor product of modules. The quivers A_∞ and $\widetilde{A_1}$ give examples with more going on but this simple example should convey the idea.

Here is how to convert right modules into functors on finitely presented left modules.

Given any right R-module M, there is the corresponding tensor functor, $(M \otimes -) = (M \otimes_R -) : R\text{-mod} \longrightarrow \mathbf{Ab}$, given by $(M \otimes -)(L) = M \otimes_R L$ on objects $L \in R\text{-mod}$ and with the obvious effect on morphisms: if $g : L \to K \in R\text{-mod}$, then $(M \otimes -)g = M \otimes_R g : M \otimes L \to M \otimes K$.

Thus from $M \in \text{Mod-}R$ we obtain $(M \otimes -) \in (R\text{-mod}, \mathbf{Ab})$.

Define the functor $\epsilon : \text{Mod-}R \longrightarrow (R\text{-mod}, \mathbf{Ab})$ by $\epsilon M = M \otimes -$ on objects and, if $M \xrightarrow{f} N$ is a morphism of right R-modules, then $\epsilon f : (M \otimes -) \to (N \otimes -)$ is the natural transformation whose component at $L \in R$-mod is defined to be $f \otimes 1_L : M \otimes L \to N \otimes L$, that is, $(- \otimes_R L)f$.

Lemma 12.1.2. *Suppose that $F, F' \in (R\text{-mod}, \mathbf{Ab})$ with F right exact and let $\tau, \tau' : F \to F'$ be natural transformations. If $\tau_R = \tau'_R$, then $\tau = \tau'$. If R has more than one object, then the hypothesis should be interpreted as $\tau_G = \tau'_G$ for each object G of R.*

Proof. Let L be a finitely presented left module, say $(R^m \to)R^n \xrightarrow{\pi} L \to 0$ is exact (if R has more than one object, replace the projective modules R^m, R^n by direct sums of projectives of the form $(-, G)$ with $G \in R$). There is a commutative diagram with the top row exact

$$\begin{array}{ccccc} FR^n & \xrightarrow{F\pi} & FL & \longrightarrow & 0 \\ \downarrow{\scriptstyle \tau^{(\prime)}_{R^n}} & & \tau_L \Vert \tau'_L & & \\ F'R^n & \xrightarrow[F'\pi]{} & F'L & \longrightarrow & 0 \end{array}$$

and with vertical maps being $\tau_{R^n} = (\tau_R)^n = (\tau'_R)^n = \tau'_{R^n}$ and either τ_L or τ'_L for the other. It follows that $\tau_L = \tau'_L$ since $\tau_L \cdot F\pi = F'\pi \cdot \tau_{R^n} = F'\pi \cdot \tau'_{R^n} = \tau'_L \cdot F\pi$ and $F\pi$ is an epimorphism. $\square$

The basic results about $M \mapsto M \otimes -$ are stated (for rings) in [254] and some more detail may be found in [256, §1] (also see the exposition in [323, B16]).

Theorem 12.1.3. *Let R be a small preadditive category. The functor $\epsilon : \text{Mod-}R \longrightarrow (R\text{-mod}, \mathbf{Ab})$ given on objects by $M \mapsto M \otimes -$ is a full embedding and is left adjoint to the functor "evaluation at R" from $(R\text{-mod}, \mathbf{Ab})$ to Mod-R: so, if R is a ring, $(M \otimes -, F) \simeq (M, F(R))$.*

A sequence $0 \to M \to N \to N' \to 0$ of right R-modules is pure-exact iff the image $0 \to \epsilon M \to \epsilon N \to \epsilon N' \to 0$ is exact. Furthermore, ϵ commutes with direct limits and products.

Proof. If $(M \otimes -) \simeq (N \otimes -)$, then, evaluating at ${}_RR$, we obtain $M \simeq M \otimes_R R \simeq N \otimes_R R \simeq N$. If $\tau : (M \otimes -) \to (N \otimes -)$ is a morphism, then its component, τ_R, at ${}_RR$ is a morphism from $M \simeq M \otimes_R R$ to $N \simeq N \otimes_R R$. The natural transformations τ and $\tau_R \otimes -$ agree at R so, since $M \otimes -$ and $N \otimes -$ are right exact, 12.1.2 yields $\tau = \tau_R \otimes -$. Therefore ϵ is a full embedding.

For the adjointness, first note that if $F \in (R\text{-mod}, \mathbf{Ab})$, then, since $\text{End}({}_RR) = R$ (acting on the right), $F({}_RR)$ has the structure of a right R-module: if $a \in F({}_RR)$ and $s \in R$, then set $as = F(- \times s) \cdot a$ and note that $a(st) =$

$F(- \times st) \cdot a = F((- \times t)(- \times s)) \cdot a = (as)t$. The natural isomorphism $(M \otimes -, F) \simeq (M, FR)$ takes $\tau \in (M \otimes -, F)$ to τ_R. By 12.1.2 this map is monic. To define the inverse map, given $g : M \to FR$, let τ_g be the natural transformation from $M \otimes -$ to F the component of which at $L \in R$-mod is defined by taking $m \otimes l \in M \otimes L$ to $Fl \cdot g(m)$, where Fl denotes the value of F at the morphism from ${}_R R$ to L which takes 1 to l. One may check that τ_g is a natural transformation, that $(\tau_g)_R = g$ and that these processes do define an adjunction. If R is not a ring, then the "evaluation of F at R" means the right R-module, that is the functor from R^{op} to **Ab**, which takes an object G of R to $F(G, -)$ and which has the obvious effect on morphisms (recall, 10.1.8, that the embedding of R into R-Mod $= (R, \mathbf{Ab})$ is contravariant).

The assertion concerning pure exact sequences is 2.1.29 (and 10.1.6).

For every $L \in R$-mod the functor $- \otimes L$, being a left adjoint, commutes with direct limits and, from this, the first part of the final statement follows easily.[1]

The assertion about products follows directly from 1.3.24. □

Example 12.1.4. The natural embedding, j, of the $\mathbb{Z}$-module $\mathbb{Z}$ into the $\mathbb{Z}$-module $\mathbb{Q}$ is not pure: consider the equation $x2 = 1$, where $2 \in \mathbb{Z}$ ($\mathbb{Z}$ as a ring) and $1 \in \mathbb{Z}$ ($\mathbb{Z}$ as a module). By 12.1.3 the natural transformation $j \otimes - : (\mathbb{Z} \otimes -) \to (\mathbb{Q} \otimes -)$ is not monic: we demonstrate this explicitly.

A morphism $\tau : F \to G$ in $(\mathbf{Ab}, \mathbf{Ab})$ is monic iff for every $L \in \mathbf{Ab}$ the component $\tau_L : FL \to GL$ is monic (10.1.5). Set $L = \mathbb{Z}_2$ (that is, $\mathbb{Z}/2\mathbb{Z}$). Then $\mathbb{Z} \otimes \mathbb{Z}_2 \simeq \mathbb{Z}_2$ but $\mathbb{Q} \otimes \mathbb{Z}_2 = 0$ because every element of $\mathbb{Q} \otimes \mathbb{Z}_2$ is a linear combination of elements of the form $a \otimes 1_2$, and $a \otimes 1_2 = \frac{1}{2}a.2 \otimes 1_2 = \frac{1}{2}a \otimes 2.1_2 = \frac{1}{2}a \otimes 0 = 0$. Thus we see that $j \otimes L : \mathbb{Z} \otimes \mathbb{Z}_2 \to \mathbb{Q} \otimes \mathbb{Z}_2$ is not monic.

The next result is immediate from the fact that ϵ is full and faithful, hence preserves and reflects idempotents (that is, projections to direct summands).

Corollary 12.1.5. *A module M is indecomposable iff the corresponding functor $M \otimes -$ in $(R\text{-mod}, \mathbf{Ab})$ is indecomposable.*

[1] I really should draw attention to the increasingly frequent sleight of hand by which all the concepts that we have discussed for modules over rings are now being applied to modules over small preadditive categories. In Part I, R was said to be a ring. Logically, we should have made R a small preadditive category from the outset, but, pedagogically, this might well have been a mistake since it would have made the material less accessible. I am now making the sweeping statement that everything which worked for rings works in the more general context and the only modifications needed are the obvious ones. Scattered around there are comments addressing this point but there is by now a large logical gap in the exposition (at least, if one cares about the more general case). It is, however, left to the reader to check these points. Usually rather little needs to be done: categorical formulations most clearly generalise but so do formulations involving elements (see Sections 10.2.4 and B.1). In particular, the various definitions and results around purity from Section 2.1.1 make sense and are valid in these functor categories. An example where such extension (of results from [174] and [631], see Section 3.4.3) is carried out in detail is [335].

Theorem 12.1.6. ([254], [256, §1]) *An exact sequence* $0 \to M \to N \to N' \to 0$ *in* Mod-R *is a pure exact sequence iff the sequence* $0 \to \epsilon M \to \epsilon N \to \epsilon N' \to 0$ *is a pure exact sequence in* (R-mod, **Ab**).

If M is a right R-module, then $\epsilon M = M \otimes -$ is an absolutely pure object of (R-mod, **Ab**), *indeed every absolutely pure functor is isomorphic to one of this form.*

Furthermore, $M \otimes -$ is injective iff M is pure-injective.

Proof. The given pure exact sequence is, 2.1.4, a direct limit of split exact sequences. Since ϵ commutes with direct limits (12.1.3) the image sequence is a direct limit of split exact sequences, hence, 2.1.3, is pure exact. By 12.1.3 we also have the converse.

Next we show that $Q \in (R\text{-mod}, \mathbf{Ab})$ is absolutely pure iff Q is a right exact functor.

Let $F \in (R\text{-mod}, \mathbf{Ab})^{\mathrm{fp}}$ have projective presentation $0 \to (C, -) \to (B, -) \to (A, -) \to F \to 0$, where $A \to B \to C \to 0$ is an exact sequence in R-mod (see the proof of 10.2.5). Then the homology groups of the chain complex $0 \to ((A, -), Q) \to ((B, -), Q) \to ((C, -), Q) \to 0$, that is, (by Yoneda) $0 \to QA \to QB \to QC \to 0$, are, by definition (see Appendix E, p. 706), precisely (F, Q), $\mathrm{Ext}^1(F, Q)$ and $\mathrm{Ext}^2(F, Q)$. Therefore Q is right exact iff for all finitely presented functors F we have $\mathrm{Ext}^1(F, Q) = 0 = \mathrm{Ext}^2(F, Q)$. We have $\mathrm{Ext}^2(F, -) \simeq \mathrm{Ext}^1(F', -)$, where, with the notation above, F' is the kernel of $(A, -) \to F$. Since (10.2.3) (R-mod, **Ab**) is locally coherent F' also is finitely presented. So the condition on Q reduces to $\mathrm{Ext}^1(F, Q) = 0$ for all finitely presented F, that is (2.3.1), Q is an absolutely pure functor.

It follows that every functor of the form $M \otimes -$ is absolutely pure.

For the converse, suppose that Q is absolutely pure, hence, as shown above, right exact. Note that $Q(_RR)$ is a right ($\mathrm{End}(_RR) \simeq$)R-module. Define the natural transformation $(Q(R) \otimes -) \to Q$ to have component at $L \in R$-mod the map $Q(R) \otimes L \to QL$ defined by taking $m \otimes l$ ($m \in Q(R)$, $l \in L$) to $Q(l : {_RR} \to L) \cdot m$ (a special case of that in the proof of adjointness in 12.1.3). It is straightforward to check that this is, indeed, a natural transformation and that at $L = R$ it is an isomorphism. So, by 12.1.2, this is an isomorphism from Q to $Q(R) \otimes -$, as required.

Given $M \in$ Mod-R, there is the pure-exact sequence $0 \to M \to H(M) \to H(M)/M \to 0$, where $H(M)$ is the pure-injective hull (Section 4.3.3) of M, hence, by the first part, the sequence of functors $0 \to (M \otimes -) \to (H(M) \otimes -) \to (H(M)/M \otimes -) \to 0$ is pure-exact. If $M \otimes -$ is injective, then this sequence is split and so, therefore (by 12.1.3, ϵ is full), is the first, whence M is a direct summand of $H(M)$, hence is pure-injective (and equal to $H(M)$). For the converse, take an injective hull, $E(M \otimes -)$, of $M \otimes -$. Since any injective is absolutely pure, $E(M \otimes -) \simeq (N \otimes -)$ for some $N \in$ Mod-R. So there is an exact sequence $0 \to (M \otimes -) \to (N \otimes -) \to (N \otimes -)/(M \otimes -) \simeq (N/M \otimes -) \to 0$ (the isomorphism by right exactness of $\otimes$) and, therefore, by the first part, there is the pure-exact sequence $0 \to M \to N \to N/M \to 0$ in Mod-R. If M is pure-injective this sequence is split, so the sequence of functors is split and, therefore, $M \otimes -$ is already injective. □

We record the following point which was established in the course of the proof above.

Proposition 12.1.7. *An object of* (R-mod, **Ab**) *is absolutely pure iff it is a right exact functor.*

Corollary 12.1.8. *The embedding $M \to N$ is a pure-injective hull in* Mod-R *iff $(M \otimes -) \to (N \otimes -)$ is an injective hull in* (R-mod, **Ab**)*: $E(M \otimes -) \simeq (H(M) \otimes -)$.*

Corollary 12.1.9. *There is a bijection between isomorphism classes of indecomposable pure-injective right R-modules, N, and isomorphism classes of indecomposable injective objects, Q, in the functor category* (R-mod, **Ab**)*, given by $N \mapsto (N \otimes -)$ and $Q \mapsto Q({}_R R)$.*

Corollary 12.1.10. *If N is pure-injective and $S = \mathrm{End}(N)$, then $S/J(S)$ is von Neumann regular, self-injective and* **idempotents lift modulo** *$J(S)$ (that is, if $s \in S$ is such that $s^2 - s \in J(S)$, then there is $t \in S$ with $t^2 = t$ and $s - t \in J(S)$).*

Proof. By 12.1.3 and 12.1.6 the endomorphism ring of a pure-injective is the endomorphism ring of an injective object in a Grothendieck category and such a ring has, by, for example, [187, 19.27], the stated properties. □

Corollary 12.1.11. *The bijection above induces a bijection between:*

isomorphism classes of indecomposable Σ-pure-injective right R-modules; and

isomorphism classes of indecomposable Σ-injective objects in the functor category.

It also induces a bijection between:

isomorphism classes of indecomposable right R-modules of finite endolength; and

isomorphism classes of indecomposable injective objects of finite endolength in the functor category.

Proof. The first statement follows from the fact, 12.1.3, that the embedding preserves direct sums and direct products and from the definition of Σ-pure-injective (and 4.4.16). For the second statement we may, for example, use that fact that an indecomposable Σ-pure-injective is of finite endolength iff it is a closed point of the Ziegler spectrum (5.3.17 and 5.3.23), together with the fact, which follows from 12.1.12 below, that Zg_R is homeomorphically embedded as a closed subset of the Ziegler spectrum of the functor category (that is, the set of isomorphism classes of indecomposable pure-injective functors, endowed with the Ziegler topology). □

The embedding ϵ has been used in Part I to give efficient proofs of many results on pure-injectivity, sometimes by reducing to standard results on injective objects in Grothendieck abelian categories.

Proposition 12.1.12. Mod-R *is a definable subcategory of* (R-mod, **Ab**).

Proof. The class of absolutely pure objects is always closed under products and pure subobjects (2.3.5), so, by 3.4.7, closure under direct limits has to be established. But that follows by 12.1.6 and 12.1.3, since, by those results, every directed system of absolutely pure objects of (R-mod, **Ab**) is, up to isomorphism, the image of a directed system in Mod-R and, also, the embedding of Mod-R into (R-mod, **Ab**) commutes with direct limits.

Alternatively, we may say that, since (R-mod, **Ab**) is locally coherent, this follows by 12.1.6 and 3.4.24 (at least, the latter generalised to small preadditive categories in place of R).

A further alternative proof is to show directly that this class is definable by pp-pairs in any suitable language for the functor category (Sections 10.2.4 and B.1) that is, using "scalars" from any generating subcategory of $(R\text{-mod}, \mathbf{Ab})^{\mathrm{fp}}$. We do this; let us take the language with a sort for every isomorphism type of finitely presented projective functor $(L, -)$, $L \in R$-mod. By 12.1.7 it has to be shown that right exactness of a functor is expressible by pp conditions. An object Q of the functor category is right exact as a functor on R-mod if for every exact sequence $K \xrightarrow{f} L \xrightarrow{g} M \to 0$ in R-mod the sequence $QK \xrightarrow{Qf} QL \xrightarrow{Qg} QM \to 0$, that is, by 10.1.7, $\big((K, -), Q\big) \xrightarrow{((f,-),Q)} \big((L, -), Q\big) \xrightarrow{((g,-),Q)} \big((M, -), Q\big) \to 0$, is exact. This is so iff the following pp-pairs are closed on Q, where Q considered as a structure for our chosen language:

$$\big(x_{(M,-)} = x_{(M,-)}\big) \,/\, \big(\exists y_{(L,-)}\, (x_{(M,-)} = y_{(L,-)}(g, -)\big),$$
$$\big(y_{(L,-)}(f, -) = 0\big) \,/\, \big(\exists z_{(K,-)}\, (y_{(L,-)} = z_{(M,-)}(g, -)\big).$$

□

There is another natural embedding of Mod-R into a functor category on finitely presented modules, namely the Yoneda embedding $M \mapsto (-, M) \in ((\text{mod-}R)^{\text{op}}, \mathbf{Ab})$. If R is an artin algebra, then the duality between mod-R and R-mod induces an equivalence between $((\text{mod-}R)^{\text{op}}, \mathbf{Ab})$ and $(R\text{-mod}, \mathbf{Ab})$. Herzog generalised this, [286, Thm 23] (also see [711]), by showing that there is an equivalence, given by $N \mapsto \text{Ext}^1(-, N)$, between the category of pure-injective right R-modules modulo the ideal of morphisms which factor through an injective and the category of injective objects in $((\underline{\text{mod}}\text{-}R)^{\text{op}}, \mathbf{Ab})$. Here $\underline{\text{mod}}$-R is the category mod-R modulo the ideal of morphisms which factor through a projective, see the example of stable module categories in Section 17.1. In particular, this gives a natural bijection between $\text{pinj}_R \setminus \text{inj}_R$ and $\text{inj}((\underline{\text{mod}}\text{-}R)^{\text{op}}, \mathbf{Ab})$ which extends to one between pinj_R and $\text{inj}((\text{mod-}R)^{\text{op}}, \mathbf{Ab})$. Furthermore, [286, Thm 31], $M \to N$ is a pure-injective hull of the module M iff the induced map $\text{Ext}^1(-, M) \to \text{Ext}^1(-, N)$ is an injective hull in the functor category. Compare with 12.1.9 and 12.1.8.

The natural map mod-$R \to \underline{\text{mod}}$-$R$ induces a full and faithful embedding $((\underline{\text{mod}}\text{-}R)^{\text{op}}, \mathbf{Ab}) \to (\text{mod-}R)^{\text{op}}, \mathbf{Ab})$ with image those $G : \text{mod-}R \to \mathbf{Ab}$ with $GR_R = 0$. It is the case, [286, Thm 4], that a module N is pure-injective iff the flat functor $(-, N) \in ((\text{mod-}R)^{\text{op}}, \mathbf{Ab})$ is cotorsion, where (for example, using 16.1.3) a functor $F \in ((\text{mod-}R)^{\text{op}}, \mathbf{Ab})$ is cotorsion iff $\text{Ext}^1((-, M), F) = 0$ for every $M \in \text{Mod-}R$.

It is also the case, [286, Thm 29], that the Auslander–Bridger transpose from R-$\underline{\text{mod}}$ to $(\underline{\text{mod}}\text{-}R)^{\text{op}}$ induces an equivalence between $((\underline{\text{mod}}\text{-}R)^{\text{op}}, \mathbf{Ab})$ and $(R\text{-}\underline{\text{mod}}, \mathbf{Ab})$.

For further results clarifying the relations between these various functor categories see [286].

12.1.2 Injectives in the category of finitely presented functors

The category of finitely presented functors has enough injectives (12.1.13), though there are injective hulls iff the ring is Krull–Schmidt (12.1.15). Over such rings a pp condition is irreducible iff it has a free realisation in a strongly indecomposable finitely presented module (12.1.16).

Proposition 12.1.13. [256, 5.5] *The category* $(R\text{-mod}, \mathbf{Ab})^{\text{fp}}$ *of finitely presented functors has enough injectives (Appendix E, p. 709) and the injective objects are the functors (isomorphic to one) of the form* $A \otimes -$ *with* $A \in \text{mod-}R$.

Proof. Any functor of the form $A \otimes_R -$ with $A \in \text{mod-}R$ is finitely presented (10.2.36) and is, by 12.1.6, absolutely pure in $(R\text{-mod}, \mathbf{Ab})$, hence is injective in $(R\text{-mod}, \mathbf{Ab})^{\text{fp}}$; for, every pure embedding in the latter category is split, since it has finitely presented cokernel – see 2.1.18.

Conversely, if $G \in (R\text{-mod}, \mathbf{Ab})^{\text{fp}}$ is injective in this category, $\text{Ext}^1(F, G) = 0$ for all $F \in (R\text{-mod}, \mathbf{Ab})^{\text{fp}}$, that is, G is absolutely pure in $(R\text{-mod}, \mathbf{Ab})$, so, by 12.1.6, G is isomorphic to a functor of the form $A \otimes -$ with $A \in \text{Mod-}R$. By 10.2.36, $A \in \text{mod-}R$.

That proves the second statement: to see the first, let $F \in (R\text{-mod}, \mathbf{Ab})^{\text{fp}}$. There is, 10.1.13, an epimorphism $(A, -) \to dF$ for some $A \in \text{mod-}R$, where $dF \in (\text{mod-}R, \mathbf{Ab})^{\text{fp}}$ denotes the dual (in the sense of Section 10.3) of F. This dualises (10.3.4) to an embedding $F \to d(A, -) \simeq (A \otimes_R -)$ (by 10.3.1), as required. □

So, cf. 10.2.1 from which this also follows by duality (§10.3), for every finitely presented functor $F \in (R\text{-mod}, \mathbf{Ab})$ there is a morphism $K \to L$ in R-mod and an exact sequence $0 \to F \to (- \otimes K) \to (- \otimes L)$. The resulting (from 10.2.32) characterisation of definable subcategories is noted by Bazzoni at [58, 5.1].

Corollary 12.1.14. *A subcategory $\mathcal{X}$ of* Mod-R *is definable iff there is a set $(g_\lambda : K_\lambda \to L_\lambda)_\lambda$ of morphisms in* R-mod *such that* $\mathcal{X} = \{M \in \text{Mod-}R : 1_M \otimes g_\lambda : M \otimes K_\lambda \to M \otimes L_\lambda \text{ is monic}\}$.

In general $(R\text{-mod}, \mathbf{Ab})^{\text{fp}}$ does not have injective hulls (that is, minimal injective extensions), for $(R\text{-mod}, \mathbf{Ab})^{\text{fp}}$ having these is, by 10.3.4, equivalent to $(\text{mod-}R, \mathbf{Ab})^{\text{fp}}$ having projective covers, hence, by 4.3.69, to R being Krull–Schmidt.

Proposition 12.1.15. *Let R be a ring. The category* $(R\text{-mod}, \mathbf{Ab})^{\text{fp}}$ *has injective hulls iff R is Krull–Schmidt.*

Proposition 12.1.16. *Suppose that R is Krull–Schmidt (Section 4.3.8). Then a pp condition is irreducible iff it has a free realisation in a finitely presented module which has local endomorphism ring.*

Proof. One half is proved at 1.2.32, so suppose that ϕ is irreducible and let $(C, \overline{c})$ be a free realisation of ϕ. By assumption C is a direct sum of modules with local endomorphism rings. So, by 12.1.13 and 12.1.3, $C \otimes -$ is a direct sum of indecomposable injectives in $(R\text{-mod}, \mathbf{Ab})^{\text{fp}}$. Since ϕ is irreducible the finitely presented functor $({}_RR^n, -)/F_{D\phi}$ is uniform (12.2.3). By 10.3.9 this functor is a subobject of $C \otimes -$. It follows, using the exchange property (E.1.25), that $({}_RR^n, -)/F_{D\phi}$ is a subfunctor of an indecomposable direct summand of $(C \otimes -)$. Such a summand has the form $D \otimes -$ for some indecomposable direct summand, D, of C and, by 10.3.9, if $\overline{d} : R^n \to D$ is the corresponding morphism, then $(D, \overline{d})$ is a free realisation of ϕ and D has local endomorphism ring, as required. □

12.1.3 The Ziegler spectrum revisited yet again

The Ziegler spectrum is realised, via *the tensor embedding, as a space of indecomposable injective functors (12.1.17). Subquotients of (localised) finitely presented functors are related to inclusions of basic open sets (12.1.19, 12.1.20).*

The results of this chapter give a fresh route to the Ziegler spectrum of R. It has been shown already, 12.1.9, that there is a bijection between isomorphism types of indecomposable pure-injective R-modules and indecomposable injective objects of the functor category (R-mod, **Ab**): that gives us the points of the spectrum. For each finitely presented functor $F \in (\text{mod-}R, \mathbf{Ab})^{\text{fp}}$ define the set $(F) = \{N \in \text{Zg}_R : \overrightarrow{F} N \neq 0\}$ and declare this to be open.[2] By 10.2.31 and 5.1.8 this gives exactly the Ziegler topology.

By 10.3.5, for $F \in (\text{mod-}R, \mathbf{Ab})^{\text{fp}}$ one has $(F) = \{N \in \text{Zg}_R : (dF, N \otimes -) \neq 0\}$. So this gives yet another way of defining the spectrum which refers just to the functor category (R-mod, **Ab**). Namely, the Ziegler spectrum of Mod-R is the set, inj(R-mod, **Ab**), of isomorphism classes of indecomposable injective objects of (R-mod, **Ab**) and the basic open sets are those of the form $(G)^{\not\perp} = \{E \in \text{inj}(R\text{-mod}, \mathbf{Ab}) : (G, E) \neq 0\}$, so, up to identification, $\{N \in \text{Zg}_R : (G, N \otimes -) \neq 0\}$, as G ranges over $(R\text{-mod}, \mathbf{Ab})^{\text{fp}}$. That is, if $F \in (\text{mod-}R, \mathbf{Ab})^{\text{fp}}$, then $(F) = (dF)^{\not\perp}$ as open subsets of Zg_R. In practice we will usually (we do it already in 12.1.19 below) drop the $^{\not\perp}$ since it will normally be clear from the context which meaning the (otherwise ambiguous) symbol (F) should have.

Theorem 12.1.17. (for example, [497, p. 201]) *The space defined above (in either way) is homeomorphic to, indeed is canonically identified with, the Ziegler spectrum,* Zg_R, *of* R.

Example 12.1.18. Let $A \in \text{mod-}R$. Then $(A, -)$ is a finitely presented functor in $(\text{mod-}R, \mathbf{Ab})^{\text{fp}}$ so $\big(d(A, -)\big)^{\not\perp} = \{N \in \text{Zg}_R : (A, -).N = (A, N) \neq 0\} = \{N \in \text{Zg}_R : (A \otimes -, N \otimes -) \neq 0\}$ is an open set. By 10.3.1 , $d(A, -) = A \otimes -$, so this illustrates directly that $(A, -)$ and $A \otimes -$ define (in dual ways) the same subset of Zg_R.

Proposition 12.1.19. (for example, [278, 3.2]) *If F and F' are finitely presented functors and F' is a subquotient of F, then $(F') \subseteq (F)$ as subsets of* Zg_R.

Indeed, if $0 \to F' \to F \to F'' \to 0$ is an exact sequence of finitely presented functors, then $(F) = (F') \cup (F'')$.

[2] Recall, Section 10.2.8, that $\overrightarrow{F}$ is the extension of F to a functor on all of Mod-R which commutes with direct limits.

Proof. If F, F' are understood as being in $(R\text{-mod}, \mathbf{Ab})^{\text{fp}}$, then $N \in (F')$ means $(F', N \otimes -) \neq 0$, so, since $N \otimes -$ is injective, there is a non-zero morphism from F to $N \otimes -$, hence $N \in (F)$, as required.

If F, F' are understood as being in $(\text{mod-}R, \mathbf{Ab})^{\text{fp}}$, then, by 10.3.4, dF' is a subquotient of dF, so we arrive at the same conclusion.

The proof of the second statement is similar. □

It is reasonable, when convenient, to extend the above notation to finitely presented functors, G, in localisations, $(R\text{-mod}, \mathbf{Ab})_{DX}$, of $(R\text{-mod}, \mathbf{Ab})$, where X a closed subset of Zg_R (see Section 12.3) writing

$$\begin{aligned}(G) &= \{N \in \mathrm{Zg}_R : (N \otimes -) \text{ is } \tau_{DX}\text{-torsionfree and } \big(G, (N \otimes -)_{DX}\big) \neq 0\} \subseteq X \\ &= \{N \in \mathrm{Zg}_R : (N \otimes -) \in \mathrm{inj}(R\text{-mod}, \mathbf{Ab})_{DX} \text{ and } (G, N \otimes -) \neq 0\}.\end{aligned}$$

The localisation indicated by subscript ${}_{DX}$ is defined just before 12.3.12.

Lemma 12.1.20. *Suppose that X is a closed subset of Zg_R, $G \in (R\text{-mod}, \mathbf{Ab})^{\text{fp}}_{DX}$ (see 11.1.33) and let $F \in (R\text{-mod}, \mathbf{Ab})^{\text{fp}}$. If G is isomorphic to a subquotient of the localised functor F_{DX}, then $(G) \subseteq (F)$. More generally, if $G \in (R\text{-mod}, \mathbf{Ab})^{\text{fp}}$ and G_{DX} is isomorphic to a subquotient of F_{DX}, then $(G) \cap X \subseteq (F)$.*

Proof. By definition $N \otimes -$ is τ_{DX}-torsionfree iff $N \in X$. By assumption and injectivity of $(N \otimes -)_{DX}$ we have, cf. the proof of 12.1.19, if $N \in X$ and $(G, N \otimes -) \neq 0$, then $(F, (N \otimes -)_{DX}) \simeq (F_{DX}, N \otimes -) \simeq \big(F_{DX}, (N \otimes -)_{DX}\big) \neq 0$ (the first isomorphism by 11.1.5 since $N \otimes -$ is τ_{DX}-torsionfree and τ_{DX}-injective, the second by the adjunction in 11.1.5), so $(G) \subseteq (F)$. The second statement follows since clearly $(G_{DX}) = (G) \cap X$. □

12.2 Pp-types, subfunctors of $({}_R R^n, -)$ and finitely generated functors

Pp-types dualise to arbitrary subfunctors of the forgetful functor, giving an isomorphism between the respective lattices (12.2.1), with irreducible pp-types corresponding to couniform functors (12.2.3). Every finitely generated functor is the quotient of a pp condition by a pp-type (12.2.2). Morphisms from an arbitrary finitely generated functor are described in terms of realisations of types modulo pp conditions (12.2.4).

The kernel of the morphism induced by a tuple is the functor dual to the pp-type of that tuple (12.2.5).

A functorial version of a technical lemma, used to produce irreducible pp-types, is given (12.2.7).

Functors which commute with ultraproducts of pure-injectives must be finitely generated (12.2.8).

Let p be a pp-type in n free variables. Each pp condition $\phi \in p$ defines the finitely generated (10.2.29) subfunctor, F_ϕ, of the nth power, $(R^n, -)$, of the forgetful functor and $\psi \le \phi$ iff $F_\psi \le F_\phi$. So to each pp-type p corresponds a filter in the lattice of finitely generated subfunctors of $(R^n, -)$. Since, 10.2.16, every finitely generated subfunctor of $(R^n, -)$ has the form F_ϕ for some pp condition ϕ, it follows that the pp-types with n free variables are in bijection with the filters in the lattice, $\mathrm{Latt}^{\mathrm{f}}(R^n, -)$, of finitely generated subfunctors. Applying elementary duality (see after 10.3.6) gives something more directly useful: a filter of finitely generated subfunctors of $(R^n, -)$ dualises to an ideal (that is, a set closed under finite sums and downwards closed) of finitely generated subfunctors of $d(R^n_R, -) = (R^n_R \otimes -) \simeq ({}_R R^n, -)$ (10.3.1) and this may be replaced by the sum of these subfunctors. By the AB5 condition, $F \cap \sum_{\to\lambda} F_\lambda = \sum_{\to\lambda} F \cap F_\lambda$ (see E.1.3), these are equivalent data. Indeed, we can think of this sum as something like an ideal in the usual algebraic sense.

Therefore, given a pp-n-type, p, for right R-modules, set[3]

$$F_{Dp} = \sum_{\to} \{F_{D\phi} : \phi \in p\} \le (R^n \otimes -).$$

This notation fits with our duality notation, $\phi \mapsto D\phi$, for pp conditions since if p is finitely generated (Section 1.2.2), by ϕ, then clearly $F_{Dp} = F_{D\phi}$.

The lattice of pp-types was introduced at the end of Section 3.2.1. The lattice structure can be defined directly, as at [495, §8.1], or we can just carry over the structure from the lattice of subfunctors of $(R^n \otimes -)$.

Proposition 12.2.1. *For each n, the map $p \mapsto F_{Dp}$ is an anti-isomorphism between the lattice of pp-n-types for right R-modules and the lattice of all subfunctors of $({}_R R^n, -) \simeq (R^n \otimes -)$.*

Proof. Let G be any subfunctor of $(R \otimes -)$. Set $p = \{\phi : F_{D\phi} \le G\}$: we claim that $G = F_{Dp}$. Certainly $F_{Dp} = \sum_{\phi \in p} F_{D\phi} \le G$. If these were different there would be a finitely generated functor, $F_{D\psi}$ say (by 10.2.16 and 1.3.1), with $F_{D\psi} \le G$ but $F_{D\psi}$ not contained in F_{Dp} – impossible by definition of p.

To see that the correspondence is bijective it is enough to check that if $F_{D\psi} \le F_{Dp}$, then $\psi \in p$. Since F_{Dp} is the sum of the $F_{D\phi}$ with $\phi \in p$, if $F_{D\psi} \le F_{Dp}$, then

[3] In [278, p. 532] the notation $T_M(a)$ is used for F_{Dp}; this is similar to the notation used by Zimmermann, for example, in [735].

$F_{D\psi} \le F_{D\phi_1} + \cdots + F_{D\phi_k}$, say, with $\phi_1, \ldots, \phi_k \in p$. But $F_{D\phi_1} + \cdots + F_{D\phi_k} = F_{D(\phi_1 \wedge \cdots \wedge \phi_k)}$, so $\psi \ge \phi_1 \wedge \cdots \wedge \phi_k$ (by 1.3.1 and 1.3.3) and, since p is a filter of pp conditions, $\psi \in p$ as required. The correspondence is clearly order-reversing so the result is proved. □

There is an obvious version which is relative to a definable subcategory and which follows directly from 12.3.16, cf. 12.3.18.[4]

We extend the notation $F_{D\psi/Dp}$ to allow the possibility that F_{Dp} is not contained in $F_{D\psi}$, that is, define $F_{D\psi/Dp} = (F_{D\psi} + F_{Dp})/F_{Dp}$ $(\simeq F_{D\psi}/(F_{D\psi} \cap F_{Dp}))$.

Corollary 12.2.2. ([102, 5.2]) *Every finitely generated functor in* (R-mod, **Ab**) *has the form $F_{D\psi/Dp}$ for some pp-type p and pp condition ψ.*

Proof. If G is finitely generated, then (10.1.13) there is an epimorphism $F \to G$ with F finitely presented. By 10.2.31 and elementary duality (1.3.1) $F \simeq F_{D\psi/D\phi}$ for some pp-pair $\phi \ge \psi$ for right modules. The kernel of the epimorphism $F_{D\psi} (\le ({}_RR^n, -)$ for some $n) \to G$ is a subfunctor of $({}_RR^n, -)$, so is, by 12.2.1, of the form F_{Dp}, as required. □

One may, in 12.2.2, take ψ to be quantifier-free, since, by 10.1.13, it may be supposed that F, as in the proof of 12.2.2, has the form $(L, -)$ for some $L \in R$-mod and then 10.2.13 applies.

Recall (Section 4.3.6) that a pp-type p is irreducible if whenever ψ_1, ψ_2 are pp conditions not in p there is $\phi \in p$ such that $\phi \wedge \psi_1 + \phi \wedge \psi_2$ is not in p.

Corollary 12.2.3. *A pp-n-type p (for right R-modules) is irreducible iff the functor $({}_RR^n, -)/F_{Dp} = (R^n \otimes -)/F_{Dp}$ is uniform. In particular, a pp condition ϕ is irreducible in the sense of Section 1.2.3 iff the functor $({}_RR^n, -)/F_{D\phi}$ is uniform.*

Proof. ($\Rightarrow$) Let $F_1 = F_{D\psi_1/Dp}$ (that is, $= (F_{D\psi_1} + F_{Dp})/F_{Dp}$) and $F_2 = F_{D\psi_2/Dp}$ be non-zero finitely generated subfunctors of $(R^n, -)/F_{Dp}$: so $\psi_1, \psi_2 \notin p$. Let $\phi \in p$ be such that $\psi_1 \wedge \phi + \psi_2 \wedge \phi \notin p$, that is, using 1.3.1, $F_{D(\psi_1 \wedge \phi + \psi_2 \wedge \phi)} = F_{(D\psi_1 + D\phi) \wedge (D\psi_2 + D\phi)} = (F_{D\psi_1} + F_{D\phi}) \cap (F_{D\psi_2} + F_{D\phi}) \nleq F_{Dp}$. The image of this in $(R^n, -)/F_{Dp}$ is a non-zero subfunctor of $F_1 \cap F_2$, as required.

($\Leftarrow$) Given $\psi_1, \psi_2 \notin p$, we have $F_{D\psi_1}, F_{D\psi_2} \nleq F_{Dp}$, so, by assumption, $(F_{D\psi_1} + F_{Dp}) \cap (F_{D\psi_2} + F_{Dp}) \nleq F_{Dp}$, hence some finitely generated subfunctor of $(F_{D\psi_1} + F_{Dp}) \cap (F_{D\psi_2} + F_{Dp})$ is not contained in F_{Dp}. Such a finitely generated subfunctor has, by definition of F_{Dp}, the form $(F_{D\psi_1} + F_{D\phi_1}) \cap (F_{D\psi_2} + F_{D\phi_2})$ for

[4] It is also obvious from the above that the $\wedge$-compact elements of the lattice of pp-types, that is, those p such that $p \ge \bigwedge_i p_i$ implies $p \ge p_{i_1} \wedge \ldots \wedge p_{i_n}$ for some $i_1, \ldots, i_n$, are those which are finitely generated, that is, equivalent to a single pp condition. For, by 12.2.1, $p \ge \bigwedge_i p_i$ is equivalent to $F_{Dp} \le \sum_i F_{Dp_i}$ and this sum reduces to a finite one iff F_{Dp} is a finitely generated object, that is, iff Dp, hence p, is finitely generated.

some $\phi_1, \phi_2 \in p$ and this (by 1.3.1) equals $F_{D(\psi_1\wedge\phi_1+\psi_2\wedge\phi_2)}$. We may replace $D\phi_1$ and $D\phi_2$ by $D\phi_1 + D\phi_2 = D(\phi_1 \wedge \phi_2)$ and, taking $\phi = \phi_1 \wedge \phi_2$ ($\in p$), we deduce that p is irreducible.

(We remark that the statement for pp conditions/finitely generated pp-types is already a corollary of 10.2.17.) □

The following generalises 10.3.5 and 10.3.8.

Proposition 12.2.4. ([101, 6.1.7, 6.1.9]) *If F is any finitely generated functor in (R-mod, **Ab**), say (12.2.2) $F \simeq F_{D\psi/Dp}$, and M is any right R-module, then $(F_{D\psi/Dp}, M \otimes -) \simeq p_\psi(M)$, where (see Section 4.3.9) $p_\psi(M) = (p(M) + \psi(M))/\psi(M)$.*

In particular $p(M) \simeq ((R^n, -)/F_{Dp}, M \otimes -)$ if p is a pp-type with n free variables.

Proof. By definition $F_{Dp} = \overrightarrow{\sum}\{F_{D\phi} : \phi \in p\}$, so $F_{D\psi}/F_{Dp}$, that is, $(F_{D\psi} + F_{Dp})/F_{Dp}$, is $\varinjlim_{\phi\in p}(F_{D\psi} + F_{D\phi})/F_{D\phi}$, since a direct limit of exact sequences in (R-mod, **Ab**) is exact. Therefore $(F_{D\psi/Dp}, M \otimes -) \simeq \big(\varinjlim_{\phi\in p}((F_{D\psi} + F_{D\phi})/F_{D\phi}), M \otimes -\big) \simeq \varprojlim_{\phi\in p}\big((F_{D\psi+D\phi})/F_{D\phi}, M \otimes -\big)$ (by definition of $\varinjlim$) $= \varprojlim_{\phi\in p}\phi(M)/(\phi(M) \cap \psi(M))$ (by 10.3.8) $\simeq \varprojlim_{\phi\in p}\big((\phi(M) + \psi(M))/\psi(M)\big) = \bigcap_{\phi\in p}\big((\phi(M) + \psi(M))/\psi(M)\big) = \big(\bigcap_{\phi\in p}(\phi(M) + \psi(M))\big)/\psi(M) = \big((\bigcap_{\phi\in p}\phi(M)) + \psi(M)\big)/\psi(M) = (p(M) + \psi(M))/\psi(M)$, as required.[5] □

The next result generalises 10.3.9 and is an immediate consequence of that result and the various definitions.

Proposition 12.2.5. *Let $\overline{a}$ be an n-tuple from the module M. Then the morphism $\overline{a} \otimes - : (R^n \otimes -) \to (M \otimes -)$ has kernel F_{Dp}, where $p = \mathrm{pp}^M(\overline{a})$ is the pp-type of $\overline{a}$ in M.*

Corollary 12.2.6. ([102, 5.6]) $E(F_{D\psi/Dp}) \simeq H(p_\psi) \otimes -$

Proof. This follows from the discussion before 4.3.76. □

The next result is a functorial version of 4.3.50 (to compare the results: H below is $\sum_{\phi\in\Phi} F_{D\phi}$, G is $\sum_{\psi\in\Psi} F_\psi$ and H' is F_{Dp}).

Lemma 12.2.7. *Let X be a closed subset of Zg_R. Suppose that G is a subfunctor of the localised nth power of the forgetful functor $(R_R^n, -)_X$ and H is a subfunctor of*

[5] Similarly, if $F = \sum\{F_{D\phi} : F_{D\phi} \leq F\}$ is any subfunctor of some $(R, -)^n$, then $(F, M \otimes -) = \varprojlim_\phi M/\phi(M)$ for any module M, [101, p. 101].

the dual functor $(R^n \otimes -)_{DX}$, such that, for every finitely generated subfunctor H_0 of H, we have $DH_0 \nleq G$ (thus "DH" is a filter of finitely generated subfunctors of $(R^n, -)_X$, none of which is contained in G). Let H' be a subfunctor of $(R^n \otimes -)_{DX}$ maximal containing H and still satisfying the condition that if F is a finitely generated subfunctor of H', then $DF \nleq G$. Then $(R^n \otimes -)/H'$ is a uniform functor.

Proof. Let F_0, F_1 be finitely generated functors subfunctors of $(R^n \otimes -)$, neither of which is contained in H'. By maximality of H' there are finitely generated functors $H_0, H_1 \leq H'$ such that $D(F_0 + H_0), D(F_1 + H_1) \leq G$. Therefore $D(F_0 + H_0) + D(F_1 + H_1) \leq G$. So, by definition of H', $D\big(D(F_0 + H_0) + D(F_1 + H_1)\big) \nleq H'$, that is, $(F_0 + H_0) \cap (F_1 + H_1)$ is not contained in H' (by 10.3.7 for $X = \mathrm{Zg}_R$, 12.3.18 for the general case). So the image of $F_0 \cap F_1$ in $(R^n \otimes -)/H'$ is non-zero, as required. □

Theorem 12.2.8. *Suppose that $\mathcal{X}$ is a definable subcategory of* Mod-R *and suppose that $F : \mathcal{X} \longrightarrow$* **Ab** *is the restriction to $\mathcal{X}$ of a functor from* Mod-R *to* **Ab** *of the form $F = F_p/F_\psi : M \mapsto (p(M) + \psi(M))/\psi(M)$ for some pp-type p and pp condition ψ. If p/ψ is not a finitely generated pp-type modulo $\mathcal{X}$, that is, if $F \restriction$* mod-R *is not a finitely generated functor in the localisation (Section 12.3)* (mod-R, **Ab**)$_{\mathcal{X}}$, *then there is a pure-injective object $N \in \mathcal{X}$ and a pure-injective ultrapower (Section 3.3.1) $N^* = N^J/\mathcal{F}$ of N such that $F(N^*) > (FN)^J/\mathcal{F}$. In particular, if F commutes with ultraproducts (of pure-injective objects of $\mathcal{X}$), then F must be finitely generated modulo $\mathcal{X}$.*

Proof. Suppose that p/ψ is not finitely generated modulo $\mathcal{X}$ ("modulo $\mathcal{X}$" means that essentially we are working in the lattice of pp conditions $\mathrm{pp}(\mathcal{X})$). For each pp condition $\phi \in p$ there is $M = M_\phi \in \mathcal{X}$ and there is $\overline{a} = \overline{a}_\phi$ from M such that the pp-type, $\mathrm{pp}^M(\overline{a}/\psi)$, of its image modulo ψ (Section 4.3.9) is generated modulo $\mathcal{X}$ by ϕ. Since p/ψ is not finitely generated modulo $\mathcal{X}$, $\overline{a}/\psi \notin F(M)$. We may replace M by its pure-injective hull and also make a uniform choice of M over all $\phi \in p$ simply by taking the direct product of all the M_ϕ obtained from individual $\phi \in p$. The resulting pure-injective module we denote by N.

Now, let J be the set of all pp conditions $\phi \in p$ and let $\mathcal{F}$ be an ultrafilter on J which contains all the sets of the form $\langle\phi\rangle = \{\phi' : \phi' \geq \phi\}$ as ϕ ranges over p. Let $\overline{a}^\star$ from $N^\star = N^J/\mathcal{F}$ be the image in $N^\star$ of the tuple in the product constructed from those tuples $\overline{a}_\phi$ chosen above. A direct application of Łoś' Theorem in the form 3.3.3 gives $\overline{a}^\star/\psi \in p(N^\star)/\psi(N^\star) = F(N^\star)$. But, by construction, $\overline{a}^\star/\psi$ is not the image, under the canonical projection from the power to the ultrapower, of any tuple in $p(N^J)/\psi(N^J) = F(N^J)$ for, if it were, then at least one (in fact many) of its component tuples would have to satisfy p modulo ψ. Finally to ensure that the ultrapower is pure-injective we replace $N^\star$ by a suitable ultrapower of it (4.2.18). □

12.3 Definable subcategories again

Under the tensor-embedding of modules into the functor category, definable subcategories of the module category correspond to definable subcategories of the functor category consisting of those absolutely pure functors which are torsionfree for some finite-type torsion theory (12.3.2). After appropriate localisation, the objects of a definable category of modules correspond to the absolutely pure objects of the localised category (12.3.3). The closed subsets of the Ziegler spectrum correspond to the sets of indecomposable injective functors which are torsionfree with respect to some finite-type torsion theory (12.3.5) and this gives a torsion-theoretic description of Ziegler-closure (12.3.7).

The torsion theory cogenerated by a set of indecomposable injectives in the functor category is of finite type iff the corresponding indecomposable pure-injectives together give an elementary cogenerator (12.3.8).

There are various ways of describing the torsion theory corresponding to a definable subcategory (12.3.10) and of describing the corresponding injective objects (12.3.12).

Finitely presented functors give a basis of open sets for any closed subset of the spectrum (12.3.14) and there is a localised version of the equivalence between evaluation at pp-pairs and forming hom-sets from the dual functor (12.3.15).

A criterion is given for equivalence of pp conditions under localisation (12.3.16) and it is shown that the lattice of pp conditions relative to a closed subset of the spectrum is isomorphic to the lattice of finitely generated subfunctors of the corresponding localised forgetful functor (12.3.18). Equivalents to a pp-pair being closed on a definable subcategory are listed in 12.3.19.

The category of pp-pairs relative to a definable subcategory is equivalent to the category of finitely presented objects of the corresponding localisation of the functor category, and each is anti-isomorphic to the relevant dual (12.3.20).

Some of the results in this section are essentially restatements of general results in this particular context.

This section describes how to relativise to a closed subset of the Ziegler spectrum, equivalently how to work in definable subcategories of Mod-R. It describes what happens when a set of pp-pairs is collapsed. The main point is that the functor categories (both direct and dual) are replaced by their localisations at the corresponding torsion theories.

By 12.1.6 the image of the category of right R-modules under the embedding Mod-$R \longrightarrow (R\text{-mod}, \mathbf{Ab})$ is the class of absolutely pure (= right exact = fp-injective) functors and these form a definable subcategory of $(R\text{-mod}, \mathbf{Ab})$ (12.1.12). In 12.3.2 we characterise the images of the definable subcategories of Mod-R in $(R\text{-mod}, \mathbf{Ab})$ as exactly the classes of absolutely pure τ-torsionfree objects as τ varies over torsion theories of finite type on $(R\text{-mod}, \mathbf{Ab})$. For torsion theories of finite type see Section 11.1.2, and also the subsequent sections, which deal with finite-type localisation in locally coherent categories such as $(R\text{-mod}, \mathbf{Ab})$.

Remark 12.3.1. The images of the definable subcategories of Mod-R are exactly the definable (in the functor category) subclasses of the class, Abs$(R\text{-mod}, \mathbf{Ab})$ of absolutely pure functors. For, each such image satisfies the conditions (3.4.7) for being definable by 12.1.3 (for products and direct limits) and 12.1.6 (which implies that any pure subobject of $M \otimes -$ is of the form $M' \otimes -$ for some pure subobject M' of M).

Proposition 12.3.2. (for example, [355, 4.5]) *A subcategory $\mathcal{X}$ of* Mod-R *is definable iff its image under the embedding $M \mapsto M \otimes -$ of* Mod-R *into* $(R\text{-mod}, \mathbf{Ab})$ *has the form $\mathcal{F} \cap$ Abs, where $\mathcal{F}$ is the torsionfree class for some torsion theory, $\tau_{D\mathcal{X}}$, of finite type on* $(R\text{-mod}, \mathbf{Ab})$ *and* Abs = Abs$(R\text{-mod}, \mathbf{Ab})$ *denotes the class of absolutely pure objects of the functor category.*

Proof. Any class of the form $\mathcal{F} \cap$ Abs is closed under products and pure subobjects (2.3.5). It is also closed under direct limits since this is true of Abs (by 10.2.3 and 3.4.24 (generalised), alternatively by 12.1.12) and of $\mathcal{F}$ (11.1.12) since the torsion theory is assumed to be of finite type. It follows that the corresponding class of R-modules has these closure conditions (12.1.3 for direct limits and products, 12.1.6 for pure submodules) hence, 3.4.7, is definable.

For the converse, if $\mathcal{X}$ is definable, then, by 12.3.1, the corresponding subcategory, $\epsilon\mathcal{X}$, of $(R\text{-mod}, \mathbf{Ab})$ is definable. Let $\mathcal{F}$ be the class of subobjects of members of $\epsilon\mathcal{X}$. Clearly (from 11.1.1) $\mathcal{F}$ is a hereditary torsionfree class, so, by 11.1.20, it will enough to show that $\mathcal{F}$ is definable. Closure under direct limits seems not so obvious (there is not a canonical way of lifting a direct system in $\mathcal{F}$ to one in $\epsilon\mathcal{X}$) but closure under reduced (or ultra-) products is immediate since a reduced power of embeddings is an embedding (this follows directly from 3.3.3 applied to the pp condition $x = 0$) and since, by 3.4.7, $\epsilon\mathcal{X}$ is closed under ultraproducts. So $\mathcal{F}$ is definable, hence is the torsionfree class for a torsion theory of finite type and clearly $\epsilon\mathcal{X}$ consists of those absolutely pure functors which lie in $\mathcal{F}$. □

The next result is direct from 11.1.31.

Corollary 12.3.3. *Suppose that $\mathcal{X}$ is a definable subcategory of* Mod-R, *let* $X \subseteq \mathrm{Zg}_R$ *be the corresponding (5.1.1) closed set and let* $(R\text{-mod}, \mathbf{Ab})_{D\mathcal{X}}$ *be the corresponding (in the above sense) localisation of* $(R\text{-mod}, \mathbf{Ab})$. *Then* $M \mapsto (M \otimes -)_{D\mathcal{X}}$ *gives an equivalence between the categories* $\mathcal{X}$ *and* $\mathrm{Abs}\big((R\text{-mod}, \mathbf{Ab})_{D\mathcal{X}}\big)$. *In particular,* $N \mapsto (N \otimes -)_{D\mathcal{X}} \simeq N \otimes -$ *is a bijection between* X *and the set of isomorphism types of indecomposable injectives of* $(R\text{-mod}, \mathbf{Ab})_{D\mathcal{X}}$.

Corollary 12.3.4. *Let* $\mathcal{X}$ *be a definable subcategory of* Mod-R. *Then* $\mathcal{X}$ *is closed under pure-injective hulls.*

Proof. This is immediate from 12.3.2, 12.1.6, 4.3.14 and the fact (Section 11.1.1) that the torsionfree class for any hereditary torsion theory is closed under injective hulls. (Alternative proofs of this fact are at 4.3.21.) □

This gives yet another characterisation of the closed subsets of the Ziegler spectrum which uses the correspondence, $N \mapsto (N \otimes -)$ (12.1.9), between indecomposable pure-injective R-modules and indecomposable injective functors.

Theorem 12.3.5. (the second part essentially [726, 4.10]) *A subset* $X \subseteq \mathrm{Zg}_R$ *of the Ziegler spectrum is closed iff* $\{N \otimes - : N \in X\}$ *is, up to isomorphism, the set of indecomposable* τ*-torsionfree injectives for some hereditary torsion theory* τ *of finite type on* $(R\text{-mod}, \mathbf{Ab})$, *in which case this set of injectives cogenerates the torsionfree class for* τ.

Proof. If X is Ziegler-closed, then 12.3.2 gives the conclusion, the last statement being 11.1.29.

Suppose, conversely, that τ is of finite type. By 12.1.6 the τ-torsionfree indecomposable injectives are those functors isomorphic to one of the form $N \otimes -$, where N is an indecomposable pure-injective such that $(F, N \otimes -) = 0$ for every torsion functor F. Since τ is of finite type it is enough, by 11.1.14, to consider just finitely presented torsion functors F. Each such functor has, by 10.2.31, the form $F_{\phi/\psi}$ for some pp-pair ϕ/ψ (for left modules), so, since $(F_{\phi/\psi}, N \otimes -) \simeq D\psi(N)/D\phi(N)$ (10.3.8), the resulting set of indecomposable pure-injectives is $\bigcap\{[D\psi/D\phi] : F_{\phi/\psi} \text{ is } \tau\text{-torsion}\}$ and this is, by 5.1.7, a closed subset of Zg_R. □

This makes clear that 5.1.4 is a corollary of 11.1.29.

In the corollary below the second statement follows by 11.2.10.

Corollary 12.3.6. *There is a bijection between definable subcategories of* Mod-R, *closed subsets of* Zg_R, *and torsion theories of finite type on* $(R\text{-mod}, \mathbf{Ab})$, *hence also with torsion theories of finite type on* $(\text{mod-}R, \mathbf{Ab})$.

It is not clear whether there is, in general, a (natural) bijection of Zg_R with inj(mod-R, **Ab**): this is the same as the question of whether Zg_R and ${}_R\mathrm{Zg}$ are (naturally) homeomorphic (see 5.4.8).

Let X be an arbitrary subset of Zg_R. The indecomposable injectives $N \otimes -$, for $N \in X$, together cogenerate (see 11.1.7) a hereditary torsion theory, let us write it as $\mathrm{cog}(X \otimes -)$, on ($R$-mod, **Ab**), but this torsion theory need not be of finite type. By 11.1.37 there is a largest hereditary torsion theory of finite type contained in[6] $\mathrm{cog}(X \otimes -)$. It follows from 12.3.5 that this is the torsion theory, $\mathrm{cog}(\overline{X} \otimes -)$, which is cogenerated by the $N \otimes -$ with $N \in \overline{X}$, the Ziegler-closure of X.

Corollary 12.3.7. *Suppose that $X \subseteq \mathrm{Zg}_R$. Then the Ziegler-closure of X is the set of indecomposable pure-injectives N such that $N \otimes -$ is τ-torsionfree for the largest hereditary torsion theory, τ, of finite type contained in the torsion theory cogenerated by $\{N' \otimes - : N' \in X\}$.*

Even if $X \subseteq \mathrm{Zg}$ is not closed it might cogenerate a torsion theory of finite type; by 11.1.7 (and 12.1.3) this is so exactly if every point N in the closure of X is a direct summand of a direct product of copies of modules in X. That, in the terminology of Section 5.3.5, is exactly the condition that the pure-injective hull of $\bigoplus_{N\in X} N$ be an elementary cogenerator. We state this.

Remark 12.3.8. Let X be any subset of Zg_R. Then the torsion theory cogenerated by $\{(N \otimes -) : N \in X\}$ is of finite type iff $H(\bigoplus_{N\in X} N)$ is an elementary cogenerator.

For instance, taking $R = \mathbb{Z}$, the torsion theory cogenerated by $(\mathbb{Z}_{p^\infty} \otimes -)$ is of finite type and the indecomposable torsionfree injectives are $(\mathbb{Z}_{p^\infty} \otimes -)$ and $(\mathbb{Q} \otimes -)$. In contrast, the torsion theory cogenerated by $(\overline{\mathbb{Z}_{(p)}} \otimes -)$ is not of finite type and there is no other indecomposable torsionfree injective. (These comments follow from 5.3.51.)

We can deduce similar statements for subclasses of Mod-R.

Remark 12.3.9. Let $\mathcal{X} \subseteq$ Mod-R be definable and suppose that $\mathcal{Y} \subseteq \mathcal{X}$.

(a) If $\mathcal{X}$ is the closure of $\mathcal{Y}$ under products and pure submodules, then the torsion theory cogenerated by the $M \otimes -$ with $M \in \mathcal{Y}$ coincides with that corresponding (in the sense of 12.3.2) to $\mathcal{X}$, in particular it is of finite type.
(b) If $\mathcal{X}$ is the definable subcategory generated by $\mathcal{Y}$, then the finite type torsion theory corresponding to $\mathcal{X}$ is the largest torsion theory of finite type for which all the $M \otimes -$ with $M \in \mathcal{Y}$ are torsionfree.

[6] Recall, 11.1.9, that "contained in" above refers to containment of torsion classes, so τ has fewer torsion objects, therefore more torsionfree objects, than $\mathrm{cog}(X \otimes -)$.

Given a definable subcategory, $\mathcal{X}$, of Mod-R we are writing $\tau_{D\mathcal{X}}$ for the hereditary torsion theory of finite type on (R-mod, **Ab**) which corresponds to $\mathcal{X}$ and we write $\tau_{\mathcal{X}}$ for the dual torsion theory (in the sense of 11.2.10) on (mod-R, **Ab**). Extend the notation in the usual way, using X in place of $\mathcal{X}$ if X is the closed subset of Zg_R defined (5.1.1) by $\mathcal{X}$, or using M in place of $\mathcal{X}$ if the latter is the definable subcategory of Mod-R generated (Section 3.4.1) by M. Sometimes we might drop the "D" since the context determines whether a torsion theory or its dual is being referred to.

It might be noticed that the notation $\tau_{D\mathcal{X}}$ is potentially ambiguous: recall from Section 3.4.2 that $D\mathcal{X}$ denotes the definable subcategory of R-Mod dual to $\mathcal{X}$. By the next result there is, in fact, no ambiguity. Given a definable subcategory $\mathcal{X}$ of Mod-R there are three ways to associate to $\mathcal{X}$ a torsion theory of finite type on the functor category, (mod-R, **Ab**). They all lead to the same result.

Proposition 12.3.10. *Let $\mathcal{X} \subseteq$ Mod-R be definable. Then the following hereditary torsion theories of finite type on* (mod-R, **Ab**) *coincide:*

(i) the dual, $\tau_{\mathcal{X}}$, in the sense of Section 11.2.2, of the torsion theory, $\tau_{D\mathcal{X}}$, cogenerated by the image of $\mathcal{X}$ in (R-mod, **Ab**)*;*
(ii) the torsion theory cogenerated by the image of the dual definable subcategory, $D\mathcal{X} \subseteq R$-Mod (Section 3.4.2), of $\mathcal{X}$ in (mod-R, **Ab**)*, that is, $\tau_{D(D\mathcal{X})}$;*
(iii) the torsion theory generated, in the sense of 11.1.36, by those finitely presented functors $F_{\phi/\psi}$ in (mod-R, **Ab**) *with $\overrightarrow{F_{\phi/\psi}}M = 0$ for every $M \in \mathcal{X}$.*

Proof. The equivalences follow from the definitions and the formula $(F_{D\psi/D\phi}, M \otimes -) \simeq F_{\phi/\psi}(M)$ (10.3.8). □

For the localisation of (R-mod, **Ab**) at $\tau_{D\mathcal{X}}$ we use any appropriate subscript, for example, $(R\text{-mod}, \mathbf{Ab})_{D\mathcal{X}}$, $(R\text{-mod}, \mathbf{Ab})_{\tau_{D\mathcal{X}}}$, $(R\text{-mod}, \mathbf{Ab})_{\mathcal{X}}$, $(R\text{-mod}, \mathbf{Ab})_X$ etc. could all be used for the same category (the latter two if the context were clear enough that we could drop the "D"). The next statement is immediate from 12.3.5 and 11.1.34.

Proposition 12.3.11. *Let X be a closed subset of Zg_R. Then the localised functor category* $(R\text{-mod}, \mathbf{Ab})_{DX}$ *is locally coherent and* $\left((R\text{-mod}, \mathbf{Ab})_{DX}\right)^{\mathrm{fp}} = \left((R\text{-mod}, \mathbf{Ab})^{\mathrm{fp}}\right)_{DX}$.

By 11.1.5 the indecomposable injectives of this localised category may be identified with the indecomposable $\tau_{D\mathcal{X}}$-torsionfree injectives in the original functor category, so, by 12.3.3, one has the following.

Corollary 12.3.12. *Let $\mathcal{X}$ be a definable subcategory of* Mod-R *and let* $X = \mathcal{X} \cap \mathrm{Zg}_R$ *be the corresponding closed subset of the Ziegler spectrum. Then there are natural bijections between isomorphism classes of:*

(i) pure-injective objects in $\mathcal{X}$*;*
(ii) $\tau_{D\mathcal{X}}$*-torsionfree injectives in* $(R\text{-mod}, \mathbf{Ab})$*;*
(iii) injective objects of the localised category $(R\text{-mod}, \mathbf{Ab})_{D\mathcal{X}}$*,*

with the points of X in bijection with the isomorphism classes of indecomposable objects in each of (ii) and (iii).

Corollary 12.3.13. *Suppose that* $X \supseteq Y$ *are closed subsets of* Zg_R*. Let* σ *be the torsion theory on* $(R\text{-mod}, \mathbf{Ab})_{DX}$ *such that the composition of localisations* $Q_\sigma Q_{\tau_{DX}}$ *equals* $Q_{\tau_{DY}}$ *(cf. 11.1.10). Then* σ *is of finite type and is cogenerated by the set of* $(N \otimes -)_{DX}$ *with* $N \in Y$*.*

Corollary 12.3.14. *If X is a closed subset of* Zg_R*, then a basis of open sets for the induced topology on X is given by the sets* $(G) = \{N \in X : (G, N \otimes -) \neq 0\}$*, where G ranges over finitely presented objects of* $(R\text{-mod}, \mathbf{Ab})_{DX}$ *and where we identify* $N \otimes -$ *with* $(i)(N \otimes -)_{DX}$ *(in the notation of 11.1.5).*

If R is a ring, the subquotients, G, of $(R \otimes -)_{DX}$ *suffice to give a basis.*

Proof. The basis for the topology induced on X by that on Zg_R consists of those sets of the form $(F) \cap X = \{N \in X : (F, N \otimes -) = 0\}$ as F ranges over the finitely presented functors in $(R\text{-mod}, \mathbf{Ab})$ (12.1.17). By 11.1.34, as F varies, the corresponding localisations, F_{DX}, vary over the finitely presented objects of the localised category. If $N \in X$, then $N \otimes -$ is τ_{DX}-torsionfree (12.3.2), so, by 11.1.5, $(N \otimes -) \simeq (N \otimes -)_{DX}$, therefore, where i denotes the right adjoint of the localisation functor, the adjunction gives $(F, (i)(N \otimes -)_{DX}) \simeq (F_{DX}, (N \otimes -_{DX})$, which is enough.

The last statement follows by 11.1.39. □

In the proof above we saw a relativised/localised version of 10.3.8. Recall (Section 5.1.1) that $\mathrm{supp}(M) \subseteq X$ is equivalent to $\langle M \rangle \subseteq \mathcal{X}$, where $\mathcal{X}$ is the definable subcategory corresponding to X.

Proposition 12.3.15. *If M is any right R-module,* $F_{D\psi/D\phi}$ *is any functor in* $(R\text{-mod}, \mathbf{Ab})^{\mathrm{fp}}$ *and X is any closed subset of* Zg_R *such that* $\mathrm{supp}(M) \subseteq X$*, then* $\big((F_{D\psi/D\phi})_{DX}, M \otimes -\big) \simeq \phi(M)/\psi(M)$*.*

Proposition 12.3.16. (for example, [278, 6.2]) *Let* ϕ, ϕ' *be pp conditions for right R-modules, each in n free variables, and let* $\mathcal{X} \subseteq$ Mod-R *be a definable subcategory. Then the following are equivalent:*

(i) $\phi(M) = \phi'(M)$ *for every* $M \in \mathcal{X}$*;*
(ii) $(F_{D\phi})_{D\mathcal{X}} = (F_{D\phi'})_{D\mathcal{X}}$ *in* $(R\text{-mod}, \mathbf{Ab})_{D\mathcal{X}}$*;*
(iii) $(F_{\phi})_{\mathcal{X}} = (F_{\phi'})_{\mathcal{X}}$ *in* $(\text{mod-}R, \mathbf{Ab})_{\mathcal{X}}$.

Here equality as subfunctors of $(R^n \otimes -)_{D\mathcal{X}}$*, respectively* $(R^n_R, -)_{\mathcal{X}}$ *is meant.*

Proof. Equality of the functors, say in the case of $(F_\phi)_{\mathcal{X}}$ and $(F_{\phi'})_{\mathcal{X}}$, is, by 11.1.2, equivalent to each of F_ϕ and $F_{\phi'}$ being $\tau_{\mathcal{X}}$-dense in $F_\phi + F_{\phi'} \leq (R^n, -)$. That is equivalent to each of $(F_\phi + F_{\phi'})/F_\phi$ and $(F_\phi + F_{\phi'})/F_{\phi'}$ being $\tau_{\mathcal{X}}$-torsion, that is (by 12.3.2), to the elementary dual, $(-)$, of each functor satisfying $((-), M \otimes -) = 0$ for every $M \in \mathcal{X}$, and that, by 10.3.8, is equivalent to condition (i). □

Corollary 12.3.17. *Let* $\mathcal{X}$ *be a definable subcategory of* Mod-R*. Then the kernel of the localisation* $(R\text{-mod}, \mathbf{Ab}) \longrightarrow (R\text{-mod}, \mathbf{Ab})_{D\mathcal{X}}$*, respectively of* $(\text{mod-}R, \mathbf{Ab}) \longrightarrow (\text{mod-}R, \mathbf{Ab})_{\mathcal{X}}$*, restricted to the finitely presented functors*[7] *consists of those functors of the form* $F_{D\psi/D\phi}$*, resp.* $F_{\phi/\psi}$*, such that* ϕ/ψ *is closed on every member of* $\mathcal{X}$*, equivalently which are closed on any generating set of* $\mathcal{X}$*.*

Corollary 12.3.18. *Let* $\mathcal{X}$ *be a definable subcategory of* Mod-R*, where* R *is a ring. Then the lattice,* $\text{pp}^n(\mathcal{X})$*, of* $\mathcal{X}$*-equivalence classes of pp conditions is isomorphic to the lattice of finitely generated subfunctors of the localisation* $(R^n, -)_{\mathcal{X}}$*, which, in turn, is isomorphic to the opposite of the lattice of finitely generated subfunctors of* $(R^n \otimes -)_{D\mathcal{X}}$*.*

This follows from the above, 10.2.17 and 10.3.7.

A similar statement statement holds for any small preadditive category R: the lattice of pp conditions modulo $\mathcal{X}$, with free variables of a certain sort (Section 10.2.4), G, say, is isomorphic to the lattice of finitely generated subfunctors of the $\mathcal{X}$-localisation of the object $(G, -)$ and that is anti-isomorphic to the lattice of finitely generated subfunctors of the $D\mathcal{X}$-localisation of $G \otimes -$.

Lemma 12.3.19. *Let* $\phi_\lambda/\psi_\lambda$ $(\lambda \in \Lambda)$ *and* ϕ/ψ *be pp-pairs and let* $\mathcal{X}$ *be the definable subcategory of* Mod-R *defined, in the sense of Section 3.4.1, by* $\{\phi_\lambda/\psi_\lambda\}_{\lambda\in\Lambda}$*. Then the following are equivalent:*

(i) ϕ/ψ *is closed on* $\mathcal{X}$*;*
(ii) $F_{D\psi/D\phi}$ *belongs to the Serre subcategory of* $(R\text{-mod}, \mathbf{Ab})^{\text{fp}}$ *generated by the* $F_{D\psi_\lambda/D\phi_\lambda}$*;*
(iii) $F_{\phi/\psi}$ *belongs to the Serre subcategory of* $(\text{mod-}R, \mathbf{Ab})^{\text{fp}}$ *generated by the* $F_{\phi_\lambda/\psi_\lambda}$*;*
(iv) $(\phi/\psi) \subseteq \bigcup_{\lambda\in\Lambda}(\phi_\lambda/\psi_\lambda)$ *(inclusion of Ziegler-open sets).*

[7] That is, the Serre subcategory which generates the localisation in the sense of 11.1.36.

Proof. This follows by 12.3.2, 11.1.38 and 12.3.17. For the last, use, say, 12.1.19. □

Corollary 12.3.20. ([274, 6.2] for the outer equivalence[8]) *Let $\mathcal{X}$ be a definable subcategory of* Mod-R. *Then* $(\mathbb{L}_R^{\mathrm{eq}+})_{\mathcal{X}} \simeq (\text{mod-}R, \mathbf{Ab})_{\mathcal{X}}^{\mathrm{fp}} \simeq \left((R\text{-mod}, \mathbf{Ab})_{D\mathcal{X}}^{\mathrm{fp}}\right)^{\mathrm{op}} \simeq \left(({}_R\mathbb{L}^{\mathrm{eq}+})_{D\mathcal{X}}\right)^{\mathrm{op}}$.

Proof. Let $\tau = \tau_{\mathcal{X}}$ be the finite-type localisation corresponding (12.3.6) to $\mathcal{X}$. By 10.2.30 and 10.3.4 the absolute categories (no subscript $\mathcal{X}$) are equivalent. The restriction of the localisation functor $Q_\tau : (\text{mod-}R, \mathbf{Ab}) \longrightarrow (\text{mod-}R, \mathbf{Ab})_{\mathcal{X}}$ to $(\text{mod-}R, \mathbf{Ab})^{\mathrm{fp}}$ is (11.1.42) the localisation $(\text{mod-}R, \mathbf{Ab})^{\mathrm{fp}} \longrightarrow (\text{mod-}R, \mathbf{Ab})_{\mathcal{X}}^{\mathrm{fp}}$ of $(\text{mod-}R, \mathbf{Ab})^{\mathrm{fp}}$ at the Serre subcategory $\mathcal{S}_{\mathcal{X}} = \mathcal{T}_{\mathcal{X}} \cap (\text{mod-}R, \mathbf{Ab})^{\mathrm{fp}}$. By definition of τ (12.3.10), the Serre subcategory $\mathcal{S}_{\mathcal{X}}$ consists exactly of those finitely presented functors F such that $\overrightarrow{F} M = 0$ for every $M \in \mathcal{X}$. These, in turn, correspond exactly to the pp-pairs which form the kernel of the functor $\mathbb{L}_R^{\mathrm{eq}+} \longrightarrow (\mathbb{L}_R^{\mathrm{eq}+})_{\mathcal{X}}$ (3.2.15). Since that functor is an exact functor to an abelian category, it must, by 11.1.40, be equivalent to the Serre localisation of the functor category, as required for the first (and third) equivalence of the statement.

By definition (Section 11.2.2) of the dual, $d(\mathcal{S}_{\mathcal{X}})$ (alternatively from the definition (Section 3.4.2) of the dual $D\mathcal{X}$) the equivalence with the opposite localised category follows. □

Corollary 12.3.21. *Let $\mathcal{X}$ be a definable subcategory of* Mod-R. *Then the objects $(F_\phi/F_\psi)_{\mathcal{X}}$, respectively $(F_{D\psi}/F_{D\phi})_{D\mathcal{X}}$, are, up to isomorphism, the finitely presented objects of the localisation* $(\text{mod-}R, \mathbf{Ab})_{\mathcal{X}}$, *resp.* $(R\text{-mod}, \mathbf{Ab})_{D\mathcal{X}}$. *This category of finitely presented objects may be regarded as the category of pp-defined functors evaluated on objects of $\mathcal{X}$, resp. $D\mathcal{X}$.*

We will refer to $(\text{mod-}R, \mathbf{Ab})_{\mathcal{X}}^{\mathrm{fp}}$ as the **functor category** of $\mathcal{X}$ and denote it $\mathrm{fun}(\mathcal{X})$, also using the notation $\mathrm{Fun}(\mathcal{X})$ for the larger category $(\text{mod-}R, \mathbf{Ab})_{\mathcal{X}}$. The category $(R\text{-mod}, \mathbf{Ab})_{D\mathcal{X}}^{\mathrm{fp}}$ will be denoted by $\mathrm{fun}^{\mathrm{d}}(\mathcal{X})$ and referred to as the **dual functor category**, and $\mathrm{Fun}^{\mathrm{d}}(\mathcal{X})$ will denote $(R\text{-mod}, \mathbf{Ab})_{D\mathcal{X}}$. Thus $\mathrm{fun}(\mathcal{X}) \simeq \left(\mathrm{fun}^{\mathrm{d}}(\mathcal{X})\right)^{\mathrm{op}} \simeq \mathrm{fun}(D\mathcal{X})$. Loosely we use the term "the functor category of $\mathcal{X}$" to denote any of these, allowing context or a descriptor such as "finitely presented" or "dual" to determine exactly which is meant.

12.4 Ziegler spectra and Serre subcategories: summary

In summary, there are natural bijections as follows.

[8] Also see [161], [146] for some cases.

Theorem 12.4.1. *For every small preadditive category R there are natural bijections between the following:*

(i) definable subcategories of Mod-R;
(i)′ definable subcategories of R-Mod;
(ii) closed subsets of Zg_R;
(ii)′ closed subsets of ${}_R\mathrm{Zg}$;
(iii) hereditary torsion theories of finite type (equivalently elementary torsion theories) on (R-mod, **Ab**);
(iii)′ hereditary torsion theories of finite type (equivalently elementary torsion theories) on (mod-R, **Ab**);
(iv) Serre subcategories of (R-mod, **Ab**)^{fp};
(iv)′ Serre subcategories of (mod-R, **Ab**)^{fp};

All the bijections are easily described; we recall some of them.

Suppose that $\mathcal{X}$ is a definable subcategory of Mod-R, defined by the set $(\phi_\lambda/\psi_\lambda)_{\lambda\in\Lambda}$, of pp-pairs, $\mathcal{X} = \{M : \phi_\lambda(M)/\psi_\lambda(M) = 0\ \forall\lambda \in \Lambda\}$. The finitely presented functors $F_\lambda = F_{\phi_\lambda/\psi_\lambda}$ generate a torsion theory, $\tau_{\mathcal{X}}$, of finite type on (mod-R, **Ab**), and a finitely presented functor $F = F_{\phi/\psi}$ is $\tau_{\mathcal{X}}$-torsion iff $\overrightarrow{F}M = 0$ (that is, $\phi(M)/\psi(M) = 0$) for every $M \in \mathcal{X}$, that is, iff ϕ/ψ is closed on $\mathcal{X}$. The intersection of the torsion class, $\mathcal{T}_{\mathcal{X}}$, with the class of finitely presented functors is a Serre subcategory $\mathcal{S}_{\mathcal{X}} = \mathcal{T}_{\mathcal{X}} \cap (\text{mod-}R, \mathbf{Ab})^{\mathrm{fp}}$ of (mod-R, **Ab**)^{fp}. The dual torsion theory on (R-mod, **Ab**) is generated by the dual Serre subcategory, $d\mathcal{S}_{\mathcal{X}} = \{dF : F \in \mathcal{S}_{\mathcal{X}}$ (Section 11.2.2), of $\mathcal{S}_{\mathcal{X}}$: $d\mathcal{S}_{\mathcal{X}}$ is a Serre subcategory of (mod-R, **Ab**)^{fp} and $d^2\mathcal{S}_{\mathcal{X}} = \mathcal{S}_{\mathcal{X}}$. By 12.3.10 this dual torsion theory coincides with $\tau_{D\mathcal{X}}$, the finite-type torsion theory corresponding to the dual definable subcategory, $D\mathcal{X}$ (Section 3.4.2), of R-Mod. A module M belongs to $\mathcal{X}$ iff $M \otimes -$ is $\tau_{D\mathcal{X}}$-torsionfree (and $\tau_{D\mathcal{X}}$-injective since, see 11.1.31, $M \otimes -$ is absolutely pure and $\tau_{D\mathcal{X}}$ is of finite type), indeed $\mathcal{X}$ is recovered from $\tau_{D\mathcal{X}}$ as the class of modules M such that $M \otimes -$ is $\tau_{D\mathcal{X}}$-torsionfree. So $\mathcal{X}$ may be identified with the class of absolutely pure $\tau_{D\mathcal{X}}$-torsionfree functors (12.3.3) and X (the closed subset, $\mathcal{X} \cap \mathrm{Zg}_R$, of Zg_R corresponding to $\mathcal{X}$) may (12.3.5) be identified with the set of indecomposable $\tau_{D\mathcal{X}}$-torsionfree injectives up to isomorphism.

Starting from a closed subset X of Zg_R, the corresponding definable subcategory $\mathcal{X}$ is that which it generates. The functors in $\mathcal{S}_X$ are those $F \in (\text{mod-}R, \mathbf{Ab})^{\mathrm{fp}}$ such that $\overrightarrow{F}N = 0$ for each $N \in X$ and those in $d\mathcal{S}_X$ are those $G \in (R\text{-mod}, \mathbf{Ab})^{\mathrm{fp}}$ such that $(G, N \otimes -) = 0$ for each $N \in X$.

Going from hereditary torsion theories of finite type to Serre subcategories is simply intersecting with the subcategory of finitely presented functors and the reverse process is closure under direct limit.

From a Serre subcategory $\mathcal{S}$ of $(\text{mod-}R, \mathbf{Ab})^{\text{fp}}$ we obtain the corresponding definable subcategory of Mod-R, respectively of R-Mod, as $\{M : \overrightarrow{F}M = 0 : F \in \mathcal{S}\}$, resp. as $\{L : (F, - \otimes L) = 0 \ \forall F \in \mathcal{S}\} = \{L : \overrightarrow{dF}L = 0 \ \forall F \in \mathcal{S}\}$, and the corresponding closed subset of Zg_R consists of all those N such that $N \otimes -$ is $d\mathcal{S}$-torsionfree.

The other cases are similarly easily deduced.

These are also in bijection with certain classes of morphisms in mod-R, characterised in [358, §2.2]; namely, to the definable subcategory $\mathcal{X} \subseteq$ Mod-R associate the set of morphisms f such that every member of $\mathcal{X}$ is injective over f. This is the set of morphisms f such that the corresponding closed subset $\mathcal{X} \cap \text{Zg}_R$ is contained in $(f)^{\text{c}}$ (notation as in Section 5.1.3).

Corollary 12.4.2. *Let $\mathcal{C}$ be a locally finitely presented abelian category. Then the equivalents of 12.4.1 are valid for $\mathcal{C}$ in place of* Mod-R, *with $\mathcal{C}^{\text{fp}}$ replacing* mod-R *and $(\mathcal{C}^{\text{d}})^{\text{fp}}$ replacing R*-mod, *where $\mathcal{C}^{\text{d}}$ is the elementary dual of $\mathcal{C}$ (Section 16.1.3).*

Proof. Various details have to be checked but they all follow fairly directly from 12.4.1, since, by 11.1.27 and 11.1.35, $\mathcal{C}$ is itself a definable subcategory (and a finite-type localisation) of a functor category Mod-R. □

Indeed, the equivalents are valid for any definable category (Section 18.1.1) $\mathcal{D}$ with its dual $\mathcal{D}^{\text{d}}$, with the various associated functor categories, fun($\mathcal{D}$), Fun($\mathcal{D}$) and their duals, in the obvious places. Since every definable category is a definable subcategory of a functor category, and since iterated localisation is just localisation, this follows from 12.4.1 using the correspondence between definable subcategories and finite type localisations.

If R is a von Neumann regular ring, then it has been shown already (3.4.29) that the definable subcategories of Mod-R are in bijection with the two-sided ideals of R. Also recall, 10.2.38, that R_R and $(R_R, -)$ may be essentially identified if R is von Neumann regular. It follows that, for such rings, to the list in 12.4.1 one may add "two-sided ideals in R". In particular one has the following, which may be found as [282, Prop. 8].

Corollary 12.4.3. *Suppose that R is von Neumann regular. Then there is a natural bijection between two-sided ideals, I, of R and Serre subcategories, $\mathcal{S}$, of* $(\text{mod-}R, \mathbf{Ab})^{\text{fp}}$, *which is given explicitly by:*

$$I \mapsto \{F : \textit{for all } M \in \text{Mod-}R,\ MI = 0 \textit{ implies } \overrightarrow{F}M = 0\};$$
$$\mathcal{S} \mapsto \tau_{\mathcal{S}}(R, -).R\}.$$

In particular, if R is a simple von Neumann regular ring, then the only definable subcategories of Mod-R are the whole category and the zero category, a result already seen as 8.2.93.

For artin algebras one may, by 15.3.4, add to the list in 12.4.1 the fp-idempotent ideals of mod-R and those of R-mod.

12.5 Hulls of simple functors

Every simple functor gives a point of the Ziegler spectrum, which is isolated by a minimal pair iff the simple functor is finitely presented (12.5.1, 12.5.2). The version relative to a closed subset is 12.5.4. Arbitrary simple functors give the neg-isolated points (12.5.5, 12.5.6).

Elementary cogenerators correspond to injective cogenerators of finite-type torsion theories (12.5.7). The approach of this section is based on Crawley-Boevey's [141] and, mostly subsequent, work of Burke, Krause, Herzog and Prest.

Much of what is said in this section is said, sometimes in different terms, in Sections 5.3.1 and 5.3.5.

Throughout R is a small preadditive category.

Every simple functor in the category (R-mod, **Ab**) has, by 10.1.13, the form $S = (A, -)/G$, where $A \in R$-mod and G is a maximal subfunctor, of $(A, -)$. There is a fundamental division into two cases: if G is a finitely generated functor, then the quotient S is (by E.1.16) a finitely presented functor; otherwise S, though finitely generated, does not lie in $(R\text{-mod}, \mathbf{Ab})^{\text{fp}}$. In the case that S is finitely presented, there is a projective presentation of the form $(B, -) \xrightarrow{(f,-)} (A, -) \to S \to 0$, where $A \xrightarrow{f} B$ is a morphism in R-mod. Also in this case the dual, $dS \in (\text{mod-}R, \mathbf{Ab})^{\text{fp}}$, is, by 10.3.4, a finitely presented simple functor. In this case, where $S = F_f$, note that $(S) = (f)$ in the notation of Section 5.1.3.

Recall, Section 3.2.1, that a pair $\phi > \psi$ of pp conditions is said to form a minimal pair if for every pp condition θ with $\phi \geq \theta \geq \psi$ either $\theta = \phi$ or $\theta = \psi$. By 10.2.17 it is immediate that $\phi > \psi$ is a minimal pair iff the finitely presented functor $F_{\phi/\psi} = F_\phi / F_\psi$ is simple. By the equivalence, 10.2.30, every finitely presented simple functor may be represented in this form.

Proposition 12.5.1. *Suppose that $S \in (R\text{-mod}, \mathbf{Ab})^{\text{fp}}$ is simple. Then there is a unique indecomposable pure-injective $N \in$ Mod-R such that $(S, N \otimes -) \neq 0$. If S is finitely presented, then $(S) = \{N\} \subseteq \mathrm{Zg}_R$ and N is isolated by a minimal pair.*

Proof. If $E \in (R\text{-mod}, \mathbf{Ab})$ is indecomposable injective and $(S, E) \neq 0$, then S, being simple, must embed in E, so E must be the injective hull of S. That, with 12.1.9, proves the first statement and the second then follows immediately from 12.1.17, by the above observations and the definition (Section 5.3.1) of being isolated by a minimal pair. □

In the above situation we also say that N is **isolated** in Zg_R **by** the finitely presented simple functor S.

Corollary 12.5.2. *Let $N \in \mathrm{Zg}_R$. The following are equivalent:*

(i) N is isolated by a finitely presented simple functor;
(ii) N is isolated by a minimal pair;
(iii) $N \otimes -$ is the injective hull of a finitely presented simple functor in $(R\text{-mod}, \mathbf{Ab})$.

Next we give the local version of this. Let X be a closed subset of Zg_R, with corresponding (12.3.10) finite-type torsion theory τ_{DX} on $(R\text{-mod}, \mathbf{Ab})$ and let $N \in X$. By 11.1.34 every finitely presented simple functor of the localised category $(R\text{-mod}, \mathbf{Ab})_{DX}$ is the localisation of a finitely presented functor, hence, by 12.3.20 and the definition (Section 5.3.1), has the form $(F_{\phi/\psi})_{DX}$ for some X-minimal pair ϕ/ψ. So, with 12.3.20, we have the following.

Remark 12.5.3. Let X be a closed subset of Zg_R. Then the finitely presented simple objects of $(R\text{-mod}, \mathbf{Ab})_{DX}$, respectively of $(\text{mod-}R, \mathbf{Ab})_X$, are those of the form $(F_{D\psi/D\phi})_{DX}$, resp. $(F_{\phi/\psi})_X$, where ϕ/ψ is an X-minimal pair.

Say that N **is isolated by** a finitely presented simple functor **within** (or **modulo**) X if there is a functor $F \in (R\text{-mod}, \mathbf{Ab})^{\mathrm{fp}}$ such that $(F) \cap X = \{N\}$ and such that the localisation $F_{DX} \in (R\text{-mod}, \mathbf{Ab})^{\mathrm{fp}}_{DX}$ is simple.

The next result is proved just as 12.5.2, bearing in mind that $N \in X$ iff $N \otimes -$ is τ_{DX}-torsionfree (12.3.5) and in that case $(N \otimes -) \simeq (N \otimes -)_{DX}$ is an indecomposable injective object of the localised category (11.1.2(1), 11.1.5).

Proposition 12.5.4. *Let X be a closed subset of Zg_R and let $N \in X$. Then the following conditions are equivalent:*

(i) N is isolated by a finitely presented simple functor within X;
(ii) N is isolated by an X-minimal pair;
(iii) $N \otimes -$ is the injective hull of a functor $F \in (R\text{-mod}, \mathbf{Ab})^{\mathrm{fp}}$ whose localisation, F_{DX}, at X is simple;
(iv) the localisation, $(N \otimes -)_{DX}$, of $N \otimes -$ at τ_{DX} is the injective hull of a finitely presented simple functor in $(R\text{-mod}, \mathbf{Ab})_{DX}$.

Now we turn to those indecomposable pure-injectives N such that, either in the absolute or in the relative situation, $N \otimes -$ is the injective hull of a simple, not necessarily finitely presented functor. Recall the description, 12.2.2, of arbitrary finitely generated functors in terms of pp conditions and pp-types, and also the notion ofneg-isolation (5.3.5).

Proposition 12.5.5. *Let p be a pp-type and let ψ be a pp condition. Then:*

(a) p is neg-isolated by ψ iff $F_{D\psi}/F_{Dp}$ is a simple functor, and every simple functor in $(R\text{-mod}, \mathbf{Ab})$ has this form for some choice of p and ψ;
(b) if X is a closed subset of Zg_R, then p is X-neg-isolated by ψ iff the localisation $(F_{D\psi}/F_{D\phi})_{DX}$ is a simple object of $(R\text{-mod}, \mathbf{Ab})_{DX}$, and every simple functor of the localised category has this form for some choice of p and ψ.

Proof. (a) By definition, p is neg-isolated by ψ if $\psi \notin p$ and if p is maximal, as a filter of pp conditions, subject to $\psi \notin p$. By 10.3.7 this is equivalent to $F_{D\psi} \not\leq F_{Dp}$ and F_{Dp} being maximal such, hence to $(F_{D\psi} + F_{Dp})/F_{Dp} \simeq F_{D\psi}/F_{D\psi} \cap F_{Dp}$, that is what in Section 12.2 we have written as $F_{D\psi/Dp}$, being a simple functor. The fact that every simple functor has this form is by 12.2.2.

(b) The relativised version follows by 11.1.15 and (the comment after) 12.3.18. □

Corollary 12.5.6. *(a) $N \in \mathrm{Zg}_R$ is neg-isolated iff $N \otimes -$ is the injective hull of a simple functor in $(R\text{-mod}, \mathbf{Ab})$.*
(b) Let X be a closed subset of Zg_R, let τ_{DX} be the corresponding finite-type torsion theory on $(R\text{-mod}, \mathbf{Ab})$ and suppose that $N \in X$. Then N is X-neg-isolated iff $(N \otimes -)_{DX}$ is the injective hull of a simple functor in the localisation $(R\text{-mod}, \mathbf{Ab})_{DX}$ and this is so iff $N \otimes -$ is the injective hull of a functor in $(R\text{-mod}, \mathbf{Ab})$ whose localisation under τ_{DX} is simple.

Proof. (a) If N is neg-isolated, that is, if $N = H(p)$ for some neg-isolated (by ψ say) pp-type p, then, by 12.2.4, $(F_{D\psi/Dp}, N \otimes -) \neq 0$ so $N \otimes -$ is the injective hull of the simple functor $F_{D\psi/Dp}$. The converse follows from 12.5.5(a).

(b) The relative version follows similarly. □

In the above situation we may say that N is **neg-isolated (relative to X) by** the simple functor S (resp. the simple functor S_{DX}) where S is $F_{D\psi/Dp}$. Then N is the only point of Zg_R (resp. X) such that $(F_{D\psi/Dp}, N \otimes -) \neq 0$, that is, such that $p(N)/p(N) \cap \psi(N) \neq 0$.

Corollary 12.5.7. *Let N be pure-injective (not necessarily indecomposable) and let $\tau = \tau_{DN}$ be the torsion theory on $(R\text{-mod}, \mathbf{Ab})$ corresponding to the definable subcategory of* Mod-R *generated by N. Then N is an elementary cogenerator iff $(N \otimes -)_{DN}$ is an injective cogenerator of the localised category $(R\text{-mod}, \mathbf{Ab})_{DN}$, equivalently if every simple functor in $(R\text{-mod}, \mathbf{Ab})_{DN}$ embeds in $(N \otimes -)_{DN}$.*

Proof. The first statement follows from the definition of elementary cogenerator, 12.1.6 and 12.3.3. Since an injective object of a locally finitely generated category is an elementary cogenerator iff it embeds every simple object, the second statement is immediate. □

As deduced in 5.3.50, this is so iff N realises every $\overline{\{N\}}$-neg-isolated pp-type, where $\overline{\{N\}}$ denotes the Ziegler-closure of N.

Finally in this section, in connection with (non-)existence of superdecomposable pure-injectives, 5.3.24 has been extended by Burke as follows. Say that a locally coherent category $\mathcal{C}$ has the **isolation property** if, for every torsion theory τ of finite type on $\mathcal{C}$, if there is a finitely presented object C such that there is a unique τ-torsionfree indecomposable injective E with $(C, E) \neq 0$, then there is such an object C with C_τ simple in $\mathcal{C}_\tau$. Taking $\mathcal{C} = (R\text{-mod}, \mathbf{Ab})$ gives, by 12.3.2, the isolation condition as we defined it earlier.

Theorem 12.5.8. ([104, 2.5]) *If $\mathcal{C}$ is a locally coherent abelian category with no superdecomposable injective object (then $\mathcal{C}$ is said to be* **locally coirreducible***), then $\mathcal{C}$ satisfies the isolation condition.*

A corollary is 5.3.25.

12.6 A construction of pp-types without width

It was shown in 7.3.4 that, if there is a superdecomposable pure-injective right R-module, then the lattice, pp_R, of pp conditions for right R-modules does not have width, $\mathrm{w}(\mathrm{pp}_R) = \infty$; for width see Section 7.3.1. It is not known whether the converse is true in general but Ziegler did prove the converse under the assumption that the ring is countable (12.6.1), more generally, a relative version, 12.6.2, under the assumption that there are only countably many pp conditions. The proof is quite technical, so the absolute version is established first and the proof of the version relative to a closed subset of the Ziegler spectrum is indicated after that. By 11.1.39 and 13.4.4 it would make no difference whether we make the width assumption for pp conditions in one free variable or in $m > 1$ free variables.[9]

Theorem 12.6.1. ([726, 7.8(2)]) *Suppose that R is a countable ring and that $\mathrm{w}(\mathrm{pp}_R) = \infty$. More generally, suppose that the lattice, $\mathrm{Latt}^{\mathrm{f}}(R^m \otimes -)$, of finitely generated subfunctors of $(R^m \otimes -) \in (R\text{-mod}, \mathbf{Ab})$ is countable and that $\mathrm{w}\big(\mathrm{Latt}^{\mathrm{f}}(R^m \otimes -)\big) = \infty$. Then there is a subfunctor $K < (R^m \otimes -)$ such that $(R^m \otimes -)/K$ has no uniform subfunctor, hence such that the injective hull*

[9] But the number of free variables can make a difference to the countability condition: for example, the lattice of pp-1-types over a field k has just two elements but the lattice pp_R^2 has card(k) elements if k is infinite.

$E\big((R^m \otimes -)/K\big)$ is superdecomposable. In particular, there is a superdecomposable pure-injective R-module.

Proof. Enumerate the finitely generated subfunctors of $R^m \otimes -$ as $(F_n)_{n\geq 0}$. A sequence of sets of intervals in the lattice $\mathrm{Latt}^{\mathrm{f}}(R^m \otimes -)$ of finitely generated subfunctors of $(R^m \otimes -)$ is defined as follows:

$$\mathcal{I}_{-1} = \{[R^m \otimes -, 0]\}.$$

Having defined $\mathcal{I}_n = \{[G_{n1}, G'_{n1}], \dots, [G_{nt_n}, G'_{nt_n}]\}$ with, inductively, $\mathrm{w}[G_{ni}, G'_{ni}] = \infty$ for each i, consider, at **stage** $n+1$, the functor F_{n+1}.

For each $i = 1, \dots, t_n$:

case (a) $\mathrm{w}[F_{n+1} + G_{ni}, F_{n+1} + G'_{ni}] = \infty$ – then put this interval, $[F_{n+1} + G_{ni}, F_{n+1} + G'_{ni}]$, into $\mathcal{I}_{n+1}$;

case (b) $\mathrm{w}[F_{n+1} + G_{ni}, F_{n+1} + G'_{ni}] < \infty$. Choose a diamond as shown (the double lines in the diagram are allowed to represent equalities, the single lines represent proper inclusions)

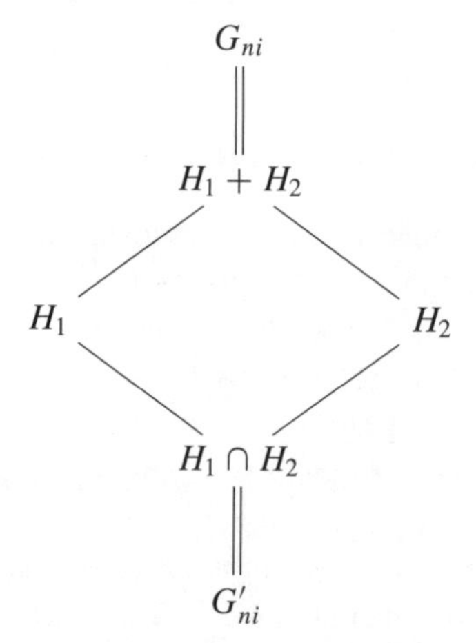

with both $\mathrm{w}[H_1 + H_2, H_j] = \infty$ $(j = 1, 2)$ and put the intervals $[H_1 + H_2, H_1]$ and $[H_1 + H_2, H_2]$ into $\mathcal{I}_{n+1}$. Note that the width of each interval in $\mathcal{I}_{n+1}$ is ∞ so the induction continues. Noting that each interval in $\mathcal{I}_n$ has either one or two descendants – those formed from it – in $\mathcal{I}_{n+1}$, we define, for ease of expression, "descendence" to be the transitive closure of this relation. Observe that for each descendant interval $[G, G']$ of $[G_{ni}, G'_{ni}]$ we have, for any $F \in \mathrm{Latt}^{\mathrm{f}}(R^m \otimes -)$, $\mathrm{w}[F + G, F + G'] \leq \mathrm{w}[F + G_{ni}, F + G'_{ni}]$, because the first interval is, inductively, isomorphic to a subquotient of the second interval.

Set $K_n = \bigcap_1^{t_n} G'_{ni}$. By construction, for all n and all j we have $G'_{n+1,j} \geq G'_{ni}$ for some i, so $K_{n+1} \geq K_n$.

At the end of the process set $K = \sum_n K_n$.

The fate of F_{n+1}, that is whether or not $F_{n+1} \leq K$, is determined at stage $n+1$: if case (a), then $F_{n+1} \leq K$, and if case (b), then $F_{n+1} \nleq K$. For, if case (a) holds at that stage for each $i = 1, \dots, t_n$, then, by definition and construction, $F_{n+1} \leq K_{n+1}$. If case (b) holds for some i, then for each descendant interval $[G, G']$ of $[G_{ni}, G'_{ni}]$ we have $\mathrm{w}[F_{n+1} + G, F_{n+1} + G'] \leq \mathrm{w}[F_{n+1} + G_{ni}, F_{n+1} + G'_{ni}] < \infty$. Therefore $F_{n+1} \nleq G'$ (because $\mathrm{w}[G, G'] = \infty$), hence $F_{n+1} \nleq K$ (because F_{n+1} is finitely generated so, if contained in K, is contained in some K_l).

The following is claimed. If, at stage $n+1$, case (b) holds for i, then $F_{n+1} \cap H_j \nleq K$ $(j = 1, 2)$, where H_1, H_2 are as before.

To see this, taking $j = 1$:

$$(F_{n+1} \cap H_1) + H_2 = (F_{n+1} \cap H_1) + G'_{ni} + H_2 = H_1 \cap (F_{n+1} + G'_{ni}) + H_2$$

by modularity, and

$$(F_{n+1} \cap H_1) + H_1 + H_2 = H_1 + H_2 = H_1 \cap (F_{n+1} + G_{ni}) + H_2.$$

Therefore the interval

$$[(F_{n+1} \cap H_1) + H_1 + H_2, (F_{n+1} \cap H_1) + H_2]$$

is identical to

$$[H_1 \cap (F_{n+1} + G_{ni}) + H_2, H_1 \cap (F_{n+1} + G'_{ni} + H_2)],$$

which is isomorphic to a subquotient of $[F_{n+1} + G_{ni}, F_{n+1} + G'_{ni}]$, which, since we are in case (b), has width $< \infty$.

Now, $F_{n+1} \cap H_1 = F_k$ for some k (10.2.7).

case $k \leq n$. If case (a) held at stage k, then, as observed already, $F_k \leq K_k \leq K_{n+1} \leq G'_{ni}$, which is impossible since $\mathrm{w}[H_1 + H_2, H_2] = \infty$ and $G'_{ni} \leq H_2$. Therefore case (b) held at stage k, so, as observed above, $F_{n+1} \cap H_1 = F_k \nleq K$, as claimed.

case $k = n+1$. If $F_{n+1} \cap H_1 = F_{n+1}$, that is, $F_{n+1} \leq H_1$, then we could not have case (b) for i since already $\mathrm{w}[G_{ni}, H_1] = \infty$ and case (a) would have applied, so this is impossible.

case $k > n$. By an argument made above, case (b) will hold for F_k and any descendant interval of $[H_1 + H_2, H_2]$. For, note that the interval $[H_1, H_1 \cap F_{n+1} + G'_{ni}] = [H_1, H_1 \cap (F_{n+1} + G'_{ni})]$ is isomorphic to

$[H_1 + F_{n+1}, G'_{ni} + F_{n+1}]$, which is contained in $[G_{ni} + F_{n+1}, G'_{ni} + F_{n+1}]$, hence has width $< \infty$. Therefore $\mathrm{w}[H_1 + H_2, H_1 \cap F_{n+1} + H_2] < \infty$, that is, $\mathrm{w}[F_k + H_1 + H_2, F_k + H_2] < \infty$. So, as remarked before, it follows that $F_k \not\leq K$.

Thus the claim is established.

Now suppose, for a contradiction, that $(R^m \otimes -)/K$ had a non-zero uniform subobject. Then there would be a finitely generated subfunctor, F, of $R^m \otimes -$ such that $(F + K)/K \simeq F/F \cap K$ is non-zero and uniform. Take n such that F is F_{n+1}.

Since $F \not\leq K$, case (b) holds at stage $n + 1$ for some interval in $\mathcal{I}_n$. Say, reindexing $\mathcal{I}_n$ if necessary, case (b) holds for $i = 1, \ldots, s\ (\geq 1)$ and case (a) holds for $i = s + 1, \ldots, t_n$. For $i = 1, \ldots, s$ and $j = 1, 2$ we have, as shown above, $F \cap H_{ij} \not\leq K$ (writing H_{ij} for what before was H_j). Now

$$K_{n+1} = \bigcap_{i=1}^{s} H_{i1} \cap H_{i2} \cap \bigcap_{i=s+1}^{t_n} (F + G'_{ni})$$

so

$$F \cap K_{n+1} = \bigcap_{i=1}^{s} (F \cap H_{i1}) \cap (F \cap H_{i2}).$$

Thus we have finitely many non-zero subfunctors of $F/F \cap K$ with zero intersection; a contradiction, as required.[10] □

This is Ziegler's proof set in the functorial context. For the original proof see [726, 7.8(2)], also [495, 10.13].[11]

The proof could be simplified marginally, as in [495, 10.13], by working in the quotient category obtained by factoring out all finitely presented functors with width $< \infty$ (see Section 13.4). This is equivalent to working in the quotient lattice which results from identifying those points ϕ, ψ where $\mathrm{w}[\phi + \psi, \phi \wedge \psi] < \infty$, as done in [495].

Corollary 12.6.2. *Suppose that X is a closed subset of Zg_R and suppose that the lattice $\mathrm{pp}^m(X)$, equivalently the lattice, $\mathrm{Latt}^{\mathrm{f}}((R^m \otimes -)_{DX})$, of finitely generated*

[10] The rough idea of the proof is that in case (a) F_{n+1} is "low", so can go into K: enough complexity can still be seen above F_{n+1}. Otherwise F_{n+1} is "high" (at least for some interval), so it is kept out of K and the complexity between it and K (that is, non-uniformity of the factor) is expressed through the intervals created, and retained, in some sense, between it and K.

[11] In the latter, I modified the proof slightly, by requiring that each finitely generated functor appear infinitely often in the list, in order to avoid proving directly some things about the construction.

subfunctors of $(R^m \otimes -)_{DX}$, *is countable and has width undefined. Then there is a superdecomposable pure-injective* R*-module with support contained in* X.

Proof. The proof of 12.6.1 applies also in the lattice of finitely generated subfunctors of $(R^m \otimes -)_{DX}$. (An alternative would be to take the F_n as in the proof there to be finitely generated and τ_{DX}-dense in the pullbacks to $(R, -)$ of the finitely generated subfunctors of $(R, -)_{DX}$ and then to check that $(R^m \otimes -)/\overline{K}$ has no uniform subobject, where $\overline{K}$ is the τ_{DX}-closure of K in $R^m \otimes -$, yielding that its injective hull is a superdecomposable injective of the localised category.) □

12.7 The full support topology again

Finitely generated functors give a basis for the full support topology (5.3.65, 12.7.3), the closed sets of which are exactly the sets of torsionfree injectives for some hereditary torsion theory (12.7.2). The closed set/torsion theory correspondence is, however, not in general a bijection (12.7.5). The isolated points are exactly the injective hulls of simple functors and they are dense (12.7.4).

Burke (see [102, §6]) defined, on any locally finitely generated Grothendieck additive category $\mathcal{C}$, a topology on the set, inj($\mathcal{C}$), of (isomorphism classes of) indecomposable injective objects of $\mathcal{C}$ by taking, for a basis of open sets, those of the form $(C) = \{E \in \text{inj}(\mathcal{C}) : (C, E) \neq 0\}$, where C ranges over arbitrary objects of $\mathcal{C}$. Since $(C) = \bigcup\{(F) : F \leq C \text{ is finitely generated}\}$ (because $\mathcal{C}$ is locally finitely generated) we can be more restrained in the definition and take, for a basis, only the sets F with F finitely generated. First we check that this does give a basis for a topology.

This topology is, by 12.7.3 below, that already seen in Section 5.3.7 and is called the **full support**, or just **fs**, topology, in order to contrast it with the topology defined in terms only of finitely presented functors.

Lemma 12.7.1. *The sets* $(F) = \{E \in \text{inj}(\mathcal{C}) : (F, E) \neq 0\}$, *as* F *ranges over finitely generated objects of* $\mathcal{C}$, *form a basis for a topology.*

Proof. If $E \in (F) \cap (G)$, say $f : F \to E$ and $g : G \to E$ are non-zero, then let H be any non-zero finitely generated subfunctor of $\text{im}(f) \cap \text{im}(g)$ (that is non-zero since E is uniform). Then $E \in (H) \subseteq (F) \cap (G)$: to see that, say, $(H) \subseteq (F)$; if $E' \in (H)$, say $h' : H \to E'$ is non-zero, then, by injectivity of E', this extends to $h : E \to E'$, hence hf is a non-zero map from F to E', as required. □

Proposition 12.7.2. *The fs-closed subsets of* inj($\mathcal{C}$) *are precisely those of the form* $\mathcal{F} \cap \text{inj}(\mathcal{C})$, *where* τ *is a hereditary torsion theory (Section 11.1.1) on* $\mathcal{C}$.

Proof. The complement of an open set (C), where $C \in \mathcal{C}$, consists of the indecomposable τ-torsionfree injectives, where τ is the torsion theory generated by C, that is, which is determined by declaring C to be torsion. A typical closed set is an intersection of such complements, hence consists of the indecomposable injectives of some torsionfree class.

If, conversely, $\mathcal{F} \subseteq \mathcal{C}$ is a hereditary torsionfree class, then, just from the definitions and 11.1.1, $\mathcal{F} \cap \text{inj}(\mathcal{C})$ is the intersection of the $(C) \cap \text{inj}(\mathcal{C})$ as C ranges over (finitely generated) torsion (for $\mathcal{F}$) objects of $\mathcal{C}$, hence is closed. □

So this topology, denote it $\text{fs}(\mathcal{C})$, is, cf. 12.4.1, finer than the Ziegler topology on this set since the latter uses only the torsion theories of finite type.

Burke notes that, in the case where $\mathcal{C}$ has no superdecomposable injectives ($\mathcal{C}$ is "locally coirreducible" in the terminology of [486]), this topology coincides with that defined by Popescu in [486, 5.3.9].

If $\mathcal{C}$ has the form $(R\text{-mod}, \mathbf{Ab})$, then regard this, via 12.1.9, as a topology on pinj_R (which is also the underlying set of the Ziegler topology, Zg_R) and denote it by fs_R. The fact that this is the topology seen already in Section 5.3.7 is immediate from the description, 12.2.2, of arbitrary finitely generated functors in terms of pp-types and pp conditions.

Proposition 12.7.3. ([102, 7.2]) *The topology* fs_R *is the full support topology in the sense defined in Section 5.3.7.*

Corollary 12.7.4. *A point of* $fs(\mathcal{C})$ *is isolated iff it is the injective hull of a simple object.*

The isolated points of $fs(\mathcal{C})$ *are dense in* $fs(\mathcal{C})$.

Proof. Certainly if $E = E(S)$ and S is simple, then $(S) = \{E\}$. Conversely, if F is finitely generated and is such that $(F) = \{E\}$ for some $E \in \text{inj}(\mathcal{C})$, then let S be a simple quotient of F (since F is finitely generated it has a maximal subfunctor). Clearly $(S) = (F) = \{E\}$, so $E \simeq E(S)$. This argument also establishes density of points of the form $E(S)$ with S simple. □

Unless every hereditary torsion theory on $\mathcal{C}$ is cogenerated by a set of indecomposable injectives, the map $\mathcal{F} \mapsto \mathcal{F} \cap \text{inj}(\mathcal{C})$ from the set of hereditary torsion theories to the set of fs-closed subsets will not be injective.

Example 12.7.5. The following is an example of a hereditary torsion theory which is not cogenerated by a set of indecomposable injectives. Let R be the free associative algebra in two generators over a field (or over $\mathbb{Z}$). Then the injective hull, E, of R is a superdecomposable injective module (4.4.3). Consider the torsion theory, τ, cogenerated by E. Any τ-torsionfree module embeds in a power of E, therefore has a submodule isomorphic to R, so cannot be an indecomposable injective module, as required.

So $\mathcal{F}_\tau \cap \mathrm{inj}_R = \emptyset$.

It follows from all this and 12.2.4 that if τ is a hereditary torsion theory on $(R\text{-mod}, \mathbf{Ab})$, then there is a set $(p_\lambda, \psi_\lambda)_{\lambda\in\Lambda}$ with each p_λ a pp-type and each ψ_λ a pp condition, such that, for all $M \in \text{Mod-}R$, $M \otimes -$ is τ-torsionfree iff for all λ we have $p_\lambda(M) = p_\lambda(M) \cap \psi_\lambda(M)$.

12.8 Rings of definable scalars again

The functor category over a ring of definable scalars is compared with the corresponding localised functor category (12.8.1). Epimorphisms of rings are characterised in terms of this comparison (12.8.7).

The ring of definable scalars may be realised as the endomorphism ring of the localised forgetful functor (12.8.2). Similarly for endomorphism rings of objects in the category of relative pp-pairs (12.8.3). These rings may be realised as biendomorphism rings (12.8.4, 12.8.5).

If every object of the quotient category is projective, equivalently (12.8.9) if the lattice of closed subobjects of the forgetful functor is complemented, then the ring of definable scalars is von Neumann regular (12.8.8).

The Gabriel–Zariski spectrum of $(R\text{-mod}, \mathbf{Ab})$ is the underlying space of an analogue, indeed a generalisation, of the structure sheaf of an algebraic variety, see Sections 14.1.2, 14.1.4. Here we describe the localisations involved.

Let R be a ring (or small preadditive category) and let $\mathcal{S}$ be a Serre subcategory of $(R\text{-mod}, \mathbf{Ab})^{\mathrm{fp}}$. Write $R_{\mathcal{S}}$ for $\mathrm{End}\big(({}_RR, -)_{\mathcal{S}}\big)$, the endomorphism ring of the image of the forgetful functor $({}_RR, -)$ in the localised category $(R\text{-mod}, \mathbf{Ab})^{\mathrm{fp}}/\mathcal{S}$. In view of 12.8.2 we refer to the map $R \longrightarrow R_{\mathcal{S}}$ induced by $R \simeq \mathrm{End}({}_RR, -) \to \mathrm{End}\big(({}_RR, -)_{\mathcal{S}}\big)$ as a ring of definable scalars. The inclusion of $R_{\mathcal{S}}$, regarded as a one-point category mapped to $({}_RR, -)_{\mathcal{S}}$, into $(R\text{-mod}, \mathbf{Ab})^{\mathrm{fp}}/\mathcal{S}$ induces, by 10.2.37, an exact functor $p : (R_{\mathcal{S}}\text{-mod}, \mathbf{Ab})^{\mathrm{fp}} \longrightarrow (R\text{-mod}, \mathbf{Ab})^{\mathrm{fp}}/\mathcal{S}$ or, in more economical notation and using 10.2.37, $\mathrm{Ab}(R_{\mathcal{S}}) \longrightarrow \mathrm{Ab}(R)/\mathcal{S}$. Also, the composition of the ring morphism $R \to R_{\mathcal{S}}$ with the canonical morphism $R_{\mathcal{S}} \to \mathrm{Ab}(R_{\mathcal{S}})$ induces an exact functor $\mathrm{Ab}(R) \longrightarrow \mathrm{Ab}(R_{\mathcal{S}})$. Since the composition of this with $\mathrm{Ab}(R_{\mathcal{S}}) \longrightarrow \mathrm{Ab}(R)/\mathcal{S}$ is an exact functor and agrees on R with the localisation functor $\mathrm{Ab}(R) \longrightarrow \mathrm{Ab}(R)/\mathcal{S}$, which is also exact, the uniqueness in the definition (Section 10.2.7) of free abelian category implies that these are equal up to natural equivalence. We state this.

Proposition 12.8.1. ([282, p. 253]) *Let $\mathcal{S}$ be a Serre subcategory of $(R\text{-mod}, \mathbf{Ab})^{\mathrm{fp}}$ and let $f : R \to R_{\mathcal{S}}$ be the corresponding ring of definable scalars. Then there is a diagram as shown, which is commutative up to natural equivalence.*

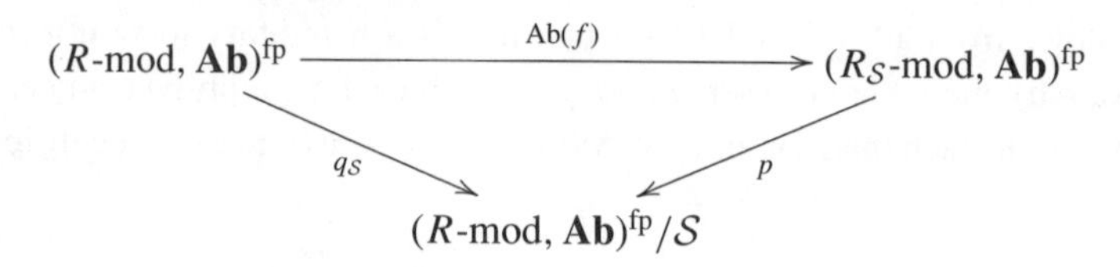

that is,

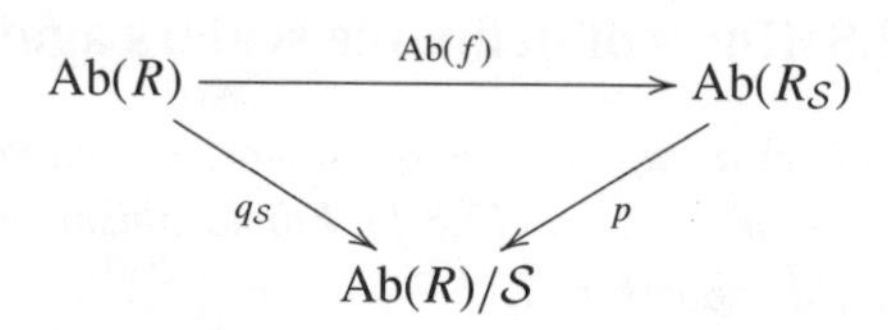

where q_s is the localisation functor (Section 11.1.4) and where p is the functor induced by the inclusion of $R_{\mathcal{S}}$ in $(R\text{-mod}, \mathbf{Ab})^{\mathrm{fp}}/\mathcal{S}$. The functor p is exact, full and dense, so is the quotient of $(R_{\mathcal{S}}\text{-mod}, \mathbf{Ab})^{\mathrm{fp}}$ by the Serre subcategory $\mathcal{S}' = \{G \in (R_{\mathcal{S}}\text{-mod}, \mathbf{Ab})^{\mathrm{fp}} : pG = 0\}$. In particular, $\mathrm{Ab}(R)/\mathcal{S} \simeq \mathrm{Ab}(R_{\mathcal{S}})/\mathcal{S}'$.

The functor p is an equivalence, $\mathrm{Ab}(R)_{\mathcal{S}} \simeq \mathrm{Ab}(R_{\mathcal{S}})$, iff the kernel of p (in the sense of objects) is 0.

The fact that p is exact and a localisation at a Serre subcategory is by 11.1.40; it is full and dense since $q_{\mathcal{S}}$ already has these properties.

Note that the definable subcategory, $\mathcal{X}_{\mathcal{S}}$ of Mod-R, corresponding to $\mathcal{S}$, if regarded as a subcategory of Mod-$R_{\mathcal{S}}$ via the natural action of its ring of definable scalars $R_{\mathcal{S}}$, is definable. For, clearly it is closed under products and direct limits and, if M' is a pure $R_{\mathcal{S}}$-submodule of $M \in \mathcal{X}_{\mathcal{S}}$ with the natural $R_{\mathcal{S}}$-structure, then certainly M' is a pure submodule of M_R, hence is in $\mathcal{X}$. Since $\mathrm{Ab}(R)/\mathcal{S}$ is the functor category of $\mathcal{X}_{\mathcal{S}}$ (12.3.20) the Serre subcategory, $\mathcal{S}' \subseteq \mathrm{Ab}(R_{\mathcal{S}})$, appearing above is that which corresponds to this definable subcategory of Mod-$R_{\mathcal{S}}$.

Theorem 12.8.2. ([498, A4.2], see [106, 4.6]) *Let $\mathcal{X} \subseteq$ Mod-R be a definable subcategory and let $\mathcal{S} = \mathcal{S}_{\mathcal{X}}^{\mathrm{d}} = \tau_{D\mathcal{X}} \cap (R\text{-mod}, \mathbf{Ab})^{\mathrm{fp}}$ be the Serre subcategory of $(R\text{-mod}, \mathbf{Ab})^{\mathrm{fp}}$ corresponding, in the sense of 12.4.1, to $\mathcal{X}$. Then the natural map $R \to \mathrm{End}\big(({}_RR, -)_{\mathcal{S}}\big)$, that is, $R \to \mathrm{End}\big(({}_RR, -)_{DX}\big)$, where $X = \mathcal{X} \cap \mathrm{Zg}_R$, is the ring of definable scalars, $R \to R_{\mathcal{X}}$, of $\mathcal{X}$.*

If $\mathcal{S}' = \mathcal{S}_{\mathcal{X}}$ is the, dual, Serre subcategory of $(\text{mod-}R, \mathbf{Ab})^{\mathrm{fp}}$ corresponding to $\mathcal{X}$, then the canonical map $R \to \big(\mathrm{End}\big((R_R, -)_{\mathcal{S}'}\big)\big)^{\mathrm{op}}$ also is the ring of definable scalars of $\mathcal{X}$.

Proof. The ring of definable scalars of $\mathcal{X}$ (Section 6.1.1) is, by 3.2.15, the opposite of the endomorphism ring of the image of the object $x = x/x = 0$ in the category, $(\mathbb{L}_R^{\mathrm{eq}+})_{\mathcal{X}}$, of definable scalars of $\mathcal{X}$. So this follows directly from the equivalence 12.3.20. □

As stated, this is for a ring R; if R is a small preadditive category the slightly modified statement and proof are valid. More generally, the proof applies to give the following (using 12.4.2), which we state in terms of finite-type torsion theories, rather than Serre subcategories, and in terms of objects, rather than definable subcategories (recall, 3.4.12, that every definable subcategory has a "generator"). For the case $\mathcal{C} = \text{Mod-}R$ this is from [106].

Theorem 12.8.3. *Let $\mathcal{C}$ be a locally finitely presented category. Take $C \in \mathcal{C}$ and let τ_C be the hereditary torsion theory of finite type on* Fun($\mathcal{C}$) *corresponding to the definable subcategory generated by C. Let ϕ/ψ be a pp-pair. Then the natural map* $\mathrm{End}(F_{\phi/\psi}) \to \mathrm{End}\big((F_{\phi/\psi})_{\tau_C}\big)$ *coincides with the map from the ring of pp-definable endomorphisms of the sort ϕ/ψ to the ring of pp-definable endomorphisms of the image of ϕ/ψ in the category $\mathbb{L}(\mathcal{C})_C^{\mathrm{eq}+}$ (see the end of Section 3.2.2 for the latter), that is, to the natural map between the (opposites of the) corresponding rings of definable scalars.*

Similarly, from 6.1.18 the following is obtained.

Corollary 12.8.4. ([498, A4.1], [106, 3.5]) *Let $\mathcal{A}$ be a small preadditive category and let $M \in$* Mod-$\mathcal{A}$. *Suppose that N is an elementary cogenerator for the definable subcategory generated by M, equivalently, 12.5.7, N is a module such that $N \otimes -$ is an injective cogenerator for the finite-type hereditary torsion theory τ_M^{d} on* $\mathrm{Fun}^{\mathrm{d}}$-$\mathcal{A}$ *corresponding to M. Then there is an index set I such that the biendomorphism ring of N^I is the ring of definable scalars of M: $\mathcal{A}_M \simeq \mathrm{Biend}(N^I)$.*

More generally, it is straightforward to prove the following (it is done for the case $\mathcal{C} = \text{Mod-}R$ in [498] and [106]).

Corollary 12.8.5. *Let $\mathcal{C}$ be a locally finitely presented abelian category and suppose that $C \in \mathcal{C}$. Then there is a pure-injective object N in the definable subcategory of $\mathcal{C}$ generated by C, and an index set I, such that, for every pp-pair, ϕ/ψ, one has $R_C^{\phi/\psi} \simeq \mathrm{End}\big(_{\mathrm{End}(N)}(\phi/\psi)(N^I)\big)$, where the action of* End(N) *on $(\phi/\psi)(N^I)$ is that induced by the diagonal action on N.*

We may rephrase this in terms of functors, as follows.

Corollary 12.8.6. *Let $\mathcal{C}$ be a locally finitely presented abelian category and suppose that $C \in \mathcal{C}$. Then there is a pure-injective object N in the definable subcategory of $\mathcal{C}$ generated by C, and an index set I, such that, for every $F \in$*

$(\mathcal{C}^{\mathrm{fp}}, \mathbf{Ab})^{\mathrm{fp}}$ *one has* $\big(\mathrm{End}(F_C)\big)^{\mathrm{op}} \simeq \mathrm{End}\big({}_{\mathrm{End}(N)}F(N^I)\big)$, *where the action of* $\mathrm{End}(N)$ *on* $F(N^I)$ *is that induced by the diagonal action on* N.

That is, if $\mathcal{C}$ is a locally finitely presented abelian category and if $F = F_{\phi/\psi}$ is a finitely presented functor in $\mathrm{Fun}(\mathcal{C}) = (\mathcal{C}^{\mathrm{fp}}, \mathbf{Ab})$ and if $\mathcal{X}$ is a definable subcategory of $\mathcal{C}$, then the ring of definable scalars $R_{\mathcal{X}}^{\phi/\psi}$ is, where we regard it as acting on the right, naturally identified with the opposite of the endomorphism ring of the localisation of F at the hereditary finite-type torsion theory $\tau_{\mathcal{X}}$ on (mod-R, $\mathbf{Ab}$) which corresponds to $\mathcal{X}$. In particular, any endomorphism of F is pp-definable. We call $(\mathcal{C}^{\mathrm{fp}}, \mathbf{Ab})_{\tau_{\mathcal{X}}}^{\mathrm{fp}} \simeq (\mathcal{C}^{\mathrm{fp}}, \mathbf{Ab})^{\mathrm{fp}}/\mathcal{S}_{\mathcal{X}}$, where $\mathcal{S}_{\mathcal{X}} = \mathcal{T}_{\mathcal{X}} \cap (\mathcal{C}^{\mathrm{fp}}, \mathbf{Ab})^{\mathrm{fp}}$ (11.1.33, 11.1.42), the **category of definable scalars** of $\mathcal{C}$ **at** $\mathcal{X}$. We have the notation $\mathrm{fun}(\mathcal{X})$ for this category: it is shown at 18.1.19 that this depends only on the category $\mathcal{X}$ – it is independent of representation of $\mathcal{X}$ as a definable subcategory of some locally finitely presented category. By 12.3.20 this skeletally small abelian category can be regarded as the category $\mathbb{L}(\mathcal{X})^{\mathrm{eq+}}$ of pp-imaginaries, that is, the category of pp-pairs, on $\mathcal{X}$.

Let $f : R \to S$ be any morphism of rings. Let $\mathcal{X}_f$ denote the definable subcategory of Mod-R generated by Mod-S, the latter regarded as a subcategory of Mod-R via f. Just from the definitions and 6.1.5, the ring of definable scalars for $\mathcal{X}_f$ is the ring R_f defined just before 6.1.12 and, by that result, R_f is contained in the dominion of f (the "epimorphic hull" of fR in S). Let f_0 denote the corestriction of f to R_f and let f_1 denote the inclusion of R_f in S. So $\mathcal{X}_f$ is also the closure of the image of Mod-S in Mod-R_f induced by the inclusion $R_f \to S$. These morphisms combined with those in the diagram in 12.8.1 give the following diagram, where $\mathcal{S}$ and $\mathcal{S}'$ are as in 12.8.1, in particular $\mathcal{S}$ is the Serre subcategory corresponding to $\mathcal{X}_f$, so $R_f = R_{\mathcal{S}}$. The outer triangle is obtained as the factorisation of $\mathrm{Ab}(f)$ through its kernel; note that $\mathcal{S} = \mathcal{S}_f = \{F_{D\psi/D\phi} : \phi(M_R) = \psi(M_R)\ \forall M_S\}$. All three functors in the outer triangle are exact, the embedding of $\mathrm{Ab}(R)/\mathcal{S}_f$ in $\mathrm{Ab}(S)$ being exact by 11.1.40.

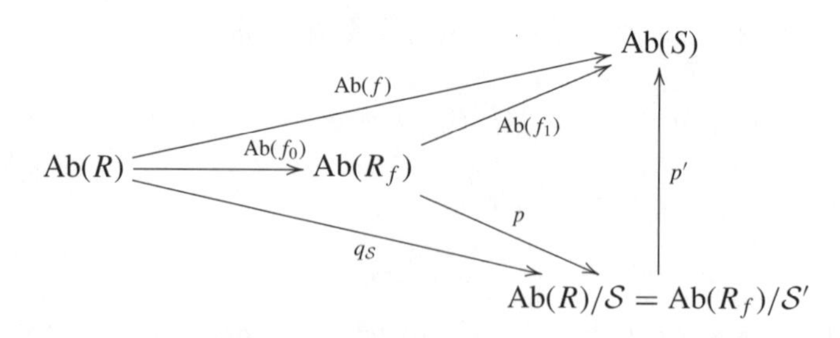

The right-most triangle commutes (at least, up to natural equivalence and one may adjust p'), since both routes are exact functors which take the one-point category

R_f to S, and so the uniqueness clause in the definition of free abelian category applies.

For any ring morphism f one has this commutative diagram of exact functors. In the special case that f is an epimorphism, $R_f = S$, so $\mathrm{Ab}(f_1)$ is an equivalence, hence the kernel of p is zero and so, by 12.8.1, p is an equivalence. That proves one half of the next result, which is [282, Thm 5] (and is presented there as a gloss on 6.1.8). This result can be seen as 6.1.9 extended from pp conditions to the categories $\mathbb{L}^{\mathrm{eq}+}$ and then expressed in terms of the equivalent (by 10.2.30) categories of finitely presented functors.

Theorem 12.8.7. *A morphism $R \to S$ of rings is an epimorphism of rings iff the induced functor $(R\text{-mod}, \mathbf{Ab})^{\mathrm{fp}} \longrightarrow (S\text{-mod}, \mathbf{Ab})^{\mathrm{fp}}$ is the quotient of $(R\text{-mod}, \mathbf{Ab})^{\mathrm{fp}}$ by the Serre subcategory $\mathcal{S}_f = \{F_{D\psi/D\phi} : \phi(M_R) = \psi(M_R)\ \forall M_S\}$, that is, iff this is an isomorphism $\mathrm{Ab}(S) \simeq \mathrm{Ab}(R)/\mathcal{S}_f$.*

Proof. For the half not proved above, if $\mathrm{Ab}(S)$ is this localisation, that is, if the canonical map from $\mathrm{Ab}(R)/\mathcal{S}_f \longrightarrow \mathrm{Ab}(S)$ which, note, takes $(R, -)_{\mathcal{S}_f}$ to $(S, -)$, is an equivalence, then $R_f = \mathrm{End}\big((R, -)_{\mathcal{S}_f}\big) \simeq \mathrm{End}(S, -) = S$. As noted before 6.1.12, that implies that f is an epimorphism, as required. □

Proposition 12.8.8. ([282, Prop. 9]) *Let $\mathcal{S}$ be a Serre subcategory of $(R\text{-mod}, \mathbf{Ab})^{\mathrm{fp}}$. If every object of the quotient category $(R\text{-mod}, \mathbf{Ab})^{\mathrm{fp}}/\mathcal{S}$ is projective, then $(R\text{-mod}, \mathbf{Ab})^{\mathrm{fp}}/\mathcal{S} \simeq (R_{\mathcal{S}}\text{-mod}, \mathbf{Ab})^{\mathrm{fp}}$ and $R_{\mathcal{S}}$ is von Neumann regular.*

Proof. If $r \in R_{\mathcal{S}}$, then the image of the induced morphism $(r, -) : (R_{\mathcal{S}}, -) \to (R_{\mathcal{S}}, -)$ is finitely generated, hence a projective object. Therefore, $(r, -)$ is a projection map, so there is $f \in \mathrm{End}(R_{\mathcal{S}}, -)$ such that $(r, -)f(r, -) = (r, -)$; via the identification of $\mathrm{End}((R, -)_{\mathcal{S}})$ with $R_{\mathcal{S}}$, we may regard f as $(s, -)$ for some $s \in R_{\mathcal{S}}$, hence $rsr = r$. So $R_{\mathcal{S}}$ is indeed von Neumann regular. For the other part, the canonical map $R_{\mathcal{S}} \longrightarrow (R\text{-mod}, \mathbf{Ab})^{\mathrm{fp}}/\mathcal{S}$ induces $(R_{\mathcal{S}}\text{-mod}, \mathbf{Ab})^{\mathrm{fp}} \longrightarrow (R\text{-mod}, \mathbf{Ab})^{\mathrm{fp}}/\mathcal{S}$, which, by 12.8.1, is full and dense and which, since R is regular so, 10.2.38, $(R_{\mathcal{S}}\text{-mod}, \mathbf{Ab})^{\mathrm{fp}} \simeq R_{\mathcal{S}}\text{-Mod}$, has zero kernel by 3.4.30. □

Proposition 12.8.9. ([282, Prop. 7]) *Suppose that $\mathcal{S}$ is a Serre subcategory of $(R\text{-mod}, \mathbf{Ab})^{\mathrm{fp}}$. Then every object of $(R\text{-mod}, \mathbf{Ab})^{\mathrm{fp}}/\mathcal{S}$ is projective iff the lattice of finitely generated subobjects of $({}_R R, -)_{\mathcal{S}}$ is complemented.*

Proof. If every object of $(R\text{-mod}, \mathbf{Ab})^{\mathrm{fp}}/\mathcal{S}$ is projective, then every exact sequence in this category splits, so every finitely generated = finitely presented subobject of $(R, -)_{\mathcal{S}}$ is a direct summand.

For the converse, if $\mathrm{Latt}^{\mathrm{f}}(R_{\mathcal{S}}, -)$ is complemented, then so, one may check, is $\mathrm{Latt}^{\mathrm{f}}(R_{\mathcal{S}}, -)^n$. Therefore, since each representable functor, $(A, -)$, with $A \in R\text{-mod}$ embeds in some $(R, -)^n$, the embedding being induced by any epimorphism $R^n \to A$, each lattice $\mathrm{Latt}^{\mathrm{f}}(A, -)_{\mathcal{S}}$ also is complemented. For every object, F, of $(R\text{-mod}, \mathbf{Ab})^{\mathrm{fp}}/\mathcal{S}$ there is an epimorphism $(A, -)_{\mathcal{S}} \to F$ with $A \in R$-mod, and it follows that $\mathrm{Latt}^{\mathrm{f}} F$ also is complemented. □

Corollary 12.8.10. *If the module M is such that for every pp condition ϕ in one free variable the subgroup $\phi(M)$ of M has a pp-definable complement, meaning that there is a pp condition ψ such that $M = \phi(M) \oplus \psi(M)$, then the ring of definable scalars, R_M, of M is von Neumann regular and $R \to R_M$ is an epimorphism.*

A more direct proof of von Neumann regularity is given before 6.1.15.

13
Dimensions

A general framework for defining dimensions on Grothendieck categories is presented. The link with the lattice dimensions of Chapter 7 is made. The special cases, Krull–Gabriel dimension and uniserial dimension, are considered.

13.1 Dimensions

In Section 13.1.1 a general notion of dimension for Grothendieck categories, based on collapsing objects of a given kind, is defined. This is specialised in Section 13.1.2 to collapsing those objects of locally coherent categories whose lattice of finitely generated subobjects is of a given kind.

13.1.1 Dimensions via iterated localisation

Let Θ be any property of objects of abelian categories. Let $\mathcal{C}$ be any Grothendieck abelian category. Define $\mathcal{T}_\Theta$ to be the hereditary torsion subclass (see Section 11.1.1) of $\mathcal{C}$ generated by those objects of $\mathcal{C}$ which have property Θ. Form the quotient category $\mathcal{C}/\mathcal{T}_\Theta$. This is the localisation of $\mathcal{C}$ obtained by forcing every object with property Θ to become zero. Since $\mathcal{C}/\mathcal{T}_\Theta$ is a Grothendieck category (11.1.5) the process may be repeated: set $\mathcal{C}_0 = \mathcal{C}$ and $\mathcal{C}_1 = \mathcal{C}/\mathcal{T}_\Theta$, then continue inductively by setting $\mathcal{C}_{\alpha+1} = (\mathcal{C}_\alpha)_1$. This process may be continued transfinitely by setting, whenever λ is a limit ordinal, $\mathcal{C}_\lambda = \varinjlim_{\alpha<\lambda} \mathcal{C}_\alpha$, where the localisations $\mathcal{C}_\alpha \longrightarrow \mathcal{C}_{\alpha+1}$ give the directed system and where the direct limit is taken in a suitable category of categories. By 11.1.10 and induction each $\mathcal{C}_\alpha$ is a localisation of $\mathcal{C}$ at a hereditary torsion theory. Denote by $\mathcal{T}_\alpha$ the torsion subcategory of $\mathcal{C}$ which is the kernel of this localisation functor $Q_{\alpha+1} : \mathcal{C} \longrightarrow \mathcal{C}_{\alpha+1}$ (see 11.1.3) in case α is finite, and the kernel of $Q_\alpha : \mathcal{C} \longrightarrow \mathcal{C}_\alpha$ if $\alpha \geq \omega$ (since at limit ordinals we are going to denote $\bigcup_{\alpha<\lambda} \mathcal{T}_\alpha$ by $\mathcal{T}_\lambda$ rather than $\mathcal{T}_{<\lambda}$). In particular, $\mathcal{T}_\Theta = \mathcal{T}_0$.

At limit ordinals, λ, set $\mathcal{T}_\lambda = \bigcup_{\alpha<\lambda} \mathcal{T}_\alpha$: then $\mathcal{T}_\lambda$ is the kernel of the localisation $Q_\lambda : \mathcal{C} \longrightarrow \mathcal{C}_\lambda \longrightarrow 0$. Since every Grothendieck category has just a set of localisations (because it has a well-powered generating object, see Appendix E, p. 708) we may define the limit of this sequence of localisations to be the localisation $Q_\infty : \mathcal{C} \longrightarrow \mathcal{C}_\infty = \varinjlim_\alpha \mathcal{C}_\alpha$ with kernel $\mathcal{T}_\infty = \bigcup_\alpha \mathcal{T}_\alpha$.

Define the Θ-**dimension** of the category $\mathcal{C}$ to be the least ordinal α such that $\mathcal{C}_{\alpha+1}$ is the trivial category with just one object up to isomorphism or, if there is no such ordinal, ∞ (“**undefined**”).

If $C \in \mathcal{C}$, then we define the Θ-**dimension** of C to be the least ordinal α such that $C \in \mathcal{T}_\alpha$, that is, such that the image, $Q_{\alpha+1}C$, of C in $\mathcal{C}_{\alpha+1}$ (for $\alpha < \omega$; $Q_\alpha C \in \mathcal{C}_\alpha$ for $\alpha \geq \omega$) is zero or to be ∞ if there is no such ordinal. Applied both to categories and to objects, write Θ-dim for this dimension. Thus $\mathcal{T}_\Theta = \{C \in \mathcal{C} : \Theta\text{-dim}(C) = 0\}$.

Example 13.1.1. ([202, §IV.1]) Let Θ be the property of being a simple object. The Serre subcategory generated by the simple objects of $\mathcal{C}$ is clearly the full subcategory, $\mathcal{C}^{\mathrm{fl}}$, of all objects of finite length. The localising subcategory (that is, hereditary torsion class), $\mathcal{T}_\Theta$, generated by this consists of all **semiartinian** objects: those $C \in \mathcal{C}$ with the property that every factor object has non-zero socle (see [663, §VIII.2], [486, §5.6]). So $\mathcal{C}/\mathcal{T}_\Theta$ is obtained from $\mathcal{C}$ by “collapsing to zero all the simple objects of $\mathcal{C}$”. The corresponding dimension is called **Gabriel dimension**, for example, [486, §5.5] but note that Popescu calls this “Krull–Gabriel dimension”, which is the term used here for the result of the iterative process of making *finitely presented* simple objects 0, see Section 13.2.1. It does, however, turn out, 13.2.9, that for locally coherent categories, Gabriel dimension is defined iff Krull–Gabriel dimension is defined. A Grothendieck category with Gabriel dimension 0 is said to be **semiartinian**: this is equivalent (see the above references) to every non-zero object having a simple subobject. If R is a ring, then R is said to be **semiartinian** if Mod-R is semiartinian: this is equivalent (for example, [663, VIII.2.5]) to the (transfinite) Loewy series, $\mathrm{soc}(R)$, $\mathrm{soc}^2(R)$ (the full inverse image in $R/\mathrm{soc}(R)$ of $\mathrm{soc}(R/\mathrm{soc}(R))$, $\dots$, reaching R. A Grothendieck category with Gabriel dimension $< \infty$ is said to be **seminoetherian** ([486, §5.5]). It is equivalent ([486, 5.5.2]) that every non-trivial quotient category have at least one simple object. If $\mathcal{C} = $ Mod-R for some ring R, then the **Gabriel dimension** of R, $\mathrm{Gdim}(R)$, is set to be that of its module category.

A Grothendieck category $\mathcal{C}$ has $\mathcal{T}_\Theta = 0$ iff it has no simple object, which is equivalent to its having no non-zero finitely generated object: for if $C \in \mathcal{C}$ is a non-zero finitely generated object, then Zorn's Lemma may be applied to the set of proper subobjects of C (this is a set since the category is Grothendieck hence well powered) in order to obtain a maximal proper subobject C'; then C/C' is simple.

There are rings without Gabriel dimension, for instance an atomless boolean algebra, regarded as a ring, since a von Neumann regular ring is seminoetherian iff it is semiartinian (13.2.12), so the category of modules over such a ring has some localisation without any simple objects.

A ring has Gabriel dimension iff it has Krull dimension (7.1.3) ([245, §§1, 2]) although the values, if infinite, may differ by 1 for indexing reasons ([245, 2.4]).

Other examples of the general process are given below.

13.1.2 Dimensions on lattices of finitely presented subfunctors

Given a class of lattices, the finitely presented objects which are collapsed by the corresponding localisation are characterised (13.1.2). It is shown that the inductive collapsing of lattices of finitely generated subobjects exactly matches the inductive sequence of localisations (13.1.4). A couple of lemmas on computing this dimension are given. The approach in this section comes from [357, §1].

Given a class, $\mathcal{L}$, of modular lattices as in Section 7.1, that is, one closed under sublattices and quotients, we can define various corresponding properties of objects in abelian categories. For instance, if $\mathcal{L}$ consists just of two-point and trivial lattices and if the corresponding property, Θ, of objects is taken to be that the lattice of all subobjects should belong to $\mathcal{L}$, that is, the object should be simple or zero, then we obtain the notion of Gabriel dimension seen above. Here we want something that is defined in terms of finitely presented functors: say that $C \in \mathcal{C}$ **has property** $\Theta_{\mathcal{L}}$ iff C is a coherent object and the lattice of finitely generated (= finitely presented) subobjects of C is in $\mathcal{L}$. For $\mathcal{L}$ as just above this would correspond to the property of being a *finitely presented* simple functor or 0.

Given a locally coherent abelian category $\mathcal{C}$ denote by $\mathcal{S}_{\mathcal{L}}$ (or $\mathcal{S}_{\Theta}$) the Serre subcategory of $\mathcal{C}^{\mathrm{fp}}$ consisting of finitely presented objects generated by those with property $\Theta = \Theta_{\mathcal{L}}$. The corresponding torsion class will be denoted by $\mathcal{T}_{\mathcal{L}}$. By 11.1.36 this is the closure, $\overrightarrow{\mathcal{S}_{\mathcal{L}}}$, of $\mathcal{S}_{\mathcal{L}}$ under direct limits. For example, if $\mathcal{L}$ is the class of two-point and trivial lattices, then every finitely presented subobject of an object of $\mathcal{T}_{\mathcal{L}}$ must be of finite length (by 11.1.36). Any semiartinian non-artinian von Neumann regular (hence coherent) ring illustrates the difference between this and 13.1.1, that is, between $\mathcal{T}_{\mathcal{L}} = \mathcal{T}_{\Theta_{\mathcal{L}}}$ and $\mathcal{T}_{\Theta}$, the point being that, for a general hereditary torsion theory, even on a locally coherent category, the torsion class is not simply the closure of the class of finitely presented torsion objects under direct limits.

Lemma 13.1.2. *Let $\mathcal{L}$ be a class of modular lattices which is closed under sublattices and quotient lattices. Let $\mathcal{C}$ be a locally coherent abelian category and let $F \in \mathcal{C}^{\mathrm{fp}}$.*

Then $F \in \mathcal{S}_\mathcal{L}$ iff F has a finite chain of finitely generated subfunctors $F = F_0 > F_1 > \cdots > F_n = 0$ such that for each i the lattice of finitely generated subfunctors of F_i/F_{i+1} belongs to $\mathcal{L}$.

Proof. Since $\mathcal{C}$ is locally coherent each subfactor, F_i/F_{i+1}, is finitely presented and its lattice of finitely presented subfunctors is a subquotient of that of F (for example, using E.1.19). Therefore, if F is in $\mathcal{S}_\mathcal{L}$, so must be F_i/F_{i+1}. Conversely it is easily checked that the class of functors with such filtrations is closed under extensions, subobjects and quotient objects, hence is a Serre subcategory of $\mathcal{C}^{\mathrm{fp}}$ containing $\mathcal{S}_\mathcal{L}$. □

Lemma 13.1.3. *Let $\mathcal{L}$ be a class of modular lattices which is closed under sublattices and quotient lattices. Let $\mathcal{C}$ be a locally coherent abelian category. Then the hereditary torsion theory with torsion class $\mathcal{T}_\mathcal{L}$ is of finite type.*

Proof. This is immediate from 11.1.14 and the fact that $\mathcal{T}_\mathcal{L}$ is generated by a class of finitely presented objects. □

It follows by 11.1.33 and induction that, for any $\mathcal{L}$, each corresponding quotient category $\mathcal{C}_\alpha$ is locally coherent and, moreover, using 11.1.18, each localisation $\mathcal{C} \longrightarrow \mathcal{C}_\alpha$ is of finite type.

The proposition which follows can be extracted from [495, §10.2], though pp conditions rather than finitely presented functors appear there. Krause, in [355, §12], gives a treatment of these dimensions in terms of Serre subcategories. The quotients of lattices referred to in the proposition are those defined in Section 7.1.

Proposition 13.1.4. *Let $\mathcal{L}$ be a class of modular lattices which is closed under sublattices and quotient lattices and let $\Theta_\mathcal{L}$ be the corresponding property of coherent functors. Let $\mathcal{C}$ be a locally coherent category. Take $F \in \mathcal{C}^{\mathrm{fp}}$ and let $\mathrm{Latt}^{\mathrm{f}}(F)$ denote the lattice of finitely generated (= finitely presented) subfunctors of F. Let α be an ordinal or ∞.*

Then $(\mathrm{Latt}^{\mathrm{f}}(F))_\alpha \simeq \mathrm{Latt}^{\mathrm{f}}(F_\alpha)$, where F_α denotes the image of F under the localisation map from $\mathcal{C}$ to $\mathcal{C}_\alpha$, the isomorphism, θ_α, being given by taking $\pi_\alpha G$ to $G_\alpha (\leq F_\alpha$ by exactness of localisation, 11.1.5) for G any finitely generated subobject of F. Here $\pi_\alpha : \mathrm{Latt}^{\mathrm{f}}(F) \to \mathrm{Latt}^{\mathrm{f}}(F)_\alpha$ is the canonical quotient map of lattices (Section 7.1).

In particular $\Theta_\mathcal{L}$-$\dim(F) = \mathcal{L}$-$\dim(\mathrm{Latt}^{\mathrm{f}}(F))$.

Proof. The proposition is proved by induction on α.

Case $\alpha = 0$. First we show that θ_0 is well defined. So suppose that G, H are finitely generated subobjects of F with $G \sim H$, where $\sim = \sim_\mathcal{L}$ (notations as in

Section 7.1) Then (7.1.1) there exists a chain of finitely generated objects $G + H = K_0 \geq K_1 \geq \cdots \geq K_n = G \cap H$ with each successive factor having lattice of finitely generated subfunctors in $\mathcal{L}$. Hence $(G+H)/G \cap H \in \mathcal{S}_0$. Therefore both $(G+H)/G$ and $(G+H)/H$ are in $\mathcal{S}_0$, so (11.1.2(2)) $G_0 = (G+H)_0 = H_0$.

Now we compute the kernel of θ_0 (by 11.1.34 each θ_α is surjective). Suppose that $\theta_0 G = \theta_0 H$, that is, $G_0 = H_0$, so both equal $(G+H)_0$ and also equal $(G \cap H)_0$. Then $(G+H)/G \cap H \in \mathcal{S}_0$, so (13.1.2) there is a chain, $\{F_i\}_i$, of finitely generated subobjects of $(G+H)/G \cap H$ as in 7.1.1, which may then be pulled back to a chain, $G + H = K_0 \geq K_1 \geq \cdots \geq K_n = G \cap H$, of finitely generated objects (since $G \cap H$ is finitely generated) with successive quotients having lattice of finitely presented subfunctors in $\mathcal{L}$; for short say each interval or gap is "in $\mathcal{L}$". Thus $G \sim H$ as required.

Case α to $\alpha + 1$. Suppose we have the result for α: we show that $\theta_{\alpha+1}$ is well defined and monic.

Let $G, H \in \mathrm{Latt}^{\mathrm{f}} F$ with $G \sim_{\alpha+1} H$ (Section 7.1 for the definition of $\sim_{\alpha+1}$). That is, $\pi_\alpha G \sim \pi_\alpha H$, so, in $(\mathrm{Latt}^{\mathrm{f}} F)_\alpha$, there is a chain, $\pi_\alpha(G+H) = \pi_\alpha K_0 \geq \pi_\alpha K_1 \geq \cdots \geq \pi_\alpha K_n = \pi_\alpha(G \cap H)$, with the successive intervals in $\mathcal{L}$, where $K_i \in \mathrm{Latt}^{\mathrm{f}} F$, and, without loss of generality, with $G + H = K_0 \geq K_1 \geq \cdots \geq K_n = G \cap H$ (replace any original choice K by $(G+H) \cap K + (G \cap H)$). Now apply the isomorphism θ_α to obtain $(G+H)_\alpha = (K_0)_\alpha \geq (K_1)_\alpha \geq \cdots \geq (K_n)_\alpha = (G \cap H)_\alpha$ with, since θ_α is an isomorphism, each gap in $\mathcal{L}$. So $\big((G+H)/G \cap H\big)_\alpha = (G+H)_\alpha/(G \cap H)_\alpha \in \mathcal{S}_0(\mathcal{C}_\alpha)$ (that is, $\mathcal{S}_0$ in $\mathcal{C}_\alpha$), hence $(G+H)/(G \cap H) \in \mathcal{S}_{\alpha+1}$ (in $\mathcal{C}$) and so, as before, $G_{\alpha+1} = H_{\alpha+1}$.

To compute the kernel of $\theta_{\alpha+1}$, suppose that $\theta_{\alpha+1}(G) = \theta_{\alpha+1}(H)$, that is, $G_{\alpha+1} = H_{\alpha+1}$. So G_α is equivalent to H_α modulo $\mathcal{S}_0(\mathcal{C}_\alpha)$, therefore $(G+H)_\alpha/(G \cap H)_\alpha \in \mathcal{S}_0(\mathcal{C}_\alpha)$. It follows, by 13.1.2, that there is a chain $(G+H)_\alpha = F_0 \geq F_1 \geq \cdots \geq F_n = (G \cap H)_\alpha$ in $\mathcal{C}$ with F_i finitely presented and each gap in $\mathcal{L}$. Since θ_α is an isomorphism, and by the case $\alpha = 0$, it follows that $G + H \sim_{\alpha+1} G \cap H$, so $G \sim_{\alpha+1} H$ as required.

Case λ a limit. Finally, if λ is a limit ordinal, then $(\mathrm{Latt}^{\mathrm{f}} F)/(\bigcup_{(1\leq)\alpha<\lambda} \sim_\alpha) = \varinjlim_{\alpha<\lambda} \big((\mathrm{Latt}^{\mathrm{f}} F)/\sim_\alpha\big)$ and $F/(\bigcup_{\alpha<\lambda} \mathcal{S}_\alpha) = \varinjlim_{\alpha<\lambda} F/\mathcal{S}_\alpha$ (writing $F/\mathcal{S}$ for the localisation of F at $\mathcal{S}$). Therefore $\mathrm{Latt}^{\mathrm{f}}\big(F/(\bigcup_{\alpha<\lambda} \mathcal{S}_\alpha)\big) = \mathrm{Latt}^{\mathrm{f}}(\varinjlim_{\alpha<\lambda} F_\alpha) = \varinjlim_{\alpha<\lambda} \mathrm{Latt}(F_\alpha)$ (since we are dealing with the lattice of finitely presented objects) and then the isomorphisms θ_α induce an isomorphism, $\theta_\lambda = \varinjlim_{\alpha<\lambda} \theta_\alpha$, from $(\mathrm{Latt}^{\mathrm{f}} F)/(\bigcup_{\alpha<\lambda} \sim_\alpha) = (\mathrm{Latt}^{\mathrm{f}} F)_\lambda$ to $\mathrm{Latt}^{\mathrm{f}}\big(F/(\bigcup_{\alpha<\lambda} \mathcal{S}_\alpha)\big) = F_\lambda$, as required. □

Corollary 13.1.5. *With assumptions as in 13.1.4, for each α, $\mathcal{S}_\alpha \cap \mathcal{C}^{\mathrm{fp}}$ is the collection of finitely presented objects of $\mathcal{C}$ with $\Theta_{\mathcal{L}}$-dim $\leq \alpha$.*

Lemma 13.1.6. *Again with assumptions as in 13.1.4, suppose that* $0 \to F \to G \to H \to 0$ *is an exact sequence of finitely presented objects of* $\mathcal{C}$. *Then* $\Theta_{\mathcal{L}}\text{-dim}(G) = \max\{\Theta_{\mathcal{L}}\text{-dim}(F), \Theta_{\mathcal{L}}\text{-dim}(H)\}$.

Proof. If $\mathcal{S}$ is a Serre subcategory of $\mathcal{C}^{\mathrm{fp}}$ then $G \in \mathcal{S}$ iff $F, H \in \mathcal{S}$ so this is immediate from 13.1.5. □

Lemma 13.1.7. *Let* R *be a small preadditive category. Then* $\Theta_{\mathcal{L}}\text{-dim}(R) = \sup\{\Theta_{\mathcal{L}}\text{-dim}(G, -) : G \in R\}$.

Proof. In $\mathcal{C} = (\text{mod-}R, \mathbf{Ab})$ every object of $\mathcal{C}^{\mathrm{fp}}$ is a subquotient of a coproduct of finitely many representable functors $(G, -)$ with G an object of R (see the comment after 10.2.6). Therefore the least α such that $\mathcal{C}_\alpha = 0$, if there is such, is also the least α such that every $(G, -)$ is in the kernel, $\mathcal{T}_{\alpha-1}$ or $\mathcal{T}_\alpha$ if $\alpha \geq \omega$, of the localisation map $\mathcal{C} \longrightarrow \mathcal{C}_\alpha$, as required. □

In particular, if $\mathcal{C}$ is a module category over a ring, then the least ordinal α such that $\mathcal{C}_\alpha = 0$ cannot be a limit ordinal.

13.2 Krull–Gabriel dimension

Krull–Gabriel dimension is defined, and examples are given, in Section 13.2.1. In Section 13.2.2 it is shown that, for a locally coherent category, the Krull–Gabriel dimension is $< \infty$ *iff the Gabriel dimension is* $< \infty$.

13.2.1 Definition and examples

Krull–Gabriel dimension is defined via collapsing finitely presented simple objects. The Krull–Gabriel dimension of a finitely presented object in a locally coherent category is equal to the m-dimension of its lattice of finitely generated subobjects (13.2.1). This is applied to localisations of functor categories (13.2.1, 13.2.3).

This dimension is considered over various kinds of rings: hereditary artin algebras; string algebras and canonical algebras; serial rings. In particular this dimension cannot take the value 1 *for serial rings (13.2.5) and if a serial ring has Krull dimension* 1, *then its Krull–Gabriel dimension is* 2 *(13.2.6). For uniserial ring this extends to arbitrary values of Krull dimension (13.2.7).*

Let $\mathcal{C}$ be a locally coherent abelian (hence, 16.1.11, Grothendieck) category.

Let $\mathcal{L}$ be the collection of two-point and trivial lattices, so $\Theta_{\mathcal{L}}$ is the property of being a finitely presented simple object or zero. The corresponding Serre subcategory of $\mathcal{C}^{\mathrm{fp}}$ is, by 13.1.2, that of all finitely presented objects of $\mathcal{C}$ of finite length. Define, as in Section 13.1.1, the sequence of (finite type) localisations of $\mathcal{C}$ based on this. The notion of dimension so obtained is called **Krull–Gabriel dimension**, written $\mathrm{KGdim}(\mathcal{C})$ for the category and $\mathrm{KGdim}(C)$ for the objects C of $\mathcal{C}$. This was introduced in [224].[1] If $\mathcal{C}$ is the functor category $(\text{mod-}R, \mathbf{Ab})$, where R is a ring or small preadditive category, then we write $\mathrm{KG}(R)$ for $\mathrm{KGdim}(\text{mod-}R, \mathbf{Ab})$. In [495, p. 266] the next result, which is a special case of 13.1.4, was glimpsed: it became clear when the connection, see 10.2.30, between pp conditions and finitely presented functors was better understood. For the definition of m-dimension see Section 7.2.

Proposition 13.2.1. ([101], see [104, 4.5]) *Let $\mathcal{C}$ be a locally coherent Grothendieck category and let $F \in \mathcal{C}^{\mathrm{fp}}$. Let $\mathrm{Latt}^{\mathrm{f}}(F)$ denote the lattice of finitely generated (= finitely presented) subfunctors of F. Then $\mathrm{KGdim}(F) = \mathrm{mdim}(\mathrm{Latt}^{\mathrm{f}}(F))$.*

Corollary 13.2.2. *Let $\mathcal{X}$ be a definable subcategory of a module category $\text{Mod-}R$ and let $\mathcal{S}_{\mathcal{X}}$ be the corresponding (12.4.1) Serre subcategory of $(\text{mod-}R, \mathbf{Ab})$. Then the Krull–Gabriel dimension of the "functor category" of $\mathcal{X}$ – the localisation of $(\text{mod-}R, \mathbf{Ab})$ at the finite type torsion theory corresponding to $\mathcal{X}$ – equals the m-dimension of the lattice of pp conditions for $\mathcal{X}$: $\mathrm{KGdim}\big((\text{mod-}R, \mathbf{Ab})_{\mathcal{X}}\big) = \mathrm{mdim}(\mathrm{pp}(\mathcal{X}))$. In particular $\mathrm{KG}(R) = \mathrm{mdim}(\mathrm{pp}_R)$.*

That follows directly from 13.1.4 and 12.3.18. The common value is also that of the dual objects, since this is true, 7.2.7, for the pp lattices. It also follows from the next result, which is immediate from the correspondence, 11.2.9, between Serre subcategories of $(\text{mod-}R, \mathbf{Ab})^{\mathrm{fp}}$ and $(R\text{-mod}, \mathbf{Ab})^{\mathrm{fp}}$ and the definition of KGdim via localisation.

Corollary 13.2.3. *For any small preadditive category R and definable subcategory, $\mathcal{X}$, of $\text{Mod-}R$, $\mathrm{KGdim}\big((\text{mod-}R, \mathbf{Ab})_{\mathcal{X}}\big) = \mathrm{KGdim}\big((R\text{-mod}, \mathbf{Ab})_{D\mathcal{X}}\big)$.*

Now we discuss this dimension over various types of ring and summarise some earlier results.

Dimension 0 $\mathrm{KG}(R) = 0$ iff R is of finite representation type (7.2.8).

Dimension 1 This value is not attained by artin algebras (15.4.5), nor by serial rings (13.2.5 below) but is attained by, for instance, some von Neumann regular rings, see 8.2.26 and 8.2.92 – an example to which the latter applies, giving the value 1, is the boolean algebra of all finite or cofinite subsets of a countably infinite set.

[1] But the terminology does clash with that used in [486, p. 344].

Dimension 2 Rings with Krull–Gabriel dimension 2 include PI Dedekind domains (5.2.6) and tame hereditary algebras (8.1.12).

Dimension defined and > 2 The domestic string algebras Λ_n from Section 8.1.3, have $\mathrm{KG}(\Lambda_n) = n + 1$. It is conjectured that any domestic string algebra has finite Krull–Gabriel dimension (8.1.22).

Dimension ∞ Wild finite-dimensional algebras have Krull–Gabriel dimension ∞: the argument at [495, pp. 281, 2] applied to, say, the example at 7.2.14 (or see [358, 8.15]) shows this.[2]

Hereditary artin algebras It follows from 7.2.18 that $\mathrm{KG}(R) = \infty$ whenever R is the path algebra of a wild quiver ([32, 4.3], also [492, Pt. 1]). If R is a tame hereditary finite-dimensional algebra, then $\mathrm{KG}(R) = 2$ by [224, 4.3] (see [323, 8.58] or [35, §4]). This also follows from 8.1.12 and 5.3.60 (plus 5.3.21 if R is countable and by the details of the proof otherwise).

String algebras/canonical algebras From 8.1.20 we have examples of domestic string algebras with any finite Krull–Gabriel dimension ≥ 2. Exactly what values of Krull–Gabriel dimension can occur for domestic (string) algebras is a subject of conjecture (9.1.15). On the other hand, the method of 7.2.15 can be used to show that no non-domestic string algebra has Krull–Gabriel dimension; of course, this now follows also by Puninsky's result 7.3.28. As discussed in Section 8.1.4, canonical tubular algebras fail to have Krull–Gabriel dimension.

Serial rings As in Section 8.2.7, if R is a serial ring, then denote by $e_1, \dots, e_n$ a set of primitive orthogonal idempotents with sum 1, so each $e_i R$ and each Re_i is a uniserial right, resp. left, indecomposable projective module and each ring $e_i R e_i$ is uniserial.

Define $J_0(R) = J(R)$ and, inductively, set $J_{\alpha+1}(R) = \bigcap_n J_\alpha(R)^n$ and $J_\lambda(R) = \bigcap_{\alpha<\lambda} J_\alpha(R)$. By [450, 5.5, 5.6], if R is a serial ring, then $\mathrm{Kdim}(R) = \alpha$ iff α is the least ordinal such that $J_\alpha(R)$ is nilpotent. In particular $\mathrm{Kdim}(R_R) = \mathrm{Kdim}({}_R R)$.

In [570, 5.1] Puninski showed that if e is a primitive idempotent of a serial ring R, then the following are all equal:

(i) the Krull dimension of the lattice of all submodules of eR;
(ii) the Krull dimension of the lattice of all cyclic submodules of eR;
(iii) the m-dimension of the lattice of all cyclic submodules of eR.

[2] The argument at [495, pp. 281, 2] is for width/uniform dimension and is not complete but it certainly works for m-dimension/Krull–Gabriel dimension.

Recall, 8.2.60, that if R is a serial ring, then $\mathrm{KG}(R) < \infty$ iff $\mathrm{Kdim}(R) < \infty$ and, in that case, $\mathrm{KG}(R) \leq \mathrm{Kdim}(R) \oplus \mathrm{Kdim}(R)$ (13.2.7). Also, 8.2.58, every serial ring satisfies the isolation condition, so, for a serial ring R, $\mathrm{CB}(\mathrm{Zg}_R) = \mathrm{KG}(R) = \mathrm{mdim}(\mathrm{pp}_R)$.

If R is a commutative valuation domain of Krull dimension α, then, by [560, 3.6], the Cantor–Bendixson rank of Zg_R, hence the Krull–Gabriel dimension of R, is equal to $\alpha \oplus \alpha$.

Let e be a primitive idempotent of a serial ring R. Let L be a linearly ordered non-trivial set. The functions $f : L \to Re$ and $g : L \to eR$ form what Puninski calls a **boundary pair** if for every $a, b \in L$ with $a < b$ we have $Rf(a) < Rf(b)$, $g(a)R < g(b)R$ and $Rf(a)g(a)R < Rf(b)g(b)R$.

Proposition 13.2.4. ([570, 5.5]) *Let R be a serial ring and let $f : L \to Re$ and $g : L \to eR$ be a boundary pair. If* $\mathrm{mdim}(L) = \alpha$ *then* $\mathrm{KG}(R) \geq \alpha \oplus \alpha$.

Proposition 13.2.5. ([570, 5.6]) *Let R be a serial ring. Then* $\mathrm{CB}(\mathrm{Zg}_R) \neq 1$, *hence* $\mathrm{KG}(R) \neq 1$.

Compare this with the similar result, 15.4.5, for artin algebras. The result here is somewhat stronger since for artin algebras it has not been shown (without some additional condition) that the Cantor–Bendixson rank cannot be 1.

If R is a right noetherian serial ring, then, [703, 5.11], R has Krull dimension 1, so by 13.2.5, the next result follows.

Theorem 13.2.6. ([570, 5.7], [593, 3.19] for two-sided noetherian rings) *If R is a serial ring of Krull dimension* 1, *then* $\mathrm{KG}(R) = 2$. *In particular, if R is serial and right noetherian, but not artinian, then* $\mathrm{KG}(R) = 2$.

A serial ring R is **semiduo** if for every $r \in e_iRe_j$ where e_i, e_j are primitive idempotents, either $e_iRr \subseteq rR$ or $rRe_j \subseteq Rr$. A serial ring is semiduo iff every finitely presented indecomposable (right or left) module has local endomorphism ring (see [564, 2.13]), so this is equivalent to the ring being Krull–Schmidt.

Theorem 13.2.7. ([570, 5.8, 5.9]) *(a) If R is a Krull–Schmidt uniserial ring of Krull dimension α, then* $\mathrm{KG}(R) = \alpha \oplus \alpha$.

(b) If R is a Krull–Schmidt serial ring of finite Krull dimension n, then $\mathrm{KG}(R) = 2n$.

Puninski conjectures, [570, 5.10], the following.

Question 13.2.8. Suppose that R is a serial ring with finite Krull dimension. Is the Krull–Gabriel dimension of R even?

13.2.2 Gabriel dimension and Krull–Gabriel dimension

It is proved that Krull–Gabriel dimension is coexistent with Gabriel dimension (13.2.9). The Krull–Gabriel dimensions of a locally coherent category and its dual category are equal, similarly for definable subcategories (13.2.11).

The Gabriel analysis of a Grothendieck category proceeds inductively, making simple objects zero at every stage. The Krull–Gabriel analysis of a locally coherent Grothendieck category is similar but makes only finitely presented simple objects zero at each stage. Therefore, the latter grows more slowly than the former. Burke showed that, nevertheless, these dimensions coexist in a locally coherent category: if one is defined, so is the other.

Theorem 13.2.9. ([104, 5.1]) *Suppose that $\mathcal{C}$ is a locally coherent category which is seminoetherian (that is, the Gabriel dimension of $\mathcal{C}$ is defined). Then the Krull–Gabriel dimension of $\mathcal{C}$ is defined, and is bounded above by the Gabriel dimension of $\mathcal{C}$.*

Proof. Let $\mathcal{S}$ be the subcategory of $\mathcal{C}^{\mathrm{fp}}$ consisting of all objects with KG dimension: by 13.1.6 this is a Serre subcategory of $\mathcal{C}^{\mathrm{fp}}$. Let τ be the (finite type, 11.1.38) hereditary torsion theory on $\mathcal{C}$ generated by $\mathcal{S}$. Assume that KG($\mathcal{C}$) is not defined: then there is $A \in \mathcal{C}^{\mathrm{fp}} \setminus \mathcal{S}$, that is, such that $A/\tau A \neq 0$. Recall (Section 11.1.1) that for an object $C \in \mathcal{C}$ the notation τC denotes the largest τ-torsion subobject of C.

Let $\mathcal{T}_\alpha$ be the sequence of hereditary torsion subcategories of $\mathcal{C}$ corresponding to the filtration for Gabriel dimension, see 13.1.1. Since $\mathcal{C}$ is seminoetherian every object of $\mathcal{C}$ belongs to some $\mathcal{C}_\alpha$. Choose α minimal such that there is some $A \in \mathcal{C}^{\mathrm{fp}}$ with $A/\tau A \in \mathcal{T}_{\alpha+1} \setminus \mathcal{T}_\alpha$. By definition of the Gabriel filtration the object $A/\tau A$ has finite length in $\mathcal{C}/\mathcal{T}_\alpha$, so it contains a simple object of that category. That object has the form $B/\tau B$ for some subobject B of A. Since B is the sum of its finitely generated subobjects, these cannot all be contained in τB, so, replacing B' with one of its finitely generated subobjects which is not contained in τB, we may suppose, without loss of generality, that B is finitely generated, hence, being contained in the coherent object A, finitely presented.

We claim that $B/\tau B$ is $\mathcal{T}_\alpha$-torsionfree. For otherwise there would be a finitely generated, hence finitely presented, subobject B' of B not contained in τB and with $B'/\tau B' = B' \cap \tau B$ being $\mathcal{T}_\alpha$-torsion. Since $B'/\tau B' \neq 0$, this would contradict minimality of α.

We claim that there is $C < B$ finitely generated with $C/\tau C \neq 0$ and KGdim$(B/C) = \infty$. If there were not, then every such $C \leq B$ either would be contained in τB or would satisfy KGdim$(B/C) \leq \beta$ for some ordinal β (since there is just a set of subobjects of B). Therefore the image of B under the finite-type torsion theory which sends to 0 every object of Krull–Gabriel dimension $\le \beta$ would be simple or 0; this would imply $\mathrm{KGdim}(B) \le \beta + 1$ – contrary to assumption. So, indeed, we may choose $C < B$ finitely generated with $\mathrm{KGdim}(B/C) = \infty$.

Since $B/\tau B$ is $\mathcal{T}_\alpha$-torsionfree, so is $(C + \tau B)/\tau B$, which, since the localisation of $B/\tau B$ in $\mathcal{C}/\mathcal{T}_\alpha$ is simple, must therefore be $\mathcal{T}_\alpha$-dense in $B/\tau B$. Thus B/C is an object of $\mathcal{C}^{\mathrm{fp}} \backslash \mathcal{S}$ which belongs to $\mathcal{T}_\alpha$ and which, therefore, is such that $(B/C)/\tau(B/C) \in \mathcal{T}_{\gamma+1} \setminus \mathcal{T}_\gamma$ for some $\gamma < \alpha$, contradicting choice of α. Contradiction, as required. □

Corollary 13.2.10. *(*[104, 5.3]) If $\mathcal{C}$ is locally coherent and seminoetherian, then so is its dual category $\mathcal{C}^{\mathrm{d}}$ (dual in the sense of Section 16.1.3).

This is immediate from 13.2.9 and the next result which follows from the bijection, 18.1.2, between Serre subcategories of the functor category and its dual, cf. 13.2.3.

Proposition 13.2.11. *Let $\mathcal{C}$ be a locally coherent category and let $\mathcal{C}^{\mathrm{d}}$ denote its dual. Then* $\mathrm{KGdim}(\mathcal{C}) = \mathrm{KGdim}(\mathcal{C}^{\mathrm{d}})$. *More generally, if $\mathcal{D}$ is a definable category and $\mathcal{D}^{\mathrm{d}}$ is its dual (Section 18.1.1), then* $\mathrm{KGdim}(\mathcal{D}) = \mathrm{KGdim}(\mathcal{D}^{\mathrm{d}})$.

We refer to the next result a couple of times and include a proof here for ease of reference.

Proposition 13.2.12. *A von Neumann regular ring is seminoetherian iff it is semiartinian.*

Proof. If R is semiartinian, then every non-zero module has non-zero socle and it follows that the localisation of Mod-R at the torsion theory generated by the simple modules is already zero, hence R is seminoetherian.

Suppose R is von Neumann regular and seminoetherian, so Mod-R is locally coherent and seminoetherian. By 13.2.9 the Krull–Gabriel dimension of Mod-R is defined. Let τ be the torsion theory generated by the finitely presented simple objects. By 11.1.36 this is of finite type. It follows by [663, XI.3.3, XI.3.4] that Mod-R_τ is equivalent to the localised category $(\text{Mod-}R)_\tau$ and hence that the functor Mod-$R_\tau \longrightarrow$ Mod-R induced by the map from R to its ring of quotients R_τ is full. Therefore $R \to R_\tau$ is an epimorphism of rings. Since R is von Neumann regular that implies (see [663, p. 226]) that R_τ is a factor ring of R. Since the Krull–Gabriel dimension of Mod-R_τ is strictly less than that of Mod-R we finish by induction. □

13.3 Locally simple objects

A dimension based on collapsing "locally simple" objects is defined. It is defined for a finitely presented functor iff the isolation condition holds for the support of every point in the corresponding Ziegler-open set (13.3.1).

Here is another example, due to Herzog, of the general construction of dimensions by localisation (Section 13.1.1).

Let R be a small preadditive category. Define a functor $F \in (\text{mod-}R, \mathbf{Ab})^{\text{fp}}$ to be **locally simple**[3] if for every irreducible closed subset X of the Ziegler spectrum Zg_R the localisation F_X of F at X is a simple or zero object of $(\text{mod-}R, \mathbf{Ab})^{\text{fp}}_X$. An example of such a functor is $(\mathbb{Z}_p, -) \in (\text{mod-}\mathbb{Z}, \mathbf{Ab})^{\text{fp}}$. The definition could be made more generally for finitely presented functors in the functor category of a definable additive category (§18.1.1).

If R is countable, then, by 5.4.6, every irreducible closed subset of Zg_R has a generic point so the condition becomes (and in all cases implies) that FN is, for every $N \in \mathrm{Zg}_R$, either 0 or a simple $\mathrm{End}(N)$-module.

Herzog defined the relation $\sqsubseteq$ on the objects of any abelian category by $A \sqsubset B$ if there is a chain of subobjects $0 \leq B_1 < B_2 \leq B_3 < B_4 \leq B$ such that $B_2/B_1 \simeq B_4/B_3 \simeq A$, that is, if A appears at least twice in some "composition series" of B. This relation is transitive. He proved in [281] that $F \in (\text{mod-}R, \mathbf{Ab})^{\text{fp}}$ is $\sqsubseteq$-minimal (that is, is minimal among non-zero objects in the $\sqsubseteq$-preordering) iff F is locally simple.

Herzog then defined a dimension, **local Krull–Gabriel dimension**, LKGdim, following the route in Section 13.1.1 (though in the context of Serre subcategories of abelian categories), the localisation at each stage being that which makes 0 the locally simple objects. Since any simple object is locally simple this grows faster than Krull–Gabriel dimension, which, recall, can be tracked by localising the abelian category of finitely presented functors at the Serre subcategory generated by the finitely presented simple functors. In particular $\mathrm{LKGdim}(F) \leq \mathrm{KGdim}(F)$ for every finitely presented functor F.

Theorem 13.3.1. [281] *For* $F \in (\text{mod-}R, \mathbf{Ab})^{\text{fp}}$ *the following conditions are equivalent:*

(i) for every $N \in (F)$*, the support* $\mathrm{supp}(N)$ *satisfies the isolation condition;*
(ii) the set of finitely generated subobjects of F *has the descending chain condition under the relation* $\sqsubseteq$*;*
(iii) $\mathrm{LKGdim}(F) < \infty$.

[3] Herzog noticed that such functors play an important role in [107] and [529] which considered Ziegler spectra, Krull–Gabriel and uniserial dimensions for string algebras.

For the definition of support see Section 5.1.1; the isolation condition is defined in Section 5.3.2. The condition that supp(N) satisfy the isolation condition is clearly equivalent to N being isolated in supp(N) by a minimal pair (Section 5.3.1) which, in turn, is equivalent to the existence of a finitely presented functor G such that $GN \neq 0$ and such that the localisation, G_N, at the finite type torsion theory corresponding to N be simple. This, in turn, is exactly the condition that $(\text{mod-}R, \mathbf{Ab})_N$ contain a finitely presented simple functor.[4]

13.4 Uniserial dimension

Another dimension is defined, this time based on collapsing uniserial finitely presented objects; the objects in the corresponding Serre subcategory are characterised (13.4.1). The uniserial dimension for the category of modules over a ring is not a limit ordinal and has the same value for right and for left modules (13.4.2). For any definable subcategory, the uniserial dimension of the functor category is equal to the breadth of the lattice of pp conditions (13.4.3), hence it is defined iff the width is defined (13.4.4). Uniserial dimension is bounded above by Krull–Gabriel dimension (13.4.5).

The uniserial dimension for modules over a PI Dedekind domain which is not a division ring is 1 *(13.4.7). For von Neumann regular rings this dimension equals the Krull–Gabriel dimension (13.4.8).*

Let $\mathcal{C}$ be a locally coherent abelian category.

Let $\mathcal{L}$ be the collection of all chains (totally ordered posets). Define the corresponding sequence of congruences and quotient lattices based on $\mathcal{L}$ as in Section 7.1.

An object of $\mathcal{C}$ is **uniserial** if its lattice of subobjects is a chain. Set $\mathcal{S}_u = \mathcal{S}_u(\mathcal{C})$ to be the Serre subcategory of $\mathcal{C}^{\text{fp}}$ generated by the uniserial finitely presented functors and let $\mathcal{T}_u = \overrightarrow{\mathcal{S}_u}$ (11.1.36) denote the corresponding hereditary torsion class of finite type. Note that an object is uniserial if its finitely generated subobjects are linearly ordered, so this is another example of the construction of Section 13.1.2. By 13.1.2 the objects of $\mathcal{S}_u$ are as follows.

Lemma 13.4.1. *If $\mathcal{C}$ is locally coherent and $F \in \mathcal{C}^{\text{fp}}$, then $F \in \mathcal{S}_u$ iff F has a finite chain of finitely generated subfunctors $F = F_0 > F_1 > \cdots > F_n = 0$ such that each quotient F_i/F_{i+1} is uniserial.*

[4] At which point one sees that there is a link with the seminoetherian condition and the fact, 5.3.17, that existence of Krull–Gabriel dimension (= m-dimension, 13.2.1) implies that the isolation condition holds.

The general definitions of Section 13.1.2 specialise as follows, where notation is as in that section. The least ordinal α such that $\mathcal{C}_{\alpha+1} = 0$, if such exists, is the **uniserial dimension** of $\mathcal{C}$, $\mathrm{Udim}(\mathcal{C})$. If there is no such α, set $\mathrm{Udim}(\mathcal{C}) = \infty$. For any object F of $\mathcal{C}$, $\mathrm{Udim}(F)$ is the least α such that $F \in \mathcal{T}_\alpha$. If $\mathcal{C} = (\text{mod-}R, \mathbf{Ab})$ set $\mathrm{UD}(R) = \mathrm{Udim}(\mathcal{C})$.

Lemma 13.4.2. *Let R be any ring. Then* $\mathrm{UD}(R)$ *is not a limit ordinal. Also* $\mathrm{UD}(R) = \mathrm{UD}(R^{\mathrm{op}})$.

Proof. Since $(R, -)$ cannot be expressed as a proper direct limit of finitely presented objects, the supremum in 13.1.7 is attained. The bijection, 11.2.9, between Serre subcategories of $(\text{mod-}R, \mathbf{Ab})^{\mathrm{fp}}$ and those of $(R\text{-mod}, \mathbf{Ab})^{\mathrm{fp}}$, plus the observation that the dual of a uniserial functor is uniserial, gives the second statement. □

Also note that the statements of 13.1.4 and 13.1.5 apply. The next result has a proof like that of 13.2.2. Breadth, the lattice property based on factoring out chains, was defined in Section 7.3.1, as is the notion, appearing in the corollary, of having width $< \infty$.

Corollary 13.4.3. *Let $\mathcal{X}$ be a definable subcategory of the category* Mod-R. *Then the breadth of* $\mathrm{pp}(\mathcal{X})$ *equals* $\mathrm{Udim}\big((\text{mod-}R, \mathbf{Ab})^{\mathrm{fp}}_{\mathcal{X}}\big)$ *– we denote this (with little danger of ambiguity) by* $\mathrm{UD}(\mathcal{X})$*; this is also the value for* $\mathrm{UD}(D\mathcal{X})$*, where $D\mathcal{X}$ is the elementary dual category of $\mathcal{X}$ (cf. 13.2.3).*

In particular $\mathrm{Udim}\big((\text{mod-}R, \mathbf{Ab})^{\mathrm{fp}}\big)$ *equals the breadth of* pp_R.

Corollary 13.4.4. *Let $\mathcal{X}$ be a definable subcategory of the category* Mod-R. *Then* $\mathrm{UD}(\mathcal{X}) < \infty$ *iff* $\mathrm{w}(\mathrm{pp}(\mathcal{X})) < \infty$.

Proposition 13.4.5. *Let $\mathcal{C}$ be a locally coherent category and let $F \in \mathcal{C}^{\mathrm{fp}}$. Then* $\mathrm{Udim}(F) \le \mathrm{KGdim}(F)$*. In particular, if* $\mathrm{KGdim}(\mathcal{C}) < \infty$*, then so is* $\mathrm{Udim}(\mathcal{C}) < \infty$.

Proof. The congruence which identifies points a, b such that the interval $[a, b]$ is a chain certainly identifies points a, b such that $[a, b]$ is the two-point interval, therefore this follows by induction on dimension. □

Now let $\mathcal{X}$ be any definable subcategory of Mod-R, where R is a small preadditive category, and consider the uniserial dimension process applied to the corresponding localised, locally coherent, category $(\text{mod-}R, \mathbf{Ab})_{\mathcal{X}}$. Let $\mathcal{T}_\alpha$ be the torsion subcategory of $(\text{mod-}R, \mathbf{Ab})_{\mathcal{X}}$ which is the kernel of the localisation corresponding to $UD \le \alpha$, and let $\mathcal{S}_\alpha = \mathcal{T}_\alpha \cap (\text{mod-}R, \mathbf{Ab})^{\mathrm{fp}}_{\mathcal{X}}$ be the corresponding Serre subcategory of finitely presented objects. Denote by $U^{\alpha+1}(\mathcal{X})$ the Ziegler-closed subset of $\mathrm{supp}(\mathcal{X})$ corresponding, in the sense of 12.4.2, to $\mathcal{S}_\alpha$;

that is, $U^{\alpha+1}(\mathcal{X}) = \{N \in \mathrm{Zg}(\mathcal{X}) : \overrightarrow{F} N = 0 \ \forall F \in \mathcal{S}_\alpha\}$. Recall, 12.3.14, that the basic open sets for the relative topology on $\mathrm{Zg}(\mathcal{X}) = \mathcal{X} \cap \mathrm{Zg}_R$ have the form $(F) = \{N \in \mathrm{Zg}_R : (F, N \otimes -) \neq 0\}$ as F ranges over $(R\text{-mod}, \mathbf{Ab})^{\mathrm{fp}}_{D\mathcal{X}}$ equivalently, by 11.1.34, as F ranges over $(R\text{-mod}, \mathbf{Ab})^{\mathrm{fp}}$. Another form of the typical basic open set is $(F) = \{N \in \mathrm{Zg}_R : \overrightarrow{F} N \neq 0\}$ as F ranges over $(R\text{-mod}, \mathbf{Ab})^{\mathrm{fp}}_{\mathcal{X}}$ by, for example, 12.3.15.

Lemma 13.4.6. *Let $\mathcal{X}$ be a definable category and let $N \in \mathrm{Zg}(\mathcal{X})$. Then $N \notin U^1(\mathcal{X})$ iff there is $F \in (R\text{-mod}, \mathbf{Ab})^{\mathrm{fp}}$ such that $N \in (F)$ and such that $F_{D\mathcal{X}}$ is uniserial. More generally, $N \notin U^{\alpha+1}(\mathcal{X})$ iff there is a neighbourhood, (F), of N such that $F \in \mathcal{S}_\alpha$, where $\mathcal{S}_\alpha$ denotes the Serre subcategory of $(R\text{-mod}, \mathbf{Ab})^{\mathrm{fp}}_{D\mathcal{X}}$ defined above; furthermore, such an F may be chosen so that its image in $\left((R\text{-mod}, \mathbf{Ab})^{\mathrm{fp}}_{D\mathcal{X}}\right)_\alpha$ is zero or uniserial.*

Proof. We have $N \notin U^{\alpha+1}$ iff the indecomposable injective functor $N \otimes -$ is not $\mathcal{S}_\alpha$-torsionfree, that is, iff $(F, N \otimes -) \neq 0$ for some functor F in $\mathcal{S}_\alpha$, which may be taken to be as described. The condition $(F, N \otimes -) \neq 0$ is equivalent to $N \in (F)$. □

We give some examples, also see [529].

Theorem 13.4.7. *Let R be a PI Dedekind domain which is not a division ring. Then* $\mathrm{UD}(R) = 1$.

Proof. Write U^α for $U^\alpha(\text{Mod-}R)$.
$\mathrm{UD}(R) \geq 1$: Let P be a non-zero prime ideal of R. By 6.1.8 Mod-$R_{(P)}$ is a definable subcategory of Mod-R. It follows directly that $\mathrm{UD}(R) \geq \mathrm{UD}(R_{(P)})$, so it will be enough to consider the case that R is local with maximal ideal P, generated by, say, π. If $\mathrm{UD}(R) = 0$, then $U^1 = \emptyset$ and the division ring of quotients Q of R is a point of Zg_R, so, by 13.4.6, there is a neighbourhood, (F), of Q with F uniserial. Now, a neighbourhood basis of Q is given (Section 5.2.1) by neighbourhoods of the form $(P^k \mid x) / (xP^t = 0)\ (k \geq 0, t \geq 1)$: in terms of functors, these neighbourhoods are defined by functors of the form $(R, -)/\mathrm{im}(f, -)$, where $f : R \to R/P^{k+t}$ is the morphism taking $1 \in R$ to $\pi^k + P^{k+t}$. We show that no such functor is uniserial. Consider the two factorisations of f: first, the natural epimorphism $g : R \to R/P^t$ followed by the natural embedding into R/P^{k+t}; second, the endomorphism h "multiplication by π^k" of R, followed by the natural epimorphism $R \to R/P^{k+t}$. Clearly neither of g, h factors through the other, hence (cf. 10.2.2) there are two incomparable subfunctors of $(R, -)/\mathrm{im}(f, -)$, as required.

$\mathrm{UD}(R) \leq 1$: Let P be a non-zero prime ideal of R and let N be the P-adic point of Zg_R (5.2.2). Let F be the functor $(R, -)/\mathrm{im}(f, -)$, where $f : R \to R/P$ is the canonical map. We claim that F is uniserial. The set of finite-length points of the form R/P^n is (5.2.3) dense in (F) so let $M = \bigoplus_n \{R/P^n\}$. Let G be a finitely generated subfunctor of F and consider $\overrightarrow{G}M$. Since $F(R/P^{n+1}) = R/P$ and R/P is a simple End(R/P)-module, if $G(R/P^{n+1}) \neq 0$, then $G(R/P^{n+1}) = F(R/P^{n+1})$. Let $g : R/P^{n+1} \to R/P^n$ be the natural projection and note that $Fg : F(R/P^{n+1}) \to F(R/P^n)$ is an isomorphism. It follows that if $G(R/P^{n+1}) \neq 0$, then $G(R/P^n) = F(R/P^n)$ and hence $G(R/P^m) = F(R/P^m)$ for all $m \leq n+1$. Therefore the evaluations of proper subfunctors of F on M are all of finite length and are linearly ordered. If $M' = \bigoplus\{R/P_1^n : P_1 \neq P \text{ is a non-zero prime }, n \geq 1\}$, then $\overrightarrow{G}(M') = 0$, so it is also the case that the subfunctors of $\overrightarrow{G}$, evaluated on $M \oplus M'$, are uniserial. Therefore, since the definable subcategory generated by $M \oplus M'$ is all of Mod-R (by 5.2.3) it follows easily, using 12.3.16 for example, that F is uniserial, as required.

Therefore, by 13.4.6, $N \notin U^1$.

Next, suppose that N is a Prüfer point. Then N is the elementary dual (Section 5.4) of an adic left module L. By what was shown above, there is a uniserial functor $F' \in (R\text{-mod}, \mathbf{Ab})^{\mathrm{fp}}$ such that $L \in (F')$. Hence $N \in (DF')$. But DF' is uniserial, so $N \notin U^1$ by 13.4.6.

We deduce that U^1 contains at most the point $\{Q\}$, where Q is the division ring of quotients of R. But R has Krull–Gabriel dimension (5.2.6), so, if $\mathcal{C}$ denotes the localisation of (mod-R, **Ab**) at $\mathcal{S}_0$, then KGdim$(\mathcal{C}) = 0$ (by 5.3.60) and hence, by 13.4.5, Udim$(\mathcal{C}) = 0$. Therefore UD$(R) \leq 1$, as required. □

Recall, 10.2.38, that if R is a von Neumann regular ring, then the category (mod-R, **Ab**) is equivalent to the module category Mod-R. Moreover, if R is von Neumann regular, then any interval in the lattice of finitely generated ideals of R is easily seen to be a chain iff it is simple. Therefore, noting 3.4.30 for inductive steps, the dimensions KGdim(mod-R, **Ab**) and Udim(mod-R, **Ab**) coincide.

Proposition 13.4.8. *If R is a von Neumann regular ring, then* KG(R) = UD(R), *and similarly for each definable subcategory of* Mod-R.

There are commutative von Neumann regular rings of arbitary Krull–Gabriel dimension (by 8.2.92, since it is easy to construct spectral spaces, hence commutative von Neumann regular rings, with spectrum of arbitrary Cantor–Bendixson rank), hence there are commutative von Neumann regular rings of arbitrary uniserial dimension.

For artin algebras the possible values are unknown, as is the case with Krull–Gabriel dimension (Section 9.1.4).

Question 13.4.9. What are the possible values of UD(R) for R an artin algebra?

14

The Zariski spectrum and the sheaf of definable scalars

The Gabriel–Zariski topology associated to a locally finitely presented additive category is introduced. This topology for the functor category is shown to be equivalent to the dual-Ziegler topology for the module category. The "structure sheaf" formed by rings of definable scalars over this topology is defined and computed for certain types of ring.

14.1 The Gabriel–Zariski spectrum

In Section 14.1.1 we see how to define the Zariski spectrum of a commutative noetherian ring purely in terms of its category of modules. This results in a definition that makes good sense in any locally finitely presented abelian category and yields, in Section 14.1.2, a spectrum based on indecomposable injectives. This space carries a presheaf of localisations which we also define.

The coincidence of various topologies over various types of ring is considered in Section 14.1.3 and it is shown in that section that the dual-Ziegler and rep-Zariski topologies coincide.

We are going to define a presheaf of rings over the set, pinj_R, of isomorphism types of indecomposable pure-injective right R-modules equipped with the dual-Ziegler topology. That topology was defined in Section 5.6 and we will see another way of defining it here. The rings which occur as sections and stalks are rings of definable scalars; these can be seen as localisations of R, but localisations in a sense broader than usual.

Localisation in the sense of rings of fractions of non-commutative rings goes back to Ore, [474]; it was developed much further, and in different directions, by Gabriel, [202], (torsion theory, see Section 11.1.1) and Cohn, [127] (universal localisation, see 5.5.5). The 'localisations' here will include the latter and the finite-type case of the former.

We can, perhaps rather pretentiously, say that we are going to define a kind of "non-commutative geometry", which is briefly described as follows.[1] Take a commutative noetherian ring R; describe $\mathrm{Spec}(R)$, not in terms of prime ideals but in terms of the category, Mod-R, of representations of R; now use the same description with any small pre-additive category, $\mathcal{A}$, in place of R. The result is a topological space, with points being some kind of "primes" of $\mathcal{A}$, and with associated presheaf of localisations of $\mathcal{A}$.

We may, in particular, apply this definition with $\mathcal{A} = \text{mod-}R$ to obtain a new ringed space associated to R. Admittedly this has "moved us up one level of representation" since it is the spectrum of mod-R rather than that of R. Nevertheless the "usual spectrum" of R, if it has one, sits inside this richer structure (more accurately, inside that with $(R\text{-mod})^{\mathrm{op}}$ for $\mathcal{A}$). For example, if R is commutative noetherian, then the usual Zariski spectrum of R is a subspace of the larger space, 14.1.5, and the usual structure sheaf is a part of the larger presheaf, see 14.1.10 and 14.1.11.

14.1.1 The Zariski spectrum through representations

The Zariski spectrum – space and structure sheaf – is defined in terms of the category of modules, 14.1.1. The key, going back to [202], is to replace a prime ideal P by the injective hull of R/P.

Recall the definition of the **Zariski spectrum**, $\mathrm{Spec}(R)$, of a commutative ring R. The points are the prime ideals of R and a basis of open sets for the topology is given by the sets $D(r) = \{P \in \mathrm{Spec}(R) : r \notin P\}$ for $r \in R$. Thus the open sets have the form $D(I) = \{P \in \mathrm{Spec}(R) : I \nleq P\} = \bigcup_{r \in I} D(r)$ and their complements are denoted $V(I) = \{P \in \mathrm{Spec}(R) : I \leq P\} = D(I)^{\mathrm{c}}$. If R is the coordinate ring of an affine variety, so in particular is noetherian, then the maximal

[1] This space was introduced in [497] but my route to it was not as direct as it should have been. I had been led by earlier work ([487], [488] for example) on the model theory of injective modules to expect that there would be a natural topology on the set of indecomposable pure-injectives over any ring, but my early attempts to define such a topology, which would fit well with the model theory, failed. With the benefit of hindsight, I can see that the waters were muddied by the fact that there are *two* natural topologies on the set of indecomposable pure-injectives. Ziegler defined the topology which is the "right" one for model theory. In [495, p. 105] it is noted that when Ziegler's topology is restricted to the (Ziegler-closed by 5.1.11) set of indecomposable injectives over a commutative noetherian ring one obtains a topology which is in some sense dual to the Zariski topology, the latter being regarded as a topology on this same set via the natural bijection between prime ideals and indecomposable injectives. There, I ask whether there is, in general, a more "algebraic-geometric" topology on the underlying set of the Ziegler spectrum. Later, [497], I realised that the functorial approach leads directly to a positive answer to this question, though my route to this (dual) topology was through seeing how far Hochster's duality, [294], for spectral spaces could be applied to the Ziegler topology. Only then did I realise that this space could be presented as a natural and literal generalisation of the Zariski spectrum.

primes of R, we denote the subspace of these by maxspec(R), correspond to the usual, that is, closed, points of the variety. The other primes correspond to, indeed are, generic points of, irreducible subvarieties (for example, [269, §II.2]).

Assuming now that R is commutative noetherian, we show how to define the spectrum purely in terms of the category, Mod-R, of R-modules. This is essentially Gabriel's approach [202, §V.4, Chpt. VI] and in part that builds on earlier work of Matlis [438, §3].

Each point $P \in \text{Spec}(R)$ is replaced by the injective hull (Appendix E, p. 708), $E_P = E(R/P)$, of the corresponding quotient module R/P. Because P is prime, hence $\cap$-irreducible in the lattice of right ideals, E_P is uniform hence indecomposable (this does use commutativity of R). Furthermore, each indecomposable injective R-module E has this form. To see this, let a and b be non-zero elements of E and let I, J be their respective annihilators in R. Since E is indecomposable injective, hence uniform, the intersection $aR \cap bR$ is non-zero, so choose a non-zero element c in this intersection. Then $c(I + J) = 0$. Since R is noetherian it follows that there is a unique maximal annihilator, P say, of non-zero elements of E. A line of calculation (which can be found at 14.4.2) shows that this is a prime ideal and so, since E contains an element with annihilator exactly P, E is the injective hull of R/P. This ideal P is the **associated prime** of E and is denoted ass(E). So we can, and will, take the underlying set of Spec(R) to be the set, inj_R, of isomorphism classes of indecomposable injective R-modules.

As for the topology, $D(r) = \{E \in \text{inj}_R : \text{Hom}(R/rR, E) = 0\}$. For, if $r \in R \setminus P$ and if $f : R/rR \to E_P$, then $\text{ann}_R\big(f(1 + rR)\big) \geq \text{ann}_R(1 + rR) = rR$, so, since P is the unique maximal annihilator of non-zero elements of E_P, it must be that $f(1 + rR) = 0$, hence $f = 0$. For the converse, if $r \in P$, then the canonical surjection from R/rR to R/P, followed by inclusion, is a non-zero morphism from R/rR to E_P.

Thus a basis of open sets for the Zariski topology is given by those sets of the form $[M] = \{E \in \text{inj}_R : \text{Hom}(M, E) = 0\}$ as M ranges over modules of the form R/rR. This, however, is not yet a description of the topology in terms of the category Mod-R because the property of being of this form, namely cyclic-projective modulo cyclic, is not invariant under equivalence of categories. We will show that if M is allowed instead to range over arbitrary finitely presented modules, then the topology does not change.

First consider the case that $M = R/I$ is cyclic. The argument used to reinterpret $D(r)$ applies equally well to show that, for any ideal I, the set $[R/I] = \{E \in \text{inj}_R : \text{Hom}(R/I, E) = 0\}$ coincides with $\{E_P : P \in \text{Spec}(R),\ I \nleq P\}$ and, since I is finitely generated, say $I = \sum_1^n r_i R$, we have $[R/I] = \bigcup_1^n [R/r_i R]$ (the inclusion "$\subseteq$" uses commutativity of R, see 14.4.3) $= \bigcup_1^n D(r_i)$, which is indeed open. But this is not yet enough because the property of being cyclic is not Morita-invariant.

So consider the case that M is finitely presented, in particular, is finitely generated, by $b_1, \dots, b_n$ say. Set $M_k = \sum_{j \leq k} b_j R$, where $M_0 = 0$. Each factor $C_j = M_j/M_{j-1}$ is finitely presented cyclic and, we claim, $[M] = [C_1] \cap \cdots \cap [C_n]$. For, if there is a non-zero morphism from C_j to E, then, by injectivity of E, this extends to a morphism from M/M_{j-1} to E, hence there is induced a non-zero morphism from M to E. Conversely, if $f : M \to E$ is non-zero let j be minimal such that the restriction of f to M_j is non-zero. Then f induces a non-zero morphism from C_j to E.

This is summarised in the following result, which comes from [508] (but some, perhaps all, of it is folklore). Beyond the commutative situation, the same result holds if R is fully bounded noetherian, see 14.1.9. Beware that if R is not noetherian, then $\mathrm{Spec}(R)$ might correspond to a proper subset of inj_R and hence the identification of $D(I)$ with $\{E \in \mathrm{inj}_R : (R/I, E) = 0\}$ cannot be made. See Section 14.4 for more on this.

Theorem 14.1.1. *Let R be a commutative noetherian ring. Then the space whose points are the isomorphism classes of indecomposable injective R-modules, E, and which has, for a basis of open sets, those of the form $[M] = \{E : \mathrm{Hom}(M, E) = 0\}$ as M ranges over finitely presented R-modules, is naturally homeomorphic to* $\mathrm{Spec}(R)$. *The bijection is given by $P \mapsto E(R/P)$ and $E \mapsto \mathrm{ass}(E)$.*

The term “naturally homeomorphic” in the statement of the result may be taken in the categorical sense. If $\alpha : R \to S$ is a morphism of commutative rings, then there is induced the map $\mathrm{Spec}(S) \to \mathrm{Spec}(R)$ which takes the prime ideal Q of S to $\alpha^{-1}Q = \{r \in R : \alpha r \in Q\}$. For instance, the projection $k[X] \to k \simeq k[X]/\langle X \rangle$ induces the embedding $\{0\} = \mathrm{Spec}(k) \to \{\langle X \rangle\} \subseteq \mathrm{Spec}(k[X])$. In terms of indecomposable injectives it is equally direct: the map from inj_S to inj_R takes $E_S(S/Q)$ to $E_R(R/\alpha^{-1}Q)$. It is easy to check that, assuming the rings are noetherian, the two topologies on the image of inj_S in inj_R, namely the quotient topology induced from inj_S and the subspace topology induced from inj_R, coincide.

The statement of 14.1.1 rather begs the question of whether it applies to arbitrary commutative rings and we do consider these in Section 14.4 but now we proceed to the next stage, which is to apply this definition more widely.

14.1.2 The Gabriel–Zariski and rep-Zariski spectra

These spaces are defined as topologies on sets of indecomposable injectives, and their associated presheaves are defined using finite type localisations.

The reformulated definition of the Zariski spectrum that was obtained in the previous section may be applied to any abelian category but surely makes most sense when the category is Grothendieck (so has enough injectives) and is locally finitely presented (so has enough finitely presented objects). Throughout this section, therefore, $\mathcal{C}$ will be a locally finitely presented abelian, hence, 16.1.11, Grothendieck, category. Recall, 11.1.27, that every such category is a nice localisation of a functor category $(\mathcal{A}^{(\mathrm{op})}, \mathbf{Ab})$ with $\mathcal{A}$ a skeletally small preadditive category.

We have particularly in mind the special case where $\mathcal{A}$ is a small preadditive category and $\mathcal{C} = \text{Mod-}\mathcal{A}$ and the even more particular case where R is a ring and $\mathcal{C}$ is the category of right R-modules. At some places we will impose the condition of local coherence on $\mathcal{C}$; recall that the functor category $(\mathcal{A}\text{-mod}, \mathbf{Ab})$ is locally coherent (10.2.3).

Denote by $\mathrm{inj}(\mathcal{C})$ the set of isomorphism types of indecomposable injective objects of $\mathcal{C}$; so, in this notation, $\mathrm{inj}_R = \mathrm{inj}(\text{Mod-}R)$. That this is a set follows directly from the assumption that $\mathcal{C}$ has a generating set of finitely generated objects and the fact, see Appendix E, p. 708, that each object of such a category has only a set of subobjects.

Equip $\mathrm{inj}(\mathcal{C})$ with the topology which has, for a basis of open sets, those of the form $[F] = \{E \in \mathrm{inj}(\mathcal{C}) : (F, E) = 0\}$ as F ranges over the category $\mathcal{C}^{\mathrm{fp}}$ of finitely presented objects of $\mathcal{C}$. Since $[F] \cap [G] = [F \oplus G]$ these sets are closed under finite intersection, so an arbitrary open set will have the form $\bigcup_\lambda [F_\lambda]$ with $F_\lambda \in \mathcal{A}^{\mathrm{fp}}$. Denote the resulting space by $\mathrm{Zar}(\mathcal{C})$ and refer to this as the **Gabriel–Zariski topology** on $\mathrm{inj}(\mathcal{C})$.

In the case $\mathcal{C} = \text{Mod-}\mathcal{A} = (\mathcal{A}^{\mathrm{op}}, \mathbf{Ab})$ for some small preadditive category $\mathcal{A}$, we write $\mathrm{inj}_{\mathcal{A}}$ for $\mathrm{inj}(\text{Mod-}\mathcal{A})$ and call this space the **Gabriel–Zariski spectrum** of $\mathcal{A}$, writing $\mathrm{GZspec}(\mathcal{A})$. If $\mathcal{C} = \mathcal{A}\text{-Mod}$, then we refer to the *left* Gabriel–Zariski spectrum of $\mathcal{A}$.

If $\mathcal{A}$ is itself of the form $(R\text{-mod})^{\mathrm{op}}$ for some ring or small preadditive category R, that is, if we are topologising $\mathrm{inj}(R\text{-mod}, \mathbf{Ab})$, then we will refer to this as the **rep-Zariski spectrum** of R (the "rep" indicating the change of representation level) and denote it Zar_R. We will use the notation ${}_R\mathrm{Zar}$ for $\mathrm{Zar}_{R^{\mathrm{op}}}$. In summary, our notation is as follows:

$\mathrm{GZspec}(\mathcal{A}) = \mathrm{Zar}(\text{Mod-}\mathcal{A})$ and
$\mathrm{Zar}_{\mathcal{A}} = \mathrm{GZspec}((\mathcal{A}\text{-mod})^{\mathrm{op}}) = \mathrm{Zar}(\mathcal{A}\text{-mod}, \mathbf{Ab})$.

If R is commutative noetherian, then, by the previous section, we may identify $\mathrm{GZspec}(R)$ with the usual spectrum, $\mathrm{Spec}(R)$, so the latter may be seen as a subset indeed, see 14.1.1 and 14.1.6, a subspace, of Zar_R.

In the case that $\mathcal{C} = (R\text{-mod}, \mathbf{Ab})$ the indecomposable injectives are in bijection (12.1.9) with the indecomposable pure-injective objects of Mod-R, so we may regard Zar_R as a topology on pinj_R. It will be seen (14.1.7) that this is the dual-Ziegler topology which was introduced earlier.

Example 14.1.2. Let $\mathcal{A}$ be the path category, over a field k, of the quiver, A_∞, shown below (see 10.1.2).

$$1 \longrightarrow 2 \longrightarrow 3 \longrightarrow \cdots$$

Recall, 10.1.4, that $\mathcal{C} = \mathcal{A}\text{-Mod} \simeq \text{Mod-}kA_\infty$ is the category of left $\mathcal{A}$-modules and this may equally be thought of as the category of right modules over the path algebra or as the category of k-representations of A_∞. Also recall, 10.1.15, that the indecomposable injective objects of $\mathcal{C}$ are: the E_n, where, using the representation-of-A_∞ description, $E_n(i) = k$ if $i \leq n$, $= 0$ if $i > n$, and where each non-zero morphism is an isomorphism; together with E_∞, which is one-dimensional at each vertex and has all maps isomorphisms. Furthermore, E_n is the injective hull of the simple representation S_n which is 0 everywhere except at n, where it is one-dimensional. These are the points of $\text{inj}(\mathcal{C})$.

Dually, the indecomposable projective objects are the P_n $(n \geq 1)$, where P_n is one-dimensional at each vertex $m \geq n$ and 0 elsewhere (so $P_1 = E_\infty$). Clearly the indecomposable representation $M_{[n,m]}$ which is one-dimensional at i for $n \leq i \leq m$ and 0 elsewhere is finitely presented.

For any representation, M, one has $(M, E_n) = 0$ iff M does not have S_n as a subquotient, that is, iff $M(n) = 0$. Therefore each indecomposable injective E_n is an isolated point in the Gabriel–Zariski topology on $\mathcal{A}$-inj $(= \text{GZspec}(\mathcal{A}^{\text{op}}))$, being isolated by $[S_1] \cap \cdots \cap [S_{n-1}] \cap [P_{n+1}]$. Also, a basis of open neighbourhoods of E_∞ consists of the cofinite sets which contain that point: clearly the latter are open and, from the description of the P_n, it is easily seen that every infinite-dimensional finitely presented object is eventually > zero-dimensional, so there are no other open neighbourhoods of E_∞.

Therefore $\text{Zar}(\mathcal{A}\text{-Mod})$ is the one-point compactification, by E_∞, of the discrete set $\{E_n : n \geq 1\}$.

The category Mod-$\mathcal{A}$ is the category of representations of the opposite quiver, which we regard as the same quiver but with arrows reversed. In this case there is a finite-dimensional indecomposable projective, P_n', for each n and, also for each n, an indecomposable injective, E_n'. The dimension vector of P_n' is as for E_n above and that of E_n' is as for P_n above.

A similar check shows that the open sets of $\text{Zar}(\text{Mod-}\mathcal{A})$ are, apart from the empty set, the cofinite sets.

If $\mathcal{C}$ is locally coherent, then there is a presheaf of localisations of $\mathcal{C}^{\text{fp}}$ over $\text{Zar}(\mathcal{C})$. If $\mathcal{C}$ has the form $(\mathcal{A}\text{-mod}, \mathbf{Ab})$, then there is also a presheaf of "localisations" of $\mathcal{A}$ over $\text{Zar}_{\mathcal{A}}$. Let us define the first of these.

Suppose that $\mathcal{C}$ is locally coherent. A presheaf of localisations over $\text{Zar}(\mathcal{C})$ is defined on a basis for the topology, just as for the usual definition of the structure sheaf of an affine algebraic variety: that is enough data for the sheafification process (for example, [674, 4.2.6]). Unlike the classical case, what we define here is, even on the basis, a presheaf rather than a sheaf: one may have a basic open set $U = V \cup W$, where V, W are basic open and disjoint, yet with the set of sections over U not equal to the product of that over V with that over W (see after 14.3.4).

Let $F \in \mathcal{C}^{\mathrm{fp}}$, so $[F] = \{E \in \mathrm{inj}(\mathcal{C}) : (F, E) = 0\}$ is basic open in $\mathrm{Zar}(\mathcal{C})$. Let $\tau_{[F]} = \mathrm{cog}([F])$ denote the hereditary torsion theory on $\mathcal{C}$ cogenerated by this set of injectives (Section 11.1.1). This torsion theory is, since the torsion class is generated by the finitely presented object F, of finite type (11.1.14). If $G \in \mathcal{C}^{\mathrm{fp}}$ with $[G] \subseteq [F]$, then $\mathcal{T}_{[G]} \supseteq \mathcal{T}_{[F]}$ so the localisation $\mathcal{C} \to \mathcal{C}_{[G]}$ factors through $\mathcal{C} \to \mathcal{C}_{[F]}$. Thus we obtain the presheaf (on a basis) of finite-type localisations of $\mathcal{C}$.

This is a presheaf of large categories. By 11.1.34 there is induced a presheaf of localisations of $\mathcal{C}^{\mathrm{fp}}$; the presheaf of quotients of $\mathcal{C}^{\mathrm{fp}}$ by the Serre subcategories $\mathcal{T}_{[F]} \cap \mathcal{C}^{\mathrm{fp}}$. We will refer to this latter as the **finite-type presheaf on** $\mathcal{C}$, and denote it by $\mathrm{FT}(\mathcal{C})$. Denote its sheafification by $\mathrm{LFT}(\mathcal{C})$.

If $\mathcal{C}$ has the form Mod-$\mathcal{A}$ for some skeletally small preadditive category $\mathcal{A}$ which is right coherent, then we also obtain a presheaf of localisations of $\mathcal{A}$. For the Yoneda embedding, 10.1.7, which takes an object A of $\mathcal{A}$ to the functor $(-, A)$, embeds $\mathcal{A}$ in mod-$\mathcal{A}$ as the "module" $\mathcal{A}_{\mathcal{A}}$, so we may consider the image of the latter in any localisation of the form $(\text{Mod-}\mathcal{A})_{\tau_{[F]}}$ with $F \in$ mod-$\mathcal{A}$. This image, which we denote $\mathcal{A}_{\tau_{[F]}}$, is the localisation of $\mathcal{A}$ which we assign to the basic open set $[F]$. With the natural restriction maps, this again gives a presheaf (defined on a basis) which clearly is a subpresheaf of the first-defined presheaf of localisations of mod-$\mathcal{A}$. We denote this subpresheaf by $\mathrm{FT}_{\mathcal{A}}$ and refer to it as the **finite-type presheaf**[2] **of** $\mathcal{A}$. Its sheafification we denote by $\mathrm{LFT}_{\mathcal{A}}$.

In particular, if we start with $\mathcal{A} = R$ a ring and $\mathcal{C} =$ Mod-R, then we obtain a presheaf of rings, the stalks of which are finite-type localisations of R (14.1.12). It will be seen, 14.1.11, that if R is commutative noetherian, then $\mathrm{FT}_R = \mathcal{O}_R$, the usual structure sheaf on the Zariski spectrum of R.

Next, suppose that $\mathcal{C} = (\mathcal{A}\text{-mod}, \mathbf{Ab})$, the category we have also denoted $\mathrm{Fun}^{\mathrm{d}}$-$\mathcal{A}$; we define a subpresheaf of "localisations" of $\mathcal{A}$ over $\mathrm{Zar}_{\mathcal{A}}$. Note the change in representation level.

Let us first examine $\mathrm{FT}((\mathcal{A}\text{-mod}, \mathbf{Ab}))$ in a little more detail. The localisations $\tau_{[F]}$ for $F \in \mathrm{fun}^{\mathrm{d}}\text{-}\mathcal{A} = (\mathcal{A}\text{-mod}, \mathbf{Ab})^{\mathrm{fp}}$ yield a presheaf of skeletally small abelian categories; localisations, $(\mathcal{A}\text{-mod}, \mathbf{Ab})^{\mathrm{fp}}_{[F]}$, of $(\mathcal{A}\text{-mod}, \mathbf{Ab})^{\mathrm{fp}}$. By 12.3.20 these categories can be regarded as categories of imaginaries for definable subcategories of

[2] I remark that we can make a similar definition for Mod-$\mathcal{A}$ without any assumption of coherence on $\mathcal{A}$ but making use of 11.1.37 to give finite-type torsion theories.

$\mathcal{A}$-Mod. By moving to the dual, that is (11.2.9), taking opposite functor categories, one obtains a presheaf of localisations of fun-$\mathcal{A} = (\text{mod-}\mathcal{A}, \mathbf{Ab})^{\text{fp}}$, equivalently (12.3.20) a presheaf of categories of pp imaginaries for the definable subcategories of Mod-$\mathcal{A}$, where to $F \in (\text{mod-}\mathcal{A}, \mathbf{Ab})^{\text{fp}}$ corresponds (by 10.2.31 and 10.3.8)

$$\{M \in \text{Mod-}\mathcal{A} : (dF, M \otimes -) = 0\} = \{M \in \text{Mod-}\mathcal{A} : \overrightarrow{F} M = 0\}. \qquad (\star\star)$$

Also note that these basic open sets of $\text{Zar}_{\mathcal{A}}$ – the $[F]$ with $F \in (\mathcal{A}\text{-mod}, \mathbf{Ab})^{\text{fp}}$ – are closed sets in the Ziegler topology on $\text{pinj}_{\mathcal{A}}$, since $[F]$ is the complement of the basic Ziegler-open set $(F) = \{N \in \text{pinj}_{\mathcal{A}} : (F, N \otimes -) = 0\}$ (Section 12.1.3).

The presheaf-on-a-basis of skeletally small abelian categories of the form $(\mathcal{A}\text{-mod}, \mathbf{Ab})^{\text{fp}}_{[F]}$ we denote by $\mathbb{D}\text{ef}_{\mathcal{A}}$ and call the **presheaf of categories of definable scalars** of $\mathcal{A}$. This, rather its opposite – the corresponding presheaf of localisations of fun$(\mathcal{A})$ – was considered briefly in [510, §13] and is referred to in [106] as the "presheaf of definable scalars in all sorts". Within this presheaf of localisations of $\text{fun}^{\text{d}}(\mathcal{A})$ we may pick out the presheaf of images of $\mathcal{A}$ itself – two applications of the Yoneda lemma embed $\mathcal{A}$ in $(\mathcal{A}\text{-mod}, \mathbf{Ab})^{\text{fp}}$. This is the presheaf of "localisations" of $\mathcal{A}$ that we have been aiming for. Denote it by $\text{Def}_{\mathcal{A}}$.

In the case that R is a ring we refer to this last as the presheaf of rings of definable scalars. The justification for the name is that the section over a basic open set $[F]$ is, indeed (12.8.2), the ring of definable scalars of the corresponding definable subcategory of Mod-R, see $(\star\star)$ above. Explicitly, this presheaf takes the open set $[F]$ defined by the finitely presented functor $F \in (R\text{-mod}, \mathbf{Ab})$ to the endomorphism ring, $\text{End}\big(({}_RR, -)_{[F]}\big)$, of the localisation of the forgetful functor at $\tau_{[F]}$.

We generally use the prefix "L" to denote the corresponding sheaf. In the case of definable scalars we refer to the sheafification as the **sheaf of locally definable scalars** of $\mathcal{A}$.

It should be emphasised that, over most types of ring, Zar_R will be a presheaf (defined on a basis), rather than a sheaf, though, 14.1.13, it will always be separated (p. 620) and, in many cases, interesting parts of it (for example, see 14.3.10) will have the glueing property, hence will be subsheaves.

We end this section by elaborating a few points: first, how this looks for a small preadditive category, a "ring with many objects", $\mathcal{A}$.

We have (12.1.3) the category Mod-$\mathcal{A}$ embedded via $M \mapsto (M \otimes - : \mathcal{A}\text{-mod} \longrightarrow \mathbf{Ab})$ as a full subcategory of $(\mathcal{A}\text{-mod}, \mathbf{Ab})$. Contained in the image of Mod-$\mathcal{A}$ is the Yoneda-image of the Yoneda-image of $\mathcal{A}$ in $\mathcal{A}$-mod. That is, map $A \in \mathcal{A}$ to $(A, -) \in \mathcal{A}$-mod and then map this to the representable functor $((A, -), -) \simeq (-, A) \otimes -$ in $(\mathcal{A}\text{-mod}, \mathbf{Ab})$. This latter does not look quite so bad in the case where $\mathcal{A}$ is a ring, R, since the usual practice is to denote the projective left R-module $(R, -)$ also by R (or ${}_RR$) and then the image of this in the functor category is denoted $({}_RR, -)$, which is isomorphic to $R_R \otimes -$ and which is just the forgetful functor. If $\mathcal{A}$ has more than one object, then the natural equivalence $((A, -), \sim) \simeq (-, A) \otimes \sim$ is seen as follows. Both $(-, A) \otimes \sim$ and $((A, -), \sim)$ are right exact, for $(A, -)$ is projective, so it is enough to check on the projective generators, 10.1.13, $(B, -)$, $B \in \mathcal{A}$, of $\mathcal{A}$-mod. By Yoneda, $((A, -), (B, -)) \simeq (B, A)$ and the isomorphism $(-, A) \otimes (B, -) \simeq (B, A)$ is part of the definition of $\otimes_{\mathcal{A}}$ (§12.1.1).

Next, in the case where $\mathcal{A}$ is a ring R, which may be identified with the endomorphism ring of the forgetful functor $({}_RR, -)$ sitting in (R-mod, **Ab**), the localisation at F is again a one-point category, that is, a ring which, *qua* ring, is the endomorphism ring $R_{[F]} = \mathrm{End}\big(({}_RR, -)_{[F]}\big)$. We emphasise that the ring morphism $R \to R_{[F]}$ is not necessarily a localisation in the usual sense since we used a torsion theory on the functor category rather than on the module category, but, as we have seen, 6.1.17, this notion of localisation in the functor category includes all finite type localisations in the module category, so, at least for a right noetherian ring, this is literally a more general notion of localisation since over such a ring all hereditary torsion theories are of finite type by 11.1.14.

We also reiterate that the construction of the Gabriel–Zariski spectrum can be made at different levels. For instance, given a ring R, it can be applied to Mod-R, giving a topology on the space of (isomorphism types of) indecomposable injective R-modules. Or it can be applied to the functor category (R-mod, **Ab**) and then this topology on the set of indecomposable injective functors may be viewed as a topology on the set of indecomposable pure-injective right R-modules. One may go further: the next stage would give a topology on the set of indecomposable injective functors on the functor category, that is, on the set of indecomposable pure-injective functors, and so on. But here at most three layers, and the corresponding two spaces, will concern us: a ring R or small preadditive category; its module category Mod-R and the functor category (R-mod, **Ab**).

14.1.3 Rep-Zariski = dual-Ziegler

If R is any ring, then inj_R *is a subset of* pinj_R, *so we may compare the Gabriel–Zariski topology on* inj_R, *that is,* GZspec(R), *with the topology inherited from the larger space* Zar_R. *It is shown in this section that if R is right coherent, then, 14.1.5, these do coincide, so in this case the injective spectrum really is part of the larger pure-injective spectrum when the latter is endowed with the rep-Zariski topology.*

The rep-Zariski topology coincides with the dual-Ziegler topology (14.1.6, 14.1.7).

Throughout, R, which we will treat as a ring, could be replaced by a small preadditive category.

First, observe the following (the proof is just as before 14.1.1).

Lemma 14.1.3. *For any ring R the sets $[M] = \{E \in \mathrm{inj}_R : (M, E) = 0\}$, with M a finitely presented R-module, form a basis of open sets of a topology on inj_R which coincides with that obtained by taking just those sets of the form $[R/I]$, where I is a finitely generated right ideal of R, as sub-basic open.*

The following also will be used.

Lemma 14.1.4. *Let $I \leq J$ be right ideals of a ring R and let E be an injective right R-module. Then $\mathrm{ann}_E(I)/\mathrm{ann}_E(J) \simeq \mathrm{Hom}(J/I, E)$.*

Proof. Just apply $(-, E)$ to the short exact sequence $0 \to J/I \to R/I \to R/J \to 0$ and note that $\mathrm{Hom}(R/I, E) \simeq \mathrm{ann}_E(I)$ (and similarly for J) and $\mathrm{Ext}^1(R/J, E) = 0$. □

Proposition 14.1.5. *Let R be right coherent. Then the Gabriel–Zariski topology on inj_R coincides with that induced from the rep-Zariski topology on pinj_R. That is, we may regard $\mathrm{GZspec}(R)$ as a subspace of Zar_R.*

Proof. One direction needs no assumption on R: given $M \in \text{mod-}R$, the basic Gabriel–Zariski-open set $[M] = \{E \in \mathrm{inj}_R : (M, E) = 0\}$ is just the intersection of the basic rep-Zariski-open set $[(M, -)]$ with $\mathrm{inj}_R \subseteq \mathrm{pinj}_R$; here we use the dual form of the definition of the topology at (⋆⋆) in the previous section and recall, 10.2.35, that $(M, -)$ is finitely presented.

For the other direction, given $F \in (R\text{-mod}, \mathbf{Ab})^{\mathrm{fp}}$ there is the basic rep-Zariski-open set $[F] = \{N \in \mathrm{pinj}_R : (F, N \otimes -) = 0\} = \{N \in \mathrm{pinj}_R : \overrightarrow{dF}N = 0\}$ (10.3.8). Since every finitely presented functor is the quotient of two finitely generated subfunctors of some power of the forgetful functor (10.2.31) we have $dF \simeq DG/DH$ for some finitely generated $G \leq H \leq ({}_RR^n, -)$ for some n. It follows from 2.3.3 that for $E \in \mathrm{inj}_R$, $\overrightarrow{dF}E = \mathrm{ann}_E(I)/\mathrm{ann}_E(J)$ for some right ideals I, J, which, by 2.3.19 are finitely generated. So, by 14.1.4, $\{E \in \mathrm{inj}_R : \overrightarrow{dF}E = 0\} = [J/I]$ (in the notation established in Section 14.1.1). Since J/I is finitely presented this is a basic Gabriel–Zariski open set, as required. □

In [497], see Section 5.6, the underlying set of Zg_R was retopologised by taking, for a basis of open sets, the complements of the compact open sets, that is, and in functorial notation, those of the form $[F] = \mathrm{Zg}_R \setminus (F) = \{N \in \mathrm{Zg}_R : \mathrm{Hom}(F, N \otimes -) = 0\}$, where $F \in (R\text{-mod}, \mathbf{Ab})^{\mathrm{fp}}$. Clearly this, dual-Ziegler, topology, extended to allow R to be any skeletally small preadditive category, coincides with the rep-Zariski topology, Zar_R, defined in Section 14.1.2 and the notation $[F]$ here has the same meaning as there.

More generally, one may prove the following extension of 14.1.5 ([525]).

Theorem 14.1.6. *Suppose that $\mathcal{C}$ is locally coherent. Then the Gabriel–Zariski topology on* $\mathrm{inj}(\mathcal{C})$ *coincides with the restriction of the dual-Ziegler topology on* $\mathrm{pinj}(\mathcal{C})$ *to* $\mathrm{inj}(\mathcal{C})$.

Take $\mathcal{C} = (\mathcal{A}\text{-mod}, \mathbf{Ab})$.

Corollary 14.1.7. *Suppose that $\mathcal{A}$ is a small preadditive category. Then the dual-Ziegler topology on* $\mathrm{pinj}_{\mathcal{A}}$ *coincides with the rep-Zariski topology, via the identification of* $\mathrm{pinj}_{\mathcal{A}}$ *with* $\mathrm{inj}(\mathcal{A}\text{-mod}, \mathbf{Ab})$.

Corollary 14.1.8. *For any ring, or small preadditive category R, there is a homeomorphism at the level of topology between* Zar_R *and* ${}_R\mathrm{Zar}$.

In particular, if $\mathrm{KG}(R) < \infty$*, then* ${}_R\mathrm{Zar} \simeq \mathrm{Zar}_R$ *(at the level of points).*

The first statement follows from 5.4.1 and the fact that the dual-Ziegler (= rep-Zariski by 14.1.8) topology can be defined directly from the Ziegler topology (Section 5.6).[3]

The second statement of 14.1.8 follows from 5.4.20. Even without any assumption on R, there is a homeomorphism at the level of topology (in the sense of 5.4.1) between ${}_R\mathrm{Zar}$ and Zar_R.

Observe that general inj_R is neither an open nor closed subset of Zar_R. Take $R = \mathbb{Z}$. To see that $\mathrm{inj}_{\mathbb{Z}}$ is not closed, refer to the list in Section 14.3.1 or use the list after 5.2.4 and the fact that $\mathbb{Q} \in \text{Zg-cl}(\overline{\mathbb{Z}_p})$ to see that $\overline{\mathbb{Z}_p} \in \text{Zar-cl}(\mathbb{Q})$. To see that inj_R is not open, again just check the list in Section 14.3.1 and note that it is not the case that every injective point has an open neighbourhood completely contained in $\mathrm{inj}_{\mathbb{Z}}$. The same will be true for any PI Dedekind domain with infinitely many primes.

If R is not noetherian, then there is the question of whether it is more appropriate to use all right ideals rather than only the finitely generated ones to define a topology on the set of indecomposable injectives. We could, for contrast, refer to the Gabriel–Zariski topology, that is, that defined using finitely presented modules, also as the **fg-ideals** topology and refer to the space with basis of open sets those of the form $\{E \in \mathrm{inj}_R : (R/I, E) = 0\}$, where now I is *any* ideal, as the **ideals** topology. By the same sort of argument that we used with the Gabriel–Zariski topology, this is the same topology as that with basis of open sets the sets $[M]$, where now M is finitely *generated*. This is Burke's full support topology from Section 12.7 (12.7.1).

[3] The converse is false: the Ziegler topology cannot, in general, be recovered from its dual, see Section 14.3.5.

Commutativity can be weakened in the module-category description of the prime spectrum.

Proposition 14.1.9. *Let R be a fully bounded noetherian (FBN) ring (see for example, [441, 6.4.7]). Then the following spaces are naturally homeomorphic:*

(i) $\mathrm{Spec}(R)$ *(defined exactly as in the commutative case);*
(ii) $\mathrm{GZspec}(R) = \mathrm{Zar}(\mathrm{Mod}\text{-}R)$.

Proof. The main difference between this and the commutative case is that, if P is a prime ideal, then $E(R/P)$ need not be indecomposable. It will, however, be a direct sum of finitely many copies of a unique indecomposable injective (for example, see [243, p. 139]) which, in this context, we denote by E_P. Then everything goes more or less as in the commutative case (cf. [490] which can be regarded as dealing with the, dual, Ziegler spectrum). □

For other cases where there is a bijection between (certain) primes and (certain) indecomposable injectives, see [582] and references there.

14.1.4 The sheaf of locally definable scalars

Extending the result 14.1.5 from the previous section on restrictions of topologies it is shown (14.1.10) that over a right coherent ring the restriction of the presheaf of definable scalars to injective points coincides with the presheaf of torsion-theoretic localisations. In particular for a commutative noetherian ring we recover the usual structure sheaf (14.1.11).

The stalk of the presheaf of definable scalars at an indecomposable pure-injective is the ring of definable scalars of that module (14.1.12). The centre of this presheaf is, 14.1.14, a presheaf of commutative local rings.

In this section our starting point is again, for simplicity, a ring R, but this could be replaced by any skeletally small preadditive category. Indeed, Mod-R could be replaced with any definable additive category (Section 18.1.1). For the results in this generality, and for the proofs, we refer to [525].

We have, from Section 14.1.2, the presheaf, $\mathbb{D}\mathrm{ef}_R$, of small abelian categories (localisations of the associated functor category) defined on a basis of Zar_R and the associated sheaf, $\mathrm{L}\mathbb{D}\mathrm{ef}_R$.

Running through this is Def_R: the thread, described earlier, of endomorphism rings of localisations of $({}_RR, -)$ – rings of definable scalars of the various definable subcategories of Mod-R. We may restrict the latter to the topology induced on $\mathrm{inj}_R \subseteq \mathrm{Zar}_R$.

We also have the presheaf FT_R of localisations of R at finite-type torsion theories, also over inj_R.

If R is right coherent, then these coincide. Part of this has been seen already: recall, 6.1.17, that if $\tau = (\mathcal{T}, \mathcal{F})$ is a hereditary torsion theory of finite type on Mod-R and if $\mathcal{E} = \mathcal{F} \cap \mathrm{inj}_R \subseteq \mathrm{Zar}_R$ is the corresponding set of indecomposable torsionfree injectives (by 11.1.29 there is a cogenerating set of indecomposable injectives), then the torsion-theoretic localisation, R_τ, of R coincides, as an R-algebra, with the ring of definable scalars, $R_{\mathcal{E}}$, of R at the Ziegler-closed set $\mathcal{E}$.

Theorem 14.1.10. ([583, 2.4.2], see [525]) *Suppose that R is a right coherent ring. Then the restriction to inj_R of the presheaf of rings of definable scalars coincides with the presheaf of finite type localisations FT_R. Indeed, there is an inclusion of ringed spaces $(\mathrm{inj}_R, \mathrm{FT}_R) \to (\mathrm{Zar}_R, \mathrm{Def}_R)$, where inj_R carries the Gabriel–Zariski topology.*

A similar statement holds with any locally coherent category in place of Mod-R ([525]).

Next we see that this really does extend the usual definition of the structure sheaf over a commutative noetherian ring.

In that classical case, associated to an affine variety, V, equivalently to its coordinate ring R, is a sheaf, written $\mathcal{O}_{V=\mathrm{Spec}(R)}$, of rings which is defined on the standard basis sets $D(r)$ (r not nilpotent) of $\mathrm{Spec}(R)$ by $\mathcal{O}_V \cdot D(r) = R[r^{-1}]$ – the localisation of R obtained by inverting $r \in R$ – and the restriction maps are just the canonical localisation maps. Any open set is already a basic open set (note that $D(r) \cap D(s) = D(rs)$), so this presheaf defined on a basis already is a presheaf, indeed, see, for example, [269, §II.2], it is a sheaf of local rings with the stalk at a prime P being the localisation, $R_{(P)}$, of R at P. Let us rewrite this sheaf in terms of Mod-R, as has been done already in Section 14.1.1 for the underlying space $\mathrm{Spec}(R)$.

Given $r \in R$ (commutative, noetherian), consider the corresponding open set $D(r) = \{E \in \mathrm{inj}_R : (R/rR, E) = 0\}$ and consider the hereditary torsion theory on Mod-R which this set of injective modules cogenerates. This is the torsion theory previously denoted, in Section 14.1.2, by $\tau_{[R/rR]}$ and it is easy to check that the localisation of R at this torsion theory is precisely $R[r^{-1}]$. Thus $\mathrm{Spec}(R)$ and FT_R agree on a basis and hence, just from the definition of the process, the sheafification of FT_R equals $\mathrm{Spec}(R)$.

Proposition 14.1.11. *If R is commutative noetherian, then (the sheafification of) the finite-type presheaf FT_R coincides with the usual structure sheaf $\mathcal{O}_{\mathrm{Spec}(R)}$.*

Recall that, given any presheaf, F, on a topological space, T, and given any point, $t \in T$, the stalk (Appendix E, p. 716) of F at t is $F_t = \varinjlim\{F(U) : t \in U \text{ and } U \text{ is open}\}$. If $\mathcal{U}_0$ is a basis for the topology, then clearly $F_t = \varinjlim\{F(U) : t \in U \in \mathcal{U}_0\}$ so a presheaf defined on a basis is enough to determine the stalks of any extension to a sheaf and, since the topology is defined locally, to determine the sheaf. The next result is from [498, C1.1], see [510, 3.1], for rings and [525] in the general case.

Proposition 14.1.12. *Let $\mathcal{A}$ be a small preadditive category and let $E \in \mathrm{inj}_{\mathcal{A}}$. Then the stalk of the presheaf $\mathrm{FT}_{\mathcal{A}}$ at E is the localisation of $\mathcal{A}$ at the torsion theory of finite type corresponding to E (in the sense of 11.1.37).*

In particular if R is a ring and if $N \in \mathrm{Zar}_R$, then the stalk of the presheaf, Def_R, of definable scalars at N is the ring of definable scalars at N: $(\mathrm{Def}_R)_N = R_N$.

Remark 14.1.13. The canonical map from a ring R to the ring of global sections of LDef_R is, by 6.1.3, an embedding.

The centre of the sheaf LDef_R is a sheaf of local rings. This follows from the fact (4.3.44) that if N is any indecomposable pure-injective, then the endomorphisms (in particular, the multiplications by elements of the centre of R) of N which are not automorphisms form a prime ideal, so there is a prime ideal, P, of $C(R)$ such that N is a module over the localisation of R at $C(R) \setminus P$. The same applies to the stalk, R_N of the presheaf of definable scalars at N, so there is the following.

Theorem 14.1.14. ([498, D1.1], see [510, 6.1]) *The centre of the presheaf of definable scalars is a presheaf of commutative local rings.*

There is also the following, which generalises 14.1.12.

Proposition 14.1.15. *Let R be any ring and let Y be any subset of Zar_R. Then $\varinjlim\{R_{[F]} : [F] \supseteq Y \text{ and } F \in (R\text{-mod}, \mathbf{Ab})^{\mathrm{fp}}\} \simeq R_{\mathrm{Zg\text{-}cl}(Y)}$, where the latter is the ring of definable scalars associated, in the sense of 1.1.13, with the closure, $\mathrm{Zg\text{-}cl}(Y)$, of Y in the Ziegler spectrum.*

The duality between right and left Ziegler spectra and right and left functor categories also, using 6.1.22 and 5.4.1, extends to these ringed spaces, see [525].

Clearly one may also define, in an entirely similar way, the presheaf of type-definable scalars (Section 6.1.6) over the full support topology (Section 5.3.7). Also see [358, Chpts. 13, 15].

14.2 Topological properties of Zar_R

Miscellaneous general points about this space are noted: Zg_R need not be recoverable from Zar_R; Zar_R need not be compact; N is in the Ziegler-closure of N' iff N' is in the Zariski-closure of N. An irreducible basic Zariski-closed set has a generic point (14.2.6).

First, although the rep-Zariski topology can be defined in terms of the Ziegler topology, the reverse is not true. This was shown in [107, 3.1], where one sees a homeomorphism of Zar_{Λ_2} (see Section 14.3.5 and Section 8.1.3) which is not a homeomorphism with respect to the Ziegler topology.

Example 14.2.1. Let $R = \mathbb{Z}_{(p)}$ for some prime p. Then

$$\mathrm{Zar}_R = \{\overline{\mathbb{Z}_{(p)}}, \mathbb{Q}\} \cup \{\mathbb{Z}_{p^\infty}, \mathbb{Q}\} \cup \bigcup_n \{\mathbb{Z}_{p^n}\}$$

is a union of Zariski-open sets with no finite subcover. The first two sets are defined by the conditions "p acts monomorphically", "p acts epimorphically" respectively, hence are Zariski-open, and the finite-length points are Zariski-open, 14.3.1.

Remark 14.2.2. $N' \in \text{Zar-cl}(N)$ iff $N \in \text{Zg-cl}(N')$. (That is straight from the definition of the dual-Ziegler topology.)

Proposition 14.2.3. *Suppose that* $X \subseteq \mathrm{Zar}_R$. *If* $N \in \text{Zar-cl}(X)$, *then there is* $N' \in \text{Zg-cl}(X)$ *such that* $N \in \text{Zar-cl}(N')$.

Proof. Let $\mathcal{F}$ be the filter of basic Zariski-open neighbourhoods of N. By assumption, for each $[F] \in \mathcal{F}$, $X \cap [F] \neq \emptyset$. Since Zg_R is compact (5.1.23) so is its closed subset $\text{Zg-cl}(X)$, so there is $N' \in \text{Zg-cl}(X) \cap \bigcap \mathcal{F}$, as required. □

This does not characterise Zariski-closure: to see this, take $R = \mathbb{Z}$, $X = \{\mathbb{Z}_{2^\infty}\}$ and $N = \overline{\mathbb{Z}_{(2)}}$. So $N \notin \text{Zar-cl}(X)$ but $\mathbb{Q} \in \text{Zg-cl}(X)$ and $N \in \text{Zar-cl}(\mathbb{Q})$.

Corollary 14.2.4. *Suppose that X is Ziegler-closed. Then*

$$\text{Zar-cl}(X) = \bigcup \{\text{Zar-cl}(N') : N' \in X\}.$$

Remark 14.2.5. An indecomposable pure-injective module N is Zariski-closed iff for every indecomposable pure-injective, N', if $N \in \langle N' \rangle$ (the definable subcategory generated by N'), then $\langle N \rangle = \langle N' \rangle$; this is straight from the definitions.

Proposition 14.2.6. *If the basic Zariski-closed set (F) is irreducible, then it has a generic point.*

Proof. Suppose, for a contradiction, that (F) does not have a Zariski-generic point. Then for each $N \in (F)$ there is $N' \in (F)$ such that $N' \notin \text{Zar-cl}(N)$, that is, such that $N \notin \text{Zg-cl}(N')$, so there is a Ziegler-open neighbourhood (F_N) of N which is a proper subset of (F). Thus we obtain a representation of (F) as a union of proper Ziegler-open subsets, so, since (F) is Ziegler-compact, there is a finite subcover. But this gives a covering of (F) by finitely many Zariski-closed sets. By Zariski-irreducibility of (F), there is a single one of these, (F_N) say, which is equal to (F), a contradiction, as required. □

Question 14.2.7. Does every irreducible Zariski-closed set have a generic point?

14.3 Examples

In any example where the Ziegler spectrum has been computed one may, in principle, compute the Zariski spectrum and perhaps also have some chance of computing the presheaf of definable scalars.

We begin with PI Dedekind domains (Sections 14.3.1, 14.3.2), generalise this to PI hereditary noetherian prime rings (Section 14.3.3) and use this for tame hereditary artin algebras (Section 14.3.4). There are brief comments on other examples (finite-dimensional algebras, generalised Weyl algebras) in Section 14.4.

14.3.1 The rep-Zariski spectrum of a PI Dedekind domain

The spectrum of the title is computed. If the ring has infinitely many primes, then the isolated points are exactly the points of finite length (14.3.1) and these points together are dense in the space (14.3.2)

Let R be a PI Dedekind domain. As commented in the computation of the Ziegler spectrum of a PI Dedekind domain (Section 5.2.1), since any Dedekind prime ring is Morita equivalent to a Dedekind domain ([441, 5.2.12]) the results here apply equally well to PI Dedekind prime rings, because everything involved is Morita-invariant.

The topology on Zar_R is described by listing a basis of open neighbourhoods at each point. Of course, to do this we need to know something about the finitely presented functors, equivalently the pp conditions.

Over these rings all finitely presented functors are, see 2.4.10, built up from annihilator and divisibility conditions. As in Section 5.2.1, a fairly obvious notation, based on that after 5.2.3, is used, with M as dummy variable, in referring to such functors and their quotients.

- R/P^n: This point is isolated by the open set $[MP^n] \cap [\mathrm{ann}(P) / (MP^{n-1} \cap \mathrm{ann}(P))]$; we go through the details. The open set $[MP^n]$ contains exactly those indecomposable pure-injectives N satisfying $NP^n = 0$, namely $R/P, R/P^2, \ldots, R/P^n$. The open set $[\mathrm{ann}(P) / (MP^{n-1} \cap \mathrm{ann}(P))]$ contains exactly those N with $\mathrm{ann}_N(P) \leq NP^{n-1}$ and, on consulting the list at 5.2.2, one sees that this defines the set $\{R/P^n, R/P^{n+1}, \ldots, R_{P^\infty}\}$. The intersection of these two open sets is exactly $\{R/P^n\}$, as claimed. We leave similar checks largely to the reader and say more only where some new feature arises.

- $\overline{R}_P$: First, there is a neighbourhood, namely $[\mathrm{ann}(P)]$, which excludes all points associated to the prime P apart from $\overline{R}_P$ itself. Then, given finitely many non-zero primes $Q_1, \dots, Q_k$ different from P, there is a neighbourhood of $\overline{R}_P$ which excludes all points associated to those primes, namely $\bigcap_{i=1}^k \big([M/MQ_i] \cap [\mathrm{ann}(Q_i)]\big)$. We cannot exclude points associated to more than finitely many primes since, otherwise, looking at the complementary rep-Zariski-closed = Ziegler-open set, we could express a basic (so, 1.1.7, compact) Ziegler-open set as a union of infinitely many proper open subsets (one for each of the excluded primes). Therefore a basis consists of the sets given by "finite localisation", that is, removing all trace of finitely many other primes, then removing all other points associated to P.

  If R has only finitely many primes, then there is a minimal neighbourhood, $\{\overline{R}_P, Q\}$.
- R_{P^∞}: The comments for $\overline{R}_P$ apply here also; alternatively use elementary duality. The sets $[M/MP] \cap \bigcap_{i=1}^k \big([M/MQ_i] \cap [\mathrm{ann}(Q_i)]\big)$, where $Q_1, \dots, Q_k$ are any non-zero primes of R different from P, form a basis of open neighbourhoods.
- Q: Again, 'finite localisation' allows us to remove all trace of any finitely many non-zero primes but, for the same reasons as before, no more.

So, if R has only finitely many primes, then Q is a Zariski-open point.

Observe that Zar_R, provided R is not a division ring, is not compact: it is, for any non-zero prime P, the union of the sets $[M/MP]$, $[\mathrm{ann}(P)]$ and the $[MP^n] \cap [\mathrm{ann}(P)/\big(MP^{n-1} \cap \mathrm{ann}(P)\big)]$ for $n \geq 1$, and there is no finite subcover.

With this description to hand one may check the following.

Proposition 14.3.1. *Let R be a PI Dedekind domain. The isolated, that is, open, points of* Zar_R *are precisely the points of finite length, except in the case where R has only finitely many primes, in which case the generic point Q also is isolated. Every point of* Zar_R*, apart from Q, is closed.*

Taking the simplest example, that of a local ring, say $k[X]_{\langle X \rangle}$ or $k[[X]]$, one has the maximal ideal, which is closed, together with the generic point, which is open and has closure itself plus the closed point $\langle X \rangle$. The 'extra' (compared to the usual spectrum) points in Zar_R, the finite length modules $k[X]/\langle X^n \rangle$, are all clopen.

The Zariski spectrum in the usual sense is embedded in Zar_R via the indecomposable injective modules: these rings are PI, hence fully bounded, so there is a bijection between indecomposable injectives and prime ideals, see Section 14.1.3.

One may also draw the following conclusions where $\mathrm{Zar}_R^{\mathrm{f}}$ denotes the set of points of Zar_R of finite length: by 5.2.3 this is an open set.

Lemma 14.3.2. *If R is a PI Dedekind domain with infinitely many primes, then* $\mathrm{Zar}_R^{\mathrm{f}}$ *is Zariski-dense in* Zar_R. *In particular,* $\mathrm{Zar}_R^{\mathrm{f}}$ *is exactly the set of isolated points of* Zar_R.

Proof. Check the list above to see that every open neighbourhood of every infinite length point contains a point of finite length. □

Denote by Zar_R^1 the set, $\mathrm{Zar}_R \setminus \mathrm{Zar}_R^{\mathrm{f}}$, of points of infinite length and endow this with the topology inherited from Zar_R. Since $\mathrm{Zar}_R^{\mathrm{f}}$ coincides with the set of all isolated points, Zar_R^1 also equals the first Cantor–Bendixson derivative (Section 5.3.6) of Zar_R. From the description of open neighbourhoods we have the following.

Lemma 14.3.3. *Let R be a PI Dedekind domain, with division ring of quotients Q. Then the non-empty rep-Zariski-open subsets of* Zar_R^1 *are exactly the cofinite sets which contain the generic point Q.*

Also see 14.3.7 for a comparison with the prime spectrum.

14.3.2 The sheaf of locally definable scalars of a PI Dedekind domain

The ring of definable scalars of the first Cantor–Bendixson derivative is just the original ring (14.3.6) and the quotient of this derivative by the natural involutive homeomorphism (14.3.7) gives the structure sheaf of the affine line.

To every point, N, of Zar_R, where R is a PI Dedekind domain, is associated a prime ideal, $P(N)$, of R, which may be obtained by taking the unique maximal ideal of the local ring $\mathrm{End}(N)$, intersecting this with the canonical copy of the centre, $C(R)$, of R which lies in $\mathrm{End}(N_R)$, and then taking $P(N)$ to be the prime ideal of R which is generated by this prime ideal of $C(R)$.

The rings of definable scalars of individual points (see 14.1.12) are as follows.

- The ring of definable scalars of the module R/P^n (P a non-zero prime) is the ring R/P^n: for, the module has finite endolength, hence (6.1.20) its ring of definable scalars coincides with its biendomorphism ring, which is R/P^n.
- If N is the P-adic or P-Prüfer module, then the ring of definable scalars, R_N, is the localisation, $R_{(P)}$, of R at P: this follows by 14.1.12 for the Prüfer module and then directly, or using duality, [274, 6.3], and 5.4.1, for the adic module.
- If N is the generic point, Q, the quotient division ring of R, then R_N is the ring Q, again by 6.1.20 since N has finite endolength and $\mathrm{Biend}(Q_R) = Q$ (because $R \to Q$ is an epimorphism of rings).

For any subset, $\mathcal{P}$, of the set, $\mathrm{maxspec}(R)$, of maximal ideals of R, let $U(\mathcal{P}) = \{N \in \mathrm{Zar}_R : P(N) \notin \mathcal{P}\}$. Denote by $R[\mathcal{P}^{-1}]$ the localisation of R at

$\mathrm{maxspec}(R)\backslash\mathcal{P}$. Since the canonical map $R \to R[\mathcal{P}^{-1}]$ is an epimorphism, it follows, see 6.1.8, that $U(\mathcal{P})$ is a Ziegler-closed subset of Zg_R. In the case that $\mathcal{P}$ is a finite subset of $\mathrm{maxspec}(R)$, then $U(\mathcal{P})$ is Zariski-open being, in the notation used in the previous section, $\bigcap_{P\in\mathcal{P}}\big([\mathrm{ann}(P)]\cap[M/MP]\big)$.

Now we compute the presheaf of definable scalars. The following result is immediate from 6.1.17.

Lemma 14.3.4. *Let R be a PI Dedekind domain and let $\mathcal{P}$ be a finite subset of* $\mathrm{maxspec}(R)$. *Then the ring of definable scalars over the corresponding Zariski-open subset, $U(\mathcal{P})$, of Zar_R is the localisation $R[\mathcal{P}^{-1}]$ of R. Furthermore, if $\mathcal{P}\subseteq\mathcal{P}'$, then the restriction map from $R_{U(\mathcal{P})}$ to $R_{U(\mathcal{P}')}$ is the natural embedding between these localisations of R.*

Lemma 14.3.4 gives all the information needed to compute the sheaf, LDef_R, of locally definable scalars. We describe, for illustration, the ring of definable scalars and the ring of sections of LDef_R, that is, the ring of locally definable scalars, over some basic open subsets of Zar_R.

- If $U=\{R/P_1^{n_1},\dots,R/P_t^{n_t}\}$, where the primes $P_1,\dots,P_t$ are all distinct, then $\mathrm{LDef}_R(U)=R/P_1^{n_1}\times\cdots\times R/P_t^{n_t}=R/\langle P_1^{n_1}\dots P_t^{n_t}\rangle$.
- If $U=\{R/P^m,R/P^n\}$ with $n\geq m$, then $R_U=R/P^n$: by 6.1.20, R_U is $\mathrm{Biend}(R/P^m\oplus R/P^n)_R$ and, noting that there is the endomorphism of this module projecting the second component on to the first, one easily computes that this is R/P^n. Since the set U has the discrete topology, $\mathrm{LDef}_R(U)$ is the direct product $R/P^m\times R/P^n$. Therefore the presheaf of definable scalars is not a sheaf.
- R_U, for U an arbitrary finite set of finite length points, is computed by combining the above observations.
- Let $U=\mathrm{Zar}_R\backslash\{N_1,\dots,N_t\}$, where each N_i is an adic or Prüfer module: then $R_U=R_V$, where V is the smallest set of the form $U(\mathcal{P})$ which contains U; for, provided at least one of the P-adic or P-Prüfer is in U, then P cannot be inverted over U. The presence in U of infinitely many finite-length, hence isolated, points means that the ring of locally definable scalars, $\mathrm{LDef}_R(U)$, is rather large. It makes sense, therefore, to throw away these isolated points (see 14.3.6 below).

Example 14.3.5. Consider the special (but illustrative) case, $R=k[X]$, $U=U\big(\langle X\rangle\big)\cup\{R/\langle X\rangle\}$. A module with support U, which is basic Zariski-open so also Ziegler-closed, is $k[X,X^{-1}]\oplus(k[X]/\langle X\rangle)$, so it will be enough to compute the definable scalars of this module. By 1.1.13(iii) this is the biendomorphism ring of a module of the form $M\oplus M'$, where M, respectively M', is a "large enough" module in the definable subcategory generated by $k[X,X^{-1}]$, resp. by $k[X]/\langle X\rangle$. Since $\mathrm{Hom}(M, M') = 0 = \mathrm{Hom}(M', M)$ the endomorphism ring of this module is just the block-diagonal matrix ring $\begin{pmatrix} \mathrm{End}(M_R) & 0 \\ 0 & \mathrm{End}(M'_R) \end{pmatrix}$. Therefore the biendomorphism ring is block-diagonal and one may check that it is $\begin{pmatrix} k[X, X^{-1}] & 0 \\ 0 & k[X]/\langle X\rangle \end{pmatrix}$, that is, the direct product of these rings. For instance, to see directly that the scalar $(1, 0) \in k[X, X^{-1}] \times (k[X]/\langle X\rangle)$ is pp-definable, let $\rho(x, y)$ be the condition $(y - x)X = 0 \wedge \exists z\,(y = zX)$.

Now we compute the sheaf of locally definable scalars restricted to the set Zar^1_R obtained by throwing away the finite length points. Since Zar^1_R is not Zariski-open, we need the following observations concerning restriction. Define LDef^1_R to be the inverse image sheaf of LDef_R under the inclusion of Zar^1_R in Zar_R: by definition (for example, see [269, p. 65]) this is the sheaf associated to the presheaf which assigns to a relatively open subset, $U \cap \mathrm{Zar}^1_R$, of Zar^1_R, where U is a Zariski-open subset of Zar_R, the direct limit of the rings $\mathrm{LDef}_R(V)$ as V ranges over Zariski-open subsets of Zar_R with $V \supseteq U \cap \mathrm{Zar}^1_R$. If $U \cap \mathrm{Zar}^1_R = \mathrm{Zar}^1_R \setminus \{N_1, \dots, N_t\}$, then let V be the set of all points of Zar_R except those which belong to a prime P such that both the P-adic and P-Prüfer modules appear among $N_1, \dots, N_t$: that is, V is the smallest set of the form $U(\mathcal{P})$ which contains U. Then, by the computations above, this limit is already equal to R_V, hence this presheaf is already a sheaf. Thus one has the following description of LDef^1_R.

Proposition 14.3.6. *Let R be a PI Dedekind domain and let Zar^1_R be the set of infinite-length points, regarded as a subspace of Zar_R. Let LDef_R denote the sheaf of locally definable scalars over Zar_R. Then the inverse image sheaf, LDef^1_R, on Zar^1_R may be computed as follows. Given a Zariski-open subset, U, of Zar_R, let V be smallest set of the form $U(\mathcal{P})$ which contains U. Then $\mathrm{LDef}^1_R(U \cap \mathrm{Zar}^1_R) = \mathrm{LDef}_R(V) = R[\mathcal{P}^{-1}]$ and the restriction maps are those of LDef_R, that is, the canonical localisation maps.*

In particular, the ring of definable scalars, $R_{\mathrm{Zar}^1_R}$, of Zar^1_R is R itself.

The sheaf LDef^1_R is "unseparated" in the sense that it contains points N, N' such that U is an open neighbourhood of N iff $(U \setminus \{N\}) \cup \{N'\}$ is an open neighbourhood of N', namely the Prüfer and adic associated to any maximal prime. In order to recover the 'classical' situation such points should be identified.

Therefore let $\alpha : \mathrm{Zar}^1_R \to \mathrm{Zar}^1_R$ be the map which interchanges the P-adic and P-Prüfer point for every P and which fixes the generic point.

Lemma 14.3.7. *The map $\alpha : \mathrm{Zar}^1_R \to \mathrm{Zar}^1_R$ is a homeomorphism of order 2 and $\mathrm{LDef}^1_R \simeq \alpha^\star \mathrm{LDef}^1_R \simeq \alpha_\star \mathrm{LDef}^1_R$, where $\alpha^\star, \alpha_\star$ denote the inverse image and direct image sheaves respectively (see [269], [318] or [674]).*

Proof. From the description of the topology it is clear that α is a homeomorphism. For any basic Zariski-open set, U, of Zar^1_R we have, again by what has been said above, $R_U \simeq R_{\alpha U}$ so the isomorphisms are direct from the definitions. $\square$

One may, therefore, form the quotient space, Zar^1_R/α, of α-orbits and the corresponding sheaf, LDef^1_R/α, over this space, to obtain a ringed space with centre isomorphic, via the identification of $\mathrm{maxspec} R$ with $\mathrm{maxspec}(C(R))$, to the structure sheaf over the commutative Dedekind domain $C(R)$.

14.3.3 The presheaf of definable scalars of a PI HNP ring

In 8.1.13 the Ziegler spectra of PI hereditary noetherian prime rings were described. Here we describe their rep-Zariski spectra and associated rings of definable scalars. In brief, and as in the previous section, there is little to check since the description of the Ziegler spectrum, points and topology, and hence of the rep-Zariski topology, is just as in the case of a Dedekind domain. In particular, to every point, N, of Zar_R is associated a prime ideal, $P(N)$, of R. The only significant difference is that if the ring R is not a Dedekind prime ring, then the map from $\mathrm{Spec}(R)$ to $\mathrm{Spec}(C(R))$ given by intersecting a prime ideal with the centre is not 1–1. So it is essential here to use $\mathrm{Spec}(R)$, rather than $\mathrm{Spec}(C(R))$, to parametrise the primes.

Example 14.3.8. Let R be the ring $\begin{pmatrix} \mathbb{Z} & 2\mathbb{Z} \\ \mathbb{Z} & \mathbb{Z} \end{pmatrix}$ – a non-maximal order in the simple artinian ring $A = \begin{pmatrix} \mathbb{Q} & \mathbb{Q} \\ \mathbb{Q} & \mathbb{Q} \end{pmatrix}$. For each non-zero prime $p \in \mathbb{Z}$, $p \neq 2$, there is the corresponding prime ideal $P_p = \begin{pmatrix} p\mathbb{Z} & 2p\mathbb{Z} \\ p\mathbb{Z} & p\mathbb{Z} \end{pmatrix}$, the corresponding p-adic and p-Prüfer right R-modules, which may be regarded as $(\overline{\mathbb{Z}}_{(p)}, \overline{\mathbb{Z}}_{(p)})$ and $(\mathbb{Z}_{p^\infty}, \mathbb{Z}_{p^\infty})$ respectively, as well as the finite length indecomposable modules, R/P^n, associated to P.

Corresponding to the prime $p = 2$, there are two prime ideals of R: $P_1 = \begin{pmatrix} \mathbb{Z} & 2\mathbb{Z} \\ \mathbb{Z} & 2\mathbb{Z} \end{pmatrix}$ and $P_2 = \begin{pmatrix} 2\mathbb{Z} & 2\mathbb{Z} \\ \mathbb{Z} & \mathbb{Z} \end{pmatrix}$ with corresponding simple modules $S_i = R/P_i$ and $\mathrm{Ext}(S_i, S_{3-i}) \neq 0$ for $i = 1, 2$. These two simple modules are at the mouth of a tube (see Section 15.1.3) of rank 2. The P_1-adic module is the inverse limit of the coray beginning with S_1 and has a unique infinite descending chain of submodules with simple composition factors S_1,S_2 alternating and starting with S_1. Dually, the P_1-Prüfer module is the direct limit of the ray beginning with S_1 and is the injective hull of S_1; similarly for P_2.

Next we compute rings of definable scalars. These are obtained for primes P belonging to singleton cliques (in the sense of [140], see Section 8.1.2) by localising, just as in the Dedekind prime case. For the other primes we use universal localisation, as in [140] (cf. 8.1.7), alternatively, as mentioned there, Goodearl's localisation from [241], to obtain the corresponding Prüfer and adic modules. An example is given below for illustration. Beyond this, the description of the presheaf of definable scalars and the corresponding sheaf, both on Zar_R and on Zar^1_R, is as in the PI Dedekind case (Section 14.3.1), using the fact, 5.5.5, that every universal localisation is an epimorphism of rings and that if $R \to S$ is an epimorphism of rings, then, in addition to the induced homeomorphic embedding of Zar_S into Zar_R (6.1.8), there is induced an embedding of LDef_S, "up to Morita equivalence" into LDef_R: this follows from the argument of [510, §8] which is developed in [525] in the general context of interpretation functors, see Section 18.2.1.

By [140, 3.2], which is an analogue of 8.1.7 for PI HNP rings, there are finitely many universal localisations of R to hereditary orders (in fact, even to PI Dedekind prime rings) such that the inverse images cover $\mathbf{R} \cup \mathrm{Zar}^1_R$. Therefore Zar^1_R, and the restriction, LDef^1_R, of LDef_R to it, can be described using the results of Sections 14.3.1 and 14.3.4.

For illustration, we continue 14.3.8 by computing the various rings of definable scalars. We retain the notation of that example.

Example 14.3.9. Corresponding to the prime P_p there is the ring of definable scalars $\begin{pmatrix} \mathbb{Z}_{(p)} & \mathbb{Z}_{(p)} \\ \mathbb{Z}_{(p)} & \mathbb{Z}_{(p)} \end{pmatrix}$, which is a maximal order in A.

The rings of definable scalars corresponding to P_1 and P_2 may be computed using [241, first paragraphs of §2]. Explicitly, and adopting the notation of that paper, the simple module S_1 is removed by localising at $X_1 = \{P_2\} \cup \{P_p : p \neq 2\}$. Let $\mathcal{S}_1$ be the set of essential right ideals, I, of R such that none of R/P_2, R/P_p ($p \neq 2$) occurs as a composition factor of R/I. Note that $P_1 \in \mathcal{S}_1$. Let $R_{(1)}$ denote the localisation of R at the hereditary torsion theory which has $\mathcal{S}_1$ as dense set of right ideals (see Section 11.1.1). Then, [241, §2], $R_{(1)} = \{a \in M_2(\mathbb{Q}) : aI \leq R$ for some $I \in \mathcal{S}_1\}$. One checks that $R_{(1)} = \begin{pmatrix} \mathbb{Z} & \mathbb{Z} \\ \mathbb{Z} & \mathbb{Z} \end{pmatrix}$. Similarly, if $R_{(2)}$ denotes the ring obtained by localising away S_2 then one checks that $R_{(2)} = \begin{pmatrix} \mathbb{Z} & 2\mathbb{Z} \\ (1/2)\mathbb{Z} & \mathbb{Z} \end{pmatrix}$.

Therefore the ring of definable scalars at P_1 is $\begin{pmatrix} \mathbb{Z}_{(2)} & \mathbb{Z}_{(2)} \\ \mathbb{Z}_{(2)} & \mathbb{Z}_{(2)} \end{pmatrix}$ and that at P_2 is

$\begin{pmatrix} \mathbb{Z}_{(2)} & 2\mathbb{Z}_{(2)} \\ (1/2)\mathbb{Z}_{(2)} & \mathbb{Z}_{(2)} \end{pmatrix}$. Notice that, as rings, though not as R-algebras, these are isomorphic (by the map taking $\begin{pmatrix} a & b \\ c & d \end{pmatrix}$ to $\begin{pmatrix} a & 2b \\ c/2 & d \end{pmatrix}$).

One way of regarding this is that we have two epimorphisms from R to the maximal order $M_2(\mathbb{Z})$. The corresponding Ziegler-closed and, one may check, Zariski-open sets cover Zar_R. So LDef_R is covered by two (very much) overlapping copies of '$M_2(\mathrm{LDef}_{\mathbb{Z}})$'.

Lemma 14.3.10. *Let R be a PI HNP ring. Then the presheaf of definable scalars over Zar^1_R is already a sheaf.*

Proof. Take an open cover, $\{U_i\}_i$, of Zar^1_R, say $U_i = \mathrm{Zar}^1_R \setminus Y_i$ (where Y_i may be any finite subset of Zar^1_R which does not contain the generic Q) and elements $s_i \in R_i = R_{U_i}$ such that on $U_i \cap U_j = \mathrm{Zar}^1_R \setminus \{Y_i \cup Y_j\}$ we have $s_i = s_j = s$, say; we identify all the rings R_i with subrings of the full, simple artinian, quotient ring of R, so this equality makes sense. We have $s \in R_i \cap R_j$ and this ring equals $R_{U_i \cup U_j}$ since a prime P satisfies $P \cdot R_i \cap R_j = R_i \cap R_j$ iff both the P-Prüfer and P-adic modules lie in both Y_i and Y_j, and this is so iff $P \cdot R_{U_i \cup U_j} = R_{U_i \cup U_j}$. So now, taking any finite subcover, say $U_1, \dots, U_n$, it follows that $s_1 = \cdots = s_n = s \in R = R_{\mathrm{Zar}^1_R}$. Thus s is a global section which restricts on each U_i to s_i and is already in the presheaf, as required. □

Proposition 14.3.11. *Suppose that R is a hereditary order. Let $\mathrm{Spec}\,R$ denote the space of prime ideals of R with the Zariski topology and let $\pi : \mathrm{Zar}^1_R \to \mathrm{Spec}\,R$ be the map which sends $N \in \mathrm{Zar}^1_R$ to $P(N)$, the prime ideal of R associated with N. Then the direct image, $\pi_\star \mathrm{LDef}^1_R$, of LDef^1_R is a sheaf on $\mathrm{Spec}\,R$ which sends an open set $U = \mathrm{Spec}\,R \setminus Y$, where Y is a finite subset of $\mathrm{maxspec}\,R$, to the localisation, in the sense of [241], of R at U. Also $\pi_\star \mathrm{LDef}^1_R$ may be identified with LDef^1_R/α, where α is the homeomorphism interchanging corresponding Prüfer and adic points (cf. 14.3.7).*

Proof. The direct image of a sheaf is always a sheaf (for example, [674, 3.7.3]) and the description of the sheaf $\pi_\star \mathrm{LDef}\,R^1$ and its identification with $\mathrm{LDef}\,R^1/\alpha$ follows from the previous discussion. □

The underlying space, $\mathrm{Spec}\,R$, of this sheaf may also be identified with the space $\mathrm{Spec}_s R$ (based on the set of simple modules) from [140, §6], via $S \mapsto \mathrm{ass}(E(S))$, the associated prime of $E(S)$. The sheaf $\mathrm{Spec}_s R$ in [140] may, therefore be identified with the centre of LDef^1_R/α.

14.3.4 The presheaf of definable scalars of a tame hereditary artin algebra

Throughout, let R be a tame hereditary artin algebra which is not of finite representation type. The points and topology of Zar_R were described in Section 8.1.2. As before, let Zar^1_R denote the set or space of infinite-dimensional (= non-isolated) points.

From the description of the spectrum in Section 8.1.2, to each point N in either $\mathbf{R}$ (the set of indecomposable regular modules) or Zar^1_R, apart from the generic point, there is an associated quasi-simple module which we denote $S(N)$. In this context the quasi-simples play the role that primes did in previous sections.[4] As in the previous examples, the process of 'finite localisation', that is, removing all trace of finitely many primes/quasi-simples, lies behind the description of neighbourhood bases. Given a set, $\mathcal{S}$, of quasisimple modules, let $U(\mathcal{S})$ denote the set consisting of the generic point and all points of $\mathbf{R} \cup \mathrm{Zar}^1_R$ which are associated to some quasi-simple *not* in $\mathcal{S}$. The following is easily deduced.

Theorem 14.3.12. ([507]) *Let R be a tame hereditary artin algebra. A basis of open sets for Zar_R is as follows.*

(a) As for every artin algebra (5.3.33), the finite-dimensional points are open.
(b) If N is S-adic or S-Prüfer module, then the sets of the form $\{N\} \cup U(\mathcal{S})$, where $\mathcal{S}$ is a finite set of quasi-simples, form a basis of open neighbourhoods for N.
(c) The sets of the form $U(\mathcal{S})$, where $\mathcal{S}$ is a finite set of quasi-simples, form a basis of open neighbourhoods for the generic point G.

In particular, the sets $\mathbf{P}$ and $\mathbf{I}$, of preprojective, resp. preinjective, modules (Section 8.1.2), are rep-Zariski-closed, as well as rep-Zariski-open, so do not figure in the description of neighbourhood bases of the infinite-dimensional points.

It follows from 6.1.8 that the intersection of Mod-$R_{\mathcal{S}}$ with Zg_R is the closed subset $U(\mathcal{S})$, which, if $\mathcal{S}$ is finite, is basic Zariski-open, being defined by the conditions $\mathrm{Ext}^1(S, -) = 0 = \mathrm{Hom}(S, -)$ for $S \in \mathcal{S}$, and the ring of definable scalars for this rep-Zariski-open set is just $R_{\mathcal{S}}$. This is used to deduce the next result.

Proposition 14.3.13. *Let R be a tame hereditary artin algebra. Then the restriction, $\mathrm{Def}_R \restriction (\mathbf{R} \cup \mathrm{Zar}^1_R)$, of the presheaf of definable scalars is given by sending a set of the form $U(\mathcal{S})$ to the localisation $R \to R_{\mathcal{S}}$ and sending any open subset of $U(\mathcal{S})$ to the ring of definable scalars of the corresponding subset, regarded as an open subset of the rep-Zariski spectrum of the PI HNP ring $R_{\mathcal{S}}$ (these were discussed in Section 14.3.3).*

[4] This is more than an analogy, see [396], [140].

In particular, the ring of definable scalars of any member, N, of $\mathbf{R} \cup \mathrm{Zar}_R^1$ may be computed as follows. Choose a set, $\mathcal{S}$, of quasi-simples which contains all quasi-simples from at least one tube and which does not contain the quasi-simple, $S(N)$, corresponding to N; localise R at $\mathcal{S}$ to obtain the hereditary order $R_{\mathcal{S}}$ (which, by choosing $\mathcal{U}$ to contain all but at most one quasi-simple from each inhomogeneous tube, may be assumed to be a Dedekind prime ring). Then compute the ring of definable scalars of N, regarded as an $R_{\mathcal{S}}$-module.

Example 14.3.14. Let R be the Kronecker algebra $\widetilde{A_1}(k)$ (1.1.4). The explicit description of the sheaf LDef_R^1 as a sheaf of hereditary orders, in fact maximal orders, is as follows.

The full quotient ring, in the sense of 8.1.7, of $\widetilde{A_1}(k)$ is the ring, $M_2(k(X))$, of 2×2 matrices over the function field $k(X)$. In order to maintain the symmetry between the arrows α and β of $\widetilde{A_1}(k)$ we represent $k(X)$ in the form $k(X_0, X_1)_0$, where the subscript denotes the 0-grade part of the quotient field of the graded ring $k[X_0, X_1]$ (with the usual grading). Then there is a natural embedding of $\widetilde{A_1}(k)$ into $M_2\big(k(X_0, X_1)_0\big)$ which takes e_1 to $\begin{pmatrix} 1 & 0 \\ 0 & 0 \end{pmatrix}$, e_2 to $\begin{pmatrix} 0 & 0 \\ 0 & 1 \end{pmatrix}$, α to $\begin{pmatrix} 0 & X_0 \\ 0 & 0 \end{pmatrix}$ and β to $\begin{pmatrix} 0 & X_1 \\ 0 & 0 \end{pmatrix}$ and, under this embedding, the quotient ring A may be identified with $M_2\big(k(X_0, X_1)_0\big)$.

Let S_0 (respectively S_1) be the quasi-simple module which satisfies $S_0 X_1 = 0$ (resp. $S_1 X_0 = 0$). Let R_i denote the localisation of R at S_i $(i = 0, 1)$. Let D_i be the Zariski-open subset of Zar_R^1, $D_i = \mathrm{Zar}_R^1 \cap [M/MX_{1-i}] \cap [\mathrm{ann}(X_{1-i})]$. Then the localisation map $R \to R_i$ identifies D_i with $\mathrm{Zar}_{R_i}^1$ and $\mathrm{LDef}_R^1 \restriction D_i$ with $\mathrm{LDef}_{R_i}^1$. Each of R_0, R_1 is isomorphic as a ring to the polynomial ring over k in one indeterminate and the localisation, $R_{0,1}$, of R at $\{S_0, S_1\}$ (which corresponds to the intersection $D_0 \cap D_1$) is a ring isomorphic to $k[T, T^{-1}]$. It is straightforward to compute these as subalgebras of $M_2\big(k(X_0, X_1)_0\big)$ and one obtains:

$$R_0 = \begin{pmatrix} k[X_0 X_1^{-1}] & kX_1 \oplus X_0 k[X_1^{-1} X_0] \\ X_1^{-1} k[X_0 X_1^{-1}] & k[X_1^{-1} X_0] \end{pmatrix};$$

$$R_1 = \begin{pmatrix} k[X_1 X_0^{-1}] & kX_0 \oplus X_1 k[X_0^{-1} X_1] \\ X_0^{-1} k[X_1 X_0^{-1}] & k[X_0^{-1} X_1] \end{pmatrix};$$

$$R_{0,1} = \begin{pmatrix} k[X_0 X_1^{-1}, X_1 X_0^{-1}] & X_1 k[X_0^{-1} X_1] \oplus X_0 k[X_1^{-1} X_0] \\ X_1^{-1} k[X_0 X_1^{-1}] \oplus X_0^{-1} k[X_1 X_0^{-1}] & k[X_1^{-1} X_0, X_0^{-1} X_1] \end{pmatrix}$$

$$= \begin{pmatrix} k(X_0, X_1)_0 & k(X_0, X_1)_1 \\ k(X_0, X_1)_{-1} & k(X_0, X_1)_0 \end{pmatrix},$$

where subscript denotes degree in the graded ring $k(X_0, X_1)$.

Also see [358, Chpt. 14].

14.3.5 Other examples

If R is any ring of finite representation type, then it follows from 6.1.20, 5.3.26 and 4.5.22 that the ring of global sections of the sheaf LDef_R of locally definable scalars is the product of the biendomorphism rings of the distinct indecomposable modules.

Let R be the k-path algebra of one of the quivers Λ_n $(n \geq 2)$, see Section 8.1.3. In that section the description of the Ziegler spectum, computed in [107], is given. In that paper the Gabriel–Zariski spectrum is computed [107, §3]. Roughly, after removing the finite-dimensional points, there are n double-except-for-generics copies of the projective line over k (cf. 14.3.7) with some $\mathbb{N}$-parametrised families of points linking them into a chain. The computation of rings of definable scalars, and the corresponding (pre)sheaf, is not done there but should not be difficult.

It is shown, in particular, that, over Λ_2, every point except for the two generics is closed ([107, 3.2]). Since the Ziegler-closure of every infinite-dimensional point contains at least one generic, the Zariski-closure of the two generics is exactly the set, $\mathrm{Zar}^1_{\Lambda_n}$, of non-isolated = infinite-dimensional points. Therefore the Gabriel–Zariski spectrum has "geometric" dimension 1.

It is also shown that there is a homeomorphism of the Gabriel–Zariski spectrum to itself (of order 2) which is not a homeomorphism of that set endowed with the Ziegler topology ([107, 3.1]).[5] Therefore, although (Section 5.6) the Gabriel–Zariski topology may be defined from the Ziegler topology, the converse is not the case.

It is conjectured (though based on rather few examples) that the "geometric" dimension of the Gabriel–Zariski spectrum of any finite-dimensional algebra is 0 (if the algebra is of finite representation type), 1 (perhaps if the algebra is of domestic type), or ∞, the last meaning that algebraic varieties of arbitrarily high dimension embed. It is shown in [510, §9] that wild algebras do have algebraic-geometric dimension ∞ in this sense.

Question 14.3.15. If R is a tame finite-dimensional algebra what is the "geometric" dimension of Zar_R?

The first Weyl algebra, $A_1(k)$, where k is a field of characteristic zero, is wild ([342, §2], see Section 7.3.2). One may, however, look at parts of the spectrum, for example, 8.2.42, and obtain some information on the Gabriel–Zariski spectrum. For example, in [522, §8] the relative topologies on inj_R, and on the set of torsionfree indecomposable pure-injective modules are described for a certain class

[5] Most of the points have a parameter belonging to the projective line over the base field; the homeomorphism arises by interchanging those with parameter 0 with those with parameter ∞.

of generalised Weyl algebras, as are the rings of definable scalars, the stalks being those already seen in 8.2.44.

Recall, 8.2.2, that if R is a generalised Weyl algebra satisfying the conditions in 8.2.42, with division ring of quotients D, and if M is a divisible R-module, then the ring of definable scalars of M is $\{sr^{-1} : \mathrm{ann}_M(r) \leq \mathrm{ann}_M(s)\} \subseteq D$.

14.4 The spectrum of a commutative coherent ring

Throughout this section R will be a commutative ring. The results here are taken from [508].

The bijection between prime ideals and indecomposable injectives over a commutative noetherian ring breaks down for commutative coherent rings (14.4.1) but each indecomposable injective is topologically equivalent to an injective which comes from a prime ideal (14.4.5). Injectives of the latter sort are those which cogenerate torsion theories of finite type (14.4.10). This gives a description of irreducible Ziegler-closed sets of injectives (14.4.11).

There is also, over coherent rings, a separation of topologies which coincide over noetherian rings and which agree when restricted to the classical Zariski spectrum (14.4.4).

The closure of any point in the fg-ideals = rep-Zariski topology is described (14.4.8), as are irreducible closed sets (14.4.12, 14.4.13).

In the reinterpretation of Spec(R) as a topology on the set, inj_R, of indecomposable injective R-modules when R is commutative noetherian (Section 14.1.1) the noetherian hypothesis was used in the pairing up of indecomposable injectives with prime ideals (and also in that it was not necessary to choose between using finitely presented or finitely generated modules to define the topology).

In the general commutative case the association $P \mapsto E(R/P)$ gives only an injection of Spec(R) into inj_R.

The first example (which was pointed out to me by T. Kucera) shows that $\mathrm{Spec}(R) \to \mathrm{inj}_R$ need not be surjective.

Example 14.4.1. Let $R = k[X_n\ (n \in \omega)]$ be a polynomial ring over a field k in infinitely many commuting indeterminates. It is easily checked that R is coherent (see 2.3.18). Let $I = \langle X_n^{n+1} : n \in \omega \rangle$. Clearly I is not prime but $E = E(R/I)$ is an indecomposable injective. To see this, it is enough to show that R/I is uniform and this may be proved as follows. First note that a polynomial $\sum a_\nu X^\nu$ (each multi-index ν occurring at most once) is in I iff each of its monomial factors is in I. Let x_i denote the image in R/I of X_i. Let $p \in R \backslash I$. A short inductive (on the number of monomials) argument shows that there is a multiple of p whose image in R/I has the form $x_1x_2^2\dots x_n^n \neq 0$ for some n. Hence any two non-zero elements of R/I have a common multiple of this form so R/I is uniform and $E(R/I)$ is indeed indecomposable.

On the other hand, E does not have the form $E(R/P)$ for any prime P. This follows from 14.4.2 below because it is easy to see that $P(E)$, as defined below, is the maximal ideal $\langle X_n : n \in \omega\rangle$ and that the injective hull, $E\big(R/\langle X_n : n \in \omega\rangle\big)$, of the corresponding factor of R has non-zero socle, whereas $E(R/I)$ has zero socle; for, if $p \in R\setminus I$, say $p \in k[X_0, \dots, X_n]$, then $(p+I)X_{n+1}$ generates a non-zero proper submodule of the submodule generated by $p+I$. Hence E is not isomorphic to $E(R/P(E))$, so, by 14.4.2, E does not have the form E_P for any prime ideal P.

Let E be any indecomposable injective R-module. Set $P = P(E) = \{r \in R : er = 0 \text{ for some } e \in E,\ e \neq 0\}$ to be the sum of annihilator ideals of non-zero elements, equivalently non-zero submodules, of E. Since E is uniform the set of annihilator ideals of non-zero elements of E is closed under finite sum (see the proof below) so the only issue is whether the sum, $P(E)$, of them all is itself an annihilator ideal.

As before we use the notation E_P to denote $E(R/P)$.

Lemma 14.4.2. *If $E \in \mathrm{inj}_R$, then $P(E)$ is a prime ideal. The module E has the form E_P for some prime ideal P iff the set of annihilator ideals of non-zero elements of E has a maximal member, namely $P(E)$, in which case $E = E_{P(E)}$.*

Proof. Suppose that $rs \in P(E)$. By definition of $P(E)$ there is $a \in E, a \neq 0$ such that $ars = 0$. Then either $ar = 0$, in which case $r \in P(E)$, or $ar \neq 0$, hence $s \in P(E)$. This shows that $P(E)$ is prime. So, if $P(E)$ is an annihilator ideal, then $E = E_{P(E)}$.

If $E = E(R/P)$, then $P \leq P(E)$, by definition of the latter. Suppose there were $r \in P(E)\setminus P$. Let $b \in E$ be non-zero with $br = 0$ and let $a \in E$ be such that $\mathrm{ann}_R(a) = P$. By uniformity of E there is a non-zero element $c \in aR \cap bR$, say $c = at$ with $t \in R$. Since $cr = 0$ we have $atr = 0$, hence $tr \in P$ and hence $t \in P$ (impossible since $c = at \neq 0$) or $r \in P$, a contradiction. Therefore $P = P(E)$. □

Before examining the relation between $E \in \mathrm{inj}_R$ and $E_{P(E)} \in \mathrm{inj}_R$ we address the issue of which topology we should be using on inj_R.

Extending the notation of Section 14.1.1, if W a subset of R set $D(W) = \{P \in \mathrm{spec}(R) : W \nsubseteq P\}$. Since $D(W) = \bigcup_{r\in W} D(r)$, this is a Zariski-open subset of $\mathrm{Spec}(R)$.

For I an ideal of R set $D^{\mathrm{m}}(I) = \{E \in \mathrm{inj}_R : (R/I, E) = 0\}$ ("m" for "morphism"). Since[6] $D^{\mathrm{m}}(I) \cap D^{\mathrm{m}}(J) = D^{\mathrm{m}}(I \cap J)$ these form a basis for what we earlier (Section 14.1.3) called the fg-ideals topology on inj_R and which is a special case of the full support topology defined in Section 12.7. Note, however, that if $I = \sum_\lambda I_\lambda$, then clearly $D^{\mathrm{m}}(I) \supseteq \bigcup_\lambda D^{\mathrm{m}}(I_\lambda)$, but, as illustrated by Example 14.4.1, the inclusion may be proper: take E as there, take I to be $P(E)$ and take the I_λ to be the annihilators of non-zero elements of E.

Lemma 14.4.3. *If I is an ideal of R and $I = \sum_1^n I_i$ then $D^{\mathrm{m}}(I) = \bigcup_1^n D^{\mathrm{m}}(I_i)$.*

Proof. Suppose that $E \notin \bigcup_1^n D^{\mathrm{m}}(I_i)$. Then for each i there is a non-zero morphism $f_i : R/I_i \to E$. The intersection of the images of these morphisms is non-zero. Since R is commutative, any element in this intersection is annihilated by each I_i, hence by I, that is, $(R/I, E) \neq 0$. Thus $E \notin D^{\mathrm{m}}(I)$, as required. □

In Section 14.1.1 we argued the (noetherian version of the) following, though there we could identify $\mathrm{Spec}(R)$ and inj_R; here we have only a containment of the former in the latter.

Corollary 14.4.4. *For any ideal I we have $D^{\mathrm{m}}(I) \cap \mathrm{Spec}(R) = D(I)$.*

Therefore the restriction of both the ideals topology and the, in general coarser, fg-ideals topology[7] on inj_R to $\mathrm{spec}(R)$ gives the usual Zariski topology. By the results at the beginning of Section 14.1.3 the topology induced on inj_R by the rep-Zariski = dual-Ziegler (14.1.7) (= induced rep-Zariski, by 14.1.5) topology is intermediate between these two. For, it uses just the (right) ideals of the form $D\phi({}_RR)$, where ϕ is a pp condition for right R-modules and this, see Section 1.1.4, includes all finitely generated ideals, but not necessarily all ideals. If R is coherent, then, by 2.3.19, this third topology coincides with the fg-ideals one. Of course, all four spaces, inj_R with its various topologies and $\mathrm{spec}(R)$ with the Zariski topology, coincide if R is noetherian.

In the remainder of this section we will mostly assume that R is coherent and concentrate on the fg-ideals = rep-Zariski (in this case) and Ziegler topologies.

Note that for I any right ideal of a ring R and $r \in R$, there is an isomorphism $R/(I : r) \simeq (rR + I)/I$, where $(I : r) = \{s \in R : rs \in I\}$; it is induced by sending $1 + (I : r)$ to $r + I$.

Theorem 14.4.5. *Let R be commutative coherent, let E be an indecomposable injective module and let $P(E)$ be the (prime) ideal of annihilators of non-zero*

[6] For the non-immediate direction, note that any morphism from $R/(I \cap J)$ to E extends, by injectivity of E, to one from $R/I \oplus R/J$.

[7] As remarked just before 14.1.9, the full support topology, when restricted to inj_R, coincides with this topology.

elements of E. Then E and $E_{P(E)}$ are topologically indistinguishable in Zg_R*, hence also in* Zar_R.

Proof. Let I be such that $E = E(R/I)$. For each $r \in R \setminus I$, by the remark just above, the annihilator of $rR + I \in E$ is $(I : r)$, so, by definition of $P(E)$, $(I : r) \le P(E)$. The natural projection $(rR + I)/I \simeq R/(I : r) \to R/P(E)$ extends to a morphism from E to $E_{P(E)}$ which is non-zero on $r + I$. Forming the product of these morphisms as r varies over $R \setminus I$, we obtain a morphism from E to a product of copies of $E_{P(E)}$ which is monic on R/I, hence is monic. Therefore E is a direct summand of a direct product of copies of $E_{P(E)}$, so, by 3.4.7, E is in the definable subcategory generated by $E_{P(E)}$. Therefore (5.1.1) $E \in \mathrm{Zg\text{-}cl}(E_{P(E)})$ (this conclusion required no assumption on R beyond commutativity).

For the converse, take a basic Ziegler-open neighbourhood of $E_{P(E)}$: by (the proof of) 14.1.5 this has the form $[J/I]^{\mathrm{c}}$ for a pair, $I < J$, of finitely generated ideals of R (the notation $[M]$ is as at 14.1.5). Now, $E_{P(E)} \in [J/I]^{\mathrm{c}}$ means that there is a non-zero morphism $f : J/I \to E_{P(E)}$. Since $R/P(E)$ is essential in $E_{P(E)}$ the image of f has non-zero intersection with $R/P(E)$, so there is an ideal J', without loss of generality finitely generated, with $I < J' \le J$ and such that the restriction, f', of f to J'/I is non-zero (and contained in $R/P(E)$). Since $R/P(E) = \varinjlim R/I_\lambda$, where I_λ ranges over the annihilators of non-zero elements of E, and since J'/I is finitely presented, f' factorises through one of the maps $R/I_\lambda \to R/P(E)$. In particular, there is a non-zero morphism $J'/I \to E$, hence, by injectivity of E, an extension to a morphism $J/I \to E$, showing that $E \in [J/I]^{\mathrm{c}}$. Therefore $E_{P(E)} \in \mathrm{Zg\text{-}cl}(E)$, as required. □

Remark 14.4.6. If E is not isomorphic to $E_{P(E)}$, then E is not in the closure of $E_{P(E)}$ with respect to the ideals topology, for there is the open neighbourhood $[R/P(E)]$ of $E_{P(E)}$ which does not contain E. Therefore, the ideals topology is strictly finer than the fg-ideals topology whenever inj_R is strictly larger than $\mathrm{Spec}(R)$.

Example 14.4.7. Coherence of R is not necessary (nor, as we saw in 14.4.1, sufficient) for equality of the prime and injective spectra. Take, for instance, $R = k[x_i \, (i \ge 1) : x_i x_j = 0 \, (i, j \ge 1)]$ where k is a field. The Jacobson radical, $J = \sum_{i \ge 1} x_i R$, is the only $\cap$-irreducible ideal, so $E = E(R/J) = E(k)$ is the only point of inj_R; thus we do have just the one indecomposable injective, corresponding to the single point of $\mathrm{Spec}(R)$. Since R/J embeds in R but is not finitely presented, R is not coherent. We continue this example, showing that inj_R is not a closed subset of Zg_R.

Since R is not coherent there is, by 3.4.24, a non-absolutely pure module M in the definable subcategory generated by E and hence, by 2.3.2, the pure-injective hull of M is not injective. We claim that there are are no superdecomposable pure-injectives in this definable subcategory and hence that it contains an indecomposable pure-injective, non-injective module.

Consider the dual definable subcategory. By 3.4.26 this is the definable subcategory generated by R. By the analysis of 5.3.62, the m-dimension of this definable subcategory, and hence of its dual, is defined and so there are no superdecomposable pure-injectives (7.3.5). Therefore the pure-injective hull of M as above is, by 4.4.2, the pure-injective hull of a direct sum of indecomposable pure-injectives, not all of which can be injective (by 4.3.12).

In fact, since duality (Section 5.4) clearly interchanges R_R and $E(R/J)$, it follows from 5.3.62 that this pure-injective, non-injective point is just R/J.

Corollary 14.4.8. *Let R be a commutative coherent ring and let $P \in \mathrm{Spec}(R)$. Then the closure of E_P in the fg-ideals = rep-Zariski topology on inj_R is $\{E \in \mathrm{inj}_R : P(E) \geq P\}$.*

Proof. For every prime Q, one has $E_Q \in \text{Zar-cl}(E_P)$ iff $Q \geq P$ (14.4.4). Therefore, by 14.4.5, $E \in \text{Zar-cl}(E_P)$ iff $P(E) \geq P$. □

If E is an injective module, then $\mathrm{cog}(E)$ denotes the hereditary torsionfree class cogenerated by E, that is, the class of those modules which embed in a power of E, see Section 11.1.1. If E' is an indecomposable injective in $\mathrm{cog}(E)$, then, since it is a direct summand of a direct product of copies of E, it is (by 3.4.7) in the definable subcategory generated by E, hence it is a member of $\mathrm{supp}(E) \subseteq \mathrm{Zg}_R$ (Section 5.1.1); in particular, if E is indecomposable, then $E' \in \text{Zg-cl}(E)$, hence (14.2.2) $E \in \text{Zar-cl}(E')$. The first half of the proof of 14.4.5 shows that $E \in \text{Zg-cl}(E_{P(E)})$ whether or not R is coherent. It also shows the following.

Lemma 14.4.9. *If $I \leq J$ are (right) ideals of R, then $E(R/I) \in \mathrm{cog}(E(R/J))$.*

Proposition 14.4.10. *Let E be an indecomposable injective module over the commutative coherent ring R. Then the torsion theory cogenerated by E is of finite type iff $E = E_P$ for some prime P.*

Proof. ($\Leftarrow$) By 14.4.4, for any ideal, I, of R we have $E_P \in D^{\mathrm{m}}(I)$ iff $E_P \in D(I)$, that is, iff $(R/I, E_P) = 0$ (that is, if R/I is E_P-torsion) iff $I \nleq P$. The latter is so iff some finitely generated ideal $I' \leq I$ satisfies $I' \nleq P$. Therefore each E_P-dense ideal contains a finitely generated E_P-dense ideal, so, by 11.1.14, the torsion theory cogenerated by E is of finite type.

($\Rightarrow$) If E cogenerates a torsion theory of finite type, then, by the proof of the second half of 14.4.5, $E_{P(E)} \in \mathrm{cog}(E)$. For there, taking $J = R$, it is shown that if I is a finitely generated ideal with $\mathrm{Hom}(R/I, E) = 0$, that is, with R/I E-torsion, then $\mathrm{Hom}(R/I, E_{P(E)}) = 0$, so, by finite type, $E_{P(E)} \in \mathrm{cog}(E)$. Therefore there is an embedding $R/P(E) \to E^\kappa$ for some index set κ. It follows that $\mathrm{ann}_E(P(E)) \neq 0$ so, by definition of $P(E)$, $E \simeq E_{P(E)}$, as required. □

Thus $\mathrm{Spec}(R)$ may be identified within inj_R as consisting of those points which cogenerate torsion theories of finite type.

Recall, 5.1.11, that if R is any right coherent ring, then a subset of inj_R is Ziegler-closed iff it has the form $\mathcal{F} \cap \mathrm{inj}_R$, where $\mathcal{F}$ is the torsionfree class for some torsion theory of finite type.

Theorem 14.4.11. *Let R be commutative coherent and let $X \subseteq \mathrm{inj}_R$ be Ziegler-closed. Then X is irreducible in the Ziegler topology iff $X = \mathrm{cog}(E_P) \cap \mathrm{inj}_R$ for some prime ideal, P, of R.*

Proof. By 5.1.11 there is a torsionfree class, $\mathcal{F}$, of finite type with $\mathcal{F} \cap \mathrm{inj}_R = X$. Let $\mathcal{I}$ be the set of annihilators of non-zero element of members of $\mathcal{F}$. If $\{I_\lambda\}_\lambda$ is a chain of members of $\mathcal{I}$ with their union = sum equal to I, say, then, since $\mathcal{F}$ is closed under direct limits (11.1.12), there is $M \in \mathcal{F}$ and $a \in M$ with $a \neq 0$ and $aI = 0$. So, by Zorn's Lemma, every $I \in \mathcal{I}$ is contained in a maximal member of $\mathcal{I}$. Denote the set of these maximal members by $\mathcal{P}$. The argument used in 14.4.2 shows that all ideals in $\mathcal{P}$ are prime.

Choose $P_0 \in \mathcal{P}$, set $E_0 = E_{P_0}$ and $E' = \bigoplus\{E_P : P \in \mathcal{P}, P \neq P_0\}$. By 14.4.9 (and comments before that) $\mathrm{supp}(E_0) \cup \mathrm{supp}(E') = X$ (because R is coherent the support of any injective module will, by 3.4.24, consist of injective modules). By irreducibility of X, either $X = \mathrm{supp}(E_0)$, which equals $\mathrm{cog}(E_0) \cap \mathrm{inj}_R$ by 14.4.10 and 5.1.11, as required, or $X = \mathrm{supp}(E')$. But, in the latter case we would have $E_0 \in \mathrm{cog} E'$, so there would be an embedding of the form $f : R/P_0 \to \prod\{E_P^{\kappa(P)} : P \in \mathcal{P}, P \neq P_0\}$ with, say $(1 + P_0) \to (e_\lambda)_\lambda$. Some e_λ would be non-zero so P_0 would be (properly!) contained in $\mathrm{ann}_R(e_\lambda)$ – contradicting $P_0 \in \mathcal{P}$. □

Proposition 14.4.12. *Let R be commutative coherent. A subset, V, of inj_R is rep-Zariski-closed and irreducible in the rep-Zariski topology iff there is a prime ideal Q of R such that $V = \{E : P(E) \geq Q\}$.*

Proof. This is just the usual description of irreducible closed subsets of $\mathrm{Spec}(R)$ combined with 14.4.5. □

Corollary 14.4.13. *Let R be commutative coherent. Then there are natural bijections between the following:*

(i) the set of irreducible closed sets in the Ziegler topology on inj_R;
(ii) the set of irreducible closed sets in the Zariski topology on $\mathrm{Spec}(R)$;
(iii) the points of $\mathrm{Spec}(R)$;
(iv) the set of irreducible (Gabriel-/induced rep-)Zariski-closed subsets of inj_R.

The bijections are given by $\{E : P(E) \leq Q\} \sim \{P : P \geq Q\} \sim Q \sim \{E : P(E) \geq Q\}$.

The duality between the rep-Zariski and Ziegler topologies was used by Garkusha and Prest, [221], to explain some results deriving from homotopy theory. Thomason, [675], extending a result of Hopkins, [300], and Neeman, [457], showed that for a commutative ring R there is a natural bijection between thick subcategories of the derived category of perfect complexes of R-modules (see Chapter 17 for these notions) and certain subsets of the prime spectrum, Spec(R), of R. The subsets of Spec(R) which appear are those subsets which are unions of complements of compact open subsets. Hovey, [302], showed that, for certain commutative rings, there is also a bijection between these thick subcategories and Serre subcategories of mod-R. He asked for a direct connection between such Serre subcategories and those subsets of Spec(R), expecting that this connection would also be more general. In [221] this is shown: the above subsets of Spec(R) are, under the bijection Spec(R) $\leftrightarrow$ inj_R, identified as exactly the Ziegler-open subsets and the connection is shown for all coherent commutative rings. In [223] the connection is extended to arbitrary commutative rings, though torsion theories of finite type on Mod-R must replace Serre subcategories of mod-R. It is also shown in these papers how the ringed space structure over Spec(R) may be recovered from the category of modules and corresponding results for projective schemes are given in [222], using a certain quotient category of the category of graded modules in place of Mod-R. There is overlap in result but not in method with [616]. These results are extended to quasi-compact, quasi-separated schemes in [216].

Also see [99], [41]–[43], [671], [672], [369].

15
Artin algebras

After some basic definitions around quivers and their representations, the Ziegler closure of a tube is described. Then results, particular to artin algebras, on duality, ideals of mod-R and m-dimension, are presented. Modules over group rings are considered briefly.

Useful background references: [27], [16], [142].

15.1 Quivers and representations

Representations of quivers and path algebras are defined in Section 15.1.1. Almost split sequences and the Auslander–Reiten quiver are defined in Section 15.1.2.

Tubes – special types of components of Auslander–Reiten quivers – generalisations of these, and their Ziegler-closures are described in Section 15.1.3.

15.1.1 Representations of quivers

Basic definitions and examples are given. For artin algebras there is a duality (15.1.3) between finitely presented right and left modules.

A **quiver**, Q, is a directed graph. A **representation**, M, of Q **in the category** of R-modules, or an R-**representation of** Q, where R is any ring, is given by the following data: for each vertex i of Q, an R-module $M(i)$ and, for each arrow $i \xrightarrow{\alpha} j$ of Q, an R-linear map $M(i) \xrightarrow{M(\alpha)} M(j)$. The category of R-representations of Q, is obtained by defining morphisms in the natural way: a morphism, f, from a representation M to a representation N is given by the following data: for each vertex i, an R-linear map $f_i : M(i) \to N(i)$, such that for every arrow $i \xrightarrow{\alpha} j$ we have $N(\alpha) f_i = f_j M(\alpha)$.

$$\begin{array}{ccc} M(i) & \xrightarrow{M(\alpha)} & M(j) \\ {\scriptstyle f_i}\downarrow & & \downarrow{\scriptstyle f_j} \\ N(i) & \xrightarrow[N(\alpha)]{} & N(j) \end{array}$$

The **path algebra** over R of the quiver Q is denoted $R(Q)$ or just RQ and is a free R-module with basis the set of all finite directed paths in the quiver. A **path** is a walk along the directed arrows of the quiver: formally, a "composable" sequence, possibly empty, of directed arrows. Staying at vertex i is a path of length 0, denoted e_i. Multiplication is defined on $R(Q)$ by setting the product, pq, of paths p and q to be 0 if the end of p is not the starting point of q and to be the composition of the paths, p then q, otherwise (this way round because we usually deal with right modules). This multiplication of paths is extended to a multiplication on $R(Q)$ by linearity and by declaring that each arrow commutes with every element of R. The arrows of Q are identified with paths of length 1 in $R(Q)$. If there are only finitely many vertices in Q, then this ring has a 1 which is the sum of all the idempotents, e_i, for i a vertex of Q. In any case it has enough local identities (finite sums of the e_i).

The category of right $R(Q)$-modules is equivalent to the category of R-representations of Q. For instance given an $R(Q)$-module, M, the R-modules which form the representation built from M are the $M(i) = Me_i$ and $M(\alpha)$ is defined to be the restriction to $M(i)$/corestriction to $M(j)$ of right multiplication by α. Conversely, given an R-representation, M, of the quiver Q, the corresponding module is, as an R-module, $\bigoplus\{M(i) : i \text{ is a vertex of } Q\}$ and right multiplication by $\alpha : i \to j$ is defined to be zero on every component except $M(i)$, where it is the map $M(\alpha)$ from $M(i)$ to $M(j)$. One checks that these processes are inverse, and similarly for the morphisms of the categories.

Examples 15.1.1. (a) If Q is the quiver with one vertex and one arrow (that is, Q is a loop), then $R(Q)$ is isomorphic to the polynomial ring $R[T]$ (T corresponds to the arrow).

(b) The path algebra $R = k\widetilde{A_1}$, over a field k, of the quiver $\widetilde{A_1}$ $1 \underset{\beta}{\overset{\alpha}{\rightrightarrows}} 2$ (see 1.1.4) is a four-dimensional algebra which can be represented as the matrix ring $\begin{pmatrix} k & k \oplus k \\ 0 & k \end{pmatrix}$ where the actions on the bimodule $k \oplus k$ both are the diagonal ones.

A representation, V, of this quiver is given by two vector spaces, V_1, V_2 and two linear maps $V_\alpha, V_\beta : V_1 \to V_2$. For instance, R_R is the direct sum of the three-dimensional indecomposable projective, P_1, generated by $e_1 = \begin{pmatrix} 1 & 0 \\ 0 & 0 \end{pmatrix}$ and the one-dimensional projective, P_2, generated by $e_2 = \begin{pmatrix} 0 & 0 \\ 0 & 1 \end{pmatrix}$. These modules have dimension vectors $(1, 2)$ and $(0, 1)$ respectively and R is the direct sum of one copy of each.

A morphism, f, from the representation V to, say, $W = (W_1, W_2, W_\alpha, W_\beta)$ is given by a pair of linear maps, $f_1 : V_1 \to W_1$, $f_2 : V_2 \to W_2$ such that $W_\alpha f_1 = f_2 V_\alpha$ and $W_\beta f_1 = f_2 V_\beta$.

$$\begin{array}{ccccccc} V_1 & \xrightarrow{V_\alpha} & V_2 & & V_1 & \xrightarrow{V_\beta} & V_2 \\ {\scriptstyle f_1}\downarrow & & \downarrow{\scriptstyle f_2} & & {\scriptstyle f_1}\downarrow & & \downarrow{\scriptstyle f_2} \\ W_1 & \xrightarrow[W_\alpha]{} & W_2 & & W_1 & \xrightarrow[W_\beta]{} & W_2 \end{array}$$

For example, a morphism from P_2 to P_1 is given by sending $e_2 \in P_2$ to $e_1\alpha$. Another is given by sending e_2 to $e_1\beta$ and it is easy to see that these form a basis for the two-dimensional space $\mathrm{Hom}_R(P_2, P_1)$; on the other hand clearly $\mathrm{Hom}(P_1, P_2) = 0$.

(c) The path algebra of the quiver $\widetilde{D_4}$ (see 1.1.3) is nine-dimensional and, as a right module over itself, has four non-isomorphic two-dimensional indecomposable projective summands and a single one-dimensional summand.

If k is a field, if the quiver Q is finite in the sense of having just finitely many vertices and finitely many arrows, and if there are no oriented cycles in the quiver, then the path algebra $k(Q)$ will be a finite-dimensional algebra, so, in particular, an artin algebra. An example of an infinite quiver is 10.1.2.

These definitions are extended to **quivers with relations**: given a quiver Q and a fixed coefficient ring R (not necessary if the relations to be imposed only have integer coefficients) a **relation** is an R-linear combination of paths declared to be 0. Then the category of those R-representations of Q which satisfy a given set of relations is clearly equivalent to the category of modules over the ring $R(Q)/I$, where I is the ideal of $R(Q)$ generated by the relations.

Example 15.1.2. The first Weyl algebra (1.1.1(c)) is obtained as the path algebra of a quiver with one vertex and two loops, α, β, by imposing the relation $\beta\alpha - \alpha\beta = 1$.

Another example is the Gelfand–Ponomarev quiver $\mathrm{GP}_{n,m}$, which is the quiver with one vertex and two arrows, α, β with relations $\alpha\beta = \beta\alpha = 0$ and $\alpha^n = 0 = \beta^m$. For further examples see 1.2.16 and those in Section 8.1.3.

If R is any ring and k is a commutative ring contained in the centre of R, then each $r \in k$ acts as an endomorphism, $r \times -$, on each R-module; furthermore, every morphism group in Mod-R has the k-module structure defined by $fr : a \mapsto f(ar) = fa \cdot r$ (that is, $f(r \times -) = (r \times -)f$) for $f : A \to B$, $a \in A$, $r \in k$. Then, if F is any additive functor from Mod-R to **Ab**, each group FA, for $A \in$ Mod-R, has the k-module structure defined by $mr = F(r \times -)m$ for $m \in FA$. Furthermore, for $A, B \in$ Mod-R, F is a k-linear map from (A, B) to (FA, FB), since, if $g : A \to B$, then the diagrams below are commutative.

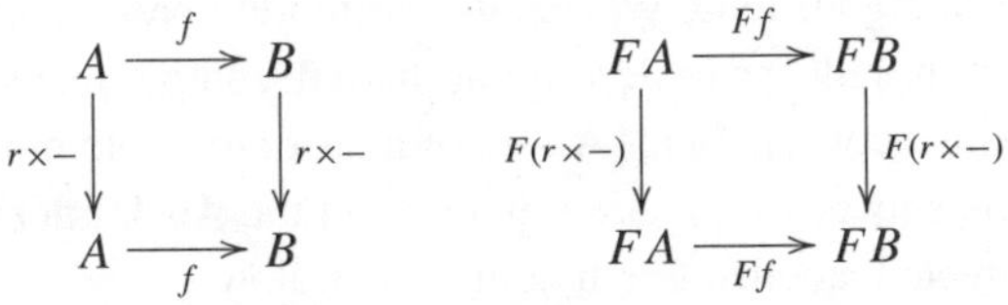

$$\begin{array}{ccc} A & \xrightarrow{f} & B \\ {\scriptstyle r\times-}\downarrow & & \downarrow{\scriptstyle r\times-} \\ A & \xrightarrow[f]{} & B \end{array} \qquad\qquad \begin{array}{ccc} FA & \xrightarrow{Ff} & FB \\ {\scriptstyle F(r\times-)}\downarrow & & \downarrow{\scriptstyle F(r\times-)} \\ FA & \xrightarrow[Ff]{} & FB \end{array}$$

Thus every additive functor from Mod-R to **Ab** may be regarded as a k-linear functor from Mod-R to Mod-k. Therefore, when dealing with artin k-algebras, we may use categories, such as (mod-R, Mod-k), of k-linear functors in place of naturally equivalent categories (see, for example, [467, p. 92]) such as (mod-R, **Ab**).

For every artin algebra there is a natural duality between the category of finitely presented right modules and the category of finitely presented left modules (also cf. Section 1.3.3).

Proposition 15.1.3. (for example, [27, II.3.3]) *Let R be an artin algebra with centre $C = C(R)$ and let E be a minimal injective cogenerator for* Mod-C *(the injective hull of $C/J(C)$ if C is basic). Then* $\mathrm{Hom}_C(-, E_C) : (\text{mod-}R)^{\mathrm{op}} \longrightarrow$ R-mod *is an equivalence, as is* $\mathrm{Hom}_C(-, {}_CE) : (R\text{-mod})^{\mathrm{op}} \longrightarrow (\text{mod-}R)$.

If R is a finite-dimensional algebra over a field k, then $M \mapsto$ $\mathrm{Hom}_k(M, k)$ *(the latter equipped with the natural left R-module structure) is equivalent to the functor above* (for example, [27, §II.3]).

An artin algebra R is **basic** if R/J ($J = J(R)$ the Jacobson radical of R) is a product of division rings (as opposed to a product of matrix rings over division rings). For instance, the path algebra, over a field, of a finite quiver without oriented cycles is a basic artin algebra. Every artin algebra R is Morita-equivalent to one which is basic: each simple component, $M_{n_i}(D_i)$, of R/J lifts to a direct sum, $P_i^{n_i}$, of copies of an indecomposable projective P_i (so the $S_i = P_i/\mathrm{rad}(P_i)$ are the simple R-modules). Let $S = (\mathrm{End}(\bigoplus_i P_i))^{\mathrm{op}}$ (just one copy of each indecomposable projective). Then S is basic and Morita-equivalent to R (for example, [477, §6.6], [27, II.2.5]).

15.1.2 The Auslander–Reiten quiver of an artin algebra

The Auslander–Reiten quiver is defined and the existence theorem for almost split (= Auslander–Reiten) sequences is stated (15.1.4).

Throughout this subsection R is an artin algebra. Recall that this means that the centre of R is artinian and R is finitely generated as a module over its centre.

Since every finitely generated R-module is of finite length, it is a direct sum of indecomposable R-modules. These indecomposables have local endomorphism rings (4.3.58), so, E.1.24, this decomposition is unique to isomorphism. The task of describing all finitely generated = finitely presented = finite-length R-modules therefore reduces to describing the indecomposable such R-modules and the morphisms between them. All these indecomposable modules and some of the morphisms between them are organised in the **Auslander–Reiten quiver** of R: this is a directed labelled graph defined as follows.

The vertices of the graph are the isomorphism classes of indecomposable finite length R-modules. Given two such modules, A and B, consider the $(\mathrm{End}(B)/J\mathrm{End}(B), \mathrm{End}(A)/J\mathrm{End}(A))$-bimodule[1] $\mathrm{rad}(A, B)/\mathrm{rad}^2(A, B)$. If the dimension on the left of this bimodule is d and on the right is e, then we put an arrow from A to B with label (e, d) (see [27, VII.1.6]).

If R is a finite-dimensional algebra over an algebraically closed field, k, then replace the labelled arrow just described by $\dim_k(\mathrm{rad}(A, B)/\mathrm{rad}^2(A, B))$ arrows going from A to B and $\dim_k(\mathrm{rad}(B, A)/\mathrm{rad}^2(A, B))$ arrows going from B to A. There is an example at 15.1.5 below and there are more in Section 8.1.2.

The next result is central to understanding the structure of mod-R when R is an artin algebra.

Theorem 15.1.4. (see, for example, [27, V.1.14, V.1.15]) *Suppose that R is an artin algebra and let A be any indecomposable finitely generated module which is not injective. Then there is, up to isomorphism, a unique non-split exact sequence $0 \to A \xrightarrow{f} B \xrightarrow{g} C \to 0$ with the property that every morphism $A \to B'$ in* Mod-R *which is not a split embedding factors through f. This sequence is also determined by the property that every morphism $B'' \to C$ in* Mod-R *which is not a split epimorphism factors through g. Moreover, B, C are finitely generated and C is indecomposable. Furthermore, for every indecomposable finitely generated module C which is not projective there is a unique to isomorphism exact sequence as above,[2] in which A, B are finitely generated and A is indecomposable.*

[1] The radical, $\mathrm{rad}(A, B)$, of (A, B) and its powers are defined at 9.1.1.

[2] More generally, if R is semiperfect and C is finitely presented with local endomorphism ring and

An exact sequence of the above form is said to be **Auslander–Reiten** or **almost split**; we tend to use the second term because it is so well chosen: such a sequence is a non-zero element of the simple socle of the $(\mathrm{End}(A), \mathrm{End}(C))$-bimodule $\mathrm{Ext}^1_R(C, A)$ (see, for example, [27, V.2.1]). Set $C = \tau^{-1}A$ and $A = \tau C$: τ is referred to as the **Auslander–Reiten translate** and may be regarded as an operation on the set ind-R of finitely generated indecomposable modules, or on the Auslander–Reiten quiver of R, which is defined at all but the projective modules. If R is a k-algebra and $R/J(R)$ is a product of copies of k, then the connection with the Auslander–Reiten quiver is as follows. With the notation of 15.1.4, let $B = B_1 \oplus \cdots \oplus B_n$ be a decomposition of B into indecomposables: then $B_1, \ldots, B_n$ are the immediate successors (counting multiplicity) of A in the Auslander–Reiten quiver of R and are also the immediate predecessors of C. This result is of immense importance both in proving general results and in the actual computation of Auslander–Reiten quivers.

A morphism f in mod-R is **irreducible** if it is neither a split monomorphism nor a split epimorphism and if it has no non-trivial factorisation, where, by a trivial factorisation of f is meant a factorisation $f = gh$ with $g, f \in$ mod-R and where either h is a split monomorphism or g is a split epimorphism. The irreducible morphisms between indecomposables are those which occur as components of the morphisms of an almost split sequence, together with the components of the inclusion of $\mathrm{rad}(P)$ into P, where P is an indecomposable projective, and the components of the projection from E to $E/\mathrm{soc}(E)$, where E is an indecomposable injective. Each irreducible morphism is, in particular, left almost split (see Section 5.3.3). See [27, §V.5] for all this.

Many examples of almost split sequences can be read off from the diagrams in Section 8.1.2 which show Auslander–Reiten **components** (connected components of the Auslander–Reiten quiver): a dashed line from A on the right to C on the left corresponds to an almost split sequence $0 \to A \to B \to C \to 0$, where B is the direct sum of the modules B_i such that there is an arrow from A to B_i, and hence also an arrow from B_i to C. For example, from the preprojective component for $\widetilde{A_3}$ on p. 331 one may read off, among others, the almost split sequence $0 \to \begin{smallmatrix} & 1 & \\ 0 & & 0 \\ & 0 & \end{smallmatrix} \to \begin{smallmatrix} & 1 & \\ 1 & & 0 \\ & 1 & \end{smallmatrix} \oplus \begin{smallmatrix} & 1 & \\ 0 & & 1 \\ & 1 & \end{smallmatrix} \to \begin{smallmatrix} & 1 & \\ 1 & & 1 \\ & 2 & \end{smallmatrix} \to 0$. A more familiar example, which may be taken to be over the ring of integers or over a suitable quotient, such as $\mathbb{Z}_4$ (which is an artin algebra) is $0 \to \mathbb{Z}_2 \to \mathbb{Z}_4 \to \mathbb{Z}_2 \to 0$.

not projective, then there is, [21, 3.9], such a sequence, but in Mod-R, not necessarily in mod-R. Zimmermann, [733, Thms 1, 2], gave a necessary and sufficient condition for such a sequence to exist in mod-R and also for the corresponding result starting with A. Also see [275, §2].

Example 15.1.5. (continuing 4.5.19) Let R be the path algebra of the quiver $A_3 = 1 \to 2 \to 3$ (see 4.5.19). The inclusions of P_2 into P_1 and of P_3 into P_2 are in $\mathrm{rad}(P_2, P_1)$, resp. $\mathrm{rad}(P_3, P_2)$, but have no further non-trivial factorisations, so are not in rad^2. Indeed, $\mathrm{rad}(P_2, P_1)/\mathrm{rad}^2(P_2, P_1)$ is easily checked to be one-dimensional, and similarly for (P_3, P_2). The inclusion of P_3 into P_1 is the composition of the other two inclusions, so is in $\mathrm{rad}^2(P_3, P_1)$, indeed there is no arrow between P_3 and P_1 in the Auslander–Reiten quiver of R. The full Auslander–Reiten quiver of this algebra of finite representation type is, with notation as in 4.5.19, as follows, where the lower version shows the dimension vectors.

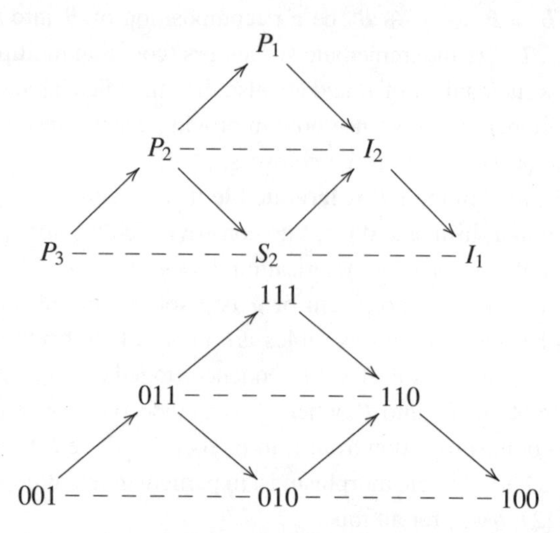

The dashed lines show the three Auslander–Reiten sequences for this algebra. They are as follows.

- $0 \to P_3 \to P_2 \to S_2 \to 0$, that is, $0 \to 001 \to 011 \to 010 \to 0$
- $0 \to P_2 \to P_1 \oplus S_2 \to I_2 \to 0$, that is, $0 \to 011 \to 111 \oplus 010 \to 110 \to 0$
- $0 \to S_2 \to I_2 \to I_1 \to 0$, that is, $0 \to 010 \to 110 \to 100 \to 0$

So the Auslander–Reiten translates are:

$S_2 = \tau^{-1} P_3 \quad I_1 = \tau^{-1} S_2 = \tau^{-2} P_3 \quad I_2 = \tau^{-1} P_2$ and $P_3 = \tau S_2$ etc.

For many examples of Auslander–Reiten quivers see, for instance, [601].

15.1.3 Tubes and generalised tubes

Tubes in the Auslander–Reiten quiver are described and generalised tubes are defined. The Ziegler-closure of a generalised tube is given (15.1.8), with more precise results for the closure of a tube (15.1.10, 15.1.12).

Over a tame artin algebra it seems that "most" indecomposable modules of finite length lie in components of the Auslander–Reiten quiver called tubes. A precise expression of this is the result, [138, Cor. E], of Crawley-Boevey, that if R is a finite-dimensional algebra over an algebraically closed field, then, for each dimension d, all but finitely many d-dimensional indecomposable modules lie in tubes. We describe this type of component; for a definition see [602, p. 113], also see, for example, [27].

A familiar example of a tube, outside the context of artin algebras, is given by the modules of the form $\mathbb{Z}_{p^n}$, for a fixed prime p, as n varies over the set of positive integers. For each $n \geq 1$ there is the canonical monomomorphism $\mathbb{Z}_{p^n} \to \mathbb{Z}_{p^{n+1}}$ and, for $n > 1$, the canonical epimorphism $\mathbb{Z}_{p^n} \to \mathbb{Z}_{p^{n-1}}$. Putting these together we obtain an almost split sequence (see 15.1.4) $0 \to \mathbb{Z}_{p^n} \to \mathbb{Z}_{p^{n+1}} \oplus \mathbb{Z}_{p^{n-1}} \to \mathbb{Z}_{p^n} \to 0$ for $n > 1$ and $0 \to \mathbb{Z}_p \to \mathbb{Z}_{p^2} \to \mathbb{Z}_p \to 0$ at $n = 1$. This quiver looks more like a line of modules than a tube but one can imagine the periodic planar structure of morphisms being rolled up into a tube which is, in this case, degenerate, being just a line.

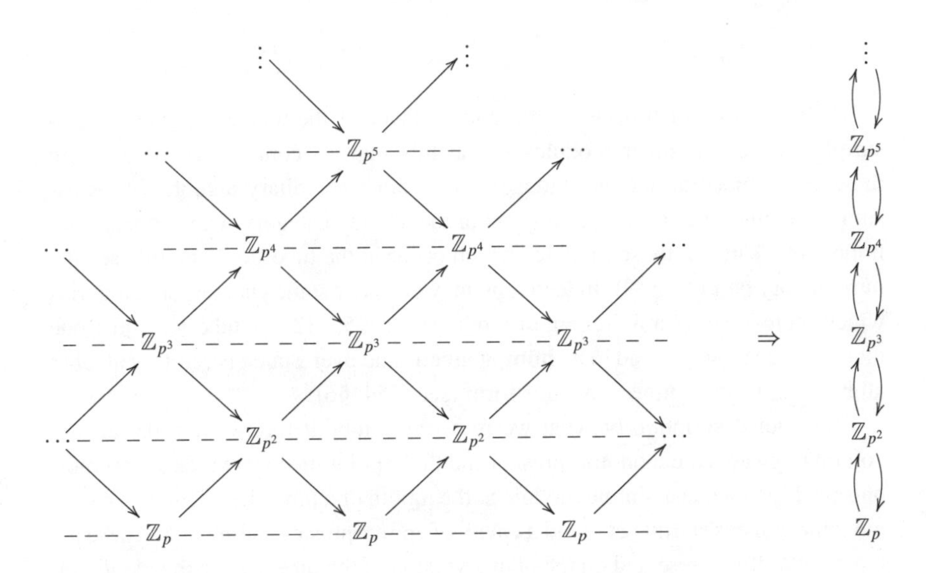

More general tubes are obtained similarly (see, for example, [602, §3.1]), but allowing for period possibly greater than 1. For example, rolling up a diagram of the shape below gives a tube of rank 2 (an example can be found in Section 8.1.2). The Auslander–Reiten translates are not marked but the pattern is as in the diagram above.

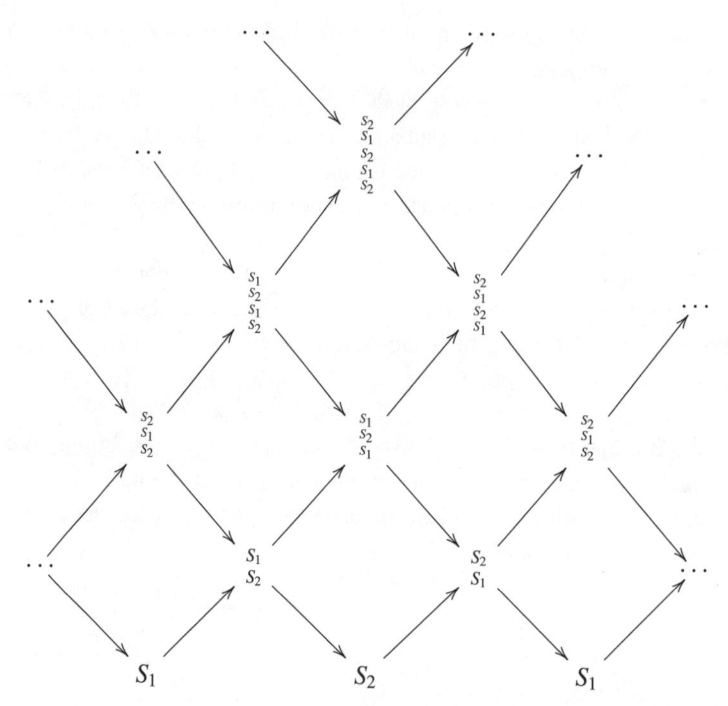

The modules which occur at the bottom layer of the tube are termed **quasi-simple** and all the other modules in the tube have a composition series with these as composition factors. Indeed, over a tame hereditary algebra the quasi-simple modules are the simple objects in the abelian category whose objects are finite direct sums of those modules which occur in the tube (for example, see the introductory part of [599]). Indeed, one may consider somewhat larger categories which include some infinite-length modules (cf. 15.1.12). A tube with just one module at its mouth is said to be **homogeneous** and over a tame hereditary algebra all but at most three tubes have this form (see [154, §6]).

From our description of what we mean by a tube it follows that the arrows pointing upwards are monomorphisms and those pointing downwards are epimorphisms. If S is a quasi-simple module at the mouth of a tube, then there is a **ray** of monomorphisms starting at S and a **coray** of epimorphisms ending at S. These are more naturally represented on the planar version of the tube, before the identifications have been made. In the abelian groups example the single ray of monomorphisms is the sequence of natural embeddings $\mathbb{Z}_p \to \mathbb{Z}_{p^2} \to \cdots \to \mathbb{Z}_{p^n} \to \cdots$ and the single coray of epimorphisms is the sequence of natural surjections $\cdots \to \mathbb{Z}_{p^n} \to \cdots \to \mathbb{Z}_{p^2} \to \mathbb{Z}_p$. Under fairly general circumstances (see 15.1.9 below) the direct limit of a ray of monomorphisms in a tube is an indecomposable, "Prüfer"-type, pure-injective module and the inverse limit of such a coray is an indecomposable, "adic"-type, pure-injective module (cf. Sections 5.2.1, 8.1.2). As in the abelian groups example above, both these will be in the Ziegler-closure of the modules forming the tube (see 15.1.9, 15.1.10). The first results along these lines were by Ringel, [599, p. 326]. The most general results are due to Krause, who considered situations where one has a similar structure of morphisms, though not necessarily the almost split sequences. We describe this now.

A **generalised tube**, [357, §8], is $(M_n, f_n, g_n)_{n\geq 0}$ with $f_n : M_{n+1} \to M_n$ and $g_n : M_n \to M_{n+1}$, $M_0 = 0$ such that for every n the sequence

$$0 \to M_n \xrightarrow{\begin{pmatrix} f_{n-1} & g_n \end{pmatrix}} M_{n-1} \oplus M_{n+1} \xrightarrow{\begin{pmatrix} g_{n-1} \\ -f_n \end{pmatrix}} M_n \to 0$$

is exact (interpret M_{-1} as 0). It follows (inductively) that each morphism f_n is surjective, with kernel isomorphic to M_1 and each morphism g_n is injective with cokernel isomorphic to M_1.

Let $0 \to M_1 \to M_\infty \xrightarrow{f} M_\infty \to 0$ be the (exact) direct limit of the exact sequences $0 \to M_1 \to M_n \xrightarrow{f_n} M_{n+1} \to 0$. Thus one obtains a module, M_∞, and an epimorphic endomorphism, f, of M_∞. Conversely, from such a pair, (M, f), one defines a generalised tube by setting $f_n = f^n$, $M_n = \ker(f^n)$ and taking g_n to be the inclusion of M_n in M_{n+1}. Indeed, [357, 8.1], there is a bijection between generalised tubes and pairs (M, f), where $f \in \operatorname{End}(M)$ is an epimorphism such that $M = \varinjlim_n \ker(f^n)$.

Proposition 15.1.6. ([357, 8.3]) *Let R be an artin algebra and let $f : M \to M$ be an epimorphism of an R-module M. Set $M_i = \ker(f^i)$. Suppose that $M = \varinjlim M_i$ and that M_1 has finite length. Then M is artinian over its endomorphism ring, in particular is Σ-pure-injective, and has m-dimension ≤ 1. If infinitely many of the M_i are indecomposable, then every non-zero direct summand of M is of infinite length.*

Also let $0 \to \widehat{M} \xrightarrow{g} \widehat{M} \to M_1 \to 0$ be the (exact, by, for example, [708, 3.5.7]) inverse limit of the sequences $0 \to M_n \xrightarrow{g_n} M_{n+1} \to M_1 \to 0$. Then the directed system of exact sequences $0 \to \widehat{M} \xrightarrow{g^n} \widehat{M} \to M_n \to 0$ has direct limit the (exact) sequence $0 \to \widehat{M} \to Q \to M_\infty \to 0$, say. Thus to a generalised tube there corresponds a canonical ("universal" in the terminology of [357]) exact sequence $0 \to \widehat{M} \to Q \to M_\infty \to 0$, where $\widehat{M}$ is the inverse limit of the tube and M_∞ is its direct limit, and, with further conditions as in 15.1.8, the module Q turns out to be generic in the sense of Section 4.5.5.

Proposition 15.1.7. ([357, 8.6]) *Let R be an artin algebra and suppose that $(M_n, f_n, g_n)_n$ is a generalised tube of finite length R-modules, with associated pair*

(M_∞, f). Suppose that each morphism $g_n : M_n \to M_{n+1}$ is in the radical (Section 9.1.1) of mod-R. *Then there is a generic module in the support,* $\mathrm{supp}(M_\infty) \subseteq \mathrm{Zg}_R$ *(Section 5.1.1), of M_∞, hence also in* $\mathrm{supp}(\{M_n\}_n)$ *(which contains* $\mathrm{supp}(M_\infty)$ *since M_∞ is a direct limit of the M_n).*

In particular, if R is an artin algebra and there is a generalised tube in the radical of mod-R, then there is a generic R-module.

In the case that R is a finite-dimensional algebra over a field a more detailed analysis may be made, showing that $\widehat{M}$ is a $(k[[T]], R)$-bimodule which is free over $k[[T]]$ of rank equal to $\dim_k M_1$ and the canonical exact sequence is $0 \to k[[T]] \otimes_{k[[T]]} \widehat{M} \to k((T)) \otimes_{k[[T]]} \widehat{M} \to \big(k((T))/k[[T]]\big) \otimes_{k[[T]]} \widehat{M} \to 0$ ([357, 8.8]). Then one obtains the following more precise result.

Theorem 15.1.8. ([357, 8.10]) *Let R be a finite-dimensional k-algebra and suppose that $(M_n, f_n, g_n)_n$ is a generalised tube of finite-dimensional R-modules. Let $0 \to \widehat{M} \to Q \to M_\infty \to 0$ be the associated canonical exact sequence. Let N be any infinite-dimensional module in the Ziegler-closure of the set $\{M_i\}_i$. Then:*

(a) Q is of endolength at most $\dim_k M_1$*;*
(b) N is a direct summand of either $\widehat{M}$, Q or M_∞;
(c) $\mathrm{mdim}(N) \le 1$*;*
(d) $\mathrm{End}(N)/\mathrm{JEnd}(N)$ *is a PI ring and, if N is of finite endolength, then* $\mathrm{End}(N)$ *is a PI ring;*[3]
(e) every module in the Ziegler-closure of M_∞, respectively $\widehat{M}$, is a direct summand of M_∞ or Q, resp. $\widehat{M}$ or Q.

Thus the Ziegler-closures of tubes over PI Dedekind domains (5.2.2) and hereditary artin algebras (Section 8.1.2) are typical in many respects, though also compare 8.2.42.

Returning to tubes, rather than generalised tubes, the following can be deduced. It first appeared, within the description of the Ziegler spectra of certain tame hereditary algebras, in [492]. The general result is [357, 9.1].

Theorem 15.1.9. *Let R be an artin algebra and suppose that $M_1 \to M_2 \to \cdots$ is a ray of (irreducible) monomorphisms in a tube, with M_1 quasi-simple. Let $M_\infty = \varinjlim M_i$. Then:*

(a) M_∞ is indecomposable, Σ-pure-injective, but not of finite endolength;
(b) if N_1 is a quasi-simple module starting a ray of irreducible monomorphisms $N_1 \to N_2 \to \cdots$ and $N_\infty = \varinjlim N_i$, then $N_\infty \simeq M_\infty$ only if $N_1 \simeq M_1$;

[3] Krause, [357, §7], considered preservation of this property (cf. Section 4.5.5) by functors which, see Section 18.2, commute with direct limits and products.

(c) the Ziegler-closure of M_∞ consists of M_∞ together with finitely many generic modules (the distinct direct summands of Q in the notation above).

See [606] for a direct construction of generic modules in this situation.

Using duality (as in Sections 1.3.3, 3.4.2) and, say, 5.4.14, one deduces the next result.

Theorem 15.1.10. *Let R be an artin algebra and suppose that $\cdots \to M_2 \to M_1$ is a coray of (irreducible) epimorphisms in a tube, with M_1 quasi-simple. Let $\widehat{M} = \varprojlim M_i$. Then:*

(a) $\widehat{M}$ is indecomposable, pure-injective, not Σ-pure-injective;
(b) if N_1 is a quasi-simple module ending a coray of irreducible epimorphisms $\cdots \to N_2 \to N_1$ and $\widehat{N} = \varprojlim N_i$, then $\widehat{N} \simeq \widehat{M}$ only if $N_1 \simeq M_1$;
(c) the Ziegler-closure of $\widehat{M}$ consists of $\widehat{M}$ together with finitely many generic modules (the distinct direct summands of Q).

Example 15.1.11. Let $R = k[x, y : xy = 0]_{(x,y)}$, where k is a field, be the localisation of $k[x, y : xy = 0]$ at the maximal ideal generated by x and y. Thus R is the local ring at the origin $(0, 0)$ of the union of the two coordinate lines $X = 0$, $Y = 0$ in the affine plane. The unique simple module $S = R/(x, y)$ and its self-extensions form a tube. In the Ziegler-closure of this tube are the S-Prüfer and S-adic modules and *two* generics, namely the quotient fields of the domains $R/(x)$ and $R/(y)$ corresponding to the two lines which have this point as their intersection.

Corollary 15.1.12. (see [357, 9.3]) *Let R be a finite-dimensional algebra over a field k and let $\{M_i\}_i$ be a homogeneous tube, where M_1 is the quasi-simple at the mouth of the tube. Then the Ziegler-closure of this tube consists of: the modules M_i; the corresponding adic module $\widehat{M}$; the corresponding Prüfer module M_∞; a finite number of generic modules, the sum of whose endolengths is no more than* $\dim_k M_1$.

15.2 Duality over artin algebras

Here we prove the result, used in Section 15.3, that, if R is an artin algebra, if F is a finitely presented functor from mod-R *to* **Ab** *and if G is any subfunctor of F, then G is the intersection of the finitely generated subfunctors of F which contain G (15.2.3). On the way, duality for finitely presented functors is extended (15.2.1) and it is shown that, over an artin algebra, the lattice of* all *subfunctors of a finitely presented functor is anti-isomorphic to that of the dual functor (15.2.2).*

Suppose throughout this section that R is an artin k-algebra, where k is a commutative artinian ring (not necessarily a field) which acts centrally on R.

Following [22, p. 131] consider the category, (R-mod, mod-k), of k-linear functors from R-mod to the category, mod-k, of *finite length* k-modules. The forgetful functor from k-Mod to **Ab** induces an equivalence of (R-mod, Mod-k) with (R-mod, **Ab**) (see comments in Section 15.1.1) and we identify (R-mod, mod-k) with its image in (R-mod, **Ab**). Clearly, as a subcategory of (R-mod, **Ab**), (R-mod, k-mod) is closed under kernels, cokernels and extensions.

We follow [358, §5.1] in extending the duality for finitely presented functors in (R-mod, Mod-k) to those possibly infinitely presented functors whose values on finitely presented R-modules are finite length k-modules: certainly every finitely presented functor in (R-mod, Mod-k) satisfies this condition.

Let $F \in$ (mod-R, mod-k). Define $F^\star \in$ (R-mod, mod-k) by: if $L \in R$-mod set $F^\star L = F(L^*)^*$, where * on the right-hand side is the duality $(-)^* = \mathrm{Hom}_k(-, E)$, E being a minimal injective cogenerator of Mod-k as in 15.1.3 (or 1.3.15), between mod-R and R-mod. Note that $\star$ extends to a functor which is exact: if $0 \to F \to G \to H \to 0$ is exact, then so is $0 \to F(L^*) \to G(L^*) \to H(L^*) \to 0$, hence so is $0 \to H(L^*)^* \to G(L^*)^* \to F(L^*)^* \to 0$.

Proposition 15.2.1. ([358, 5.3] also cf. [22, p. 132]) *Let R be an artin k-algebra.*

(a) If $F \in (\text{mod-}R, \mathbf{Ab})^{\mathrm{fp}}$, then $F^\star \simeq dF$.
(b) If F is a subquotient of a finitely presented functor, then so is $F^\star$.
(c) If $F \in$ (mod-R, mod-k), then $F^\star \in$ (R-mod, mod-k) and $F^{\star\star} \simeq F$.

Proof. (a) First suppose that $F = (X, -)$, so $dF = X \otimes -$, hence $dF \cdot L = X \otimes L$ for any finitely presented L. Also $F^\star L = \mathrm{Hom}_k\big(\mathrm{Hom}_R\big(X, \mathrm{Hom}_k(L, E)\big)\big) \simeq \mathrm{Hom}_k\big(\mathrm{Hom}_k(X \otimes_R L, E)\big) \simeq (X \otimes_R L)^{**} \simeq X \otimes_R L$. Thus $F^\star$ and dF agree on representable functors, so, since both functors $d(-)$ and $(-)^\star$ are exact, they agree on the objects of $(\text{mod-}R, \text{Mod-}k)^{\mathrm{fp}}$. The isomorphism $dF \cdot L \simeq F^\star L$ is functorial in L, hence $dF \simeq F^\star$.

(b) Say $0 \to F \to G$ is exact, where $F_1 \to G \to 0$ is exact and F_1 is finitely presented. Then there are (since * is a duality and exact) exact sequences $G^\star \to F^\star \to 0$ and $0 \to G^\star \to F_1^\star$, with $F_1^\star$ finitely presented by part (a), hence $F^\star$ is of the form claimed.

(c) The first part is clear from the definition of $\star$ and, for the second part,

$$F^{\star\star}X = (F^\star(X^*))^* = \mathrm{Hom}_k(F^\star(X^*), E) = \mathrm{Hom}_k(F(X^{**})^*, E)$$
$$\simeq \mathrm{Hom}_k\big(\mathrm{Hom}_k(FX, E), E\big) \simeq (FX)^{**} \simeq FX$$

since FX is of finite length over k. □

In particular * induces a duality between subquotients of finitely presented functors. Let F be any object of (mod-R, Mod-k)^{fp}. Recall that Latt(F) denotes the modular lattice of *all* subfunctors of F. If G is a subfunctor of F, then denote by DG the kernel of $(G \xrightarrow{i} F)^{\star} = F^{\star} \xrightarrow{i^{\star}} G^{\star}$, where i is the inclusion. Note that DG is a subfunctor of $dF = F^{\star}$. This notation is consistent with our earlier notation: if ϕ is a pp condition then $D(F_{\phi}) = F_{D\phi}$ (comment after 10.3.6).

Proposition 15.2.2. ([515, 3.2]) *Let R be an artin k-algebra and let $F \in$* (mod-R, Mod-k)^{fp}. *Then the map from* Latt(F) *to* Latt(dF) *which takes a subfunctor, G, of F to $DG = \ker\big((G \xrightarrow{i} F)^{\star}\big)$, where i is the inclusion of G in F, induces a duality* $(\mathrm{Latt}(F))^{\mathrm{op}} \simeq \mathrm{Latt}(dF)$ *which commutes with arbitrary intersections and sums:*

$$D(\textstyle\bigcap_{\lambda} G_{\lambda}) = \sum_{\lambda} DG_{\lambda};$$
$$D(\textstyle\sum_{\lambda} G_{\lambda}) = \bigcap_{\lambda} DG_{\lambda}.$$

Proof. The exact sequence $0 \to G \to F \to H \to 0$ (say) dualises to $0 \to H^{\star} \to F^{\star} \to G^{\star} \to 0$, so $H^{\star} = DG = \big(\mathrm{coker}(G \xrightarrow{i} F)\big)^{\star}$. Applying the same construction to $H^{\star}$ and using that $^{\star\star} \simeq \mathrm{Id}$ for subquotients of finitely presented functors (15.2.1) we see that, since the isomorphism $^{\star\star} \simeq \mathrm{Id}$ is, one may check, functorial, we may regard D^2 as Id (modulo a fixed identification of F with $F^{\star\star}$).

Clearly the map D is order-reversing.

Set $G = \bigcap_{\lambda} G_{\lambda}$, so $DG \geq DG_{\lambda}$ for all λ, hence $DG \geq \sum_{\lambda} DG_{\lambda} \geq DG_{\mu}$ for all μ. Therefore $G_{\mu} \geq D\big(\sum_{\lambda} DG_{\lambda}\big) \geq D^2G = G$ for all μ and hence $G = \bigcap_{\lambda} G_{\lambda} = D(\sum_{\lambda} DG_{\lambda})$. Therefore $DG = D^2(\sum_{\lambda} DG_{\lambda}) = \sum_{\lambda} DG_{\lambda}$, that is, $D(\bigcap_{\lambda} G_{\lambda}) = \sum_{\lambda} DG_{\lambda}$, as required.

The proof for the other part is similar. □

Note that if $F_1 \leq F$ are finitely presented, then DF_1 is finitely presented since it is the kernel of $F^{\star} \to F_1^{\star}$, which, by 15.2.1(a), is the map $dF \to dF_1$ (and these are finitely presented by 10.3.3).

Corollary 15.2.3. *Let R be an artin k-algebra and let $G \leq F \in$* (mod-R, **Ab**)^{fp}. *Then*

$$G = \bigcap\{F' \leq F : F' \text{ is finitely presented and } G \leq F'\}.$$

Proof. We have $DG = \sum\{F'' : F'' \leq DG (\leq dF),\ F'' \text{ finitely presented}\}$ so $G = D\big(\sum\{F'' : F'' \leq DG(\leq dF),\ F'' \text{ finitely presented}\}\big) = \bigcap\{DF'' : DF'' \geq G,\ F'' \leq F,\ F'' \text{ finitely presented}\}$ and each DF'' is finitely presented (as noted above), as required. □

15.3 Ideals in mod-R when R is an artin algebra

Throughout R is an artin algebra.

The Galois correspondence between ideals of mod-R *and collections of finitely presented functors gives a bijection between fp-idempotent ideals of* mod-R *and Serre subcategories of the category of finitely presented functors (15.3.4).*

It is shown that if (the extension to all modules of) a functor does not annihilate a given module, then it does not annihilate some morphism in mod-R *which factors through that module (15.3.2). The set of morphisms in* mod-R *which factor through some member of a class of modules which is closed under direct sums is not increased by allowing factorisation through the definable subcategory generated by that class (15.3.3). The morphisms annihilated by a Serre subcategory of the category of finitely presented functors are those which factor through some module in the corresponding definable subcategory (15.3.5).*

Any morphism in the ω-radical over an artin algebra factors through a finite direct sum of infinite-length points of the Ziegler spectrum (15.3.8).

Recall (Sections 9.2, 11.2.1) that if $\mathcal{I}$ is a collection of morphisms in mod-R, then ann($\mathcal{I}$) denotes the collection of finitely presented functors F with $Ff = 0$ for every $f \in \mathcal{I}$. Similary if $\mathcal{A}$ is a collection of functors in $(\text{mod-}R, \mathbf{Ab})^{\text{fp}}$, then ann($\mathcal{A}$) denotes the set of morphisms $f \in$ mod-R such that $Ff = 0$ for every $F \in \mathcal{A}$.

Proposition 15.3.1. ([358, 5.7]) *Let R be an artin algebra and let $\mathcal{I}$ be an ideal of* mod-R. *Then* $\mathcal{I} = \text{ann}(\text{ann}(\mathcal{I}))$.

Proof. Set $\overline{\mathcal{I}} = \text{ann}(\text{ann}(\mathcal{I}))$; so $\mathcal{I} \subseteq \overline{\mathcal{I}}$. Let $A \in$ mod-R. Since R is an artin algebra, by 15.2.3 each of $\mathcal{I}(A, -)$ and $\overline{\mathcal{I}}(A, -)$ is the intersection of the finitely generated subfunctors of $(A, -)$ containing it. So it will suffice to show that if $F \leq (A, -)$ is finitely generated and $F \geq \mathcal{I}(A, -)$, then $F \geq \overline{\mathcal{I}}(A, -)$. We may suppose (10.2.2) that $F = \text{im}(f, -)$ for some $A \xrightarrow{f} B \in$ mod-R. If $F \geq \mathcal{I}(A, -)$ then, by 11.2.5, $F_f(\mathcal{I}) = 0$, so $F_f \in \text{ann}(\mathcal{I})$. Hence, by definition of $\overline{\mathcal{I}}$, $F_f(\overline{\mathcal{I}}) = 0$ and so, again by 11.2.5, $F = \text{im}(f, -) \geq \overline{\mathcal{I}}(A, -)$, as required. □

Thus if $\mathcal{I}$ is an fp-idempotent ideal of mod-R, then, with the notation of 11.2.7, $r_{\text{ann}(\mathcal{I})} = \mathcal{I}$.

Lemma 15.3.2. ([515, 4.4]) *Let R be an artin k-algebra. Take a functor $F \in$* $(\text{mod-}R, \text{mod-}k)^{\text{fp}}$ *and let $\overrightarrow{F}$ denote the extension of F to a functor in* (Mod-R, Mod-k) *which commutes with direct limits (Section 10.2.8). Let $M \in$* Mod-R. *If $\overrightarrow{F}M \neq 0$, then $Fh \neq 0$ for some $h \in$* mod-R *which factors through M.*

Proof. Represent F as F_f for some $A \xrightarrow{f} B \in$ mod-R: so, see 10.2.42, the same representation serves to define $\overrightarrow{F}$ in the larger functor category. Since $\overrightarrow{F} M \neq 0$ there is $X \xrightarrow{g} M$ which does not factor through f.

Since R is an artin algebra there is a pure embedding $M \xrightarrow{i} \prod_\lambda N_\lambda$ with the N_λ of finite length (5.3.35). Let $\pi_\mu : \prod_\lambda N_\lambda \to N_\mu$ be the projection and set $g_\mu = \pi_\mu i g$.

If, for each λ, the morphism g_λ factors through f, say $g_\lambda = k_\lambda f$ for some $B \xrightarrow{k_\lambda} N_\lambda$, then the product, $B \xrightarrow{k} \prod_\lambda N_\lambda$, of these morphisms satisfies $ig = kf$. Since i is a pure embedding it follows by 2.1.7 that there is $B \xrightarrow{k'} M$ such that $g = k'f$, a contradiction.

So there is λ such that g_λ does not factor through f. Thus we have, taking $h = g_\lambda$, a morphism $h \in$ mod-R which factors through M and with $F_f(h) \neq 0$, as required. □

Suppose that $\mathcal{Y}$ is any subclass of Mod-R. Following [358, §5] consider and, extending the notation of Section 9.1.1, denote by $\mathcal{I}_\mathcal{Y}$ the class of morphisms in mod-R which factor through add($\mathcal{Y}$) (the closure of $\mathcal{Y}$ under direct summands of finite direct sums). Then $\mathcal{I}_\mathcal{Y}$ is an ideal of mod-R. For clearly $\mathcal{I}_\mathcal{Y}$ is closed under right and left composition. Also, if $A \xrightarrow{f} Y \xrightarrow{g} B$, $A \xrightarrow{f'} Y' \xrightarrow{g'} B$ with $Y, Y' \in \mathcal{Y}$, then $Y \oplus Y' \in$ add($\mathcal{Y}$) and $gf + g'f = A \xrightarrow{\binom{f}{f'}} Y \oplus Y' \xrightarrow{(g,g')} B$, so $\mathcal{I}_\mathcal{Y}$ is also closed under addition.

Theorem 15.3.3. ([515, 4.5]) *Let R be an artin algebra and let $\mathcal{Y} \subseteq$ Mod-R be any class of modules. Let $\mathcal{I}_\mathcal{Y}$ denote the class of morphisms in* mod-R *which factor through* add($\mathcal{Y}$). *Let $F \in ($mod-R, $\mathbf{Ab})^{\mathrm{fp}}$. Then $\overrightarrow{F}\mathcal{Y} = 0$ iff $F \in$ ann($\mathcal{I}_\mathcal{Y}$).*

Therefore $\mathcal{I}_\mathcal{Y} = \mathcal{I}_{\langle\mathcal{Y}\rangle}$, where $\langle\mathcal{Y}\rangle$ denotes the definable subcategory of Mod-R *generated by $\mathcal{Y}$.*

Proof. Let $F \in$ ann($\mathcal{I}_\mathcal{Y}$). By 15.3.2, $\overrightarrow{F}\mathcal{Y} = 0$, hence $\overrightarrow{F}\langle\mathcal{Y}\rangle = 0$ (see 10.2.44), so, clearly, $F \in$ ann($\mathcal{I}_{\langle\mathcal{Y}\rangle}$). Therefore ann($\mathcal{I}_\mathcal{Y}$) $\subseteq$ ann($\mathcal{I}_{\langle\mathcal{Y}\rangle}$). Since $\mathcal{Y} \subseteq \overline{\mathcal{Y}}$, the opposite inclusion is clear, so ann($\mathcal{I}_\mathcal{Y}$) = ann($\mathcal{I}_{\overline{\mathcal{Y}}}$). But then (both $\mathcal{I}_\mathcal{Y}$ and $\mathcal{I}_{\langle\mathcal{Y}\rangle}$ are ideals) by 15.3.1 we have $\mathcal{I}_\mathcal{Y} = \mathrm{ann}(\mathrm{ann}(\mathcal{I}_\mathcal{Y})) = \mathrm{ann}(\mathrm{ann}(\mathcal{I}_{\langle\mathcal{Y}\rangle})) = \mathcal{I}_{\langle\mathcal{Y}\rangle}$. □

Over any ring, each Serre subcategory of (mod-R, $\mathbf{Ab}$)^{fp} is the annihilator of a suitable set of modules (12.4.1). Over an artin algebra, the above result allows such a Serre subcategory to be represented as the annihilator of a set, not of arbitrary *modules*, but of *morphisms between* finitely presented modules.

Corollary 15.3.4. ([358, 5.10]) *Let R be an artin k-algebra. Then annihilation induces a bijection between Serre subcategories of* (mod-R, mod-k)^{fp} *and fp-idempotent ideals of* mod-R.

Proposition 15.3.5. ([358, 5.2]) *Let R be an artin k-algebra and let $\mathcal{S}$ be a Serre subcategory of* $(\text{mod-}R, \text{Mod-}k)^{\text{fp}}$. *Then* $\text{ann}(\mathcal{S}) = \mathcal{I}_{\mathcal{X}_\mathcal{S}}$, *where $\mathcal{X}_\mathcal{S}$ denotes the definable subcategory of* Mod-R *corresponding, in the sense of 12.4.1, to $\mathcal{S}$.*

Proof. If $h \in \mathcal{I}_{\mathcal{X}_\mathcal{S}}$, then $h = h''1_M h'$ for some $M \in \mathcal{X}_\mathcal{S}$ $(= \text{add}\mathcal{X}_\mathcal{S})$, so, if $F \in \mathcal{S}$, then, since $\overrightarrow{F}1_M = 1_{\overrightarrow{F}M} = 0$, we have $Fh = \overrightarrow{F}h = \overrightarrow{F}h''\overrightarrow{F}1_M\overrightarrow{F}'_h = 0$, so $h \in \text{ann}(\mathcal{S})$. That is, $\mathcal{I}_{\mathcal{X}_\mathcal{S}} \subseteq \text{ann}(\mathcal{S})$.

Therefore $\text{ann}(\mathcal{I}_{\mathcal{X}_\mathcal{S}}) \supseteq \text{ann}(\text{ann}(\mathcal{S})) = \mathcal{S}$ by 15.3.2. But now, if $F \in \text{ann}(\mathcal{I}_{\mathcal{X}_\mathcal{S}})$, then, by 2.1.7, $\overrightarrow{F}M = 0$ for every $M \in \mathcal{X}_\mathcal{S}$. Therefore, by 12.4.1, $F \in \mathcal{S}$. So $\text{ann}(\mathcal{I}_{\mathcal{X}_\mathcal{S}}) \subseteq \mathcal{S}$ and these must, therefore, be equal.

So $\mathcal{I}_{\mathcal{X}_\mathcal{S}} = \text{ann}(\text{ann}(\mathcal{I}_{\mathcal{X}_\mathcal{S}}))$ (by 15.3.1), which equals $\text{ann}(\mathcal{S})$, as required. □

Corollary 15.3.6. *Let R be an artin algebra. Let $\mathcal{X}$ be a definable subcategory of* Mod-R *and let* $\text{ann}(\mathcal{S}_\mathcal{X})$ *$(= \mathcal{I}_\mathcal{X}$ by 15.3.5) be the associated ideal of* mod-R. *Then, given any dense subset Y of $\mathcal{X} \cap \text{Zg}_R$, every morphism in this ideal factors through a finite direct sum of copies of members of Y.*

For, the definable closure of the set Y of modules is $\mathcal{X}$, so 15.3.3 applies.

Corollary 15.3.7. ([515, 4.7]) *Let R be an artin algebra. Let $\mathcal{I}$ be an fp-idempotent ideal of* mod-R *with associated Serre subcategory* $\mathcal{S} = \text{ann}(\mathcal{I})$ *and let Y be a dense subset of the closed subset of* Zg_R *associated to $\mathcal{S}$. Then each morphism $h \in \mathcal{I}$ factors through a finite direct sum of copies of members of Y.*

This also is immediate from 15.3.3 and 15.3.4.

This may be applied to the infinite radical, rad^ω (Section 9.1.2), of the category mod-R, where R is an artin algebra; as seen in 9.2.3(a) this is an fp-idempotent ideal of mod-R. Let $\mathcal{S}_0$ be the Serre subcategory of finite length functors. By 5.3.37 and 5.3.57 localising at $\mathcal{S}_0$ corresponds to removing the isolated = finite-length points of the Ziegler spectrum and so the ideal of mod-R which corresponds to $\mathcal{S}_0$ in the sense of 15.3.4 is that consisting of morphisms which factor through a module in the definable subcategory generated by the infinite-length indecomposable pure-injectives. Therefore, we obtain, as a special case of the above, the following conclusion, originally obtained in [502] by a very different proof which involved more analysis of lattices of pp-definable subgroups.

Theorem 15.3.8. ([515, 4.9]) *Let R be an artin algebra and let* $f \in \text{rad}^\omega(\text{mod-}R)$. *Then there is a factorisation of f through a finite direct sum of indecomposable, infinite length, pure-injective modules. Therefore, by 9.1.6,* rad^ω *consists exactly of the morphisms which have such a factorisation.*

15.4 mdim $\neq 1$ for artin algebras

Let R be an artin algebra over $C = C(R)$ and let $\mathcal{S}_0$ be the Serre subcategory of finite-length objects in $(\text{mod-}R, \mathbf{Ab})^{\text{fp}}$.

Proposition 15.4.1. ([279, 3.1]) *Let R be an artin C-algebra, let $\mathcal{C} \subseteq \text{mod-}R$ be covariantly finite (Section 3.4.4) and let $F \in (\mathcal{C}, \mathbf{Ab})^{\text{fp}}$. Then $\text{End}(F)$ is an artin C-algebra.*

Proof. By 10.1.13 there is an epimorphism $(A, -) \to F$ for some $A \in \mathcal{C}$. Evaluating at A gives an epimorphism $\text{End}(A) \to FA$. Since C is central in R this is an epimorphism of C-modules, so FA is a finite-length C-module. Also, applying $(-, F)$ to the first sequence gives an embedding, $\text{End}(F) \to ((A, -), F) \simeq FA$, of C-modules, so the result follows. □

Say that a C-algebra R' is a **locally artin C-algebra** if every finitely generated C-subalgebra of R' is artinian; this will be so if R' is a direct limit of artin C-algebras.

Theorem 15.4.2. ([279, 3.2]) *Let R be an artin C-algebra and let $\mathcal{S}_0$ be the Serre subcategory of finite-length functors on* mod-R. *Let $F \in (\text{mod-}R, \mathbf{Ab})^{\text{fp}}$. Then $\text{End}(F_{\mathcal{S}_0})$, where $F_{\mathcal{S}_0}$ is the localisation of F at $\mathcal{S}_0$, is a locally artin C-algebra.*

Proof. The key point of the proof, which is just indicated here, is that $\mathcal{S}_0$ is the direct limit of its Serre subcategories which are generated by finite sets of simple objects and, therefore, hom groups in the localisation are correspondingly direct limits of hom groups at these "finite" localisations. These finitely generated Serre subcategories $\mathcal{S}'$ are, by a result of Auslander and Smalø (see [26, 2.9]), all of the form $\text{ann}(\mathcal{C}) = \{F \in (\text{mod-}R, \mathbf{Ab})^{\text{fp}} : FC = 0 \ \forall C \in \mathcal{C}\}$ for some covariantly finite subcatgories $\mathcal{C} = \mathcal{C}_{\mathcal{S}'}$ of mod-R. Then a result of Krause applies to give that the endomorphism ring of the localisation of F at $\mathcal{S}'$ is isomorphic to the endomorphism ring of its restriction to $\mathcal{C}_{\mathcal{S}'}$ which, by the above result, is an artin C-algebra. Thus the endomorphism ring of $F_{\mathcal{S}_0}$ is a direct limit of artin C-algebras, as required. □

Corollary 15.4.3. ([279, 3.4]) *If S is a simple object of the localised category $(\text{mod-}R, \mathbf{Ab})^{\text{fp}}_{\mathcal{S}_0}$, then the division ring $\text{End}(S)/\text{JEnd}(S)$ is algebraic over the field $k_S = C/C \cap \text{ann}_C(S)$.*

Proposition 15.4.4. ([279, 3.5]) *If R is any ring and $\text{KG}(R) = 1$, then the simple objects of $(\text{mod-}R, \mathbf{Ab})^{\text{fp}}_{\mathcal{S}_0}$ correspond to the generic modules.*

Proof. By 5.3.5 these simple objects correspond to the isolated points of the first Cantor–Bendixson derivative, Zg_R^1 of the Ziegler spectrum of R. By 5.3.60 and assumption these are all the points of Zg_R^1, there are only finitely many such points and they are all of finite endolength, as required. □

The following result was proved by Krause, [357, 11.4], for the case where R is a finite-dimensional algebra over an algebraically closed field (using his results on tubes, see Section 15.1.3) and by Herzog, [279, 3.6], in general.

Theorem 15.4.5. *If R is an artin C-algebra, then* $\mathrm{KG}(R) \neq 1$.

Proof. The above results imply that for every generic R-module N the division ring $\mathrm{End}(N)/\mathrm{JEnd}(N)$ is algebraic over the corresponding factor field of C (namely, k_S, where S is the simple functor corresponding to N). This contradicts (the proof of) [141, 9.6], see 4.5.38. If the base field is algebraically closed, then 4.5.41 is enough to finish the proof. □

15.5 Modules over group rings

We gather together some results specific to modules over group rings.

Throughout this section G is a finite group and k is a field, the interesting case being when the characteristic of k divides the order of G. If M is any kG-module, then it has a projective cover $P_M \to M$ and an injective hull $M \to E(M)$. The kernel of the first map is the **first syzygy** of M and is denoted ΩM. The cokernel of the second map is denoted $\Omega^{-1}M$ and is the **first cosyzygy** of M. Thus we have exact sequences $0 \to \Omega M \to P \to M \to M \to 0$ and $0 \to M \to E(M) \to \Omega^{-1}M \to 0$: the beginning of a projective, respectively injective, resolution of M. Note that, if M is finitely generated, then so are all the other modules appearing.

Suppose that $0 \to A \to B \to C \to 0$ is an exact sequence of kG-modules. We have an exact sequence $0 \to \Omega A \to P_1 \to A \to 0$, where P_1 is a projective cover of A and similarly an exact sequence $0 \to \Omega C \to P_3 \to C \to 0$. Lifting $P_3 \to C$ over $B \to C$ and combining these gives a projective resolution $0 \to K \to P_1 \oplus P_3 \to B \to 0$ of which the minimal projective resolution of B must be a direct summand. Thus the initial exact sequence $0 \to A \to B \to C \to 0$ "translates under Ω" to an exact sequence of the form $0 \to \Omega A \to \Omega B \oplus P \to \Omega C \to 0$ for some projective module P. Working in the derived category, see Section 17.1, tends to give smoother proofs since this allows one to ignore pesky projectives such as P.

Theorem 15.5.1. ([73, 4.2.1]) *An exact sequence $0 \to A \to B \to C \to 0$ of kG-modules is pure-exact iff the translated sequence $0 \to \Omega A \to \Omega B \oplus P \to \Omega C \to 0$, where P is projective, is pure-exact.*

A kG-module M is pure-projective iff ΩM is pure-projective iff $\Omega^{-1}M$ is pure-projective.

A kG-module M is pure-injective iff ΩM is pure-injective iff $\Omega^{-1}M$ is pure-injective.

Proof. The last statement, about pure-injectivity, is direct from 17.3.10 and since pure-injectivity in a triangulated category is preserved by shift.

For the second statement, suppose that M is pure-projective, so, by 2.1.26 and the comment after that, because R is a Krull–Schmidt ring M is actually a direct sum of finitely presented modules: $M = \bigoplus_i M_i$. For each i there is an exact sequence $0 \to L_i \to P_i \to M_i \to 0$ with P_i a finitely generated projective, hence with L_i finitely presented. Put these together to obtain an exact sequence $0 \to \bigoplus_i L_i \to \bigoplus_i P_i \to M \to 0$. Since ΩM is a direct summand of $\bigoplus_i L_i$ it is pure-projective, as required. The argument for $\Omega^{-1}M$ is similar, using that the injective hull $E(M_i)$, and hence $\Omega^{-1}M_i$, is finitely presented.

The first statement also follows from the results and methods described in Section 17.3, for example, from 17.3.13. □

Also see [373, 2.6] for a somewhat related result.

We quote some further results from [73].

Proposition 15.5.2. ([73, 4.3.1]) *Let H be a subgroup of the finite group G. Then both induction and restriction between kH-modules and kG-modules take pure-projective modules to pure-projective modules, pure-injective modules to pure-injective modules and pure-exact sequences to pure-exact sequences.*

For modules of finite endolength one also has the bounds $\mathrm{el}(M|_H) \le \mathrm{el}(M)$ and $\mathrm{el}(N|^G) \le [G : H]\,\mathrm{el}(N)$, where M, N are respectively a kG- and a kH-module; it follows that if k has characteristic p, then M_{kG} is of finite endolength iff its restriction to a Sylow p-subgroup is of finite endolength ([75, 2.2, 2.3]), since then M is a direct summand of the induction of its restriction.

If M and N are right kG-modules, then their tensor product over k, $M \otimes_k N$, has the structure of a right kG-module, defined by $m \otimes n \cdot g = mg \otimes ng$ for $m \in M, n \in N, g \in G$. Benson and Gnacadja noted, [73, 4.3.2], that, if M and N are pure-projective kG-modules, then $M \otimes_k N$ is a pure-projective kG-module.

They also showed, [73, 4.3.4], that if k is an algebraically closed field of characteristic $p > 0$ and G is a finite group and if $0 \to L \to M \to N \to 0$ is a pure-exact sequence of kG-modules, then $\mathcal{V}_G(M) = \mathcal{V}_G(L) \cup \mathcal{V}_G(N)$. Here the $\mathcal{V}_G(M)$ are the varieties mentioned at the end of Section 17.4.

In general, the process of induction from a subgroup $H \le G$ of finite index coprime to the characteristic of k is a mild process: for any H-module M the induced module M^G is interpretable, in the model-theoretic sense, in M, that is,

"M^G lives in $M^{\mathrm{eq}+}$", see [425, §1] for instance. This is also formulated in [577, 4.8], where it is shown that the lattice $\mathrm{pp}(M^G)$ of pp-definable subgroups of M^G embeds naturally in the lattice $\mathrm{pp}^n(M)$, where n is the index of H in G and, further, that, if G is finite and $\mathrm{char}(k)$ does not divide the order of G, then there is a natural embedding of lattices $\mathrm{pp}_{kG} \to \mathrm{pp}^n_{kH}$. Among many consequences there is the following one, concerning Krull–Gabriel dimension.

Corollary 15.5.3. ([577, 4.9]) *If H is a subgroup of index n in the finite group G and k is a field whose characteristic does not divide n, then* $\mathrm{KG}(kH) = \mathrm{KG}(kG)$.

16

Finitely accessible and presentable additive categories

This book began with categories of modules but almost everything works in the more general context of locally finitely presented abelian categories. This broader applicability is illustrated by categories of sheaves and by categories subgenerated by modules. In both contexts conditions are given under which such a category is locally finitely presented. Finitely accessible categories give a yet more general context. Dual and conjugate categories are defined.

16.1 Finitely accessible additive categories

Section 16.1.1 gives the representation of finitely accessible additive categories as categories of flat functors as well as refinements of this. Purity in finitely accessible additive categories is described in Section 16.1.2. Two different notions of "dual" of a finitely accessible additive category are given in Section 16.1.3 and their equivalence under coherence assumptions noted.

16.1.1 Representation of finitely accessible additive categories

This section, which mostly summarises results rather than giving their proofs, is based on [516]. More details can be found there and in the original references.

The context, that of finitely accessible categories, is set up. Flat objects are described (16.1.2) and finitely accessible additive categories characterised as the flat functors on their finitely presented objects (16.1.3). Conditions on the category of finitely presented objects which are equivalent to "finiteness" conditions on the additive category are stated (16.1.4).

A finitely accessible additive category has products iff the category of finitely presented objects has pseudocokernels (16.1.6) and then the former is a definable subcategory of a functor category (16.1.7).

A finitely accessible additive category has cokernels iff the category of finitely presented objects has cokernels iff the category is locally finitely presented (16.1.8).

Any finitely accessible abelian category is Grothendieck (16.1.11).

Various equivalents to coherence of a functor category are given (16.1.12, 16.1.14).

A category $\mathcal{C}$ is said to be **finitely accessible** if it has direct limits, if every object is a direct limit, equivalently filtered colimit, of finitely presented objects and if the full subcategory, $\mathcal{C}^{\mathrm{fp}}$, of finitely presented objects is skeletally small.[1] A finitely accessible category is **locally finitely presented** if it is complete, equivalently [3, 2.47] cocomplete. For abelian categories there is no conflict with other definitions of locally finitely presented which are often used in the additive context: accessible additive categories have arbitrary direct sums (see, for example, [144, 1.1]), so, since an abelian category has cokernels, a finitely accessible abelian category is locally finitely presented. In fact, rather less is needed (16.1.9).

Examples 16.1.1. The category of torsion abelian groups is locally finitely presented; the category of divisible abelian groups is accessible (drop the requirement that the generators be finitely presented) but not finitely accessible (it contains no finitely presented objects apart from 0, see 18.1.1). Both are examples of the kind of category $\sigma[M]$ discussed in Section 16.2, as is the category of comodules over a coalgebra over a field (a language is set up and pp conditions described for these in [147]). The category of flat right modules over a ring which is not left coherent is, by 16.1.3, finitely accessible but, by 16.1.5 and 2.3.21, not locally finitely presented.

A significant part of the broad theory here is an additive analogue of that which is known in, and what I will refer to as, the "non-additive" (or "general") context. By that I mean functor categories of the form $(\mathcal{A}, \mathbf{Set})$, that is, "presheaves on $\mathcal{A}^{\mathrm{op}}$" see, for example, [15, Expose I]. The additive results are usually not literally specialisations of the non-additive ones, because of the replacement of **Set** by **Ab**, although the proofs often are very similar. There does not (yet) seem to be a general, well-developed theory which specialises to both the additive and non-additive cases, though there are partial theories along these lines (see, for example, [338], [85]) but also seeming limitations, for example, see [585] where the correspondence between finitely presented functors and appropriate imaginaries which holds in both the additive case (10.2.30) and the **Set**-based case, see [584] is shown to fail in the affine case.

[1] In the additive context such categories have been called "locally finitely presented" by various authors but this clashes with the standard terminology.

In the general context an object is said to be **flat** if it is a direct limit of representable functors. In the additive context the definition of flatness is usually given in terms of a tensor functor, but these are equivalent; in the additive context this is due to Lazard and Govorov for modules, and Oberst and Rohrl for the general case.[2] Here we work entirely within the additive context, so, even where not said explicitly, **all categories in this section are preadditive**.

Let $\mathcal{A}$ be a small preadditive category. Within Mod-$\mathcal{A} = (\mathcal{A}^{\mathrm{op}}, \mathbf{Ab})$ one has the full subcategory, Flat-$\mathcal{A}$ = Flat(Mod-$\mathcal{A}$), of flat functors, where $M \in$ Mod-$\mathcal{A}$ is **flat** if the tensor functor (Section 12.1.1) $M \otimes_{\mathcal{A}} - : \mathcal{A}$-Mod $\longrightarrow \mathbf{Ab}$ is exact, see, for example, 12.1.1. Recall (10.1.14) that a finitely generated $\mathcal{A}$-module is projective iff it is a direct summand of a finite direct sum of representable functors.

Theorem 16.1.2. ([391, 1.1, 1.2], [246], [467, 3.2], [661, Thm 3], also see [144, 1.3]) *Let $\mathcal{A}$ be a skeletally small preadditive category and let $M \in$ Mod-$\mathcal{A}$. Then the following are equivalent:*

(i) M is flat;
(ii) M is a direct limit (equivalently a filtered colimit, see Appendix E, p. 715) of finitely generated projective $\mathcal{A}$-modules;
(iii) each morphism from a finitely presented $\mathcal{A}$-module to M factors through a finitely generated projective module, equivalently, through a direct sum of representable functors $(-, A)$, $A \in \mathcal{A}$.

Denote by flat-$\mathcal{A}$ = flat(Mod-$\mathcal{A}$) the category of flat functors which are finitely presented in Mod-$\mathcal{A}$.

The following representation theorem is due to Crawley-Boevey and is a generalisation of results of Gabriel [202, II.4, Thm 1], Roos [614, 2.2] and Breitsprecher [89, 2.7]. There are related results in the non-additive case, see, for example, [3, 2.36].

Theorem 16.1.3. ([144, 1.4])

(a) Let $\mathcal{A}$ be a skeletally small preadditive category. Then Flat-$\mathcal{A}$ *is finitely accessible, with* (Flat-$\mathcal{A}$)^{fp} *consisting of the direct summands of finite direct sums $\bigoplus_{i=1}^{n}(-, A_i)$ of representable functors. The Yoneda embedding $\mathcal{A} \longrightarrow$* Flat-$\mathcal{A}$*, $A \mapsto (-, A)$ is a full embedding of $\mathcal{A}$ as a generating subcategory of finitely presented objects of* Flat-$\mathcal{A}$ *and is an equivalence of $\mathcal{A}$ with* flat-$\mathcal{A}$ *iff $\mathcal{A}$ is additive with split idempotents.*

[2] In fact, in the general context there is also a definition of tensor product and a corresponding result, see [419, p. 381 ff.].

(b) *If $\mathcal{C}$ is an additive finitely accessible category, then the full subcategory, $\mathcal{C}^{\mathrm{fp}}$, of finitely presented objects is skeletally small, additive, has split idempotents and $\mathcal{C} \longrightarrow$ Flat-$\mathcal{C}^{\mathrm{fp}}$, $C \mapsto (-, C) \restriction \mathcal{C}^{\mathrm{fp}}$ is an equivalence.*

Corollary 16.1.4. and Theorems ([144, 1.4]) *There is a bijection between skeletally small additive categories $\mathcal{A}$ with split idempotents and finitely accessible categories $\mathcal{C}$. Under this correspondence:*

(Gabriel [202, II.4, Thm 1]) *$\mathcal{C}$ is locally noetherian iff $\mathcal{A} = \mathcal{C}^{\mathrm{fp}}$ is abelian and every object of $\mathcal{C}^{\mathrm{fp}}$ is noetherian (that is, has the acc on subobjects);*
(Roos [614, 2.2]) *$\mathcal{C}$ is locally coherent iff $\mathcal{A} = \mathcal{C}^{\mathrm{fp}}$ is abelian;*
(essentially Breitsprecher [89, 2.7]) *$\mathcal{C}$ is abelian iff $\mathcal{A} = \mathcal{C}^{\mathrm{fp}}$ has cokernels, every epimorphism of $\mathcal{C}^{\mathrm{fp}}$ is a cokernel and, for every right exact sequence $A \xrightarrow{f} B \xrightarrow{g} C \to 0$ in $\mathcal{C}^{\mathrm{fp}}$, for every morphism $h : B' \to B$ in $\mathcal{C}^{\mathrm{fp}}$ with $gh = 0$, there are $A' \in \mathcal{C}^{\mathrm{fp}}$ and morphisms $k : A' \to A$ and $l : A' \to B'$ with l epi and $hl = fk$.*

$$\begin{array}{ccccccc} A & \xrightarrow{f} & B & \xrightarrow{g} & C & \longrightarrow & 0 \\ {\scriptstyle k}\uparrow & & \uparrow{\scriptstyle h} & & & & \\ A' & \dashrightarrow_{l} & B' & & & & \end{array}$$

Sometimes an additive category is defined to be locally coherent if it is finitely accessible and if every finitely presented object is coherent in the sense of Section 2.3.3. It follows from 16.1.4 that such a category is abelian and, from 16.1.5, that it has products. Therefore, by 16.1.11, such a category is actually Grothendieck and it is equivalent to say that a category is locally coherent if it is Grothendieck and has a generating set (in the sense of Appendix E, p. 707) of coherent objects.

There has been recent work of Guil Asensio and Herzog (see towards the end of Section 4.6) on the context where existence of products is not assumed but, although much can be developed in that context, in this book it is finitely accessible categories $\mathcal{C}$ with products, and definable subcategories of these, that are considered. The required condition on $\mathcal{A}$, resp. on $\mathcal{C}^{\mathrm{fp}}$, for existence of products is below (16.1.5, 16.1.6).

By $\mathcal{A}^+$ we denote the category generated by $\mathcal{A}$ under finite direct sums and by $\mathcal{A}^{++}$ the category generated by "closing under direct summands" – the smallest additive category containing $\mathcal{A}$, which has finite direct sums and which is such that all idempotents split (see any of [195, pp. 60, 61], [467, p. 92], [486, p. 22], [516]).

Theorem 16.1.5. ([467, §4], [144, 2.1], also cf. [116, 2.1] for rings) *Let $\mathcal{A}$ be a skeletally small preadditive category. Then the following are equivalent:*

(i) $\mathcal{A}^+$ *has pseudocokernels;*
(ii) $\mathcal{A}^{++}$ *has pseudocokernels;*
(iii) Flat-$\mathcal{A}$ *has products;*
(iv) Flat-$\mathcal{A}$ *is closed in* Mod-$\mathcal{A}$ *under products.*

Corollary 16.1.6. *An additive category $\mathcal{C}$ which is finitely accessible has products iff $\mathcal{C}^{\text{fp}}$ has pseudocokernels.*

Corollary 16.1.7. *Let $\mathcal{C}$ be a finitely accessible category with products. Then $\mathcal{C}$ is a definable subcategory of* Mod-$\mathcal{C}^{\text{fp}}$.

That follows immediately from 3.4.7, 16.1.3 and the fact (see 2.3.11 or, more generally, [516]) that a pure subobject of a flat object is flat.

With a somewhat stronger condition on $\mathcal{A}$ one obtains local finite presentation of $\mathcal{C}$.

Theorem 16.1.8. ([144, 2.2], [359, 6.3] *for (v)⇒(i)) For a skeletally small preadditive category $\mathcal{A}$ the following are equivalent:*

(i) $\mathcal{A}^+$ *has cokernels;*
(ii) $\mathcal{A}^{++}$ *has cokernels;*
(iii) Flat-$\mathcal{A}$ *has cokernels, hence is complete and cocomplete.*

Corollary 16.1.9. ([144, 2.2], [359, 5.7]) *If $\mathcal{C}$ is finitely accessible, then $\mathcal{C}$ has cokernels iff $\mathcal{C}^{\text{fp}}$ has cokernels iff $\mathcal{C}$ is locally finitely presented.*

In that case one has equivalent formulations in terms of left exact functors. If $\mathcal{B}$ is an abelian category, denote by Lex$(\mathcal{B}, \mathbf{Ab})$ the full subcategory of $(\mathcal{B}, \mathbf{Ab})$ whose objects are the left exact functors.

Proposition 16.1.10. ([191, 2.4, 2.9], [467, 3.4]) *If $\mathcal{A}$ is a skeletally small preadditive category and $\mathcal{A}^{+(+)}$ has cokernels, then* Flat-$\mathcal{A} \simeq \text{Lex}\big((\mathcal{A}^+)^{\text{op}}, \mathbf{Ab}\big) \simeq \text{Lex}\big((\mathcal{A}^{++})^{\text{op}}, \mathbf{Ab}\big)$.

In many cases of interest $\mathcal{C}$ is actually abelian: it was shown by Crawley-Boevey that in this case the Grothendieck condition is automatic.

Theorem 16.1.11. ([144, 2.4]) *For an additive category $\mathcal{C}$ the following are equivalent:*

(i) $\mathcal{C}$ is finitely accessible and abelian;
(ii) $\mathcal{C}$ is locally finitely presented and abelian;
(iii) $\mathcal{C}$ is locally finitely presented Grothendieck.

If $\mathcal{C}$ is a finitely accesssible category with products, then by a **definable subcategory** of $\mathcal{C}$ we mean, cf. 3.4.7, a full subcategory closed under products, direct limits and pure subobjects. By 16.1.7 this does not actually extend the class of definable subcategories and, also by that result, all the usual characterisations of, and results about, definable subcategories hold. That includes those which involve the notion of pp condition, since a pp condition for $\mathcal{C}$ may be taken to mean simply a pp condition for Mod-$\mathcal{C}^{\mathrm{fp}}$ restricted to $\mathcal{C}$ (there is, however, an equivalent "internal" definition of pp condition for finitely accessible categories, see Section 16.1.2).

Theorem 16.1.12. *Let $\mathcal{A}$ be skeletally small preadditive and let $\mathcal{C}$ be finitely accessible. Then the following are equivalent.*

(a) (in terms of $\mathcal{A}$)

- *(i) $\mathcal{A}^{+(+)}$ has pseudocokernels;*
- *(ii) ($\mathcal{C}$=)* Flat-$\mathcal{A}$ *has products;*
- *(iii)* Flat-$\mathcal{A}$ *is closed under products in* Mod-$\mathcal{A}$*;*
- *(iv)* $\mathcal{A}$-Mod *is locally coherent;*
- *(v)* Flat-$\mathcal{A}$ *is a definable subcategory of* Mod-$\mathcal{A}$*;*
- *(vi)* $\mathcal{A}$-Abs *is a definable subcategory of* $\mathcal{A}$-Mod.

(b) (in terms of $\mathcal{C}$)

- *(i) ($\mathcal{A}$=) $\mathcal{C}^{\mathrm{fp}}$ has pseudocokernels;*
- *(ii) $\mathcal{C}$ has products;*
- *(iii) $\mathcal{C}$ is closed under products in* Mod-$\mathcal{C}^{\mathrm{fp}}$*;*
- *(iv)* $\mathcal{C}^{\mathrm{fp}}$-Mod *is locally coherent;*
- *(v) $\mathcal{C}$ is a definable subcategory of* Mod-$\mathcal{C}^{\mathrm{fp}}$*;*
- *(vi)* $\mathcal{C}^{\mathrm{fp}}$-Abs *is a definable subcategory of* $\mathcal{C}^{\mathrm{fp}}$-Mod.

Equivalence of the first three conditions is 16.1.8, the equivalence of (iii) and (v) is from the fact that a pure subobject of a flat object is flat and since direct limits of flat objects are flat (Section 2.3.2). The equivalence of (iii) and (iv) is 2.3.21 (for $\mathcal{A}$ a ring, though the proof generalises, or see [467, 4.1] for the general case). The equivalence of the last three conditions is 3.4.24 (again extended to rings with many objects). For a generalisation to the non-additive context, see [62].

The next result was proved directly at 10.2.3. It is also immediate from 16.1.12 since each of mod-$\mathcal{A}$ and $\mathcal{A}$-mod has cokernels.

Corollary 16.1.13. *For any skeletally small preadditive category $\mathcal{A}$ the categories* (mod-$\mathcal{A}$, **Ab**) *and* ($\mathcal{A}$-mod, **Ab**) *are locally coherent.*

Theorem 16.1.14. *Let $\mathcal{A}$ be skeletally small preadditive. The following are equivalent:*

(i) $\left((\text{mod-}\mathcal{A})^{\text{op}}, \mathbf{Ab}\right)^{\text{fp}}$ *is abelian;*
(ii) $\left((\text{mod-}\mathcal{A})^{\text{op}}, \mathbf{Ab}\right)$ *is locally coherent;*
(iii) mod-$\mathcal{A}$ *has pseudokernels;*
(iv) mod-$\mathcal{A}$ *has kernels;*
(v) mod-$\mathcal{A}$ *is abelian;*
(vi) Mod-$\mathcal{A}$ *is locally coherent;*
(vii) $\mathcal{A}^{+(+)}$ *has pseudokernels.*

Proof. The only part which is not immediate from what is above is (iii)⇒(iv). So suppose that $f : L \to M$ is a morphism in mod-$\mathcal{A}$, let $(i : K \to L) = \ker(f)$ in Mod-$\mathcal{A}$ and let $g : K_0 \to L$ be a pseudokernel of f in mod-$\mathcal{A}$. Since $K = \ker(f)$ there is $h : K_0 \to K$ such that $g = ih$.

If h were not an epimorphism there would be (since Mod-$\mathcal{A}$ is locally finitely presented) $K_1 \in$ mod-$\mathcal{A}$ and $k : K_1 \to K$ such that $\text{im}(k) \nleq \text{im}(h)$. Now $f(ik) = 0$ so there is $k' : K_1 \to K_0$ such that $ik = gk'$, which equals ihk'. Since i is monic $k = hk'$ – contrary to choice of K_1 and k.

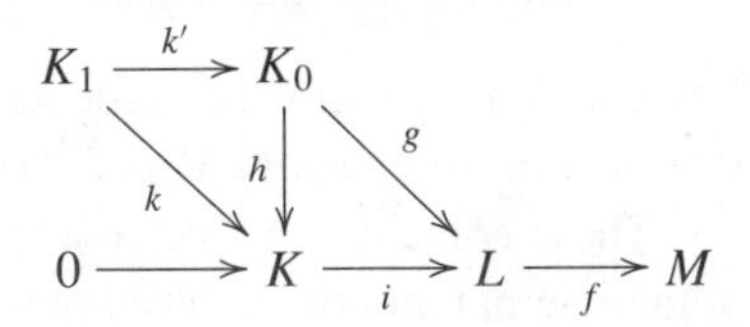

Therefore h is an epimorphism and K is finitely generated. That is, the kernel in Mod-$\mathcal{A}$ of each map in mod-$\mathcal{A}$ is finitely generated, which is enough, for then $\ker(h) = \ker(ih)$ is finitely generated so K is finitely presented. □

16.1.2 Purity in finitely accessible categories

Equivalents to purity in this generality are given (16.1.15), including one in terms of ultrapowers (16.1.16).

If $\mathcal{C}$ is a finitely accessible additive category, then an exact sequence $0 \to L \to M \to N \to 0$ is **pure(-exact)** if for every finitely presented $A \in \mathcal{C}$ the induced sequence $0 \to (A, L) \to (A, M) \to (A, N) \to 0$ of abelian groups is exact; in that case, say that $L \to M$ is a **pure monomorphism** and $M \to N$ is a **pure epimorphism**. The following equivalents, which have been proved already for module categories in Section 2.1.1 (2.1.4, 2.1.7, 2.1.16), can be proved just as in that case. Those proofs which mention elements can be reformulated purely categorically; alternatively one may interpret elements to be morphisms from objects of $\mathcal{C}^{\text{fp}}$, cf. Sections 10.2.4 and B.1. For purity in additive categories developed in this generality see, for example, [144], [232].

Theorem 16.1.15. ([391, 2.2, 2.3] for modules) *If* $0 \to L \xrightarrow{f} M \xrightarrow{g} N \to 0$ *is an exact sequence in the finitely accessible additive category* $\mathcal{C}$, *then the following are equivalent:*

(i) the sequence is pure-exact;
(ii) for every $A \in \mathcal{C}^{\text{fp}}$ *and morphism* $h : A \to N$ *there is* $k : A \to M$ *such that* $gk = h$;
(iii) for every morphism $h : A \to B$ *in* $\mathcal{C}^{\text{fp}}$ *and commutative diagram as shown*

$$\begin{array}{ccc} A & \xrightarrow{h} & B \\ {\scriptstyle k}\downarrow & & \downarrow{\scriptstyle k'} \\ L & \xrightarrow[f]{} & M \end{array}$$

there is $l : B \to L$ *with* $lh = k$;
(iv) as (iii) but requiring of $h \in \mathcal{C}$ *only that* $\text{coker}(h)$ *be in* $\mathcal{C}^{\text{fp}}$.
(v) the sequence is a direct limit of split exact sequences in $\mathcal{C}$.

This is an "internal" notion of purity. If $\mathcal{C}$ has products, then, by 16.1.7, $\mathcal{C}$ is a definable subcategory of the functor category Mod-$\mathcal{C}^{\text{fp}}$ so it inherits a notion of purity from that category. These coincide. In fact, in any representation of $\mathcal{C}$ as a definable subcategory, the internal and the induced notions of purity coincide, even if the objects of $\mathcal{C}^{\text{fp}}$ are not finitely presented objects of the larger category. That follows immediately from the next characterisation of purity. For module categories this was already stated as 4.2.18 and the proof of this generalisation is by the same result from model theory.

Theorem 16.1.16. *Suppose that* $0 \to L \xrightarrow{f} M \xrightarrow{g} N \to 0$ *is an exact sequence in the finitely accessible category* $\mathcal{C}$ *which has products. Then the following are equivalent:*

(i) the sequence is pure-exact;
(ii) some ultrapower of the sequence is split, that is, there is an index set, I, *and an ultrafilter,* $\mathcal{F}$, *on* I, *both of which may be chosen independently of the sequence and with the cardinality of* I *being bounded in terms of the number of arrows in a skeletal verison of* $\mathcal{C}^{\text{fp}}$, *such that*

$$0 \to L^I/\mathcal{F} \xrightarrow{f^I/\mathcal{F}} M^I/\mathcal{F} \xrightarrow{g^I/\mathcal{F}} N^I/\mathcal{F} \to 0$$

is split.

Corollary 16.1.17. *Suppose that* $\mathcal{C}$ *is a finitely accessible category with products and suppose also that* $\mathcal{C}$ *is a definable subcategory of the locally finitely presented* *abelian category $\mathcal{B}$. Let $0 \to L \to M \to N \to 0$ be a sequence of morphisms in $\mathcal{C}$. Then this is a pure-exact sequence in $\mathcal{C}$ iff it is a pure-exact sequence in $\mathcal{B}$.*

One may then define pure-injectivity, algebraic compactness (using the internal notion of pp condition) and pure-injective hulls as in Chapter 4. It is straightforward to check that the various equivalences and characterisations go through. Because we may simply treat finitely accessible categories as definable subcategories of functor categories, with the inherited notion of purity, we do not give the details here but these, including an embedding which turns pure-injectives into injectives, can be found in [144, §3].

As pointed out at [144, p. 1658], the embedding used in that paper is not the Gruson–Jensen embedding seen in Section 12.1.1. Certainly one has a Gruson–Jensen type embedding: because $\mathcal{C}$ is a definable subcategory of a functor category, there is the corresponding localisation of the functor category (12.4.1), and a full embedding of $\mathcal{C}$ into a locally coherent abelian category which turns pure-exact sequences into exact sequences and under which the pure-injectives of $\mathcal{C}$ correspond to the injectives of the locally coherent category (12.3.3).

Crawley-Boevey, [144, 3.3], used a different embedding. He considered the category $(\mathcal{C}^{\mathrm{fp}}, \mathbf{Ab})$-Flat, which equals $\mathrm{Lex}\big((\mathcal{C}^{\mathrm{fp}}, \mathbf{Ab})^{\mathrm{fp}}, \mathbf{Ab}\big)$ if $\mathcal{C}^{\mathrm{fp}}$ has cokernels. This category also is locally coherent, its absolutely pure objects are the exact functors from $\mathcal{C}^{\mathrm{fp}}$-mod to $\mathbf{Ab}$ and there is an embedding of $\mathcal{C}$ into this which has the properties of Gruson and Jensen's embedding ([144, 3.3]). Since the absolutely pure objects of this category are precisely the exact functors, this representation of $\mathcal{C}$ is as $\mathrm{Ex}\big((\mathcal{C}^{\mathrm{fp}}, \mathbf{Ab})^{\mathrm{fp}}, \mathbf{Ab}\big)$. This is a representation of $\mathcal{C}$ as an exactly definable category (see Section 18.1.2) compared with the Gruson–Jensen representation (Section 12.1.1) of $\mathcal{C}$ as a definable subcategory (Section 18.1.1).

16.1.3 Conjugate and dual categories

If $\mathcal{C}$ is finitely accessible, then, 16.1.3, $\mathcal{C} \simeq \text{Flat-}\mathcal{C}^{\mathrm{fp}}$. The **conjugate** category is defined to be $\widetilde{\mathcal{C}} = \mathcal{C}^{\mathrm{fp}}\text{-Flat}$ (Roos, [614], at least for $\mathcal{C}$ locally coherent). By 16.1.3, $\widetilde{\mathcal{C}}$ is again finitely accessible and, by 16.1.4, it is locally coherent iff $\mathcal{C}$ is. Clearly $\widetilde{\widetilde{\mathcal{C}}} \simeq \mathcal{C}$.

If $\mathcal{C} = \text{Mod-}R$, where R is a skeletally small preadditive category, and it happens that there is a duality, $(\text{mod-}R)^{\mathrm{op}} \simeq R\text{-mod}$, between mod-$R$ and R-mod, then $\widetilde{\mathcal{C}} = (\text{mod-}R)\text{-Flat} \simeq \text{Flat-}(R\text{-mod}) \simeq R\text{-Mod}$. That is, if there is such a duality, then $\widetilde{\text{Mod-}R} \simeq R\text{-Mod}$. Note that the existence of a duality $(\text{mod-}R)^{\mathrm{op}} \simeq R\text{-mod}$ implies that mod-R and R-mod are abelian, hence that R is left and right coherent.

One may check, see [516, §9], that, without any condition on R, we have $(\widetilde{\text{mod-}R}, \mathbf{Ab}) \simeq (R\text{-mod}, \mathbf{Ab})$.

Another category associated to a finitely accessible category $\mathcal{C}$ is its **(elementary) dual** category: $\mathcal{C}^{\mathrm{d}} \simeq \mathcal{C}^{\mathrm{fp}}$-Abs (cf. 16.1.12). This definition will be extended in Section 18.1.1 to arbitrary definable categories.

For example, if $\mathcal{C} = \text{Mod-}R$, with R skeletally small preadditive, then $\mathcal{C}^{\mathrm{d}} = \text{Abs}(\text{mod-}R, \mathbf{Ab}) \simeq R\text{-Mod}$ (12.1.6). That is, for module categories, $\mathcal{C}^{\mathrm{d}}$ always is the module category over the opposite ring(oid) (for this also see [161]).

Proposition 16.1.18. *Suppose that $\mathcal{A}$ is a small preadditive category. Then* $(\text{Mod-}\mathcal{A})^{\mathrm{d}} \simeq \mathcal{A}\text{-Mod}$.

Again, if there is a duality $(\text{mod-}R)^{\mathrm{op}} \simeq R\text{-mod}$, then one may check (see [516, §9]) that $(\text{mod-}R, \mathbf{Ab})^{\mathrm{d}} \simeq (R\text{-mod}, \mathbf{Ab})$ and so existence of a duality between finitely presented right and left modules implies that conjugate and dual coincide at both "levels" of representation (modules, and functors on finitely presented modules).

The connection between a finitely accessible category $\mathcal{C}$ and its dual is also strong in that there is a natural bijection, see the more general 18.1.2, between the definable subcategories of $\mathcal{C}$ and those of $\mathcal{C}^{\mathrm{d}}$ which extends that which we know already, 12.4.1, for module, more generally functor, categories. In this sense $\mathcal{C}$ and $\mathcal{C}^{\mathrm{d}}$ are dual to each other in the way that Mod-R and R-Mod are dual.

16.2 Categories generated by modules

A condition is given for a cyclic module to be finitely presented in the category subgenerated by a module (16.2.2); then there is a finiteness condition on right ideals characterising when such a category is locally finitely presented (16.2.3) and a generating set of finitely presented objects is described (16.2.4).

There is an example of a non-trivial category which is subgenerated by a module and which contains no non-zero finitely presented object (16.2.5).

Then there is discussion of the, in general tenuous, relation between purity in the category subgenerated by a module and purity in the whole module category.

Given an R-module M, let

$$\sigma[M] = \{N \in \text{Mod-}R : N \text{ embeds in some epimorphic image of } M\}$$

denote the category **subgenerated** by M, [713, §15]. This is an exact subcategory of Mod-R (a short exact sequence in the subcategory is exact in the subcategory iff it is exact in the whole category) and it is Grothendieck ([713, p. 118 ff.]). In general, $\sigma[M]$, though locally finitely generated (see just below), need not be locally finitely presented, although it is easy to see that it is α-presentable for some α. In fact, see 16.2.5, $\sigma[M]$ need not contain any non-zero finitely presented object. Here a necessary and sufficient condition is given for a category of the kind $\sigma[M]$ to be locally finitely presented. The criterion is one which is often easily checkable and so we have a useful source of examples of locally finitely presented categories. The results in this section are taken from [530].

The inclusion of $\sigma[M]$ in Mod-R has a right adjoint, see just before 16.2.7, hence preserves colimits. It follows that if $A \in \sigma[M]$, then A is finitely generated as an object of $\sigma[M]$ iff it is finitely generated as an R-module. Since every module is a sum of its finitely generated subobjects and since $\sigma[M]$ is cocomplete, the category $\sigma[M]$ is therefore determined by the finitely generated, hence by the cyclic, modules in it. Therefore $\sigma[M]$ is determined by the set $\mathcal{F}_M = \{I \leq R_R : R/I \in \sigma[M]\}$, easily seen to be a filter, of right ideals of R. Say that $J \in \mathcal{F}_M$ is **$\mathcal{F}_M$-finitely generated**, **$\mathcal{F}_M$-fg** for short, if for all $J' \leq J$ with $J' \in \mathcal{F}_M$ it is the case that J/J' is finitely generated.

Lemma 16.2.1. ([530, 1.1]) *Let M be any module. A right ideal $J \in \mathcal{F}_M$ is $\mathcal{F}_M$-finitely generated if and only if whenever $J = \sum_\lambda J_\lambda$ with $J_\lambda \in \mathcal{F}_M$, then $J = J_{\lambda_1} + \cdots + J_{\lambda_n}$ for some $\lambda_1, \ldots, \lambda_n$.*

Proof. If J is $\mathcal{F}_M$-finitely generated and $J = \sum_\lambda J_\lambda = J_{\lambda_1} + \sum_{\lambda \neq \lambda_1} J_\lambda$, then, since J/J_{λ_1} is finitely generated, there are $\lambda_2, \ldots, \lambda_n$ such that $J/J_{\lambda_1} = \sum_2^n (J_{\lambda_i} + J_{\lambda_1})/J_{\lambda_1}$, hence $J = \sum_1^n J_{\lambda_i}$.

Conversely if $J \geq J' \in \mathcal{F}_M$ and if J/J' were not finitely generated, then there would be $(J_\lambda)_\lambda$ with $J_\lambda \geq J'$, hence $J_\lambda \in \mathcal{F}_M$, and $J/J' = \sum_\lambda J_\lambda/J'$, but with no finite subsum equal to J/J'. Then we would have $J = \sum J_\lambda$ with no finite subsum equal to J. □

Proposition 16.2.2. ([530, 1.2]) *Given a module M and $J \in \mathcal{F}_M$ we have $R/J \in \sigma[M]^{\mathrm{fp}}$ iff J is $\mathcal{F}_M$-finitely generated.*

Proof. Suppose that J is $\mathcal{F}_M$-fg. Let $\big((L_\lambda)_\lambda, (g_{\lambda\mu} : L_\lambda \to L_\mu)_{\lambda \leq \mu}\big)$ be a directed system in $\sigma[M]$ with limit $\big(L, (g_{\lambda\infty} : L_\lambda \to L)_\lambda\big)$ and suppose that $f : R/J \to L$. It must be shown that f factors through some $g_{\lambda\infty}$. Set $a = f(1 + J)$. For each λ and each $b \in L_\lambda$ such that $g_{\lambda\infty}(b) = a$ (if there is such in L_λ) set $I_{\lambda,b} = \mathrm{ann}_R(b)$. So $I_{\lambda,b} \in \mathcal{F}_M$ and $\mathrm{ann}_R(b) \leq \mathrm{ann}_R(a)$. From the construction of direct limits (E.1.21) $\mathrm{ann}_R(a) = \sum_{\lambda,b} \mathrm{ann}_R(b)$, so $J \leq \sum_{\lambda,b} I_{\lambda,b}$, hence $J = \sum_{\lambda,b} J \cap I_{\lambda,b}$. Note that $J \cap I_{\lambda,b} \in \mathcal{F}_M$. By 16.2.1, $J = \sum_1^n J \cap I_{\lambda_i,b_i}$ for some λ_i, b_i. For each $i, j \in \{1, \ldots, n\}$ we have $g_{\lambda_i\infty}b_i = g_{\lambda_j\infty}b_j$, so there is $\lambda \geq \lambda_1, \ldots, \lambda_n$ such that $g_{\lambda_i\lambda}b_i = g_{\lambda_j\lambda}b_j = b_0$ say, for all i, j. Clearly $J = J \cap I_{\lambda,b_0}$. Thus $\mathrm{ann}_R(b_0) \geq J$ and f factors through $g_{\lambda\infty}$, as required.

For the converse let $J \in \mathcal{F}_M$ be such that R/J is finitely presented in $\sigma[M]$. Then for any $J' \in \mathcal{F}_M$ with $J' \leq J$, the kernel of $R/J' \to R/J$ is, by E.1.17, finitely generated and is equal to J/J', that is, J is $\mathcal{F}_M$-fg. □

Say that $\mathcal{F}_M$ is **cofinally $\mathcal{F}_M$-finitely generated** if for every $I \in \mathcal{F}_M$ there is some $\mathcal{F}_M$-finitely generated $J \in \mathcal{F}_M$ with $J \leq I$.

Theorem 16.2.3. ([530, 1.3]) *For any module M the category $\sigma[M]$ is locally finitely presented iff $\mathcal{F}_M$ is cofinally $\mathcal{F}_M$-finitely generated.*

Proof. Suppose first that $\sigma[M]$ is locally finitely presented. Let $I \in \mathcal{F}_M$. Then $R/I \in \sigma[M]$, so (E.1.15) there is an epimorphism $f : F \to R/I$ with $F \in \sigma[M]^{\mathrm{fp}}$. Let $a_1, \ldots, a_n$ be a finite set of generators for F where, without loss of generality, $f(a_1) = 1 + I$. Say $f(a_i) = r_i + I$, $i = 2, \ldots, n$. Set $F' = F/\langle a_1 r_i - a_i : i = 2, \ldots, n\rangle$ and let $p : F \to F'$ be the projection. Then F' is cyclic and also finitely presented (see E.1.12). There is a factorisation of f through p, say $f' : F' \to R/I$, such that $f'p = f$.

Now, F' is cyclic, isomorphic to R/J with $J = \mathrm{ann}_R(pa_1)$, and finitely presented, so by 16.2.2, J is $\mathcal{F}_M$-finitely generated. Furthermore, J is contained in I, as required.

For the converse, if $\mathcal{F}_M$ is cofinally $\mathcal{F}_M$-finitely generated, then the R/J with J a $\mathcal{F}_M$-fg member of $\mathcal{F}_M$ form a generating (by cofinality of these in $\mathcal{F}_M$) set of finitely presented (by 16.2.2) objects of $\sigma[M]$, as required. □

Also note, in a special case, the related [284, Prop. 11].

Corollary 16.2.4. ([530, 1.4]) *If M is a module such that $\sigma[M]$ is locally finitely presented, then the R/J, as J ranges over the $\mathcal{F}_M$-finitely generated right ideals in $\mathcal{F}_M$, form a generating set of finitely presented objects of $\sigma[M]$.*

The condition of 16.2.3 is often readily checkable and one can recover known conditions for $\sigma[M]$ to be locally finitely presented quite easily. For example, if R is right noetherian or if M is a locally coherent module, then $\sigma[M]$ is locally finitely presented. To see the latter, note that the annihilators of elements of finite powers M^n of M are cofinal in $\mathcal{F}_M$ and that, by 2.3.16, these annihilators are finitely generated. In particular the category of comodules over a k-coalgebra C, where k is a field (and more generally, see, for example, [714]), is locally finitely presented since it is equivalent to $\sigma[{}_{C^*}C]$, where C^* is the dual algebra to C. If $\mathcal{F}_M$ has a minimal element, then $\sigma[M]$ is locally finitely presented, indeed, it is a module category.

It may also be checked, [530, 1.6], that the category $\sigma[M]$ is locally finitely presented iff for every $F \in \text{mod-}R$ and every morphism $f : F \to A \in \sigma[M]$ there is a factorisation of f through a member of $\sigma[M]^{\mathrm{fp}}$.

For contrast, here is an example of a category of the form $\sigma[M]$ where the only finitely presented object is the zero object. In particular, this is a Grothendieck category $\mathcal{C}$ with $\mathcal{C}^{\mathrm{fp}} = \{0\}$.

Example 16.2.5. ([530, 1.7]) Let $R = K[X_n : n \geq 0]$ be the polynomial ring over a field K in countably many indeterminates. Set $I_n = \langle X_{k2^n} : k \geq 1 \rangle$. So $I_0 > I_1 > \cdots$ is a decreasing sequence of ideals with each factor I_n/I_{n+1} an infinitely generated R-module. Let $\mathcal{F}$ be the filter of ideals generated by the I_n. Therefore, if $M = \bigoplus\{R/I : I \in \mathcal{F}\}$, then $\mathcal{F} = \mathcal{F}_M$ (because R is commutative, $a \in R/I$ implies $\mathrm{ann}_R(a) \geq I$). Then there is no finitely presented object in $\sigma[M]$ other than 0; we prove this.

Proof. If there is a non-zero finitely presented object, then there is one which is cyclic (see the end of the proof), so, for a contradiction and noting 16.2.4, suppose that there is $I \in \mathcal{F}_M$, $I \neq R$ such that I is $\mathcal{F}_M$-finitely generated. Since $I \in \mathcal{F}_M$, $I \geq I_n (> I_{n+1})$ for some n so, since I is $\mathcal{F}_M$-finitely generated, $I = I_{n+1} + \sum_1^t a_i R$ for some $a_i \in R$.

Let m be of the form $m = k2^n$ with k odd and be chosen such that for all X_j appearing in $a_1, \ldots, a_t$ we have $j < m$. Then $X_m \in I \setminus I_{n+1}$ and we assume, for a contradiction, that there is a representation $X_m = f + \sum_i a_i g_i$ with $f \in I_{n+1}$ and the $g_i \in R$. Consider

$$I' = I \cap K[X_0, \ldots, X_{m-1}] = \left(I_{n+1} + \sum_{i=1}^{t} a_i R \right) \cap K[X_0, \ldots, X_{m-1}].$$

Since $I \neq R$, I' is a proper ideal in $K[X_0, \ldots, X_{m-1}]$ so there is a maximal ideal, J, of $K[X_0, \ldots, X_{m-1}]$ with $I' \subseteq J$. Let $L = K[X_0, \ldots, X_{m-1}]/J$ and regard this as an extension field of K. Consider the projection from $R = K[X_0, \ldots, X_{m-1}][X_m, \ldots]$ to $L[X_m, \ldots]$ with kernel $J \cdot R$, followed by the projection to L with kernel $\langle X_m - 1 \rangle + \langle X_n : n > m \rangle$. Denote the composite morphism by $\theta : R \to L$.

Since $f \in I_{n+1}$ and $X_m \notin I_{n+1}$, we have

$$f = f_0 + f_1,$$

where

$$f_0 \in \big(K[X_0, \ldots, X_{m-1}] \cap I_{n+1} \big) \cdot R$$

(that is, every monomial of f_0 is divisible by some $X_j \in I_{n+1}$ with $j < m$) and where every monomial of f_1 is divisible by some X_j with $j > m$. Then $\theta(f_1) = 0$ and $\theta(f_0) = 0$, since $f_0 \in I'$. Moreover, each $a_i \in I'$, hence $\theta(\sum_i a_i g_i) = 0$. But this is a contradiction because $\theta\big(f + \sum_i a_i g_i = X_m\big) = 1$.

Therefore there is no non-zero finitely presented cyclic object. Now suppose that A were a non-zero finitely presented object of $\sigma[M]$. Choose some minimal generating set $a_1, \ldots, a_n$ for A. Then $A/\sum_2^n a_i R$ is a non-zero cyclic object in $\sigma[M]$ and is finitely presented.

We conclude that $\sigma[M]^{\mathrm{fp}}$ has only the zero object. □

In general, $\sigma[M]$ does not "sit nicely" within Mod-R with respect to purity. Nevertheless, there is some relation between purity and pure-injectivity in these categories, and this is investigated in [530]. For example, [530, 2.1], if $\sigma[M]$ is locally finitely presented, then a pure-exact sequence in $\sigma[M]$ also is pure-exact in Mod-R. That the converse fails is shown by the next example.

Example 16.2.6. ([530, 2.2]) Let R be a von Neumann regular ring which is not semisimple. Suppose that R has simple modules S, T (possibly isomorphic) such that $\mathrm{Ext}^1(S, T) \neq 0$ (so, because every exact sequence of R-modules is pure, S cannot, by 2.1.18, be finitely presented), say M is a non-split extension of S by T. Both S and T are clearly finitely presented objects of $\sigma[M]$ and they generate this category, which is, therefore, locally finitely presented. The non-split exact sequence $0 \to T \to M \to S \to 0$ cannot, therefore, be pure in $\sigma[M]$ – otherwise (2.1.18) it would split. On the other hand, every short exact sequence in Mod-R is pure, because R is von Neumann regular (2.3.22).

For such examples see [242, 3.18].

On the other hand, [530, 2.5], if $\sigma[M]$ is locally finitely presented and $\sigma[M]^{\mathrm{fp}} \subseteq$ mod-R, then an exact sequence in $\sigma[M]$ is pure in $\sigma[M]$ iff it is pure in Mod-R. Note that the condition that $\sigma[M]^{\mathrm{fp}}$ be contained in mod-R is, by 16.2.2, equivalent to the condition that every $\mathcal{F}_M$-finitely generated right ideal be finitely generated. For instance, $\sigma[M]^{\mathrm{fp}} \subseteq$ mod-R if M is locally coherent because every finitely presented object of $\sigma[M]$ has the form A/B for some finitely generated modules $B \leq A \leq M^n$. The combined conditions that $\sigma[M]$ be locally finitely presented and that $\sigma[M]^{\mathrm{fp}}$ be contained in mod-R are equivalent to there being a cofinal set of finitely generated right ideals in $\mathcal{F}_M$, therefore they are satisfied for any module M if R is right noetherian.

Although there is more on this in [530] it is not clear that there are natural conditions on $\mathcal{F}_M$, necessary and sufficient for purity in Mod-R and $\sigma[M]$ to coincide.

Define a functor, T_M, from Mod-R to $\sigma[M]$, by $T^M(N) = \sum\{N' \leq N : N' \in \sigma[M]\}$ and to be restriction on morphisms: so $T_M(N)$ is the largest submodule of N which is in $\sigma[M]$. Then, see [713, 45.11], T_M is right adjoint to the inclusion of $\sigma[M]$ in Mod-R. It is known, [713, 17.9], that an object $N \in \sigma[M]$ is injective in $\sigma[M]$ iff $N = T^M E(N)$. In [530] there are similar, though weaker, results for pure-injective objects. In particular, [530, 3.1], if $\sigma[M]$ is locally finitely presented and $N \in$ Mod-R is pure-injective, then $T^M N$ is a pure-injective object of $\sigma[M]$. But the converse is false.

Example 16.2.7. ([530, 3.2]) Take R to be the first Weyl algebra over a field of characteristic 0 (1.1.1(3)) and let S be a simple R-module. Since R is (right) noetherian, $\sigma[S]$ is locally finitely presented and $\sigma[S]^{\mathrm{fp}} \subseteq \text{mod-}R$. One may check that the category $\sigma[S]$ is semisimple and that S is an injective, in particular pure-injective, object of $\sigma[S]$. But, as an R-module, S is not pure-injective (8.2.35).

It is the case, [530, 3.3], that if $\sigma[M]$ is locally finitely presented and $\sigma[M]^{\mathrm{fp}} \subseteq \text{mod-}R$, then, for every $A \in \sigma[M]$, the pure-injective hull of A in $\sigma[M]$ is a direct summand of $T^M H(A)$. It is not, however, known whether these are equal in general; it is true, [530, 3.5], if $T^M H(A)$ is pure in $H(A)$, but that certainly is not always the case: take $M = \mathbb{Z}_{p^n}$ and $N = \mathbb{Z}_{p^\infty}$. Nevertheless, with that assumption, along with the assumptions that $\sigma[M]$ is locally finitely presented and that $\sigma[M]^{\mathrm{fp}} \subseteq \text{mod-}R$, one does get, for example, a homeomorphic embedding of $\mathrm{Zg}(\sigma[M])$ into Zg_R ([530, 3.8]).

16.3 Categories of presheaves and sheaves

The category Mod-$\mathcal{O}_X$ *of sheaves of modules over a ringed space* $\mathcal{O}_X$ *is Grothendieck abelian (for example, [663, pp. 214-216], [486, §4.7]). Here we look at the question of when it is locally finitely presented and, when it is, we make some observations on the form of pp conditions.*

Basic definitions are given in Appendix E, pp. 715, 716 but, for more details on sheaves and presheaves, see, for instance, [269], [318], [332], [674].

If X is a topological space and $\mathcal{O}_X$ is a (pre-)sheaf of rings on X, then PreMod-$\mathcal{O}_X$ *denotes the category of $\mathcal{O}_X$-presheaves and* Mod-$\mathcal{O}_X$ *denotes the category of $\mathcal{O}_X$-sheaves. Both are abelian Grothendieck categories: the former since it is a functor category (and by 10.1.3); the latter by 11.1.5 since it is a localisation of a functor category (see [486, §4.7], also [415, §5]). We stay at this level of generality but much of what is here would apply if X were replaced by a locale.*

The algebraic context of presheaves is considered in Section 16.3.1 and some conditions under which the sheafification functor (a torsion-theoretic localisation) yields a locally finitely presented category of sheaves are given in Section 16.3.2.

The question of when the category of sheaves is locally finitely presented is considered in Section 16.3.3 and the corresponding question for local finite generation in Section 16.3.4.

Section 16.3.5 gives some information on pp conditions on categories of sheaves.

This section is largely based on [526].

16.3.1 Categories of presheaves

Every category of presheaves is locally finitely presented (16.3.1) and a generating set of finitely presented objects may be obtained from the collection of open sets (16.3.2). An algebraic description of finitely generated presheaves is given as 16.3.3 and local coherence of the category of presheaves of modules is considered (16.3.4).

Theorem 16.3.1. (for example, [15, §I.3]) *For any space, X, and presheaf, $\mathcal{O}_X$, of rings on X the category* PreMod-$\mathcal{O}_X$ *is a locally finitely presented abelian category.*

A generating set of finitely presented presheaves is as follows.

Let $U \subseteq X$ be open. Our generic notation for the inclusion, $U \to U'$, of one subset in another is j, and hence $j_!$, j^* etc. denote associated direct and inverse image functors between categories of (pre)sheaves; for these see [318, §§2.4, 2.6]. Let $G \in$ PreMod-$\mathcal{O}_U$, where $\mathcal{O}_U$ denotes the restriction, $\mathcal{O}_X \upharpoonright U$, of the structure presheaf $\mathcal{O}_X$ to U. Define the presheaf $j_0 G \in$ PreMod-$\mathcal{O}_X$ by

$$j_0 G \cdot V = \begin{cases} G(V) & \text{if } V \subseteq U \\ 0 & \text{otherwise} \end{cases}$$

and in the obvious way on inclusions. It is easy to check that j_0 extends to an exact functor from PreMod-$\mathcal{O}_U$ to PreMod-$\mathcal{O}_X$.

Proposition 16.3.2. (for example, [87, p. 7 Prop. 6] or [526, 2.14]) *The $j_0\mathcal{O}_U$, with $U \subseteq X$ open, form a generating set of finitely presented objects of* PreMod-$\mathcal{O}_X$.

Note that $j_0\mathcal{O}_U \leq \mathcal{O}_X$. We describe the finitely generated presheaves.

Let $F \in$ PreMod-$\mathcal{O}_X$, let $U \subseteq X$ be open and let $a \in FU$. Define the presheaf $\langle a \rangle$ by

$$V \mapsto \begin{cases} \mathrm{res}_{UV}(a) \cdot R_V & \text{if } V \subseteq U \\ 0 & \text{otherwise} \end{cases},$$

where res_{UV} denotes the restriction map from FU to FV. Then $\langle a \rangle \in$ PreMod-$\mathcal{O}_X$ and there is a natural inclusion $\langle a \rangle \leq F$.

Recall from Appendix E, p. 710 the definition of (an object, in particular a sheaf) F being finitely generated: whenever $F = \sum_\lambda F_\lambda$ with the F_λ subsheaves of F, then there are $\lambda_1, \dots, \lambda_n$ such that $F = F_{\lambda_1} + \cdots + F_{\lambda_n}$.

Lemma 16.3.3. ([526, 2.16]) *The presheaf $F \in$* PreMod-$\mathcal{O}_X$ *is finitely generated, iff there exist open subsets $U_1, \dots, U_n$ of X and sections $a_i \in FU_i$ such that $F = \sum_1^n \langle a_i \rangle$.*

Proof. ($\Rightarrow$) On comparing at each open set U, one has $F = \sum_{U \subseteq X} \sum_{a \in FU} \langle a \rangle$, so if F is finitely generated, then F is a sum of finitely many of these.

($\Leftarrow$) If F has the given form and $F = \sum_{\overrightarrow{\lambda}} F_\lambda$ (a directed sum) then choose, for each $i = 1, \dots, n$, some λ_i such that $a_i \in F_{\lambda_i} U_i$; since $FU_i = \sum_\lambda F_\lambda U_i$ this exists. Then $\langle a_i \rangle \leq F_{\lambda_i}$, so $F \leq \sum_1^n F_{\lambda_i}$. □

The next result gives conditions under which PreMod-$\mathcal{O}_X$ is a locally coherent category.

Theorem 16.3.4. ([526, 2.18]) *Let $\mathcal{O}_X$ be a presheaf of rings on the space X. If, for each open subset U of X, the ring, $\mathcal{O}_X(U)$, of sections over U is a right coherent ring, then the category,* PreMod-$\mathcal{O}_X$, *of presheaves over $\mathcal{O}_X$ is locally coherent.*

If, for each inclusion $V \subseteq U$ of open sets, the ring $\mathcal{O}_X(V)$ is a flat left $\mathcal{O}_X(U)$-module, via the corresponding morphism of rings, then the converse also holds.

16.3.2 Finite-type localisation in categories of presheaves

Finite-type localisations on presheaves are characterised (16.3.5) and density described in terms of open covers (16.3.7). The separated presheaves are those which are torsionfree with respect to the sheafification localisation (16.3.8). The sheafification localisation is of finite type iff it is elementary (16.3.12) iff there is a basis of compact open sets (that is, the space is noetherian) (16.3.10), iff the class of separated presheaves is definable (16.3.11). Therefore the category of sheaves of modules on a noetherian ringed space is locally finitely presented (16.3.13), whatever the sheaf of rings.

By 11.1.11, 16.3.1 and 16.3.2, if τ is a hereditary torsion theory on PreMod-$\mathcal{O}_X$, then τ is determined by the collection of Gabriel filters: $\mathcal{U}_\tau(j_0\mathcal{O}_U) = \{\mathcal{I} \leq j_0\mathcal{O}_U : j_0\mathcal{O}_U/\mathcal{I} \in \mathcal{T}\}$, where U is an open subset of X and $\mathcal{T}$ denotes the class of τ-torsion objects. Then 11.1.14 gives the following.

Proposition 16.3.5. ([526, 3.2]) *A hereditary torsion theory, τ, on* PreMod-$\mathcal{O}_X$ *is of finite type iff for all open $U \subseteq X$ the filter $\mathcal{U}_\tau(j_0\mathcal{O}_U)$ has a cofinal set of finitely generated presheaves.*

The category, Mod-$\mathcal{O}_X$, of sheaves sits inside the category of presheaves as a localisation of the latter. There is (see the references suggested) the canonical process "sheafification", which, given a presheaf, forms a sheaf and which is, [486, §4.7], just localisation at a certain hereditary torsion theory (the torsion objects are described below). The sheafification process is, first, to make zero any section which locally is zero (that is, factor out the torsion subsheaf), then to patch together compatible sets of sections to make sections over larger open sets (and that is the second process, of completing to the τ-injective hull, described in Section 11.1.1).

For the remainder of this section suppose that $\mathcal{O}_X$ is a sheaf of rings, let τ be the torsion theory corresponding to the sheafification functor and let $\mathcal{T}, \mathcal{F}$ denote the corresponding torsion and torsionfree classes respectively. Thus $\mathcal{T}$ is the class of presheaves with sheafification equal to 0.

Lemma 16.3.6. ([526, 3.3]) *Let $F \in$ PreMod-$\mathcal{O}_X$. Then $F \in \mathcal{T}$ iff for every open $U \subseteq X$ and for every $a \in FU$ there is an open cover, $\{U_i\}_i$, of U such that, for all i, we have $\mathrm{res}^F_{U,U_i} a = 0$.*

Proof. Clearly such a presheaf is torsion (its sheafification must be 0). Conversely, if $F \in \mathcal{T}$ and if U, a are as given, then a must be identified with $0 \in FU$ by the sheafification process and this can happen only if there is a cover of U as described. □

Corollary 16.3.7. ([526, 3.4]) *Let $G \le F \in$ PreMod-$\mathcal{O}_X$. Then G is τ-dense in F (that is, F/G is τ-torsion) iff for every open $U \subseteq X$ and for every $a \in FU$ there is an open cover, $\{U_i\}_i$, of U such that, for all i, we have $\mathrm{res}^F_{U,U_i} a \in GU_i$.*

Recall that the stalk of a (pre)sheaf F at a point x is $F_x = \varinjlim_{U \ni x} FU$.

Lemma 16.3.8. ([526, 3.5]) *Let $F \in$ PreMod-$\mathcal{O}_X$. Then $F \in \mathcal{T}$ iff for every $x \in X$ we have $F_x = 0$.*

Proof. ($\Rightarrow$) $F \in \mathcal{T}$ implies $qF = 0$, where q is the sheafification functor, that is, the localisation functor Q_τ associated to τ, and this implies that $(qF)_x = 0$ for all $x \in X$, which, in turn, implies, since stalks are unchanged by sheafification (see, for example, [318, II.2.4]), that $F_x = 0$ for all $x \in X$.

($\Leftarrow$) If every stalk of F is 0, then the same is true for qF, hence $qF = 0$, so $F \in \mathcal{T} = \ker(q)$. □

A presheaf is said to be **separated** (or a **monopresheaf**) if it is torsionfree for the sheafification torsion theory.

Proposition 16.3.9. ([526, 3.7]) *Let $U \subseteq X$ be open. Then $j_0\mathcal{O}_U$ has a cofinal family of finitely generated τ-dense subpresheaves iff U is compact.*

Proof. ($\Leftarrow$) Suppose that $G \le F = j_0\mathcal{O}_U$ is τ-dense. Then there is an open cover, $\{U_i\}_i$, of U which, by compactness of U, may be taken to be a finite cover $U_1, \dots, U_n$, such that for each i, $\mathrm{res}^F_{U,U_i}(1) \in G(U_i)$, where $1 \in FU$. That is, $G \upharpoonright U_i = F \upharpoonright U_i = \mathcal{O}_{U_i}$ for each i.

Let $G_i = j_0\mathcal{O}_{U_i}$. Note that $G_i \le G$. So $G_1 + \dots + G_n \le G$ is a finitely generated subobject of G and, by construction and by 16.3.7, is τ-dense, as required.

($\Rightarrow$) If U is not compact, then choose an open cover, $\{U_i\}_i$, of U which has no finite subcover and define $G \in \text{PreMod-}\mathcal{O}_X$ by $GV = FV$ if $V \subseteq U_i$ for some i, $GV = 0$ otherwise. So $G \leq F$ and G is τ-dense in F because $\text{res}_{U,U_i}(1) \in GU_i$ for every i.

If there were $G' \leq G$ with G' finitely generated and τ-dense in G, then there would be, in particular, an open cover, $\{V_j\}_j$, of U with $\text{res}_{U,V_j}(1) \in G'V_j$ for every j. Suppose, using 16.3.3, that $G' = \sum_1^m \langle a_l \rangle$ with $a_l \in G'V_l$ $(l = 1, \dots, m)$. Since $G' \leq G$, if $a_l \neq 0$ we have $V_l \subseteq U_{i_l}$ for some i_l. But then, if V is any open set not contained in $U_{i_1} \cup \dots \cup U_{i_m}$ (a proper subset of U) it must be that $G'V = 0$. So, choosing V_j not contained in this union, we see that $\text{res}_{U,V_j}(1)$, which is non-zero, cannot be in $G'V_j$, a contradiction. $\square$

Corollary 16.3.10. ([526, 3.8]) *Let $\mathcal{O}_X$ be a sheaf of rings. The localisation "sheafification" from the category of $\mathcal{O}_X$-presheaves to the category of $\mathcal{O}_X$-sheaves is of finite type iff every open subset of X is compact.*

The corollary is immediate by 16.3.9 and 16.3.5. The next corollary is by 11.1.20.

Corollary 16.3.11. ([526, 3.9]) *Let $\mathcal{O}_X$ be a sheaf of rings. The class of separated presheaves is a definable subcategory of* PreMod-$\mathcal{O}_X$ *iff every open subset of X is compact.*

Notice that every open subset of X is compact iff the space X is **noetherian** – has the ascending chain condition on open subsets, equivalently the descending chain condition on closed subsets.

Theorem 16.3.12. ([526, 3.10, 3.12]) *Let $\mathcal{O}_X$ be a sheaf of rings on the space X and let τ denote the sheafification torsion theory on* PreMod-$\mathcal{O}_X$. *Then the following are equivalent:*

(i) τ is an elementary torsion theory (Section 11.1.3);
(ii) τ is of finite type;
(iii) X is a noetherian space;
(iv) Mod-$\mathcal{O}_X$ *is a definable subcategory of* PreMod-$\mathcal{O}_X$.

Proof. For the equivalence of the first three conditions, it remains to be shown that if every open subset is compact, then τ is elementary. Let $F \leq G = j_0\mathcal{O}_U$ with G/F torsion. It will be enough, by 11.1.22, to show that F contains a finitely presented τ-dense subobject. Since F is τ-dense in G, there is an open cover, $\{U_\lambda\}_\lambda$, of U such that for each λ, $\text{res}_{U,U_\lambda} 1 = 1_\lambda \in FU_\lambda$. Since U is compact, it may be supposed that the index set is finite, say $F \geq F' = \sum_1^n j_0\langle 1_{U_i} \rangle$, where $U_1, \dots, U_n$ cover U. Then F' is τ-dense in G, hence in F. But clearly F' is finitely presented,[3] the relations between its generators being generated by the $\mathrm{res}_{U_i,U_i\cap U_j}(1_{U_i}) = \mathrm{res}_{U_j,U_i\cap U_j}(1_{U_j})$, where $1 \leq i < j \leq n$.

Equivalence with the last condition is by 11.1.21. □

The next result therefore follows from 16.3.1, 16.3.12 and 11.1.26.

Theorem 16.3.13. ([526, 3.11]) *Let $\mathcal{O}_X$ be any sheaf of rings on a noetherian space X. Then* Mod-$\mathcal{O}_X$ *is locally finitely presented.*

It will be seen that this condition is certainly not necessary for Mod-$\mathcal{O}_X$ to be locally finitely presented; X being locally noetherian will suffice (16.3.18).

16.3.3 The category Mod-$\mathcal{O}_X$: local finite presentation

Any basis of open sets gives a generating family for the category of sheaves (16.3.14), with finitely presented generators corresponding to compact open sets (16.3.16). If there is a basis of compact open sets, then the category of sheaves is locally finitely presented (16.3.17).

Every category of additive sheaves, being an additive Grothendieck category, is locally presentable[4] but it is by no means always the case that the category, Mod-$\mathcal{O}_X$, of sheaves over a ringed space is locally *finitely* presented. This is in contrast (16.3.1) with the presheaves, which are genuinely algebraic objects so (see, for example, [3, §3.A]) form a locally finitely presented category. But the sheaf property is not an algebraic one and does not fit well with notions like "finitely presented" and "finitely generated" unless the base space has a strong compactness property.

Recall from Sections 10.2.4 and 11.1.5 that the "elements" of an object C in a locally finitely presented category are the morphisms from finitely presented objects to C. Such elements have finitary character, unlike the sections of a sheaf over an arbitrary open set. So the aim of this section, only partially realised, is to discover when a sheaf is determined by its (finitary) elements.

A key difference between presheaves and sheaves lies in the fact that in the category of presheaves direct limits are computed at the level of sections: if G is the presheaf direct limit of a directed system, $(G_\lambda)_\lambda$, of presheaves, then, since this is how one defines the direct limit of functors, for each open set, U, GU is the direct limit of the $G_\lambda U$. Even if all the G_λ are sheaves this recipe will yield a presheaf rather than a sheaf in general.

[3] That "finitely presented" has the usual algebraic meaning in categories of presheaves follows from [3, 3.5(3), 3.10] for example.

[4] Also in the non-additive case, for example, [84, 3.4.2, 3.4.16].

If $j : U \to X$ is the inclusion of an open set in X, then $j_!$ is the functor from Mod-$\mathcal{O}_U$ to Mod-$\mathcal{O}_X$ defined on objects, $G \in$ Mod-$\mathcal{O}_U$, by

$$j_! G \cdot V = \{s \in G(V \cap U) : \mathrm{supp}(s) \text{ is closed in } V\}$$

for $V \subseteq X$ open; this is the sheafification of, $j_0 G$, the presheaf extension by zero of G defined in Section 16.3.1. Then, for example, [318, pp. 108–9], $j_!$ is left adjoint to j^* : for every $G \in$ Mod-$\mathcal{O}_U$ and $F \in$ Mod-$\mathcal{O}_X$ we have $(j_! G, F) \simeq (G, F \mid_U)$.

By 16.3.2 and 11.1.5 the sheaves $j_! \mathcal{O}_U$, where $U \subseteq X$ is open, together generate Mod-$\mathcal{O}_X$. Because sheaves are locally determined, those corresponding to a basis of open sets suffice to generate the category.

Proposition 16.3.14. ([526, 5.2]) *If $\mathcal{U}$ is a basis of open sets for X, then the $j_! \mathcal{O}_U$ for $U \in \mathcal{U}$ together generate* Mod-$\mathcal{O}_X$.

Proof. (This does not follow via localisation of PreMod-$\mathcal{O}_X$ since it is easy to see that the corresponding result for that category is false.)

Let $F \in$ Mod-$\mathcal{O}_X$, let $x \in X$ and take $a \in F_x$. Since $F_x = \varinjlim_{x \in U} FU$ there is U open and $b = b_{U,a} \in FU$ such that the canonical map from FU to F_x takes b to a. Without loss of generality $U \in \mathcal{U}$. Define $f' : \mathcal{O}_U \to F \restriction U$ by $1 \in \mathcal{O}_U U \mapsto b_{U,a} \in FU$ (this is enough to define a presheaf, hence sheaf, morphism).

Since $j_!$ is left adjoint to $(-)_U$, there is, corresponding to $f' \in (\mathcal{O}_U, F \restriction U)$, a morphism $f_{x,U,a} \in (j_! \mathcal{O}_U, F)$ with $(f_{x,U,a})_U : 1_U \mapsto b_{U,a}$. Consider the image of 1_U in $(j_! \mathcal{O}_U)_x$, that is, $1_{\mathcal{O}_{X,x}}$. Then $(f_{x,U,a})_x : (j_! \mathcal{O}_U)_x \to F_x$ (the stalkwise at x map induced by $f_{x,U,a}$) maps $1_{\mathcal{O}_{X,x}}$ to $a \in F_x$.

Therefore $\prod_{x \in X} \prod_{a \in F_x} f_{x,U,a} : \bigoplus_x \bigoplus_a j_! \mathcal{O}_{U = U(x,a)} \to F$ is an epimorphism on stalks, hence (for example, [318, II.2.6]) is an epimorphism in the category of sheaves. □

Proposition 16.3.15. ([165, 2.4], [526, 5.3]) *Suppose $\varinjlim_\lambda G_\lambda = G$ in* Mod-$\mathcal{O}_X$ *and suppose that $U \subseteq X$ is compact open. Then the canonical map $g : \varinjlim_\lambda (G_\lambda U) \to GU$ is an isomorphism.*

Proof. The sheaf G is the sheafification of the presheaf direct limit, G', of the G_λ and this is given by $G'V = \varinjlim (G_\lambda V)$ for $V \subseteq X$ open. So the proposition asserts that G' and G agree at all compact open sets. Let $g : G' \to G$ be the sheafification map.

Let $s \in G'U$ and suppose that $gs = 0$. Then there must be an open cover $\{U_i\}_i$ of U such that, for all i, we have $\mathrm{res}^{G'}_{U,U_i} s = 0$. Since U is compact the cover may be taken to be finite: $U_1, \dots, U_n$ say. By construction of direct limit, for each i there are λ_i and $a_i \in G_{\lambda_i} U$ such that $(g'_{\lambda_i,\infty})_U a_i = s$ (where $g'_{\lambda_i,\infty} : G_{\lambda_i} \to G'$ is the canonical map) and $(g'_{\lambda_i,\infty})_{U_i} \mathrm{res}^{G_{\lambda_i}}_{U,U_i} a_i = 0$. Since there are just finitely many U_i one may take λ with $\lambda \geq \lambda_1, \ldots, \lambda_n$ and also such that $(g_{\lambda_i,\lambda})_U a_i = (g_{\lambda_j,\lambda})_U a_j = b$, say, for all i, j (since the a_i all map to the same element in the limit) and, furthermore, such that $(g_{\lambda_i,\lambda})_{U_i}\mathrm{res}^{G_{\lambda_i}}_{U,U_i}(a_i) = 0$ for all i (since each $\mathrm{res}^{G_{\lambda_i}}_{U,U_i} a_i$ maps to 0 in the limit). But then $\mathrm{res}^{G_\lambda}_{U,U_i}(b) = (g_{\lambda_i,\lambda})_{U_i}\mathrm{res}^{G_{\lambda_i}}_{U,U_i}(a_i) = 0$ for each i, so, since G_λ is a sheaf, $b = 0$. Therefore $s = (g'_{\lambda,\infty})_U(b) = 0$ and g is monic.

To see that g is onto, take any $t \in GU$. For each $x \in U$ there is an open neighbourhood, U_x, of x and $t_x \in G'U_x$ such that $gt_x = \mathrm{res}^G_{UU_\lambda}(t)$. Since U is compact, there are finitely many open sets $U_1, \ldots, U_n$ say, with corresponding $t_i \in G'U_i$, which cover U.

For each i there is λ_i and $s_i \in G_{\lambda_i}U_i$ with $g'_{\lambda_i\infty}s_i = t_i$. Since there are only finitely many λ_i, one may take $\lambda \geq \lambda_1, \ldots, \lambda_n$ and it may be supposed, without loss of generality, that $s_i \in G_\lambda U_i$ for each i.

Furthermore, for each pair i, j of indices, $\mathrm{res}^{G'}_{U_i,U_i\cap U_j}(t_i) = \mathrm{res}^{G'}_{U_j,U_i\cap U_j}(t_j)$, that is, $g'_{\lambda\infty}\mathrm{res}^{G_\lambda}_{U_i,U_i\cap U_j}(s_i) = g'_{\lambda\infty}\mathrm{res}^{G_\lambda}_{U_j,U_i\cap U_j}(s_j)$, hence there is $\mu_{ij} \geq \lambda$ such that $g'_{\lambda\mu_{ij}}\mathrm{res}^{G_\lambda}_{U_i,U_i\cap U_j}(s_i) = g'_{\lambda\mu_{ij}}\mathrm{res}^{G_\lambda}_{U_j,U_i\cap U_j}(s_j)$. So, choosing $\mu \geq \mu_{ij}$ for each i, j and setting $s'_i = g_{\lambda\mu}(s_i)$ it may be supposed that for all i, j we have $\mathrm{res}^{G_\mu}_{U_i,U_i\cap U_j}(s'_i) = \mathrm{res}^{G_\mu}_{U_j,U_i\cap U_j}(s'_j)$. Since G_μ is a sheaf, there is $s \in G_\mu U$ such that $\mathrm{res}^{G_\mu}_{U,U_i}(s) = s_i$ for each $i = 1, \ldots, n$. Then $g'_{\mu\infty}s \in G'U$ (since G is separated) and $gg'_{\mu\infty}s = t$, as required. □

Corollary 16.3.16. ([526, 5.8]) *Let $\mathcal{O}_X$ be a sheaf of rings and let $U \subseteq X$ be open. Then $j_!\mathcal{O}_U$ is finitely presented in* Mod-$\mathcal{O}_X$ *iff U is compact.*

Proof. Suppose that U is compact and let $(G_\lambda)_\lambda$ be a directed system in Mod-$\mathcal{O}_X$ with direct limit G. We have the adjuction, $(j_!\mathcal{O}_U, G) \simeq (\mathcal{O}_U, G \upharpoonright U)$ and also $(\mathcal{O}_U, G \upharpoonright U) \simeq (G \upharpoonright)U = GU$. Similarly $\varinjlim_\lambda (j_!\mathcal{O}_U, G_\lambda) \simeq \varinjlim(G_\lambda U)$ – and these coincide by 16.3.15, as required.

If, on the other hand, U is not compact, let $\{U_\lambda\}_\lambda$ be an open cover with no finite subcover. It may be supposed that $\{U_\lambda\}_\lambda$ is closed under finite union. Set $G_\lambda = j_!\mathcal{O}_{U_\lambda}$ and $G = \varinjlim j_!\mathcal{O}_{U_\lambda}$. Then $G = j_!\mathcal{O}_U$ since, from the canonical inclusions $j_!\mathcal{O}_{U_\lambda} \to j_!\mathcal{O}_U$, one obtains a map $G = \varinjlim j_!\mathcal{O}_{U_\lambda} \to j_!\mathcal{O}_U$ which locally, and hence stalkwise, is an isomorphism and which is, therefore (for example, [318, II.2.2]), an isomorphism.

The identity map of G would factor through some $j_!\mathcal{O}_{U_i}$ if $G = j_!\mathcal{O}_U$ were finitely presented but, since $U_i \neq U$, there can be no such factorisation, since if $x \notin U_i$, then $(j_!\mathcal{O}_{U_i})_x = 0$ (for example, by [318, II.6.4]). □

Theorem 16.3.17. ([587] see [526, 5.6]) *If X has a basis of compact open sets, then* Mod-$\mathcal{O}_X$ *is locally finitely presented, with the $j_!\mathcal{O}_U$ for $U \subseteq X$ compact open* *(or with the U from any basis of such sets) a generating set of finitely presented objects of* Mod-$\mathcal{O}_X$.

Corollary 16.3.18. ([526, 5.7]) *If X is* **locally noetherian**, *that is, a union of open subsets each of which is noetherian, and $\mathcal{O}_X$ is any sheaf of rings on X, then* Mod-$\mathcal{O}_X$ *is locally finitely presented.*

Concerning stronger conditions, one has, for example, that if $(X, \mathcal{O}_X)$ is a locally noetherian prescheme (a condition on the rings, not just the topology) then, [268, 7.8], Mod-$\mathcal{O}_X$ even has a noetherian set of generators, however, [268, p. 135], the category of quasi-coherent sheaves over such a prescheme need not be locally noetherian.

There are some results in [526] giving conditions for sheaves of the form $j_!\mathcal{O}_K$, with K locally closed, to be finitely presented. The condition for such a sheaf to be finitely generated is 16.3.21 below.

Question 16.3.19. When is Mod-$\mathcal{O}_X$ locally finitely presented?

Certainly some conditions are needed for local finite presentation of the category of sheaves: by 16.3.22 if X is not locally compact, then the category of sheaves is not even locally finitely generated. The results are compatible with the possibility that the answer depends only on the underlying space, not on the ringed space per se and, in [526] we asked if this is so.[5]

16.3.4 The category Mod-$\mathcal{O}_X$: local finite generation

Any finitely generated sheaf has compact support (16.3.20). If the category of sheaves is locally finitely generated, then the space must be locally compact (16.3.22).

Here we consider the weaker condition that Mod-$\mathcal{O}_X$ be locally finitely generated.

Suppose that $F \in$ Mod-$\mathcal{O}_X$ and $s \in FX$. As in Section 16.3.1, define a subpresheaf, $\langle s\rangle^0$, of F by setting: $\langle s\rangle^0 X = sR_X$, $\langle s\rangle^0 U = \mathrm{res}^F_{X,U}(s) \cdot \mathcal{O}_X(U)$ and with the restriction maps coming from F. This is a subpresheaf of F, hence is separated, but is not necessarily a sheaf. Let $\langle s\rangle$ denote the sheafification of $\langle s\rangle^0$: this is a subsheaf of F.

More generally, if $U \subseteq X$ is open and $s \in FU$, then define the subsheaf of F **generated by** s to be $j_!\langle s\rangle$, where $j : U \to X$ is the inclusion and $\langle s\rangle \leq F \restriction_U$ is defined as above.

[5] Subsequently Tibor Beke has outlined an argument, since carried through by Philip Bridge, that this should, indeed, be the case, over more general Grothendieck sites.

The adjunction between $j_!$ and restriction yields the inclusion $j_! F \mid_U \to F$, so, since $j_!$ is (left) exact, $j_! \langle s \rangle$ is indeed a subsheaf of F. Although we call it the sheaf generated by s it need not be a finitely generated sheaf: unless U is compact s might not be a "finitary element" of F but note that if $G \leq F$ is a subsheaf such that $s \in GU$, then $j_! \langle s \rangle \leq G$ (this is immediate from the definitions).

It also follows easily ([526, 6.2]) that if $F \in \text{Mod-}\mathcal{O}_X$, then $F = \sum \{ j_! \langle s \rangle : U \subseteq X$ is open and $s \in FU\}$, where we write $j_!$ for $(j_U)_!$ where $j_U : U \to X$ is the inclusion. Infinite sums in Mod-$\mathcal{O}_X$ are obtained by first forming the presheaf sum, that is, the algebraic sum of U-sections at each open $U \subseteq X$, and then sheafifying (see, for example, [318, p. 90]). Furthermore ([526, 6.3]), if $s \in FX$ and $(V_\lambda)_\lambda$ is an open cover of X, then $\langle s \rangle = \sum_\lambda j_! \langle \text{res}^F_{X, V_\lambda} s \rangle$.

The **support** of $s \in FU$ is $\text{supp}(s) = \{x \in X : s_x \neq 0\}$ and this is always a closed subset of U (by the construction of stalks). Set $\text{supp}(F) = \{x \in X : F_x \neq 0\}$.

Proposition 16.3.20. ([526, 6.5]) *Let $\mathcal{O}_X$ be a sheaf of rings and let $F \in$ Mod-$\mathcal{O}_X$ be finitely generated. Then* supp(F) *is compact.*

A subset of X is **locally closed** if it has the form $U \cap C$ for some open $U \subseteq X$ and closed $C \subseteq X$.

Proposition 16.3.21. ([526, 6.7]) *Let $K \subseteq X$ be locally closed. Then $j_! \mathcal{O}_K$ is finitely generated iff K is compact.*

Say that a topological space X is **locally compact** if for every $x \in X$ and for every open set U containing x there is an open set V containing x and a compact locally closed set K with $x \in V \subseteq K \subseteq U$.

Theorem 16.3.22. ([526, 6.11]) *Suppose that* Mod-$\mathcal{O}_X$ *is locally finitely generated. Then X is locally compact.*

Proof. Given $x \in X$ and $U \subseteq X$, consider $j_! \mathcal{O}_U$. By assumption there is an epimorphism from a direct sum of finitely generated sheaves to $j_! \mathcal{O}_U$ and this must be surjective on stalks. Hence there is a finitely generated sheaf, $F \in$ Mod-$\mathcal{O}_X$, and $f : F \to j_! \mathcal{O}_U$ such that $f_x : F_x \to (j_! \mathcal{O}_U)_x = \mathcal{O}_x$ has $1_x \in \mathcal{O}_x$ in its image and which, therefore, is surjective at x. It follows that there is an open set, V', with $x \in V' \subseteq U$ and sections $s' \in FV'$ and $t' \in j_! \mathcal{O}_U \cdot V'$ such that $f_{V'} s' = t'$ and $t' \mapsto 1_x$ under the canonical map $\mathcal{O}_U V' \to \mathcal{O}_x$. Since $(t' - 1_{V'})_x = 0$ there is an open set, V, with $x \in V \subseteq V'$ and with $\text{res}_{V'V}(t') = \text{res}_{V'V}(1_{V'}) = 1_V$. So there is an open set V with $x \in V \subseteq U$ and a section s $(= \text{res}^F_{V'V}(s'))$ with $f_V s = 1_V \in \mathcal{O}_V V$. Thus $V \supseteq \text{supp}(s) \supseteq \text{supp}(f_V s) = V$, that is, $\text{supp}(s) = \text{supp}(f_V s) = V$.

Since F is finitely generated, fF is a finitely generated subsheaf of $j_!\mathcal{O}_U$, so, by 16.3.21, $\mathrm{supp}(fF) \subseteq \mathrm{supp}\, j_!\mathcal{O}_U = U$ is a compact locally closed subset of X which contains V. □

The property of Mod-$\mathcal{O}_X$ being locally finitely generated depends on the sheaf $\mathcal{O}_X$ of rings, not just on X. For example, if X is the closed interval $[0, 1]$ and if $\mathcal{O}_X$ is the sheaf of continuous functions from X to $\mathbb{R}$, then ([526, 6.12]) the $j_!\mathcal{O}_K$ with $K = [e, f], 0 \le e < f \le 1$ form a generating set of finitely presented objects, so Mod-$\mathcal{O}_X$ is locally finitely generated. On the other hand, if $\mathcal{O}_K$ is the (locally) constant sheaf on $X = [0, 1]$ with values in, say, a field K, then ([526, 6.13]) Mod-$\mathcal{O}_K$ is not locally finitely generated.

16.3.5 Pp conditions in categories of sheaves

We point out some relations, and lack thereof, between pointwise, local and global pp-definability. Most of this comes from [519] (which also has some small examples based on quivers).

Suppose that Mod-$\mathcal{O}_X$ is locally finitely presented. Appropriate many-sorted languages for such a category, and the corresponding notions of pp condition, are described in Sections 10.2.4 and 11.1.5. To each choice of generating subcategory, $\mathcal{G} \subseteq$ mod-$\mathcal{O}_X$ there is a corresponding language, with a sort for each isomorphism type of object in that subcategory, but they are all equivalent: a condition written in one language may be rewritten as a condition in any other one.

By 16.3.17 the objects of $\mathcal{G}$ may be taken to have the form $j_!\mathcal{O}_U$ for U ranging over the compact open sets (or just a basis of these). For purposes of illustrating the pp conditions of the corresponding language we give a direct proof of the following result which is a consequence of 11.1.21 and 16.3.12.

Proposition 16.3.23. ([519, 1.3]) *Suppose that X has a basis $\mathcal{B}$ of compact open sets. Then the category* Mod-$\mathcal{O}_X$ *is a definable subcategory of the functor category* $(\mathcal{G}^{\mathrm{op}}, \mathbf{Ab})$, *where $\mathcal{G}$ is the full subcategory of* Mod-$\mathcal{O}_X$ *on the $j_!\mathcal{O}_U$ with $U \in \mathcal{B}$.*

Proof. We need the fact, seen in the proof of 16.3.16, that for any sheaf, $F \in$ Mod-$\mathcal{O}_X$, and open set, $U \subseteq X$, one has $(j_!\mathcal{O}_U, F) \simeq FU$. Since the open sets corresponding to objects of $\mathcal{G}$ are compact, it is then clear that the presheaves on which the following pp-pairs are closed are exactly the sheaves:

$$\left(\bigwedge_{i=1}^{n} \mathrm{res}_{U,U_i} x_U = 0\right) \Big/ (x_U = 0),$$

$$\left(\bigwedge_{i,j=1}^{n} \mathrm{res}_{U_i,U_i\cap U_j} x_{U_i} = \mathrm{res}_{U_j,U_i\cap U_j} x_{U_j}\right) \Big/ \left(\exists x_U \left(\bigwedge_{i=1}^{n} \mathrm{res}_{U,U_i} x_U = x_{U_i}\right)\right),$$

where U ranges over compact open sets and $\{U_1, \ldots, U_n\}$ ranges over finite open covers of such U by compact open sets. These, therefore, form a subcategory which is definable according to the definition in Section 3.4.1.

The paragraphs which follow should clarify why these are pp conditions in the language based on $\mathcal{G}$. $\square$

The form of a general pp condition $\phi(x_1, \ldots, x_n)$ where the sort of x_i is V_i (meaning the sort corresponding to $j_!\mathcal{O}_{V_i}$) is $\exists y_1, \ldots, y_m \bigwedge_k \Sigma_i x_i r_{ik} + \Sigma_j y_j s_{jk} = 0$. Here, if the sort of y_j is W_j, then for each k there is a sort U_k such that $r_{ik} : j_!\mathcal{O}_{U_k} \to j_!\mathcal{O}_{V_i}$ and $s_{jk} : j_!\mathcal{O}_{U_k} \to j_!\mathcal{O}_{W_j}$: each term $x_i r_{ik}$, $y_j s_{kj}$ represents, when the variables x_i, y_j are given values, an element of sort U_k.

In order to understand the language beyond this formalism it is necessary to consider the morphism groups of the form $(j_!\mathcal{O}_U, j_!\mathcal{O}_V)$, where U, V are compact open. Recall, from the proof of 16.3.14, that $j_!$ is left adjoint to the restriction functor. Therefore there is the immediate reinterpretation of the "elements" of a sheaf $F \in \text{Mod-}\mathcal{O}_X$, namely $s_U F = (j_!\mathcal{O}_U, F) \simeq (\mathcal{O}_U, F\,|_U) \simeq FU$. That is, the elements of F of sort U are precisely the sections of F over U, where U is compact open.[6]

Now consider the function symbols of the language. Let $r \in (j_!\mathcal{O}_U, j_!\mathcal{O}_V)$; here "$j$" is used for both embeddings $U \to X$, $V \to X$ without, I hope, causing confusion. Then $r \in j_!\mathcal{O}_V \cdot U = \{s \in \mathcal{O}_V(U \cap V) : \text{supp}(s) \text{ is closed in } U\} = \{s \in \mathcal{O}_X(U \cap V) : \text{supp}(s) \text{ is closed in } U\}$. That is, $(j_!\mathcal{O}_U, j_!\mathcal{O}_V)$ may be identified with the set of those sections of the structure sheaf, $\mathcal{O}_X$, over $U \cap V$ whose support is closed in U. In particular, if $U \subseteq V$, then $(j_!\mathcal{O}_U, j_!\mathcal{O}_V) = \mathcal{O}_X(U)$ (recall that the support of a section over an open set is always closed in that open set).

Lemma 16.3.24. ([519, 1.4]) *The action of $r \in (j_!\mathcal{O}_U, j_!\mathcal{O}_V)$ on $F \in \text{Mod-}\mathcal{O}_X$, regarded as a map from FV to FU, is restriction from FV to $F(U \cap V)$, followed by multiplication by r, regarded as an element of $\mathcal{O}_X(U \cap V)$, followed by inclusion in FU.*

Proof. Note that, since $\text{supp}(r)$, r regarded as an element of $\mathcal{O}_X(U \cap V)$, is closed in U, any section of $F(U \cap V)$ which is a multiple of r can be extended to a section of FU, by defining it to be 0 on the, open, complement of $\text{supp}(r)$ in U.

That the action is exactly as described can be deduced by following through the adjunction isomorphism at the level of stalks ([318, Chpt. 2] gives such details). $\square$

[6] I remark that if we wished sections of F over arbitrary open sets to be considered as "elements", then an infinitary language, a multi-sorted $L_{\kappa\infty}$ where κ is the least cardinal such that every open cover of an open set has a subcover of cardinality less than κ, would be appropriate.

In particular, if $U \subseteq V$, then the action of $r \in (j_!\mathcal{O}_U, j_!\mathcal{O}_V) = \mathcal{O}_X(U)$ is restriction to U followed by multiplication by r. Note that the element $1 \in \mathcal{O}_X U$ corresponds, for each $V \supseteq U$ to the element in $(j_!\mathcal{O}_U, j_!\mathcal{O}_V)$ whose action on any sheaf $F \in \text{Mod-}\mathcal{O}_X$ is just restriction, $\text{res}_{V,U}$, from V to U; so the restriction maps between compact opens are function symbols of the language.

For example, consider the annihilator condition $xr = 0$, where x has sort U and $r \in (j_!\mathcal{O}_V, j_!\mathcal{O}_U)$, that is, $r : s_U \to s_V$. Then $a \in FU$ satisfies $xr = 0$ iff $\text{res}^F_{U,U\cap V}(a) \cdot r = 0$, where in the latter equation we regard $r \in \mathcal{O}_X(U \cap V)$.

For another example, consider the divisibility condition $r \mid x$, that is, $\exists y\,(x = yr)$, where x has sort U, $r \in (j_!\mathcal{O}_U, j_!\mathcal{O}_V)$ and y has sort V. Then $a \in FU$ satisfies $r \mid x$ iff $\exists b \in FV$ such that $\text{res}^F_{V,U\cap V}(b) \cdot r = \text{res}^F_{U,U\cap V}(a)$.

In particular, it is clear that any pp condition for $\mathcal{O}_X(U)$-modules may be "translated" to one for $\mathcal{O}_X$-modules whenever U is compact open. For, each $r \in R$ may be regarded as corresponding to a function symbol in the language for sheaves since $r \in (\mathcal{O}_X(U), \mathcal{O}_X(U)) \simeq \mathcal{O}_X(U) \simeq (j_!\mathcal{O}_U, j_!\mathcal{O}_U)$, noting that $(j_!\mathcal{O}_U, j_!\mathcal{O}_U) \simeq \big(\mathcal{O}_U, (j_!\mathcal{O}_U) \restriction_U\big) \simeq \big((j_!\mathcal{O}_U) \restriction_U\big)(U) \simeq (j_!\mathcal{O}_U)(U) \simeq \mathcal{O}_U(U) \simeq \mathcal{O}_X(U)$.

Then it is easy to prove the following.

Lemma 16.3.25. ([519, 1.5]) *If $U \subseteq X$ is compact open and $\phi(\overline{x})$ is a pp condition for $\mathcal{O}_X(U)$-modules, then there is a pp condition, $\phi'(\overline{x}')$, for $\mathcal{O}_X$-modules with all variables (free, and bound by an existential quantifier) of sort U, such that for every $F \in$ Mod-$\mathcal{O}_X$ and for every tuple $\overline{a}$ in the $\mathcal{O}_X(U)$-module FU, we have $\overline{a} \in \phi(FU)$ iff $\overline{a} \in \phi'(F)$.*

Note that the "$\overline{a}$" on the two sides of this equivalence refer, strictly speaking, to different, but equivalent, objects.

Corollary 16.3.26. ([519, 1.6]) *If $U \subseteq X$ is compact open, $F \in$ Mod-$\mathcal{O}_X$ and if H is a pp-definable subgroup of the $\mathcal{O}_X(U)$-module FU, then H is also a pp-definable subgroup of the sheaf F in sort U.*

The converse to the corollary is far from being true, as can be seen from various examples presented in [519, §2]. Also see [165, §§3, 4] where purity and pure-injectivity of sheaves are considered. A different sense of pure-injectivity of sheaves is considered in [668].

17

Spectra of triangulated categories

Purity and associated ideas may be defined in compactly generated triangulated categories via their associated functor categories and also internally. Basic definitions and results on purity are presented, mostly without proof. The Ziegler spectrum of a ring is related to that of the derived category of its modules and, in certain contexts, the rep-Zariski spectrum is related to the Zariski spectrum of the cohomology ring.

Triangulated categories (Section 17.1), such as the derived category of an abelian category and the stable module category of a group ring, often arise from abelian categories but, though additive, they are almost never themselves abelian. Nevertheless, compactly generated triangulated categories (Section 17.2) are roughly analogous to locally finitely presented abelian categories and it turns out that the theory around purity can be developed for such categories (Section 17.3). This can be done via an associated functor category but can also be done directly, without reference to that associated abelian category. Krause defined the Ziegler spectrum of a compactly generated triangulated category in terms of the associated functor category and the usual connections, for instance with definable subcategories, are there (Section 17.3.1). The relevant notion of localisation is described in Section 17.4. There is some relation between the Ziegler spectrum of a ring and that of its derived category (Section 17.3.2).

Benson and Gnacadja [73] linked triangulated categories and purity in the context of stable module categories of group rings. Krause reformulated and investigated the Telescope Conjecture of homotopy theory from the purity/functor category point of view. Benson and Krause showed, [76], that the spectrum of the Tate cohomology ring of a finite group is homeomorphic to a part of the rep-Zariski spectrum of the group ring (Section 17.5). These authors, Beligiannis and others have built up a considerable body of work developing ideas and results around purity in the context of triangulated categories (and more general contexts, for example, [64]). Brown representability (Section 17.2.1) plays a central technical role.

Most of the key examples of triangulated categories are compactly generated and we stay in that "finitary" context. Just as with abelian categories and, more generally, accessible categories, infinitary languages seem most appropriate for more general contexts (see especially [462], also, for example, [367]).

For background and more detail on this topic see, for example, [708, Chpt. 10], [318, Chpt. 1], [343], [264], [229], [336], [462], [252] and, from a different viewpoint, Strickland's survey [667].

This chapter gives definitions, statements of some results and examples, but few proofs (and those just to give some flavour).

17.1 Triangulated categories: examples

Basic definitions and examples, and a few illustrative arguments, are given.

A **triangulated category** is an additive category, $\mathcal{C}$, equipped with an autoequivalence, variously called **shift** or **translation**, and a class of **distinguished triangles**, such that the axioms listed below are satisfied. By an autoequivalence is meant a functor $T : \mathcal{C} \longrightarrow \mathcal{C}$ such that there is a functor $T^{-1} : \mathcal{C} \longrightarrow \mathcal{C}$ with both $T^{-1}T$ and TT^{-1} naturally equivalent to the identity functor of $\mathcal{C}$. We will use the notation $C \mapsto C[1]$, $f \mapsto f[1]$ to denote shift and $(-) \mapsto (-)[-1]$ to denote its inverse, so $(C[1])[-1]$ is naturally isomorphic to, and in practice identified with, C.

A **triangle** is a triple, (f, g, h), of morphisms of $\mathcal{C}$ of the form $A \xrightarrow{f} B \xrightarrow{g} C \xrightarrow{h} A[1] (\xrightarrow{f[1]} B[1] \cdots)$.

A **morphism** $(f, g, h) \to (f', g', h')$ of triangles is a triple, (a, b, c), of morphisms such that the diagram shown commutes

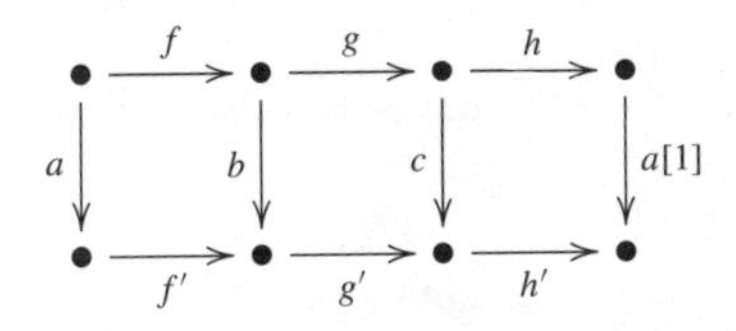

(a) Every morphism $f \in \mathcal{C}$ is contained in a distinguished triangle of the form (f, g, h).
(b) For every object $A \in \mathcal{C}$ the triangle $(1_A, 0_{A\to 0}, 0)$ is distinguished.
(c) Every triangle isomorphic to a distinguished triangle is distinguished.

(d) If (f, g, h) is a distinguished triangle, then so are $(g, h, -f[1])$ and $(-h[-1], f, g)$. (Therefore any, appropriately signed, "rotation"[1] of a distinguished triangle is distinguished.)

(e) If (f, g, h) and (f', g', h') are distinguished triangles and morphisms a, b of $\mathcal{C}$ are given such that $bf = f'a$, then there is a (not necessarily unique) completion c to a morphism of triangles as shown.

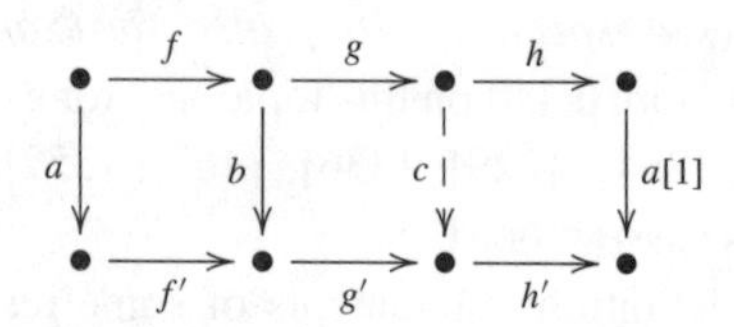

(f) (Octahedral Axiom) Given objects A, B, C, A', B', C' of $\mathcal{C}$ suppose that there are three distinguished triangles $A \overset{f}{\to} B \overset{g}{\to} C \overset{h}{\to} A[1]$, $B \overset{a}{\to} C' \overset{b}{\to} A' \overset{c}{\to} B[1]$ and $A \overset{af}{\to} C' \overset{d}{\to} B' \overset{e}{\to} A[1]$. Then there is a fourth distinguished triangle $C \overset{k}{\to} B' \overset{l}{\to} A' \xrightarrow{g[1]c} C'[1]$ with $ek = h, ld = b$, $da = kg$, $f[1]e = cl$. Refer to [708, p. 375] (or [332, p. 244], or [336, p. 74], ...) for a diagram illustrating the name. Also see the alternative visualisation on p. 376 of [708] which makes it clearer that this is more or less one of the standard isomorphism theorems.

Here are some of the principal examples.

Derived categories From each abelian category, $\mathcal{A}$, a triangulated category is defined as follows (again, for details, see the references). First form the category, $\mathrm{Ch}(\mathcal{A})$, of cochain complexes:[2] the objects are $\mathbb{Z}$-indexed complexes $\cdots \to C^n \xrightarrow{d^n(=d^n_C)} C^{n+1} \to \cdots$ of objects and morphisms of $\mathcal{A}$. Write $C^\bullet$ or $(C^n, d^n)_n$ for such a complex. A **morphism** (of degree 0), $f^\bullet$ from $C^\bullet$ to $D^\bullet$ is given by, for each n, a morphism $f^n : C^n \to D^n$ such that $d^n_D f^n = f^{n+1} d^n_C$. We often drop the $^\bullet$ in the notation for morphisms.

A **homotopy**, s, from $f^\bullet$ to $g^\bullet$, where these are morphisms from $C^\bullet$ to $D^\bullet$, is given by, for each n, a morphism $s^n : C^n \to D^{n-1}$ such that $d^{n-1}_D s^n - s^{n+1} d^n_C =$

[1] This, and the term "triangle", refer to the way of representing a triangle as shown where h is actually to $A[1]$.

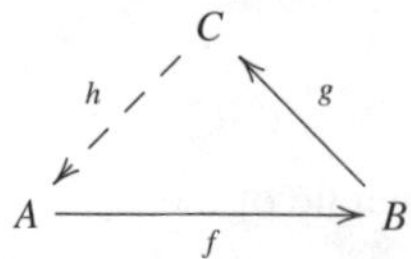

[2] "Cochain" as opposed to "chain" is used to indicate that we are using superscripted indices which increase to the right.

$f^n - g^n$. If such s exists, then say that $f^\bullet$ and $g^\bullet$ are **homotopic**. Define the **homotopy category**, $K\mathcal{A}$, to have the same objects as $\mathrm{Ch}(\mathcal{A})$ but to have morphisms replaced by their homotopy equivalence classes. That is, $K\mathcal{A}$ is $\mathrm{Ch}(\mathcal{A})$ modulo the ideal of morphisms homotopic to 0.

The category $K\mathcal{A}$ is triangulated, as are versions of this built respectively from bounded above, bounded below or bounded (above and below) complexes (superscripts $+, -, b$ are used to denote these variants). The translation functor is $C^\bullet \mapsto C^\bullet[1]$, where $C^\bullet[1]$ is $C^\bullet$ shifted one place to the left, that is, $(C^\bullet[1])^n = C^{n+1}$. The triangulated structure is given by cones on morphisms. The **cone** on a morphism $f^\bullet : C^\bullet \to D^\bullet$ is the complex, cone$(f^\bullet)$, with nth term $C^{n+1} \oplus D^n$ and differential given by (multiplication on the left by) $\begin{pmatrix} -d_C^{n+1} & 0 \\ f^{n+1} & d_D^n \end{pmatrix}$. This fits into an exact sequence (with obvious maps, though note that $\delta^\bullet$ has alternating signs) $0 \to D^\bullet \xrightarrow{g^\bullet} \mathrm{cone}(f^\bullet) \xrightarrow{\delta^\bullet} C^\bullet[1] \to 0$ (for example, [318, p. 23], [708, p. 18]). The distinguished triangles are those triples isomorphic to a triple of the form $(f^\bullet, g^\bullet, \delta^\bullet)$.

The (co)**homology**, $H^*(C^\bullet)$, of a complex $C^\bullet$ consists of the (co)homology objects (in $\mathcal{A}$), $H^n C^\bullet = \ker(d^n)/\mathrm{im}(d^{n-1})$, of the complex. Any morphism $f^\bullet : C^\bullet \to D^\bullet$ of chain complexes induces morphisms (in $\mathcal{A}$) $H^n f : H^n C^\bullet \to H^n D^\bullet$ for each n and $f^\bullet$ is a **quasi-isomorphism** if each $H^n f$ is an isomorphism. The **derived category**, $D\mathcal{A}$, of $\mathcal{A}$ is obtained from the homotopy category by a process of localisation which inverts the quasi-isomorphisms in $K\mathcal{A}$, that is, invert those morphisms in the category $K\mathcal{A}$ which induce an isomorphism of (co)homology. The localisation process, which is forming a "category of fractions", has much in common with torsion-theoretic localisation (Section 11.1.1) and universal localisation (5.5.5). See, for example, [206], [332], [486, §4.1] or [708], for details (including the question of existence). The result is again a triangulated category, with the triangulated structure induced directly from that on $K\mathcal{A}$. We will also write $\mathcal{D}(R)$ for $D(\text{Mod-}R)$.

There is a full embedding $\text{Mod-}R \longrightarrow \mathcal{D}(R)$ which takes $M \in \text{Mod-}R$ to the complex which has M in degree 0 and zeroes elsewhere. Such a complex is said to be **concentrated in degree** 0.

An important point is that Ext groups between objects of $\mathcal{A}$ are realised as hom groups in the derived category: $\mathrm{Ext}^n_{\mathcal{A}}(C, D) \simeq \mathrm{Hom}_{D(\mathcal{A})}(C, D[n]) \simeq \mathrm{Hom}_{D(\mathcal{A})}(C[-n], D)$, where C and D are regarded as objects, concentrated in degree 0, of the derived category (e.g. [318, pp. 60, 63] or [708, §10.7]).

The category $D\mathcal{A}$ is hardly every abelian (it is so iff all short exact sequences in $\mathcal{A}$ split, for example, see [252, 6.12]).

Stable module categories Let R be a **quasi-Frobenius** ring; that is, R is right and left artinian and is injective as a right and left module over itself. Every group ring kG, where k is a field and G is a finite group, is quasi-Frobenius, hence so is every block (ring direct factor) of such a group ring. If R is quasi-Frobenius, then every finite projective resolution is a split exact sequence (since every projective module is injective) hence every module either is projective = injective or has projective dimension ∞.

For M, N in Mod-R denote by $\mathcal{P}(M, N)$ the subgroup of (M, N) consisting of those morphisms which factor through a projective module. The **stable module category** $\underline{\text{Mod}}$-R is the quotient of Mod-R by the ideal (Section 9.1.1) $\mathcal{P}$ formed by these, that is, the ideal generated by the (identity maps of) projective modules. Thus $\underline{\text{Mod}}$-R has the same objects as Mod-R but the morphism group from M to N, written $\underline{\text{Hom}}(M, N)$, in this new category is $(M, N)/\mathcal{P}(M, N)$. See [24] and, for example, [596, §3] for more detail on the relation between the two categories.

Then $\underline{\text{Mod}}$-R is a triangulated category with translation being the functor which takes M to $\Omega^{-1}M$, where $\Omega^{-1}M$ is the cokernel of the inclusion of M into an injective envelope: $0 \to M \to E(M) \to \Omega^{-1}M \to 0$. The inverse translation, Ω, is given by the kernel of a projective cover, that is, the beginning of a *minimal* projective resolution $P_0 \to M$ of M (by 4.3.65 and E.1.23 such exist over quasi-Frobenius rings). For $M, N \in$ Mod-R one has $\text{Ext}^1_R(M, \Omega N) \simeq \underline{\text{Hom}}(M, N) \simeq \text{Ext}^1_R(\Omega^{-1}M, N)$ (for example, [708, §10.7]).

The distinguished triangles are those isomorphic to one of the form $X \to Y \to Z \to \Omega^{-1}X$ (equivalently, to those of the form $\Omega Z \to X \to Y \to Z$), where $0 \to X \to Y \to Z \to 0$ is an exact sequence in Mod-R and the morphism $Z \to \Omega^{-1}X$ is that which corresponds to the given extension of Z by X under the isomorphism $\text{Ext}(Z, X) \simeq \underline{\text{Hom}}(Z, \Omega^{-1}X)$ (see, for example, [69, 2.5.7]).

Stable module categories of more general rings are considered in [326].

Homotopy categories The homotopy category of CW spectra is triangulated, with suspension giving the translation functor and mapping cones giving the triangles, see, for example, [708, p. 409ff.].

I just point, for example, see [301], [303], to somewhat related ideas concerning model categories.

The following basic lemmas are included for illustration.

Lemma 17.1.1. *Suppose that* $A \xrightarrow{f} B \xrightarrow{g} C \xrightarrow{h} A[1]$ *is a distinguished triangle of the triangulated category* $\mathcal{C}$. *Then* $0 = gf = hg = f[1]h$.

Proof. It is enough "by rotating" (applying some power of the shift functor) to prove this for one composition. A dashed arrow making the diagram (and its rotation) commutative exists by a rotation of the fifth axiom, so $gf = 0$ by commutativity of the left-hand square.

$$
\begin{array}{ccccccc}
A & \xrightarrow{f} & B & \xrightarrow{g} & C & \xrightarrow{h} & A[1] \\
\downarrow & & \downarrow{\scriptstyle b} & & \| & & \downarrow \\
0 & \longrightarrow & C & = & C & \longrightarrow & 0
\end{array}
$$

□

Also every morphism $A \xrightarrow{f} B$ has a pseudocokernel (Appendix E, p. 704), namely if $A \xrightarrow{f} B \xrightarrow{g} C \to A[1]$ is a distinguished triangle (existence by the first axiom), then g is a pseudocokernel for f. For, given $B \xrightarrow{h} B'$ with $hf = 0$ consider the diagram

$$
\begin{array}{ccccccc}
A & \xrightarrow{f} & B & \xrightarrow{g} & C & \longrightarrow & A[1] \\
\downarrow & & \downarrow{\scriptstyle h} & & \downarrow{\scriptstyle k} & & \\
0 & \longrightarrow & B' & = & B' & \longrightarrow & 0
\end{array}
$$

where the morphism $C \xrightarrow{k} B'$ with $kg = h$ exists by the fifth axiom.

Lemma 17.1.2. *Suppose that $A \xrightarrow{f} B \xrightarrow{g} C \xrightarrow{h} A[1]$ is a distinguished triangle of the triangulated category $\mathcal{C}$. Then for every object $D \in \mathcal{C}$ the induced chain complex $\cdots \to (D, C[-1]) \to (D, A) \to (D, B) \to (D, C) \to (D, A[1]) \to \cdots$ is exact, as is $\cdots \to (A[1], D) \to (C, D) \to (B, D) \to (A, D) \to (C[-1], D) \to \cdots$.*

Proof. By 17.1.1 these are chain complexes. By rotating, it is enough to check exactness at just one point, say at (D, B). Let $D \xrightarrow{a} B \in \ker\big((D, B) \to (D, C)\big)$. Consider the diagram shown: by assumption the middle square commutes, so the dotted arrow exists by (a rotation of) the fifth axiom.

$$
\begin{array}{ccccccc}
D & = & D & \longrightarrow & 0 & \longrightarrow & D[1] \\
\downarrow & & \downarrow{\scriptstyle a} & & \downarrow & & \downarrow \\
A & \xrightarrow[f]{} & B & \xrightarrow[g]{} & C & \xrightarrow[h]{} & A[1]
\end{array}
$$

This shows that $a \in \operatorname{im}\big((D, A) \to (D, B)\big)$.

The opposite of a triangulated category clearly is triangulated so exactness of the second long exact sequence follows from that of the first. □

17.2 Compactly generated triangulated categories

Compact objects are somewhat analogous to finitely presented objects and the compactly generated triangulated categories are reminiscent of locally finitely presented abelian categories.

Suppose that $\mathcal{C}$ is a triangulated category with coproducts.

An object, C, of $\mathcal{C}$ is **compact** if for all sets, $(D_\lambda)_\lambda$, of objects in $\mathcal{C}$ the canonical morphism $\bigoplus_\lambda (C, D_\lambda) \to (C, \bigoplus_\lambda D_\lambda)$ is an isomorphism. Denote by $\mathcal{C}^c$ the full subcategory of compact objects of $\mathcal{C}$.

A **triangulated subcategory** of a triangulated category is a full subcategory which is closed under shift and its inverse and which is such that if $A \to B \to C \to A[1]$ is a distinguished triangle and two of A, B, C are in the subcategory, then so is the third. Such a subcategory, with the induced structure, is itself triangulated.

Lemma 17.2.1. *Let $\mathcal{C}$ be triangulated. Then the full subcategory, $\mathcal{C}^c$, of compact objects is a triangulated subcategory of $\mathcal{C}$.*

Proof. The shift of a compact object is compact since shift is an automorphism of $\mathcal{C}$. If $A \to B \to C \to A[1]$ is a distinguished triangle and $(C_i)_i$ are objects of $\mathcal{C}$, then, for each i, the sequence $\cdots \to (A[1], C_i) \to (C, C_i) \to (B, C_i) \to (A, C_i) \to (C[-1], C_i) \to \cdots$ is exact (by 17.1.2), as is the sequence with $\bigoplus_i C_i$ replacing C_i. On comparing the latter sequence with $\cdots \to \bigoplus_i (A[1], C_i) \to \bigoplus_i (C, C_i) \to \bigoplus_i (B, C_i) \to \bigoplus_i (A, C_i) \to \bigoplus_i (C[-1], C_i) \to \cdots$, we see that if two of A, B, C are compact then so is the third. □

Say that $\mathcal{C}$ is **compactly generated** (**algebraic** in the terminology of [304]) if $\mathcal{C}$ has coproducts, $\mathcal{C}^c$ is skeletally small and if for every non-zero object $D \in \mathcal{C}$ there is a compact object C of $\mathcal{C}$ and a non-zero morphism from C to D. This is weaker than the usual categorical notion of generation which, see Appendix E, p. 707, is "sees every morphism" rather than "sees every object, that is, identity morphism", and, in fact, the difference is highly significant (see, for example, 17.3.3). If $\mathcal{C}$ is compactly generated, then a morphism, $f : M \to N$, of $\mathcal{C}$ is said to be **phantom** if $(C, f) : (C, M) \to (C, N)$ is the zero map for every compact object $C \in \mathcal{C}$. Informally, f is phantom if f is not seen by $\mathcal{C}^c$. "Most" compactly generated triangulated categories have non-zero phantom maps, see [63, §9] for those which do not.

Say that a subcategory, $\mathcal{G}$, of a triangulated category, $\mathcal{C}$ is **generating** provided that for every $C \in \mathcal{C}$ if $(\mathcal{G}, C) = 0$, then $C = 0$.

Example 17.2.2. If R is any ring, then it is straightforward to check that the derived category, $\mathcal{D}(R)$, of unbounded complexes of right R-modules is compactly generated, with the compact objects being the **perfect** complexes – those isomorphic to a bounded complex consisting of finitely generated projectives. For example, a module of projective dimension n which is also FP_n is, if regarded as a complex concentrated in degree 0, isomorphic to a perfect complex, namely any finite projective resolution by finitely generated projectives.

Proposition 17.2.3. ([119, 1.1]) *The following are equivalent for a ring R:*

(i) R is right coherent and each finitely presented right R-module has finite projective dimension;
(ii) every finitely presented right R-module is compact as an object of $\mathcal{D}(R)$;
(iii) a complex, $M^\bullet$, is compact iff each of its cohomology groups, $H^n M^\bullet$, is a finitely presented R-module and $H^n M^\bullet = 0$ for all but finitely many n.

Example 17.2.4. (see [597, §3]) Suppose that R is a quasi-Frobenius ring. Then it is easily checked that the stable module category $\underline{\mathrm{Mod}}$-R is compactly generated, with the image of mod-R being a generating set (up to isomorphism) of compact objects, indeed $(\underline{\mathrm{Mod}}\text{-}R)^{\mathrm{c}}$ may be identified[3] with $\underline{\mathrm{mod}}$-R.

It is a theorem of Rickard, [595, 2.1], that if R is quasi-Frobenius, then $\underline{\mathrm{Mod}}$-R is the quotient of $\mathcal{D}^b(R)$ by the localising subcategory (see Section 17.4) generated by the perfect complexes.

If $\mathcal{C}$ is the homotopy category of CW spectra, then $\mathcal{C}^{\mathrm{c}}$ is the subcategory of finite spectra (that is, shifts of finite CW complexes). Over suitable schemes X the derived category of chain complexes of quasi-coherent sheaves over $\mathcal{O}_X$ and the derived category of quasi-coherent sheaves over the sheaf, $\mathcal{D}_X$, of differential operators on X are compactly generated, see [458]. Other examples can be found in [327]. For some non-examples see [463].

17.2.1 Brown representability

In the abelian context the links between projective, flat and representable functors are clear. In this context the identification of representable functors is non-trivial.

For more detail see the introduction to [119] and Section 9 of [667], on which I have based this summary.

Say that a contravariant additive functor, F, from the triangulated category $\mathcal{C}$ to an abelian category $\mathcal{A}$ is **homological** (or **exact**) if it sends triangles to exact sequences. That is, if $A \to B \to C \to A[1]$ is a triangle in $\mathcal{C}$, then applying F to this and its rotations gives a (doubly) long exact sequence $\cdots \to F(A[1]) \to FC \to FB \to FA \to F(C[-1]) \to \cdots$. For example, any representable functor

[3] But the corresponding statement is not true of arbitrary quotients of Mod-R, see [251].

$(-, C)$ is homological by 17.1.2. The same terminology is used for covariant functors.

For any subcategory, $\mathcal{C}'$, of $\mathcal{C}$ denote by $Y_{\mathcal{C}'}$ the restricted Yoneda functor $\mathcal{C} \longrightarrow ((\mathcal{C}')^{\mathrm{op}}, \mathbf{Ab})$ given by $C \mapsto (-, C) \upharpoonright \mathcal{C}'$. This kind of functor underlies much of what follows.

In [2] Adams, extending a result of Brown [95], showed that every contravariant homological functor from the category of finite spectra (that is, from $\mathcal{C}^{\mathrm{c}}$, where $\mathcal{C}$ is the homotopy category of spectra) is representable by an object of $\mathcal{C}$, that is, is of the form $Y_{\mathcal{C}^{\mathrm{c}}}(C)$ for some $C \in \mathcal{C}$. Moreover, every natural transformation between such functors on $\mathcal{C}^{\mathrm{c}}$ also is representable. In other words, the restricted Yoneda functor $Y_{\mathcal{C}^{\mathrm{c}}}$ induces, for this particular category, an equivalence between $\mathcal{C}$ and the category of homological functors from $(\mathcal{C}^{\mathrm{c}})^{\mathrm{op}}$ to $\mathbf{Ab}$. Brown also showed the following result for the special case of the homotopy category of spectra.

Theorem 17.2.5. *Let $\mathcal{C}$ be a triangulated category with coproducts and suppose that $F : \mathcal{C}^{\mathrm{op}} \longrightarrow \mathbf{Ab}$ is a homological functor which takes coproducts to products. Then F is representable.*

This is sometimes referred to as "Brown representability for cohomology", and it holds for any compactly generated triangulated category $\mathcal{C}$, [462, 8.4.2, 8.4.3], [193, 2.4]. It is a direct consequence that any such triangulated category also has products, see [462, 8.4.6]; for, given a set $(C_i)_i$ of objects of $\mathcal{C}$, the product functor $D \mapsto \prod_i (D, C_i)$ must be representable – by an object which, by definition of product, is $\prod_i C_i$.

The result for functors on the smaller category, that every homological functor from $(\mathcal{C}^{\mathrm{c}})^{\mathrm{op}}$ to $\mathbf{Ab}$ is representable by an object of $\mathcal{C}$, does not hold in general, even for $\mathcal{C}$ of the form $\mathcal{D}(R)$, see [119, 2.11, 2.12], and is referred to as "Brown representability for homology", the reason according to [119] being that in many cases $\mathcal{C}^{\mathrm{c}}$ is self-dual (see [667, pp. 77, 80]) and hence such contravariant functors can also be regarded as covariant functors on $\mathcal{C}^{\mathrm{c}}$. These, in turn, cf. 17.2.16 below, extend canonically to functors on $\mathcal{C}$ which are homological and commute with coproducts.

In [119] the following terminology and notations are introduced. Say that a triangulated category, $\mathcal{C}$, with coproducts satisfies **Brown representability for objects, BRO**, if every contravariant homological functor from $\mathcal{C}^{\mathrm{c}}$ to $\mathbf{Ab}$ is represented by some object of $\mathcal{C}$ (sic), that is, iff the closure under isomorphism of the image of the restricted Yoneda functor, $Y_{\mathcal{C}^{\mathrm{c}}}$, is the class of homological functors. Say that $\mathcal{C}$ satisfies **Brown representability for morphisms, BRM**, if every morphism between functors of the form $(-, C) \upharpoonright \mathcal{C}^{\mathrm{c}}$ with $C \in \mathcal{C}$ is representable, that is, if $Y_{\mathcal{C}^{\mathrm{c}}}$ is full.

Theorem 17.2.6. ([63, 11.8]) *For every compactly generated triangulated category BRM implies BRO.*

Theorem 17.2.7. ([459]) *For a compactly generated triangulated category, $\mathcal{C}$, which satisfies certain countability conditions (for example, if $\mathcal{C} = \mathcal{D}(R)$ for R a countable ring), then BRM, hence also BRO, holds.*

Following Beligiannis [63], define the **pure global dimension** (cf. Section 2.2), pgldim($\mathcal{C}$), of $\mathcal{C}$ to be the supremum over objects $M \in \mathcal{C}$ of the projective dimension of $Y_{\mathcal{C}^c}(M) = (-, M) \restriction \mathcal{C}^c$. This turns out to be equal to the supremum of projective dimensions over all homological contravariant functors even though these classes of functors need not coincide.

Theorem 17.2.8. *(a)* ([63, 11.2]) *If $\mathcal{C}$ is a compactly generated triangulated category, then*[4] $\text{pgldim}(\mathcal{C}) = \sup\{\text{projdim}(F) : F \textit{ is a homological functor on } \mathcal{C}\}$.

(b) ([459], also (i)⇒(ii) in [120]) *The compactly generated triangulated category $\mathcal{C}$ satisfies both BRO and BRM iff* $\text{pgldim}(\mathcal{C}) \leq 1$.

(c) [63, 11.12] *If $\mathcal{C}$ is a compactly generated triangulated category with* $\text{pgldim}(\mathcal{C}) \leq 2$, *then $\mathcal{C}$ satisfies BRO.*

It is shown in [119, §2] that if $R = k[X, Y]$ and $|k| \geq \aleph_2$, then BRM fails for $\mathcal{D}(R)$, and if $|k| \geq \aleph_3$, then BRO also fails. Therefore, as remarked there, the question whether $\mathcal{D}(\mathbb{C}[X, Y])$ satisfies BRO is independent of the usual, ZFC, set theory, cf. 2.2.1(4).

The first example of failure of BRM in a category of the form $\mathcal{D}(R)$ was given by Keller; it is presented in [459]. Also, in [95] there is an example which shows that BRO can fail in such a category.

The relationship between the pure global dimension of R and that of $\mathcal{D}(R)$ is investigated by Christensen, Keller and Neeman [119].

Theorem 17.2.9. ([119, 1.4]) *(a) If R is right coherent and if all finitely presented right R-modules are of finite projective dimension (cf. 17.2.3), then* $\text{pgldim}(\mathcal{D}(R)) \geq \text{pgldim}(R)$.
(b) If R is right hereditary then $\text{pgldim}(\mathcal{D}(R)) = \text{pgldim}(R)$.

The inequality in part (a) can be strict: two finite-dimensional algebras with equivalent derived categories but different pure global dimension are exhibited in [119, 1.5].

[4] This may be compared with the abelian case, see [647, Thm 2.7].

Theorem 17.2.10. ([119, 2.11]) *Suppose that R is right coherent and that every finitely presented right R-module has finite projective dimension. Suppose that there is an R-module M such that* $\mathrm{pinjdim}(M) - \mathrm{injdim}(M) \geq 2$. *Then BRO fails for $\mathcal{D}(R)$.*

Theorem 17.2.11. ([119, 2.13] (cf. 17.2.8)) *Suppose that R is hereditary.*

(1) $\mathcal{D}(R)$ satisfies BRM iff $\mathrm{pgldim} R \leq 1$;
(2) $\mathcal{D}(R)$ satisfies BRO iff $\mathrm{pgldim} R \leq 2$.

For the stable module category BRM is characterised in the following result.

Theorem 17.2.12. ([73, 4.7.1]) *Suppose that G is a finite group and k is a field. Then the following are equivalent.*

(i) BRM holds for $\underline{\mathrm{Mod}}$-kG;
(ii) $\mathrm{pgldim}(kG) \leq 1$;
(iii) either k is countable or the Sylow p-subgroups of G are cyclic, where $p = \mathrm{char}(k)$ *(the latter is equivalent to kG being of finite representation type).*

Brown representability for certain stable (with respect to certain ideals) module categories is proved by Jørgensen [325]. Also see [64], [67], [77], [78] [366] for more on Brown representability.

17.2.2 The functor category of a compactly generated triangulated category

Coherent functors on compactly generated triangulated categories are characterised (17.2.13), as are the flat functors (17.2.15). Extensions of functors from compact objects to all objects are considered (17.2.16, 17.2.17).

In this book, purity in Mod-R has been investigated by looking at associated functor categories, (mod-R, **Ab**) and (R-mod, **Ab**). Some more use could have been made of ((mod-R)^{op}, **Ab**), see comments at the end of Section 16.1.2 and also near the beginning of Section 12.1.1. Investigating a compactly generated triangulated category $\mathcal{C}$ via the associated functor category $((\mathcal{C}^{\mathrm{c}})^{\mathrm{op}}, \mathbf{Ab})$ goes back at least to [120], [460]. Krause used this to define the Ziegler spectrum for such categories.

First consider functors from all of $\mathcal{C}$ to **Ab**. The general results of Chapter 10 apply, though some care is needed since $\mathcal{C}$ is not skeletally small. Say that $F : \mathcal{C} \longrightarrow \mathbf{Ab}$ is **coherent** if there is an exact sequence $(D, -) \to (C, -) \to F \to 0$ induced by a morphism $C \to D$ in $\mathcal{C}^{\mathrm{c}}$. The restriction of F to $\mathcal{C}^{\mathrm{c}}$ is, note, simply a finitely presented object of $(\mathcal{C}^{\mathrm{c}}, \mathbf{Ab})$ (Section 10.2.1).

A **homology colimit** of a diagram in $\mathcal{C}$ composed of objects $(C_i)_i$ and morphisms $(g_k)_k$ is a cocone on the diagram, that is (Appendix E, p. 704), an object C, and compatible (with the g_k) maps $f_i : C_i \to C$, with the following property. Let $X \in \mathcal{C}^c$ and consider the diagram in **Ab** induced by $(X, -)$: this has objects the (X, C_i), and morphisms the (X, g_k). The requirement is that, for all $X \in \mathcal{C}^c$, the map, induced by the $(X, f_i) : (X, C_i) \to (X, C)$, from the colimit of this diagram to (X, C), is an isomorphism. If this is so, then write $C = \text{hocolim}\, C_i$. For the definition of the terms involving purity which appear in the next result, see Section 17.3.

Theorem 17.2.13. ([362, 5.1, 5.2, 5.3]) *Suppose that $\mathcal{C}$ is a compactly generated triangulated category and that $F : \mathcal{C} \longrightarrow$ **Ab** is an additive functor. The following are equivalent.*

(i) F is coherent.
(ii) F commutes with direct products and sends every homology colimit in $\mathcal{C}$ to a colimit in **Ab**.
(iii) F commutes with direct sums and direct products and for every pure triangle, $A \to B \to C \to A[1]$, the sequence $0 \to FA \to FB \to FC \to 0$ is exact.
(iv) F commutes with direct sums of pure-projectives and direct products of pure-injectives and for every pure triangle, $A \to B \to C \to A[1]$, the sequence $0 \to FA \to FB \to FC \to 0$ is exact.
(v) F commutes with direct sums and direct products of compact objects and for every pure triangle $A \to B \to C \to A[1]$ the sequence $0 \to FA \to FB \to FC \to 0$ is exact.

Next consider functors defined on the compact objects.

Remark 17.2.14. Suppose that $\mathcal{C}$ is a compactly generated triangulated category. The restricted Yoneda functor, $Y_{\mathcal{C}^c}$, given by $C \mapsto (-, C)$, $f \mapsto (-, f)$, from $\mathcal{C}$ to $((\mathcal{C}^c)^{\text{op}}, \textbf{Ab})$, is, just from the definitions, injective on objects. A morphism belongs to the kernel of this functor iff it is a phantom morphism.

In general $Y_{\mathcal{C}^c}$ is not full; BRM from the previous section is the condition that it is full.

One has the following characterisation of homological functors. Recall that a functor $F : (\mathcal{C}^c)^{\text{op}} \longrightarrow \textbf{Ab}$ is said to be flat if the functor $F \otimes_{\mathcal{C}^c} - : (\mathcal{C}^c, \textbf{Ab}) \longrightarrow \textbf{Ab}$ is exact, equivalently, 16.1.2, if F is a direct limit of representable functors.

Proposition 17.2.15. (see, for example, [362, 1.5]) *Let $\mathcal{C}$ be a compactly generated triangulated category. The following are equivalent for a functor $F : (\mathcal{C}^c)^{\text{op}} \longrightarrow$* **Ab***:*

(i) F is flat;
(ii) F is homological;
(iii) F is an absolutely pure object of $((\mathcal{C}^c)^{op}, \mathbf{Ab})$.

Indeed, the result is true with any skeletally small triangulated category in place of $(\mathcal{C}^c)^{op}$.

Therefore all representable functors are absolutely pure and the converse is exactly the condition BRO.

Concerning the connection between functors on $(\mathcal{C}^c)^{op}$ and on $\mathcal{C}$ there is the following.

Theorem 17.2.16. ([361, 2.3, 2.4, 2,8, Thm E]) *Suppose that $\mathcal{C}$ is compactly generated. Then for every homological functor $F : (\mathcal{C}^c)^{op} \longrightarrow \mathcal{A}$, where $\mathcal{A}$ is Grothendieck, there is a unique homological functor $F' : \mathcal{C} \longrightarrow \mathcal{A}$ which commutes with coproducts and satisfies $FY_{\mathcal{C}^c} = F'$; indeed, the category of homological functors from $(\mathcal{C}^c)^{op}$ to* $\mathbf{Ab}$ *is equivalent to the category of those homological functors from $\mathcal{C}$ to* $\mathbf{Ab}$ *which commute with coproducts. Furthermore, F' commutes with products iff F, hence F', is representable by an object of $\mathcal{C}^c$.*

The proof uses the fact, [361, 2.1, 2.2], that every functor $F : \mathcal{C}^c \longrightarrow \mathcal{A}$, where $\mathcal{A}$ is abelian, has a unique extension to a right exact functor $((\mathcal{C}^c)^{op}, \mathbf{Ab}) \longrightarrow \mathcal{A}$ which is exact iff F is homological.

Proposition 17.2.17. ([361, 2.9]) *Suppose that the triangulated category $\mathcal{C}$ is compactly generated and suppose that $F : \mathcal{C} \longrightarrow$* $\mathbf{Ab}$ *commutes with coproducts. Then F commutes with products iff F commutes with products of objects from $(\mathcal{C}^c)^{op}$ iff $F = (C, -)$ for some object $C \in \mathcal{C}^c$.*

17.3 Purity in compactly generated triangulated categories

Pure-exact triangles are characteristed (17.3.1, 17.3.2), as are pure-injective objects (17.3.3, 17.3.4). The Jensen–Lenzing characterisation of pure-injectives remains valid (17.3.7). Pure-injectives correspond to injectives of the functor category (17.3.5) and hulls of strongly indecomposable compact objects are as for modules (17.3.6). Indecomposable pure-injectives are purely cogenerating (17.3.8).

Pure-injectivity and phantom maps are linked (17.3.9, 17.3.11, 17.3.12). The pure-injectives in the stable category are the images of the pure-injective modules (17.3.10).

The definable subcategory generated by a collection of objects is described (17.3.15).

The notions of purity, pure-injectivity and the Ziegler spectrum of a compactly generated triangulated category $\mathcal{C}$ are defined, following [361], via the restricted Yoneda functor $Y_{\mathcal{C}^c} : \mathcal{C} \longrightarrow ((\mathcal{C}^c)^{\mathrm{op}}, \mathbf{Ab})$. It has been remarked already that, in contrast with the case where $\mathcal{C}$ is Mod-R and where $\mathcal{C}^c$ is replaced by mod-R, this functor is, in general, neither faithful on morphisms nor full.

Some of the results of this section first appeared, in the context of the stable module category of a group ring, in Benson and Gnacadja [73] (which begins with a useful "homological" account of purity). The results in their form here come mainly from Krause [361], [362], also see [63]. Benson and Gnacadja pointed, for their inspiration, to various works in homotopy theory, in particular [304], [120], [460] (Neeman, in his review, [461], of [73], credits Keller with first noticing a connection between phantom maps and purity). Indeed, the general case is essentially as described in the first few pages of [460].

In a general compactly generated triangulated category, as opposed to the stable module category, there is no abelian category to refer back to, so it is necessary to give intrinsic definitions of purity and associated notions, either within the category or referring to the associated functor category.

Let $\mathcal{C}$ be a compactly generated triangulated category. A morphism $f : A \to B$ of $\mathcal{C}$ is a **pure-monomorphism** if $Y_{\mathcal{C}^c} f = (-, f) : (-, A) \to (-, B)$ is a monomorphism in the functor category $((\mathcal{C}^c)^{\mathrm{op}}, \mathbf{Ab})$; f is a **pure-epimorphism** if the induced map $(-, f)$ is an epimorphism. A triangle $A \to B \to C \to A[1]$ is **pure exact** if the induced sequence, $0 \to (-, A) \to (-, B) \to (-, C) \to 0$, of functors in $((\mathcal{C}^c)^{\mathrm{op}}, \mathbf{Ab})$ is exact; note that this is consistent with the terminology for module categories, 16.1.15. A pure-monomorphism need not be a monomorphism(!) and similarly for epimorphisms. The definition of pure-monomorphism is that, *as far as compact objects are concerned*, the map looks like a monomorphism.

Lemma 17.3.1. ([361, 1.3]) *The following are equivalent for a distinguished triangle $A \xrightarrow{f} B \xrightarrow{g} C \xrightarrow{h} A[1]$ in the compactly generated triangulated category $\mathcal{C}$:*

(i) the triangle is pure-exact;
(ii) f is a pure-monomorphism;
(iii) g is a pure-epimorphism;
(iv) h is a phantom map.

There is an analogue of 2.1.4.

Proposition 17.3.2. ([362, 2.8]) *Suppose that $A \xrightarrow{f} B \xrightarrow{g} C \xrightarrow{h} A[1]$ is a distinguished triangle in the compactly generated triangulated category $\mathcal{C}$. Then the following are equivalent:*

(i) the triangle is is pure-exact;
(ii) the triangle is a homology colimit of a filtered diagram of split triangles;
(iii) the triangle is a homology colimit of a filtered diagram of split triangles in $\mathcal{C}^c$.

An object N of $\mathcal{C}$ is **pure-injective** if every pure-monomorphism with source N splits.

Proposition 17.3.3. ([361, 1.4], also see [73, 4.2.5]) *The following are equivalent for an object N of a compactly generated triangulated category $\mathcal{C}$:*

(i) N is pure-injective;
(ii) if $f : M \to N$ is phantom, then $f = 0$;
(iii) N is injective over all pure-monomorphisms, that is, given any pure-monomorphism $g : A \to B$ and any morphism $f : A \to N$ there is a factorisation of f through g.

Indeed, [361, 2.5] generalising [73, 4.2.2], phantom maps are characterised by: $g : A \to B$ is phantom iff for every indecomposable pure-injective N and every morphism $f : B \to N$ the restriction, fg, of f along g is 0.

Theorem 17.3.4. ([361, 1.7]) *An object N of a compactly generated triangulated category $\mathcal{C}$ is pure-injective iff for every object, M, of $\mathcal{C}$ the Yoneda map from (M, N) to $(Y_{\mathcal{C}^c}M, Y_{\mathcal{C}^c}N)$ is a bijection.*

Compare the next result with 12.1.9.

Proposition 17.3.5. ([361, 1.7, 1.9]) *Let $\mathcal{C}$ be a compactly generated triangulated category. Then every injective object of $((\mathcal{C}^c)^{op}, \mathbf{Ab})$ is isomorphic to a functor of the form $(-, N)$ for some pure-injective object $N \in \mathcal{C}$. The restricted Yoneda functor, $Y_{\mathcal{C}^c}$, restricts to a natural equivalence between the category of pure-injective objects of $\mathcal{C}$ and the category of injective objects of $((\mathcal{C}^c)^{op}, \mathbf{Ab})$.*

It follows, for example, that an indecomposable pure-injective object of $\mathcal{C}$ has local endomorphism ring.

The proof of the next result is just like that of 4.3.55; existence of pure-injective hulls (see below) in $\mathcal{C}$ is via the existence of injective hulls in $((\mathcal{C}^c)^{op}, \mathbf{Ab})$, cf. Section 4.3.3.

Theorem 17.3.6. ([220, 2.3]) *If A is a compact object of the compactly generated triangulated category $\mathcal{C}$, then its pure-injective hull, $H(A)$, is indecomposable iff $\mathrm{End}(A)$ is local. If B is compact and $H(B) \simeq H(A)$, then $B \simeq A$.*

There are the expected further equivalents to pure-injectivity, including the criterion of Jensen and Lenzing (see 4.3.6).

Theorem 17.3.7. ([361, 1.8]) *For an object, N, of a compactly generated triangulated category $\mathcal{C}$ the following are equivalent:*

(i) N is pure-injective;
(ii) for every index set, I, the summation map $N^{(I)} \to N$ factors through the pure-monomorphism $N^{(I)} \to N^I$.

Say that an object N of a compactly generated triangulated category is of **finite endolength** if for every compact object C the $\mathrm{End}(N)$-module (C, N) is of finite length. There is, [360, 1.2], a decomposition theorem like 4.4.19 and there are characterisations and uses of these objects in homotopy categories in [370].

The next result should be compared with Ziegler's [726, 6.9], see 5.1.4, which states that every module is elementarily equivalent to a direct sum of indecomposable pure-injectives or, better, the fact that every module purely embeds in a direct product of indecomposable pure-injectives (5.3.53).

Proposition 17.3.8. ([361, 1.10]) *Let $\mathcal{C}$ be a compactly generated triangulated category. For every object $C \in \mathcal{C}$ there is a pure-monomorphism $C \to \prod_i N_i$ where the N_i are indecomposable pure-injective objects of $\mathcal{C}$.*

A **pure-injective hull** of an object M of a triangulated category $\mathcal{C}$ is a pure-monomorphism $M \to N$ with N pure-injective and which is such that the composition with any morphism $f : N \to N'$ in $\mathcal{C}$ with domain N is a pure-monomorphism iff f is a pure-monomorphism. As remarked already, existence and the usual properties are obtained via the functor category ([361, 1.12]). We write $H(M)$ for a pure-injective hull of M.

Proposition 17.3.9. ([361, 1.13, 1.14], also [73, 4.2.2]) *Let $\mathcal{C}$ be a compactly generated triangulated category. Suppose that $M' \xrightarrow{f} M \xrightarrow{g} H(M) \to M'[1]$ is a distinguished triangle where g is the canonical pure-monomorphism of an object into its pure-injective hull. Then a morphism $C \to M$ is phantom iff it factors through f.*

In the context of stable module categories these properties are closely related to those in the original category.

Proposition 17.3.10. ([361, 1.16]) *Let R be quasi-Frobenius. Then $N \in$ Mod-R is pure-injective iff its image in $\underline{\mathrm{Mod}}$-R is pure-injective.*

Proof. If N is pure-injective, then so is its image in $\underline{\mathrm{Mod}}$-R, by the characterisation of pure-injectivity in 17.3.7 and by the Jensen–Lenzing criterion, 4.3.6. The converse also follows from that characterisation since if N is a module whose image in $\underline{\mathrm{Mod}}$-R is pure-injective, and if I is any index set, then, by 17.3.7, there is $f : N^I \to N$ in Mod-R such that $fi = \Sigma$ in $\underline{\mathrm{Mod}}$-R, where $i : N^{(I)} \to N^I$ is the canonical inclusion and $\Sigma : N^{(I)} \to N$ is the summation map. That is, $fi - \Sigma$ factors through a projective module, say $fi - \Sigma = hg$ where $g : N^{(I)} \to P$ and P is projective. Since R is quasi-Frobenius, P is also injective, so $g = g'i$ for some $g' : N^I \to P$. Therefore $\Sigma = fi + hg = (f + g')i$ does factor through i and so (4.3.6) N is pure-injective. □

Suppose that G is a finite group and k is a field. Every morphism in the stable module category, $\underline{\text{Mod}}\text{-}kG$, is the image of a morphism in Mod-kG so one may say, as in [73], that a morphism in Mod-kG is **phantom** if its image in $\underline{\text{Mod}}\text{-}kG$ is phantom. Therefore, by 17.2.4, a morphism $f : M \to N$ is phantom iff for every finitely generated (= finitely presented) submodule M' of M the restriction of f to M' factors through a projective module. The next results show that the intrinsic definitions agree with these which refer back to Mod-kG.

Proposition 17.3.11. ([73, 4.2.2]) *Let $f : M \to N$ be in* Mod-kG. *Then the following are equivalent:*

(i) f is phantom;
(ii) the corresponding extension, $0 \to \Omega N \to X \to M \to 0$ in $\text{Ext}^1_{kG}(M, \Omega N) \simeq \underline{\text{Hom}}_{kG}(M, N)$ is pure;
(iii) the composition of f with the natural inclusion of N into its double dual N^{**} (as in Section 1.3.3) factors through a projective module.

Theorem 17.3.12. ([73, 4.2.4, 4.2.5]) *A kG-module M is pure-injective iff whenever $f : N \to M$ is a phantom map in $\underline{\text{Mod}}\text{-}kG$, then $f = 0$.*

A kG-module M is pure-projective iff whenever $f : M \to N$ is a phantom map in $\underline{\text{Mod}}\text{-}kG$, then $f = 0$.

There are similar results for more general rings. Herzog, [287, Thm 7], showed that for any module M over any ring there is a short exact sequence $0 \to K \to \text{Ph}(M) \to M \to 0$, where Ph($M$) is a phantom cover of M and K is pure-injective. The definition of phantom cover is like that in Section 4.6 for $\mathcal{F}$-cover but with the class of maps from $\mathcal{F}$ replaced by the class of phantom maps. He also showed ([287, Cor. 11]) that if R is right coherent and N is pure-injective, then every phantom morphism to N is zero. The category used here is Mod-R modulo the ideal of morphisms which factor through a *flat* module.

Nilpotence of the ideal of phantom maps has been investigated. Write $\underline{\text{Ph}}_{kG}(M, N)$ for the group of phantom maps between M and N in $\underline{\text{Mod}}\text{-}kG$. As M, N vary these form an ideal of $\underline{\text{Mod}}\text{-}kG$.

Proposition 17.3.13. ([73, 4.2.3]) *There are natural isomorphisms $\underline{\text{Ph}}_{kG}(M, N) \simeq \text{Pext}^1_{kG}(M, \Omega N) \simeq \text{Pext}^1_{kG}(\Omega^{-1} M, N)$ (for* Pext *see Section 2.2).*

It is shown ([73, 4.6.1]) that the image of the natural map $\mathrm{Pext}^n_{kG}(-, \Omega^n -) \to \mathrm{Ext}^n_{kG}(-, \Omega^n -)$ is naturally isomorphic to $\underline{\mathrm{Ph}}^n_{kG}(-, -)$, the nth power of the ideal of phantom maps and, when $n = 1$, the map $\mathrm{Pext}^1_{kG}(M, \Omega N) \to \mathrm{Ext}^1_{kG}(M, \Omega N)$ is injective, so this gives the isomorphism $\mathrm{Pext}^1_{kG}(M, \Omega N) \simeq \underline{\mathrm{Ph}}_{kG}(M, N)$ above.

Theorem 17.3.14. ([73, 4.6.2]) *If k is a field of cardinality $\aleph_t$ ($t \in \mathbb{N}$), then the composite of any $t + 2$ phantom maps in $\underline{\mathrm{Mod}}$-kG is zero.*

This bound is not achieved in all cases. If k is a field of characteristic 2 and $G = \mathbb{Z}_2 \times \mathbb{Z}_2$ (more generally, if the Sylow 2-subgroups of G are of this form), then ([73, 4.6.3]):

(a) if k is countable, the composite of any two phantom maps in $\underline{\mathrm{Mod}}$-kG is zero;
(b) if k is uncountable, the composite of any three phantom maps in $\underline{\mathrm{Mod}}$-kG is zero.

Also see [286] for more on spectra and stable module categories.

Just as (Section 16.1.1) for finitely accessible additive categories which, in contrast to most compactly generated triangulated categories,[5] have arbitrary direct limits, say that a subclass or subcategory $\mathcal{X}$ of the compactly generated category $\mathcal{C}$ is **definable** if it is an intersection of kernels of coherent functors: $\mathcal{X} = \{M \in \mathcal{C} : FM = 0 \text{ for all } F \in \mathcal{T}\}$ for some set, $\mathcal{T}$, of coherent functors from $\mathcal{C}$ to **Ab**.

There is a characterisation of definable subcategories analogous to 3.4.7. Of course, in this context there is the issue of existence of (suitable) colimits (also see [83]) and we refer to [362, 2.4, 2.6, 3.2, §7] for details.

Theorem 17.3.15. ([362, 7.5]) *Let $\mathcal{C}$ be a compactly generated triangulated category which satisfies BRO. Let $\mathcal{X}$ be a class of objects of $\mathcal{C}$ and let $C \in \mathcal{C}$. Then C belongs to the definable subcategory, $\langle \mathcal{X} \rangle$, generated by $\mathcal{X}$ iff there is a pure-exact triangle $C \to D \to E \to C[1]$ such that D is a reduced product of objects of $\mathcal{X}$.*

Krause remarks, in the introduction to [362], that Freyd's Generating Hypothesis[6] may be formulated in these terms, as saying that the definable subcategory of the stable homotopy category generated by the sphere together with its loops and suspensions is the whole category.

An appendix in [362] gives a way of making the analogy between finitely presented objects in abelian categories and compact objects in triangulated categories precise.

[5] One kind of exception is, [350, §6], the stable module category of a quasi-Frobenius ring of finite representation type. In fact, [361, 2.10], a compactly generated triangulated category has filtered colimits iff it is pure-semisimple; also see [63, §9].

[6] In connection with this I also point out the paper [305].

17.3.1 The Ziegler spectrum of a compactly generated triangulated category

The Ziegler spectrum of a compactly generated triangulated category is defined and, 17.3.16, its closed subsets shown to be in bijection with definable subcategories and with Serre subcategories of the functor category.

Let $\mathcal{C}$ be a compactly generated triangulated category. Define the **Ziegler spectrum** of $\mathcal{C}$, $\mathrm{Zg}(\mathcal{C})$, to have for its points the isomorphism classes of indecomposable pure-injective objects of $\mathcal{C}$. As in the case of abelian categories (Sections 5.1, 12.1.3) the topology may be defined in a number of equivalent ways: via the functor category; in terms of internal purity using pp conditions; that given below. It turns out that a great deal may be lifted from the abelian context to this one. Here we mention only a few examples. For instance Beligiannis' [66] has analogues of various of the results in Chapter 5 on isolated points and Krull–Schmidt rings.

Say that a subset of $\mathrm{Zg}(\mathcal{C})$ is **closed** if it has the form $\mathcal{D} \cap \mathrm{Zg}(\mathcal{C})$ for some definable subcategory $\mathcal{D}$ of $\mathcal{C}$. It is easily checked that this does give a topology.

A **cohomological ideal** of $\mathcal{C}^{\mathrm{c}}$ is one which has the form $\{f : Ff = 0\}$ where F is a homological functor (Section 17.2.1). For any $M \in \mathcal{C}$ its **annihilator**, $\mathrm{ann}(M) = \{A \xrightarrow{f} B \in \mathcal{C}^{\mathrm{c}} : (B, M) \xrightarrow{(f,M)} (A, M) \text{ is } 0\}$ is, therefore, a cohomological ideal, being defined by the coherent functor which is the cokernel of $(B, -) \to (A, -)$.

If $\mathcal{C}$ is a compactly generated triangulated category, then, [362, 7.4], every cohomological ideal of $\mathcal{C}^{\mathrm{c}}$ has the form $\mathrm{ann}(N)$ for some pure-injective $N \in \mathcal{C}$. Compare the next result with 12.4.1.

Theorem 17.3.16. ([362, §7]) *Let $\mathcal{C}$ be a compactly generated triangulated category. Then there is a bijection between:*

(i) closed subsets of the Ziegler spectrum of $\mathcal{C}$;
(ii) definable subcategories of $\mathcal{C}$;
(iii) Serre subcategories of the category, $\mathrm{Coh}(\mathcal{C})$, *of coherent functors from $\mathcal{C}$ to* **Ab***;*
(iv) cohomological ideals of $\mathcal{C}^{\mathrm{c}}$.

Just from the categorical nature of the various definitions it follows that if $\mathcal{C}$ and $\mathcal{D}$ are compactly generated triangulated categories and if $f : \mathcal{C} \longrightarrow \mathcal{D}$ is an equivalence of triangulated categories, meaning an equivalence of categories which preserves the triangulated structure, then f induces a homeomorphism between the Ziegler, respectively rep-Zariski, spectra of $\mathcal{C}$ and $\mathcal{D}$ (the latter with the obvious definition, cf. Section 14.1.2). More generally, representation embeddings (in an appropriate sense) induce homeomorphic embeddings of Ziegler spectra, [66, 6.13] cf. 5.5.9.

One may check, [220, 6.1], that if R is a quasi-Frobenius ring, then $\mathrm{Zg}(\underline{\mathrm{Mod}}\text{-}R)$ is homeomorphic to a clopen subset of Zg_R, its complement being the finite set of indecomposable projective (= injective) R-modules.

The restricted Yoneda functor from a compactly generated triangulated category, $\mathcal{C}$, to $((\mathcal{C}^{\mathrm{c}})^{\mathrm{op}}, \mathbf{Ab})$ gives one way of setting up the notion of pp condition via the kind of language for $\mathcal{C}$ that is described in Section B.1: that is, apply the functor and then use the language of the functor category. Setting up the language may, however, be done directly (cf. Section 10.2.4), using the compact objects in place of finitely presented objects to give the sorts, and introducing function symbols corresponding to the morphisms of $\mathcal{C}^{\mathrm{c}}$. This is done in [220, §3]. It is shown, in particular, that every pp condition is equivalent to one of the form $f|x$, where f is a morphism of $\mathcal{C}^{\mathrm{c}}$ and x is a free variable of an appropriate sort, equivalently to one of the form $xg = 0$; thus we have elimination of quantifiers. The proof uses the fact, noted after 17.1.1, that a triangulated category has pseudocokernels. For details see [220, 3.2, 3.1].

17.3.2 The Ziegler spectrum of $\mathcal{D}(R)$

Purity in a module category and its derived category are linked (17.3.17, 17.3.18, 17.3.19). Shifts of the Ziegler spectrum of the ring lie inside the Ziegler spectrum of the derived category (17.3.20) and, over hereditary (17.3.22) and von Neumann regular (17.3.23) rings, cover the latter but do not do so over arbitrary rings (17.3.21). The results here mainly come from [220].

Recall that $\mathcal{D}(R)$ denotes the derived category of a ring R.

Proposition 17.3.17. ([220, 2.4]) *If $A \to B \to C \to A[1]$ is a pure-exact sequence in $\mathcal{D}(R)$, then, for each n, the sequence $0 \to H^n(A) \to H^n(B) \to H^n(C) \to 0$ of R-modules is pure-exact.*

Theorem 17.3.18. ([220, 5.2, 5.4]) *If C is an absolutely pure object of $\mathcal{D}(R)$, then each cohomology group $H^n(C)$ is an absolutely pure R-module. If R is right coherent, then the converse is true.*

Theorem 17.3.19. ([220, 7.1]) *If C is a pure-injective object of $\mathcal{D}(R)$, then each cohomology group $H^n(C)$ is a pure-injective R-module. If N is a pure-injective module, then, for each n, the complex $N[-n]$, with N concentrated in degree n, is a pure-injective object of $\mathcal{D}(R)$.*

It follows that if $\mathcal{D}(R)$ is pure-semisimple, in the sense that every object is pure-injective, then R is pure-semisimple but the converse is false: the example 17.3.21 below shows this. More generally, if R is an algebra of finite representation type and with infinite global dimension, then the derived category, $\mathcal{D}(R)$, is not pure-semisimple ([63, 12.16]).

Denote by $\mathrm{Zg}_R[-n]$ the set of isomorphism classes of those objects of $\mathcal{D}(R)$ which are isomorphic to a complex of the form $N[-n]$ where $N \in \mathrm{Zg}_R$.

Theorem 17.3.20. ([220, 7.3]) *Let R be any ring.*

(a) Each $\mathrm{Zg}_R[n]$ is a closed subset of $\mathrm{Zg}(\mathcal{D}(R))$.
(b) If R is right coherent and every finitely presented R-module has finite projective dimension, then $N \mapsto N[n]$ induces a homeomorphism between Zg_R and $\mathrm{Zg}_R[n]$, where the latter has the subspace topology inherited from $\mathrm{Zg}(\mathcal{D}(R))$.
(c) $\bigcup_n \mathrm{Zg}_R[n]$ is a closed subset of $\mathrm{Zg}(\mathcal{D}(R))$ and its complement consists of those indecomposable pure-injective objects of $\mathcal{D}(R)$ with at least two non-zero cohomology groups.

Example 17.3.21. The union $\bigcup_n \mathrm{Zg}_R[n]$ is, in general, a proper subset of $\mathrm{Zg}(\mathcal{D}(R))$ [220, p. 19]. Let R be the path algebra, over a field k, of the quiver

$$1 \xrightarrow{\alpha_1} 2 \xrightarrow{\alpha_2} \cdots \xrightarrow{\alpha_8} 9 \xrightarrow{\alpha_9} 10\,,$$

with relation $\alpha_8 \dots \alpha_1 = 0$ (sic). Let S be the path algebra of the quiver

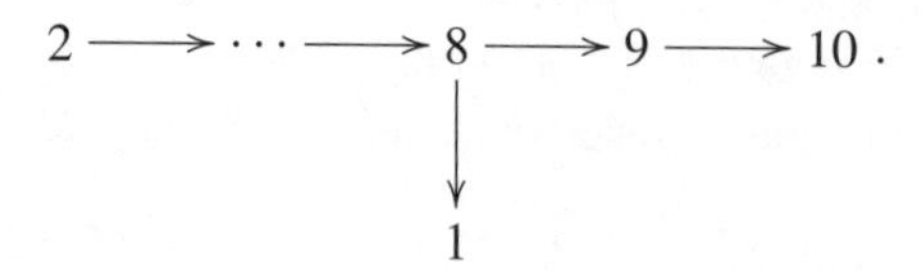

Then R is of finite representation type so $\bigcup_n \mathrm{Zg}_R[n]$ is a countable set. On the other hand, if k is uncountable, then Zg_S, hence $\mathrm{Zg}(\mathcal{D}(R))$, is uncountable (there are families of non-isomorphic indecomposable modules parametrised by k, see Section 8.1.2). But, [119, pp. 1349–50], $\mathcal{D}(R)$ and $\mathcal{D}(S)$ are equivalent as triangulated categories, hence, as commented after 17.3.16, they have homeomorphic Ziegler spectra, so $\bigcup_n \mathrm{Zg}_R[n]$ cannot exhaust $\mathrm{Zg}(\mathcal{D}(R))$.

On the other hand, 17.3.22 below shows that $\mathrm{Zg}(\mathcal{D}(R)) \simeq \mathrm{Zg}(\mathcal{D}(S)) \simeq \bigcup_n \mathrm{Zg}_S[n]$.

In the following cases every indecomposable pure-injective has cohomology concentrated in just one degree.

Theorem 17.3.22. ([220, 8.1]) *Suppose that R is right hereditary. An object, N, of $\mathcal{D}(R)$ is pure-injective iff it is isomorphic to $\bigoplus_n H^n(N)[-n]$ with each $H^n(N)$ a pure-injective R-module. In particular, $\mathrm{Zg}(\mathcal{D}(R)) \simeq \bigcup_n \mathrm{Zg}_R[n]$.*

The key point is a result of Neeman, [455, 6.7], which says that, for a right hereditary ring R, every object in $\mathcal{D}(R)$ is isomorphic to the coproduct of its cohomology modules.

Theorem 17.3.23. ([220, 8.2, 8.3, 8.5]) *The ring R is von Neumann regular iff every pure-injective object of $\mathcal{D}(R)$ is injective iff every object of $\mathcal{D}(R)$ is absolutely pure. In this case, an object, N, of $\mathcal{D}(R)$ is pure-injective iff it is isomorphic to $\bigoplus_n H^n(N)[-n]$ with each $H^n(N)$ a pure-injective R-module. In particular, $\mathrm{Zg}(\mathcal{D}(R)) \simeq \bigcup_n \mathrm{Zg}_R[n]$. Furthermore, the open subsets of $\mathrm{Zg}(\mathcal{D}(R))$ are exactly those of the form $\bigcup_n \{N : (I_n[-n], N) \neq 0\}$, where $(I_n)_{n\in\mathbb{Z}}$ is any sequence of two-sided ideals of R.*

The last part follows from the description, 3.4.30, of definable subsets of Mod-R when R is von Neumann regular.

17.4 Localisation

In triangulated categories there is a process analogous to torsion-theoretic localisation.

A **thick subcategory**, $\mathcal{S}$, of a triangulated category, $\mathcal{C}$, is a triangulated subcategory which is closed under direct summands. This is somewhat analogous to a Serre subcategory of an abelian category (see 11.1.1). If $\mathcal{C}$ has arbitrary coproducts, then say that a **localising subcategory** of $\mathcal{C}$ is a triangulated subcategory which is closed under coproducts. Such a subcategory, $\mathcal{S}$, must be thick: if $A \oplus B \in \mathcal{S}$, then consider a distinguished triangle of the form $(A \oplus B)^{\aleph_0} \to A \oplus (B \oplus A)^{\aleph_0} \to A \to (A \oplus B)^{\aleph_0}[1]$. These subcategories are comparable to hereditary torsionfree classes in a Grothendieck category.

Set $\Sigma_{\mathcal{S}} = \{A \xrightarrow{f} B : \text{ if } A \xrightarrow{f} B \to C \to A[1]$ is a distinguished triangle then $C \in \mathcal{S}$ (hence also $C[-1] \in \mathcal{S}$)$\}$. It may be checked that $\Sigma_{\mathcal{S}}$ is closed under composition and, ignoring possible set-theoretic difficulties, one may form the category of fractions, $\mathcal{C}[\Sigma_{\mathcal{S}}^{-1}]$ by inverting the morphisms in $\Sigma_{\mathcal{S}}$ (see, for example, [486, §4.1], [206, Chpt. 1]). This has the same objects as $\mathcal{C}$ but the morphisms from A to B in $\mathcal{C}[\Sigma_{\mathcal{S}}^{-1}]$ are given by certain equivalence classes of diagrams of the form shown below, where $f \in \Sigma_{\mathcal{S}}$.

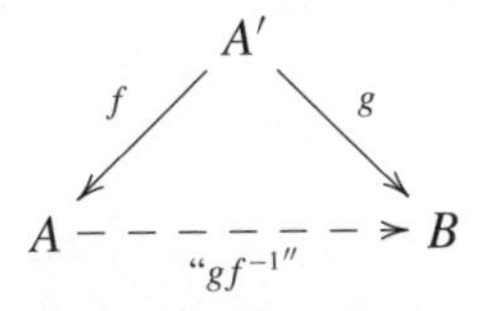

Then (see, for example, [597, 5.2]) the category $\mathcal{C}[\Sigma_{\mathcal{S}}^{-1}]$ is triangulated and has the universal property that if $F : \mathcal{C} \longrightarrow \mathcal{D}$ is a triangulated functor between triangulated categories and $F\mathcal{S} = 0$, then there is an essentially unique factorisation of F through the obvious and canonical functor $\mathcal{C} \longrightarrow \mathcal{C}[\Sigma_{\mathcal{S}}^{-1}]$. The notation $\mathcal{C}/\mathcal{S}$ is more commonly used for $\mathcal{C}[\Sigma_{\mathcal{S}}^{-1}]$.

For example, if $\mathcal{C} = K\mathcal{A}$ is the homotopy category of (co)chain complexes from an abelian category $\mathcal{A}$ and $\mathcal{S}$ is the class of **acyclic** complexes (those with zero (co)homology in every degree), then $\Sigma_{\mathcal{S}}$ is the class of quasi-isomorphisms and $K\mathcal{A}[\Sigma_{\mathcal{S}}^{-1}]$ is the derived category $D\mathcal{A}$ defined in Section 17.1.

For $\mathcal{Y} \subseteq \mathcal{C}$ set $\mathcal{Y}^{\perp} = \{C \in \mathcal{C} : (\mathcal{Y}, C) = 0\}$ and ${}^{\perp}\mathcal{Y} = \{C \in \mathcal{C} : (C, \mathcal{Y}) = 0\}$. If $\mathcal{S}$ is a thick subcategory of the triangulated category $\mathcal{C}$ and $C \in {}^{\perp}\mathcal{S}$, then it may be checked that for every $D \in \mathcal{C}$ the natural map $\mathcal{C}(C, D) \to (\mathcal{C}/\mathcal{S})(C, D)$ is an isomorphism.

Theorem 17.4.1. (see [456], also [597, §3] and references there) *If $\mathcal{S}$ is a localising subcategory of the triangulated category $\mathcal{C}$ and if $\mathcal{S}$ is compactly generated, with $\mathcal{S}^{c} \subseteq \mathcal{C}^{c}$, then $\mathcal{S}$ is contravariantly finite (§3.4.4) in $\mathcal{C}$.*

Denoting a "universal" map from $\mathcal{S}$ to $C \in \mathcal{C}$ by $EC \to C$ and completing to a distinguished triangle, $EC \to C \to FC \to EC[1]$, the map $C \to FC$ is universal for maps from C to $\mathcal{S}^{\perp}$.

The assignment $C \mapsto FC$ extends to a functor $F : \mathcal{C} \longrightarrow \mathcal{S}^{\perp}$ which is left adjoint to the inclusion $\mathcal{S}^{\perp} \longrightarrow \mathcal{C}$.

Moreover $\mathcal{C}/\mathcal{S} \simeq \mathcal{S}^{\perp}$.

Objects of $\mathcal{S}^{\perp}$ are referred to as $\mathcal{S}$-**local**. A construction of $C \to FC$, taken from [596, §5], is as follows. Given $C \in \mathcal{C}$, we set $C_0 = C$ and define $C_k \to C_{k+1}$ inductively. Given C_k, choose S_k to be a direct sum of objects of $\mathcal{S}^{c}$ together with a morphism, $S_k \to C_k$, such that every morphism from an object of $\mathcal{S}^{c}$ to C_k factors through this morphism (cf. 2.1.25). Define C_{k+1} to be such that $S_k \to C_k \to C_{k+1} \to S_k[1]$ is a distinguished triangle. So we have the chain of morphisms $C \to C_1 \to \cdots \to C_k \to C_{k+1} \to \cdots$. Let $FC = \mathrm{hocolim}(C_k)$ be the homology colimit (§17.2.2) of this sequence and let $C \to FC$ be the colimit map. To see that $FC \in \mathcal{S}^{\perp}$, take $S \in \mathcal{S}^{c}$. Then, by the definition (Section 17.2.2), $(S, FC) = \varinjlim(S, C_k)$, which is zero by construction since each induced $(S, C_k) \to (S, C_{k+1})$ is zero. Therefore $FC \in (\mathcal{S}^{c})^{\perp}$ and the latter is easily seen to equal $\mathcal{S}^{\perp}$.

Theorem 17.4.2. (for example, [597, 5.1]) *Suppose that $\mathcal{C}$ is a compactly generated triangulated category with coproducts and let $\mathcal{S} \subseteq \mathcal{C}$ be a localising subcategory. Then the inclusion functor $\mathcal{S} \longrightarrow \mathcal{C}$ has a right adjoint, $C \mapsto EC$, iff $\mathcal{C}/\mathcal{S}$ exists.*

Now we just point to some results in the context of the stable module category of the group ring, kG, of a finite group over field, where, for example, in the case of a p-group there is a classification, due to Benson, Carlson and Rickard

[72], of the thick subcategories of $\underline{\text{Mod}}\text{-}kG$. The tensor product in the module category induces a symmetric monoidal product on the stable category and this is used to give a definition of "idempotent" object. Such objects arise from the distinguished triangle seen in 17.4.1 in connection with localisations defined in terms of certain subvarieties of the maximal ideal spectrum of the cohomology ring $H^*(G, k)$, see [596]. There are connections with some of the infinite-dimensional pure-injective modules over string algebras appearing in Section 8.1.3, see [70], and the question of pure-injectivity of these modules has arisen ([73, §5.6]) and been answered in part in [75], where it is also shown, for instance, that a kG-module is of finite endolength as a module iff it is so as an object in the stable module category.

Finally we mention [6, §5] where it is shown, roughly, that localisation, in the sense of this section and Section 11.1.1, commutes with forming derived categories.

Also see [68].

17.5 The spectrum of the cohomology ring of a group ring

If G is a finite group and k is a field, then the injective modules over the Tate cohomology ring correspond to the pure-injective modules in the definable subcategory generated by the trivial module and its syzygies (17.5.1); this gives a homeomorphic embedding of Zariski spectra (17.5.3).

In [76] Benson and Krause show that certain pure-injective kG-modules, where G a finite group and k is a field of finite characteristic, correspond to injective modules over the Tate cohomology ring of kG, hence correspond to points of the (ordinary) Zariski spectrum of that graded-commutative ring. The setting is somewhat more general: they assume that R is a finite-dimensional cocommutative Hofp algebra over a field k of finite characteristic which, for various of their results, should satisfy certain conditions, namely that Tate cohomology is periodic or that negative Tate cohomology is nilpotent (the fact that kG satisfies these conditions is [76, 2.4]).

First they show that the injective modules over the Tate cohomology ring, $\widehat{H}^*(R, k)$ (k now denoting the trivial module!), all are coinduced from injective modules over the ordinary cohomology ring, $H^*(R, k)$, that is, have the form $\text{Hom}^*_{H^*(R,k)}(\widehat{H}^*(R, k), E)$ for some injective $H^*(R, k)$-module E ([76, 2.8]). The Tate cohomology ring is defined like the ordinary cohomology ring but using a complete resolution of the trivial module k: this is the result of splicing together a projective and an injective resolution of this module. Equivalently, $\widehat{H}^*(R, k)$ is $\text{End}^*(k) = \bigoplus_{n\in\mathbb{Z}}(k, k[-n])$ computed in the stable module category; the ordinary cohomology ring $H^*(R, k)$ is the non-negatively graded part of this ring. This correspondence between injectives induces an equivalence, [76, 6.3], between the respective spectral categories of injective modules: in particular, indecomposable injectives correspond.

If E is an indecomposable injective $\widehat{H}^*(R, k)$-module, then the functor $M \mapsto \mathrm{Hom}_{H^*(R,k)}(\widehat{H}^*(R, M), E)$ takes triangles in $\underline{\mathrm{Mod}}$-R to exact sequences and coproducts to products, so, by Brown representability (17.2.5), there is an object, $T(E)$ say, of $\underline{\mathrm{Mod}}$-R such that this functor is equivalent to that given by $M \mapsto \underline{\mathrm{Hom}}_R(M, T(E))$.

Proposition 17.5.1. ([76, 4.6, 4.7, 5.1]) *Let k be a field of finite characteristic and let R be a finite-dimensional cocommutative Hofp algebra over k for which either Tate cohomology is periodic or negative Tate cohomology is nilpotent. For example, let $R = kG$, where G is a finite group. Let $H^*(R, k)$ denote the cohomology ring of R and let E be an indecomposable injective $H^*(R, k)$-module. Then the corresponding object, $T(E) \in \underline{\mathrm{Mod}}$-$R$ is a direct summand of a direct product of modules of the form $\Omega^n k$ $(n \in \mathbb{Z})$, where k denotes the trivial module. In particular, $T(E)$ is pure-injective. Furthermore, every indecomposable direct summand of such a product has the form $T(E)$ for some indecomposable injective $H^*(R, k)$-module E. The statements obtained by deleting the term "indecomposable" also are true.*

It is also the case, [76, 6.5], that the correspondence $E \mapsto T(E)$ is one-to-one.

Any module of the form $T(E)$ therefore belongs to the Ziegler-closure of the set of syzygies, $\Omega^n k$ $(n \in \mathbb{Z})$, of the trivial module and these latter are, [76, 3.1], exactly the $T(E)$ where E is a shift of the trivial module k. (Note that the trivial module corresponds to the maximal, "inessential" ideal, $P_0 = \bigoplus_{n>0} H^n(R, k)$, generated by all homogeneous elements of positive degree.) Since $H^*(R, k)$ is noetherian, [197, 1.1], the indecomposable injectives over it form a closed subset of its Ziegler spectrum, 5.1.11. In particular, $E\big(\bigoplus_n k[n]\big)$, where $[n]$ now denotes shift in the graded sense, is an elementary cogenerator. By [76, 3.9 and §6] the functor $E \mapsto T(E)$ preserves the elementary cogeneration property, so we have the following.

Corollary 17.5.2. *Let R be as in 17.5.1. Then the Ziegler-closure of the set $\{\Omega^n k : n \in \mathbb{Z}\}$ of points of Zg_R consists of those modules of the form $T(E)$ for E an indecomposable injective $H^*(R, k)$-module.*

Let $\mathrm{Proj}(H^*(R, k))$ denote the set of homogeneous prime ideals, P, of $H^*(R, k)$ apart from the maximal ideal P_0. This set carries the usual Zariski topology. Then $E = E(R/P)$ is an indecomposable injective module (and the injective hull of R/P_0 is the trivial module k).

Theorem 17.5.3. ([76, 8.3]) *Let R be as in 17.5.1. Then the map $E \mapsto T(E)$ is a homeomorphic embedding from the homogeneous prime ideal spectrum,* $\mathrm{Proj}(H^*(R,k))$*, of the cohomology ring (equipped with the Zariski topology) into the set of non-isolated points of the Ziegler-closure of $\{\Omega^n k : n \in \mathbb{Z}\}$ (equipped with the rep-Zariski topology).*

If we replace the set of homogeneous, non-maximal, prime ideals by the set of all indecomposable injectives, excluding just the translates of the injective hull of the trivial module, then we obtain a homeomorphism with the first Cantor–Bendixson derivative, that is, the set of non-isolated points, of the Ziegler closure of $\{\Omega^n k : n \in \mathbb{Z}\}$. Or we may, alternatively, say that the latter set modulo the action induced by translate/syzygy and equipped with the rep-Zariski = dual-Ziegler (Section 14.1.3) topology is homeomorphic to $\mathrm{Proj}(H^*(R,k))$.

These results can be generalised, see [219], by replacing $\underline{\mathrm{Mod}}\text{-}R$ by any compactly generated triangulated category $\mathcal{C}$, replacing the trivial module, k, by any compact object, X, of $\mathcal{C}$ and replacing the Tate cohomology ring by $S = \mathrm{End}^*(X) = \bigoplus_n (S, S[-n])$. Then, [219, 4.1, 4.4], Brown representability yields, as above, an equivalence between injective S-modules and certain pure-injective objects of $\mathcal{C}$. To obtain the strongest results one should assume that X and its shifts generate the triangulated subcategory, $\mathcal{C}^{\mathrm{c}}$, of compact objects (as is the case in the original situation). In the case that S is right noetherian one obtains analogues, [219, 4.3, 4.6], of some of the results of Benson and Krause (though note that the Tate cohomology ring is not usually noetherian, hence the work required in [76] to relate the injective modules over the Tate and the, noetherian, ordinary cohomology rings). Just with the assumption that S is right noetherian one obtains, [219, 5.1, 5.2], the analoguous results giving a homeomorphism between the Zariski spectrum of S, interpreted as a topology on indecomposable injectives if S is not near enough to commutative, and the rep-Zariski topology on the corresponding subset of $\mathrm{Zg}(\mathcal{C})$.

Appendix B Languages for definable categories

In this appendix to Part II we define languages, in particular pp conditions, for finitely accessible categories as well as the enriched, "imaginaries", language for functor categories and their localisations.

B.1 Languages for finitely accessible categories

The language for a finitely accessible additive category is defined and the objects of the category are regarded as structures for this language.

This book began with the notion of a pp condition for modules. In Section 10.2.4 this was extended from the category of modules over a ring to functor categories, that is, categories of modules over any small preadditive category. Further extensions were described/indicated: to locally finitely presented categories in Section 11.1.5; to finitely accessible categories in Section 16.1.2 (with a detailed treatment for categories of sheaves in Section 16.3.5); to compactly generated triangulated categories in Section 17.3.1. Indeed, any definable category, $\mathcal{D}$, may be represented as a definable subcategory of a locally coherent functor category (see Section 18.1.1), so the notion of pp condition for the latter category may be applied to $\mathcal{D}$. To obtain something intrinsic (18.1.19) for $\mathcal{D}$, however, one should pass to its functor category, fun($\mathcal{D}$), equivalently (12.3.20), to the category of pp-pairs for $\mathcal{D}$.

Quite a few steps have now been taken from the naïve origins: from pp conditions to the functors they define; then to quotients of such functors; to noting the theorem, 12.3.20, that the category of pp-pairs is equivalent this category of functors; to making these pairs the objects of a category by including the pp-defined maps; then to seeing a module as a functor (evaluation) on this category of pp-pairs; then to regarding restriction to a definable subcategory as the replacement of the category of pp-pairs, equivalently category of finitely presented functors, by a suitable localisation.

Although one may regard everything in terms of functors, we have seen that it is still possible to think in terms of pp conditions provided one allows multi-sorted languages, in particular languages which build in "imaginaries", always, however, staying in the "pp" context, since we don't want to lose the abelian group structure. Recall, see Appendix A, that a general language will have more than function symbols and a symbol for 0; also general formulas involve negations, disjunctions and universal quantifiers. Pp-elimination of quantifiers (A.1.1) does, however, mean that we have not really lost anything by restricting to pp conditions.

Let us review what has been done already.

Suppose that $\mathcal{C}$ is a finitely accessible additive category, and let $\mathcal{G} \subseteq \mathcal{C}^{\text{fp}}$ be a generating set of finitely presented objects; so every object of $\mathcal{C}$ is the colimit of a directed diagram of objects in $\mathcal{G}$. If $G \in \mathcal{G}$ and $C \in \mathcal{C}$, then a G-**element** of C is a morphism from G to C. Since $\mathcal{G}$ is generating, every object is determined by its elements: the elements of the (G, C) as G ranges over $\mathcal{G}$. But an object is not just a set of elements: it has some structure, and that is given by the morphisms of $\mathcal{G}$. Namely, if $f : G \to H$ is a morphism between objects of $\mathcal{G}$, then, for every $C \in \mathcal{C}$ there is the induced map $(f, C) : (H, C) \to (G, C)$.

In effect then, an object $C \in \mathcal{C}$ has been replaced by the restricted representable functor $(-, C) : \mathcal{G}^{\text{op}} \longrightarrow \mathbf{Ab}$. Equivalently, and this is the model-theoretic view, the object C has been replaced, not by a functor, but by a multi-sorted structure for a certain language.

In the model-theoretic view we define a language $\mathcal{L}^{\mathcal{G}}(\mathcal{C})$ which has a sort for each $G \in \mathcal{G}$ and a function symbol for each morphism in $\mathcal{G}$. Then the collection of sets (G, C) for $G \in \mathcal{G}$, and the induced morphisms, (f, C), for f in $\mathcal{G}$, is the $\mathcal{L}^{\mathcal{G}}(\mathcal{C})$-structure on an object C of $\mathcal{C}$.

These equivalent views, and this idea of what an "element" is, are commonplace in some areas, particularly in topos theory, but not really in algebra or even in "ordinary" model theory. (Note, though, that a representation of a quiver is a natural example of a multi-sorted structure and the identification of Me_i (using the notation of Section 15.1.1) with (S_i, M) is just this view of elements.)

Almost everything that was done in the first part for categories of modules applies to functor categories and much, with the statements modified as necessary, also applies in finitely accessible categories. In most cases "the same" proof works. That includes proofs which involve working with elements, provided one interprets "elements" in the way described above.

The language associated to $\mathcal{C}$ does depend on the choice of generating set $\mathcal{G}$, though not in any essential way: if $\mathcal{H} \subseteq \mathcal{C}^{\text{fp}}$ is another choice of generating set, then for each pp condition in the language built from $\mathcal{H}$ there is a pair of pp conditions in the language built from $\mathcal{G}$ which is equivalent in the sense that these conditions define, uniformly (over $\mathcal{C}$) and definably (by pp conditions), isomorphic groups in every object of $\mathcal{C}$. That is, from each language one may define the category of pp-pairs, as in Section 3.2.2, and, by 10.2.30, each of these categories is naturally equivalent to the category of finitely presented functors on $\mathcal{C}^{\text{op}}$, hence they are equivalent to each other. At least, that is the case when $\mathcal{C}$ is a functor category. For the general case, the language for $\mathcal{C}$ based on $\mathcal{G}$ may be regarded as the language of Mod-$\mathcal{G}$ restricted to $\mathcal{C}$ and the category of pp-pairs for $\mathcal{C}$ is a localisation of $\mathbb{L}_{\mathcal{G}}^{\text{eq}+}$ (12.3.20). Then use the fact, 18.1.14 below, that this localised category depends only on $\mathcal{C}$, not on the choice of $\mathcal{G}$. More direct proofs of "inter-interpretability" of the languages based on different choices of generating sets, may be given (for instance, it is done for locally finitely presented abelian categories in [488, 2.22]).

Let us give the formal definitions.

Let $\mathcal{C}$ be a finitely accessible category. Let $\mathcal{G}$ be a subcategory of $\mathcal{C}^{\text{fp}}$ which is generating. The language $\mathcal{L}_{\mathcal{G}}$ has, for each object G of $\mathcal{G}$, a sort, denoted s_G, and, for each morphism $r : G \to H$ in $\mathcal{G}$, a function symbol, which we denote as right multiplication by r, with domain s_H and codomain s_G. There are also, for each object G, symbols $+_G$ and 0_G (usually we drop the G) for the abelian group structure on elements of sort G (so $+_G$ is a 2-ary function symbol with domain $s_G \times s_G$ and codomain s_G, and 0_G is a constant symbol of sort s_G).

Then pp conditions (and more general formulas) are formed in the usual way (see Appendix A). A linear equation has the form $\sum_i v_i r_i = 0$, where v_i is a variable of sort H_i (each variable of the language has a specified sort) and r_i is a (function symbol corresponding to the) morphism from a fixed (as i varies) sort G to H_i. As in the case of modules we tend to ignore the difference between morphisms of $\mathcal{G}$ and the corresponding symbols of the language. A pp condition is just an existentially quantified system of linear equations: $\exists v_{k+1}, \dots, v_n \bigwedge_{j=1}^m \sum_{i=1}^n v_i r_{ij} = 0$, where $r_{ij} : H_i \to G_j$.

An (**additive**) $\mathcal{L}_{\mathcal{G}}$**-structure** M is given by: for each sort s_G, an abelian group M_G; for each function symbol r from s_H to s_G a morphism of abelian groups, $M_r : M_H \to M_G$. Of course, $\mathcal{G}$ was a category with structure so it is natural to insist that $M_{1_G} = 1_{M_G}$ and $M_{sr} = M_r M_s$; in model-theoretic terms, we impose, as axioms, the conditions $x_G = x_G \to x_G 1_G = x_G$ and $x_K = x_K \to (x_K s)r = x_K(rs)$, where $G \xrightarrow{r} H \xrightarrow{s} K$. In this way we obtain precisely the contravariant additive functors from $\mathcal{G}$ to **Ab** as the models for these axioms. Thus Mod-$\mathcal{G}$ is a definable class of structures for the language $\mathcal{L}_{\mathcal{G}}$; and note that it was defined by a set of pp-pairs being closed, cf. Section 3.4.1. As for the original category $\mathcal{C}$, recall, 11.1.27, that it is a definable subcategory of Mod-$\mathcal{G}$ so it is obtained by imposing further axioms saying that further pp-pairs are closed. Different choices of $\mathcal{G}$ correspond to representations of $\mathcal{C}$ as definable subcategories of different functor categories.

The possibility of localising a functor category to get the general case means that it is sufficient, for many purposes, to treat the case $\mathcal{C} = \text{Mod-}\mathcal{G}$.

If $\phi(\overline{x})$ is a pp condition, then the solution set in any $\mathcal{G}$-module is a subgroup of the product of the relevant sorts of elements of M. For example, if $r, s : G \to H$, then the condition $zr = x \wedge zs = y$ defines, ordering the variables as (x, y, z), a subgroup of $MG \times MG \times MH$. Just as for modules over a ring, one has pp-elimination of quantifiers (A.1.1) reducing general conditions to boolean combinations of pp conditions. The proof for functor categories is hardly different from that for rings,[1] indeed, since any formula involves only finitely many sorts, the former can be reduced to the latter.

We also mention Fisher's work, [190], on "abelian structures" – certain categories of many-sorted additive structures, where function, relation and constant symbols are allowed in the language (so pp-definable sets will be affine sets rather than subgroups). He developed a good deal of theory about these, for example proving existence of pure-injective hulls and the more general hulls of sets/pp-types (Section 4.3.5). There is some description of what is in [190] in [495, §3A] and some discussion of how to relate this to the approach *via* pp-imaginaries in [382, §1].

The equivalence between the category of pp-pairs and the "functor category", $(\text{mod-}\mathcal{G}, \mathbf{Ab})^{\text{fp}}_{\mathcal{C}}$, of $\mathcal{C}$ has been extended to the non-additive case by Rajani, [583] see [584], where it becomes essentially a description of the coherent objects of the classifying topos of a coherent theory. The surrounding ideas – purity, interpretations, Ziegler spectra – are further investigated in these references and in [585] and [586].

[1] So similar that it has, so far as I know, never appeared in print.

B.1.1 Languages for modules

The various natural languages for R-modules are compared.

All this applies to our original context, the category of modules over a ring. In particular, different choices for generating subcategory $\mathcal{G} \subseteq \text{mod-}R$ give rise to different languages, which we compare.

(1a) $\mathcal{G}$ consists just of the full one-object subcategory R_R. Elements of an R-module M may be regarded as the morphisms from R_R to M so this is the single sort of M in this language. Also, multiplication by $r \in R$ on M, $a \in M \mapsto ar$, is the composition of left multiplication, $s \mapsto rs$, by r on R (the endomorphisms of R_R are exactly the left multiplications by elements of R), with the morphism $a \in (R, M)$, since this composition takes $s \in R$ to ars and this is the morphism $ar \in (R, M)$. This is the usual view of a module.

(1b) $\mathcal{G}$ consists of one copy of each R^n and all maps (rectangular matrices over R) between these. Under this view n-tuples from a module really are elements of the module and the actions are multiplications by matrices. Various arguments, from the first couple of chapters, which involve tuples and look slightly messy become more streamlined if one regards a module as a structure for this language. A simple example is the notation for a pp condition. In this view the usual existentially quantified system of linear equations becomes simply $\exists y\,(xA = yB)$ for some matrices, that is, function symbols in the language, A, B.

(2) $\mathcal{G} = \text{mod-}R$ (rather, a small version thereof). The elements of sort A ($A \in \text{mod-}R$) of an R-module M are the morphisms from A to M: $s_A(M) = (A, M)$. Thus the elements of M of a given sort are finite tuples which satisfy some finite system of equations.

Therefore, if regarding R-modules as structures for the language $\mathcal{L}_R$ is taking the view that a module is an additive functor from the one-object category R to **Ab**, then regarding R-modules as structures for the language $\mathcal{L}_{\text{mod-}R}$ is viewing modules as certain contravariant additive functors from mod-R to **Ab** (namely, the flat ones, 16.1.3).

Clearly a module regarded as an $\mathcal{L}_{\text{mod-}R}$-structure is, in some sense, richer than that same module regarded as an $\mathcal{L}_R$-structure. Nevertheless, the richer structure is definable, in a fairly obvious sense, from the $\mathcal{L}_R$-structure. This is an example of the general fact, discussed in the previous section, that the choice of generating subcategory of $\mathcal{C}^{\text{fp}}$ is relevant to the precise language but not to the "expressive power" of that language.

(3) There is an even richer language for R-modules, which is described in the next section. All still is definable in terms of $\mathcal{L}_R$, hence this richer language is no more expressive, though it is more elegant in expression. This language is not one based on finitely presented objects of Mod-R since with (2) we have gone as far as we can with that in that direction. It is, rather, based on regarding Mod-R as a category of (exact) functors on its "functor category" $(\text{mod-}R, \mathbf{Ab})^{\text{fp}}$, via evaluation or, if one prefers to use the dual functor category, *via* $M \mapsto (-, M \otimes -)$.

The diagrams in Section 18.1.3 show the views at these three levels ((1a/b), (2), (3)).

The view of Part 1 of this book was, at least in statements of results, if not in all their proofs, largely that based on (1a) and (1b). The view of a module in Part 2 was that based on (2).[2] The view of Part 3 is based on the enriched language described next.

B.2 Imaginaries

The enriched "(pp-)imaginaries" language for R-modules is defined. A richer language, suited to restricting to a limited collection of sorts is described.

The enriched language for R-modules that I am going to describe now is constructed from the usual one, $\mathcal{L}_R$, using a modification of a process which was introduced by Shelah, [644], and which is now standard in model theory. Shelah's idea was that we should count, as "elements" of a structure, finite tuples of elements of the structure and even equivalence classes of such tuples modulo any equivalence relation which is definable in the language. For instance, if M is any structure, then, as well as the ordered pairs of elements of M, the unordered pairs of elements of M are of this kind, being simply the elements of $M^2/\sim$, where $\sim$ is the equivalence relation on M^2 defined by $(x, y) \sim (x', y')$ iff $(x = x' \wedge y = y') \vee (x = y' \wedge y = x')$ (read "$\vee$" as "or"). To be precise, one forms a multi-sorted language by introducing a sort for every $n \geq 1$ and every definable equivalence relation on n-tuples of elements. In the standard version one adds these sorts together with the projection functions[3] $M^n \to M^n/\sim$.

For our purposes some modifications are required. First, we will factor only by equivalence relations which are defined by pp conditions. This is because the intention is that every sort should inherit a natural structure of an abelian group (for example, the sort of unordered pairs does not have such a structure). Second, all definable subgroups, in all sorts, should be sorts in their own right: then the category of sorts has, for instance, kernels. Third, the language should contain not just the standard projection and injection functions but all those functions between sorts which are definable by pp conditions; again we insist on pp-definability since we want these functions to be group homomorphisms between the sorts. The result is the language $\mathcal{L}_R^{\mathrm{eq}+}$, and every R-module gives rise to (may be "enriched" to) a structure for this language in an obvious way. Indeed, if we regard the collection of sorts and pp-definable functions between them as a category, then, of course, this category is nothing other than the category, $\mathbb{L}_R^{\mathrm{eq}+}$, of pp-pairs (Section 3.2.2) and replacing a module by the corresponding enrichment to an $\mathbb{L}_R^{\mathrm{eq}+}$-structure (the model-theoretic view) is equivalent to replacing the module by the evaluation functor on this category, that is, is equivalent to making it an $\mathbb{L}_R^{\mathrm{eq}+}$-module.

More precisely, let $\mathcal{G}$ be a small preadditive category and denote by $\mathcal{L}_{\mathcal{G}}$ its associated language, that is, that appropriate for $\mathcal{G}$-modules, as defined in Section B.1.

For every pp condition $\phi = \phi(\overline{x})$, where $\overline{x} = (x_1, \dots, x_n)$, x_i having sort G_i, and each pp condition $\psi \leq \phi$, add a sort, $s_{\phi/\psi}$, and a function symbol $\pi_{\phi/\psi}$ from sort $s_\phi = s_{\phi/\overline{x}=\overline{0}}$ to sort $s_{\phi/\psi}$. (Different but equivalent pairs of conditions will give different but isomorphic sorts.) If M is a $\mathcal{G}$-module set $s_{\phi/\psi}(M) = \phi(M)/\psi(M)$ and interpret the function symbol $\pi_{\phi/\psi}$ as

[2] Well, strictly speaking, it was the dual view, with modules viewed as functors on R-mod, not mod-R.

[3] Also the standard injections and projections between the various finite powers of a structure are there, though this is not always made explicit.

the projection from $\phi(M)$ to $\phi(M)/\psi(M)$. Also add, given ϕ as above, a function symbol $i_{\phi,G_1,\dots,G_n}$ from sort s_ϕ to the product sort $G_1 \times \cdots \times G_n$ which may be identified with the sort s_θ, where θ is $x_1 = x_1 \wedge \cdots \wedge x_n = x_n$. This function symbol will be interpreted as the inclusion of $\phi(M)$ in $G_1(M) \times \cdots \times G_n(M)$. Our convention concerning free variables in formulas (viz., that writing $\phi = \phi(x_1, \dots, x_n)$ indicates that the set of free variables of ϕ is a *subset* of $\{x_1, \dots, x_n\}$) means that we have, in effect, also added function symbols which correspond to the canonical inclusions and projections between arbitrary product sorts. For each sort, s, add symbols, $+_s$ and 0_s, to express the abelian group structure of that sort.

One could stop at this point (and in model theory one often does) but, from the functorial view it is clear that we should continue by adding all pp-definable functions between sorts. That is, given any two sorts s, t as above and any pp condition, $\rho(x_s, y_t)$, such that, in any $\mathcal{G}$-module M, ρ defines a function from $s(M)$ to $t(M)$, add a function symbol corresponding to ρ.

This gives the language $\mathcal{L}_\mathcal{G}^{\mathrm{eq}+}$. Since all this new structure is definable in terms of the original $\mathcal{L}_\mathcal{G}$-structure, every $\mathcal{L}_\mathcal{G}$-structure has a unique extension to an $\mathcal{L}_\mathcal{G}^{\mathrm{eq}+}$-structure. This follows from 10.2.37 (up to ^{op}), since, by definition, any module, that is, any functor from the one-object category R, has a unique extension to an exact functor on the free abelian category on R and that result identifies the latter with the functor category, that is (10.2.30), with the category of pp-pairs.

Notice that what we are doing in going from $\mathcal{L}_\mathcal{G}$ to $\mathcal{L}_\mathcal{G}^{\mathrm{eq}+}$ is to add, explicitly, structure which was implicit.

Thus every $\mathcal{G}$-module M has a canonical extension, which we denote $M^{\mathrm{eq}+}$, to a structure for this enriched language.

Note that 18.1.4 says that the categories of additive $\mathcal{G}$-structures (that is, $\mathcal{G}$-modules) and *exact* $\mathcal{L}_\mathcal{G}^{\mathrm{eq}+}$-structures (meaning exact functors from the category, $\mathbb{L}_\mathcal{G}^{\mathrm{eq}+}$, which underlies the language $\mathcal{L}_\mathcal{G}^{\mathrm{eq}+}$ to **Ab**) are equivalent.

We also mention pp-elimination of imaginaries which, for a definable category $\mathcal{X}$, of R-modules, is the statement/result that every pp-pair (ϕ/ψ) in $\mathbb{L}^{\mathrm{eq}+}(\mathcal{D})$ is (pp-)definably isomorphic to a subsort (ϕ' say) of some power of the "home" sort; that is, every object of fun($\mathcal{D}$) is isomorphic to a subobject of a finite power of the forgetful functor. For $\mathcal{D} = \text{Mod-}R$ this is equivalent to R being von Neumann regular (10.2.40).

There is one further step which may be taken. We consider this because we will sometimes detach sorts from the whole structure (see Section 18.2.1) and then it will be necessary to detach, with that sort, or collection of sorts, *all* the induced structure. It is not, in general, the case (see 18.2.13) that the pp-defined functions we have added are enough to define all structure "locally". Therefore we extend $\mathcal{L}_R^{\mathrm{eq}+}$ to obtain a language, we will denote $\mathcal{L}_R^{\mathrm{eq}++}$, by adding, for any pp-definable n-ary relation between sorts, an n-ary relation symbol of the appropriate sort. Explicitly, if θ is a pp condition of sort $s_1 \times \cdots \times s_n$, then we introduce a corresponding relation symbol R_θ which will be interpreted in every structure as the subgroup defined by θ. In particular, for every pp-definable subgroup of a sort, we have added a corresponding predicate (= 1-ary relation) symbol.

Each $\mathcal{L}_\mathcal{G}^{\mathrm{eq}+}$-structure, M, has a unique extension to this new language, defined by setting $\theta(M)$ to be the group consisting of all $\overline{a}$ from M such that $\theta(\overline{a})$ holds in M. Note that this larger language, as opposed to $\mathcal{L}_\mathcal{G}^{\mathrm{eq}+}$ is not a purely functional one.

A useful observation is the following.

Proposition B.2.1. (for example, [106, 4.2]) *Suppose that N is a pure-injective module and that ϕ/ψ is a pp-pair. Then every endomorphism of $\phi(N)/\psi(N)$ regarded as an $\mathcal{L}_R^{\mathrm{eq}++}$-structure extends to an endomorphism of N.*

By restricting the language of a functor category, Mod-$\mathcal{G}$, to a definable subcategory, all this extends to definable subcategories, in particular, to finitely accessible additive categories.

Appendix C A model theory/functor category dictionary

This is a short "dictionary" giving translations between model-theoretic and functor-theoretic concepts.

1a. pp formula (in n free variables)
1b. finitely generated subfunctor of a power (the nth power) of the forgetful functor $(R_R, -)$ (10.2.15, 10.2.14, Section 1.1.2)

2a. pp-type (in n free variables)
2b. (via duality) arbitrary subfunctor of a power (the nth power) of the forgetful functor $({}_RR, -) \simeq (R \otimes_R -)$ (12.2.1)

3a. pp-type of $\overline{a} = (a_1, \dots, a_n)$ in M
3b. kernel of the morphism $\overline{a} \otimes -$ from $R^n \otimes -$ to $M \otimes -$ induced by the morphism $\overline{a} : R^n \to M$ which takes $(r_1, \dots, r_n)$ to $\sum_{i=1}^n a_i r_i$ (10.3.9, 12.2.5, Section 1.2.1)

4a. pp-type in a sort
4b. finitely generated functor (12.2.2, Section 4.3.9)

5a. irreducible pp-type p in pp_R^n
5b. subfunctor G of $R^n \otimes -$ such that $(R^n \otimes -)/G$ is uniform ($G = F_{Dp} = \sum_{\phi \in p} F_{D\phi}$) (12.2.3, Section 4.3.6)

6a. pp-pair; corresponding quotient sort
6b. pair of finitely generated subfunctors of some $(R^n, -)$, one contained in the other; arbitrary finitely presented functor in (mod-R, **Ab**) (10.2.30, 10.2.31, Section 3.2.2)

7a. hull, $H(p)$, of a pp-n-type p
7b. injective hull, $E(R^n \otimes -/F_{Dp})$, of corresponding finitely generated functor (Section 4.3.5 esp. 4.3.34, also 12.2.6)

8a. lattice of pp formulas (in n free variables)
8b. lattice of finitely generated subfunctors of (the nth power of) the forgetful functor (10.2.17, Section 3.2.1)

9a. category, $\mathbb{L}_R^{\mathrm{eq}+}(M)$, of pp-pairs (relative to M)
9b. category, $(\text{mod-}R, \mathbf{Ab})^{\mathrm{fp}}_{\tau_M}$, of finitely presented functors (modulo the Serre subcategory of functors which vanish, at least whose $\varinjlim$-commuting extensions vanish, on M) (10.2.30, Section 3.2.2)

10a. subclass of modules axiomatised by closure of certain pp-pairs/elementary subclass of Mod-R closed under products and direct summands
10b. definable subcategory of Mod-R/subcategory of Mod-R closed under products, direct limits and pure submodules (Section 3.4.1, p. 413, also 12.3.2)

11a. restriction to a definable subcategory
11b. localisation of the functor category at a finite type torsion theory (Section 3.4.1, 5.1.1, Section 12.3 esp. 12.3.6, 12.3.10, 12.3.12, 12.3.16, 12.3.19 and 12.3.20)

12a. elementary embedding/elementary submodule
12b. pure embedding between modules which have the same support in the Ziegler spectrum, equivalently which generate the same definable subcategory (A.1.4, A.1.3)

13a. ring of definable scalars of M
13b. endomorphism ring of forgetful functor localised at torsion theory corresponding to M (12.8.2, Section 6.1.1)

14a. $M^{\mathrm{eq}+}$: M_R extended to a structure for the pp-imaginaries language
14b. the (image of the) functor evaluation-at-M from the free abelian category on R (Ab(R), equivalently the functor category (R-mod, **Ab**)^{fp}) to the category of abelian groups (3.2.11, p. 450, Section B.2)

15a. reduced product/ultraproduct
15b. a certain direct limit of products (3.3.1)

16a. (isolated by) minimal pair
16b. (isolated by) finitely presented simple functor (5.3.2, 12.5.2, 12.5.4, §3.2.1)

17a. neg-isolated pp-type/indecomposable pure-injective
17b. (isolated by) simple functor (5.3.45, 12.5.5, 12.5.6)

18. m-dimension and Krull–Gabriel dimension:
$\mathrm{mdim}([\phi, \psi]_M) = \mathrm{KGdim}((F_{\phi/\psi})_{\tau_M}) = \mathrm{KGdim}((F_{D\psi/D\phi})_{\tau_{DM}})$ (13.2.1, 13.2.2)

19. breadth (also see width) and uniserial dimension:
$\mathrm{w}([\phi, \psi]_M) = \mathrm{Udim}((F_{\phi/\psi})_{\tau_M}) = \mathrm{Udim}((F_{D\psi/D\phi})_{\tau_{DM}})$ (13.4.3)

20. linking of pp conditions (an important ingredient in the model-theoretic approach) said in terms of functors (proof of 4.3.74)

21. various descriptions of the Ziegler spectrum (Section 5.1.1 esp. 5.1.1, 5.1.6, 5.1.7 and 5.1.8, 5.1.25, 10.2.45, 12.1.17, Section 12.4, 18.1.19)

Part III

Definable categories

Here we examine definable subcategories per se, cutting them free from context. That is, we study their intrinsic properties, showing that most of these do not depend on any particular category within which they are definable. We also investigate the natural functors between definable categories.

Since this material is still developing I have kept this third part quite short, providing just an introduction.

18

Definable categories and interpretation functors

Definable subcategories and categories of exact functors on small abelian categories are shown to be equivalent. It turns out that the functor category of a definable category has an intrinsic definition. Functors which preserve products and direct limits are introduced as the "appropriate" functors between definable categories and these are mirrored by exact functors between the associated functor categories/categories of imaginaries. Those functors between definable categories which preserve direct limits and products are shown to be exactly the model-theoretic interpretation functors. A variety of examples is presented.

18.1 Definable categories

Definable subcategories are represented as categories of exact functors in Section 18.1.1 and the converse is shown in Section 18.1.2. The dual of a definable category is defined in Section 18.1.1.

One way of obtaining the functor category of a definable category intrinsically is given in Section 18.1.3; a simpler description is in Section 18.1.4.

18.1.1 Definable subcategories

The elementary dual of a definable subcategory is defined: the respective functor categories are dual (18.1.3). Every definable category is the category of exact functors on its functor category (18.1.4).

Let $\mathcal{C}$ be a finitely accessible additive category with products (Section 16.1.1, esp. 16.1.12) and let $\mathcal{D}$ be a full subcategory of $\mathcal{C}$. As in the case of subcategories of module categories in Section 3.4.1, say that $\mathcal{D}$ is a **definable subcategory** of $\mathcal{C}$ if $\mathcal{D}$ is closed under arbitrary products, direct limits and pure subobjects; note, from Section 16.1.2, that it follows that purity in $\mathcal{C}$ is defined and has properties as usual. There are equivalent sets of conditions exactly as in 3.4.7. The proofs are as there if $\mathcal{C}$ is actually a functor category and, in the general case, either one may use these proofs, with the generalised notion of element (that is, a morphism from a finitely presented object, see Sections 10.2.4 and B.1), or one may use the fact, 16.1.7, that $\mathcal{C}$ is a definable subcategory of a functor category and restrict from the functor category to $\mathcal{C}$.

In particular, $\mathcal{D}$ may be defined by the closure of a set of pp-pairs, $\mathcal{D} = \mathrm{Mod}(T)$ (that notation is defined in Section 3.4.1), where T is a set of pp-pairs. Pp conditions are defined for functor categories in Section 10.2.4 and for general locally finitely presented categories in Section 11.1.5.

By 16.1.16 every definable subcategory, $\mathcal{D}$, of a finitely accessible category $\mathcal{C}$ with products has an internal notion of purity which coincides with any natural "external" notion of purity, so every definable subcategory of $\mathcal{D}$ is actually a definable subcategory of $\mathcal{C}$. Since, by 16.1.7, every definable subcategory in the above sense is a definable subcategory of a functor category this definition has not, in fact, extended the notion of definable subcategory. Indeed, by 12.1.12 every definable subcategory is a definable subcategory of a locally coherent functor category. In view of this and of what we will show, we will say that $\mathcal{D}$ is a **definable category** if it is a definable subcategory of some (locally coherent functor) category.

Recall that we limit ourselves here to finitary definability. For categories definable in infinitary languages see, for instance, [3], [205], [462], [367, Thm 2] (for a generalisation of 3.4.7). We note that definable categories are certainly more general than finitely accessible categories.

Example 18.1.1. Consider the definable subcategory (3.4.28) Inj-$\mathbb{Z}$ of injective abelian groups. Any $\varinjlim$-generating set for Inj-$\mathbb{Z}$ must contain a module with $\mathbb{Q}$ as a direct summand (since $(\mathbb{Z}_{p^\infty}, \mathbb{Q}) = 0$) but $\mathbb{Q}$ is not a finitely presented object. To see this, regard $M = \mathbb{Z}_{p^\infty}^{(\aleph_0)}$ as a direct limit $\mathbb{Z}_{p^\infty} \longrightarrow \mathbb{Z}_{p^\infty} \oplus \mathbb{Z}_{p^\infty} \longrightarrow \cdots \longrightarrow \mathbb{Z}_{p^\infty}^n \longrightarrow \cdots$. Define a morphism from $\mathbb{Q}$ to M by sending 1 to $(a_{11}, 0, \dots)$, where $a_{11} \neq 0$ and $a_{11}p = 0$, sending $1/p$ to $(a_{12}, a_{21}, 0, \dots)$, where $a_{12}p = a_{11}$, $a_{21} \neq 0$ and $a_{21}p = 0$, etc., then extending to all of $\mathbb{Q}$. Clearly any such morphism factors through no finite subsum. This argument applies to any indecomposable injective, showing that $(\text{Inj-}\mathbb{Z})^{\mathrm{fp}} = 0$.

Suppose that $\mathcal{D}$ is a definable subcategory, of Mod-$\mathcal{A}$ say. By 12.4.1 there is a corresponding, dual, definable subcategory of $\mathcal{A}$-Mod. Namely, if $\mathcal{S}_{\mathcal{D}}$ is the Serre subcategory of $(\text{mod-}\mathcal{A}, \mathbf{Ab})^{\mathrm{fp}}$ consisting of those functors F which annihilate $\mathcal{D}$, in the sense that $\overrightarrow{F}(\mathcal{D}) = 0$, where $\overrightarrow{F}$ is as in Section 10.2.8, then set $\mathcal{S}_{\mathcal{D}}^{\mathrm{d}} = \{dF : F \in \mathcal{S}_{\mathcal{D}}\}$ to be the dual Serre subcategory (11.2.9) of $(\mathcal{A}\text{-mod}, \mathbf{Ab})^{\mathrm{fp}}$, and then define $\mathcal{D}^{\mathrm{d}} = \{L \in \mathcal{A}\text{-Mod} : \overrightarrow{G}\, L = 0 \text{ for every } G \in \mathcal{S}^{\mathrm{d}}_{\mathcal{D}}\}$ to be the (**elementary**) **dual** of $\mathcal{D}$. This extends the notion of dual category from Section 16.1.3.[1]

Corollary 18.1.2. *For any definable category $\mathcal{D}$ there is an inclusion-preserving bijection between definable subcategories of $\mathcal{D}$ and definable subcategories of $\mathcal{D}^{\mathrm{d}}$.*

This is immediate from 12.4.1 since a definable subcategory of $\mathcal{D}$ is simply a definable subcategory of Mod-$\mathcal{A}$ which is contained in $\mathcal{D}$. We extend the notation introduced after 12.3.21, writing fun-$\mathcal{A}$ for fun(Mod-$\mathcal{A}$). Note that in all cases, since Fun($\mathcal{D}$) is locally coherent (12.3.11), the category fun($\mathcal{D}$) is abelian (E.1.19).

It is easily checked that the functor category, in the sense defined just after 12.3.21, of the elementary dual, $\mathcal{D}^{\mathrm{d}}$, of $\mathcal{D}$ is $\mathrm{Fun}(\mathcal{D}^{\mathrm{d}}) = (\mathcal{A}\text{-mod}, \mathbf{Ab})_{\tau^{\mathrm{d}}_{\mathcal{D}}}$, where $\tau^{\mathrm{d}}_{\mathcal{D}}$ is the hereditary finite type torsion theory generated by $\mathcal{S}^{\mathrm{d}}_{\mathcal{D}}$. By 12.3.21, $\mathrm{fun}(\mathcal{D}^{\mathrm{d}}) \simeq (\mathrm{fun}(\mathcal{D}))^{\mathrm{op}}$; model-theoretically $\mathbb{L}(\mathcal{D})^{\mathrm{eq}+} \simeq (\mathbb{L}(\mathcal{D}^{\mathrm{d}})^{\mathrm{eq}+})^{\mathrm{op}})$. We record this for reference.

Proposition 18.1.3. *Let $\mathcal{D}$ be a definable category. Then* $\mathrm{fun}(\mathcal{D}^{\mathrm{d}}) \simeq (\mathrm{fun}(\mathcal{D}))^{\mathrm{op}}$.

The discussion above begs some questions. Namely, does the dual of $\mathcal{D}$ depend only on $\mathcal{D}$, as opposed to the particular representation of $\mathcal{D}$ as a definable subcategory? For certainly one may have naturally equivalent definable subcategories of quite different functor categories. More generally, is the "definable" structure on a definable subcategory intrinsic? It was shown in Section 16.1.2 that such a category does have an intrinsic theory of purity but does, for example, the associated functor category depend only on $\mathcal{D}$? And, if so, how do we obtain this functor category from $\mathcal{D}$? A (positive) answer to these questions was given by Krause in [355] and we discuss this, and a more general solution, at 18.1.14 and 18.1.19 below.

By the "definable structure" on a definable subcategory we include, as well as the associated category of functors and the category of pp-definable sorts and maps to which this is equivalent (12.3.20), the Ziegler spectrum and the sheaf of locally definable scalars. Since all these can be defined from the functor category it will follow that if $\mathcal{D} \subseteq \mathcal{C}$ and $\mathcal{D}' \subseteq \mathcal{C}'$ are definable subcategories of finitely accessible categories $\mathcal{C}$, $\mathcal{C}'$ with products, and if $\mathcal{D} \simeq \mathcal{D}'$ just as categories, then Zg($\mathcal{D}$) will be homeomorphic to Zg($\mathcal{D}'$) (the Ziegler spectrum of $\mathcal{D}$ is defined as in Section 12.3) and a naturally homeomorphic space is obtained by restricting the Ziegler topology of $\mathcal{C}$, respectively $\mathcal{C}'$ to the closed subset corresponding to $\mathcal{D}$, resp. $\mathcal{D}'$.

[1] It also introduces notational redundancy since $\mathcal{D}^{\mathrm{d}} = D\mathcal{D}$ in the notation of Section 3.4.2 (each notation fits well with others, depending on the context).

Krause's construction of the functor category from $\mathcal{D}$ will be presented in Section 18.1.3. A more direct construction is then given in Section 18.1.4. The second construction allows us to remove a restriction from Krause's characterisation, [355, 7.2], of the key functors between definable categories, see Section 18.2.

We also show that $\mathcal{D}$ may be recovered from its functor category as, up to equivalence, the category of exact functors from fun($\mathcal{D}$) to **Ab**: this was shown to me for definable subcategories of module categories by Herzog,[2] can be found for locally finitely presented categories as [144, 3.3, Lemma 1], and the general case was dealt with by Krause, see [355, §4]. In fact, this is the additive, finitary version of a very general result of Hu ([307, 5.10(ii)], also see [420, 5.1]), from which it could be derived using Makkai's [420, §6].

One thing that we have not done is to give an entirely intrinsic characterisation of definable categories (in the sense of their being exactly the categories satisfying some list of conditions that make no reference to other categories). Some necessary conditions are discussed in Section 18.1.4.

Theorem 18.1.4. *Let $\mathcal{D}$ be a definable subcategory. Then $\mathcal{D} \simeq \mathrm{Ex}(\mathrm{fun}(\mathcal{D}), \mathbf{Ab})$, the map from left to right being $D \mapsto \mathrm{ev}_D$, where the latter is evaluation at D, defined by $\mathrm{ev}_D(F) = \overrightarrow{F} D$.*

Proof. Suppose first that $\mathcal{D} = \text{Mod-}R$, where R is a ring. For each object $D \in \mathcal{D}$ the functor, evaluation at D, $\mathrm{ev}_D : \mathrm{fun}(\mathcal{D}) \longrightarrow \mathbf{Ab}$, certainly is exact (a sequence of functors is exact iff it is exact under each evaluation).

For the converse, suppose that $e : \mathrm{fun}(\mathcal{D}) \longrightarrow \mathbf{Ab}$ is exact. Set $M = e(R, -)$: this is a right R-module under the action $r(\in R) \mapsto \big(a(\in M) \mapsto e(r, -)a\big)$. We claim that $e = \mathrm{ev}_M$. For each $A \in \text{mod-}R$ there is an exact sequence $R^m \to R^n \to A \to 0$ hence an exact sequence $0 \to (A, -) \to (R^n, -) \to (R^m, -)$ in the functor category. Since e and ev_M are exact and agree on $(R, -)$ they therefore agree on A. For each object of fun($\mathcal{D}$), that is, for each finitely presented functor, F, on mod-R, there is (10.2.1) an exact sequence $(B, -) \to (A, -) \to F \to 0$ so, again, since each functor is exact and since they agree on representable functors, they agree on F and $e = \mathrm{ev}_M$, as required.

More generally, if $R = \mathcal{A}$ is a small preadditive category, then M is replaced by a functor from R, viewed via the Yoneda embedding (10.1.8) as a category of projective modules, to **Ab**, and the proof goes as before, with functors of the form $(G, -)$, where G is an object of R replacing $(R, -)$.

For the general case, regard $\mathcal{D}$ as a definable subcategory of a category of the form Mod-$\mathcal{A}$. Then fun($\mathcal{D}$) is a localisation of fun-$\mathcal{A}$. Localisation is exact (11.1.40), so any exact functor on fun($\mathcal{D}$) may be regarded, composing with the

[2] In 1993 at the ASL meeting in Notre Dame.

localisation functor, as an exact functor on fun-$\mathcal{A}$. Therefore, by the above, every exact functor e on fun($\mathcal{D}$) is evaluation, ev_M, at some $\mathcal{A}$-module M, which must be in $\mathcal{D}$ by the bijection (12.4.1) between Serre subcategories and definable subcategories since $\overrightarrow{F} M = 0$ for each functor $F \in \mathcal{S}_{\mathcal{D}}$ (the kernel of the localisation). □

Corollary 18.1.5. *Let $\mathcal{D}$ be a definable category. Then* $\mathcal{D} \simeq \mathrm{Ex}(\mathrm{fun}(\mathcal{D}), \mathbf{Ab}) \simeq \mathrm{Ex}((\mathrm{fun}(\mathcal{D}^{\mathrm{d}}))^{\mathrm{op}}, \mathbf{Ab})$*, where $\mathcal{D}^{\mathrm{d}}$ is the elementary dual of $\mathcal{D}$.*

Recall (after 12.3.11) that $\mathrm{fun}^{\mathrm{d}}(\mathcal{D}) = \mathrm{fun}(\mathcal{D}^{\mathrm{d}})$ is referred to as the dual functor category of $\mathcal{D}$. By 18.1.5, $\mathcal{D} \simeq \big((\mathrm{fun}^{\mathrm{d}}(\mathcal{D}))^{\mathrm{op}}, \mathbf{Ab}\big)$, and it follows from 12.3.15 that this equivalence is given by $D \mapsto (G \mapsto (G, D \otimes_{\mathcal{A}} -)$, if $\mathcal{D}$ is a definable subcategory of Mod-$\mathcal{A}$.

It is the categories of *finitely presented* functors which are dual: note that $\mathrm{Fun}^{\mathrm{d}}(\mathcal{D})$ is, in general, quite different from $\big(\mathrm{Fun}(\mathcal{D})\big)^{\mathrm{d}}$ – take $\mathcal{D} = \mathrm{Mod}\text{-}\mathcal{A}$, so $\mathrm{Fun}^{\mathrm{d}}(\mathcal{D}) = (\mathcal{A}\text{-mod}, \mathbf{Ab})$, whereas $\big(\mathrm{Fun}(\mathcal{D})\big)^{\mathrm{d}} = (\mathrm{mod}\text{-}\mathcal{A}, \mathbf{Ab})^{\mathrm{d}} = \big((\mathrm{mod}\text{-}\mathcal{A})^{\mathrm{op}}, \mathbf{Ab}\big)$ and, in the absence of a duality $\mathcal{A}\text{-mod} \simeq (\mathrm{mod}\text{-}\mathcal{A})^{\mathrm{op}}$, these are different.

The model-theoretic view of the equivalence in 18.1.4 is, given an object $D \in \mathcal{D}$, to regard it, not as a functor (on $\mathrm{fun}(\mathcal{D}) \simeq \mathbb{L}(\mathcal{D})^{\mathrm{eq}+}$) per se but rather as a structure consisting of the image of this functor. It is exactly the view which regards a module as a set with structure, as opposed to a functor from a one-object category to $\mathbf{Ab}$. Since $\mathbb{L}(\mathcal{D})^{\mathrm{eq}+}$ is skeletally small we may consider the image of D regarded as a functor on $\mathbb{L}(\mathcal{D})^{\mathrm{eq}+}$ to be a set of objects of $\mathbf{Ab}$ together with various morphisms between those objects. This is the structure which is denoted by $D^{\mathrm{eq}+}$, see Section B.2 (in [347] it is called the endocategory of D – though, strictly speaking, Krause uses this term for this structure regarded as an object over the endomorphism ring of D).

18.1.2 Exactly definable categories

Definable subcategories coincide with exactly definable categories (the categories of exact functors on small abelian categories, 18.1.6). The dual is the category of exact functors on the opposite abelian category (18.1.8). The relation between exact (absolutely pure), left exact (flat) and all functors is described (before 18.1.7).

A definable category is finitely accessible iff its dual functor category has enough injectives (18.1.12).

A category is **exactly definable** in the terminology of Krause ([358], [355]) if $\mathcal{D} \simeq \mathrm{Ex}(\mathcal{A}^{\mathrm{op}}, \mathbf{Ab})$ for some skeletally small abelian category[3] $\mathcal{A}$. Since $\mathcal{A}$ is

[3] Of course, since the opposite of an abelian category is abelian, the ^{op} is a choice rather than a necessity.

abelian, Mod-$\mathcal{A}$ is locally coherent (E.1.19), hence (16.1.12 and cf. proof of 12.1.6) $\mathcal{D} \simeq \mathrm{Ex}(\mathcal{A}^{\mathrm{op}}, \mathbf{Ab}) \simeq \mathrm{Abs}\text{-}\mathcal{A}$ is a definable subcategory of Mod-$\mathcal{A}$.

Conversely, if $\mathcal{D}$ is a definable subcategory (of some finitely accessible category with products), then, by 18.1.5, $\mathcal{D} \simeq \mathrm{Ex}(\mathrm{fun}(\mathcal{D}^{\mathrm{d}})^{\mathrm{op}}, \mathbf{Ab})$ and, 12.3.11, $\mathrm{fun}(\mathcal{D}^{\mathrm{d}})^{\mathrm{op}}$ is abelian so $\mathcal{D}$ is exactly definable. Thus the concepts of definable subcategory and exactly definable category are equivalent. Note that both, however, depend on a representation of the category being given, so fail to be "intrinsic" definitions.

Proposition 18.1.6. (see [355, 2.2]) *A category $\mathcal{D}$ is a definable subcategory (of a finitely accessible category with products) iff it is exactly definable iff it is equivalent to the full subcategory of absolutely pure (= fp-injective) objects of a locally coherent category.*

That follows from the above discussion and (a general version of) 3.4.24. In the next diagram of inclusions $\mathcal{A}$ is assumed to be abelian.

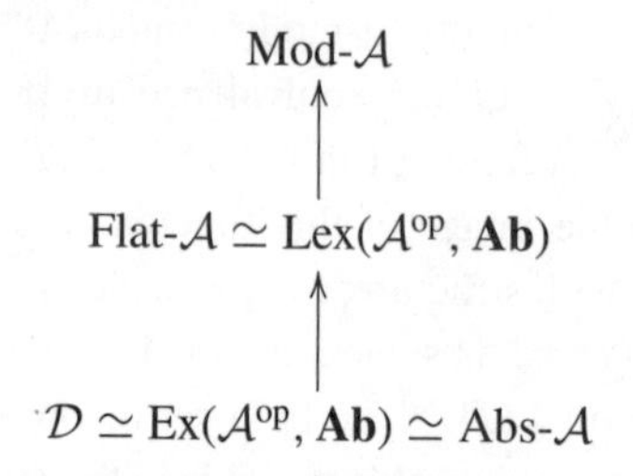

If $\mathcal{D} = \mathrm{Ex}(\mathcal{A}^{\mathrm{op}}, \mathbf{Ab})$ with $\mathcal{A}$ abelian, let $\mathcal{C} = \mathrm{Flat}\text{-}\mathcal{A}$ denote the locally coherent category corresponding, in the sense of 16.1.4, to $\mathcal{A}$, so $\mathcal{A} \simeq \mathcal{C}^{\mathrm{fp}}$. Then, since $\mathrm{Abs}(\mathcal{C}) \simeq \mathrm{Ex}\big((\mathcal{C}^{\mathrm{fp}})^{\mathrm{op}}, \mathbf{Ab}\big)$, the above diagram may be reexpressed as follows.

$$\mathrm{Mod}\text{-}\mathcal{C}^{\mathrm{fp}}$$

$$\uparrow$$

$$\mathcal{C} \simeq \mathrm{Flat}\text{-}\mathcal{C}^{\mathrm{fp}} \simeq \mathrm{Lex}\big((\mathcal{C}^{\mathrm{fp}})^{\mathrm{op}}, \mathbf{Ab}\big)$$

$$\uparrow$$

$$\mathcal{D} \simeq \mathrm{Ex}\big((\mathcal{C}^{\mathrm{fp}})^{\mathrm{op}}, \mathbf{Ab}\big) = \mathrm{Abs}(\mathcal{C})$$

Proposition 18.1.7. *In the above diagram $\mathcal{C} \simeq \mathrm{Fun}^{\mathrm{d}}(\mathcal{D})$. That is, if $\mathcal{D} \simeq \mathrm{Ex}(\mathcal{A}^{\mathrm{op}}, \mathbf{Ab})$ with $\mathcal{A}$ abelian, then $\mathrm{Fun}^{\mathrm{d}}(\mathcal{D}) \simeq \mathrm{Lex}(\mathcal{A}^{\mathrm{op}}, \mathbf{Ab})$. Hence $\mathrm{fun}^{\mathrm{d}}(\mathcal{D}) \simeq$ flat-$\mathcal{A} \simeq \mathcal{A}$.*

Proof. Calling on 18.1.15 below, 18.1.4 gives $\mathcal{C}^{\mathrm{fp}} \simeq \mathrm{fun}^{\mathrm{d}}(\mathcal{D})$, therefore the corresponding finitely accessible categories (16.1.3), $\mathrm{Fun}^{\mathrm{d}}(\mathcal{D})$ and $\mathcal{C}$, are equivalent.

(As presented, this argument is circular, but circularity can be avoided by a direct proof of this result, for which see [516, 11.2, 11.3].) □

Corollary 18.1.8. *If* $\mathcal{D} = \mathrm{Ex}(\mathcal{A}^{\mathrm{op}}, \mathbf{Ab})$*, then the elementary dual is* $\mathcal{D}^{\mathrm{d}} \simeq \mathrm{Ex}(\mathcal{A}, \mathbf{Ab})$.

This follows from 18.1.5, given 18.1.15. One may check that there is a canonical tensor functor $\mathcal{D}^{\mathrm{d}} \times \mathcal{D} \longrightarrow \mathbf{Ab}$ determined by $(D, D') \mapsto D \otimes_{\mathcal{A}} D'$ and, that in the diagram after 18.1.6, and using 18.1.7, the equivalence $\mathcal{D} \simeq \mathrm{Abs}(\mathrm{Fun}(\mathcal{D}^{\mathrm{d}}))$ is given by $M(\in \mathcal{D}) \mapsto (M \otimes_{\mathcal{A}} -)_{\mathcal{D}}$. Here $M \otimes_{\mathcal{A}} -$ is the functor in $\mathcal{C} = \mathrm{Fun}^{\mathrm{d}}(\mathcal{D}) = \mathrm{Fun}(\mathcal{D}^{\mathrm{d}})$ given by $L \mapsto M \otimes_{\mathcal{A}} L$ for $L \in (\mathcal{A} = \mathcal{C}^{\mathrm{fp}})$-mod and the subscript denotes its localisation at the torsion theory on $\mathrm{Fun}^{\mathrm{d}}(\mathcal{D})$ corresponding (12.3.6) to $\mathcal{D}^{(\mathrm{d})}$.

Corollary 18.1.9. *Suppose that* $0 \to L \to M \to N \to 0$ *is an exact sequence in the definable category* $\mathcal{D}$*. Then the sequence is pure-exact iff for every* $F \in \mathrm{fun}(\mathcal{D})$ *the sequence* $0 \to FL \to FM \to FN \to 0$ *of abelian groups is exact.*

Proof. Evaluation of F at an object $M \in \mathcal{D}$ is equivalent to computing $(dF, M \otimes -)$ (10.3.8) so this follows since $(M \otimes_{\mathcal{A}} -)_{\mathcal{D}}$ is absolutely pure (12.3.3). □

We also note the following results on the relation between $\mathcal{D}$ and $\mathrm{fun}(\mathcal{D})$.

Proposition 18.1.10. ([355, 2.3(1)]) *If* $\mathcal{A}$ *is skeletally small abelian and* $\mathcal{D} = \mathrm{Ex}(\mathcal{A}^{\mathrm{op}}, \mathbf{Ab})$*, then the Yoneda embedding* $\mathcal{A} \longrightarrow \text{Mod-}\mathcal{A}$*, given by* $A \mapsto (-, A)$*, induces an equivalence* $\mathrm{Inj}(\mathcal{A}) \simeq \mathcal{D}^{\mathrm{fp}}$.

Proof. For one direction, if $A \in \mathrm{Inj}(\mathcal{A})$, then the functor $(-, A)$ is, being exact, in $\mathcal{D}$. It is also a finitely presented object of Mod-$\mathcal{A}$, so, since direct limits in $\mathcal{D}$ are direct limits in Mod-$\mathcal{A}$ ($\mathcal{D}$ being a definable subcategory of Mod-$\mathcal{A}$), $(-, A) \in \mathcal{D}^{\mathrm{fp}}$.

The other direction takes more work: suppose that $D \in \mathcal{D}^{\mathrm{fp}}$ and identify $\mathcal{D}$ with Abs(Mod-$\mathcal{A}$). So $(D, -)$ commutes with direct limits of absolutely pure objects of Mod-$\mathcal{A}$. Breitsprecher showed ([89, 1.8]) that if $\mathcal{C}$ is a Grothendieck category, $C \in \mathcal{C}$ and $(C, -)$ commutes with direct limits of injective objects,[4] then $(C, -)$ commutes with direct limits of arbitrary objects, that is, C is finitely presented. Therefore $D \in$ mod-$\mathcal{A}$. Since D also is in Flat(Mod-$\mathcal{A}$) it must be a representable functor $(-, A)$ for some $A \in \mathcal{A}$ with, since $(-, A)$ is exact, $A \in \mathrm{Inj}(\mathcal{A})$, as required. □

[4] That is, if the natural map $(C, \varinjlim E_\lambda) \to \varinjlim(C, E_\lambda)$ is a bijection whenever $(E_\lambda)_\lambda$ is a direct system with the E_λ injective (hence, in this case of a locally coherent category, with $\varinjlim E_\lambda$ absolutely pure).

Example 18.1.11. ([256, 5.5]) Take $\mathcal{D} = \text{Mod-}R \simeq \text{Ex}\big(((R\text{-mod}, \mathbf{Ab})^{\text{fp}})^{\text{op}}, \mathbf{Ab}\big) \leq \text{Mod-}\big((R\text{-mod}, \mathbf{Ab})^{\text{fp}}\big)^{\text{op}}$. Under the Yoneda correspondence (10.1.8) $\text{mod-}R \simeq \text{Inj}\big((R\text{-mod}, \mathbf{Ab})^{\text{fp}}\big)$ and this is explicitly given by $M \mapsto M \otimes -$; recall, 12.1.13, that $M \otimes -$ is only absolutely pure in the full functor category although it is injective in the category of finitely presented functors.

Corollary 18.1.12. ([355, 2.3]) *A definable category $\mathcal{D}$ is finitely accessible iff* $\text{fun}^{\text{d}}(\mathcal{D})$ *has enough injectives.*

Equivalently, the definable category $\mathcal{D} = \text{Ex}(\mathcal{A}, \mathbf{Ab})$ is finitely accessible iff the abelian category $\mathcal{A}$ has enough projectives.

Note that in our typical situation of Part II, where we have a locally finitely presented category $\mathcal{C}$, a definable subcategory, $\mathcal{X}$ and the corresponding localisation $\text{fun}(\mathcal{C}) \longrightarrow \text{fun}(\mathcal{X})$ with kernel the Serre subcategory $\mathcal{S}_{\mathcal{X}} = \{F \in \text{fun}(\mathcal{D}) : \overrightarrow{F}\mathcal{X} = 0\}$, there is another definable category around, namely $\mathcal{D} = \text{Ex}(\mathcal{S}, \mathbf{Ab})$. Krause ([355]) referred to this as a **definable quotient category** of $\mathcal{C}$ and investigated these, showing that there is an equivalence between $\text{Pinj}(\mathcal{D})$ and the stable category $\text{Pinj}(\mathcal{C})/\mathcal{P}_{\mathcal{X}}$ where $\mathcal{P}_{\mathcal{X}}$ is the ideal of morphisms of $\mathcal{C}$ which factor through an object of $\text{Pinj}(\mathcal{X})$ ([355, 5.1]).

18.1.3 Recovering the definable structure

Still, it must be shown that the definable structure on a definable category is unique for, without that, it does not even make sense to talk about $\text{fun}(\mathcal{D})$. *This is done at 18.1.14. As a consequence there is a bijection between definable categories and skeletally small abelian categories (18.1.15).*

Let $\mathcal{D}$ be a definable category. Since there is (Section 16.1.2) an intrinsic theory of purity on $\mathcal{D}$, which coincides with that induced by any representation of $\mathcal{D}$ as a definable subcategory, the class, $\text{Pinj}(\mathcal{D})$, of pure-injective objects of $\mathcal{D}$ may be defined as consisting of those $D \in \mathcal{D}$ such that every pure monomorphism in $\mathcal{D}$ with domain D is split. Alternatively this class may be defined directly, as in [355, §2], using the Jensen–Lenzing test (4.3.6) for pure-injectivity.

We make use of the fact (see the second diagram in Section 18.1.2) that if $\mathcal{D}$ is exactly definable, then $\mathcal{D} \simeq \text{Ex}\big((\mathcal{C}^{\text{fp}})^{\text{op}}, \mathbf{Ab}\big)$ for some locally coherent category $\mathcal{C}$ and, via the embedding of $\text{Ex}\big((\mathcal{C}^{\text{fp}})^{\text{op}}, \mathbf{Ab}\big)$ into $\text{Lex}\big((\mathcal{C}^{\text{fp}})^{\text{op}}, \mathbf{Ab}\big) \simeq \mathcal{C}$, one has the identification (18.1.6) of $\mathcal{D}$ with $\text{Abs}(\mathcal{C})$, hence the identification of $\text{Pinj}(\mathcal{D})$ with $\text{Inj}(\mathcal{C})$.

Lemma 18.1.13. ([355, 2.7]) *Suppose that $\mathcal{C}$ is abelian with enough injectives. Then the Yoneda map* $\mathcal{C}^{\text{op}} \longrightarrow (\text{Inj}(\mathcal{C}), \mathbf{Ab})^{\text{fp}}$ *given by* $C \mapsto (C, -) \upharpoonright \text{Inj}(\mathcal{C})$ *is an equivalence.*

Proof. If $F \in (\mathrm{Inj}(\mathcal{C}), \mathbf{Ab})^{\mathrm{fp}}$, then there is $f : E \to E'$ in $\mathrm{Inj}(\mathcal{C})$ such that $(E', -) \xrightarrow{(f,-)} (E, -) \to F \to 0$ is exact. Let $K = \ker(f)$, so $0 \to K \to E \xrightarrow{f} E'$ is exact. Since we are evaluating at injectives, the induced sequence $(E', -) \to (E, -) \to (K, -) \to 0$ also is exact, therefore $F \simeq (K, -)$.

Conversely, if $K \in \mathcal{C}$, then there is an exact sequence $0 \to K \to E \xrightarrow{f} E'$, where E and E' are injective, so, as above, $(K, -) = \mathrm{coker}(f, -)$ and $(K, -) \upharpoonright \mathrm{Inj}(\mathcal{C})$ is finitely presented, as required. □

Putting this together with the equivalence $\mathrm{Pinj}(\mathcal{D}) \simeq \mathrm{Inj}(\mathcal{C})$ and 18.1.7 one obtains a description of $\mathrm{fun}(\mathcal{D})$ from $\mathcal{D}$.

Theorem 18.1.14. *Given a definable category* $\mathcal{D} = \mathrm{Ex}(\mathcal{A}^{\mathrm{op}}, \mathbf{Ab})$, *where* $\mathcal{A}$ *is abelian, set* $\mathcal{C} = \text{Mod-}\mathcal{A}$, *so* $\mathcal{C} = \mathrm{Fun}^{\mathrm{d}}(\mathcal{D})$ *is locally coherent and* $\mathcal{D} = \mathrm{Ex}\big((\mathcal{C}^{\mathrm{fp}})^{\mathrm{op}}, \mathbf{Ab}\big)$. *Then* $\big(\mathrm{Pinj}(\mathcal{D}), \mathbf{Ab}\big)^{\mathrm{fp}} \simeq \big(\mathrm{Inj}(\mathcal{C}), \mathbf{Ab}\big)^{\mathrm{fp}} \simeq \mathcal{C}^{\mathrm{op}}$, *hence* $\mathrm{Fun}^{\mathrm{d}}(\mathcal{D}) \simeq \big((\mathrm{Pinj}(\mathcal{D}), \mathbf{Ab})^{\mathrm{fp}}\big)^{\mathrm{op}}$ *and so* $\mathrm{fun}^{\mathrm{d}}(\mathcal{D}) \simeq \big(((\mathrm{Pinj}(\mathcal{D}), \mathbf{Ab})^{\mathrm{fp}})^{\mathrm{op}}\big)^{\mathrm{fp}}$ *and* $\mathrm{fun}(\mathcal{D}) \simeq \Big(\big(((\mathrm{Pinj}(\mathcal{D}), \mathbf{Ab})^{\mathrm{fp}})^{\mathrm{op}}\big)^{\mathrm{fp}}\Big)^{\mathrm{op}}$.

Thus $\mathrm{fun}(\mathcal{D})$ may be recovered from the category $\mathcal{D}$.

Theorem 18.1.15. ([355, 2.9]) *There is a bijection between natural equivalence classes of skeletally small abelian categories* $\mathcal{A}$ *and definable categories* $\mathcal{D}$, *given by* $\mathcal{A} \mapsto \mathrm{Ex}(\mathcal{A}^{\mathrm{op}}, \mathbf{Ab})$, *with inverse* $\mathcal{D} \mapsto \big(((\mathrm{Pinj}(\mathcal{D}), \mathbf{Ab})^{\mathrm{fp}})^{\mathrm{op}}\big)^{\mathrm{fp}}$.

In fact, there is an equivalence of suitable 2-categories, see 18.2.3 below.

This leaves us with the questions: is there a simpler characterisation of $\mathrm{fun}(\mathcal{D})$, and is there a category-theoretic characterisation of definable categories? We give a positive answer to the first question in 18.1.19 but do not offer an answer to the second.

18.1.4 Definable categories

A simpler way of deriving the functor category of a definable category, namely as the category of functors which commute with direct products and direct limits, is given (18.1.19). Since this area is still somewhat in flux, I mainly just report results and refer to [516] for proofs and to [525], [584] for related developments.

Concerning the possibility of giving a set of category-theoretic conditions which determine exactly the definable categories $\mathcal{D}$, part of such a characterisation should be that the category has direct limits and products. There should also be some "smallness" condition, reflecting the fact that any definable subcategory is in some sense determined by a small subcategory. For example, we might require that there is a skeletally small subcategory, $\mathcal{D}_0$, of $\mathcal{D}$ such that every object of $\mathcal{D}$

is a direct limit of objects from $\mathcal{D}_0$. Any definable subcategory satisfies this (the smallness condition is immediate by, for example, the downwards Löwenheim–Skolem theorem, A.1.10). This is just the requirement that $\mathcal{D}$ should be accessible, have products and arbitrary direct limits.

There are other natural possibilities for the "smallness" condition. For example, one could require that there be a small subcategory, $\mathcal{D}_0$, such that every object of $\mathcal{D}$ is a direct limit of pure subobjects from $\mathcal{D}_0$: certainly this is true in every definable category. As said in the previous section, in view of 16.1.16 one may define an internal notion of purity for such categories: say that a morphism, f, of $\mathcal{D}$ is a **pure embedding** if some ultrapower of f is a split monomorphism. Such a map f must be monic since if an ultrapower, $f^I/\mathcal{U}$, of f is monic and if $gf = hf$ for some maps g, h, then $g^I/\mathcal{U} \cdot f^I/\mathcal{U} = (gf)^I/U = (hf)^I/\mathcal{U} = h^I/\mathcal{U} \cdot f^I/\mathcal{U}$, so $g^I/\mathcal{U} = h^I/\mathcal{U}$, hence (restricting to the diagonal) $g = h$.

Observe that, because a direct limit of a set of objects is a pure subobject of a reduced product of these objects, 3.3.2, every object of such a category is a pure subobject of a reduced product of objects from a generating subcategory $\mathcal{D}_0$.

A characterisation of fun($\mathcal{D}$), neater than 18.1.14, for a definable category $\mathcal{D}$, is obtained through considering the (large) abelian category $(\mathcal{D}, \mathbf{Ab})$. Certainly, there is a set-theoretic issue here because $\mathcal{D}$ is not skeletally small so some care is necessary. In particular, a functor in $(\mathcal{D}, \mathbf{Ab})$ might not be determined by its action on any particular *set* of objects of $\mathcal{D}$, hence will not have a presentation by representable functors, since there will be no epimorphism from the direct sum of a *set* of representable functors to it. Denote by $(\mathcal{D}, \mathbf{Ab})^{\to\prod}$ the full subcategory of $(\mathcal{D}, \mathbf{Ab})$ consisting of those functors which commute with direct products and direct limits.

Lemma 18.1.16. *Let $\mathcal{D}$ be a definable category. Then $(\mathcal{D}, \mathbf{Ab})^{\to\prod}$ is abelian.*

Proof. Since $(\mathcal{D}, \mathbf{Ab})$ is Abelian it is enough to check that if $f : F \to G$ is in the smaller category and if $0 \to F_0 \to F \xrightarrow{f} G \to G_1 \to 0$ is exact in $(\mathcal{D}, \mathbf{Ab})$, then $F_0, G_1 \in (\mathcal{D}, \mathbf{Ab})^{\to\prod}$. So take a collection, $\{D_i\}_i$, of objects of $\mathcal{D}$. Each sequence $0 \to F_0 D_i \to FD_i \to GD_i \to G_1 D_i \to 0$ is exact and, since products of exact sequences in $\mathbf{Ab}$ are exact, so is $0 \to \prod_i F_0 D_i \to \prod_i FD_i \to \prod_i GD_i \to \prod_i G_1 D_i \to 0$. But also $0 \to F_0 \prod_i D_i \to F \prod_i D_i = \prod_i FD_i \to G \prod_i D_i = \prod_i GD_i \to G_1 \prod_i D_i \to 0$ is exact. Therefore $F_0 \prod_i D_i \simeq \prod_i F_0 D_i$ and $G_1 \prod_i D_i \simeq \prod_i G_1 D_i$.

The argument showing that F_0 and G_1 commute with direct limits is similar, using that $\varinjlim$ is exact in $\mathbf{Ab}$. □

For the proofs of the next results we refer to [516]. Say that a functor $G \in (\mathcal{D}, \mathbf{Ab})$ **has a presentation** if there are index sets I, J and an exact sequence $\bigoplus_{j \in J} (E_j, -) \to \bigoplus_{i \in I} (D_i, -) \to G \to 0$ in $(\mathcal{D}, \mathbf{Ab})$ with the $E_j, D_i \in \mathcal{D}$.

Proposition 18.1.17. (see [516, 12.6]) *Suppose that $\mathcal{D}$ is a category with products and let $G \in (\mathcal{D}, \mathbf{Ab})$ be a functor with a presentation. Then G is finitely presented as an object of $(\mathcal{D}, \mathbf{Ab})$ iff G commutes with products.*

Proposition 18.1.18. ([516, 12.8]) *If $\mathcal{D}$ is a definable category, then we have $(\mathrm{Pinj}(\mathcal{D}), \mathbf{Ab})^{\prod} \simeq (\mathrm{Fun}^{\mathrm{d}}(\mathcal{D}))^{\mathrm{op}}$, the former being the category of functors from the category of pure-injective objects of $\mathcal{D}$ to $\mathbf{Ab}$ which commute with products.*

Theorem 18.1.19. ([516, 12.10]) *Let $\mathcal{D}$ be a definable category. Then $\mathrm{fun}(\mathcal{D}) \simeq (\mathcal{D}, \mathbf{Ab})^{\to\prod}$: the category of functors from $\mathcal{D}$ to $\mathbf{Ab}$ which commute with direct limits (equivalently filtered colimits) and products.*

If $\mathcal{D}$ is locally finitely presented, then the proof is essentially as for module categories and fairly straightforward (for example, [355, 7.2]) but the general case seems to require more (ultraproducts play a central role in the proof in [516]). This is essentially a special case of a general result of Makkai, [420, 5.1] also see [307, esp. 5.10(ii)] for the 2-category equivalence (cf. 18.2.3 below).

Corollary 18.1.20. *Let $\mathcal{D}$ be a definable subcategory of $\text{Mod-}\mathcal{A}$ ($\mathcal{A}$ a small preadditive category) and let $G \in (\mathcal{D}, \mathbf{Ab})$. Then G commutes with products and direct limits iff there is a pp-pair $\phi > \psi$ for $\mathcal{A}$-modules such that $G = F_{\phi/\psi} \upharpoonright \mathcal{D}$.*

18.2 Functors between definable categories

There are three "natural" types of functors between definable categories: those which preserve products and direct limits; those which are induced by exact functors between their respective functor categories; and the interpretation functors from model theory (Section 18.2.1). The latter two are easily seen to be equivalent (18.2.9) and, from the characterisation of the functor category in 18.1.19, it follows that all three types of functor coincide; this is stated as 18.2.2.

The effect of such functors on purity and definable subcategories is considered (18.2.4), more satisfactory results being obtained if fullness on pure-injectives is assumed (18.2.7, 18.2.8).

Let $\mathcal{C}$ and $\mathcal{D}$ be definable categories. If $I_0 : \mathrm{fun}(\mathcal{D}) \longrightarrow \mathrm{fun}(\mathcal{C})$ is an exact functor, then I_0 induces, by composition, a functor $I = (I_0)^* : \mathcal{C} = \mathrm{Ex}(\mathrm{fun}(\mathcal{C}), \mathbf{Ab}) \longrightarrow \mathcal{D} = \mathrm{Ex}(\mathrm{fun}(\mathcal{D}), \mathbf{Ab})$ (18.1.4). So if $G \in \mathrm{fun}(\mathcal{D})$ and $C \in \mathcal{C}$, then (see 18.1.4) $G(I_0^*C) = \mathrm{ev}_{I_0^*C} G = \mathrm{ev}_C(I_0 G) = I_0 G \cdot C$.

Theorem 18.2.1. ([516, 13.1], the first part in [355, 7.2]) *Let $\mathcal{C}$, $\mathcal{D}$ be definable categories. If $I_0 : \mathrm{fun}(\mathcal{D}) \longrightarrow \mathrm{fun}(\mathcal{C})$ is an exact functor, then $(I_0)^* : \mathcal{C} \longrightarrow \mathcal{D}$ commutes with direct limits and products.*

Conversely, if $I : \mathcal{C} \longrightarrow \mathcal{D}$ commutes with direct limits and products, then it induces an exact functor $I_0 : \mathrm{fun}(\mathcal{D}) \longrightarrow \mathrm{fun}(\mathcal{C})$.

These processes are inverse: $(I_0)^*$ is naturally equivalent to I and $((I_0)^*)_0$ is naturally equivalent to I_0.

Proof. Given I_0, take a set, $\{C_\lambda\}_\lambda$, of objects in $\mathcal{C}$. Then $I_0^*(\prod_\lambda C_\lambda)$ is the functor from $\mathcal{D}$ to **Ab** which takes G to $I_0G(\prod_\lambda C_\lambda) = \prod_\lambda I_0GC_\lambda$ since all functors in $\mathrm{fun}(\mathcal{C})$ commute with products. This is the value of $\prod_\lambda I_0^*C_\lambda$ on G. Argue similarly for direct limits.

The second paragraph is immediate from 18.1.5 and 18.1.19, and the last statement is easily checked. □

Corollary 18.2.2. ([516, 13.2], [355, 7.2] for the case where $\mathcal{C}$ is locally finitely presented) *Let $\mathcal{C}$, $\mathcal{D}$ be definable categories. Then a functor from $\mathcal{C}$ to $\mathcal{D}$ commutes with direct limits and products iff it has the form I_0^* for some exact functor I_0 from* $\mathrm{fun}(\mathcal{D})$ *to* $\mathrm{fun}(\mathcal{C})$.

In fact there is an equivalence of 2-categories. We refer to [525] for the relevant definitions and proof.

Denote by $\mathbb{DEF}$ the 2-category whose objects are definable additive categories and whose morphisms are those which preserve direct products and direct limits, and by $\mathbb{ABEX}$ denote the 2-category of skeletally small abelian categories and exact functors. In each case the 2-arrows are the natural transformations (for 2-categories see, for instance, [421], [393]).

Theorem 18.2.3. *The assignments $\mathcal{D} \mapsto$* $\mathrm{fun}(\mathcal{D})$ *and $\mathcal{A} \mapsto$* $\mathrm{Ex}(\mathcal{A}, \mathbf{Ab})$ *on objects, $I \mapsto I_0$ and $I_0 \mapsto (I_0)^*$ on functors extend to inverse natural equivalences of the 2-categories $\mathbb{DEF}$ and $\mathbb{ABEX}$.*

A functor $I : \mathcal{C} \longrightarrow \mathcal{D}$ which is of the form I_0^*, that is (18.2.2), which commutes with products and direct limits, we will refer to as a **definable** functor. This terminology is in agreement with that used in [355] and, as said already, it does turn out, see 18.2.10, to be equivalent to being "pp-definable", more precisely, to being an interpretation functor in the sense of Section 18.2.1.

The next couple of results can be found variously in [355, §7] (in full generality), in [323, esp. 7.35] and, said in terms of interpretation functors, in [503, §3].

Proposition 18.2.4. ([503, 3.16] for module categories, [355, 7.8] in general) *Let $\mathcal{C}$, $\mathcal{D}$ be definable categories and let $I_0 : \mathrm{fun}(\mathcal{D}) \longrightarrow \mathrm{fun}(\mathcal{C})$ be exact. Then $I = I_0^* : \mathcal{C} \longrightarrow \mathcal{D}$ preserves pure exact sequences and pure-injectivity. Also $\ker(I) = \{C \in \mathcal{C} : IC = 0\}$ is a definable subcategory of $\mathcal{C}$. Indeed, for any definable subcategory, $\mathcal{D}'$, of $\mathcal{D}$ the inverse image, $I^{-1}\mathcal{D}'$, is a definable subcategory of $\mathcal{C}$.*

Proof. If $0 \to A \to B \to C \to 0$ is a pure exact sequence in $\mathcal{C}$, then, by 18.1.9, for any $F \in \text{fun}(\mathcal{C})$ the sequence $0 \to FA \to FB \to FC \to 0$ is exact. Apply this to functors of the form $F = I_0G$ for $G \in \text{fun}(\mathcal{D})$, then recall that $G(IC') = I_0G.C'$ for $C' \in \mathcal{C}$ to deduce, using 18.1.9 again, that the image under I of the sequence is pure exact.

The fact that I preserves pure-injectivity is direct from the Jensen–Lenzing criterion, 4.3.6, and 18.2.1. Also by 18.2.1 the last statement will follow once it has been shown that $I^{-1}(\mathcal{D}')$ is closed under pure subobjects. So suppose that B is a pure subobject of $C \in \mathcal{C}$, where $IC \in \mathcal{D}'$. Some ultrapower, $B^J/\mathcal{U} \to C^J/\mathcal{U}$, of the inclusion is split and, since I commutes with ultrapowers (it commutes with products and direct limits), $(IB)^J/\mathcal{U}$ is a direct summand of $(IC)^J/\mathcal{U}$, which is in $\mathcal{D}'$. Therefore $(IB)^J/\mathcal{U} \in \mathcal{D}'$. But every object is a pure subobject of each of its ultrapowers (3.3.4), so $IB \in \mathcal{D}'$, as required. □

An alternative argument for the first part is to use that a sequence is pure-exact iff some ultrapower of it is split.

Corollary 18.2.5. *Let $\mathcal{C}$, $\mathcal{D}$ be definable categories and let $I : \mathcal{C} \longrightarrow \mathcal{D}$ commute with products and direct limits. Then I preserves pure exact sequences and pure-injectivity.*

If $C \in \mathcal{C}$ is an indecomposable pure-injective, then it may well be that IC is decomposable: take the forgetful functor from Mod-R, where R is a k-algebra (k a field), to Mod-k to obtain examples. To see that I need not preserve pure-injective hulls, take R to be a finite-dimensional algebra not of finite representation type, so, see Section 4.5.4, there is a module which is a proper submodule of its pure-injective hull).

Example 18.2.6. Let $R \to S$ be a morphism of rings and let I : Mod-$S \longrightarrow$ Mod-R be the induced restriction of scalars functor. Of course, this commutes with products and direct limits. If $\mathcal{D}'$ is any definable subcategory of Mod-R, then, by the result above, the subcategory of Mod-S with objects those modules M_S such that $M_R \in \mathcal{D}'$ is definable. To obtain explicitly the set of pp-pairs for S-modules defining $I^{-1}\mathcal{D}'$, take a set, T, of pp-pairs such that $\mathcal{D}' = \text{Mod}(T)$ and, in each pp condition appearing in T, replace each occurrence of each element r of R by the element fr of S (cf. 3.2.7 ff.).

If I is a definable functor, then we know (18.2.13) that the image, IC' of a definable subcategory $\mathcal{C}'$ of Mod-S need not be a definable subcategory of Mod-R since it need not be closed under pure submodules.

Say that a functor $I : \mathcal{C} \longrightarrow \mathcal{D}$ between definable categories is **full on pure-injectives** if, restricted to Pinj$(\mathcal{C})$, it is full; this is equivalent, see 18.2.11, to the model-theoretic notion of preserving all induced structure. Clearly, it is enough to assume that for every $N \in \mathrm{Pinj}(\mathcal{C})$ the induced map $\mathrm{End}(N) \to \mathrm{End}(IN)$ is surjective (since $N \oplus N'$ is pure-injective if both N and N' are).

Corollary 18.2.7. ([503, 3.16, 3.17]) *Suppose that the functor $I : \mathcal{C} \longrightarrow \mathcal{D}$ between definable categories commutes with products and direct limits and is full on pure-injectives. Then I preserves indecomposability of pure-injectives in the weak sense that if $N \in \mathcal{C}$ is an indecomposable pure-injective, then either $IN = 0$ or IN is an indecomposable pure-injective. Also, for every $C \in \mathcal{C}$, one has $IH(C) = H(IC)$, where H denotes pure-injective hull.*

Proof. That I preserves indecomposability is immediate from that fact that it is full. For the statement about pure-injective hulls note, by 18.2.5, that IC is pure in $IH(C)$, which is pure-injective, so $H(IC)$ is a direct summand of $IH(C)$. If it were a proper direct summand, then there would, by fullness, be an endomorphism of $IH(C)$ which was the identity on IC but not an automorphism, so there would, by fullness, be a proper direct summand of $H(C)$ strictly between C and $H(C)$, contradicting the definition of pure-injective hull. □

For further consequences of fullness on pure-injectives see Section 18.2.5.

The next result gives a set of conditions under which the image of a definable subcategory is definable. The proof does seem to require model theory[5] in the sense that it uses a result of model theory which appears not to translate in any natural way to something algebraic. That the countability hypothesis cannot be dropped is shown in by an example of Herzog ([503, 3.11]).

Theorem 18.2.8. ([503, 3.8]) *Suppose that the functor $I : \mathcal{C} \longrightarrow \mathcal{D}$ between definable categories commutes with products and direct limits and is full on pure-injectives. Suppose that* fun($\mathcal{C}$) *has just countably many objects up to isomorphism. Then for every definable subcategory, $\mathcal{C}'$, of $\mathcal{C}$ the image $I\mathcal{C}'$ is a definable subcategory of $\mathcal{D}$.*

18.2.1 Interpretation functors

It is shown (18.2.10) that interpretation functors between definable categories coincide with those discussed in the previous section. The relation between preserving all induced structure and fullness on pure-injectives is given (18.2.11). We assume rather more acquaintance with model theory in this section.

The notion of a structure, or class of structures, being interpreted in another comes from model theory. Roughly, a class, $\mathcal{D}$, of structures is interpretable in

[5] Existence of locally atomic models under a countability hypothesis.

another class, $\mathcal{C}$, of structures if $\mathcal{D}$ "can be found definably within" $\mathcal{C}$. The model-theoretic definition is given, then it is reformulated in functorial terms. Some examples and properties are presented.

To say that M is interpreted in N is to say that M can "be found definably within" N, not necessarily as a subset of N, rather, as a definable section, that is, a definable subset of some quotient of a finite power of N by a definable equivalence relation. Of course, it is not only a copy of the underlying subset of M which can be found definably within N: all the functions, relations and constants that make up the structure on M also should be definable on that "definable section" which is in bijection with the underlying set of M. This idea extends in the obvious way to that of one class of structures being interpreted in another.

A simple example (which does not involve definable quotients) is to take a module M over the Kronecker quiver with, in the notation of 1.1.4, α being an isomorphism from Me_1 to Me_2. There are a couple of obvious isomorphic $k[T]$-modules sitting definably within M: one has underlying set Me_1 and the action of T on this set is given by $\beta\alpha^{-1}$.

For a similar example see 7.2.18.

A variety of examples of interpretations can be found in the following sections. Of course, in our context, all interpretations should be defined using pp conditions in order to retain the underlying abelian group structures ([106, 2.1]).

Consider two structures M and N, usually structures for different languages, suppose that M is an $\mathcal{L}$-structure and N is an $\mathcal{L}_1$-structure. An **interpretation of** M **in** N is given by the following data.

- A sort γ/ϵ of[6] $\mathcal{L}_1^{\mathrm{eq}}$ modulo the theory of N; that is, similar to what is seen in Section B.2 but with γ an *arbitrary*, not necessarily pp, formula and ϵ a formula defining, modulo the theory of N, an equivalence relation on the solution set of γ. We require that there be a bijection $t : M \to \gamma(N)/\epsilon(N)$, and
- for every basic (function, relation, constant) symbol f, R, or c of $\mathcal{L}$ there be a corresponding formula η_f, ζ_R or κ_c of $\mathcal{L}_1^{\mathrm{eq}}$ which defines the appropriate kind of function, relation or constant (of the correct arity) on $\gamma(N)/\epsilon(N)$ in such a way that, if $\gamma(N)/\epsilon(N)$ is thus regarded as an $\mathcal{L}$-structure, then t is an isomorphism.

In this definition it was assumed, for simplicity, that $\mathcal{L}$ is a 1-sorted language. The many-sorted case is an obvious modification and can be found in [516, §25].

The phrasing of that definition emphasises just the two structures involved but a more functorial formulation can be extracted easily and can be found in [516].

[6] This is defined like $(-)^{\mathrm{eq}+}$ but using any formulas in place of pp ones.

Clearly the above data define, sitting within[7] N^{eq}, an $\mathcal{L}$-structure: the $\mathcal{L}$-structure so **interpreted in** $\mathcal{N}$. If $\mathcal{N}$ is a class of $\mathcal{L}_1$-structures, then we may consider the class, $\mathcal{M}$, of $\mathcal{L}$-structures which arise in this way (fixed γ, ϵ, η_f etc.) as N ranges over $\mathcal{N}$. Simple examples (see 18.2.15) show that this is, in general, not an axiomatisable class, but no matter: we say that there is an **interpretation of $\mathcal{M}$ in $\mathcal{N}$**. Note, however, that the functor involved actually goes the other way: given $N \in \mathcal{N}$ the interpretation gives an object of $\mathcal{M}$. This motivates the form of the definition, below, of interpretation functor, where now we specialise again to the additive case, so use $^{\mathrm{eq+}}$.

Suppose that $\mathcal{C}$ and $\mathcal{D}$ are definable additive categories. An **interpretation functor** $I : \mathcal{C} \longrightarrow \mathcal{D}$ is a functor which is given by the following data. These will be said in terms of the canonical languages, $\mathcal{L}(\mathcal{C})$, $\mathcal{L}(\mathcal{D})$ (Section B.1), of these categories: for other languages one makes the obvious modifications.

(i) For each sort, A, of $\mathcal{L}(\mathcal{D})$ a pp-pair, ϕ_A/ψ_A, in $\mathcal{L}(\mathcal{C})$, each of ϕ_A, ψ_A having free variable of sort $B = B(A)$, say.[8]
(ii) For each function symbol, f, of $\mathcal{L}(\mathcal{C})$ from sort A to sort A_1, a pp condition, $\rho_f(x, y)$, where x has sort $B(A)$ and y has sort $B_1 = B(A_1)$ (the sort of the free variable of ϕ_{A_1}).

Then we require that for every $M \in \mathcal{C}$ each ρ_f actually well define a function from $\phi_A(M)/\psi_A(M)$ to $\phi_{A_1}(M)/\psi_{A_1}(M)$ and that every atomic formula in the theory, Th($\mathcal{D}$), of $\mathcal{D}$ translate, via $f \mapsto \rho_f$, to a formula true in $\mathcal{C}$ – in other words, the "addition and multiplication tables" for (the morphisms of) $\mathcal{D}^{\mathrm{fp}}$ should be preserved. The next result may be found at [516, 25.2] and we refer there for the proof (and for the precise definitions necessary before giving a proof).

Theorem 18.2.9. *Let $\mathcal{C}$, $\mathcal{D}$ be definable categories. Then there is a natural bijection between interpretation functors, I, from $\mathcal{C}$ to $\mathcal{D}$ and exact functors I_0 from $\mathbb{L}(\mathcal{D})^{\mathrm{eq+}}$ to $\mathbb{L}(\mathcal{C})^{\mathrm{eq+}}$, that is, from* fun($\mathcal{D}$) *to* fun($\mathcal{C}$).

In the terminology used above, this gives an **interpretation of** the (not necessarily definable, see comment after 18.2.5) subcategory $I\mathcal{C}$ of $\mathcal{D}$ **in** $\mathcal{C}$.

The next result then follows from 18.2.1.

Corollary 18.2.10. *Let $\mathcal{C}$, $\mathcal{D}$ be definable categories. Then a functor from $\mathcal{C}$ to $\mathcal{D}$ is an interpretation functor iff it is a definable functor iff it commutes with products and direct limits.*

[7] Like $N^{\mathrm{eq+}}$ but using $\mathcal{L}_1^{\mathrm{eq}}$ in place of an $(-)^{\mathrm{eq+}}$ language.
[8] If $\mathcal{C}$ is finitely accessible, then, recall, the sorts correspond to objects of $\mathcal{C}^{\mathrm{fp}}$ and, even in the general case, we may suppose these categories to be definable subcategories of finitely accessible categories, hence we may suppose that the language is that of a finitely accessible category.

Any interpretation functor $I : \mathcal{C} \longrightarrow \mathcal{D}$ induces a map on the enriched languages, from $\mathcal{L}(\mathcal{D})^{\mathrm{eq}+}$ to $\mathcal{L}(\mathcal{C})^{\mathrm{eq}+}$, namely that defined by the data of the interpretation functor, extended to the categories,[9] $\mathbb{L}^{\mathrm{eq}+}$, underlying the eq+ languages and then extended to pp conditions (and even to arbitrary formulas). That is, to each sort, σ, of $\mathbb{L}(\mathcal{D})^{\mathrm{eq}+}$ corresponds the sort $I_0\sigma$ of $\mathbb{L}(\mathcal{C})^{\mathrm{eq}+}$ (where $I_0 : \mathrm{fun}(\mathcal{D}) = \mathbb{L}(\mathcal{D})^{\mathrm{eq}+} \longrightarrow \mathbb{L}(\mathcal{C})^{\mathrm{eq}+} = \mathrm{fun}(\mathcal{C})$ is such that, in the notation of 18.2.1, $I_0^* = I$) and to each function symbol, f, corresponds $I_0 f$ (thought of now as a function symbol of a language rather than a morphism). The extension to pp conditions is simply replacement of function symbols (f by $I_0 f$) and replacement of variables of sorts in $\mathbb{L}(\mathcal{D})^{\mathrm{eq}+}$ by variables of the corresponding sorts in $\mathbb{L}(\mathcal{C})^{\mathrm{eq}+}$. A simple example is seen in 18.2.6. A slightly less simple example is that of $R = kA_2$-modules (18.2.13, 18.2.15) being interpreted as k-representations of the quiver A_2: the language for k-representations of A_2 has two basic sorts, γ_1 and γ_2 say, corresponding to the two vertices of the quiver A_2, so variables of the "home" sort for R-modules must be replaced by variables of the product sort $\gamma_1 \times \gamma_2$; also the function symbol (from the home sort to itself) in the language for R-modules corresponding to multiplication by α, where $\alpha \in R$ is the element representing the arrow of A_2, is replaced by a function symbol from sort γ_1 to sort γ_2 in the language of A_2-representations.

Note also that the action of I_0 on formulas commutes with the propositional connectives $\wedge, \vee, \neg$ and also with $+$ on pp formulas.

If $I : \mathcal{C} \longrightarrow \mathcal{D}$ is an interpretation functor, then elements of IC ($C \in \mathcal{C}$) may be seen as elements of C in the following way. If $a \in \tau(IC)$, where τ is a sort of $\mathbb{L}(\mathcal{D})^{\mathrm{eq}+}$, so $a \in F_\tau(IC)$ where F_τ is the functor in $\mathrm{fun}(\mathcal{D})$ corresponding to τ in $\mathbb{L}(\mathcal{D})^{\mathrm{eq}+}$, then (see just before 18.2.1) $F_\tau(IC) = I_0 F \cdot C$. Therefore a may be regarded as an element, a' say, of C of the sort $\sigma(\in \mathbb{L}(\mathcal{C})^{\mathrm{eq}+})$ corresponding to $I_0 F_\tau(\in \mathrm{fun}(\mathcal{C}))$. Also note the relation between $\mathrm{pp}^{IC}(a)$ and $\mathrm{pp}^{C}(a')$, namely if $a \in \phi(IC)$, that is, $a \in F_\phi(IC)$, then $a' \in I_0 F_\phi(C)$, that is, $a' \in (I_0\phi)(C)$, where we have denoted by $I_0\phi$ the "translation" of $\phi \in \mathcal{L}(\mathcal{D})^{\mathrm{eq}+}$ to a pp condition in $\mathcal{L}(\mathcal{C})^{\mathrm{eq}+}$, as described in the previous paragraph.

Given an interpretation functor $I : \mathcal{C} \longrightarrow \mathcal{D}$ and the corresponding exact functor (see the beginning of Section 18.2) $I_0 : \mathbb{L}(\mathcal{D})^{\mathrm{eq}+} = \mathrm{fun}(\mathcal{D}) \longrightarrow \mathrm{fun}(\mathcal{C}) = \mathbb{L}(\mathcal{C})^{\mathrm{eq}+}$, say that the interpretation functor **preserves all induced structure** if I_0 is full. An explanation of this, in particular its model-theoretic significance and an explanation of why it is the same as the concept given that name in [503, 3.7], can be found in [516, §25].

Recall that I is said to be full on pure-injectives if it is full when restricted to $\mathrm{Pinj}(\mathcal{C})$.

[9] [355, 8.4] expresses the functorial version of this.

Theorem 18.2.11. ([503, 3.17], also see [516, 25.9]) *Let $I : \mathcal{C} \longrightarrow \mathcal{D}$ be an interpretation functor between definable categories. Then I preserves all induced structure iff I is full on pure-injectives. If this is so then I_0 is full.*

Corollary 18.2.12. *Let $I : \mathcal{C} \longrightarrow \mathcal{D}$ be an interpretation functor between definable additive categories. If I is full, then I preserves all induced structure.*

Example 18.2.13. Consider the path algebra, $R = kA_2$, over a field k, of the quiver $A_2 = \bullet \to \bullet$. Consider the sort which corresponds to the second component (Me_2 in the obvious notation) of a module M. One may check that the only endomorphisms of, that is, pp-definable functions on, this sort are the multiplications by the scalars of k, hence the only structure on the restriction of any module to this sort is that of a k-vectorspace. In particular, the image of the arrow from the first to the second vertex has been lost as a definable subset of Me_2. Thus we have an interpretation functor from Mod-R to Mod-k which does not preserve all induced structure. This example is continued below at 18.2.15.

In practice, an interpretation functor is, as will be seen from the examples I give, usually defined on a generating subset of the functor category. More precisely, an intrepretation functor from the definable category $\mathcal{C}$ to the definable category $\mathcal{D}$ may be given by the following data:

- a (not necessarily full) subcategory $\mathcal{B}'$ of fun($\mathcal{D}$) which is generating in a suitable sense;
- a functor $F' : \mathcal{B}' \longrightarrow$ fun($\mathcal{C}$) which can be extended to an exact functor from fun($\mathcal{D}$) to fun($\mathcal{C}$).

The second condition implies, in particular, that whenever $A \xrightarrow{f} B \xrightarrow{g} C$ is a sequence in $\mathcal{B}'$ which is exact as a sequence in fun($\mathcal{D}$), then the image must be exact in fun($\mathcal{C}$).

In the original definition of interpretation functor between module categories over rings ([503]) the second condition was not seen because it was not necessary: the category fun-S, where S is a ring, is the free abelian category on S (10.2.37) so any functor defined on S extends (uniquely) to an exact functor on fun-S. But once we have a pp-generating subcategory for which fun-S is not the free abelian category, then some conditions must be imposed on the functor F above. The next result generalises the first observation since fun-$\mathcal{A}$ is the free abelian category on $\mathcal{A}$.

Corollary 18.2.14. *Suppose that $\mathcal{C}$ is a definable category and that $\mathcal{A}$ is a small preadditive category. Then any functor $F' : \mathcal{A} \longrightarrow$ fun($\mathcal{C}$) induces a unique interpretation functor from $\mathcal{C}$ to* Mod-$\mathcal{A}$.

Theorems about, and examples of, functors which share at least some of the properties of interpretation functors can be found in, for example, [395], [323].

18.2.2 Examples of interpretations

Example 18.2.15. Let R be the path algebra over the field k of the quiver $A_2 = (\bullet \to \bullet)$. Let $\mathcal{D}$ be the full subcategory of Mod-R whose objects are those in which the linear map corresponding to the arrow of A_2 is an isomorphism. Let S be the field k and consider the forgetful functor from Mod-R to Mod-S induced by the ring inclusion of $S = k$ into R. This forgetful functor is an example of an interpretation functor: it is defined on the "home" sort $(x = x)/(x = 0)$ and simply restricts to a non-full subcategory of the category of modules over a subring of the ring of definable scalars (Section 6.1.1). Note that the image of this functor is not a definable subcategory of Mod-S since it consists of the even-dimensional vectorspaces.

It is shown in [501, Cors. 2,3] that any strictly wild k-algebra interprets the category of modules over any finite-dimensional k-algebra. More generally, see Section 6.1.2, any epimorphism $R \to S$ of rings induces an interpretation functor.[10]

Example 18.2.16. An interpretation of the category of $k[T]$-modules within $k\widetilde{A}_1$-modules was used for illustration at the beginning of Section 18.2.1. If $\mathcal{C}$ is the full subcategory of those $k\widetilde{A}_1$-modules M in which α is an isomorphism from Me_1 to Me_2 and $I : \mathcal{C} \longrightarrow \mathcal{D} = \text{Mod-}k[T]$ is the functor described there (the action on morphisms is the obvious one), then I is an interpretation functor. The corresponding functor $I_0 : \text{fun}(\mathcal{D}) \longrightarrow \text{fun}(\mathcal{C})$ is determined by its action on the home sort/forgetful functor $(k[T], -)$, which is to take this to the sort σ defined by $(x = xe_1)/(x = 0)$, and to take the map multiplication-by-T to the map from σ to itself defined by a pp condition $\rho(x, y)$, say $y\alpha = x\beta$, which defines the action of $\beta\alpha^{-1}$.

Example 18.2.17. Let k be any field and let $R = k[x, y, z : x^2 = y^2 = z^2 = 0 = xy = xz = yz]$ – a four-dimensional k-algebra with k-basis $\{1, x, y, z\}$. Let $S = k\langle X, Y\rangle$ be the free associative k-algebra on two generators. Let $\mathcal{C}$ be the (clearly) definable subcategory of Mod-R consisting of all R-modules M such that $\text{ann}_M(z) = Mz$. An interpretation functor I from $\mathcal{C}$ to $\mathcal{D} = \text{Mod-}S$ is defined as follows.

Consider the sort σ which is the pp-pair $(u = u) / (z|u)$: so $\sigma(M) = M/Mz$. Let $\rho_X(u, v)$ be the pp condition $ux - vz = 0$ and let $\rho_Y(u, v)$ be the pp condition

[10] There is a sort of converse at [501, Prop. 6].

$uy - vz = 0$. Then it is easy to check that for any $M \in \mathcal{C}$, ρ_X, and similarly ρ_Y, defines functions from $\sigma(M)$ to itself. For instance, if $a + Mz \in \sigma(M)$, then $ax \in \mathrm{ann}_M(z)$ since $xz = 0$, so, since $M \in \mathcal{C}$, there is $b \in M$ such that $ax = bz$. If also $b'z = ax$, then $b - b' \in \mathrm{ann}_M(z) = Mz$, so the coset $b + Mz$ is uniquely defined. Also, if $a - a' \in Mz$, then $ax - a'x = (a - a')x = 0$ because $zx = 0$, so the value of b modulo Mz depends only on the value of a modulo Mz.

The functor I is defined on objects by sending $M \in \mathcal{C}$ to $\sigma(M)$ equipped with an action of $X \in S$ defined by ρ_X and an action of Y defined by ρ_Y. Again note that the corresponding functor from fun-S to fun-R is defined on a subcategory which is pp-generating in a natural sense.

The image of I is actually the whole of Mod-S. For, if C is any S-module, then construct the R-module $M(C)$ which has underlying set $C \oplus C$ and where the actions of x, y and z are given by $(c, d)x = (0, cX)$, $(c, d)y = (0, cY)$, $(c, d)z = (0, c)$. It is easy to check that $M(C) \in \mathcal{C}$ and that $I(M(C) \simeq C$. Thus Mod-S is interpreted in the definable subcategory $\mathcal{D}$ of Mod-R.

For variety, we also give the definition of I_0 in terms of the realisation of fun$(\mathcal{C})$ as the category, $(\text{mod-}R, \mathbf{Ab})^{\text{fp}}_{\mathcal{C}}$, of functors (rather than as the category, $\mathbb{L}(\mathcal{C})^{\text{eq}+}$, of pp sorts). It is enough to define the relevant functors in $(\text{mod-}R, \mathbf{Ab})^{\text{fp}}$. Let $F \in (\text{mod-}R, \mathbf{Ab})^{\text{fp}}$ be the functor with projective presentation $(R/zR, -) \xrightarrow{(\pi,-)} (R, -) \xrightarrow{p} F \to 0$, where $\pi : R \to R/zR$ is the projection and where p is the cokernel of $(\pi, -)$. So $FA = A/Az$ for any (finitely presented) R-module A. Define the natural transformation ρ_X from F to itself by $\rho_{X,A} : FA \to FA$ where $\rho_{X,A}$ is the map from A/Az to itself defined by taking $a + Az$ to the unique $b + Az$ such that $ax = bz$. Checking that this is a well-defined map is the argument as in the paragraph above. One must also check that this does give a natural transformation, hence an endomorphism of F: this is easy. Similarly define the natural transformation ρ_Y.

We note the particular form that the notion of interpretation functor takes when the definable categories are subcategories of module categories over rings. Suppose that $\mathcal{C}$ is a definable subcategory of Mod-R for some ring R and let S be a ring. Then an interpretation functor from $\mathcal{C}$ to (a subcategory of) Mod-S is given by the following data:

- a sort, ϕ/ψ, of $\mathbb{L}(\mathcal{C})^{\text{eq}+}$;
- a ring morphism $S \to R_{\mathcal{C}}^{\phi/\psi}$, where the latter is, recall (12.8.3), the ring of definable scalars for $\mathcal{C}$ in the sort ϕ/ψ.

Then I is the functor from $\mathcal{C}$ to Mod-S which takes $M \in \mathcal{C}$ to $\phi(M)/\psi(M)$ regarded as an S-module *via* the given morphism $S \to R_{\mathcal{C}}^{\phi/\psi}$.

Proposition 18.2.18. *Let F be a functor from* Mod-R *to* **Ab** *which commutes with direct limits and which is such that the restriction, F_0, of F to* mod-R *is finitely presented in* (mod-R, **Ab**). *Let $S = \text{End}(F) = \text{End}(F_0)$: so for every R-module M, FM is an S-module. Then the functor F from* Mod-R *to* Mod-S *is an interpretation functor.*

Proof. First note that if two endomorphisms of F agree on all finitely presented modules, then, since F commutes with direct limits, they agree on all modules, so indeed $\text{End}(F) = \text{End}(F_0)$.

By 10.2.43 F has the form $F_{\phi/\psi}$ for some pp conditions $\phi > \psi$. Also, see 10.2.30, corresponding to each element $s \in S$ there is a pp condition ρ_s which defines the action of s on FM for every finitely presented R-module M. Again using that F commutes with direct limits, it follows that ρ_s defines the action of s on any R-module. □

There are examples of such functors in Section 10.2.6, there are some simple examples in [503] and one can find many further examples, often rather complicated, in various papers, such as those of Baur, Toffalori *et al.*, on (un)decidability, (see the Bibliography).

18.2.3 Tilting functors

We illustrate these ideas by applying them to the process of tilting. The notion of tilting arose in the representation theory of finite-dimensional algebras ([79], [93], [266]) but is now applied in much more general contexts, see, for instance, [11]. For more details see that reference and, for instance, [602, §4.1].

Tilting allows us to compare, often very closely, certain categories of modules over different rings. For instance, if one takes two quivers without relations with the same underlying directed graph but possibly differently-oriented arrows, then the categories of modules over their respective path algebras are almost the same and this relationship can be said precisely in terms of Coxeter functors, [79], which are particular kinds of tilting functors. Geisler showed, [227, 4.2], that Coxeter functors are interpretation functors. The general case described here is given in [503].

Let R be any ring. An R-module T is a **tilting module** (in the terminology of [133] a classical tilting module[11]) if T is a finitely presented module of projective dimension ≤ 1, if $\text{Ext}^1_R(T, T) = 0$ and if there is an embedding of R into a direct

[11] Note, for example, [11], that now the term "tilting module" is used for a much more general concept.

sum of copies of T such that the factor module is in add(T) – the category of direct summands of finite direct sums of copies of T.

If T_R is a tilting module and if $S = \mathrm{End}(T_R)$, then ${}_ST$ is tilting and $R = \mathrm{End}({}_ST)$ (see 1.1 of [131] for the general case and [133, 2.15b] for the fact that ${}_ST$ is finitely presented). Moreover, the classes of R-modules generated by T (that is, epimorphic images of direct sums of copies of T) and of R-modules ext-orthogonal to T (that is, the kernel of $\mathrm{Ext}^1_R(T, -)$) coincide.[12]

Associated with any tilting module are certain categories of modules and equivalences between these. The theorem below is proved in [93] for finite-dimensional algebras and it is generalised to arbitrary rings in [131].

Given a module T_R, define the following subcategories of Mod-R:

$\mathcal{F}(T) = \{X_R : \mathrm{Hom}_R(T, X) = 0\}$;
$\mathcal{G}(T) = \{X_R : \mathrm{Ext}^1_R(T, X) = 0\}$

and the following subcategories of Mod-S where $S = \mathrm{End}(T_R)$:

$\mathcal{Y}(T) = \{N_S : \mathrm{Tor}^R_1({}_ST, N_S) = 0\}$;
$\mathcal{X}(T) = \{N_S : N_S \otimes T = 0\}$.

Theorem 18.2.19. (for example, [11, p. 10]) *If T is a tilting module, then $(\mathcal{G}(T), \mathcal{F}(T))$, respectively $(\mathcal{X}(T), \mathcal{Y}(T))$, is a not necessarily hereditary torsion theory on* Mod-R, *resp. in* Mod-S.

Furthermore, the functor $\mathrm{Hom}_R({}_ST_R, -)$ *from $\mathcal{G}(T)$ to $\mathcal{Y}(T)$ is an equivalence of categories, with inverse $- \otimes_S T_R$, as is the functor* $\mathrm{Ext}^1_R({}_ST_R, -)$ *from $\mathcal{F}(T)$ to $\mathcal{X}(T)$, with inverse* $\mathrm{Tor}^R_1({}_ST_R, -)$.

The next result follows from Section 10.2.6, 10.2.35 and 10.2.36 in particular.

Theorem 18.2.20. ([503, 4.4]) *Let R be any ring and let T_R be an R-module with endomorphism ring S. Suppose that T_R is* FP_2 *and that ${}_ST$ is* FP_2. *Then the subclasses $\mathcal{F}(T)$ and $\mathcal{G}(T)$ of* Mod-R, *respectively the subclasses $\mathcal{X}(T)$ and $\mathcal{Y}(T)$ of* Mod-S, *are definable.*

Theorem 18.2.21. ([503, 4.5]) *Let R be any ring and let T_R be an R-module with endomorphism ring S. Suppose that T_R is* FP_2 *and that ${}_ST$ is* FP_2. *Then the functors* $\mathrm{Hom}_R({}_ST_R, -)$, *respectively* $\mathrm{Ext}^1_R({}_ST_R, -)$, *restrict to interpretation functors from $\mathcal{G}(T)$, resp. from $\mathcal{F}(T)$, to* Mod-S, *and the functors ${}_ST \otimes_R -$ and* $\mathrm{Tor}^R_1({}_ST_R, -)$ *restrict to interpretation functors from $\mathcal{Y}(T)$, resp. from $\mathcal{X}(T)$, to* Mod-R.

Proof. It must be shown that the actions of the elements of the relevant ring, R or S, are pp-definable. By the equivalence, 10.2.30, of $\mathbb{L}_R^{\mathrm{eq}+}$ and $(\text{mod-}R, \mathbf{Ab})^{\mathrm{fp}}$ it

[12] See 1.3 of [133] where a converse is also given.

will be sufficient to show that S acts on elements of $\mathcal{Y}(T)$ and $\mathcal{X}(T)$ as a subring of $\text{End}\big((T,-) \upharpoonright \text{mod-}R\big)$ and $\text{End}\big(\text{Ext}^1_R(T,-) \upharpoonright \text{mod-}R\big)$ respectively and that R acts on elements of $F(T)$ and $G(T)$ as a subring of $\text{End}\big(\text{Tor}_1({}_ST,-) \upharpoonright \text{mod-}S\big)$ and $\text{End}\big((- \otimes_S T) \upharpoonright \text{mod-}R\big)$ respectively. For Hom, this is by the Yoneda Lemma and for tensor, Ext and Tor it is straightforward to check (and is a consequence of the functoriality of these bifunctors). □

Corollary 18.2.22. ([503, 4.8]) *Let R be any ring and let T_R be a tilting module with endomorphism ring S. Then the subclasses $\mathcal{F}(T)$ and $\mathcal{G}(T)$ of* Mod-R, *respectively the subclasses $\mathcal{X}(T)$ and $\mathcal{Y}(T)$ of* Mod-S, *are definable. Furthermore, the functors* $\text{Hom}_R({}_RT_S,-)$, *respectively* $\text{Ext}^1_R({}_ST_R,-)$, *restrict to interpretation functors from $\mathcal{G}(T)$ to $\mathcal{Y}(T)$, resp. from $\mathcal{F}(T)$ to $\mathcal{X}(T)$, which are equivalences and whose respective inverses ${}_ST \otimes_T -$ and* $\text{Tor}^R_1({}_ST_R,-)$ *are also interpretation functors.*

Proof. Since any tilting module has projective dimension less than or equal to 1, from the fact that it is finitely presented it follows that it is actually FP_∞. As remarked after the definition of tilting module, ${}_ST$ is tilting if T_R is, so the results above apply to give the conclusion. □

With 18.2.28, the next result follows.

Corollary 18.2.23. ([503, 4.9]) *Let R be any ring and let T_R be a tilting module with endomorphism ring S. Then the functors* $\text{Hom}_R({}_ST_T,-)$, *respectively* $\text{Ext}^1_R({}_ST_T,-)$, *induce homeomorphisms from* $\text{Zg}(\mathcal{G}(T)) = \mathcal{G}(T) \cap \text{Zg}_R$ *to* $\text{Zg}(\mathcal{Y}(T))$, *resp. from* $\text{Zg}(\mathcal{F}(T))$ *to* $\text{Zg}(\mathcal{X}(T))$.

The conditions on a tilting module T imply that the classes $\mathcal{F}(T)$ and $\mathcal{G}(T)$ have only the zero module in common, as have the classes $\mathcal{X}(T)$ and $\mathcal{Y}(T)$ (see 18.2.19). Therefore the corresponding closed subsets of the relevant Ziegler spectrum are disjoint. In some cases, such as the earlier-mentioned Coxeter functors, they almost cover the respective spectra.

These results allow a good deal of information to be transferred between the module categories, for example, [503, 4.11], if S is an artin algebra which is tilted from tame, then it follows from these results and the description of the Ziegler spectra of tame hereditary algebras, 8.1.12 in particular, that the Cantor–Bendixson rank of Zg_S is 2 or 0. Another immediate consequence is that if an algebra is tilted from an indecomposable hereditary algebra and is not of finite representation type, then it has exactly one generic module (in the sense of Section 4.5.5) since, by the description of the spectrum in Section 8.1.2, an indecomposable tame hereditary algebra has just one such module (this also follows from [602, p. 189 ff], [139, 5.7]).

There is a kind of dual notion to tilting, which we only mention here, just pointing to the literature. Recall that $\mathrm{cog}(M)$ denotes the class of modules which are cogenerated by M – which embed in some power of M. The module M is said to be **cotilting**[13] if $\mathrm{cog}(M) = {}^{\perp}M$, where the latter denotes $\{A \in \text{mod-}R : \mathrm{Ext}^1(A, M) = 0\}$. These modules arise, for example, in the context of cotorsion theories, which are defined rather like not necessarily hereditary torsion theories but using Ext^1-(or Ext^∞)orthogonality rather than Hom-orthogonality, see [687]. It was observed that all known examples of cotilting modules were pure-injective and for a while it was open whether this is always true; that this *is* so was proved by Bazzoni [56]. She also proved, [56, 3.2], that if M is cotilting, then ${}^{\perp}M$ is a definable subcategory. There are also generalisations of these notions, and this question, and answer, [665], by Šťovíček, using Ext^n, see [57]. Purity, definable categories, etc. are relevant here[14] and I just point the reader to the survey papers on this currently-developing topic, [688], [658], [132], the book [234], and the extensive lists of references that these contain.

A result which we have not made use of here but which is important in this work as well as that of Guil Asensio and Herzog is a theorem of Auslander which states that if N is pure-injective and $(C_i)_i$ is a directed system of modules then $\mathrm{Ext}^1(\varinjlim_i C_i, N) \simeq \varprojlim_i (\mathrm{Ext}^1(C_i, N)$ ([21, 10.1]).

18.2.4 Another example: lattices over groups

This example is rather more involved than those described earlier and is based on an interpretation functor which is developed in [431] and modified and used further in, for instance, [45], [681]. The functor is based on a construction of Butler, [108], which relates lattices over a group ring to certain configurations of vector spaces.

We will work with the definable class of torsionfree modules, which is, since every torsionfree R-module is a direct limit of lattices, the definable closure of the class of lattices ([144, 5.2]). So, to define an interpretation functor on the class of torsionfree modules it will be enough to define, using pp conditions, a functor on the class of lattices.

Let R be the group ring $\mathbb{Z}_{(2)}G$, where $\mathbb{Z}_{(2)}$ is the localisation of the ring of integers at the prime 2 and where G is the four-group $\mathbb{Z}_2 \times \mathbb{Z}_2$ with generators, say, x and y. Let $\mathcal{C}'$ denote the definable class of $\mathbb{Z}$-torsionfree R-modules: those which

[13] As with tilting, there is a much more general notion of cotilting, meaning n-cotilting for some n, where the notion just defined is 1-cotilting; see, for example, [688].

[14] There is also a version of the Telescope Conjecture, mentioned at the beginning of Chapter 17, for module categories, see [374], [12].

are finitely generated as $\mathbb{Z}$-modules are the G-**lattices**, so every $\mathbb{Z}$-torsionfree G-module is the sum of its submodules which are lattices. One may check that R_R is indecomposable. Let $L_1, \ldots, L_4$ be the four (non-isomorphic, indecomposable) lattices which are obtained by taking the abelian group $\mathbb{Z}_{(2)}$ and allowing each of x, y to act as $\pm$the identity. Say that a lattice M is **reduced** if none of $L_0 = R, L_1, \ldots, L_4$ occurs as a direct summand of M. Let $\mathcal{C} \subseteq \mathcal{C}'$ be the definable closure of the class of reduced lattices.

Consider the group ring $\mathbb{Q}G$: this is a semisimple artinian ring and is easily seen to split as the direct product of four simple rings, namely $\mathbb{Q}$ with the four possible actions of G given by having x, y act as $\pm$the identity. The corresponding primitive idempotents of $\mathbb{Q}G$ are denoted $e_1, \ldots, e_4$. It is easily checked that for each i, $e_i 4 \in \mathbb{Z}G$: for example the idempotent corresponding to the action of x as 1 and of y as -1 is $\frac{1}{4}(1 + x - y - xy)$.

It is shown in [108, 1.5] that a lattice M is reduced iff for each i, $M \cap Me_i = M2e_i$. There is an obvious obstacle to using this: these are not conditions in terms of the R-module structure since $e_i \in \mathbb{Q}G \backslash R$. But, because these modules are $\mathbb{Z}$-torsionfree, the above condition is equivalent to $M4 \cap M4e_i = M2 \cdot 4e_i$, which *is* a condition that can be expressed in terms of pp conditions over R. Therefore $\mathcal{C}$ is exactly the class of $\mathbb{Z}$-torsionfree R-modules which satisfy these five conditions. Let $C = \mathrm{Zg}_R \cap \mathcal{C}$ be the corresponding closed subset of the Ziegler spectrum. By [428, 2.1], C consists of the reduced (in the above sense) indecomposable $\mathbb{Z}$-torsionfree modules together with the four simple $\mathbb{Q}G$-modules.

Now we claim that a $\mathbb{Z}$-torsionfree R-module M is in $\mathcal{C}$ iff none of $L_0, \ldots, L_4$ is a pure submodule of M. We outline one argument here; an alternative would be to modify the proof of [108, 1.5] or, for the whole argument, to take the route in [431, §3].

First, as noted in [108, p. 200], $R \cap Re_1 < R2e_1$, hence the pp-pair $(8e_1 \mid x)/(4 \mid x \wedge 4e_1 \mid x)$ is open on R. If M is $\mathbb{Z}$-torsionfree and this pair is open on M, then any element of M in the gap may be seen to have the same pp-type as 1 in R, so it follows that $R = L_0$ purely embeds in M. Also, [108, p. 199], for each $i = 1, \ldots, 4$, one has $L_i \cap L_i e_i = L_i > L_i 2e_i$, hence the pp-pair $(4 \mid x \wedge 4e_i \mid x)/(8e_i \mid x)$ is open in L_i. It follows similarly (cf. 5.3.15, though that result is not directly applicable) that L_i embeds purely in any module which opens this pair. Since $\mathcal{C}$ is defined by closure of all these pairs it follows that if M is $\mathbb{Z}$-torsionfree and not in $\mathcal{C}$, then at least one of $L_0, \ldots, L_4$ is pure in M.

Now, given $M \in \mathcal{C}$ set $V(M) = Me_*/M4$, where $e_* = e_1 + \cdots + e_4$ and $V_i(M) = (M4e_i + M4)/M4$ for $i = 1, \ldots, 4$. Notice that each of these five $\mathbb{F}_2$-vectorspaces is the value of M at a sort of $\mathbb{L}_R^{\mathrm{eq}+}$ so these vectorspaces are part of $M^{\mathrm{eq}+}$. There are also the obvious, pp-definable (hence in $(-)^{\mathrm{eq}+}$), inclusions of $V_i(M)$ into $V(M)$. This gives, as is easily checked, a functor from the definable category $\mathcal{C}$ to the definable category whose objects are $\mathbb{F}_2$-vectorspaces with four specified subspaces. Thus we have an interpretation functor from $\mathcal{C}$ to the category of representations of the quiver $\widetilde{D_4}$ with image that definable subcategory corresponding to the closed subset of the Ziegler spectrum which is the complement of the four simple non-subspace representations. Indeed, in [108] this functor is shown to be a representation equivalence between these categories, so the classification of finite-dimensional four-subspace configurations (see Section 8.1.2) yields a classification of $\mathbb{Z}_{(2)}G$-lattices. In [431], also see [433], decidability of the former is then used to deduce decidability of $\mathcal{C}$ and hence of the class $\mathcal{C}'$ of all torsionfree modules.

Observe that the above functor is given on objects by replacing an object M by $M^{\mathrm{eq}+}$, then restricting this object (that is, functor on $\mathbb{L}^{\mathrm{eq}+}(\mathcal{C})$) to a subcategory $\mathcal{A}'$ of $\mathbb{L}^{\mathrm{eq}+}(\mathcal{C})$ consisting of five objects, four morphisms and the $\mathbb{F}_2$-action on each sort, thus regarding the image as belonging to a certain definable category of 4-subspace configurations, that is, $\mathbb{F}_2(\widetilde{D_4})$-modules.

18.2.5 Definable functors and Ziegler spectra

Definable functors induce continuous maps of Ziegler spectra (18.2.25), a homeomorphic embedding at the level of points if the functor is full on pure-injectives (18.2.26, 18.2.27) and a homeomorphism if the functor is also an equivalence (18.2.28).

Many of the results of this section were first proved in [503] for the case where $\mathcal{D}$ is a definable subcategory of a module category, using the model-theoretic view, as discussed after 18.1.5 and in Section 18.2.1. These same techniques are adequate to deal with the general case, which was established using the functorial approach in [355].

The first result generalises 4.3.54.

Proposition 18.2.24. ([500, Prop. 2], [355, 7.8(3)]) *Suppose that $I : \mathcal{C} \longrightarrow \mathcal{D}$ is a definable functor between definable categories, that is, one which commutes with products and direct limits. Let $\mathcal{C}'$ be a definable subcategory of $\mathcal{C}$ and suppose that N is an indecomposable pure-injective direct summand of IC for some $C \in \mathcal{C}'$. Then there is $N' \in \mathrm{Zg}(\mathcal{C}') = \mathcal{C}' \cap \mathrm{Zg}(\mathcal{C})$ such that N is a direct summand of IN'.*

Proof. Represent $\mathcal{D}$ as a definable subcategory of Mod-$\mathcal{B}$ for some $\mathcal{B}$. Let a be a non-zero element of N, say $a \in ((-, B), N)$ is of sort $(-, B)$ for some $B \in \mathcal{B}$. Set $p = \mathrm{pp}^N(a)$ to be the pp-type of a in N. Consider the map $I_0 : \mathrm{fun}(\mathcal{D}) \longrightarrow \mathrm{fun}(\mathcal{C})$ such that $I = I_0^*$ (existence by 18.2.2). Identifying these functor categories with the categories of imaginaries (12.3.20, Section B.2), the image under I_0 of the sort $(-, B)$ is a sort, σ say, of $\mathbb{L}(\mathcal{C})^{\mathrm{eq}+}$ and the image of the pp-type p under the induced map (see after 18.2.10) from $\mathcal{L}(\mathcal{D})^{\text{eq}+}$ to $\mathcal{L}(\mathcal{C})^{\text{eq}+}$ is a pp-type, p_1 say, of sort σ. Also set Ψ to be the image under I_0 of $p^- = \{\psi \in \mathcal{L}(\mathcal{D})^{\text{eq}+} : \psi$ is pp of sort $(-, B)$ and $\psi \notin p\}$.

Consider the lattice of pp conditions of $\mathcal{L}(\mathcal{C})^{\text{eq}+}$ with free variable of sort σ. The filter generated by p_1 and the ideal generated by Ψ have empty intersection, since, regarding a as an element, a', of C of sort σ, a' satisfies each formula in the filter and none in the ideal (note that $\text{pp}^{IC}(a) = \text{pp}^N(a)$ since N is pure in IC).

The necessary adaptations having been made, the proof now continues as in [500, Prop. 2], cf. the proof of 4.3.54. Let q be a pp-type of sort σ maximal with respect to containing p_1 and missing Ψ. We check Ziegler's criterion (4.3.49) to show that q is irreducible.

So let η_1, η_2 be pp conditions of sort σ which are not in q. By maximality of q there are $\phi_1, \phi_2 \in p_1$ and $\psi_1, \psi_2 \in \Psi$ such that $\eta_i \wedge \phi_i \leq \psi_i$ $(i = 1, 2)$. Without loss of generality $\phi_1 = \phi_2 = \phi$ say (replace each by $\phi_1 \wedge \phi_2$). Let $\phi' \in p$ and $\psi_1', \psi_2' \in p^-$ be such that $\phi = I_0\phi'$ and $\psi_i = I_0\psi_i'$, $i = 1, 2$, (notation as after 18.2.10). Since p is irreducible, by Ziegler's criterion there is $\phi_0 \in p$ with $\phi_0 \leq \phi'$ such that $\phi_0 \wedge \psi_1' + \phi_0 \wedge \psi_2' \notin p$ and hence such that $(I_0\phi_0) \wedge \psi_1 + (I_0\phi_0) \wedge \psi_2 \in \Psi$ (since the action of I_0 on formulas commutes with $\wedge$ and $+$). Since $I_0\phi_0 \in p_1 \subseteq q$ this is as required for 4.3.49, so q is irreducible.

Therefore, see 4.3.49, there is an indecomposable pure-injective object $N' \in \mathcal{C}$ and an element b' of $\sigma(N')$ such that $\text{pp}^{N'}(b') = q$. Consider $IN' \in \mathcal{D}$ and the element, b say, of IN' of sort $(-, B)$ such that b in IN' corresponds to b' in N'. Clearly $\text{pp}^{IN'}(b) = p$. Since N is indecomposable it is the hull, $H(p)$, of p and hence, by 4.3.35, N is isomorphic to a direct summand of IN', as required. □

Proposition 18.2.25. *Suppose that $I : \mathcal{C} \longrightarrow \mathcal{D}$ is a functor between definable categories which commutes with products and direct limits. Then I induces a closed and continuous map from $\text{Zg}(\mathcal{C})$ to $\text{Zg}(\mathcal{D})$ at the level of topology: that is, it induces a map, I_*, from the lattice of closed sets of $\text{Zg}(\mathcal{C})$ to that of $\text{Zg}(\mathcal{D})$, which preserves finite union and arbitrary intersection and it induces a map, I^{-1}, from the lattice (in fact, the complete Heyting algebra) of open subsets of $\text{Zg}(\mathcal{D})$ to that of $\text{Zg}(\mathcal{C})$, which preserves arbitrary union and finite intersection.*

Proof. Given a closed subset, X, of $\text{Zg}(\mathcal{C})$, let $\mathcal{X}$ be the corresponding definable subcategory of $\mathcal{C}$. Then the closure of $I\mathcal{X}$ under pure subobjects is a definable subcategory of $\mathcal{D}$ and determines a closed subset, let us denote it I_*X, of $\text{Zg}(\mathcal{D})$.

If Y is another closed subset of $\text{Zg}(\mathcal{C})$ and $\mathcal{Y}$ is the corresponding definable subcategory, then the definable subcategory corresponding to the closed set $X \cup Y$ is (3.4.9) $\{L : L \text{ is pure in } M' \oplus M'' \text{ for some } M' \in \mathcal{X}, M'' \in \mathcal{Y}\}$, so the definable subcategory of $\mathcal{D}$ generated by the image of this set under I is

$\{L : L$ is pure in $IM' \oplus IM''$ for some $M' \in \mathcal{X}, M'' \in \mathcal{Y}\}$. By 18.2.24 any indecomposable pure-injective in that set must already be in I_*X or I_*Y; thus I_* does commute with finite unions.

Next, suppose that X_λ $(\lambda \in \Lambda)$ are closed subsets of $\mathrm{Zg}(\mathcal{C})$ and let X be their intersection. Denote the corresponding definable subcategories of $\mathcal{C}$ by $\mathcal{X}_\lambda$, respectively $\mathcal{X}$. Certainly $I_*X \subseteq \bigcap_\lambda I_*X_\lambda$. Suppose that $N \notin I_*X$. Then there is $F \in \mathrm{fun}(\mathcal{D})$ such that $FN \neq 0$ but $FIC = 0$ for every $C \in X$, since the open sets (G) for $G \in \mathrm{fun}(\mathcal{D})$ form a basis of open sets. Hence $F(I\mathcal{X}) = 0$. Therefore $I_0F \cdot C = 0$ for every $C \in X$, that is, the intersection of the compact (5.1.23) open set (I_0F) with X is empty. Therefore, for some λ, $(I_0F) \cap X_\lambda = \emptyset$. That is, $F(IC) = I_0F \cdot C = 0$ for every $C \in X_\lambda$, hence $F(I\mathcal{X}_\lambda) = 0$. Therefore, $N \notin I\mathcal{X}_\lambda$, as required.

Regarding continuity, the sets $(G) \cap \mathrm{im}\big(\mathrm{Zg}(\mathcal{C})\big)$ form a basis of open sets for the induced topology on $\mathrm{im}\big(\mathrm{Zg}(\mathcal{C})\big)$ and the formula $G(IN) = I_0G \cdot N$ shows that the inverse image of this set under I is $(I_0G) \subseteq \mathrm{Zg}(\mathcal{C})$, which is open. So I is continuous at the level of open sets hence I^{-1} is as described. □

Corollary 18.2.26. (see [503, §3], [355, 7.8]) *Suppose that $I : \mathcal{C} \longrightarrow \mathcal{D}$ is a functor between definable categories which commutes with products and direct limits. If I is full on pure-injectives, then I induces a map $\mathrm{Zg}(\mathcal{C}) \to \mathrm{Zg}(\mathcal{D})$. This map is continuous and closed.*

Proof. Since I is full on pure-injectives, $N \in \mathrm{Zg}(\mathcal{C})$ implies IN also is indecomposable, hence, if non-zero, is in $\mathrm{Zg}(\mathcal{D})$. Moreover, it follows from 18.2.24 that the image of a closed set is closed, so I_* (as in 18.2.25) is just the usual direct image map. □

Actually, we should be more careful in the statement since we do not usually allow the zero module to be a point of the spectrum and it could well be that $IN = 0$ for some $N \in \mathrm{Zg}(\mathcal{C})$. So, if K denotes the closed (by 18.2.4) subset of $\mathrm{Zg}(\mathcal{C})$ corresponding to the kernel of I, then the correct statement is that I induces a map from $\mathrm{Zg}(\mathcal{C})\backslash K$ to $\mathrm{Zg}(\mathcal{D})$.

Therefore, in the general case, if $I : \mathcal{C} \longrightarrow \mathcal{D}$ is a functor between definable categories which commutes with products and direct limits and if $K \subseteq \mathrm{Zg}(\mathcal{C})$ is the closed subset corresponding to the kernel of I, then the maps in 18.2.25 factor through the (lattices of the) space $\mathrm{Zg}(\mathcal{C})\backslash K$ where this is given the relative topology.

Theorem 18.2.27. (mainly from [503, 3.19]) *Suppose that $I : \mathcal{C} \longrightarrow \mathcal{D}$ is a definable functor between definable categories which is full on pure-injectives and which has kernel, K, as above. Then the induced map $I_* : \mathrm{Zg}(\mathcal{C})\backslash K \to \mathrm{Zg}(\mathcal{D})$ is a homeomorphism of its domain with its image; the latter is the closed subset of*

$\mathrm{Zg}(\mathcal{D})$ corresponding to the definable subcategory of $\mathcal{D}$ generated, under taking pure subobjects, by the image of I.

Proof. If $N, N' \in \mathrm{Zg}(\mathcal{C})\backslash K$, then any isomorphism $IN \simeq IN'$ would, by fullness, be the image of an isomorphism $N \simeq N'$, so I_* is an embedding of $\mathrm{Zg}(\mathcal{C})\backslash K$. It remains to be seen that the topology on $\mathrm{Zg}(\mathcal{C})\backslash K$ is no finer than that induced by $\mathrm{Zg}(\mathcal{D})$ on its image. But that follows by the discussion before 18.2.11. □

Theorem 18.2.28. ([503, 3.10] for the modules case) *Let $F : \mathcal{C} \longrightarrow \mathcal{D}$ and $G : \mathcal{D} \longrightarrow \mathcal{C}$ be definable functors between definable categories such that GF and FG are naturally equivalent to the identity functors of $\mathcal{C}$ and $\mathcal{D}$ respectively. Then:*

(a) F and G are full;
(b) GF and FG are definable functors;
(c) F and G induce homeomorphisms between $\mathrm{Zg}(\mathcal{C})$ and $\mathrm{Zg}(\mathcal{D})$.

Proof. The first point is immediate, the second follows since the compositions of the exact functors between $\mathcal{C}^{\mathrm{eq}+}$ and $\mathcal{D}^{\mathrm{eq}+}$ are exact and induce the compositions of F and G. The third statement is immediate from 18.2.27 since the kernels of F and G are zero. □

For further results linked to the presheaves of definable scalars, see [525].

Appendix D Model theory of modules: an update

For most of the model theory of modules per se, and certainly for model-theoretic stability theory in the context of modules, my earlier book, [495], rather than this one, is the place to look, although for a model-theoretic view of some more recent developments one may look at the later parts of the paper [516]. At the time [495] was written it seemed that all the basic results were in place and most of the general development had been carried through, with a limited number of hard model-theoretic problems still open. This was not correct: elementary duality at the level of theories and Ziegler spectra was still to come. Also, with the internal development seemingly complete, the way forward was clearly to applications and perhaps analogous developments in other contexts. That was largely correct, though the equivalence of the category of imaginaries with the category of finitely presented functors, and related functor-category reformulations was still a major theme to come.

These new features, elementary duality and equivalence with functor categories, have been dealt with in the main body of this text. Progress on the major purely model-theoretic open questions (Vaught's conjecture, various questions still open in the absence of a countability hypothesis) has been significant, in that they have often been settled for particular types of ring, but disappointing in that all the main questions left open in Ziegler's paper [726] still are open in the general case. The situation is similar for some of the more "applied" questions, such as that about the equivalence of decidability and tame representation type: though there has been impressive progress in some contexts the general question remains open. It may be, of course, that the general theorems are not there; that there do exist counterexamples and that positive results in particular contexts are all that can be achieved.

Apart from work on the long-standing questions, some fresh avenues have been explored; here I give some signposts. It is not my intention to summarise what has been done, only to point the reader to the relevant work.

To anyone who is looking at this but not at the main body of the text I should certainly point out the book, [323], of Jensen and Lenzing which was published not long after [495]. These two books, though both mainly about the model theory of modules, are largely complementary rather than overlapping, both in material covered and in approach. The approach of this book is somewhere in between.

The right/left duality of Ziegler spectra, expressed as a homeomorphism, 5.4.1, at the level of topology, between Zg_R and ${}_R\mathrm{Zg}$ has a more precise expression as a duality, [274,

6.6],[1] between complete theories of right modules and complete theories of left modules. By pp-elimination of quantifiers, A.1.1, a complete theory T of right R-modules is determined by a set of sentences of the form $\sigma =$"card($\phi(-)/\psi(-)$) $\sim k$" where ϕ/ψ is a pp-pair, where $\sim$ is one of $\geq$, $=$, $\leq$ and where k is a positive integer (A.1.2). Set $D\sigma$ to be the sentence "card($D\psi(-)/D\phi(-)$) $\sim k$", regarded as a sentence for left R-modules. If T is any complete theory of right modules, then $DT = \{D\sigma : \sigma \in T\}$ is a complete theory for left modules and, since D^2T is equivalent to T this gives a natural bijection between complete theories of right and left modules. Note that 5.4.1 is equivalent to the restriction of this bijection to those complete theories T with all invariants, card($\phi(-)/\psi(-)$), either 1 or infinite.[2] Of course, if R is an algebra with an infinite central subfield, then this, by 1.1.8, is no restriction.

Let M be any module and let ϕ/ψ be a pp-pair. Let M^* be any chosen model of the dual, $D\mathrm{Th}(M)$, of the complete theory of M. We may consider $\phi(M)/\psi(M)$ as a right module over its ring, $R_M^{\phi/\psi}$ of definable scalars (this is defined after 6.1.7) and, by 12.3.20, $D\psi(M^*)/D\phi(M^*)$ is a left module over the same ring. Then, [274, 6.8], these $R_M^{\phi/\psi}$-modules have dual theories.

Nedelmann, [454], defined an analogue of the Ziegler spectrum for simple theories. The points are the domination-equivalence classes of types of weight 1 and the topology is induced by that on the Stone space of all types. Herzog and Cherlin had previously observed that the Ziegler spectrum for modules could be regarded this way. With some additional assumptions on the simple theory Nedelmann obtained ([454, §2.3, Chpt. 3]) a generalisation of the correspondence between closed subsets of the spectrum and certain theories.

Strong minimality is an important concept in model theory; a module M is **strongly minimal** if every proper pp-definable subgroup of M is finite. Herzog, [274, §7], gave an analysis of the Ziegler spectrum of such modules. Puninskya and Puninski have a number of papers (for example, [535], [544]) on this and Puninskaya gave, [543], with new results and including many proofs, an extensive survey.

Strongly minimal sheaves of modules are considered in [519, §3].

A module is **weakly minimal** if every pp-definable subgroup is either finite or of finite index. Herzog and Rothmaler, [273, §3, §4], [290], [291], investigated the model theory of these and other modules with a regular generic type, as did Loveys, [413], [414]. The non-modular models of a unidimensional theory of modules are characterised as the elementary submodels of the prime pure-injective model in [480]; also see [479]. Vaught's conjecture is proved by Buechler, [100], for modules satisfying certain weak minimality conditions, also see [465], [466]. Classification of modules which are minimal in one of these, or some related, senses are given in particular cases in [401], [407], [408], [409].

A positive version of Deissler rank (a layering by definability) is considered in [377], [378], [115]. A layering using regular types is defined and applied to injective modules in [380].

The content for modules of some general stability-theoretic concepts is described quite explicitly in [383], along with various (counter)examples. The topics discussed include regular types, (hereditary) orthogonality, p-simplicity, internality and analysability.

[1] The proof of which needs 6.1.22 and a result, [274, 6.4], along the lines of 5.3.10.

[2] That result gives a bijection between definable subcategories; the complete theory corresponding to a definable subcategory is the theory of $M^{(\aleph_0)}$ for any module M which generates the definable subcategory.

The introduction to that paper grew into a paper in its own right: [382] begins with a description of how Fisher's "abelian structures" may be subsumed under the model theory of modules and objects of more general abelian categories. Categories of pp-imaginaries are introduced and canonical bases and weak elimination of imaginaries, including results of Poizat, Evans and Pillay ([485], [180]), are discussed. Various results on hulls in the imaginary category are given and some of the usual model theory of modules is lifted to that context (the "endocategory" of Krause, [347]).

The pure-injective modules are the positively saturated modules. At the other extreme are the positively atomic and positively constuctible modules; these are shown to be the Mittag–Leffler, respectively pure-projective, modules, by Kucera and Rothmaler ([621], [384]).

The question of the relation between decidability and representation type was asked in [495] and further investigated in the general case in [496] and [501], where it is shown that any strictly wild algebra has undecidable theory of modules. For general results around decidability for modules see [483], [495, Chpt. 17].

The decidability question has been much investigated over group rings. Let G be a finite group. First, suppose that k is a (decidable) field, the characteristic of which divides the order of G (otherwise kG is semisimple and the problem is trivial). We may ask when the theory of all kG-modules is decidable: it is conjectured that the decidability/undecidability split coincides with that between tame (or finite)/wild representation type. Most cases are settled in [425] and the unsettled ((semi)dihedral and quaternionic in characteristic 2) cases are part of larger unsettled questions such as whether or not the theory of modules over a non-domestic string algebra is decidable.

Alternatively we may consider the theory of $\mathbb{Z}G$-modules. Here the question becomes whether this is decidable iff the order of G is square-free (see [676]). There are various partial results, for instance undecidability in some wild cases, [676], [677], and decidability, using the representation of $\mathbb{Z}C_p$ as a pullback ring (Section 8.2.9), in some tame cases, [678], [679]. More generally, one may consider the question over RG, where R is a (decidable) Dedekind domain of characteristic 0. The theory of $\mathbb{Z}G$-modules links to that of abelian-by-finite groups ([425], [427], [430]).

Similar questions about modules over rings such as $\mathbb{Z}_{p^n}G$ have been considered (for example, [426], [429]).

A third related question is the decidability of the theory of $\mathbb{Z}G$-lattices, that is, those $\mathbb{Z}G$-modules which are free as $\mathbb{Z}$-modules. Considering the definable closure of the underlying abelian groups, one may check that the definable closure of this class is that of $\mathbb{Z}G$-modules which are torsionfree as $\mathbb{Z}$-modules, so it is the decidability of (the theory of) this class which is being considered. (The theories of these classes are, however, not the same – see [121]). The class of $\mathbb{Z}G$-lattices is not wild iff either G is cyclic of prime or prime-squared order, or G has order 8, or G is the Klein 4-group. Toffalori, [681], proved undecidability for $\mathbb{Z}G$-lattices if G has no Sylow subgroup of the kinds listed; that is, wild implies undecidability for $\mathbb{Z}G$-lattices. Baratella and Toffalori, [45], proved decidability when G is the Klein 4-group. In the same paper they proved various results on the model theory of $\mathbb{Z}G$-lattices. The relevant part of the Ziegler spectrum of $\mathbb{Z}G$ is investigated in [428]. See also other papers by these authors, for instance at [431, p. 192] it is shown that decidability of $\mathbb{Z}_{(p)}G$-modules for each prime p implies decidability of $\mathbb{Z}G$-modules (similarly for stability, see [271]). Baratella and Toffalori showed, under some assumptions, that lattices over integral group rings are elementarily equivalent iff they are in the same genus ([45]).

In [123] classification and (un)decidability for pairs of free abelian groups are considered.

Puninski and Toffalori, [580], went much of the way towards establishing the dividing line between having decidable or undecidable theory of modules for finite commutative rings; in particular they showed that wildness implies undecidability for such rings. That paper also gives a very careful and clear account of a number of general, folklorish, results on proving decidability, especially concerning how to use an explicit description of the Ziegler spectrum to achieve this.

Completely different methods are used in [521, §6] to prove undecidability of the theory of modules of certain generalised Weyl algebras, for instance the quantum Weyl algebras $A_1(q)$, where q is not a root of unity, the universal enveloping algebra, $U\mathrm{sl}_2(k)$ over a field of characteristic 0, certain differential polynomial rings.

Interpretability between definable subcategories of module categories is given a somewhat uniform, somewhat category-theoretic treatment in [503]. Note the counterexample [503, 3.11] of Herzog to the generalisation of a result on extending structure from a restricted collection of sorts. That result, [503, 3.8] (18.2.8 here), uses existence of locally atomic models which in general needs countability. A much more category-based, and broader, treatment which draws on work ([355], [359]) of Krause, is in [516]

Vaught's Conjecture is that if T is a countable set of sentences in some first-order language (such as those we have used here), then the number of countable models of T up to isomorphism is finite, countably infinite or $2^{\aleph_0}$. Of course, assuming the continuum hypothesis gives an easy positive answer but the point is to show it without this assumption, part of the motivation being that, in order to prove it, one likely will have to develop a reasonably strong structure theorem that will allow a count. It is the case that large parts of the general question reduce to the question for theories of modules. The conjecture has been verified in particular cases; over particular sorts of rings and for particular sorts of modules. The related property of having few, that is, less than continuum many, types is usually the key. See, for instance, [273], [537] for the kinds of arguments which can be used.[3] Completely general results on Vaught's Conjecture, however, are lacking. I refer to [541] for background on the problem and a survey of cases in modules where it has been settled. Also see [546].

Properties such as ω-stability and superstability for modules have general characterisations, which can be found, with references, in [495]. Explicit descriptions and classifications of modules with these properties have been made in various situations, see, for instance, [425], [426], [271], [518].

See the comments towards the end of Section 8.2.3 for results of Herzog from [282] on the theory of pseudofinite-dimensional representations of the Lie algebra $\mathrm{sl}_2(k)$; pseudofiniteness is a general and fruitful model-theoretic theme.

The view of the lattice pp_R as a generalisation of the lattice of right ideals of R, particularly the relation between regular types and primes, is pursued in [376].

Kucera and Rothmaler showed, [384, 3.1], that the positively constructible modules are precisely the pure-projective modules: contrast this with the fact that the positively saturated modules are exactly the pure-injective modules.

[3] Herzog's characterisation, reported in [495, pp. 161 ff.], of countable theories of modules with few types was published as [273, 1.2].

Kamensky examined the theory of modules over variable rings, [328], [329]. By restricting the rings to n-generated commutative algebras over fields and adding predicates for Grassmannians he obtains a model-completion which, of course, specialises to that of Eklof and Sabbagh in [174].

Katsov, [335], used multi-sorted languages to extend results of Eklof and Sabbagh ([174], [631]) to functor categories. As seen in the second part of this book, one can go further.

Villemaire, [694]–[696], investigated abelian groups $\aleph_0$-categorical over a subgroup and Hodges, [297], investigated general categoricity over a subgroup.

Dellunde, Delon and Point considered separably closed fields of characteristic p as modules over a ring which retains information on p-bases. They examined the theory of the resulting modules, obtaining an axiomatisation and showing elimination of quantifiers after p-component functions had been added, [149]; they described the corresponding closed subset of the Ziegler spectrum, proved that the width is undefined, obtained information about pp-types and ranks and proved an elimination of imaginaries result, [150]. Also see [82].

A somewhat similar approach was taken for a decidability result of Pheidas and Zahidi, [476], where they regarded perfect fields of finite characteristic as modules over a ring which includes the action of the Frobenius map.

A somewhat analogous investigation of difference rings is undertaken in [306].

If T is a complete theory in a language $\mathcal{L}$, let $\mathcal{L}_\sigma$ denote the language obtained by adding a new unary function symbol σ and let T_0 denote T together with the set of sentences which say that σ is an automorphism. If T_0 has a model-companion, then denote this by T_A. It was shown by Chatzidakis and Pillay, [117], that if T is the theory of an ω-stable module (that is, a Σ-pure-injective over a countable ring, Section 4.4.2), then T_A exists. They asked whether there is an example of non-ω-stable module whose theory T is such that T_A exists. More generally, they asked what is the condition on the theory T of a module for T_A to exist. Granger, [248, §4.5], gave an example as asked for in their first question and also obtained information on the more general question. Since his work was not prepared for publication I give some details here.

First we recall that Eklof and Sabbagh proved that the common theory of all right R-modules has a model-companion iff R is right coherent and they also described this theory quite explicitly, [174]. If, with notation as above, (M, σ^M) is a model of T_0, then we think of this as an $R[\sigma, \sigma^{-1}]$-module. So, to give an example of the kind asked for, it is enough to find a ring R such that common theory of non-zero R-modules is complete (that is, an indiscrete ring in the sense of Section 8.2.12), which is not ω-stable (so, in this context, not a simple artinian ring) and such that $R[\sigma, \sigma^{-1}]$ is not right coherent. By 8.2.93 it will be enough to find a simple, non-artinian, von Neumann regular ring R such that the Laurent polynomial ring $R[T, T^{-1}]$ is right coherent. By a result of Dicks and Schofield, [151], the last condition is equivalent to $R[T]$ being right coherent. Despite results in particular cases, there seems to be no general necessary and sufficient condition on a von Neumann regular ring R for $R[T]$ to be coherent, so this indicates that the second question from [117] might not have an easy answer.

In any case, we now have a guide to an example which will answer the first question and Granger constructed such. Let $R_n = M_{2^n}(K)$ be the ring of $2^n \times 2^n$ matrices over a field K. Each embeds in the next by the block-diagonal map $A \mapsto \begin{pmatrix} A & 0 \\ 0 & A \end{pmatrix}$. Let R be the direct limit of these embeddings. It is easy to see that R is simple, non-artinian and von Neumann regular. It is also true, though it takes some work to show this [248, 4.5.10], that $R[T]$ is coherent, as required.

It is also noted that R being simple, non-artinian and von Neumann regular does not guarantee that $R[T]$ will be coherent.

Baldwin, Eklof and Trlifaj, [38], investigated when classes of modules of the form ${}^{\perp}N = \{A : \mathrm{Ext}^i(A, N) = 0 \ \forall i > 0\}$ are abstract elementary classes; such classes arise from cotilting and cotorsion theories. They showed that if ${}^{\perp}N$ is an abstract elementary class, then N must be cotorsion and, in the other direction, if N satisfies the stronger condition of being pure-injective, then ${}^{\perp}N$ is an abstract elementary class. They obtained results on when ${}^{\perp}N$ is a cotilting class and on when it is definable. They showed that ${}^{\perp}N$ always has amalgamation and, under some conditions, has closures (minimal special submodels over sets). Also see [689], where it is shown that every cotilting class gives an abstract elementary class of finite character.

They also defined an appropriate notion of type in terms of automorphisms and defined corresponding notions of stability and locality.

The model theory around tensor product of modules is investigated in [509]. The tensor product of pp-types is defined and, although elementary equivalence is not in general preserved by tensor product, it is preserved in some cases and, in particular, the tensor product of complete theories is well-defined.

If two rings are elementarily equivalent, even more if one is an elementary substructure of the other, then, in view of 1.1.13, one would expect there to be some correspondence between the model theory of their respective module categories. Indeed ([506]), in the latter case there is a natural inclusion of lattices of pp conditions. From this one can derive various corollaries, for example the result, [272, 2.5], of Herrmann, Jensen and Lenzing that if R is an elementary subring of S, then R has finite representation type iff S has. There are various results in [506] on induction and restriction of theories of modules when $R \prec S$.

Zayed, [722], showed that an ultrapower of a pure-semisimple ring must be pure-semisimple, answering a question of Jensen and Lenzing, [323, 13.16].

Various axiomatisability questions for rings, though not the concern in this book, are quite closely related to the themes considered. See [323], [333], [334].

The existence of a complete variety which is not model-complete follows from the existence of an indiscrete ring, that is, one whose non-zero modules all are elementarily equivalent, which is not von Neumann regular, that is (by 2.3.22 and A.1.4), such that not every embedding is an elementary embedding. This answers a question of Baldwin and Lachlan, [39, Q.1].

The elementary duality of pp formulas (Section 1.3.2) is extended to certain types of infinitary formulas in [528, §§3,4], and classes axiomatised by implications between these, in particular the classes of flat and absolutely pure modules outside the case where the ring is coherent are considered.

As normally presented, comodules over a coalgebra are not structures in the sense of model theory: part of the structure of a comodule M over a coalgebra C (over a field k) is a map from M to $M \otimes_k C$. But the category of C-comodules *is* locally finitely presented, so does have a model theory, some basics of which are presented in [147].

Moving away from the "additive" situation which has been our concern throughout this book, work with Rajani, [584], also see [307], [420], [422], shows that the "functor category", $(\text{mod-}R, \mathbf{Ab})^{\mathrm{fp}} \simeq \mathbb{L}_R^{\mathrm{eq}+}$, is the additive analogue of the category of coherent objects of the classifying topos of a coherent theory, and a description of both finitely presented and coherent objects in the latter (unlike the additive case, they are different!) is given in terms of positive existential formulas. It is shown that this commonality, essentially between **Ab**-valued and **Set**-valued categories of models of coherent theories, is not shared with models taking values in the category of affine spaces: in that context there are finitely presented functors which are not in any sense definable, [585]. An analogue of the Ziegler spectrum is developed for G-sets (G a group) in [586].

Appendix E Some definitions

Formulas Essentially, but not literally, what we have called a condition in this book. We use the latter term since it de-emphasises the syntactic view of a formula as a string of symbols in a formal language and encourages one to focus on the solution set of such a formula and on the functor that it defines; a formula per se is merely a representative of an equivalence class, so, though useful to manipulate, is not in itself the object of interest. Formal languages are discussed in Appendices A and B.

In practice, the term "condition" is used in this book with some ambiguity: sometimes it does refer to a particular formula (in the strict syntactic sense) but, if pressed, I would formally define a condition to be a certain type of functor; then a formula is a kind of presentation of this functor. A pp-type would, therefore, literally be a set of functors: it can be represented by a set of formulas and it is often convenient to identify it with such a set of formulas.

Model theory Basic definitions have been given in Appendix A. I would not be so brave (or foolish) as to attempt a definition here; it is a rather wide-ranging subject. But I can certainly point to some arguments in the book which have a model-theoretic flavour (1.3.26, 1.3.27, 2.1.21, 2.4.5, 3.3.6, part of 3.4.24, 4.1.4, 4.2.1, 4.2.2, 4.3.11, 4.3.21, 7.3.2, for instance). For the model theory of modules per se, see Appendices A and D as well as [495].

(Skeletally) Small category A category $\mathcal{C}$ is **small** if the class of objects is a set. It is **skeletally small** if the class of isomorphism classes of objects is a set. The default assumption is that the collection of morphisms between any two given objects of a category is a set. If, however, $\mathcal{C}$ is not skeletally small, then a functor category $(\mathcal{C}, \mathcal{D})$ may fail to satisfy this assumption: the collection of natural transformations between two functors may be a proper class. There are various ways of dealing with potential problems caused by this: one can simply be careful (for example, not to take limits over proper classes); one can use hierarchies (of "universes") that allow proper classes; one can decide that one does not really care that much about objects of unimaginably large cardinality and just cut off categories in an ad hoc way so that they are always skeletally small.

In this book, in cases where $\mathcal{C}$ is not skeletally small, we will never need to consider the entirety of a functor category $(\mathcal{C}, \mathcal{D})$. In case $\mathcal{C}$ is skeletally small we will sometimes tacitly replace $\mathcal{C}$ by a small version of $\mathcal{C}$ – that is, by a subcategory of $\mathcal{C}$ which contains an isomorphic copy of each object of $\mathcal{C}$ and which has just a set of objects.

Morphisms An epimorphism in a category of modules is surjective: in the category of rings it need not be. For instance, the embedding of the ring, $\mathbb{Z}$, of integers into the field, $\mathbb{Q}$, of rationals is an epimorphism because every element of $\mathbb{Q}$ is the unique solution of a certain equation over $\mathbb{Z}$. (For precise expression of this see, say, [315]–[317].)

A morphism $f \in (C, D)$ is a **split monomorphism** if there is a morphism $g \in (D, C)$ such that $gf = 1_C$; since this forces f to be monic, f may be regarded as the inclusion of a direct summand into D. Dually, f is a **split epimorphism** if there is a morphism $h \in (D, C)$ such that $hf = 1_D$; f may be thought of as a projection of D onto a direct summand. In either case the morphism f is said to **split**. If $\mathcal{C}$ is any category, then **idempotents split** in $\mathcal{C}$ if whenever $e \in (C, C)$ is an **idempotent** morphism of $\mathcal{C}$ (that is, $e^2 = e$), then there is an object B and there are morphisms $i \in (B, C)$ and $p \in (C, B)$ such that $pi = 1_B$ and $ip = e$; then e may be regarded as the projection of C onto a direct summand B, followed by the split inclusion of this direct summand into C.

The additive category $\mathcal{C}$ **has pseudocokernels** if every morphism $f \in (C, D)$ in $\mathcal{C}$ has a **pseudocokernel**, meaning a morphism $p \in (D, E)$ such that $pf = 0$ and such that for every $F \in \mathcal{C}$ and every morphism $g \in (D, F)$, if $gf = 0$, then there is $g' \in (E, F)$ (perhaps not unique) such that $g = g'p$. The notion of **pseudokernel** is defined dually.

Limits and colimits Suppose that $(C_i)_i$ is a set of objects and that $(\alpha_k)_k$ is a set of morphisms between various of those objects. This collection of objects and morphisms is refered to as a **diagram** in $\mathcal{C}$ (the term is also used for the corresponding functor to $\mathcal{C}$ from a category with objects and morphisms corresponding to the given data). A **cocone** on this diagram is given by an object $C \in \mathcal{C}$ and, for each i, a morphism $f_i : C_i \to C$ such that, for each arrow, $\alpha : C_i \to C_j$ of the diagram, $f_i = \alpha f_j$ (that is, the f_i are compatible with any arrows in the diagram). A **colimit** of this diagram is a cocone as above such that, for any cocone $(g_i : C_i \to D)_i$ on the diagram, there is a unique arrow $h : C \to D$ with $g_i = hf_i$ for all i. If this exists it is unique up to isomorphism. There are many familiar special cases: a coproduct of a set of objects is a colimit of the diagram consisting of those objects and no maps; the definitions of pushout and direct limit also are special cases.

Cones and limits are defined dually.

If the indexing category is a directed (upwards) poset (Appendix E, p. 712), then the colimit is referred to as a directed colimit or, more usually but perhaps confusingly, as a **direct limit**. An **inverse limit** is a limit over a poset which is directed downwards. For the more general filtered (co)limits see Appendix E, p. 715.

A category is **(co)complete** if it has arbitrary (co)limits. A preadditive category is (co)complete iff it has (co)products and (co)kernels. Indeed (see, for example, [663, IV.8.2, IV.8.4] or almost any category theory text, such as [418]), (co)limits may be constructed explicitly via (co)products and (co)kernels.

Functors If $F : \mathcal{C} \longrightarrow \mathcal{D}$ is a functor, then its **image** – the collection of all objects of the form FC with $C \in \mathcal{C}$ and all morphisms of the form Ff with $f \in \mathcal{C}$ – is a, not necessarily full, subcategory of $\mathcal{C}$ (a subcategory $\mathcal{B}$ of $\mathcal{C}$ is **full** if for every $A, B \in \mathcal{B}$ every $f \in \mathcal{C}(A, B)$ is in $\mathcal{B}$). If the image of F is full say that F is a **full** functor.

A subcategory, $\mathcal{B}$, of $\mathcal{C}$ is **dense** if every object of $\mathcal{C}$ is isomorphic to an object of $\mathcal{B}$. A functor is **dense** if its image is dense.

If $\mathcal{C}, \mathcal{D}$ are preadditive categories, then a functor $F : \mathcal{C} \longrightarrow \mathcal{D}$ is **additive** if, for all $C, C' \in \mathcal{C}$ and for all $f, g \in \mathcal{C}(C, C')$, we have $F(f + g) = Ff + Fg$.

Suppose that $\mathcal{C}$ and $\mathcal{D}$ are preadditive and that F is an additive functor. The **kernel** of F is used, ambiguously (though the context should make the meaning clear) to refer to the collection of all morphisms $f \in \mathcal{C}$ such that $Ff = 0$ or to the class (or full subcategory) of objects $C \in \mathcal{C}$ such that $FC = 0$.

A functor F is **exact** if it sends (short) exact sequences to (short) exact sequences, and is **right**, respectively **left exact**, if it takes each short exact sequence $0 \to A \to B \to C \to 0$ to an exact sequence $A \to B \to C \to 0$, resp. $0 \to A \to B \to C$.

If $\mathcal{D}$ is, say, the category of sets or the category of abelian groups and $\mathcal{C}$ is a category of sets, respectively abelian groups, with extra structure, then the **forgetful functor** from $\mathcal{C}$ to $\mathcal{D}$ is that which takes each object of $\mathcal{C}$ to its underlying set or group and which takes each morphism of $\mathcal{C}$ to the same map, regarded as a map of underlying sets. If $\mathcal{C} = \text{Mod-}R$ for a ring R, then the representable functor $(R, -)$ is naturally isomorphic to the forgetful functor from Mod-R to **Ab**.

Adjoint functors Suppose that $F : \mathcal{C} \longrightarrow \mathcal{D}$ and $G : \mathcal{D} \longrightarrow \mathcal{C}$ are functors. Say that F is **left adjoint** to G and G is **right adjoint** to F (and that (F, G) is an **adjoint pair**) if there is a natural bijection $\mathcal{D}(F(-), *) \simeq \mathcal{C}(-, G(*))$, meaning that for every $C \in \mathcal{C}$ and $D \in \mathcal{D}$ there is a natural bijection $\beta_{C,D} : \mathcal{D}(FC, D) \to \mathcal{C}(C, GD)$. By "natural" we mean that for all $f : C \to C'$ in $\mathcal{C}$ and $g : D \to D'$ in $\mathcal{D}$ the obvious diagrams commute.

Theorem E.1.1. (for example, [663, IV.9.4] or a general text, such as [418]) *Suppose that $F : \mathcal{C} \longrightarrow \mathcal{D}$ and $G : \mathcal{D} \longrightarrow \mathcal{C}$ are such that (F, G) is an adjoint pair. Then F is preserves colimits and G preserves limits. In particular, if the categories are abelian and the functors are additive, then F is right exact and G is left exact.*

Natural transformations and categories of functors If $\mathcal{C}, \mathcal{D}$ are categories, then $(\mathcal{C}, \mathcal{D})$ denotes the category whose objects are the functors from $\mathcal{C}$ to $\mathcal{D}$ (thus $(\mathcal{C}^{\text{op}}, \mathcal{D})$ will denote the category whose objects are the contravariant functors from $\mathcal{C}$ to $\mathcal{D}$) and whose morphisms are the natural transformations between functors. If $F, G : \mathcal{C} \longrightarrow \mathcal{D}$ are functors, then a **natural transformation**, $\tau : F \to G$, from F to G is given by: for each object $C \in \mathcal{C}$, a morphism $\tau_C : FC \to GC$ in $\mathcal{D}$ (the **component** of τ at C) such that, for every arrow $h \in \mathcal{C}(C, C')$ one has $\tau_{C'}.Fh = Gh.\tau_C$.

Composition of natural transformations is defined in the obvious way and is easily seen to give the structure of a category to $(\mathcal{C}, \mathcal{D})$; for instance, the identity natural transformation of a functor F has, for its component at $C \in \mathcal{C}$, just 1_C. So (F, G) is the collection of natural transformations from F to G.

In the case that both $\mathcal{C}$ and $\mathcal{D}$ are preadditive, then we use $(\mathcal{C}, \mathcal{D})$ to denote the category of *additive* functors from $\mathcal{C}$ to $\mathcal{D}$. In this case there is an obvious definition of addition on (F, G), giving $(\mathcal{C}, \mathcal{D})$ the structure of a preadditive category.

Projective objects Let C be an object of the abelian category $\mathcal{C}$. Then C is **projective** if the functor $(C, -) : \mathcal{C} \longrightarrow \textbf{Ab}$ is exact. Since $(C, -)$ is in any case left exact, it follows that C is projective iff whenever $p : A \to A''$ is an epimorphism in $\mathcal{C}$ and $f : C \to A''$ is a morphism there is a morphism $g : C \to A$ such that $f = pg$.

The **projective dimension**, $\text{pdim}(C)$, of an object C of $\mathcal{C}$ is the least n such that there is a **projective resolution** of C of length n, that is, an exact sequence $0 \to P_n \to P_{n-1} \to \cdots \to P_0 \to C \to 0$ with each P_i projective; set $\text{pdim}(C) = \infty$ if there is no such n. If

$D \in \mathcal{D}$, then the **homology** groups (kernel of one map modulo the image of the previous one) of the chain complex $0 \to (P_0, D) \to (P_1, D) \to (P_2, D) \to \cdots$ are, respectively, (C, D), $\mathrm{Ext}^1(C, D)$, $\mathrm{Ext}^2(C, D), \ldots$.

The **global dimension** of a category is defined to be the supremum of the projective dimensions of its objects, a natural number or ∞.

Representable and flat functors Important examples of functors are the **representable functors**. If $C \in \mathcal{C}$, then there is a (covariant, representable) functor $(C, -) : \mathcal{C} \longrightarrow \mathbf{Set}$, also denoted $\mathcal{C}(C, -)$, which, on objects $D \in \mathcal{C}$, is given by $(C, -)D = \mathcal{C}(C, D)$, the set of morphisms from C to D, and which is given by composition on arrows, that is, if $g : D \to E$ is an arrow of $\mathcal{C}$, then $(C, -)g : (C, -)D \to (C, -)E$ is the mapping, also denoted (C, g), from (C, D) to (C, E) given by $(C, g) : h \mapsto gh$ for $h \in (D, E)$.

Since composition in a preadditive category is bilinear all representable functors $(C, -)$ will be additive, hence may, and will, be regarded as taking values in the category, **Ab**, of abelian groups: $(C, -) : \mathcal{C} \longrightarrow \mathbf{Ab}$.

Dually there are the representable functors from the **opposite** category $\mathcal{C}^{\mathrm{op}}$ (this category has the same objects as $\mathcal{C}$ but "all the arrows are turned round"), which we also regard as contravariant functors from $\mathcal{C}$: given $C \in \mathcal{C}$ define the contravariant representable functor $(-, C) : \mathcal{C}^{\mathrm{op}} \longrightarrow \mathbf{Set}$ by $(-, C)D = (D, C)$ for objects D of $\mathcal{C}$ and, on morphisms $g \in (D, E)$ of $\mathcal{C}$, set $(-, C)g = (g, C) : (E, C) \to (D, C)$ where the latter is given by $(g, C)h = hg$ for $h \in (E, C)$.

The Yoneda Lemma (see 10.1.7) is the key result about representable functors. Also note their projectivity in the functor category (10.1.12).

Preadditive category A category $\mathcal{C}$ is **preadditive** if each morphism set, (C, D), is endowed with the structure of an abelian group in such a way that composition is bilinear: $h(f + g) = hf + hf$ and $(f + g)h = fh + gh$ (when one side is defined so is the other and they are equal).

A preadditive category with a single object is really just the same thing as a ring, see Section 10.1.1.

Additive categories A preadditive category $\mathcal{C}$ is **additive** if it has a **zero object** – an object 0 (unique to isomorphism when it exists) such that $\mathrm{card}(0, C) = 1 = \mathrm{card}(C, 0)$ for every object C of $\mathcal{C}$ – and if every finite set of objects of $\mathcal{C}$ has a direct sum (see, for example, [195, p. 60]).

Abelian categories An additive category is **abelian** if every morphism has a kernel and a cokernel, if every monomorphism is the kernel of its cokernel and if every epimorphism is the cokernel of its kernel. In an abelian category a morphism is an isomorphism iff it is both a monomorphism and epimorphism [195, 2.12] (this does not hold, consider $\mathbb{Z} \to \mathbb{Q}$, in the category of rings which is, therefore, not abelian).

In any abelian category there is a well-defined notion of intersection and sum of subobjects of an object and, under these operations, the collection of subobjects of any object forms a modular lattice, possibly, though not if the category has a generating set ([195, 3.35]), a proper class. If the category has limits and colimits, then the lattice of subobjects has intersections and sums over arbitrary index sets.

A subcategory (or subclass of objects), $\mathcal{G}$, of the additive category $\mathcal{C}$ is said to **generate** $\mathcal{C}$ if for every non-zero morphism $f : A \to B$ in $\mathcal{C}$ there is $G \in \mathcal{G}$ and a morphism $g : G \to A$ such that $fg \neq 0$.

Lemma E.1.2. (see [195, 3.36, 3.34]) *Let $\mathcal{C}$ be an abelian category and let $\mathcal{G}$ be a subcategory of $\mathcal{C}$.*

(a) If $\mathcal{C}$ has arbitrary coproducts hence is cocomplete, then $\mathcal{G}$ generates $\mathcal{C}$ iff every object of $\mathcal{C}$ is an epimorphic image of a coproduct of copies of objects from $\mathcal{G}$.
(b) If every object of $\mathcal{G}$ is projective, then $\mathcal{G}$ generates $\mathcal{C}$ iff, given any non-zero object A of $\mathcal{C}$, there is a non-zero morphism from some object of $\mathcal{G}$ to A.

Let $\mathcal{C}$ be an abelian category. A (not necessarily full) subcategory, $\mathcal{D}$, of $\mathcal{C}$ is an **abelian subcategory** if $\mathcal{D}$ is abelian and if the inclusion functor from $\mathcal{D}$ to $\mathcal{C}$ is exact.

Pushouts and pullbacks Given morphisms $A \xrightarrow{f} M$, $A \xrightarrow{f'} M'$ in an abelian category their **pushout** (if it exists) is (g, g', N) in the diagram shown (but often we say just that N is the pushout)

$$\begin{array}{ccc} A & \xrightarrow{f} & M \\ {\scriptstyle f'}\downarrow & & \downarrow{\scriptstyle g} \\ M' & \xrightarrow[g']{} & N \end{array}$$

where, for any pair of morphisms $h : M \to N'$ and $h' : M' \to N'$ such that $h'f' = hf$, there is a unique $k : N \to N'$ with $kg = h$ and $kg' = h'$. If the pushout exists it is unique to isomorphism and, where this makes sense, $N = (M \oplus M')/\{(fa, -f'a) : a \in A\}$ and g, g' are the embeddings into $M \oplus M'$ followed by the projection to N.

Turning all the arrows round gives the notion of **pullback** (and that is realised as $N = \{(a, a') \in M \oplus M' : fa = f'a'\}$).

Grothendieck categories The abelian category $\mathcal{C}$ is **AB5** if it is cocomplete and if **direct limits are exact**, meaning the following. Consider the category with objects the chain complexes of the form $0 \to A \to B \to C \to 0$ of objects of $\mathcal{C}$ and with morphisms from $0 \to A \xrightarrow{f} B \xrightarrow{g} C \to 0$ to $0 \to A' \xrightarrow{f'} B' \xrightarrow{g'} C' \to 0$ being those triples, $(h_1, h_2, h_3) \in (A, A') \times (B, B') \times (C, C')$, of morphisms such that $f'h_1 = h_2 f$ and $g'h_2 = h_3 g$.

One may check that since $\mathcal{C}$ has direct limits so does this category: the direct limit of a directed system of chain complexes $0 \to A_\lambda \xrightarrow{f_\lambda} B \xrightarrow{g_\lambda} C_\lambda \to 0$ is obtained by taking the direct limits, A, B, C say, in $\mathcal{C}$ of the three directed sets, $(A_\lambda)_\Lambda, (B_\lambda)_\Lambda, (C_\lambda)_\Lambda$, of objects and then using the universal property of direct limits to obtain the "direct limit morphisms" $f \in \mathcal{C}(A, B)$ and $g \in \mathcal{C}(B, C)$. The condition is that, if each of the sequences $0 \to A_\lambda \xrightarrow{f_\lambda} B_\lambda \xrightarrow{g_\lambda} C_\lambda \to 0$ is exact, then so is the sequence $0 \to A \to B \to C \to 0$, which we would write as $0 \to \varinjlim A_\lambda \xrightarrow{\varinjlim f_\lambda} \varinjlim B_\lambda \xrightarrow{\varinjlim g_\lambda} \varinjlim C_\lambda \to 0$; it is enough, see, for example, [486, 2.8.6], that the morphism from A to B be a monomorphism.

Lemma E.1.3. (for example, [486, 2.8.6]) *The cocomplete abelian category $\mathcal{C}$ is AB5 iff whenever $C \in \mathcal{C}$, D is a subobject of C, Λ is a directed set and $C_\lambda \leq C$ ($\lambda \in \Lambda$) is a directed* *system of subobjects of* C *(that is, the morphisms in the directed system are the inclusions), then* $D \cap \sum_{\lambda}^{\to} C_\lambda = \sum_{\lambda}^{\to} (D \cap C_\lambda)$. *(We have written* $\underset{\to}{\sum}$ *instead of* $\sum$ *just for emphasis.)*

A **Grothendieck** category is an AB5 category with a generator, equivalently (given cocompleteness) a generating set of objects. For many results AB5 is enough but one notable result which does need the generator is the following.

Theorem E.1.4. (Gabriel–Popescu, [204]), *Let* $\mathcal{C}$ *be a Grothedieck category with* $\{G_i\}_i$ *for a generating set. Set* $G = \bigoplus_i G_i$ *and let* $T = \mathrm{End}(G)$. *Then the functor* $\mathcal{C} \longrightarrow T\text{-Mod}$ *which is defined on objects by* $C \mapsto \mathcal{C}(G, C)$ *is an embedding of* $\mathcal{C}$ *as a* **Giraud subcategory** *of* T-Mod. *That is, there is a hereditary torsion theory on* T-Mod *such that* $\mathcal{C}$ *is the corresponding quotient (sub)category (*$i\mathcal{C}_\tau$ *in the notation of 11.1.5).*

Theorem E.1.5. (see, for example, proof of [195, 5.21] [486, 3.4.2]) *If* $\mathcal{A}$ *is a small preadditive category and* $\mathcal{B}$ *is Grothendieck, then the functor category* $(\mathcal{A}, \mathcal{B})$ *is Grothendieck.*

Sometimes we use the fact (see [195, 3.3.5]) that any abelian category with a generator, in particular any Grothendieck category, is **well powered**; that is, the subobjects of any given object form a set, rather than a proper class.

Injective objects One of the main features of Grothendieck abelian categories is the existence of injective hulls. An object C of the abelian category $\mathcal{C}$ is **injective** if the contravariant representable functor $(-, C) : \mathcal{C} \longrightarrow \mathbf{Ab}$ is exact. That is, C is injective iff whenever $A \xrightarrow{i} B$ is a monomorphism, and $A \xrightarrow{f} C$ is a morphism, then there is a morphism $B \xrightarrow{g} C$ such that $gi = f$.

Lemma E.1.6. *An object* E *of an abelian category* $\mathcal{C}$ *is injective iff every embedding* $E \xrightarrow{j} D$ *from* E *is a split embedding.*

Proof. The direction $\Rightarrow$ is immediate from the definition. For the other direction, given a monomorphism $A \to B$ and a morphism $A \to E$ form the pushout and finish in the obvious way. □

An **injective hull** of C is an object, denoted $E(C)$, together with an embedding $C \xrightarrow{i} E(C)$ such that any embedding $C \xrightarrow{j} E'$ of C into an injective object E' factors through i, meaning that there is $E(C) \xrightarrow{k} E'$ such that $j = ki$. Existence of injective hulls of modules is due to Eckmann and Schopf, [168].

A subobject A' of an object A is said to be **essential in** A if, for every non-zero subobject A'' of A we have $A' \cap A'' \neq 0$. An object A is **uniform** if every non-zero subobject of A is essential in A, that is, if every two non-zero subobjects of A have non-empty intersection.

Proposition E.1.7. ([486, §3.10], [663, §5.2]) *Let* $C \xrightarrow{i} E$ *be a monomorphism in an abelian category* $\mathcal{C}$. *Then the following conditions are equivalent:*

(i) $C \xrightarrow{i} E$ *is an injective hull of* C;
(ii) E *is injective and* C *is essential in* E;
(iii) if $C \xrightarrow{j} E'$ *is any monomorphism of* C *into an injective object* E', *then there is a split embedding* $E \xrightarrow{k} E'$ *such that* $ki = j$.

In particular, any indecomposable injective object is uniform.

It is easy to check that an injective hull, $C \xrightarrow{i} E(C)$, of C, if it exists, is **unique to isomorphism over** C, meaning that if $C \xrightarrow{j} E'$ is also an injective hull of C, then there is an isomorphism $E(C) \xrightarrow{k} E'$ such that $ki = j$.

Theorem E.1.8. ([486, 3.10.10], [663, X.4.3]) *Let $\mathcal{C}$ be a Grothendieck abelian category. Then every object of $\mathcal{C}$ has an injective hull.*

A category $\mathcal{C}$ **has enough injectives** if every object of $\mathcal{C}$ embeds into an injective object. This is weaker than having injective hulls, for example, see the comment after 12.1.13.

Say that an object is **superdecomposable** (the term **continuous** was used in [495] and in some other references) if it is non-zero and has no indecomposable summand.

Theorem E.1.9. ([486, p. 335]) *Let $\mathcal{C}$ be a Grothendieck category and let $E \in \mathcal{C}$ be an injective object. Then $E = E\big(\bigoplus_{\lambda\in\Lambda} E_\lambda\big) \oplus E_c$, where each E_λ is indecomposable injective and where E_c is a superdecomposable injective.*

If also $E \simeq E\big(\bigoplus_{\mu\in M} E_\mu\big) \oplus E'$ with the E_μ indecomposable and E' superdecomposable, then $E' \simeq E_c$ and there is a bijection $\sigma : M \to \Lambda$ with $E_{\sigma\lambda} \simeq E_\lambda$ for each $\lambda \in \Lambda$.

Lemma E.1.10. *If $\mathcal{C}$ is a locally finitely presented abelian category, then $\mathcal{C}$ has just a set of indecomposable injective objects.*

Proof. Every indecomposable injective is the injective hull of a finitely generated object, so it will be enough to show that $\mathcal{C}^{\mathrm{fg}}$ is skeletally small. By E.1.13 every finitely generated object is an epimorphic image of a morphism from a finitely presented object. Since the kernel of an epimorphism is a subobject which determines the epimorphism, it is enough to show that every finitely presented object has just a set of subobjects. For this we need just that $\mathcal{C}$ is abelian and has a generating set, $\mathcal{G}$, of objects, since, if C is any object of $\mathcal{C}$ and if C' is a subobject of C, then C' is determined by the collection of morphisms f from objects G of $\mathcal{G}$ such that f factors through the inclusion of C' in C and, since $\bigcup\{(G, C) : G \in \mathcal{G}\}$ is a set, we are finished. □

An **injective cogenerator** for a category $\mathcal{C}$ is an injective module E such that every object of $\mathcal{C}$ embeds in some power of E.

Lemma E.1.11. *If $\mathcal{C}$ is a locally finitely generated Grothendieck category, then $\bigoplus_S E(S)$, where S runs over the isomorphism types of simple objects of $\mathcal{C}$ is an injective cogenerator for $\mathcal{C}$.*

The argument at [663, p. 21] applies to give this since every object in a locally finitely generated category is a sum of its finitely generated subobjects and since every finitely generated object has, by Zorn's Lemma, a simple factor object.

Finitely presented objects The familiar idea is that a structure is finitely presented if it can be generated by a finite set of elements such that the set of relations which are satisfied by that generating tuple are generated by a finite number of them. In situations where it makes sense to talk about "elements (of finite character)", for instance, in finitely accessible categories with cokernels (Section 16.1.1), the definition that we are about to give is, indeed, equivalent to this; we will prove this for modules.

The general definition is as follows. An object A of a category $\mathcal{C}$ is **finitely presented** if the representable functor $(A, -)$ commutes with direct limits, meaning that if $((C_\lambda)_\lambda, (f_{\lambda\mu} :$

$C_\lambda \to C_\mu)_{\lambda<\mu})$ is any directed system in $\mathcal{C}$ and if $(C, (f_\lambda : C_\lambda \to C)_\lambda)$ is the direct limit of this system, then the direct limit of the (clearly) directed system $\big((A, C_\lambda)_\lambda, (A, f_{\lambda\mu} : (A, C_\lambda) \to (A, C_\mu))_{\lambda<\mu}\big)$ in **Set** is $\big((A, C), (A, f_\lambda)_\lambda\big)$. We will use this definition only in categories which have arbitrary direct limits.

Example E.1.12. (also see E.1.20.) Usually one says that a module $M \in \text{Mod-}R$ is finitely presented if it has a finite generating set $a_1, \ldots, a_n$ and if the right R-module (a submodule of R^n) of tuples $\overline{r}$, such that the linear relations $\overline{a} \cdot \overline{r} = \sum_i a_i r_i = 0$ hold between these elements, is finitely generated, say by $\overline{r}_1, \ldots, \overline{r}_m$. This does amount to the same as the above definition.

To see this suppose, first, that M is finitely generated and finitely related as just above and let $(C_\lambda, f_{\lambda\mu})$ be a directed system in Mod-R with direct limit (C, f_λ). We must show that $\big((M, C), (M, f_\lambda)\big)$ has the universal property for the directed system (in **Ab**) $\big((M, C_\lambda), (M, f_{\lambda\mu})\big)$. So take an object D of **Ab** and a collection $g_\lambda : (M, C_\lambda) \to D$ of morphisms such that for every $\lambda \le \mu$ we have $g_\lambda = g_\mu(M, f_{\lambda\mu})$. It must be shown that there is a morphism $h : (M, C) \to D$ such that $g_\lambda = h(M, f_\lambda)$ for each $\lambda \in \Lambda$. So let $k \in (M, C)$: say $k(a_i) = c_i \in C$. Recalling the construction (see E.1.20) of C and the f_λ, it is clear that, for each i, there is some C_{λ_i} and some $c_{\lambda_i} \in C_{\lambda_i}$ such that $c_i = f_{\lambda_i}(c_{\lambda_i})$. Since there are only finitely many a_i, there is λ and some $c_{\lambda i} \in C_\lambda$ such that $c_i = f_\lambda(c_{\lambda i})$ for each $i = 1, \ldots, n$. Furthermore, since $k(a_i) = c_i$, it must be that $\overline{c} \cdot \overline{r}_j = \sum_i c_i r_{ij} = 0$ for $j = 1, \ldots, m$. So for each j there is, by construction of C, some $\mu_j \ge \lambda$ such that $\sum_i f_{\lambda\mu_j}(c_{\lambda i}) r_{ij} = 0$. Again, since there are just finitely many such relations, there is $\mu \ge \lambda$ such that, setting $c_{\mu i} = f_{\lambda\mu}(c_{\lambda i})$, we have $\sum_i c_{\mu i} r_{ij} = 0$ for each $j = 1, \ldots, m$. Since the $c_{\mu i}$ satisfy all the relations satisfied by the a_i there is a morphism $k' : M \to C_\mu$ defined by sending a_i to $c_{\mu i}$ and, therefore, such that $k = f_\mu k'$. Finally, set $h(k) = g_\mu(k')$. The verification that this defines a map h with the required properties and that this is the only possible way to define such a map is straightforward.

For the converse, suppose that $(M, -)$ commutes with direct limits. Let $(M_\lambda)_\Lambda$ be the directed system of finitely generated submodules of M (with morphisms the inclusions). The limit of this system is just M together with the inclusions i_λ of its finitely generated submodules M_λ. Then $1_M \in (M, M) = \varinjlim_\lambda (M, M_\lambda)$: therefore, from the construction, there is λ and $k' \in (M, M_\lambda)$ such that $1_M = (M, i_\lambda)(k') = i_\lambda k'$ from which it follows that $M = M_\lambda$ is finitely generated. We also know, from E.1.20, that M is the direct limit of a directed system of finitely presented modules and, by essentially the same argument, it follows that M is also finitely presented.

Say that an object C of a category $\mathcal{C}$ (with direct limits) is **finitely generated** if whenever $(C_\lambda)_\lambda$ is an upwards directed system of subobjects of C such that $C = \varinjlim_\lambda C_\lambda$, then there is λ_0 such that $C = C_{\lambda_0}$. Clearly any finitely presented object is finitely generated. It is equivalent (for example, [486, 3.5.7]) that $\varinjlim(C, C_\lambda) \to (C, \varinjlim C_\lambda)$ be a bijection for every directed system $(C_\lambda, f_{\lambda\mu})$ of monomorphisms $f_{\lambda\mu}$. The next lemma follows from the fact that the direct limit of a directed system $(C_\lambda)_\lambda$ may be represented as a epimorphic image of $\bigoplus_\lambda C_\lambda$.

Lemma E.1.13. *If $\mathcal{C}$ is an abelian category with a generating set $\mathcal{G}$ of finitely generated objects, then an object C is finitely generated iff there is a finite direct sum A of objects of $\mathcal{G}$ and an epimorphism $A \to C$. Hence any epimorphic image of a finitely generated object is finitely generated.*

It also follows that any finitely generated projective object is finitely presented.

The next couple of results are direct consequences of the definitions.

Lemma E.1.14. *Suppose that $\mathcal{C}$ is a locally finitely presented abelian category and let $C \in \mathcal{C}$. Then C is finitely presented iff whenever C is represented as a direct limit, $\varinjlim C_\lambda = C$, of a direct system, $\big((C_\lambda)_\lambda, (f_{\lambda\mu})_{\lambda<\mu}\big)$, of finitely presented objects C_λ, then there is λ such that the limit morphism $f_\lambda : C_\lambda \to C$ is an isomorphism.*

Lemma E.1.15. *If $\mathcal{C}$ is an additive category with a generating set, $\mathcal{G}$, of finitely presented objects, then an object C of $\mathcal{C}$ is:*

(a) finitely generated iff there is an exact sequence $G \to C \to 0$ with G a finite direct sum of objects in $\mathcal{G}$;
(b) finitely presented iff there is an exact sequence $H \to G \to C \to 0$ with G, H each a finite direct sum of objects in $\mathcal{G}$.

Lemma E.1.16. *Let $\mathcal{C}$ be a locally finitely presented abelian category and let $0 \to A \to B \xrightarrow{\pi} C \to 0$ be an exact sequence. Suppose that B is finitely presented. Then C is finitely presented iff A is finitely generated.*

Proof. Suppose first that A is finitely generated. Let $D = \varinjlim_\lambda D_\lambda$ be the direct limit of a directed system $((D_\lambda)_\lambda, (g_{\lambda\mu} : D_\lambda \to D_\mu)_{\lambda<\mu})$ and let $f : C \to D$ be a morphism. Since B is finitely presented the composition $f\pi$ factors through some D_λ; say $f' : B \to D_\lambda$ is such that $g_\lambda f' = f\pi$, where $g_\lambda : D_\lambda \to D$ is the map to the limit. Since the composition $A \to B \xrightarrow{f'} D_\lambda \xrightarrow{g_\lambda} D$ is 0 and A is finitely generated, there is some $\mu \geq \lambda$ such that $g_{\lambda\mu} f'.A = 0$, hence such that there is an induced $f'' : C \to D_\mu$ with $f''\pi = g_{\lambda\mu} f'$. It follows that $g_\mu f''\pi = f\pi$, hence that $g_\mu f'' = f$, as required.

For the converse suppose that C is finitely presented. Since $\mathcal{C}$ is locally finitely presented we have $A = \varinjlim_\lambda A_\lambda$ for some directed system of finitely presented objects A_λ. For each λ we have an exact sequence $0 \to A'_\lambda \to B \to C_\lambda \to 0$, say, where A'_λ is the image of $A_\lambda \to A$. Since, by 16.1.11, $\mathcal{C}$ is Grothendieck, the direct limit $0 \to A \to B \to \varinjlim C_\lambda \to 0$, of this directed system of exact sequences is also exact. Hence $C = \varinjlim C_\lambda$. Since C is finitely presented already some map $C_\lambda \to C$ to the limit is an isomorphism, so the inclusion of A'_λ in A is the identity and $A = A'_\lambda$ is indeed finitely generated, being the image of a finitely presented object. □

Lemma E.1.17. *If $\mathcal{C}$ is a locally finitely presented abelian category and if, in the exact sequence $0 \to A \to B \xrightarrow{g} C \to 0$, C is finitely presented and B is finitely generated, then A is finitely generated.*

To see this, choose an epimorphism $\pi : G \to B$ with G finitely presented. Then, by E.1.16, $K = \ker(g\pi)$ is finitely generated. Hence $A = \pi K$ is finitely generated.

Accessible/locally finitely presented/finitely generated/coherent categories If $\mathcal{C}$ is any category, then $\mathcal{C}^{\mathrm{fp}}$ denotes the full subcategory consisting of finitely presented objects of $\mathcal{C}$. The category $\mathcal{C}$ is **finitely accessible** if $\mathcal{C}^{\mathrm{fp}}$ is skeletally small and if every object in $\mathcal{C}$ is a direct limit of a directed system of objects from $\mathcal{C}^{\mathrm{fp}}$. If $\mathcal{C}$ also is complete (has products and kernels), equivalently [3, 2.47] is cocomplete (has arbitrary direct sums and cokernels), then it is **locally finitely presented**. In the additive situation some authors (including this one) have used the term "locally finitely presented" to mean just finitely accessible but it is better to use the terminology which is standard outside the additive case. It is easy to check (cf. E.1.12) that every category Mod-R, with R a small preadditive category, is locally finitely presented (see Sections 10.1.1, 10.1.3)

Say that $\mathcal{C}$ is **locally finitely generated** if $\mathcal{C}^{\mathrm{fg}}$ is skeletally small and if every object in $\mathcal{C}$ is a direct limit of a directed system of objects from $\mathcal{C}^{\mathrm{fg}}$, where $\mathcal{C}^{\mathrm{fg}}$ denotes the full subcategory consisting of finitely generated objects of $\mathcal{C}$. Since the image of a finitely generated object is finitely generated, in such a category every object is the directed sum of its finitely generated subobjects. Note that this implies that the finitely generated objects are generating in the sense of Appendix E, p. 707 and, indeed conversely, if the finitely generated objects are generating in that sense then they are also $\underrightarrow{\lim}$-generating.

Many of the basic results on these categories can be found in [89].

A finitely presented object C of $\mathcal{C}$ is **coherent** if every finitely generated subobject is finitely presented. Let $\mathcal{C}^{\mathrm{coh}}$ denote the full subcategory consisting of coherent objects of $\mathcal{C}$. The category $\mathcal{C}$ is **locally coherent** if $\mathcal{C}^{\mathrm{coh}}$ is skeletally small and if every object in $\mathcal{C}$ is a direct limit of a directed system of objects from $\mathcal{C}^{\mathrm{coh}}$; for an equivalent definition see after 16.1.4. A module category Mod-R, where R is a ring, is locally coherent iff R is a coherent ring (2.3.18). The statement for R a small preadditive category is [467, 4.1].

If $\mathcal{C}$ has a generating set of **noetherian** objects (that is, objects with the ascending chain condition on subobjects), then $\mathcal{C}$ is said to be **locally noetherian**.

Lemma E.1.18. *If $\mathcal{C}$ is an abelian category which is locally coherent, then every finitely presented, object of $\mathcal{C}$ is coherent.*

Proof. If C is finitely presented, then there is an epimorphism $f : D \to C$ with D coherent. By E.1.16, $K = \ker(f)$ is finitely generated. If $A \leq C$ is finitely generated, then (easily) there is a finitely generated, hence finitely presented, subobject, B say, of G with $fB = A$. Since (cf. 2.3.15) $K \cap B$ is finitely generated, by E.1.16 again, $A \simeq B/(K \cap B)$ is finitely presented. □

Proposition E.1.19. ([614, 2.2]) *The locally finitely presented abelian category $\mathcal{C}$ is locally coherent iff the full subcategory $\mathcal{C}^{\mathrm{fp}}$ of finitely presented objects is an abelian subcategory.*

Also see 16.1.4

It is also the case, see 16.1.11, that any locally finitely presented abelian category is Grothendieck.

Posets Let $(P, \leq)$ be a partially ordered set; a **poset**, P, for short. This may be regarded as a (small) category whose objects are the elements of P and with $\mathrm{card}(p, q) = 1$ if $p \leq q$ and $(p, q) = \emptyset$ if $p \nleq q$.

For $a, b \in P$ a **join** (or least upper bound) for a and b is an element $d \in P$ with $a \leq c$, $b \leq c$ and such that every element $d \in P$ with $a \leq d$ and $b \leq d$ satisfies $c \leq d$. Clearly, if this exists it is unique and is denoted $a \vee b$ or often, in an additive context, by $a + b$. Dually, a **meet**, $a \wedge b$, of a and b is an element c' with $c' \leq a$, $c' \leq b$ and with $d \leq a, b$ implying $d \leq c'$. A **lattice** is poset in which every pair of elements, a, b, has a join and a meet.

A lattice is **modular** if it satisfies the identity $a \wedge (c \vee b) = (a \wedge c) \vee b$ – in "additive" notation $a \wedge (c + b) = a \wedge c + b$ – whenever $a \geq b$. In a modular lattice, for all a, b, the intervals $[a + b, a]$ and $[b, a \wedge b]$ are isomorphic, via $c \in [a + b, a] \mapsto c \wedge b$. In, for example, the (modular) lattice of submodules of a module, this corresponds to one of the standard isomorphism theorems.

A lattice is **distributive** if it satisfies the identities $a + (b \wedge c) = (a + b) \wedge (a + c)$ and $a \wedge (b + c) = (a \wedge b) + (a \wedge c)$ (in fact, one is enough). This implies modularity and is a very strong condition in the additive context but does occur.

Both these conditions are preserved by surjective homomorphisms: a quotient of a modular, respectively distributive, lattice is modular, resp. distributive. By a **(homo)morphism** of lattices we mean one which preserves both meets and joins (and hence the partial order). The **kernel** of a homomorphism f with domain a lattice L, say, is the equivalence relation on L defined by $a \sim b$ iff $fa = fb$. This is a **congruence** on L, that is, $a \sim b$ implies $a \wedge c \sim b \wedge c$ and $a + c \sim b + c$ for all $a, b, c \in L$. Conversely, if $\sim$ is any congruence on L, then the set, $L/\sim$, of equivalence classes is a lattice under the (well-defined) induced operations and $a \mapsto a/\sim : L \to L/\sim$ is a surjective homomorphism of lattices.

A lattice is modular, respectively distributive, iff it does not contain a sublattice isomorphic to the first, resp. second, diagram below (see pretty well any book on lattice theory, for instance [249]).

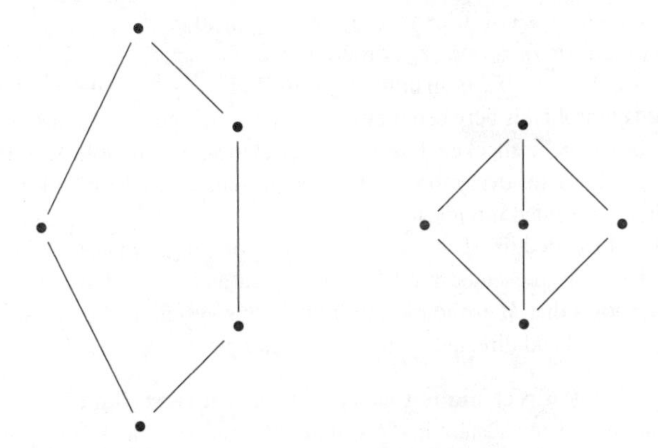

A lattice L with meet, $\wedge$, join, $+$, top, 1_L, and bottom, 0_L, is **complemented** if for every $a \in L$ there is some $a' \in L$ such that $a + a' = 1_L$ and $a \wedge a' = 0_L$.

I remark that almost always we will be dealing with lattices having a top, 1 (so $1 \geq a$ for all a) and bottom 0 (so $0 \leq a$ for all a).

Cantor sum of ordinals Any ordinal $\alpha > 0$ may be written in the form $n_1\omega^{\alpha_1} + n_2\omega^{\alpha_2} + \cdots + n_k\omega^{\alpha_k}$, where $n_1, \ldots, n_k$ are natural numbers and $\alpha_1 > \cdots > \alpha_k$ are ordinals. This expression is unique if the n_i all are non-zero. If α and β are ordinals (> 0), then write them in comparable forms, allowing zero coefficients: $\alpha = m_1\omega^{\gamma_1} + m_2\omega^{\gamma_2} + \cdots + m_k\omega^{\gamma_k}$, $\beta = n_1\omega^{\gamma_1} + n_2\omega^{\gamma_2} + \cdots + n_k\omega^{\gamma_k}$. The **Cantor sum** of α and β, written $\alpha \oplus \beta$, is the ordinal $(m_1 + n_1)\omega^{\gamma_1} + (m_2 + n_2)\omega^{\gamma_2} + \cdots + (m_k + n_k)\omega^{\gamma_k}$. This differs from the ordinary ordinal sum, $\alpha + \beta$, in which the leading term of β will swallow any smaller terms in the tail of α (possibly all α, for example, $1 + \omega = \omega$). For instance, $(\omega^\omega + 5\omega^2 + 2\omega + 1) + (\omega^3 + \omega) = \omega^\omega + \omega^3 + \omega$ but the Cantor sum of these ordinals is $\omega^\omega + \omega^3 + 5\omega^2 + 3\omega + 1$. For finite ordinals, both sums reduce to the usual sum.

Directed limits and colimits A partially ordered set $(\Lambda, \leq)$ is **directed (upwards)** if, for every $\lambda, \mu \in \Lambda$, there is $\nu \in \Lambda$ such that $\lambda \leq \nu$ and $\mu \leq \nu$, and is **directed downwards** if, for every $\lambda, \mu \in \Lambda$, there is $\rho \in \Lambda$ such that $\rho \leq \lambda$ and $\rho \leq \mu$. A subset Λ' is **cofinal** in Λ if for every $\lambda \in \Lambda$ there is $\mu \in \Lambda'$ with $\lambda \leq \mu$.

If Λ is any poset, a Λ-**indexed system** in the category $\mathcal{C}$ is given by specifying, for every $\lambda \in \Lambda$, and object $C_\lambda \in \mathcal{C}$ and, for every $\lambda \leq \mu$ in Λ, a morphism $f_{\lambda\mu} : C_\lambda \to C_\mu$ in $\mathcal{C}$ such that, whenever $\lambda \leq \mu \leq \nu$, we have $f_{\lambda\nu} = f_{\mu\nu} f_{\lambda\mu}$ and such that $f_{\lambda\lambda} = 1_{C_\lambda}$ for each $\lambda \in \Lambda$; in other words, a functor from the poset Λ, regarded as a category, to $\mathcal{C}$. In case Λ is directed we refer to an (**upwards** or **downwards**) (Λ-)**directed system** in $\mathcal{C}$.

For example, if M is an R-module, then consider the directed system in Mod-R consisting of all finitely generated submodules of M together with the inclusion morphisms between them. Then M is the direct limit of this system.

Example E.1.20. (also see E.1.12) Let M be any module. For each finite tuple, $\overline{a} = (a_1, \dots, a_n)$, of elements of M, and for each finite set, $\overline{r}_1, \dots, \overline{r}_m$, of linear relations $\sum_i a_i r_{ij} = 0$ $(\overline{r}_j = (r_{1j}, \dots, r_{nj}))$ satisfied by $\overline{a}$ in M, let $M_{\overline{a}, \overline{r}_1, \dots, \overline{r}_m}$ be the module $R^n / \sum_j \overline{r}_j R$ which is generated by the images, $d_1, \dots, d_n$, of the elements $e_1, \dots, e_n \in R^n$, where $e_i \in R^n$ has a 1 in the ith position and 0 elsewhere. Let $f_{\overline{a}, \overline{r}_1, \dots, \overline{r}_m} : M_{\overline{a}, \overline{r}_1, \dots, \overline{r}_m} \to M$ be the morphism which is (well) defined by sending d_i to a_i $(i = 1, \dots, n)$. Let Λ be the set of tuples of the form $\overline{a}, \overline{r}_1, \dots, \overline{r}_m$, ordered by $\overline{a}, \overline{r}_1, \dots, \overline{r}_m \leq \overline{b}, \overline{s}_1, \dots, \overline{s}_l$ iff $\overline{a}$ is an initial tuple of $\overline{b}$ and $\overline{r}_1, \dots, \overline{r}_m$ is an initial tuple of $\overline{s}_1, \dots, \overline{s}_l$. Then the modules $M_{\overline{a}, \overline{r}_1, \dots, \overline{r}_m}$, with the obvious morphisms between them, form a Λ-directed system of finitely presented modules which, as may be checked, has, for its direct limit, the module M together with the morphisms $f_{\overline{a}, \overline{r}_1, \dots, \overline{r}_m}$. Thus every module may be presented as the direct limit of a directed system of finitely presented modules.

One can avoid the details of this construction if one takes instead a small version of the category of finitely presented modules, regards it as a filtered system (see below for the definition), shows that M is the filtered limit of this system and then uses the essential equivalence of filtered and directed systems, for example, [3, pp. 13ff.].

A category $\mathcal{C}$ **has direct limits** (respectively **has inverse limits**) if every upwards (resp. downwards) directed system in $\mathcal{C}$ has a direct limit (resp. an inverse limit).

Example E.1.21. The categories **Set** and **Ab** have both direct and inverse limits, as has the category, Mod-R, of modules over any ring R. Sometimes the explicit description of direct limits in **Set** (equally in **Ab**) is used, so we give it here.

Let $\big((C_\lambda)_\lambda, (f_{\lambda\mu})_{\lambda\leq\mu}\big)$ be a directed system of sets and maps between them. Let C' be the disjoint union, $\bigcup_\lambda C_\lambda$, of the sets C_λ and let $\sim$ be the equivalence relation on C' which is generated by $c_\lambda (\in C_\lambda) \sim c_\mu (\in C_\mu)$ iff $f_{\lambda\mu} c_\lambda = c_\mu$. Let C be the quotient of C' by this equivalence relation and define the map $f_\lambda : C_\lambda \to C$ by $f_\lambda c_\lambda = c_\lambda / \sim$. Then it is easily checked that $\big(C, (f_\lambda)_\lambda\big)$ is the direct limit of this system.

This construction is also the basis of the construction of direct limits in many categories of sets-with-structure. In particular, the direct limit of a directed family of modules is just the direct limit of the underlying sets equipped with the inherited (and well-defined) module structure; one may use the direct sum in place of the union of sets and obtain the same result.

Filtered colimits This is a generalisation of the notion of direct limit. A **filtered category** is one which satisfies the conditions that, given any two objects, λ, μ, there is an object ν and there are morphisms $\lambda \to \nu$ and $\mu \to \nu$, and also that, given any parallel pair of morphisms $\alpha, \beta : \lambda \to \mu$, there is ν' and a morphism $\gamma : \mu \to \nu$ such that $\gamma\alpha = \gamma\beta$. So a directed category is just a filtered category where there is at most one morphism between any two objects. The definition of a filtered diagram in a category is the obvious generalisation of a directed diagram and a **filtered colimit** is the direct limit of such a diagram. It is the case that a category has filtered colimits iff it has direct limits, indeed, given a filtered diagram in any category, there is a directed diagram involving some of the objects and morphisms of the diagram which has, as its direct limit, the filtered colimit of the first diagram (see, for example, [3, pp. 13ff.]). There are occasions when a filtered diagram is the object which arises naturally; for example, in the representation, as at 16.1.2, of a flat object as a filtered colimit of projective objects.

Reduced products and ultraproducts (also see Section 3.3.1) Let $(C_\lambda)_{\lambda\in\Lambda}$ be a collection of objects of the category $\mathcal{C}$. Consider the **power set**, $\mathcal{P}(\Lambda)$, the set of all subsets of the index set Λ, as a boolean algebra under intersection, union and complement. Let $\mathcal{F}$ be a **filter** in $\mathcal{P}(\Lambda)$ (one also says a filter **on** Λ). That is, $\mathcal{F} \subseteq \mathcal{P}(\Lambda)$ is non-empty, is not all of $\mathcal{P}(\Lambda)$, is closed under finite intersection and is **upwards closed** in $\mathcal{P}(\Lambda)$, meaning that if $\Lambda' \in \mathcal{F}$ and $\Lambda' \subseteq \Lambda'' \subseteq \Lambda$, then $\Lambda'' \in \mathcal{F}$. Let $\mathcal{I} = \{\Lambda' \subseteq \Lambda : (\Lambda')^c \in \mathcal{F}\}$. Then $\mathcal{I}$ is a proper ideal of the boolean ring $\mathcal{P}(\Lambda)$ (where multiplication is intersection and addition is symmetric difference). Assuming that $\mathcal{C}$ has products, consider the directed (by canonical inclusions) system of products of the form $\prod_{\lambda\in\Lambda'} C_\lambda$ as Λ' ranges over $\mathcal{I}$; this system is directed since $\mathcal{I}$ is closed under finite union. All these are subobjects of $P = \prod_{\lambda\in\Lambda} C_\lambda$, hence so is their direct limit, K, which, if we are dealing with modules, consists of all elements, $(c_\lambda)_\lambda$, of the product P which are zero "$\mathcal{F}$-almost everywhere", meaning that $\{\lambda \in \Lambda : c_\lambda = 0\} \in \mathcal{F}$. Then the **reduced product** of the C_λ ($\lambda \in \Lambda$) by the filter $\mathcal{F}$ is the quotient object P/K. If $\mathcal{F}$ is an **ultrafilter**, that is, a maximal filter, equivalently a filter, $\mathcal{F}$, with the property that for every $\Lambda' \subseteq \Lambda$ either $\Lambda' \in \mathcal{F}$ or $\Lambda'^c \in \mathcal{F}$, then we refer to P/K as an **ultraproduct**. If all C_λ are isomorphic to some fixed object C, then the terms **reduced (ultra)power** of C are used. Łoś' Theorem, A.1.7, also see 3.3.3, relates the properties of an ultraproduct to the properties of its components C_λ.

Presheaves and sheaves Let X be a topological space. A **presheaf**, F, of sets over X is given by assigning, to each open subset U of X, a set, FU (called the set of **sections** of F over U), and to each inclusion of open subsets, $V \subseteq U$, a **restriction** map, $\mathrm{res}^F_{U,V} : FU \to FV$; in many classical examples of sheaves FU is some ring of functions on the set U and the map r_{UV} really is just restriction. It is required that $\mathrm{res}^F_{U,U} = 1_U$ and that if $W \subseteq V \subseteq U$, then $\mathrm{res}_{V,W}\mathrm{res}_{U,V} = \mathrm{res}_{U,W}$.

In other words, if the poset of open subsets of X is regarded as a category in the usual way (Appendix E, p. 712), then a presheaf is just a contravariant functor from this category to the category of sets. Replacing the category of sets by other categories one obtains presheaves of abelian groups, of rings, etc. A **morphism** between presheaves is just a natural transformation of functors (that is, for each open set one has a morphism, $f_U : FU \to GU$, in the image category such that, for every $V \subseteq U$, $\mathrm{res}^G_{UV} f_U = f_V \mathrm{res}^F_{UV}$). A **global section** is an element of FX.

A **sheaf** is a presheaf in which sections are determined locally. That is, given an open set U and sections $s, t \in FU$, then, if there is an open cover, $(U_i)_i$, of U such that the restrictions, $\mathrm{res}_{U,U_i}(s)$ and $\mathrm{res}_{U,U_i}(t)$ of s and t to each open set U_i are equal, $s = t$ (of course, in the additive case one may take $t = 0$). Moreover, given such an open cover of an open set and a corresponding collection, $(s_i \in FU_i)_i$, of **compatible** sections (that is, $\mathrm{res}_{U_i,U_i\cap U_j}(s_i) = \mathrm{res}_{U_j,U_i\cap U_j}(s_j)$ for all i, j), it is required that these "glue together" to give a section over U, meaning that there exists $s \in FU$ such that $\mathrm{res}_{U,U_i}(s) = s_i$ for each i.

A topological space X together with a sheaf, $\mathcal{O}_X$ of rings on it is called a **ringed space**. If the space has just one point, then we have simply a ring and, in the general case, a sheaf of rings can be viewed as a "continuously varying" family of rings. A **premodule** over a presheaf of rings is a presheaf, M, of abelian groups such that, for each open set $U \subseteq X$, $M(U)$ is a right R_U-module, where $R_U = \mathcal{O}_X(U)$ and such that, for every inclusion $V \subseteq U \subseteq X$ of open subsets of X, the restriction map, $\mathrm{res}^M_{U,V} : M(U) \to M(V)$, is a homomorphism of R_U-modules, where $M(V)$ is regarded as an R_U-module via $\mathrm{res}^{\mathcal{O}_X}_{U,V} : R_U \to R_V$. Such objects will be referred to as presheaves even though the image category is now varying. There is a natural notion of morphism (in the notation used a couple of paragraphs above, each f_U should be R_U-linear) so one obtains the category, PreMod-$\mathcal{O}_X$ of $\mathcal{O}_X$-premodules and, at least when $\mathcal{O}_X$ itself is a sheaf of rings, the full subcategory, Mod-$\mathcal{O}_X$, of $\mathcal{O}_X$-modules, the objects of which are those premodules which are sheaves. Such an object will be referred to as a **sheaf of modules** or as an **$\mathcal{O}_X$-module**.

Both PreMod-$\mathcal{O}_X$ and Mod-$\mathcal{O}_X$ are Grothendieck abelian categories, see, for example, [15, §I.3], [663, pp. 214–16], [486, §4.7].

Example E.1.22. The category of presheaves of k-vectorspaces over a space X, where k is a field, is equivalent to PreMod-$\mathcal{O}_X$, where $\mathcal{O}_X$ is taken to be the constantly k presheaf defined by $\mathcal{O}_X(U) = k$, $\mathrm{res}^{\mathcal{O}_X}_{U,V} = \mathrm{id}_k$ for $U \subseteq X$ open, $U \neq \emptyset$.

If F is a presheaf over X and $x \in X$, then the **stalk** of F at x is the direct limit of the FU, taken over the poset, directed by reverse inclusion, of open neighbourhoods, U, of x: $F_x = \varinjlim_{U \ni x} FU$.

Local endomorphism rings and Krull–Remak–Schmidt decomposition of objects A ring R is **local** if it has a unique maximal two-sided ideal, which must be the Jacobson radical, $J(R)$, of R and if the factor ring $R/J(R)$ is a division ring; equivalently if for each $r \in R$ either r or $1 - r$ is (say) right invertible.

In particular, if R is the endomorphism ring of an additive structure, then R is local iff the set of non-isomorphisms is closed under addition.

If $C \in \mathcal{C}$ is any non-zero object and if End(C) is local, then C must be indecomposable: for, otherwise, let $e = e^2 \in \mathrm{End}(C)$, $e \neq 1$ be projection to a non-zero proper factor of C. Then neither e nor $1 - e$ is invertible, so End(C) is not local.

Theorem E.1.23. *Suppose that $\mathcal{C}$ is an abelian category and that $E \in \mathcal{C}$ is an injective object of $\mathcal{C}$. Then E is indecomposable iff* End(E) *is a local ring.*

Proof. One direction has just been seen. For the other, suppose, conversely, that E is indecomposable and that $f \in \mathrm{End}(C)$ is not invertible. Then f cannot be monic, since then $\mathrm{im}(f) \simeq E$ would be a non-zero proper direct summand of E. Therefore $\ker(f)$ is essential in E. The "fixed point set" of f, the largest subobject, M, of E such that $f \upharpoonright M = \mathrm{id}_M$

must, therefore be 0. That is, $\ker(1 - f) = 0$, so $1 - f$ is monic, hence (by the first part of this argument) also an epimorphism. □

Theorem E.1.24. (Krull–Remak–Schmidt–Azumaya Theorem; for example, [183, 2.12], [486, 5.1.3], [626, 2.9.17]) *If $M = \bigoplus_I M_i$, where each M_i has local endomorphism ring, and if also $M = \bigoplus_J N_j$, where each N_j is indecomposable, then there is a bijection, $\sigma : I \to J$, such that $M_i \simeq N_{\sigma i}$ for all $i \in I$.*

Theorem E.1.25. (Exchange Property; for example, [183, 2.8], [486, 5.1.1], [626, 2.9.15]) *Suppose $M = \bigoplus_I M_i$ where each M_i has local endomorphism ring.*

If $M = N \oplus N'$, then some M_i is isomorphic to a direct summand of N.

If N is indecomposable, then there exists i such that $M_i \simeq N$ and $N \oplus \bigoplus_{j \neq i} M_j = M = M_i \oplus N'$.

Theorem E.1.26. (Crawley–Jonsson–Warfield Theorem; for example, [183, 2.55]) *Let $M = \bigoplus_{i \in I} M_i$, where each M_i is countably generated and has local endomorphism ring. Then any direct summand of M is isomorphic to $\bigoplus_{i \in J} M_i$ for some $J \subseteq I$.*

Beyond this, things can go wrong: for example, see the introduction to Section 2.5.

Theorem E.1.27. ([486, 5.3.4], [700, 4.2], [453, Prop. 1]) *Suppose that $\mathcal{C}$ is Grothendieck and that $E = E(\bigoplus_I C_i) \simeq E(\bigoplus_J D_j)$, where the C_i, D_j are indecomposable injective objects of $\mathcal{C}$. Then there is a bijection, $\sigma : I \to J$, such that $C_i \simeq D_{\sigma i}$ for all $i \in I$.*

If F is a direct summand of E, then there is $I' \subseteq I$ such that $E = F \oplus E(\bigoplus_{i \in I'} C_i)$.

There is a corresponding statement, 4.4.1, for pure-injective hulls of direct sums of indecomposable pure-injective objects.

Main examples

$A_2 = 1 \xrightarrow{\alpha} 2$: after 3.3.4; after 3.4.14; 12.1.1; 18.2.13; 18.2.15

$A_3 = 1 \xrightarrow{\alpha} 2 \xrightarrow{\beta} 3$: 4.5.19; 5.1.2; 5.1.9; 15.1.5

A_∞, the quiver $1 \longrightarrow 2 \longrightarrow 3 \longrightarrow \cdots$: 10.1.2; 10.1.4; 10.1.15; 14.1.2

Kronecker quiver $\widetilde{A_1}$: 1.1.4; 5.5.8; Section 8.1.2; 14.3.14; 15.1.1(b); 18.2.16

$\widetilde{A_3}$: Section 8.1.2

$k\widetilde{D_4}$: 1.1.3; 15.1.1(c)

$k[a, b : a^2 = 0 = b^2 = ab]$: 3.4.23; 4.3.73

Over $\mathbb{Z}$: p. xxv; 1.1.12; after 1.3.15; 2.1.1; after 2.1.5; after 2.1.28; 2.4.12; 3.4.13; 4.3.3; 4.3.22; 5.2.7; 5.3.51; 5.3.68; before 6.1.29; 6.1.36; 11.1.4; 12.1.4; after 12.3.8; 18.1.1

$\mathbb{Z}_n$ and similar: 1.1.1(1); 1.1.18

At p, $\mathbb{Z}_{(p)}$: p. xxv; 2.4.12; 4.1.2; after 4.2.14; 4.2.7; 4.4.40; 5.3.51; 6.1.6; 6.1.32; 14.2.1

At p, completed, $\overline{\mathbb{Z}_{(p)}}$: p. xxv; p. 22; after 3.4.7

$\begin{pmatrix} \mathbb{Z}_{(p)} & \mathbb{Q} \\ 0 & \mathbb{Q} \end{pmatrix}$: before 2.4.20; 8.2.49; 8.2.56

$\begin{pmatrix} \mathbb{Z}_{(p)} & p\mathbb{Z}_{(p)} \\ \mathbb{Z}_{(p)} & \mathbb{Z}_{(p)} \end{pmatrix}$: 8.2.49; 14.3.8; 14.3.9

$\begin{pmatrix} \mathbb{Z} & 2\mathbb{Z} \\ \mathbb{Z} & \mathbb{Z} \end{pmatrix}$: 14.3.8

Λ_2, Λ_n: Section 8.1.3

Canonical algebras: Section 8.1.4

Bibliography

[1] G. Abrams, F-rings need not be $\prod$-homogeneous, *Comm. Algebra*, **17(6)** (1989), 1495–1504.

[2] J. F. Adams, A variant of E. H. Brown's representability theorem, *Topology*, **10(3)** (1971), 185–198.

[3] J. Adámek and J. Rosický, *Locally Presentable and Accessible Categories*, London Math. Soc. Lect. Note Ser., Vol. 189, Cambridge University Press, 1994.

[4] M. Adelman, Abelian categories over additive ones, *J. Pure Appl. Algebra*, **3(2)** (1973), 103–117.

[5] S. Tempest Aldrich, E. E. Enochs, J. R. García Rozas and L. Oyonarte, Covers and envelopes in Grothendieck categories: flat covers of complexes with applications, *J. Algebra*, **243(2)** (2001), 615–630.

[6] L. Alonso Tarrío, A. Jeremías López, and M. J. Souto Salorio, Localization in categories of complexes and unbounded resolutions, *Canad. J. Math.*, **52(2)** (2000), 225–247.

[7] F. W. Anderson and K. R. Fuller, *Rings and Categories of Modules*, Grad. Texts in Math., Vol. 13, Springer, 1974.

[8] L. Angeleri Hügel, Direct summands of products, *Arch. Math.*, **78(1)** (2002), 12–23.

[9] L. Angeleri Hügel, S. Bazzoni and D. Herbera, A solution to the Baer splitting problem, *Trans. Amer. Math. Soc.*, **360(5)** (2008), 2409–2421.

[10] L. Angeleri Hügel, A key module over pure-semisimple hereditary rings, *J. Algebra*, **307(1)** (2007), 361–376.

[11] L. Angeleri Hügel, D. Happel and H. Krause, *Handbook of Tilting Theory*, London Math. Soc. Lect. Note Ser., Vol. 332, Cambridge University Press, 2007.

[12] L. Angeleri Hügel, J, Šaroch and J. Trlifaj, On the telescope conjecture for module categories, *J. Pure Appl. Algebra*, **212(2)** (2008), 297–310.

[13] L. Angeleri Hügel and J. Trlifaj, Direct limits of modules of finite projective dimension, pp. 27–44 in A. Facchini *et al.* (Eds), *Rings, Modules, Algebras and Abelian Groups*, Dekker, 2004.

[14] D. Arnold and R. Laubenbacher, Finitely generated modules over pullback rings, *J. Algebra*, **184(1)** (1996), 304–332.

[15] M. Artin, A. Grothendieck and J. L. Verdier, *Théorie des Topos et Cohomologie Etale des Schémas (SGA4)*, Lecture Notes in Mathematics, Vol. 269, Springer-Verlag, 1972.

[16] I. Assem, D. Simson and A. Skowroński, *Elements of the Representation Theory of Associative Algebras. 1: Techniques of Representation Theory*, London Math. Soc. Student Texts, Vol. 65, Cambridge University Press, 2006.

[17] M. Auslander, Coherent functors, pp. 189–231 in S. Eilenberg *et al.* (Eds.) *Proceedings of the Conference on Categorical Algebra, La Jolla, 1965*, Springer-Verlag, 1966.

[18] M. Auslander, *Representation Dimension of Artin Algebras*, Queen Mary College Mathematics Notes, Queen Mary College, University of London (1971).

[19] M. Auslander, Representation theory of Artin algebras II, *Comm. Algebra*, **1(4)** (1974), 269–310.

[20] M. Auslander, Large modules over Artin algebras, pp. 1–17 in A. Heller and M. Tierney (Eds.), *Algebra, Topology and Category Theory*, Academic Press, 1976.

[21] M. Auslander, Functors and objects determined by objects, pp. 1–244 in R. Gordon (Ed.), *Representation Theory of Algebras, Philadelphia 1976*, Marcel Dekker, 1978.

[22] M. Auslander, A functorial approach to representation theory, pp. 105–179 in *Representations of Algebras, Puebla 1980*, Lecture Notes in Mathematics, Vol. 944, Springer-Verlag, 1982.

[23] M. Auslander, Isolated singularities and existence of almost split sequences, (notes by Louise Unger), pp. 194–242 in *Representation Theory II, Groups and Orders, Ottawa 1984*, Lecture Notes in Mathematics, Vol. 1178, Springer-Verlag, 1986.

[24] M. Auslander and M. Bridger, Stable module theory, *Mem. Amer. Math. Soc.*, **94** (1969).

[25] M. Auslander and I. Reiten, Representation theory of Artin algebras III. Almost split sequences, *Comm. Algebra*, **3(3)** (1975), 239–294.

[26] M. Auslander and I. Reiten, Homologically finite subcategories, pp. 1–42 in H. Tachikawa and S. Brenner (Eds.) *Representations of Algebras and Related Topics*, London Mathematical Society Lecture Notes Series, Vol. 168, Cambridge University Press, 1992.

[27] M. Auslander, I. Reiten, S. Smalø, *Representation Theory of Artin Algebras*, Cambridge Studies in Advanced Mathematics, Vol. 36, Cambridge University Press, 1995.

[28] M. Auslander and S. Smalø, Preprojective modules over artin algebras, *J. Algebra*, **66(1)** (1980), 62–122.

[29] G. Azumaya, Locally pure-projective modules, *Contemp. Math.*, **124** (1992), 17–22.

[30] G. Azumaya and A. Facchini, Rings of pure global dimension zero and Mittag-Leffler modules, *J. Pure Appl. Algebra*, **62(2)** (1989), 109–122.

[31] D. Baer, Zerlegungen von Moduln und Injektive über Ringoiden, *Arch. Math.*, **36(1)** (1981), 495–501.

[32] D. Baer, Homological properties of wild hereditary Artin algebras, pp. 1–12 in V. Dlab *et al.* (Eds.), *Representation Theory I*, Lecture Notes in Mathematics, Vol. 1177, Springer-Verlag, 1986.

[33] D. Baer and H. Lenzing, A homological approach to representations of algebras I: the wild case, *J. Pure Appl. Algebra*, **24(3)** (1982), 227–233.

[34] D. Baer, H. Brune and H. Lenzing, A homological approach to representations of algebras II: tame hereditary algebras, *J. Pure Appl. Algebra*, **26(2)** (1982), 141–153.

[35] D. Baer, W. Geigle and H. Lenzing, The preprojective algebra of a tame hereditary Artin algebra, *Comm. Algebra*, **15(1–2)** (1987), 425–457.

[36] S. Balcerzyk, On the algebraically compact groups of I. Kaplansky, *Fund. Math.*, **44** (1957), 91–93.

[37] S. Balcerzyk, On factor groups of some subgroups of the complete direct sum of infinite cyclic groups, *Bull. Acad. Polon. Sci.*, **7** (1959), 141–142.

[38] J. T. Baldwin, P. C. Eklof and J. Trlifaj, ${}^{\perp}N$ as an abstract elementary class, *Ann. Pure Appl. Logic*, **149(1–3)** (2007), 25–39.

[39] J. T. Baldwin and A. H. Lachlan, On universal Horn classes categorical in some infinite power, *Algebra Universalis*, **3(1)** (1973), 98–111.

[40] J. T. Baldwin and R. N. McKenzie, Counting models in universal Horn classes, *Algebra Universalis*, **15** (1982), 359–384.

[41] P. Balmer, Presheaves of triangulated categories and reconstruction of schemes, *Math. Ann.*, **324(3)** (2002), 557–580.

[42] P. Balmer, The spectrum of prime ideals in tensor triangulated categories, *J. reine angew. Math.*, **588** (2005), 149–168.

[43] P. Balmer, Supports and filtrations in algebraic geometry and modular representation theory, *Amer. J. Math.*, **129(5)** (2007), 1227–1250.

[44] S. Baratella and M. Prest, Pure-injective modules over the dihedral algebras, *Comm. Algebra*, **25(1)** (1997), 11–31.

[45] S. Baratella and C. Toffalori, The theory of $\mathbb{Z}C(2)^2$-lattices is decidable, *Arch. Math. Logic*, **37(2)** (1998), 91–104.

[46] H. Bass, Finitistic dimension and a homological generalization of semi-primary rings, *Trans. Amer. Math. Soc.*, **95(3)** (1960), 466–488.

[47] W. Baur, Decidability and undecidability of theories of abelian groups with predicates for subgroups, *Compos. Math.*, **31(1)** (1975), 23–30.

[48] W. Baur, Undecidability of the theory of abelian groups with a subgroup, *Proc. Amer. Math. Soc.*, **55(1)** (1976), 125–128.

[49] W. Baur, Elimination of quantifiers for modules, *Israel J. Math.*, **25(1–2)** (1976), 64–70.

[50] W. Baur, On the elementary theory of quadruples of vector spaces, *Ann. Math. Logic*, **19(3)** (1980), 243–262.

[51] R. Bautista, On some tame and discrete families of modules, pp. 321–330 in [371].

[52] V. Bavula, Generalised Weyl algebras and their representations (in Russian), *Algebra and Analiz*, **4(1)** (1992), 75–97; English transl. in *St. Petersburg Math. J.*, **4(1)** (1993), 71–92.

[53] V. Bavula, The extension group of the simple modules over the first Weyl algebra, *Bull. London Math. Soc.*, **32(2)** (2000), 182–190.

[54] V. V. Bavula and D. A. Jordan, Isomorphism problems and groups of automorphisms for generalized Weyl algebras, *Trans. Amer. Math. Soc.*, **353(2)** (2000), 769–794.

[55] V. Bavula and F. van Oystaeyen, The simple modules of certain generalised crossed products, *J. Algebra*, **194(2)** (1997), 521–566.

[56] S. Bazzoni, Cotilting modules are pure-injective, *Proc. Amer. Math. Soc.*, **131(12)** (2003), 3665–3672.

[57] S. Bazzoni, n-cotilting modules and pure-injectivity, *Bull. London Math. Soc.*, **36(5)** (2004), 599–612.

[58] S. Bazzoni, When are definable classes tilting and cotilting classes?, *J. Algebra*, **320(12)** (2008), 4281–4299.

[59] S. Bazzoni and D. Herbera, One dimensional tilting modules are of finite type, *Algebras and Representation Theory*, **11(1)** (2008), 43–61.

[60] S. Bazzoni and J. Šťovíček, All tilting modules are of finite type, *Proc. Amer. Math. Soc.*, **135(12)** (2007), 3771–3781.

[61] K. I. Beidar, A. V. Mikhalev and G. E. Puninski, Logical aspects of ring and module theory, *Fundam. Prik. Mat.*, **1** (1995), 1–62.

[62] T. Beke, P. Karazeris and J. Rosický, When is flatness coherent?, *Comm. Algebra*, **33(6)** (2005), 1903–1912.

[63] A. Beligiannis, Relative homological algebra and purity in triangulated categories, *J. Algebra*, **227(1)** (2000), 268–361.

[64] A. Beligiannis, Purity and almost split sequences in abstract homotopy categories: a unified approach *via* Brown representability, *Algebras and Representation Theory*, **5(5)** (2002), 483–525.

[65] A. Beligiannis, On the Freyd categories of an additive category, *Homology, Homotopy and Apps.*, **2(1)** (2000), 147–185.

[66] A. Beligiannis, Auslander–Reiten triangles, Ziegler spectra and Gorenstein rings, *K-theory*, **32(1)** (2004), 1–82.

[67] A. Beligiannis and H. Krause, Realizing maps between modules over Tate cohomology rings, *Contribut. Algebra. Geom.*, **44(2)** (2003), 451–466.

[68] A. Beligiannis and I. Reiten, Homological and homotopical aspects of torsion theories, *Mem. Amer. Math. Soc.*, **188** (2007).

[69] D. J. Benson, Representations and Cohomology I: Basic Representation Theory of Finite Groups and Associative Algebras, Cambridge Studies in Advanced Math., Vol. 30, Cambridge University Press, 1991.

[70] D. Benson, Infinite dimensional modules for finite groups, pp. 251–272 in [371].

[71] D. J. Benson, J. F. Carlson and J. Rickard, Complexity and varieties for infinitely generated modules, *Math. Proc. Camb. Phil. Soc.*, **118** (1995), 223–243.

[72] D. Benson, J. Carlson and J. Rickard, Thick subcategories of the stable module category, *Fund. Math.*, **153** (1997), 59–80.

[73] D. Benson and G. Gnacadja, Phantom maps and purity in modular representation theory I, *Fund. Math.*, **161** (1999), 27–91.

[74] D. Benson and G. Gnacadja, Phantom maps and purity in modular representation theory II, *Algebras and Representation Theory*, **4(4)** (2001), 395–404.

[75] D. Benson and H. Krause, Generic idempotent modules for a finite group, *Algebras and Representation Theory*, **3(4)** (2000), 337–346.

[76] D. Benson and H. Krause, Pure injectives and the spectrum of the cohomology ring of a finite group, *J. reine angew. Math.*, **542** (2002), 23–51.

[77] D. Benson, H. Krause and S. Schwede, Realizability of modules over Tate cohomology, *Trans. Amer. Math. Soc.*, **356(9)** (2004), 3621–3668.

[78] D. Benson, H. Krause and S. Schwede, Introduction to realizability of modules over Tate cohomology, pp. 81–97 in *Representation of Algebras and Related Topics*, Fields Inst. Comms., Vol. 45, The Fields Institute, 2005.

[79] I. N. Bernstein, I. M. Gelfand and V. A. Ponomarev, Coxeter functors and Gabriel's theorem (in Russian), *Uspehi Mat. Nauk*, **28(2)** (1973) 19–33, translated in *Russian Math. Surveys*, **28** (1973), 17–32.

[80] K. Bessenrodt, H. H. Brungs and G. Törner, Right chain rings, Part 1, Schriftenreihe des Fach. Math., Universität Duisburg, 1990.

[81] L. Bican, R. El Bashir and E. Enochs, All modules have flat covers, *Bull. London Math. Soc.*, **33(4)** (2001), 385–390.

[82] T. Blossier, Subgroups of the additive group of a separably closed field, *Ann. Pure Appl. Logic*, **134(2–3)** (2005), 169–216.

[83] M. Bökstedt and A. Neeman, Homotopy limits in triangulated categories, *Compos. Math.*, **86(2)** (1993), 209–234.

[84] F. Borceux, *Handbook of Categorical Algebra*, (3 Vols.) Cambridge University Press, 1994.

[85] F. Borceux and C. Quinteiro, A theory of enriched sheaves, *Cahiers Topologie Géom. Différentielle Catég.*, **37** (1996), 145–162.

[86] F. Borceux and J. Rosický, Purity in algebra, *Algebra Universalis*, **56(1)** (2007), 17–35.

[87] F. Borceux and G. Van den Bossche, *Algebra in a Localic Topos with Applications to Ring Theory*, Lecture Notes in Mathematics, Vol. 1038, Springer-Verlag, 1983.

[88] F. Borceux and B. Veit, On the left exactness of orthogonal reflections, *J. Pure Appl. Algebra*, **49(1)** (1987), 33–42.

[89] S. Breitsprecher, Lokal endlich präsentierbare Grothendieck–Kategorien, *Mitt. math. Sem. Giessen*, **85** (1970), 1–25.

[90] S. Brenner, Endomorphism algebras of vector spaces with distinguished sets of subspaces, *J. Algebra*, **6(1)** (1967), 100–114.

[91] S. Brenner, Some modules with nearly prescribed endomorphism rings, *J. Algebra*, **23(2)** (1972), 250–262.

[92] S. Brenner, Decomposition properties of some small diagrams of modules, *Symp. Math. Ist. Naz. Alta. Mat.*, **13** (1974), 127–141.

[93] S. Brenner and M. C. R. Butler, Generalisations of the Bernstein–Gel'fand–Ponomarev reflection functors, pp. 103–169 in *Representation Theory II*, Lecture Notes in Mathematics, Vol. 832, Springer, 1980.

[94] S. Brenner and C. M. Ringel, Pathological modules over tame rings, *J. London Math. Soc. (2)*, **14(2)** (1976), 207–215.

[95] E. H. Brown, Cohomology theories, *Annals of Math.*, **75(3)** (1962), 467–484.

[96] K. S. Brown, Homological criteria for finiteness, *Comment. Math. Helvetici*, **50** (1975), 129–135.

[97] H. Brune, Some left pure semisimple ringoids which are not right pure semisimple, *Comm. Algebra*, **7(17)** (1979), 1795–1803.

[98] H. Brune, On the global dimension of the functor category ((mod-R)^{op}, Ab) and a theorem of Kulikov, *J. Pure Appl. Algebra*, **28(1)** (1983), 31–39.

[99] A. B. Buan, H. Krause and Ø. Solberg, Support varieties: an ideal approach, *Homology, Homotopy and Apps.*, **9(1)** (2007), 45–74.

[100] S. Buechler, The classification of small weakly minimal sets, Part III, *J. Symbolic Logic*, **53(3)** (1988), 975–979.

[101] K. Burke, Some model-theoretic properties of functor categories for modules, Doctoral Thesis, University of Manchester, 1994.

[102] K. Burke, Some connections between finitely generated functors and pp-types, University of Manchester, preprint, 1997.

[103] K. Burke, A generalised character theory for modules, *Comm. Algebra*, **28(1)** (2000), 265–297.

[104] K. Burke, Co-existence of Krull filtrations, *J. Pure Appl. Algebra*, **155(1)** (2001), 1–16.

[105] K. Burke, Pure-injective hulls of expanding string modules, University of Manchester, preprint, 1998.

[106] K. Burke and M. Prest, Rings of definable scalars and biendomorphism rings, pp. 188–201 in *Model Theory of Groups and Automorphism Groups*, London Math. Soc. Lect. Note Ser., Vol. 244, Cambridge University Press, 1997.

[107] K. Burke and M. Prest, The Ziegler and Zariski spectra of some domestic string algebras, *Algebras and Representation Theory*, **5(3)** (2002), 211–234.

[108] M. C. R. Butler, The 2-adic representations of Klein's four group, pp. 197–203 in *Proceedings of the 2nd International Conference on Theory of Groups*, Lecture Notes in Mathematics, Vol. 372, Springer-Verlag, 1974.

[109] M. C. R. Butler and G. Horrocks, Classes of extensions and resolutions, *Phil. Trans. Roy. Soc. London Ser. A*, **254(1039)** (1961), 155–222.

[110] M. C. R. Butler and C. M. Ringel, Auslander–Reiten sequences with few middle terms and applications to string algebras, *Comm. Algebra*, **15(1–2)** (1987), 145–179.

[111] A. I. Cárceles and J. L. García, Pure semisimple finitely accessible categories and Herzog's criterion, *J. Algebra Appl.*, **6(6)** (2007), 1001–1025.

[112] H. Cartan and S. Eilenberg, *Homological Algebra*, Princeton University Press, 1956.

[113] R. Carter, G. Segal and I. Macdonald, *Lectures on Lie Groups and Lie Algebras*, London Math. Soc. Student Texts, Vol. 32, Cambridge University Press, 1995.

[114] C. C. Chang and H. J. Keisler, *Model Theory*, North-Holland, Amsterdam, 1973 (and later editions).

[115] R. Chartrand and T. Kucera, Deissler rank complexity of powers of indecomposable injective modules, *Notre Dame J. Formal Logic*, **35(3)** (1994), 398–402.

[116] S. U. Chase, Direct products of modules, *Trans. Amer. Math. Soc.*, **97(3)** (1960), 457–473.

[117] Z. Chatzidakis and A. Pillay, Generic structures and simple theories, *Ann. Pure Appl. Logic*, **95(1–3)** (1998), 71–92.

[118] T. Cheatham and E. Enochs, Injective hulls of flat modules, *Comm. Algebra*, **8(20)** (1980), 1989–1998.

[119] J. D. Christensen, B. Keller and A. Neeman, Failure of Brown representability in derived categories, *Topology*, **40(6)** (2001), 1339–1361.

[120] J. D. Christensen and N. P. Strickland, Phantom maps and homology theories, *Topology*, **37(2)** (1998), 339–364.

[121] S. Cittadini and C. Toffalori, Comparing first order theories of modules over group rings, *Math. Log. Quart.*, **48(1)** (2002), 147–156.

[122] S. Cittadini and C. Toffalori, Comparing first order theories of modules over group rings II: decidability, *Math. Log. Quart.*, **48(4)** (2002), 483–498.

[123] S. Cittadini and C. Toffalori, On pairs of free modules over a Dedekind domain, *Arch. Math. Logic*, **45(1)** (2006), 75–95.

[124] F. U. Coelho, E. N. Marcos, H. A. Merklen and A. Skowroński, Module categories with infinite radical square zero are of finite type, *Comm. Algebra*, **22(11)** (1994), 4511–4517.

[125] F. U. Coelho, E. N. Marcos, H. A. Merklen and A. Skowroński, Module categories with infinite radical cube zero, *J. Algebra*, **183(1)** (1996), 1–23.

[126] P. M. Cohn, On the free product of associative rings, *Math. Z.*, **71(1)** (1959), 380–398.

[127] P. M. Cohn, *Free Rings and their Relations*, London Math. Soc. Monographs, Vol. 2, Academic Press, 1971.

[128] P. M. Cohn, Progress in free associative algebras, *Israel J. Math*, **19** (1974), 109–151.

[129] P. M. Cohn, The affine scheme of a general ring, pp. 197–211 in *Applications of Sheaves*, Lecture Notes in Mathematics, Vol. 753, Springer-Verlag, 1979.

[130] P. M. Cohn, *Free Ideal Rings and Localization in General Rings*, New Mathematical Monographs, Vol. 3, Cambridge University Press, 2006.

[131] R. R. Colby and K. R. Fuller, Tilting, cotilting and serially tilted rings, *Comm. Algebra*, **18(5)** (1990), 1595–1615.

[132] R. Colpi and K. R. Fuller, Cotilting dualities, pp. 345–358 in [11].

[133] R. Colpi and J. Trlifaj, Tilting modules and tilting torsion theories, *J. Algebra*, **178(2)** (1995), 614–634.

[134] F. Couchot, Sous-modules purs et modules de type cofini, pp. 198–208 in *Séminaire d'Algebre P. Dubriel*, Lecture Notes Mathematics, Vol. 641, Springer-Verlag, 1978.

[135] F. Couchot, Pure-injective hulls of modules over valuation rings, *J. Pure Appl. Algebra*, **207(1)** (2006), 63–76.

[136] F. Couchot, RD-flatness and RD-injectivity, *Comm. Algebra*, **34(10)** (2006), 3675–3689.

[137] S. C. Coutinho and M. P. Holland, Differential operators on smooth varieties, pp. 201–219 in *Séminaire d'Algèbra Paul Dubriel et Marie-Paul Malliavin*, Lecture Notes in Mathematics, Vol. 1404, Springer, 1989.

[138] W. Crawley-Boevey, On tame algebras and bocses, *Proc. London Math. Soc.*, **56(3)** (1988), 451–483.

[139] W. Crawley-Boevey, Tame algebras and generic modules, *Proc. London Math. Soc.*, **63(2)** (1991), 241–265.

[140] W. Crawley-Boevey, Regular modules for tame hereditary algebras, *Proc. London Math. Soc.*, **62(3)** (1991), 490–508.

[141] W. Crawley-Boevey, Modules of finite length over their endomorphism rings, pp. 127–184 in *Representations of Algebras and Related Topics*, London Math. Soc. Lect. Note Ser. Vol. 168, Cambridge University Press, 1992.

[142] W. Crawley-Boevey, Lectures on representations of quivers, (1992), at http://www.amsta.leeds.ac.uk/~pmtwc/.

[143] W. Crawley-Boevey, Additive functions on locally finitely presented Grothendieck categories, *Comm. Algebra*, **22(5)** (1994), 1629–1639.

[144] W. Crawley-Boevey, Locally finitely presented additive categories, *Comm. Algebra*, **22(5)** (1994), 1641–1674.

[145] W. Crawley-Boevey, Infinite-dimensional modules in the representation theory of finite-dimensional algebras, pp. 29–54 in I. Reiten, S. Smalø and Ø. Solberg (Eds.), *Algebras and Modules I*, Canadian Math. Soc. Conf. Proc., Vol. 23, American Mathematical Society, 1998.

[146] S. Crivei and J. L. García, Gruson–Jensen duality for idempotent rings, *Comm. Algebra*, **33(11)** (2005), 3949–3966.

[147] S. Crivei, M. Prest and G. Reynders, Model theory of comodules, *J. Symbolic Logic*, **69(1)** (2004), 137–142.

[148] A. P. Dean and F. Okoh, Extensionless modules of infinite rank, pp. 121–124 in I. Reiten *et al.* (Eds.), *Algebra and Modules II*, Canad. Math. Soc. Conf. Proc., Vol. 24, Canadian Mathematical Society, 1998.

[149] P. Dellunde, F. Delon and F. Point, The theory of modules of separably closed fields 1, *J. Symbolic Logic*, **67(3)** (2002), 997–1015.

[150] P. Dellunde, F. Delon and F. Point, The theory of modules of separably closed fields 2, *Ann. Pure Applied Logic*, **129(1–3)** (2004), 181–210.

[151] W. Dicks and A. H. Schofield, On semihereditary rings, *Comm. Algebra*, **16(6)** (1988), 1243–1274.

[152] J. Dixmier, *Algèbres Enveloppantes*, Gauthier-Villars, 1974, Engl. transl. *Enveloping Algebras*, North-Holland, 1977.

[153] V. Dlab and C. M. Ringel, Decomposition of modules over right uniserial rings, *Math. Z.*, **129(3)** (1972), 207–230.

[154] V. Dlab and C. M. Ringel, Indecomposable representations of graphs and modules, *Mem. Amer. Math. Soc.*, **173** (1976).

[155] P. Donovan and M. R. Freislich, *The Representation Theory of Finite Graphs and Associated Algebras*, Carleton Math. Lect. Notes, Vol. 5, Carleton University, 1973.

[156] P. Dowbor and A. Skowroński, On the representation type of locally bounded categories, *Tsukuba J. Math.*, **10(1)** (1986), 63–72.

[157] Yu. Drozd, On generalised uniserial rings, *Math. Zametki*, **18(5)** (1975), 705–710.

[158] Du, X., Generic modules for tilted algebras, *Chinese Sci. Bull.*, **42(3)** (1997), 177–180.

[159] N. Dubrovin, The rational closure of group rings of left-ordered groups, Schriftenreihe des Fach. Math., Universität Duisburg, 1994.

[160] N. Dubrovin and G. Puninski, Classifying projective modules over some semilocal rings, *J. Algebra Appl.*, **6(5)** (2007), 839–865.

[161] N. V. Dung and J. L. García, Additive categories of locally finite representation type, *J. Algebra*, **238(1)** (2001), 200–238.

[162] N. V. Dung and J. L. García, Copure semisimple categories and almost split maps. *J. Pure Appl. Algebra*, **188(1)** (2004), 73–94.

[163] N. V. Dung and J. L. García, Endofinite modules and pure semisimple rings, *J. Algebra*, **289(2)** (2005), 574–593.

[164] N. V. Dung and J. L. García, Endoproperties of modules and local duality, *J. Algebra*, **316(1)** (2007), 368–391.

[165] M. Mehdi Ebrahimi, Equational compactness of sheaves of algebras on a Noetherian locale, *Algebra Universalis*, **16** (1983), 318–330.

[166] S. Ebrahimi Atani, On pure-injective modules over pullback rings, *Comm. Algebra*, **28(9)** (2000), 4037–4069.

[167] S. Ebrahimi Atani, On secondary modules over pullback rings, *Comm. Algebra*, **30(6)** (2002), 2675–2685.

[168] B. Eckmann and A. Schopf, Über injektive Moduln, *Arch. Math.*, **4(2)** (1953), 75–78.

[169] S. Eilenberg, Abstract description of some basic functors, *J. Indian Math. Soc.*, **24** (1960), 231–234.

[170] D. Eisenbud and J. C. Robson, Modules over Dedekind prime rings, *J. Algebra*, **16(1)** (1970), 67–85.

[171] D. Eisenbud and J. C. Robson, Hereditary noetherian prime rings, *J. Algebra*, **16(1)** (1970), 86–104.

[172] P. Ç. Eklof and E. Fisher, The elementary theory of abelian groups, *Ann. Math. Logic*, **4(2)** (1972), 115–171.

[173] P. C. Eklof and I. Herzog, Model theory of modules over a serial ring, *Ann. Pure Appl. Logic*, **72(2)** (1995), 145–176.

[174] P. Eklof and G. Sabbagh, Model-completions and modules, *Ann. Math. Logic*, **2(3)** (1971), 251–295.

[175] P. Eklof and J. Trlifaj, How to make Ext vanish, *Bull. London Math. Soc.*, **33(1)** (2001), 41–51.

[176] H. B. Enderton, *Elements of Set Theory*, Academic Press, 1977.

[177] E. Enochs, A note on absolutely pure modules, *Canad. Math. Bull.*, **19** (1976), 361–362.

[178] E. Enochs, Flat covers of flat cotorsion modules, *Proc. Amer. Math. Soc.*, **92(2)** (1984), 179–184.

[179] E. Enochs and L. Oyonarte, Flat covers and cotorsion envelopes of sheaves, *Proc. Amer. Math. Soc.*, **130(5)** (2002), 1285–1292.

[180] D. Evans, A. Pillay and B. Poizat, Le groupe dans le groupe, (English transl.), *Algebra and Logic*, **29(3)** (1990), 244–252.

[181] A. Facchini, Decompositions of algebraically compact modules, *Pacific J. Math.*, **116(1)** (1985), 25–37.

[182] A. Facchini, Relative injectivity and pure-injective modules over Prüfer rings, *J. Algebra*, **110(2)** (1987), 380–406.

[183] A. Facchini, *Module Theory: Endomorphism Rings and Direct Sum Decompositions in Some Classes of Modules*, Progress in Math., Vol. 167, Birkhäuser, 1998.

[184] A. Facchini and G. Puninksi, Σ-pure-injective modules over a serial ring, pp. 145–162 in *Abelian Groups and Modules*, Kluwer, 1995.

[185] C. Faith, Rings with ascending chain condition on annihilators, *Nagoya Math. J.*, **27(1)** (1966), 179–191.

[186] C. Faith, *Algebra I, Ring Theory*, Springer, 1976.

[187] C. Faith, *Algebra II, Ring Theory*, Springer, 1976.

[188] C. Faith and E. A. Walker, Direct sum representations of injective modules, *J. Algebra*, **5(2)** (1967), 203–221.

[189] S. N. Fedin, The pure dimension of rings and modules, *Russ. Math. Surv.*, **37** (1982), 170–171.

[190] E. Fisher, Abelian Structures, Yale University, 1974/5, partly published as Abelian Structures I, pp. 270–322 in *Abelian Group Theory*, Lecture Notes in Mathematics, Vol. 616, Springer-Verlag, 1977.

[191] J. Fisher, The tensor product of functors, satellites and derived functors, *J. Algebra*, **8(3)** (1968), 277–294.

[192] D. J. Fieldhouse, Pure theories, *Math. Ann.*, **184(1)** (1969), 1–18.

[193] J. Franke, On the Brown representability theorem for triangulated categories, *Topology*, **40(4)** (2001), 667–680.

[194] B. Franzen, Algebraic compactness of filter quotients, pp. 228–241 in *Abelian Group Theory*, Lectures Notes in Math., Vol. 874, Springer-Verlag, 1981.

[195] P. Freyd, *Abelian Categories*, Harper and Row, 1964.

[196] P. Freyd, Representations in abelian categories, pp. 95–120 in *Proceedings of the Conference on Categorical Algebra, La Jolla, 1965*, Springer-Verlag, 1966.
[197] E. M. Friedlander and A. Suslin, Cohomology of finite group schemes over a field, *Invent. Math.*, **127(2)** (1997), 209–270.
[198] L. Fuchs, Infinite *Abelian Groups*, Vol. I, Academic Press. 1970.
[199] L. Fuchs and L. Salce, *Modules over Valuation Domains*, Lecture Notes in Pure and Appl. Math., Vol. 97, Marcel Dekker, 1985.
[200] K. R. Fuller, On rings whose left modules are direct sums of finitely generated modules, *Proc. Amer. Math. Soc.*, **54(1)** (1976), 39–44.
[201] W. Fulton and J. Harris, *Representation Theory*, Grad. Texts in Math., Vol. 129, Springer-Verlag, 1991.
[202] P. Gabriel, Des catégories abéliennes, *Bull. Soc. Math. France*, **90** (1962), 323–448.
[203] P. Gabriel, Unzerlegbare Darstellungen I, *Manus. Math.*, **6(1)** (1972), 71–103.
[204] P. Gabriel and N. Popescu, Caractérisation des catégories abéliennes avec générateurs et limites inductives exactes, *C. R. Acad. Sci. Paris*, **258** (1964), 4188–4191.
[205] P. Gabriel and F. Ulmer, *Lokal präsentierbare Kategorien*, Lecture Notes in Mathematics, Vol. 221, Springer-Verlag, 1971.
[206] P. Gabriel and M. Zisman, *Calculus of Fractions and Homotopy Theory*, Springer-Verlag, 1967.
[207] S. Garavaglia, Direct product decomposition of theories of modules, *J. Symbolic Logic*, **44(1)** (1979), 77–88.
[208] S. Garavaglia, Decomposition of totally transcendental modules, *J. Symbolic Logic*, **45(1)** (1980), 155–164.
[209] S. Garavaglia, Dimension and rank in the model theory of modules, preprint, University of Michigan, 1979, revised 1980.
[210] J. L. García and J. Martínez, Purity through Gabriel's functor rings, *Bull. Soc. Math. Belgique*, **45** (1993), 115–141.
[211] J. L. García and D. Simson, On rings whose flat modules form a Grothendieck category, *Colloq. Math.*, **73** (1997), 115–141.
[212] G. A. Garkusha, Grothendieck categories (in Russian), pp. 1–68 in *Algebra and Analysis (2)*, **13(2)** (2001), English transl. in *St. Petersburg Math. J.*, **13(2)** (2002).
[213] G. A. Garkusha, FP-injective and weakly quasi-Frobenius rings (in Russian), *Zap. Nauchn. Sem. POMI*, **265** (1999), 110–129, English transl. in *J. Math. Sci.*, **112(3)** (2002), 4303–4312.
[214] G. Garkusha, Relative homological algebra for the proper class ω_f, *Comm. Algebra*, **32(10)** (2004), 4043–4072.
[215] G. A. Garkusha, A note on almost regular group rings (in Russian), *Zap. Nauchn. Sem. POMI*, **281** (2001), 128–132, English transl. in *J. Math. Sci.*, **120(4)** (2004), 1561–1562.
[216] G. Garkusha, Classifying finite localizations of quasi-coherent sheaves, Algebra i Analiz, to appear.
[217] G. A. Garkusha and A. I. Generalov, Grothendieck categories as quotient categories of (R-mod, **Ab**) (in Russian), *Fund. Appl. Math.*, **7(4)** (2001), 983–992, English language version of preprint available.

[218] G. A. Garkusha and A. I. Generalov, Duality for categories of finitely presented modules (in Russian), *Algebra Analiz.*, **11(6)** (1999), 139–152, English transl. in *St. Petersburg Math. J.*, **11(6)** (2000), 1–11.

[219] G. Garkusha and M. Prest, Injective objects in triangulated categories, *J. Algebra Appl.*, **3(4)** (2004), 367–389.

[220] G. Garkusha and M. Prest, Triangulated categories and the Ziegler spectrum, *Algebras and Representation Theory*, **8(4)** (2005), 499–523.

[221] G. Garkusha and M. Prest, Classifying Serre subcategories of finitely presented modules, *Proc. Amer. Math. Soc.*, **136(3)** (2008), 761–770.

[222] G. Garkusha and M. Prest, Reconstructing projective schemes from Serre subcategories, *J. Algebra*, **319(3)** (2008), 1132–1153.

[223] G. Garkusha and M. Prest, Torsion classes of finite type and spectra, pp. 393–412 in G. Cortiñas *et al.*, *K-theory and Noncommutative Geometry* European Math. Soc., 2006.

[224] W. Geigle, The Krull–Gabriel dimension of the representation theory of a tame hereditary artin algebra and applications to the structure of exact sequences, *Manus. Math.*, **54(1–2)** (1985), 83–106.

[225] W. Geigle and H. Lenzing, A class of weighted projective lines arising in representation theory of finite-dimensional algebras, pp. 265–297 in *Singularities, Representations of Algebras and Vector Bundles*, Lecture Notes in Mathematics, Vol. 1273, Springer-Verlag, 1987.

[226] W. Geigle and H. Lenzing, Perpendicular categories with applications to representations and sheaves, *J. Algebra*, **144(2)** (1991), 273–343.

[227] G. Geisler, Zur Modelltheorie von Moduln, Doctoral dissertation, Universität Freiburg, 1994.

[228] C. Geiss and H. Krause, On the notion of derived tameness, *J. Algebra and Apps.*, **1(2)** (2002), 133–157.

[229] S. I. Gelfand and Yu. I. Manin, *Homological Algebra*, Springer, 1994.

[230] I. M. Gelfand and V. A. Ponomarev, Indecomposable representations of the Lorenz group (in Russian), *Uspehi Mat. Nauk*, **23(2)** (1968), translated in *Russian Math. Surveys*, **23(2)** (1968), 1–58.

[231] O. Gerstner, Algebraische Kompaktheit bei Faktorgruppen von Gruppen ganzzahliger Abbildungen, *Manus. Math.*, **11(2)** (1974), 103–109.

[232] J. M. Hernández Gil, Pureza en Categorías de Grothendieck, Thesis de Licenciatura, Universidad de Murcia, 1994.

[233] S. Glaz, *Commutative Coherent Rings*, Lecture Notes in Math., Vol. 1371, Springer, 1989.

[234] R. Göbel and J. Trlifaj, *Approximations and Endomorphism Algebras of Modules*, W. de Gruyter, 2006.

[235] J. S. Golan, Finiteness conditions on filters of left ideals, *J. Pure Appl. Algebra*, **3(3)** (1973), 251–259.

[236] J. S. Golan, *Structure Sheaves over a Non-commutative Ring*, Lecture Notes in Pure Appl. Math., Marcel Dekker, 1975.

[237] J. L. Gómez Pardo and P. A. Guil Asensio, Endomorphism rings of completely pure-injective modules, *Proc. Amer. Math. Soc.*, **124(8)** (1996), 2301–2309.

[238] J. L. Gómez Pardo and P. A. Guil Asensio, Indecomposable decompositions of finitely presented pure-injective modules, *J. Algebra*, **192(1)** (1997), 200–208.

[239] J. L. Gómez Pardo and P. A. Guil Asensio, Chain conditions on direct summands and pure quotient modules, pp. 195–203 in *Interactions between Ring Theory and Representations of Algebras (Murcia)*, Lecture Notes in Pure Appl. Math, 210, Marcel Dekker, 2000.

[240] K. R. Goodearl, *Von Neumann Regular Rings*, Pitman, London, 1979, second edn, Krieger Publishing, 1991.

[241] K. R. Goodearl, Localisation and splitting in hereditary noetherian prime rings, *Pacific J. Math.*, **53(1)** (1974), 137–151.

[242] K. R. Goodearl, Artinian and noetherian modules over regular rings, *Comm. Algebra*, **8(5)** (1980), 477–504.

[243] K. R. Goodearl and R. B. Warfield, *An Introduction to Noncommutative Noetherian Rings*, London Math. Soc. Student Texts, Vol. 16, Cambridge University Press, 1989.

[244] R. Gordon and J. C. Robson, *Krull Dimension*, Mem. Amer. Math. Soc., Vol. 133, American Mathematical Society, 1973.

[245] R. Gordon and J. C. Robson, The Gabriel dimension of a module, *J. Algebra*, **29(3)** (1974), 459–473.

[246] V. E. Govorov, On flat modules (in Russian), *Sibirsk. Mat. Z.*, **6** (1965), 300–304.

[247] N. Granger, Transfer Report, University of Manchester, 1998.

[248] N. Granger, Stability, simplicity and the model theory of bilinear forms, Doctoral Thesis, University of Manchester, 1999.

[249] G. Grätzer, *General Lattice Theory*, Birkhäuser, 1978.

[250] P. A. Griffith, *Infinite Abelian Group Theory*, University of Chicago Press, 1970.

[251] M. Grime, Finite dimensional modules and perpendicular subcategories, preprint, University of Bristol, 2007.

[252] P.-P. Grivel, Catégories dérivées et foncteurs dérivés, pp. 1–108 in A. Borel *et al.*, *Algebraic D-Modules, Perspectives in Mathematics*, Vol. 2, Academic Press, 1987.

[253] L. Gruson, Simple coherent functors, pp. 156–159 in *Representations of Algebras*, Lecture Notes in Mathematics, Vol. 488, Springer-Verlag, 1975.

[254] L. Gruson and C. U. Jensen, Modules algébriquement compact et foncteurs $\varprojlim^{(i)}$, *C. R. Acad. Sci. Paris*, **276** (1973), 1651–1653.

[255] L. Gruson and C. U. Jensen, Deux applications de la notion de L-dimension, *C. R. Acad. Sci. Paris*, **282** (1976), 23–24.

[256] L. Gruson and C. U. Jensen, Dimensions cohomologiques reliées aux foncteurs $\varprojlim^{(i)}$, pp. 243–294 in M.-P. Malliavin (Ed.) *Sèminaire d'algébre Paul Dubreil et Marie-Paule Malliavin* Lecture Notes in Mathematics, Vol. 867, Springer-Verlag, 1981.

[257] P. A. Guil Asensio and I. Herzog, Left cotorsion rings, *Bull. London Math. Soc.*, **36(3)** (2004), 303–309.

[258] P. A. Guil Asensio and I. Herzog, Sigma-cotorsion rings, *Adv. Math.*, **191(1)** (2005), 11–28.

[259] P. A. Guil Asensio and I. Herzog, Indecomposable flat cotorsion modules, *J. London Math. Soc.*, **76(3)** (2007), 797–811.

[260] P. A. Guil Asensio and I. Herzog, Model-theoretic aspects of Σ-cotorsion modules, *Ann. Pure Appl. Logic*, **146(1)** (2007), 1–12.

[261] P. A. Guil Asensio and I. Herzog, Pure-injectivity in the category of flat modules, pp. 155–166 in Contemp. Math, Vol. 419, American Math. Soc., 2006.

[262] P. A. Guil Asensio and D. Simson, Indecomposable decompositions of pure-injective objects and the pure-semisimplicity, *J. Algebra*, **244(2)** (2001), 478–491.
[263] Y. Han, Controlled wild algebras, *Proc. London Math. Soc. (3)*, **83(2)** (2001), 279–298.
[264] D. Happel, *Triangulated Categories in the Representation Theory of Finite Dimensional Algebras*, London Math. Soc. Lect. Note Ser., Vol. 119, Cambridge University Press, 1988.
[265] D. Happel and B. Huisgen-Zimmermann, Viewing finite dimensional representations through infinite dimensional ones, *Pacific J. Math.*, **187(1)** (1999), 65–89.
[266] D. Happel and C. M. Ringel, Tilted algebras, *Trans. Amer. Math. Soc.*, **274(2)** (1982), 399–443.
[267] M. Harada and Y. Sai, On categories of indecomposable modules I, *Osaka J. Math.*, **7(2)** (1970), 323–344.
[268] R. Hartshorne, *Residues and Duality*, with an appendix by P. Deligne, Lecture Notes in Mathematics, Vol. 20 Springer-Verlag, 1966.
[269] R. Hartshorne, *Algebraic Geometry*, Graduate Texts in Math., Vol. 52, Springer-Verlag, 1977.
[270] R. Hartshorne, Coherent functors, *Adv. in Math.*, **140(1)** (1998), 44–94.
[271] F. Haug, On preservation of stability for finite extensions of abelian groups, *Math. Logic, Q.*, **40** (1994), 14–26.
[272] C. Herrmann, C. U. Jensen and H. Lenzing, Applications of model theory to representations of finite-dimensional algebras, *Math. Z.*, **178(1)** (1981), 83–98.
[273] I. Herzog, Modules with few types, *J. Algebra*, **149(2)** (1992), 358–370.
[274] I. Herzog, Elementary duality of modules, *Trans. Amer. Math. Soc.*, **340(1)** (1993), 37–69.
[275] I. Herzog, The Auslander–Reiten translate, *Contemp. Math.*, **130** (1992), 153–165.
[276] I. Herzog, A test for finite representation type, *J. Pure Appl. Algebra*, **95(2)** (1994), 151–182.
[277] I. Herzog, Finitely presented right modules over a left pure-semisimple ring, *Bull. London Math. Soc.*, **26(4)** (1994), 333–338.
[278] I. Herzog, The Ziegler spectrum of a locally coherent Grothendieck category, *Proc. London Math. Soc.*, **74(3)** (1997), 503–558.
[279] I. Herzog, The endomorphism ring of a localised coherent functor, *J. Algebra*, **191(1)** (1997), 416–426.
[280] I. Herzog, Model theory of modules, pp. 66–72 in *Logic Colloquium '95 (Haifa)*, Lecture Notes Logic, Vol. 11, Springer, 1998.
[281] I. Herzog, Locally simple objects, pp. 341–351 in P. Eklof and R. Göbel (Eds.), *Proceedings of the International Conference on Abelian Groups and Modules, Dublin*, Birkhäuser, 1999.
[282] I. Herzog, The pseudo-finite dimensional representations of $sl(2, k)$, *Selecta Mathematica*, **7(2)** (2001), 241–290.
[283] I. Herzog, Pure injective envelopes, *J. Algebra Appl.*, **2(4)** (2003), 397–402.
[284] I. Herzog, Finite matrix topologies, *J. Algebra*, **282(1)** (2004), 157–171.
[285] I. Herzog, Applications of duality to the pure-injective envelope, *Algebras and Representation Theory*, **10(2)** (2007), 135–155.
[286] I. Herzog, Contravariant functors on the category of finitely presented modules, *Israel J. Math.*, **167(1)**, (2008), 347–410.

[287] I. Herzog, The phantom cover of a module, *Adv. Math.*, **215(1)** (2007), 220–249.

[288] I. Herzog and S. L'Innocente, The nonstandard quantum plane, *Ann. Pure Appl. logic*, **156(1)** (2008), 78–85.

[289] I. Herzog and V. Puninskaya, The model theory of divisible modules over a domain (in Russian), *Fundamentalnaya i Prikladnaya Matematika*, **2(2)** (1996), 563–594.

[290] I. Herzog and Ph. Rothmaler, Modules with regular generic types, I–III, pp. 138–176 in A. Nesin and A. Pillay (Eds.), *The Model Theory of Groups*, University of Notre Dame Press, 1989.

[291] I. Herzog and Ph. Rothmaler, Modules with regular generic types, Part IV, *J. Symbolic Logic*, **57(1)** (1992), 193–199.

[292] I. Herzog and Ph. Rothmaler, Pure projective approximations, *Math. Proc. Camb. Philos. Soc.*, **146** (2009), 83–94.

[293] I. Herzog and Ph. Rothmaler, when cotorsion modules are pure injective, preprint.

[294] M. Hochster, Prime ideal structure in commutative rings, *Trans. Amer. Math. Soc.*, **142** (1969), 43–60.

[295] T. J. Hodges, Noncommutative deformations of type-A Kleinian singularities, *J. Algebra*, **161(2)** (1993), 271–290.

[296] W. Hodges, *Model Theory, Encyclopedia of Mathematics*, Vol. 42, Cambridge University Press, 1993.

[297] W. Hodges, Relative categoricity in abelian groups pp. 157–168 in *Models and Computability*, London Math. Soc. Lect. Note Ser., Vol. 259, Cambridge University Press, 1999.

[298] I. Hodkinson and S. A. Shelah, A construction of many uncountable rings using SFP domains and Aronszajn trees, *Proc. London Math. Soc. (3)*, **67(3)** (1993), 449–492.

[299] H. Holm and P. Jørgensen, Covers, precovers and purity, *Ill. J. Math.*, to appear.

[300] M. J. Hopkins, Global methods in homotopy theory, pp. 73–96 in *Homotopy Theory (Durham, 1985)*, London Math. Soc. Lect. Note Ser., Vol. 117, Cambridge University Press, 1987.

[301] M. Hovey, *Model Categories*, Math. Surv. Monographs, Vol. 63, American Mathematical Society, 1999.

[302] M. Hovey, Classifying subcategories of modules, *Trans. Amer. Math. Soc.*, **353(8)** (2001), 3181–3191.

[303] M. Hovey, Cotorsion pairs, model category structures and representation theory, *Math. Z.*, **241(3)** (2002), 553–592.

[304] M. Hovey, J. H. Palmieri and N. P. Strickland, Axiomatic stable homotopy theory, *Mem. Amer. Math. Soc.*, **128(610)**, 1997.

[305] M. Hovey, K. Lockridge, G. Puninski, The generating hypothesis in the derived category of a ring, *Math. Z.*, **256(4)** (2007), 789–800.

[306] E. Hrushovski and F. Point, On von Neumann regular rings with an automorphism, *J. Algebra*, **315(1)** (2007), 76–120.

[307] H. Hu, Dualities for accessible categories, pp. 211–242 in R. A. G. Seely (Ed.) *Category Theory 1991*, Canad. Math. Soc. Conf. Proc., Vol. 13, Canadian Mathematical Society, 1992.

[308] B. Huisgen-Zimmermann, Purity, algebraic compactness, direct sum decompositions and representation type, pp. 331–367 in [371].

[309] B. Huisgen-Zimmermann, The finitistic dimension conjectures – a tale of 3.5 decades, pp. 501–517 in A. Facchini and C. Menini (Eds.), *Abelian Groups and Modules*, Kluwer, 1995.

[310] B. Huisgen-Zimmermann and F. Okoh, Direct products of modules and the pure semisimplicity conjecture, *Comm. Algebra*, **29(1)** (2001), 271–276.

[311] B. Huisgen-Zimmermann and M. Saorín, Direct sums of representations as modules over their endomorphism rings, *J. Algebra*, **250(1)** (2002), 67–89.

[312] A. Hulanicki, The structure of the factor group of an unrestricted sum by the restricted sum of abelian groups, *Bull. Acad. Polon. Sci.*, **10** (1962), 77–80.

[313] J. E. Humphreys, *Introduction to Lie Algebras and Representation Theory*, Graduate Texts in Math., Vol. 9, Springer, 1972.

[314] J. E. Humphreys, Highest weight modules for semisimple Lie algebras, pp. 72–103 in V. Dlab and P. Gabriel (Eds.) *Representation Theory I*, Lecture Notes in Mathematics, Vol. 831, Springer, 1980.

[315] J. R. Isbell, Epimorphisms and dominions, pp. 232–246 in S. Eilenberg *et al.* (Eds.) *Proceedings of the Conference on Categorical Algebra, La Jolla, 1965*, Springer-Verlag, 1966.

[316] J. R. Isbell, Epimorphisms and dominions III, *Amer. J. Math.*, **90(4)** (1968), 1025–1030.

[317] J. R. Isbell, Epimorphisms and dominions IV, *J. London Math. Soc.*, **1(1)** (1969), 265–273.

[318] B. Iversen, *Cohomology of Sheaves*, Springer-Verlag, 1986.

[319] A. V. Jategaonkar, *Localization in Noetherian Rings*, London Math. Soc. Lect. Note Ser., Vol. 98, Cambridge University Press, 1986.

[320] C. U. Jensen, *Les Foncteurs Dérivés de* $\varinjlim$ *et leurs Applications en Théorie des Modules*, Lecture Notes in Mathematics, Vol. 254, Springer, 1972.

[321] C. U. Jensen and H. Lenzing, Homological dimension and representation type of algebras under base field extension, *Manus. Math.*, **39(1)** (1982), 1–13.

[322] C. U. Jensen and H. Lenzing, Algebraic compactness of reduced products and applications to pure global dimension, *Comm. Algebra*, **11(3)** (1983), 305–325.

[323] C. U. Jensen and H. Lenzing, *Model Theoretic Algebra; with Particular Emphasis on Fields, Rings and Modules*, Gordon and Breach, 1989.

[324] C. U. Jensen and B. Zimmermann-Huisgen, Algebraic compactness of ultrapowers and representation type, *Pacific J. Math.*, **139(2)** (1989), 251–265.

[325] P. Jørgensen, Brown representability for stable categories, *Math. Scand.*, **85(2)** (1999), 195–218.

[326] P. Jørgensen, Spectra of modules, *J. Algebra*, **244(2)** (2001), 744–784.

[327] P. Jørgensen, The homotopy category of complexes of projective modules, *Adv. Math.*, **193(1)** (2005), 223–232.

[328] M. Kamensky, Applications of model theory to fields and modules, Doctoral Thesis, Hebrew University, 2006.

[329] M. Kamensky, The model completion of the theory of modules over finitely generated commutative algebras, *J. Symbolic Logic*, to appear.

[330] I. Kaplansky, *Infinite Abelian Groups*, University of Michigan Press, 1954. Also revised edition, Ann Arbor, 1969.

[331] I. Kaplansky, *Commutative Rings*, University of Chicago Press, 1970.

[332] M. Kashiwara and P. Schapira, *Categories and Sheaves*, Grund. math. Wiss., Vol. 332, Springer, 2006.

[333] S. Kasjan, On the problem of axiomatization of tame representation type, *Fund. Math.*, **171** (200), 53–67.

[334] S. Kasjan and J. A. de la Pe na, Galois coverings and the problem of axiomatization of the representation type of algebras, *Extracta Math.*, **20(2)** (2005), 137–150.

[335] Y. Katsov, Model theory of functors: axiomatizability problems, *J. Pure Appl. Algebra*, **119(3)** (1997), 285–296.

[336] B. Keller, Derived categories and their uses, pp. 671–702 *in* M. Hazewinkel (Ed.), *Handbook of Algebra*, Vol. 1, Elsevier, 1996.

[337] G. M. Kelly, On the radical of a category, *J. Austral. Math. Soc.*, **4** (1964), 299–307.

[338] G. M. Kelly, *Basic Concepts in Enriched Category Theory*, London Math. Soc. Lect. Note Ser., Vol. 64, Cambridge University Press, 1982.

[339] O. Kerner and A. Skowroński, On module categories with nilpotent infinite radical, *Compos. Math.*, **77(3)** (1991), 313–333.

[340] R. Kielpinski, On Γ-pure injective modules, *Bull. Polon. Acad. Sci. Math.*, **15** (1967), 127–131.

[341] R. Kielpinski and D. Simson, On pure homological dimension, *Bull. Acad. Pol. Sci.*, **23** (1975), 1–6.

[342] L. Klingler and L. Levy, Wild torsion modules over Weyl algebras and general torsion modules over HNPs, *J. Algebra*, **172(2)** (1995), 273–300.

[343] S. König and A. Zimmermann, *Derived Equivalences for Group Rings*, Lecture Notes in Mathematics, Vol. 1685, Springer, 1998.

[344] M. Ja. Komarnitski, Axiomatisability of certain classes of modules connected with a torsion (in Russian), *Mat. Issled.*, **56** (1980), 92–109 and 160–161.

[345] G. Krause and T. H. Lenagan, Transfinite powers of the Jacobson radical, *Comm. Algebra*, **7(1)** (1979), 1–8.

[346] H. Krause, Dualizing rings and a characterization of finite representation type, *Compt. Rend. Acad. Sci. Paris*, **332** (1996), 507–510.

[347] H. Krause, The endocategory of a module, pp. 419–432 *in* R. Bautista *et al.* (Eds.), *Representation Theory of Algebras*, Canad. Math. Soc. Conf. Proc., Vol. 18, Canadian Mathematical Society, 1996.

[348] H. Krause, An axiomatic description of a duality for modules, *Adv. Math.*, **130(2)** (1997), 280–286.

[349] H. Krause, The spectrum of a locally coherent category, *J. Pure Applied Algebra*, **114(3)** (1997), 259–271.

[350] H. Krause, Stable equivalence preserves representation type, *Comm. Math. Helv.*, **72(2)** (1997), 266–284.

[351] H. Krause, Constructing large modules over artin algebras, *J. Algebra*, **187(2)** (1997), 413–421.

[352] H. Krause, On the nilpotency of the Jacobson radical for noetherian rings, *Arch. Math.*, **70(6)** (1998), 435–437.

[353] H. Krause, Stable equivalence and representation type, pp. 387–391 in I. Reiten *et al.* (Eds.), *Algebra and Modules II*, Canad. Math. Soc. Conf. Proc., Vol. 24, 1998.

[354] H. Krause, Representation type and stable equivalence of Morita type for finite dimensional algebras, *Math. Z.*, **229(4)** (1998), 601–606.

[355] H. Krause, Exactly definable categories, *J. Algebra*, **201(2)** (1998), 456–492.

[356] H. Krause, Finitistic dimension and Ziegler spectrum, *Proc. Amer. Math. Soc.*, **126(4)** (1998), 983–987.

[357] H. Krause, Generic modules over artin algebras, *Proc. London Math. Soc. (3)*, **76(2)** (1998), 276–306.

[358] H. Krause, *The spectrum of a module category*, Habilitationsschrift, Universität Bielefeld, 1997, published as *Mem. Amer. Math. Soc.*, **707** (2001).

[359] H. Krause, Functors on locally finitely presented additive categories, *Colloq. Math.*, **75** (1998), 105–132.

[360] H. Krause, Decomposing thick subcategories of the stable module category, *Math. Ann.*, **313(1)** (1999), 95–108.

[361] H. Krause, Smashing subcatgories and the telescope conjecture – an algebraic approach, *Invent. Math.*, **139(1)** (2000), 99–133.

[362] H. Krause, Coherent functors in stable homotopy theory, *Fundamenta Math.*, **173** (2002), 33–56.

[363] H. Krause, Finite versus infinite dimensional representations – a new definition of tameness, pp. 393–403 *in* [371].

[364] H. Krause, Brown representability and flat covers, *J. Pure Appl. Algebra*, **157(1)** (2001), 81–86.

[365] H. Krause, On Neeman's well generated triangulated categories, *Documenta Math.*, **6** (2001), 119–125.

[366] H. Krause, A Brown representability theorem via coherent functors, *Topology*, **41(4)** (2002), 853–861.

[367] H. Krause, Coherent functors and covariantly finite subcategories, *Algebras and Representation Theory*, **6(5)** (2003), 475–499.

[368] H. Krause, Cohomological quotients and smashing localizations, *Amer. J. Math.*, **127(6)** (2005), 1191–1246.

[369] H. Krause, with an appendix by S. Iyengar, Thick subcategories of modules over commutative noetherian rings, *Math. Ann.*, **340(4)** (2008), 733–747.

[370] H. Krause and U. Reichenbach, Endofiniteness in stable homotopy theory, *Trans. Amer. Math. Soc.*, **353(1)** (2001), 157–173.

[371] H. Krause and C. M. Ringel (Eds.), *Infinite Length Modules*, Birkhäuser, 2000.

[372] H. Krause and M. Saorín, On minimal approximations of modules, pp. 227–236 in E. Green and B. Huisgen-Zimmermann (Eds.), *Proceedings of the Conference on Representations of Algebras, Seattle 1997*, Contemp. Math., Vol. 229, American Mathematical Society, 1998.

[373] H. Krause and Ø. Solberg, Filtering modules of finite projective dimension, *Forum Math.*, **15(3)** (2003), 377–393.

[374] H. Krause and Ø. Solberg, Applications of cotorsion pairs, *J. London Math. Soc.*, **68(3)** (2003), 631–650.

[375] H. Krause and G. Zwara, Stable equivalence and generic modules, *Bull. London Math. Soc.*, **32(5)** (2000), 615–618.

[376] T. G. Kucera, Totally transcendental theories of modules: decompositions of modules and types, *Ann. Pure Appl. Logic*, **39(3)** (1988), 239–272.

[377] T. G. Kucera, Generalizations of Deissler's minimality rank, *J. Symbolic Logic*, **53(1)** (1988), 269–283.

[378] T. G. Kucera, Positive Deissler rank and the complexity of injective modules, *J. Symbolic Logic*, **53(1)** (1988), 284–293.

[379] T. G. Kucera, Explicit descriptions of injective envelopes: generalizations of a result of Northcott, *Comm. Algebra*, **17(11)** (1989), 2703–2715.

[380] T. G. Kucera, The regular hierarchy in modules, *Comm. Algebra*, **21(12)** (1993), 4615–4628.

[381] T. G. Kucera, Explicit descriptions of the indecomposable injective modules over Jategaonkar's rings, *Comm. Algebra*, **30(12)** (2002), 6023–6054.

[382] T. G. Kucera and M. Prest, Imaginary modules, *J. Symbolic Logic*, **57(2)** (1992), 698–723.

[383] T. G. Kucera and M. Prest, Four concepts from "geometrical" stability theory in modules, *J. Symbolic Logic*, **57(2)** (1992), 724–740.

[384] T. G. Kucera and Ph. Rothmaler, Pure-projective modules and positive constructibility, *J. Symbolic Logic*, **65(1)** (2000), 103–110.

[385] A. Laradji, Algebraic compactness of reduced powers over commutative perfect rings, *Arch. Math.*, **64(4)** (1995), 299–303.

[386] A. Laradji, A generalization of algebraic compactness, *Comm. Algebra*, **23(10)** (1995), 3589–3600.

[387] A. Laradji, On duo rings, pure semisimplicity and finite representation type, *Comm. Algebra*, **25(12)** (1997), 3947–3952.

[388] A. Laradji, On a weak form of equational compactness, *Algebra Universalis*, **39(1)** (1998), 71–80.

[389] A. Laradji, Inverse limits of algebras as retracts of their direct products, *Proc. Amer. Math. Soc.*, **131(4)** (2003), 1007–1010.

[390] D. Lazard, Sur les modules plats, *C. R. Acad. Sci. Paris*, **258** (1964), 6313–6316.

[391] D. Lazard, Autour de la platitude, *Bull. Soc. Math. France*, **97** (1969), 81–128.

[392] S. Lefschetz, *Algebraic Topology*, Colloquium Publications, 1942.

[393] T. Leinster, *Higher Operads, Higher Categories*, London Math. Soc. Lect. Note Ser., Vol. 298, Cambridge University Press, 2004.

[394] H. Lenzing, Endlich präsentierbare Moduln, *Arch. Math.*, **20(3)** (1969), 262–266.

[395] H. Lenzing, Homological transfer from finitely presented to infinite modules, pp. 734–761 in *Abelian Group Theory*, Lecture Notes in Math., Vol. 1006, Springer-Verlag, 1983.

[396] H. Lenzing, Curve singularities arising from the representation theory of tame hereditary algebras, pp. 199–231 in *Representation Theory I: Finite Dimensional Algebras*, Lecture Notes in Mathematics, Vol. 1177, Springer-Verlag, 1986.

[397] H. Lenzing, Generic modules over tubular algebras, pp. 375–385 in R. Göbel (Ed.) *Advances in Algebra and Model Theory*, Gordon and Breach, 1997.

[398] H. Lenzing, Auslander's work on artin algebras, pp. 83–105 in I. Reiten *et al.* (Eds.) *Algebras and Modules I (Workshop, 1996, Trondheim)*, Canad. Math. Soc. Conf. Procs., Vol. 23, American Mathematical Society, 1998.

[399] H. Lenzing, Invariance of tameness under stable equivalence: Krause's theorem, pp. 405–418 in [371].

[400] H. Lenzing, Hereditary categories, pp. 105–146 in [11].

[401] S. Leonesi, S. L'Innocente and C. Toffalori, Weakly minimal modules over integral group rings and over related classes of rings, *Math. Log. Quart.*, **51(6)** (2005), 613–625.

[402] L. S. Levy, Modules over pullbacks and subdirect sums, *J. Algebra*, **71(1)** (1981), 50–61.

[403] L. S. Levy, Mixed modules over $\mathbb{Z}G$, G cyclic of prime order, and over related Dedekind pullbacks, *J. Algebra*, **71(1)** (1981), 62–114.

[404] L. S. Levy, Modules over Dedekind-like rings, *J. Algebra*, **93(1)** (1985), 1–116.

[405] S. L'Innocente and A. J. Macintyre, Towards decidability of the theory of pseudo-finite dimensional representations of $\mathrm{sl}_2(k)$, pp. 235–260 in A. Ehrenfeucht *et al.*, Andrej Mostowski and Foundation Studies, IOS Press, 2008.

[406] S. L'Innocente and M. Prest, Rings of definable scalars of Verma modules, *J. Algebra Appl.*, **6(5)** (2007), 779–787.

[407] S. L'Innocente, V. Puninskaya and C. Toffalori, Strongly minimal modules over group rings, *Comm. Algebra*, **33(7)** (2005), 2089–2107.

[408] S. L'Innocente, V. Puninskaya and C. Toffalori, Minimalities and modules over Dedekind-like rings, *Comm. Algebra*, **34(7)** (2006), 2453–2466.

[409] S. L'Innocente, V. Puninskaya and C. Toffalori, Minimal modules over serial rings, *Comm. Algebra*, **36(7)** (2008), 2750–2763.

[410] J. Łoś, Quelques remarques, théorèmes et problèmes sur les classes définissables d'algèbres, pp. 98–113 in L. E. J. Brouwer *et al.* (Eds.), *Mathematical Interpretation of Formal Systems*, North-Holland, 1955.

[411] J. Łoś, Abelian groups that are direct summands of every abelian group which contains them as pure subgroups, *Bull. Acad. Polon. Sci.*, **4** (1956), 73 and *Fund. Math.*, **44** (1957), 84–90.

[412] J. Łoś, Linear equations and pure subgroups, *Bull. Acad. Polon. Sci.*, **7** (1959), 13–18.

[413] J. Loveys, Weakly minimal groups of unbounded exponent, *J. Symbolic Logic*, **55(3)** (1990), 928–937.

[414] J. Loveys, Abelian groups with modular generic, *J. Symbolic Logic*, **56(1)** (1991), 250–259.

[415] W. Lowen, A generalization of the Gabriel–Popescu theorem, *J. Pure Appl. Algebra*, **190(1–3)** (2004), 197–211.

[416] S. Lubkin, Imbedding of abelian categories, *Trans. Amer. Math. Soc.*, **97(3)** (1960), 410–417.

[417] A. J. Macintyre, On ω_1-categorical theories of abelian groups, *Fund. Math.*, **70** (1971), 253–270.

[418] S. Mac Lane, *Categories for the Working Mathematician*, Springer-Verlag, 1971.

[419] S. Mac Lane and I. Moerdijk, *Sheaves in Geometry and Logic: A First Introduction to Topos Theory*, Springer-Verlag, 1992.

[420] M. Makkai, A theorem on Barr-exact categories with an infinitary generalization, *Ann. Pure Appl. Logic*, **47** (1990), 225–268.

[421] M. Makkai and R. Paré, *Accessible Categories: The Foundations of Categorical Model Theory*, Contemp. Math., Vol. 104, American Mathematical Society, 1989.

[422] M. Makkai and G. Reyes, *First Order Categorical Logic*, Lecture Notes in Mathematics, Vol. 611, Springer-Verlag, 1977.

[423] P. Malcolmson, Construction of universal matrix localisations, pp. 117–132 in P. J. Fleury (Ed.) *Advances in Noncommutative Ring Theory (Plattsburgh 1981)*, Lecture Notes in Math., Vol. 951, Springer-Verlag, 1982.

[424] J. M. Maranda, On pure subgroups of abelian groups, Arch. Math., **11(1)** (1960), 1–13.

[425] A. Marcja, M. Prest and C. Toffalori, The stability classification for abelian-by-finite groups and modules over a group ring, *J. London Math. Soc. (2)*, **47(2)** (1993), 212–226.

[426] A. Marcja, M. Prest and C. Toffalori, Classification theory for abelian groups with an endomorphism, *Arch. Math. Logic*, **31(2)** (1991), 95–104.

[427] A. Marcja, M. Prest and C. Toffalori, On the undecidability of some classes of abelian-by-finite groups, *Ann. Pure Appl. Logic*, **62(2)** (1993), 167–173.

[428] A. Marcja, M. Prest and C. Toffalori, The torsionfree part of the Ziegler spectrum of RG when R is a Dedekind domain and G is a finite group, *J. Symbolic Logic*, **67(3)** (2002), 1126–1140.

[429] A. Marcja and C. Toffalori, Decidability for modules over a group ring, *Comm. Algebra*, **21(7)** (1993), 2251–2264.

[430] A. Marcja and C. Toffalori, Abelian-by-G groups, for G finite, from the model theoretic point of view, *Math. Logic Q.*, **40** (1994), 125–131.

[431] A. Marcja and C. Toffalori, Decidable representations, *J. Pure Appl. Algebra*, **103(2)** (1995), 189–203.

[432] A. Marcja and C. Toffalori, On the elementarity of some classes of abelian-by-finite groups, *Studia Logica*, **62(2)** (1999), 201–213.

[433] A. Marcja and C. Toffalori, Decidability for $\mathbb{Z}_2G$-lattices when G extends the noncyclic group of order 4, *Math. Logic Q.*, **48** (2002), 203–212.

[434] A. Marcja and C. Toffalori, *A Guide to Classical and Modern Model Theory*, Trends in Logic, Studia Logica Library, Vol. 19, Kluwer, 2003.

[435] D. Marker, *Model Theory: an Introduction*, Graduate Texts in Mathematics, Vol. 217, Springer, 2002.

[436] H. Marubayashi, Modules over bounded Dedekind prime rings II, Osaka *J. Math.*, **9(3)** (1972), 427–445.

[437] H. Marubayashi, Pure injective modules over hereditary noetherian prime rings with enough invertible ideals, *Osaka J. Math.*, **18(1)** (1981), 95–107.

[438] E. Matlis, Injective modules over noetherian rings, *Pacific J. Math.*, **8(3)** (1958), 511–528.

[439] H. Matsumura, *Commutative Ring Theory*, Studies in Advanced Math., Vol. 8, Cambridge University Press, 1980 (1989 printing).

[440] J. C. McConnell and J. C. Robson, Homomorphisms and extensions of modules over certain differential polynomial rings, *J. Algebra*, **26(2)** (1973), 319–342.

[441] J. C. McConnell and J. C. Robson, *Noncommutative Noetherian Rings*, John Wiley and Sons, 1987.

[442] W. W. McGovern, G. Puninski and Ph. Rothmaler, When every projective module is a direct sum of finitely generated modules, *J. Algebra*, **315(1)** (2007), 454–481.

[443] R. McKenzie and S. Shelah, The cardinals of simple models for universal theories, pp. 53–74 in L. Henkin *et al.* (Eds.), *Proceedings of the Tarski Symposium*, American Mathematical Society, 1974.

[444] D. McKinnon and M. Roth, Curves arising from endomorphism rings of Kronecker modules, *Rocky Mt J. Math.*, **37(3)**, 2007, 879–892.

[445] P. Menal and R. Raphael, On epimorphism-final rings, *Comm. Algebra*, **12(15)** (1984), 1871–1876.

[446] P. Menal and P. Vámos, Pure ring extensions and self FP-injective rings, *Math. Proc. Camb. Phil. Soc.*, **105** (1989), 447–458.

[447] B. Mitchell, *Theory of Categories*, Academic Press, 1965.

[448] B. Mitchell, Rings with several objects, *Adv. in Math.*, **8** (1972), 1–161.

[449] E. Monari-Martinez, On Σ-pure-injective modules over valuation domains, *Arch. Math.*, **46(1)** (1986), 26–32.

[450] B. J. Müller and S. Singh, Uniform modules over serial rings, *J. Algebra*, **144(1)** (1991), 94–109.

[451] J. Mycielski, Some compactifications of general algebras, *Colloq. Math.*, **13** (1964), 1–9.

[452] J. Mycielski and C. Ryll-Nardzewski, Equationally compact algebras II, *Fund. Math.*, **61** (1968), 271–281.

[453] C. Năstăsescu and N. Popescu, Sur la structure des objects de certaines catégories abéliennes, *C. R. Acad. Sci. Paris*, **262** (1966), 1295–1297.

[454] J. Nedelmann, The spectrum of simple modular theories, Dissertation, Universität Darmstadt, Logos Verlag, Berlin, 2001.

[455] A. Neeman, The Brown representability theorem and phantomless triangulated categories, *J. Algebra*, **151(1)** (1992), 118–155.

[456] A. Neeman, The connection between the K-theory localization theorem of Thomason, Trobaugh and Yao and the smashing subcategories of Bousfield and Ravenel, *Ann. Scinet. Éc. Norm. Sup. (4)*, **25** (1992), 547–566.

[457] A. Neeman, with an appendix by M. Bökstedt, The chromatic tower for $D(R)$, *Topology*, **31(3)** (1992), 519–532.

[458] A. Neeman, The Grothendieck duality theorem via Bousfield's techniques and Brown representability, *J. Amer. Math. Soc.*, **9(1)** (1996), 205–236.

[459] A. Neeman, On a theorem of Brown and Adams, *Topology*, **36(3)** (1997), 619–645.

[460] A. Neeman, Brown representability for the dual, *Invent. Math.*, **133(1)** (1998), 97–105.

[461] A. Neeman, review of [73], *Mathematical Reviews*, 2000k:20013.

[462] A. Neeman, *Triangulated Categories*, Annals of Mathematics Studies, Vol. 148, Princeton University Press, 2001.

[463] A. Neeman, On the derived category of sheaves of a manifold, *Documenta Math.*, **6** (2001), 483–488.

[464] A. Neeman, A. Ranicki and A. Schofield, Representations of algebras as universal localisations, *Math. Proc. Camb. Phil. Soc.*, **136** (2004), 105–117.

[465] L. Newelski, A proof of Saffe's conjecture, *Fund. Math.*, **134** (1990), 143–155.

[466] L. Newelski, On U-rank 2 types, *Trans. Amer. Math. Soc.*, **344(2)** (1994), 553–581.

[467] U. Oberst and H. Rohrl, Flat and coherent functors, *J. Algebra*, **14(1)** (1970), 91–105.

[468] F. Okoh, Hereditary algebras that are not pure-hereditary, pp. 432–437 in V. Dlab and P. Gabriel (Eds.) *Representation Theory II*, Lecture Notes in Mathematics, Vol. 832, Springer, 1980.

[469] F. Okoh, Indecomposable pure-injective modules over hereditary Artin algebras of tame type, *Comm. Algebra*, **8(20)** (1980), 1939–1941.

[470] F. Okoh, Cotorsion modules over tame finite-dimensional hereditary algebras, pp. 263–269 in M. Auslander and E. Luis (Eds.), *Representations of Algebras*, Lecture Notes in Mathematics, Vol. 903, Springer, 1981.

[471] F. Okoh, Pure-injective modules over path algebras, *J. Pure Appl. Algebra*, **75(1)** (1991), 75–83.

[472] F. Okoh, Direct sums and direct products of finite-dimensional modules over path algebras, *J. Algebra*, **151(2)** (1992), 487–501.

[473] F. Okoh, The structure of the reduced product of preinjective modules over a hereditary right pure semisimple ring, *Comm. Algebra*, **34(1)** (2006), 235–250.

[474] Ø. Ore, Linear equations in non-commutative fields, *Ann. Math.*, **32(3)** (1931), 463–477.

[475] B. L. Osofsky, The subscript of $\aleph_n$, projective dimension, and the vanishing of $\varprojlim^{(n)}$, *Bull. Amer. Math. Soc.*, **80(1)** (1974), 8–26.

[476] Th. Pheidas and K. Zahidi, Elimination theory for addition and the Frobenius map in polynomial rings, *J. Symbolic Logic*, **69(4)** (2004), 1006–1026.

[477] R. S. Pierce, Modules over commutative regular rings, *Mem. Amer. Math. Soc.*, **70**, 1967.

[478] A. Pillay and M. Prest, Modules and stability theory, *Trans. Amer. Math. Soc.*, **300(2)** (1987), 641–662.

[479] A. Pillay and Ph. Rothmaler, Non-totally transcendental unidimensional theories, *Arch. Math. Logic*, **30(2)** (1990), 93–111.

[480] A. Pillay and Ph. Rothmaler, Unidimensional modules: uniqueness of maximal non-modular submodels, *Ann. Pure Appl. Logic*, **62(2)** (1993), 175–181.

[481] A. Pillay and M. Ziegler, On a question of Herzog and Rothmaler, *J. Symbolic Logic*, **69(2)** (2004), 478–481.

[482] F. Piron, Doctoral thesis, Universität Bonn (1987).

[483] F. Point and M. Prest, Decidability for theories of modules, *J. London Math. Soc.*, **38(2)** (1988), 193–206.

[484] F. Point, Decidability questions for modules, pp. 266–280 in J. Oikkonen and J. Väänänen, *Logic Colloquium 1990*, Lecture Notes in Logic, Vol. 2, Springer, 1993.

[485] B. Poizat, Une théorie de Galois imaginaire, *J. Symbolic Logic*, **48(4)** (1983), 1151–1170.

[486] N. Popescu, *Abelian Categories with Applications to Rings and Modules*, Academic Press, 1973.

[487] M. Y. Prest, Some model-theoretic aspects of torsion theories, *J. Pure Applied Alg.*, **12(3)** (1978), 295–310.

[488] M. Prest, Applications of logic to torsion theories in abelian categories, Doctoral Thesis, University of Leeds, 1979.

[489] M. Prest, Elementary torsion theories and locally finitely presented categories, *J. Pure Applied Algebra*, **18(2)** (1980), 205–212.

[490] M. Prest, Elementary equivalence of Σ-injective modules, *Proc. London Math. Soc. (3)*, **45(1)** (1982), 71–88.

[491] M. Prest, Rings of finite representation type and modules of finite Morley rank, *J. Algebra*, **88(2)** (1984), 502–533.

[492] M. Prest, Tame categories of modules and decidability, preprint, University of Liverpool, 1985.

[493] M. Prest, Model theory and representation type of algebras, pp. 219–260 in F. R. Drake and J. K. Truss (Eds.) *Logic Colloquium '86 (Proceedings of the European Meeting of the Association for Symbolic Logic, Hull, 1986)*, North-Holland, Amsterdam, 1988.

[494] M. Prest, Duality and pure-semisimple rings, *J. London Math. Soc. (2)*, **38** (1988), 403–409.

[495] M. Prest, *Model Theory and Modules*, London Math. Soc. Lect. Note Ser., Vol. 130, Cambridge University Press, 1988.

[496] M. Prest, Wild representation type and undecidability, *Comm. Algebra*, **19(3)** (1991), 919–929.

[497] M. Prest, Remarks on elementary duality, *Ann. Pure Applied Logic*, **62(2)** (1993), 183–205.

[498] M. Prest, The (pre-)sheaf of definable scalars, University of Manchester, preprint, 1994.

[499] M. Prest, A note concering the existence of many indecomposable infinite-dimensional pure-injective modules over some finite-dimensional algebras, preprint, University of Manchester, 1995, revised 1996.

[500] M. Prest, Representation embeddings and the Ziegler spectrum, *J. Pure Appl. Algebra*, **113(3)** (1996), 315–323.

[501] M. Prest, Epimorphisms of rings, interpretations of modules and strictly wild algebras, *Comm. Algebra*, **24(2)** (1996), 517–531.

[502] M. Prest, Maps in the infinite radical of mod-R factor through large modules, University of Manchester, preprint, 1997.

[503] M. Prest, Interpreting modules in modules, *Ann. Pure Applied Logic*, **88(2–3)** (1997), 193–215.

[504] M. Prest, Morphisms between finitely presented modules and infinite-dimensional representations, pp. 447–455 in Reiten *et al.* (Eds.), *Algebras and Modules II*, Canad. Math. Soc. Conf. Proc., Vol. 24, Canadian Mathematical Society, 1998.

[505] M. Prest, Ziegler spectra of tame hereditary algebras, *J. Algebra*, **207(1)** (1998), 146–164.

[506] M. Prest, The representation theories of elementarily equivalent rings, *J. Symbolic Logic*, **63(2)** (1998), 439–450.

[507] M. Prest, Sheaves of definable scalars over tame hereditary algebras, University of Manchester, preprint, 1998.

[508] M. Prest, The Zariski spectrum of the category of finitely presented modules, University of Manchester, preprint, 1998, revised, 2004, 2006, 2007, final version at http://eprints.ma.man.ac.uk/1049/

[509] M. Prest, Tensor product and theories of modules, *J. Symbolic Logic*, **64(2)** (1999), 617–628.

[510] M. Prest, The sheaf of locally definable scalars over a ring, pp. 339–351 in S. B. Cooper and J. K. Truss (Eds.), *Models and Computability*, London Math. Soc. Lect. Note Ser., Vol. 259, Cambridge University Press, 1999.

[511] M. Prest, Topological and geometric aspects of the Ziegler spectrum, pp. 369–392 in H. Krause and C. M. Ringel (Eds.), *Infinite Length Modules*, Birkhäuser, 2000.

[512] M. Prest, with input from T. Kucera and G. Puninski, Notes based on a working seminar on infinite-dimensional pure-injectives over canonical algebras, University of Manchester, 2002.

[513] M. Prest, Model theory for algebra, pp. 199–226 in M. Hazewinkel (Ed.), *Handbook of Algebra*, Vol. 3, Elsevier, 2003.

[514] M. Prest, Model theory and modules, pp. 227–253 in M. Hazewinkel (Ed.), *Handbook of Algebra*, Vol. 3, Elsevier, 2003.

[515] M. Prest, Ideals in mod-R and the ω-radical, *J. London Math. Soc.*, **71(2)** (2005), 321–334.

[516] M. Prest, Definable additive categories: purity and model theory, *Mem. Amer. Math. Soc.*, to appear.
[517] M. Prest and V. Puninskaya, Vaught's Conjecture for modules over a commutative Prüfer ring, *Algebra and Logic*, **38(4)** (1999), 228–236.
[518] M. Prest and V. Puninskaya, Modules with few types over some finite-dimensional algebras, *J. Symbolic Logic*, **67(2)** (2002), 841–858.
[519] M. Prest, V. Puninskaya and A. Ralph, Some model theory of sheaves of modules, *J. Symbolic Logic*, **69(4)** (2004), 1187–1199.
[520] M. Prest and G. Puninski, Σ-pure-injective modules over a commutative Prüfer ring, *Comm. Algebra*, **27(2)** (1999), 961–971.
[521] M. Prest and G. Puninski, Some model theory over hereditary noetherian domains, *J. Algebra*, **211(1)** (1999), 268–297.
[522] M. Prest and G. Puninski, Pure injective envelopes of finite length modules over a Generalized Weyl Algebra, *J. Algebra*, **251(1)** (2002), 150–177.
[523] M. Prest and G. Puninski, One-directed indecomposable pure injective modules over string algebras, *Colloq. Math.*, **101(1)** (2004), 89–112.
[524] M. Prest and G. Puninski, Krull–Gabriel dimension of 1-domestic string algebras, *Algebras and Representation Theory*, **9(4)** (2006), 337–358.
[525] M. Prest and R. Rajani, Structure sheaves of definable additive categories, preprint, University of Manchester, 2008.
[526] M. Prest and A. Ralph, Locally finitely presented categories of sheaves of modules, preprint, University of Manchester, 2001.
[527] M. Prest, Ph. Rothmaler and M. Ziegler, Absolutely pure and flat modules and "indiscrete" rings, *J. Algebra*, **174(2)** (1995), 349–372.
[528] M. Prest, Ph. Rothmaler and M. Ziegler, Extensions of elementary duality, *J. Pure Appl. Algebra*, **93(1)** (1994), 33–56.
[529] M. Prest and J. Schröer, Serial functors, Jacobson radical and representation type, *J. Pure Appl. Algebra*, **170(2–3)** (2002), 295–307.
[530] M. Prest and R. Wisbauer, Finite presentation and purity in categories $\sigma[M]$, *Colloq. Math.*, **99(2)** (2004), 189–202.
[531] P. Příhoda, On uniserial modules that are not quasi-small, *J. Algebra*, **299(1)** (2006), 329–343.
[532] P. Příhoda, Projective modules are determined by their radical factors, *J. Pure Appl. Algebra*, **210(3)** (2007), 827–835.
[533] H. Prüfer, Untersuchungen über die Zerlegbarkeit der abzählbaren primären abelschen Gruppen, *Math. Z.*, **17(1)** (1923), 35–61.
[534] V. Puninskaya, Injective minimal modules (in Russian), *Algebra i Logika*, **33** (1994), 211–226, English transl. in Algebra and Logic, **33(2)** (1994), 120–128.
[535] V. Puninskaya, Strongly minimal modules over right distributive rings, *Algebra and Logic*, **35(3)** (1996), 196–203.
[536] V. Puninskaya, Vaught's conjecture for modules over a Dedekind prime ring, *Bull. London Math. Soc.*, **31(2)** (1999), 129–135.
[537] V. Puninskaya, Vaught's conjecture for modules over a serial ring, *J. Symbolic Logic*, **65(1)** (2000), 155–163.
[538] V. Puninskaya, Modules with few types over a commutative valuation ring, *J. Math. Sci.*, **102** (2000), 4652–4661.

[539] V. A. Puninskaya, Modules with few types over semichain rings (in Russian), *Uspekhi Mat. Nauk*, **56** (2001), 209–210, English transl. in *Russ. Math. Surv.*, **56** (2001), 41–420.

[540] V. Puninskaya, Modules with few types over a hereditary noetherian prime ring, *J. Symbolic Logic*, **66(1)** (2001), 271–280.

[541] V. Puninskaya, Vaught's Conjecture, *J. Math. Sci.*, **109** (2002), 1649–1668 (transl. from Russian original).

[542] V. Puninskaya, Modules with few types over a serial ring and over a commutative Prüfer ring, *Comm. Algebra*, **30(3)** (2002), 1227–1240.

[543] V. A. Puninskaya, Strongly minimal modules, *J. Math. Sci.*, **114** (2003), 1127–1156 (transl. from Russian original, 2001).

[544] V. A. Puninskaya and G. E. Puninski, Minimal modules over Prüfer rings (in Russian), *Algebra i Logika*, **30(5)** (1991), 557–567, English transl. in *Algebra and Logic*, **30(5)** (1991), 361–368.

[545] V. Puninskaya, G. Puninski and C. Toffalori, Decidability of the theory of modules over commutative valuation domains, *Ann. Pure Appl. Logic*, **145(3)** (2007), 258–275.

[546] V. Puninskaya and C. Toffalori, Vaught's conjecture and group rings, *Comm. Algebra*, **33(11)** (2005), 4267–4281.

[547] V. Puninskaya and C. Toffalori, Superdecomposable pure injective modules over commutative Noetherian rings, *J. Algebra Appl.* **7(6)** (2008), 809–830.

[548] G. E. Puninski, Boolean powers of modules (in Russian), *Vestnik Mosk. Univ.*, **1** (1983), 28–31.

[549] G. E. Puninski, The pure-injective and RD-injective hulls of rings (in Russian), *Mat. Zametki*, **52(6)** (1992), 81–88, English transl. in *Math. Notes*, **52(6)** (1992), 1224–1230.

[550] G. E. Puninski, Superdecomposable pure-injective modules over commutative valuation rings (in Russian), *Algebra i Logika*, 31 (1992), 655–671, English transl. in *Algebra and Logic*, **31(6)** (1992), 377–386.

[551] G. E. Puninski, RD-formulas and W-rings (in Russian), *Mat. Zametki*, **53(1)** (1993), 95–103, English transl. in *Math. Notes*, **53(1)** (1993), 66–71.

[552] G. E. Puninski, Endodistributive and pure-injective modules over uniserial rings (in Russian), *Uspekhi Mat. Nauk*, **48** (1993), 201–202 English transl. in *Russ. Math. Surv.*, **48(3)** (1993), 213–214.

[553] G. E. Puninski, Rings defined by purities (in Russian), *Uspekhi Mat. Nauk*, **48** (1993), 169–170 English transl. in *Russ. Math. Surv.*, **48(6)** (1993), 181–182.

[554] G. E. Puninski, On Warfield rings (in Russian), *Algebra i Logika*, **33(3)** (1994), 264–285, English transl. in *Algebra and Logic*, **33(3)** (1994) 147–159.

[555] G. E. Puninski, A remark on pure-injective modules over a semichain ring (in Russian), *Uspekhi Mat. Nauk*, **49** (1994), 171–172 English transl. in *Russ. Math. Surv.*, **49(5)** (1994), 185–186.

[556] G. E. Puninski, Lattices of *pp*-defined subgroups (in Russian) pp. 193–203, in *Abelian Groups and Modules*, No. 11, 12, 258, Tomsk. University Press, 1994.

[557] G. E. Puninski, Indecomposable purely injective modules over chain rings (in Russian), *Trudy Moskov. Mat. Obshch.*, **56** (1995), English transl. in *Proc. Moscow Math. Soc.*, **56** (1995), 176–191.

[558] G. E. Puninski, Pure-injective modules over right noetherian serial rings, *Comm. in Algebra*, **23(4)** (1995), 1579–1591.

[559] G. E. Puninski, Serial Krull-Schmidt rings and pure injective modules, *Fund. Prikl. Mat.*, **1** (1995), 471–489.

[560] G. Puninski, Cantor–Bendixson rank of the Ziegler spectrum over a commutative valuation domain, *J. Symbolic Logic*, **64(4)** (1999), 1512–1518.

[561] G. Puninski, Left almost split morphisms and generalised Weyl algebras (in Russian), *Mat. Zam.*, **66(5)** (1999), 734–740, English transl. in *Math. Notes*, **66** (1999) 608–612.

[562] G. Puninski, Finite length and pure-injective modules over a ring of differential operators, *J. Algebra*, **231(2)** (2000), 546–560.

[563] G. Puninski, Pure-injective and finite length modules over certain rings of differential polynomials, pp. 171–185 in Y. Fong *et al.* (Eds.), *Proceedings of the Moscow–Taiwan Workshop*, Springer-Verlag, 2000.

[564] G. Puninski, *Serial Rings*, Kluwer, 2001.

[565] G. Puninski, Some model theory over an exceptional uniserial ring and decompositions of serial modules, *J. London Math. Soc. (2)*, **64(2)** (2001), 311–326.

[566] G. Puninski, Some model theory over a nearly simple uniserial domain and decompositions of serial modules, *J. Pure Applied Algebra*, **163(3)** (2001), 319–337.

[567] G. E. Puninski, Pure projective modules over an exceptional uniserial coherent ring (in Russian), *St Petersburg Math. J.*, **13(6)** (2001, English translation 2002), 1–13.

[568] G. Puninski, Model theory of rings and modules, *J. Math. Sci.*, **109** (2002), 1641–1648.

[569] G. Puninski, Pure-injective modules over a commutative valuation domain, *Algebras and Representation Theory*, **6(3)** (2003), 239–250.

[570] G. Puninski, The Krull-Gabriel dimension of a serial ring, *Comm. Algebra*, **31(12)** (2003), 5977–5993.

[571] G. Puninski, Superdecomposable pure-injective modules exist over some string algebras, *Proc. Amer. Math. Soc.*, **132(7)** (2004), 1891–1898.

[572] G. Puninski, When a superdecomposable pure injective module over a serial ring exists, pp. 449–463 in A. Facchini, E. Houston and L. Salce (Eds.), *Rings, Modules and Abelian Groups*, Marcel Dekker, 2004.

[573] G. Puninski, Band combinatorics of domestic string algebras, *Colloq. Math.*, **108(2)** (2007), 285–296.

[574] G. Puninski, How to construct a 'concrete' superdecomposable pure-injective module over a string algebra, *J. Algebra*, **212(4)** (2008), 704–717.

[575] G. Puninski, M. Prest and Ph. Rothmaler, Rings described by various purities, *Comm. Algebra*, **27(5)** (1999), 2127–2162.

[576] G. Puninski, V. Puninskaya and C. Toffalori, Superdecomposable pure-injective modules and integral group rings, *J. London Math. Soc. (2)*, **73(1)** (2006), 48–64.

[577] G. Puninski, V. Puninskaya, and C. Toffalori, Krull–Gabriel dimension and the model-theoretic complexity of the category of modules over group rings of finite groups, *J. London Math. Soc.*, **78(1)** (2008), 125–142.

[578] G. Puninski and Ph. Rothmaler, When every finitely generated flat module is projective, *J. Algebra*, **277(2)**, (2004), 542–558.

[579] G. Puninski and Ph. Rothmaler, Pure-projective modules, *J. London Math. Soc.*, **71(2)** (2005), 304–320.

[580] G. Puninski and C. Toffalori, Towards the decidability of the theory of modules over finite commutative rings, *Ann. Pure Appl. Logic*, to appear.

[581] G. E. Puninski and A. A. Tuganbaev, *Rings and Modules*, Soyuz, 1998.

[582] G. Puninski and R. Wisbauer, Σ-injective modules over left duo and left distributive rings, *J. Pure Appl. Algebra*, **113(1)** (1996), 55–66.

[583] R. Rajani, Model-theoretic imaginaries and Ziegler spectra in general categories, Doctoral Thesis, University of Manchester, 2007.

[584] R. Rajani and M. Prest, Model-theoretic imaginaries and coherent sheaves, *Applied Categorical Structures, to appear.*

[585] R. Rajani and M. Prest, Positive imaginaries and coherent affine functors, *Appl. Categorical Structures*, to appear.

[586] R. Rajani and M. Prest, Pure-injectives and model theory of G-sets, *J. Symbolic Logic*, to appear.

[587] A. Ralph, An approach to the model theory of sheaves, Dissertation, University of Manchester, 2000.

[588] M. Raynaud and L. Gruson, Critères de platitude et de projectvité, Seconde partie, *Invent. Math.*, **13(1–2)** (1971), 52–89.

[589] I. Reiner, *Maximal Orders*, Academic Press, 1975.

[590] I. Reiten and C. M. Ringel, Infinite dimensional representations of canonical algebras, *Canad. J. Math.*, **58(1)** (2006), 180–224.

[591] R. Rentschler and P. Gabriel, Sur la dimension des anneaux et ensembles ordonnés, *C. R. Acad. Sci. Paris, Ser.*, **A5** (1967), 712–715.

[592] G. Reynders, Ziegler Spectra over serial rings and coalgebras, Doctoral Thesis, University of Manchester, 1998.

[593] G. Reynders, Ziegler spectra of serial rings with Krull dimension, *Comm. Algebra*, **27(6)** (1999), 2583–2611.

[594] J. Rickard, Morita theory for derived categories, *J. London Math. Soc.*, **39(3)** (1989), 436–456.

[595] J. Rickard, Derived categories and stable equivalence, *J. Pure Appl. Algebra*, **61(3)** (1989), 303–317.

[596] J. Rickard, Idempotent modules in the stable category, *J. London Math. Soc.*, **56(1)** (1997), 149–170.

[597] J. Rickard, Bousfield localisation for representation theorists, pp. 273–283 in [371].

[598] C. M. Ringel, The indecomposable representations of the dihedral 2-groups, *Math. Ann.*, **214(1)** (1975), 19–34.

[599] C. M. Ringel, Infinite-dimensional representations of finite-dimensional hereditary algebras, *Symp. Math. inst. Naz. Alta. Mat.*, **23** (1979), 321–412.

[600] C. M. Ringel, The spectrum of a finite dimensional algebra, pp. 535–597 in F. Van Oystaeyen (Ed.), *Ring Theory*, Lect. Notes in Pure Appl. Math., Marcel Dekker, 1979.

[601] C. M. Ringel, Tame algebras, pp. 137–287 in V. Dlab and P. Gabriel (Eds.), *Representation Theory* I, Lecture Notes in Math., Vol. 831, Springer-Verlag, 1980.

[602] C. M. Ringel, *Tame Algebras and Integral Quadratic Forms*, Lecture Notes in Math., Vol. 1099, Springer-Verlag, 1984.

[603] C. M. Ringel, The canonical algebras (with an appendix by W. W. Crawley-Boevey), pp. 407–432 in S. Balcerzyk *et al.* (Eds.), *Topics in Algebra*, Banach Centre Publications, Vol. 26(1), Polish Scientific Publishers, 1990.

[604] C. M. Ringel, Recent advances in the representation theory of finite dimensional algebras, pp. 141–192 in *Representation Theory of Finite Groups and Finite-dimensional Algebras*, Birkhauser, 1991.

[605] C. M. Ringel, Some algebraically compact modules I, pp. 419–439 in A. Facchini and C. Menini (Eds.), *Abelian Groups and Modules*, Kluwer, 1995.

[606] C. M. Ringel, A construction of endofinite modules, pp. 387–399 in M. Droste and R. Göbel (Eds.), *Advances in Algebra and Model Theory*, Gordon and Breach, 1997.

[607] C. M. Ringel, The development of the representation theory of finite-dimensional algebras 1968–1975, pp. 89–115 in A. Martsinkovsky and G. Todorov (Eds.), *Representation Theory and Algebraic Geometry*, London Math. Soc. Lect. Note Ser., Vol. 238, Cambridge University Press, 1997.

[608] C. M. Ringel, The Ziegler spectrum of a tame hereditary algebra, *Colloq. Math.*, **76** (1998), 105–115.

[609] C. M. Ringel, Tame algebras are Wild, *Algebra Coll.*, **6** (1999), 473–480.

[610] C. M. Ringel, Infinite length modules: some examples as introduction, pp. 1–73 in [371].

[611] C. M. Ringel, On generic modules for string algebras, *Bol. Soc. Mat. Mexicana (3)*, **7** (2001), 85–97.

[612] C. M. Ringel, Algebraically compact modules arising from tubular families: a survey, *Algebra Colloq.*, **11** (2004), 155–172.

[613] J.-E. Roos, Sur les dérivés de lim. Applications, *C.R. Acad. Sci. Paris*, **252** (1961), 3702–3704.

[614] J.-E. Roos, Locally Noetherian categories and generalised strictly linearly compact rings: Applications, pp. 197–277 in *Category Theory, Homology Theory and their Applications*, Lecture Notes in Mathematics, Vol. 92, Springer-Verlag, 1969.

[615] A. L. Rosenberg, *Noncommutative Algebraic Geometry and Representations of Quantized Algebras*, Math. and Appl., Vol. 330. Kluwer, 1995.

[616] A. L. Rosenberg, The spectrum of abelian categories and reconstruction of schemes, pp. 257–274 in *Rings, Hopf Algebras, and Brauer Groups*, Lect. Notes Pure Appl. Math., Vol. 197, Marcel Dekker, 1998.

[617] Ph. Rothmaler, Some model theory of modules II: on stability and categoricity of flat modules, *J. Symbolic Logic*, **48(4)** (1983), 970–985.

[618] Ph. Rothmaler, Some model theory of modules III: on infiniteness of sets definable in modules, *J. Symbolic Logic*, **49(1)** (1984), 32–46.

[619] Ph. Rothmaler, A trivial remark on purity, p. 127 in H. Wolter (Ed.), *Proceedings of the 9th Easter Conference on Model Theory, Gosen 1991*, Seminarber. 112, Humboldt University, 1991.

[620] Ph. Rothmaler, Mittag–Leffler modules and positive atomicity, Habilitationsschrift, Universität Kiel, 1994.

[621] Ph. Rothmaler, Mittag–Leffler modules, *Ann. Pure Applied Logic*, **88(2–3)** (1997), 227–239.

[622] Ph. Rothmaler, Purity in model theory, pp. 445–469 in R. Göbel and M. Droste (Eds.), *Advances in Algebra and Model* Theory, Gordon and Breach, 1997.

[623] Ph. Rothmaler, *Introduction to Model Theory*, transl./revised from German original (1995), Algebra, Logic and Appl., Vol. 15., Gordon and Breach, 2000.

[624] Ph. Rothmaler, When are pure-injective envelopes of flat modules flat?, *Comm. Algebra*, **30(6)** (2002), 3077–3085.

[625] Ph. Rothmaler, Elementary epimorphisms, *J. Symbolic Logic*, **70(2)** (2005), 473–487.

[626] L. H. Rowen, *Ring Theory*, Student Edition, Academic Press, 1991.

[627] G. Sabbagh, Sur la pureté dans les modules, *C. R. Acad. Sci. Paris*, **271** (1970), 865–867.

[628] G. Sabbagh, Aspects logiques de la pureté dans les modules, *C. R. Acad. Sci. Paris*, **271** (1970), 909–912.

[629] G. Sabbagh, Sous-modules purs, existentiellement clos et élementaires, *C. R. Acad. Sci. Paris*, **272** (1971), 1289–1292.

[630] G. Sabbagh, Catégoricité et stabilité: quelques exemples parmi les groupes et anneaux, *C. R. Acad. Sci. Paris*, **280** (1975), 531–533.

[631] G. Sabbagh and P. Eklof, Definability problems for rings and modules, *J. Symbolic Logic*, **36(4)** (1971), 623–649.

[632] L. Salce, Valuation domains with superdecomposable pure injective modules, pp. 241–245 in L. Fuchs and R. Göbel (Eds.), *Abelian Groups*, Lect. Notes Pure Appl. Math., Vol. 146, Marcel Dekker, 1993.

[633] J. Šaroch and J. Šťovíček, The countable telescope conjecture for module categories, *Adv. Math.*, **219(3)** (2008), 1002–1036.

[634] K. Schmidt, The endofinite spectrum of a tame algebra, *J. Algebra*, **279(2)** (2004), 771–790.

[635] M. Schmidmeier, Auslander–Reiten–Köcher für artinische Ringe mit Polynomidentität, Dissertation, Universität Munchen, 1996.

[636] M. Schmidmeier, The local duality for homomorphisms and an application to pure semisimple PI-rings, *Colloq. Math.*, **77(1)** (1998), 121–132.

[637] M. Schmidmeier, Endofinite modules over hereditary artinian PI-rings, pp. 497–511 in I. Reiten *et al.* (Eds.), *Algebras and Modules II*, Canad. Math. Soc. Conf. Proc., Vol. 24, Canadian Mathematical Society, 1998.

[638] A. H. Schofield, *Representations of Rings over Skew Fields*, London Math. Soc. Lect. Note Ser., Vol. 92, Cambridge University Press, 1985.

[639] A. Schofield, Universal localisation for hereditary rings and quivers, pp. 149–164 in F. M. J. Van Ostaeyen (Ed.), *Ring Theory, Proceedings Antwerp 1985*, Lecture Notes in Math., Vol. 1197, Springer-Verlag, 1986.

[640] J. Schröer, Hammocks for string algebras, Doctoral Thesis, Universität Bielefeld, 1997.

[641] J. Schröer, On the Krull–Gabriel dimension of an algebra, *Math. Z.*, **233(2)** (2000), 287–303.

[642] J. Schröer, On the infinite radical of a module category, *Proc. London Math. Soc. (3)*, **81(3)** (2000), 651–674.

[643] J. Schröer, The Krull–Gabriel dimension of an algebra – open problems and conjectures, pp. 419–424 in [371].

[644] S. Shelah, *Classification Theory and the Number of Non-Isomorphic Models*, North-Holland, 1978.

[645] S. Shelah and J. Trlfaj, Spectra of the Γ-invariant of uniform modules, *J. Pure Appl. Algebra*, **162(2–3)** (2001), 367–379.

[646] L. Silver, Noncommutative localizations and applications, *J. Algebra*, **7(1)** (1967), 44–76.

[647] D. Simson, On pure global dimension of locally finitely presented Grothendieck categories, *Fund. Math.*, **96** (1977), 91–116.

[648] D. Simson, Pure semisimple categories and rings of finite representation type (corrigendum), *J. Algebra*, **67(1)** (1980), 254–256.

[649] D. Simson, Partial Coxeter functors and right pure semisimple hereditary rings, *J. Algebra*, **71(1)** (1981), 195–218.

[650] D. Simson, On representation types of module subcategories and orders, *Bull. Pol. Acad. Sci. Math.*, **41** (1993), 77–93.

[651] D. Simson, An Artin problem for division ring extensions and the pure semisimplicity conjecture I, *Arch. Math.*, **66(2)** (1996), 114–122.

[652] D. Simson, A class of potential counterexamples to the pure semisimplicity conjecture, pp. 345–373 in M. Droste and R. Göbel (Eds.), *Advances in Algebra and Model Theory*, Gordon and Breach, 1997.

[653] D. Simson, An Artin problem for division ring extensions and the pure semisimplicity conjecture, II, *J. Algebra*, **227(2)** (2000), 670–705.

[654] D. Simson and A. Skowroński, The Jacobson radical power series of module categories and the representation type, *Bol. Soc. Mat. Mexicana (3)*, **5** (1999), 223–236.

[655] Z. Skoda, Noncommutative localization in noncommutative geometry, pp. 220–313 in A. Ranicki (Ed.), *Noncommutative Localization in Algebra and Topology*, London Math. Soc. Lect. Note Ser., Vol. 330, Cambridge University Press, 2006.

[656] A. Skowroński, Module categories over tame algebras, pp. 281–313 in R. Bautista *et al.* (Eds.) *Representation Theory of Algebras and Related Topics*, CMS Conf. Procs. Vol. 19, Canadian Mathematical Society, 1996.

[657] S. O. Smalø, The inductive step of the second Brauer–Thrall conjecture, *Canad. J. Math.*, **32** (1980), 342–349.

[658] Ø. Solberg, Infinite dimensional tilting modules over finite dimensional algebras, pp. 323–344 in [11].

[659] J. T. Stafford, Stable structure of noncommutative Noetherian rings, *J. Algebra*, **47(2)** (1977), 244–267.

[660] B. Stenström, Pure submodules, *Arkiv Mat.*, **7(2)** (1967), 159–171.

[661] B. Stenström, Purity in functor categories, *J. Algebra*, **8(3)** (1968), 352–361.

[662] B. Stenström, Coherent rings and FP-injective modules, *J. London Math. Soc. (2)*, **2(2)** (1970), 323–329.

[663] B. Stenström, *Rings of Quotients*, Springer-Verlag, 1975.

[664] W. Stephenson, Modules whose lattice of submodules is distributive, *Proc. London Math. Soc.*, **28(2)** (1974), 291–310.

[665] J. Šťovíček, All n-cotilting modules are pure-injective, *Proc. Amer. Math. Soc.*, **134(7)** (2006), 1891–1897.

[666] J. Šťovíček and J. Trlifaj, All tilting modules are of countable type, *Bull. London Math. Soc.*, **39(1)** (2007), 121–132.

[667] N. P. Strickland, Axiomatic stable homotopy, a survey, pp. 69–98 in J. P. C. Greenlees (Ed.), *Axiomatic, Enriched and Motivic Homotopy Theory*, Kluwer, 2004.

[668] R. G. Swan, Cup products in sheaf cohomology, pure injectives, and a substitute for projective resolutions, *J. Pure Appl. Algebra*, **144(2)** (1999), 169–211.

[669] W. Szmielew, Elementary properties of abelian groups, *Fund. Math.*, **41** (1954), 203–271.

[670] H. Tachikawa, *Quasi-Frobenius Rings and Generalisations*, Lecture Notes in Math., Vol. 351, Springer-Verlag, 1973.

[671] R. Takahashi, Classifying subcategories of modules over a commutative noetherian ring, *J. London Math. Soc.*, **78(3)**, (2008), 767–782.

[672] R. Takahashi, On localizing subcategories of derived categories, preprint, 2007.

[673] W. Taylor, Some constructions of compact algebras, *Ann. Math. Logic*, **3(4)** (1971), 395–435.

[674] B. R. Tennison, *Sheaf Theory*, London Math. Soc. Lecture Notes Ser., Vol. 20, Cambridge University Press, 1975.

[675] R. W. Thomason, The classification of triangulated subcategories, *Compos. Math.*, **105(1)** (1997), 1–27.

[676] C. Toffalori, Decidability for modules over a group ring II, *Comm. Algebra*, **22(2)** (1994), 397–405.

[677] C. Toffalori, Decidability for modules over a group ring III, *Comm. Algebra*, **24(1)** (1996), 331–344.

[678] C. Toffalori, Decidability for $\mathbb{Z}G$-modules when G is cyclic of prime order, *Math. Logic. Quart.*, **42(1)** (1996), 369–378.

[679] C. Toffalori, Some decidability results for $\mathbb{Z}G$-modules when G is cyclic of square-free order, *Math. Logic. Quart.*, **42(1)** (1996), 433–445.

[680] C. Toffalori, Two questions concerning decidability for lattices over a group ring, *Boll. Un. Mat. Ital., (7)*, **10-B** (1996), 799–814.

[681] C. Toffalori, Wildness implies undecidability for lattices over group rings, *J. Symbolic Logic*, **62(4)** (1997), 1429–1447.

[682] C. Toffalori, An decidability theorem for lattices over group rings, *Ann. Pure Appl. Logic*, **88(2–3)** (1997), 241–262.

[683] C. Toffalori, The decision problem for $\mathbb{Z}C(p^3)$-lattices with p prime, *Arch. Math. Logic*, **37(2)** (1998), 127–142.

[684] J. Trlifaj, Two problems of Ziegler and uniform modules over regular rings, pp. 373–383 in D. M. Arnold and K. M. Rangaswamy (Eds.), *Abelian Groups and Modules*, M. Dekker, 1996.

[685] J. Trlifaj, Modules over non-perfect rings, pp. 471–492 in M. Droste and R. Göbel (Eds.), *Advances in Algebra and Model Theory*, Gordon and Breach, 1997.

[686] J. Trlifaj, Uniform modules, Γ-invariants, and Ziegler spectra of regular rings, pp. 327–340 in P. C. Eklof and R. Göbel (Eds.), *Abelian Groups and Modules*, Trends in Mathematics, Birkhäuser, 1999.

[687] J. Trlifaj, Covers, envelopes and cotorsion theories, Lecture Notes from the Workshop "Homological Methods in Module Theory", Cortona, 2000.

[688] J. Trlifaj, Infinite dimensional tilting modules and cotorsion pairs, pp. 279–321 in [11].

[689] J. Trlifaj, Abstract elementary classes induced by tilting and cotilting modules have finite character, *Proc. Amer. Math. Soc.*, to appear.

[690] L. V. Tyukavkin, Model completeness of certain theories of modules (in Russian), *Algebra i Logika*, **21(1)** (1982), 73–83, English transl. in *Algebra and Logic*, **21(1)** (1982), 50–57.

[691] F. Van Oystaeyen and L. Willaert, Grothendieck topology, coherent sheaves and Serre's theorem for schematic algebras, *J. Pure Appl. Algebra*, **104(1)** (1995), 109–122.

[692] A. Verschoren, Sheaves and localization, *J. Algebra*, **182(2)** (1996), 341–556.

[693] A. Verschoren and L. Willaert, Noncommutative algebraic geometry: from pi-algebras to quantum groups, *Bull. Belg. Math. Soc.*, **4(5)** (1997), 557–588.

[694] R. Villemaire, Catégoricité relative et groupes abéliens, *C. R. Acad. Sci. Paris Sér. I Math.*, **309** (1989), 747–749.

[695] R. Villemaire, Abelian groups $\aleph_0$-categorical over a subgroup, *J. Pure Appl. Algebra*, **69(2)** (1990), 193–204.

[696] R. Villemaire, Completely decomposable abelian groups $\aleph_0$-categorical over a subgroup. *Arch. Math. Logic*, **31(4)** (1992), 263–275.

[697] R. Villemaire, Theories of modules closed under direct products, *J. Symbolic Logic*, **57(2)** (1992), 515–521.

[698] R. Villemaire and M. Hébert, Theories of abelian groups and modules preserved under extensions, pp. 239–254 in F. Haug *et al.* (Eds.) *Algebra, Logic and Set Theory, Festschrift für Ulrich Felgner*, College Publications, 2007.

[699] R. B. Warfield, Purity and algebraic compactness for modules, *Pacific J. Math.*, **28(3)** (1969), 699–719.

[700] R. B. Warfield, Decompositions of injective modules, *Pacific J. Math.*, **31(1)** (1969), 263–276.

[701] R. B. Warfield, Decomposability of finitely presented modules, *Proc. Amer. Math. Soc.*, **25(1)** (1970), 167–172.

[702] R. B. Warfield, Exchange rings and decomposition of modules, *Math. Ann.*, **199(1)** (1972), 31–36.

[703] R. B. Warfield, Serial rings and finitely presented modules, *J. Algebra*, **37(2)** (1975), 187–222.

[704] R. B. Warfield, Large modules over artinian rings, pp. 451–463 in R. Gordon (Ed.), *Representation Theory of Algebras (Proceedings)*, Lect. Notes in Pure Appl. Math., Vol. 37, Marcel-Dekker, 1978.

[705] C. E. Watts, Intrinsic characterisation of some additive functors, *Proc. Amer. Math. Soc.*, **11(1)** (1960), 5–8.

[706] B. Węglorz, Equationally compact algebras (I), *Fund. Math.*, **59** (1966), 289–298.

[707] B. Węglorz, Equationally compact algebras (III), *Fund. Math.*, **60** (1966), 89–93.

[708] C. A. Weibel, *An Introduction to Homological Algebra*, Cambridge University Press, 1994.

[709] V. Weispfenning, Quantifier elimination for modules, *Arch. Math. Logik Grund. Math.*, **25(1)** (1985), 1–11.

[710] G. H. Wenzel, Equational compactness, Appendix 6 (pp. 417–447) in G. Grätzer, *Universal Algebra*, second edn, Springer-Verlag, 1979.

[711] E. Wilkins, An equivalence induced by Ext and Tor applied to the finitistic weak dimension of coherent rings, *Glasgow Math. J.*, **40(2)** (1998), 167–176.

[712] R. Wisbauer, On modules with the Kulikov property and pure semisimple modules and rings, *J. Pure Appl. Algebra*, **70(3)** (1991), 315–320.

[713] R. Wisbauer, *Foundations of Module and Ring Theory*, Gordon and Breach, 1991.

[714] R. Wisbauer, *Module and Comodule Categories – A Survey, Proceedings of the Mathematics Conference (Birzeit University 1998)*, World Scientific (2000), 277–304.

[715] R. Wisbauer, Correct classes of modules, *Algebra Discrete Math.*, **4(4)** (2004), 106–118.

[716] J. Xu, *Flat Covers of Modules*, Lecture Notes in Mathematics, Vol. 1634, Springer, 1996.

[717] M. Zayed, Ultraproduits et modules sans facteurs directs de type fini, pp. 312–324 in M.-P. Malliavin (Ed.), *Sèminaire d'algébre Paul Dubreil et Marie-Paule Malliavin (Paris, 1986)*, Lecture Notes in Math., Vol. 1296, Springer, 1987.

[718] M. Zayed, Indecomposable modules over right pure semisimple rings, *Monatsh. Math.*, **105(2)** (1988), 165–170.

[719] M. Zayed, Ultrapuissances et algèbres de représentation finie, *Comm. Algebra*, **16(10)** (1988), 2125–2132.

[720] M. Zayed, An application of a theorem of Ziegler, *Monatsh. Math.*, **109(4)** (1990), 327–331.

[721] M. Zayed, A characterization of algebras of finite representation type, *Pure Math. Appl. Ser. A*, **3** (1992), 295–297.

[722] M. Zayed, Pure-semisimplicity is preserved under elementary equivalence, *Glasgow Math. J.*, **36** (1994), 345–346.

[723] M. Zayed, On C-pure-injective modules, *Comm. Algebra*, **25(2)** (1997), 347–355.

[724] M. Zayed and A. A. Abdel-Aziz, On modules which are subisomorphic to their pure-injective envelopes. *J. Algebra Appl.*, **1(3)** (2002), 289–294.

[725] D. Zelinsky, Linearly compact rings and modules, *Amer. J. Math.*, **75(1)** (1953), 79–90.

[726] M. Ziegler, Model theory of modules, *Ann. Pure Appl. Logic*, **26(2)** (1984), 149–213.

[727] M. Ziegler, Ein stabiles Modell mit der finite cover property aber ohne Vaughtsche Paare, pp. 179–185 in B. Dahn and H. Wolter (Eds.), *Proceedings of the 6th Easter Conference on Model Theory (Wendisch Rietz, 1988)*, Seminarberichte, 98, Humboldt University, 1998.

[728] M. Ziegler, Decidable modules, pp. 255–274 in *Algebraic Model Theory (Toronto, 1996)*, NATO Adv. Sci. Inst. Ser. C Math. Phys. Sci., Vol. 496, Kluwer, 1997.

[729] B. Zimmermann-Huisgen and W. Zimmermann, Algebraically compact rings and modules, *Math. Z.*, **161(1)** (1978), 81–93.

[730] B. Zimmermann-Huisgen and W. Zimmermann, On the sparsity of representations of rings of pure global dimension zero, *Trans. Amer. Math. Soc.*, **320(2)** (1990), 695–711.

[731] W. Zimmermann, Rein injektive direkte Summen von Moduln, *Comm. Algebra*, **5(10)** (1977), 1083–1117.

[732] W. Zimmermann, (Σ-)algebraic compactness of rings, *J. Pure Appl. Algebra*, **23(3)** (1982), 319–328.

[733] W. Zimmermann, Auslander-Reiten sequences over artinian rings, *J. Algebra*, **119(2)** (1988), 366–392.

[734] W. Zimmermann, Auslander–Reiten sequences over derivation polynomial rings, *J. Pure Appl. Algebra*, **74(3)** (1991), 317–332.

[735] W. Zimmermann, Modules with chain conditions for finite matrix subgroups, *J. Algebra*, **190(1)** (1997), 68–87.

[736] W. Zimmermann, Extensions of three classical theorems to modules with maximum condition for finite matrix subgroups, *Forum Math.*, **10(3)** (1998), 377–392.

[737] W. Zimmermann, On locally pure-injective modules, *J. Pure Applied Algebra*, **166(3)** (2002), 337–357.

[738] G. Zwara, Tame algebras and degenerations of modules, pp. 311–319 in [371].

Index